SOLID STATE PHYSICS LITERATURE GUIDES
Volume 8

BIBLIOGRAPHY OF MICROWAVE OPTICAL TECHNOLOGY

Solid State Physics Literature Guides

Prepared under the auspices of the Research Materials Information Center
Oak Ridge National Laboratory

General Editor: T. F. Connolly

Solid State Division
*Oak Ridge National Laboratory** *
Oak Ridge, Tennessee

---

*Oak Ridge National Laboratory is operated by Union Carbide Corporation for the U.S. Energy Research and Development Administration.

SOLID STATE PHYSICS LITERATURE GUIDES
Volume 8

BIBLIOGRAPHY OF MICROWAVE OPTICAL TECHNOLOGY

Compiled by

A. F. Harvey

Royal Signals and Radar Establishment
Malvern, England

IFI/PLENUM • NEW YORK-WASHINGTON-LONDON

Library of Congress Cataloging in Publication Data

Harvey, Arthur Frank.
  Bibliography of microwave optical.

  (Solid state physics literature guides; v. 8)
  Includes indexes.
  1. Microwaves — Bibliography. 2. Optics — Bibliography. I. Title. II. Series.
Z5838.M521I26 [TK7876]   016.621381'3   76-24892

ISBN-13: 978-1-4684-6218-0   e-ISBN-13: 978-1-4684-6216-6
DOI: 10.1007/978-1-4684-6216-6

PREFACE

Although microwaves and coherent optics, being two of the largest and most useful branches of electrical engineering to emerge technologically, are usually considered as distinct subjects, many of the underlying fundamental principles, scientific achievements, and practical applications have common features. Following the evolvment of the initial principles and techniques during the closing decade of the last century, microwave engineering has long matured to a stage of ready availability of components, automation and accuracy of measurement, economical manufacturing methods, and application of sophisticated systems. Further, this development of electromagnetic phenomena having spatial and temporal coherence has, based on several centuries of study and practice of noncoherent light, in the last two decades reached the optical region. Hence, it is now practicable to consider a comprehensive treatment of these two fields, division being made by subject matter rather than by the artificial distinctions of frequency and/or wavelength ranges. However, a full text on the combined subjects would be very large and unwieldy and, thus, this Bibliography is presented in the hope that it will prove useful as a compact reference source to a large body of workers and, by putting forward the latest scientific and technical advances, stimulate a multi-disciplinary approach.

The material of the book commences with the fundamentals of radiation and matter, progressing through components and devices, amplification and generation, transmission, reception and processing of information, and methods of measurement to conclude with a wide range of applications. The spectral range covered extends from the centimetric region, through the millimetre, submillimetre, far-to-near infrared, visible, and ultraviolet bands to the X-ray region. As such, this range is bounded, on the high-frequency side, by regions where devices are geometrically small and subject to close mechanical tolerances, where physical phenomena are associated with bound charges, and where quantum effects are dominant and, on the low-frequency side, by regions that exhibit most strongly the wave nature of radiation, where diffraction effects are important, and where operation is associated with mobile charges. As examples of combined technology it will be noted that optical integrated circuits can provide compact modulators, filters, and directional couplers that work alongside microwave components and systems. Moreover, interfacing of optical and microwave devices is practicable and thus an overall design can be selected on the basis of costs and technical advantages. The work also covers, both as a byproduct and partly by intent, recent advances in associated acoustic devices and technology.

As research and development builds up and then, as production and application take over, decreases, so scientific activity in a given field initially rises strongly but eventually subsides. Hence, to achieve an adequate and representative selection of disciplines without including obsolescent material, the publication period covered has been chosen to be the last ten years, with the first two or three years restricted more to well established subjects rather than the beginnings of new ones. The work contains about fifteen thousand references to original technical articles carefully selected from the world archival literature, these being intended also to follow on from the extensive bibliographies previously given in the author's High Frequency Thermionic Tubes (1943), Microwave Engineering (1963), and Coherent Light (1970).

Although the Bibliography generally includes the full original title and authors, for reasons like easy search through the Indexes, adequate and accurate description of content, economy in space, and limitation of a uniform-style typescript to reproduce

faithfully the results of a variety of publication methods, some minor changes have
occasionally been made. These modifications include unification of spelling of authors'
names, adopting 'et al' where there are more than three authors, writing Greek letters
out in full, using current standard nomenclature, omission of inessential prepositions
and subtitles, adding in parentheses additional description, and using accepted
abbreviations for long words and/or phrases. Thus, although the references are
certainly accurate enough for search, use, and listing, it is recommended that, when
the precise and full details are required, the original article or even the manuscript,
in English or foreign language and with possible later corrections, be consulted.

The references are arranged in six Parts, each of which is broken down according
to subject matter into Sections that are sequentially numbered through the book. The
individual references are arranged and numbered in approximate chronological order
while, in addition, extensive author and subject Indexes are provided. Hence, despite
the large size and high information density of the work, it is hoped that this four-
way retrieval system will enable the reader to locate rapidly the required material.

                                                                    A.F.HARVEY

Malvern, Worcs., England
April 1976

CONTENTS

CONTENTS

Part II. COMPONENTS AND DEVICES

Part III. AMPLIFICATION AND GENERATION

Part IV. TRANSMISSION AND RECEPTION

Part V. MEASUREMENT PRINCIPLES AND TECHNIQUES

Part VI. APPLICATIONS

REFERENCED JOURNALS

In the individual references of this Bibliography the final numbers represent, in sequence, years of publication, volume, issue, and pages. The full titles of the journals referenced are given below, in alphabetical order of abbreviations employed, with town and country of publication in parentheses.

ABU Tech. Rev.	ABU Technical Review (Tokyo, Japan)
Acqua	Acqua (Rome, Italy)
Acta Cient.	Acta Cientifica (Buenos Aires, Argentine)
Acta Cryst.	Acta Crystallographica (Copenhagen, Denmark)
Acta Electron.	Acta Electronica (Limeil-Brevannes, France)
Acta Electron. Sin.	Acta Electronica Sinica (Peking, China)
Acta Entomol. Bohemoslov.	Acta Entomologica Bohemoslovaca (Prague, Czechoslovakia)
Acta Geophys. Pol.	Acta Geophysica Polonica (Warsaw, Poland)
Acta Odontol. Scand.	Acta Odontologica Scandinavaca (Stockholm, Sweden)
Acta Phys. Acad. Sci. Hung.	Acta Physica Acadamiae Scientiarum Hungaricae (Budapest, Hungary)
Acta Phys. Austriaca	Acta Physica Austriaca (Vienna, Austria)
Acta Phys. Chem. Szeged.	Acta Physica et Chemica Universitatis Szegediensis (Szeged, Hungary)
Acta Phys. Hung.	Acta Physica Hungaricae (Budapest, Hungary)
Acta Phys. Pol.	Acta Physica Polonica (Warsaw, Poland)
Acta Phys. Sin.	Acta Physica Sinica (Peking, China)
Acta Polytech. Scand.	Acta Polytechnica Scandinavica (Stockholm, Sweden)
Acta Radiol.	Acta Radiologica (Stockholm, Sweden)
Acta Tech. Acad. Sci. Hung.	Acta Technica Academiae Scientiarum Hungaricae (Budapest, Hungary)
Acta Tech. CSAV	Acta Technica Ceskoslovenska Akademie Ved (Prague, Czechoslovakia)
Acta Tech. Hung.	Acta Technica Hungaricae (Budapest, Hungary)
Acustica	Acustica (Stuttgart, Germany)
Adv. Cyclic Nucleotide Res.	Advances in Cyclic Nucleotide Research (New York, USA)
Adv. Phys.	Advances in Physics (London, England)
AECL Res. Dev. Eng.	AECL Research and Development in Engineering (Chalk River, Canada)
Aerosp. Med.	Aerospace Medicine (Washington, USA)
Aerotech. Missili Spazio	Aerotecnica Missili e Spazio (Milan, Italy)
AIAA J.	AIAA Journal (New York, USA)
Aircr. Eng.	Aircraft Engineering (London, England)
Akust. Zh.	Akusticheskii Zhurnal (Moscow, USSR)
Albiswerk-Ber.	Albiswerk-Berichte (Zurich, Switzerland)
Alta Freq.	Alta Frequenza (Milan, Italy)
Am. Ceram. Soc. Bull.	American Ceramic Society Bulletin (Columbus, USA)
Am. J. Ophthal.	American Journal of Ophthalmology (St. Louis, USA)
Am. J. Optom.	American Journal of Optometry (Minneapolis, USA)
Am. J. Phys.	American Journal of Physics (New York, USA)
Am. J. Phys. Med.	American Journal of Physical Medicine (Baltimore, USA)

Am. J. Physiol.	American Journal of Physiology (Boston, USA)
Am. Surg.	American Surgeon (Baltimore, USA)
An. Fis.	Anales de Fisica (Madrid, Spain)
An. Stiint. Cuza Iasi	Analele Stiintife Ale Universitatu Al. I. Cuza din Iasi (Iasi, Rumania)
Anal. Chem.	Analytical Chemistry (Washington, USA)
Ann. Astrophys.	Annales d'Astrophysique (Paris, France)
Ann. Geophys.	Annales de Geophysique (Paris, France)
Ann. Inst. Poincare	Annales de l'Institut Henri Poincare (Paris, France)
Ann. N. Y. Acad. Sci.	Annals of the New York Academy of Science (New York, USA)
Ann. Phys. (Leipzig)	Annalen der Physik (Leipzig, Germany)
Ann. Phys. (Paris)	Annales de Physique (Paris, France)
Ann. Phys. Chem.	Annalen der Physik und Chemie (Leipzig, Germany)
Ann. Radioelectr.	Annales de Radioelectricite (Paris, France)
Ann. Soc. Sci. Brux.	Annales de la Societe Scientifique de Bruxelles (Namur, Belgium)
Ann. Telecommun.	Annales des Telecommunications (Paris, France)
Annee Biol.	Annee Biologique (Paris, France)
Annu. Rep. Eng. Res. Inst. Univ. Tokyo	Annual Report of the Engineering Research Institute, University of Tokyo (Tokyo, Japan)
Annu. Rep. Sofia Fac. Phys.	Annuaire de l'Universite de Sofia Faculte de Physique (Sofia, Bulgaria)
Anritsu Tech. Bull.	Anritsu Technical Bulletin (Tokyo, Japan)
Antenna	Antenna (Milan, Italy)
APL Tech. Dig.	Applied Physics Laboratory Technical Digest (Silver Springs, USA)
Appl. Phys. Lett.	Applied Physics Letters (New York, USA)
Appl. Sol. Energy	Applied Solar Energy (New York, USA). Translation of Geliotekhnika
Appl. Spect.	Applied Spectroscopy (New York, USA)
Arch. Biochem. Biophys.	Archives of Biochemistry and Biophysics (New York, USA)
Arch. Elektr. Ubertrag.	Archiv der Elektrischen Ubertragung (Stuttgart, Germany)
Arch. Elektrotech.	Archiv fur Electrotechnik (Berlin, Germany)
Arch. Elektrotech. (Warsaw)	Archiwum Elektrotechniki (Warsaw, Poland)
Arch. Ophthal.	Archives of Ophthalmology (Chicago, USA)
Arch. Sci.	Archives des Sciences (Geneva, Switzerland)
Arch. Surg.	Archives of Surgery (London, England)
Arch. Tech. Mess.	Archiv fur Technisches Messen und Messtechnische Praxis (Munich, Germany)
Ark. Astron.	Arkiv for Astronomi (Stockholm, Sweden)
Ark. Fys.	Arkiv for Fysik (Stockholm, Sweden)
Ark. Kemi	Arkiv for Kemi (Stockholm, Sweden)
Astron. Astrophys.	Astronomy and Astrophysics (Berlin, Germany)
Astron. J.	Astronomical Journal (New York, USA)
Astron. Zh.	Astronomicheskii Zhurnal (Moscow, USSR)
Astrophys. J.	Astrophysical Journal (Chicago, USA)
Astrophys. Lett.	Astrophysical Letters (London, England)
At. Energ.	Atomnaya Energiya (Moscow, USSR)
Atmos. Envir.	Atmospheric Environment (Oxford, England)
Atomkernenergie	Atomkernenergie (Munich, Germany)
Atti Accad. Sci. Torino	Atti della Accademia delle Scienze di Torino (Turin, Italy)
Atti Fond. Giorgio Ronchi	Atti della Fondazione Giorgio Ronchi (Arcetri-Firenze, Italy)
Atti Semin. Mat. Fis. Univ. Modena	Atti del Seminario Matematico e Fisico dell' Universita di Modena (Modena, Italy)
Aust. J. Phys.	Australian Journal of Physics (Mellbourne, Australia)
Aust. J. Sci.	Australian Journal of Science (Sydney, Australia)
Aust. Telecommun. Res.	Australian Telecommunication Research (Melbourne, Australia)
Autom. Strum.	Automazione e Strumentazione (Milan, Italy)

Autom. Weld.	Automatic Welding (Cambridge, England). Translation of Avtom. Svarka
Avtom. Svarka	Avtomaticheskaya Svarka (Kiev, USSR)
Avtom. Telemekh. Svyaz	Avtomatika, Telemekhanika, Svyaz' (Moscow, USSR)
Avtometriya	Avtometriya (Novosibirsk, USSR)
Behav. Res. Meth.	Behaviour Research Methods and Instrumentation (Austin, USA)
Beitr. Plasma Phys.	Beitrage aus der Plasma Physik (Berlin, Germany)
Bell Lab. Rec.	Bell Laboratories Record (Murray Hill, USA)
Bell Syst. Tech. J.	Bell System Technical Journal (New York, USA)
Ber. Bunsenges. Phys. Chem.	Berichte der Bunsengesellschaft fur Physikalische Chemie (Weinheim, Germany)
Biofizika	Biofizika (Moscow, USSR)
Biokhimiya	Biokhimiya (Moscow, USSR)
Biol. Kozlem.	Biologiai Kozlemenyek (Budapest, Hungary)
Biol. Nauk	Biologicheskie Nauk (Moscow, USSR)
Biomed.Tech.	Biomedizinsche Technik (Berlin, Germany)
Br. J. Ind. Med.	British Journal of Industrial Medicine (London, England)
Brain Res.	Brain Research (Amsterdam, Netherlands)
Brown Boveri Rev.	Brown Boveri Review (Baden, Switzerland)
Budavox Telecommun. Rev.	Budavox Telecommunication Review (Budapest, Hungary)
Bul. Inst. Politeh. Iasi	Buletinul Institutului Politehnic din Iasi (Iasi, Rumania)
Bull. Acad. Pol. Sci.	Bulletin de l'Academie Polonaise des Sciences (Warsaw, Poland)
Bull. Acad. Sci. USSR, Phys. Ser.	Bulletin of the Academy of Sciences of the USSR, Physical Series (New York, USA). Translation of Izv. Akad. Nauk SSSR, Ser. Fiz.
Bull. Am. Meteorol. Soc.	Bulletin of the American Meteorological Society (Boston, USA)
Bull. Assoc. Suisse Electr.	Bulletin de l'Association Suisse des Electriciens (Zurich, Switzerland)
Bull. Astron.	Bulletin Astronomique (Paris, France)
Bull. Electrotech. Lab.	Bulletin of the Electrotechnical Laboratory (Tokyo, Japan)
Bull. Fac. Eng. Yokohama Natl. Univ.	Bulletin of the Faculty of Engineering of Yokohama National University (Yokohama, Japan)
Bull. Jap. Soc. Precis. Eng.	Bulletin of the Japan Society of Precision Engineering (Tokyo, Japan)
Bull. Nagoya Inst. Technol.	Bulletin of the Nagoya Institute of Technology (Nagoya, Japan)
Bull. Res. Inst. Appl. Electr.	Bulletin of the Research Institute of Applied Electricity (Sapporo, Japan)
Bull. Tokyo Inst. Technol.	Bulletin of Tokyo Institute of Technology (Tokyo, Japan)
Bull. Yamagata Univ.	Bulletin of the Yamagata University (Yamagata, Japan)
Byull. Eksp. Biol. Med.	Byulleten' Eksperimentalnoi Biologii i Meditsiny (Moscow, USSR)
C. R. Acad. Bulg. Sci.	Comptes Rendus de l'Academie Bulgare des Sciences (Sofia, Bulgaria)
C. R. Acad. Sci. (Paris)	Comptes Rendus Hebdomadaires des Seances de l'Academie des Sciences (Paris, France)
Cables Transm.	Cables et Transmission (Paris, France)
Can. J. Chem.	Canadian Journal of Chemistry (Ottowa, Canada)
Can. J. Phys.	Canadian Journal of Physics (Ottowa, Canada)
Cesk. Cas. Fis.	Ceskoslovensky Casopis pro Fisiku (Prague, Czechoslovakia)
Chart. Mech. Eng.	Chartered Mechanical Engineer (London, England)
Chem. Phys. Lett.	Chemical Physics Letters (Amsterdam, Netherlands)
Chin. J. Phys.	Chinese Journal of Physics (Taiwan, China)
Cienc. Tec.	Ciencia y Tecnica (Buenos Aires, Argentine)
Combust. Explos.	Combustion, Explosion, and Shock Waves (New York, USA). Translation of Fiz. Goreniya Vzryva
Combust. Flame	Combustion and Flame (New York, USA)

Comments Astrophys. Space Phys.	Comments on Astrophysics and Space Physics (London, England)
Comments At. Mol. Phys.	Comments on Atomic and Molecular Physics (London, England)
Comments Solid State Phys.	Comments on Solid State Physics (London, England)
Commun. Equip. Sys. Des.	Communications Equipment and Systems Design (Valley Stream, USA)
Commun. Math. Phys.	Communication in Mathematical Physics (Heidelberg, Germany)
Component Technol.	Component Technology (Ilford, England)
COMSAT Tech. Rev.	COMSAT Technical Review (Washington, USA)
Contemp. Phys.	Contemporary Physics (London, England)
Copeia	Copeia (New York, USA)
Cosmic Res.	Cosmic Research (New York, USA). Translation of Kosm. Issled.
Cryogenics	Cryogenics (Guildford, England)
Czech. J. Phys.	Czechoslovak Journal of Physics (Prague, Czechoslovakia)
Dainichi-Nippon Cables Rev.	Dainichi-Nippon Cables Review (Osaka, Japan)
Def. Sci. J.	Defence Science Journal (New Delhi, India)
Des. Electron.	Design Electronics (London, England)
Des. News	Design News (Denver, USA)
DISA Inf.	Danske Industri Syndikat A/S Information (Herlev, Denmark)
Dokl. Akad. Nauk SSSR	Doklady Akademii Nauk SSSR (Moscow, USSR)
Electr. Commun.	Electrical Communication (New York, USA)
Electr. Commun. Lab. Tech. J.	Electrical Communication Laboratories Technical Journal (Tokyo, Japan)
Electr. Eng. Jap.	Electrical Engineering in Japan (Washington, USA). Translation of Trans. IEE Jap.
Electr. Eng. Rev.	Electrical Engineering Review (Lahore, Pakistan)
Electr. Wld	Electrical World (New York, USA)
Electro-Opt. Syst. Des.	Electro-Optical Systems Design (Chicago, USA)
Electro-Technol.	Electro-Technology (Bangalore, India)
Electron. Appl. Bull.	Electronic Applications Bulletin (Eindhoven, Netherlands)
Electron. Aust.	Electronics Australia (Sydney, Australia)
Electron Commun. Jap.	Electronics and Communications in Japan (Washington, USA). Translation of Trans. Inst. Electron. Commun. Eng. Jap., A-D
Electron. Compon.	Electronic Components (London, England)
Electron. Des.	Electronic Design (Rochelle Park, USA)
Electron. Eng.	Electronic Engineering (London, England)
Electron. Equip. News	Electronic Equipment News (London, England)
Electron. Fis. Apl.	Electronica y Fisica Aplicada (Madrid, Spain)
Electron. Ind.	Electronic Industry and Tele-Tech (Philadelphia, USA)
Electron. Lett.	Electronics Letters (London, England)
Electron. Mainland China	Electronics of Mainland China (New York, USA). Translation of Acta Electron. Sin.
Electron.Microelectron. Ind.	Electronique et Microelectronique Industrielles (Paris, France)
Electron Technol.	Electron Technology (Warsaw, Poland)
Electron. Wld	Electronics World (New York, USA)
Electronics	Electronics (New York, USA)
Electroplat. Met. Finish.	Electroplating and Metal Finishing (Watford, England)
Electrotech. J.	Electrotechnical Journal (Tokyo, Japan)
Elektr. Masch.	Elektrische Maschinen (Heidelberg, Germany)
Elektro-Anz.	Elektro-Anzeiger (Essen, Germany)
Elektro Prom. Prib.	Elektro Promishlenost i Priborostroene (Sofia, Bulgaria)
Elektro-Tech. Z	Elektrotechnische Zeitschrift (Berlin, Germany)
Elektron Int.	Elektron International (Linz-Donau, Austria)
Elektron Wiss. Tech.	Elektron in Wissenschaft und Technik (Munich, Germany)

Elektron. Datenverarb.	Elektronische Datenverarbeitung (Braunschweig, Germany)
Elektron. Ind.	Elektronik Industrie (Heidelberg, Germany)
Elektron. J.	Elektronik Journal (Munich, Germany)
Elektron. Rdsch.	Elektronische Rundschau (Berlin, Germany)
Elektronika	Elektronika (Warsaw, Poland)
Elektronikka	Elektronikka (Helsinki, Finland)
Elektrosvyaz	Elektrosvyaz (Moscow, USSR)
Elektrotech. Cas.	Elektrotechnicky Casopis (Senicka, Czechoslovakia)
Elektrotech. Maschin.	Elektrotechnik und Maschinenbau (Vienna, Austria)
Elektrotech. Obz.	Elektrotechnicky Obzor (Prague, Czechoslovakia)
Elektrotechniek	Elektrotechniek (Doetinchem, Netherlands)
Elektrotechnika	Elektrotechnika (Budapest, Hungary)
Elektroteh. Vestn.	Elektrotehniski Vestnik (Ljubljana, Yugoslavia)
Elektrotek. Tidsskr.	Elektroteknisk Tidsskrift (Oslo, Norway)
Elteknik	Elteknik (Stockholm, Sweden)
Eng. Cybern.	Engineering Cybernetics (Washington, USA). Translation of Tekh. Kibern.
Eng. J.	Engineering Journal (Montreal, Canada)
Entwick. Siemens Halske	Entwicklungsberichten Siemens und Halske (Berlin, Germany)
Environ. Lett.	Environmental Letters (New York, USA)
Ericsson Tech.	Ericsson Technics (Stockholm, Sweden)
Exp. Cell Res.	Experimental Cell Research (New York, USA)
Exp. Tech. Phys.	Experimentelle Technik der Physik (Berlin, Germany)
Federation Proc.	Federation Proceedings (Washington, USA)
Feingerate Tech.	Feingerate Technik (Berlin, Germany)
Fernmelde-Ing.	Fernmelde-Ingenieur (Bad Windsheim, Germany)
Fernmelde-Praxis	Fernmelde-Praxis (Berlin, Germany)
Fernmeldetech. Z.	Fernmeldetechnische Zeitschrift (Brunswick, Germany)
Ferroelectrics	Ferroelectrics (London, England)
Finommech.	Finommechanika Mikrotechnika (Budapest, Hungary)
Fiz. Goreniya Vzryva	Fizika Goreniya i Vzryva (Moscow, USSR)
Fiz. Khim. Obrab. Mater.	Fizika Khimiya Obrabotki Materialov (Moscow, USSR)
Fiz. Plazmy	Fizika Plazmy i Problemy Upravlyaemogo Termoyadernogo Sinteza (Kiev, USSR)
Fiz. Tekh. Poluprov.	Fizika i Tekhnika Poluprovodnikov (Moscow, USSR)
Fiz. Tver. Tela	Fizika Tverdogo Tela (Moscow, USSR)
Fizika	Fizika (Zagreb, Yugoslavia)
Fiziol. Zh.	Fiziologicheskii Zhurnal (Moscow, USSR)
Food Technol.	Food Technology (London, England)
Forskning	Forskning (Oslo, Norway)
Frequenz	Frequenz (Berlin, Germany)
Fujitsu Sci. Tech. J.	Fujitsu Scientific and Technical Journal (Kawasaki, Japan)
Funktech. Mh.	Funktechnische Monatshefte (Berlin, Germany)
Funk-Tech.	Funk-Technik (Berlin, Germany)
GEC-AEI Telecommun.	GEC-AEI Telecommunications (London, England)
GEC J. Sci. Technol.	GEC Journal of Science and Technology (London, England)
Geofiz. Sb.	Geofizicheskii Sbornik (Leningrad, USSR)
Geliotekhnika	Geliotekhnika (Tashkent, USSR)
Geomagn. Aeronomiya	Geomagnetizm i Aeronomiya (Moscow, USSA)
Geomagn. Aeron.	Geomagnetism and Aeronomy (Washington, USA). Translation of Geomagn. Aeronomiya
Geophys. Prospect.	Geophysical Prospecting (Hague, Netherlands)
Glass Technol.	Glass Technology (Sheffield, England)

Hautarzt	Hautarzt (Berlin, Germany)
Health Phys.	Health Physics (London, England)
Heat Transfer - Jap. Res.	Heat Transfer - Japanese Research (Washington, USA)
Helv. Phys. Acta	Helvetica Physica Acta (Basle, Switzerland)
Hewlett-Packard J.	Hewlett-Packard Journal (Palo Alto, USA)
High Temp.	High Temperature (New York, USA). Translation of Teplofiz. Vys. Temp.
High Temp. Pressures	High Temperatures - High Pressures (London, England)
High Temp. Sci.	High Temperature Science (New York, USA)
Hitachi Rev.	Hitachi Review (Tokyo, Japan)
Hiradastechnika	Hiradastechnika (Budapest, Hungary)
Hochfreq. Elektroak.	Hochfrequenztechnik und Elektroakustik (Leipzig, Germany)
Hosp. Manage.	Hospital Management (Chicago, USA)
Houille Blanche	Houille Blanche (Grenoble, France)
Hydrospace	Hydrospace (Bromley, England)
IBM J. Res. Dev.	IBM Journal of Research and Development (Yorktown Heights, USA)
IBM Tech. Disclosure Bull.	IBM Technical Disclosure Bulletin (Yorktown Heights, USA)
Icarus	Icarus (New York, USA)
Image Technol.	Image Technology (Los Angeles, USA)
Ind. Electron.	Industrial Electronics (London, England)
Ind. Ital. Elettrotec.	Industria Italiana Elettrotecnica ed Elettronica (Milan, Italy)
Ind. Lab.	Industrial Laboratory (New York, USA). Translation of Zavod. Lab.
Indian J. Meteorol. Geophys.	Indian Journal of Meteorology and Geophysics (New Delhi, India)
Indian J. Phys.	Indian Journal of Physics (Calcutta, India)
Indian J. Pure Appl. Phys.	Indian Journal of Pure and Applied Physics (New Delhi, India)
Indian J. Technol.	Indian Journal of Technology (New Delhi, India)
INE Rev. Electronica	INE Revista de Informacion Electronica (Madrid, Spain)
Infrared Phys.	Infrared Physics (Oxford, England)
Ingegnere	Ingegnere (Rome, Italy)
Ingenieur	Ingenieur (Hague, Netherlands)
Inorg. Mater.	Inorganic Materials (New York, USA). Translation of Izv. Akad. Nauk SSSR, Neorg. Mater.
Inst. Eng. Aust.	Institution of Engineers, Australia (Sydney, Australia)
Instrum. Control Syst.	Instruments and Control Systems (Philadelphia, USA)
Int. Conv. Rec. IEEE	International Convention Record of the Institute of Electrical and Electronic Engineers (New York, USA)
Int. Elektron. Rundsch.	International Elektronische Rundschau (Berlin, Germany)
Int. J. Circuit Theory Appl.	International Journal of Circuit Theory and Applications (Chichester, England)
Int. J. Electron.	International Journal of Electronics (London, England)
Int. J. Eng. Sci.	International Journal of Engineering Science (Oxford, England)
Int. J. Mech. Sci.	International Journal of Mechanical Sciences (Oxford, England)
Int. J. Radiat. Biol.	International Journal of Radiation Biology (London, England)
Inter Electron.	Inter Electronique (Paris, France)
Invest. Ophthal.	Investigative Ophthalmology (St. Louis, USA)
Inzh-Fiz. Zh.	Inzhenerno-Fizicheskii Zhurnal (Minsk, USSR)
ISA Trans.	ISA Transactions (New York, USA)
Israel J. Technol.	Israel Journal of Technology (Jerusalem, Israel)
Izmer. Tekh.	Izmeritelnaya Tekhnika (Moscow, USSR)
Izv. Acad. Sci. USSR, Atmos. Oceanic Phys.	Izvestiya Academy of Sciences USSR, Atmospheric and Oceanic Physics (Washington, USA). Translation of Izv. Akad. Nauk SSSR, Fiz. Atmos. Okeana
Izv. Acad. Sci. USSR, Phys. Solid Earth	Izvestiya Academy of Sciences USSR, Physics of the Solid Earth (Washington, USA). Translation of Izv. Akad. Nauk SSSR, Fiz. Zemli

Izv. Akad. Nauk Arm. SSR Fiz.	Izvestiya Akademia Nauk Armyanskoi SSR, Fizika (Barekamutyan, USSR)
Izv. Akad. Nauk SSSR, Fiz. Atmos. Okeana	Izvestiya Akademii Nauk SSSR, Fizika Atmosfery i Okeana (Moscow, USSR)
Izv. Akad. Nauk SSSR, Fiz. Zemli	Izvestiya Akademii Nauk SSSR, Fizika Zemli (Moscow, USSR)
Izv. Akad. Nauk SSSR, Neorg. Mater.	Izvestiya Akademii Nauk SSSR, Neorganicheskie Materialy (Moscow, USSR)
Izv. Akad. Nauk SSSR, Ser. Fiz.	Izvestiya Akademii Nauk SSSR, Seriya Fizicheskaya (Moscow, USSR)
Izv. Fiz. Inst. ANEB	Izvestiya na Fizicheskiya Instituts ANEB (Sofia, Bulgaria)
Izv. Gl. Astron. Obs. Pulkove	Izvestiya Glavnoi Astronomicheskoi Observatorii v Pulkove (Leningrad, USSR)
Izv. Inst. Elektron.	Izvestiya na Instituta po Elektronika (Sofia, Bulgaria)
Izv. VUZ Elektromekh.	Izvestiya Vysshikh Vchebnykh Zavedenii Elektromekhanika (Moscow, USSR)
Izv. VUZ Fiz.	Izvestiya VUZ Fizika (Tomsk, USSR)
Izv. VUZ Gornyi Zh.	Izvestiya VUZ Gornyi Zhurnal (Moscow, USSR)
Izv. VUZ Mashinostr.	Izvestiya VUZ Mashinostroenie (Moscow, USSR)
Izv. VUZ Prib.	Izvestiya VUZ Priborostroenie (Leningrad, USSR)
Izv. VUZ Radioelektron.	Izvestiya VUZ Radioelektronika (Vepikoi, USSR)
Izv. VUZ Radiofiz.	Izvestiya VUZ Radiofizika (Lyadova, USSR)
Izv. VUZ Radiotekh	Izvestiya VUZ Radiotekhnika (Kiev, USSR)
J. Acoust. Soc. Am.	Journal of the Acoustical Society of America (New York, USA)
J. Acoust. Soc. Jap.	Journal of the Acoustical Society of Japan (Tokyo, Japan)
J. Aerosol Sci.	Journal of Aerosol Science (Oxford, England)
J. AEU	Journal of the Asia Electronics Union (Tokyo, Japan)
J. Am. Ceram. Soc.	Journal of the American Ceramic Society (Columbus, USA)
J. Am. Chem. Soc.	Journal of the American Chemical Society (Washington, USA)
J. Am. Dental Assoc.	Journal of the American Dental Association (Chicago, USA)
J. Am. Diet.	Journal of the American Dietetic Association (Chicago, USA)
J. Am. Med. Assoc.	Journal of the American Medical Association (Chicago, USA)
J. Appl. Mech. Tech. Phys.	Journal of Applied Mechanics and Technical Physics (New York, USA). Translation of Zh. Prikl. Mekh. Tekh. Fiz.
J. Appl. Meteorol.	Journal of Applied Meteorology (Boston, USA)
J. Appl. Phys.	Journal of Applied Physics (New York, USA)
J. Atmos. Sci.	Journal of Atmospheric Sciences (Boston, USA)
J. Atmos. Terr. Phys.	Journal of Atmospheric and Terrestrial Physics (Oxford, England)
J. Belge Radiol.	Journal Belge de Radiologie (Brussels, Belgium)
J. Br. Interplan. Soc.	Journal of the British Interplanetary Society (London, England)
J. Cell Biol.	Journal of Cell Biology (Baltimore, USA)
J. Chem. Phys.	Journal of Chemical Physics (New York, USA)
J. Chim. Phys.	Journal de Chimie Physique (Paris, France)
J. Cryst. Growth	Journal of Crystal Growth (Amsterdam, Netherlands)
J. Dental Res.	Journal of Dental Research (Baltimore, USA)
J. Electrochem. Soc.	Journal of the Electrochemical Society (Princeton, USA)
J. Electron. Mater.	Journal of Electronic Materials (New York, USA)
J. Eng. Phys.	Journal of Engineering Physics (New York, USA). Translation of Inzh. Fiz. Zh.
J. Fac. Eng. Chiba Univ.	Journal of the Faculty of Engineering, Chiba University (Chiba, Japan)
J. Fac. Eng. Univ. Tokyo	Journal of the Faculty of Engineering, University of Tokyo (Tokyo, Japan)
J. Fluid Mech.	Journal of Fluid Mechanics (London, England)
J. Franklin Inst.	Journal of the Franklin Institute (Oxford, England)
J. Geophys. Res.	Journal of Geophysical Research (Richmond, USA)
J. Indian Inst. Sci.	Journal of the Indian Institute of Science (Bangalore, India)
J. Inorg. Nucl. Chem.	Journal of Inorganic and Nuclear Chemistry (Oxford, England)

J. Inst. Electr. Commun. Eng. Jap.	Journal of the Institute of Electrical Communication Engineers of Japan (Tokyo, Japan)
J. Inst. Electron. Commun. Eng. Jap.	Journal of the Institute of Electronics and Communication Engineers of Japan (Tokyo, Japan)
J. Inst. Eng.	Journal of the Institute of Engineers (Calcutta, India)
J. Inst. Eng. Aust.	Journal of the Institution of Engineers, Australia (Sydney, Australia)
J. Inst. Navig.	Journal of the Institute of Navigation (London, England)
J. Inst. Telecommun. Eng.	Journal of the Institution of Telecommunication Engineers (New Delhi, India)
J. Inst. Telev. Eng. Jap.	Journal of the Institute of Television Engineers of Japan (Tokyo, Japan)
J. Invest. Derm.	Journal of Investigative Dermatology (Baltimore, USA)
J. Jap. Soc. Precis. Eng.	Journal of the Japan Society of Precision Engineering (Tokyo, Japan)
J. Korean Inst. Electron. Eng.	Journal of Korean Institute of Electronic Engineers (Seoul, Korea)
J. Korean Phys. Soc.	Journal of the Korean Physical Society (Seoul, Korea)
J. Less-Common Metals	Journal of the Less-Common Metals (Amsterdam, Netherlands)
J. Low-Temp. Phys.	Journal of Low-Temperature Physics (New York, USA)
J. Mater. Sci.	Journal of Materials Science (London, England)
J. Math. Phys.	Journal of Mathematical Physics (New York, USA)
J. Mech. Eng. Sci.	Journal of Mechanical Engineering Science (London, England)
J. Meteorol.	Journal of Meteorology (Boston, USA)
J. Microwave Power	Journal of Microwave Power (Edmonton, Canada)
J. Mol. Spectrosc.	Journal of Molecular Spectroscopy (New York, USA)
J. Natl. Cancer Inst.	Journal of the National Cancer Institute (Washington, USA)
J. Nematol.	Journal of Nematology (Beltville, USA)
J. Nucl. Energy	Journal of Nuclear Energy (Oxford, England)
J. Opt. Soc. Am.	Journal of the Optical Society of America (New York, USA)
J. Photogr. Soc.	Journal of the Photographic Society (London, England)
J. Phys.	Journal of Physics A-D (London, England)
J. Phys. Chem.	Journal of Physical Chemistry (Washington, USA)
J. Phys. Chem. Solids	Journal of Physics and Chemistry of Solids (Oxford, England)
J. Phys. (Paris)	Journal de Physique (Paris, France)
J. Phys. Soc. Jap.	Journal of the Physical Society of Japan (Tokyo, Japan)
J. Plasma Phys.	Journal of Plasma Physics (London, England)
J. Polym. Sci.	Journal of Polymer Science (New York, USA)
J. Quant. Spectrosc.	Journal of Quantitative Spectroscopy and Radiative Transfer (Oxford, England)
J. Quantum Electron. IEEE	Journal of Quantum Electronics of the IEEE (New York, USA)
J. Rech. Atmos.	Journal de Recherche Atmospheriques (Clemont - Ferrand, France)
J. Res. Natl. Bur. Stand.	Journal of Research of the National Bureau of Standards (Washington, USA)
J. Roy. Astron. Soc. Can.	Journal of the Royal Astronomical Society of Canada (Toronto, Canada)
J. Sci. Food Agr.	Journal of the Science of Food and Agriculture (London, England)
J. Sci. Ind. Res.	Journal of Scientific and Industrial Research (New Delhi, India)
J. Sci. Instrum. Control Eng.	Journal of the Society of Instrument and Control Engineers (Tokyo, Japan)
J. Solid State Circuits IEEE	Journal of Solid State Circuits of the IEEE (New York, USA)
J. Spacecr. Rockets	Journal of Spacecraft and Rockets (New York, USA)
J. Strain Anal.	Journal of Strain Analysis (London, England)
J. Surg. Res.	Journal of Surgical Research (New York, USA)
J. Test. Eval.	Journal of Testing and Evaluation (Philadelphia, USA)
J. Vac. Sci. Technol.	Journal of Vacuum Science and Technology (New York, USA)
Jap. J. Appl. Phys.	Japanese Journal of Applied Physics (Tokyo, Japan)
Jap. Telecommun. Rev.	Japan Telecommunications Review (Tokyo, Japan)

Jemna Mech. Opt.	Jemna Mechanika a Optika (Prerov, Czechoslovakia)
Jena Rev.	Jena Review (Berlin, Germany)
JETP Lett.	JETP Letters (New York, USA). Translation of Zh. Eksp. Teor. Fiz. Pis'ma
JPL Q. Tech. Rev.	JPL Quarterly Technical Review (Pasadena, USA)
K. Dan. Vidensk. Selsk. Mat.-Fys. Skr	Kangelige Danske Videnskabernes Selskab Matematisk - Fysiske Skrifter (Copenhagen, Denmark)
K. Tekn. Hogsk. Handl.	Kongliga Tekniska Hogskolans Handligar (Stockholm, Sweden)
KDD Tech. J.	KDD Technical Journal (Tokyo, Japan)
KFKI Kozl.	KFKI Kozlemenyek (Budapest, Hungary)
Kibern. Vychisl. Tekh.	Kibernetika i Vychislitel'naya Tekhnika (Kiev, USSR)
Kiserl. Orvostud.	Kiserletes Orvostudomany (Budapest, Hungary)
Klin. Oczna	Klinika Oczna (Warsaw, Poland)
Kogaku Giyutsu	Kogaku to Giyutsu (Osaka, Japan)
Kosm. Issled.	Kosmicheske Issledovaniya (Leningrad, USSR)
Kristall Tech.	Kristall und Technik (Berlin, Germany)
Kristallogr.	Kristallografiya (Moscow, USSR)
Kvantovaya Elektron.	Kvantovaya Elektronika (Moscow, USSR)
Kvantovaya Elektron. (Kiev)	Kvantovaya Elektronika (Kiev, USSR)
Lab. Prac.	Laboratory Practice (London, England)
Labdev J. Sci. Technol.	Labdev Journal of Science and Technology (Kampur, India)
Laser Angew. Strahlentech.	Laser und Angewandte Strahlentechnik (Aarau, Switzerland)
Laser Focus	Laser Focus (Newtonville, USA)
Laser Unconv. Opt. J.	Laser and Unconventional Optics Journal (Gothenberg, Sweden)
Latv. PSR Zinat. Akad. Vestis	Latvijas PSR Zinatnu Akademijas Vestis (Riga, USSR)
Lek. Wojsk.	Lekarz Wojskowy (Warsaw, Poland)
Lett. Nuovo Cim.	Lettre el Nuovo Cimento (Bologna, Italy)
Lichttechnik	Lichttechnik (Berlin, Germany)
Life Sci.	Life of Science (Krakow, Poland)
Litov. Fiz. Sb.	Litovskii Fizicheskii Sbornik (Vil'nyus, USSR)
Magn. Gidrodin.	Magnitnaya Gidrodinamika (Riga, USSR)
Magy. Fiz. Foly.	Magyar Fizikai Folyoirat (Budapest, Hungary)
Mar. Technol. Soc. J.	Marine Technology Society Journal (Washington, USA)
Marconi Instrum.	Marconi Instrumentation (St. Albans, England)
Marconi Rev.	Marconi Review (Chelmsford, England)
Mater. Eval.	Materials Evaluation (Evanston, USA)
Mater. Res. Bull.	Materials Research Bulletin (Elmsford, USA)
Mater. Res. Stand.	Materials Research and Standards (Philadelphia, USA)
Meas. Control	Measurement and Control (London, England)
Meas. Tech.	Measurement Techniques (New York, USA)
Mech. Eng.	Mechanical Engineering (New York, USA)
Med. Biol. Eng.	Medical and Biological Engineering (Stevenage, England)
Med. Biol. Illus.	Medical and Biological Illustration (London, England)
Mekh. Avtom. Proiz.	Mekhanizatsiya i Avtomatizatsiya Proizvodstva (Moscow, USSR)
Mekh. Polim.	Mekhanika Polimerov (Riga, USSR)
Mem. Chubu Inst. Technol.	Memoirs of the Chubu Institute of Technology (Nagoya, Japan)
Mem. Def. Acad.	Memoirs of the Defence Academy (Yokusuka, Japan)
Mem. Fac. Eng. Hokkaido Univ.	Memoirs of the Faculty of Engineering, Hokkaido University (Sapporo, Japan)
Mem. Fac. Eng. Kyoto Univ.	Memoirs of the Faculty of Engineering, Kyoto University (Kyoto, Japan)
Mem. Fac. Eng. Nagoya Univ.	Memoirs of the Faculty of Engineering, Nagoya University (Nagoya, Japan)
Mem. Fac. Eng. Osaka City Univ.	Memoirs of the Faculty of Engineering, Osaka City University (Osaka, Japan)

Mem. Fac. Ind. Arts Kyoto Tech. Univ.	Memoirs of the Faculty of Industrial Arts, Kyoto Technical University (Kyoto, Japan)
Mem. Fac. Sci. Kyoto Univ.	Memoirs of the Faculty of Science, Kyoto University (Kyoto, Japan)
Mem. Fac. Sci. Kyushu Univ.	Memoirs of the Faculty of Science, Kyushu University (Kyushu, Japan)
Mem. Inst. Sci. Ind. Res. Osaka Univ.	Memoirs of the Institute of Scientific and Industrial Research, Osaka University (Osaka, Japan)
Mem. Res. Inst. Sci. Eng. Ritsumeikan Univ.	Memoirs of the Research Institute of Science and Engineering, Ritsumeikan University (Kyoto, Japan)
Mem. Soc. Astron. Ital.	Memoirie della Societa Astronomica Italiana (Milan, Italy)
Mem. Soc. Roy. Sci. Liege	Memoires de la Societe Royal des Sciences de Liege (Liege, Belgium)
Mes. Regul. Autom.	Mesures, Regulation, Automatisme (Paris, France)
Mess. Pruef.	Messen und Pruefen (Bad Worishofen, Germany)
Messtechnik	Messtechnik (Brunswick, Germany)
Meteorol. Gidrol.	Meteorologiya i Gidrologiya (Moscow, USSR)
Meteorol. Mag.	Meteorological Magazine (Bracknell, England)
Methods Enzymol.	Methods in Enzymology (New York, USA)
Microelectron. Reliab.	Microelectronics and Reliability (Oxford, England)
Microscope	Microscope (London, England)
Microwave J.	Microwave J. (Dedham, USA)
Microwaves	Microwaves (New York, USA)
Mitsubishi Denki Lab. Rep.	Mitsubishi Denki Laboratory Reports (Amagasaki, Japan)
Mitt. AGEN	Mitteilumgen der Arbeitgemeinschaft fur Elektrische Nachrichtentechnick der Stiftung Hasler-Werke, Bern (Zurich, Switzerland)
Mol. Phys.	Molecular Physics (London, England)
Mon. Not. Roy. Astron. Soc.	Monthly Notices of the Royal Astronomical Society (London, England)
Monatsber. Dtsch. Akad. Wiss. Berlin	Monatsberichte der Deutschen Akademie der Wissenschaften zu Berlin (Berlin, Germany)
Monogr. Res. Inst. Appl. Electr.	Monograph of the Research Institute of Applied Electricity (Sapporo, Japan)
Mullard Tech. Commun.	Mullard Technical Communications (Mitcham, England)
N. Z. Eng.	New Zealand Engineering (Wellington, New Zealand)
Nachr. Tech.	Nachrichtentechnik (Berlin, Germany)
Nachr. Tech. Elektron.	Nachrichtentechnik Elektronik (Berlin, Germany)
Nachr. Tech. Fachber.	Nachrichtentechnische Fachberichte (Berlin, Germany)
Nachr. Tech. Z.	Nachrichtentechnische Zeitschrift (Berlin, Germany)
Natl. Conv. Rec. IEEE	National Convention Record of the IEEE (New York, USA)
Natl. Tech. Rep.	National Technical Report (Osaka, Japan)
Naturwiss.	Naturwissenschaften (Berlin, Germany)
Naturwiss. Rundsch.	Naturwissenschaftliche Rundschau (Stuttgart, Germany)
NEC Res. Dev.	NEC Research and Development (Tokyo, Japan)
Ned. Tijdschr. Natuurk.	Nederlands Tijdschrift voor Natuurkunde (Hague, Netherlands)
News Rohde Schwarz	News from Rohde and Schwarz (Munich, Germany)
NHK Lab. Note	NHK Laboratory Note (Tokyo, Japan)
NHK Tech. J.	NKH Technical Journal (Tokyo, Japan)
Non-Destr. Test.	Non-Destructive Testing (Guildford, England)
Non-Ioniz. Radiat.	Non-Ionizing Radiation (Guildford, England)
Note Recens. Not.	Note Recensioni e Notizie (Rome, Italy)
Nouv. Rev. Opt. Appl.	Nouvelle Revue d'Optique Appliquee (Paris, France)
Nucl. Fusion	Nuclear Fusion (Vienna, Austria)
Nucl. Instrum. Methods	Nuclear Instruments and Methods (Amsterdam, Netherlands)
Nukleonika	Nukleonika (Warsaw, Poland)
Nuovo Cim.	Nuovo Cimento (Bologna, Italy)

Nutr. Rep. Int.   Nutrition Reports International (Los Altos, USA)

Oftal'mol. Zh.   Oftal'mologicheskii Zhurnal (Moscow, USSR)
Oki Rev.   Oki Review (Tokyo, Japan)
Onde Electr.   Onde Electrique (Paris, France)
Opt. Acta   Optica Acta (London, England)
Opt. Commun.   Optics Communications (Amsterdam, Netherlands)
Opt. Eng.   Optical Engineering (Redondo Beach, USA)
Opt. Laser Technol.   Optics and Laser Technology (Guildford, England)
Opt.-Mekh. Prom.   Optiko-Mekhanicheskaya Promyshlennost (Moscow, USSR)
Opt. Pura Apl.   Optico Pura y Aplicada (Madrid, Spain)
Opt. Spectra   Optical Spectra (Pittsfield, USA)
Opt. Spectrosc.   Optics and Spectroscopy (New York, USA). Translation of
                              Opt. Spekt.
Opt. Spekt.   Optika i Spektroskopiya (Moscow, USSR)
Optik   Optik (Stuttgart, Germany)
Opto-Electron.   Opto-Electronics (London, England)
Oral Surg. Med. Pathol.   Oral Surgery, Medicine and Pathology (St. Louis, USA)
Otbor Peredacha Inf.   Otbor i Peredacha Informatsii (Kiev, USSR)
Otolaryngol. Pol.   Otolaryngologia Polska (Warsaw, Poland)
Oyo Buturi   Oyo Buturi (Tokyo, Japan)

Pak. J. Sci. Ind. Res.   Pakistan Journal of Scientific and Industrial Research
                              (Karachi, Pakistan)
Part. Accel.   Particle Accelerators (London, England)
Patol. Fiziol. Eksp. Ter. Patologicheskaya Fiziologiya i Eksperimental' naya Terapiya
                              (Moscow, USSR)
Period. Math.-Phys.   Periodicum Mathematico-Physicum et Astronomicum
   Astron.   (Zagreb, Yugoslavia)
Period. Polytech. Electr. Periodica Polytechnica. Electrical Engineering (Budapest,
   Eng.   Hungary)
Philips Electron. Meas.   Philips Electronic Measurement and Microwave Notes
   Microwave Notes   (Eindhoven, Netherlands)
Philips Res. Rep.   Philips Research Reports (Eindhoven, Netherlands)
Philips Test Meas. Notes   Philips Test and Measurement Notes (Eindhoven, Netherlands)
Philos. Trans.   Philosophical Transactions of the Royal Society
                              (London, England)
Photochem. Photobiol.   Photochemistry and Photobiology (Oxford, England)
Photogramm. Eng.   Photogrammetric Engineering (Falls Church, USA)
Phys. Fenn.   Physica Fennica (Helsinki, Finland)
Phys. Fluids   Physics of Fluids (New York, USA)
Phys. Lett.   Physics Letters (Amsterdam, Netherlands)
Phys. Norv.   Physica Norvegica (Oslo, Norway)
Phys. Rev.   Physical Review A-D (New York, USA)
Phys. Rev. Lett.   Physical Review Letters (New York, USA)
Phys. Scr.   Physica Scripta (Stockholm, Sweden)
Phys. Status Solidi   Physica Status Solidi (Berlin, Germany)
Phys. Teach.   Physics Teacher (New York, USA)
Phys. Z.   Physikalische Zeitschrift (Leipzig, Germany)
Physica   Physica (Amsterdam, Netherlands)
Planet. Space Sci.   Planetary and Space Science (Oxford, England)
Plasma Phys.   Plasma Physics (Oxford, England)
Point-to-Point   Point-to-Point Telecommunications (Chelmsford, England)
   Telecommun.
Pol. Med. J.   Polish Medical Journal (Warsaw, Poland)
Poluprov. Tekh.   Poluprovnikovaya Tekhnika i Mikroelectronica (Kiev, USSR)
   Mikroelectron.
Polym. Mech.   Polymer Mechanics (New York, USA). Translation of Mekh. Polim.

Polytech. Tijdschr. Elektrotech. Elektron.	Polytechnisch Tijdschrift Elektrotechniek Elektronika (New Parklaan, Netherlands)
Polytech. Tijdschr. Werk.	Polytechnisch Tijdschrift Werktingbouw (New Parklaan, Netherlands)
Pomiary Autom. Kont.	Pomiary Automatyka Kontrola (Warsaw, Poland)
Post Office Electr. Eng. J.	Post Office Electrical Engineers' Journal (London, England)
Post Office Telecommun. J.	Post Office Telecommunications Journal (London, England)
Posta Telecom.	Posta si Telecomunicatii (Bucarest, Rumania)
Postepy Fiz.	Postepy Fizyki (Kracow, Poland)
Powder Technol.	Powder Technology (Lausanne, Switzerland)
Pr. Inst. Lacznosci	Prace Instytutu Lacznosci (Warsaw, Poland)
Pr. PIE	Prace Przemyslowego Instytutu Electroniki (Warsaw, Poland)
Pr. PIT	Prace Przemyslowego Instytutu Telekomunikacji (Warsaw, Poland)
Pramana	Pramana (Bangalore, India)
Prib. Sist. Up.	Pribory i Sistemy Upravleniya (Moscow, USSR)
Prib. Tekh. Eksp.	Pribory i Tekhnika Experimenta (Moscow, USSR)
Probl. Peredachi Inf.	Problemy Peredachi Informatsii (Moscow, USSR)
Problems Inf. Transm.	Problems of Information Transmission (New York, USA). Translation of Probl. Peredachi Inf.
Proc. Br. Acoust. Soc.	Proceedings of the British Acoustical Society (London, England)
Proc. Br. Ceram. Soc.	Proceedings of the British Ceramic Society (Stoke, England)
Proc. Fla. State Hort. Soc.	Proceedings of the Florida State Horticultural Society (Lake Alfred, USA)
Proc. IEE	Proceedings of the Institution of Electrical Engineers (London, England)
Proc. IEEE	Proceedings of the Institute of Electrical and Electronic Engineers (New York, USA)
Proc. Indian Acad. Sci.	Proceedings of the Indian Academy of Sciences (Bangalore, India)
Proc. Inst. Mech. Eng.	Proceedings of the Institution of Mechanical Engineers (London, England)
Proc. Iowa Acad. Sci.	Proceedings of the Iowa Academy of Science (Iowa, USSR)
Proc. IREE Aust.	Proceedings of the Institution of Radio and Electronic Engineers (Sydney, Australia)
Proc. Natl. Acad. Sci.	Proceedings of the National Academy of Science (Washington, USA)
Proc. Natl. Inst. Sci. India	Proceedings of the National Institute of Sciences of India (Calcutta, India)
Proc. Res. Inst. Atmos. Nagoya Univ.	Proceedings of the Research Institute of Atmospherics, Nagoya University (Nagoya, Japan)
Proc. Roy. Soc.	Proceedings of the Royal Society (London, England)
Proc. Soc. Inf. Disp.	Proceedings of the Society for Information Display (Los Angeles, USA)
Proc. West. Pharmocol. Soc.	Proceedings of the Western Pharmocological Society (Seattle, USA)
Process Eng.	Process Engineering (London, England)
Prog. Nucl. Phys.	Progress in Nuclear Physics (London, England)
Przegl. Elektron.	Przeglad Elektroniki (Warsaw, Poland)
Przegl. Elektrotech.	Przeglad Elektrotechniczny (Warsaw, Poland)
Pregl. Telekomun.	Przeglad Telekomunikacynjy (Warsaw, Poland)
PTT Bedr.	PTT - Bedriff (Hague, Netherlands)
Publ. Astron. Soc. Jap.	Publications of the Astronomical Society of Japan (Tokyo, Japan)
Publ. Astron. Soc. Pac.	Publications of the Astronomical Society of the Pacific (San Francisco, USA)
Q. Appl. Math.	Quarterly of Applied Mathematics (Providence, USA)
Q. J. Math.	Quarterly Journal of Mathematics (Oxford, England)

Q. J. Mech. Appl. Math.	Quarterly Journal of Mechanics and Applied Mathematics (London, England)
Q. J. Roy. Meteorol. Soc.	Quarterly Journal of the Royal Meteorological Society (London, England)
Q. Prog. Rep. Res. Lab. Electron. MIT	Quarterly Progress Report. Research Laboratory of Electronics MIT (Massachusetts, USA)
Radiat. Res.	Radiation Research (New York, USA)
Radio Electron. Eng.	Radio and Electronic Engineer (London, England)
Radio Eng. Electron. Phys.	Radio Engineering and Electronic Physics (New York, USA). Translation of Radiotekh. Elektron.
Radio Fernsehen Elektron.	Radio Fernsehen Elektronik (Berlin, Germany)
Radio Mentor Electron.	Radio Mentor Electronic (Berlin, Germany)
Radio Sci.	Radio Science (Richmond, USA)
Radiophys. Quantum Electron.	Radiophysics and Quantum Electronics (New York, USA). Translation of Izv. VUZ Radiofiz.
Radiotekh. Elektron.	Radiotekhnika i Elektronika (Moscow, USSR)
Radiotekh. (Kharkov)	Radiotekhnika (Kharkov, USSR)
Radiotekhnika	Radiotekhnika (Moscow, USSR)
Rafena Tech. Commun.	Rafena Technical Communications (Radeberg, Germany)
RCA Rev.	RCA Review (Princeton, USA)
RCA Tech. Notes	RCA Technical Notes (Princeton, USA)
Rec. Electr. Commun. Eng. Conversaz. Tohoku Univ.	Record of Electrical and Communication Engineering Conversazione Tohoku University (Sendai, Japan)
Rech. Aerosp.	Recherche Aerospatiale (Chatillon, France)
Rech. Spat.	Recherche Spatiale (Paris, France)
Remote Sensing Envir.	Remote Sensing of Environment (New York, USA)
Rend. Accad. Naz. Lincei	Rendiconti Accademia Nazionale dei Lincei (Rome, Italy)
Rend. Ist. Lomb.	Rendiconti Istituto Lombardo (Milan, Italy)
Rend. Riun. Assoc. Elettrotec. Ital.	Rendiconti della Riunione Annuale dell' Associazione Elettrotecnica Italiana (Milan, Italy)
Rep. Electr. Commun. Lab.	Reports of the Electrical Communication Laboratory (Tokyo, Japan)
Rep. Fac. Eng. Shizuoka Univ.	Reports of the Faculty of Engineering Shizuoka University (Hamamatsu, Japan)
Rep. Himeji Inst. Technol.	Reports of Himeji Institute of Technology (Himeji, Japan)
Rep. Inst. Phys. Chem. Res.	Reports of the Institute of Physical and Chemical Research (Saitama, Japan)
Rep. NRL Prog.	Report of NRL Progress (Springfield, USA)
Rep. Prog. Phys.	Reports on Progress in Physics (London, England)
Rep. Res. Inst. Electr. Commun. Tohoku Univ.	Reports of the Research Institute of Electrical Communication Tohoku University (Sendai, Japan)
Rep. Univ. Electro-Commun.	Reports of the University of Electro-Communications (Tokyo, Japan)
Res. Electrotech. Lab.	Researches of the Electrotechnical Laboratory (Tokyo, Japan)
Res. Rep. Fac. Eng. Meiji Univ.	Research Report of the Faculty of Engineering Meiji University (Meiji, Japan)
Res. Rep. Fac. Eng. Niigata Univ.	Research Report of the Faculty of Engineering, Niigata University (Nagaoka, Japan)
Rev. Cethedec	Revue du Cethedec (Paris, France)
Rev. Cienc. Apl.	Revista de Ciencia Aplicada (Madrid, Spain)
Rev. Eleotr. Commun. Lab.	Review of the Electrical Communication Laboratory (Tokyo, Japan)
Rev. Electrotec.	Revista Electrotecnica (Barcelona, Spain)
Rev. Esp. Electron.	Revista Espanola de Electronica (Barcelona, Spain)
Rev. FITCE	Revue FITCE (Brussels, Belgium)

Rev. Fr. Electr.	Revue Francaise de l'Electricite (Paris, France)
Rev. Gen. Elect.	Revue Generale de l'Electricite (Malakoff, France)
Rev. Geophys. Space Phys.	Reviews of Geophysics and Space Physics (Richmond, USA)
Rev. HF	Revue HF (Brussels, Belgium)
Rev. MBLE	Revue MBLE (Brussels, Belgium)
Rev. Mex. Fis.	Revista Mexicana de Fisica (Mexico, Mexico)
Rev. Mod. Phys.	Reviews of Modern Physics (New York, USA)
Rev. Phys. Appl.	Revue de Physique Appliquee (Orsay, France)
Rev. Radio Res. Lab.	Review of the Radio Research Laboratories (Tokyo, Japan)
Rev. Roum. Phys.	Revue Roumaine de Physique (Bucarest, Rumania)
Rev. Roum. Sci. Tech.	Revue Roumaine des Science Techniques (Bucarest, Rumania)
Rev. Sci. Instrum.	Review of Scientific Instruments (New York, USA)
Rev. Tech. Thomson - CSF	Revue Technique Thomson - CSF (Paris, France)
Rev. Telecomun.	Revista de Telecomunicacion (Madrid, Spain)
Ric. Sci.	Ricerca Scientifica (Rome, Italy)
Robotron Tech. Commun.	Robotron Technical Communications (Radeberg, Germany)
Rozpr. Electrotech.	Rozprawy Electrotechniczne (Warsaw, Poland)
Saehkoe	Sahko Electricity in Finland (Helsinki, Finland)
Schweissen Schneiden	Schweissen und Schneiden (Dusseldorf, Germany)
Sci. Electr.	Scientia Electrica (Zurich, Switzerland)
Sci. Eng. Rep. Saitama Univ.	Science and Engineering Reports of Saitama University (Saitama, Japan)
Sci. Eng. Rev. Doshisha Univ.	Science and Engineering Review of Doshisha University (Kyoto, Japan)
Sci. Light	Science of Light (Tokyo, Japan)
Sci. Pap. Inst. Phys. Chem. Res.	Scientific Papers of the Institute of Physical and Chemical Research (Saitama, Japan)
Sci. Rep. Res. Inst. Tohoku Univ.	Science Reports of the Research Institutes, Tohoku University (Sendai, Japan)
Sci. Sin.	Scientia Sinica (Peking, China)
Science	Science (Washington, USA)
Sdelovaci Tech.	Sdelovaci Technika (Prague, Czechoslovakia)
SEL Nachr.	Standard Elektrik Lorenz Nachrichten (Stuttgart, Germany)
Siemens Forsch. Entwick.	Siemens Forschungs - und Entwicklungs-berichte (Berlin, Germany)
Sky Telesc.	Sky and Telescope (Cambridge, USA)
Slab. Obz.	Slaboproudy Obzor (Prague, Czechoslovakia)
Soc. Photo-Opt. Instrum. Eng. J.	Society of Photo-Optical Instrumentation Engineers Journal (Redondo Beach, USA)
Sol. Phys.	Solar Physics (Dordrecht, Netherlands)
Solid State Commun.	Solid State Communications (New York, USA)
Solid State Electron.	Solid State Electronics (Oxford, England)
Solid State Technol.	Solid State Technology (Washington, USA)
Sound Vision Broadcast.	Sound and Vision Broadcasting (Chelmsford, England)
Sov. Astron. AJ	Soviet Astronomy - AJ (New York, USA). Translation of Astron. Zh.
Sov. J. Opt. Technol.	Soviet Journal of Optical Technology (Washington, USA). Translation of Opt.-Mekh. Prom.
Sov. J. Quantum Electron.	Soviet Journal of Quantum Electronics (New York, USA). Translation of Kvantovaya Elektron.
Sov. Phys. J.	Soviet Physics Journal (New York, USA). Translation of Izv. VUZ Fiz.
Sov. Phys. — Acoust.	Soviet Physics —Acoustics (New York, USA). Translation of Akust. Zh.
Sov. Phys.-Crystallogr.	Soviet Physics-Crystallography (New York, USA). Translation of Kristallogr.

Sov. Phys.-Dokl.	Soviet Physics—Doklady (New York, USA). Translation of Dokl. Akad. Nauk SSSR
Sov. Phys.-JETP	Soviet Physics-JETP (New York, USA). Translation of Zh. Eksp. Teor. Fiz.
Sov. Phys.-Semicond.	Soviet Physics-Semiconductors (New York, USA). Translation of Fiz. Tekh. Poluprov.
Sov. Phys.-Solid State	Soviet Physics-Solid State (New York, USA). Translation of Fiz. Tver. Tela
Sov. Phys.-Tech. Phys.	Soviet Physics-Technical Physics (New York, USA). Translation of Zh. Tekh. Fiz.
Sov. Phys.-Usp.	Soviet Physics-Uspekhi (New York, USA). Translation of Usp. Fiz. Nauk
Spectrochim. Acta	Spectrochimica Acta (Oxford, England)
Spectrosc. Lett.	Spectroscopy Letters (New York, USA)
Sperry Rand Eng. Rev.	Sperry Rand Engineering Review (New York, USA)
State Inst. Tech. Res. Rep.	State Institute for Technical Research : Reports (Helsinki, Finland)
Stomatologiya	Stomatologiya (Moscow, USSR)
Stud. Biophys.	Studia Biophysica (Berlin, Germany)
Stud. Cercet. Fiz.	Studii si Cercetari de Fizica (Bucarest, Rumania)
Surf. Sci.	Surface Science (Amsterdam, Netherlands)
Surg. Forum	Surgical Forum (Chicago, USA)
Surg. Gynec. Obstet.	Surgery, Gynecology, and Obstetrics (Chicago, USA)
Surgery	Surgery (St. Louis, USA)
Svar. Proizvod.	Svarochnoye Proizvodstra (Moscow, USSR)
Tec. Regul. Mando Autom.	Tecnica de la Regulacion y Mando Automatico (Barcelona, Spain)
Tech. Chron.	Technika Chronika (Athens, Greece)
Tech. Dig.	Technical Digest (New York, USA)
Tech. J. Jap. Broadcast. Corp.	Technical Journal of Japan Broadcasting Corporation (Tokyo, Japan)
Tech. Mitt. AEG-Telefunken	Technische Mitteilungen AEG-Telefunken (Berlin, Germany)
Tech. Mitt. BRF	Technische Mitteilungen BRF (Berlin, Germany)
Tech. Mitt. Krupp Forsch.	Technische Mitteilungen Krupp Forschungberichte (Essen, Germany)
Tech. Mitt. PTT	Technische Mitteilungen PTT (Berne, Switzerland)
Tech. Mitt. RFZ	Technische Mitteilungen RFZ (Berlin, Germany)
Tech. Mod.	Technique Moderne (Paris, France)
Tech. Rundsch.	Technische Rundschau (Berne, Switzerland)
Tech.-Wiss. Abh. Osram-Ges.	Technisch-Wissenschaftliche Abhandlungen aus der Osram-Gesellschaft (Berlin, Germany)
Technol. Rep. Kyushu Univ.	Technology Reports of the Kyushu University (Fukuoka, Japan)
Technol. Rep. Osaka Univ.	Technology Reports of the Osaka University (Osaka, Japan)
Technol. Rep. Seikei Univ.	Technology Reports of the Seikei University (Tokyo, Japan)
Technol. Rep. Tohoku Univ.	Technology Reports of the Tohoku University (Sendai, Japan)
Tecnica	Tecnica (Lisbon, Portugal)
Tek. Tidskr.	Teknisk Tidskrift (Stockholm, Sweden)
Tek. Ukebl.	Teknisk Ukeblad (Oslo, Norway)
Tekh. Kibern.	Tekhnicheskaya Kibernetika (Moscow, USSR)
Tekh. Kino Telev.	Tekhnika Kino i Televideniya (Moscow, USSR)
Telecommun. Radio Eng. Pt 1	Telecommunications and Radio Engineering, Part 1 : Telecommunications (New York, USA). Translation of Elektrosvyaz
Telecommun. Radio Eng. Pt 2	Telecommunications and Radio Engineering, Part 2 : Radio Engineering (New York, USA). Translation of Radiotekhnika

Telecommunications	Telecommunications (Dedham, USA)
Telecomunicatii	Telecomunicatii (Bucarest, Rumania)
Telefunken-Rohre	Telefunken-Rohre (Berlin, Germany)
Telefunken Z.	Telefunken Zeitung (Berlin, Germany)
Telegr. Telef.	Telegraafen Telefoon (Hague, Netherlands)
Telektronikk	Telektronikk (Oslo, Norway)
Teleph. Eng. Management	Telephone Engineer and Management (Chicago, USA)
Teleteknik	Teleteknik (Aarhus, Denmark)
Tellus	Tellus (Stockholm, Sweden)
Teor. Eksp. Khim.	Teoreticheskaya i Eksperimental'naya Khimiya (Moscow, USSR)
Teor. Mat. Fiz.	Teoreticheskaya Matematicheskaya Fizika (Moscow, USSR)
Teor. Veroyatn. Primen.	Teoriya Veroyatnostei i Primeneniya (Moscow, USSR)
Teplofiz. Vys. Temp.	Teplofizika Vysokikh Temperatur (Moscow, USSR)
TESLA Electron.	TESLA Electronics (Prague, Czechoslovakia)
Theor. Exp. Chem.	Theoretical and Experimental Chemistry (New York, USA). Translation of Teor. Eksp. Khim.
Theor. Math. Phys.	Theoretical and Mathematical Physics (New York, USA). Translation of Teor. Mat. Phys.
Theory Probab. Appl.	Theory of Probability and Applications (Philadelphia, USA). Translation of Teor. Veroyatn. Primen.
Thin Solid Films	Thin Solid Films (Lausanne, Switzerland)
Throm. Diath. Haemorrh.	Thrombosis et Diathesis Haemorrhagica (Stuttgart, Germany)
Tijdschr. Ned. Elektron. Radiogenoot.	Tijdschrift van het Nederlands Elektronica en Radiogenootschap (Leidschendam, Netherlands)
Toim. Eesti NSV Tead. Akad. Fuus. Mat.	Eesti NSV Teaduste Akadeemia Toimetised Fuusika Matemaatika (Tallinn, Estonia)
Tokyo Astron. Bull.	Tokyo Astronomical Bulletin (Tokyo, Japan)
Tokyo Astron. Obs. Rep.	Tokyo Astronomical Observatory Report (Tokyo, Japan)
Toshiba Rev.	Toshiba Review (Kawasaki, Japan)
Toute Electron.	Tout l'Electronique (Paris, France)
Tr. Mosk. Obshchest. Ispyt. Prir.	Trudy Moskovskogo Obshchestva Ispytatelei Prirody (Moscow, USSR)
Trans. Am. Acad. Ophthal. Otol.	Transactions of the American Academy of Ophthalmology and Otology (St. Louis, USA)
Trans. Am. Geophy. Union	Transactions of the American Geophysical Union (Washington, USA)
Trans. ASME	Transactions of the American Society of Mechanical Engineers (New York, USA)
Trans. IEEE	Transactions of the Institute of Electrical and Electronic Engineers (New York, USA)
Trans. Inst. Electron. Commun. Eng. Jap.	Transactions of the Institute of Electronics and Communication Engineers of Japan (Tokyo, Japan)
Trans. S. Afr. IEE	Transactions of the South African Institute of Electrical Engineers (Johannesburg, South Africa)
Trans. Soc. Instrum. Control. Eng.	Transactions of the Society of Instrument and Control Engineers (Tokyo, Japan)
Ukr. Fiz. Zh.	Ukrayins'kyi Fizychnyi Zhurnal (Kiev, USSR)
Umsche. Wiss. Tech.	Umschau in Wissenschaft und Technik (Frankfurt, Germany)
Usp. Fiz. Nauk	Uspekhi Fizicheskii Nauk (Moscow, USSR)
Usp. Sovrem. Biol.	Uspekhi Sovremennoi Biologii (Moscow, USSR)
VDE Fachber.	VDE Fachberichte (Berlin, Germany)
Vestn. Leningr. Univ. Fiz. Khim.	Vestnik Leningradskogo Universiteta, Fizika i Khimiya (Leningrad, USSR)

Vestn. Mosk. Univ. Fiz. Astron.	Vestnik Moskovskogo Universiteta, Fizika, Astronomiya (Moscow, USSR)
Vestn. Mosk. Univ. Mat. Mekh.	Vestnik Moskovskogo Universiteta, Matematika, Mekhanika (Moscow, USSR)
Vestsi Akad. Navuk BSSR	Vestsi Akademii Navuk BSSR (Minsk, USSR)
Vojen. Zdrav. Listy	Vojenske Zdravotnicke Listy (Prague, Czechoslovakia)
Vop. Fiziol.	Voprosy Fiziologii (Kiev, USSR)
Vop. Kurortol.	Voprosy Kurortologii (Moscow, USSR)
Vop. Pitan.	Voprosy Pitaniya (Moscow, USSR)
Vrash. Pril. Deform. Zemli	Vrashchenie i Prilivnye Deformatsii Zemli (Kiev, USSR)
Vychisl Metody Program.	Vychislitel' nye Metody i Programmirovanie (Moscow, USSR)
Weed Sci.	Weed Science (Geneva, USA)
Weld. Prod.	Welding Production (Cambridge, England). Translation of Svar. Proizvod.
Werk. Betr.	Werkstatt und Betrieb (Munich, Germany)
Werkstattstechnik	Werkstattstechnik (Berlin, Germany)
Western Electr. Eng.	Western Electric Engineer (New York, USA)
Wireless Wld	Wireless World (London, England)
Wiss. Ber. AEG-Telefunken	Wissenschaftliche Berichte AEG-Telefunken (Berlin, Germany)
Wiss. Z. Friedrich-Schiller Univ. Jena	Wissenschaftliche Zeitschrift der Friedrich-Schiller Universitaet Jena (Jena, Germany)
Wiss. Z. Tech. Hochsch. Ilmenau	Wissenschaftliche Zeitschrift der Technischen Hochschule Ilmenau (Ilmenau, Germany)
Z. Angew. Math. Phys.	Zeitschrift fur Angewandte Mathematik und Physik (Basle, Switzerland)
Z. Angew. Phys.	Zeitschrift fur Angewandte Physik (Heidelberg, Germany)
Z. Astrophys.	Zeitschrift fur Astrophysik (Berlin, Germany)
Z. Elektr. Inf. Energietech.	Zeitschrift fur Elektrische Informatins und Energietechnik (Leipzig, Germany)
Z. Flugwiss.	Zeitschrift fur Flugwissenschaften (Brunswick, Germany)
Z. Geophys.	Zeitschrift fur Geophysik (Heidelberg, Germany)
Z. Naturforsch.	Zeitschrift fur Naturforschung (Oberkochen, Germany)
Z. InstrumKde	Zeitschrift fur Instrumentenkunde (Berlin, Germany)
Z. Phys.	Zeitschrift fur Physik (Berlin, Germany)
Z. Phys. Chem. (Frankfurt)	Zeitschrift fur Physikalische Chemie (Frankfurt, Germany)
Z. Tech. Phys.	Zeitschrift fur Technische Physik (Leipzig, Germany)
Z. Vermessungswes.	Zeitschrift fur Vermessungswesen (Stuttgart, Germany)
Z. Zellforsch.	Zeitschrift fur Zellforschung und Mikroskopische Anatomie (Berlin, Germany)
Zavod. Lab.	Zavodskaya Laboratoriya (Moscow, USSR)
Zeiss Mitt.	Zeiss Mitteilungen (Wurttemberg, Germany)
Zh. Eksp. Teor. Fiz.	Zhurnal Eksperimentalnoi i Teoreticheskoi Fiziki (Moscow, USSR)
Zh. Eksp. Teor. Fiz. Pis'ma	Zhurnal Eksperimentalnoi i Teoreticheskoi Fizika Pis'ma v Redaktsiyu (Moscow, USSR)
Zh. Nauchn. Pribl. Fotogr. Kinematogr.	Zhurnal Nauchnoi i Prikladnoi Fotografii i Kinematografii (Moscow, USSR)
Zh. Prikl. Mekh. Tekh. Fiz.	Zhurnal Prikladnoi Mekhaniki i Tekhnicheskoi Fiziki (Novosibirsk, USSR)
Zh. Prikl. Spekt.	Zhurnal Prikladnoi Spektroskopii (Minsk, USSR)
Zh. Tekh. Fiz.	Zhurnal Tekhnikeskoi Fiziki (Moscow, USSR)
Zh. Vychisl. Mat. Fiz.	Zhurnal Vychislitel'noi Matematiki i Matematicheskoi Fiziki (Moscow, USSR)

Part I

RADIATION AND MATTER

1. PROPERTIES OF MEDIA

1.1. Shah, M.A., Hasted, J.B., and Moore, L.: "Microwave Absorption by Water in Building Materials. Aerated Concrete", Br. J. Appl. Phys., 1965, 16, p.1747.

1.2. Hasty, T.E., and Wisseman, W.R.: "Effect of High Electric Fields on Propagation of 24-GHz Waves along the Surface of InSb", J. Appl. Phys., 1965, 36, p.3617.

1.3. Sinha, B., Roy, S.B., and Kastha, G.S.: "Absorption of 7.7-mm Waves by Solutions of Substituted Nitrobenzenes in Different Nonpolar Solvents", Indian J. Phys., 1965, 39, p.328.

1.4. Carr, E.F.: "Influence of Electric and Magnetic Fields on the Microwave Permittivity of a Liquid Crystal with Positive Anisotropy", J. Chem. Phys., 1965, 42, p.738.

1.5. Syamalamba, K., and Premaswarup, D.: "Dielectric Dispersion of Pure Liquids. Acrylates", Indian J. Pure Appl. Phys., 1965, 3, p.267.

1.6. Dykman, I.A.: "Microwave Conductivity of Semiconductors with Carriers Heated by a DC Field", Fiz. Tver. Tela, 1965, 7, p.414, and Sov. Phys.-Solid State, 1965, 7, p.332.

1.7. Deb, S., and Roy, S.K.: "Microwave Conductivity of a Thin Semiconductor Slab in the Presence of Surface Effects", J. Phys. Soc. Jap., 1965, 20, p.879.

1.8. Hamaguchi, C., Ishida, A., and Inuishi, Y.: "Microwave Attenuation by Supersonic Electrons in CdS", Proc. IEEE, 1965, 53, p.1259.

1.9. Sokoloski, M.M., and Fang, P.H.: "Dielectric Constant of Lead Salts", Phys. Lett., 1965, 16, p.222.

1.10. Bagguley, D.M.S., and Liesegang, J.: "Microwave Absorption in Rare-Earth Metals", Phys. Lett., 1965, 7, p.95.

1.11. Das, P.: "Microwave Hot-Electron Conduction in Many-Valley Semiconductors", Phys. Rev., 1965, 138, p.A590.

1.12. Sun, S.F.,: "Magnetoresistance of InSb-NiSb at Microwave Frequencies", Solid-State Electron., 1965, 8, p.344.

1.13. Wong, A.Y.: "Local Heating of a Cs Plasma by Microwave Irradiation at the Upper Hybrid Frequency", Appl. Phys. Lett., 1965, 6, p.147.

1.14. Kataoka, S., and Fujisada, H.: "Magneto-Resistance Effect in InSb at 34 GHz", Proc. IEEE, 1965, 53, p.178.

1.15. Szymczak, H., and Tharowski, J.: "Microwave Orientation of Ruby Crystals", Przegl. Elektron., 1965, 6, p.246.

1.16. Fante, R.L., and Mullin, C.R.: "Effective Field for Microwave Breakdown", Proc. IEEE, 1965, 53, p.484.

1.17. Fanelli, R., and Meissner, H.,: "Microwave Absorption in Normal-Conducting Films on Superconducting Substrates", Phys. Rev., 1966, 147, p.227.

1.18. Sheppard, A.P.: "Microwave Attenuation with Photoconductive Materials", Trans. IEEE, 1966, IM-15, p.44.

1.19. Bhattacharyya, J., Roy, S.B., and Kastha, G.S.: "Dielectric Absorption of 3-cm Waves in Polar Liquids", Indian J. Technol., 1966, 40, p.187.

1.20. Frenkel, L., and Woods, D.: "Microwave Absorption by H_2O Vapour and its Mixtures with other Gases at 100-300 GHz", Proc. IEEE, 1966, 54, p.498.

1.21. Felsenthal, P.: "Nanosecond-Pulse Microwave Breakdown in Air", J. Appl. Phys., 1966, 37, p.4557.

1.22. Yazytskii, B.Ya, and Poplavko, Yu. M.: "SHF Dispersion in $BaTiO_3$ Above the Curie Temperature", Fiz. Tver. Tela, 1966, 8, no.12, pp.3639-41, and Sov. Phys.-Solid State, 1967, 8, no.12, pp.2906-7.

1.23. Eru, I.I.: "Dielectric Properties of Crystals in the Millimetre Range", Zh. Tekh. Fiz., 1966, 36, p.1315, and Sov. Phys.-Tech. Phys., 1967, 11, no.7, pp.979-80.

1.24. Geilikman, B.T.: "Different Mechanisms for Superconductivity", Elektrotech. Cas., 1967, 18, p.563.

1.25. Mashkovich, M.D., and Smelyanskaya, E.N.: "Nature of Microwave Dielectric Losses in Glasses", Fiz. Tver. Tela, 1967, 9, no.4, pp.1249-50, and Sov. Phys.-Solid State, 1967.

1.26. Poszler, L.: "Microwave Losses of Polycrystalline Ferrites as a Function of the Crystal Structure", Hiradastechnika, 1967, 18, no.1, pp.15-9.

1.27. Dagg, I.R., Reesor, G.E., and Doble, R.P.: "Permittivity of Bromine at Microwave Frequencies", Can. J. Phys., 1967, 45, no.7, pp.2349-54.

1.28. Bagguley, D.M.S., and Liesegang, J.: "Microwave Absorption Phenomena in Rare-Earth Metals", Proc. Roy, Soc., 1967, 300A, no. 1463, pp.497-518.

1.29. Bennemann, K.H.: "Anomalous Micro-wave Absorption of Superconducting Alloys in a Static Magnetic Field", Phys. Lett., 1967, 24A, no.7, pp.357-8.

1.30. Ramasastry, C., and Ramaiah, C.V.: "Permittivity of $NaClO_3$ Crystals for Micro-waves", Phys. Status Solidi, 1967, 19, no.1, pp.K15-7.

1.31. Blinc, R., and Zeks, B.: "Pressure Dependence of the Ferroelectric Properties of KH_2PO_4 and KD_2PO_4", Helv. Phys. Acta, 1968, 41, p.700.

1.32. Tsandoulas, G.N.: "Doppler Effect in Plane-Wave Scattering by Moving Conducting Bodies", Proc. IEEE, 1968, 56, p.1749.

1.33. Potekhin, V.A., et al.: "Statistical Properties of an Arbitrary Partially Polarized Wave", Radiotekh. Elektron., 1968, 13, p.1443, and Radio Eng. Electron. Phys., 1968, 13.

1.34. Fleury, P.: "Relationships between Electricity and Optics", Rev. Franc. Electr., 1969, 42, p.17.

1.35. Briganti, E.: "Application of Principles of Relativity to Propagation of Electromagnetic Fields", Note Recens. Not., 1969, 18, no.2, p.256.

1.36. Hess, K., Nimtz, G., and Seeger, K.: "Nonohmic Microwave Conductivity in Semiconductor Posts", Solid-State Electron., 1969, 12, no.2, p.79.

1.37. D'Aiello, R.V., and Freedman, S.J.: "Microwave Conductivity of Granular Superconducting Aluminium Films", J. Appl. Phys., 1969, 40, p.2156.

1.38. Mayhan, J.T.: "Nonlinear Microwave Breakdown Systems Study", Trans. IEEE, 1969, AP-17, p.251.

1.39. Nishino, T., Okuyama, M., and Hamakawa, Y.: "Electro-Reflectance of p-GaAs", J. Phys. Chem. Solids, 1969, 30, p. 2671.

1.40. Afinogenov, V.M., et al.: "Photoconductivity of p-CdSb in the Millimetre Band", Zh. Eksp. Teor. Fiz. Pis'ma, 1969, 10, no.8, p.370, and JETP Lett., 1969, 10, no. 8, pp.234-7.

1.41. Braginski, A., and Buck, D.C.: "Polycrystalline Ferrite Films for Microwave Applications", Trans. IEEE, 1969, MAG-5, pp.484 and 924.

1.42. Auzel, F.: "Contribution to the Spectroscopic Study of Er^{3+}-Doped Glasses for Obtaining Laser Effect", Ann. Telecommun., 1969, 24, no.5-6, p.199.

1.43. Edelson, D., Jaeger, R.E., and Williams, J.C.: "Transient Effects of Nuclear Radiation on Dielectric Properties of Refractory Low-Loss Ceramics at Microwave Frequencies", J. Am. Ceram. Soc., 1969, 52, no.7, p.359.

1.44. Hosaya, M., and Nakamura, E.: "Dielectric Properties of SbSI at Microwave Frequencies", Jap. J. Appl. Phys., 1970, 9, no.5, p.552.

1.45. Luther, G., and Muser, H.E.: "Microwave Dispersion of Ferroelectrics near Transition Point", Z. Angew. Phys., 1970, 29, no.4, p.237.

1.46. Kell, R.C., et al.: "Novel Temperature-Stable Ceramics for Microwave Dielectric Resonators and Microstrip Substrates", Electron. Lett., 1970, 6, p.614.

1.47. Mavaddat, R.: "Millimetre-Wave Attenuation through Illuminated Semiconductor Panel", Trans. IEEE, 1970, 18, p.360.

1.48. Sonnenberg, H.: "$InAsP-Cs_2O$. High Efficiency IR Photocathode", Appl. Phys. Lett., 1970, 16, p.245.

1.49. Andrew, D., et al.: "GaAs-Cs-0 Transmission Photocathode", J. Phys. D, 1970, 3, p.320.

1.50. Gaver, R.L., and Seguin, H.J.: "High Quality Sputtered Multilayer Coatings for IR Laser Applications", Rev. Sci. Instrum., 1970, 41, p.427.

1.51. Loudon, R.: "Propagation of Electromagnetic Energy through an Absorbing Dielectric", Proc. Phys. Soc., 1970, 3, p.233.

1.52. Enderlein, R.: "Franz-Keldysh Effect and Hot-Electron Effects in the Interband Absorption of Semiconductors in an External Electric Field", Phys. Status Solidi, 1970, 41, no.1, pp.107-16.

1.53. Brescia, G., and Grossetti, E.: "Complex Dielectric Constant of Mixtures of Liquids at 5 GHz", RC Accad. Naz. Lincei, 1970, 48, no.6, pp.619-24.

1.54. Lau, B.W., Chung, J.S., and Rhee, J.R.: "Properties of Thin Metal Films at Microwave Frequencies", J. Korean Phys. Soc., 1970, 3, no.1, pp.15-20.

1.55. Rhee, J.R., Chung, J.S., and Lau, B.W.: "Power Reflectance of Thin Films at Microwave Frequencies", J. Korean Phys. Soc., 1970, 3, no.2, pp.61-2.

1.56. Grosser, P.: "Microwave Absorption of Polycrystalline Barium Ferrite between Room Temperature and Curie Point", Z. Angew. Phys., 1970, 30, no.1, pp.133-8.

1.57. Zohta, Y.: "Surface Resistance of Tin under Superconducting Condition", Bull. Electrotech. Lab., 1970, 34, no.4, pp.11-21.

1.58. Wu, S.M., Bridges, E., and Kao, K.C.: "Microwave Complex Permittivities of Si and GaAs in the Presence of High Steady Electric Fields", Int. J. Electron., 1971, 31, no.3, pp.233-41.

1.59. Glover, G.H.: "High-Field Microwave Permittivity of Electrons in Bulk GaAs", J. Appl. Phys., 1971, 42, no.13, pp.5590-5.

1.60. Bodi, A., Baican, R., and Barbur, I.: "Microwave Dielectric Properties of Gamma-Irradiated Ferroelectric $(NH_4)_2SO_4$", Acta Phys. Pol., 1971, 39A, no.1, pp.39-44.

1.61. Mercier, M.: "Property of Magnetic Materials. Magnetoelectric Effect", Rev. Gen. Electr., 1971, 18, no.2, pp.143-52.

1.62. Maki, K., and Sakurai, M.: "Microwave Response of Superconducting Thin Films in Magnetic Fields", J. Low Temp. Phys., 1971, 4, no.5, pp.515-23.

1.63. Strogryn, A.: "Equations for Calculating the Permittivity of Saline Water", Trans. IEEE, 1971, MTT-19, no.8, pp.733-6.

1.64. Bratescu, G.G., and Tudor, T.: "Modification of Spectral Structure of Light by Modulation", An. Univ. Bucur. Fiz., 1972, 21, pp.9-28.

1.65. Ollendorff, F.: "Reflection of an EM Wave by a Moving Mirror", Arch. Elektrotech., 1972, 54, no.5, pp.262-8.

1.66. Landstorfer, F., et al.: "Energy Flow in EM Wavefields", Nachr. Tech. Z., 1972, 25, no.5, pp.225-31.

1.67. Bussemer, P.: "Remarks on the Coherent States of the Radiation Field", Wiss. Z. Friedrich-Schiller Univ. Jena, 1972, 21, no.1, pp.13-7.

1.68. Asakura, T.: "Coherence and Fluctuation Properties of Optical Fields. II", J. Inst. Electron. Commun. Eng. Jap., 1972, 55, no.8, pp.1011-8.

1.69. Morokuma, T.: "Coherence (of Light)", J. Soc. Instrum. Control. Eng., 1972, 11, no.7, pp.624-34.

1.70. Laxpati, S.R.: "Closed-Form Solutions for Photocount Statistics of Superposed Coherent and Chaotic Radiation", J. Appl. Phys., 1972, 43, no.11, pp.4773-6.

1.71. Delisle, C., and Zardecki, A.: "Moments of the Sum of Photocounts in Gaussian Light", Phys. Rev., 1972, 6A, no.6, pp.2237-42.

1.72. Imbert, C.: "Photon's Inertial Spin Effect. Theory and Experimental Proof", Nouv. Rev. Opt. Appl., 1972, 3, no.4, pp.199-208.

1.73. Polukhin, A.T.: "Absorption of Microwave Radiation in a Polar Liquid", Radiotekh. Elektron., 1972, 16, no.9, pp.1691-7, and Radio Eng. Electron. Phys., 1972, 16, no.9, pp.1534-9.

1.74. Pancharatnam, S.: "Theory of Dispersion in Relation to Light Shifts", Proc. Roy. Soc., 1972, 330A, no.1581, pp.281-9.

1.75. Baldini, G., Cottini, M., and Grilli, E.: "UV Reflection and Absorption of KH_2PO_2 and $(NH_4)H_2PO_4$ Crystals", Solid-State Commun., 1972, 11, no.9, pp.1257-60.

1.76. Strukov, B.A., et al.: "Electrical and Thermal Properties of Mixed Crystal $K(HD)PO_4$", Fiz. Tver. Tela, 1972, 14, no.4, pp.1034-9, and Sov. Phys.-Solid State, 1972, 14, no.4.

1.77. Gerharz, R.: "Photon as a Fundamental Phenomenon", Int. J. Electron., 1972, 32, no.3, pp.335-45.

1.78. Anisimov, V.Ya., and Sotskii, B.A.: "Higher-Order Coherence of Electromagnetic Field", Opt. Spektr., 1972, 33, no.1, pp.172-5, and Opt. Spectrosc., 1972, 33, no.1, pp.94-5.

1.79. Abe, M., and Kaneda, S.: "Millimetre-Wave Frequency Response of Hot Electrons in n-GaAs", Electron. Commun. Jap., 1972, 55, no.6, pp.113-8.

1.80. Kumar, A., and Kothari, P.C.: "Microwave Conductivity and Permittivity in Ge, Si, and GaAs", Indian J. Pure Appl. Phys., 1972, 10, no.10, pp.740-1.

1.81. Hardy, J.R., and Agrawal, B.S.: "Determination of Origin of 10.6-micron Absorption in Laser Window Materials", Appl. Phys. Lett., 1973, 22, no.5, pp.236-7.

1.82. Anisimov, V.Ya., and Sotskii, B.A.: "Relation between Coherence and Entropy of Radiation", Theor. Math. Phys., 1973, 14, no.2, pp.211-3.

1.83. Davidson, F., and Amoss, J.: "Sequential Photon-Counting Statistics and Maximum-Likelihood Estimation Techniques for Gaussian Optical Fields", J. Opt. Soc. Am., 1973, 63, no.1, pp.30-7.

1.84. Hattori, T., et al.: "Indices of Refraction of ZnS, ZnSe, ZnTe, CdS, and CdTe, in the Far-IR", Opt. Commun., 1973, 7, no.3, pp.229-32.

1.85. Shaw, D.E., Hones, M.J., and Wunderlich, F.J.: "Measurements of Reflected and Transmitted Energies near the Critical Angle", Am. J. Phys., 1973, 41, no.4, pp.561-2.

1.86. Moles, M., and Vigier, J.P.: "Possible Physical Consequencies of the Existence of a Nonzero Photon Mass", C.R. Acad. Sci. (Paris), 1973, 276B, no.17, pp.697-700.

1.87. Schmutzer, E.: "Maxwell's Theory (in Media) and Quantum Mechanics in a Rotating Reference Frame", Ann. Phys. (Leipzig), 1973, 29, no.1, pp.75-95.

1.88. Keck, D.B., Maurer, R.D., and Schultz, P.C.: "Ultimate Lower Limit of Attenuation in Glass Optical Waveguides", Appl. Phys. Lett., 1973, 22, no.7, pp.307-9.

1.89. Thornber, K.K.: "Transverse Force on Light Refracted by Matter", Phys. Lett., 1973, 43A, no.6, pp.501-3.

1.90. Cole, K.D.: "Finite Rest Masses of Wave Quanta in Inhomogeneous Material Media", Aust. J. Phys., 1973, 26, no.3, pp.359-67.

1.91. Riccius, H.D., and Siemsen, K.J.: "Far-IR Optical Properties of Proustite", Opt. Commun., 1973, 8, no.3, pp.207-9.

1.92. Rich, T.C., and Pinnow, D.A.: "Optical Absorption in Fused Silica at 1060nm", Appl. Opt., 1973, 12, no.10, p.2234.

1.93. Armington, A.F., Posen, H., and Lipson, H.: "Strengthening of Halides for IR Windows", Electron. Mater., 1973, 2, no.1, pp.127-36.

1.94. Deryugin, I.A., and Kurashov, V.N.: "Coherence and Phase Relations in Quantum Optics", Izv. Akad. Nauk. SSSR, Ser. Fiz., 1973, 37, no.10, pp.2032-45.

1.95. Saleh, B.E.A.: "Photon Arrival. Time between Consecutive Photons and the Moment Generating Function", J. Phys. A, 1973, 6, no.12, pp.L161-4.

1.96. Saleh, B.E.A., and Minkowski, J.M.: "Spatial Properties of Quasi-Stationary Gaussian Optical Fields", J. Phys. A, 1973, 6, no.12, pp.L165-8.

1.97. Bishop, P.J., and Gibson, A.F.: "Absorption Coefficient of Ge at 10.6 micron", Appl. Opt., 1973, 12, no.11, pp.2549-50.

1.98. Bendow, B., and Gianino, P.D.: "Theoretical Lower Bound on the Absorption Coefficient of IR Transmitting Materials", Opt. Commun., 1973, 9, no.3, pp.306-10.

1.99. Belskii, A.M., and Khapalyruk, A.P.: "Reflection of a Laser Beam from the Interface of Isotropic Dielectrics", Opt. Spekt., 1973, 35, no.1, pp.117-9, and Opt. Spectrosc., 1973, 35, no.1, pp.67-80.

1.100. Agrawal, G.P.: "Higher-Order Angular Coherence Functions", Nuovo Cim., 1973, 18B, no.2, pp.265-76.

1.101. Muuss, H., and Scholz, V.: "Investigation of Coherence of Pulsed Argon-Ion Laser", Optik, 1973, 37, no.1, pp.26-30.

1.102. Bazhenov, S.V., Brazovskii, V.E., and Telegin, G.G.: "Statistical Phenomena in the Transient of a He-Ne Laser with Given Initial Photon Distribution", Opt. Spekt., 1973, 35, no.1, pp.106-16, and Opt. Spectrosc., 1973, 35, no.1, pp.62-6.

1.103. Vize, L., Pinter, F., and Gati, L.: "First-Order Coherence of Radiation of a Dye Laser", Acta Phys. Chem. Szeged., 1973, 19, no.4, pp.417-21.

1.104. Deryugin, I.A., Vishenskii, A.A., and Kurashov, V.N.: "Phase Variables in Quantum Optics", Kvantovaya Elektron., 1973, no.7, pp.152-67.

1.105. Deutsch, T.F.: "Absorption Coefficient of IR Laser Window Materials", J. Phys. Chem. Solids, 1973, 34, no.12, pp.2091-104.

1.106. Bendow, B.: "Optical Properties of IR Transmitting Materials", J. Electron. Mater., 1974, 3, no.1, pp.101-35.

1.107. Funke, J.: "N-th Order Correlation Function for Superposition of Coherent and Chaotic Fields", Czech. J. Phys., 1974, 24B, no.3, pp.245-8.

1.108. Perina, J., and Mista, L.: "Definitions of Coherence Time, Area, and Volume, for Superposed Coherent and Chaotic Radiation", Opt. Acta, 1974, 21, no.4, pp.329-40.

1.109. Dutriaux, L., and Huard, S.: "Force Exerted by a Fresnel Evanescent Wave on a Perfectly Absorbing Sphere", C.R. Acad. Sci. (Paris), 1974, 278B, no.14, pp.639-42.

1.110. Corti, M., and Degiorgio, V.: "Third-Order Intensity Correlations in Laser Light", Opt. Commun., 1974, 11, no.1, pp.1-4.

1.111. Antar, Y.M., and Boerner, W.M.: "Gaussian Beam Interaction with a Planar Dielectric Interface", Can. J. Phys., 1974, 52, no.11, pp.962-72.

1.112. Pescetti, D., and Piano, E.: "Interference between Polarized Radiation Beams", Atti Fond. Giorgio Ronchi, 1974, 29, no.1, pp.1-6.

1.113. Sparks, M., and Chow, H.C.: "High-Power 2- to 6-micron Window Material Figures of Merit with Edge Cooling and Surface Absorption Included", J. Appl. Phys., 1974, 45, no.4, pp.1510-7.

1.114. Duthler, C.J.: "Extrinsic Absorption in 10.6-micron Laser-Window Materials due to Molecular-Ion Impurities", J. Appl. Phys., 1974, 45, no.6, pp.2668-71.

1.115. Antar, Y.M., and Boerner, W.M.: "Generation of Complex Gaussian Beam Modes in Beam Interaction with a Planar Dielectric Interface", Trans. IEEE, 1974, AP-22, no.6, pp.837-9.

1.116. Mishina, H., and Asakura, T.: "Two-Gaussian Beam Interference", Nouv. Rev. Opt., 1974, 5, no.2, pp.101-7.

1.117. Kononenko, V.L.: "Submillimetre Absorption Spectra and Phase Transitions of Indirect Excitons in Ge", Trans. IEEE, 1974, MTT-22, no.12, pp.1086-9.

1.118. Szmolszhaja, T.I., et al.: "First Order Coherence of Radiation of a Pulsed Dye Laser", Acta Phys. Chem. Szeged., 1974, 20, no.3, pp.305-13.

2. LINES WITH PURELY TRANSVERSE FIELDS

2.1. Burton, R.W., and King, R.W.P.: "Experimental Investigation of a Two-Slot Transmission Line of Nonplanar Surfaces", Trans. IEEE, 1965, MTT-13, p.303.

2.2. Kovalev, I.S.: "Field Calculations for Symmetric Air-Filled Stripline under Static Conditions", Izv. VUZ Radiotekh., 1965, 8, p.78, and Sov. Radio Eng., 1965, 8, p.56.

2.3. Green, H.E., and Pyle, J.R.: "Characteristic Impedance and Velocity Ratio of Dielectric-Supported Stripline", Trans. IEEE, 1965, MTT-13, p.135.

2.4. Jutzi, W.: "Magnetic Field of Unsymmetric Striplines", Arch. Elektr. Ubertrag., 1965, 19, p.119.

2.5. Duncan, J.W.: "Characteristic Impedance of Multiconductor Striplines", Trans. IEEE, 1965, MTT-13, p.107.

2.6. Wheeler, H.A.: "Transmission-Line Properties of Parallel Strips Separated by Dielectric Sheet", Trans. IEEE, 1965, MTT-13, p.172.

2.7. Metcalf, W.S.: "Characteristic Impedance of Rectangular Transmission Lines", Proc. IEE, 1965, 112, p.2033.

2.8. Kartashev, V.G.: "Fundamental Wave Propagation in a Coaxial Line with Nonuniform Dielectric Filling", Radiotekh. Elektron., 1965, 10, p.1057, and Radio Eng. Electron. Phys., 1965, 10, p.904.

2.9. Seshagiri, N.: "Regular Polygon Coaxial Transmission Line", Proc. IEEE, 1965, 53, p.1549.

2.10. Krank, W.: "Influence of Production Tolerances on Characteristic Impedance of Shielded Striplines", Telefunken Z., 1966, 39, no.6, pp.215-22.

2.11. Sato, R.: "Theory of Parallel-Wire Line Covered with Multilayer Dielectric. II", Technol. Rep. Tohoku Univ., 1966, 31, no.1, pp.41-9.

2.12. Laura, P.A.: "Characteristic Impedance of Rectangular Transmission Lines", Proc. IEE, 1966, 113, no.10, pp.1595-6.

2.13. Howe, H.G.: "Dielectrically Loaded Stripline at 18 GHz", Microwave J., 1966, 9, no.1, p.52.

2.14. Ginzburg, V.M., and Yakovleva, E.A.: "Field in a Coaxial Line which Contains an Inhomogeneous Isotropic Dielectric", Radiotekh. Elektron., 1966, 11, no.5, pp.850-9, and Radio Eng. Electron. Phys., 1966, 11, no.5, pp.734-42.

2.15. Carson, C.T.: "Numerical Solution of TEM-Mode Transmission Lines with Curved Boundaries", Trans. IEEE, 1967, MTT-15, no.4, pp.269-70.

2.16. Okean, H.C.: "Properties of a TEM Transmission Line Used in Microwave Integrated Circuit Applications", Trans. IEEE, 1967, MTT-15, no.5, pp.327-8.

2.17. Brooke, R.L., Hoer, C.A., and Love, C.H.: "Inductance and Characteristic Impedance of a Strip Transmission Line", J. Res. Natl. Bur. Stand., 1967, 71C, no.1, pp.59-67.

2.18. Brenner, H.E.: "Numerical Solution of TEM-Line Problems Involving Inhomogeneous Media", Trans. IEEE, 1967, MTT-15, no.8, pp.485-7.

2.19. Lesik, I.I.: "Wave Impedance of an Asymmetrical Stripline with Infinitely Thin Conductor", Radiotekh. Elektron., 1967, 12, no.10, pp.1817-9, and Radio Eng. Elektron. Phys., 1967, 12, no.10.

2.20. Kammler, D.W.: "Calculation of Characteristic Admittances and Coupling Coefficients for Striplines", Trans. IEEE, 1968, MTT-16, p.925.

2.21. Yamashita, E., and Yamazaki, S.: "Parallel-Strip Line Embedded in or Printed on a Dielectric Sheet", Trans. IEEE, 1968, MTT-16, p.972.

2.22. Graf, H., and Krank, W.: "Calculation of Properties of Striplines of Various Cross-Sectional Forms", Frequenz, 1968, 22, p.237.

2.23. Groll, H., and Koffler, H.: "Triplate Line with Two Inner Conductors of Various Dimensions", Nachr. Tech. Z., 1968, 21, p.384.

2.24. Earle, M.A., and Benedek, P.: "Characteristic Impedance of Dielectric-Supported Stripline", Trans. IEEE, 1968, MTT-16, p.884.

2.25. Lesik, I.I., Sedykh, V.M., and Kondratyev, B.V.: "Coupled Asymmetric Striplines with Infinitely Thin Boards", Izv. VUZ Radioelektron., 1968, 11, p.1027.

2.26. Hornsby, J.S., and Gopinath, A.: "Fourier Analysis of a Dielectric-Loaded Waveguide with a Microstrip Line", Electron. Lett., 1969, 5, p.265.

2.27. Hellman, M.E., and Palocz, I.: "Effect of Neighbouring Conductors and Fields in Plane Parallel Transmission Lines", Trans. IEEE, 1969, MTT-17, p.254.

2.28. Hilberg, W.: "Exact Relations for Characteristic Impedances", Trans. IEEE, 1969, MTT-17, p.229.

2.29. Bolle, D.M.: "Microstrip Modal Spectrum", Trans. IEEE, 1969, MAG-5, p.483.

2.30. Hamasaki, J.: "Microwave Stripline. I and II", J. Inst. Electron. Commun. Eng. Jap., 1969, 52, no.11, pp.1350 and 1426.

2.31. Yamashita, E.: "Design Methods for Striplines", J. Inst. Electron. Commun. Eng. Jap., 1969, 52, no.11, p.1426.

2.32. Gunston, M.A.R.: "Transmission Characteristics of Microstrip", Marconi Rev., 1969, 32, no.174, p.226.

2.33. Yamashita, E., and Atsuki, K.: "Design of Transmission-Line Dimensions for a Given Characteristic Impedance", Trans. IEEE, 1969, MTT-17, p.638.

2.34. Gupta, R.R.: "Accurate Impedance Determination of Coupled TEM Conductors", Trans. IEEE, 1969, MTT-17, p.479.

2.35. Schneider, M.V.: "Dielectric Loss in Integrated Microwave Circuits", Bell Syst. Tech. J., 1969, 48, p.2325.

2.36. Weiss, J.A.: "Parameters of Microstrip", Electron. Lett., 1969, 5, p.517.

2.37. Gunston, M.A.R., and Weale, J.R.: "Variation of Microstrip Impedance with Strip Thickness", Electron. Lett., 1969, 5, p.697.

2.38. Hill, Y.M., Reckord, N.O., and Winner, D.R.: "Method for Impedance and Coupling Characteristics of Microstrip and Triplate Configurations", IBM J. Res. Dev., 1969, 13, p.314.

2.39. Smith, R.M.: "Microwave Striplines", Des. Electron., 1969, 6, no.7, p.44.

2.40. Kondrat'ev, B.V., et al.: "Asymmetric Stripline with Plate of Arbitrary Thickness", Radiotekh. Elektron., 1969, 14, p.524, and Radio Eng. Electron. Phys., 1969, 14.

2.41. Cohn, S.B.: "Slot Line on a Dielectric Substrate", Trans. IEEE, 1969, MTT-17, p.768.

2.42. Hornsby, J.S., and Gopinath, A.: "Numerical Analysis of a Dielectric-Loaded Waveguide with Microstrip Line. I", Trans. IEEE, 1969, MTT-17, p.684.

2.43. Eaves, R.E., and Bolle, D.M.: "Guided Waves in Parallel-Plate Structures", Trans. IEEE, 1970, MTT-18, p.66.

2.44. Farrar, A., and Adams, A.T.: "Characteristic Impedance of Microstrip by Method of Moments", Trans. IEEE, 1970, MTT-18, p.65.

2.45. Gish, D.L., and Graham, O.: "Characteristic Impedance and Phase Velocity of a Dielectric-Supported Air Stripline with Sidewalls", Trans. IEEE, 1970, MTT-18, p.131.

2.46. Mittra, R., and Itoh, I.: "Charge and Potential Distributions in Shielded Striplines", Trans. IEEE, 1970, MTT-18, p.149.

2.47. Kuny, W.: "Calculation of Distributed Capacitance and Inductance of Planar Striplines by Conformal Mapping", Frequenz, 1970, 24, no.4, p.113.

2.48. Weiss, J.A., and Bryant, T.G.: "Dielectric Green's Function for Parameters of Microstrip", Electron. Lett., 1970, 6, p.462.

2.49. Delogne, P.: "On Wu's Theory of Microstrip", Electron. Lett., 1970, 6, p.541.

2.50. Hatsuda, T., and Shimada, S.: "Calculation of Stripline Characteristics", J. Inst. Electr. Commun. Eng. Jap., 1970, 53, no.12, pp.1720-4.

2.51. Atsuki, K., and Yamashita, E.: "Analytical Method for Transmission Lines with Thick-Strip Conductor, Multidielectric Layers, and Shielding Conductor", Electron. Commun. Jap., 1970, 53, no.6, pp.85-90.

2.52. Andresciani, V., and Pesamosca, G.: "Calculation of Characteristic Impedance and Phase Velocity of Striplines", RC Riun. Ass. Elettrotec. Ital., 1970, 45, no.202, pp.1-13.

2.53. Ivanov, S.A., and Lyapunov, N.V.: "Calculation of Parameters of Symmetrical Stripline by Magnetic Vector Potential Method", Annu. Univ. Sofia Fac. Phys., 1970-2, 64-5, pp.257-69.

2.54. Eaves, R.E., and Bolle, D.M.: "Guided Waves in Limit Cases of Microstrip", Trans. IEEE, 1970, MTT-18, p.231.

2.55. Yamashita, E., and Atsuki, K.: "Stripline with Rectangular Outer Conductor and Three Dielectric Layers", Trans. IEEE, 1970, MTT-18, p.238.

2.56. Gelder, D.: "Numerical Determination of Microstrip Properties Using Transverse Field Components", Proc. IEE, 1970, 117, p.699.

2.57. Davies, J.B., and Corr, D.G.: "Computer Analysis of Fundamental and Higher-Order Modes in Single and Coupled Microstrip", Electron. Lett., 1970, 6, no.25, p.806.

2.58. Grunberger, G.K., Keine, V., and Meinke, H.H.: "Longitudinal Field Components and Frequency-Dependent Phase Velocity in Microstrip", Electron. Lett., 1970, 6, no.21, pp.683-5.

2.59. Daly, P.: "Hybrid-Mode Analysis of Microstrip by Finite-Element Methods", Trans. IEEE, 1971, MTT-19, no.1, pp.19-25.

2.60. Denlinger, E.J.: "Frequency-Dependent Solution for Microstrip", Trans. IEEE, 1971, MTT-19, no.1, pp.30-9.

2.61. Mittra, R., and Itoh, T.: "Technique for Analysis of Dispersion Characteristics of Microstrip Lines", Trans. IEEE, 1971, MTT-19, no.1, pp.47-56.

2.62. Yamashita, E., and Atsuki, K.: "Analysis of Thick-Strip Transmission Lines", Trans. IEEE, 1971, MTT-19, no.1, pp.120-2.

2.63. Bryant, T.G., and Weiss, J.A.: "Programme for Microstrip Parameters", Trans. IEEE, 1971, MTT-19, no.4, pp.418-9.

2.64. Schmitt, H.J., and Sarges, K.H.: "Wave Propagation in Microstrip", Nachr. Tech. Z., 1971, 24, no.5, pp.260-4.

2.65. Kowalski, G., and Pregla, R.: "Dispersion Characteristics of Shielded Microstrip with Finite Thickness", Arch. Elektr. Ubertrag., 1971, 25, no.4, pp.193-6.

2.66. Jain, O.P., and Makios, V.: "Coupled-Mode Model of Dispersion in Microstrip", Electron. Lett., 1971, 7, no.14, pp.405-7.

2.67. Balint, L.: "Computer Simulation of Two- and Three-Conductor Transmission Lines", Hiradastechnika, 1971, 22, no.5, pp.135-46.

2.68. Itoh, T., and Mittra, R.: "Dispersion Characteristics of Slot Lines", Electron. Lett., 1971, 7, no.13, pp.364-5.

2.69. Vorob'ev, P.A., Matyutin, N.D., and Solomonik, I.Sh.: "Capacitance of Unsymmetric Zigzag Screened Stripline", Izv. VUZ Radioelektron., 1971,14,no.5,pp.489-93.

2.70. Cohn, S.B.: "Sandwich Slot Line", Trans. IEEE, 1971, MTT-19, pp.773-4.

2.71. Chudobiak, W.J.: "Dispersion in Microstrip", Trans. IEEE, 1971, MTT-19, pp.783-4.

2.72. Grunberger, G.K., and Meinke, H.H.: "Proof of Longitudinal Field Components in the Dominant Mode of Microstrip", Nachr. Tech. Z., 1971, 24, no.7, pp.364-8.

2.73. Foster, K., and Anderson, R.: "Characteristic Impedance of TEM Lines by Variational Methods", Proc. IEE, 1971, 118, no.8, pp.280-2.

2.74. Belousova, L.I., and Sal'nikova, L.P.: "Higher Modes in a Shielded Nonsymmetrical Stripline", Radiotekh. Elektron., 1971, 16, no.7, pp.1262-6, and Radio Eng. Electron. Phys., 1971, 16, no.7,pp.1199-202.

2.75. Avdeev, Ye.V., and Potapova, V.I.: "Determination of Parameters of Open Strip Waveguides", Radiotekhnika, 1971, 26, no.8, pp.56-64, and Telecommun. Radio Eng. Pt 2, 1971, 26, no.8, pp.98-103.

2.76. Belousova, L.I., and Sal'nikova, L.P.: "Cutoff Frequencies of Higher Modes in a Screened Stripline", Radiotekh. (Kharkov), 1971, no.17, pp.72-6.

2.77. Hatsuda, T., and Kimura, T.: "Computation of Stripline Characteristics by Relaxation Method", Electron. Commun. Jap., 1971, 54, no.9, pp.48-56.

2.78. Hatsuda, T.: "Computation of Coplanar Stripline Characteristics by Relaxation Method", Electron. Commun. Jap., 1971, 54, no.11, pp.76-81.

2.79. Iwakura, H., and Arakawa, T.: "Analysis of Asymmetrical Stripline", Rep. Univ. Electro-Commun., 1971, 22, no.1, pp.1-10.

2.80. Minor, J.C., and Bolle, D.M.: "Propagation in Shielded Microslot on Ferrite Substrate", Electron. Lett., 1971, 7, no.17, pp.502-4.

2.81. Horton, R., Easter, B., and Gopinath, A.: "Variation of Microstrip Losses with Thickness of Strip", Electron. Lett., 1971, 7, no.17, pp.490-1.

2.82. Hasegawa, H., Furukawa, M., and Yanai, H.: "Properties of Microstrip Line on Si-SiO$_2$ System", Trans. IEEE, 1971, MTT-19, no.11, pp.869-81.

2.83. Horton, R., and Easter B.: "Impedance and Guide Wavelength on Arbitrary Ferromagnetic Substrate", Electron. Lett., 1971, 7, no.27, pp.642-3.

2.84. Okoshi, T., and Miyoshi, T.: "Analysis of a Planar Circuit. I and II", Annu. Rep. Eng. Res. Inst. Univ. Tokyo, 1971, 30, pp.153-63 and 165-70.

2.85. Atechian, J., and Crampagne, R.: "Determination of Charge Density on the Central Conductor of a Stripline", C.R. Acad. Sci. (Paris), 1972, 274B, no.10, pp.677-9.

2.86. Foster, K., and Anderson, R.: "Characteristic Impedance of TEM Lines by Variational Methods", Int. J. Electron., 1972, 32, no.6, pp.715-7.

2.87. Cohn, S.B.: "Slot-Line Field Components", Trans. IEEE, 1972, MTT-20, no.2, pp.172-4.

2.88. Schneider, M.V.: "Microstrip Dispersion", Proc. IEEE, 1972, 60, no.1, pp.144-6.

2.89. Ivashka, V.P., and Shugurov, V.K.: "Boundary Effects in Symmetric Microstrip Line", Litov. Fiz. Sb., 1972, 12, no.5, pp.781-8.

2.90. Gal'chenko, N.A., and Mikhalevskii, V.S.: "Use of Schwartz' Method for Computing Electrical Characteristics of Striplines", Radiotekh. Elektron., 1972, 17, no.2, pp.240-6, and Radio Eng. Electron. Phys., 1972, 17, no.2, pp.183-8.

2.91. Iwakura, H., and Arakawa, T.: "Fringing Capacitances of Shielded Rectangular Bar", Electron. Commun. Jap., 1972, 55, no.4, pp.51-8.

2.92. Wiesbeck, W.: "Calculation of Attenuation of Microstrips", Wiss. Ber. AEG-Telefunken, 1972, 45, no.4, pp.162-6.

2.93. Hatsuda, T.: "Computation of Characteristics of Coplanar-Type Striplines by Relaxation Method", Electr. Commun. Lab. Tech. J., 1972, 21, no.1, pp.15-26, and Trans. IEEE, 1972, MTT-20, no.6, pp.413-6.

2.94. Benedek, P., and Silvester, P.: "Capacitance of Parallel Rectangular Plates Separated by a Dielectric Sheet", Trans. IEEE, 1972, MTT-20, no.8, pp.504-10.

2.95. Pal, S., and Babu, G.R.: "Characteristic Impedance of Microstrip", J. AEU, 1972, 5, no.1, pp.21-7.

2.96. Wharton, R.P., and Rodrigue, G.P.: "Dominant-Mode Analysis of Microstrip", Trans. IEEE, 1972, MTT-20, no.8, pp.552-5.

2.97. Riblet, H.J.: "Exact Dimensions of a Family of Rectangular Coaxial Lines with Given Impedance", Trans. IEEE, 1972, MTT-20, no.8, pp.538-41.

2.98. Shimasaki, M., and Kiyono, T.: "Analysis of Microstrip by Integral-Equation Approach", Electron. Commun. Jap., 1972, 54, no.2, pp.80-8.

2.99. Hatsuda, T., and Kimura, T.: "Computation of Stripline Characteristics by Relaxation Method", Rev. Electr. Commun. Lab., 1972, 20, no.1-2, pp.36-46.

2.100. Malyutin, N.D., and Vorob'ev, P.A.: "Linear Capacitance and Characteristic Impedance of Planar Striplines", Izv. VUZ Radioelektron., 1972, 15, no.5, pp.662-3.

2.101. Jain, O.P., Makios, V., and Chudobiak, W.J.: "Open-End and Edge Effect in Microstrip", Trans. IEEE, 1972, MTT-20, no.9, pp.626-30.

2.102. Okugawa, S.: "Calculation of Characteristics of Embedded Coupled Microstrip Lines with Finite Thickness", Electron. Lett., 1972, 8, no.20, pp.494-5.

2.103. Essayag, G., and Sauve, B.: "Effects of Geometrical Parameters of a Microstrip on its Dispersive Properties", Electron. Lett., 1972, 8, no.21, pp.529-30.

2.104. Itoh, T., Mittra, R., and Ward, R.D.: "Method for Computing Edge Capacitance of Finite and Semi-Infinite Microstrip", Trans. IEEE, 1972, MTT-20, no.12, pp.847-9.

2.105. Essayag, G., and Sauve, B.:"Study of Higher-Order Modes in a Microstrip Structure", Electron. Lett., 1972, 8, no.23, pp.564-6.

2.106. Dudley, D.G., and Quintenz, J.P.: "Multimodal Transient Excitation Effects in an Infinite, Parallel-Plate, Waveguide", Can. J. Phys., 1972, 50, no.22, pp.2826-35.

2.107. Getsinger, W.J.: "Microstrip Dispersion Model", Trans. IEEE, 1973, MTT-21, no.1, pp.34-9.

2.108. Costamagna, E.: "Fast Parameters Calculation of Dielectric-Supported Air-Strip Transmission Line", Trans. IEEE, 1973, MTT-21, no.3, pp.155-6.

2.109. Deryugin, L.N., Kurdyumov, O.A., and Sotin, V.Ye.: "Fundamental and Parasitic Waves in Microstrip", Izv. VUZ Radiofiz., 1973, 16, no.1, pp.118-28.

2.110. Arndt, F.: "Application of Similarity Transformation to Microstrip", Nachr. Tech. Z., 1973, 26, no.2, pp.46-9.

2.111. Lagerlof, R.O.E.: "Ridged Waveguide for Planar Microwave Circuits", Trans. IEEE, 1973, MTT-21, no.7, pp.499-501.

2.112. Kuzora, I., and Bajorek, J.: "Analogue Method to Calculate Capacitance of Microstrip", Rozpr. Electrotech., 1973, 19, no.1, pp.165-77.

2.113. Riblet, H.J.: "Limiting Value of Interaction between Symmetrical Fringing Capacitances", Trans. IEEE, 1973, MTT-21, no.10, pp.644-7.

2.114. Van de Capelle, A.R., and Paert, P.J.L.: "Fundamental and Higher-Order Modes in Open Microstrip", Electron. Lett., 1973, 9, no.15, pp.345-6.

2.115. Farrar, A., and Adams, A.T.: "Potential-Theory Method for Covered Microstrip", Trans. IEEE, 1973, MTT-21, no.7, pp.494-6.

2.116. Itoh, T., and Mittra, R.: "Spectral-Domain Approach for Calculating Dispersion Characteristics of Microstrip Lines", Trans. IEEE, 1973, MTT-21, no.7, pp.496-9.

2.117. Golubev, V.I., and Kovalev, I.S.: "Attenuation in a Step Waveguide with Central Conductor of Two Equipotential Strips", Radiotekhnika, 1973, 28, no.7, pp.46-51, and Telecommun. Radio Eng. Pt 2, 1973, 28, no.7, pp.90-3.

2.118. Avdeev, Ye.V., and Potapova, V.I.: "Nomograms for Coupled Open Microstrip Lines", Radiotekhnika, 1973, 28, no.9, pp.54-9, and Telecommun. Radio Eng. Pt 2, 1973, 28, no.9, pp.89-93.

2.119. Yashin, A.A.: "Capacitance of a Coplanar Microstrip Line Allowing for Effect of a Screening Plane", Radiotekhnika, 1973, 28, no.11, pp.85-6, and Telecommun. Radio Eng. Pt 2, 1973, 28, no.11, pp.114-5.

2.120. Kitasawa, T., et al.: "Slot Line with Thick Metal Coating", Trans. IEEE, 1973, MTT-21, no.9, pp.580-2.

2.121. Carlin, H.J.: "Simplified Circuit Model for Microstrip", Trans. IEEE, 1973, MTT-21, no.9, pp.589-91.

2.122. Atechian, J., and Crampagne, R.: "Studies for Microstrip Lines", Electron. Fis. Apl., 1973, 16, no.2, pp.131-5.

2.123. Costamagna, E.: "Observations on Calculation of (Stripline) Fringing Capacitance", Alta Freq., 1973, 42, no.7, pp.332-4.

2.124. Dupuis, P.A., and Campbell, C.K.: "Characteristic Impedance of Surface-Strip Coplanar Waveguides", Electron. Lett., 1973, 9, no.16, pp.354-5.

2.125. Crampagne, R., Guirand, J.L., and de Robert, A.M.: "Generalization of Method of Images for Microstrip and Microaperture Structures", C. R. Acad. Sci. (Paris), 1973, 277B, no.20, pp.591-4.

2.126. Akhunlar, A., and Idemen, M.: "Calculation of the Characteristic Impedance of Stripline with Elliptical Outer Conductor", Arch. Elektrotech., 1973, 55, no.6, pp.310-1.

2.127. Ivashka, V.P., and Shugurov, V.K.: "Symmetric Stripline with any Number of Central Strips", Litov. Fiz. Sb., 1973, 13, no.5, pp.709-22.

2.128. Zamansky, P.: "Study of Parallel-Plate Microwave Lines and their Discontinuities", Rev. Tech. Thomson-CSF, 1973, 5, no.3, pp.505-39.

2.129. Hofmann, H.: "Relative Convergence in Mode-Matching Solutions of Microstrip Problems", Electron. Lett., 1974, 10, no.8, pp.126-7.

2.130. John, S., and Arlett, P.: "Simple Method for Calculation of Characteristic Impedance of Microstrip", Electron. Lett., 1974, 10, no.10, pp.188-90.

2.131. Albrey, I.J., and Gunn, M.W.: "Reduction of Attenuation Coefficient of Microstrip", Trans. IEEE, 1974, MTT-22, no.7, pp.739-42.

2.132. Farrar, A., and Adams, A.T.: "Multilayer Microstrip Lines", Trans. IEEE, 1974, MTT-22, no.10, pp.889-91.

2.133. Itoh, T., and Mittra, R.: "Technique for Computing Dispersion Characteristics of Shielded Microstrip Lines", Trans. IEEE, 1974, MTT-22, no.10, pp.896-8.

2.134. Bianco, B., et al.: "Frequency Dependence of Microstrip Parameters", Alta Freq., 1974, 43, no.7, pp.413-6.

2.135. Linner, L.J.P.: "Method for Computation of Characteristic Immittance Matrix of Multiconductor Striplines with Arbitrary Widths", Trans. IEEE, 1974, MTT-22, no.11, pp.930-7.

2.136. Riblet, H.J.: "Determination of Excess (Stripline) Capacitance", Trans. IEEE, 1974, MTT-22, no.4, p.467.

2.137. Ivashka, V.P., and Shugurov, V.K.: "Microstrip Line with Conductors of Any Section Shape", Litov. Fiz. Sb., 1974, 14, no.3, pp.507-15.

2.138. Ivashka, V.P., and Shugurov, V.K.: "Microstrip Line with Inhomogeneous Dielectric Filling in TEM Approximation", Litov. Fiz. Sb., 1974, 14, no.3, pp.517-23.

2.139. Zolotov, E.M., Kiselev, V.A., and Sychugov, V.A.: "Optical Phenomena in Thin-Film Waveguide", Usp. Fiz. Nauk, 1974, 112, no.1-2, pp.231-73, and Sov. Phys.-Usp., 1974, 17, no.1, pp.64-84.

3. HOLLOW CONDUCTING TUBES

3.1. Yee, H.Y., and Audeh, N.F.: "Cutoff Frequencies of Waveguides with Arbitrary Cross Sections", Proc. IEEE, 1965, 53, p.637.

3.2. Zelby, L.W.: "Unified Approach to Theory of Guided Waves", Int. J. Electr. Eng. Educ., 1965, 3, no.1, p.39.

3.3. Yashkin, A.Ya., Golubev, A.N., and Kalashnikov, V.G.: "Calculation of Transmission Band of Waveguides with Stepped Cross Section", Radiotekh. Elektron., 1965, 10, p.1038, and Radio Eng. Electron. Phys., 1965, 10, p.888.

3.4. Chatterjee, S.K., and Chatterjee, R.: "Dielectric-Loaded Waveguides. Review of Theoretical Solutions. I-III", Radio Electron. Eng., 1965, 30, pp.145,195,and 353.

3.5. Hao-Ming, S.: "Solution of Field Equations in Regular Waveguide", Acta Electron. Sin., 1965, no.1, p.85.

3.6. Yeh, C., Casey, K.F., and Kaprielian, Z.A.: "TM-Wave Propagation in Sinusoidally Stratified Dielectric Media", Trans. IEEE, 1965, MTT-13, p.297.

3.7. Yamanada, R., and Watanabe, K.: "Propagation in Cylindrical Waveguide Containing Inhomogeneous Dielectric", Trans. IEEE, 1965, MTT-13, p.716.

3.8. Smorgonskii, V.Ya., and Kovshov, A.I.: "Transmittance of Elliptical Waveguide for H_{11} Modes", Radiotekh. Elektron., 1965, 10, p.945, and Radio Eng. Electron. Phys., 1965, 10, p.804.

3.9. Yee, H.Y., and Audeh, N.F.: "Uniform Waveguides with Arbitrary Cross Section Considered by the Point-Matching Method", Trans. IEEE, 1965, MTT-13, p.847.

3.10. Laura, P.A.: "Determination of Cutoff Frequencies of Waveguides with Arbitrary Cross Sections by Point Matching", Proc. IEEE, 1965, 53, p.1660.

3.11. Prache, P.M.: "Propagation Parameters of a Rectangular Waveguide Containing a Dielectric Plate Parallel with One of the Walls", Cables Transm., 1966, 20, p.11.

3.12. Pyle, J.R.: "Cutoff Wavelength of TE_{10} Mode in Ridged Guide of any Aspect Ratio", Trans. IEEE, 1966, MTT-14, p.175.

3.13. Davies, J.B., and Muilwyk, C.A.: "Numerical Solution of Uniform Hollow Waveguides with Boundaries of Arbitrary Shape", Proc. IEE, 1966, 113, p.277.

3.14. Yee, H.Y., and Audeh, N.F.: "Attenuation Coefficient of Waveguide with General Cross Sections", Trans. IEEE, 1966, MTT-14, p.252.

3.15. Brown, J.: "Electromagnetic Momentum Associated with Waveguide Modes", Proc. IEE, 1966, 113, p.27.

3.16. Yee, H.Y.: "Cutoff Frequencies of Waveguides with Arbitrary Cross Section", Proc. IEEE, 1966, 54, p.64.

3.17. Laura, P.A.: "Conformal Mapping and Determination of Cutoff Frequencies of Waveguides with Arbitrary Cross Section", Proc. IEEE, 1966, 54, p.1078.

3.18. Yee, H.Y., and Audeh, N.F.: "Cutoff Frequencies of Eccentric Waveguides", Trans. IEEE, 1966, MTT-14, p.487.

· 3.19. Laura, P.A.: "Simple Method for Determination of Cutoff Frequencies of Waveguides with Arbitrary Cross Sections", Proc. IEEE, 1966, 54, p.1495.

3.20. Upstain, S.T., and Audeh, N.F.: "Transverse Resonance Solutions of Uniform Trapezoidal Waveguides", Trans. IEEE, 1966, MTT-14, p.158.

3.21. Seckelmann, R.: "Propagation of TE Modes in Dielectric-Loaded Waveguides", Trans. IEEE, 1966, MTT-14, p.518.

3.22. Smorgonskii, V.Ya, and Kovshov, A.T.: "Cutoff Conditions in a Circular Waveguide with Dielectric Insert", Radiotekh. Elektron., 1966, 11, p.752, and Radio Eng. Electron. Phys., 1966, 11, p.647.

3.23. Lyapunov, N.V., Dmitreev, V.M., and Sedykh, V.M.: "Calculation of Cutoff Frequencies of Double- and Single-Ridge Waveguides", Radiotekh. Elektron., 1966, 11, p.345, and Radio Eng. Electron. Phys., 1966, 11, p.285.

3.24. Kikina, N.G., and Merkulov, V.V.: "Use of Airy's Function for Design of Waveguides of Complicated Shape", Radiotekh. Elektron., 1966, 11, p.1493, and Radio Eng. Electron. Phys., 1966, 11, p.1295.

3.25. Sam, E.Ah.: "Propagation in Cylindrical Waveguide Containing Inhomogeneous Dielectric", Trans. IEEE, 1967, MTT-15, p.60.

3.26. Laura, P.A., and Faulstich, A.J.: "Cutoff Frequencies of Uniform Waveguides of Regular Polygonal Cross Section", Proc. IEEE, 1967, 55, no.3, pp.410-1.

3.27. Eberhardt, N.: "Propagation in the Off-Centre E-Plane Dielectrically Loaded Waveguide", Trans. IEEE, 1967, MTT-15, no.5, pp.282-9.

3.28. Davies, J.B.: "Propagation in Rectangular Waveguide Filled with Skew Uniaxial Dielectric", Trans. IEEE, 1967, MTT-15, no.6, pp.372-6.

3.29. Golubev, A.N.: "Calculation of Field in Stepped Waveguides", Radiotekh. Elektron., 1967, 12, no.2, pp.350-2, and Radio Eng. Electron. Phys., 1967, 12, no.2.

3.30. Ricard, J.: "Maxwell Tensor of a Guided Wave in a Tube of Square Section", C. R. Acad. Sci. (Paris), 1967, 264B, no.13, pp.959-62.

3.31. Ricard, J.: "Propagation of Energy and Poynting's Vector for a Square Waveguide", C. R. Acad. Sci. (Paris), 1967, 264B, no.14, pp.995-8.

3.32. Casarella, M.J., and Laura, P.A.: "Determination of Cutoff Frequencies of Grooved Waveguides", Proc. IEEE, 1967, 55, no.6, pp.1096-7.

3.33. Gruner, L.: "Higher-Order Modes in Rectangular Coaxial Waveguides", Trans. IEEE, 1967, MTT-15, no.8, pp.483-5.

3.34. Waldron, R.A.: "Characteristic Impedances of Waveguides", Marconi Rev., 1967, 30, pp.125-36.

3.35. Fujioka, H., Nihei, F., and Kumagai, N.: "Propagation Characteristics of Waveguides Filled with a Moving Medium", J. Inst. Electr. Commun. Eng. Jap., 1967, 50, no.5, pp.920-1.

3.36. Jiang, C.L.: "Transformation Theory of Guided Waves in Moving Media", J. Appl. Phys., 1967, 38, no.9, pp.3692-6.

3.37. Nagelberg, E.R.: "Phase Progression in Conical Waveguides", Bell Syst. Tech. J., 1967, 46, no.10, pp.2453-66.

3.38. Yashkin, A.Ya.: "Calculation Data of the Lower Cutoff Frequencies for Stepped Waveguides", Radiotekhnika, 1967, 22, no.5, pp.103-5, and Telecommun. Radio Eng. Pt 2, 1967, 22, no.5, pp.144-6.

3.39. Baier, W.: "Modes in Waveguides Consisting of Conductors of Rectangular Cross Section", Arch. Elektr. Ubertrag., 1968, 22, no.4, pp.179-85.

3.40. Hord, W.E., and Rosenbaum, F.J.: "Approximation Technique for Dielectric-Loaded Waveguides", Trans. IEEE, 1968, MTT-16, no.4, pp.228-33.

3.41. Gardiol, F.E.: "Higher-Order Modes in Dielectric-Loaded Waveguides", Trans. IEEE, 1968, MTT-16, p.919.

3.42. Sedin, V.A., et al.: "Calculation of Fields in Waveguides of Arbitrary Cross Section Using Quasi-Static Field Models", Radiotekh. Elektron., 1968, 13, p.1931, and Radio Eng. Electron. Phys., 1968, 13, no.11, pp.1694-1702.

3.43. Berger, M.N.: "Design of Waveguide Devices Partly Filled with Dielectric", Radiotekh. Elektron., 1968, 13, no.10, p.1910, and Radio Eng. Electron. Phys., 1968, 13, no.10, pp.1681-3.

3.44. Burrell, R., and Dmitrevsky, S.: "Approximate Cutoff Frequency of Dielectric Ridge Waveguide", Proc. IEEE, 1968, 56, p.2183.

3.45. Young, L.: "Characteristic Impedance of Waveguide", Trans. IEEE, 1968, MTT-16, p.975.

3.46. Van der Vorst, A.S., and Govaerts, R.J.M.: "Accuracy Obtained when Using Variational Techniques for Asymmetrically Loaded Waveguides", Trans. IEEE, 1969, MTT-17,p.51.

3.47. Gardiol, F.E., and Van der Vorst, A.S.: "Wave Propagation in a Rectangular Waveguide Loaded with an H-Plane Dielectric Slab", Trans. IEEE, 1969, MTT-17, p.56.

3.48. Fuller, J.A., and Audeh, N.F.: "Point-Matching Solution of Uniform Nonsymmetric Waveguides", Trans. IEEE, 1969, MTT-17, p.114.

3.49. Gonzalez, G., and Stinson, D.C.: "Propagation in Rectangular Waveguide Partially Filled with an Inhomogeneous Dielectric", Trans. IEEE, 1969, MTT-17, p.284.

3.50. Silvester, P.: "High-Order Finite-Element Waveguide Analysis Programme", Trans. IEEE, 1969, MTT-17, p.204.

3.51. Molodkin, V.A.: "Nonhomogeneous Waveguide", Radiotekhnika, 1969, 24, no.7, p.99, and Telecommun. Radio Eng. Pt 2, 1969, 24, no.7.

3.52. Kao, T.W.: "Reflection and Transmission of Electromagnetic Waves in Inhomogeneous Dielectric-Filled Rectangular Waveguide", Trans. IEEE, 1969, MTT-17, p.639.

3.53. Yurov, Yu.Ya.: "Theory of Thin Wires in Waveguides", Izv. VUZ Radioelektron., 1969, 12, no.6, p.603.

3.54. Bulley, R.M., and Davies, J.B.: "Computation of Approximate Polynomial Solutions to TE Modes in an Arbitrarily Shaped Waveguide", Trans. IEEE, 1969,MTT-17, p.440.

3.55. Smorgonskii, V.Ya.: "Dispersion Properties of Round Waveguides with Dielectric Inserts", Radiotekh. Elektron., 1969, 14, no.6, p.1099, and Radio Eng. Electron. Phys., 1969, 14, no.6, pp.949-52.

3.56. Jull, E.V., Bleakley, W.J., and Steen, M.M.: "Design of Waveguides with Symmetrically Placed Double Ridges", Trans. IEEE, 1969, MTT-17, p.397.

3.57. Bruscantini, S., and Colaviti, C.: "Calculation of Attenuation Coefficient in Elliptical Waveguide", Note Recens. Not., 1969, 18, p.909.

3.58. Bartling, J.Q.: "Propagation in an Infinite Rectangular Dielectric Waveguide", J. Franklin Inst., 1969, 287, no.5, p.389.

3.59. Kretzschmar, J.G.: "Difference between TM_{11} Mode in a Circular and an Elliptical Waveguide", Electron. Lett., 1969, 5, p.602.

3.60. Kibler, L.V.: "Cutoff Region of a Rectangular Waveguide with Losses", Bell Syst. Tech. J., 1969, 48, p.2221.

3.61. Krohne, M.: "Contribution to Wave Propagation in a Circular Waveguide Containing a Coaxial Dielectric Rod", Nachr. Tech. Z., 1969, 22, no.11, p.655.

3.62. Oleinik, I.S., and Tereshchenko, A.I.: "Symmetric and Asymmetric Modes in a Conical Waveguide", Izv. VUZ Radioelektron., 1969, 12, no.9, p.1077.

3.63. Ahmed, S., and Daly, P.: "Waveguide Solutions by Finite-Difference Method", Radio Electron. Eng., 1969, 38, no.4, p.217.

3.64. Filippov, Yu.F.: "Theory of Propagation of H-Waves with Finite Amplitude in a Circular Waveguide", Izv. VUZ Radiofiz., 1969, 12, no.4, p.593.

3.65. Watson, A.: "Parabolic Cylindrical Waveguides", Aust. Telecommun. Res., 1969, 3, no.1, p.3.

3.66. Robinson, J.E., Mohsen, A., and Hamid, M.A.K.: "Application of Ray-Optical Technique to Dielectric-Loaded Rectangular Waveguides", Electron. Lett., 1969, 5, p.380.

3.67. Van der Vorst, A.S., and Govaerts, R.J.M.: "Application of a Variation-Iteration Method to Inhomogeneously Loaded Waveguides", Trans. IEEE, 1970, MTT-18, p.468.

3.68. Kretzschmar, J.G.: "Wave Propagation in Hollow Conducting Elliptical Waveguides", Trans. IEEE, 1970, MTT-18, p.547.

3.69. Baier, W.: "Waves and Evanescent Fields in Rectangular Waveguides Filled with a Transversely Inhomogeneous Dielectric", Trans. IEEE, 1970, MTT-18, p.696.

3.70. Mohsen, A., and Hamid, M.A.K.: "Propagation in Circular Waveguide with an Absorbing Wall", J. Appl. Phys., 1970, 41, p.433.

3.71. Veselov, G.I., and Semenov, S.G.: "Theory of Circular Waveguide with Eccentrically Placed Metallic Conductor", Radiotekh. Elektron., 1970, 15, no.4, and Radio Eng. Electron. Phys., 1970, 15, no.4, pp.687-90.

3.72. Veselov, G.I., and Semenov, S.G.: "Critical Parameters of a Shielded Three-Layer Waveguide", Radiotekhnika, 1970, 25, no.2, p.61, and Telecommun. Radio Eng. Pt 2, 1970, 25, no.2.

3.73. Gal'chenko, N.A., and Mikhalevskii, V.S.: "Application of Schwartz' Method to Calculation of Electrical Parameters of Single- and Double-Ridged Waveguides", Radiotekh. Elektron., 1970, 15, p.51, and Radio Eng. Electron. Phys., 1970, 15, no.1,pp.38-44.

3.74. Raevskii, S.B., and Smorgonskii, V.Ya.: "Method of Computation of Critical Frequencies of Elliptical Waveguide", Radiotekh. Elektron., 1970, 15, no.9, and Radio Eng. Electron. Phys.,1970,15,no.9,pp.1702-5.

3.75. Veselov, G.I., and Gaydar, V.I.: "Analysis of Circular Waveguide with Internal Cross-Shaped Conductor", Radiotekhnika, 1970, 25, no.11, and Telecommun. Radio Eng. Pt 2, 1970, 25, no.11, pp.147-9.

3.76. Gal'chenko, N.A., and Mikhalevskii, V.S.: "Electrical Parameters of a Rectangular Waveguide with T-Shaped Pedestal", Radiotekh. Elektron., 1970, 15, no.12, and Radio Eng. Electron. Phys., 1970, 15, no.12, pp.2210-7.

3.77. Csendes, Z.J., and Silvester, P.: "Numerical Solution of Dielectric-Loaded Waveguides. I and II", Trans. IEEE, 1970, MTT-18, no.12, pp.1124-31, and 1971, MTT-19, no.6, pp.504-9.

3.78. English, W.J.: "Vector Variational Solutions of Inhomogeneously Loaded Cylindrical Waveguide Structures", Trans. IEEE, 1971, MTT-19, no.1, pp.9-18.

3.79. English, W.J., and Young, F.J.: "Vector Variational Formulation of Maxwell's Equations for Cylindrical Waveguide Problems", Trans. IEEE, 1971, MTT-19, no.1, pp.40-6.

3.80. Yee, H.Y.: "Dispersion Relations for Guided Waves in a Simple Moving Medium", Trans. IEEE, 1971, MTT-19, no.4, pp.400-1.

3.81. Montgomery, J.P.: "Complete Eigenvalue Solution of Ridged Waveguide", Trans. IEEE, 1971, MTT-19, no.6, pp.547-55.

3.82. de Gouvenain, A.: "(Nomogram for) Waveguides and Resonant Cavities", Toute Electron., 1971, no.355, pp.47-52.

3.83. Sedykh, V.M.: "Analysis of Cutoff Frequencies of Single- and Double-Ridged Waveguides", Izv. VUZ Radioelektron., 1971, 14, no.7, pp.830-1.

3.84. Lerer, A.M., Gal'chenko, N.A., and Mikhalevskii, V.S.: "Cutoff Wavelength in Guides with Semicircular Ridges", Izv. VUZ Radioelektron., 1971, 14, no.7, pp.837-8.

3.85. Bui, V.R., and Gagne, R.R.J.: "Cutoff Frequencies in Rectangular Waveguides Containing Layers of Dielectric in the H Plane", Proc. IEE, 1971, 118, no.12, pp.1751-3.

3.86. Mann, J.E.: "Propagation of Electromagnetic Waves in Tubes with Large Wall Conductivity", Radio Sci., 1971, 6, no.11, pp.997-1001.

3.87. Raevskii, S.B., and Smorgonskii, V.Ya.: "Analysis of Dispersion Characteristics of Elliptical Waveguide with Dielectric Rod for the HE_{11} Wave", Radiotekh. Elektron., 1971, 16, no.6, pp.941-6, and Radio Eng. Electron. Phys., 1971, 16, no.6, pp.939-43.

3.88. Ilarionov, Yu.A., and Smorgonskii, V.Ya.: "Shift of Cutoff Frequencies in a Round Partially Filled Waveguide", Radiotekh. Elektron., 1971, 16, no.7, pp.1128-32, and Radio Eng. Electron. Phys., 1971, 16, no.7, pp.1093-7.

3.89. Bareukov, Yu.K.: "Complete System of Normal Modes of a Waveguide", Radiotekh. Elektron., 1971, 16, no.12, pp.2167-74, and Radio Eng. Electron. Phys., 1971, 16, no.12, pp.2000-6.

3.90. Smorgonskii, V.Ya.: "Computation of Critical Frequencies in a Partially Filled Elliptical Waveguide", Radiotekh. Elektron., 1971, 16, no.11, pp.2032-8, and Radio Eng. Electron. Phys., 1971, 16, no.11, pp.1812-7.

3.91. Ruiz-Aguirre, R.D., and Deus, L.F.P.: "Propagation in Rectangular Waveguide Partially Filled with an Isotropic Medium", Electron. Fis. Apl., 1971, 14, no.2, pp.181-7.

3.92. Rhodes, J.D.: "General Constraints on Propagation Characteristics of Electromagnetic Waves in Uniform Inhomogeneous Waveguides", Proc. IEE, 1971, 118, no.7, pp.849-56.

3.93. Kretzschmar, J.G.: "Field Configuration of the TM_{001} Mode in an Elliptical Waveguide", Proc. IEE, 1971, 118, no.9, pp.1187-9.

3.94. Heeren, R.G., and Baird, J.A.: "Inhomogeneously Filled Rectangular Waveguide Capable of Supporting TEM Propagation", Trans. IEEE, 1971, MTT-19, no.11, pp.884-5.

3.95. Smorgonskii, V.Ya., and Ilarionov, Yu.A.: "Cutoff Frequencies in a Round Waveguide with a Dielectric Insert", Izv. VUZ Radioelektron., 1971, 14, no.7, pp.736-42.

3.96. Nevesely, M.: "Dispersion Equation of a Waveguide Filled with Two Media", Elektrotech. Cas., 1972, 23, no.2, pp.93-104.

3.97. Kretzschmar, J.G.: "Attenuation Characteristics of Hollow Conducting Elliptical Waveguides", Trans. IEEE, 1972, MTT-20, no.4, pp.280-3.

3.98. Laura, P.A.: "Solution of Helmholtz' Equation in Elliptical Domains", Trans. IEEE, 1972, MTT-20, no.4, p.292.

3.99. Lagasse, P., and Van Bladel, J.: "Square and Rectangular Waveguides with Rounded Corners", Trans. IEEE, 1972, MTT-20, no.5, pp.331-7.

3.100. Laura, P.A., Romanelli, E., and Maurizi, M.J.: "Determination of Cutoff Frequencies of a Square Waveguide with a Concentric Inner Circular Boundary", Proc. IEEE, 1972, 60, no.4, pp.451-2.

3.101. Philipp, W.: "TE_{10} Mode in Rectangular Waveguide", Fernmelde-Praxis, 1972, 49, no.3, pp.119-34, and no.5, pp.207-22.

3.102. Larsen, T.: "Relation between Modes in Rectangular Elliptical and Parabolic Waveguides and a Mode-Classifying System", Trans. IEEE, 1972, MTT-20, no.6, pp.379-84.

3.103. Davies, J.B., and Kretzchmar, J.G.: "Analysis of Hollow Elliptical Waveguides by Polygon Approximation", Proc. IEE, 1972, 119, no.5, pp.519-22.

3.104. Arnold, R.M., and Rosenbaum, F.J.: "Approximate Analysis of Dielectric Ridge Loaded Waveguide", Trans. IEEE, 1972, MTT-20, no.10, pp.699-701.

3.105. Ng, F.L., and Bates, R.H.T.: "Null-Field Method for Waveguides of Arbitrary Cross Section", Trans. IEEE, 1972, MTT-20, no.10, pp.658-62.

3.106. Spielman, B.E., and Harrington, R.F.: "Waveguides of Arbitrary Cross Section by Solution of a Nonlinear Integral Eigenvalue Equation", Trans. IEEE, 1972, MTT-20, no.9, pp.578-85.

3.107. Pfaler, C.E.: "Fundamental System of Solution of Maxwell's Equations for a Circular-Cylindrical Tube", Siemens Forsch. Entwick., 1972, 1, no.2, pp.197-203.

3.108. Khadzhinov, V.D.: "Diffraction of E-Polarized Plane Waves on an Infinitely Thin, Longitudinally Slotted, Cylindrical Waveguide with Coaxial Metal Inner Conductor", Radiotekh. (Kharkov), 1972, no.20, pp.109-15.

3.109. Johns, P.B.: "Application of Transmission-Line-Matrix Method to Homogeneous Waveguides of Arbitrary Cross Section", Proc. IEE, 1972, 119, no.8, pp.1086-91.

3.110. Kiselev, A.P.: "Normal Modes in a Waveguide Formed by Branches of a Hyperbola", Radiotekh. Elektron., 1972, 17, no.11, pp.2445-7, and Radio Eng. Electron. Phys., 1972, 17, no.11, pp.1966-8.

3.111. Belozerov, Yu. S., et al.: "Computation of Critical Frequencies of Waves of Higher Types in a Hollow Elliptical Waveguide", Radiotekh. Elektron., 1972, 17, no.6, pp.1300-1, and Radio Eng. Electron. Phys., 1972, 17, no.6, pp.1009-11.

3.112. Smorgonskii, V.Ya.: "Analysis of Dispersion Equation of a Two-Layer Elliptic Waveguide in Critical Regime", Radiotekh. Elektron., 1972, 17, no.6, pp.1302-4, and Radio Eng. Electron. Phys., 1972, 17, no.6, pp.1011-3.

3.113. Davies, J.B.: "Methods for Numerical Solution of Hollow-Waveguide Problem", Proc. IEE, 1972, 119, no.1, pp.33-7.

3.114. Bartashevskii, Ye, L., et al.: "Analysis of Transversely Irregular Waveguides", Izv. VUZ Radioelektron., 1972, 15, no.7, pp.839-41.

3.115. Yanson, Z.A.: "Asymptotic Properties of Solutions of the Wave Equation for a Waveguide", Izv. Akad. Nauk SSSR, Fiz. Zemli, 1972, no.3, pp.15-24, and Izv. Acad. Sci. USSR, Phys. Solid Earth, 1972, no.3, pp.137-41.

3.116. Parekh, S.V.: "Mode-Matching Method as Applied to Ridged Waveguides", Int. J. Electron., 1973, 34, no.2, pp.285-7.

3.117. Laura, P.A., Davies, J.B., and Kretzchmar, J.G.: "Analysis of Hollow Elliptical Waveguides by Polygon Approximation", Proc. IEE, 1973, 120, no.3, p.336.

3.118. Rollin, R.E.: "Incompleteness of E and H Modes in Waveguides", Can. J. Phys., 1973, 51, no.11, pp.1135-40.

3.119. Garlichs, G.: "Approximation Method for Calculating Attenuation of Dielectric-Filled Circular Waveguides", Arch. Elektron. Ubertrag., 1973, 27, no.7-8, pp.355-6.

3.120. Parekh, S.V.: "Eigenvalue Solutions of Asymmetric-Ridge Waveguides Using the Mode-Matching Method", Int. J. Electron., 1973, 35, no.3, pp.413-20.

3.121. Vuorinen, P.A.: "Comparison of the Conformal Mapping Solution and the Exact Solution for TE_{0n} Modes in Sector Waveguide", J. Microwave Power, 1973, 8, no.2, pp.179-83.

3.122. Casey, K.F.: "Inhomogeneously Filled Rectangular Waveguides", Trans. IEEE, 1973, MTT-21, no.8, pp.566-7.

3.123. Akhtarzad, S., and Johns, P.B.: "Transmission-Line-Matrix Solution of Waveguides with Wall Losses", Electron. Lett., 1973, 9, no.15, pp.335-6.

3.124. Gal'chenko, N.A., Mikhalevskii, V.S., and Sinyavskii, G.P.: "E-Modes in Waveguides of Complex Cross Section", Izv. VUZ Radioelektron., 1973, 16, no.8, pp.12-7.

3.125. Ilarionov, Yu.A.: "Ambiguity of Dispersion Characteristics of Waves in a Circular Partially Filled Waveguide", Izv. VUZ Radioelektron., 1973, 16, no.8, pp.30-7.

3.126. Johns, P.B., and Slater, G.F.: "Transient Analysis of Waveguides with Curved Boundaries", Electron. Lett., 1973, 9, no.21, pp.486-7.

3.127. Lee, C.Q., and Christian, R.: "High-Order Modes in a Square Coaxial Waveguide", Proc. IEEE, 1973, 61, no.12, pp.1754-5.

3.128. Lewin, L.: "Attenuation in an Elliptical Waveguide", Electron. Lett., 1973, 9, no.17, pp.388-9.

3.129. Bird, T.S.: "Evaluation of Attenuation from Lossless Triangular-Finite-Element Solutions for Inhomogeneously Filled Guiding Structures", Electron. Lett., 1973, 9, no.25, pp.590-2.

3.130. Snyder, H.H.: "Guided-Wave Propagation in Cylindrical Domains with Non-Orthogonal Sets of Normal Modes", Proc. Iowa Acad. Sci., 1973, 80, no.2, pp.91-102.

3.131. Kuhn, E.: "Mode-Matching Method for Solving Field Problems in Waveguide and Resonator Circuits", Arch. Elektron. Ubertrag. 1973, 27, no.12, pp.511-8.

3.132. Schoonaert, D.H., and Luypaert, P.J.: "Use of Nonseparable Solutions of Helmholtz' Wave Equation in Waveguides and Cavities", Electron. Lett., 1973, 9, no.26, pp.617-8.

3.133. Radionov, A.A., and Smorgonskii, V.Ya.: "Bandwidth Properties of a Two-Layer Elliptical Waveguide", Radiotekh. Elektron., 1973, 18, no.1, pp.179-82, and Radio Eng. Electron. Phys., 1973, 18, no.1, pp.129-32.

3.134. Smorgonskii, V.Ya., Belozerov, Yu.S., and Ilarionov, Yu.A.: "Calculation of Cutoff Wavelengths of TE Waves in Elliptical Guide", Radiotekh. Elektron., 1973, 18, no.1, pp.182-4, and Radio Eng. Electron. Phys., 1973, 18, no.1, pp.132-4.

3.135. Raevskii, S.B., et al.: "Transverse Distribution of the HE_{11}-Mode Field in a Two-Layer Elliptical Waveguide", Izv. VUZ Radioelektron., 1973,16,no.12,pp.100-3.

3.136. Raevskii, S.B., Simkina, L.G., and Smorgonskii, V.Ya.: "Distribution of Electromagnetic Field of the H_{11p} Mode in Transverse Cross Section of a Double-Layer Elliptical Guide", Radiotekh. Elektron., 1973, 18, no.7, pp.1335-41, and Radio Eng. Electron. Phys., 1973, 18, no.7, pp.985-90.

3.137. Kalmyk, V.A., and Raevskii, S.B.: "Graphical Analysis of Complex Wave Numbers in a Two-Layer Round Waveguide", Radiotekh. Elektron., 1973, 18, no.2, pp.385-91, and Radio Eng. Electron. Phys., 1973, 18, no.2, pp.272-8.

3.138. DeVore, R., Toth, J.F., and Caldecott, R.: "Dielectric Coaxial Waveguide", J. Appl. Phys., 1973, 44, no.10, pp.4488-500.

3.139. Vol'man, V.I.: "Analysis of EH Inhomogeneities and Singular H Inhomogeneities in a Rectangular Waveguide", Radiotekhnika, 1973, 28, no.3, pp.34-41, and Telecommun. Radio Eng. Pt 2, 1973, 28, no.3, pp.90-5.

3.140. Lewin, L., and Al-Hariri, A.M.B.: "Effect of Cross-Section Curvature on Attenuation in Elliptical Waveguides and a Basic Correction to Previous Formulae", Trans. IEEE, 1974, MTT-22, no.5, pp.504-9.

3.141. Mahmoud, S.F., and Wait, J.R.: "Theory of Wave Propagation Along a Thin Wire Inside a Rectangular Waveguide", Radio Sci., 1974, 9, no.3, pp.417-20.

3.142. Ferdinandov, E.S.: "Parameters of Transmission in a Two-Cylinder Waveguide", Elektro Prom. Prib., 1974, 9, no.1, pp.12-5.

3.143. Chen, M.H., Tsandoulas, G.N., and Willwerth, F.G.: "Modal Characteristics of Quadruple-Ridged Circular and Square Waveguides", Trans. IEEE, 1974, MTT-22, no.8, pp.801-4.

3.144. Aginskii, A.L., and Dakhno, V.N.: "Perturbed H_{10} Modes in a Rectangular Waveguide with Finite Conductivity of the Walls", Izv. VUZ Radioelektron., 1974,17,no.5,pp.70-5.

3.145. Akhtarzad, S., and Johns, P.B.: "Numerical Solution of Lossy Waveguides. TLM Computer Programme", Electron. Lett., 1974, 10, no.15, pp.309-11.

3.146. Rabinowitz, I.I., and Staniforth, J.A.: "Propagation in Two Circular Coaxial Waveguides", Electron. Lett., 1974, 10, no.20, pp.415-7.

3.147. Iskander, M.F., and Hamid, M.A.K.: "Analysis of Triangular Waveguides of Arbitrary Dimensions", Arch. Elektron. Ubertrag., 1974, 28, no.11, pp.455-61.

3.148. Narasimhan, M.S., and Balasubramanya, K.S.: "Transmission Characteristics of Spherical TE and TM Modes in Conical Waveguides", Trans. IEEE, 1974, MTT-22, no.11, pp.965-70.

3.149. Lerer, A.M.: "Allowance for Edge Discontinuity in Calculation of Cutoff Frequencies and Fields in a Rectangular Waveguide with a T-Step", Izv. VUZ Radioelektron., 1974, 17, no.9, pp.90-3.

3.150. Iskander, M.F., and Hamid, M.A.K.: "Eigenvalues for a Trapezoidal Waveguide", Radio Electron. Eng., 1974, 44, no.11, pp.593-6.

3.151. Shiau, Y., and Yang, R.F.H.:"Ridged Circular Waveguide", Trans. IEEE, 1974, MTT-22, no.2, pp.130-2.

3.152. Somlo, P.L.: "Exact Numerical Evaluation of Complex Dielectric Constant of a Dielectric Partially Filling a Waveguide", Trans. IEEE, 1974, MTT-22, no.2, pp.150-1, and no.4, pp.468-9.

3.153. Daly, P.: "Polar Geometry Waveguides by Finite-Element Methods", Trans. IEEE, 1974, MTT-22, no.3, pp.202-9.

3.154. Findakly, T., and Haskal, H.: "Attenuation and Cutoff Frequencies of Double-Ridged Waveguides", Microwave J., 1974, 17, no.12, pp.49-50.

3.155. Eung, Y.I., and Ryul, K.B.: "Cutoff Mode Characteristics in Step-Ridged Waveguide", J. Korean Inst. Electron. Eng., 1974, 11, no.5, pp.212-20.

3.156. Zorkin, A.F., Kartavtseva, L.F., and Ivanishina, Z.V.: "Graphical Method of Calculating the Range of Single-Wave Mode of a Three-Layered Symmetrical Rectangular Waveguide", Radiotekh.(Kharkov),1974,no.29,pp.83-6.

4. RESONANT INTERACTION

4.1. Severin, H.: "Ferrites at High Microwave Power Levels", Nachr. Tech. Z., 1965, 18, p.7.

4.2. Damon, R.W., and van de Haart, H.: "Propagation of Magnetostatic Spin Waves at Microwave Frequencies in a Normally Magnetized Disc", J. Appl. Phys., 1965, 36, p.3453.

4.3. Lam, Y.W.: "Magnetostatic-Mode Linewidth in YIG", Solid-State Electron., 1965, 8, p.923.

4.4. Bodway, G.E., and Wang, S.: "Instabilities in the Magnetization Vector in YIG at High Microwave Power Levels", J. Appl. Phys., 1965, 36, p.2566.

4.5. Brand, H., and Fieweger, H.W.: "Ferromagnetic Resonance Linewidth and g-Factor in Ferrites at 2-18 GHz", Trans. IEEE, 1965, MTT-13, p.712.

4.6. Brown, I.M., Weissman, S.I., and Snyder, L.C.: "Triplet-State ESR of Metal Chelate Compounds", J. Chem. Phys., 1965, 42, p.1105.

4.7. Hebel, L.C., Blount, E.I., and Smith, G.E.: "Theory of Spin Resonance in Bismuth", Phys. Rev., 1965, 138, p.A1636.

4.8. Turoff, R.D.: "Spin-Lattice Relaxation in Caesium Chrome Alum", Phys. Rev., 1965, 138, p.A1524.

4.9. Drumheller, J.E., Locher, K., and Waldner, F.: "Electron Paramagnetic Resonance of Cr^{3+} and Fe^{3+} in Synthetic $ZnAl_2O_4$ Spinel", Helv. Phys. Acta, 1965, 38, p.172.

4.10. Galt, J.K., Merritt, F.R., and Klauder, J.R.: "Cyclotron Resonance in Cd", Phys. Rev., 1965, 139, p.A823.

4.11. Geist, D.: "Cyclotron and Electron-Spin Resonance in Diamagnetic Semiconductors", Phys. Status Solidi,1965,5,p.217.

4.12. Chin-chi, C., and Wu-yan, L.: "Influence of Rare-Earth Ions on the Ferromagnetic Resonance at High Power Levels", Acta Phys. Sin., 1965, 21, p.1188.

4.13. Huiszoon, C., and Dymanus, A.: "Stark Effect of Millimetre-Wave Transitions. I. H_2S", Physica, 1965, 31, p.1049.

4.14. Solov'ev, V.I., and Gurevich, A.G.: "Ferromagnetic Resonance in Yttrium Garnet Doped with Terbium", Fiz. Tver. Tela, 1965, 7, p.1761, and Sov. Phys.-Solid State, 1965, 7, p.1420.

4.15. Itoh, K., and Miyata, N.: "Effect of Eddy Currents on Ferromagnetic Resonance of Spherical Speciments", J. Phys. Soc. Jap., 1965, 20, p.1528.

4.16. Polivanov, K.M., and Pollak, B.P.: "Resonance Characteristics of Magnetically Uniaxial Polycrystalline Ferrites", Arch. Elektrotech., 1965, 14, p.213.

4.17. Miyamoto, S., Tanaka, N., and Iida, S.: "Ferromagnetic Resonance in Single Crystals of Cobalt-Substituted Nickel Ferrite", J. Phys. Soc. Jap., 1965, 20, p.753.

4.18. Maguire, E.A., and Green, J.J.: "Magnetic Properties of Gd-Y-Fe-In Garnets", J. Am. Ceram. Soc., 1965, 48, p.369.

4.19. Schirmer, W., and Kempel, K.A.: "Ferrimagnetic Resonance in Ellipsoids", Phys. Kondens Materie, 1965, 3, p.187.

4.20. Seiden, P.E., and Sparks, M.: "Frequency Dependence of Two-Magnon Ferrimagnetic Resonance Linewidth", Phys. Rev., 1965, 137, p.A1278.

4.21. Lobachev, V.P., and Kichigin, D.A.: "Cyclotron Resonance of Hot Electrons", Fiz. Tver. Tela, 1965, 7, p.470, and Sov. Phys.-Solid State, 1965, 7, p.375.

4.22. Comstock, R.L., et al.: "Spin-Wave Spectrum of GdIG", Appl. Phys. Lett., 1966, 9, p.274.

4.23. Orton, J.W., Fruin, A.S., and Walling, J.C.: "Spin-Lattice Relaxation of Cr^{3+} in Single Crystals of $ZnWO_4$", Proc. Phys. Soc., 1966, 81, p.703.

4.24. Demirdjioghlou, S.F., and Pointon, A.J.: "Microwave Properties of Low-Saturation-Magnetization Ni-Ga-Al Ferrites", Proc. IEE, 1966, 113, p.525.

4.25. Bagguley, D.M.S., and Liesegang, J.: "Microwave Absorption in Rare-Earth Metals", J. Appl. Phys., 1966, 37, p.1220.

4.26. Holzer, W.C., Perry, B.W., and Portis, A.M.: "Microwave Transmission through Permalloy Films", J. Appl. Phys., 1966, 37, p.1222.

4.27. Duneav, N.M., and Pil'shchikov, A.I.: "Surface Magnetostatic Modes in a Hollow Ferrite Cylinder", Radiotekh. Elektron., 1966, 11, no.6, pp.1019-29, and Radio Eng. Electron. Phys., 1966, 11, no.6, pp.886-95.

4.28. Lewis, M.F., and Lacklison, D.E.: "Mode in Axially Magnetised YIG Rods", Appl. Phys. Lett., 1966, 9, no.11, pp.414-5.

4.29. Bar'yakhtar, V.G., Savchenko, M.A., and Stepanov, K.N.: "Interaction of Electromagnetic Waves and Spin Waves in Helicoidal Magnetic Structures", Fiz. Tver. Tela, 1966, 8, no.7, pp.2168-72, and Sov. Phys.-Solid State, 1967, 8, no.7, pp.1723-5.

4.30. Bell, M.D., and Leivo, W.J. "Electron Spin Resonance in Semiconducting Diamonds", J. Appl. Phys., 1967, 38, no.1, pp.337-9.

4.31. Wolff, I.: "Material Investigation of Ferrites in the Millimetre-Wave Region", Frequenz, 1967, 21, no.5, pp.161-5.

4.32. Rusov, G.I.: "Ferromagnetic and Spin-Wave Resonances in Fe-Ni Alloy Thin Films", Fiz. Tver. Tela, 1967, 9, no.1, pp.196-9, and Sov. Phys.-Solid State, 1967, 9, no.1, pp.146-8.

4.33. de Santis, P.: "Magnetostatic Waves at Singular Turning Points", Electron. Lett., 1967, 3, no.12, pp.555-6.

4.34. Riches, E.E.: "Advances in Ferrite Materials for Microwave Devices", GEC J. Sci. Technol., 1968, 35, no.3, p.94.

4.35. Schott, W.F., and Tao, T.F.: "Classification of Electromagnetic Waves in Ferrite Rods", Trans. IEEE, 1968, MTT-16, p.959.

4.36. Kriz, T.A., and Ishii, T.K.: "Statistical Analysis of Initial Microwave Susceptibility Tensor of a Polycrystalline Ferrite", J. Appl. Phys.,1968,39,p.5029.

4.37. Vylegzhanin, D.N.: "Calculations of the Relative Probability of Cross Relaxation in Ruby from the Inversion Coefficients", Radiotekh. Elektron., 1968, 13, p.2198, and Radio Eng. Electron. Phys., 1968, 13, no.12, pp.1926-32.

4.38. Shevchenko, Yu, A., and Mikhalevskii, V.S.: "Study of Magnetostatic Oscillations in Cylindrical YIG Monocrystals", Radiotekh. Elektron., 1968, 13, p.1429, and Radio Eng. Electron. Phys., 1968, 13, no.8, pp.1241-4.

4.39. Deryugin, I.A., Yemlin, P.G., and Sugakov, V.I.: "Effect of Waveguide Walls on Magnetostatic Modes Excited in a Ferrite Sphere", Radiotekh. Elektron., 1968, 13, p.2075, and Radio Eng. Electron. Phys., 1968, 13, no.11, pp.1818-20.

4.40. Green, J.J., Patton, C.E., and Stern, E.: "Threshold Microwave-Field Amplitude for Unstable Growth of Spin Waves under Oblique Pumping", J. Appl. Phys., 1969, 40, p.172.

4.41. Young, P.: "Effect of Boundary Conditions on Propagation of Surface Magnetostatic Waves in a Transversely Magnetized Thin YIG Slab", Electron. Lett., 1969, 5, p.429.

4.42. Deschamps, A.: "Ferrites for Use at Millimetre Wavelengths", Z. Angew. Phys., 1969, 26, p.190.

4.43. Helszajn, J., and McStay, J.: "External Susceptibility Tensor of Magnetized Ferrite Ellipsoid in Terms of Uniform-Mode Ellipticity", Proc. IEE, 1969, 116,p.2088.

4.44. Winkler, G., and Hansen, P.: "Ca-V-In Substituted YIG with Very Narrow Resonance Lines", Mater. Res. Bull., 1969, 4, no.11, p.825.

4.45. Deryugin, I.A., Yemlin, P.G., and Sugakov, V.I.: "Excitation of Magnetostatic Modes in a Ferrite Sphere in a Waveguide", Radiotekh. Elektron., 1969, 14, no.12, p.2167, and Radio Eng. Electron. Phys., 1969, 14, no.12, pp.1876-9.

4.46. Seshadri, S.R.: "Walker Modes of Ferrite Column Magnetized in an Arbitrary Direction", Proc. IEEE, 1969,57,no.8,p.1438.

4.47. Seshadri, S.R.: "Resonance of an Axially Magnetized Hollow Ferrite Cylinder", Proc. IEEE, 1969, 57, no.8, p.1454.

4.48. Patton, C.E.: "Microwave Properties of Fine-Grain Polycrystalline YIG", J. Appl. Phys., 1970, 41, p.1355.

4.49. Mann, M.M., and DeShazer, L.G.: "Energy Levels and Spectral Broadening of Nd^{3+} Ions in Laser Glass", J. Appl. Phys., 1970, 41, p.2951.

4.50. Silber, L.M.: "Ferromagnetic Resonance in Planar Ferrites", Trans. IEEE, 1970, MTT-18, p.68.

4.51. de Santis, P.: "Magnetoelastic Damping of Microwaves in Ferrite Single Crystals", Trans. IEEE, 1970, MAG-6, p.84.

4.52. Dixon, S., Weiner, M., and Coin, T.R.A.: "Ferromagnetic Resonance and Nonlinear Effects in Ferrites with Uniaxial Anisotropy", J. Appl. Phys., 1970, 41, p.1357.

4.53. Seshadri, S.R.: "Surface Magnetostatic Modes of a Ferrite Slab", Proc. IEEE, 1970, 58, p.506.

4.54. Morotani, T., and Nisida, Y.: "Cyclotron Resonance in n-InSb by HCN Laser", Solid-State Commun., 1970, 8, p.755.

4.55. Brunet, H.: "Saturation of IR Absorption in SF_6", J. Quantum Electron. IEEE, 1970, QE-6, no.11, pp.678-84.

4.56. Uehara, Y., and Kobayashi, S.: "(MgMn) Ferrite for 6-11 GHz", Fujitsu Sci. Tech. J., 1971, 7, no.2, pp.171-91.

4.57. Scotter, D.G.: "Microwave Loss in Polycrystalline Ferrites. Role of Spin Waves", J. Sci. Technol., 1971, 38, no.2, pp.55-62.

4.58. Strozyk, J.W. "Lutetium Effects on UV Absorption Spectra of Nd^{3+}:YAG", J. Quantum Electron. IEEE, 1971, QE-7, no.9, pp.467-9.

4.59. Kraftmakher, G.A., and Meriakri, V.V.: "Investigation of Ferrites in the Submillimetre Band", Radiotekh. Elektron., 1971, 16, no.11, pp.2221-2, and Radio Eng. Electron. Phys., 1971, 16, no.11, pp.1967-8.

4.60. Masuda, M., Chang, N.S., and Matsuo, Y.: "Azimuthally Dependent Magnetostatic Modes in Cylindrical Ferrites", Trans. IEEE, 1971, MTT-19, no.10, pp.834-6.

4.61. Gurkut, B., Schermann, J.P., and Audion, C.: "Frequency Shift of Hyperfine Transition of Atomic Hydrogen by Two Coherent Perturbations", J. Quantum Electron. IEEE, 1972, QE-8, no.5, pp.459-60.

4.62. Danishevskii, A.M., Kochegarov, S.F., and Subashiev, V.K.: "Nonlinear Absorption of IR Radiation in InSb", Fiz. Tver. Tela, 1972, 14, no.11, pp.3233-9, and Sov. Phys.-Solid State, 1973,14,no.11,pp.2754-8.

4.63. Kurbatov, L.N., et al.: "Luminescence of Semiconductors in the 0.5-11 micron Band", Izv. Akad. Nauk SSSR, Ser. Fiz., 1973, 37, no.2, pp.363-8.

4.64. Yamanoi, M.: "Spectral Intensity of Thermal Radiation in a Medium Near an Absorption Line", Phys. Lett., 1973, 44A, no.1, pp.23-4.

4.65. Doring, H., and Wolff, I.: "Investigations of Properties of Ferrites in the Millimetre-Wave Region", Elektr. Masch., 1973, 90, no.5, pp.233-40.

4.66. Boka, A., and Gilanyi, T.: "Research on Microwave Ferrites Suitable for Frequencies Higher than 10 GHz", Hiradastechnika, 1973, 24, no.5, pp.146-9.

4.67. Murmuzhev, B.A.: "Propagation of Magnetostatic Waves in a Ferrite-Teflon-Copper Layered Medium", Radiotekh. Elektron., 1973, 18, no.7, pp.1499-505, and Radio Eng. Electron. Phys., 1973, 18, no.7, pp.1107-9.

4.68. Dotsch, H., and Schmitt, H.J.: "Interaction of Microwaves with Ring Domains in Magnetic Garnet Films", Appl. Phys. Lett., 1974, 24, no.9, pp.442-4.

4.69. James, D.A.: "Reflection Factor of Magnetized Ferrite Sphere in Rectangular Waveguide", Radio Electron. Eng., 1974, 44, no.9, pp.481-90.

5. ANISOTROPIC MEDIA

5.1. Mansfield, R., and Borst, M.R.: "Microwave Faraday Effect in InSb", Br. J. Appl. Phys., 1965, 16, p.570.

5.2. Nag, B.R., and Engineer, M.H.: "Experimental Observation of Faraday Rotation in Artificial Dielectrics", J. Appl. Phys., 1965, 36, p.3388.

5.3. Mort, J., Luty, F., and Brown, F.C.: "Faraday Rotation and Spin-Orbit Splitting of the F-Centre in Alkali Halides", Phys. Rev., 1965, 137, p.A566.

5.4. Dascola, G., Giori, D.C., and Varacca, V.: "Cotton-Mouton Effect Associated with ESR in a Birefringent Medium", Nuovo Cim., 1965, 37, p.382.

5.5. Nag, B.R., and Engineer, M.H.: "Faraday Rotation in Artificial Dielectrics", J. Appl. Phys., 1965, 36, p.192.

5.6. Gandhi, O.P.: "Birefringence in Semiconductor Magnetoplasmas for Ultramicrowave Gyratory Devices", Proc. IEE, 1965, 112, p.483.

5.7. Servant, Y.: "EPR Faraday Effect of MnF_2 in 3-GHz Band", C. R. Acad. Sci. (Paris), 1965, 260B, p.5494.

5.8. Dong, N.Q.: "Scattering of Light by a Plasma", Phys. Lett., 1966, 21, pp.159-60.

5.9. Kibble, T.W.B.: "Radiative Corrections to Thomson Scattering from Laser Beams", Phys. Lett., 1966, 20, pp.627-8.

5.10. Vorob'ev, L.E.: "Influence of a Strong Electric Field on the Faraday Effect in InSb", Fiz. Tekh. Poluprov., 1967, 9, no.1, pp.145-6, and Sov. Phys.-Semicond., 1967, 9, no.1.

5.11. Furdyna, J.K.: "Microwave Faraday Rotation in Semiconductor Plasmas in the High-Magnetic-Field Limit", Solid-State Commun., 1967, 5, no.7, pp.539-42.

5.12. Potzl, H.W., and Richter, K.: "Microwave Conductivity Anisotropy of Hot Electrons in n-InSb at 77°K", Proc. IEEE, 1967, 55, no.8, pp.1497-8.

5.13. Mashukov, Yu.P.: "Features of the Faraday Effect in GaAs", Fiz. Tekh. Poluprov., 1967, 1, no.7, pp.990-2, and Sov. Phys.-Semicond., 1968, 1, no.7.

5.14. Judy, J.H., et al.: "Large Longitudinal Kerr Rotations and Factors of Merit in Thin Iron Films", Trans. IEEE, 1968, MAG-4, p.401.

5.15. Krinchik, G.S.: "Exchange Resonance and Optical Faraday Effect in Rare-Earth Iron Garnets", Zh. Eksp. Teor. Fiz. Pis'ma, 1968, 8, no.9, p.462, and JETP Lett., 1968, 8, no.9, p.284.

5.16. Zubkov, V.I., and Monosov, Ya.A.: "Instability of Nonlinear Ferromagnetic Resonance", Radiotekh. Elektron., 1968, 13, no.10, p.1897, and Radio Eng. Electron. Phys., 1968, 13, no.10, pp.1665-8.

5.17. Kaganov, M.I., and Yankelevich, R.P.: "Propagation of EM Waves in Gyro-Anisotropic Media", Fiz. Tver. Tela, 1968, 10, p.2771, and Sov. Phys.-Solid State, 1969, 10, no.9, pp.2181-5.

5.18. Zanmarchi, G., and Bongers, P.F.: "IR Faraday Rotation in Ferrites", J. Appl. Phys., 1969, 40, p.1230.

5.19. Crossley, W.A., Cooper, R.W., and Page, J.L.: "Faraday Rotation in Rare-Earth Iron Garnets", J. Appl. Phys.,1969,40,p.1497.

5.20.   Borrelli, N.F., and Layton, M.M.:
"Electrooptic Properties of Transparent
Ferroelectric Glass-Ceramic Systems",
Trans. IEEE, 1969, ED-16, p.511.

5.21.   Thacher, P.D., and Land, C.E.:
"Ferroelectric Electrooptic Ceramics with
Reduced Scattering", Trans. IEEE, 1969,
ED-16, p.515.

5.22.   Land, C.E., and Thacher, P.D.:
"Ferroelectric Ceramic Electrooptic Materials and Devices", Proc. IEEE, 1969,57,p.751.

5.23.   Watson, K.M.: "Multiple Scattering
of Waves in an Underdense Plasma", J. Math.
Phys., 1969, 10, p.688.

5.24.   Blum, F.A., et al.: "Microwave
Scattering and Noise Emission from After-
glow Plasmas in a Magnetic Field", Phys.
Fluids, 1969, 12, p.1018.

5.25.   Hussein, A.M., and Pakhomov, V.I.:
"Study of EM Surface Waves in Anisotropic
Media", Physica, 1969, 45, p.321.

5.26.   Bagaev, V.S., et al: "Electrooptic
Effect in GaAs and CdTe", Fiz. Tekh. Polu-
prov., 1969, 3, no.11, p.1687, and Sov.
Phys.-Semicond., 1970, 3, no.11, pp.1418-20.

5.27.   Pinnow, D.A.: "Guide Lines for Selection of Acoustooptic Materials", J.
Quantum Electron. IEEE, 1970, QE-6, p.223.

5.28.   Tada, K., et al.: "Linear Electrooptic Effect in ZnTe", J. Fac. Eng. Univ.
Tokyo, 1970, ser.A, no.8, pp.34-5.

5.29.   Stoffel, A.M., and Schneider, J.:
"Magnetooptic Properties of MnAs Films",
J. Appl. Phys., 1970, 41, p.1405.

5.30.   Baranovskii, S.N., et al.: "Anisotropy of Faraday Rotation in n-Ge Subjected
to Strong Electric Fields", Fiz. Tekh.
Poluprov., 1970, 4, p.589, and Sov. Phys.-
Semicond., 1970, 4, no.3, pp.492-5.

5.31.   Abdurakhmanov, A.A.: "Theory of
Magnetooptic Phenomena in Thin Ferromagnetic Films", Izv. VUZ Fiz., 1970, no.1, p.60.

5.32.   Buhrer, C.F.: "Enhancement of Faraday Rotation of Iron Garnets by Bismuth and
Rare-Earth Ions", J. Appl. Phys., 1970, 41,
p.1393.

5.33.   Asche, M.: "Faraday Effect of Hot
Electrons in Many-Valley Semiconductors",
Phys. Status Solidi, 1970,41,no.1,pp.67-73.

5.34.   Ermet, H.: "Gaussian Beams in Anisotropic Media", Electron. Lett., 1970, 6,
no.22, pp.720-1.

5.35.   Dixon, R.W.: "Acoustooptic Interactions and Devices", Trans. IEEE, 1970,
ED-17, p.229,

5.36.   Hoffman, A.S., Hagemeier, W.R., and
Cushner, S.H.: "Enhancement of the Longitu-
dinal Kerr Magnetooptic Effect in Thin
Films", J. Appl. Phys., 1970, 41, p.1407.

5.37.   Chen, D., Aagard, R.L., and Liu,
T.S.: "Magnetooptic Properties of Quenched
Thin Films of MnBi", J. Appl. Phys., 1970,
41, p.1395.

5.38.   Lee, R.E., and White, R.M.: "Elastic-
Wave Generation by a Gunn-Effect Oscillator
Coupled to a Piezoelectric", Appl. Phys. Lett.,
1970, 16, p.343.

5.39.   Cummins, S.E.: "Electrical, Optical,
and Mechanical Behaviour of Ferroelectric
$Gd_2(MoO_4)_3$", Ferroelectrics, 1970, 1, p.11.

5.40.   Luther-Davies, B., et al.: "Sign of
the Electrooptic Coefficients for $LiTaO_3$",
J. Phys.C, 1970, 3, p.L106.

5.41.   Buts, V.A.: "Excitation of a Plane
Gyrotropic Waveguide", Izv. VUZ Radiofiz.,
1970, 13, no.3, p.422.

5.42.   Pisarev, R.V., Berdennikova, E.V.,
and Petrov, R.A.: "Faraday Effect and Saturation Magnetization of Substituted Ferrite
Garnets", Fiz. Tver. Tela, 1970, 12, no.5,
p.1547, and Sov. Phys.-Solid State, 1970, 12,
no.5, pp.1218-9.

5.43.   Chu, R.S., and Tamir, T.: "Guided-
Wave Theory of Light Diffraction by Acoustic
Microwaves", Trans. IEEE, 1970, MTT-18,p.486.

5.44.   Adrion, R.F.: "Doppler Shift in
Bragg Cells", Proc. IEEE, 1970, 58, p.1391.

5.45.   Sueta, T., et al.: "Modulation of
10.6-micron Laser Radiation by CuCl", Proc.
IEEE, 1970, 58, p.1378.

5.46.   Oguchi, T.: "Scattering of a Plane
Wave by a Ferrite Sphere", J. Radio Res. Lab.,
1970, 17, no.90, pp.103-35.

5.47.   Datta, A.N., and Nag, B.R.: "Faraday
Rotation in Media Stratified Parallel with
the Direction of Propagation", Radio Sci.,
1970, 5, no.8-9, pp.1185-9.

5.48.   Auth, D.C.: "Equivalence of Doppler
and Quantum Frequency Shifting in Bragg Cells",
Proc. IEEE, 1971, 59, no.3, pp.413-4.

5.49.   Hudson, A.S., Sneider, J., and
Dorlejn, J.W.E.: "Mn Substitution in Garnets
for Remanent Phaseshifters", Trans. IEEE,
1971, MTT-19, no.1, pp.119-20.

5.50.   Fox, A.J.: "Longitudinal Quadratic
Electrooptic Effect in KTN", Electron. Lett.,
1971, 7, no.5-6, pp.139-40.

5.51.   Arunasalam, V., Heald, M.A., and
Sinnis, J.: "Microwave Scattering from Unstable Electron Plasma Waves", Phys. Fluids,
1971, 4, no.6, pp.1194-203.

5.52.   Schadt, M., and Helfrich, W.: "Voltage Dependent Optical Activity of a Twisted
Nematic Liquid Crystal", Appl. Phys. Lett.,
1971, 18, no.4, pp.127-8.

5.53.   Ermet, H.: "Propagation of Gaussian
Beams in Gyrotropic Media", Arch. Elektr.
Ubertrag., 1971, 25, no.1, pp.17-24.

5.54. Tada, K., and Aoki, M.: "Linear Electrooptic Properties of ZnTe at 10.6 micron", Jap. J. Appl. Phys., 1971, 10, no.8, pp.998-1001.

5.55. Bogdanov, S.V., Zubrinov, I.I., and Sheloput, D.V.: "Investigation of Materials for Acoustooptic Devices", Izv. Akad. Nauk SSSR, Ser. Fiz., 1971, 35, no.5, pp.1013-4.

5.56. Zhashkov, A.A., et al.: "Longitudinal Electrooptic Effect in Oblique Sections of LiNbO$_3$", Kvantovaya Elektron., 1971, no.1, pp.147-8.

5.57. Furuhata, Y.: "Guidelines for Single-Crystal Material for Optoelectronics", Solid-State Phys., 1971, 6, no.7, pp.36-9.

5.58. Angert, N.B., et al.: "Phase-Matching Angles and Temperatures of LiNbO$_3$ Crystals of Different Stoichiometries", Kvantovaya Elektron., 1971, no.5, pp.128-9.

5.59. Goto, K.: "Methods of Using Non-linear Optical Materials. LiNbO$_3$", Oyo Buturi, 1971, 40, no.10, pp.1101-4.

5.60. Kolokolov, A.A., and Skrotskii, G.V.: "Self-Focusing Kinetics of Short Light Pulses", Opt. Spekt., 1971, 31, no.4, pp.650-2, and Opt. Spectrosc., 1971, 31, no.4, pp.342-3.

5.61. Mirovskaya, Ye.A.: "Four-Wave Parametric Interaction in a Nonlinear Medium", Radiotekh. Elektron., 1971, 16, no.6, pp.1058-60, and Radio Eng. Electron. Phys., 1971, 16, no.6, pp.1035-7.

5.62. Kludzin, V.V., et al.: "Ultrasonic Light Modulation in Dense Flints", Opt.-Mekh. Prom., 1972, 39, no.1, pp.3-5, and Sov. J. Opt. Technol.,1972,39,no.1, pp.1-3.

5.63. Roland, G.W., Gottlieb, M., and Feichtner, J.D.: "Acoustooptic Properties of Tl$_3$AsS$_4$", Appl. Phys. Lett., 1972, 21, no.2, pp.52-4.

5.64. Ohmachi, Y., and Uchida, N.: "Acoustooptic Property of Single-Crystal Pb$_5$Ge$_3$O$_{11}$", J. Appl. Phys., 1972, 43, no.8, pp.3583-4.

5.65. Shimizu, F.: "Effect of Dispersion on Pulse Distortion in Optical Filaments", J. Quantum Electron. IEEE, 1972, QE-8, no.11, pp.851-2.

5.66. Gaisser, T.K.: "Quantum Treatment of Thomson Scattering in a Magnetic Field", Am. J. Phys., 1972, 40, no.12, pp.1782-6.

5.67. Ward, G., and Pechacek, R.E.: "Scattering of Light by Relativistic Electrons", Phys. Fluids, 1972, 15, no.12, pp.2202-10.

5.68. Ehlotzky, F.: "Effect of Laser Light Polarization on Multiphonon Scattering Processes", Acta Phys. Austriaca, 1972, 36, no.3, pp.243-7.

5.69. Brodskii, Yu.Ya., Kondrat'ev, I.G., and Miller, M.A.: "Electromagnetic Beams in Anisotropic Media. II", Izv. VUZ Radiofiz., 1972, 15, no.4, pp.592-600.

5.70. Thacher, P.D.: "Linear Electrooptic Effect in Ferroelectric Ceramics. PLZT", Ferroelectrics, 1972, 3, no.2-4, pp.147-56.

5.71. Bichard, V.M., Davies, P.H., and Hulme, K.F.: "1060-nm Absorption Coefficients of KDP with 70-100 % Deuteration", Electron. Lett., 1972, 8, no.6, pp.147-8.

5.72. Ninomiya, Y.: "Method of Damping Piezoelectric Resonance of Electrooptic Devices", Electron. Commun. Jap., 1972, 55, no.6, pp.106-12.

5.73. Lisitsa, M.P., et al.: "Two-Photon Absorption in Cd(SSe) Single Crystals", Kvantovaya Elektron., 1972, no.5, pp.53-7, and Sov. J. Quantum Electron., 1972, 2, no.5, pp.432-4.

5.74. Ternov, I.M., Khalilov, V.R., and Kholomai, B.V.: "Frequency Drift in a High Intensity (Electron) Combinatorial Scattering", Vestn. Mosk. Univ., Fiz. Astron., 1972, 13, no.5, pp.621-2.

5.75. Chen, S.H., and Tartaglia, P.: "Light Scattering from N Non-Interacting Particles", Opt. Commun., 1972,6,no.2,pp.119-24.

5.76. Welsch, D.: "Quantum Theory of Spontaneous Many-Photon Raman Scattering", Acta Phys. Pol., 1972, 42A, no.5, pp.467-74.

5.77. Borcherds, P.H.: "Scattering of Light by Elastic Waves in Cubic Crystals", Opt. Acta, 1973, 20, no.2, pp.147-59.

5.78. Walls, D.F.: "Master-Equation Approach to the Raman Effect", J. Phys. A, 1973, 6, no.4, pp.496-505.

5.79. Piekarra, A.H., and Gustafson, T.K.: "Interaction of Vibrational Enhancement and Orientational Kerr Effects in Liquids", Opt. Commun., 1973, 7, no.3, pp.197-200.

5.80. Veklenko, B.A.: "Theory of Scattering of an Electromagnetic Field in Gases", Izv. VUZ Fiz., 1973, no.4, pp.81-6.

5.81. Kogelnik, H., Sosnowski, T.P., and Welser, H.P.: "Ray-Optical Analysis of Thin-Film Polarization Converters", J. Quantum Electron. IEEE, 1973, QE-9, no.8,pp.795-800.

5.82. Wettling, W., et al.: "Optical Absorption and Faraday Rotation in YIG", Phys. Status Solidi, 1973, 59, no.1, pp.63-70.

5.83. Krivoshchekov, G.V., et al.: "Special Features of Vector Synchronous Interaction of Light Waves in Anisotropic Crystals", Opt. Spekt., 1973, 34, no.2, pp.347-50, and Opt. Spectrosc., 1973, 34, no.2, pp.196-7.

5.84. Godomskii, O.N., and Nagibarov, V.R.: "Scattering of Light by Coherent Systems", Opt. Spekt., 1973, 34, no.2, pp.387-92, and Opt. Spectrosc., 1973,34,no.2,pp.218-20.

5.85. Esayan, S.Kh., et al.: "Elastooptic Properties of Proustite", Fiz. Tver. Tela, 1973, 15, no.3, pp.907-8, and Sov. Phys.- Solid State, 1973, 15, no.3, pp.627-8.

5.86. Cheo, P.K.: "Electrooptic Properties of Reverse Biased GaAs Epitaxial Thin Film at 10.6 micron", Appl. Phys. Lett., 1973, 23, no.8, pp.439-41.

5.87. Shaldin, Yu.V., and Belogurov, D.A.: "Anisotropy of Nonlinear Refractive Index in Acentric Crystals", Opt. Spekt., 1973, 35, no.4, pp.693-701, and Opt. Spectrosc., 1973, 35, no.4, pp.403-7.

5.88. Choudhury, B.J., and Bhakar, B.S.: "Potential Scattering of Electrons in the Presence of an Electromagnetic Field", J. Phys. B., 1974, 7, no.4, pp.L137-40.

5.89. Turner, E.H., Kaminow, I.P., and Schwab, C.: "Temperature Dependence of Raman-Scattering,Electrooptic, and Dielectric, Properties of CuBr", Phys. Rev., 1974, 9B, no.6, pp.2524-9.

5.90. Shen, Y.R.: "Distinction between Resonance Raman Scattering and Hot Luminescence", Phys. Rev., 1974, 9B, no.2,pp.622-6.

5.91. Brannon, P.J., et al.: "Faraday Rotator Using ED4 Glass", Appl. Opt., 1974, 13, no.7, pp.1555-7.

5.92. Stogryn, A.: "Electromagnetic Scattering by Random Dielectric Constant Fluctuations in a Bounded Medium", Radio Sci., 1974, 9, no.5, pp.509-18.

5.93. Hepner, G., and Desormiere, B.: "Influence of the Quadratic Magnetooptical Effect on Light Propagation in Garnet Films", Appl. Opt., 1974, 13, no.9, p.2007.

5.94. Gottlieb, M., et al.: "Acoustooptic Properties of Chalcogenide Crystals", J. Appl. Phys., 1974, 45, no.12, pp.5145-51.

5.95. Sussman, A.: "Electrooptic Transfer Function in Nematic Liquids", RCA Rev., 1974, 35, no.2, pp.176-97.

5.96. Little, V.I., et al.: "Scattering of Laser Light from Light-Induced Periodic Structures", Contemp. Phys., 1974, 15, no.3, pp.271-98.

5.97. Stith, J.H., et al.: "Hypersound Speeds in CS_2, Acetone, and Benzene", J. Acoust. Soc. Am., 1974, 55, no.4, pp.785-9.

6. NONLINEAR EFFECTS

6.1. Kurtz, S.K.: "Nonlinear Optical Materials", J. Quantum Electron. IEEE, 1968, QE-4, p.578.

6.2. Murmuzhev, B.A., and Tarazenko, V.V.: "Nonlinear Theory of Frequency Conversion in Ferrites", Fiz. Tver. Tela, 1968, 10, no.11, p.3443.

6.3. Abramov, A.A., Lugovoi, V.N., and Prokhorov, A.M.: "Self-Focusing of Ultrashort Laser Pulses", Zh. Eksp. Teor. Fiz. Pis'ma, 1969, 9, no.12, p.675, and JETP Lett., 1969, 9, no.12, p.419.

6.4. Kielich, S.: "Laser-Beam-Induced Nonlinear Effects in Optically Active Media", Opto-Electron., 1969, 1, no.2, p.75.

6.5. Rao, B.R.: "Parametric Interaction in a Dielectric Medium with Second-Harmonic Space-Time Pump Modulation", Proc. IEEE, 1969, 57, p.2173.

6.6. Barsukov, Yu.K.: "Dispersion in Parametric Waveguides", Izv. VUZ Radiotekh., 1969, 12, no.6, p.610.

6.7. Taran, J.P.E., and Gustafson, T.K.: "Self-Focusing of Short Light Pulses", J. Quantum Electron. IEEE, 1969, QE-5, p.381.

6.8. Byer, R.L., et al.: "Nonlinear Optical Properties of $Ba_2NaNb_5O_{15}$ in the Tetragonal Phase", J. Appl. Phys., 1969,40,p.444.

6.9. Sukhorukov, A.P., and Tomov, I.V.: "Trebling of Optical Frequencies. I", Opt. Spekt., 1969, 27, p.58, and Opt. Spectrosc., 1969, 27, p.119.

6.10. Chmela, P.: "Synchronization of Phases of First and Second Harmonics in Nonlinear Crystals Under the Interaction of Two Waves", Optik, 1969, 28, p.346.

6.11. Nath, G.: "Strong SHG of a Ruby Laser in $LiIO_3$", Phys. Lett., 1969, 29A, p.91.

6.12. Tucker, J., and Walls, D.F.: "Quantum Theory of the TW Frequency Converter", Phys. Rev. 1969, 178, p.2036.

6.13. Meadors, J.G.: "Steady-State Theory of BW Parametric Interactions", J. Appl. Phys., 1969, 40, p.2510.

6.14. Gorlich, P., and Hofman, C.: "Aspects of Nonlinear Optics. I", Feingerate Tech., 1970, 19, no.2, p.49.

6.15. Goldberg, L.S., and Schnur, J.M.: "Optical Harmonic Generation in Liquid Crystals", Radio Electron. Eng., 1970, 39, p.279.

6.16. Gustafson, T.K., et al.: "Self-Modulation of Picosecond Pulses in Electrooptic Crystals", Opt. Commun., 1970, 2, no.1, p.17.

6.17. Kerr, E.L.: "Transient and Steady-State Electrostrictive Laser Beam Trapping", J. Quantum Electron. IEEE,1970,QE-6, p.616.

6.18. Kielich, S.: "Frequency Doubling of Light in an Isotropic Medium with Electrically Destroyed Centre of Inversion", Opto-Electron., 1970, 2, no.1, p.5.

6.19. Shvartsburg, A.B.: "Self-Focusing of a Beam of Electromagnetic Waves with an Ellipsoidal Phase Front", Izv. VUZ Radiofiz., 1970, 13, no.12, pp.1775-8.

6.20. Butyagin, O.F., et al.: "Observation of Optical Rectification Effect of Ultrashort Light Pulses", Radiotekh. Elektron., 1970, 15, no.7, and Radio Eng. Electron. Phys., 1970, 15, no.7, pp.1252-5.

6.21. Levkov, I.G., Minakova, I.I., and Semenova, T.A.: "Effect of System Parameters on Output Power for Three-Photon Frequency Multiplication in NH_3", Radiotekh. Elektron., 1970, 15, no.8, and Radio Eng. Electron. Phys., 1970, 15, no.8, pp.1439-43.

6.22. Orlov, R.Yu., Sukhorukov, A.P., and Tomov, I.V.: "Simultaneous Generation of Second and Third Optical Harmonics", Annu. Univ. Sofia Fac. Phys., 1970, 64-5,pp.283-8.

6.23. Armstrong, J.A., Jha, S.S., and Shiren, N.S.: "Effects of Group-Velocity Dispersion on Parametric Interactions", J. Quantum Electron. IEEE, 1970, QE-6, p.123.

6.24. Cassedy, E.S., and Evans, C.R.: "Optical Parametric BW Instabilities", Proc. IEEE, 1970, 58, p.164.

6.25. Rozzi, T.E.: "Modal Analysis for Nonlinear Processes in Optical Waveguides", J. Quantum Electron. IEEE, 1970, QE-6,p.539.

6.26. Singh, S., et al.: "Nonlinear Optical Susceptibility of Lithium Formate", Appl. Phys. Lett., 1970, 17, no.7,pp.292-4.

6.27. Kielich, S.: "Doubling and Mixing of Laser-Light Frequencies in Crossed Electric and Magnetic Fields", Opt. Commun., 1970, 2, no.5, pp.197-9.

6.28. Jerphagnon, J.: "Optical SHG in Isocyclic and Heterocyclic Organic Compounds", J. Quantum Electron. IEEE, 1971, QE-7, no.1, pp.42-3.

6.29. Jungling, K.C., and Gaddy, O.L.: "Self-Focusing of 10.6-micron Radiation in Liquid CS_2", J. Quantum Electron. IEEE, 1971, QE-7, no.2, pp.97-8.

6.30. Chemla, D.S., et al.: "Nonlinear Properties of Cuprous Halides", J. Quantum Electron. IEEE, 1971, QE-7, no.3, pp.126-32.

6.31. Gandrud, W.B.: "Possibility of Phase-Matched IR Mixing Using Induced Birefringence", J. Quantum Electron. IEEE, 1971, QE-7, no.3, pp.132-3.

6.32. Semenova, T.A.: "Theory of Multi-Photon Frequency Multiplication in Gases", Izv. VUZ Radiofiz., 1971, 14, no.4,pp.585-91.

6.33. Kielich, S.: "Frequency Mixing of Laser Waves on Electrically Polarized Media", Opto-Electron., 1971, 3, no.1, pp.5-11.

6.34. Topp, M.R., Jones, R.P., and Rentzepis, P.M.: "Optical Third-Harmonic Generation in Organic Liquids", Opt. Commun., 1971, 3, no.4, pp.264-8.

6.35. Andrews, R.A.: "Crystal Symmetry Effects on (Phase Matching) Processes in Optical Waveguides", J. Quantum Electron. IEEE, 1971, QE-7, no.11, pp.523-9.

6.36. Kerr, E.L.: "Electrostrictive Self-Focusing of Picosecond Laser Pulse Trains", J. Quantum Electron. IEEE, 1971, QE-7, no.11, pp.532-3.

6.37. Boyd, G.D., Kasper, H.M., and McFee, J.H.: "Nonlinear Optical Properties of $AgGaS_2$, $CuGaS_2$, and $CuInS_2$, Using Wedge Technique for Measurement", J. Quantum Electron. IEEE, 1971, QE-7, no.12, pp.563-73.

6.38. Courtens, E.: "Formation of Trapped Filaments by Self-Focusing", J. Quantum Electron. IEEE, 1971, QE-7, no.12, pp.578-9.

6.39. Tomov, I.V., and Chirkin, A.S.: "Efficiency of Generation of Higher-Order Optical Harmonics and Many-Quantum Processes in a Multimode Radiation Field", Kvantovaya Elektron., 1971, no.1, pp.110-5.

6.40. Deserno, U.: "Optical-Frequency Conversion in Mercury Thiocyanate with Phase Matching", Siemens Forsch. Entwick., 1971, 1, no.1, pp.16-9.

6.41. Okada, M., and Ieiro, S.: "Influence of Self-Induced Thermal Effects on SHG", J. Quantum Electron. IEEE, 1971, QE-7, no.9, pp.469-70.

6.42. Mikhina, T.V., Sukhorukov, A.P., and Tomov, I.V.: "Effect of Thermal Self-Interaction on Coherent Nonlinear Optical Flow Processes", Zh. Prikl. Spekt., 1971, 15, no.6, pp.1001-7.

6.43. Ito, H., Inaba, H., and Naito, H.: "Characteristics of Nonlinear Crystal HIO_3. I and II", Rec. Electr. Commun. Eng. Conversaz. Tohoku Univ., 1971, 40, no.3, pp.174-9 and 180-5.

6.44. Sodha, M.S., et al.: "Self-Focusing of Laser Beams in Inhomogeneous Dielectrics", Opto-Electron., 1971, 3, no.4, pp.157-61.

6.45. Tewari, D.P., et al.: "Waveguide Propagation of Laser Beams in Media with Saturating Nonlinearity", Opto-Electron., 1971, 3, no.4, pp.171-5.

6.46. Luk'yanov, D.P.: "Nonlinear Luminous Interactions in Electrooptic Media Excited by a Circular Modulating Field", Radiotekh. Elektron., 1971, 16, no.10, pp.1859-64, and Radio Eng. Electron. Phys., 1971, 16, no.10, pp.1667-71.

6.47. Darznek, S.A., and Suchkov, A.F.: "Estimate of Limiting Diameter of a Self-Focusing Channel in a Medium with Cubic Nonlinearity", Kvantovaya Elektron., 1971, no.4, pp.109-12.

6.48. LaCour, B., and Pocholle, J.P.: "Optical Kerr Constant Measurement in Liquids and Glasses", J. Quantum Electron. IEEE, 1972, QE-8, no.5, pp.456-7.

6.49. Singh, S., et al.: "Second-Harmonic Generation in d-Threonine", Opt. Commun., 1972, 5, no.2, pp.131-3.

6.50. Vyshkind, S.Ya., and Rabinovich, M.I.: "Parametric Wave Transformation in Active Media", Izv. VUZ Radiofiz., 1972, 15, no.10, pp.1502-8.

6.51. Virmani, C.J.K., and Dubey, P.K.: "Self-Focusing of Laser Beams in Inhomogeneous Nonlinear Dielectrics", Atti Fond Giorgio Ronchi, 1972, 27, no.3, pp.337-42.

6.52. Crosignani, B., Di Porto, P., and Solimeno, S.: "Quantum Effects in SHG", J. Phys. A, 1972, 5, no.11, pp.1119-21.

6.53. Butyagin, O.F.: "Influence of Linear Inhomogeneity of Refractive Index of Nonlinear Crystals on SHG", Kvantovaya Elektron., 1972, no.7, pp.26-32, and Sov. J. Quantum Electron., 1972, 2, no.1, pp.18-22.

6.54. Butyagin, O.F.: "Temperature and Angular Widths of Phase-Matching Curve of $LiNbO_3$", Kvantovaya Elektron., 1972, no.7, pp.76-8, and Sov. J. Quantum Electron., 1972, 2, no.1, pp.66-8.

6.55. Boyd, G.D., et al.: "Linear and Nonlinear Optical Properties of Ternary Selenides", J. Quantum Electron. IEEE, 1972, QE-8, no.12, pp.900-8.

6.56. Abdullaev, G.B., et al.: "GaSe for Nonlinear Optics", Eksp. Teor. Fiz. Pis'ma, 1972, 16, no.3, pp.130-3, and JETP Lett., 1972, 16, no.3, pp.90-2.

6.57. Shaldin, Yu.V., et al.: "Nonlinear Optical Properties of Cubic Ti-Sillenite Crystals", Kristallogr., 1972, 17, no.3, pp.674-7, and Sov. Phys.-Crystallogr., 1972, 17, no.3, pp.588-90.

6.58. Guha, S., and Tripathi, V.K.: "Laser Focusing in GaAs Intervalley Transfer Mechanism", Phys. Status Solidi, 1972, 13, no.2, pp.681-4.

6.59. Krivoshchekov, G.V., et al.: "Cascade Frequency Conversion of Laser Radiation in Nonlinear Crystals", Avtometriya, 1972, no.5, pp.106-12.

6.60. Kamiura, Y., and Kawabe, K.: "Temperature Dependence of Index-Matching Angle for SHG in KDP", J. Phys. Soc. Jap., 1972, 33, no.6, pp.1643-4.

6.61. Schenzle, A., and Haken, H.: "Self-Induced Transparency of Excitons", Opt. Commun., 1972, 6, no.2, pp.96-7.

6.62. Hermann, J.P., and Ducuing, J.: "Dispersion of Two-Photon Cross Section in Rhodamine Dyes", Opt. Commun., 1972, 6, no.2, pp.101-5.

6.63. Thek-de, Im., et al.: "Influence of Laser-Field Polarization on Nonlinear Interference Effects", Zh. Eksp. Teor. Fiz., 1972, 62, no.5, pp.1661-5, and Sov. Phys.-JETP, 1972, 35, no.5, pp.865-6.

6.64. Giuliani, J.F.: "Enhanced Optical Four-Photon Parametric Interaction in a Low-Dispersive Liquid", J. Appl. Phys., 1972, 43, no.10, pp.4266-7.

6.65. Mizuno, J.: "Pseudo-Particle Method for Nonlinear Polarizabilities and Multi-Photon Processes", J. Phys. B, 1972, 5, no.10, pp.1875-83.

6.66. Leviev, G.I.: "(Microwave) SHG in Bismuth During Anomalous Skin Effect", Zh. Eksp. Teor. Fiz., 1972, 62, no.3, pp.1031-6, and Sov. Phys.-JETP, 1972, 35, no.3, pp.544-6.

6.67. Somekh, S., and Yariv, A.: "Phase Matching by Periodic Modulation of Nonlinear Optical Properties", Opt. Commun., 1972, 6, no.3, pp.301-4.

6.68. Ohi, K., Takagi, N., and Hanawa, Y.: "Optical Rectification in alpha-HIO_3", Acta Crystallogr., 1972, 28A, pt 4, p.S239.

6.69. Danishevskii, A.M., et al.: "Dependence of Two-Photon Absorption Coefficient on Light Polarization in Cubic Semiconductors", Zh. Eksp. Teor. Fiz. Pis'ma, 1972, 16, no.11, pp.625-8, and JETP Lett., 1972, 16, no.11, pp.440-3.

6.70. Muzart, J., et al.: "Noncollinear and Collinear SHG in ZnS", Opt. Commun., 1972, 6, no.4, pp.329-32.

6.71. Kiselev, V.A., et al.: "Investigation of Spontaneous Parametric Radiation in alpha-HIO_3", Zh. Eksp. Teor. Fiz. 1972, 62, no.4, pp.1291-301, and Sov. Phys.-JETP, 1972, 35, no.4, pp.682-6.

6.72. Sukhorukov, A.P., et al.: "Steady-State Thermal Self-Focusing of Laser Beams", Kvantovaya Elektron., 1972, no.2, pp.53-60, and Sov. J. Quantum Electron., 1972, no.2, pp.133-7.

6.73. Gorbunov, L.M.: "Transient Processes in Parametrically Unstable Nonlinear Media", Zh. Eksp. Teor. Fiz., 1972, 62, no.6, pp.2141-6, and Sov. Phys.-JETP, 1972, 35, no.6, pp.1119-21.

6.74. Sokolovskii, R.I.: "Multiphoton Molecular Scattering of Light", Opt. Spekt., 1972, 33, no.3, pp.586-8, and Opt. Spectrosc., 1972, 33, no.3, pp.317-9.

6.75. Makhlin, A.N., and Skrotskii, G.V.: "Ponderomotive Forces in Optics of Nonlinear Media", Kvantovaya Elektron., 1972, no.3, pp.18-23, and Sov. J. Quantum Electron., 1972, 2, no.3, pp.211-4.

6.76. V'yukov, L.A., Lokhov, Yu.N., and Fiveiskii, Yu.D.: "Problems in the Theory of Transient Self-Focusing", Kvantovaya Elektron., 1972, no.4, pp.91-4, and Sov. J. Quantum Electron., 1972, 2, no.4, pp.367-9.

6.77. Mercier, A., and Levy, F.: "Optical SHG in GaSe", Helv. Phys. Acta, 1972, 45, no.6, pp.872-4.

6.78. Strizhevskii, V.L.: "Statistical Effects in Nonlinear Scattering of Light", Kvantovaya Elektron., 1972, no.6, pp.165-71, and Sov. J. Quantum Electron., 1972, 2, no.6.

6.79. Lyakhov, G.A.: "Stratification of a Gaussian Laser Beam in a Cubic Medium", Opt. Spekt., 1972, 33, no.5, pp.969-74, and Opt. Spectrosc., 1972, 33, no.5, pp.530-2.

6.80. Mirovskaya, Ye.A.: "Four-Photon Interaction in a Nonlinear Medium of Anisotropic Molecules", Radiotekh. Elektron., 1972, 17, no.10, pp.2159-64, and Radio Eng. Electron. Phys., 1972, 17, no.10, pp.1724-8.

6.81. Andreev, R.B., Volosov, V.D., and Kalintsev, A.G.: "Features of SHG in $LiNbO_3$", Kvantovaya Elektron., 1972, no.6, pp.44-9, and Sov. J. Quantum Electron., 1972, 2, no.6, pp.529-32.

6.82. Marange, C., et al.: "Two-Photon Absorption in CuBr", Opt. Commun., 1972, 6, no.2, pp.138-41.

6.83. Bivas, A., et al.:"Two-Photon Absorption in CuCl", Opt. Commun., 1972, 6, no.2, pp.142-5.

6.84. Eichler, H., Fery, H., and Hermann, F.: "Noncollinear Parametric Four-Photon Interaction in CdS and $SiTiO_3$", Opt. Commun., 1972, 6, no.2, pp.152-5.

6.85. Miller, R.C., Nordland, W.A., and Ballman, A.A.: "Nonlinear Optical Properties of Ferroelectric Pb_5GeO_{11}", Opt. Commun., 1972, 6, no.2, pp.210-2.

6.86. Meisner, L.B.: "Nonlinear Optical Properties of $BaTiO_3$, $LiNbO_3$, and $LiIO_3$ Crystals in a Model of Polarizable Ions", Fiz. Tver. Tela, 1972, 14, no.8, pp.2220-4, and Sov. Phys.-Solid State, 1973, 14, no.8, pp.1922-5.

6.87. Bairamov, B.Kh., et al.: "Self-Focusing of Argon-Laser Radiation in $Bi_{12}SiO_{20}$ Crystals", Fiz. Tver. Tela, 1972, 14, no.9, pp.2730-6, and Sov. Phys.-Solid State, 1973, 14, no.9, pp.2357-62.

6.88. Zil'berberg, V.V.: "Dispersion of Nonlinear Polarizability of InSb", Fiz. Tekh. Poluprov., 1972, 6, no.11, pp.2202-5, and Sov. Phys.-Semicond., 1973, 6, no.11, pp.1856-8. See also Al'tshuler, Yu.G., et al.: "Submillimetre Propagation in a Waveguide Filled With n-InSb in a Magnetic Field", Izv. VUZ Radioelektron., 1972, 15, no.8, pp.938-44.

6.89. Karov, D.D., and Koikov, S.N.: "Theoretical Analysis of Polarization and of Nonlinear Optical Properties of Single Crystals", Fiz. Tver. Tela, 1972, 14, no.12, pp.3659-64, and Sov. Phys.-Solid State, 1973, 14, no.12, pp.3064-7.

6.90. Zanadvorov, P.N., et al.: "Optical Rectification in $LiIO_3$ Crystals", Fiz. Tver. Tela, 1972, 14, no.9, pp.2794-5, and Sov. Phys.-Solid State, 1973, 14, no.9, pp.2424-5.

6.91. Arumov, G.P., et al.: "Mixing of Krypton- and CO_2-Laser Radiation in Proustite", Kvantovaya Elektron., 1973, no.5, pp.95-9, and Sov. J. Quantum Electron., 1973, 3, no.5, pp.421-3.

6.92. Kamal, A.K., and Agrawal, M.C.: "Second-Order Nonlinear Coefficient of Crystal Class 3m with DC Polarization", J. Inst. Electron. Telecommun. Eng., 1973, 19, no.12, pp.684-5.

6.93. Naito, H., and Inaba, H.: "Measured Refractive Indices of Crystalline Lithium Formate", Opto-Electron., 1973, 5, no.3, pp.256-9.

6.94. Loy, M.M.T., and Shen, Y.R.: "Study of Self-Focusing and Small-Scale Filaments of Light in Nonlinear Media", J. Quantum Electron. IEEE, 1973, QE-9, no.3, pp.409-22.

6.95. Sherman, G.H., and Coleman, P.D.: "Measurement of SHG Nonlinear Susceptibilities of Se, CdTe, and InSb, at 28 micron by Comparison with Te", J. Appl. Phys., 1973, 44, no.1, pp.238-41.

6.96. Crane, G.R.: "Determination of Relative Signs of Nonlinear Optical Coefficients", J. Appl. Phys., 1973, 44, no.2, pp.915-6.

6.97. Bates, H.E.: "Analysis of Noncollinear-Phase-Matching Effects in Uniaxial Crystals", J. Opt. Soc. Am., 1973, 63, no.2, pp.146-51.

6.98. Hofmann, B., and Vogt, H.: "Directional Pattern of Phase-Matched SHG in Biaxial $NaNO_2$", J. Phys. C, 1973, 6, no.3, pp.543-50.

6.99. Doviak, J.M., et al.: "Two-Photon Absorption in InSb at 10.6 micron", J. Phys. C, 1973, 6, no.3, pp.593-600.

6.100. Burns, W.K., and Andrews, R.A.: "Noncritical Phase Matching in Optical Waveguides", Appl. Phys. Lett., 1973, 22, no.4, pp.143-5.

6.101. Owen, T.C., Coleman, L.W., and Burgess, T.J.: "Ultrafast Optical Kerr Effect in CS_2 at 10.6 micron", Appl. Phys. Lett., 1973, 22, no.6, pp.272-3.

6.102. Sodha, M.S., et al.: "Nonlinear Mechanisms for Self-Focusing of Microwaves in Semiconductors", J. Appl. Phys., 1973, 44, no.4, pp.1699-705.

6.103. Chemla, D.S., Kupecek, Ph. J., and Schwartz, C.A.: "Redetermination of Nonlinear Optical Coefficients of Proustite by Comparison with Pyrargyrite and GaAs", Opt. Commun., 1973, 7, no.3, pp.225-8.

6.104. Levine, B.F., et al.: "Nonlinear Optical Properties of Zn_3AgInS_5 and Zn_5AgInS_7", J. Quantum Electron. IEEE, 1973, QE-9, no.2, pp.258-9.

6.105. Sherman, G.H., and Coleman, P.D.: "Absolute Measurement of SHG Nonlinear Susceptibilities of Te at 28 micron", J. Quantum Electron. IEEE, 1973,QE-9,no.3,pp.403-8.

6.106. Sobolev, V.V., Synakh, V.S., and Zakharov, V.E.: "Numerical Investigations in Nonlinear Optics", Comput. Phys. Commun., 1973, 5, no.1, pp.48-50.

6.107. Levine, B.F., Nordland, W.A., and Shiever, J.W.: "Nonlinear Optical Susceptibility of AgI", J. Quantum Electron. IEEE, 1973, QE-9, no.4, pp.468-70.

6.108. Glodz, M., and Krasinski, J.: "Two-Photon Absorption of 633-nm Radiation by Benzopyrene Solid Solution in Perspex", Lett. Nuovo Cim., 1973, 6, no.14, pp.566-8.

6.109. Morosin, B., Bergman, J.G., and Crane, G.R.: "Crystal Structure, Linear and Nonlinear Optical Properties of $Ca(IO_3)_2$", Acta Cryst., 1973, 29B,pp.1067-72.

6.110. Adhav, R.S., and Wallace, R.W.: "SHG in 90°-Phase-Matched KDP Isomorphs", J. Quantum Electron. IEEE, 1973, QE-9, no.8, pp.855-6.

6.111. Jain, R.K., and Gustafson, T.K.: "SHG of Several Argon-Ion Laser Lines", J. Quantum Electron. IEEE, 1973, QE-9, no.8, pp.859-61.

6.112. Karlov, N.V., et al.: "Self-Focusing of CO_2-Laser Radiation in Resonantly Absorbing Gases", Zh. Eksp. Teor. Fiz. Pis'ma, 1973, 17, no.7, pp.337-40, and JETP Lett., 1973, 17, no.7, pp.239-41.

6.113. Bemecker, O., Matthes, H., and Marshall, A.: "Optical Properties of $Ba_2NaNb_5O_{15}$", Phys. Status Solidi, 1973, 17A, no.2, pp.453-8.

6.114. Sodha, M.S., Khanna, R.K., and Tripathi, V.K.: "Self-Focusing of Electromagnetic Beams in InSb", Appl. Phys., 1973, 2, no.1, pp.39-42.

6.115. Desemo, U., and Haussuhl, S.: "Phase-Matchable Optical Nonlinearity in Strontium Formate", J. Quantum Electron. IEEE, 1973, QE-9, no.6, pp.586-601.

6.116. Tang, C.L.: "Simple Molecular-Orbital Theory of Nonlinear Optical Properties of III-V and II-VI Compounds", J. Quantum Electron. IEEE, 1973, QE-9, no.7, pp.755-62.

6.117. Boyd, G.D., Kasper, H.M., and McFee, J.H.: "Linear and Nonlinear Optical Properties of $LiInS_2$", J. Appl. Phys., 1973, 44, no.6, pp.2809-12.

6.118. Danishevskii, A.M., et al.: "Linear/Circular Two-Photon Dichroism in Degenerate InSb", Zh. Eksp. Teor. Fiz. Pis'ma, 1973, 17, no.4, pp.181-3, and JETP Lett., 1973, 17, no.4, pp.129-30.

6.119. Belyaev, L.M., et al.: "Dispersion of Nonlinear Susceptibility of TeO_2 in the Optical Region", Zh. Eksp. Teor. Fiz. Pis'ma, 1973, 17, no.4, pp.201-4, and JETP Lett., 1973, 17, no.4, pp.142-4.

6.120. Boyd, G.D., and Pollack, M.A.: "Microwave Nonlinearities in Anisotropic Dielectrics", Phys. Rev., 1973, 7B, no.12, pp.5345-59.

6.121. Stolen, R.H., and Ashkin, A.: "Optical Kerr Effect in Glass Waveguide", Appl. Phys. Lett., 1973, 22, no.6, pp.294-6.

6.122. Kielich, S., and Zawodny, R.: "Optical Nonlinear Phenomena in Magnetized Crystals and Isotropic Media", Acta Phys. Pol., 1973, A43, no.4, pp.579-602.

6.123. Vlasenko, N.A., et al.: "Phase Matching in Molecular Biaxial Methylnitroaniline and Anaesthesin Monocrystals", Izv. VUZ Radiofiz., 1973, 16, no.3, pp.363-8.

6.124. Iio, K., and Yanagi, Y.: "Nonlinear Optical Properties of $NaNO_2$. I and II", J. Phys. Soc. Jap., 1973, 34, no.1, pp.138-47 and 35, no.5, pp.1465-71.

6.125. Adonts, G.G., et al.: "Polarization Effects on Self-Focusing of Light", Izv. Akad. Nauk Arm. SSR Fiz., 1973, 8, no.1, pp.28-32.

6.126. Askar'yan, G.A., et al.: "Experimental Investigation of Waveguide Concentration of Radiation in a Nonlinear Medium", Zh. Eksp. Teor. Fiz. Pis'ma, 1973, 17, no.9, pp.504-8, and JETP Lett., 1973, 17, no.9, pp.363-5.

6.127. Lu, E.Y.C.: "Critical and Oscillatory Behaviour in Parametric Amplification with Time-Dependent Pump Amplitude and Phase", Lett. Nuovo Cim., 1973, 7, no.14, pp.646-8.

6.128. Dubey, P.K., and Paranjape, V.V.: "Self-Focusing of Laser Beams in Semi-conductors", Phys. Rev., 1973, 8B, no.4, pp.1514-22.

6.129. Khallik, M.: "Sum-Frequency Generation in a Nonlinear Film", Toim. Eesti NSV Tead. Akad. Fuus. Mat., 1973, 22, no.2, pp.139-46.

6.130. Hermann, J.P., Ricard, D., and Ducuing, J.: "Optical Nonlinearities in Conjugated Systems. beta-Carotene", Appl. Phys. Lett., 1973, 23, no.4, pp.178-80.

6.131. Chmela, P.: "Generation of Second Harmonic in a Gaussian Beam", Czech. J. Phys., 1973, 23B, no.9, pp.884-7.

6.132. Levine, B.F.: "Magnitude and Dispersion of Kleinman Forbidden Nonlinear Optical Coefficients", J. Quantum Electron. IEEE, 1973, QE-9, no.9, pp.946-54.

6.133. Miller, R.C., et al.: "Nonlinear Optical Properties of Cuprous Halides", J. Appl. Phys., 1973, 44, no.8, pp.3700-2.

6.134. Anderson, R.W., et al.: "Treatment of Nonlinear Absorption of Laser Radiation by a Monte Carlo Method", J. Chem. Phys., 1973, 59, no.2, pp.977-8.

6.135. Eilbeck, J.C., et al.: "Solitons in Nonlinear Optics. I", J. Phys. A, 1973, 6, no.9, pp.1337-47.

6.136. Uematsu, Y.: "Nonlinear Optical Coefficients of $KNbO_3$", Jap. J. Appl. Phys., 1973, 12, no.8, pp.1257-8.

6.137. Shigorin, V.D., and Shipulo, G.P.: "Laser SHG in Organic Crystals", Opt. Spekt., 1973, 34, no.1, pp.151-6, and Opt. Spectrosc., 1973, 34, no.1, pp.83-6.

6.138. Kelley, P.L., and Gustafson, T.K.: "Backward Stimulated Light Scattering and Limiting Diameter of Self-Focused Beams", Phys. Rev., 1973, 8A, no.1, pp.315-8.

6.139. Asher, I.M., Hopf, F.A., and Kie, L.K.: "Temperature-Dependent Phase Memory of Ruby Determined by Self-Induced Transparency", Phys. Rev., 1973, 8, no.1, pp.396-401.

6.140. Hasegawa, A., and Tappert, F.: "Transmission of Stationary Nonlinear Optical Pulses in Dispersive Dielectric Fibres. I and II", Appl. Phys. Lett., 1973, 23, no.3, pp.142-4, and no.4, pp.171-2.

6.141. Kolokolov, A.A., and Skrotskii, G.V.: "Self-Focusing and -Trapping of Light Beams in a Nonlinear Medium", Acta Phys. Austriaca, 1973, 38, no.2, pp.216-24.

6.142. Haueisen, D.C.: "Description of Nonlinear Optics on an Atomic Scale", Am. J. Phys., 1973, 41, no.11, pp.1251-4.

6.143. Chen, W.Y., and Chiang, M.F.: " "Two-Dimensional Time-Dependent Self-Curving Effect", Chin. J. Phys., 1973, 11, no.1, pp.33-40.

6.144. Vakhitov, N.G., and Kolokolov, A.A.: "Stationary Solutions of the Wave Equation with Nonlinearity Saturation", Izv. VUZ Radiofiz., 1973, 16,no.7,pp.1020-8.

6.145. Wang, C.S., Chen, J.M., and Bower, J.R.: "Second-Harmonic Generation from Alkali Metals", Opt. Commun., 1973, 8, no.4, pp.275-9.

6.146. Lippmann, B.A.: "Macroscopic Theory of Self-Induced Transparency", Opt. Commun., 1973, 8, no.4, pp.394-6.

6.147. Freund, I.: "Two-Photon Absorption of Optical and X-Ray Quanta", Opt. Commun., 1973, 8, no.4, pp.401-3.

6.148. Shaldin, Yu.V., and Belogurov, D.A.: "Dispersion of the Nonlinear Refractive Index in Noncentrosymmetric Crystals", Fiz. Tver. Tela, 1973, 15, no.2, pp.483-5, and Sov. Phys.-Solid State, 1973, 15, no.2, pp.339-40.

6.149. Zverev, G.M., Naumov, V.S., and Pashkov,V.A.: "Self-Focusing of Ultrashort Laser Pulses in Solid Dielectrics", Fiz. Tver. Tela, 1973, 15, no.2, pp.575-6, and Sov. Phys.-Solid State, 1973, 15, no.2, p.399.

6.150. Kizel', V.A., et al.: "SHG in Lead Dithionate", Fiz. Tver. Tela, 1973, 15, no.3, pp.905-7, and Sov. Phys.-Solid State, 1973, 15, no.3, pp.625-6.

6.151. Bushnev, V.A.: "Nonlinear Scattering of Light in Crystals", Vestn. Mosk. Univ., Fiz. Astron., 1973, 14, no.2, pp.177-85.

6.152. Frahm, J., and Fischer, R.: "Theory of Parametric Four-Photon Exchange Effect. II", Ann. Phys. (Leipzig), 1973, 30, no.1, pp.6-36.

6.153. Fischer, R.: "Generation of Third Optical Harmonic with High Conversion Efficiency", Ann. Phys. (Leipzig), 1973, 30, no.1, pp.93-6.

6.154. Feldman, A., Horowitz, D., and Waxler, R.M.: "Mechanisms for Self-Focusing in Optical Glasses", J. Quantum Electron. IEEE, 1973, QE-9, no.11, pp.1054-61.

6.155. Owyoung, A.: "Ellipse Rotation Studies in Laser Host Materials", J. Quantum Electron. IEEE, 1973, QE-9,no.11, pp.1064-9.

6.156. Gyuzalyan, R.N., Karmenyan, K.V., and Chilingaryan, Yu.S.: "Nonlinear Optical Effects for Picosecond Pumping", Izv. Akad. Nauk Arm. SSR Fiz., 1973, 8, no.2, pp.125-32.

6.157. Bridenbaugh, P.M.: "Nonlinear Optical Properties of $InPS_4$", Mater. Res. Bull., 1973, 8, no.9, pp.1055-60.

6.158. Wright, N., and Newstein, M.C.: "Self-Focusing of Coherent Pulses", Opt. Commun., 1973, 9, no.1, pp.8-13.

6.159. Hermann, J.P.: "Absolute Measurements of Third-Order Susceptibilities", Opt. Commun., 1973, 9, no.1, pp.74-9.

6.160. Gelbart, W.M.: "SHG by Atomic Fluids", Chem. Phys. Lett., 1973, 23, no.1, pp.53-5.

6.161. Moskalenko, S.A., Sinyak, V.A., and Khadzhi, P.I.: "Vortex-Type Solutions for Self-Focusing of Laser Radiation", Izv. Akad. Nauk SSSR, Ser. Fiz., 1973, 37, no.10, pp.2152-4.

6.162. Davis, L.W., and Lin, Y.S.: "Propagation of Optical Pulses in a Saturable Absorber", J. Quantum Electron. IEEE, 1973, QE-9, no.12, pp.1135-8.

6.163. Kielich, S., and Zawodny, R.: "Nonlinear Magnetooptical Phenomena in Crystals and Liquids", Opt. Acta, 1973, 20, no.11, pp.867-77.

6.164. Chen, J.M., Bower, J.R., and Wang, C.S.: "Optical SHG from Submonolayer Na-Covered Ge Surfaces", Opt. Commun., 1973, 9, no.2, pp.132-4.

6.165. Kato, K.: "Efficient Second-Harmonic Generation in CDA", Opt. Commun., 1973, 9, no.3, pp.249-51.

6.166. Kildal, H., and Mikkelsen, J.C.: "Nonlinear Optical Coefficient, Phasematching, and Optical Damage in the Chalcopyrite $AgGaSe_2$", Opt. Commun., 1973, 9, no.3, pp.315-8.

6.167. Mitsui, T., et al.: "Optical SHG in Ammonium Tartrate", Opt. Commun., 1973, 9, no.3, pp.322-4.

6.168. Fischer, R.: "Theory of Optical Third-Harmonic Generation by Optimal Focusing", Opto-Electron., 1973, 5, no.6, pp.503-8.

6.169. Stefanovich, S.Yu., and Venevtsev, Yu.N.: "Determination of Parameters of Ferroelectric Crystals by SHG", Phys. Status Solidi, 1973, 20A, no.1, pp.K49-52.

6.170. Rao, D.V.G.L.N., and Jayaraman, S.: "Self-Focusing of Laser Light in the Isotropic Phase of a Nematic Liquid Crystal", Appl. Phys. Lett., 1973, 23, no.10, pp.539-40.

6.171. Kalintsev, A.G., Volosov, V.D., and Andreev, R.B.: "Optical SHG in $LiIO_3$ Crystal under (oee)-Wave Interaction", Opt. Spekt., 1973, 35, no.1, pp.167-8, and Opt. Spectrosc., 1973, 35, no.1, p.96.

6.172. Shaldin, Yu.V., et al.: "Nonlinear Optical Properties of Para-Tellurite Crystals", Kristallogr., 1973, 18, no.3, pp.570-2, and Sov. Phys.-Crystallogr., 1973, 18, no.3, pp.358-9.

6.173. Campillo, A.J., Shapiro, S.L., and Suydam, B.R.: "Periodic Breakup of Optical Beams due to Self-Focusing", Appl. Phys. Lett., 1973, 23, no.11, pp.628-30.

6.174. Shimizu, F.: "Numerical Calculation of Self-Focusing and Trapping of a Short Light Pulse in Kerr Liquids", IBM J. Res. Dev., 1973, 17, no.4, pp.286-98.

6.175. Sodha, M.S., Tripathi, V.K., and Nayyar, V.P.: "Cross-Focusing of Two Coaxial Spatial Modes of a Single Frequency of a Laser Beam in a Nonlinear Dielectric", Opt. Commun., 1973, 9, no.4, pp.381-4.

6.176. Bjorkholm, J.E., and Ashkin, A.: "CW Self-Focusing and Self-Trapping of Light in Sodium Vapour", Phys. Rev. Lett., 1973, 32, no.4, pp.129-32.

6.177. Fossum, H.J., and Chang, D.B.: "Two-Photon Excitation Rate in InSb", Phys. Rev., 1973, 8B, no.6, pp.2842-9.

6.178. Haveisen, D.C., and Mahr, H.: "Noncollinear-Beam SHG in CuCl", Phys. Rev., 1973, 8B, no.6, pp.2969-74.

6.179. Freund, I., and Levine, B.F.: "Surface Effects in Nonlinear Interaction of X-Ray and Optical Fields", Phys. Rev., 1973, 8B, no.6, pp.3059-60.

6.180. Gorelik, V.S., et al.: "Temperature Dependence of Intensity of SHG in $BaTiO_3$", Fiz. Tver. Tela, 1973, 15, no.6, pp.1688-92, and Sov. Phys.-Solid State, 1973, 15, no.6, pp.1133-5.

6.181. Akhundov, G.A., et al.: "SHG in III-VI Compounds", Fiz. Tekh. Poluprov., 1973, 7, no.6, pp.1229-31, and Sov. Phys.-Semicond., 1973, 7, no.6, pp.826-7.

6.182. Ablova, L.A., and Bobrysheva, A.I.: "Influence of Uniaxial Deformation and of a Static Electric Field on Two-Photon Exciton Absorption of Cu_2O", Fiz. Tekh. Poluprov., 1973, 7, no.6, pp.712-6.

6.183. Dmitriev, V.G., et al.: "Frequency Multiplication in Nonlinear $LiIO_3$", Kvantovaya Elektron., 1973, no.2, pp.64-6, and Sov. J. Quantum Electron., 1973, 3, no.2, pp.126-7.

6.184. Berezovskii, V.V., et al.: "Two-Photon Absorption in Proustite", Kvantovaya Elektron., 1973, no.2, pp.74-5, and Sov. J. Quantum Electron., 1973, 3, no.2, pp.134-5.

6.185. Anderson, R.J.: "Theory of Interaction of Light Beams via Polariton-Polariton Scattering", Phys. Rev., 1973, 8B no.8, pp.3861-4.

6.186. Tripathi, B.P., Laloraya, R.K., and Srivastava, S.L.: "Calculation of Linear and Nonlinear Polarizability Coefficient of Xenon at Ruby-Laser Frequency", Indian J. Pure. Appl. Phys., 1973, 11, no.8, pp.594-6.

6.187. Popolitov, V.I., and Lobachev, A.N.: "Synthesis and Properties of Single Crystals of CuI", Izv. Akad. Nauk SSSR, Neorg. Mater., 1973, 9, no.6, pp.1062-3, and Inorg. Mater., 1973, 9, no.6, pp.949-50.

6.188. Kolokolov, A.A., and Skrotskii, G.V.: "Self-Constriction of Spherically Symmetric Pulses in a Nonlinear Medium", Opt. Spekt., 1973, 35, no.5, pp.898-901, and Opt. Spectrosc., 1973, 35, no.5, pp.520-1.

6.189. Sokolovskii, R.I.: "Nonlinear Interaction of Intersecting Light Beams in a Gas", Opt. Spekt., 1973, 35, no.5, pp.972-3, and Opt. Spectrosc., 1973, 35, no.5, pp.563-4.

6.190. Kabelka, V.I., Piskarskas, A.S., and Stabinis, A.Yu.: "Frequency Dependence of Group Velocities and Dispersion Spreading of Wave Packets in a KDP Crystal", Kvantovaya Elektron., 1973, no.5, pp.135-7, and Sov. J. Quantum Electron., 1973, 3, no.5, p.453.

6.191. Bairamov, B. Kh., et al.: "Nonlinear Rotation of Plane of Polarization of Light in $Bi_{12}GeO_{20}$ Crystals", Fiz. Tver. Tela, 1973, 15, no.6, pp.1868-73, and Sov. Phys.-Solid State, 1973, 15,no.6,pp.1245-8.

6.192. Golyaev, Yu.D., et al.: "Efficient Frequency Doubler Utilizing CDA", Kvantovaya Elektron., 1973, no.7, pp.122-3, and Sov. J. Quantum Electron., 1973, 3, no.1, pp.72-3.

6.193. Kerherve, F.: "Application of Nonlinear Optical Methods to Study of Plastic Films", Opt. Commun., 1973, 9, no.4,pp.420-3.

6.194. Lisitsa, M.P., and Fekeshgazi, I.V.: "Angular Dependence of Second Harmonic in Nonlinear Crystals with Tetragonal Symmetry", Zh. Tekh. Fiz., 1973, 43, no.10, pp.2153-7, and Sov. Phys.-Tech. Phys., 1974, 18, no.10, pp.1352-4.

6.195. Christensen, C.P., and Schwarz, S.E.: "Nonlinear-Optical-Device Characteristics of Gaseous Saturable Absorbers", J. Quantum Electron. IEEE, 1974, QE-10, no.3, pp.338-46.

6.196. Inoue, K., and Ishidate, T.: "Determination of Relative Signs of Optical Nonlinear Susceptibility Components of $NaNO_2$", J. Phys. Soc. Jap., 1974, 36, no.1, p.304.

6.197. Chmela, P.: "Theory of SHG in Focused Light Beams with (ooe)-Wave Interaction", Czech. J. Phys., 1974, 24B, no.1, pp.1-23.

6.198. Levenson, M.D.: "Feasibility of Measuring the Nonlinear Index of Refraction by Third-Order Frequency Mixing", J. Quantum Electon. IEEE, 1974, QE-10, no.2, pp.110-5.

6.199. Chemla, D.S., Begley, R.F., and Byer, R.L.: "Experimental and Theoretical Studies of Third-Harmonic Generation in Chalcopyrite $CdGeAs_2$", J. Quantum Electron. IEEE, 1974, QE-10, no.1, pp.71-80.

6.200. Thomas, P., Jares, A., and Stoicheff, B.P.: "Nonlinear Refractive Index and DC Kerr Constant of Liquid CS_2 at 10.6 micron", J. Quantum Electron. IEEE, 1974, QE-10, no.5, pp.493-4.

6.201. Campillo, A.J., Shapiro, S.L., and Saydam, B.R.: "Relationship of Self-Focusing to Spatial Instability Modes", Appl. Phys. Lett., 1974, 24, no.4,pp.178-80.

6.202. Wong, G.K.L., and Shen, Y.R.: "Transient Self-Focusing in a Nematic Liquid Crystal in the Isotropic Phase", Phys. Rev. Lett., 1974, 32, no.10, pp.527-30.

6.203. Agrawal, G.P., and Mehta, C.L.: "Dynamics of Parametric Processes with a Trilinear Hamiltonian", J. Phys. A, 1974, 7, no.5, pp.607-16.

6.204. McNeil, K.J., and Walls, D.F.: "Master-Equation Approach to Nonlinear Optics", J. Phys. A, 1974, 7, no.5, pp.617-31.

6.205. Byer, R.L., et al.: "SHG and IR Mixing in $AgGaSe_2$", Appl. Phys. Lett., 1974, 24, no.2, pp.65-8.

6.206. Venkataraman, R., and Rivlin, R.S.: "Harmonic Generation (in Nonlinear Dielectrics)", Z. Angew. Math. Phys., 1974, 24, no.5, pp.661-75.

6.207. Averbukh, B.B., Krivoshchekov, G.V., and Nikulin, N.G.: "Excitation of Second Harmonic by Random Series of Ultrashort Light Pulses", Avtometriya, 1974,no.1,pp.97-9.

6.208. Sodha, M.S., and Tripathi, V.K.: "Self-Focusing of a Laser Beam in Anisotropic Crystals", Optik, 1974, 40, no.2, pp.121-9.

6.209. Chmela, P.: "Method of SHG by (oee)-Wave Interaction", Optik, 1974, 40, no.3, pp.310-4.

6.210. Ostrovskii, L.A.: "Approximate Methods in Nonlinear Wave Theory", Izv. VUZ Radiofiz., 1974, 17, no.4, pp.454-76.

6.211. Akhmanov, S.A.: "Interaction of Random Waves in Nonlinear Media", Izv. VUZ Radiofiz., 1974, 17, no.4, pp.541-69.

6.212. Yakubovich, E.I.: "Nondegenerate Interaction of Parametrically Coupled Waves", Izv. VUZ Radiofiz., 1974, 17, no.4, pp.627-9.

6.213. Kupecek, Ph. J., Schwartz, C.A., and Chemla, D.S.: "Silver Thiogallate $(AgGaS_2)$. I. Nonlinear Optical Properties", J. Quantum Electron. IEEE, 1974, QE-10, no.7, pp.540-5.

6.214. Miller, R.C., and Nordland, W.A.: "Optical SHG Coefficients for Guanidinium Aluminium Sulphate", J. Appl. Phys., 1974, 45, no.2, pp.898-900.

6.215. Finn, R.S., and Ward, J.F.: "Measurements of Hyperpolarizabilities for Halogenated Methanes", J. Chem. Phys., 1974, 60, no.2, pp.454-8.

6.216. Levenson, M.D., and Bloembergen, N.: "Dispersion of Nonlinear Optical Susceptibilities of Organic Liquids and Solutions", J. Chem. Phys., 1974, 60, no.4, pp.1323-7.

6.217. Brancus, D., and Dorobantu, I.A.: "Kinetic Theory of Nonlinear Magnetooptic Phenomena. II", Rev. Roum. Phys., 1974, 19, no.2, pp.137-47.

6.218. Chmela, P.: "Theory of SHG in Guassian Beams for (oee)-Wave Interaction", Czech. J. Phys., 1974, 24B, no.5, pp.506-21.

6.219. Zakharov, V.E.: "Hamiltonian Formalism for Waves in Nonlinear Media with Dispersive Phase", Izv. VUZ Radiofiz., 1974, 17, no.4, pp.431-53.

6.220. De Martini, F., and Giuliani, G.: "Optical Shocks in Nonlinear Two-Photon Interactions", Opt. Commun., 1974, 11, no.1, pp.42-5.

6.221. Cotter, D., et al: "BW Medium-IR Downconversion in Proustite", Opt. Commun., 1974, 11, no.1, pp.54-6.

6.222. Chemla, D.S., et al.: "Optical BW Mixing in NaNO$_2$", Opt. Commun., 1974, 11, no.1, pp.57-61.

6.223. Wherret, B.S., and Pidgeon, C.R.: "Electric-Dipole Contributions to Resonant Far-IR Difference-Frequency Mixing in InSb", Phys. Rev., 1974, 9B, no.2, pp.711-5.

6.224. Kramer, S.D., Parsons, F.G., and Bloembergen, N.: "Interference of Third-Order Mixing and Exciton SHG in CuCl", Phys. Rev., 1974, 9B, no.4, pp.1853-6.

6.225. Tulub, A.V.: "Derivation of Non-linear Optics Equations", Dokl. Akad. Nauk SSSR, 1973, 212, no.1-3, pp.584-7, and Sov. Phys.-Dokl., 1974, 18, no.9, pp.610-1.

6.226. Glaze, J.A., and Bliss, E.S.: "Self-Induced Beam Steering of Intense Laser Pulses in Nonlinear Media", Appl. Opt., 1974, 13, no.6, pp.1288-9.

6.227. Kato, K.: "SHG in CDA and CD*A", J. Quantum Electron. IEEE, 1974, QE-10, no.8, pp.616-8.

6.228. Kato, K.: "High Efficient UV Generation at 347.2nm in RDA", J. Quantum Electron. IEEE, 1974, QE-10, no.8, pp.622-4.

6.229. Chakravarti, A.K., and Phadke, U.P.: "Optical SHG in Ba$_2$NaNb$_5$O$_{15}$", J. Appl. Phys., 1974, 45, no.3, pp.1461-2.

6.230. Bergman, J.G., and Crane, G.R.: "Structural Aspects of Nonlinear Optics", J. Chem. Phys., 1974, 60, no.6, pp.2470-4.

6.231. Tomov, I.V.: "Four-Photon Parametric Processes in Metal Vapours and Inert Gases", Phys. Lett., 1974, 48A, no.2, pp.153-4.

6.232. Zawadzki, W.: "Nonlinear Optical Response of Mobile Holes in Tellurium", Phys. Rev. Lett., 1974, 32, no.24, pp.1373-6.

6.233. Rahn, O., and Maier, M.: "Raman-Limited Beam Diameters in Self-Focusing of Laser Light", Phys. Rev., 1974, 9A, no.3, pp.1427-37.

6.234. Biswas, S.N., Hague, S.N., and Mohan, M.: "Two-Raman-Photon Emission in an Intense EM Field", Phys. Rev., 1974, 9A, no.2, pp.631-6.

6.235. Piekarra, A.H., Moore, J.S., and Feld, M.S.: "Analysis of Self-Trapping Using the Wave Equation with High-Order Nonlinear Electric Permittivity", Phys. Rev., 1974, 9A, no.3, pp.1403-7.

6.236. Weinmann, D., and Vogt, H.: "Second-Harmonic Light Scattering by Laminar Ferroelectric Domains", Phys. Status Solidi, 1974, 23A, no.2, pp.463-72.

6.237. Gundersen, M.: "Conversion of 28-micron Radiation to Visible Light Using Bound Excitons in CdS", Appl. Phys. Lett., 1974, 24, no.12, pp.591-2.

6.238. Cornolti, F., et al.: "Two-Photon Absorption in CdS by Low-Power CW Laser", J. Lumin., 1974, 8, no.6, pp.462-70.

6.239. Kupecek, Ph.J., Batifol, E., and Kuhn, A.: "Optical Frequency Conversion in GaSe", Opt. Commun., 1974, 11, no.3, pp.291-5.

6.240. Ralford, M.T.: "Degenerate Parametric Amplification with Time-Dependent Pump Amplitude and Phase", Phys. Rev., 1974, 9A, no.5, pp.2060-9.

6.241. Leung, K.M., Ward, J.F., and Orr, B.J.: "Two-Photon Resonant Optical, Third-Harmonic, Generation in Caesium Vapour", Phys. Rev., 1974, 9A, no.6, pp.2440-8.

6.242. Klyshko, D.N., and Polkovnikov, B.F.: "Phase- and Self-Modulation of Light in Three-Photon Processes", Kvantovaya Elektron., 1974, no.4, pp.81-5, and Sov. J. Quantum. Electron., 1974, 3, no.4, pp.324-6.

6.243. Dolgopolov, V.V., El-Siragy, N.M., and Sayed, Y.A.: "Generation and Radiation of Second Harmonics by s-Polarized Waves Incident on a Narrow Inhomogeneous Plasma Layer", J. Plasma Phys., 1974, 12, pt 1, pp.15-20.

6.244. Saydam, B.R.: "Self-Focusing of Very Powerful Laser Beams. II", J. Quantum Electron. IEEE, 1974, QE-10, no.11, pp.837-43.

6.245. Uematsu, Y.: "Nonlinear Optical Properties of KNbO$_3$ Single Crystal in the Orthorhombic Phase", Jap. J. Appl. Phys., 1974, 13, no.9, pp.1362-8.

6.246. Brancus, D.: "Resonant Magnetooptical Phenomena and Self-Induced Effects of Laser Beam in n-Type Semiconductors", Int. J. Electron., 1974, 37, no.4, pp.469-80.

6.247. Zhidko, Yu.M.: "Maximum Pulse Compression in a Uniform Medium with Dispersion", Izv. VUZ Radiofiz., 1974, 17, no.7, pp.959-63.

6.248. Kondratenko, A.N., Shaptala, V.G., and Kuklin, V.M.: "SH Generation and Emission in Wave Conversion at a Plasma Boundary", Zh. Tekh. Fiz., 1974, 44, no.1, pp.38-42, and Sov. Phys.-Tech. Phys., 1974, 19, no.1, pp.22-4.

6.249. Kato, K., and Nakao, S.: "Frequency Doubling of Nd^{3+}:YAG Laser Radiation in RDP", Jap. J. Appl. Phys., 1974, 13, no.10, pp.1681-2.

6.250. Hsu, H., and Yu, C.: "Parametric Amplification, Oscillation, and Mixing, in Nonlinear Backward Scattering", Oyo Buturi, 1974, 43, pp.75-80.

6.251. Kato, K.: "Efficient UV generation at 354.7 nm in RDP", Appl. Phys. Lett., 1974, 25, no.6, pp.342-3.

6.252. Gupta, S.C.: "Edge Structure Patterns in LiNbO₃ Using Ruby Laser", Indian J. Pure Appl. Phys., 1974,12,no.1,pp.57-9.

6.253. Levine, B.F., Bethea, C.G., and Kasper, H.M.: "Nonlinear Optical Susceptibility of Thiogallate CdGa₂S₄", J. Quantum Electron. IEEE, 1974, QE-10,no.12, pp.904-6.

6.254. Diels, J.C., and Schafer, F.P.: "Phase-Matched Third-Harmonic Generation in Dye Solutions", Appl. Phys., 1974, 5, no.3, pp.197-202.

6.255. Vedenov, A.A., et al.: "Generation of Far-IR Radiation by Mixing in GaAs at Room Temperature", Phys. Lett., 1974, 50A, no.2, pp.79-80.

6.256. Voronov, V.V., et al.: "Nonlinear Properties of Ba₂NaNb₅O₁₅", Zh. Eksp. Teor. Fiz. Pis'ma, 1974, 20, no.1, pp.26-7, and JETP Lett., 1974, 20, no.1.

6.257. Andreev, R.B., Volosov, V.D., and Kalintsev, A.G.: "Optical Characteristics of Various Nonlinear Crystals in Second- and Fourth-Harmonic Generation", Opt.Spekt., 1974, 37, no.2, pp.294-9, and Opt. Spectrosc., 1974, 37, no.2, pp.169-71.

6.258. Brukner, F., et al.: "Self-Induced Transparency in a Semiconductor", Zh. Eksp. Teor. Fiz., 1974, 67, no.6, pp.2219-26, and Sov. Phys.-JETP.

6.259. Kabelka, V.I., et al.: "Features of Parametric Interaction of Ultrashort Light Packets in LiIO₃ Crystals", Zh.Prikl. Spekt., 1974, 21, no.5, pp.947-50.

6.260. Marie, P.: "Self-Focusing of Waves having Circular Symmetry", Ann. Telecommun., 1974, 29, no.9-10, pp.471-80.

6.261. Oberman, C., and Auer, G.: "General Theory of Enhanced Induced Emission in Plasmas", Phys. Fluids, 1974, 17, no.11, pp.1980-2.

6.262. Varnavskii, O.P., Veduta, A.P., and Kirsanov, B.P.: "Upconversion of an Image by Stimulated Parametric Scattering in Isotropic Media", Kvantovaya Elektron., 1974, 1, no.2, pp.410-3, and Sov. J. Quantum Electron., 1974, 4, no.2, pp.230-1.

6.263. Reutov, A.T., and Tarashchenko, P.P.: "Frequency Multiplication in an Optical Waveguide with Nonlinear Layer of LiNbO₃", Opt. Spekt., 1974, 37, no.4, pp.786-7, and Opt. Spectrosc., 1974, 37, no.4, pp.447-8.

6.264. Brukner, F., Dneprovskii, V.S., and Koshchug, D.G.: "Self-Induced Exciton Transmittance", Zh. Eksp. Teor. Fiz. Pis'ma, 1974, 20, no.1, pp.10-3, and JETP Lett., 1974, 20, no.1.

6.265. Chemla, D.S.: "Nonlinear Optical Properties of Crystals with Chalcopyrite Structure", Ann. Telecommun., 1974, 29, no.11-12, pp.565-76.

6.266. Vlasov, S.N., Gurbatov, S.N., and Piskunova, L.V.: "Self-Focusing of Beams with Elliptical Cross Section", Izv. VUZ Radiofiz., 1974, 17, no.12, pp.1805-11.

6.267. Wherrett, B.S., et al.: "Electric-Dipole Contributions to Resonant Far-IR Difference-Frequency Mixing in InSb", Trans. IEEE, 1974, MTT-22, no.12, pp.1100-3.

6.268. Yeleonskii, V.M., Oganesyants, L.G., and Silin, V.P.: "Dispersion Relationships and Energy Flux of Self-Focusing Waveguides", Izv. VUZ Radiofiz., 1974, 17, no.12, pp.1812-6.

6.269. Stroganov, V.I., and Samarin, V.I.: "Interference Effects during Excitation of Optical Harmonics", Opt. Spekt., 1974, 37, no.2, pp.300-2, and Opt. Spectrosc., 1974, 37, no.2, pp.172-3.

7. RADIATION TECHNIQUES

7.1. Faulkner, T.R.: "Diffraction of a Plane Wave by a Metallic Strip", J. Inst. Math. Applic., 1965, 1, p.149.

7.2. Borison, S.L.: "Diagonal Representation of Radar Scattering Matrix for an Axially Symmetric Body", Trans. IEEE, 1965, AP-13, p.176.

7.3. Barakat, R.: "Diffraction of Cylindrical EM Waves by Ribbons and Slits", J. Opt. Soc. Am., 1965, 55, p.992.

7.4. Rheinstein, J.: "Scattering of Short Pulses of EM Waves", Proc. IEEE, 1965, 53, p.1069.

7.5. Lewis, B.L.: "Extended Coverage Corner Reflector", Proc. IEEE, 1965, 53, p.734.

7.6. Baumler, P., and Horn, M.: "Effect of Curvature of a Screen Edge on Diffraction of Centimetre Waves", Z. Phys., 1965, 187, p.386.

7.7. Richmond, J.H.: "Scattering by an Arbitrary Array of Parallel Wires", Trans. IEEE, 1965, MTT-13, p.408.

7.8. Moffatt, D.L., and Kennaugh, E.M.: "Axial Echo Area of a Perfectly Conducting Prolate Spheroid", Trans. IEEE, 1965, AP-13, p.401.

7.9. Tricoles, G., and Rope, E.L.: "Scattering of Microwaves by Dielectric Slabs and Hollow Dielectric Wedges", J. Opt. Soc. Am., 1965, 55, p.1479.

7.10. Litvinenko, L.N., and Shestopalov, V.P.: "Diffraction Properties of Two-Element Asymmetric Metal Grids", Radiotekh. Elektron., 1965, 10, p.1135, and Radio Eng. Electron. Phys., 1965, 10, no.6, pp.968-71.

7.11. Richmond, J.H.: "Scattering by a Dielectric Cylinder of Arbitrary Cross-Section Shape", Trans. IEEE, 1965, AP-13, p.334.

7.12. Blore, W.E., and Musal, H.M.: "Radar Cross Section of Metal Hemispheres, Spherical Segments, and Partially Capped Spheres", Trans. IEEE, 1965, AP-13, p.478.

7.13. Crispin, J.W., and Maffett, A.L.: "Radar-Cross-Section Estimation for Simple Shapes", Proc. IEEE, 1965, 53, p.833.

7.14. Crispin, J.W., and Maffett, A.L.: "Radar-Cross-Section Estimation for Complex Shapes", Proc. IEEE, 1965, 53, p.972.

7.15. Yeh, C.: "Backscattering Cross Section of a Dielectric Elliptic Cylinder", J. Opt. Soc. Am., 1965, 55, p.309.

7.16. Gruber, H.: "Intensity and Anomalous Phase in the Image Space of Microwave Lenses", Ann. Phys. (Paris), 1965, 15, p.225.

7.17. Schrank, H.E.: "Spherical Radar Reflectors with High-Gain Omnidirectional Response", Proc. IEEE, 1965, 53, p.1117.

7.18. Arbuthnot, R.S., and Badcoe, S.R.: "Enhancement of Radar Echoing Area of Gliders at S and X Bands", Radio Electron. Eng., 1965, 30, p.123.

7.19. Ross, R.A.: "Radar Cross Section of Rectangular Flat Plates as a Function of Aspect Angle", Trans. IEEE, 1966, AP-14, p.329.

7.20. Moreland, J., and Peters, L.: "Specular Radar Cross Section of Absorber-Coated Bodies", Trans. IEEE, 1966, AP-14, no.6, pp.799-800.

7.21. Weiner, S.D., and Borison, S.L.: "Radar Scattering from Blunted Cone Tips", Trans. IEEE, 1966, AP-14, no.6, pp.774-81.

7.22. DiCaudo, V.J., and Martin, W.W.: "Approximate Solution to Bistatic Radar Cross Section of Finite Length, Infinitely Conducting, Cylinder", Trans. IEEE, 1966, AP-14, no.6, pp.782-6.

7.23. Mayer, A.: "Mean Radar Cross Section of Finite Cylindrical Wires. Dependence on Conductivity and Frequency", Proc. IEEE, 1967, 55, no.8, pp.1502-4.

7.24. Weiner, S.D.: "Scattering by Edges. Hemisphere", Trans. IEEE, 1967, AP-15, no.5, pp.709-10.

7.25. Panchenko, B.A.: "Diffraction by a Flat Screen Having a Finite Thickness and Evenly Distributed Apertures", Radiotekh. Elektron., 1967, 12, no.4, pp.719-21, and Radio Eng. Electron. Phys., 1967, 12, no.4, pp.667-70.

7.26. Pimenov, Yu.V., and Kravtsov, V.A.: "Reflection of a Plane Wave from a Perfectly Conducting Ellipsoid", Radiotekhnika, 1967, 22, no.2, pp.93-5, and Telecommun. Radio Eng. Pt 2, 1967, 22, no.2, pp.141-3.

7.27. Bowman, J.J., and Senior, T.B.A.: "Diffraction of a Dipole Field by a Perfectly Conducting Half-Plane", Radio Sci., 1967, 2, no.11, pp.1339-45.

7.28. Lee, S.W., and Mittra, R.: "Scattering by Moving Cylinder in Free Space", Can. J. Phys., 1967, 45, no.9, pp.2999-3008.

7.29. Kattawar, G.W., and Plass, G.N.: "Scattering from Absorbing Spheres", Appl. Opt., 1967, 6, no.8, pp.1377-82.

7.30. Hong, S., and Borison, S.L.: "Short-Pulse Scattering by a Cone. Direct and Inverse", Trans. IEEE, 1968, AP-16, no.1, pp.98-102.

7.31. Ross, R.A., Freeny, C.C., and Cleary, J.C.: "Bistatic Scattering Matrix for a Finite Right-Circular Cylinder", Electron. Lett., 1968, 4, no.8, pp.148-9.

7.32. Hamid, M.A.K.: "Diffraction by a Conical Horn", Trans. IEEE, 1968, AP-16, p.520.

7.33. Restrick, R.C.: "Scattering by a Moving Conducting Sphere", Radio Sci., 1968, 3, p.1144.

7.34. Hyde, G.: "Studies of the Focal Region of a Spherical Reflector", Trans. IEEE, 1968, AP-16, p.646.

7.35. Whitford, B.G., and Pavlasek, T.J.F.: "Focal-Region Fields of Annular and Sectoral Microwave Apertures", J. Opt. Soc. Am., 1968, 58, p.1591.

7.36. Witt, H.R., and Price, E.L.: "Scattering from Hollow Conducting Cylinders", Proc. IEE, 1968, 115, no.1, pp.94-9.

7.37. Bruning, J.H., and Lo, Y.T.: "Electromagnetic Scattering by Two Spheres", Proc. IEEE, 1968, 56, no.1, pp.119-20.

7.38. Tan, H.S.: "Electromagnetic Diffraction by a Plane Slit Aperture", Aust. J. Phys., 1968, 21, no.1, pp.35-41.

7.39. Weiss, M.R.: "Numerical Evaluation of Geometrical-Optics Radar Cross Section", Trans. IEEE, 1969, AP-17, p.239.

7.40. Thomas, B.M., Minnett, H.C., and Bao, V.T.: "Fields in the Focal Region of a Spherical Reflector", Trans. IEEE, 1969, AP-17, p.229.

7.41. Malushkov, G.D.: "Scattering of a Plane Wave from Cones", Izv. VUZ Radiofiz., 1969, 12, p.1862.

7.42. Censor, D.: "Scattering by a Cylinder Moving Along its Axis", Trans. IEEE, 1969, MTT-17, p.154.

7.43. Bem, D.J.: "Electric-Field Distribution in the Focal Region of an Offset Paraboloid", Proc. IEE, 1969, 116, p.679.

7.44. Banes, P.: "Connection of Geometrical Optics with Propagation of Gaussian Beams and Theory of Resonators", Opto-Electron., 1969, 1, no.2, p.103.

7.45. Mitchell, R.L.: "RCS Statistics of Randomly Oriented Discs and Rods", Trans. IEEE, 1969, AP-17, p.370.

7.46. Pace, J.R.: "Asymptotic Formulae for Coupling between Two Antennas in the Fresnel Region", Trans. IEEE, 1969, AP-17, p.285.

7.47. Einarrson, O.: "Scattering by a Thin Finite Wire", Acta Polytech. Scand., 1969, EL23, p.3.

7.48. Sedov, G.S., and Smorchkova, S.A.: "Transformation of Spatial Structure of Laser Radiation by a Lens", Opt. Spekt., 1969, 26, no.3, p.398, and Opt. Spectrosc., 1969, 26, no.3, p.218.

7.49. Rudge, A.W.: "Focal-Plane Field Distribution of Parabolic Reflectors", Electron. Lett., 1969, 5, p.510.

7.50. Alexopoulos, N.G.: "Radar Cross Section of Perfectly Conducting Spheres Coated with Radially Inhomogeneous Dielectrics", Trans. IEEE, 1969, AP-17, p.667.

7.51. Ivanov, E.A., and Lukhanina, G.M.: "Diffraction of a Plane EM Wave by Two Circular Cylinders of Finite Length", Izv. VUZ Radiofiz., 1969, 12, p.1849.

7.52. Lavanchy, J.P., and Odone, G.B.: "Automatic Rangefinder and Focus-Control System", J. Soc. Motion Picture Telev. Eng., 1969, 78, p.32.

7.53. Eichmann, G.: "Propagation of Light Beams in Continuously Focusing Lenslike Media", Proc. IEEE, 1970, 58, p.837.

7.54. Thomas, B.M.: "Matching Focal-Region Fields with Hybrid Modes", Trans. IEEE, 1970, AP-18, p.404.

7.55. Bem, D.J.: "Poynting's Vector in the Focal Region of a Paraboloidal Reflector", Arch. Elektrotech., 1970, 19, no.2, p.203.

7.56. Crispin, J.W., and Maffett, A.L.: "Estimating the Radar Cross Section of a Cavity", Trans. IEEE, 1970, AES-6, p.672.

7.57. Peters, W.N., and Ledger, A.M.: "Techniques for Matching Laser TEM_{00} Mode to Obscured Circular Aperture", Appl. Opt., 1970, 9, p.1435.

7.58. Uchida, T., et al.: "Optical Characteristics of a Light-Focusing Fibre Guide", J. Quantum Electron. IEEE, 1970, QE-6, p.606.

7.59. Rigard, J.: "Elementary Method of Calculating the Diffraction of a Plane EM Wave by a Perfectly Conducting Half-Plane", Ann. Telecommun., 1970, 25, p.22.

7.60. Felsen, L.B., and Whitman, G.M.: "Wave Propagation in Time-Varying Media", Trans. IEEE, 1970, AP-18, p.242.

7.61. Karnishin, V.V., et al.: "Study of EM Wave Scattering in Resonant Conditions", Radiotekh. Elektron., 1970, 15, p.14, and Radio Eng. Electron. Phys., 1970, 15, no.1, pp.10-4.

7.62. Juranek, H.J.: "Scattering of EM Waves from a Rough Surface", Z. Phys., 1970, 233, p.324.

7.63. Bodnar, D.G.: "Variational Principle in Electromagnetics", Trans. IEEE, 1970, AP-18, p.213.

7.64. Palocz, I., and Oliner, A.A.: "Equivalent Network of a Multimode Planar Grating", Trans. IEEE, 1970, MTT-18, p.244.

7.65. Yoneyama, T., and Nishida, S.: "Method of Wave Beam Analysis", Rep. Res. Inst. Electr. Commun. Tohoku Univ., 1970, 22, no.1, pp.11-26.

7.66. Moshinskii, A.V.: "Scattering of a Plane TEM Wave on Two Parallel Elliptic Cylinders, Metallic Strips, and their Combinations", Radiotekh. Elektron., 1970, 15, no.7, and Radio Eng. Electron. Phys., 1970, 15, no.7, pp.1158-64.

7.67. Kvavadze, D.K., et al.: "Diffraction of an Infinite Array Placed above a Dielectric Layer of Finite Thickness", Radiotekh. Elektron., 1970, 15, no.7, and Radio Eng. Electron. Phys., 1970, 15, no.7, pp.1184-8.

7.68. Burckhardt, C.B.: "Laser Speckle Pattern. Narrowband Noise Model", Bell Syst. Tech. J., 1970, 49, p.309.

7.69. Kevanishvili, G.Sh., and Tsagareyshvili, O.P.: "Theory of Diffraction of a Plane Wave at an Array of Cylinders", Radiotekh. Elektron., 1970, 15, no.7, and Radio Eng. Electron. Phys., 1970, 15, no.7, pp.1288-92.

7.70. Sen, A.K., Fedotowsky, A., and Rouillard, M.: "Microwave Scattering from Dielectric Structures", Can. J. Phys., 1970, 48, no.1, pp.2611-22.

7.71. Chen, C.C.: "Scattering by a Two-Dimensional Periodic Array of Conducting Plates", Trans. IEEE, 1970, AP-18, no.5, pp.660-5.

7.72. Bucci, O.M., and Franceschetti, G.: "Scattering from Wedge-Shaped Absorbers", Trans. IEEE, 1971, AP-19, no.1, pp.96-104.

7.73.   Carr, J.F., and Charschan, S.S.:
"Obtaining Near-Gaussian Intensity Dis-
tributions from Multimode Pulsed Ruby
Lasers", Appl. Opt., 1971,10,no.3,pp.684-6.

7.74.   Knott, E.F., and Senior, T.B.A.:
"Polarization Characteristics of Scattered
Fields", Electron. Lett., 1971, 7, no.8,
pp.183-4.

7.75.   Chen, C.C.: "Diffraction by a Conducting Screen Perforated Periodically with
Circular Holes", Trans. IEEE, 1971, MTT-19,
no.5, pp.475-81.

7.76.   Yamamoto, S., and Makimoto, T.:
"Ray-Transfer Matrix of a Tapered Lenslike
Medium", Proc. IEEE,1971,59,no.8,pp.1254-5.

7.77.   Lewin, L.: "Rayleigh Distance in
Relation to Reflection from Curved Surfaces",
Electron. Lett., 1971, 7, no.25, pp.744-5.

7.78.   Nemoto, S., and Makimoto, T.: "Reflection and Transmission of Two-Dimensional
Gaussian Beam at the Plane Interface of
Dielectrics", Electron. Commun. Jap., 1971,
54, no.12, pp.30-6.

7.79.   O'Kelly, P.D., and Kharadly, M.M.Z.:
"Backscattering of EM Waves from a Rough
Surface Model", Can. J. Phys., 1972, 50,
no.23, pp.2928-38.

7.80.   Shigematsu, H., Okumura, K., and
Fukumitsu, O.: "Experimental Study of Mode
Matching of a Laser Beam", Technol. Rep.
Kyushu Univ., 1972, 45, no.6, pp.890-4.

7.81.   Zook, J.D., and Lee, T.C.: "f-Number
(in Optics)", Appl. Opt., 1972, 11, no.10,
pp.2140-5.

7.82.   Gaponov, S.V., Salashchenko, N.N.,
and Khanin, Ya.I.: "Method for Improving
Homogeneity of Spatial Distribution of Laser
Radiation", Kvantovaya Elektron., 1972,no.7,
pp.48-53, and Sov. J. Quantum Electron.,
1972, 2, no.1, pp.34-8.

7.83.   Harrache, R.J.: "Electric-Field
Components in a Focused Laser Beam", Appl.
Opt., 1973, 12, no.1, pp.133-5.

7.84.   Rossing, T.D., Stadum, R., and
Lang, D.: "Bragg Diffraction of Microwaves",
Am. J. Phys., 1973, 41, no.1, pp.129-30.

7.85.   Chen, C.C.: "Transmission of Microwaves through Perforated Flat Plates of
Finite Thickness", Trans. IEEE, 1973, MTT-21,
no.1, pp.1-6.

7.86.   Antes, G.P.: "Prefocused Objective-
Pinhole Unit for Beam Expanding and Spatial
Filtering", Appl. Opt.,1973,12,no.3,pp.493-5.

7.87.   Williams, C.S.: "Gaussian-Beam
Formulae from Diffraction Theory", Appl.
Opt., 1973, 12, no.4, pp.872-6.

7.88.   Carter, W.H.: "Anomalies in the
Field of a Gaussian Beam near Focus", Opt.
Commun., 1973, 7, no.3, pp.211-8.

7.89.   Peng, S.T., Tarrier, T., and Bertoni,
H.L.: "Leaky-Wave Analysis of Optical Periodic
Couplers", Electron. Lett., 1973, 9, no.6,
pp.150-2.

7.90.   Solymar, L.: "Scattering Cross Section of Passive Linear Arrays", Trans. IEEE,
1973, AP-21, no.3, pp.391-2.

7.91.   Nagata, K., and Umehara, T.: "Spatial
Correlation of Gaussian Beam in Moving Ground
Glass", Jap. J. Appl. Phys., 1973, 12, no.5,
pp.694-705.

7.92.   Kas, G.: "(Results of) Measurements
on Radar Reflectors", Int. Elektron. Rundsch.,
1973, 27, no.6, pp.127-9.

7.93.   Ito, M.: "Theory of Ellipsoidal
Waves and Seidel Aberrations of Gaussian
Beams", Jap. J. Appl. Phys., 1973, 12, no.6,
pp.856-75.

7.94.   Barakat, R.: "Brightness Distribution of the Sum of Two Correlated Speckle
Patterns", Opt. Commun., 1973, 8, no.1,
pp.14-6.

7.95.   Hayes, C.L., and Brandewie, R.A.:
"Reflection Factors for Wires and Cables at
10.6 micron", Appl. Opt., 1973, 12, no.7,
pp.1564-9.

7.96.   Jaiswal, A.K., Agrawal, G.P., and
Mehta, C.L.: "Coherence Functions in the
Far-Field Diffraction Plane", Nuovo Cim.,
1973, 15B, no.2, pp.295-307.

7.97.   Carter, W.H.: "Longitudinal Electric
Field in a Gaussian Beam near Focus", Appl.
Opt., 1973, 12, no.8, p.1732.

7.98.   Barakat, R.: "First-Order Probabi-
lity Densities of Laser Speckle Patterns
Observed through Finite Scanning Apertures",
Opt. Acta, 1973, 20, no.9, pp.729-40.

7.99.   Roscoe, B.J., and Banas, J.F.:
"Cross Section of Blunted Cones", Proc. IEEE,
1973, 61, no.11, p.1646.

7.100.   Tanaka, K., and Fukumitsu, O.:
"Transmission of an Optical Wave Beam through
a System of Two Aperture Stops", Trans. IEEE,
1973, MTT-22, no.2, pp.81-6.

7.101.   Richmond, J.H., Schwab, L.M., and
Wickliff, R.G.: "Tumble-Average Radar Back-
scatter of Thin-Wire Chaff Elements", Trans.
IEEE, 1974, AP-22, no.1, pp.124-6.

7.102.   Sambasivan, R.: "Diffraction of
Focused Laser Beams", Opt. Acta, 1974, 21,
no.4, pp.323-8.

7.103.   Matsumura, M.: "Speckle Noise
Reduction by Random Phaseshifters", Jap. J.
Appl. Phys., 1974, 13, no.3, pp.557-8.

7.104.   Frediani, D.J., Lang, K.C., and
LaPage, B.F.: "Comparison of Measured Radar
Cross Section of Solid and Mesh Ogives",
Trans. IEEE, 1974, AP-22, no.3, pp.427-32.

7.105. Prudky, J.: "Focusing of a Laser Beam by an Optical System", Jemna Mech. Opt., 1974, 19, no.1, pp.12-4.

7.106. Hogge, C.B., Butts, R.R., and Burlakoff, M.: "Characteristics of Phase-Aberrated Nondiffraction-Limited Laser Beams", Appl. Opt., 1974, 13, no.5, pp.1065-70.

7.107. Tavis, M.T.: "Radar Cross Section of Thin Wires at All Angles of Incidence", J. Appl. Phys., 1974, 45, no.3, pp.1179-82.

7.108. Aagard, R.L.: "Methods for Optimizing the Beam Shape in a Focused Coherent Optical System", Appl. Opt., 1974, 13, no.7, pp.1633-8.

7.109. Liggitt, R.C., and Liang, C.S.: "Novel Target Designed for Calibration of Radars", Trans. IEEE, 1974, AP-22, no.5, pp.730-2.

7.110. Ramm, A.G.: "Scalar Scattering on Small Bodies of Arbitrary Shape", Izv. VUZ Radiofiz., 1974, 17, no.7, pp.1062-8.

7.111. Vrazakov, E.I.: "Scattering of Waves in a Waveguide by Bounded Isotropic Media", Vestn. Mosk. Univ. Fiz. Astron., 1974, 15, no.4, pp.386-94.

7.112. Rusch, W.V.T.: "Physical-Optics Diffraction Coefficients for a Paraboloid", Electron. Lett., 1974, 10, no.17, pp.358-60.

7.113. Ishino, K., Watanabe, T., and Hashimoto, Y.: "Electromagnetic Wave Absorbers and Applications", Oyo Buturi, 1974, 43, no.11, pp.1157-63.

7.114. Elmoazzen, Y.E., and Shafai, L.: "Mutual Coupling Between Two Circular Waveguides", Trans. IEEE, 1974, AP-22, no.6, pp.751-60.

8. PRACTICAL ASPECTS

8.1. Piefke, G., and Strube, R.: "Reflection and Transmission of a TE_{01} Mode Incident on a Step Discontinuity in the E Plane of a Rectangular Waveguide", Arch. Elektr. Ubertrag., 1965, 19, p.231.

8.2. Smith, W.F., and Sokolowski, T.: "Effects of Corrosion on Waveguide Insertion Loss", Int. Conv. Rec. IEEE, 1965, 13, pt 5, p.209.

8.3. MacKenzie, T.E.: "Advances in Design of Precision Coaxial Standards and Components", Int. Conv. Rec. IEEE, 1965, 13, pt 5, p.190.

8.4. Kaminskii, F.: "Synthesis of Stepped Waveguide Transitions with a Large Number of Sections", Radiotekh. Elektron., 1965, 10, p.2134, and Radio Eng. Electron. Phys., 1965, 10, p.1822.

8.5. Gans, M., Kajfez, D., and Rumsey, V.H.: "Frequency-Independent Baluns", Proc. IEEE, 1965, 53, p.647.

8.6. Pyle, J.R.: "Circular-to-Rectangular Waveguide Transition Maintaining a Constant Cutoff Wavelength", Proc. IREE Aust., 1965, 26, p.338.

8.7. Reitzig, R.: "H-Plane Bends in Overmoded Rectangular Waveguide Systems", Int. J. Electron. 1965, 19, p.509.

8.8. Nefedev, E.I.: "Dielectric Prism in a Bend of a Plane Wide Waveguide", Radiotekh. Elektron., 1965, 10, p.879, and Radio Eng. Electron. Phys., 1965, 10, p.749.

8.9. Ehrreich, J.E., and Nimoy, M.: "Shielding Performance of Reinforced, Metal-Filled, Conductive-Plastic, Flat Gaskets", Trans. IEEE, 1965, EMC-7, no.1, p.50.

8.10. Cullen, A.L., Reitzig, R., and Robson, P.N.: "Considerations of Overmoded Rectangular Waveguide for High-Power Transmission", Proc. IEE, 1965, 112, p.1301.

8.11. Young, L., and Schiffman, B.M.: "Pulse Power Capacity of Short-Slot Couplers", Trans. IEEE, 1965, MTT-13, p.133.

8.12. Matsumoto, T., Suzuki, M., and Funatsu, C.: "Equivalent Width of Waveguide", J. Inst. Electr. Commun. Eng. Jap., 1965, 48, p.1046.

8.13. Krasnushkin, P.Ye.: "Method of Design of a Nonhomogeneous Septate Waveguide of Finite Length", Dokl. Akad. Nauk SSSR, 1965, 160, p.1285, and Sov. Phys.-Dokl.,1965.

8.14. Kotik, I.P., and Sivov, A.N.: "Diffraction at a Plane Mirror in a Waveguide Bend", Radiotekh. Elektron., 1965, 10, p.175, and Radio Eng. Electron. Phys.,1965,10,p.140.

8.15. Quine, J.P.: "E- and H-Plane Bends for High-Power Oversize Rectangular Waveguide", Trans. IEEE, 1965, MTT-13, p.54.

8.16. Bottcher, F.: "Thermal Stresses in Waveguide Apertures", Frequenz,1965,19,p.320.

8.17. Yashkin, A.Ya., Golubev, A.N., and Kalashnikov, V.G.: "Calculation of Passband of Straight Waveguide with Stepped Section", Radiotekh. Elektron., 1965, 10, p.1038, and Radio Eng. Electron. Phys., 1965, 10.

8.18. Tereshchenko, A.I., and Dolzhikov, V.V.: "Selecting the Optimum Shape for a Multimode Rectangular-Waveguide Transition", Izv. VUZ Radiotekh., 1965, 8, p.48, and Sov. Radio Eng., 1965, 8, p.34.

8.19.   Nowak, W.: "Rapidly Converging Approximate Calculation Methods for Waveguides with Internal Volume Perturbations", Hochfreq. Elektroak., 1965, 74, p.94.

8.20.   Young, I.: "Incremental Phase Shift due to Changes in Cross Section of Rectangular Waveguides", Microwave J., 1966, 9, no.5, p.45.

8.21.   de Ronde, F.C.: "Claw Flange. International Standardized Millimetre Waveguide Flange", Microwave J., 1966, 9, no.5, p.55.

8.22.   Udelson, B.J., and McDonald, D.W.: "Design of Two-Section Transformers having Variable Length", Microwave J., 1966, 9, no.4, p.68.

8.23.   Stegen, R.J.: "Simplified Determination of Impedances of Chebyshev Transformers", Trans. IEEE, 1966, MTT-14, p.354.

8.24.   Cochran, J.A., and Pecina, R.G.: "Mode Propagation in Continuously Curved Waveguides", Radio Sci., 1966, 1, no.6, p.679.

8.25.   Hall, A.H.: "Impedance Matching by Tapered or Stepped Transmission Lines", Microwave J., 1966, 9, no.3, p.109.

8.26.   Potter, B.E.: "Phase Modulation in Waveguide due to Mechanical Vibration", Trans. IEEE, 1966, AP-14, p.667.

8.27.   Waldron, R.A.: "Theory of Reflections in a Tapered Waveguide", Radio Electron. Eng., 1966, 32, p.245.

8.28.   Lucas, I.: "Reflection Factors at Offsets in Rectangular Waveguides", Arch. Elektr. Ubertrag., 1966, 20, p.683.

8.29.   Somlo, P.I.: "Computation of Coaxial-Line Step Capacitances", Trans. IEEE, 1967, MTT-15, p.48.

8.30.   Cole, W.J., Nagelberg, E.R., and Nagel, C.M.: "Iterative Solution of Waveguide Discontinuity Problems", Bell Syst. Tech. J., 1967, 46, no.3, pp.649-722.

8.31.   Mazzola, V., and Becker, J.E.: "Coupler-Type Bend for Pillbox Antennas", Trans. IEEE, 1967, MTT-15, no.8, pp.462-8.

8.32.   Kashiwagi, H.: "Rectangular Waveguide Taper for Oversized Waveguide", Bull. Electrotech. Lab., 1967, 31, no.8, pp.963-71.

8.33.   Reitzig, R.: "Use of the Variational Principle to Determine the Effect of Small Discontinuities in Circular Waveguides", Frequenz, 1968, 22, no.2, pp.45-51.

8.34.   Paul, D.K., and Paul, A.: "Power-Carrying Capacity of Dielectric Rod Waveguide", Electron. Commun., 1968, 3, no.3, p.9.

8.35.   Bahar, E., and Crain, G.E.: "Synthesis of Multimode Waveguide Transition Sections", Proc. IEE, 1968, 115, p.1395.

8.36.   Knetsch, H.D.: "Theory of Abrupt Cross-Sectional Changes in Waveguides", Arch. Elektr. Ubertrag., 1968, 22, p.591.

8.37.   French, G.N., and Fooks, E.H.: "Design of Stepped Transmission-Line Transformers", Trans. IEEE, 1968, MTT-16, p.885.

8.38.   Toyonaga, T., et al.: "Waveguide Devices for Prevention of Leakage of High-Power Microwaves. I", Sci. Eng. Rev. Doshisha Univ., 1968, 8, p.217.

8.39.   Cross, P.H.: "Broadband Waveguide Window for High-Power Microwave Tubes", Nachr. Tech. Fachber., 1968, 35, p.248.

8.40.   Tomiyasu, K.: "Conversion of TE_{11} Mode by a Large-Diameter Conical Junction", Trans. IEEE, 1969, MTT-17, p.277.

8.41.   Poulton, G., Karbowiak, A.E., and Tang, C.C.H.: "Mode Conversion in Overmoded Waveguide Tapers", Trans. IEEE, 1969, MTT-17, p.288.

8.42.   Virgile, L.: "Supporting Waveguide under Vibration", Microwaves, 1969, 8, p.58.

8.43.   Napoli, L.S., and Hughes, J.J.: "High-Frequency Behaviour of Microstrip Lines", RCA Rev., 1969, 30, p.268.

8.44.   Schneider, M.V.: "Microstrip Lines for Integrated Circuits", Bell Syst. Tech. J., 1969, 48, no.5, p.1421.

8.45.   Brady, M.M.: "Cutoff Wavelengths and Frequencies of Standard Rectangular Waveguides", Electron. Lett., 1969, 5, p.410.

8.46.   Bandler, J.W.: "Computer Optimization of Inhomogeneous Waveguide Transformers", Trans. IEEE, 1969, MTT-17, p.563.

8.47.   Tresselt, C.P.: "Design and Computed Theoretical Performance of Three Classes of Equal-Ripple Nonuniform Line Couplers", Trans. IEEE, 1969, MTT-17, p.218.

8.48.   Bates, C.P.: "Intermodal Coupling at Junction between Straight and Curved Waveguides", Bell Syst. Tech. J., 1969, 48, p.2259.

8.49.   Brady, M.M.: "Discrepancies in Waveguide Attenuation Expressions", Electron. Lett., 1969, 5, p.718.

8.50.   Knetsch, H.D.: "Rotary-Symmetrical Discontinuities in Cross Section in a Circular Waveguide", Nachr. Tech. Z., 1969, 22, p.501.

8.51.   Makarov, T.V.: "Propagation of TE_{01}-Mode Analogue in the Bend of a Cross-Shaped Waveguide", Izv. VUZ Radiofiz., 1969, 12, no.10, p.1557.

8.52.   Shein, A.G., and Shavorykin, Yu.V.: "Theory of Inclined Dielectric Windows in Rectangular Waveguide", Izv. VUZ Radioelektron., 1969, 12, no.10, p.1215.

8.53.   Kulikov, E.L., and Soshestvenskaya, L.A.: "Design of Waveguide Transformers", Izv. VUZ Radiofiz., 1969, 12, no.9, p.1398.

8.54.   Veselov, G.I., and Platonov, N.I.: "Analysis of Corner Waveguides", Radiotekhnika, 1969, 24, no.12, p.98, and Telecommun. Radio Eng. Pt 2, 1969, 24, no.12.

8.55. Lewin, L., and Crain, G.E.: "Theoretical Performance of an Overmoded Right-Angled Corner", Proc. IEE, 1969, 116, p.667.

8.56. Iveland, T.D.: "Computer Design of Inhomogeneous Quarter-Wave Transformers", Trans. IEEE, 1969, MTT-17, p.120.

8.57. Bramanti, M., and Calamia, M.: "Tapered Waveguide-to-Waveguide Adaptor", Trans. IEEE, 1969, MTT-17, p.116.

8.58. Tomiyasu, K.: "Multiconical 5-m-Long Taper for TE_{01} Mode in 0.75-m-Diameter Waveguide at X Band", Proc. IEE, 1969, 116, p.373.

8.59. Knetsch, H.D.: "Axial Offsets of Circular Waveguides of Arbitrary Radii", Arch. Elektr. Ubertrag., 1969, 23, p.23.

8.60. Freylikher, V.D., and Fuks, I.M.: "Attenuation of Mean Field in a Waveguide at the Critical Frequency", Izv. VUZ Radiofiz., 1970, 13, p.128.

8.61. DeLoach, B.C.: "Coaxial Line v. Rectangular Waveguides for Solid-State Microwave Circuits", Proc. IEEE, 1970, 58, p.505.

8.62. Wardrop, B.: "Electrical Lengths of Stripline Bends", Electron. Lett., 1970, 6, p.494.

8.63. Toyonaga, T., and Ochi, H.: "Waveguide Devices for Leakage Prevention of Microwave High Power. II", Sci. Eng. Rev. Doshiska Univ., 1970, 10, no.4, p.297.

8.64. Tang, C.H.: "Orthogonal Coordinate System for Curved Pipes", Trans. IEEE, 1970, MTT-18, p.69.

8.65. Bava, G.P., and Naldi, C.: "Discussion of Design Methods for Dielectric Steps in Rectangular Waveguides", Trans. IEEE, 1970, MTT-18, p.167.

8.66. Koslov, I.A.: "Reflections Inside Chebyshev Matching Transitions", Radiotekh. Elektron., 1970, 15, no.7, and Radio Eng. Electron. Phys., 1970, 15, no.7, pp.1293-4.

8.67. Kozlov, I.A.: "Synthesis of Chebyshev Stepped Junctions", Radiotekh. Elektron., 1970, 15, no.9, and Radio Eng. Electron. Phys., 1970, 15, no.9, pp.1705-7.

8.68. Maeda, T., Sato, Y., and Ueda, Y.: "Aluminium Elliptical Waveguide. VII", Dainichi-Nippon Cables Rev., 1970, no.46, pp.15-9.

8.69. Wen, C.P.: "Attenuation Characteristics of Coplanar Waveguides", Proc. IEEE, 1970, 58, p.141.

8.70. Gopinath, A., Horton, R., and Easter, B.: "Microstrip Loss Calculations", Electron. Lett., 1970, 6, p.40.

8.71. van Heuven, J.H.C., and van Nie, A.G.: "Properties of Microstrip Lines on Fused Quartz", Trans. IEEE, 1970, MTT-18, p.113.

8.72. Cherry, W., et al.: "60- to 90-GHz High-Temperature Pressure Window", Trans. IEEE, 1970, ED-17, p.85.

8.73. Crain, G.E.: "Symmetries of the Overmoded Right-Angle Corner", Proc. IEE, 1970, 117, p.713.

8.74. Lewin, L.: "Inadequacy of Discrete Mode-Matching Techniques in Wave Discontinuity Problems", Trans. IEEE, 1970, MTT-18, p.364.

8.75. Phelan, H.R.: "Wideband Parallel-Connected Balun", Trans. IEEE, 1970, MTT-18, p.259.

8.76. Shimada, Y.: "Input Impedance Analysis of 1:1 Balun", Trans. IEEE, 1970, MTT-18, p.264.

8.77. Rulea, Gh.: "Design Curves for Uniform Waveguide Matching with a Dielectric Sheet", Telecomunicattii, 1970, 14, no.6, pp.244-6.

8.78. Vidrascu, T.: "Rectangular Waveguides with Improved Aspect Ratio", An. Stiint. Univ. Cuza Iasi, 1970, 16, no.2b, pp.185-8.

8.79. Marchini, C.: "Attenuation for a Waveguide with Absorbing Walls. I and II", Atti. Semin. Mat. Fiz. Univ. Modena, 1970, 19, no.1, pp.42-52 and 65-9.

8.80. Ferrari, L.A., and Zucker, M.S.: "Lowloss Waveguide for Low-Temperature Applications", Rev. Sci. Instrum., 1971, 42, no.1, p.167.

8.81. Gopinath, B., and Sondhi, M.M.: "Inversion of the Telegraph Equation and Synthesis of Nonuniform Lines", Proc. IEEE, 1971, 59, no.3, pp.383-92.

8.82. King, H.E., and Wong, J.L.: "Characteristics of Oversize Circular Waveguides and Transitions at 3-mm Wavelength", Trans. IEEE, 1971, MTT-19, no.1, pp.116-9.

8.83. Christensen, P.S.: "Design of a Tapered Transition in a Circular Waveguide", Trans. IEEE, 1971, MTT-19, no.1, pp.99-100.

8.84. Kaffenberger, E.: "(Rectangular/ Elliptical) Fittings for Flexwell Waveguides", Nachr. Tech. Z., 1971, 24, no.3, pp.146-51.

8.85. Kazantsev, Yu.N., and Kharlashkin, O.A.: "Wide Rectangular Low-Loss Waveguides", Radiotekh. Elektron., 1971, 16, no.6, pp.1063-5, and Radio Eng. Electron. Phys., 1971, 16, no.6, pp.1040-2.

8.86. Piefke, G.: "Reflection Problems and their Solutions with Special Line Joints", Nachr. Tech. Z., 1971, 24, no.5, pp.245-7.

8.87. Bava, G.P., and Naldi, C.: "Analysis of Dielectric Steps in Partially Filled Rectangular Waveguides", Alta Freq., 1971, 40, no.5, pp.435-41.

8.88. Schilder, D.: "Calculating the Optimum Continuous Transition between Striplines of Various Characteristic Impedances", Nachr. Tech., 1971, 21, no.10, pp.342-6.

8.89. James, D.S., and Tse, S.H.: "Microstrip End Effects", Electron. Lett., 1972, 8, no.2, pp.46-7.

8.90. Lit, J.W.Y., and Boulay, R.: "Characteristic Parameters of a Gaussian Beam", Can. J. Phys., 1972,50,no.3,pp.301-2.

8.91. Bergqvist, A.: "Wave Propagation on Nonuniform Transmission Lines", Trans. IEEE, 1972, MTT-20, no.8, pp.557-8.

8.92. Ashkinazi, L.A., and Kisilev, A.B.: "Effect of Substrate on Resistance of a Thin Metal Film", Izv. VUZ Radioelektron., 1972, 15, no.5, pp.664-5.

8.93. Nalbandian, V., and Steenaart, W.: "Discontinuities in Symmetric Striplines due to Impedance Steps", Trans. IEEE, 1972, MTT-20, no.9, pp.573-8.

8.94. Bui, V.R., and Gagne, R.R.J.: "Dielectric Losses in H-Plane-Loaded Rectangular Waveguide", Trans. IEEE, 1972, MTT-20, no.9, pp.621-3.

8.95. Kaffenberger, E., and Schuttloffel, E.: "Bendable Oval Section Aluminium Waveguide with Good Characteristic Impedance", Frequenz, 1972, 26, no.9, pp.252-8.

8.96. Benedek, P., and Silvester, P.: "Equivalent Capacitances for Microstrip Gaps and Steps", Trans. IEEE, 1972, MTT-20, no.11, pp.729-33.

8.97. Hecken, R.P.: "Near-Optimum Matching Section without Discontinuities", Trans. IEEE, 1972, MTT-20, no.11, pp.734-9.

8.98. Montgomery, J.P., and Lewin, L.: "Note on E-Plane Waveguide Step with Simultaneous Change of Media", Trans. IEEE, 1972, MTT-20, no.11, pp.763-4.

8.99. Acampora, A.S., and Sproul, P.T.: "Waveguide Breakdown Effects at High Average Power and Long Pulses", Bell Syst. Tech. J., 1972, 51, no.9, pp.2065-91.

8.100. Kozlov, I.A.: "Design of a Smooth Transition with Maximally-Flat Frequency Response", Radiotekh. Elektron., 1972, 17, no.8, pp.1739-40, and Radio Eng. Electron. Phys., 1972, 17, no.8, pp.1375-7.

8.101. Kravchenko, N.I., et al.: "Temperature Effect of H_{10} Wave on the Absorbing Wall of a Rectangular Waveguide", Inzh.-Fiz. Zh. 1972, 22, no.1, p.158, and J. Eng. Phys., 1972, 22, no.1, p.116.

8.102. Reisdorf, F., and Knetsch, H.D.: "Wave Propagation in a Bent Rectangular Waveguide", Nachr. Tech. Z., 1972, 25, no.7, pp.312-7.

8.103. Michaelides, M.: "Tapered Transmission Lines for Impedance Matching Sections", Mullard Tech. Commun., 1972, 12, no.116, pp.170-6.

8.104. Shavorykin, Yu.V., and Bondarenko, B.N.: "Propagation in Waveguides Containing a Longitudinally Nonuniform Dielectric", Radiotekh. (Kharkov), 1972, no.22, pp.3-9.

8.105. Hamid, M.A.K.: "Reflection Factor of Dielectric-Loaded Waveguide Bends", Electron. Lett., 1973, 9, no.2, pp.37-8.

8.106. Bova, N.T., Sorokovoi, P.I., and Tolstikov, Yu.V.: "Parameter Tolerances in a Microstrip Line", Izv. VUZ Radioelektron., 1973, 16, no.2, pp.28-33.

8.107. Silvester, P., and Benedek, P.: "Microstrip Discontinuity Capacitances for Right-Angle Bends, T-Junctions, and Crossings", Trans. IEEE, 1973, MTT-21, no.5, pp.341-6.

8.108. De Jong, G., and Offringa, W.: "Reflection and Transmission by a Slant Interface between Two Media in a Rectangular Waveguide", Int. J. Electron., 1973, 34, no.4, pp.453-63.

8.109. Reisdorf, F.: "Scattering Matrix of a Double Truncated Corner in a Rectangular Waveguide", Nachr. Tech. Z., 1973, 26, no.4, pp.176-9.

8.110. Chow, Y.L., and Wu, S.C.: "Moment Method with Mixed Basis Functions for Scatterings by Waveguide Junctions", Trans. IEEE, 1973, MTT-21, no.5, pp.333-40.

8.111. Hecken, R.P., and Anuff, A.: "Optimum Design of Tapered Waveguide Transitions", Trans. IEEE, 1973, MTT-21, no.6, pp.374-80.

8.112. Horton, R.: "Electrical Characterization of a Right-Angled Bend in Microstrip Line", Trans. IEEE,1973,MTT-21,no.6,pp.427-9.

8.113. Gardiol, F.E., and Parriaux, O.: "Excess Losses in H-Plane Loaded Waveguides", Trans. IEEE, 1973, MTT-21, no.7, pp.457-61.

8.114. Bommas, G., and Bergmann, R.: "Losses in Microstrip with Respect to Circuit Design", Frequenz, 1973, 27, no.6, pp.159-63.

8.115. Goyal, I.C., Sodha, M.S., and Ghatak, A.K.: "Propagation in a Medium with Random Radial Permittivity Gradient", J. Opt. Soc. Am., 1973, 63, no.8, pp.940-3.

8.116. Horton, R.: "Equivalent Representation of an Abrupt Impedance Step in Microstrip Line", Trans. IEEE, 1973, MTT-21, no.8, pp.562-4.

8.117. Knorr, J.B., and Saenz, J.: "Effect of Surface Metal Adhesive on Slot-Line Wavelength", Trans. IEEE, 1973, MTT-21, no.10, pp.642-4.

8.118. Pitschi, F.: "Determination of Thermal Load Limit of Coaxial Transmission Lines", Nachr. Tech. Z., 1973, 26, no.11, pp.487-9.

8.119. Komisarczuk, J.: "Propagation Problems in Very Lossy Lines", Bull. Acad. Pol. Sci. Tech., 1973,21,no.6,pp.493-7.

8.120. Bahar, E., and Govindarajan, G.: "Rectangular and Annular Model Analyses of Multimode Waveguide Bends", Trans. IEEE, 1973, MTT-21, no.12, pp.819-24.

8.121. Brinson, J.R.: "Automation of Smith-Chart Plots", Electron. Des., 1973, 21, no.18, pp.130-6.

8.122. Ferdinandov, E.S.: "Two-Cylinder Waveguide", Elektro Prom. Prib., 1973, 8, no.9, pp.323-5.

8.123. Metrikin, A.A., Nadenenko, B.S., and Polushin, G.P.: "Feeder Line of Bimetallic and Elliptical Waveguides for Radio-Relay Communication", Elektrosvyaz, 1973, 27, no.6, pp.1-9, and Telecommun. Radio Eng. Pt 1, 1973, 27, no.6, pp.1-7.

8.124. Zhurav, S.M., and Losev, V.S.: "Diffraction at a Junction of Circular Waveguides. General Solution", Antenny, 1973, no.18, pp.114-24.

8.125. Vaganov, R.B.: "Focusing Waveguide Transition", Radiotekh. Elektron., 1973, 18, no.6, pp.1272-6, and Radio Eng. Electron. Phys., 1973, 18, no.6, pp.933-6.

8.126. Vaganov, R.B.: "Phase Corrector in the Bend of a Wide Waveguide", Radiotekh. Elektron., 1973, 18, no.2, pp.235-42, and Radio Eng. Electron. Phys., 1973, 18, no.2, pp.170-5.

8.127. Kaminow, I.P., Mammel, W.L., and Weber, H.P.: "Metal-Clad Optical Waveguides", Appl. Opt., 1974, 13, no.2, pp.396-405.

8.128. Crow, J.D.: "Power-Handling Capability of Glass-Fibre Lightguides", Appl. Opt., 1974, 13, no.3, pp.467-8.

8.129. Johns, P.B.: "Solution of Inhomogeneous Waveguide Problems Using a Transmission-Line Matrix", Trans. IEEE, 1974, MTT-22, no.3, pp.209-15.

8.130. McDonald, B.H., Friedman, M., and Wexler, A.: "Variational Solution of Integral Equations", Trans. IEEE, 1974, MTT-22, no.3, pp.237-48.

8.131. Henke, H.: "Wave Propagation in Waveguides of Varying Cross Section", Arch. Elektron. Ubertrag., 1974, 28, no.3, pp.146-7.

8.132. Mur, G.: "Finite Difference Method for Solution of Waveguide Discontinuity Problems", Trans. IEEE, 1974, MTT-22, no.1, pp.54-7.

8.133. Rahmat-Samii, Y., Itoh, T., and Mittra, R.: "Spectral Domain Analysis for Solving Microstrip Discontinuity Problems", Trans. IEEE, 1974, MTT-22, no.4, pp.372-8.

8.134. Checcacci, P.F., Falciai, R., and Scheggi, A.M.: "Circular Bends in Dielectric Frame-Beam Waveguides", Trans. IEEE, 1974, MTT-22, no.5, pp.576-8.

8.135. Tischer, F.J.: "Excess Surface Resistance due to Surface Roughness at 35 GHz", Trans. IEEE, 1974, MTT-22, no.5, pp.566-9.

8.136. Tischer, F.J.: "Effect of Roughness on Resistance of Plane Copper Surfaces at Millimetre Waves", Proc. IEE, 1974, 121, no.5, pp.333-6.

8.137. Hu, A.S., Lam, F.W., and Lin, C.: "Recursive Formulae for Multi-Sectioned (Waveguide) Line", Trans. IEEE, 1974, CAS-21, no.5, pp.640-2.

8.138. Van Heuven, J.H.C.: "Conduction and Radiation Losses in Microstrip", Trans. IEEE, 1974, MTT-22, no.9, pp.841-4.

8.139. Sargent, G.A.: "Reflection Factors of Offset Rectangular Waveguides at 56 GHz", Trans. IEEE, 1974, IM-23, no.3, pp.246-7.

8.140. Gopinath, A., and Easter, B.: "Moment Method of Calculating Discontinuity Inductance of Microstrip Right-Angled Bends", Trans. IEEE, 1974, MTT-22, no.10, pp.880-3.

8.141. Lynch, A.C.: "Transmission Loss in Waveguide Partially Filled with Dielectric", Proc. IEE, 1974, 121, no.10, pp.1057-8.

8.142. Costamagna, E.: "Method of Calculating Gap Capacitances in Microstrip Structures", Alta Freq., 1974, 43, no.6, pp.362-4.

8.143. Rao, K.N.S., and Kosta, S.P.: "Tapered Microstrip Transmission Lines", J. AEU, 1974, 7, no.1, pp.34-6.

8.144. Nayfeh, A.H., and Asfar, O.R.: "Propagation in Conducting Waveguides having Slowly Varying Cross Sections", Radio Sci., 1974, 9, no.10, pp.867-71.

8.145. McRitchie, W.K.: "Properties of Interface between Homogeneous and Inhomogeneous Waveguides", Proc. IEE, 1974, 121, no.11, pp.1367-74.

8.146. Arnold, R.M.: "Transmission-Line Impedance Matching Using the Smith Chart", Trans. IEEE, 1974, MTT-22, no.11, pp.977-8.

8.147. Unrau, U.: "Design of Mitred H-Plane Elbows for Dominant Mode Waveguide", Arch. Elektron. Ubertrag., 1974, 28, no.10, pp.441-2.

8.148. Kravchenko, N.I., and Kukush, V.D.: "Investigation into the Thermal Effect of the Dominant Wave on the Walls of Rectangular Waveguides", Radiotekh. (Kharkov), 1974, no.28, pp.111-20.

8.149. Nishihara, H., Inoue, T., and Kayama, J.: "Lowloss Parallel-Plate Waveguide at 10 micron", Appl. Phys. Lett., 1974, 25, no.7, pp.391-3.

Part II

COMPONENTS AND DEVICES

9. FREE-SPACE COMPONENTS

9.1. Schunk, E.: "Wedge-Ferrite Absorber Combinations for Electromagnetic Waves", Z. Angew. Phys., 1965, 19, p.420.

9.2. Degenford, J.E.: "Power-Handling Capability of a Reflecting Beam Waveguide", Proc. IEEE, 1965, 53, p.493.

9.3. Grassler, L.E.: "Electromagnetic-Wave Absorber with Stepwise Attenuation Characteristic. I", Nachr. Tech., 1965, 15, p.391.

9.4. Semenov, G.F.: "Design of Resonators and Waveguides with Multilayer Dielectric", Izv. VUZ Radiofiz., 1965, no.2, p.407, and Sov. Radiophys., 1965, no.2, p.293.

9.5. Deutsch, J., Jung, H.J., and Volhardt, G.: "Wideband Absorbers for Electromagnetic Waves", Z. Angew. Phys., 1966, 20, p.511.

9.6. Deutsch, J., Jung, H.J., and Volhardt, C.: "Investigations on Wideband Absorbers with Low Reflectance", Z. Angew. Phys., 1966, 20, p.540.

9.7. Kumagai, N., Yoshida, K., and Nakahara, T.: "Reflecting Beam Waveguide", J. Inst. Electr. Commun. Eng. Jap., 1966, 49, p.1099.

9.8. Laurie, R.E., and Peters, L.: "Control of Echo Area of Ogives by Cutoff Corrugated Surfaces", Trans. IEEE, 1966, AP-14, p.798.

9.9. Chen, K.M., and Vincent, M.: "Method of Minimizing Radar Cross Section of a Sphere", Proc. IEEE, 1966, 54, p.1629.

9.10. Sueta, T., Kumagai, N., and Kurazono, S.: "Wideband Quasi-Optic Prism Components", J. Inst. Electr. Commun. Eng. Jap., 1966, 49, no.3, p.23.

9.11. Corti, E., et al.: "Two-Dimensional Artificial Dielectrics", Alta Freq., 1966, 35, p.E882.

9.12. Yoshida, K.: "Light Rays in Reflecting Beam Waveguides", J. Inst. Electr. Commun. Eng. Jap., 1967, 50, no.3, pp.370-6.

9.13. Vaganov, R.B., and Matveev, R.F.: "Theoretical Study of Extended Slightly Irregular Beamguides", Radiotekh. Elektron., 1968, 13, no.2, pp.232-42, and Radio Eng. Electron. Phys., 1968, 13, no.2.

9.14. Kazantsev, Yu.N.: "Natural Modes in Gas-Filled Beamguides", Radiotekh. Elektron., 1968, 13, p.1227, and Radio Eng. Electron. Phys., 1968, 13.

9.15. Shevchenko, V.V.: "Nonreflecting Lenses for Quasi-Optical Lines", Radiotekh. Elektron., 1969, 14, no.10, p.1764, and Radio Eng. Electron. Phys., 1969, 14, no.10.

9.16. Mink, J.W.: "Experimental Investigation with an Iris Beam Waveguide", Trans. IEEE, 1969, MTT-17, p.48.

9.17. Christian, J.R., et al.: "Diffractional Distortions in Beam Waveguides with Off-Axis Beams", Proc. IEEE, 1969, 57, p.829.

9.18. Hongo, K.: "Beam-Wave Transmission in a Circular Hollow Cylinder", Electron. Commun. Jap., 1969, 52, no.2, p.102.

9.19. Vaganov, R.B.: "Dielectric Prism in a Beam Waveguide", Radiotekh. Elektron., 1969, 14, no.3, p.393, and Radio Eng. Electron. Phys., 1969, 14, no.3, pp.340-6.

9.20. Abrams, R.L., and Gandrud, W.B.: "Variable 10.6-micron Attenuator", J. Quantum Electron. IEEE, 1969, QE-5, p.212.

9.21. Gebhardt, F.G., McCoy, J.H., and Smith, D.C.: "Variable Attenuator for CO_2-Laser Radiation", J. Quantum Electron. IEEE, 1969, QE-5, p.471.

9.22. Kahan, W.: "Birefringent Laser Mirrors", Appl. Opt., 1969, 8, p.985.

9.23. Ono, M., Yokokawa, S., and Suzuki, M.: "Design Method of Narrowband Multilayer Absorber for Microwaves", Bull. Yamagata Univ., 1969, 10, no.2, pp.317-31.

9.24. Nestrizhenko, Yu.A.: "Lummer-Gehrcke Plate Polarizer for Nd^{3+}:Glass Laser", Opt. Spekt., 1969, 26, p.1000, and Opt. Spectrosc., 1969, 26, p.542.

9.25. Chatterjee, S.K., et al.: "Two-Dimensional Array Absorber for Microwaves", J. Indian Inst. Sci., 1969, 51, p.103.

9.26. Carson, J.W.: "Type of Quasi-Optical Waveguide Component", Trans. IEEE, 1970, MTT-18, p.57.

9.27. Weisman, D.L.: "Laser-Beam Shaping for Streak Interferometry", Appl. Opt., 1970, 9, p.1213.

9.28. Barlow, H.E.M.: "High-Frequency Impedance of a Practical Metal Surface and Effect of a Thin Coating of Dielectric", Electron. Lett., 1970, 6, p.413.

9.29. Sincerbox, G.T.: "Laser Beam (Birefringent) Combining", IBM Tech. Disclosure Bull., 1970, 12, p.1663.

9.30. Shimizu, Y., and Suetake, K.: "Practical Wideband Absorbing Wall Using Dielectric Material", Electron. Commun. Jap., 1970, 53, no.3, pp.14-5.

9.31. Daly, J.C.: "Analogy of Beam Waveguides to Electric Transmission Lines", Trans. IEEE, 1970, MTT-18, p.657.

9.32. Suzuki, M., et al.: "Theoretical Analysis of Discrete Reflecting-Beam Waveguide with Parabolic Cylindrical Reflectors", Trans. IEEE, 1970, MTT-18, p.338.

9.33. Kamimura, M., et al.: "Experiments on a Discrete Reflecting-Beam Waveguide with Parabolic Cylindrical Reflectors", Trans. IEEE, 1970, MTT-18, p.348.

9.34. Zyatitskii, V.A.: "Theory of Generalized Regular Beamguides", Radiotekh. Elektron., 1970, 15, no.9, and Radio Eng. Electron. Phys., 1970, 15, no.9, pp.1574-82.

9.35. Edmonds, H.D., DePalma, C., and Harris, E.P.: "Preparation and Properties of SiO Antireflection Coatings for GaAs Injection Lasers with External Resonators", Appl. Opt., 1971, 10, no.7, pp.1591-6.

9.36. Mallozzi, P.J., et al.: "Method for Isolating Q-Switched Lasers", J. Appl. Phys., 1971, 42, no.11, pp.4531-2.

9.37. Naito, Y., and Suetake, K.: "Application of Ferrite to EM-Wave Absorber and Characteristics", Trans. IEEE, 1971, MTT-19, no.1, pp.65-72.

9.38. Adonina, A.I., et al.: "Prismatic Polarizers", Izv. VUZ Radioelektron., 1971, 14, no.1, pp.61-6.

9.39. Shcherbov, V.A.: "Matching (Wire-Array) Transformer for Quasi-Optical Line", Izv. VUZ Radioelektron., 1971, 14, no.4, pp.379-83.

9.40. Cheo, P.K., and Bass, C.D.: "Efficient Wire-Grid Diplexer Polarizer for CO_2 Lasers", Appl. Phys. Lett., 1971, 18, no.12, pp.565-7.

9.41. Voronkov, L.: "Method of (Diffuse) Attenuating Radiation from a CO_2 Laser", I er. Tekh., 1971, 14, no.9, pp.27-8, and Meas. Tech., 1971, 14, no.9, pp.1327-9.

9.42. Kuleshov, Ye.M., and Litvinov, D.D.: "Beam Diplexing in Quasi-Optical Microwave Channels", Radiotekh. (Kharkov), 1971, no.18, pp.98-104.

9.43. Ono, M., and Suzuki, M.: "Improved Synthesis of Multilayer Absorber", Electron. Commun. Jap., 1971, 54, no.4, pp.64-9.

9.44. Katsenelenbaum, B.Z.: "Theory of Quasi-Optical Elements in Wide Waveguides", Radiotekh. Elektron., 1971, 16, no.10, pp.1797-804, and Radio Eng. Electron. Phys., 1971, 16, no.10, pp.1619-25.

9.45. Wardrop, B.: "Quasi-Optical Directional Coupler", Marconi Rev., 1972, 35, no.185, pp.159-69.

9.46. Shcherbov, V.A.: "Application of Polarization Gratings for Matching Beam-guides", Izv. VUZ Radioelektron., 1972, 15, no.6, pp.739-44.

9.47. Varnado, S.G., and Smith W.D.: "Electrooptic Ceramics as Wavelength-Selection Devices in Dye Lasers", J. Quantum Electron. IEEE, 1972, QE-8, no.2, pp.88-9.

9.48. Lashmore, D.S., and Baldwin, K.M.: "Laboratory Construction of Multilayer Dielectric Mirrors for He-Ne Lasers", Am. J. Phys., 1972, 40, no.2, pp.294-7.

9.49. Yoneyama, T., and Nishida, S.: "Optimum Design of Periscope-Type Beam Waveguides", Electron. Commun. Jap., 1972, 55, no.7, pp.69-75.

9.50. Sellner, J.: "Theory of Beam Waveguides", Acta Phys. Austriaca, 1972, 35, no.4, pp.378-88.

9.51. Checcacci, P.F., Falciai, R., and Scheggi, A.: "Phase-Step Beam Waveguide", Trans. IEEE, 1972, MTT-20, no.9, pp.608-13.

9.52. Bangert, H., Theron, E., and Eigner, G.: "Design of Laser Mirrors with Intermediate Reflectances", Opt. Commun., 1972, 6, no.4, pp.399-401.

9.53. Checcacci, P.F., and Scheggi, A.M.: "Metallic Frame Beam Waveguide", Trans. IEEE, 1973, MTT-21, no.10, pp.649-51.

9.54. McClain, W.M.: "Mounting a Pellin-Broca Prism for Laser Work", Appl. Opt., 1973, 12, no.1, p.153.

9.55. Siegman, A.E.: "Antiresonant Ring (Hybrid-T) Interferometer for Coupled Laser Cavities", J. Quantum Electron. IEEE, 1973, QE-9, no.2, pp.247-50.

9.56. Johnson, M.M.: "Continuously Tunable Resonant Ruby-Laser Reflector", Appl. Opt., 1973, 12, no.3, pp.510-8.

9.57. Voronkov, G.L.: "Quartz Attenuator of IR Radiation", Izmer. Tekh., 1973, 16, no.9, pp.69-70, and Meas. Tech., 1973, 16, no.9, pp.1386-7.

9.58. Mikoshiba, S., and Ahlborn, B.: "Laser Mirror with Variable Focal Length", Rev. Sci. Instrum., 1973, 44, no.4, pp.508-11.

9.59. Bin-Nun, E., and Dothan-Deutsch, F.: "Mirror with Adjustable Radius of Curvature", Rev. Sci. Instrum., 1973, 44, no.4, pp.512-3.

9.60. Hill, N.: "Microwave Transmission through a Series of Inclined Gratings", Proc. IEE, 1973, 120, no.4, pp.407-12.

9.61. Loomis, J.S.: "Absorption in Coated Laser Windows", Appl. Opt., 1973, 12, no.4, pp.877-8.

9.62. Wilson, M.J.F., and Teh, G.A.: "Improved Tolerance in Optical Directional Couplers", Electron. Lett., 1973, 9, no.19, pp.453-5.

9.63. Orr, B.J.: "Constant Deviation Laser Tuning Device", J. Phys. E, 1973, 6, no.5, pp.426-8.

9.64. Taylor, M.J.: "Compound TIR Prism for Polarization-Selective Laser Resonators", Opto-Electron., 1973, 5, no.3, pp.255-6.

9.65. Bronifin, F.B., et al.: "Focusing Optical Elements with Regularly Distributed Refractive Index", Zh. Prikl. Spekt., 1973, 18, no.3, pp.523-49.

9.66. Checcacci, P.F., Falciai, R., and Scheggi, A.M.: "Beam Waveguides with Minimized Dielectric Structures", Trans. IEEE, 1973, MTT-21, no.5, pp.362-3.

9.67. Mahlein, H.F.: "Graphs for the Design of Laser Mirrors at Normal Incidence", Opt. Laser Technol., 1973, 5, no.2, pp.60-8.

9.68. Mahlein, H.F.: "Properties of Laser Mirrors at Non-Normal Incidence", Opt. Acta, 1973, 20, no.9, pp.687-97.

9.69. Firester, A.H., Heller, M.E., and Wittke, J.P.: "Inexpensive Laser Mirrors", Am. J. Phys., 1973, 41, no.10, pp.1202-3.

9.70. Mott, L.P.: "Call for Standardized (Optical) Coatings", Laser Focus, 1973, 9, no.8, pp.54-6.

9.71. Spiller, E.: "High-Quality Fabry-Perot Mirrors for the UV", Optik, 1973, 39, no.2, pp.118-25.

9.72. Keilmann, F., et al.: "Optical Isolation Using a Doppler-Broadened Molecular Absorber", Appl. Phys. Lett., 1973, 23, no.11, pp.612-4.

9.73. Dreyfus, R.W., von Gutfeld, R.J., and Wallace, S.C.: "Aluminium Mirror Degradation in a Vacuum-UV Laser", Opt. Commun., 1973, 9, no.4, pp.342-5.

9.74. Bogomolov, A.M., and Pankratov, V.M.: "IR Radiation Polarizer", Prib. Tekh. Eksp., 1973, 16, no.3, pt 2, p.271, and Instrum. Exp. Tech., 1973, 16, no.3, pt 2, p.982.

9.75. Vlasov, S.N., and Orlova, I.M.: "Quasi-Optical Converter of Waves in a Circular-Section Waveguide into a Narrow-Directed Wave Beam", Izv. VUZ Radiofiz., 1974, 17, no.1, pp.148-54.

9.76. Leeb, W.: "Variable (Double-Prism) Beam Attenuator for the IR", Appl. Opt., 1974, 13, no.1, pp.17-9.

9.77. King, W.B.: "Unobscured Laser-Beam-Expander Pointing System with Tilted Spherical Mirrors", Appl. Opt., 1974, 13, no.1, pp.21-2.

9.78. Hockman, G.A., and Sharpe, A.B.: "Influence of Gap between (Semiconductor) Laser and Optical Waveguide", Electron. Lett., 1974, 10, no.8, pp.135-7.

9.79. Lynk, E.T., and Major, L.B.: "Restrahlen Crystals as Wavelength-Selective Laser Reflectors", Rev. Sci. Instrum., 1974, 45, no.1, pp.132-3.

9.80. Austin, R.R.: "Thin-Film Polarizing Devices", Electro-Opt. Syst. Des., 1974, 6, no.2, pp.30-5.

9.81. MacPherson, R.W.: "Variable Attenuator for TEA CO_2 Lasers", Rev. Sci. Instrum., 1974, 45, no.2, p.316.

9.82. Holtom, G.: "Design of a Birefringent Filter for High-Power Dye Lasers", J. Quantum Electron. IEEE, 1974, QE-10, no.8, pp.577-9.

9.83. Mahlein, H.F.: "Nonpolarizing Beamsplitters", Opt. Acta, 1974, 21, no.7, pp.577-83.

9.84. Simmons, W.W., Leppelmeier, G.W., and Johnson, B.C.: "Optical Beam-Shaping Devices Using Polarization Effects", Appl. Opt., 1974, 13, no.7, pp.1629-32.

9.85. Peng, S.T., and Tamir, T.: "Directional Blazing of Waves Guided by Asymmetrical Dielectric Gratings", Opt. Commun., 1974, 11, no.4, pp.405-9.

9.86. Uehara, S., Izawa, T., and Nakagome, H.: "Optical Waveguiding Polarizer", Appl. Opt., 1974, 13, no.8, pp.1753-4.

9.87. Rudisill, J.E., Braunstein, M., and Braunstein, A.I.: "Optical Coatings for High Energy ZnSe Laser Windows", Appl. Opt., 1974, 13, no.9, pp.2075-80.

9.88. Shegai, V.V.: "Case of Double-Layer Matching", Vestn. Mosk. Univ. Fiz. Astron., 1974, 15, no.4, pp.419-23.

9.89. Bar-Isaac, C., and Hardy, A.: "Grid Attenuators in Laser Machining Systems", Appl. Phys., 1974, 5, no.3, pp.251-7.

9.90. Shafer, D.R.: "Anamorphic Laser-Beam Shaper with Phase Step", Opt. Eng., 1974, 13, no.4, pp.331-4.

9.91. Nestrizhenko, Yu.A.: "Total-Internal-Reflection Polarizer", Opt. Spektr., 1974, 37, no.2, pp.326-32, and Opt. Spectrosc., 1974, 37, no.2, pp.186-9.

9.92. Starostina, G.P., et al.: "Silica Light Filters Doped with Eu, Yb, and Sm", Opt.-Mekh. Prom., 1974, 41, no.1, pp.43-8, and Sov. J. Opt. Technol., 1974, 41, no.1, pp.39-42.

9.93. Westwood, W.D.: "Sectioned Thin-Film Grating", J. Opt. Soc. Am., 1974, 64, no.12, pp.1631-5.

10. GUIDED-WAVE COMPONENTS

10.1. Levinson, I.B., and Fridberg, P.Sh.: "Coupling between Two Volumes by a Narrow Slot", Radiotekh. Elektron., 1965, 10,p.260, and Radio Eng. Electron. Phys.,1965,10,p.215.

10.2. Berner, H.: "Waveguide Components of Low VSWR for Antenna Feeds", SEL Nachr., 1965, 13, p.44.

10.3. Arora, R.K.: "Bifurcation of a Parallel-Plate Waveguide by a Unidirectionally Conducting Screen", Proc. Roy. Soc. Edin., 1965, 67, pt 1, p.50.

10.4. Karjala, D.S., and Mittra, R.: "Scattering by a Semi-Infinite Impedance Strip in a Waveguide", Appl. Sci. Res., 1965, 12B, p.157.

10.5. Lavendol, L., and Taub, J.J.: "Reentrant Directional Coupler Using Strip Transmission Line", Trans. IEEE, 1965, MTT-13, p.700.

10.6. Itakura, K., Yamamoto, S., and Azakami, T.: "Coupled Stripline with Three Centre Conductors", Technol. Rep. Osaka Univ., 1965, 16, p.413.

10.7. Itakura, K., Azakami, T., and Koyama, M.: "Optimum Design of Stripline Hybrid-T", Technol. Rep. Osaka Univ., 1965, 15, p.95.

10.8. Cullen, A.L.: "Convergence of the EH Tuner", Electron. Lett., 1965, 1, p.55.

10.9. Atiya, F.S.: "Thin Metal Sheet Dividing Broad Sides of a Rectangular Waveguide", Arch. Elektr. Ubertrag., 1965, 19, p.43.

10.10. Levy, R.: "Transmission-Line Directional Couplers for Very Broadband Operation", Proc. IEE, 1965, 112, p.469.

10.11. Sangster, A.J.: "Variational Method for Analysis of Waveguide Coupling", Proc. IEE, 1965, 112, p.2171.

10.12. Didenko, A.N.: "Use of Normal-Mode Method for Calculation of Rectangular Waveguides With Diaphragm", Zh. Tekh. Fiz., 1965, 35, p.967, and Sov. Phys.-Tech. Phys., 1965, 10, no.5, pp.744-5.

10.13. Brown, J.: "Design of Waveguide Components", J. Inst. Telecommun. Eng., 1965, 11, p.156.

10.14. Janke, R.: "Natural Reflection Factor of a Directional Line", Nachr. Tech., 1965, 15, p.144.

10.15. Boronski, S.: "Multi-Channel Waveguide Rotatable Joint", Microwave J., 1965, 8, no.6, p.102.

10.16. Zabolotikov, Yu.P.: "Determination of Reflection Factor of an Unsymmetrical Diaphragm in a Rectangular Waveguide Containing a Dielectric Plate", Radiotekh. Elektron., 1965, 10, p.642, and Radio Eng. Electron. Phys., 1965, 10, p.546.

10.17. Seo, W.Y.: "Stripline Phaseshifter", Proc. IEEE, 1965, 53, p.208.

10.18. Shestopalov, V.P., and Shcherbak, V.V.: "Discontinuities in Rectangular Waveguides. Capacitive and Inductive Obstacles", Radiotekh. Elektron., 1965, 10, pp.1043 and 1202, and Radio Eng. Electron. Phys., 1965, 10, pp.892 and 1032.

10.19. Sablin, Ye.S.: "Use of Mirror-Image Method to Evaluate Scattering Matrix of a Slot Waveguide Bridge", Izv. VUZ Radiotekh., 1965, 8, p.111, and Sov. Radio Eng., 1965, 8, p.81.

10.20. Gekker, I.R., et al.: "Study of a Corrugated H_{01}-E_{11} Mode Converter in Circular Waveguide", Radiotekh. Elektron., 1965, 10, p.756, and Radio Eng. Electron. Phys., 1965, 10, p.642.

10.21. Stuchly, S., and Kraszewski, A.: "Wideband Rectangular-to-Circular Waveguide Mode and Impedance Transformer", Arch. Elektrotech., 1965, 14, p.597, and Trans. IEEE, 1965, MTT-13, p.379.

10.22. Paul, M.: "Possibility of Altering the Transmission Factor of Microwave Mode-Transforming Elements", Solid-State Electron. 1965, 8, no.2, p.137.

10.23. Toulios, P.P., and Todd, A.C.: "Synthesis of Symmetrical TEM-Mode Directional Couplers", Trans. IEEE, 1965, MTT-13,p.536.

10.24. Cristal, E.G., and Young, L.: "Theory and Tables of Optimum Symmetrical TEM-Mode Coupled-Transmission-Line Directional Couplers", Trans. IEEE, 1965,MTT-13,p.544.

10.25. Sablin, Ye.S.: "Analytical Determination of Network Transfer Constants for Three Waveguide Directional Couplers in Cascade", Izv. VUZ Radiotekh., 1965, 8, p.93, and Sov. Radio Eng., 1965, 8, p.66.

10.26. Raicu, D.: "Novel Hybrid for Microwaves", Telecomunicatti, 1965, 9, p.489.

10.27. Parad, L.I., and Moynihan, R.L.: "Split-Tee Power Divider", Trans. IEEE, 1965, MTT-13, p.91.

10.28. Mushiake, Y., and Ishida, T.: "Characteristics of Loaded Rectangular Waveguides", Trans. IEEE, 1965, MTT-13, p.451.

10.29. Meyer, E., Helberg, H.W., and Macrander, C.: "Wedge Absorber with Magnetic Losses", Z. Angew. Phys., 1965, 20, p.51.

10.30. Gandhi, O.P., and Khandelwal, D.D.: "Wideband Stripline Directional Coupler", J. Inst. Telecommun. Eng., 1965, 11, p.58.

10.31. Schroeder, K.G., and Davis, J.C.: "Averaging of Reflection Factors in Broadband Hybrid Tees", Proc. IEEE, 1965, 53, p.626.

10.32. Gekker, I.R., Luk'yanchikov, G.S., and Sergeichev, K.F.: "Combined Slot Exciter of H_{01} and E_{11} Waves in a Round Waveguide", Radiotekh. Elektron., 1965, 10, p.1138, and Radio Eng. Electron. Phys., 1965, 10, no.6, pp.974-6.

10.33. Shimada, S.: "Rectangular Waveguide TE_{10}-TE_{20} Mode Transducer", J. Inst. Electr. Commun. Eng. Jap., 1965,48,p.1127.

10.34. Rulea, Gh.: "Theoretical and Experimental Methods for Study of Waveguide Discontinuities", Telecomunicatti, 1965, 9, no.2, p.40.

10.35. Zabolotikov, Yu.P.: "Determination of Amplitudes of Higher Modes for H_{01} Wave Incident on an Unsymmetrical Diaphragm in Rectangular Waveguide", Radiotekh. Elektron., 1965, 10, p.762, and Radio Eng. Electron. Phys., 1965, 10, p.648.

10.36. Kulikov, E.L.: "Variational Methods of Designing Microwave Networks", Radiotekh. Elektron., 1965, 10, p.559, and Radio Eng. Electron. Phys., 1965, 10, p.476.

10.37. Kirillov, L.E., and Yavich, L.R.: "Design of Stepped Transition and Directional Couplers with an Arbitrary Number of Elements", Radiotekh. Elektron., 1965, 10, p.1153, and Radio Eng. Electron. Phys., 1965, 10, p.991.

10.38. Cooper, D.C.: "Waveguide Directional Couplers Using Inclined Slots", Microwave J., 1966, 9, no.8, p.97.

10.39. Lind, W.R.: "TE_{11}-Mode-Selective, Coaxial, Directional Coupler", Microwave J., 1966, 9, no.10, p.91.

10.40. von Dall'Armi, G.: "Wideband and Heavily Loadable Coupling between Waveguide and Coaxial Line", Frequenz, 1966,20,p.270.

10.41. Sablin, Ye.S.: "Application of Method of Images to Calculation of Input Impedance of a Thin Transverse Post Exciting a Rectangular Waveguide", Radiotekhnika, 1966, 21, no.2, p.12, and Telecommun. Radio Eng. Pt 2, 1966, 21, no.2, p.71.

10.42. Cristal, E.G.: "Coupled-Transmission-Line Directional Couplers with Lines of Unequal Characteristic Impedances", Trans. IEEE, 1966, MTT-14, p.337.

10.43. Keep, D.N., and Porter, N.E.: "Bandwidth Properties of Arrays of Shunt Slots in Ridged Waveguide", Microwave J., 1966, 9, no.9, p.97.

10.44. Shestopalev, V.P., and Shcherbak, V.V.: "Discontinuities in Rectangular Waveguides. Higher-Mode Waves", Radiotekh. Elektron., 1966, 11, p.675, and Radio Eng. Electron. Phys., 1966, 11, p.577.

10.45. Martini, W.: "Directional Couplers with Coupled Lines", Ann. Radioelectr., 1966, 21, p.279.

10.46. Model', A.M.: "Propagation in Two Coupled Waveguides of Different Cross Sections", Radiotekhnika, 1966, 21, no.3, and Telecommun. Radio Eng. Pt 2, 1966, 21, no.3, p.90.

10.47. Rzepecka, M., and Stuchly, S.: "Rectangular Waveguide Short-Circuit with Cylindrical Slugs", Trans. IEEE, 1966, MTT-14, p.161.

10.48. Yanovskii, M.S., and Knyaz'kov, B.N.: "Possibility of Reduction of Spectral Distortions and Range Expansion of Continuous Waveguide Phaseshifters", Radiotekhnika, 1966, 21, no.7, p.69, and Telecommun. Radio Eng. Pt 2, 1966, 21, no.7.

10.49. Dorfman, L.G., and Filatov, V.V.: "Scattering Factors Caused by Abrupt Change in Electrical Properties of Narrow Wall of a Waveguide", Radiotekh. Elektron., 1966, 11, p.202, and Radio Eng. Electron. Phys., 1966, 11, p.170.

10.50. Clarricoats, P.J.B., Green, P.E., and Oliner, A.A.: "Slot-Mode Propagation in Rectangular Waveguide", Electron. Lett., 1966, 2, p.307.

10.51. Meriakri, V.V.: "Tunable Multimode Waveguide Coupler", Prib. Tekh. Eksp., 1966, no.2, p.204, and Instrum. Exp. Tech., 1966, no.2, p.479.

10.52. Levinson, I.B., and Fridberg, P.Sh.: "Slot Couplings of Rectangular H_{01}-Mode Waveguides. I and II", Radiotekh. Elektron., 1966, 11, pp.831 and 1076, and Radio Eng. Electron. Phys., 1966, 11, pp.717 and 937.

10.53. Shimada, S., Miyamoto, K., and Suzuki, Y.: "Rectangular TE_{10}-TE_{20} Mode Transducer for 7GHz", J. Inst. Electr. Commun. Eng. Jap., 1966, 49, p.1182.

10.54. Clarricoats, P.J.B., and Slinn, K.R.: "Numerical Method for Solution of Waveguide-Discontinuity Problems", Electron. Lett., 1966, 2, p.226.

10.55. Munier, J., and Munier, P.: "Thin-Conducting-Diaphragm Susceptances in Circular TE_{11} and TM_{01} Waveguides", Electron. Lett., 1966, 2, p.38.

10.56. Plonus, M.A.: "Impedance of a Finite Slot", Trans. IEEE, 1966, MTT-14, p.48.

10.57. Meriakri, V.V.: "H-Mode Slot Filters in Circular Waveguide", Radiotekh. Elektron. 1966, 11, p.934, and Radio Eng. Electron. Phys., 1966, 11, p.807.

10.58. Fjerstad, R.L.: "Compact, Wide-Tuning-Range, Dual TE_{111} Mode Preselector", Trans. IEEE, 1966, MTT-14, p.424.

10.59. Nolan, P.E.: "Evaluation of Micro-Wave Phaseshifters", ISA Trans., 1966, 5, no.1, p.74.

10.60. Schuegraf, E.: "Wideband Transformers for Converting H_{10}-Rectangular to H_{01}-Circular Mode", Nachr. Tech. Z., 1966, 19, p.31.

10.61. Kerns, D.M., and Grandy, W.T.: "Perturbation Theorems for Waveguide Junctions", Trans. IEEE, 1966, MTT-14, p.85.

10.62. Shimada, S.: "Coupled-Wave-Type Rectangular-TE_{01} to Circular-TE_{02} Mode Transducer", J. Inst. Electr. Commun. Eng. Jap., 1966, 49, no.3, p.29.

10.63. Yamamoto, S., Azakami, T., and Itakura, K.: "Slit-Coupled Strip Transmission Lines", Trans. IEEE, 1966, MTT-14, p.542.

10.64. Pace, J.R., and Mittra, R.: "Trifurcated Waveguide", Radio Sci., 1966, 1, no.1, p.117.

10.65. Felsen, L.B., and Ren, C.L.: "Scattering by Obstacles in a Multimode Waveguide", Proc. IEE, 1966, 113, p.16.

10.66. Cristal, E.G.: "Reentrant Directional Couplers Having Direct-Coupled Centre Conductors", Trans. IEEE, 1966, MTT-14, p.207.

10.67. Schiffman, B.M.: "Multisection Microwave Phase-Shift Network", Trans. IEEE, 1966, MTT-14, p.209.

10.68. Kraszewski, A., et al.: "Absorbing Material for Microwave Applications", Arch. Elektrotech., 1966, 15, p.525.

10.69. Bolster, M.F.: "High-Power Microwave Load, with Uniform Power Absorption", Microwave J., 1966, 9, no.4, p.56.

10.70. Shimada, S.: "Resonant-Cavity-Type Mode Transducer", Trans. IEEE, 1966, MTT-14, p.384.

10.71. Braccesi, A., et al.: "Simple and Accurate, Liquid-Dielectric, Variable-Length Line", Proc. IEEE, 1966, 54, p.69.

10.72. Shestopalov, V.P., and Shcherbak, V.V.: "Discontinuities in Rectangular Waveguides. Double-Strip Obstacles", Radiotekh. Elektron., 1966, 11, p.1066, and Radio Eng. Electron. Phys., 1966, 11, p.928.

10.73. DuHamel, R.H., and Armstrong, M.E.: "Log-Periodic Transmission-Line Circuits. I", Trans. IEEE, 1966, MTT-14, p.264.

10.74. Dorfman, L.G., and Filatov, V.V.: "Parameters of a Waveguide Directional Coupler", Radiotekh. Elektron., 1966, 11, p.1933, and Radio Eng. Electron. Phys., 1966, 11, no.11, pp.1698-1706.

10.75. Dangl, J.R., and Steele, K.P.: "Use of Stripline to Design Microwave Circuits. I and II", Electronics, 1966, 39, Feb.7, p.72, and Feb.21, p.90.

10.76. Dybdal, R.B., Rudduck, R.C., and Tsai, L.L.: "Mutual Coupling between TEM and TE_{01} Parallel-Plate Waveguide Apertures", Trans. IEEE, 1966, AP-14, p.574.

10.77. Wlodek, J.: "Elements of Microwave Systems in Microstrip", Przegl. Telekomun., 1966, no.6, p.167.

10.78. Singletary, J.: "Fringing Capacitance in Stripline Coupler Design", Trans. IEEE, 1966, MTT-14, p.398.

10.79. Palais, J.C.: "Complete Solution of Inductive Iris with TE_{k0} Incidence in Rectangular Waveguide", Trans. IEEE, 1967, MTT-15, p.156.

10.80. Querido, H., Frank, J., and Cheston, T.C.: "Wideband Phaseshifters", Trans. IEEE, 1967, AP-15, p.300.

10.81. Mori, S., et al.: "Properties of Short-Slot Hybrid Junction with Dielectric Slab", Bull. Univ. Osaka Prefecture, 1967, 16A, p.265.

10.82. Sharp, E.D.: "Exact Calculation for a T-Junction of Rectangular Waveguides Having Arbitrary Cross Sections", Trans. IEEE, 1967, MTT-15, p.109.

10.83. Roberts, F.B.: "High-Power Microwave Loads", Electron. Compon., 1967, 8, no.4, pp.374-8.

10.84. Hancock, K.E.: "Design and Manufacture of Waveguide Chebyshev Directional Couplers", Electron. Eng., 1967, 39, no.5, pp.292-7.

10.85. Flachenecker, G., Lange, K.P., and Meinke, H.H.: "Waveguide Irises of any Longitudinal Section", Nachr. Tech. Z., 1967, 20, no.2, pp.70-6.

10.86. Barker, L.R.: "Stripline Coupler Design Chart", Microwaves, 1967, 6, no.2, pp.46-8.

10.87. Bouthinon, N., and Coumes, A.: "Broadband Hybrids", Trans. IEEE, 1967, MTT-15, no.7, pp.431-2.

10.88. Kumagai, N., Kurazono, S., and Shiomi, S.: "Circular TE_{0n} Mode Filter", J. Inst. Electr. Commun. Eng. Jap., 1967, 50, no.3, pp.463-4.

10.89. Gillitzer, E.: "Dominant-Mode Reflection at Circular Horn Junctions", Proc. IEEE, 1967, 55, no.7, pp.1210-1.

10.90. Muilwyk, C.A., and Davies, J.B.: "Numerical Solution of Rectangular Waveguide Junctions and Discontinuities of Arbitrary Cross Section", Trans. IEEE, 1967, MTT-15, no.8, pp.450-4.

10.91. Lucas, I.: "General Theory of the Short-Slot Directional Coupler", Arch. Elektr. Ubertrag., 1967, 21, no.7, pp.339-44.

10.92. Gewartowski, J.B., and Swan, C.B.: "Constructive Coupling in Directional Couplers", Trans. IEEE, 1967, MTT-15, no.2, p.134.

10.93. Wexler, A.: "Solution of Waveguide Discontinuities by Modal Analysis", Trans. IEEE, 1967, MTT-15, no.9, pp.508-16.

10.94.   Johnson, R.C., Cain, F.L., and
Bone, E.N.: "Dual-Mode Coupler", Trans.
IEEE, 1967, MTT-15, no.11, pp.651-2.

10.95.   Hall, M.B., and Little, W.E.:
"Directional Coupler with a Readily Calcul-
able Coupling Ratio", Trans. IEEE, 1967,
MTT-15, no.11, pp.598-603.

10.96.   Quine, J.P.: "Finite-Length Metal
Septa for Suppression of Higher-Order Modes",
Trans. IEEE, 1968, MTT-16, p.879.

10.97.   Bryant, T.G., and Weiss, J.A.:
"Parameters of Microstrip Lines and Coupled
Pairs", Trans. IEEE, 1968, MTT-16, p.1021.

10.98.   Vernon, R.J.: "General Theorem on
Symmetry-Imposed Decoupling of Ports of a
Waveguide Junction", Electron. Lett., 1968,
4, p.586.

10.99.   Kraus, A.: "Equivalent Circuit of
the Unterminated Directional Coupler",
Nachr. Tech. Z., 1968, 21, p.471.

10.100.   Orleanskaya, E.V.: "Broadband
Device for Rotation of Polarization", Radiotekhnika, 1968, 23, no.8, p.23, and Telecommun. Radio Eng. Pt 2, 1968, 23, no.8.

10.101.   Mouw, R.B.: "Broadband Hybrid
Junction", Trans. IEEE, 1968, MTT-16,p.911.

10.102.   Miller, S.E.: "Solutions of Two
Waves with Periodic Coupling", Bell Syst.
Tech. J., 1968, 47, p.1801.

10.103.   Muehe, C.E.: "High-Power Waveguide
Tuner", Trans. IEEE, 1968, MTT-16, p.882.

10.104.   Levy, R.: "Analysis and Synthesis
of Waveguide Multiaperture Directional
Couplers", Trans. IEEE, 1968, MTT-16, p.995.

10.105.   Carson, C.T.: "Numerical Solution
of Waveguide Problems by Fast Fourier Transforms", Trans. IEEE, 1968, MTT-16, p.955.

10.106.   Franceschetti, G.: "Thin Resistive Layers in Waveguides. Theory and
Experiment", Arch. Elektrotech., 1969, 18,
no.4, p.655.

10.107.   Masterman, P.H., Clarricoats,
P.J.B., and Hannaford, C.D.: "Computer
Method of Solving Waveguide Iris Problems",
Electron. Lett., 1969, 5, p.23.

10.108.   Luzzato, G.: "180° Ring Couplers",
Proc. IEEE, 1969, 57, p.82.

10.109.   Mohr, C.: "Mode Conversion in
Discontinuities of Rectangular Waveguides",
Arch. Elektr. Ubertrag., 1969, 23, p.147.

10.110.   Levy, R., and Lind, L.F.: "Synthesis of Asymmetrical Branch-Guide, Directional-Coupler, Impedance Transformers",
Trans. IEEE, 1969, MTT-17, p.45.

10.111.   Chikunov, L.I.: "Limits of Physical Realisability of Circuits in Coupled
Striplines", Radiotekhnika, 1969, 24, no.1,
p.51, and Telecommun. Radio Eng. Pt 2, 1969,
24, no.1.

10.112.   Levy, R.: "Analysis of Practical
Branch-Guide Directional Couplers", Trans.
IEEE, 1969, MTT-17, p.289.

10.113.   Tan, P.C.: "Dual Adjacent Directional Coupler", Electron. Lett., 1969, 5,
p.283.

10.114.   Rhinewine, M.: "Linear Polarization Diplexer at 34 GHz", Rev. Sci. Instrum.,
1969, 40, p.951.

10.115.   Vendik, O.G., Mironenko, I.G., and
Yavnoshan, F.V.: "Calculation of Complex
Waveguide Junctions", Zh. Tekh. Fiz., 1969,
39, no.3, p.483, and Sov. Phys.-Tech. Phys.,
1969, 14, no.3, pp.357-62.

10.116.   Sinnott, D.H., et al.: "Finite
Difference Solution of Microwave Circuit
Problems", Trans. IEEE, 1969, MTT-17, p.464.

10.117.   Gupta, R.R.: "Fringing Capacitance
Curves for Coplanar Rectangular Coupled Bars",
Trans. IEEE, 1969, MTT-17, p.637.

10.118.   Yurov, Yu.Ya., and Sablin, Ye.S.:
"Coupling in Rectangular Waveguides through
Inclined Slots in Common Broad Wall", Izv.
VUZ Radioelektron., 1969, 12, no.6, p.598.

10.119.   Rozzi, T.E.: "Wideband 180° Hybrid
Junctions Using Phase-Corrected Asymmetric
Directional Couplers", Proc. IEE, 1969, 116,
p.661.

10.120.   Bisio, G.R., and Millanta, L.:
"Practical Realization of the Dumbbell Short-
Circuiting Plunger", Alta Freq., 1969, 38,
no.8, p.643.

10.121.   Kammler, D.W.: "Design of Discrete
N-Section and Continuously Symmetrical TEM
Directional Couplers", Trans. IEEE, 1969,
MTT-17, p.577.

10.122.   Felsen, L.B.: "Ray Method for
Scattering by Small Discontinuities in a
Waveguide", Electron. Lett., 1969, 5, p.542.

10.123.   Daly, P.: "Finite-Element Coupling
Matrices", Electron. Lett., 1969, 5, p.613.

10.124.   Green, R.B.: "Grating Formulation
for Problems Involving Cylindrical Discontinuities in Rectangular Waveguides", Trans.
IEEE, 1969, MTT-17, p.760.

10.125.   Yee, H.Y., and Felsen, L.B.:
"Ray-Optical Analysis of Scattering in Waveguides", Trans. IEEE, 1969, MTT-17, p.671.

10.126.   Marin, L.: "Coupling between Two
Adjacent Rectangular Waveguides through a
Slot in the Common Wall", Acta Polytech.
Scand., 1969, no.EL20, p.1.

10.127.   Millican, G.L., and Wales, R.C.:
"Practical Stripline Microwave Circuit
Design", Trans. IEEE, 1969, MTT-17, p.696.

10.128.   Ryzhkova, L.V.: "Relation between
S and T Matrices of a Waveguide Multiterminal
Network", Izv. VUZ Radioelektron., 1969, 12,
no.8, p.931.

10.129. Lind, L.F.: "Design Equations for Two-Branch Stripline Directional-Coupler/ Impedance-Transformer", Electron. Lett., 1969, 5, p.495.

10.130. Huckle, P.R., and Masterman,P.H.: "Analysis of a Rectangular-Waveguide Junction Incorporating a Row of Rectangular Posts", Electron. Lett., 1969, 5, p.559.

10.131. Lewin, L.: "Calculation of Waveguide Junction and Diaphragm Interactions", Trans. IEEE, 1969, MTT-17, p.758.

10.132. Taub, J.J., and Kurpis, G.P.: "General N-Way Hybrid Power Divider", Trans. IEEE, 1969, MTT-17, p.406.

10.133. Groendijk, H., and Versnel, W.: "Inductive Posts with Capacitive Gap in a Waveguide", Appl. Sci. Res., 1969, 21, no.5, p.309.

10.134. Lange, J.H., and Rose, B.E.: "Star-Delta Transformation of Three-Way Hybrid Junction", Trans. IEEE, 1969, MTT-17, p.789.

10.135. Dalley, J.E.: "Stripline Directional Coupler Utilizing a Nonhomogeneous Dielectric", Trans. IEEE, 1969,MTT-17,p.706.

10.136. Ishii, T.K., and Jenners, J.A.: "Transmission Characteristics of Reentrant Hybrid T", Trans. IEEE, 1969, MTT-17,p.718.

10.137. Otoshi, T.Y.: "Scattering Parameters of a Reduced Multiport", Trans. IEEE, 1969, MTT-17, p.722.

10.138. Bernardi, P.: "Improved Mode Filters for Oversized Rectangular Waveguides", Trans. IEEE, 1969, MTT-17, p.237.

10.139. Yemelin, B.F., and Mashkovtsev, B.M.: "Scattering Matrix of an Aperture in the Side of Multimode Waveguide", Radiotekhnika, 1969, 24, no.12, p.59, and Telecommun. Radio Eng. Pt 2, 1969, 24, no.12.

10.140. Guogo, V.I., Kizhlo, G., and Shugurov, V.K.: "Calculation of H_{10}-Mode Reflection by a Thin Long Pintle in Middle of a Rectangular Waveguide", Litov. Fiz. Sb., 1969, no.6, p.1059.

10.141. Stelzried, C.T.: "Effect of Capacitative-Screw Tuners on Waveguide Loss", Trans. IEEE, 1969, MTT-17, p.172.

10.142. Brady, M.M.: "Rectangular-Waveguide/ Coaxial-Line Transitions", Trans. IEEE, 1969, MTT-17, p.170.

10.143. Takeo, I., and Watanabe, K.: "Oversize-Waveguide Directional Coupler Using Two Prisms Interface-Matched by Brewster's-Angle Effect", Trans. IEEE, 1969, MTT-17, p.124.

10.144. Courbois, C., et al.: "Study of Symmetrical-Scatter Irises in Overdimensional Waveguides", C.R. Acad. Sci. (Paris), 1969, 268B, no.20, p.1286.

10.145. Mailloux, R.J.: "First-Order Solutions for Mutual Coupling between Waveguides which Propagate Two Orthogonal Modes", Trans. IEEE, 1969, AP-17, p.740.

10.146. Musil, J.: "Directional Coupler with Full Power Transfer", Slab. Obz., 1969, 30, no.11, p.489.

10.147. Braeckelmann, W., and Roeder, R.: "Characteristic Impedances of Coupled Striplines", Electron. Lett., 1970, 6, p.123.

10.148. Knetsch, H.D.: "Mode Transducers and Filters Formed by Rectangular and Circular Waveguide Elements", Nachr. Tech. Z., 1970, 23, no.2, p.57.

10.149. Ramey, R.L., Landes, H.S., and Manus, E.A.: "Microwave Properties of Thin Films with (Lossy) Apertures", Trans. IEEE, 1970, MTT-18, p.196.

10.150. Krage, M.K., and Haddad, G.I.: "Characteristics of Coupled Microstrip Lines. I and II", Trans. IEEE, 1970, MTT-18, pp.217 and 222.

10.151. Hoffswell, R.A.: "Microstrip Line Stretcher", Rev. Sci. Instrum., 1970, 41, no.9, pp.1330-1.

10.152. Yurov, Yu.Ya., and Lavrenko, K.F.: "Experimental Verification of Approximate Theory for a Thin Conductor in a Waveguide", Izv. VUZ Radioelektron., 1970, 13, no.9, pp.1047-54.

10.153. Maloratskii, L.G., and Cherne, Kh.I.: "Directional Couplers Using Lines with Imperfect Terminations", Izv. VUZ Radioelektron., 1970, 13, no.3, p.341.

10.154. Oleinikov, V.N., and Boikov, V.V.: "Broadband Waveguide Slot-Excited Launcher of Circularly Polarized TE_{11} Waves", Radiotekh. Elektron., 1970, 15, p.1080, and Radio Eng. Electron. Phys., 1970, 15, no.5, pp.907-8.

10.155. Koryakin, L.G., and Sevast'yanova, A.M.: "Arbitrarily Wide Near-Resonant Slot at the End of a Rectangular Guide", Radiotekh. Elektron., 1970, 15, no.5, p.1081, and Radio Eng. Electron. Phys., 1970, 15, no.5, pp.909-11.

10.156. Kozar', A.I., and Khizhnyak, N.A.: "Reflection from Resonant Dielectric Sphere in a Waveguide", Ukr. Fiz. Zh., 1970,5,p.847.

10.157. Kaufman, I.: "Calculation of Radiation Fields in Waveguides by Principle of Power Balance", Trans. IEEE, 1970, MTT-18, p.418.

10.158. Glaser, J.I.: "Numerical Solution of Waveguide Scattering Problems by Finite-Difference Green's Functions", Trans. IEEE, 1970, MTT-18, p.436.

10.159. Horton, M.C.: "Loss Calculations for Rectangular Coupled Bars", Trans. IEEE, 1970, MTT-18, no.10, pp.736-8.

10.160.   Someda, C.G.: "Rectangular-TE_{10} to Circular-TE_{01} Cross-Shaped Coupler", Alta Freq., 1970, 39, no.5, p.180e.

10.161.   Denlinger, E.J.: "Frequency Dependence of a Coupled Pair of Microstrip Lines", Trans. IEEE, 1970, MTT-18, p.731.

10.162.   Luzzatto, G.: "S Parameters of Ring Directional Couplers", Alta Freq., 1970, 39, no.7, p.635.

10.163.   Chao, G.: "Wideband Variable Microwave Coupler", Trans. IEEE, 1970, MTT-18, p.576.

10.164.   Arndt, F.: "Tables for Asymmetric Chebyshev High-Pass TEM-Mode Directional Couplers", Trans. IEEE, 1970, MTT-18, p.633.

10.165.   De Mesquita, D.G., and Bailey, A.G.: "Symmetrically Excited Microwave Rotary Joint", Trans. IEEE, 1970, MTT-18, p.654.

10.166.   Clarke, J.: "Transmission Characteristics of a Reentrant Hybrid Tee", Trans. IEEE, 1970, MTT-18, p.665.

10.167.   Aronson, I.: "Variational Bound Principle for Scattering by Obstacles in Waveguide", Trans. IEEE, 1970, MTT-18, no.10, pp.725-31.

10.168.   Tsandoulas, G.N., Temme, D.H., and Willwerth, F.G.: "Longitudinal Section Mode Analysis of Dielectric-Loaded Rectangular Waveguides", Trans. IEEE, 1970, MTT-18, p.88.

10.169.   Agarwal, K.K., and Nagelberg, E.R.: "Phase Characteristics of a Circularly Symmetric Dual-Mode Transducer", Trans. IEEE, 1970, MTT-18, p.69.

10.170.   Gardiol, F.E.: "Power Combiner Nomograms", Trans. IEEE, 1970, MTT-18, p.71.

10.171.   Snyder, A.W., and Davies, O.J.: "Asymptotic Solution of Coupled-Mode Equations for Sinusoidal Coupling", Proc. IEEE, 1970, 58, p.168.

10.172.   Braeckelmann, W.: "Waveguide Junctions for Rectangular Waveguide", Nachr. Tech. Z., 1970, 23, no.1, p.2.

10.173.   Chen, W.H.: "Even- and Odd-Mode Impedance of Coupled Pairs of Microstrip Lines", Trans. IEEE, 1970, MTT-18, p.55.

10.174.   Mastellari, G.A., and Someda, C.G.: "Evaluation of Coupling between Different Waveguides", Alta Freq., 1970, 39, p.200.

10.175.   Polishchuk, N.P.: "Scattering Matrix of a Directional Coupler in Stripline with Small Asymmetry", Radiotekh. Elektron., 1970, 15, p.387, and Radio Eng. Electron. Phys., 1970, 15, no.2, pp.335-6.

10.176.   Monaco, V.A., and Tiberio, P.: "Automatic Scattering Matrix Computation of Microwave Circuits", Alta Freq., 1970, 39, p.165.

10.177.   Rozzi, T.E.: "Phase-Compensated Asymmetric Directional Couplers for Design of Decade-Wide Quadrature Hybrids", Proc. IEE, 1970, 117, p.704.

10.178.   Wen, C.P.: "Coplanar-Waveguide Directional Couplers", Trans. IEEE, 1970, MTT-18, p.318.

10.179.   Okugawa, S., and Hagiwara, H.: "Analysis and Computation of Coupling between Microstrip Lines", Electron. Commun. Jap., 1970, 53, no.7, pp.30-1.

10.180.   Woodward, O.M.: "Dual-Channel Rotary Joint for High-Average-Power Operation", Trans. IEEE, 1970, MTT-18, no.12, pp.1072-7.

10.181.   Vol'man, V.I.: "Diffraction by Inhomogeneities in a Rectangular Waveguide", Radiotekhnika, 1970, 25, no.8, and Telecommun. Radio Eng. Pt 2, 1970, 25, no.8, pp.99-105.

10.182.   Rabinovich, G.I.: "Waveguide Load with Separate Control of Modulus and Phase of Reflection Factor", Radiotekhnika, 1970, 25, no.9, and Telecommun. Radio Eng. Pt 2, 1970, 25, no.9, pp.145-7.

10.183.   Oda, M., and Sato, R.: "Method of Construction of Nonuniformly Coupled Symmetrical Directional Coupler", Rec. Electr. Commun. Eng. Conversaz. Tohoku Univ., 1970, 39, no.4, pp.6-10.

10.184.   Suetake, K., et al.: "Small Waveguide Matched Load with Rubber Ferrite Sheet", Electron. Commun. Jap., 1970, 53, no.9, pp.59-62.

10.185.   Cullen, A.L., and Davies, O.J.: "Periodic Coupling of Waveguide Modes", Proc. IEEE, 1970, 117, no.11, pp.2061-8.

10.186.   Gelnovatch, V.G., and Chase, I.L I.L.: "Optimal-Seeking Computer Programme for Design of Microwave Circuits", J. Solid-State Circuits IEEE, 1970, SC-5, no.6, pp.303-9.

10.187.   Krohne, M.: "Junction of Two Circular Waveguides, One Containing a Coaxial Dielectric Rod", Nachr. Tech. Z., 1970, 23, no.12, pp.633-9.

10.188.   Stradler, E., and Ochojski, A.: "Compensated Directional Coupler with Octave Bandwidth", Frequenz, 1970, 24, no.11, pp.342-5.

10.189.   Van Bladel, J.: "Small Holes in a Waveguide Wall", Proc. IEE, 1971, 118, no.1, pp.43-50.

10.190.   Masterman, P.H., and Clarricoats, P.J.B.: "Computer Field-Matching Solution of Waveguide Transverse Discontinuities", Proc. IEE, 1971, 118, no.1, pp.51-63.

10.191. Itoh, T., and Mittra, R.: "Accurate Method for Calculating Charge and Potential Distributions in Coupled Microstrip Lines", Proc. IEEE, 1971, 59, no.2, pp.332-4.

10.192. Farrar, A., and Adams, A.T.: "Computation of Lumped Microstrip Capacities by Matrix Methods. Rectangular Sections and End Effect", Trans. IEEE, 1971, MTT-19, no.5, pp.495-7.

10.193. Knorr, J.B.: "Tensor Character of Symmetrical Waveguide Junctions", Trans. IEEE, 1971, MTT-19, no.4, pp.414-5.

10.194. Ekinge, R.B.: "Design of Impedance-Transforming Directional Couplers", Trans. IEEE, 1971, MTT-19, no.4, pp.415-6.

10.195. Katunin, V.V., and Kurilin, B.I.: "Parameter Optimization of a Cavity with a Non-Contacting Short-Circuit", Izv. VUZ Radioelektron., 1971, 14, no.3, pp.286-91.

10.196. Napoli, L.S., and Hughes, J.J.: "Foreshortening of Microstrip Open Circuits on Alumina Substrates", Trans. IEEE, 1971, MTT-19, no.6, pp.559-61.

10.197. Kashyap, S.C., and Hamid, M.A.K.: "Ray-Optical Scattering by a Thick Waveguide Diaphragm", Int. J. Electron., 1971, 30, no.6, pp.543-9.

10.198. Ekinge, R.B.: "Method of Synthesizing Matched Broadband TEM-Mode Three-Ports", Trans. IEEE, 1971, MTT-19, no.1, pp.81-8.

10.199. Fialko, Ye.I., Redkin, B.A., and Ocheret, Zh.G.: "Excitation of a Circular Waveguide by a Transverse Rod", Izv. VUZ Radioelektron., 1971, 14, no.2, pp.216-8.

10.200. Tsandoulas, G.N., and Ince, W.J.: "Modal Inversion in Circular Waveguides. I", Trans. IEEE, 1971, MTT-19, no.4, pp.386-92.

10.201. Garcia, J.A.: "Wideband Quadrature Hybrid Coupler", Trans. IEEE, 1971, MTT-19, no.7, pp.660-1.

10.202. Joines, W.T.: "Continuously Variable Dielectric Phaseshifter", Trans. IEEE, 1971, MTT-19, no.8, pp.729-32.

10.203. Schlude, F.S.: "Direct Solution of Waveguide Step with Thin Iris", Arch. Elektr. Ubertrag., 1971, 25, no.8, pp.398-400.

10.204. Johns, P.B., and Beurie, R.L.: "Numerical Solution of Two-Dimensional Scattering Problems Using a Transmission-Line Matrix", Proc. IEE, 1971, 118, no.9, pp.1203-8.

10.205. Sobel, H.: "Radiation Conductance of Open-Circuit Microstrip (Stub)", Trans. IEEE, 1971, MTT-19, no.11, pp.885-7.

10.206. Tetarenko, R.P., and Goud, P.A.: "Broadband Properties of a Class of TEM-Mode Hybrids", Trans. IEEE, 1971, MTT-19, no.11, pp.887-9.

10.207. Bryan, J.G., and Rosenbaum, F.J.: "Wideband Nearly-Constant-Susceptance Waveguide Element", Trans. IEEE, 1971, MTT-19, no.11, pp.889-91.

10.208. Nagasawa, Y., and Sato, R.: "Exact Design of Directional Coupler Using TEM-Mode Coupled Transmission Lines", Technol. Rep. Tohoku Univ., 1971, 36, no.1, pp.167-86.

10.209. Paramonov, V.K.: "Influence of Nonoptimal Parameters on the Characteristics of a Balanced Directional Coupler", Elektrosvyaz, 1971, 24, no.12, pp.30-5, and Telecommun. Radio Eng. Pt 1, 1971, 24, no.12.

10.210. Rowe, D.N.E.: "Design of 3-dB Branch-Line Coupler Centred on 1.25 GHz", Marconi Rev., 1971, 34, no.183, pp.253-8.

10.211. Minner, W.: "Components in Microstrip Technique", Nachr. Tech. Z., 1971, 24, no.10, pp.K163-7.

10.212. Ferdula, M., and Vizner, V.: "Aspects of Design of T- and Pi-Type Microwave Attenuators", Pr. PIT, 1971, 21, no.72, pp.25-36.

10.213. Vogel, R.: "Properties of Directional Couplers Employed in Microwave Integrated Circuits", Rozpr. Electrotech., 1971, 17, no.3, pp.517-40.

10.214. Lyashchenko, V.A.: "Slots in a Circular Waveguide", Izv. VUZ Radioelektron., 1971, 14, no.10, pp.1123-9.

10.215. Nicolaescu, St.: "Calculation of a 3-dB Directional Coupler", Posta Telecom., 1971, 1, no.3, pp.143-7.

10.216. Bushneva, N.N., Kurdyumov, O.A., and Sotin, V.Ye.: "Spurious Resonances in Microstrip Components", Izv. VUZ Radioelektron., 1971, 14, no.10, pp.1242-4.

10.217. Kasa, I., and Adorjan, P.: "Design of Directional Couplers Comprising Continuously Coupled Striplines. I and II", Hiradastechnika, 1971, 22, no.11, pp.327-34, and no.12, pp.368-74.

10.218. Lun'kov, A.E., and Lyakhovetskii, B.A.: "Analysis of Operation of Prism Power Divider in Multimode Waveguide", Izv. VUZ Radiofiz., 1971, 14, no.12, pp.1902-5.

10.219. Hundewadt, J., and Flabb, H.: "Increasing Coupling Factor of Microstrip by Interdigital Structure", Nachr. Tech. Z., 1971, 24, no.8, pp.424-6.

10.220. Leighton, W.H., and Milnes, A.G.: "Junction Reactance and Dimensional Tolerance Effects on X-Band 3-dB Directional Couplers", Trans. IEEE, 1971, MTT-19, no.10, pp.818-24.

10.221. Zawadzki, M., and Zakrzewski, A.: "Variable Microwave Attenuator for Waves of Arbitrary Polarization", Pr. PIT, 1971, 21, no.71, pp.55-6.

10.222. Koyama, M.: "Antisymmetric Branch-Guide Directional Couplers", Rev. Electr. Commun. Lab., 1971, 19, no.9-10, pp.1007-15.

10.223. Shinada, Y., Matsumoto, A., and Ono, K.: "Susceptance Compensation of Hybrid-Ring Circuit", J. Inst. Telev. Eng. Jap., 1971, 25, no.11, pp.873-9.

10.224. Kulinski, J.: "Microwave Directional Couplers with Inductive Loops in Rectangular Waveguide", Pr. PIT, 1971, 21, no.73, pp.21-33.

10.225. Shinada, Y., and Matsumoto, A.: "Broadbanding of Reverse-Phase Hybrid Ring", J. Inst. Telev. Eng. Jap., 1971, 25, no.11, pp.880-5.

10.226. Hoer, C.A.: "Directional-Coupler Design", Natl. Bur. Stand. Tech. News Bull., 1971, 55, no.12, pp.300-2.

10.227. Rulea, Gh.: "Calculation of Identical-Slot Directional Couplers", Posta Telecom., 1971, 1, no.6, pp.289-94.

10.228. Leikin, V.Yu.: "Coupling between Rectangular Waveguide and Stripline through a Hole", Radiotekh. (Kharkov), 1971, no.17, pp.76-80.

10.229. Hidaka, T.: "Optimum Design Theory for Absorbing Wall", Bull. Electrotech. Lab., 1971, 35, no.11, pp.931-46.

10.230. Rokushina, K., Mori, H., and Taki, I.: "Short-Slot Hybrid Junction Loaded with a Dielectric Slab", Electron. Commun. Jap., 1971, 54, no.4, pp.87-93.

10.231. Bouwkamp, C.J.: "Scattering Characteristics of a Cross-Junction of Oversized Waveguides", Philips Tech. Rev., 1971, 32, no.6-8, pp.165-77.

10.232. Bottjer, M.F., and King, H.E.: "Top-Wall and Branch-Guide Hybrids for Millimetre Waves", Trans. IEEE, 1972, MTT-20, no.2, pp.182-4.

10.233. Royer, E., and Mittra, R.: "Diffraction by Dielectric Steps in Waveguides", Trans. IEEE, 1972, MTT-20, no.4, pp.273-9.

10.234. Lacombe, D.: "Multioctave Microstrip 50-ohm Termination", Trans. IEEE, 1972, MTT-20, no.4, pp.290-2.

10.235. Wolff, I., Kompa, G., and Mehran, R.: "Calculation Method for Microstrip Discontinuities and T-Junctions", Nachr. Tech. Z., 1972, 25, no.5, pp.217-24, and Electron. Lett., 1972, 8, no.7, pp.177-9.

10.236. Meshchanov, V.P., and Kibirskii, Yu.V.: "Directional Couplers Using Coupled Lines of Different Electrical Lengths", Radiotekhnika, 1972, 27, no.2, pp.95-6, and Telecommun. Radio Eng. Pt 2, 1972, 27, no.2, pp.132-3.

10.237. Rozzi, T.E.: "Equivalent Network for Interacting Thick Inductive Irises", Trans. IEEE, 1972, MTT-20, no.5, pp.323-30.

10.238. Gould, J.W., and Talboys, E.C.: "Even- and Odd-Mode Guide Wavelengths of Coupled Lines in Microstrip", Electron. Lett., 1972, 8, no.5, pp.121-2.

10.239. Bergandt, H.G., and Pregla, R.: "Calculation of Even- and Odd-Mode Capacitance Parameters for Coupled Microstrips", Arch. Elektron. Ubertrag., 1972, 26, no.4, pp.153-8.

10.240. Maeda, M.: "Analysis of Gaps in Microstrip Lines", Trans. IEEE, 1972, MTT-20, no.6, pp.390-6.

10.241. Thong, V.K.: "Solutions for Waveguide Discontinuities by the Method of Moments", Trans. IEEE, 1972, MTT-20, no.6, pp.416-8.

10.242. Young, D.T., and Rowe, H.E.: "Optimum Coupling for Random Guides with Frequency-Dependent Coupling", Trans. IEEE, 1972, MTT-20, no.6, pp.365-72.

10.243. Uenakada, K., and Yasunaga, K.: "Microwave Equivalent Circuit of Diode Mount Using Radial-Line Structure", NHK Tech. J., 1972, 24, no.1, pp.59-67.

10.244. Singh, J., Kumar, R.C., and Singh, F.: "Rectangular-Circular Waveguide Transition", J. Inst. Telecommun. Eng., 1972, 18, no.1, pp.47-8.

10.245. Moller, E.: "Digital Microstrip 3-dB Hybrid Coupled Phaseshifters for X Band with Real Elements", Nachr. Tech. Z., 1972, 25, no.5, pp.232-7.

10.246. Hundewadt, J., and Stahlmann, R.: "Coupled Microstrip Transmission-Line Parameters", Nachr. Tech. Z., 1972, 25, no.5, pp.244-6.

10.247. Farrar, A., and Adams, A.T.: "Matrix Methods for Microstrip Three-Dimensional Problems", Trans. IEEE, 1972, MTT-20, no.8, pp.497-504.

10.248. Nagasawa, Y., and Sato, R.: "Exact Design of Directional Couplers Using TEM-Mode Coupled Transmission Lines", Electron. Commun. Jap., 1972, 54, no.1, pp.9-16.

10.249. Kowalski, G., and Pregla, R.: "Dispersion Characteristics of Single and Coupled Microstrip", Arch. Elektron. Ubertrag., 1972, 26, no.6, pp.276-80.

10.250. Schwartz, E., and Bex, H.: "Scattering Coefficients of Cyclic Symmetric Passive Three-Ports", Arch. Elektron. Ubertrag., 1972, 26, no.7-8, pp.336-42.

10.251. Brewer, T.A., and Ishii, T.K.: "Reciprocal but Nonsymmetrical Transmission Characteristics of Tapered Irises", Electron. Lett., 1972, 8, no.14, pp.362-7.

10.252. Brenner, H.: "Quarter-Wavelength Directional Coupler in an Inhomogeneous Medium", Frequenz, 1972, 26, no.6, pp.156-165.

10.253. Silvester, P., and Benedek, P.:
"Equivalent Capacitances of Microstrip Open-
Circuits", Trans. IEEE, 1972, MTT-20, no.8,
pp.511-6.
10.254. Valtonen, M.: "Synthesis Method
for Branch-Guide Directional Couplers", Acta
Polytech. Scand., 1972, no.EL30, pp.7-18.
10.255. Khac, T.Vu., and Carson, C.T.:
"Coupling by Slots in Rectangular Waveguides
with Arbitrary Wall Thickness", Electron.
Lett., 1972, 8, no.18, pp.456-8.
10.256. Corr, D.G., and Davies, J.B.:
"Computer Analysis of Fundamental and Higher-
Order Modes in Single and Coupled Microstrip",
Trans. IEEE, 1972, MTT-20, no.10, pp.669-78.
10.257. Krage, M.K., and Haddad, G.I.:
"Frequency-Dependent Characteristics of
Microstrip Lines", Trans. IEEE, 1972, MTT-20,
no.10, pp.678-88.
10.258. Gal'chenko, N.A., et al.: "Thin
Inductive Rod in Single- and Double-Ridged
Waveguides", Izv. VUZ Radioelektron., 1972,
15, no.3, pp.302-7.
10.259. McDonald, N.A.: "Electric and
Magnetic Coupling through Small Apertures
in Shield Walls of any Thickness", Trans.
IEEE, 1972, MTT-20, no.10, pp.689-95.
10.260. Gal, L.K., and Khizhnyak, N.A.:
"Scattering of a Thin Elliptical Rod in a
Waveguide", Radiotekh. (Kharkov), 1972,
no.20, pp.89-103.
10.261. Koshparenak, V.N., and Chernova,
S.V.: "Propagation in Circular Waveguide
with Axial Slots", Radiotekh. (Kharkov),
1972, no.20, pp.103-9.
10.262. Snurnikova, G.K.: "Open Rectangular and Plane-Parallel Waveguides Divided by
Diaphragm of Finite Thickness", Radiotekh.
(Kharkov), 1972, no.20, pp.185-90.
10.263. Noguchi, M., Abe, T., and Saito,
Y.: "Discontinuity Impedance in a Circular
Waveguide", Res. Rep. Fac. Eng. Niigata
Univ., 1972, no.21, pp.23-30.
10.264. Rozzi, T.E., and de Vrij, G.:
"Series Transformation for Diaphragm-Type
Discontinuities in Waveguide", Trans. IEEE,
1972, MTT-20, no.11, pp.770-1.
10.265. Sangster, A.J., and Hawkins, D.C.:
"Scattering Parameters of an Arbitrarily
Shaped Aperture in a Waveguide", Proc. IEE,
1972, 119, no.10, pp.1465-6.
10.266. Bichara, M.R.E., Arfaras, G., and
van Berchem, P.: "Variable Directional
Coupler for Microwaves", C. R. Acad. Sci.
(Paris), 1972, 275B, no.12, pp.411-3.
10.267. Gal'chenko, N.A., et al.: "Directional Couplers in Ridged Waveguides", Izv.
VUZ Radioelektron., 1972, 15, no.7, pp.842-6.

10.268. Michaelides, M.: "Microstrip
Directional Couplers", Mullard Tech. Commun.,
1972, 12, no.116, pp.177-87.
10.269. Pregla, R.: "Calculation of Distributed Capacitances and Phase Velocities
in Coupled Microstrip Lines by Conformal
Mapping Techniques", Arch. Elektron. Ubertrag., 1972, 26, no.11, pp.470-4.
10.270. Somlo, P.I., and Holloway, D.L.:
"Conductive Contacting Spheres on Centre of
Broad Wall of Rectangular Waveguides",
Electron. Lett., 1972, 8, no.20, pp.507-8.
10.271. Wu, S.C., and Chow, Y.L.: "Application of Moment Method to Waveguide Scattering Problems", Trans. IEEE, 1972, MTT-20,
no.11, pp.744-9.
10.272. Lewin, L., and Montgomery, J.P.:
"Quasi-Dynamic Method of Solution of a Class
of Waveguide Discontinuity Problems", Trans.
IEEE, 1972, MTT-20, no.12, pp.849-52.
10.273. Schaller, G.: "Directivity Improvements of Microstrip Quarter-Wave Directional
Couplers", Arch. Elektron. Ubertrag., 1972,
26, no.11, pp.508-9.
10.274. Malanchenko, V.P., and Kiezeeva,
G.M.: "Transfer Constant of a Waveguide with
Two Obstacles", Radiotekh. Elektron., 1972,
17, no.1, pp.1-6, and Radio Eng. Electron.
Phys., 1972, 17, no.1, pp.1-5.
10.275. Martin, D.: "Waveguide Load Utilizing a Multimode Cavity Partially Filled with
High-Loss Dielectric", Rev. Roum. Sci. Tech.,
1972, 17, no.4, pp.723-43.
10.276. Tripathi, V.K.: "Loss Calculations
for Coupled Transmission-Line Structures",
Trans. IEEE, 1972, MTT-20, no.2, pp.178-80.
10.277. de Jong, G.: "Scattering by a
Perfectly Conducting Cylindrical Obstacle in
Rectangular Waveguide", Int. J. Electron.,
1972, 32, no.2, pp.153-67.
10.278. Klimkiewicz, R.: "3-dB Directional
Coupler", Pr. Inst. Lacznosci, 1972, 19,
no.1, pp.25-38.
10.279. Kozalev, A.I.: "Excitation of H_{21}
Waves in Round Waveguides", Izmer. Tekh.,
1972, 15, no.7, p.91, and Meas. Tech., 1972,
15, no.7, pp.1121-2.
10.280. Cahill, L.W.: "Design Formulae for
Parallel-Coupled Striplines", Proc. IREE Aust.,
1972, 33, no.12, pp.573-4.
10.281. Nikol'skii, V.V., Izmailov, F.F.,
and Fedosaev, A.P.: "Application of Impedance
Treatment to Diffraction Problems for Rectangular Waveguide", Radiotekh. Elektron.,
1972, 17, no.6, pp.1305-8, and Radio Eng.
Electron. Phys., 1972, 17, no.6, pp.1014-8.
10.282. Koyama, M.: "Theory of Multihole
and Multislot Directional Couplers", Rev.
Electr. Commun. Lab., 1972, 20, no.11-12,
pp.1041-50, and Electron. Commun. Lab. Tech.
J., 1972, 21, no.7, pp.119-35.

10.283. Maloratskii, L.G.: "Analysis of a Six-Pole Ring-Type Power Divider", Radiotekhnika, 1972, 27, no.9, pp.58-63, and Telecommun. Radio Eng. Pt 2, 1972, 27, no.9, pp.100-4.

10.284. Yavich, L.R.: "Problems of Synthesis of Wideband Power Dividers", Radiotekh. Elektron., 1972, 17, no.8, pp.1580-5, and Radio Eng. Electron. Phys., 1972, 17, no.8, pp.1245-9.

10.285. Andreev, V.A.: "Problem of Determining the Reflection Factor of a Thin Conductor Situated Inside a Waveguide", Izv. VUZ Prib., 1972, 15, no.9, pp.5-9.

10.286. Cullen, A.L., and Byars, M.: "Mirror Corner for Use with Overmoded Circular Waveguide", Electron. Lett., 1972, 8, no.25, pp.621-3.

10.287. Lerer, A.M., and Mikhalevskii, V.S.: "Calculation of Slotted Directional Coupler for Symmetrical Stripline by Method of Integral Transformation", Radiotekh. Elektron., 1972, 17, no.5, pp.913-8, and Radio Eng. Electron. Phys., 1972, 17, no.5, pp.711-5.

10.288. Leikin, V.Yu., and Malovichko, A.A.: "Synthesis of Directional Couplers Using Lumped Elements", Radiotekh. (Kharkov), 1972, no.23, pp.102-9.

10.289. Rozzi, T.E.: "Variational Treatment of Thick Interacting Inductive Irises", Trans. IEEE, 1973, MTT-21, no.2, pp.82-8.

10.290. Levy, R.: "Zolotarev Branch-Guide Couplers", Trans. IEEE, 1973, MTT-21, no.2, pp.95-9.

10.291. Glance, B., and Trambarulo, R.: "Waveguide to Stripline Transition", Trans. IEEE, 1973, MTT-21, no.2, pp.117-8.

10.292. Kowalski, G., and Pregla, R.: "Calculation of Distributed Capacitances of Coupled Microstrip Using a Variational Integral", Arch. Elektron. Ubertrag., 1973, 27, no.1, pp.51-2.

10.293. Fritzsche, H.: "Frequency-Dependent Propagation Characteristics of Coupled Striplines with Multilayer Dielectric", Nachr. Tech. Z., 1973, 26, no.1, pp.1-8.

10.294. Getsinger, W.J.: "Dispersion of Parallel-Coupled Microstrip", Trans. IEEE, 1973, MTT-21, no.3, pp.144-5.

10.295. Kowalski, G., and Pregla, R.: "Dispersion Characteristics of Single and Coupled Microstrips with Double-Layer Substrates", Arch. Elektron. Ubertrag., 1973, 27, no.3, pp.125-30.

10.296. Rahmat-Samii, Y., Itoh, T., and Mittra, R.: "Spectral-Domain Technique for Solving Coupled Microstrip Problems", Arch. Elektron. Ubertrag., 1973, 27, no.2,pp.69-71.

10.297. Arkhangel'skii, Yu.S., and Kolomeitsev, V.A.: "Thermal Fields of Waveguide Loads in the Form of Film Absorbers on Inner Walls", Izv. VUZ Radioelektron., 1973, 16, no.1, pp.64-72.

10.298. Zaikin, B.M., and Rudeshko, G.A.: "Analysis of Impedance of Coupled Microstrip Lines", Izv. VUZ Radioelektron., 1973, 16, no.2, pp.39-42.

10.299. Horton, R.: "Loss Calculations of Coupled Microstrip Lines", Trans. IEEE, 1973, MTT-21, no.5, pp.359-60.

10.300. Torre, E.D., and Kinsner, W.: "Solution to Waveguide Problems by Successive Extrapolated Relaxation", Trans. IEEE, 1973, MTT-21, no.7, pp.490-1.

10.301. Rozzi, T.E.: "Network Analysis of Strongly Coupled Transverse Apertures in Waveguide", Int. J. Circuit Theory Appl., 1973, 1, no.2, pp.161-78.

10.302. Rizzoli, V.: "Calculation of Scattering Parameters for Coupled Microstrip Arrays of any Cross Section", Alta Freq., 1973, 42, no.4, pp.191-9.

10.303. Luebbers, R.J., and Munk, B.A.: "Analysis of Thick Rectangular Waveguide Windows with Finite Conductivity", Trans. IEEE, 1973, MTT-21, no.7, pp.461-8.

10.304. Chang, C.T.M.: "Equivalent Circuit for Partially Dielectric Filled Rectangular Waveguide Junctions", Trans. IEEE, 1973, MTT-21, no.6, pp.403-11.

10.305. Jain, A.K., and Srivastava, G.P.: "Circular Waveguide Hybrid-T and its Applications", Trans. IEEE, 1973, MTT-21, no.7, pp.482-3.

10.306. Schindler, G.: "Waveguide Transitions for Excitation of H_{11} Mode in Circular Waveguide", Arch. Elektron. Ubertrag., 1973, 27, no.7-8, pp.348-54.

10.307. Sawacki, Z., and Vogel, R.: "Three-Port Symmetrical Hybrid for Applications in MIC's", Rozpr. Electrotech., 1973, 19, no.2, pp.305-25.

10.308. Oraizi, H., and Perini, J.: "Numerical Method for the Solution of the Junction of Cylindrical Waveguides", Trans. IEEE, 1973, MTT-21, no.10, pp.640-2.

10.309. Daumas, R., et al.: "Faster Impedance Estimation for Coupled Microstrips with an Overrelaxation Method", Trans. IEEE, 1973, MTT-21, no.8, pp.552-6.

10.310. Knorr, J.B., and Saenz, J.: "End Effect in a Shorted Slot", Trans. IEEE, 1973, MTT-21, no.9, pp.579-80.

10.311. English, W.J.: "Circular-Waveguide Step-Discontinuity Mode Transducer", Trans. IEEE, 1973, MTT-21, no.10, pp.633-6.

10.312. Balint, L.: "Modelling of Coupled Multiconductor Microstrip-Like Transmission Lines", Int. J. Circuit Theory Appl., 1973, 1, no.3, pp.281-91.

10.313. Mendonca, J.T.: "Two Oversize Waveguide-Polarization Diplexers", Trans. IEEE, 1973, MTT-21, no.9, pp.586-7.

10.314. Daumas, R., et al.: "Extension of Kirchhoff's Theory to Coupled Striplines", Ann. Telecommun., 1973, 28, no.7-8,pp.325-34.

10.315. Hieber, A.L., and Vernon, R.J.: "Matching Considerations of Lossless Reciprocal Five-Port Waveguide Junctions", Trans. IEEE, 1973, MTT-21, no.8, pp.547-52.

10.316. Carr, J.W.: "Balanced-Line Microwave Hybrids", Microwave J., 1973, 16, no.5, pp.49-52.

10.317. Alger, R.P.: "Broadband 3-dB Quadrature Coupler", Microwave J., 1973, 16, no.5, p.45.

10.318. Clark, J.V., et al.: "Rapid-Switching Broadband 1:4 Waveguide Switch Matrix", Microwave J., 1973, 16, no.6, pp.39-42, and no.9, pp.47-54.

10.319. Ziermann, A.: "Bounds for Scattering Coefficients of Linear, Reciprocal, Lossless, and Symmetrical Three-Ports", Arch. Elektron. Ubertrag.,1973,27,no.9,pp.379-83.

10.320. Daumas, R., et al.: "Fast-Relaxation Method for Calculation of Microwave Couplers", C. R. Acad. Sci. (Paris), 1973, 277B, no.5, pp.119-22.

10.321. Daumas, R., et al.: "Accelerated Finite-Difference Method for Computing Coupled Microstrip Characteristics", Electron. Fis. Apl., 1973, 16, no.2, pp.167-72.

10.322. James, J.R., and Ladbrooke, P.H.: "Surface-Wave Phenomena Associated with Open-Circuited Stripline Terminations", Electron. Lett., 1973, 9, no.24, pp.570-1.

10.323. Neviere, M., et al.: "Determination of Coupling Coefficient of a Holographic Thin-Film (Grating) Coupler", Opt. Commun., 1973, 9, no.3, pp.240-5.

10.324. Richardson, J.K.: "Practical Realization of Passive Stripline Components", Aust. Telecommun. Res.,1973,7,no.1,pp.45-57.

10.325. Sveshnikov, A.G., and Repin, V.M.: "Numerical Solution of the Problem of Coupling of Rectangular Waveguides via Apertures", Vychisl. Metody Program., 1973, no.20, pp.12-21.

10.326. Il'inskii, A.S., and Galishnikova, T.N.: "Investigating Diffraction in Waveguide by the Fredholm Integral Equation Method", Vychisl. Metody Program.,1973,no.20,pp.22-37.

10.327. Sveshnikov, A.G., and Repin, V.M.: "Geometrical Optimization of Radiation via Apertures", Vychisl. Metody Program., 1973, no.20, pp.289-96.

10.328. Pal, S., and Babu, G.R.: "Coupled Microstrip Transmission Line. Even and Odd Mode Characteristic Impedance Calculations", J. AEU, 1973, 6, no.1, pp.36-9.

10.329. Stouten, P.: "Equivalent Capacitances of T Junctions", Electron. Lett., 1973, 9, no.23, pp.552-3.

10.330. Elmoazzen, Y.E., and Shafai, L.: "Mutual Coupling between Parallel-Plate Waveguides", Trans. IEEE, 1973, MTT-21, no.12, pp.825-33.

10.331. Puzanov, V.A.: "Waveguide Matching Transformer for 8-mm Band", Izmer. Tekh., 1973, 16, no.3, p.72, and Meas. Tech., 1973, 16, no.3, pp.429-30.

10.332. Yavich, L.R.: "Synthesis of Directional Couplers", Radiotekh. Elektron., 1973, 18, no.1, pp.176-8, and Radio Eng. Electron. Phys., 1973, 18, no.1, pp.126-8.

10.333. Waldron, R.A.: "Theory of Coupling in a Tapered Waveguide", Radio Electron. Eng., 1973, 43, no.12, pp.751-6.

10.334. Wolff, I.: "Static Values of Rectangular and Circular Microstrip Disc Capacitors", Arch. Elektron. Ubertrag., 1973, 27, no.1, pp.44-7.

10.335. Vogel, R.W.: "Effects of T-Junction Discontinuity on Design of Microstrip Directional Couplers", Trans. IEEE, 1973, MTT-21, no.3, pp.145-6.

10.336. Fike, G.F.: "Designing Hybrid Couplers with Uneven Power Splits", Microwaves, 1973, 12, no.11, pp.64-6.

10.337. Kozar', A.I., and Khizhnyak, N.A.: "Internal Field of a Resonant Dielectric Irregularity in a Waveguide", Radiotekh. (Kharkov), 1973, no.27, pp.152-61.

10.338. Belousova, L.I., and Sal'nikova, L.P.: "Coupling of Rectangular Waveguides by a Long Continuous Slot", Radiotekh. (Kharkov), 1973, no.27, pp.123-7.

10.339. Azumi, K., Shinada, Y., and Matsumoto, A.: "Generalization of Hybrid-Ring Circuits", Electron. Commun. Jap., 1973, 56, no.11, pp.44-52.

10.340. Bova, N.T., et al.: "Mathematical Modelling of Scattering Coefficients of a Microwave Power Divider", Izv. VUZ Radioelektron., 1973, 16, no.10, pp.137-8.

10.341. Prokhoda, I.G., and Chumachenko, V.P.: "Design of Four-Port H-Plane Junctions Using Rectangular Guide", Izv. VUZ Radioelektron., 1973, 16, no.10, pp.143-4.

10.342. Fridberg, P.Sh.: "Voltage Across a Narrow Slot in the Wall of a Waveguide", Radiotekh. Elektron., 1973, 18, no.3, pp.469-82, and Radio Eng. Electron. Phys., 1973, 18, no.3, pp.341-53.

10.343. Szczypka, Z.: "Graphical Analysis of Microwave Linear Two-Ports", Pr. PIT, 1973, 23, no.77, pp.1-14.

10.344. Hockman, G.A.: "TEM Waveguide Coupling in Presence of Short-Circuited Slot", Electron. Lett., 1974, 10, no.6, pp.62-3.

10.345. Matsuhara, M., and Kumagai, N.: "Theory of Coupled Open Transmission Lines and its Applications", Trans. IEEE, 1974, MTT-22, no.4, pp.378-82.

10.346. Chang, K., and Khan, P.J.: "Analysis of a Narrow Capacitive Strip in Waveguide", Trans. IEEE, 1974, MTT-22, no.5, pp.536-41.

10.347. Chang, T.M.: "Computation of the Equivalent-Circuit Parameters for a Junction between Empty and Side-Slab-Filled Rectangular Waveguides", Trans. IEEE, 1974, MTT-22, no.5, p.585.

10.348. Wight, J.S., et al.: "Equivalent Circuits of Microstrip Impedance Discontinuities and Launchers", Trans. IEEE, 1974, MTT-22, no.1, pp.48-52.

10.349. Daumas, R., et al.: "Determination of Microstrip Adaption Conditions for a Directional Coupler Taking into Account the Different Propagation Velocities for Each Mode", C. R. Acad. Sci. (Paris), 1974, 278B, no.11, pp.459-62.

10.350. Weirather, R.R.: "Small MIC Coupler with Good Directivity", Trans. IEEE, 1974, MTT-22, no.1, pp.70-1.

10.351. Chang, C.T.M.: "Partially Dielectric-Slab-Filled Waveguide Phaseshifter", Trans. IEEE, 1974, MTT-22, no.5, pp.481-5.

10.352. Zaentsev, V.V., and Bel'chinskii, V.V.: "Characteristics of a Hybrid Ring with Coupled Section", Izv. VUZ Radioelektron., 1974, 17, no.1, pp.49-55.

10.353. Kuhn, E.: "Improved Design and Resulting Performance of Multiple-Branch Waveguide Directional Couplers", Arch. Elektron. Ubertrag., 1974, 28, no.5, pp.206-14.

10.354. Boerner, W.M., and Chaudhuri, S.K.: "Transformation Properties of Lossless 4-Port Microwave Junctions", Arch. Elektron. Ubertrag., 1974, 28, no.5, pp.215-22.

10.355. Haupt, G., and Delfs, H.: "High-Directivity Microstrip Directional Couplers", Electron. Lett., 1974, 10, no.9, pp.142-3.

10.356. Zaentsev, V.V., and Bel'chinskii, V.V.: "Multichannel Microwave Power Divider Using Coupled Striplines", Izv. VUZ Radioelektron., 1974, 17, no.3, pp.31-7.

10.357. Gauthier, F., and Besse, M.: "Graphical Design of Coupled Microstrip Lines", Microwave J.,1974,17,no.2,pp.36-38.

10.358. Sorensen, O.H.: "Adjustable, Highly Stable, Point Contact in a 4-mm Waveguide", Cryogenics, 1974,14,no.3,pp.166-7.

10.359. Wade, W.L., and Raffalovich, A.: "Method of Forming Ferrite-Impregnated Resins for Microwave Measurements", Trans. IEEE, 1974, MAG-10, no.1, pp.96-7.

10.360. Lerer, A.M., and Mikhalevskii, V.S.: "Effect of Finite Thickness of Waveguide Walls on Parameters of a Directional Coupler", Izv. VUZ Radioelektron., 1974, 17, no.5, pp.76-9.

10.361. Bahar, E., and Govindarajan, G.: "H-Plane Multimode Waveguide Transition Sections with Large Flare Angles", Proc. IEE, 1974, 121, no.6, pp.443-9.

10.362. Bova, N.T., et al.: "Application of an Iterative Algorithm for Design of Waveguides with Irises", Izv. VUZ Radioelektron., 1974, 17, no.6, pp.136-8.

10.363. Razaz, M., and Davies, J.B.: "Admittance of a Coaxial-Line/ Circular-Waveguide Junction", Electron. Lett., 1974, 10, no.15, pp.324-6.

10.364. Carli, E., and Corzani, T.: "Existence Range of S Parameters of a Passive Two-Port Network", Trans. IEEE, 1974, MTT-22, no.9, pp.835-7.

10.365. Orth, B.J., and Speciale, R.A.: "Solution of the Coupled-Mode Equations in Terms of Redefined Even- and Odd-Mode Waves", Electron. Lett., 1974, 10, no.20, pp.423-4.

10.366. Schiek, B.: "Hybrid Branchline Couplers. Useful Class of Directional Couplers", Trans. IEEE, 1974, MTT-22, no.10, pp.864-9.

10.367. Pregla, R.: "Method for Analysis of Coupled Rectangular Dielectric Waveguides", Arch. Elektron. Ubertrag., 1974, 28, no.9, pp.349-57.

10.368. Klishch, O.M.: "Method of Calculation for Reflection Factor of a Cylindrical Post in Pi or H Guide", Teor. Elektrotech., 1974, no.17, pp.164-8.

10.369. Falciasecca, G.: "Finite-Difference Equations in Circular-Waveguide Coupling Problems", Electron. Lett., 1974, 10, no.2, pp.18-9.

10.370. Staeger, Chr.: "Development of Microwave Broadband Matching Network and Application to a Directional Coupler", Tech. Mitt. PTT, 1974, 52, no.2, pp.47-51.

10.371. Laloux, A.A., Van der Vorst, A.S., and Govaerts, R.J.M.: "Application of Variation-Iteration Method to Waveguides with Inhomogeneous Lossy Loads", Trans. IEEE, 1974, MTT-22, no.3, pp.229-36.

10.372. Linner, L.J.P.: "Explicit Expressions for Geometrical Dimensions of Symmetrical Coupled Striplines", Electron. Lett., 1974, 10, no.4, pp.45-6.

10.373. Silvester, P., and Csendes, Z.J.: "Numerical Modelling of Passive Microwave Devices", Trans. IEEE, 1974, MTT-22, no.3, pp.190-201.

10.374. Voges, E.: "Directional Couplers with Rectangular Dielectric Waveguides", Arch. Elektron. Ubertrag., 1974, 28, no.11, pp.478-9.

10.375. Vaisleib, Yu.V., Kirkheizen, E.G., and Kulikov, L.N.: "Theory of a Slot Bridge in Circular Waveguide", Izv. VUZ Radioelektron., 1974, 17, no.8, pp.33-41.

10.376. Yatsuk, D.P., Zhironkina, A.V., and Katrich, V.A.: "Parameters of Crossed Slots in a Rectangular Waveguide Allowing for Mutual Coupling between Arms", Izv. VUZ Radioelektron., 1974, 17, no.8, pp.100-3.

10.377. Belousova, L.I., and Zorya, V.D.: "Periodic Connection of Rectangular Waveguides with Dielectric Fillings", Radiotekh. (Kharkov), 1974, no.28, pp.97-104.

10.378. Belousova, L.I., and Belyavtseva, T.V.: "Natural Waves of a Rectangular Waveguide with a Specially Shaped Latticed Partition", Radiotekh. (Kharkov), 1974, no.28, pp.104-10.

10.379. Yashchuk, L.P., Zhironkina, A.V., and Katrick, V.A.: "Internal Conductance of Crossed-Slot Arms in a Rectangular Waveguide", Izv. VUZ Radioelektron., 1974, 17, no.7, pp.105-8.

10.380. Butakova, S.V., Gorobets, N.N., and Lyakhovskii, A.F.: "Band Properties of Slot Bridges with Connection through a Narrow Waveguide Wall", Radiotekh. (Kharkov), 1974, no.28, pp.94-7.

10.381. Yanovskii, M.S., Knyaz'kov, B.N., and Kuleshov, Ye.M.: "Polarizing Attenuators for Quasi-Optical Waveguide", Izv. VUZ Radioelektron., 1974, 17, no.9, pp.49-54.

10.382. Sheleg, B., and Spielman, B.E.: "Broadband Directional Couplers Using Microstrip with Dielectric Overlays", Trans. IEEE, 1974, MTT-22, no.12, pp.1216-20.

10.383. Hurst, G.J.: "Stripline for Simpler Microwave Components", Marconi Instrum., 1974, 14, no.5, pp.112-5.

10.384. Gitel'son, A.A., and Orlov, S.V.: "Ultra-Broadband Microwave Directional Couplers", Prib. Tekh. Eksp., 1974, 17, no.4, pp.112-3, and Instrum. Exp. Tech., 1974, no.4, pp.1068-9.

10.385. Nagai, N.: "TEM-Mode Hybrid Power Dividers", Bull. Res. Inst. Appl. Electr., 1974, 26, no.1-2, pp.25-41.

10.386. Lapta, S.I., and Salogub, V.G.: "Excitation of a Segment of a Circular Waveguide by a Longitudinal Dipole Located on the Waveguide Axis", Radiotekh. (Kharkov), 1974, no.30, pp.146-55.

11. RESONANT STRUCTURES

11.1. Yee, H.Y.: "Resonant Frequencies of Microwave Dielectric Resonators", Trans. IEEE, 1965, MTT-13, p.256.

11.2. Vetter, M.J.: "Linear Tuning of a Microwave Cavity", Trans. IEEE, 1965, MTT-13, p.880.

11.3. Malinov, I.A.: "Theory of Pi-Shaped Resonator", Radiotekhnika, 1965, 20, no.3, p.9, and Telecommun. Radio Eng. Pt 2, 1965, 20, no.3, p.75.

11.4. Brodwin, M.E., and Parsons, M.K.: "Perturbation of Cavity Resonators by Homogeneous Isotropic Spheres", J. Appl. Phys., 1965, 36, p.494.

11.5. Twistleton, J.R.G.: "Derivation of Cavity Coupling to Feeder from Tuning Curves", Proc. IEE, 1965, 112, p.452.

11.6. Gheorghiu, O.C.: "Transmission Lines with Losses. Resonance Criteria", Rev. Roum. Phys., 1965, 10, p.305.

11.7. Ching-Jen, C.: "Excitation of Cavities by Waveguide", Acta Electron. Sin., 1965, no.2-3, p.116.

11.8. Krasnushkin, P.Ye: "Application of Method of Functional Networks to Forced Oscillations in Volume of Complex Shape", Radiotekh. Elektron., 1965, 10, p.1214, and Radio Eng. Electron. Phys., 1965, 10, p.1043.

11.9. Hantzsche, E.: "Natural Resonances in Spherical Cavities", Hochfreq. Elektroak., 1965, 74, p.90.

11.10. Semenov, G.F.: "Design Method for Resonators with Dielectrics", Izv. VUZ Radiotekh., 1965, 8, p.55.

11.11. Rebsch, D.L., et al.: "Mode Chart for Design of Cylindrical Dielectric Resonators", Trans. IEEE, 1965, MTT-13, p.468.

11.12. Sushkov, A.D., and Vendik, I.B.: "Influence of Nonuniformity on a Ring Resonator", Zh. Tekh. Fiz., 1965, 35, p.1610, and Sov. Phys.-Tech. Phys., 1966, 10, no.9, pp.1244-8.

11.13. Stiglitz, M.R.: "Frequency Tuning of Rutile Resonators", Proc. IEEE, 1966, 54, p.413.

11.14. Sethares, J.C., and Naumann, S.J.: "Design of Microwave Dielectric Resonators", Trans. IEEE, 1966, MTT-14, p.2.

11.15. Shimada, S.: "Method of Calculating Attenuation Coefficients of Unwanted Modes in Mode Filters Using Resistive Sheets", Trans. IEEE, 1966, MTT-14, p.159.

11.16. Valentin, M.: "Fabry-Perot Resonators at 12.5-mm Wavelength", C. R. Acad. Sci. (Paris), 1966, 262B, p.1115.

11.17. Kulke, R.: "Changing the Coupling into a Microwave Cavity by a Stub Tuner", Trans. IEEE, 1967, MTT-15, p.184.

11.18. Gastine, M., Courtois, L., and Dormann, J.L.: "Electromagnetic Resonances of Free Dielectric Spheres", Trans. IEEE, 1967, MTT-15, no.12, pp.694-700.

11.19. Shatrov, A.D.: "Theory of Strip Resonators", Radiotekh. Elektron., 1967, 12, no.7, pp.1277-80, and Radio Eng. Electron. Phys., 1967, 12, no.7.

11.20. Suchkin, G.L.: "Disc Resonators", Radiotekh. Elektron., 1967, 12, no.8, and Radio Eng. Electron. Phys., 1967, 12, no.8, pp.1391-4.

11.21. Kashiwagi, H.: "Fabry-Perot Resonator in Short-Millimetre Region", Bull. Electrotech. Lab., 1968, 32, no.1, pp.100-14.

11.22. Rivier, E.: "Spheroidal Cavity. Lumped-Parameter Resonator", Proc. IEEE, 1968, 56, no.8, p.1387.

11.23. Mandel'shtam, M.Ya.: "Wideband Tuneable Rectangular-Waveguide Cavity with a Capacitive Tuning Rod", Radiotekhnika, 1968, 23, no.7, pp.9-13, and Telecommun. Radio Eng. Pt 2, 1968, 23, no.7.

11.24. Beluga, I.Sh.: "Design of Waveguides and Cavities by Methods of Partial Regions (Scalar Case)", Radiotekh. Elektron., 1968, 13, p.1357, and Radio Eng. Electron. Phys., 1968, 13, no.8, pp.1178-85.

11.25. Troughton, P.: "High-Q-Factor Resonators in Microstrip", Electron. Lett., 1968, 4, p.520.

11.26. Karp, A., Shaw, H.J., and Winslow, D.K.: "Circuit Properties of Microwave Dielectric Resonators", Trans. IEEE, 1968, MTT-16, p.818.

11.27. Popykhova, R.M.: "Influence of Intense SHF Field on the Characteristics of a Resonator with Multilayer Dielectric Walls", Izv. VUZ Fiz., 1968, no.7, p.30.

11.28. Yvandov, L.N.Ge., and Levinson, I.B.: "Variational Principle for the Admittance Matrix in Coupling of Cavities", Radiotekh. Elektron., 1968, 13, p.1249, and Radio Eng. Electron. Phys., 1968, 13, no.7, pp.1084-93.

11.29. Samson, A.M.: "Theory of Laser Resonators", Opt. Spekt., 1968, 25, p.386, and Opt. Spectrosc., 1968, 25, p.209.

11.30. Popov, M.M.: "Resonators for Lasers with Unfolded Directions of Principal Curvatures", Opt. Spekt., 1968, 25, p.394, and Opt. Spectrosc., 1968, 25, p.213.

11.31. Kontorovich, M.I., and Rogozin, V.V.: "Field Analysis in a Rectangular Cavity", Radiotekh. Elektron., 1968, 13, p.2239, and Radio Eng. Electron. Phys., 1968, 13, no.12, pp.1961-4.

11.32. Bruevich, A.N., and Evtyanov, S.I.: "Simple Formulae for Parameters of a Circuit Replacing a Coaxial Cavity", Radiotekhnika, 1968, 23, no.8, p.104, and Telecommun. Radio Eng. Pt 2, 1968, 23, no.8.

11.33. Chatterjee, S.K., and Chatterjee, R.: "Microwave Resonators", J. Indian Inst. Sci., 1968, 50, p.345.

11.34. Popov, M.M.: "Resonators for Lasers with Rotated Directions of Principal Curvatures", Opt. Spekt., 1968, 25, p.314, and Opt. Spectrosc., 1968, 25, p.170.

11.35. Uralev, G.A.: "Determination of Tolerances in Construction of Resonators with Layered Dielectric Walls", Izv. VUZ Fiz., 1968, no.8, p.148.

11.36. Engelmann, R., and Pollmann, H.: "Detailed Analysis of a TEM Transmission-Line Resonator for Solid-State Microwave Oscillator Applications", Nachr. Tech. Fachber., 1968, 35, p.487.

11.37. Fontana, J.R.: "Synthesis of TW Microwave and Optical Resonators", Nachr. Tech. Fachber., 1968, 35, p.709.

11.38. Nishimura, S., and Makimoto, T.: "Improvement of Frequency Characteristics of Preselector Using a Hybrid Circuit", Trans. IEEE, 1968, MTT-16, p.891.

11.39. Galeta, V.O., and Valitov, R.A.: "Diffraction Losses of a Coupled Resonator with Plane Mirrors", Zh. Tekh. Fiz., 1968, 38, no.12, p.2101, and Sov. Phys.-Tech. Phys., 1969, 13, no.12.

11.40. Shiryaev, V.V.: "Section of Nonuniform Line as Oscillating Structure", Radiotekhnika, 1969, 24, no.1, p.107, and Telecommun. Radio Eng. Pt 2, 1969, 24, no.1.

11.41. Ooms, G.: "Polarization State of an Open Resonator with Plane Circular Mirrors", Optik, 1969, 29, no.4, p.410.

11.42. Voitovich, N.N., and Nefedov, Ye.I.: "Open Resonators with Ring Hole in One of the Mirrors", Izv. VUZ Radiofiz., 1969, 12, no.4, p.626.

11.43. Lotsch, H.K.V.: "Fabry-Perot Resonators. II and III", Optik, 1969, 28, pp.328 and 555.

11.44. Younger, F.C., and Meads, P.F.: "Numerical Design of an Azimuthally Symmetric RF Cavity that Resonates at Given Frequencies and has Zero Response at Other Specified Frequencies", Trans. IEEE, 1969, NS-16, p.558.

11.45. McPhun, M.K.: "Comparison of TEM with Waveguide-Below-Cutoff Resonators", Electron. Lett., 1969, 5, p.425.

11.46. Ishii, T.K., and Moldovan, P.K.: "Notched Stripline Resonator", Electron. Lett., 1969, 5, p.439.

11.47. Epishin, V.A., Kamyshan, V.V., and Balitov, R.A.: "Problem of Experimental Investigation of Coupled Open Resonators", Zh. Tekh. Fiz., 1969, 39, no.5, p.954, and Sov. Phys.-Tech. Phys., 1969, 14, no.5, pp.716-8.

11.48. Kiselev, V.A.: "Modes of Open Optical Resonators with Cylindrical Mirrors", Zh. Prikl. Spekt., 1969, 11, no.1, p.48.

11.49. Checcacci, P.F.: "Open Resonators with Rimmed Mirrors", Trans. IEEE, 1969, MTT-17, p.125.

11.50. Vlasov, S.N., et al.: "Irregular Waveguides as Open Resonators", Izv. VUZ Radiofiz., 1969, 12, no.8, p.1236.

11.51. McCumber, D.E.: "Eigenmodes of Asymmetric Cylindrical Confocal Laser Resonator with a Single Output-Coupling Aperture", Bell Syst. Tech. J., 1969,48,p.1919.

11.52. Rubanov, V.S.: "Calculating Polarization Characteristics of Lasers", Zh. Prikl. Spekt., 1969, 10, no.5, p.725.

11.53. McNice, G.T., and Derr, V.E.: "Analysis of Cylindrical Confocal Laser Resonator Having a Single Circular Coupling Aperture", J. Quantum Electron. IEEE, 1969, QE-5, p.569.

11.54. Fiedziuszko, S., and Pospieszalski, M.: "Dielectric Microwave Resonators", Arch. Elektrotech., 1969, 18, p.709.

11.55. Afonin, D.G.: "Characteristics of Open Resonator with Spherical Mirrors of Different Radii of Curvature", Vestn. Mosk. Univ. Fiz. Astron., 1969, no.5, p.127.

11.56. Pellegrin, J.L.: "Filling Factor of Shielded Dielectric Resonators", Trans. IEEE, 1969, MTT-17, p.764.

11.57. Rivier, E.: "Theoretical Expressions of Lumped Parameters in a Spheroidal Cavity", Trans. IEEE, 1969, MTT-17, p.720.

11.58. Watkins, J.: "Circular Resonant Structures in Microstrip", Electron. Lett., 1969, 5, p.524.

11.59. Voitovich, N.N.: "Extremal Case in the Theory of Open Resonators", Radiotekh. Elektron., 1969, 14, no.12, p.2251, and Radio Eng. Electron. Phys., 1969, 14, no.12, pp.1943-4.

11.60. Kurilin, B.I., and Kien, L.C.: "Bandwidth Characteristics of Microwave Cavities", Izv. VUZ Radioelektron., 1969, 12, no.9, p.1084.

11.61. Tarnow, V.: "Way of Coupling a Coaxial Line to a Microwave Cavity", Proc. IEEE, 1969, 57, no.6, p.1188.

11.62. Rzepecka, M.: "Properties Analysis of Certain Microwave Hybrid Resonators", Arch. Elektrotech., 1969, 18, no.4, p.725.

11.63. Averkov, S.I., and Furashov, N.I.: "Diffraction Resonators", Izv. VUZ Radiofiz., 1969, 12, no.10, p.1532.

11.64. Sanderson, R.L., and Streifer, W.: "Unstable Laser Resonator Modes", Appl. Opt., 1969, 8, p.2129.

11.65. Kiselev, V.A.: "Asymmetrical Optical Resonators with Circular Spherical Mirrors", Zh. Prikl., Spekt., 1969, 11, no.6, p.1035.

11.66. Kiselev, V.A.: "Optical Resonators with Spherical Mirrors of Rectangular Shape", Zh. Prikl., Spekt., 1969, 11, no.6, p.1140.

11.67. Yamashita, E., Mittra, R., and Itoh, T.: "Application of the Sampling Theorem to a Multislit Coupling Structure of a Confocal-Mirror Resonator", Electron. Lett., 1969, 5, p.67.

11.68. Kolomoitsev, F.I., Aleinikov, I.N., and Abkin, Ye. B.: "Shortening of Stripline Resonators", Radiotekhnika, 1969, 24, no.2, p.72, and Telecommun. Radio Eng. Pt 2, 1969, 24, no.2.

11.69. Ross, G.F.: "Stored Energy and Bandwidth in TEM-Mode Microwave Networks", Trans. IEEE, 1969, MTT-17, p.386.

11.70. Golubev, Yu.M., et al.: "TW Mode in a Ring Resonator with Additional External Mirror", Opt. Spektr., 1969, 27, p.519, and Opt. Spectrosc., 1969, 27, p.278.

11.71. Andriyakhin, V.M., et al.: "Effect of Mismatching upon the Transmittance of TEM_{00} Waves in a Fabry-Perot Resonator", Zh. Prikl., Spekt., 1969, 11, p.464.

11.72. Hanna, D.C.: "Astigmatic Gaussian Beams Produced by Axially Asymmetric Laser Cavities", J. Quantum Electron. IEEE, 1969, QE-5, p.483.

11.73. Anan'ev, Yu.A., and Vinokurov, G.H.: "Properties of Annular Unstable Resonators with Angular Selection of Radiation", Zh. Tekh. Fiz., 1969, 39, no.7, p.1327, and Sov. Phys.-Tech. Phys., 1970, 14, no.7, pp.1000-2.

11.74. Lyubimov, V.V., and Orlova, I.B.: "Influence of Misalignment of the Mirror of a Resonator Upon Losses and Angular Distribution", Zh. Tekh. Fiz., 1969, 39, p.2183, and Sov. Phys.-Tech. Phys., 1970, 14, no.12, pp.1648-51.

11.75. Korzhenevich, I.M., and Ratner, A.M.: "Theory of Optical Resonator with Lenses", Ukr. Fiz. Zh., 1970, 15, p.191.

11.76. Burtovoi, D.P., Mironenko, V.L., and Tereshchenko, A.I.: "Frequencies of Open Cutoff Rectangular Resonator", Radiotekh. Elektron., 1970, 15, no.2, p.389, and Radio Eng. Electron. Phys.,1970,15,no.2,pp.337-9.

11.77. Cherne, Kh.I., and Maloratskii, L.G.: "Characteristics of Ring Circuits with Mismatched Loads", Radiotekhnika, 1970, 25, p.30, and Telecommun. Radio Eng. Pt 2,1970,25.

11.78. Checcacci, P.F., and Scheggi, A.M.: "Further Consideration on Open Resonators with Rimmed Mirrors", Trans. IEEE, 1970, MTT-18, p.282.

11.79. Muc, A.M., Dagg, I.R., and Reesor, G.E.: "Mode Charts for Cylindrical Microwave Resonators", Trans. IEEE, 1970,MTT-18,p.286.

11.80. Valitov, R.A., and Parkhamov, N.S.: "Coupled Signals of Open Resonators with an Arbitrary Coupling Factor", Izv. VUZ Radioelektron., 1970, 13, no.8, pp.981-6.

11.81. Cullen, A.L.: "Note on the Radiation Associated with Excitation of an Open Resonator", Electron. Lett., 1970,6,p.243.

11.82. Van Bladel, J.: "Small-Hole Coupling of Resonant Cavities and Waveguides", Proc. IEE, 1970, 117, p.1098.

11.83. Kulikov, E.L.: "Errors in Calculating the Natural Frequencies of Microwave Resonators", Radiotekh. Elektron., 1970, 15, p.593, and Radio Eng. Electron. Phys., 1970, 15, no.3, pp.503-4.

11.84. Bloom, G.H.: "Using a Pentaprism for Laser Alignment", Appl. Opt., 1970, 9, p.1210.

11.85. Banaszczyk, R., and Kowar, J.: "Equivalence of Substitute Parameters of Toroidal and Coaxial Resonators", Arch. Elektrotech., 1970, 19, no.2, p.315.

11.86. Pipiskova, A., and Lukac, P.: "Electric-Field Distribution in a Cylindrical TM_{010} Microwave Cavity with End Holes and a Glass Tube", J. Phys. D,1970,3,p.1381.

11.87. Kostychev, Yu.G.: "Analysis of Field Structure of Free Oscillations in Nonuniform Hollow Structures", Radiotekh. Elektron., 1970, 15, no.5, p.1084, and Radio Eng. Electron. Phys.,1970,15,no.5,pp.912-3.

11.88. Kretzschmar, J.G.: "Mode Charts for Elliptical Resonant Cavities", Electron. Lett., 1970, 6, p.432.

11.89. Schiffman, B.M.: "Crossed-Cylinder Microwave Resonator", Trans. IEEE, 1970, MTT-18, p.509.

11.90. Ratner, A.M.: "Theory of a Transverse Field Structure in a Plane Resonator", Radiotekh. Elektron., 1970, 15, no.2, p.394, and Radio Eng. Electron. Phys., 1970, 15, no.2, pp.342-4.

11.91. Voitovich, N.N., and Nefedov, Ye.I.: "Axisymmetric Open Resonator Having Arbitrary Quadratic Mirrors and Round Cutout in One of the Mirrors", Radiotekh. Elektron., 1970, 15, no.2, p.391, and Radio Eng. Electron. Phys., 1970, 15, no.2, pp.339-42.

11.92. Dagg, I.R., Muc, A.M., and Reesor, G.E.: "Q-Factor Calculations for Cylindrical Microwave Resonators", Trans. IEEE, 1970, MTT-18, p.669.

11.93. Raevskii, S.B.: "Natural Modes of an Elliptical Open Resonator with a Dielectric Cylinder", Izv. VUZ Radioelektron., 1970, 13, no.8, pp.987-92.

11.94. Easter, B., and Roberts, R.J.: "Radiation from Half-Wavelength Open-Circuit Microstrip Resonators", Electron. Lett., 1970, 6, p.573.

11.95. Zakharov, M.I., Troitskii, Yu.V., and Goldina, N.D.: "Investigation of Optical Resonators with a Thin-Layer Metal Diffraction Array", Izv. VUZ Radiofiz., 1970, 13, no.9, pp.1335-41.

11.96. Yamauchi, N.: "Mode Representation of Laser Resonator with Oblate Spheroidal Vector Wave Function", Electron. Commun. Jap., 1970, 53, no.3, pp.8-9.

11.97. Uenakada, K.: "Equivalent Circuit of Reentrant Cavity Resonator", Electron. Commun. Jap., 1970, 53, no.4, pp.12-13.

11.98. Ulrich, R., Bridges, T.J., and Pollack, M.A.: "Variable Metal-Mesh Coupler for Far-IR Lasers", Appl. Opt., 1970, 9, no.11, pp.2511-16.

11.99. Korzhenevich, I.M.: "General Method of Calculation the Spectrum of Optical Cavities with Spherical Mirrors", Radiotekh. Elektron., 1970, 15, no.4, and Radio Eng. Electron. Phys., 1970, 15, no.4, pp.696-8.

11.100. Bykov, V.P.: "Optical Cavity Partially Filled with an Inhomogeneous Dielectric", Radiotekh. Elektron., 1970, 15, no.4, and Radio Eng. Electron. Phys., 1970, 15, no.4, pp.594-8.

11.101. Zucker, H.: "Optical Resonators with Variable-Reflectivity Mirrors", Bell Syst. Tech. J., 1970, 49, no.9, pp.2349-76.

11.102. Bartashevskii, Ye.L., and Privalov, E.N.: "Design of Cavities with a Variable Rectangular Cross Section", Izv. VUZ Radioelektron.,1970,13,no.12,pp.1486-8.

11.103. Bogomolov, G.D., and Rusin, F.S.: "Open Cavity with Variable Quasi-Optical Coupling", Radiotekh. Elektron., 1970, 15, no.4, and Radio Eng. Electron. Phys., 1970, 15, no.4, pp.727-9.

11.104. Orlov, S.I.: "Coaxial Resonator with Parallel Capacitor Tuning", Radiotekhnika, 1970, 25, no.6, and Telecommun. Radio Eng. Pt 2, 1970, 25, no.6, pp.145-9.

11.105. Lyubimov, V.V., and Orlova, I.B.: "Approximate Calculation of Oscillations in Resonators with Concave Mirrors", Opt. Spekt., 1970, 29, no.3, pp.581-6, and Opt. Spectrosc., 1970, 29, no.3, pp.310-3.

11.106. Lugovoi, V.N., et al.: "Tunable Optical Resonator with Nearly Plane-Parallel Mirrors", Prib. Tekh. Eksp., 1970, no.5, pp.186-7, and Instrum. Exp. Tech., 1970, no.5, pp.1459-60.

11.107. Vlasov, S.N., and Talanov, V.I.: "Confocal Resonator with Slotted Mirrors", Radiotekh. Elektron., 1970, 15, no.11, pp.1095-7, and Radio Eng. Electron. Phys., 1970, 15, no.11, pp.2095-7.

11.108. Fiedziuszko, S., and Jelenski, A.: "Influence of Metallic Walls on Resonant Frequencies of a Dielectric Resonator", Arch. Elektrotech., 1970, 19, no.4, pp.883-5.

11.109. Fiedziuszko, S., and Jelenski, A.: "Double Dielectric Resonator", Arch. Elektrotech., 1970, 19, no.4, pp.879-81.

11.110. Day, W.R.: "Dielectric Resonators as Microstrip Circuit Elements", Trans. IEEE, 1970, MTT-18, no.12, pp.1175-6.

11.111. Kovbasa, A.P., and Shelamov, G.N.: "Resonant Frequencies of Coupled Dielectric Microwave Resonators", Izv. VUZ Radioelektron., 1971, 14, no.1, pp.106-9.

11.112. Bokrinskaya, A.A., and Il'chenko, M.Ye.: "Dielectric Microwave Resonators in a Transmission Line", Izv. VUZ Radioelektron., 1971, 14, no.2, pp.151-7.

11.113. Barone, S.R.: "Perturbed Open Resonators", Appl. Opt., 1971, 10, no.4, pp.935-8.

11.114. Manenkov, A.B.: "(Millimetric) Open Resonators with Dielectric Walls", Izv. VUZ Radiofiz., 1971, 14, no.4, pp.606-12.

11.115. Fong, T.T., and Lee, S.W.: "Theory of Fabry-Perot Resonators with Dielectric Medium", J. Quantum Electron. IEEE, 1971, QE-7, no.1, pp.1-11.

11.116. Wrolstad, K.H., Avizonis, P.V., and Holmes, D.A.: "Stable Resonators with Increased Fundamental-Mode Volume for CO_2 Lasers", J. Phys. E, 1971, 4, no.2, pp.143-5.

11.117. Robinson, G.H.: "Resonant-Frequency Calculations for Microstrip Cavities", Trans. IEEE, 1971, MTT-19, no.7, pp.665-6.

11.118. Popov, A.B., and Popova, M.N.: "Diffraction Losses of an Open Resonator", Radiotekh. Elektron., 1971, 16, no.12, pp.2175-81, and Radio Eng. Electron. Phys., 1971, 16, no.12, pp.2007-12.

11.119. Ogura, I.: "Gas-Laser Reflector Loss", Oyo Buturi, 1971,40,no.10,pp.1151-5.

11.120. Budanov, V.Ye., et al.: "Study of Asymmetric Single-Ring Coaxial Cylindrical Cavity", Radiotekh. (Kharkov), 1971, no.18, pp.113-8.

11.121. Kawakami, S., and Nishida, S.: "Laser Resonator with a Selector for a Higher Transverse Mode", Trans. IEEE, 1971, MTT-19, no.4, pp.403-6.

11.122. Irish, R.T.: "Elliptic Resonator and its Use in Microwave Systems", Electron. Lett., 1971, 7, no.7, pp.149-50.

11.123. Roberts, R.J., and Easter, B.: "Microstrip Resonators having Reduced Radiation Loss", Electron. Lett., 1971, 7, no.8, pp.191-2.

11.124. Bisio, G.R., Ronchi, L., and Tognetti, V.: "Considerations about Diffraction Loss of Open Resonators", Trans. IEEE, 1971, MTT-19, no.5, pp.490-1.

11.125. Yoshida, Y., Ogura, H., and Ikenoue, J.: "Fabry-Perot Resonator with Circular Apertures", Jap. J. Appl. Phys., 1971, 10, no.6, pp.754-7.

11.126. Vakhitov, N.G.: "Formation of a Natural Oscillation which Realizes a Stipulated Field Distribution on the Mirror of an Open Resonator", Dokl. Akad. Nauk SSSR, 1971, 209, no.12, pp.1154-6, and Sov. Phys.-Dokl., 1971, 15, no.12.

11.127. Bogomolov, G.D., and Manenkov, A.B.: "Interaction of Oscillations in Open Resonators with Spherical Mirrors", Izv. VUZ Radiofiz., 1971, 14, no.5, pp.748-53.

11.128. Raevskii, S.B.: "Oscillations in an Open Elliptic Cavity Containing Dielectric", Izv. VUZ Radioelektron., 1971, 14, no.5, pp.483-8.

11.129. Suematsu, Y.: "Gaussian Modes and Fabry-Perot Resonators Including Uniaxially Anisotropic Medium", Jap. J. Appl. Phys., 1971, 10, no.8, pp.1060-5.

11.130. Tanaka, T., Suzuki, M., and Matsumoto, T.: "Optimum Resonant Conditions of the Fabry-Perot Resonator Filled with Anisotropic Medium for Extraordinary Waves", Electron. Commun. Jap., 1971, 53,no.2,pp.60-4.

11.131. Fiedziuszko, S., and Jelenski, A.: "Double Dielectric Resonator", Trans. IEEE, 1971, MTT-19, no.9, pp.779-81.

11.132. Belov, Yu.G., and Raevskii, S.B.: "Natural Oscillations in a Partly Filled Resonator Containing Mirrors", Izv. VUZ Radioelektron., 1971, 14, no.6, pp.613-6.

11.133. Ctyroky, J.: "Equality of Diffraction Losses at Both Mirrors of a Confocal Laser Resonator", Acta Tech. CSAV, 1971, 16, no.4, pp.578-83.

11.134. Wolff, I.: "Open Microwave Resonators", Nachr. Tech. Z.,1971,24,no.6,pp.299-306.

11.135. Wolff, I., and Knoppik, N.: "Microstrip Ring Resonator and Dispersion Measurement", Electron. Lett., 1971, 7, no.26, pp.779-81.

11.136. Nefedov, Ye.I., and Fialkovskii, A.T.: "Open Coaxial Cylindrical Resonator", Isv. VUZ Radioelektron., 1971, 14, no.10, pp.1115-22.

11.137. Oberhettinger, F.: "Fields in Conducting Coaxial Conical Resonators", Z. Angew. Math. Phys., 1971, 22, no.5, pp.937-50.

11.138. Bolger, B.: "Multiple-Beam Interferometer with Transmission-Like Fringes in Reflection", Opt. Commun., 1971, 4, no.4, pp.313-5.

11.139. Kiselev, V.A.: "Multipass Minor Modes of Optical Cavities", Radiotekh. Elektron., 1971, 16, no.1, and Radio Eng. Electron. Phys., 1971, 16, no.1, pp.113-9.

11.140. Izmestev, A.A.: "Generating Functions of Paraxial Gaussian Beams", Radiotekh. Elektron., 1971, 16, no.1, and Radio Eng. Electron. Phys., 1971, 16, no.1, pp.149-51.

11.141. Molchanov, V.Ya., and Skrotskii, G.V.: "Matrix Method for Calculation of Polarization Eigenstates of Anisotropic Optical Resonators. Review", Kvantovaya Elektron., 1971, no.4, pp.3-26.

11.142. Furuhama, Y.: "Study on a Confocal Spherical Laser Resonator", Rev. Radio Res. Lab., 1971, 17, no.91, pp.318-91.

11.143. Anan'ev, Yu.A.: "Unstable Resonators and their Applications", Kvantovaya Elektron., 1971, no.6, pp.3-34.

11.144. Korzhenevich, I.M., Ratner, A.M., and Solov'ev, V.S.: "Selection of the Principal Transverse Mode in a Ring Resonator", Kvantovaya Elektron., 1971, no.6, pp.94-7.

11.145. Tereshchenko, A.I., et al.: "Analysis of Natural Frequencies in Irregular Below-Cutoff Rectangular Resonators", Radiotekh. (Kharkov), 1971, no.18, pp.31-5.

11.146. Enoto, T., Suzuki, M., and Matsumoto, T.: "Analysis of Fabry-Perot Resonator with a Sheet of Anisotropic Medium", Electron. Commun. Jap., 1971, 54, no.4, pp.70-7.

11.147. Yepishin, V.A., and Kiselev, V.K.: "Plane-Parallel Open Resonator with Circular Mirrors having Central Apertures", Radiotekh. Elektron., 1971, 16, no.11, pp.2027-31, and Radio Eng. Electron. Phys., 1971, 16, no.11, pp.1809-12.

11.148. Popov, M.M.: "Integral Equations of Open Resonators Filled with Inhomogeneous Medium", Opt. Spekt., 1972, 32, no.2, pp.421-4, and Opt. Spectrosc., 1972, 32, no.2, pp.220-1.

11.149. Boyd, J.T.: "Totally Internally Reflecting Thin-Film Optical Cavities", Appl. Opt., 1972, 11, no.11, pp.2635-8.

11.150. Mikaelyan, A.L., and D'yachenko, V.V.: "Conservation of Wavefront in Strongly Deformed Solid Media", Zh. Eksp. Teor. Fiz. Pis'ma, 1972, 16, no.1, pp.25-9, and JETP Lett., 1972, 16, no.1, pp.17-9.

11.151. Andrushko, L.M., and Markov, S.Ye.: "Approximate Analysis of a Cavity Resonator in Terms of n First Natural Frequencies", Izv. VUZ Radioelektron., 1972, 15, no.6, pp.793-5.

11.152. Hant, W., and Seeger, J.A.: "Modified Conical Cavity for Submillimetre-Wave Applications", Trans. IEEE, 1972, ED-19, no.1, pp.80-5.

11.153. Simonsohn, G.: "Modes of Fabry-Perot Resonators with Smooth Parallel Mirrors. I and II", Opt. Acta, 1972, 19, no.1, pp.45-58 and 59-78.

11.154. Kogelnik, H., et al.: "Astigmatically Compensated Cavities for CW Dye Lasers", J. Quantum Electron. IEEE, 1972, QE-8, no.3, pp.373-9.

11.155. Kretzschmar, J.G.: "Theoretical Results for Elliptic Microstrip Resonator", Trans. IEEE, 1972, MTT-20, no.5, pp.342-3.

11.156. Miyazaki, Y.: "Electromagnetic Theory of Open Resonators", J. Phys. Soc. Jap., 1972, 32, no.3, pp.837-44.

11.157. Fialovskii, A.T., and Chaika, V.Ye.: "Coaxial Open Resonator with Ring-Shaped Outer and Cylindrical Inner Reflector Strips", Izv. VUZ Radiofiz., 1972, 15, no.1, pp.117-25.

11.158. Kaminskaya, R.G., and Yushkov, Yu.G.: "Excitation of a TW Regime in Circular Resonators", Izv. VUZ Fiz., 1972, no.1, pp.57-62.

11.159. Chesler, R.B., and Maydan, D.: "Convex-Concave Resonators for TEM_{00} Operation of Solid-State Ion Lasers", J. Appl. Phys., 1972, 43, no.5, pp.2254-7.

11.160. Joly, J.C., and Poinsot, A.: "Eigenmodes for Straight Circular Cylinder Microwave Resonators Containing a Coaxial Dielectric Sample", Trans. IEEE, 1972, MTT-20, no.6, p.422.

11.161. Pasqualetti, F., and Ronchi, L.: "Resonances and Resonant Fields in a Fabry-Perot Resonator Illuminated by a Line Source", Appl. Opt., 1972, 11, no.5, pp.1133-42.

11.162. Furuhama, Y.: "Excitation of a Confocal Spherical Laser Resonator", Jap. J. Appl. Phys., 1972, 11, no.6, pp.874-82.

11.163. Tanaka, T., and Suzuki, M.: "Optimum Mirrors for Fabry-Perot Resonator Filled with Anisotropic Medium", Trans. IEEE, 1972, MTT-20, no.8, pp.546-7.

11.164. Wested, J., and Anderson, E.: "Resonance Splitting in Nonuniform Ring Resonators", Electron. Lett., 1972, 8, no.12, pp.301-2.

11.165. Mikhirev, V.I., and Raevskii, S.B.: "Experimental Study of an Open Resonator Partly Filled with Dielectric", Izv. VUZ Radioelektron., 1972, 15, no.5, pp.668-70.

11.166. Swain, D.W., and Mix, L.P.: "Variable Output Coupling Device for Far-IR Laser", Rev. Sci. Instrum., 1972,43,no.7,pp.1047-8.

11.167. Johnston, W.D., and Runge, P.K.: "Improved Astigmatically Compensated Resonator for CW Dye Lasers", J. Quantum Electron. IEEE, 1972, QE-8, no.8, pp.724-5.

11.168. Cullen, A.L., Nagenthiram, P., and Williams, A.D.: "Variational Approach to Theory of Open Resonator", Proc. Roy. Soc., 1972, 329A, no.1577, pp.153-69.

11.169. Schilder, D.: "Coupling a Mirror Resonator to a Waveguide", Wiss. Z. Tech. Hochsch. Ilmenau, 1972, 18, no.1, pp.87-96.

11.170. Boulay, R., et al.: "Diffraction Properties of Two Closely Spaced Apertures and Effect on Fabry-Perot Resonators", Opt. Commun., 1972, 5, no.2, pp.82-5.

11.171. Budanov, V.Ye., and Shinkarenko, V.F.: "Eigenfrequencies of Symmetric TE-Mode Signals in a Ring Coaxial Resonator", Radiotekh. (Kharkov), 1972, no.20, pp.191-6.

11.172. Kamyshan, A.V., and Kamyshan, V.V.: "Excitation of an Open Waveguide Resonator by a Coupling Aperture", Radiotekh. (Kharkov), 1972, no.20, pp.203-7.

11.173. Morawski, T.: "Analysis of Electric Field in a Rectangular Resonator with Flat Iris", Elektronika, 1972, no.7-8, pp.312-3.

11.174. Danielson, M.: "Analysis of Power Loss in the Coupling Mechanism of a Cavity Resonator", Trans. IEEE, 1972, MTT-20, no.11, pp.758-60.

11.175. Kurin, A.F., Novikov, G.P., and Orlov, V.N.: "Open Resonator with Trihedral Reflector", Izv. VUZ Radiofiz., 1972, 15, no.5, pp.766-72.

11.176. Poinsot, A., and Joly, J.C.: "Natural Frequencies of Microwave Cylindrical Cavity with Coaxial Dielectric Sample", Onde Electr., 1972, 52, no.5, pp.223-7.

11.177. Miyamoto, T.: "Output Beam from Confocal Resonator with Pinhole for Mode Selection", Appl. Opt., 1972, 11, no.9, pp.2040-6.

11.178. Chester, A.N., and Abrams, R.L.: "Mode Losses in Hollow-Waveguide Lasers", Appl. Phys. Lett., 1972, 21, no.12,pp.576-8.

11.179. Kiselev, V.A.: "Diffraction Losses in Optical Resonators with Cylindrical Mirrors", Radiotekh. Elektron., 1972, 17, no.2, pp.247-55, and Radio Eng. Electron. Phys., 1972, 17, no.2, pp.189-96.

11.180. Lyubimov, V.V., and Orlova, I.B.: "Unstable Resonator with Angular Distribution Selector", Opt. Spekt., 1972, 33, no.1, pp.138-40, and Opt. Spectrosc., 1972, 33, no.1, pp.74-5.

11.181. Sherstobitov, V.E., and Vinokurov, G.N.: "Properties of Unstable Resonators with Large Equivalent Fresnel Numbers", Kvantovaya Elektron., 1972, no.3, pp.36-44, and Sov. J. Quantum Electron., 1972, 2, no.3, pp.224-9.

11.182. Tajima, Y.: "Numerical Analysis of MIC Circular Resonators", Electron. Commun. Jap., 1972, 55, no.7, pp.46-52.

11.183. Knyazev, B.R., and Zykov, A.I.: "Coupling Device for Cryogenic Microwave Resonators", Prib. Tekh. Eksp., 1972, 15, no.5, pp.242-3, and Instrum. Exp. Tech., 1972, 15, no.5, pp.1577-8.

11.184. Korzhenvich, I.M., and Ratner, A.M.: "General Method for Finding Proper Frequencies of Optical Resonators with Caustic Surfaces", Kvantovaya Elektron., 1972, no.6, pp.88-103, and Sov. J. Quantum Electron., 1972, 2, no.6.

11.185. Korzhenvich, I.M.: "Ring Resonator with Lenses", Kvantovaya Elektron., 1972, no.6, pp.103-8, and Sov. J. Quantum Electron., 1972, 2, no.6.

11.186. Uenakada, K.: "Analysis of Admittance of a Reentrant Cavity", Electron. Commun. Jap., 1972, 55, no.8, pp.46-51.

11.187. Malanin, Yu.N., Mardanov, R.F., and Pol'skii, Yu.Ye.: "Mode Structure of Field in an Optical Resonator with Moving Mirror", Radiotekh. Elektron., 1972, 17, no.5, pp.919-25, and Radio Eng. Electron. Phys., 1972, 17, no.5, pp.715-21.

11.188. Waksberg, A., and Wood, J.: "Aligning Intracavity Modulators", Laser Focus, 1972, 8, no.11, p.43.

11.189. Shatrov, A.D.: "Expansion of Fields in Open Waveguides and Resonators", Radiotekh. Elektron., 1972, 17, no.6, pp.1153-60, and Radio Eng. Electron. Phys., 1972, 17, no.6, pp.896-902.

11.190. Mak, A.A., Mitkin, V.M., and Soms, L.N.: "Excitation of Radial and Angular Modes in Optical Cavities", Opt. Spekt., 1972, 33, no.5, pp.996-7, and Opt. Spectrosc., 1972, 33, no.5, pp.546-7.

11.191. Novikov, M.A., and Smirnov, G.T.: "Prismatic Dispersion Resonator for a Laser", Prib. Tekh. Eksp., 1972, 15, no.6, pp.172-4, and Instrum. Exp. Tech., 1972, 15, no.6, pp.1814-6.

11.192. Kiselev, V.A.: "Optical Resonators with Annular Slot Mirrors", Radiotekh. Elektron., 1972, 17, no.10, pp.2020-7, and Radio Eng. Electron. Phys., 1972, 17, no.10, pp.1609-15.

11.193. Mikaelyan, A.L., and D'yachenko, V.V.: "Waveguide-Type Optical Resonators", Kvantovaya Elektron., 1972, no.5, pp.97-9, and Sov. J. Quantum Electron., 1972, 2, no.5, pp.467-9.

11.194. Sysoev, A.S., and Tret'yakov, O.A.: "Open Resonators with a Diffraction Grating Used for One Mirror", Radiotekh. Elektron., 1972, 17, no.9, pp.1951-2, and Radio Eng. Electron. Phys., 1972, 17, no.9, pp.1556-7.

11.195. Melkov, G.A.: "Forced Oscillations in Open Dielectric Microwave Resonators", Radiotekh. Elektron., 1972, 17, no.10, pp.2027-35, and Radio Eng. Electron. Phys., 1972, 17, no.10, pp.1615-21.

11.196. Shelamov, G.N.: "Analysis of the Interaction between a Dielectric Resonator and Rectangular Waveguide", Radiotekhnika, 1973, 27, no.4, pp.61-9, and Telecommun. Radio Eng. Pt 2, 1973, 27, no.4, pp.103-9.

11.197. Safaryan, V.A.: "Calculation of Natural Frequencies and Fields of a Rectangular Resonator with a Dielectric Sphere", Izv. VUZ Radiofiz., 1973, 16, no.4, pp.640-3.

11.198. Wolff, I., and Hofmann, H.: "Three-Layer Spherical Resonators", Frequenz, 1973, 27, no.5, pp.110-9.

11.199. Il'chenko, M.Ye., and Melkov, G.A.: "Calculation of Filling Factor in a Microwave Dielectric Resonator with Waveguide", Izv. VUZ Radiofiz., 1973, 16, no.5, pp.101-3.

11.200. Ronchi, L.: "Lowloss Modes and Resonances in a Quasi-90°-Roof Mirror Resonator", Appl. Opt., 1973, 12, no.1, pp.93-7.

11.201. Uenakada, K.: "Equivalent Circuit of Reentrant Cavity", Trans. IEEE, 1973, MTT-21, no.1, pp.48-51.

11.202. Chester, A.N.: "Beam Steering in Confocal Unstable Resonators", J. Quantum Electron. IEEE, 1973, QE-9, no.2, pp.209-12.

11.203. Casperson, L.W.: "Cylindrical Laser Resonators", J. Opt. Soc. Am., 1973, 63, no.1, pp.25-9.

11.204. Bradley, H.L., and Harris, D.J.: "Open Resonator Operating at 337 micron", Electron. Lett., 1973, 9, no.4, pp.78-9.

11.205. Rensch, D.B., and Chester, A.N.: "Iterative Diffraction Calculations of Transverse Mode Distributions in Confocal Unstable Laser Resonators", Appl. Opt., 1973, 12, no.5, pp.997-1010.

11.206. Consortini, A.: "Sloped-Rim Open Resonators", Appl. Opt., 1973, 12, no.5, pp.1011-4.

11.207. Lagerlof, R.O.E.: "End Effects of Half-Wave Stripline Resonators", Trans. IEEE, 1973, MTT-21, no.5, pp.351-3.

11.208. Wu, Y.S., and Rosenbaum, F.J.: "Mode Chart for Microstrip Ring Resonators", Trans. IEEE, 1973, MTT-21, no.7, pp.487-9.

11.209. Campillo, A.J., et al.: "Fresnel Diffraction Effects in Design of High-Power Laser Systems", Appl. Phys. Lett., 1973, 23, no.2, pp.85-7.

11.210. Freiberg, R.J.: "Data on Unstable Resonators", Laser Focus, 1973, 9, no.5, pp.59-63.

11.211. Miyamoto, T.: "Approximate Expression for Beam Wave from Spherical Resonator with Arbitrarily Located Apertures", Electron. Commun. Jap., 1973, 56, no.10, pp.100-6.

11.212. Il'chenko, M.Ye., and Melkov, G.A.: "Coupling of a Dielectric Microwave Resonator with Waveguide", Izv. VUZ Radioelektron., 1973, 16, no.5, pp.101-3.

11.213. Kahn, W.K.: "Geometrical Constructions for Virtual Centres of Unstable Optical Resonators", Appl. Opt., 1973, 12, no.9, pp.2026-7.

11.214. Degnan, J.J., and Hall, D.R.: "Finite-Aperture Waveguide-Laser Resonators", J. Quantum Electron. IEEE, 1973, QE-9, no.9, pp.901-10.

11.215. Chavka, G.G.: "Effect of Losses on Broadband Matching of High-Q Loads", Izv. VUZ Radioelektron., 1973, 16, no.7, pp.79-81.

11.216. Affolter, P., and Eliasson, B.: "Resonances and Q-Factors of Lossy Dielectric Spheres", Trans. IEEE, 1973, MTT-21, no.9, pp.573-8.

11.217. Watkins, J.: "Radiation Loss from Open-Circuited Dielectric Resonators", Trans. IEEE, 1973, MTT-21, no.10, pp.636-9.

11.218. Averbakh, V.S., Vlasov, S.N., and Piskunova, L.V.: "Coupled Open Resonators", Izv. VUZ Radiofiz., 1973, 16, no.8, pp.1205-10.

11.219. Chester, A.N.: "Three-Dimensional Diffraction Calculations of Laser-Resonator Modes", Appl. Opt., 1973, 12, no.10, pp.2353-66.

11.220. Le Floch, A., and Stephen, G.: "Resonance Condition in Anisotropic Lasers Containing Birefringent Plates", C. R. Acad. Sci. (Paris), 1973, 277B, no.11, pp.265-8.

11.221. Chen, L.W., and Felsen, L.B.: "Coupled-Mode Theory of Unstable Resonators", J. Quantum Electron. IEEE, 1973, QE-9, no.11, pp.1102-13.

11.222. Jimenez, J.J., and Guijarro, J.J.: "Experimental Q-Factors of Three Types of Microstrip Resonators", Rev. Phys. Appl., 1973, 8, no.3, pp.279-82.

11.223. Yamauchi, N.: "Wave-Theoretical Analysis of Resonant Modes in Astigmatic Open Resonators", Electron. Commun. Jap., 1973, 56, no.6, pp.73-9.

11.224. Vinokurov, G.N., Lyubimov, V.V., and Orlova, I.B.: "Investigation of Selective Properties of Open Unstable Cavities", Opt. Spekt., 1973, 34, no.4, pp.741-51, and Opt. Spectrosc., 1973, 34, no.4, pp.427-32.

11.225. Itoh, T., and Mittra, R.: "Analysis of a Microstrip Disc Resonator", Arch. Elektron. Ubertrag., 1973, 27, no.11, pp.456-8.

11.226. Shlat'ko, A.A.: "Scattering Properties of Resonating Systems with Waveguide Output", Radiotekh. (Kharkov), 1973, no.26, pp.145-50.

11.227. Chen, L.W., and Felsen, L.B.: "Waveguide Analysis of Unstable Resonators", Electron. Lett., 1973, 9, no.14, pp.311-2.

11.228. Holmes, J.F., and Yamamoto, R.S.: "Numerical Calculation of Losses in Optical Resonators", Proc. IEEE, 1973, 61, no.11, pp.1652-3.

11.229. Affolter, P., and Kach, A.: "Resonance Frequencies of a Dielectric Sphere Contained within a Concentric Shield", Arch. Elektron. Ubertrag., 1973, 27, no.10, pp.423-32.

11.230. Aran'ev, Yu.A.: "Unstable Prism Resonators", Kvantovaya Elektron., 1973, no.7, pp.105-8, and Sov. J. Quantum Electron., 1973, 3, no.1, pp.58-9.

11.231. Horwitz, P.: "Asymptotic Theory of Unstable Resonator Modes", J. Opt. Soc. Am., 1973, 63, no.12, pp.1528-43.

11.232. Lax, M., Louisell, W.H., and McKnight, W.B.: "Transverse-Mode Discrimination in Three-Mirror Resonators", J. Opt. Soc. Am., 1973, 63, no.12, pp.1544-9.

11.233. Heller, H.J.: "Nomogram for Designing an Optical Resonator with Internal Thin Lenses", Z. Angew. Math. Phys., 1973, 24, no.3, pp.449-50.

11.234. Lyubimov, V.V., Peugenen, N.N., and Petrov, V.E.: "Unstable Asymmetric Resonators", Opt. Spekt., 1973, 35, no.6, pp.1132-7, and Opt. Spectrosc., 1973, 35, no.6, pp.657-9.

11.235. Ishchenko, Ye.F., and Sushkin, V.N.: "Rational Choice of Configuration of an Open Resonator with Coupling Aperture", Radiotekh. Elektron., 1973, 18, no.5, pp.930-4, and Radio Eng. Electron. Phys., 1973, 18, no.5, pp.674-7.

11.236. Slavyanov, S.Yu.: "Theory of Open Resonators", Zh. Eksp. Teor. Fiz., 1973, 64, no.3, pp.785-95, and Sov. Phys.-JETP, 1973, 37, no.3, pp.399-403.

11.237. Tereshchenko, A.I., and Grebenyuk, A.F.: "Resonant Ring Cavity Systems with E-Plane Bend", Radiotekh. (Kharkov), 1973, no.27, pp.117-23.

11.238. Consortini, A.: "Curved-Rim Open Resonators", Trans. IEEE, 1974, MTT-22, no.1, pp.60-3.

11.239. McAllister, G.L., Lacina, W.B., and Steier, W.H.: "Improved Mode Properties of Unstable Resonators with Tapered-Reflectance Mirrors and Shaped Apertures", J. Quantum Electron. IEEE, 1974, QE-10, no.3, pp.346-55.

11.240. Ter-Pogosyan, A.S.: "Calculation of Symmetrical Optical Resonators for Large Fresnel Numbers", Izv. VUZ Prib., 1974, 17, no.1, pp.112-7.

11.241. Wolff, I., and Knoppik, N.: "Microstrip Disc Resonators", Arch. Elektron. Ubertrag., 1974, 28, no.3, pp.101-8.

11.242. Nefedov, Ye.I., et al.: "Excitation of Open Coaxial Resonator through Longitudinal Slots", Izv. VUZ Radioelektron., 1974, 17, no.1, pp.121-2.

11.243. Ter-Pogosyan, A.S.: "Asymmetric Open Laser Resonators", Izv. VUZ Prib., 1974, 17, no.2, pp.97-101.

11.244. Pasqualetti, F., and Ronchi, L.: "Integral Equation for 90° Roof-Mirror Optical Resonator", J. Opt. Soc. Am., 1974, 64, no.3, pp.289-94.

11.245. Kamyshan, A.V., Tsvyk, A.I., and Shestopalov, V.P.: "Experimental Investigation of Open Resonators with Toroidal Mirrors", Izv. VUZ Radiofiz., 1974, 17, no.5, pp.727-33.

11.246. Bloom, A.L.: "Modes of a Laser Resonator Containing Tilted Birefringent Plates", J. Opt. Soc. Am., 1974, 64, no.4, pp.447-52.

11.247. Checcacci, P.F., Falciai, R., and Scheggi, A.M.: "Modes and Losses of a Four-Mirror Ring Resonator", Trans. IEEE, 1974, MTT-22, no.7, pp.751-2.

11.248. Kraszewski, A.: "One-Port TW Ring Resonator", Arch. Elektrotech., 1974, 23, no.1, pp.229-48.

11.249. Dal Pozzo, P., et al.: "Doubly-Confocal Unstable Ring Resonator", Opt. Commun., 1974, 11, no.2, pp.115-7.

11.250. Jansen, R.: "Shielded Rectangular Microstrip Disc Resonators", Electron. Lett., 1974, 10, no.15, pp.299-300.

11.251. Safaryan, V.A.: "Projection Algorithm for Rectangular Resonator with Dielectric Sphere", Izv. Akad. Nauk Arm. SSR Fiz., 1974, 9, no.2, pp.148-57.

11.252. Pekar, V.S.: "Theory of a Plane Optical Resonator and Waveguide with a Semitransparent Wall", Ukr. Fiz. Zh., 1974, 19, no.8, pp.1350-8.

11.253. Abrams, R.L., and Chester, A.N.: "Resonator Theory for Hollow Waveguide Lasers", Appl. Opt., 1974, 13, no.9, pp.2117-25.

11.254. Mur, G.: "Field Analysis and Complex Resonance Frequency of the Quasi-TE_{011} Mode in an Inhomogeneously Filled Resonator with Losses", Appl. Sci. Res., 1974, 29, no.2, pp.137-44.

11.255. Wolff, I.: "Rectangular and Circular Microstrip Disc Capacitors and Resonators", Trans. IEEE, 1974, MTT-22, no.10, pp.857-64.

11.256. De Mey, G.: "Frequency Shift in Resonant Cavities", Int. J. Electron., 1974, 37, no.3, pp.369-75.

11.257. Weiner, M.M.: "Useful Beam Quality Design Curves for Unstable Resonators", Opt. Eng., 1974, 13, no.2, pp.87-91.

11.258. Belanger, P.A., and Legare, A.: "Asymptotic Solution of the Integral Equation of a Plane Parallel Laser Resonator", Can. J. Phys., 1974, 52, no.20, pp.1981-7.

11.259. Johnson, M.M.: "Direct Application of the Fast Fourier Transform to Open Resonator Calculations", Appl. Opt., 1974, 13, no.10, pp.2326-8.

11.260. Ajvazov, J.B.: "Two-Conic Microwave Resonator", Arch. Elektrotech., 1974, 23, no.2, pp.337-48.

11.261. Garault, Y., and Guillon, P.: "Best Approximation for Design of Natural Resonance Frequencies of Microwave Dielectric Disc Resonators", Electron. Lett., 1974, 10, no.24, pp.505-7.

11.262. Itoh, T.: "Analysis of Microstrip Resonators", Trans. IEEE, 1974, MTT-22, no.11, pp.946-52.

11.263. Arnaud, J.A.: "Optical Resonators in the Gaussian Approximation", Proc. IEEE, 1974, 62, no.11, pp.1561-70.

11.264. Checcacci, P.F., Falciai, R., and Scheggi, A.M.: "Ring and 90° Roof Open Resonators", Proc. IEEE, 1974, 62, no.11, pp.1611-3.

11.265. Tereshchenko, A.I., Burtovoi, D.P., and Kanarik, G.G.: "Dependence of the Parameters of Polycylindrical Resonators on their Geometry. I and II", Radiotekh. (Kharkov), 1974, 28, pp.77-86 and 86-93.

11.266. Bel'dyugin, I.M., et al.: "Theory of Open Resonators with Cylindrical Mirrors", Kvantovaya Elektron., 1974, 1, no.4, pp. pp.881-91, and Sov. J. Quantum Electron., 1974, 4, no.4, pp.485-90.

11.267. Isaev, A.A., et al.: "Converging Beams in Unstable Telescopic Resonators", Kvantovaya Elektron., 1974, 1, no.6, pp.1379088, and Sov. J. Quantum Electron., 1974, 4, no.6, pp.761-6.

11.268. Soohoo, J., and Mevers, G.E.: "Cavity-Mode Analysis Using the Fourier-Transform Method", Proc. IEEE, 1974, 62, no.12, pp.1721-3.

11.269. Bava, E., and Bava, G.P.: "Experiments with Microwave Open Resonators Lined with Dielectric Material", Alta Freq., 1974, 43, no.11, pp.944-8.

11.270. Junghans, J., Keller, M., and Weber, H.: "Laser Resonators with Polarizing Elements", Appl. Opt., 1974, 13, no.12, pp.2793-8.

11.271. Slavyanov, S.Yu., and Farafonov, V.G.: "Aberrations in Cavities with Misadjusted Mirrors", Opt. Spekt., 1974, 37, no.1, pp.206-7, and Opt. Spectrosc., 1974, 37, no.1, pp.117-8.

11.272. Batt, R.J., et al.: "Waveguide and Open-Resonator Techniques for Submillimetre Waves", Trans. IEEE, 1974, MTT-22, no.12, pp.1089-94.

11.273. Danileiko, Yu.K., and Lobachev, V.A.: "Rotating-Field Resonator for Lasers", Kvantovaya Elektron., 1974, 1, no.3, pp.688-90, and Sov. J. Quantum Electron., 1974, 14, no.3, pp.389-90.

11.274. Kvasil, B.: "Transmission of Partially Coherent Signal through an Open Line and an Open Resonator", Acta Tech. CSAV, 1974, 19, no.6, pp.633-46.

12. PERIODIC AND GUIDING STRUCTURES

12.1. Kao, K.C.: "Electromagnetic Wave Propagation on a Double-Layer Dielectric Film", Electron. Lett., 1965, 1, p.35.

12.2. Bevensee, R.M.: "Nonuniform TEM Transmission Lines. I", Proc. IEE, 1965, 112, p.644.

12.3. Schlosser, W.: "Perturbation of Characteristic Values of Circular Dielectric Wires with Slight Elliptical Deformations", Arch. Elektr. Ubertrag., 1965, 19, p.1.

12.4. Cornet, G., and Raoult, G.: "Numerical Results on Loaded Waveguides with Rectangular Section", C.R. Acad. Sci. (Paris), 1965, 260B, p.2447.

12.5. Didenko, A.N.: "Use of Normal-Mode Method for Calculation of Rectangular Waveguides with Diaphragms", Zh. Tekh. Fiz., 1965, 35, p.967, and Sov. Phys.-Tech. Phys., 1965, 10, no.5, pp.744-5.

12.6. Mittra, R., and Jones, K.E.: "Resolution of Multimode Data in Periodic Structures and Waveguides", Trans. IEEE, 1965, AP-13, p.325.

12.7. Clarricoats, P.J.B., and Oliner, A.A.: "Transverse-Network Representation for Inhomogeneously Filled Circular Waveguide", Proc. IEE, 1965, 112, p.883.

12.8. Norton, D.E.: "Calculation and Measurement of Near Fields of a Surface Waveguide", Int. Conv. Rec. IEEE, 1965, 13, pt 5, p.200.

12.9. Laxpati, S.R., and Mittra, R.: "Energy Considerations in Open and Closed Waveguides", Int. Conv. Rec. IEEE, 1965, 13, pt 5, p.140.

12.10. Jones, K.E., and Mittra, R.: "Interpretations and Applications of the Omega-Beta Diagram", Int. Conv. Rec. IEEE, 1965, 13, pt 5, p.134.

12.11. Simonovich, D., and Ishii, T.K.: "Attenuation Coefficient of Millimetre-Wave, Two-Wire, Surface-Wave Transmission Lines", Electron. Lett., 1965, 1, p.144.

12.12. Clarricoats, P.J.B., and Slinn, K.R.: "Complex Modes of Propagation in Dielectric-Loaded Circular Waveguide", Electron. Lett., 1965, 1, p.145.

12.13. Hershenov, B.: "Paralleled Microwave Circuits, Phase Synchronized and Quasi-Statically Coupled", Proc. IEEE, 1965, 53, p.104.

12.14. Snyder, A.W.: "Excitation of Surface Modes along a Semi-Infinite Dielectric Cylinder", Electron. Lett., 1965, 1, p.208.

12.15. Timerev, N.P., and Fedorenkov, A.I.: "Unsymmetrical Wave Propagation along a Conical Helix with Variable Parameters", Radiotekh. Elektron., 1965, 10, p.760, and Radio Eng. Electron. Phys., 1965, 10, no.4, pp.646-8.

12.16. Nefedov, Ye.I.: "Thin Inhomogeneous Dielectric Prism in a Wide Waveguide", Radiotekh. Elektron., 1965, 10, p.764, and Radio Eng. Electron. Phys., 1965, 10, no.5, pp.749-57.

12.17. Gekker, I.R., et al.: "Investigation of a Corrugated Structure for Transforming H_{01} Waves into E_{11} Waves in a Round Waveguide", Radiotekh. Elektron., 1965, 10, p.756, and Radio Eng. Electron. Phys., 1965, 10, no.4, pp.642-5.

12.18. Nagao, T., and Nishida, S.: "Characteristics of Rectangular Waveguides Partially Filled with Parallel-Plate Media", Rep. Res. Inst. Electr. Commun. Tohoku Univ., 1965, 17, no.1, p.1.

12.19. Froom, J., et al.: "Ridge-Loaded Ladder Lines", Trans. IEEE, 1965, ED-12, p.411.

12.20. Bobrovnikov, M.S., Goshin, G.G., and Smirnov, V.P.: "Problem of Effective Excitation of Radial Waves on Cylindrical Surface", Radiotekh. Elektron., 1965, 10, p.1023, and Radio Eng. Electron. Phys., 1965, 10, no.6, pp.875-80.

12.21. Berceli, T.: "Optimum Design of Helix Couplers with Shielding", Trans. IEEE, 1965, MTT-13, p.471.

12.22. Bernardi, P.: "Propagation in a Rectangular Waveguide Bounded by an Impedance Wall", Alta Freq., 1965, 34, p.490.

12.23. Neumann, E.G., and Stumper, U.: "Passage of a Dielectric Transmission Line through an Iris Diaphragm", Z. Angew. Phys., 1965, 20, p.56.

12.24. Chen, M.Y., and Wu, Y.C.: "Decay in H-Shape Dielectric-Metallic Waveguide", Acta Phys. Sin., 1965, 21, p.1705.

12.25. Gillespie, E.S., and Gustincic, J.J.: "Scattering of a TM Surface Wave by a Perfectly Conducting Strip", Trans. IEEE, 1965, MTT-13, p.630.

12.26. Curnow, H.J.: "General Equivalent Circuit for Coupled-Cavity Slow-Wave Structures", Trans. IEEE, 1965, MTT-13, p.671.

12.27. Garault, Y.: "Study of a Class of Guided Waves. EH Waves", Ann. Phys. (Paris), 1965, 10, p.641.

12.28. Ruddy, J.M.: "Experimental Results in Groove Guide", Trans. IEEE, 1965, MTT-13, p.880.

12.29. Meier, P.J., and Arnow, S.: "Wideband Polarizer in Circular Waveguide Loaded with Dielectric Discs", Trans. IEEE, 1965, MTT-13, p.763.

12.30. Kaplunov, M.B.: "Bent Screened Single-Wire Surface-Wave Transmission Line", Radiotekhnika, 1965, 20, no.7, p.15, and Telecommun. Radio Eng. Pt 2, 1965, 20, no.7, p.78.

12.31. Kosta, S.P.: "Design of Surface-Wave System for 5 GHz", J. Inst. Telecommun. Eng. Jap., 1965, 11, p.123.

12.32. Tereshchenko, A.I., and Shein, A.G.: "Slow-Wave Structures Using Iris-Type Cross-Shaped Waveguide", Radiotekh. Elektron., 1965, 10, p.1029, and Radio Eng. Electron. Phys., 1965, 10, p.880.

12.33. Kotik, I.P., and Sivov, A.N.: "Analysis of Phase Characteristics of Non-symmetric Waves and of the Filtering Properties of a Helical Waveguide", Radiotekh. Elektron., 1965, 10, p.1065, and Radio Eng. Electron. Phys., 1965, 10, p.911.

12.34. Rosenbaum, F.J.: "Hybrid Modes on Anisotropic Dielectric Rods", J. Quantum Electron. IEEE, 1965, QE-1, p.367.

12.35. Rulea, Gh.: "Graphical Method for Determining the Field in Layered-Dielectric Guides", Rev. Roum. Phys., 1965, 10, p.657.

12.36. Lofy, F.J., and Ishii, T.K.: "Mode of Millimetre-Wave, Two-Wire, Surface-Wave Transmission-Line Fields", Proc. IEEE, 1965, 33, p.1652.

12.37. Kovalenko, Ye.S.: "Design of Periodically Loaded Waveguides", Izv. VUZ Radiotekh., 1965, 8, p.467, and Sov. Radio Eng., 1965, 8, p.338.

12.38.   Severin,H.: "Surface-Wave Transmission Lines for Microwave Frequencies. I and II", Philips Tech. Rev., 1965,26,p.342.

12.39.   Tereshchenko, A.I., and Shein, A.G.: "Dispersion Equations for Ring Systems Using Rectangular Iris Waveguides", Izv. VUZ Radiotekh., 1965, 8, p.535, and Soviet Radio Eng., 1965, 8, p.398.

12.40.   Chatterjee, S.K.: "Theory of Metal-Disc-Loaded Sommerfeld Surface-Wave Line", J. Inst. Telecommun. Eng., 1965, 11, p.450.

12.41.   Chatterjee, S.K., Chatterjee, R., and Zacharia, K.P.: "Theory of Open-Type Resonator with Centre Conductor", J. Inst. Telecommun. Eng., 1965, 11, p.407.

12.42.   Goncharenko, A.M.: "Properties of Rectangular Waveguide with Anisotropic Filling", Zh. Tekh. Fiz., 1965, 35, p.1444, and Sov. Phys.-Tech. Phys., 1966, 10, no.8, pp.1121-2.

12.43.   Vagin, V.A., and Kotov, V.I.: "Hybrid Waves in Circular Waveguide Partially Filled with Dielectric", Zh. Tekh. Fiz., 1965, 35, p.1273, and Sov. Phys.-Tech. Phys., 1966, 10, no.7, pp.987-91.

12.44.   Kondrat'ev, B.V.: "Power Flux in Helical Waveguide with Anisotropic Medium", Zh. Tekh. Fiz., 1965, 35, p.1447, and Sov. Phys.-Tech. Phys., 1966, 10,no.8,pp.1123-5.

12.45.   Bava, G.P., and Perona, G.: "Conformal Mapping Analysis of a Type of Groove Guide", Electron. Lett., 1966, 2, p.13.

12.46.   Zhileiko, G.I.: "Slow-Wave Structure with Variable Phase Velocity of E Wave", Radiotekhnika, 1966, 21, no.6, p.17, and Telecommun. Radio Eng. Pt 2, 1966, 21, no.6, p.79.

12.47.   Mikaelyan, A.L., and Turkov, Yu.G.: "Parasitic Internal Oscillation in Open Resonators with Dielectric Rod", Radiotekh. Elektron., 1966, 11, p.347, and Radio Eng. Electron. Phys., 1966, 11, no.2, pp.286-7.

12.48.   Haas, W., and Godtmann, H.D.: "Natural Oscillations of Dielectric Rods Arranged between Two Metallic Plates", Arch. Elektr. Ubertrag., 1966, 20, p.97.

12.49.   Waldron, R.A.: "Mode Nomenclature for Helix Waveguide", Proc. IEE, 1966, 113, p.420.

12.50.   Rosenberg, D., and Stock, D.J.R.: "Results for Thin-Iris Loaded Periodic Waveguides", Trans. IEEE, 1966, MTT-14, p.145.

12.51.   Reeder, T.M.: "Equivalent Circuit for Centipede Waveguide", Trans. IEEE, 1966, MTT-14, p.200.

12.52.   Bernardi, P.: "Waves Guided by the Most General Anisotropic Impedance Wall", Nuovo Cim., 1966, 43B, p.338.

12.53.   Paul, D.K.: "Surface-Wave Propagation through Anisotropic Circular-Cylindrical Dielectric-Rod Waveguide", J. Indian Inst. Sci., 1966, 48, no.2-3, p.102.

12.54.   Conlon, R.F.B., and Benson, F.A.: "Propagation and Attenuation in Double-Strip H-Guide", Proc. IEE, 1966, 113, p.1311.

12.55.   James, C.R., and Walker, G.B.: "Approximate Wave Equation for Axially Symmetric Periodic Waveguide", Trans. IEEE, 1966, MTT-14, p.428.

12.56.   Semenov, N.A.: "Transmitted Power in Dielectric Waveguide", Radiotekhnika, 1966, 21, no.3, and Telecommun. Radio Eng. Pt 2, 1966, 21, no.3, p.85.

12.57.   Klynev, V.M., and Prokopenko, N.I.: "Rectangular Comb in Rectangular Waveguide", Radiotekhnika, 1966, 21, no.8, p.13, and Telecommun. Radio Eng. Pt 2, 1966, 21, no.8.

12.58.   Bernard, J., et al.: "Dispersion Formula for Waveguides Loaded with Rounded-Edge Irises", C.R. Acad. Sci. (Paris), 1966, 263B, no.18, pp.1010-3.

12.59.   Nishida, S., and Nagao, T.: "Transmission Characteristics of Rectangular Waveguides Filled with Inhomogeneous Media Composed of Dielectric and Metallic Vanes", J. Inst. Electr. Commun. Eng. Jap.,1966,49,p.1173.

12.60.   Itakura, K., Azakami, T., and Harada, J.: "H-Guide Mode Superposed by Surface-Wave Mode", Technol. Rep. Osaka Univ.,1966,16,p.267.

12.61.   Barlow, H.E.M.: "Features of Wave Propagation Not Subject to Cutoff between Two Parallel Guiding Surfaces", Proc. IEE, 1967, 114, no.4, pp.421-7.

12.62.   Clarricoats, P.J.B., Oliner, A.A., and Olver, A.D.: "Propagation Behaviour of Slotted Inhomogeneous Circular Waveguides", Proc. IEE, 1967, 114, no.4, pp.457-64.

12.63.   Oliner, A.A., and Clarricoats, P.J.B.: "Transverse Equivalent Networks for Slotted Inhomogeneous Circular Waveguides", Proc. IEE, 1967, 114, no.4, pp.445-56.

12.64.   Savard, J.Y.: "Higher-Order Cylindrical Surface-Wave Modes", Trans. IEEE, 1967, MTT-15, no.3, p.151-5.

12.65.   Riblet, H.J.: "Explicit Derivation of the Relationship between Parameters of an Interdigital Structure and Equivalent Transmission-Line Cascade", Trans. IEEE, 1967, MTT-15, no.3, p.161-6.

12.66.   Shein, A.G.: "Theory of a Cross-Shaped Diaphragmed Waveguide", Radiotekh. Elektron., 1967, 12, no.6, pp.1098-1101, and Radio Eng. Electron. Phys., 1967,12,no.6.

12.67.   Sigelman, R.A.: "Surface Waves on a Grounded Dielectric Slab Covered by a Periodically Slotted Conducting Plane", Trans. IEEE, 1967, AP-15, no.5, pp.672-6.

12.68. Ol'derogge, Ye.B.: "Calculation of the Coupling and Depression Coefficients for a Slow-Wave Comb Structure", Radiotekh. Elektron., 1967, 12, no.4, pp.744-7, and Radio Eng. Electron. Phys., 1967, 12, no.4, pp.695-9.

12.69. Nilsson, O.: "Approximate Theory of Round-Wire Helix Slow-Wave Structure", Nachr. Tech. Fachber., 1968, 35, p.80.

12.70. Beal, J.C., and Dewar, W.J.: "Coaxial-Slot Surface-Wave Launcher", Electron. Lett., 1968, 4, p.557.

12.71. Shatrov, A.D.: "Integral Equations for Natural Modes of Periodic Structures", Radiotekh. Elektron., 1968, 13, p.1500, and Radio Eng. Electron. Phys., 1968, 13, no.8, pp.1303-5.

12.72. Garault, Y., and Fenelon, J.P.: "Hybrid EH Waves in a Periodic Trough Waveguide", C.R. Acad. Sci. (Paris), 1968, 267B, p.540.

12.73. Laybourn, P.J.R.: "Group Velocity of Dielectric Waveguide Modes", Electron. Lett., 1968, 4, p.507.

12.74. Teplyakov, V.A., and Stepanov, V.B.: "Study of an H Cavity", Radiotekh. Elektron., 1968, 13, p.1965, and Radio Eng. Electron. Phys., 1968, 13, no.11, pp.1724-33.

12.75. Zav'yalov, A.S.: "Waves in a Groove with an Inclined Comb", Izv. VUZ Fiz., 1968, no.10, p.138.

12.76. Subrahmaniam, V., and Chatterjee, S.K.: "Circular Cylindrical Dielectric Rod Waveguide", J. Indian Inst. Sci., 1968, 50, p.258.

12.77. Shigesawa, H., and Takiyama, K.: "Study of a Close-Grooved Guide", Sci. Eng. Rev. Doshisha Univ., 1968, 9, p.9.

12.78. Severin, H.: "Cylindrical Surface-Wave Transmission Lines", Tech. Mitt. PTT, 1968, 46, p.442.

12.79. Melekhin, V.N., and Manenkov, A.B.: "Dielectric Tubes as Waveguides with Low Losses", Zh. Tekh. Fiz., 1968, 38, no.12, p.2113, and Sov. Phys.-Tech. Phys., 1969, 13, no.12, pp.1698-9.

12.80. Barlow, H.E.M., and Sen, M.: "Two-Wire Line Resonator Supporting the Hybrid TEM-Dual Surface Wave", Electron. Lett., 1969, 5, p.68.

12.81. Cook, K.R., and Chu, T.M.: "Mode Coupling between Surface-Wave Lines", Trans. IEEE, 1969, MTT-17, p.265.

12.82. Partch, J.E.: "Dielectric Shielded G-Line", Trans. IEEE, 1969, MTT-17, p.271.

12.83. Sawa, S., and Kumagai, N.: "Surface Wave Along a Circular H-Bend of an Inhomogeneous Dielectric Thin Film", Electron. Commun. Jap., 1969, 52, no.3, p.44.

12.84. Barlow, H.E.M.: "Hybrid TEM-Dual Surface Wave in a Coaxial Cable", Proc. IEE, 1969, 116, p.489.

12.85. Vzyatyshev, V.F., and Rozhkov, G.D.: "Attenuation Criterion in Surface-Wave Transmission Lines", Izv. VUZ Radioelektron., 1969, 12, p.25.

12.86. Tischer, F.J.: "H-Guide with Laminated Dielectric", Proc. IEEE, 1969, 57, p.820.

12.87. Stringer, P.E.: "Surface-Wave Resonator Coupling", Electron. Lett., 1969, 5, p.258.

12.88. Semenov, N.A.: "Wave Parameters in Surface-Wave Coupled Waveguides", Radiotekh. Elektron., 1969, 14, p.599, and Radio Eng. Electron. Phys., 1969, 14, no.4, pp.519-22.

12.89. Paik, S.F.: "Design Formulae for Helix Dispersion Shaping", Bell Syst. Tech. J., 1969, 48, p.2189.

12.90. Yatsuk, K.P., Sapelkin, A.I., and Kagan, V.Ya.: "Study of Dispersion in Dielectric Waveguides in the Millimetre Band", Radiotekh. Elektron., 1969, 14, no.7, p.1314, and Radio Eng. Electron. Phys., 1969, 14, no.7, pp.1136-8.

12.91. Miller, S.E.: "Theory and Applications of Periodically Coupled Waves", Bell Syst. Tech. J., 1969, 48, p.2189.

12.92. Marcuse, D., and Derosier, R.M.: "Mode Conversion Caused by Diameter Changes of a Round Dielectric Waveguide", Bell Syst. Tech. J., 1969, 48, p.3217.

12.93. Marcuse, D.: "Radiation Losses of Dielectric Waveguides in Terms of Power Spectrum of Wall Distortion Function", Bell Syst. Tech. J., 1969, 48, p.3233.

12.94. Zakharov, A.A., and Sovetov, N.M.: "Equivalent Circuits for Calculating Dispersion of a Plane Convoluted Spiral Type of Delay Structure", Izv. VUZ Radioelektron., 1969, 12, no.9, p.1101.

12.95. Damestoy, E., Pouyet, J., and Rebiere, J.P.: "Coupling between Three Identical Helicoidal Lines Supporting a First-Order Wave", C. R. Acad. Sci. (Paris), 1969, 169B, p.1020.

12.96. Ivanov, V.N.: "Surface Waves in Semi-Infinite Ribbon Array", Radiotekh. Elektron., 1969, 14, no.10, p.1757, and Radio Eng. Electron. Phys., 1969, 14, no.10, pp.1520-5.

12.97. Shevchenko, V.V.: "Graphic Classification of Modes Propagated in Uniform Open Waveguides", Radiotekh. Elektron., 1969, 14, no.10, p.1768, and Radio Eng. Electron. Phys., 1969, 14, no.10, pp.1530-3.

12.98. Marcatili, E.A.J.: "Dielectric Rectangular Waveguide and Directional Coupler for Integrated Optics", Bell Syst. Tech. J., 1969, 48, p.2071.

12.99. Goell, J.E.: "Circular-Harmonic Computer Analysis of Rectangular Dielectric Waveguides", Bell Syst. Tech. J., 1969, 48, p.2133.

12.100. Shein, A.G., and Molyavko, V.I.: "Study of Two-Dimensional, Cylindrical, Cellular-Type Slow-Wave Structure", Izv. VUZ Radioelektron., 1969, 12, no.7, p.731.

12.101. Bruckelmann, W.: "Wave Propagation in a Cylinder-Loaded Waveguide", Nachr. Tech. Z., 1969, 22, no.10, p.571.

12.102. Veselov, G.I., and Krekhtunov, V.M.: "Natural Oscillations in a Set of N Open Dielectric Waveguides", Radiotekh. Elektron., 1969, 14, no.8, p.1399, and Radio Eng. Electron. Phys., 1969, 14, no.8, pp.1211-7.

12.103. Ivanov, V.N.: "Correction for Open End of Interdigital Delay System", Radiotekh. Elektron., 1969, 14, no.9, p.1679.

12.104. Snyder, A.W.: "Asymptotic Expressions for Eigenfunctions and Eigenvalues of a Dielectric Waveguide", Trans. IEEE, 1969, MTT-17, p.1130.

12.105. Wehner, D.R.: "Tailored-Response Microwave Filter", Trans. IEEE, 1969, MTT-17, p.115.

12.106. Glaser, J.I.: "Attenuation and Guidance of Modes on Hollow Dielectric Waveguides", Trans. IEEE, 1969, MTT-17, p.173.

12.107. Wu, P.R.: "Dispersion Characteristics of an Open Interdigital-Line Structure", Trans. IEEE, 1969, MTT-17, p.159.

12.108. Snyder, A.W.: "Excitation and Scattering of Modes on a Dielectric Fibre", Trans. IEEE, 1969, MTT-17, p.1138.

12.109. Sanderson, R.L., and Streifer, W.: "Unstable Laser Resonator Modes", Appl. Opt., 1969, 8, p.2129.

12.110. Poigina, M.I., and Malivanchuk, V.I.: "Theoretical and Experimental Study of a Ring-Rod Structure", Izv. VUZ Radioelektron., 1969, 12, no.6, p.618.

12.111. Gallawa, R.L., and Partch, J.E.: "Use of Nonuniform Dielectrics on Surface-Wave Structures", Electron. Lett., 1969, 5, p.177.

12.112. Vainoris, Z.A., Matseika, K.Yu., and Shtaras, S.S.: "Wave Impedance of Plane Spiral Slow-Wave Structures", Izv. VUZ Radioelektron., 1969, 12, no.9, p.1103.

12.113. Sirohi, R.S.: "Light Transmission Along a Conical Fibre", Optik, 1969, 30, no.3, p.294.

12.114. Goncharenko, A.M., and Gusak, N.A.: "Theory of Dielectric Light Guides", Zh. Prikl. Spekt., 1969, 11, no.6, p.1045.

12.115. Hashimoto, T.: "Propagation Coefficient of a Helix with Attenuation", Electron. Commun. Jap., 1969, 52, no.11, pp.77-82.

12.116. Marcuse, D.: "Mode Conversion Caused by Surface Imperfections of a Dielectric Slab Waveguide", Bell Syst. Tech. J., 1969, 48, p.3187.

12.117. Yampol'skii, I.R.: "Investigation of (2/3)-pi Mode of EH_{11} Hybrid Wave in a Diaphragm Waveguide", Zh. Tekh. Fiz., 1969, 39, no.7, p.1199, and Sov. Phys.-Tech. Phys., 1970, 14, no.7, pp.900-2.

12.118. Braeckelmann, W.: "Wave Propagation in a Rectangular Waveguide with Periodic Branches", Nachr. Tech. Z., 1970, 23, p.210.

12.119. Koroza, V.I.: "Parametric Resonance in Propagation of Waves in Periodic Delay Lines", Radiotekh. Elektron., 1970, 15, p.450, and Radio Eng. Electron. Phys., 1970, 15, no.3, pp.387-91.

12.120. Yip, G.L.: "Launching of HE_{11} Surface-Wave Mode on a Dielectric Rod", Electron. Lett., 1970, 6, no.1, p.2.

12.121. Marcuse, D.: "Modes and Pseudomodes in Dielectric Waveguides", Trans. IEEE, 1970, MTT-18, p.62.

12.122. Tischer, F.J.: "H Guide with Laminated Dielectric Slab", Trans. IEEE, 1970, MTT-18, p.9.

12.123. Cermak, I.A., Silvester, N.J.P., and Wong, S.K.: "Capacitance Determination for Infinite Interdigital Structures", Trans. IEEE, 1970, MTT-18, p.115.

12.124. Midwinter, J.E.: "Evanescent Field Coupling into a Thin-Film Waveguide", J. Quantum Electron. IEEE, 1970, QE-6, p.583.

12.125. Gunston, M.A.R., and Blunden, D.F.: "Simplified Analysis of Corrugated Waveguide Structures", Marconi Rev., 1970, 33, p.260.

12.126. Tien, P.K., and Ulrich, R.: "Theory of Prism-Film Coupler and Thin-Film Light Guides", J. Opt. Soc. Am., 1970, 60, p.1325.

12.127. Ulrich, R.: "Theory of Prism-Film Coupler by Plane-Wave Analysis", J. Opt. Soc. Am., 1970, 60, p.1337.

12.128. Snyder, A.W.: "Asymptotic Expressions for Dielectric Waveguides", J. Opt. Soc. Am., 1970, 60, p.139.

12.129. Sawatari, T., and Kapany, N.S.: "Wave Propagation along Hollow Dielectric Waveguides", J. Opt. Soc. Am., 1970, 60, p.132.

12.130. Midwinter, J.E., and Zernike, F.: "Experimental Studies of Evanescent Mode Coupling into Thin-Film Waveguide", Appl. Phys. Lett., 1970, 16, p.198.

12.131. Orleanskaya, E.V.: "Application of the Methods of Circuit Theory to Problems of Waveguides Partly Filled with Dielectric", Radiotekhnika, 1970, 25, no.3, p.37, and Telecommun. Radio Eng. Pt 2, 1970, 25, no.3.

12.132. Snyder, A.W.: "Coupling of Modes on a Tapered Dielectric Cylinder", Trans. IEEE, 1970, MTT-18, p.383.

12.133. Kevanishvili, G.Sh., and Kekeliya, V.L.: "Wave Delay in Open Waveguides", Radiotekh. Elektron., 1970, 15, p.173, and Radio Eng. Electron. Phys., 1970, 15, no.1, pp.143-5.

12.134. Mayer, N.A.: "Phenomenological Investigation of Large-Period Two-Wave Guides", Izv. VUZ Radiofiz., 1970, 13, p.133.

12.135. Snurnikova, G.K.: "Wave Propagation in a Trough Waveguide", Radiotekh. Elektron., 1970, 15, p.598, and Radio Eng. Electron. Phys., 1970, 15, no.3, pp.509-11.

12.136. Hamid, M.A.K., and Johnson, W.A.: "Ray-Optical Solution for Dyadic Green's Function in a Rectangular Cavity", Electron. Lett., 1970, 6, p.317.

12.137. Kazantsev, Yu.N.: "Waves in Dielectric Channels of Rectangular Cross Section", Radiotekh. Elektron., 1970, 15, no.6, p.1140, and Radio Eng. Electron. Phys., 1970, 15, no.6, pp.963-8.

12.138. Snyder, A.W.: "Mode Propagation in Optical Waveguides", Electron. Lett., 1970, 6, p.561.

12.139. Haidle, L.L.: "Power Transfer between Goubau Lines with Unequal Phase Velocities", Electron. Lett., 1970, 6, p.599.

12.140. Maurer, S.J., and Felsen, L.B.: "Ray Methods for Trapped and Slightly Leaky Modes in Multilayered or Multiwave Regions", Trans. IEEE, 1970, MTT-18, p.584.

12.141. Snyder, A.W., and De La Rue, R.: "Asymptotic Solution of Eigenvalue Equation for Surface Waveguide Structures", Trans. IEEE, 1970, MTT-18, p.650.

12.142. Mohsen, A., et al.: "Field Distribution in Multilayered Dielectric-Loaded Rectangular Waveguides", Proc. IEE, 1970, 117, p.709.

12.143. Vanclooster, R., and Phariseau, P.: "Coupling of Two Parallel Dielectric Fibres. I and II", Physica, 1970, 47, pp.485 and 501.

12.144. Dakss, M.L., et al.: "Grating Coupler for Efficient Excitation of Optical Guided Waves in Thin Films", Appl. Phys. Lett., 1970, 16, p.523.

12.145. Hall, D., Yariv, A., and Garmire, E.: "Observation of Propagation Cutoff and its Control in Thin Optical Waveguides", Appl. Phys. Lett., 1970, 17, p.127.

12.146. Ruiz-Aguirre, R.D., and Deus, L.F.P.: "Telegraphist's Equation for a Rectangular Helix Waveguide", Electron. Fis. Apl., 1970, 13, no.1, p.9.

12.147. Grigor'ev, A.D., and Petrov, V.: "Delay System Consisting of Cavities with External Capacitive Coupling", Izv. VUZ Radioelektron., 1970, 13, no.8, pp.1000-7.

12.148. Clarricoats, P.J.B., and Chan, K.B.: "Excitation and Propagation of Modes of a Multilayer Fibre", Electron. Lett., 1970, 6, no.23, pp.750-2.

12.149. Burke, J.J.: "Propagation Coefficients of Resonant Waves on Homogeneous Isotropic Slab Waveguides", Appl. Opt., 1970, 9, no.11, pp.2444-52.

12.150. Clarricoats, P.J.B., and Chan, K.B.: "Wave Propagation along Radially Inhomogeneous Dielectric Cylinders", Electron. Lett., 1970, 6, no.22, pp.694-5.

12.151. Al-Hakkak, M.J., and Lo, Y.T.: "Circular Waveguides with Anisotropic Walls", Electron. Lett., 1970, 6, no.24, pp.786-9.

12.152. Inamura, M.: "Effect of Curve in a Dielectric Circular Rod Waveguide", Electron. Commun. Jap., 1970, 53, no.4, p.35.

12.153. Shevchenko, V.V.: "Behaviour of Wave Numbers in Dielectric Waveguide beyond Cutoff", Izv. VUZ Radiofiz., 1970, 13, no.10, p.1528-31.

12.154. Crouzet, J.: "Experimental Study of Propagation on Helicoidal Conductors", Part. Accel., 1970, 1, no.4, pp.335-45.

12.155. Marcatili, E.A.J.: "Dielectric Guide with Curved Axis and Truncated Parabolic Index", Bell Syst. Tech. J., 1970, 49, no.8, pp.1645-63.

12.156. Marcuse, D.: "Radiation Losses of Dominant Mode in Round Dielectric Waveguides", Bell Syst. Tech. J., 1970, 49, no.8, pp.1665-93.

12.157. Marcuse, D.: "Excitation of Dominant Mode of a Round Fibre by a Gaussian Beam", Bell Syst. Tech. J., 1970, 49, no.8, pp.1695-703.

12.158. Wait, J.R.: "Launching an Azimuthal Surface Wave on a Cylindrical Impedance Boundary", Acta Phys. Austriaca, 1970, 32, no.2, pp.122-30.

12.159. Chakrabarti, N.B.: "TW Components in a Nonuniform Transmission Line", Electron. Lett., 1970, 6, no.21, pp.674-6.

12.160. Yanovskii, M.S., and Knyaz'kov, B.N.: "Polarized Quasioptical Phaseshifter", Izv. VUZ Radioelektron., 1970, 13, no.10, pp.1199-1204.

12.161. Mal'tsev, V.P., et al.: "Surface-Wave Beating in Coupled Film-Type Guides", Izv. VUZ Radioelektron., 1970, 13, no.11, pp.1381-3.

12.162. Yip, G.L.: "Launching Efficiency of HE_{11} Surface-Wave Mode on a Dielectric Rod", Trans. IEEE, 1970, MTT-18, no.12, pp.1033-41.

12.163. Deryugin, L.N., et al.: "Resonant Excitation of a Plane Dielectric Waveguide through a Below-Cutoff Layer", Izv. VUZ Radioelektron., 1970, 13, no.8, pp.973-80.

12.164. Koroza, V.I., Tragov, A.G., and Shankin, Yu.P.: "Numerical Design of Periodic Slow-Wave Structures", Radiotekh. Elektron., 1970, 15, no.10, and Radio Eng. Electron. Phys., 1970, 15, no.10,pp.1910-3.

12.165. Tikhomirov, A.A., and Kovalenko, Ye.S.: "Open Dielectric Waveguides and Cavities with Anisotropic Permittivity in the Cross Section", Radiotekh. Elektron., 1970, 15, no.12, pp.2327-30, and Radio Eng. Electron. Phys., 1970, 15, no.12.

12.166. Mirovitskii, D.I., Dubrovin, V.F., and Baskakov, V.V.: "Ring-Hybrid Couplers for Dielectric Waveguides", Radiotekh. Elektron., 1970, 15, no.12, pp.2309-11, and Radio Eng. Electron. Phys., 1970, 15, no.12.

12.167. Bhartia, P., and Hamid, M.A.K.: "Eigenvalues for a Spherical Cavity with an Impedance Wall", Trans. IEEE, 1971, MTT-19, no.1, pp.110-1.

12.168. Dybdal, R.B., Peters, L., and Peake, W.H.: "Rectangular Waveguides with Impedance Walls", Trans. IEEE, 1971, MTT-19, no.1, p.2-9.

12.169. Snyder, A.W.: "Approximate Eigenvalues for a Circular Rod of Arbitrary Relative Permittivity", Electron. Lett., 1971, 7, no.4, pp.105-6.

12.170. Cooper, D.N.: "Complex Propagation Coefficients and Step Discontinuity in Corrugated Cylindrical Waveguide", Electron. Lett., 1971, 7, no.5-6, pp.135-6.

12.171. Cooper, D.N.: "Orthogonality Relationship for Class of Waveguide with Anisotropic Walls", Electron. Lett., 1971, 7, no.5-6, p.137.

12.172. Harris, J.H., and Shubert, R.: "Variable Tunnelling Excitation of Optical Surface Waves", Trans. IEEE, 1971, MTT-19, no.3, pp.269-76.

12.173. Lewis, L.R., and Hessel, A.: "Propagation Characteristics of Periodic Arrays of Dielectric Slabs", Trans. IEEE, 1971, MTT-19, no.3, pp.276-86.

12.174. Krohne, M.: "Circular Waveguide Periodically Loaded with Coaxial Dielectric Cylinders", Nachr. Tech. Z., 1971, 24, no.1, pp.14-7.

12.175. Marcuse, D., and Marcatili, E.A.J.: "Excitation of Waveguides for Integrated Optics with Laser Beams", Bell Syst. Tech. J., 1971, 50, no.1, pp.43-57.

12.176. Anikin, V.I., Deryugin, L.N., and Sotin, V.Ye.: "Resonant Excitation of a Plane Dielectric Optical Waveguide by a Limited Beam via a Below-Cutoff Filter", Izv. VUZ Radioelektron., 1971, 14, no.4, pp.371-8.

12.177. Tischer, F.J.: "Fence Guide for Millimetre Waves", Proc. IEEE, 1971, 59, no.7, pp.1112-3.

12.178. Snyder, A.W.: "Continuous Mode Spectrum of a Circular Dielectric Rod", Trans. IEEE, 1971, MTT-19, no.8, pp.720-7.

12.179. Williams, C.G., and Campbell, G.K.: "Computation of Surface-Waveguide Modes by Use of Reactance Boundary Conditions", Electron. Lett., 1971, 7, no.12, pp.323-4.

12.180. Midwinter, J.E.: "Theory of an Ultra-Broadband Optical Dielectric Waveguide-Coupler System", J. Quantum Electron. IEEE, 1971, QE-7, no.7, pp.345-50.

12.181. Smirnov, V.P., and Shapiro, G.Ya.: "Moderately Efficient Method of Launching Surface Waves", Izv. VUZ Radioelektron., 1971, 14, no.5, pp.500-7.

12.182. Hangen, L.: "Condition for BW Propagation in a Circular Waveguide Containing a Dielectric Rod", Nachr. Tech. Z., 1971, 24, no.7, pp.361-3.

12.183. Cullen, A.L., Ozkan, O., and Jackson, L.A.: "Point-Matching Technique for Rectangular-Section Dielectric Rod", Electron. Lett., 1971, 7, no.17, pp.497-9.

12.184. Witte, H.H., and Khan, R.H.: "Novel Light Coupling Method for Fibres", Rev. Sci. Instrum., 1971, 42, no.9,pp.1374-5.

12.185. Pennington, K.S., and Kuhn, L.: "Bragg Diffraction Beamsplitter for Thin Film Optical Guided Waves", Opt. Commun., 1971, 3, no.5, pp.357-9.

12.186. Bates, R.H.T.: "Contributions to the Theory of the Azimuthal Surface Wave", Alta Freq., 1971, 40, no.8, pp.658-66.

12.187. Gillespie, E.S.: "Impedance and Scattering Properties of a Plane Annulus Surrounding a Goubau Line", Trans. IEEE, 1971, MTT-19, no.10, pp.837-9.

12.188. Knorr, J.B., and McIsaac, P.R.: "Group Theoretic Investigation of the Single-Wire Helix", Trans. IEEE, 1971, MTT-19, no.11, pp.854-61.

12.189. Lewin, L.: "Interpretation of (Stub-Loaded) Slimguide Operation in Terms of Propagating Waves", Electron. Lett., 1971, 7, no.18, pp.555-6.

12.190. Koroza, V.I., and Mokhamed, M.O.: "Analysis of Azimuthal Nonsymmetric Periodic Waveguides", Vestn. Mosk. Univ., Mat. Mekh., 1971, no.3, pp.117-23.

12.191. Takeichi, Y., Hashimoto, T., and Takeda, F.: "Ring-Loaded Corrugated Waveguide", Trans. IEEE, 1971, MTT-19, no.12, pp.947-50.

12.192. Cullen, A.L.: "Calculation of Attenuation of Dielectric-Rod Waveguides", Electron. Lett., 1971, 7, no.24, pp.707-8.

12.193. Zyatitskii, V.A., and Kazantsev, Yu.N.: "Excitation of Waveguides of Dielectric-Channel Type", Izv. VUZ Radiofiz., 1971, 14, no.10, pp.1570-3.

12.194. Andrushko, L.M., and Markov, S.Ye.: "Synthesis of Two-Stage Slow-Wave Resonator Structures", Izv. VUZ Radioelektron., 1971, 14, no.9, pp.1099-1104.

12.195. Baldwin, R.: "Attenuation in Corrugated Rectangular Waveguide", Electron. Lett., 1971, 7, no.26, pp.770-2.

12.196. Pereverzev, S.I.: "Complex Waves in Goubau's Line", Izv. VUZ Radiofiz., 1971, 14, no.12, pp.1864-8.

12.197. Veselov, G.I., and Voronina, G.G.: "Calculation of Open Dielectric Waveguide of Rectangular Section", Izv. VUZ Radiofiz., 1971, 14, no.12, pp.1891-901.

12.198. Snurnikova, G.K.: "Propagation in a Grooved Guide with Correction for the Open Side", Radiotekh. Elektron., 1971, 16, no.2, pp.411-9, and Radio Eng. Electron. Phys., 1971, 16, no.2, pp.352-4.

12.199. Mal'tsev, V.P., Mironov, V.L., and Shevchenko, V.V.: "Propagation on Surface Waveguides of Finite Width", Radiotekh. Elektron., 1971, 16, no.3, and Radio Eng. Electron. Phys., 1971, 16, no.3, pp.540-3.

12.200. Marcuse, D.: "Bending Losses of Asymmetric Slab Waveguide", Bell. Syst. Tech. J., 1971, 50, no.8, pp.2551-63.

12.201. Gordienko, V.I.: "Distribution of Longitudinal Energy Flux of Surface TM and TE Modes between Media", Otbor Peredacha Inf., 1971, no.27, pp.45-51.

12.202. Arora, R.K., and Vijayaraghavan, S.: "Excitation of Shielded Surface Wave in Parallel-Plate Waveguide with Reactive Walls", J. Inst. Eng., 1971, 52, no.3, pt ET2, pp.80-4.

12.203. Ikuta, K., and Aoki, K.: "Surface-Wave Launching on a Circular Dielectric Waveguide", Technol. Rep. Kyushu Univ., 1971, 44, no.6, pp.786-91.

12.204. Kuchikyah, L.M.: "Passage of Polarized Light through a Rectangular Lightpipe", Opt.-Mekh. Prom., 1971, 38, no.7, pp.13-6, and Sov. J. Opt. Technol., 1971, 38, no.7, pp.399-402.

12.205. Girija, H.M., and Chatterjee, S.K.: "Surface-Wave Characteristics of a Cylindrical Metallic Corrugated Structure Excited in E_0 Mode", J. Indian Inst. Sci., 1971, 53, no.4, pp.269-326.

12.206. Prabhavathi, A.S., and Chatterjee, S.K.: "Theory of Open Microwave Resonator with an Axial Corrugated Metal Rod", J. Indian Inst. Sci., 1971, 53, no.4, pp.333-54.

12.207. Mikhalev, L.A.: "Waves in a Coaxial Helical Line Containing an Inhomogeneous Dielectric", Radiotekh. Elektron., 1971, 16, no.8, pp.1486-9, and Radio Eng. Electron. Phys., 1971, 16, no.8, pp.1376-9.

12.208. Koroza, V.I., Tragov, A.G., and Shankin, Yu.P.: "Calculation of High-Frequency Characteristics of Periodic Waveguides of Complex Shape", Radiotekh. Elektron., 1971, 16, no.10, pp.1788-96, and Radio Eng. Electron. Phys., 1971, 16, no.10, pp.1612-8.

12.209. Avdeev, Ye.V., and Chegis, I.L.: "Calculation of Segments of a Meander Stripline with an Inhomogeneous Dielectric", Radiotekh. Elektron., 1971, 16, no.10, pp.1808-15, and Radio Eng. Electron. Phys., 1971, 16, no.10, pp.1628-34.

12.210. Al-Hakkak, M.J.: "Dielectric Loading of Corrugated Waveguides", Electron. Lett., 1972, 8, no.7, pp.179-80.

12.211. Shevchenko, V.V.: "Dispersion of Dielectric Waveguides Below Cutoff with Lossy Media", Izv. VUZ Radiofiz., 1972, 15, no.2, pp.257-65.

12.212. James, J.R., and Gallett, I.N.L.: "Point-Matched Solutions for Propagating Modes on Arbitrarily Shaped Dielectric Rods", Radio Electron. Eng., 1972, 42, no.3, pp.103-13.

12.213. Ihaya, A., Furuta, H., and Noda, H.: "Thin-Film Optical Directional Coupler", Proc. IEEE, 1972, 60, no.4, pp.470-1.

12.214. Clarricoats, P.J.B., and Sharpe, A.B.: "Modal Matching Applied to a Discontinuity in a Planar Surface Waveguide", Electron. Lett., 1972, 8, no.2, pp.28-9.

12.215. Arora, R.K., et al.: "Modes of Propagation in a Coaxial Waveguide with Lossless Reactive Guiding Surfaces", Trans. IEEE, 1972, MTT-20, no.3, pp.210-4.

12.216. Hockham, G.A., and Sharpe, A.B.: "Dielectric Waveguide Discontinuities", Electron. Lett., 1972, 8, no.9, pp.230-1.

12.217. Takano, T., and Hamasaki, J.: "Propagating Modes of a Metal-Clad-Dielectric Slab Waveguide for Integrated Optics", J. Quantum Electron. IEEE, 1972, QE-8, no.2, pp.206-12.

12.218. Mahan, A.I., Bitterli, C.V., and Unger, H.J.: "Reflection and Transmission Properties of Cylindrically Guided (Hybrid) Waves", J. Opt. Soc. Am., 1972, 62, no.3, pp.361-8.

12.219. Garault, Y., and Jecko, B.: "Open Terminations for Periodic Structures", Electron. Lett., 1972, 8, no.10, pp.270-1.

12.220. Falciasecca, G., and Prandi, G.M.: "Attenuation in Wall-Impedance Waveguides", Electron. Lett., 1972, 8, no.10, pp.273-4.

12.221. Chang, C.T.M.: "Circular Waveguides Lined with Artificial Anisotropic Dielectrics", Trans. IEEE, 1972, MTT-20, no.8, pp.517-23.

12.222. McRitchie, W.K., and Beal, J.C.: "Yagi-Uda Array as Surface-Wave Launcher for Dielectric Image Lines", Trans. IEEE, 1972, MTT-20, no.8, pp.493-6.

12.223. Chu, R.S., and Tamir, T.: "Wave Propagation and Dispersion in Space-Time Periodic Media", Proc. IEE, 1972, 119, no.7, pp.797-806.

12.224. Hill, K.O., Watanabe, A., and Chambers, J.G.: "Evanescent-Wave Interactions in an Optical Waveguiding Structure", Appl. Opt., 1972, 11, no.9, pp.1952-9.

12.225. Lapta, S.I.: "Excitation of a Ring Helical Conducting Waveguide by a Dipole", Radiotekh. (Kharkov), 1972, no.20, pp.58-70.

12.226. Tret'yakov, O.A., and Shmat'ko, A.A.: "Study of Resonators with Diffraction Arrays Employing Eigenmodes of Periodic Structures", Radiotekh. (Kharkov), 1972, no.20, pp.131-41.

12.227. Prandi, G.M., and Someda, C.G.: "Variation-Iteration Technique for Design of Wall-Impedance Waveguides", Alta Freq., 1972, 41, no.6, pp.439-42.

12.228. Thomas, B.M.: "Mode Conversion Using Circumferentially Corrugated Cylindrical Waveguide", Electron. Lett., 1972, 8, no.15, pp.394-6.

12.229. Snurnikova, G.K.: "Irregular Grooved Waveguide with Asymmetric Section", Radiotekh. (Kharkov), 1972, no.20,pp.178-85.

12.230. Ferdinandorff, E.: "Dielectric Waveguide with Elliptical Section", Nachr. Tech., 1972, 22, no.9, pp.302-4.

12.231. Heyke, H.J.: "Mode Excitation and Scattering in Fibre Launchers", Arch. Elektron. Ubertrag., 1972, 26, no.10, pp.456-8.

12.232. Kazantsev, Yu.N., and Udalov, V.V.: "(Dielectric-) Pipe Diaphragm Waveguide", Izv. VUZ Radiofiz., 1972, 15, no.10, pp.1561-6.

12.233. Snyder, A.W.: "Coupled-Mode Theory for Optical Fibres", J. Opt. Soc. Am., 1972, 62, no.11, pp.1267-77.

12.234. Gambling, W.A., Payne, D.N., and Matsumura, H.: "Dispersion in Lowloss Liquid-Core Optical Fibres", Electron. Lett., 1972, 8, no.23, pp.568-9.

12.235. Ogawa, K., et al.: "Theoretical Analysis of Etched Grating Couplers for Integrated Optics", J. Quantum Electron. IEEE, 1972, QE-9, no.1, pp.29-42.

12.236. Sangster, A.J.: "Higher-Order Evanescent Modes on Slow-Wave Structures", Electron. Lett., 1972, 8, no.25,pp.208-9.

12.237. Shankin, Yu.P.: "Dispersion Characteristics of Periodic Delay Systems", Radiotekh. Elektron., 1972, 17, no.3, pp.441-7, and Radio Eng. Electron. Phys., 1972, 17, no.3, pp.341-6.

12.238. Koroza, V.I.: "Dispersion Characteristics of Periodic Delay Systems by Perturbation Method", Radiotekh. Elektron., 1972, 17, no.2, pp.225-33, and Radio Eng. Electron. Phys., 1972, 17, no.2, pp.171-6.

12.239. Didenko, A.N., and Kaminskaya, R.G.: "Dispersion Characteristics of Elliptical Iris Waveguides", Radiotekh. Elektron., 1972, 17, no.2, pp.399-401, and Radio Eng. Electron. Phys., 1972, 17, no.2, pp.306-9.

12.240. Kukhmin, M.P.: "Attenuation in Comb Waveguides", Radiotekh. (Kharkov), 1972, no.22, pp.9-15.

12.241. Kukhmin, M.P.: "Critical Wavelength in Comb Waveguides", Radiotekh. (Kharkov), 1972, no.22, pp.16-9.

12.242. Dabby, F.W., Kestenbaum, A., and Paek, U.C.: "Periodic Dielectric Waveguides", Opt. Commun., 1972, 6, no.2, pp.125-30.

12.243. Labh, B.K., and Sharma, K.P.: "Effect of Metallic Discs on a Circular Rod on Transmission of Surface Waves", Indian J. Pure Appl. Phys., 1972, 10, no.5, pp.343-5.

12.244. Shatrov, A.D.: "Expansion of Fields in Open Waveguides and Resonators", Radiotekh. Elektron., 1972, 17, no.6, pp.1153-60, and Radio Eng. Electron. Phys., 1972, 17, no.6, pp.896-902.

12.245. Dzygan, V.P., Smarosmenko, V.V., and Shein, A.G.: "Investigation of a Limited Comb in a Waveguide", Radiotekh. (Kharkov), 1972, no.22, pp.19-25.

12.246. Kunz, H.: "Launching of Goubau Waves by Imperfectly Shielded Coaxial Cables", Mitt. AGEN, 1972, no.14, pp.3-38.

12.247. Raevskii, S.B., and Storgonsky, V.Ya.: "Dispersion Equation of a Corrugated Elliptical Waveguide", Radiotekh. Elektron., 1972, 17, no.6, pp.1297-9, and Radio Eng. Electron. Phys., 1972, 17, no.6, pp.1006-9.

12.248. Loshakov, L.N.: "Calculation of Parameters of Screened Helical Line with Dielectric Supports", Radiotekhnika, 1972, 27, no.8, pp.32-9, and Telecommun. Radio Eng. Pt 2, 1972, 27, no.8, pp.79-84.

12.249. Alekseychik, L.V., Gevorkyan, V.M., and Kazantsev, Yu.A.: "Excitation of Open Dielectric Resonator by a Transmission Line", Radiotekh. Elektron., 1972, 17, no.11, pp.2261-70, and Radio Eng. Electron. Phys., 1972, 17, no.11, pp.1814-21. See also Miyazaki, Y.: "Electromagnetic Theory of Open Resonators", J. Phys. Soc. Jap., 1972, 32, no.3, pp.837-44.

12.250. Babichev, R.K.: "Fast Waves in a System of Rectangular and Plane Waveguides Connected by an Array of Strips", Radiotekh. Elektron., 1972, 17, no.11, pp.2440-4, and Radio Eng. Electron. Phys., 1972, 17, no.11, pp.1961-5.

12.251. Zubovskii, V.P.: "Microwave Structure with Parallel Transverse Rods", Zh. Tekh. Fiz., 1972, 42, no.12, pp.2611-2, and Sov. Phys.-Tech. Phys., 1973, 17, no.12, pp.2029-30.

12.252. Cheo, P.K., et al.: "Optical Waveguide Structures for CO_2 Lasers", Appl. Opt., 1973, 12, no.3, pp.500-9.

12.253. Dabby, F.W., Saifi, M.A., and Kestenbaum, A.: "High-Frequency Cutoff in Periodic Dielectric Waveguides", Appl. Phys. Lett., 1973, 22, no.4, pp.190-1.

12.254. Kersten, R.Th.: "Experiments Concerning Light Propagation in Dielectric Tapered Waveguides", Arch. Elektron. Ubertrag., 1973, 27, no.3, pp.121-4.

12.255. Russo, D.P.G., and Harris, J.H.: "Wave Propagation in Anisotropic Thin-Film Optical Waveguide", J. Opt. Soc. Am., 1973, 63, no.2, pp.138-45.

12.256. Elachi, C.: "Frequency-Selective Coupler for Integrated Optics Systems", Opt. Commun., 1973, 7, no.3, pp.201-4.

12.257. Degnan, J.J.: "Waveguide Laser Mode Patterns in the Near and Far Fields", Appl. Opt., 1973, 12, no.5, pp.1026-30.

12.258. Sohler, W.: "Light-Wave Coupling to Optical Waveguides by a Tapered Cladding Medium", J. Appl. Phys., 1973, 44, no.5, pp.2343-5.

12.259. Heyke, H.J., and Kuhn, M.H.: "Dispersion Characteristics of General Gradient Fibres", Arch. Elektron. Ubertrag., 1973, 27, no.5, pp.235-8.

12.260. Noble, D.F., and Carlin, H.J.: "Circuit Properties of Coupled Dispersive Transmission Lines", Trans. IEEE, 1973, CT-20, no.1, pp.56-64.

12.261. Hessel, A., et al.: "Propagation in Periodically Loaded Waveguides with Higher Symmetries", Proc. IEEE, 1973, 61, no.2, pp.183-95.

12.262. Sukhovskii, E.S.: "Approximate Computation of Waves in a Periodic Waveguide", Radiotekh. Elektron., 1973, 17, no.2, pp.234-9, and Radio Eng. Electron. Phys., 1973, 17, no.2, pp.176-83.

12.263. Cristal, E.G.: "Meander-Line Transformers", Trans. IEEE, 1973, MTT-21, no.2, pp.69-76.

12.264. Ulrich, R., and Tacke, M.: "Submillimetre Waveguiding on Periodic Metal Structure", Appl. Phys. Lett., 1973, 22, no.5, pp.251-3.

12.265. El-Sherbiny, A.M., and Filippov, M.M.: "Dispersion Properties of Waveguides with Small Periodic Discontinuities. I and II", Vestn. Leningr. Univ. Fiz. Khim., 1973, no.3, pp.69-78 and no.4, pp.53-66.

12.266. Matsuhara, M.: "Analysis of TEM Modes in Dielectric Waveguides by Variational Method", J. Opt. Soc. Am., 1973, 63, no.12, pp.1514-7.

12.267. Stoll, H., and Yariv, A.: "Coupled-Mode Analysis of Periodic Dielectric Waveguides", Opt. Commun., 1973, 8, no.1, pp.5-8.

12.268. Neviere, M., Petit, R., and Cadilhac, M.: "Theory of Optical Grating Coupler-Waveguide Systems", Opt. Commun., 1973, 8, no.2, pp.113-7.

12.269. Saddler, N.B., and Staniforth, J.A.: "Excitation of Dielectric-Loaded Trough Waveguide by an Aperture", Radio Electron. Eng., 1973, 43, no.6, pp.385-8.

12.270. Neumann, E.G.: "Wave Propagation between Two Plane Parallel Reactive Walls", Arch. Elektron. Ubertrag., 1973, 27, no.7-8, pp.343-7.

12.271. Gallagher, W.J.: "Properties of Disc-Loaded Lines", Trans. IEEE, 1973, NS-20, no.3, pp.952-6.

12.272. Ihaya, A., Furuta, H., and Noda, H.: "Directional Coupling between Thin-Film Optical Guides", Fujitsu Sci. Tech. J., 1973, 9, no.2, pp.101-9.

12.273. Elachi, C., and Yeh, C.: "Periodic Structures in Integrated Optics", J. Appl. Phys., 1973, 44, no.7, pp.3146-52.

12.274. Laybourn, P.J.R., and Gambling, W.A.: "Bandwidths of Single- and Multi-Mode Optical Fibres", Opt. Commun., 1973, 8, no.3, pp.195-200.

12.275. Minakovic, B., and Gokgor, S.: "Attenuation and Phase-Shift Coefficients in Dielectric-Loaded Periodic Waveguides", Trans. IEEE, 1973, MTT-21, no.8, p.568.

12.276. Doswell, A., and Harris, D.J.: "Modified H-Guide for Submillimetre Wavelengths", Trans. IEEE, 1973, MTT-21, no.9, pp.587-9.

12.277. Tien, P.K., Martin, R.J., and Smolinsky, G.: "Formation of Light-Guiding Interconnections in an Integrated Optical Circuit by Composite Tapered-Film Coupling", Appl. Opt., 1973, 12, no.8, pp.1909-16.

12.278. Yariv, A.: "Coupled-Mode Theory for Guided-Wave Optics", J. Quantum Electron. IEEE, 1973, QE-9, no.9, pp.919-33.

12.279. Conwell, E.M.: "Modes in Optical Waveguides Formed by Diffusion", Appl. Phys. Lett., 1973, 23, no.6, pp.328-9.

12.280. Taylor, H.F.: "Frequency-Selective Coupling in Parallel Dielectric Waveguides", Opt. Commun., 1973, 8, no.4, pp.421-5.

12.281. Sova, A.V., Starostenko, V.V., and Shein, A.G.: "Amplitude Spectrum Calculation of Spatial Harmonics of Plane Comb-Shaped Structures", Radiotekh. (Kharkov), 1973, no.24, pp.113-8.

12.282. Belov, Yu.G., et al.: "Approximate Calculation of Cutoff Frequencies of a Corrugated Elliptical Waveguide", Izv. VUZ Radioelektron., 1973, 16, no.8, pp.44-7.

12.283. Chandezon, J., Cornet, G., and Raoult, G.: "Wave Propagation in Cylindrical Guides with Sinusoidal Generatrix", C. R. Acad. Sci. (Paris), 1973, 277B, no.14, pp.355-8, and no.15, pp.403-5.

12.284. Clarricoats, P.J.B., and Oliver, A.D.: "Low Attenuation in Corrugated Circular Waveguides", Electron. Lett., 1973, 9, no.16, pp.376-7.

12.285. Il'inskii, A.S., Al'hovskii, E.A., and Danilova, A.G.: "Method of Calculating Propagation Coefficients in Corrugated Waveguides", Izv. VUZ Radiofiz., 1973, 16, no.10, pp.1583-7.

12.286. Belanov, A.S., Ezov, G.I., and Tschernij, W.W.: "Wave Propagation in Circular Layered Dielectric Rod Transmission Lines", Arch. Elektron. Ubertrag., 1973, 27, no.11, pp.494-6.

12.287. Shankin, Yu.P.: "Theory of Periodic Waveguides Loaded by Diaphragms", Izv. VUZ Radiofiz., 1973, 16, no.11, pp.1730-5.

12.288. Il'inskii, A.S., Repin, V.M., and Starostin, L.I.: "Numerical Solution of the Two-Dimensional Problem of Propagation of H-Polarized Waves in an Aperture-Coupled System of Cavities", Vychisl. Metody Program., 1973, no.20, pp.38-49.

12.289. Clarricoats, P.J.B.: "Propagation Behaviour of Cylindrical-Dielectric-Rod Waveguides", Proc. IEE,1973,120,no.11,pp.1371-8.

12.290. Gambling, W.A., Payne, D.N., and Matsumura, H.: "Mode Excitation in a Multimode Optical-Fibre Waveguide", Electron. Lett., 1973, 9, no.18, pp.412-4.

12.291. James, J.R.: "Modal Analysis of Triangular-Cored Glass-Fibre Waveguide", Proc. IEE, 1973, 120, no.11, pp.1362-70.

12.292. Raicu, D.: "TW Resonance in Cavities", Arch. Elektron. Ubertrag., 1973, 27, no.12, pp.528-34.

12.293. Chiba, J., Sato, R., and Inouye, T.: "Fields and Phase Coefficients of Waves on a Shielded Wire Helix", Electr. Eng. Jap., 1973, 93, no.1, pp.7-13.

12.294. Shankara, K.N., and Chatterjee, S.K.: "Surface Wave Characteristics of Circular Cylindrical, Corrugated and Uniform, Dielectric Rods Excited in E_0 Mode", J. Indian Inst. Sci., 1973, 54, no.3,pp.118-38.

12.295. Ulrich, R.: "Efficiency of Optical Grating Couplers", J. Opt. Soc. Am., 1973, 63, no.11, pp.1419-31.

12.296. Shankin, Yu.P.: "Dispersion in Higher Pass and Stop Bands of Periodic Waveguides", Radiotekh. Elektron., 1973, 18, no.1, pp.68-73, and Radio Eng. Electron. Phys., 1973, 18, no.1, pp.49-53.

12.297. Marcuse, D.: "Cutoff Condition of Optical Fibres", J. Opt. Soc. Am., 1973, 63, no.11, pp.1369-71.

12.298. Marcatili, E.A.J.: "Extrapolation of Snell's Law to Optical Fibres", J. Opt. Soc. Am., 1973, 63, no.11, pp.1372-3.

12.299. Fox, A.J.: "Plane-Wave Theory for Optical Grating Waveguide", Philips Res. Rep., 1973, 28, no.4, pp.306-46.

12.300. McIntyre, P.D., and Snyder, A.W.: "Power Transfer between Optical Fibres", J. Opt. Soc. Am., 1973, 63, no.12, pp.1518-27.

12.301. Kukhtin, M.P., and Degtyar, S.V.: "Parameters of Round Corrugated Waveguide", Radiotekh. (Kharkov), 1973, no.25, pp.113-9.

12.302. Tuan, H.S., and Ou, C.H.: "Scattering of TM Surface Wave at a Guide Deformation", J. Appl. Phys., 1973, 44, no.12, pp.5522-5.

12.303. Watts, R.K.: "Evanescent Field Coupling of Thin-Film Laser and Passive Waveguide", J. Appl. Phys., 1973, 44, no.12, pp.5635-6.

12.304. Voitovich, N.N., and Shatrov, A.D.: "Excitation of an Open Waveguide with Dielectric Walls", Radiotekh. Elektron., 1973, 18, no.4, pp.687-94, and Radio Eng. Electron. Phys., 1973, 18, no.4, pp.497-503.

12.305. Yerokhin, G.A., and Kocharzhevskii, V.G.: "External Bend of an Open Waveguide with Low Radiation Losses", Radiotekh. Elektron., 1973, 18, no.4, pp.695-702, and Radio Eng. Electron. Phys., 1973, 18, no.4.

12.306. Tsuchiya, H., and Mushiake, Y.: "Characteristics of Optical Horn for Excitation of Optical Film", Electron. Commun. Jap., 1973, 56, no.4, pp.87-92.

12.307. Suematsu, Y., and Furuya, K.: "Characteristic Modes and Scattering Loss of Asymmetric Slab Optical Waveguides", Electron. Commun. Jap., 1973, 56, no.5, pp.101-6.

12.308. Mallick, A.K.: "Determination of Omega-Beta Diagram of an Inductively Loaded Periodic Structure", J. Inst. Electron. Telecommun. Eng., 1973, 19, no.11, pp.617-22.

12.309. Kukhtin, M.P.: "Influence of Geometrical Factors on Parameters of Elliptical Corrugated Waveguides", Radiotekh. (Kharkov), 1973, no.27, pp.103-10.

12.310. Dyadyuk, V.J., and Zorkin, A.F.: "Design of Meander Nonsymmetric Stripline", Radiotekh. (Kharkov),1973,no.27,pp.95-103.

12.311. Takano, T., and Hamasaki, J.: "Approximate Solutions of Propagating Modes of a Metal-Clad/ Dielectric-Slab Optical Waveguide", Electron. Commun. Jap., 1973, 56, no.7, pp.68-75.

12.312. Williams, C.G., and Cambrell, G.K.: "Numerical Solution of Surface Wave-guide Modes Using Transverse Field Components", Trans. IEEE, 1974, MTT-22, no.3, pp.329-30.

12.313. McAulay, A.D., and Charap, S.H.: "Numerical Method for Investigating Propagation of Surface Waves on Dissipative Guides", Proc. IEEE, 1974, 62, no.3, pp.402-3.

12.314. Arnaud, J.A.: "Transverse Coupling in Fibre Optics. I and II", Bell Syst. Tech. J., 1974, 53, no.2, pp.217-24, and no.4, pp.675-96.

12.315. Yip, G.L., and Ahmew, Y.H.: "Propagation Characteristics of Radially Inhomogeneous Optical Fibre", Electron. Lett., 1974, 10, no.4, pp.37-8.

12.316. Kawakami, S., and Nishida, S.: "Anomalous Dispersion of Double-Clad Optical Fibre", Electron. Lett., 1974, 10, no.4, pp.38-40.

12.317. Itoh, T., and Mittra, R.: "Resonance Conditions of Open Resonators at Microwave Frequencies", Trans. IEEE, 1974, MTT-22, no.2, pp.99-102.

12.318. Satomura, U., Hara, M.M., and Kumagai, N.: "Analysis of Modes in Anisotropic Slab Waveguide", Trans. IEEE, 1974, MTT-22, no.2, pp.86-92.

12.319. Snyder, A.W., and Mitchell, D.J.: "Whispering-Gallery Rays within Dielectric Circles and Spheres", Electron. Lett., 1974, 10, no.2, p.16.

12.320. Garault, Y., and Fray, C.: "Wave Propagation in Shielded Ring Line", Trans. IEEE, 1974, MTT-22, no.2, pp.92-9.

12.321. Brown, J.: "Impedance and Scattering Properties of a Perfectly Conducting Strip above a Plane Surface-Wave System", Trans. IEEE, 1974, MTT-22, no.2, p.152.

12.322. Arora, R.K., and Vijayaraghavan, S.: "Analysis of Reactance Discontinuities in a Coaxial Shielded Surface Waveguide", Arch. Elektron. Ubertrag., 1974, 28, no.2, pp.86-91.

12.323. Yip, G.L., and Auyeung, T.: "Launching Efficiency of HE_{11} Surface-Wave Mode on Dielectric Tube", Trans. IEEE, 1974, MTT-22, no.1, pp.6-14.

12.324. Boivin, L.P.: "Thin-Film Laser-to-Fibre Coupler", Appl. Opt., 1974, 13, no.2, pp.391-5.

12.325. Stone, J., Ramaswamy, V., and Cohen, L.G.: "Efficient End Reflector for Optical Fibres", Opto-Electron. 1974, 6, no.2, pp.181-4.

12.326. Ulrich, R.: "Frequency-Selective Beam Coupler for Integrated Optics", Appl. Phys. Lett., 1974, 24, no.1, pp.21-4.

12.327. Kharadly, M.M.Z.: "Periodically Loaded Nonreciprocal Transmission Lines for Phaseshifter Applications", Trans. IEEE, 1974, MTT-22, no.6, pp.635-40.

12.328. Nemoto, S., and Yip, G.L.: "Excitation of Self-Focusing Optical Fibre by Gaussian Beam", Electron. Lett., 1974, 10, no.9, pp.150-1.

12.329. Kogelnik, H., and Weber, H.P.: "Rays, Stored Energy, and Power Flow in Dielectric Waveguides", J. Opt. Soc. Am., 1974, 64, no.2, pp.174-85.

12.330. Weber, W.H., McCarthy, S.L., and Ford, G.W.: "Perturbation Theory Applied to Gain or Loss in an Optical Waveguide", Appl. Opt., 1974, 13, no.4, pp.715-6.

12.331. Burns, W.K., and Warner, J.: "Mode Dispersion in Uniaxial Optical Waveguides", J. Opt. Soc. Am., 1974, 64, no.4, pp.441-6.

12.332. Hudson, M.C.: "Calculation of the Maximum Optical Coupling Efficiency into Multimode Optical Waveguides", Appl. Opt., 1974, 13, no.5, pp.1029-33.

12.333. Marcatili, E.A.J.: "Slab-Coupled Waveguides", Bell Syst. Tech. J., 1974, 53, no.4, pp.645-74.

12.334. Arnaud, J.A.: "Propagation Along Contacting Dielectric Tubes", Electron. Lett., 1974, 10, no.14, pp.269-70.

12.335. Barlow, H.E.M.: "Dipole-Mode Propagation in Hollow Tubular Waveguides", Proc. IEE, 1974, 121, no.7, pp.537-40.

12.336. Brandt, G.B.: "Birefringent Coupler for Integrated Optics", Appl. Opt., 1974, 13, no.6, pp.1359-62.

12.337. Snyder, A.W., and Mitchell, D.J.: "Generalized Fresnel's Laws for Determining Radiation Loss from Optical Waveguides", Optik, 1974, 40, no.4, pp.438-59.

12.338. Fong, T.T., and Lee, S.W.: "Modal Analysis of a Planar Dielectric Strip Waveguide for Millimetre-Wave Integrated Circuits", Trans. IEEE, 1974, MTT-22, no.8, pp.776-83.

12.339. Al-Hariri, A.M.B., Olver, A.D., and Clarricoats, P.J.B.: "Low-Attenuation Properties of Corrugated Rectangular Waveguide", Electron. Lett., 1974, 10, no.15, pp.304-5.

12.340. Geshiro, M., et al.: "Analysis of Wave Modes in Slab Waveguide with Truncated Parabolic Index", J. Quantum Electron. IEEE, 1974, QE-10, no.9, pp.647-9.

12.341. Ulrich, F., and Chen, D.: "Offset Prism for Optical Waveguide Coupling", Appl. Opt., 1974, 13, no.8, pp.1850-2.

12.342. Conwell, E.M.: "Optical Wave-guiding in Graded-Index Layers", Appl. Phys. Lett., 1974, 25, no.1, pp.40-2.

12.343. Carter, W.H.: "Electromagnetic Field of a Guided Gaussian Beam", Opt. Commun., 1974, 11, no.4, pp.410-4.

12.344. Blagidze, Yu.M., et al.: "Attenuation of Light in Fibre Guides", Kvantovaya Elektron., 1974, no.4, pp.97-9, and Sov. J. Quantum Electron., 1974, 3, no.4, pp.335-6.

12.345. Zlenko, A.A., and Sychugov, V.A.: "Light-Injection Prism with a Parabolic Gap Profile", Kvantovaya Elektron., 1974, no.4, pp.101-3, and Sov. J. Quantum Electron., 1974, 3, no.4, pp.339-40.

12.346. DeVore, R.: "Coaxial Dielectric Waveguides. II", J. Appl. Phys., 1974, 45, no.7, pp.2874-9.

12.347. Kogelnik, H., and Ramaswamy, V.: "Scaling Rules for Thin-Film Optical Waveguides", Appl. Opt.,1974,13,no.8,pp.1857-62.

12.348. Pask, C., and Snyder, A.W.: "Light Acceptance Property of an Optical Fibre", Appl. Opt., 1974,13,no.8,pp.1889-92.

12.349. Imai, M., and Hara, E.H.: "Excitation of Fundamental and Low-Order Modes of Optical-Fibre Waveguides by Gaussian Beams. I", Appl. Opt., 1974, 13, no.8, pp.1893-9.

12.350. Spiller, E., and Harper, J.S.: "High-Resolution Lenses for Optical Waveguides", Appl. Opt.,1974,13,no.9,pp.2105-8.

12.351. Pruzhanovskii, V.A.: "Round Metal-Optical Waveguide Filled with a Two-Layer Dielectric", Zh. Tekh. Fiz., 1974, 44, no.1, pp.6-15, and Sov. Phys.-Tech. Phys., 1974, 19, no.1, pp.3-8.

12.352. Kumar, A., Thyagarajan, K., and Ghatak, A.K.: "Modes in Inhomogeneous Slab Waveguides", J. Quantum Electron. IEEE, 1974, QE-10, no.12, pp.902-4.

12.353. Barlow, H.E.M.: "Dipole-Type-Mode Propagation in a Rectangular Waveguide", Proc. IEE, 1974, 121, no.11, pp.1363-6.

12.354. Kawakami, S., and Nishida, S.: "Characteristics of a Doubly Clad Optical Fibre with Low-Index Inner Cladding", J. Quantum Electron. IEEE, 1974, QE-10, no.12, pp.879-87.

12.355. Elachi, C., and Yeh, C.: "Mode Conversion in Periodically Disturbed Thin-Film Waveguides", J. Appl. Phys., 1974, 45, no.8, pp.3494-9.

12.356. Ramaswamy, V.: "Ray Model of Energy and Power Flow in Anisotropic Film Waveguides", J. Opt. Soc. Am., 1974, 64, no.10, pp.1313-20.

12.357. Gedeon, A.: "Comparison between Rigorous Theory and WKB Analysis of Modes in Graded-Index Waveguides", Opt. Commun., 1974, 12, no.3, pp.329-32.

12.358. Kazantsev, Yu.N., Manenkov, A.B., and Kharlashkin, O.A.: "Hollow Dielectric and Metal-Dielectric Waveguides for Propagation of Fast H-Modes", Izv. VUZ Radiofiz., 1974, 17, no.10, pp.1529-38.

12.359. Arnaud, J.A., and Saleh, A.A.M.: "Guidance of Surface Waves by Multilayer Coatings", Appl. Opt., 1974, 13, no.10, pp.2343-5.

12.360. Galantowicz, T.A.: "Lineshape Measurement for Coupling to Waveguide Modes in the Prism-Film Coupler", Appl. Opt., 1974, 13, no.11, pp.2525-8.

12.361. Green, H.E.: "Phase Shift in a Lossless Rectangular Waveguide Pierced by a Periodic Array of Small Holes", Trans. Inst. Aust. Electr. Eng., 1974,EE-10,no.1,pp.56-8.

12.362. Ryzhenko, B.F., Timchenko, L.P., and Akulov, V.V.: "Dimensioning of Iris-Loaded Waveguides for a Given Dispersion", Izv. VUZ Radioelektron., 1974,17,no.7,pp.83-7.

12.363. Chung, P.S.: "Graphical Determination of the Coupling Efficiency by Prism-Film Couplers in Planar Optical Waveguides", J. Phys., 1974, 7, no.18, pp.2490-500.

12.364. Kiselev, V.A.: "Resonant Conversion and Reflection of Surface Waves in a Thin-Film Waveguide with a Sinusoidally Corrugated Surface", Kvantovaya Elektron., 1974, 1, no.2, pp.329-33, and Sov. J. Quantum Electron., 1974, 4, no.2, pp.182-4.

12.365. Kiselev, V.A.: "Excitation of a Thin-Film Waveguide Using a Three-Dimensional Diffraction Grating", Kvantovaya Elektron., 1974, 1, no.2, pp.320-8, and Sov. J. Quantum Electron., 1974, 4, no.2, pp.178-81.

12.366. Weiss, J.A.: "Dispersion and Field Analysis of a Microstrip Meander-Line Slow-Wave Structure", Trans. IEEE, 1974, MTT-22, no.12, pp.1194-201.

12.367. Conciauro, G.: "Group Velocity of Degenerate Modes Propagating in Lossless Periodic Structures", Alta Freq., 1974, 43, no.12, pp.998-1004.

12.368. Tiem, D.H.: "Electromagnetic Helical Waveguide", Proc. IEEE, 1974, 62, no.12, pp.1723-4.

12.369. Raevskii, S.B.: "Theory of Two-Layer Waveguides with Resistive Film between the Layers", Izv. VUZ Radiofiz., 1974, 17, no.11, pp.1703-8.

12.370. Caton, W.M.: "Propagation Coefficients in Diffused Planar Optical Waveguides", Appl. Opt., 1974, 13, no.12, pp.2755-7.

12.371. Goben, C.A., Begley, D.L., and Davarpanah, M.: "Mode Selective Filtering by a Coupling Mechanism between Glass Fibre and Thin-Film Slab Waveguide", Appl. Opt., 1974, 13, no.12, pp.2757-8.

12.372. Yamamoto, Y., Yanai, H., and Kamiya, T.: "Studies on Optical Circuit Elements. II", Annu. Rep. Eng. Res. Inst. Fac. Eng. Univ. Tokyo, 1974, 33, pp.179-84.

12.373. Zolotov, E.M., et al.: "Differential Coupling of Radiation into a Thin-Film Waveguide", Kvantovaya Elektron., 1974, 1, no.8, pp.1869-73.

12.374. Zolotov, E.M., and Logachev, F.A.: "Coupling into an Optical Waveguide through its Tapered Edge", Kvantovaya Elektron., 1974, 1, no.8, pp.1873-5.

12.375. Ushakov, V.M., and Khizhnyak, N.A.: "Periodic Array of Uniform Cylinders in a Rectangular Waveguide", Radiotekh. (Kharkov), 1974, no.30, pp.139-45.

13. SOLID-STATE DEVICES

13.1. Peppiatt, H.J., McDaniel, A.V., and Linker, J.B.: "7-GHz Narrowband Waveguide Switch Using PIN Diodes", Trans. IEEE, 1965, MTT-13, p.44.

13.2. Fisher, R.E.: "Broadbanding Microwave Diode Switches", Trans. IEEE, 1965, MTT-13, p.706.

13.3. Bonnet, D., and Roch J.: "Calculation of Coupling by Hall Effect between Two Rectangular Orthogonal Guides with a Semiconductor", C. R. Acad. Sci. (Paris), 1965, 260B, p.3029.

13.4. Rodrigue, G.P.: "Microwave Solid-State Delay Lines", Proc. IEEE, 1965, 53, p.1428.

13.5. Engineer, M.H., and Nag, B.R.: "Propagation in Rectangular Guides Filled with a Semiconductor in a Transverse Magnetic Field", Trans. IEEE, 1965, MTT-13, p.641.

13.6. Toda, M.: "Isolator Using a Solid-State Plasma Waveguide", Trans. IEEE, 1965, MTT-13, p.126.

13.7. Heinz, W.W., and Okwit, S.: "Low-Level Microwave Limiting Utilizing Impact Ionization in Bulk Ge at 4.2°K", Proc. IEEE, 1965, 53, p.1274.

13.8. Hasty, T.E., and Wisseman, W.R.: "Effect of High Electric Fields on the Propagation of 24-GHz Waves Along the Surface of InSb", J. Appl. Phys., 1965, 36, p.3617.

13.9. White, J.F.: "High-Power, PIN-Diode, Controlled Microwave Transmission Phaseshifters", Trans. IEEE, 1965, MTT-13, p.233.

13.10. Engineer, M.H., et al.: "Semiconductor Microwave Rotator Using Hot-Carrier Properties", Proc. IEEE, 1966, 54, p.429.

13.11. Arizumi, T., and Umeno, M.: "Propagation Characteristics in Waveguide Loaded with a Cylindrical Semiconductor Rod", J. Inst. Electr. Commun. Eng. Jap., 1966, 49, p.722.

13.12. Blechert, G.: "Open Microwave Resonators Using Dielectrically Anisotropic Slabs", Arch. Elektr. Ubertrag., 1966, 20, p.149.

13.13. Larrabee, R.D.: "Characteristics of Waveguides with Semiconductor Sidewall", Trans. IEEE, 1966, MTT-14, p.306.

13.14. Lontsarenko, A.M., and Lusak, N.A.: "Mode Theory of Dielectric Anisotropic Waveguides", Zh. Prikl. Spekt., 1966, 6, p.561.

13.15. Gardner, A.L., and Hawke, R.S.: "High-Speed Microwave Phaseshifters Using Varactor Diodes", Rev. Sci. Instrum., 1966, 37, p.19.

13.16. Grisson, D.: "Enhancement of Resonator Q-Factor by Superconductivity and its Usefulness", Int. Conv. Rec. IEEE, 1966, 14, pt 9, p.94.

13.17. Gabriel, G.J., and Brodwin, M.E.: "Perturbation Analysis of Rectangular Waveguide Containing Transversely Magnetized Semiconductor", Trans. IEEE, 1966, MTT-14, p.258.

13.18. Higgins, V.J., Brunton, R.H., and Hall, G.: "Semiconductor Limiters as Microwave Duplexing Devices", Microwave J., 1966, 9, no.4, p.47.

13.19. Kuno, H.J., and Hershberger, W.D.: "Observation of Microwave Faraday Rotation in a Solid-State Plasma", Proc. IEEE, 1966, 54, p.978.

13.20. Asaly, R.N.: "Designs of X-Band Diode Switches", Trans. IEEE, 1966, MTT-14, p.553.

13.21. Manenkov, A.A., et al.: "Microwave Power Switch Using Semiconductor Diodes", Radiotekh. Elektron., 1966, 11, p.1899, and Radio Eng. Electron. Phys., 1966, 11.

13.22. Gunn, M.W.: "Wave Propagation in Rectangular Guide Containing a Semiconducting Film", Proc. IEE, 1967, 114, p.207.

13.23. Kuno, H.J., and Hershberger, W.D.: "Solid-State-Plasma Controlled Nonreciprocal Microwave Device", Trans. IEEE, 1967, MTT-15, p.57.

13.24. Swartz, G.A.: "Microwave Coupling to the Helicon Mode in InSb", RCA Rev., 1967, 28, no.1, pp.64-74.

13.25. Hicinbothem, W.A., and Larrabee, R.D.: "Laminar Slow-Wave Coupler and Application to InSb", Trans. IEEE, 1967, MTT-15, no.6, pp.382-3.

13.26. Toda, M.: "Field Distribution in a Waveguide Loaded with a Thin Plate of n-InSb", Proc. IEEE, 1967, 55, no.4, pp.589-91.

13.27. Das, R.: "Thin Ferroelectric Phaseshifters", Solid-State Electron., 1967, 10, no.8, pp.857-63.

13.28. Sheikh, R.H., and Gunn, M.W.: "Wave Propagation in a Rectangular Waveguide Inhomogeneously Filled with Semiconductor", Trans. IEEE, 1968, MTT-16, no.2, pp.117-21.

13.29. Kong, J.A., and Cheng, D.K.: "Guided Waves in Moving Anisotropic Media", Trans. IEEE, 1968, MTT-16, no.2, pp.99-102.

13.30. Tripathi, V.K.: "Varactor-Loaded Ladder Lines", Nachr. Tech. Fachber., 1968, 35, p.594.

13.31. Hamaguchi, C., Moritani, A., and Naki, J.: "Microwave Propagation in Rectangular Waveguide Loaded with Semiconductor Thin Film", Technol. Rep. Osaka Univ., 1968, 18, p.191.

13.32. May, W.G., and McLeod, B.R.: "Waveguide Isolator Using InSb", Trans. IEEE, 1968, MTT-16, p.877.

13.33. Sugimoto, S.: "Ultra-High-Speed Diode Switch for 50-GHz Band Utilizing Avalanche Breakdown in Varactor Diodes", Trans. IEEE, 1968, MTT-16, p.1017.

13.34. Kulcsar, E., and Conning, S.W.: "High-Isolation PIN Diode Microwave Switch", Proc. IRE Aust., 1968, 29, p.332.

13.35. McKenna, J., and Morrison, J.A.: "Effects of Strain on Electromagnetic Modes of Anisotropic Dielectric Waveguides at p-n Junctions", Bell Syst. Tech. J., 1968, 47, p.1933.

13.36. Zepp, G.: "Attenuation Coefficient of a Wave Propagating in a Superconducting Guide", C. R. Acad. Sci. (Paris), 1969, 269B, p.1082.

13.37. Rahmann, S.A., and Gunn, M.W.: "Propagation in Rectangular Guide Filled with a Semiconductor in a Transverse Magnetic Field", Trans. IEEE, 1969,MTT-17, p.279.

13.38. Tsutsumi, M., Tomiya, T., and Aoyagi, K.: "Guided-Wave Propagation through Magnetoelectric Media", Proc. IEEE, 1969, 57, p.696.

13.39. Pedersen, N.F.: "Propagation in Longitudinally Magnetized Ge Plasma Waveguide. I and II", Phys. Status Solidi, 1969, 36, no.1, pp.157 and 165.

13.40. Ivanov, I.V., et al.: "Ferroelectric Distributed Nonlinear Elements in Microwave Circuits", Vestn. Mosk. Univ., Fiz. Astron., 1969, no.6, p.40.

13.41. Gandhi, O.P.: "Low-Loss Mode in Semiconductor Magnetoplasmas", Trans. IEEE, 1969, ED-16, p.965.

13.42. Suzuki, K.: "Room-Temperature Solid-State Plasma Nonreciprocal Microwave Devices", Trans. IEEE, 1969, ED-16, p.1018.

13.43. Zepp, G.: "Propagation of Waves in a Superconductor Guide", C. R. Acad. Sci. (Paris), 1969, 269B, p.1030.

13.44. Gandhi, O.P., and Verma, K.: "Propagation of Microwaves through InSb in a Transverse Steady Magnetic Field", Trans. IEEE, 1969, ED-16, p.964.

13.45. Nielsen, E.D.: "Scattering by a Cylindrical Post of Complex Permittivity in a Waveguide", Trans. IEEE, 1969, MTT-17, p.148.

13.46. Tsutsumi, M., and Aoyagi, K.: "Guided-Wave Propagation through Magneto-electric Media", Electron. Commun. Jap., 1969, 52, no.11, pp.83-9.

13.47. Rabinowitz, M.: "Possible Sources of Residual Power Loss in RF Superconducting Cavities", Lett. Nuovo Cim., 1970,4,p.549.

13.48. Hirota, R., and Suzuki, K.: "Field Distribution in a Magnetoplasma-Loaded Waveguide at Room Temperature", Trans. IEEE, 1970, MTT-18, p.188.

13.49. Zepp, G.: "Wave Propagation in a Superconducting Guide in Nonlocal Theory", J. Phys. (Paris), 1970, 31,no.5-6,pp.513-7.

13.50. Zepp, G.: "Wave Propagation in a Cylindrical Superconducting Guide", C. R. Acad. Sci. (Paris), 1970, 270B, p.30.

13.51. Midgley, D.: "Analysis of Helicon-Wave Propagation", Electron. Lett., 1970, 6, p.497.

13.52. Suzuki, K., and Hirota, R.: "Field Distribution in a Waveguide Loaded with a Thin Slab of InSb", Proc. IEEE, 1970, 58, p.915.

13.53. Champlin, K.S., and Glover, G.H.: "Twist Modes in Magnetoplasma-Filled Circular Waveguides", Trans. IEEE, 1970, MTT-18, p.566.

13.54. Glover, G.H., and Champlin, K.S.: "Investigation of Twist Mode Propagation in InSb at 70 GHz", Trans. IEEE, 1970, MTT-18, p.570.

13.55. Champlin, K.S., and Glover, G.H.: "Millimetre-Wave Mode Conversion by a Solid-State Magnetoplasma", Trans. IEEE, 1970, ED-17, p.637.

13.56. van Dalen, P.A.: "Propagation of Guided Waves in Semiconductors in a Longitudinal Magnetic Field", J. Appl. Phys., 1970, 41, p.3092.

13.57. Al'tschuler, Yu.G., et al.: "Application of Plasma Effects in InSb in the Design of Submillimetre Control Devices", Izv. VUZ Radioelektron., 1970, 13, no.8, pp.965-72.

13.58. Gunn, M.W., and Rahmann, S.A.: "Effect of Hall Field in a Rectangular Waveguide Containing Semiconducting Material", Proc. IEEE, 1970, 117, no.12, pp.2238-40.

13.59. Yagi, H., et al.: "Superconducting Wire-Wall Cavity for Millimetric ENDOR Experiments", Jap. J. Appl. Phys., 1970, 9, no.12, p.1534.

13.60. Arnold, R.M., and Rosenbaum, F.J.: "Nonreciprocal Propagation in Semiconductor-Loaded Waveguides with Transverse Magnetic Field", Trans. IEEE, 1971, MTT-19, no.1, pp.57-65.

13.61. Filippov, Yu.F.: "Theory of Faraday Effect in a Semiconducting Waveguide", Izv. VUZ Radiofiz., 1971, 14, no.2, pp.314-6.

13.62. Benacka, St.: "Superconductive Cavity Resonators in the 10-GHz Band", Elektrotech. Cas., 1971, 22, no.2, pp.81-92.

13.63. McLeod, B.R., and May, W.G.: "35-GHz Isolator Using a Coaxial Solid-State (InSb) Plasma in a Longitudinal Magnetic Field", Trans. IEEE, 1971, MTT-19, no.6, pp.510-6.

13.64. Combet, H.A., Chouan, Y., and Thepault, E.: "Microwave Devices Using Proximity Effect between Superconducting and Normal Metals", Cryogenics, 1971, 11, no.2, pp.102-6.

13.65. Barone, A., and Parmentier, R.D.: "Josephson Junction Transmission Lines", Alta Freq., 1971, 40, no.2, pp.166-8.

13.66. Suzuki, K., and Hirota, R.: "Nonreciprocal Millimetre-Wave Devices Using a Solid-State Plasma at Room Temperature", Trans. IEEE, 1971, ED-18, no.7, pp.408-11.

13.67. Hallford, B.R.: "90-dB Microstrip (PIN) Switch on a Plastic Substrate", Trans. IEEE, 1971, MTT-19, no.7, pp.654-7.

13.68. Toussaint, H.N., and Hoffmann, R.: "Octave-Bandwidth Adjustable SPDT Switch Using PIN Diodes", Trans. IEEE, 1971, MTT-19, no.7, pp.657-9.

13.69. Henry, R.: "PIN Diodes for Microwave Switching", Rev. Tech. Thomson-CSF, 1971, 3, no.2, pp.239-68.

13.70. Rabinowitz, M.: "Frequency Dependence of Superconducting-Cavity Q and Magnetic Breakdown Field", Appl. Phys. Lett., 1971, 19, no.3, pp.73-6.

13.71. DiNardo, A.J., Smith, F.G., and Arams, F.R.: "Superconducting Microstrip High-Q Resonators", J. Appl. Phys., 1971, 42, no.1, pp.186-9.

13.72. Mortenson, K.E., et al.: "Review of Bulk Semiconductor Microwave Control Components", Proc. IEEE, 1971, 59, no.8, pp.1191-200.

13.73. Zepp, G.: "Cylindrical Superconducting Guides for Microwaves", Rev. Phys. Appl., 1971, 6, no.2, pp.135-41.

13.74. Halama, H.J.: "Effects of Radiation on Surface Resistance of Superconducting Niobium Cavity", Appl. Phys. Lett., 1971, 19, no.4, pp.90-1.

13.75. Strongin, M.: "Sensitivity of Q-Factor of Superconducting RF Cavities to Surface Conditions", J. Appl. Phys., 1971, 42, no.10, pp.4105-7.

13.76. Channin, D.J.: "Voltage-Induced Optical Waveguide", Appl. Phys. Lett., 1971, 19, no.5, pp.128-30.

13.77. Wang, S., Crow, J.D., and Sah, M.: "Thin-Film Optical-Waveguide Mode Converters Using Gyrotropic and Anisotropic Substrates", Appl. Phys. Lett., 1971, 19, no.6, pp.187-9.

13.78. Zepp, G.: "TEM-Wave Propagation in a Cylindrical Coaxial Superconductor Guide", Onde Electr., 1971, 51, no.10, pp.869-73.

13.79. Ferrari, L.A.: "High-Power Loss Mechanism in Superconducting Microwave Cavities", Part. Accel., 1971, 2, no.4, pp.283-7.

13.80. Nozaka, M., and Nishida, S.: "Guided Waves in Bounded Nonlinear Medium. I", Rep. Res. Inst. Electr. Commun. Tohoku Univ., 1971, 23, no.1, pp.31-8.

13.81. Mironov, B.A.: "Nonlinear Penetration of a Plane Plasma Layer", Izv. VUZ Radiofiz., 1971, 14, no.9, pp.1450-2.

13.82. Al'tshuler, Yu.G., Dovzhenok, A.A., and Kats, L.I.: "Interaction of Microwave Radiation with Solid-State Plasmas", Izv. VUZ Radioelektron., 1971, 14, no.9, pp.972-96.

13.83. Max, C., and Perkins, F.: "Strong Electromagnetic Waves in Overdense Plasmas", Phys. Rev. Lett., 1971, 27, no.20, pp.1342-5.

13.84. Nozaka, M., and Nishida, S.: "Beam Modes in Open Resonator Filled with Anisotropic Uniaxial Crystals", Rep. Res. Inst. Electr. Commun. Tohoku Univ., 1971, 23, no.2, pp.79-93.

13.85. Combet, H.A.: "Superconducting Striplines", Rev. Phys. Appl., 1971, 6, no.4, pp.543-5.

13.86. Kobayashi, T., and Fujisawa, K.: "Scattering by a Rod of Anisotropic Semiconductor in a Rectangular Waveguide", Electron. Commun. Jap., 1971, 54, no.9, pp.83-93.

13.87. Dahlstrom, L.: "Passive (Optical-Kerr) Nonlinear Output Coupler for Mode-Locked High-Power Lasers", Opt. Commun., 1971, 4, no.3, pp.214-9.

13.88. Okamoto, H., and Mizushima, Y.: "Interface-Wave Instability with Two Parallel Semiconductor Sheets with Transverse Magnetic Field", Rev. Electr. Commun. Lab., 1971, 19, nos.9-10, pp.973-8.

13.89. Shah, M., Crow, J.D., and Wang, S.: "Optical-Waveguide Mode-Conversion Experiments", Appl. Phys., Lett., 1972, 20, no.2, pp.66-9.

13.90. Garwin, E.L., and Rabinowitz, M.: "Resistivity Ratio of Niobium Superconducting Cavities", Appl. Phys. Lett., 1972, 20, no.4, pp.154-6.

13.91. Brandle, R., and Sedlmair, S.: "PIN-Diode Phaseshifters in Symmetrical Stripline Technology", Frequenz, 1972, 26, no.2, pp.45-50.

13.92. Garver, R.V.: "Broadband Diode Phaseshifter", Trans. IEEE, 1972, MTT-20, no.5, pp.314-23.

13.93. Didenko, A.N., and Shiyan, V.P.: "Damping of EM Waves in Smooth Superconducting Waveguides", Radiotekh. Elektron., 1972, 17, no.11, pp.2438-40, and Radio Eng. Electron. Phys., 1972, 17, no.11, pp.1959-61.

13.94. Foggiato, G.A., and Pearson, G.L.: "High-Speed Bulk Semiconductor Microwave Switch Utilizing Ga(AsP) Mixed Crystals", Proc. IEEE, 1972, 60, no.4, pp.456-7.

13.95. Hindin, H.J.: "Phase Varied with Liquid Artificial Dielectric", Microwaves, 1972, 11, no.2, p.9.

13.96. Eschelbacher, H.C.: "Use of Superconducting Resonant Cavities at Gigahertz Frequencies", Nachr. Tech. Z., 1972, 25, no.4, pp.193-5.

13.97. Somekh, S., and Yariv, A.: "Phase-Matchable Nonlinear Optical Interactions in Periodic Thin Films", Appl. Phys. Lett., 1972, 21, no.4, pp.140-1.

13.98. Vlasov, S.N., Petrishchev, V.A., and Talanov, V.I.: "Nonlinear Quasi-Optical Systems", Izv. VUZ Radiofiz., 1972, 15, no.8, pp.1162-72.

13.99. Parris, W.J.: "PIN Variable Attenuator with Low Phase Shift", Trans. IEEE, 1972, MTT-20, no.9, pp.618-9.

13.100. Lorek, W.: "Calculation of Ridged Waveguides by Quasi-Stationary Representation (for Diode Modulation)", Nachr. Tech. Z., 1972, 25, no.11, pp.511-9.

13.101. Novikov, S.A.: "Wave Attenuation in Superconducting Elliptical Guide", Izv. VUZ Radiofiz., 1972, 15, no.12, pp.1944-6.

13.102. Nagumo, F., Naito, Y., and Suetake, K.: "Short-Circuit Controlled by PIN Diode", Electron. Commun. Jap., 1972, 54, no.5, pp.51-7.

13.103. Bychkova, N.N., and Kulik, I.O.: "Nonlinear Effects in Superconducting Resonators", Zh. Tekh. Fiz., 1972, 42, no.3, pp.584-90, and Sov. Phys.-Tech. Phys., 1972, 42, no.3, pp.461-6.

13.104. Al'tshuler, Yu.G., Kats, L.I., and Revzin, R.M.: "Submillimetre Propagation in Drift Solid-State Plasma", Izv. VUZ Radioelektron., 1972, 15, no.8, pp.932-7.

13.105. Al'tshuler, Yu.G., Kats, L.I., and Revzin, R.M.: "Submillimetre Propagation in a Waveguide Filled with n-InSb in a Magnetic Field", Izv. VUZ Radioelektron., 1972, 15, no.8, pp.938-44.

13.106. Liberman, L.S., et al.: "Semiconductor Diodes for Controlling Microwave Power", Radiotekhnika, 1972, 27, no.5, pp.9-25, and Telecommun. Radio Eng. Pt 2, 1972, 27, no.5, pp.63-76.

13.107. Somekh, S., et al.: "Channel Optical Waveguide Directional Couplers", Appl. Phys. Lett., 1973, 22, no.1, pp.46-7.

13.108. Sosnowski, T.P., and Weber, H.P.: "Polarization Conversion of Light in Thin-Film Waveguides", Opt. Commun., 1973, 7, no.1, pp.47-50.

13.109. Campbell, C.K.: "Frequency Response of a Superconducting Transmission Line", Proc. IEEE, 1973, 61, no.6, pp.799-800.

13.110. Jager, D., and Rabus, W.: "Bias-Dependent Phase Delay of Schottky Contact Microstrip Line", Electron. Lett., 1973, 9, pp.201-3.

13.111. Pinnow, D.A., et al.: "Fundamental Optical Attenuation Limits in the Liquid and Glassy State with Application to Fibre-Optical Waveguide Materials", Appl. Phys. Lett., 1973, 22, no.10, pp.527-9.

13.112. Cohen, J., and Gilden, M.: "Mathematical Model of a Varactor Package in Microstrip Line", Trans. IEEE, 1973, MTT-21, no.6, pp.412-3.

13.113. Barskii, I.V., et al.: "Application of Distributed p-n and p-i-n Structures in the Design of Integrated Circuits for Controlling Microwave Networks", Izv. VUZ Radioelektron., 1973, 16, no.4, pp.62-6.

13.114. Benacka, St.: "High-Frequency Losses and Stability of the Resonance Frequency of Superconducting Cavity Resonators", Elektrotech. Cas., 1973, 24, no.4, pp.209-16.

13.115. Williamson, A.G., and Otto, D.V.: "Analysis of a Waveguide Mounting Structure (for Semiconductor Devices)", Proc. IREE Aust., 1973, 34, no.3, pp.95-7.

13.116. Tsypkin, E.R.: "Analysis and Synthesis of Two-Channel Broadband PIN-Diode Switches", Izv. VUZ Radioelektron., 1973, 16, no.7, pp.37-40.

13.117. Mikami, O., and Ishida, A.: "Experiments on Voltage-Induced Optical Waveguide in LiNbO₃", Jap. J. Appl. Phys., 1973, 12, no.8, pp.1294-5.

13.118. Tenenholtz, R.: "Broadband MIC Multithrow PIN Diode Switches", Microwave J., 1973, 16, no.7, pp.25-8.

13.119. Reid, M.J.: "Microwave Switch and Attenuator Modules", Microwave J., 1973, 16, no.7, pp.37-40.

13.120. Brown, N.J., and Basken, U.: "Design Concepts for IFF Diode Switches", Microwave J., 1973, 16, no.7, pp.43-6.

13.121. Voges, E., and Petermann, K.: "Losses in Superconducting Coaxial Transmission Lines", Arch. Elektron. Ubertrag., 1973, 27, no.9, pp.384-8.

13.122. Reisinger, A.: "Attenuation Properties of Optical Waveguides with a Metal Boundary", Appl. Phys. Lett., 1973, 23, no.5, pp.237-9.

13.123. Askar'yan, G.A., and Tarasova, N.M.: "Production of Steep Microwave Fronts by a Laser-Flash-Evaporated Metallized Film", Zh. Eksp. Teor. Fiz. Pis'ma, 1973, 18, no.1, pp.8-10, and JETP Lett., 1973, 18, no.1, pp.3-5.

13.124. Giordano, S., et al.: "Influence of Solute Oxygen and Nitrogen on Superconducting Niobium Cavities", J. Appl. Phys., 1973, 44, no.9, pp.4185-90.

13.125. Schnitzke, K., et al.: "TE_{011} X-Band Niobium Cavity with Critical Magnetic Flux Density Higher than B_{c1}", Phys. Lett., 1973, 45A, no.3, pp.241-2.

13.126. Zykov, A.I., Knyazev, B.R., and Tereshchenko, A.I.: "Investigation of Q-Factors of Superconducting Cavities", Radiotekh. (Kharkov), 1973, no.24, pp.158-62.

13.127. Jimenez, J.J.: "Superconducting Microstrip Resonators", Electron. Fis. Apl., 1973, 16, no.2, pp.119-27.

13.128. Sato, S., and Makimoto, T.: "Stationary Expression for Scattering Coefficients at Uniform Piezoelectric Waveguide Junction", Proc. IEEE, 1973, 61, no.11, pp.1648-50.

13.129. Levin, B.J., and Weidner, G.G.: "Millimetre-Wave (PIN Diode) Phaseshifter", RCA Rev., 1973, 34, no.3, pp.489-505.

13.130. Gupta, P.N., Jain, P.K., and Tolpadi, S.K.: "Propagation Characteristics of Semiconductor-Loaded Waveguide with Transverse Magnetic Field", J. Inst. Electron. Telecommun. Eng., 1973, 19, no.12, pp.695-6.

13.131. Barsukov, Yu.K., and Makeeva, G.S.: "Voltage Dependence of Dynamic Linear Capacitance of a Nonlinear Stripline", Radiotekh. Elektron., 1973, 18, no.1, pp.32-40, and Radio Eng. Electron. Phys., 1973, 18, no.1, pp.27-33.

13.132. Zlunitsyn, E.S., Zykov, A.I., and Kushnir, V.A.: "Procedure for Investigating Parameters of Superconducting Resonators", Prib. Tekh. Eksp., 1973, 16, no.3, pt 2, pp.243-5, and Instrum. Exp. Tech., 1973, 16, no.3, pt 2, pp.947-9.

13.133. Baranov, L.I., Gamanyuk, V.B., and Usanov, D.A.: "Nonreciprocal Propagation of Waves in a Guide Partially Filled with a Semiconductor", Radiotekh. Elektron., 1973, 18, no.1, pp.73-8, and Radio Eng. Electron. Phys., 1973, 18, no.1, pp.53-7.

13.134. Didenko, A.N., and Shtein, Yu.G.: "Requirements on Stability of Parameters of Superconducting TW Resonators", Radiotekh. Elektron., 1973, 18, no.3, pp.624-6, and Radio Eng. Electron. Phys., 1973, 18, no.3, pp.447-9.

13.135. Danielson, M.: "Superconducting Lead Cavities at 35 GHz", Proc. IEEE, 1973, 61, no.1, pp.71-6.

13.136. Chiba, N., Kashiwayanagi, Y., and Mikoshiba, K.: "Frequency Characteristics of Superconductive Coaxial Lines", Proc. IEEE, 1973, 61, no.1, pp.124-5.

13.137. Weidner, G.G., and Levin, B.J.: "Distributed PIN-Diode Phaser for Millimetre Wavelengths", Microwave J., 1973, 16, no.11, pp.42-4.

13.138. Sarkisyan, Zh.A., and Simonyan, R.N.: "Broadband Metal Rotator of Plane of Polarization", Radiotekhnika, 1973, 28, no.6, pp.69-70, and Telecommun. Radio Eng. Pt 2, 1973, 28, no.6, pp.98-9.

13.139. Kononenko, V.K., and Kuleshov, Ye.M.: "Unilateral Propagation of Helicon Waves in n-InSb at Microwave Frequencies", Radiotekh. Elektron., 1973, 18, no.7, pp.1429-34, and Radio Eng. Electron. Phys., 1973, 18, no.7, pp.1051-5.

13.140. Zapol'skii, O.B., Markin, E.P., and Oraevskii, A.N.: "Inhomogeneous Medium as High-Q Resonator", Kvantovaya Elektron., 1973, no.5, pp.120-2, and Sov. J. Quantum Electron., 1973, 3, no.5, pp.438-9.

13.141. Yamamoto, S., and Makimoto, T.: "Circuit-Theoretic Treatment of Anisotropic Thin-Film Optical Waveguides", Electron. Commun. Jap., 1973, 56, no.3, pp.122-9.

13.142. Kharusi, M.S.: "Uniaxial and Biaxial Anisotropy in Thin-Film Optical Waveguides", J. Opt. Soc. Am., 1974, 64, no.1, pp.27-35.

13.143. Burns, R.W., Holden, R.L., and Tang, R.: "Low-Cost Design Techniques for Semiconductor Phaseshifters", Trans. IEEE, 1974, MTT-22, no.6, pp.675-88.

13.144. Lynes, G.D., et al.: "Design of Broadband 4-bit Loaded Switched-Line Phase-shifter", Trans. IEEE, 1974, MTT-22, no.6, pp.693-7.

13.145. Warner, J.: "Excitation of Hybrid Modes in Magnetooptic Waveguides", Appl. Opt., 1974, 13, no.5, pp.1001-4.

13.146. Matsuda, A., et al.: "Stopping Effect on Guided Light in As-S Films by a Laser Beam", Appl. Phys. Lett., 1974, 24, no.7, pp.314-5.

13.147. Obunai, T., and Sekiguchi, T.: "Observation of a Slow Surface Wave in a Millimetre-Wave Solid-State-Plasma Waveguide", Jap. J. Appl. Phys., 1974, 13, no.1, pp.93-108.

13.148. Spiller, E., and Segmuller, A.: "Propagation of X-Rays in Waveguides", Appl. Phys. Lett., 1974, 24, no.2, pp.60-1.

13.149. Timmermann, C.C.: "Material Dispersion in Optical Glass Fibres", Arch. Elektron. Ubertrag., 1974, 28, no.3, pp.144-5.

13.150. Jacobs, H., and Chrepta, M.M.: "Electronic Phaseshifter for Millimetre-Wave Semiconductor Dielectric Integrated Circuits", Trans. IEEE, 1974, MTT-22, no.4, pp.411-7.

13.151. Ramaswamy, V.: "Strip-Loaded Film Waveguide", Bell Syst. Tech. J., 1974, 53, no.4, pp.697-704.

13.152. Hu, C., and Whinnery, J.R.: "Field-Realigned Nematic-Liquid-Crystal Optical Waveguides", J. Quantum Electron. IEEE, 1974, QE-10, no.7, pp.556-62.

13.153. Hofmann, H.: "Dispersion of Ferrite-Filled Microstrip Line", Arch. Elektron. Ubertrag., 1974, 28, no.5, pp.223-7.

13.154. Smith, T.I.: "Superconducting Microwave Cavities and Josephson Junctions", J. Appl. Phys., 1974, 45, no.4, pp.1875-9.

13.155. Jager, D., Rabus, W., and Eikhoff, W.: "Bias-Dependent Small-Signal Parameters of Schottky-Contact Microstrip Lines", Solid-State Electron., 1974, 17, no.8, pp.777-83.

13.156. Hammer, J.M., and Phillips, W.: "Low-Loss Single-Mode Optical Waveguides and Efficient High-Speed Modulators of $Li(NbTa)O_3$ on $LiTaO_3$", Appl. Phys. Lett., 1974, 24, no.11, pp.545-7.

13.157. Wood, V.E., Hartman, N.F., and Verber, C.M.: "Characteristics of Effused Slab Waveguides in $LiNbO_3$", J. Appl. Phys., 1974, 45, no.3, pp.1449-51.

13.158. Finn, D., et al.: "Low-Loss Large-Area GaAs/GaAsP Heterostructure as Optical Waveguide at 10.6 micron", Opt. Commun., 1974, 11, no.2, pp.201-3.

13.159. Kneisel, P., Stoltz, O., and Halbritter, J.: "Surface Preparation and Measurement of Niobium Used in High-Frequency Cavities", J. Appl. Phys., 1974, 45, no.5, pp.2296-301.

13.160. Kneisel, P., Stoltz, O., and Halbritter, J.: "Breakdown Fields of a Superconducting Niobium Cavity at 2-4 GHz", J. Appl. Phys., 1974, 45, no.5, pp.2302-4.

13.161. de Santis, P.: "Edge-Guided Modes in Ferrite Microstrip with Curved Edges", Appl. Phys., 1974, 4, no.2, pp.67-74.

13.162. Ung, Y.I., and Kang, C.Y.: "Study of Perturbed Modes in Rectangular Waveguide Filled with a Transversely Magnetized Semiconductor", J. Korean Inst. Electron. Eng., 1974, 11, no.2, pp.12-21.

13.163. Richter, K.H.: "Transfer Characteristics of Microstrip Lines over Semiconductor Substrates", Nachr. Tech. Elektron., 1974, 24, no.1, pp.2-4.

13.164. Schnitzke, K., et al.: "Gas Exposure Tests of High-Field-Niobium X-Band Cavities", Appl. Phys., 1974, 5, no.1, pp.77-8.

13.165. Yasaitis, J.A., and Rose, R.M.: "Microwave Surface Resistance of Superconducting Mo_3Re", Appl. Phys. Lett., 1974, 25, no.6, pp.354-5.

13.166. Goncharov, V.N.: "Microwave Propagation in a Cylindrical Waveguide with Anisotropic Rod", Izv. VUZ Radioelektron., 1974, 17, no.9, pp.83-7.

13.167. Ivanov, M.M., and Matyshev, V.A.: "Power-Handling Capacity of a Diode Limiter in Microwave Transmission Line", Izv. VUZ Radioelektron., 1974, 17, no.12, pp.63-6.

13.168. Obunai, T., and Sekiguchi, T.: "Propagation Modes in Millimetre-Wave Solid-State Plasma Waveguide", Jap. J. Appl. Phys., 1974, 13, no.12, pp.2075-6.

13.169. Kang, C.Y.: "Nonreciprocal Wave Propagation in Semiconductor", J. Korean Inst. Electron. Eng., 1974, 11, no.5, pp.184-7.

13.170. Blum, F.A., Shaw, D.W., and Holton, W.C.: "Optical Striplines for Integrated Optical Circuits in Epitaxial GaAs", Appl. Phys. Lett., 1974, 25, no.2, pp.116-8.

13.171. Hu, C., and Whinnery, J.R.: "Losses of a Nematic Liquid-Crystal Optical Waveguide", J. Opt. Soc. Am., 1974, 64, no.11, pp.1424-32.

14. DEVICES WITH GYROMAGNETIC MEDIA

14.1. Damiano, R., and Kliphuis, J.: "Cryogenically Cooled, Y-Junction, Circulators for Parametric-Amplifier Applications", Proc. IEEE, 1965, 53, p.198.

14.2. Chia, H.H., and Yuan, F.D.: "Coupled-Mode Theory of Guided-Wave Propagation through a Ferrite Rod with Variable Cross Section", Sci. Sin., 1965, 14, p.378, and Acta Phys. Sin., 1965, 21, p.1653.

14.3. Deryugin, I.A., Strizhevskii, V.L., and Kuts, P.S.: "Investigation of Microwave Faraday Devices with Variable Magnetic Field", Zh. Tekh. Fiz., 1965, 35, p.546, and Sov. Phys.-Tech. Phys., 1965, 10, no.3, pp.424-32.

14.4. Wantuch, E.: "Extremely High-Power, X-Band, Three-Port Circulator", Electron. Lett., 1965, 1, p.45.

14.5. Seo, W.Y.: "Three-Port Star Waveguide Phaseshifter", Proc. IEEE, 1965, 53, p.424.

14.6. Davies, J.B.: "Theoretical Design of Wideband Waveguide Circulators", Electron. Lett., 1965, 1, p.60.

14.7. Comstock, R.L., and Fay, C.E.: "Performance and Ferrimagnetic-Material Considerations in Cryogenic Microwave Devices", J. Appl. Phys., 1965,36,pt 2,p.1253.

14.8. Mikhaelovskii, A.K., et al.: "Properties and Application of Magnetic Single-Axis Ferrites at Millimetre Wavelengths", Radiotekh. Elektron., 1965, 10, p.1739, and Radio Eng. Electron. Phys., 1965,10,p.1495.

14.9. Fay, C.E., and Comstock, R.L.: "Operation of Ferrite Junction Circulator", Trans. IEEE, 1965, MTT-13, p.15.

14.10. Weiss, J.A.: "Circulator Synthesis", Trans. IEEE, 1965, MTT-13, p.38.

14.11. Krokstad, J.: "Ferrimagnetic Microwave Power Limiter", Trans. IEEE, 1965, MTT-13, p.119.

14.12. Bolle, D.M., and Heller, G.S.: "Theoretical Considerations on Use of Circularly Symmetric TE Modes for Digital Ferrite Phaseshifters", Trans. IEEE, 1965, MTT-13, p.421.

14.13. Schlomann, E., Green, J.J., and Saunders, J.H.: "Ultimate Performance Limitations of High-Power Ferrite Circulators and Phaseshifters", Trans. IEEE, 1965, MAG-1, p.168.

14.14. Huang, H., and Fan, D.: "Coupled-Mode Theory of Guided-Wave Propagation through a Ferrite Rod with Variable Cross Section", Sci. Sin., 1965, 14, p.378.

14.15. Nakahara, S., and Kurebayashi, H.: "Directional Power Divider and Application", Mitsubishi Denki Lab. Rep.,1965,6,no.2,p.131.

14.16. Wade, W.L., Stern, R., and Collins, T.: "Millimetre Resonance Isolator Utilizing Multilayer Ni and NiZn Ferrite Films", Trans. IEEE, 1965, MTT-13, p.127.

14.17. Sedlmair, S.: "Rapid Ferrite Phaseshifter for Control of a Resonator at 35GHz", Frequenz, 1965, 19, p.348.

14.18. Ivanov, K.P.: "Helical Line with Gyrotropic Medium", Radiotekh. Elektron., 1965, 10, p.1343, and Radio Eng. Electron. Phys., 1965, 10, no.7, pp.1153-5.

14.19. Whicker, L.R., and Jones, R.R.: "Digital Latching Ferrite Stripline Phaseshifter", Trans. IEEE, 1965, MTT-13, p.781.

14.20. Helszajn, J.: "Switching Criteria for Waveguide Ferrite Devices", Radio Electron. Eng., 1965, 30, p.289.

14.21. Willoughby, R.E.: "High-Power SPDT Fast Ferrite Switch", J. Appl. Phys., 1965, 36, pt 2, p.1247.

14.22. Berger, H.: "Nonreciprocal Directional Couplers", Trans. IEEE, 1965, MTT-13, p.474.

14.23. Kraszewski, A.: "Results of Experiments with Waveguide Circulators in the 3-cm Band", Przegl. Elektron., 1965, 6, p.505.

14.24. Vol'man, V.I.: "Design of Waveguide Y-Circulators", Radiotekhnika, 1965, 20, no.3, p.21, and Telecommun. Radio Eng. Pt 2, 1965, 20, no.3, p.85.

14.25. Helszajn, J.: "Note on Tetrahedral Junction", Trans. IEEE, 1965, MTT-13, p.132.

14.26. Konishi, Y.: "Lumped-Element Y-Circulator", Trans. IEEE, 1965, MTT-13, p.852.

14.27. Siekanowicz, W.W., et al.: "Design and Performance of a 20-kW Latching Nonreciprocal X-Band Ferrite Phaseshifter", RCA Rev., 1965, 26, p.574.

14.28. Bogdanov, G.B., and Voronov, Yu.K.: "Design Method for Waveguides with Ferrite Inserts in Case of Strong Reaction", Radiotekh. Elektron., 1965, 10, p.943, and Radio Eng. Electron. Phys., 1965, 10, p.802.

14.29. Sidhu, G.S., and Gandhi, O.P.: "Design and Development of X-Band Ferrite Components", J. Inst. Telecommun. Eng., 1965, 11, no.6, p.181.

14.30. Kraszewski, A.: "Experimental Results on X-Band Junction Circulator", Trans. IEEE, 1965, MTT-13, p.382.

14.31. Sidhu, G.S., and Gandhi, O.P.: "Four-Port Waveguide Junction Circulator and Effects of Dielectric Loading on Performance", Trans. IEEE, 1965, MTT-13, p.388.

14.32. Simon, J.W.: "Broadband Stripline Y-Circulators", Trans. IEEE, 1965, MTT-13, p.335.

14.33. Bujatti, M.: "Field Measurements in Small-Cross-Section Guide Loaded with Magnetized Ferrite", Trans. IEEE, 1965, MTT-13, p.885.

14.34. Kaufman, I., and Soohoo, R.F.: "Electric Field and Wave Impedance in Spin-Wave Propagation", Trans. IEEE, 1965, MTT-13, p.703.

14.35. Veselkov, G.I.: "Autostabilization of Angle of Rotation of Polarization in a Circulator", Izv. VUZ Radiotekh., 1965, 8, p.376, and Sov. Radio Eng., 1965, 8, p.265.

14.36. Popova, G.M., and Protopopova, L.M.: "Isolators Based on Surface Waves of H_{n0} Type in Rectangular Waveguide Containing Transversely Magnetized Ferrite", Izv. VUZ Fiz., 1965, no.2, p.28, and Sov. Phys. J., 1965, no.2, p.18.

14.37. Chandra, K., et al.: "Experimental Y-Circulator for X Band", J. Inst. Telecommun. Eng., 1965, 11, p.531.

14.38. Hashimoto, T.: "Novel Design Principle to Minimize Temperature Dependence of the Resonance Isolator", Proc. IEEE, 1965, 53, p.1273.

14.39. Kislyakovskii, A.V., and Vuntesmeri, V.S.: "Phase Relationship in Interaction between a Ferrite Spheroid and the Field in a Waveguide", Izv. VUZ Radiotekh., 1965, 8, p.455, and Sov. Radio Eng., 1965, 8, p.328.

14.40. Towpik, R.: "Three-Port S-Band Ferrite Circulator", Przegl. Elektron., 1965, 6, p.609.

14.41. Tao, T.F.: "Surface Wave along a Ferrite Road Magnetized Below Saturation", Trans. IEEE, 1966, AP-14, p.375.

14.42. Bolle, D.M.: "Ferrite-Loaded Rectangular Cavities", Electron. Lett., 1966, 2, p.17.

14.43. Clarricoats, P.J.B., and Olver, A.D.: "Propagation in Anisotropic Radially Stratified Circular Waveguides", Electron. Lett., 1966, 2, p.37.

14.44. Barzilai, G., and Gerosa, G.: "Rectangular Waveguides Loaded with Magnetized Ferrite and Thermodynamic Paradox", Proc. IEE, 1966, 113, p.285.

14.45. Seckelmann, R.: "Characteristics of Helical Phaseshifters", Trans. IEEE, 1966, MTT-14, p.24.

14.46. Schlomann, E.: "Theoretical Analysis of Twin-Slab Phaseshifters in Rectangular Waveguide", Trans. IEEE, 1966, MTT-14, p.15.

14.47. Clark, W.P., Hering, K.H., and Charlton, D.A.: "TE-Mode Solutions for Partially Ferrite Filled Rectangular Waveguide Using ABCD Matrices", Int. Conv. Rec. IEEE, 1966, 14, pt 5, p.39.

14.48. Wantuch, E., and Lepore, D.: "Extremely-High-Power, X-Band, Three-Port Circulator", J. Appl. Phys., 1966, 37, p.1079.

14.49. von Aulock, W.H.: "Theory of Linear Devices for Microwave Applications", J. Appl. Phys., 1966, 37, p.939.

14.50. Gerosa, G.: "Modes in Waveguides Filled with Longitudinally Magnetized Ferrite", Trans. IEEE, 1966, AP-13, p.983.

14.51. Gerosa, G., and Ottavi, C.M.: "Experimental Verification of Theoretical Behaviour of Ferrite Structures", Electron. Lett., 1966, 2, p.132.

14.52. Helszajn, J.: "Analysis of Coupled Waveguide Circulator Using Nonreciprocal Attenuation", Radio Electron. Eng., 1966, 31, p.319.

14.53. Hagelin, S.: "Flow-Graph Analysis of 3- and 4-Port Junction Circulators", Trans. IEEE, 1966, MTT-14, p.243.

14.54. Alzetta, G., and Battaglia, A.: "Ferrite Microwave Power Levels", Nuovo Cim., 1966, 43B, p.183.

14.55. Weiner, M.M., Brodwin, M.E., and Miller, D.: "Propagation of Quasi-TEM Mode in Ferrite-Filled Coaxial Line", Trans. IEEE, 1966, MTT-14, p.49.

14.56. Samaddar, S.N.: "Modes in Waveguides Filled with Longitudinally Magnetized Ferrite", Trans. IEEE, 1966, AP-14, p.408.

14.57. Vol'man, V.I.: "Waveguide Y-Circulator with Ferrite Sphere", Radiotekhnika, 1966, 21, no.2, p.31, and Telecommun. Radio Eng. Pt 2, 1966, 21, no.2, p.85.

14.58. Stern, E., and Ince, W.J.: "Temperature Stabilization of Unsaturated Microwave Ferrite Devices", J. Appl. Phys., 1966, 37, p.1075.

14.59. Eaves, R.E., and Bolle, D.M.: "Perturbation Theoretical Calculation of Differential Phase Shifts in Ferrite Loaded Circular Waveguide in TE_{01} Mode", Electron. Lett., 1966, 2, p.275.

14.60. Whicker, L.R., and Jones, R.R.: "Digital Current-Controlled Latching Ferrite Phaseshifter", Trans. IEEE, 1966, MTT-14, p.45.

14.61. Benda, O., Novak, S., and Uher, L.: "Experimental Results on Switching Y-Circulator at X-Band", Elektrotech. Cas., 1966, 17, p.579.

14.62. Furukawa, S., and Horiguchi, S.: "Frequency Characteristics of Active Circulators", Proc. IEEE, 1966, 54, p.1615.

14.63. Buchta, G.: "Miniaturized Broadband E-Tee Circulator at X Band", Proc. IEEE, 1966, 54, p.1607.

14.64. Bonfeld, M.D., Linn, D.F., and Omori, M.: "Stripline Circulator", Trans. IEEE, 1966, MTT-14, p.98.

14.65. Helszajn, J.: "Ferrite Ring Stripline Junction Circulator", Radio Electron. Eng., 1966, 32, p.55.

14.66. Emmrich, P.: "Stripline Circulator for Low Temperatures", Arch. Elektr. Ubertrag., 1966, 20, p.237.

14.67. Passaro, W.C.: "Quasi-Single-Junction Six-Port Circulator", Proc. Natl. Electron. Conf., 1966, 22, p.45.

14.68. Antonenko, M.G.: "Study of the Operational Reliability of High-Power Ferrite Devices", Radiotekhnika, 1966, 21, no.11, p.73, and Telecommun. Radio Eng. Pt 2, 1966, 21, no.11.

14.69. Liebe, H.J., and Senitzky, B.: "Y-Circulator at 258 GHz", Trans. IEEE, 1967, MTT-15, p.190.

14.70. Hines, M.E., and Chow, K.K.: "Phaseshifter in C, X, and K Bands Using Segregated-Mode Resonance in Single-Crystal YIG", Trans. IEEE, 1967, MTT-15, p.181.

14.71. Ince, W.J., and Stern, E.: "Nonreciprocal Resonance Phaseshifters in Rectangular Waveguide", Trans. IEEE, 1967, MTT-15, no.2, p.87-95.

14.72. Bogdanov, A.G., Iretskaya, I.V., and Kartazhov, V.B.: "Experimental Study of the Field Structure in a Waveguide X-Circulator", Radiotekh. Elektron., 1967, 12, no.1, pp.153-6, and Radio Eng. Electron. Phys., 1967, 12, no.1.

14.73. Orlov, V.P.: "Calculation for a Rectangular Waveguide with a Transversely Magnetized Ferrite Rod Using the Galerkin-Ritz Method", Radiotekh. Elektron., 1967, 12, no.1, pp.41-50, and Radio Eng. Electron. Phys., 1967, 12, no.1.

14.74. Longley, S.R.: "Experimental 4-Port E-Plane Junction Circulators", Trans. IEEE, 1967, MTT-15, no.6, pp.378-80.

14.75. Kobierzycki, J.: "8-mm Ferrite Circulator", Pr. PIT, 1967, no.56, pp.69-76.

14.76. Tyukov, I.P.: "Symmetrical Y-Junctions with Ferrite-Dielectric Filling", Radiotekh. Elektron., 1967, 12, no.2, pp.367-70, and Radio Eng. Electron. Phys., 1967, 12, no.2.

14.77. Helszajn, J.: "Simplified Theory of the Three-Port Junction Ferrite Circulator", Radio Electron. Eng., 1967, 33, no.5, pp.283-8.

14.78. Ackers, J.K.: "Contribution to the Theory of Stripline Junction Circulator", Microwave J., 1967, 10, no.8, pp.57-9.

14.79. Tyukov, I.P.: "Theory of Symmetric Waveguide Circulators", Radiotekh. Elektron., 1967, 12, no.2, pp.268-77, and Radio Eng. Electron. Phys., 1967, 12, no.2.

14.80. Barsony, P.: "Stripline Y-Circulators", Hiradastechnika, 1967, 18, no.5, pp.138-43.

14.81. Bogdanov, A.G., Iretskaya, I.V., and Kartazhov, V.B.: "Experiments on the Field Structure in a Waveguide X-Circulator", Radiotekh. Elektron., 1967, 12, no.1, pp.144-6, and Radio Eng. Electron. Phys., 1967, 12, no.1, pp.140-1.

14.82. Sundarum, S., and Mani, G.S.: "Broadband 2-Port Circulators", Indian J. Technol., 1967, 5, no.7, pp.210-11.

14.83. Bosch, F., and Rupf, K.: "Wideband, Low-Field, Stripline Circulators for Use at 4GHz and $4.2^{\circ}K$", Arch. Elektr. Ubertrag., 1967, 21, no.10, pp.553-5.

14.84. Foniok, F.: "Resonant High-Power Ferrite Isolators", Pr. PIT, 1968, 18, no.62, p.21.

14.85. Hamasaki, J., and Takino, T.: "Low-Magnetic-Field Faraday Rotator Using an Iterative Optical Path", J. Quantum Electron. IEEE, 1968, QE-4, p.380.

14.86. Clark, W.P.: "Technique for Improving the Factor of Merit of a Twin-Slab Nonreciprocal Ferrite Phaseshifter", Trans. IEEE, 1968, MTT-16, p.974.

14.87. Vamberskii, M.V., and Kazantsev, V.I.: "Design of Waveguide H-Plane Y-Circulators", Radiotekhnika, 1968, 23, no.10, p.15, and Telecommun. Radio Eng. Pt 2, 1968, 23, no.10.

14.88. Vol'man, V.I., et al.: "Five-Port Waveguide Circulator", Radiotekhnika, 1968, 23, no.10, p.101, and Telecommun. Radio Eng. Pt 2, 1968, 23, no.10.

14.89. Medoks, A.G.: "Algorithm for Design of Waveguide Y-Circulators", Radiotekhnika, 1968, 23, no.9, p.21, and Telecommun. Radio Eng. Pt 2, 1968, 23, no.9.

14.90. DeMeo, E.A., and Heller, G.S.: "Nonreciprocal Polycrystalline Antiferromagnetic Devices for Millimetre Waves", Trans. IEEE, 1968, MAG-4, p.603.

14.91. Bernardi, P., and Aldoni, F.: "Impedance-Wall Concept Applied to Waveguide with Thin Ferrite Slab", Note Recens. Not., 1968, 17, p.481.

14.92. Smoczynski, L.: "Designing Broadband Stripline Ferrite Circulators", Rozpr. Elektrotech., 1968, 14, p.539.

14.93. Buck, D.C.: "Propagation in Longitudinally Magnetized-Ferrite Loaded Waveguide", Trans. IEEE, 1968, MTT-16, p.1028.

14.94. Stern, R.A., and Agrios, J.P.: "500-kW X-Band Air-Cooled Ferrite Latching Switch", Trans. IEEE, 1968, MTT-16, p.1034.

14.95. Kostychev, Yu.G.: "Calculation of Propagation Coefficients of Gyrotropic Waveguides", Radiotekh. Elektron., 1968, 13, p.2063, and Radio Eng. Electron. Phys., 1968, 13, no.11, pp.1807-9.

14.96. Bosch, F., and Moursi, Z.: "Determination of Low-Field Working Point of Stripline Circulators", Arch. Elektr. Ubertrag., 1968, 22, p.605.

14.97. Gromova, L.I., and Kurushin, P.Ye.: "Energy Distribution in Ferrite Switches with Biased Field", Radiotekh. Elektron., 1968, 13, no.9, p.1716, and Radio Eng. Electron. Phys., 1968, 13, no.9, pp.1503-5.

14.98. Hord, W.E., and Rosenbaum, F.J.: "Propagation in Longitudinally Magnetized, Ferrite-Filled, Square Waveguide", Trans. IEEE, 1968, MTT-16, p.967.

14.99. Meier, P.J.: "Latching Reciprocal Polarization-Insensitive Phaseshifter", Trans. IEEE, 1968, MTT-16, p.958.

14.100. Ford, W.E., Rosenbaum, F.J., and Boyd, C.R.: "Theory of Suppressed-Rotation Reciprocal Ferrite Phaseshifter", Trans. IEEE, 1968, MTT-16, p.902.

14.101. Nikol'skii, V.V., et al.: "Scattering Matrix of a Waveguide Transformer Containing Ferrite. Transverse Magnetization", Radiotekh. Elektron., 1968, 13, p.2172, and Radio Eng. Electron. Phys., 1968, 13, no.12, pp.1905-10.

14.102. Davis, L.E., and Bacon, J.R.: "Latched Three-Port Waveguide Circulators", Electron. Lett., 1968, 4, p.490.

14.103. Bosma, H.: "General Model for Junction Circulators. Magnetization and Bias Field", Trans. IEEE, 1968, MAG-4, p.587.

14.104. Spitalnik, R.: "Stripline Four-Port Nonreciprocal Device Using a YIG Disc", Proc. IEEE, 1968, 56, p.1747.

14.105. Feoktistov, V.G.: "Design of a Stripline Y-Circulator", Radiotekh. Elektron., 1968, 13, p.1281, and Radio Eng. Electron. Phys., 1968, 13, no.7, pp.1111-5.

14.106. Kovbasa, A.P., and Shelamov, G.N.: "Resonant Frequency of an Obliquely Oriented Ferrite Disc and a Cylinder", Izv. VUZ Radioelektron., 1969, 12, p.1452.

14.107. Kumar, R.C.: "S-Band Y-Junction Stripline Circulator with Triangular Ferrite", Electron. Eng., 1969, 41, p.214.

14.108. Gruden, M., and Trontelj, L.: "Miniaturized Stripline Ferrite Circulator for S and X Bands", Elektroteh. Vestn., 1969, 36, no.3-5, p.61.

14.109. Hagelin, S.: "Analysis of Lossy Symmetrical Three-Port Network with Circulator Properties", Trans. IEEE, 1969, MTT-17, p.328.

14.110. Kartazhov, V.B.: "Analysis of a Screened Stripline with Transversely Magnetized Ferrite Using a Variational Method", Radiotekh. Elektron., 1969, 14, p.344, and Radio Eng. Electron. Phys., 1969, 14, no.2, pp.295-8.

14.111. Carter, J.L., and McGowan, J.W.: "X-Band Ferrite Varactor Limiter", Trans. IEEE, 1969, MTT-17, p.231.

14.112. Wolff, I.: "Fields in Gyrotropic Microwave Structures", Frequenz, 1969, 23, p.172.

14.113. Gromova, L.I., and Kurushin, P.Ye.: "Temperature Stability of Waveguide Y-Circulators", Izv. VUZ Radioelektron., 1969, 12, no.6, p.648.

14.114. Kasevich, R., and Wantuch, E.: "High-Performance High-Power Analogue Phaseshifter", Proc. IEEE, 1969, 57, p.1427.

14.115. Kaneki, T.: "Analysis of Linear Microstrip Using Arbitrary Ferromagnetic Substrate", Electron. Lett., 1969, 5, p.463.

14.116. Hasheider, H.: "Ferrite Circulator Tunable over an Octave", Int. Elektron. Rdsch, 1969, 23, no.9, p.233.

14.117. Launets, V.L., Novitskas, M.M., and Shugurov, V.K.: "Exciting a Waveguide by a Small Ferrite Sphere", Litov. Fiz. Sb., 1969, no.6, p.1051.

14.118. Il'chenko, M.Ye.: "Ferrite Resonator in Square Waveguide", Izv. VUZ Radioelektron., 1969, 12, p.1443.

14.119. Kazyulin, A.F.: "Theory of Excitation of Gyrotropically Filled Waveguides", Radiotekh. Elektron., 1969, 14, no.10, p.1885, and Radio Eng. Electron. Phys., 1969, 14, no.10, pp.1627-30.

14.120. Kinsey, R.R.: "Peak-Power Threshold Approximation for Remanent Ferrite Phase Shifters", Trans. IEEE, 1969, MTT-17, p.790.

14.121. Singh, J.K.: "3-cm Nonreciprocal Circular Polarizer", Int. J. Electron., 1969, 26, p.603.

14.122. Freibergs, E.: "Half-Height-Waveguide Y- and T-Circulators", Trans. IEEE, 1969, MTT-17, p.729.

14.123. Khlystov, A.S., and Perveeva, A.I.: "Effect of Magnetic Losses on Spectrum of Natural Oscillations of a Cylindrical Cavity with a Magnetized Ferrite Rod", Radiotekh. Elektron., 1969, 14, no.9, p.1566, and Radio Eng. Electron. Phys., 1969, 14, no.9, pp.1360-4.

14.124. Siekanowicz, W.W., et al.: "Temperature-Stable, Fail-Safe, Latching Ferrite TR Switch", Trans. IEEE, 1969, MTT-17, p.712.

14.125. Kartazhov, V.B.: "Variational Calculations on a Screened Stripline with Transversely Magnetized Ferrite Inserts", Des. Electron., 1969, 7, no.3, p.54.

14.126. Grossbach, R.: "Equivalent Circuit for Spherical Planar-Ferrite Resonators", Proc. IEEE, 1969, 57, p.1236.

14.127. Lait, A.J.: "Broadband Circular Polarizers", Marconi Rev., 1969, 32, p.159.

14.128. Bogdanov, A.G.: "Design of Waveguide X-Circulators", Radiotekh. Elektron., 1969, 14, p.610, and Radio Eng. Electron. Phys., 1969, 14, no.4, pp.528-32.

14.129. Konishi, Y., et al.: "Circulator Loaded by Slow-Wave Circuit", Tech. J. Jap. Broadcast. Corp., 1969, 21, p.80.

14.130. Groll, H., Berghofer, W., and Detlefsen, J.: "Electronically Variable Phaseshifters and Attenuators in Stripline Structure", Nachr. Tech. Z., 1969,22,p.261.

14.131. Foniok, F.: "Ferrite Materials for X-Band Reciprocal Reggia-Spencer Phaseshifters", Pr. PIT, 1969, 69, no.64, p.13.

14.132. Kovbasa, A.P., and Shelamov, G.N.: "Time Response of Microwave Four-Poles with Ferrite Resonators", Izv. VUZ Radioelektron., 1969, 12, no.6, p.634.

14.133. Nelson, T., et al.: "Small Analogue Stripline X-Band Ferrite Phaseshifter", Trans. IEEE, 1970, MTT-18, p.45.

14.134. Helszajn, J., and McDermott, M.: "Junction Inductance of a Lumped-Constant Circulator", Trans. IEEE, 1970,MTT-18,p.50.

14.135. Chen, D., et al.: "Magnetooptic Mode Splitting in an Anisotropic Laser Cavity", J. Quantum Electron. IEEE, 1970, QE-6, p.259.

14.136. Il'chenko, M.Ye.: "Effect of Coupling between a Ferrite Resonator and Transmission Line on Effective Parameters above Threshold", Radiotekh. Elektron., 1970, 15, p.175, and Radio Eng. Electron. Phys., 1970, 15, no.1, pp.145-7.

14.137. Tao, T.F., and Tully, J.W.: "Non-magnetostatic Volume- and Surface-Wave Modes on Gyromagnetic YIG-Rod Waveguides", Appl. Phys. Lett., 1970, 16, no.5, p.194.

14.138. Hartwig, C.P., and Readey, D.W.: "Ferrite Film Circulator", J. Appl. Phys., 1970, 41, p.1351.

14.139. Chao, G.: "Matched Microwave Limiter", Trans. IEEE, 1970, MTT-18, p.283.

14.140. Gonzalez, G., and Johnson, V.R.: "Propagation in a Rectangular Waveguide Partially Filled with a Linearly Varying Dielectric", Trans. IEEE, 1970, MTT-18,p.404.

14.141. Hurd, R.A.: "Surface Waves at Ferrite-Metal Boundaries", Electron. Lett., 1970, 6, p.262.

14.142. O'Brien, K.C.: "Microwave Properties of Uniformly Magnetized Material Filling a Rectangular Waveguide Operating in TM_{n0} Modes", Trans. IEEE, 1970,MTT-18,p.377.

14.143. O'Brien, K.C.: "Microwave Properties of Rectangular Waveguide Semi-Infinitely Filled", Trans. IEEE, 1970, MTT-18, p.400.

14.144. Biglin, D.E.: "Nonreciprocal Junction. Circuit Element", Marconi Rev., 1970, 33, no.176, p.55.

14.145. Marriot, S.P.A.: "Microwave Devices Using Spheres of Monocrystalline Garnets", Marconi Rev., 1970, 33, no.176, p.79.

14.146. Potekhin, A.I., and Yurgenson, R.R.: "Analysis of Microwave Networks Containing Azimuthally Magnetized Ferrites", Radiotekh. Elektron., 1970, 15, p.456, and Radio Eng. Electron. Phys., 1970, 15,no.3,pp.392-9.

14.147. Maksimov, V.I.: "Experimental Study of the Mutual Coupling of Two Ferrite Resonators in a Waveguide", Radiotekh. Elektron., 1970, 15, p.601, and Radio Eng. Electron. Phys., 1970, 15, no.3, pp.512-5.

14.148. Buts, V.A.: "Theory of Resonators with Non-Mutual Ferrite Filling", Izv. VUZ Radiofiz., 1970, 13, p.426.

14.149. Moore, I.J.H.S., and Stewart, J.A.C.: "Analysis of a Magnetically Tuned Coaxial Cavity Using State-Space Techniques", Electron. Lett., 1970, 6, p.391.

14.150. Longley, S.R.: "Multioctave Tunable Three-Port Circulator Using a YIG Sphere", Electron. Lett., 1970, 6, p.406.

14.151. Skolnick, M.L., and Buczek, C.J.: "Proposed Interferometric Optical Isolator", J. Quantum Electron. IEEE,1970,QE-6,p.524.

14.152. Salmond, W.E., and Yeh, C.: "Ferrite-Filled Elliptical Waveguides. I and II", J. Appl. Phys., 1970, 41, pp.3210 and 3221.

14.153. Blanc, F., Fanguin, R., and Raoult, G.: "Propagation in Cylindrical Waveguides Filled with Anisotropic Material", Inter Electron., 1970, 25, no.6, p.40.

14.154. Khlystov, A.S., Mudrov, A.E., and Perveeva, A.I.: "Modes in a Rectangular Waveguide with a Transverse Magnetized Ferrite Plate", Izv. VUZ Fiz., 1970, no.4, p.74.

14.155. Gardiol, F.E.: "Anisotropic Slabs in Rectangular Waveguides", Trans. IEEE, 1970, MTT-18, p.461.

14.156. Helszajn, J.: "Frequency and Bandwidth of H-Plane TEM-Junction Circulator", Proc. IEE, 1970, 117, p.1235.

14.157. Holmes, W.H.: "Minimal-Sensitivity Circulator Using Differential Transconductances", Proc. IEEE, 1970, 58, p.1143.

14.158. Hammer, G.: "Analysis of a Waveguide Y-Circulator", Period. Polytech. Electr. Eng., 1970, 14, no.1, p.61.

14.159. Erberhardt, N., and Horvath, V.V.: "Plane and Quasi-Optical Wave Propagation in Gyromagnetic Media", Trans. IEEE, 1970, MTT-18, p.554.

14.160. Helszajn, J.: "Ferrite Parameters of the Junction Circulator", Electron. Compon., 1970, 11, no.7, p.785.

14.161. Castillo, J.B., and Davis, L.E.: "Computer-Aided Design of Three-Port Waveguide-Junction Circulators", Trans. IEEE, 1970, MTT-18, p.25.

14.162. Siekanowicz, W.W., Paglione, R.W., and Walsh, T.E.: "Latching Ring-and-Post Ferrite Waveguide Circulator", Trans. IEEE, 1970, MTT-18, p.212.

14.163. Panov, A.E.: "Analysis of Ferrite Y-Circulators with Lumped Constants", Radiotekh. Elektron., 1970, 15, no.2, p.298, and Radio Eng. Electron. Phys., 1970, 15, no.2, pp.258-66.

14.164. Helszajn, J.: "Adjustment of m-Port Single-Junction Circulator", Trans. IEEE, 1970, MTT-18, no.10, pp.705-11.

14.165. Hofer, W.J.R., et al.: "Input Impedance of Cavity Containing a Gyromagnetic Sample with Very Narrow Line", Nachr. Tech. Z., 1970, 23, p.121.

14.166. Kondrat'ev, B.V., and Belyaeva, M.A.: "Helical Waveguide in Anisotropic Medium", Ukr. Fiz. Zh., 1970, 15, no.10, pp.1607-10.

14.167. Janke, R.: "Termination of a Circulator with Mismatched Absorber", Nachr. Tech., 1970, 20, no.11, pp.414-21.

14.168. Nakahara, S., and Kurebayashi, H.: "Short-Slot Waveguide Latching Ferrite Switch", Trans. IEEE, 1970, MTT-18, no.12, pp.1048-51.

14.169. Nakahara, S., and Orime, N.: "Broadband Stripline Miniature Circulators", Mitsubishi Denki Lab. Rep., 1970, 11, no.1-2, pp.63-88.

14.170. Kalina, V.G., Budanov, V.N., and Belyakov, S.V.: "Resonance Characteristics of a Ferrite Sphere in an Unmatched Waveguide", Radiotekh. Elektron., 1970, 15, no.4, and Radio Eng. Electron. Phys., 1970, 15, no.4, pp.722-4.

14.171. Vol'man, V.I., and Muravtsov, A.D.: "Miniature Stripline Isolator", Radiotekhnika 1970, 25, no.7, and Telecommun. Radio Eng. Pt 2, 1970, 25, no.7, pp.152-4.

14.172. Slutskaya, V.V.: "Field-Displacement Ferrite Isolators and Influence of Dielectric", Elektrosvyaz, 1970, 24, no.8, and Telecommun. Radio Eng. Pt 1, 1970, 24, no.8, pp.51-3.

14.173. Kapilevich, B.Yu.: "Phase Characteristics of a Stripline with Two Oppositely Magnetized Ferrite Plates", Radiotekh. Elektron., 1970, 15, no.9, and Radio Eng. Electron. Phys., 1970, 15, no.9, pp.1597-600.

14.174. Bova, N.T., and Khramov, V.A.: "Design Automation of Microwave Devices", Izv. VUZ Radioelektron., 1970, 14, no.12, pp.1488-90.

14.175. Narayanan, K.G., and Sharma, G.P.: "Isolation Rating of Ferrite Components at High Pulse Powers", Trans. IEEE, 1970, MTT-18, p.322.

14.176. Owen, B., and Barnes, C.E.: "Compact Turnstile Circulator", Trans. IEEE, 1970, MTT-18, no.12, pp.1096-100.

14.177. Boyd, C.R.: "Dual-Mode Latching Reciprocal Ferrite Phaseshifter", Trans. IEEE, 1970, MTT-18, no.12, pp.1119-24.

14.178. Buck, G.J.: "Ferrite Microstrip Phaseshifters with Transverse and Longitudinal Magnetization", Trans. IEEE, 1970, MTT-18, no.12, pp.1170-3.

14.179. Konishi, Y., and Hoshino, N.: "Design of a Broadband Isolator", Trans. IEEE, 1971, MTT-19, no.3, pp.260-9.

14.180. Gardiol, F.E., and Van der Vorst, A.S.: "Computer Analysis of E-Plane Resonance Isolators", Trans. IEEE, 1971, MTT-19, no.3, pp.315-22.

14.181. Decreton, M., et al.: "Computer Optimization of E-Plane Resonance Isolators", Trans. IEEE, 1971, MTT-19, no.3, pp.322-31.

14.182. Okamoto, N., Nishioka, I., and Nakanishi, Y.: "Scattering by a Ferrimagnetic Circular Cylinder in a Rectangular Waveguide", Trans. IEEE, 1971, MTT-19, no.6, pp.521-7.

14.183. Gardiol, F.E.: "High-Power Circulators with Mismatched Terminations", Trans. IEEE, 1971, MTT-19, no.6, pp.561-2.

14.184. Ince, W.J., Temme, D.H., and Willwerth, F.G.: "Toroid Corner Chamfering as a Method of Improving the Factor of Merit of Latching Ferrite Phasers", Trans. IEEE, 1971, MTT-19, no.6, pp.563-4.

14.185. Trambarulo, R.: "30-GHz Inverted-Microstrip Circulator", Trans. IEEE, 1971, MTT-19, no.7, pp.662-4.

14.186. Kumar, R.C.: "Experimental Results on S-Band Stripline Y-Circulator", Proc. IEEE, 1971, 59, no.1, pp.109-12.

14.187. DeCamp, E.E., and True, R.M.: "1-MW Four-Port E-Plane Junction Circulator", Trans. IEEE, 1971, MTT-19, no.1, pp.100-3.

14.188. Ince, W.J., et al.: "Comparison of Two Nonreciprocal Latching Ferrite Phasers", Trans. IEEE, 1971, MTT-19, no.1, pp.105-7.

14.189. Kumar, R.C.: "Fine Tuning of Microwave Ferrite Circulators and Isolators", Electro-Technol., 1971, 15, no.1, pp.23-4.

14.190. Il'chenko, M.Ye.: "Ferrite Filter-Limiters Employing Resonant Rotation of the Plane of Polarization", Izv. VUZ Radioelektron., 1971, 14, no.12, pp.1453-9.

14.191. Fukasawa, A., Abe, M., and Goto, A.: "Microwave Nonreciprocal Circuit. II", Oki Rev., 1971, 38, no.4, pp.29-38.

14.192. Salay, S.J., and Peppiatt, H.J.: "Input Impedance Behaviour of Stripline Circulator", Trans. IEEE, 1971, MTT-19, no.1, pp.109-10.

14.193. Hines, M.E.: "Reciprocal and Non-reciprocal Modes of Propagation in Ferrite Stripline and Microstrip Devices", Trans. IEEE, 1971, MTT-19, no.5, pp.442-51.

14.194. Castillo, J.B., and Davis, L.E.: "Identification of Spurious Modes in Circulators", Trans. IEEE, 1971, MTT-19, no.1, pp.112-3.

14.195. Konishi, Y.: "Propagation Coefficient of Stripline Loaded with Arbitrarily Magnetized Ferrite", NHK Tech. J., 1971, 23, no.1, pp.68-78.

14.196. Sultan, N.B.: "Generalized Theory of Waveguide Differential-Phase Sections", Trans. IEEE, 1971, MTT-19, no.4, pp.348-57.

14.197. Bex, H., and Schwartz, E.: "Performance Limitations of Lossy Circulators", Trans. IEEE, 1971, MTT-19, no.5, pp.493-4.

14.198. Minor, J.C., and Bolle, D.M.: "Modes in Shielded Microstrip on a Ferrite Substrate Transversely Magnetized", Trans. IEEE, 1971, MTT-19, no.7, pp.570-7.

14.199. Ogasawara, N., and Kaji, M.: "Coplanar-Guide and Slot-Guide Junction Circulators", Electron. Lett., 1971, 7, no.9, pp.220-1.

14.200. Poirier, A.L.: "Integrated Microstrip Circulator", Trans. IEEE, 1971, MTT-19, no.7, pp.661-2.

14.201. Cohen, M.: "Electronically Tunable Nonreciprocal X-Band YIG Triplexer", Microwave J., 1971, 14, no.3, pp.47-52.

14.202. Andrews, R.S.: "Scattering-Matrix (Analysis) of Three-Port Circulators", Electron. Lett., 1971, 7, no.12, pp.351-3.

14.203. Pauchard, R., et al.: "Electromagnetic Surface Waves in a Metallized Ferrite Slab", Electron. Lett., 1971, 7, no.15, pp.428-30.

14.204. Hindin, H.J.: "Relationship between the Scattering Parameters of a Passive Lossy Nonreciprocal Two-Port", Trans. IEEE, 1971, MTT-19, no.9, p.781.

14.205. Ivanov, K.P., and Garichev, S.I.: "Ferrite-Loaded Helical Line with Coaxial Inner Conductor", Trans. IEEE, 1971, MTT-19, no.9, pp.784-6.

14.206. Stern, R.A.: "Fast 3-mm Ferrite Switch", Trans. IEEE, 1971, MTT-19, no.9, pp.786-8.

14.207. Wolff, I.: "Contribution to the Theory of Transversely Magnetized Ferrite", Frequenz, 1971, 25, no.8, pp.235-41.

14.208. Launets, V.L., Novitskas, M.M., and Shugurov, V.K.: "Waveguide Excitation by a Small Ferrite Sphere with Wall Effects", Litov. Fiz. Sb., 1971, 11, no.2, pp.229-38.

14.209. Taggart, D.A., and Schott, F.W.: "Ferrite-Filled Cavity Resonators", Appl. Sci. Res., 1971, 25, no.1-2, pp.35-53.

14.210. Stolyarov, A.K., and Smirnov, V.S.: "Magnetooptic Phenomena in Ferrite Microstrip", Radiotekh. Elektron., 1971, 16, no.2,pp.434-6, and Radio Eng. Electron. Phys., 1971, 16, no.2, pp.369-71.

14.211. Churkin, V.I., and Chelishchev, N.N.: "Effect of Mismatched Load on Performance of Ferrite Power Limiter", Radiotekh. Elektron., 1971, 16, no.5, pp.840-1, and Radio Eng. Electron. Phys., 1971, 16, no.5,pp.859-60.

14.212. Stolyarov, A.K., and Smirnov, V.S.: "Nonreciprocal Properties of Miniature Stripline Isolator", Radiotekh. Elektron., 1971, 16, no.10, pp.1954-5, and Radio Eng. Electron. Phys., 1971, 16, no.10, pp.1744-6.

14.213. Tsukamoto, N., Suzuki, M., and Matsumoto, T.: "Analysis of Ferrite Waveguide T-Junction", Electron. Commun. Jap., 1971, 54, no.4, pp.49-56.

14.214. Rogozin, V.V.: "Equivalent Circuit of a Two-Mode Resonator Containing a Magnetized Ferrite Ellipsoid", Radiotekh. Elektron., 1971, 16, no.12, pp.2298-9, and Radio Eng. Electron. Phys., 1971, 16, no.12, pp.2106-8.

14.215. Kaneki, T.: "Variational Analysis for a (Magnetized) Ferrite-Substrate Microstrip", Electron. Commun. Jap., 1971, 54, no.11, pp.40-5.

14.216. Helszajn, J., Walker, D., and Aitken, F.M.: "Varactor-Tuned Lumped-Element Circulators", Trans. IEEE, 1971, MTT-19, no.10, pp.825-6.

14.217. Wen, C.P.: "Symmetrical Trough Waveguide Nonresonant Ferrite Isolators and Steerable Antenna", RCA Rev., 1971, 32, no.2, pp.279-88.

14.218. Alekseev, L.V.: "Analysis of X-Circulators", Izv. VUZ Radioelektron., 1971, 14, no.8, pp.910-8.

14.219. Spaulding, W.G.: "Application of Periodic Loading to a Ferrite Phaseshifter Design", Trans. IEEE, 1971, MTT-19, no.12, pp.922-8.

14.220. Whicker, L.R.: "Reciprocal Phaser for Use at Millimetre Wavelengths", Trans. IEEE, 1971, MTT-19, no.12, pp.944-5.

14.221. Ohkawa, S.: "Analysis of Annular Ferrite-Loaded Waveguide", J. Fac. Eng. Chiba Univ., 1971, 22, no.41, pp.35-9.

14.222. de Santis, P., and Pucci, F.: "Novel Type of MIC Symmetrical 3-Port Circulator", Electron. Lett., 1972,8,no.1,pp.12-3.

14.223. Roveda, R., Borghese, C., and Cattarin, P.I.G.: "Dissipative Parameters in Ferrites and Insertion Losses in Waveguide Y-Circulators below Resonance", Trans. IEEE, 1972, MTT-20, no.2, pp.89-96.

14.224. Hord, W.E., et al.: "Reciprocal Faraday Rotation Phaseshifter", Trans. IEEE, 1972, MTT-20, no.2, pp.112-9.

14.225. Eidmann, K., et al.: "Optical Isolators for High-Power Giant-Pulse Lasers", J. Phys. E, 1972, 5, no.1, pp.56-8.

14.226. de Santis, P., et al.: "Hybrid-Frequency Cutoffs in Gyrotropic Waveguides", Trans. IEEE, 1972, MTT-20, no.3, pp.237-8.

14.227. Helszajn, J., and McDermott, M.: "Mode Chart for E-Plane Circulators", Trans. IEEE, 1972, MTT-20, no.2, pp.187-8.

14.228. Wang, S., Shah, M., and Crow, J.D.: "Gyrotropic and Anisotropic Materials for Mode Conversion in Thin-Film Optical Waveguide", J. Appl. Phys., 1972, 43, no.4, pp.1861-75.

14.229. Sinev, H., and Daskalov, R.: "Simplifying the Characteristic Control of a Gyromagnetic Coaxial Waveguide", Elektro Prom. Prib., 1972, 7, no.2, pp.43-4.

14.230. Helszajn, J.: "Three-Resonant Mode Adjustment of Waveguide Circulator", Radio Electron. Eng., 1972, 42, no.5, pp.213-6.

14.231. Konishi, Y.: "Propagation Coefficient of Stripline Loaded with Arbitrarily Magnetized Ferrite", Electron. Commun. Jap., 1972, 54, no.1, pp.82-91.

14.232. Hanfling, J.D., and Monaghan, S.R.: "Feedthrough for Digital Latching Ferrite Phasers", Trans. IEEE, 1972, MTT-20, no.8, pp.549-52.

14.233. Salay, S.J., and Peppiatt, H.J.: "Accurate Junction Circulator Design Procedure", Trans. IEEE, 1972, MTT-20, no.2, pp.192-3.

14.234. Tsandoulas, G.N., Willwerth, F.G., and Ince, W.J.: "Mode Characteristics in Phaseshifter Parametrization", Trans. IEEE, 1972, MTT-20, no.4, pp.253-8.

14.235. Hering, K.H.: "Design of an X-Band High-Power Ferrite Phaseshifter", Trans. IEEE, 1972, MTT-20, no.4, pp.284-6.

14.236. Owen, B.: "Identification of Modal Resonances in Ferrite-Loaded Waveguide Y-Junctions", Bell Syst. Tech. J., 1972, 51, no.3, pp.595-627.

14.237. Castillo, J.B., and Davis, L.E.: "Higher-Order Approximation for Waveguide Circulators", Trans. IEEE, 1972, MTT-20, no.6, pp.410-2.

14.238. Dinger, R.J., and White, D.J.: "Use of InSb Film as K-Band Field-Displacement Isolator", Proc. IEEE, 1972, 60, no.5, pp.646-7.

14.239. Pokusin, D.N.: "Analysis of Waveguide Phaseshifters Using a Ferrite with Rectangular Hysteresis Loop", Radiotekh. Elektron., 1972, 16, no.9, pp.1603-17, and Radio Eng. Electron. Phys., 1972, 16, no.9, pp.1465-77.

14.240. Konishi, Y.: "Theoretical Concept for Wideband Gyromagnetic Devices", Trans. IEEE, 1972, MAG-8, no.3, pp.505-8.

14.241. Wolff, I.: "Lowest-Order Mode and Quasi-TEM Mode in a Ferrite-Filled Coaxial Line or Resonator", Trans. IEEE, 1972, MTT-20, no.8, pp.558-60.

14.242. Kaneki, T.: "Propagation Coefficients of Stripline Using Longitudinally Magnetized Ferrite Substrate", NHK Tech. J., 1972, 24, no.2, pp.43-8.

14.243. Yoshida, T., Umeno, M., and Miki, S.: "Propagation Characteristics of a Rectangular Waveguide Containing a Cylindrical Rod of Magnetized Ferrite", Trans. IEEE, 1972, MTT-20, no.11, pp.739-43.

14.244. Helszajn, J.: "Susceptance-Slope Parameter of Junction Circulators", Proc. IEE, 1972, 119, no.9, pp.1257-61.

14.245. Khlystov, A.S., and Mudrov, A.E.: "Theory of a Ferromagnetic Resonance Isolator", Izv. VUZ Fiz., 1972, no.9, pp.158-60.

14.246. Deschamps, J., Fitaire, M., and Lagoutte, M.: "Experimental Study of Inverse Faraday Effect in Plasmas", Rev. Phys. Appl., 1972, 7, no.3, pp.155-62.

14.247. Hayashi, Y., Fujiki, Y., and Suzuki, M.: "Analytical Method for Microstrip Line with Magnetized Ferrite Slab", Electron. Commun. Jap., 1972, 54, no.11, pp.54-61.

14.248. Kostychev, Yu.G.: "Shielded Balanced Stripline with Longitudinally Magnetized Ferrite", Radiotekh. Elektron., 1972, 17, no.5, pp.926-31, and Radio Eng. Electron. Phys., 1972, 17, no.5, pp.721-5.

14.249. Churkin, V.I., Chelischev, N.N., and Izotov, V.A.: "Experimental Investigation of Interaction of a Nonlinear Ferrite Resonator with the Field of a Shortcircuited Waveguide", Radiotekh. Elektron., 1972, 17, no.5, pp.1076-7, and Radio Eng. Electron. Phys., 1972, 17, no.5, pp.836-7.

14.250. Yamamoto, S., et al.: "Normal-Mode Analysis of Gyrotropic Thin-Film Optical Waveguide", Electron. Commun. Jap., 1972, 55, no.10, pp.127-33.

14.251. Miura, T.: "Experimental High-Isolation Ferrite Substrate Circulator with Trigonally Symmetric Polepieces", Trans. IEEE, 1972, MAG-8, no.3, pp.509-10.

14.252. Nakahara, S., Kurebayashi, H., and Mizobuchi, A.: "6-GHz Rotating-Field Phaseshifter", Trans. IEEE, 1972, MAG-8, no.3, pp.544-7.

14.253. Muravtsov, A.D.: "Analysis of Variational Method of Limiting Isolator in Balanced Stripline", Radiotekhnika, 1972, 27, no.9, pp.50-7, and Telecommun. Radio Eng. Pt 2, 1972, 27, no.9, pp.93-9.

14.254. Kawabata, S., and Nakahara, S.: "Broadband Microstrip Circulators", Mitsubishi Denki Lab. Rep., 1972, 13, no.1-4, pp.97-121.

14.255. Jha, R.K., and Prasad, L.: "Stripline Ferrite Junction Circulator", Int. J. Electron., 1973, 34, no.4, pp.445-8.

14.256. Andresciani, V., and de Leo, R.: "Coherent-Light Propagation through Optical Fibres in Presence of High Magnetic Fields", Alta Freq., 1973, 42, no.4, pp.210-3.

14.257. El-Shandwily, M.E., Abdallah, E.A.F., and Kamal, A.A.: "General Field Theory Treatment of H-Plane Waveguide-Junction Circulators", Trans. IEEE, 1973, MTT-21, no.6, pp.392-403.

14.258. Libbey, W.M.: "Characteristic of a Microstrip Two-Meander Ferrite Phaseshifter", Trans. IEEE, 1973, MTT-21, no.7, pp.483-7.

14.259. Allen, J.L.: "Analysis of Lossy Ferrite Devices", Int. J. Electron., 1973, 34, no.6, pp.817-30.

14.260. Kislyakovskii, A.V., and Vodop'yanov, N.G.: "Resonant Ferrite Phaseshifter", Izv. VUZ Radioelektron., 1973, 16, no.1, pp.58-63.

14.261. Knerr, R.H.: "4-GHz Lumped-Element Circulator", Trans. IEEE, 1973, MTT-21, no.3, pp.150-1.

14.262. Kitasawa, T., et al.: "Slot Line on Magnetized Ferrite Substrate", Electron. Commun. Jap., 1973, 56, no.3, pp.50-5.

14.263. Yavnoshan, F.V., and Smirnova, A.V.: "Calculation of Losses in Waveguide Y-Circulators", Antenny, 1973, no.18, pp.135-40.

14.264. Vendik, O.G., and Yavnoshan, F.V.: "Calculating Waveguide Y-Circulators", Antenny, 1973, no.18, pp.124-35.

14.265. Kheifets, S.B.: "Calculating a Ferrite Film Microstrip Y-Circulator", Antenny, 1973, no.18, pp.140-6.

14.266. Wang, S., et al.: "Eigenmode Analysis of Propagation in Optical Waveguides on Gyrotropic Substrates", J. Appl. Phys., 1973, 44, no.7, pp.3232-9.

14.267. Markov, K.G., and Ivanov, S.A.: "Experimental Study of Coaxial-Tape S-Band Y-Circulators", Elektro Prom. Prib., 1973, 8, no.3, pp.94-6.

14.268. Helszajn, J.: "Waveguide and Stripline Four-Port Single-Junction Circulators", Trans. IEEE, 1973, MTT-21, no.10, pp.630-3.

14.269. Barr, O.C., McMahon, J.M., and Trenholme, J.B.: "Large-Aperture High-Extinction-Ratio Faraday-Rotator Isolator", J. Quantum Electron. IEEE, 1973, QE-9, no.11, pp.1124-5.

14.270. de Santis, P.: "Dispersion Characteristics for a Ferrimagnetic Plate", Appl. Phys., 1973, 2, no.4, pp.197-200.

14.271. Ushakov, V.M., and Khizhnyak, N.A.: "Rectangular Waveguide Loaded with Semi-Infinite Chain of Ferrite Spheres", Izv. VUZ Radiofiz., 1973, 16, no.6, pp.942-9.

14.272. Kapilevich, B.Yu., and Fedotova, T.N.: "Phase Characteristics of a Rectangular Waveguide with Symmetrical Transversely Magnetized Ferrite Layers", Izv. VUZ Radiofiz., 1973, 16, no.6, pp.950-5.

14.273. Vishvakarma, B.R., et al.: "X-Band Waveguide Star-Junction Circulator", J. Inst. Eng., 1973, 54, pt ET2, pp.49-51.

14.274. Muravtsov, A.D.: "Design and Experimental Investigation of Miniature Stripline Isolator", Radiotekhnika, 1973, 28, no.10, pp.47-52, and Telecommun. Radio Eng. Pt 2, 1973, 28, no.10, pp.94-7.

14.275. Allen, J.L.: "Phase and Loss Characteristics of High-Average-Power Ferrite Phasers", Trans. IEEE, 1973, MTT-21, no.8, pp.543-4.

14.276. Kapilevich, B.Yu.: "Polarization Properties of Rectangular Waveguide with Transversely Magnetized Ferrite", Radiotekhnika, 1973, 28, no.9, pp.98-100, and Telecommun. Radio Eng. Pt 2, 1973, 28, no.9, pp.125-7.

14.277. Courtois, L., Chiron, B., and Forterre, G.: "Microwave Propagation in a Magnetized Ferrite Blade", Cables Transm., 1973, 27, no.4, pp.416-35.

14.278. Modenov, V.P.: "Calculating Propagation Coefficients in Circular Waveguide with a Ferrite Rod by the Galerkin Method", Vychisl. Metody Program., 1973, no.20, pp.50-8.

14.279. Martynova, T.A.: "Propagation in a Waveguide Filled with Locally Laminar Gyrotropic Medium", Vychisl. Metody Program., 1973, no.20, pp.71-93.

14.280. Hoefer, W.J.R.: "Power/Frequency Distribution in a Cavity Containing a Resonant Ferrite Sample", Electron. Lett., 1973, 9, no.16, pp.357-9.

14.281. Stern, R.A.: "High-Power S-Band Junction Circulator", Trans. IEEE, 1973, MTT-21, no.12, pp.840-2.

14.282. Kanda, M., and May, W.G.: "Nonreciprocal Reflection-Beam Isolators for Far-IR Use", Trans. IEEE, 1973, MTT-21, no.12, pp.786-90.

14.283. Blanc, F., Fanguin, R., and Raoult, G.: "Guided-Wave Rotation and Ellipticity during Oblique Reflection on Absorbing Gyromagnetic Sample", Onde Electr., 1973, 53, no.11, pp.397-404.

14.284. Warner, J.: "Faraday Optical Isolator/Gyrator Design in Planar Dielectric-Waveguide Form", Trans. IEEE, 1973, MTT-21, no.12, pp.769-75.

14.285. Bernues, F.J., and Bolle, D.M.: "Digital Twin-Ferrite-Toroid Circular Waveguide Phaser", Trans. IEEE, 1973, MTT-21, no.12, pp.842-5.

14.286. Nagao, T.: "Circulator Action of a Stripline Y-Junction Fitted with Inhomogeneous Ferrite Rings", Mem. Def. Acad., 1973, 13, no.2, pp.215-27.

14.287. Bol'shakova, T.I.: "Miniature Waveguide Ferrite Isolator", Radiotekh. Elektron., 1973, 17, no.12, pp.2592-4, and Radio Eng. Electron. Phys., 1973, 17, no.12, pp.2079-81.

14.288. Otmakhov, Yu.A.: "Theory of a Miniature Y-Circulator with Lumped Parameters", Radiotekh. Elektron., 1973, 18, no.3, pp.461-9, and Radio Eng. Electron. Phys., 1973, 18, no.3, pp.335-41.

14.289. Babayan, G.Yu., and Vasil'yev, A.A.: "Determination of Characteristics of a Discrete Ferrite Phaseshifter in Rectangular Waveguide by the Perturbation Method", Radiotekh. Elektron., 1973, 18, no.5, pp.1063-6, and Radio Eng. Electron. Phys., 1973, 18, no.5, pp.783-5.

14.290. Korobkin, V.A., Pyatak, N.I., and Mekhed'kin, A.A.: "Electrically Controlled Wideband Polarizer Using a Circular Waveguide with a Periodic (Magnetized) Ferrite Structure", Radiotekh. Elektron., 1973, 18, no.6, pp.1270-1, and Radio Eng. Electron. Phys., 1973, 18, no.6, pp.931-2.

14.291. Chen, H.C.: "Constitutive Relations of a Moving Gyrotropic Medium", Int. J. Electron., 1974, 36, no.3, pp.319-28.

14.292. Gardiol, F.E.: "Circularly Polarized Electric Field in Rectangular Waveguide", Trans. IEEE, 1974, MTT-22, no.5, pp.563-5.

14.293. Kitlinski, M.: "Bidirectional Thin-Film (Ferrite) Lumped-Element Circulator", Electron. Lett., 1974, 10, no.6, pp.66-8.

14.294. El-Shandwily, M.E., and Abdallah, E.A.F.: "Finite-Gap Stripline Latching Circulator", Trans. IEEE, 1974, MTT-22, no.1, pp.57-60.

14.295. Bell, H.C., and Boyd, C.R.: "Optimum Filling of Ferrite Phaseshifters of Uniform Dielectric Constant", Trans. IEEE, 1974, MTT-22, no.4, pp.360-4.

14.296. Helszajn, J.: "Composite-Junction Circulators Using Ferrite Discs and Dielectric Rings", Trans. IEEE, 1974, MTT-22, no.4, pp.400-10.

14.297. Knerr, R.H.: "Lumped-Element Circulator without Crossovers", Trans. IEEE, 1974, MTT-22, no.5, pp.544-8.

14.298. Boyd, C.R.: "Comments on Design and Manufacture of Dual-Mode Reciprocal Latching Ferrite Phaseshifters", Trans. IEEE, 1974, MTT-22, no.6, pp.593-601.

14.299. Duputz, A.M., and Priou, A.C.: "Computer Analysis of Microwave Propagation in a Ferrite Loaded Circular Waveguide", Trans. IEEE, 1974, MTT-22, no.6, pp.601-13.

14.300. Moore, R.A., Kern, G.M., and Cooper, L.F.: "High-Average-Power S-Band Digital Phaseshifter", Trans. IEEE, 1974, MTT-22, no.6, pp.626-34.

14.301. Tseng, S.C.C., et al.: "Mode Conversion in Magnetooptic Waveguides Subjected to a Periodic Permalloy Structure", Appl. Phys. Lett., 1974, 24, no.6, pp.265-7.

14.302. Yamamoto, S., and Makimoto, T.: "Circuit Theory for a Class of Anisotropic and Gyrotropic Thin-Film Optical Waveguides", J. Appl. Phys., 1974, 45, no.2, pp.882-8.

14.303. Cattarin, P.I.G., and Roveda, R.: "Dissipative Parameters in Ferrites and Insertion Losses in Stripline Y-Circulators below Resonance", Trans. IEEE, 1974, MTT-22, no.7, pp.752-4.

14.304. Denlinger, E.J.: "Design of Partial Height Ferrite Waveguide Circulators", Trans. IEEE, 1974, MTT-22, no.8, pp.810-3.

14.305. Okean, H.C., and Steffek, L.J.: "Low-Loss, 3-mm, Junction Circulator", Microwave J., 1974, 17, no.4, pp.58-62.

14.306. Wu, Y.S., and Rosenbaum, F.J.: "Microwave Propagation in Magnetized Ferrite-Dielectric Composite Transmission Lines", J. Appl. Phys., 1974, 45, no.6, pp.2512-20.

14.307. Helszajn, J.: "Common Waveguide Circulator Configurations", Electron. Eng., 1974, 46, no.559, pp.66-9.

14.308. Igarashi, M., and Naito, Y.: "Theoretical Analysis of Magnetic Resonance Nonreciprocal Circuits", Trans. IEEE, 1974, MTT-22, no.9, pp.821-9.

14.309. Kanda, M., and May, W.G.: "Hollow-Cylinder Waveguide Isolators for Use at Millimetre Wavelengths", Trans. IEEE, 1974, MTT-22, no.11, pp.913-7.

14.310. Akaiwa, Y.: "Operation Modes of a Waveguide Y-Circulator", Trans. IEEE, 1974, MTT-22, no.11, pp.954-60.

14.311. Kravchenko, V.F., and Kostychev, Yu.G.: "Application of a Structural Method to Analysis of Waveguides with Homogeneous Magnetodielectric Filling", Izv. VUZ Radioelektron., 1974, 17, no.9, pp.93-6.

14.312. Zhurakhovskii, V.A.: "Mathematical Description of an Electronic Rotator Immersed in an Adiabatically Nonuniform Magnetic Field", Radiotekh. (Kharkov), 1974, no.28, pp.60-71.

14.313. Bernues, F.J., and Bolle, D.M.: "Ferrite-Loaded Waveguide Discontinuity Problem", Trans. IEEE, 1974, MTT-22, no.12, pp.1187-93.

14.314. Ikushima, I., and Maeda, M.: "Temperature-Stabilized Broadband Lumped-Element Circulator", Trans. IEEE, 1974, MTT-22, no.12, pp.1220-5.

14.315. Hord, W.E., and Rosenbaum, F.J.: "Coupled-Mode Analysis of Longitudinally Magnetized Ferrite Phaseshifters", Trans. IEEE, 1974, MTT-22, no.2, pp.135-8.

14.316. Roy, P.K., and Datta, A.N.: "Nonreciprocal Transmission in Artificial Dielectrics Using Ferromagnetic Powders", J. Phys. D, 1974, 7, no.7, pp.1053-62.

14.317. Wu, Y.S., and Rosenbaum, F.J.: "Wideband Operation of Microstrip Circulators", Trans. IEEE, 1974, MTT-22, no.10, pp.849-56.

14.318. Dinger, R.J., Waugh, T.M., and White, D.J.: "Nonreciprocal Properties of Vacuum-Deposited InSb Films at 87 GHz", Trans. IEEE, 1974, MTT-22, no.10, pp.879-80.

14.319. Akaiwa, Y., and Okazaki, T.: "Application of Hexagonal Ferrite to Millimetre-Wave Y-Circulator", Trans. IEEE, 1974, MAG-10, no.2, pp.374-8.

14.320. Leppinn, J.: "Thermal Design of Ferrite Isolators for Industrial Microwave Equipment", J. Microwave Power, 1974, 9, no.3, pp.251-61.

14.321. Jacobs, S.D., Teegarden, K.J., and Ahrenkeil, R.K.: "Faraday Rotation Optical Isolator for 10-micron Radiation", Appl. Opt., 1974, 13, no.10, pp.2313-6.

14.322. James, D.A.: "Reflection Factor of Magnetized Ferrite Sphere in Rectangular Waveguide", Radio Electron. Eng., 1974, 44, no.9, pp.481-90.

14.323. Bernardi, P., and Gerosa, G.: "Transversely Magnetized Ferrite-Filled Cylindrical Resonators", Alta Freq., 1974, 43, no.10, pp.884-8.

14.324. Noguchi, T., Akaiwa, Y., and Katoh, H.: "Edge-Guided-Mode Isolator Using Ferromagnetic-Resonance Absorption", Electron. Lett., 1974, 10, no.23, pp.501-2.

14.325. Akaiwa, Y.: "Bandwidth Enlargement of a Millimetre-Wave Y-Circulator with Half-Wavelength Line Resonators", Trans. IEEE, 1974, MTT-22, no.12, pp.1283-6.

14.326. Dotsch, H., and Schmitt, H.J.: "Interaction of Microwaves with Ring Domains in Magnetic Garnet Films", Appl. Phys. Lett., 1974, 24, no.9, pp.442-4.

15. GASEOUS-DISCHARGE DEVICES

15.1. Hess, W., Rauchle, E., and Stutzle, D.: "Change of Polarization of Microwaves on Passing through a Low-Pressure Plasma", Z. Angew. Phys., 1965, 19, p.211.

15.2. Gilson, V.A., and Johnson, C.C.: "Perturbations of Disc-Loaded Waveguide Modes by an Anisotropic Plasma", Proc. IEEE, 1965, 53, p.1639.

15.3. Kalmykova, S.S., and Kurilko, V.I.: "MHD Waveguide Excitation by Coaxial Lines", Magn. Gidrodin., 1965, no.3, p.51.

15.4. Ward, C.S., et al.: "Arc Loss of Multi-Megawatt Gas-Discharge Duplexers", Trans. IEEE, 1965, MTT-13, p.801.

15.5. Kononov, V.P., et al.: "Acceleration of Plasma by H_{11} Wave in Circular Guide", Zh. Tekh. Fiz., 1965, 35, p.51, and Sov. Phys.-Tech. Phys., 1965, 10, no.1, pp.36-9.

15.6. Nagelberg, E.R.: "Perturbation Theory of Microwave Interaction with Gyroelectric Plasmas", J. Math. Phys., 1965, 6, p.44.

15.7. Robinson, L.C.: "Reflection of Microwaves from an Anisotropic Plasma Surface", Plasma Phys., 1965, 7, no.2, p.167.

15.8. Gregory, B., and Mourier, G.: "Coupling of a Cylindrical Plasma to a Waveguide", C. R. Acad. Sci. (Paris), 1965, 260B, p.1592.

15.9. Seshadri, S.E.: "Guided Waves on a Dielectric-Coated Cylinder Immersed in a Plasma", Electron. Lett., 1965, 1, p.19.

15.10. Gekker, I.R., et al.: "Experimental Investigation of Acceleration of Plasma in a Microwave-Field Gradient", Zh. Tekh. Fiz., 1965, 35, p.577, and Sov. Phys.-Tech. Phys., 1965, 10, no.3, pp.450-3.

15.11. Lee, S.W., Lo, Y.T., and Mittra, R.: "H-Plane Bifurcation of a Waveguide with Anisotropic Plasma", Can. J. Phys., 1965, 43, p.2123.

15.12. Cullen, A.L.: "Point of Emergence of a Microwave Beam Entering a Linearly Graded Plasma", J. Res. Natl. Bur. Stand., 1965, 69D, p.177.

15.13. Cottingham, W.B., and Buchsbaum, S.J.: "Diffusion in a Microwave Plasma in Presence of Turbulent Flow", J. Appl. Phys., 1965, 36, p.2075.

15.14. Kononov, V.P.: "Resonance Inter-action between a Plasmoid and the Field in a Waveguide", Zh. Tekh. Fiz., 1965, 35, p.47, and Sov. Phys.-Tech. Phys., 1965, 10, no.1, pp.33-5.

15.15. Fehsenfeld, F.C., Evenson, K.M., and Broida, H.P.: "Microwave Discharge Cavi-ties at 2.45 GHz", Rev. Sci. Instrum., 1965, 36, p.294.

15.16. Meservey, E.B., and Schlesinger, S.P.: "Verification of Hot Plasma Theory for Microwave Propagation", Phys. Fluids, 1965, 8, p.500.

15.17. Scharer, J.E., and Trivelpiece, A.W.: "Quasistatic Analysis of Waves in a Plasma-Filled Waveguide", J. Appl. Phys., 1965, 36, p.318.

15.18. Shohet, J.L., and Moskowitz, C.: "Eigenvalues of a Microwave-Cavity/ Lossy-Plasma System", J. Appl. Phys., 1965, 36, p.1756.

15.19. Booz, J., and Erbert, H.G.: "Mean Energy Expenditure for Formation of an Elec-tron through Ionization by Collision in a Microwave Field", Z. Angew. Phys., 1965, 18, p.285.

15.20. Yeh, C., and Rusch, W.V.T.: "Inter-action of Microwaves with an Inhomogeneous and Anisotropic Plasma Column", J. Appl. Phys., 1965, 36, p.2302.

15.21. Starik, A.M.: "Cathode Field of Anomalous Glow Discharge in a Waveguide", Radiotekh. Elektron., 1965, 10, p.779, and Radio Eng. Electron. Phys., 1965, 10, no.4, pp.667-8.

15.22. Seshadri, S.R.: "Guided Waves on a Perfectly Conducting Infinite Cylinder in a Magnetoionic Medium", Proc. IEE, 1965, 112, p.1497.

15.23. Kerzar, B., and Weissglas, P.: "Plasma-Microwave Interaction", J. Appl. Phys., 1965, 36, p.2479.

15.24. Takeda, S., Minami, K., and Masumi, M.: "Microwave Absorption by a Plasma Slab", J. Phys. Soc. Jap., 1965, 20, p.845.

15.25. Starik, A.M.: "Calculation of Electron Density in a Gas-Discharge Attenu-ator with Hollow Cathode", Radiotekh. Elek-tron., 1965, 10, p.1250, and Radio Eng. Electron. Phys., 1965, 10, p.1074.

15.26. Johansen, E.L.: "Radiation Proper-ties of a Parallel-Plane Waveguide in a Transversely Magnetized, Homogeneous, Plasma", Trans. IEEE, 1965, MTT-13, p.77.

15.27. Farber, H., et al.: "DC-Triggered, High-Speed, High-Power, Microwave Spark-Gap Switch", Trans. IEEE, 1965, MTT-13, p.28.

15.28. Pinder, D.N.: "Slow-Wave Propaga-tion in a Nonuniform Plasma", Electron. Lett., 1965, 1, p.291.

15.29. Johnston, T.W., Gore, J.V., and Osborne, F.J.F.: "Isotropic Plasma Guide. TE_{11} Mode", J. Appl. Phys., 1965, 36, p.3354.

15.30. Ginzburg, V.M., and Marinov, A.: "Waves in a Plane Guide Containing Inhomo-geneous Isotropic Plasma", Radiotekh. Elek-tron., 1965, 10, p.868, and Radio Eng. Elec-tron. Phys., 1965, 10, p.739.

15.31. Caron, P.R.: "Study of Waves Sup-ported by Warm Plasma Slab", J. Res. Natl. Bur. Stand., 1965, 69D, p.729.

15.32. Giorgadze, N.P.: "Interaction of Circular Waves in Cold Plasma Guide", Zh. Tekh. Fiz., 1965, 35, p.1177, and Sov. Phys.-Tech. Phys., 1966, 10, no.7, pp.912-4.

15.33. Toda, M.: "Theory of a Microwave Plasma Instability due to Transverse Break-down", J. Appl. Phys., 1966, 37, p.37.

15.34. Wort, D.J.H.: "Microwave Trans-mission through Turbulent Plasma", Plasma Phys., 1966, 8, p.79.

15.35. Terayama, K., and Takamoto, T.: "Microwave Reflection from Striated Plasma Column", Elektr. Eng. Jap.,1966,86,no.2,p.58.

15.36. Clarricoats, P.J.B., Olver, A.D., and Wong, J.S.L.: "Propagation in Isotropic Plasma Waveguides", Proc. IEEE, 1966, 113, p.755.

15.37. Kalivoda, L., et al.: "Properties of a Plasma Waveguide", Czech. J. Phys., 1966, 16B, p.314.

15.38. Bevc, V.: "Power Flow in Plasma-Filled Waveguides", J. Appl. Phys., 1966, 37, p.3128.

15.39. Carlile, R.N.: "Quasi-TE_{11} Modes in an Anisotropic Plasma Waveguide", Trans. IEEE, 1966, MTT-14, p.350.

15.40. Ohkubo, M.: "Effect of Electron-Density Distribution in Plasma Waveguides", Electr. Eng. Jap., 1966, 86, no.5, p.9.

15.41. Webber, J., and Sims, G.D.: "Wave Propagation in Plasma Waveguides", Radio Sci., 1966, 1, p.659.

15.42. Brodskii, Yu.Ya., Vagin, V.A., and Kotov, V.I.: "Nonsymmetrical Waves in Plasma Guides", Zh. Tekh. Fiz., 1966, 36, p.453, and Sov. Phys.-Tech. Phys., 1966, 11, no.3, pp.335-9.

15.43. Tajiri, T., et al.: "Microwave Switching Tube Utilizing Glow Discharge", Toshiba Rev., 1966, no.25, p.18.

15.44. Wada, J.Y., and Rice, D.K.: "Non-linear Scattering of Microwaves by a TW Density-Modulated Plasma Column", Electron. Lett., 1966, 2, p.384.

15.45. Shohet, J.L.: "Exact Solution for the Eigenfrequencies of a Microwave Cavity Partially Filled with a Magnetized Plasma", J. Appl. Phys., 1966, 37, p.3775.

15.46. Guthart, H., Weissman, D.E., and Morita, T.: "Microwave Scattering from an Underdense Turbulent Plasma", Radio Sci., 1966, 1, p.1253.

15.47. Schaedla, W.H., and Beyer, J.B.: "Modes of a Gyroplasma-Filled Beam Waveguide", Trans. IEEE, 1966, AP-14, p.662.

15.48. Kartashev, V.G., and Shevchenko, R.F.: "Propagation of Symmetric Waves in a Cylindrical Guide Filled with Inhomogeneous Plasma", Radiotekh. Elektron., 1966, 11, p.143, and Radio Eng. Electron. Phys., 1966, 11, p.1228.

15.49. Ashkenazi, D.Ya., and Vintsents, L.M.: "Increasing Bandwidth of TR Cells while Maintaining Low Firing Power", Radiotekh. Elektron., 1966, 11, no.10, pp.1880-1, and Radio Eng. Electron. Phys., 1966, 11, no.10.

15.50. Buts, V.A., Kalmykova, S.S., and Kurilko, V.I.: "Theory of Excitation of Gyrotropic Plasma Waveguide by a Coaxial Line", Radiotekh. Elektron., 1966, 11, no.11, pp.2069-72, and Radio Eng. Electron. Phys., 1966, 11, no.11.

15.51. Ohkubo, M.: "Fields in Anisotropic Plasma-Guides Considering Effect of Electron Collision", Electr. Eng. Jap., 1966, 85, no.10, pp.79-84.

15.52. Ponomarev, V.N., and Solntsev, G.S.: "Propagation Coefficient of a Wave in a Rectangular Guide Containing Plasma in a Dielectric Tube", Zh. Tekh. Fiz., 1966, 36, p.1376, and Sov. Phys.-Tech. Phys., 1967, 11, no.8, pp.1027-31.

15.53. Lee, S.W., Liang, C., and Lo, Y.T.: "Coupling of Modal Waves in a Plasma-Filled Parallel-Plate Guide", Radio Sci., 1967, 2, p.401.

15.54. Hedvall, P.: "Observation of Nonlinear Coupling between Slow and Fast Waves in a Plasma Guide", Ark. Fys., 1967, 35, no.22, pp.283-6.

15.55. Leiba, E.: "Plasma Phaseshifter", Ann. Radioelectr.,1967,22,no.10,pp.297-310.

15.56. Katz, J.E.: "Propagation Characteristics in a Hollow Plasma Waveguide", J. Appl. Phys., 1968, 39, no.4, pp.2154-6.

15.57. Krepak, V.N., and Yakimenko, I.P.: "Symmetrical Waves in a Nonuniform Plasma Guide", Radiotekh. Elektron., 1968, 13, no.4, pp.579-85, and Radio Eng. Electron. Phys., 1968, 13, no.4.

15.58. Tsvirko, Yu.A.: "Resonance Transmission through an Opaque Boundary by a Cyclotron Plasma Wave", Ukr. Fiz. Zh., 1968, 13, p.1572.

15.59. Stewart, P.: "Propagation Normal to a Magnetic Field in a Uniform Anisotropic Maxwellian Plasma", J. Plasma Phys., 1968, 2, p.591.

15.60. Kent, G.: "Resonant Frequencies and Fields in a Cavity Containing a Magnetoplasma", J. Appl. Phys., 1968, 39, p.5919.

15.61. Stoll, I.: "Comparison of Dispersive Properties of Axial and Annular Longitudinally Magnetized Plasma Waveguides", Acta Tech. CSAV, 1968, 13, p.863.

15.62. Quine, J.P., et al.: "Spark-Gap Switch for Oversize Rectangular Waveguides", Trans. IEEE, 1968, MTT-16, p.952.

15.63. Bachynski, M.P., and Gibbs, B.W.: "Propagation of Strong Fields through Plasmas Near the Electron Cyclotron Frequency", Phys. Rev. Lett., 1969, 22, p.583.

15.64. DeCamp, E.E., True, R.M., and Edwards, E.V.: "Boron Nitride as a Dielectric Material for High-Power Duplexing Devices", Trans. IEEE, 1969, ED-16, p.209.

15.65. Tuan, H.S.: "Mode Theory of Waveguide Filled with Warm Uniaxial Plasma", Trans. IEEE, 1969, MTT-17, p.134.

15.66. Buts, V.A.: "Excitation of Magneto-active Plasma Waveguide", Ukr. Fiz. Zh., 1969, 14, p.270.

15.67. Blevin, H.A., and Reynolds, J.A.: "Resonances of a Cylindrical Cavity Containing a Coaxial Annular Plasma Column", J. Appl. Phys., 1969, 40, p.3899.

15.68. Sforza, P.F.: "Nonlinear Interaction of a Microwave Field with a Plasma Slab in a Waveguide", J. Appl. Phys., 1969, 40, p.1908.

15.69. Champlin, K.S.: "Use of Symmetry in the Variational Treatment of a Magnetoplasma-Filled Waveguide", Trans. IEEE, 1969, MTT-17, p.401.

15.70. Lee, K.F.: "Modes in Confined Plasma Waveguide", J. Appl. Phys., 1969, 40, p.428.

15.71. Krepak, V.N., and Yakimenko, I.P.: "E-Modes in a Circular Waveguide with Nonuniform Plasma", Radiotekh. Elektron., 1969, 14, p.401, and Radio Eng. Electron. Phys., 1969, 14, no.3, pp.346-9.

15.72. Goldie, H.: "26- to 40-GHz Fast-Acting Plasma Waveguide Switch", Trans. IEEE, 1969, ED-16, p.6.

15.73. Chawla, B.R., and Unz, H.: "Propagation in a Two-Stream Magnetoplasma", Trans. IEEE, 1969, AP-17, p.384.

15.74. Nandedkar, D.P., and Bhagavat, G.K.: "Analysis of Absorption in Plasma in the Region of Anomalous Dispersion", Int. J. Electron., 1969, 26, p.269.

15.75. Zaitsev, A.A., and Leonov, G.S.: "Propagation through a Waveguide Filled with a Positive Column with Travelling Striations", Izv. VUZ Radiofiz., 1969, 12, no.8, p.1256.

15.76. Morescu, M., and Zilli, E.: "Resonator Partially Filled with a Magnetized Plasma", Ric. Sci., 1969, 39,no.4-6,p.341.

15.77. Raicu, D.: "Microwave Transmission through a Plasma-Filled Waveguide of Finite Length above the Cutoff Density", Int. J. Electron., 1969, 26, no.4, p.391.

15.78. Ivanova, V.D.: "Wave Theory for Rectangular Helix in Plasma", Radiotekh. Elektron., 1969, 14, no.8, p.1495, and Radio Eng. Electron. Phys., 1969, 14, no.8, pp.1292-4.

15.79. Bulkin, P.S., and Maripov, A.: "Propagation of Symmetrical H_{01} Waves in Circular Waveguide with Plasma", Vestn. Mosk. Univ., Fiz. Astron., 1969,no.5,p.122.

15.80. Franklin, R.N., and Oldfield, M.L.G.: "Propagation in a Longitudinally Magnetized Plasma-Filled Coaxial Waveguide", Int. J. Electron., 1969, 27, p.431.

15.81. Anderson, J.M.: "Microwave Switching in Evacuated Waveguides by Metal-Vapour Arcs", Trans. IEEE, 1970, ED-17, p.939.

15.82. Yakimenko, I.P.: "Noncoherent Scattering of Waves in a Guide with Plasma", Zh. Tekh. Fiz., 1969, 39, p.2163, and Sov. Phys.-Tech. Phys., 1970, 14, no.12, pp.1632-6.

15.83. Maripov, A., et al.: "H_{0n}-Type Waves in a Cylindrical Waveguide Containing Gaseous-Discharge Plasma", Radiotekh. Elektron., 1970, 15, no.8, and Radio Eng. Electron. Phys., 1970, 15, no.8, pp.1409-16.

15.84. Ristic, V.M., and Tam, M.C.: "Class of Waveguides Suitable for Plasma Experiments", Int. J. Electron., 1970, 28, p.297.

15.85. Shohet, J.L., and Hatch, A.J.: "Eigenvalues of a Microwave Cavity Filled with a Plasma of Variable Radial Density", J. Appl. Phys., 1970, 41, p.2610.

15.86. Rozkwitalski, Z.: "TM_{010} Resonant Frequency of a Microwave Cavity Partially Filled with a Nonuniform Plasma Column", Bull. Acad. Pol. Sci., 1970,18, no.2,p.177.

15.87. Kalluri, D.: "Waveguide Modes of a Warm Drifting Uniaxial Electron Plasma", Proc. IEEE, 1970, 58, p.278.

15.88. Dorman, F.H., and McTaggart, F.K.: "Absorption of Microwave Power by Plasmas", J. Microwave Power, 1970, 5, no.1, pp.4-16.

15.89. Popovich, V.P., and Shustin, Ye.G.: "Electron-Beam Excitation of a Cavity Filled with Nonuniform Plasma", Radiotekh. Elektron., 1970, 15, p.204, and Radio Eng. Electron. Phys., 1970, 15, no.1, pp.176-8.

15.90. Ishizone, T., et al.: "Propagation along a Conducting Wire in a General Magnetoplasma", Proc. IEEE, 1970, 58, no.11, pp.1843-4.

15.91. Ishizone, T., et al.: "Measured Phase Constant along a Conducting Wire in a Magnetoplasma", Proc. IEEE, 1970, 58, no.11, pp.1852-4.

15.92. Muehe, C.E., and Browne, A.A.L.: "X-Band Gas-Tube Pulsed Attenuator for Maser Protection", Trans. IEEE, 1970, ED-17, no.12, pp.1040-7.

15.93. Fredericks, R.M., and Asmussen, J.: "Excitation of Warm-Plasma Rotationally Symmetric Resonances in Short-Gap Cavity", Proc. IEEE, 1971, 59, no.2, pp.315-7.

15.94. Buryak, V.S., and Kozyrev, Ye.N.: "Antenna Switch in Round Waveguide Propagating the TE_{01} Mode", Izv. VUZ Radioelektron., 1971, 14, no.1, pp.56-60.

15.95. Gonzalez, G.: "Propagation in a Rectangular Waveguide Partially and Completely Filled with an Isotropic Inhomogeneous Plasma", J. Appl. Phys., 1971, 42, no.5, pp.1853-6.

15.96. Kinderdijk, H.M.J., and Hagelbeuk, H.J.L.: "Propagation in a Circular Waveguide Containing a Cold Cylindrically Stratified Plasma", Physica, 1971, 52, no.2, pp.299-315.

15.97. Kinderdijk, H.M.J., and van Eck, J.: "Resonance Frequency of a Cylindrical Microwave Cavity Partially Containing a Cold Nonuniform Plasma Column", Physica, 1971, 52, no.2, pp.316-28.

15.98. Krupina, A.E., and Orlova, E.V.: "Influence of Collisions and Relativistic Effects on Wave Propagation in Magnetoactive Plasma", Izv. VUZ Radiofiz., 1971, 14, no.6, pp.817-22.

15.99. Wilcox, R.M.: "Transmission of EM Waves through a Conducting (Plasma) Slab", J. Math. Phys., 1971, 12, no.7, pp.1195-207.

15.100. Bogomolov, Yu.V.: "Absorption of a Plane Wave in a Circular Plasma Cylinder", Izv. VUZ Radiofiz., 1971, 14, no.8,pp.1168-75.

15.101. Azakami, T., Narita, H., and Thein, U.A.: "Analysis of Waveguides with Warm Plasma", Bull. Nagoya Inst. Techno., 1972, 24, pp.201-6.

15.102. Anderholm, N.C.: "Fast Gas Switch for Characterizing Laser Output Pulses", Appl. Opt., 1972, 11, no.9, pp.2057-9.

15.103. Tyutyunnik, V.B., and Tkachenko, V.M.: "Plasma Phaseshifter in the 8-mm Band with Cylindrical Hollow Cathode", Radiotekh. Elektron., 1972, 17, no.5, pp.1098-9, and Radio Eng. Electron. Phys., 1972, 17, no.5, pp.857-9.

15.104. Minami, K.: "Cyclotron Resonance of Ordinary Waves in Dense Plasmas", J. Phys. Soc. Jap., 1973, 35, no.5, pp.1509-13.

15.105. Aliev, Yu.M., Gradov, O.M., and Kirii, A.Yu.: "Anomalous Dissipation and Penetration of Strong Radiation into a Bounded Plasma", Zh. Eksp. Teor. Fiz. Pis'ma, 1973, 17, no.3, pp.177-9.

15.106. Cupini, E., Molinari, V.G., and Poli, P.: "Reflection Factor of a Plasma Column of Variable Electron Density in a Waveguide", Alta Freq., 1973, 42, no.2, pp.62-8.

15.107. Gwal, A.K., Singh, K.P., and Misra, K.D.: "Absorption of Microwaves through Laboratory Plasma", J. Inst. Telecommun. Eng., 1973, 19, no.2, pp.94-6.

15.108. Musil, J., and Zacek, F.: "Efficient Injection of High Microwave Power into an Overdense Magnetoactive Plasma in Waveguide", Czech. J. Phys., 1973, 23B, no.7, pp.736-41.

15.109. Fichet, M., and Fidone, I.: "Propagation of Electromagnetic Modes at Oblique Incidence in a Magnetized Inhomogeneous Plasma", Nuovo Cim., 1973, 16B, no.1, pp.221-39.

15.110. Miroshnichenko, V.I.: "Reflection of Waves from a Moving Plasma", Zh. Tekh. Fiz., 1973, 43, no.3, pp.467-74, and Sov. Phys.-Tech. Phys., 1973, 18, no.3, pp.299-303.

15.111. Ishida, A., and Kitao, K.: "Anomalous Penetration of an Ordinary Wave into a Magnetoplasma Slab", J. Phys. Soc. Jap., 1973, 35, no.5, pp.1514-21.

15.112. Anisimov, A.I., Vinogradov, N.I., and Poloskin, B.P.: "Anomalous Microwave Absorption at the Upper Hybrid Frequency", Zh. Tekh. Fiz., 1973, 43, no.4, pp.727-30, and Sov. Phys.-Tech. Phys., 1973, 18, no.4, pp.459-60.

15.113. Golovanivskii, K.S., and Blanco, J.C.: "Absorption of EM Waves at Harmonics of the Electron Gyrofrequency in an Inhomogeneous Magnetoactive Plasma", Zh. Tekh. Fiz., 1973, 43, no.5, pp.1068-70, and Sov. Phys.-Tech. Phys., 1973, 18, no.5, pp.673-5.

15.114. Uchida, K., and Aoki, K.: "Dispersion Relation of a Circular Waveguide Filled with an Anisotropic Plasma", Electron. Commun. Jap., 1973, 56, no.12, pp.66-73.

15.115. Alferov, V.N., and Yampol'skii, J.R.: "Controllable Spark Gap for Protection of Microwave Devices", Prib. Tekh. Eksp., 1973, 16, no.4, pp.162-4, and Instrum. Exp. Tech., 1973, 16, no.4, pp.1178-9.

15.116. Kats, L.I., and Kireev, N.N.: "Experimental Investigation of Modulation of Millimetre-Wave Radiation by a Periodically Nonstationary Magnetoplasma", Radiotekh. Elektron., 1973, 18, no.6, pp.1295-7, and Radio Eng. Electron. Phys., 1973, 18, no.6, pp.953-4.

15.117. Troitskii, S.V., and Yakimenko, I.P.: "Mode Conversion and Wave Scattering due to Fluctuations in a Rectangular Waveguide Filled with Compressible Plasma", Radiotekh. Elektron., 1973, 18, no.5, pp.1077-80, and Radio Eng. Electron. Phys., 1973, 18, no.5, pp.797-800.

15.118. Maiti, J.N., and Basu, J.: "Properties of a Dielectric-Rod Waveguide Immersed in Plasma", J. Appl. Phys., 1974, 45, no.4, pp.1650-6.

15.119. Minami, K., et al.: "Nonlinear Skin Effect of High-Power Microwaves Incident on a Collisionless Magnetized Plasma", Phys. Rev. Lett., 1974, 33, no.13, pp.740-3.

15.120. Sodha, M.S., Khanna, R.K., and Tripathi, V.K.: "Self-Focusing of Electromagnetic Beams in a Strongly Ionized Magnetoplasma", J. Phys. D, 1974, 7, no.16, pp.2188-97.

15.121. Stallcop, J.R.: "Absorption of Laser Radiation in a H-He Plasma. I", Phys. Fluids, 1974, 17, no.4, pp.751-8.

15.122. Kopecky, V., Musil, J., and Zacek, F.: "Absorption of Microwave Energy in a Plasma Column at High Magnetic Fields", Phys. Lett., 1974, 50A, no.4, pp.309-10.

15.123. Ferrari, L.A., McQuade, A.W., and LaHaye, R.J.: "Absorption of Left(-Hand) Wave near Electron Cyclotron Frequency", Phys. Fluids, 1974, 17, no.9, pp.1785-7.

15.124. Asmussen, J., et al.: "Design of a Microwave Plasma Cavity", Proc. IEEE, 1974, 62, no.1, pp.109-17.

15.125. Sodha, M.S., Sharma, R.P., and Tripathi, V.K.: "Uniform Magnetoplasma Waveguide", J. Phys. D, 1974, 7, no.6, pp.866-70.

15.126. Lax, B., and Cohn, D.R.: "Interaction of Intense Submillimetre Radiation with Plasma", Trans. IEEE, 1974, MTT-22, no.12, pp.1049-52.

15.127. Singh, S., Gupta, S.C., and Midha, J.M.: "Collisonal Damping of EM Waves in Plasma", Indian J. Pure. Appl. Phys., 1974, 12, no.11, pp.786-8.

15.128. Kurilko, V.I., et al.: "Propagation of Regular Nonlinear Waves in an Anisotropic Plasma-Filled Waveguide", Zh. Tekh. Fiz., 1974, 44, no.5, pp.985-94, and Sov. Phys.-Tech. Phys., 1974, 19, no.5, pp.622-7.

15.129. Paul, S.N., and Bondyopadhaya, R.: "Wave Propagation through Nonuniform Plasma", Indian J. Phys., 1974, 48, no.11, pp.987-1001.

15.130. Pappalardo, R., and Lempicki, A.: "Method for Detecting Transient Gain (or Loss) in Pulsed Gas Discharges", J. Quantum Electron. IEEE, 1974, QE-10, no.10, pp.816-8.

16. BEAM-TYPE ANTENNAS

16.1. Knop, C.M., and Swift, C.T.: "Radiation Conductance of an Axial Slot on a Cylinder", J. Res. Natl. Bur. Stand., 1965, 69D, p.447.

16.2. Lerner, D.S.: "Wave Polarization Converter for Circular Polarization", Trans. IEEE, 1965, AP-13, p.3.

16.3. Pozzolo, V., and Zich, R.: "Phase-Velocity of Wave Propagated over an Open Periodic Structure", Alta Freq., 1965, 34, p.188.

16.4. Takeshima, T.: "Slot-Antenna Systems with In-Phase Feed", Jap. J. Appl. Phys., 1965, 4, p.151.

16.5. Aivazyam, Yu.M., and Sedrakyan, D.M.: "Radiation from Open End of a Plane Semi-Infinite Waveguide", Zh. Tekh. Fiz., 1965, 35, p.459, and Sov. Phys.-Tech. Phys., 1965, 10, no.3, pp.358-61.

16.6. Chu, T.S., and Semplak, R.A.: "Gain of Electromagnetic Horns", Bell Syst. Tech. J., 1965, 44, p.527.

16.7. Russo, P.M., Rudduck, R.C., and Peters, L.: "Method for Computing E-Plane Patterns for Horn Antennas", Trans. IEEE, 1965, AP-13, p.219.

16.8. Blume, S., Keydel, W., and Wolter, H.: "Experimental Investigations on Plane-Surface Antennas", Z. Angew. Phys., 1965, 19, p.61.

16.9. Bates, R.H.T.: "Mode Theory Approach to Arrays", Trans. IEEE, 1965, AP-13, p.321.

16.10. Ricardi, L.J.: "Near-Field Characteristics of a Linear Array", Microwave J., 1965, 8, no.3, p.41.

16.11. Butler, J.K., and Unz, H.: "Fourier Transform Methods for Analysing Nonuniform Arrays", Proc. IEEE, 1965, 53, p.191.

16.12. Baklanov, Ye.V.: "Radiation Pattern of an Infinite Periodic Grating", Radiotekh. Elektron., 1965, 10, p.183, and Radio Eng. Electron. Phys., 1965, 10, no.1, pp.147-51.

16.13. Maclean, T.S.M., and Said, R.A.K.: "Dia-Zigzag Aerial", Proc. IEEE, 1965, 112, p.872.

16.14. Karjala, D.S., and Mittra, R.: "Radiation from Periodic Structures Excited by a Waveguide", Electron. Lett.,1965,1,p.111.

16.15. Nagai, K.: "Amplitude-Modulated End-Fire Array", Can. J. Phys., 1965,43,p.155.

16.16. Cornbleet, S.: "Simple Spherical Lens with External Foci", Microwave J., 1965, 8, no.5, p.65.

16.17. Matthews, P.A.: "Aperture Distributions, Autocorrelation, and Angular Power Spectrum for Partially Illuminated Apertures", Proc. IEE, 1965, 112, p.1492.

16.18. Zakson, M.B., and Merkulov, V.V.: "Nonuniform Antenna Arrays with Randomly Spaced Elements", Radiotekh. Elektron., 1965, 10, p.7, and Radio Eng. Electron. Phys., 1965, 10, p.4.

16.19. Maclean, T.S.M., and Williams, D.J.: "Broadband Dielectric Rod Aerial", Radio Electron. Eng., 1965, 30, p.99.

16.20. Baklanov, Ye.V.: "Antenna Pattern of a Radiator in an Infinite Periodic Array", Radiotekh. Elektron., 1965, 10, p.183, and Radio Eng. Electron. Phys., 1965, 10, p.147.

16.21. James, P.W.: "Polar Patterns of Phase-Corrected Circular Arrays", Proc. IEE, 1965, 112, p.1839.

16.22. Sletten, C.J., and Blacksmith, P.: "Paraboloidal Mirror for Microwaves", Appl. Opt., 1965, 4, p.1239.

16.23. Petrov, B.M.: "Self and Mutual Conductance of Slot Radiatiors on a Wedge", Radiotekh. Elektron., 1965, 10, p.1135, and Radio Eng. Electron. Phys., 1965, 10, no.6, pp.971-4.

16.24. Kotik, I.P., et al.: "Application of Theoretical Analysis to Circular Waveguides with Longitudinal Slots", Radiotekh. Elektron., 1965, 10, p.1138, and Radio Eng. Electron. Phys., 1965, 10, no.7, pp.1053-9.

16.25. Kales, M.L., and Brown, R.M.: "Design Considerations for Two-Dimensional Symmetric Bootlace Lenses", Trans. IEEE, 1965, AP-13, p.521.

16.26. Das, B.: "Maximum Directivity of Symmetrical Radiator Systems", Radiotekh. Elektron., 1965, 10, p.997, and Radio Eng. Electron. Phys., 1965, 10, no.6, pp.853-9.

16.27. Ivanov, E.A.: "Linear System of Spherical Radiators", Radiotekh. Elektron., 1965, 10, p.1005, and Radio Eng. Electron. Phys., 1965, 10, no.6, pp.859-66.

16.28. Maclean, T.S.M.: "Meander Line Aerial", Electron. Eng., 1965, 37, p.667.

16.29. Ma, M.T.: "Directivity of Uniformly Spaced Optimum End-Fire Arrays with Equal Sidelobes", J. Res. Natl. Bur. Stand., 1965, 69D, p.1249.

16.30. Jones, H.S., and Heinard, W.G.: "Improved Slot Array Using Post Excitation", Microwave J., 1965, 8, no.7, p.69.

16.31. Sigelmann, R.A., and Ishimaru, A.: "Radiation from Periodic Structures Excited by an Aperiodic Source", Trans. IEEE, 1965, AP-13, p.354.

16.32. Adonina, A.I.: "Design of Bi-Reflector Antenna Gratings", Radiotekh. Elektron., 1965, 10, p.190, and Radio Eng. Electron. Phys., 1965, 10, p.154.

16.33. Shered'ko, E.Yu.: "Radiation Field of a Single-Screw, Logarithmic-Elliptical, Spiral Antenna", Radiotekhnika, 1965, 20, no.6, p.13, and Telecommun. Radio Eng. Pt 2, 1965, 20, no.6.

16.34. Hanna, F., and Bishai, A.M.: "Determination of the Influence of Reflector Distance on a Horn Antenna", Slab. Obz., 1965, 26, p.85.

16.35. Gunderman, R.J., Mathis, H.F., and Zurcher, L.A.: "Two-Reflector, Non-Shadowing, Antenna", Trans. IEEE, 1965, AP-13, p.474.

16.36. Davies, D.E.N., and Fenby, R.G.: "Series-Fed Circular Array", Electron. Lett., 1965, 1, p.264.

16.37. Michelson, R.A., and Schomer, J.W.: "Three-Parameter Antenna-Pattern Synthesis Technique", Microwave J., 1965,8,no.9,p.88.

16.38. Litvinenko, L.N., and Shestopalov, V.P.: "Diffraction Properties of Two-Element Asymmetric Metal Grids", Radiotekh. Elektron., 1965, 10, p.1135, and Radio Eng. Electron. Phys., 1965, 10, no.6, pp.968-71.

16.39. Unger, H.G.: "Wave Propagation in Horns and through Horn Junctions", Arch. Elektr. Ubertrag., 1965, 19, p.459.

16.40. Blume, S., Habermehl, A., and Wolter, H.: "Investigations of Helix Antennas", Z. Angew. Phys., 1965,20,p.149.

16.41. Vasil'ev, E.N., and Seregina, A.R.: "Radiation Pattern of Slot Antenna on a Thick Cylinder of Finite Length", Radiotekhnika, 1965, 20, no.4, p.27, and Telecommun. Radio Eng. Pt 2, 1965,20,no.4,p.97.

16.42. Ronchi, L., and Russo, V.: "Diffraction Reflector Antennas", Appl. Opt., 1965, 4, p.1544.

16.43. Dyson, J.D.: "Characteristics and Design of Conical Log-Spiral Antenna", Trans. IEEE, 1965, AP-13, p.488.

16.44. Cook, J.S., Denkmann, W.J., et al.: "Open Cassegrain Antenna. I and II", Bell Syst. Tech. J., 1965, 44, pp.1255 and 1301.

16.45. Nagelberg, E.R., and Shefer, J.: "Mode Conversion in Circular Waveguide (Horn Feed)", Bell Syst. Tech. J., 1965, 44, p.1321.

16.46. Kapany, N.S., Burke, J.J., and Frame, K.: "Radiation Characteristics of Circular Dielectric Waveguides", Appl. Opt., 1965, 4, p.1534.

16.47. Eckart, G.: "Theory of Sharply Focused Directional Radiation", Arch. Elektr. Ubertrag., 1965, 19, p.581.

16.48. Gruber, H., and Koppatz, P.: "Ferrite Rod Antennas as Dielectric Radiators for Microwaves", Hochfreq. Elektroak., 1965, 74, p.83.

16.49. Vyeshnikova, I.Ye., and Yevstropov, G.A.: "Theory of Matched Slot Radiators", Radiotekh. Elektron., 1965, 10, p.1181, and Radio Eng. Electron. Phys., 1965, 10, p.1013.

16.50. Yampol'skii, V.G.: "Directional Properties of Antennas with Unequal Element Spacing", Elektrosvyaz, 1965, no.8, and Telecommun. Radio Eng. Pt 1, 1965, no.8, p.28.

16.51. Chiba, T.: "Maximum Spacing of the Array Antenna", Proc. IEEE, 1965, 53, p.1270.

16.52. Nagelberg, E.R.: "Fresnel-Region Phase Centres of Circular-Aperture Antennas", Trans. IEEE, 1965, AP-13, p.479.

16.53. Yevstropov, G.A., and Tsarapkin, S.A.: "Investigation of Slotted Waveguide Antennas with Identical Resonant Radiators", Radiotekh. Elektron., 1965, 10, p.1663, and Radio Eng. Electron. Phys., 1965, 10, p.1429.

16.54. Thust, P.: "Horn Radiator with Sector-Shaped Directional Radiation Pattern", Proc. IEEE, 1965, 53, p.1239.

16.55. Cheston, T.C.: "Matching of Phased-Array Antennas", Trans. IEEE, 1965, AP-13, p.327.

16.56. Pokras, A.M., and Sirenev, V.S.: "Compact Antenna with Patterns Omnidirectional in the Horizontal Plane and Narrow in the Vertical Plane", Radiotekhnika, 1965, 20, no.8, p.31, and Telecommun. Radio Eng. Pt 2, 1965, 20, no.8, p.99.

16.57. Balfour, M.A.: "Phased Array Simulators in Waveguide for a Triangular Arrangement of Elements", Trans. IEEE, 1965, AP-13, p.475.

16.58. Hudock, E., and Mayes, P.E.: "Near-Field Investigation of Uniform Periodic Monopole Arrays", Trans. IEEE, 1965, AP-13, p.840.

16.59. Iijima, T.: "Theory of Radiation from an Open-Ended Waveguide", J. Inst. Electr. Commun. Eng. Jap., 1965, 48, p.533.

16.60. Takeshima, T.: "Slot-Array Antennas for Simultaneous Radiation", Trans. IEEE, 1965, AP-13, p.472.

16.61. Grigor'ev, G.I., et al.: "Log-Periodic Spiral Element and Paraboloidal Reflector with Frequency Coverage of 1:7", Izv. VUZ Radiofiz., 1965, 8, p.768, and Sov. Radiophys., 1965, 8, p.549.

16.62. Takeshima, T.: "Frequency Bandwidth of Slot-Array Antenna", J. Inst. Electr. Commun. Eng. Jap., 1965, 48, p.2161.

16.63. Larson, R.W., and Powers, V.M.: "Slots in Dielectric-Loaded Waveguide", Radio Sci., 1966, 1, no.1, p.31.

16.64. Deryugin, L.N., and Kuznetsov, M.G.: "Angular Transparency Sectors of Antennas with Periodic Waveguide Channels", Radiotekh. Elektron., 1966, 11, p.187, and Radio Eng. Electron. Phys., 1966, 11, no.2, pp.158-64.

16.65. Geyer, H.: "Circular Horn Radiators with Annular Wave-Traps for Simultaneous Transmission of Two Polarization-Decoupled Waves", Frequenz, 1966, 20, p.22.

16.66. Lo, Y.T., and Lee, S.W.: "Study of Space-Tapered Arrays", Trans. IEEE, 1966, AP-14, p.22.

16.67. Brown, G.S.: "Lens Design for Azimuthally Symmetric, Split-Beam, Pattern Shaping at Millimetre Wavelengths", Trans. IEEE, 1966, AP-14, p.106.

16.68. Cullyer, W.J.: "Synthesis of Linear Arrays Using a Potential Analogue", Proc. IEE, 1966, 113, p.255.

16.69. Nyquist, D.P., and Chen, K.M.: "TW Antenna with Non-Dissipative Loading", Int. Conv. Rec. IEEE, 1966, 14, pt 5, p.200.

16.70. Stark, L.: "Radiation Impedance of a Dipole in an Infinite Planar Phased Array", Radio Sci., 1966, 1, p.361.

16.71. Breithaupt, R.W.: "Control of Undesirable Higher-Order Modes Excited at a Conical-Horn Mouth", Electron. Lett., 1966, 2, p.62.

16.72. McInnes, P.A., and Ramsay, J.F.: "Fresnel and Fraunhofer Patterns of Overmoded Feeds and Reflector Antennas", Int. Conv. Rec. IEEE, 1966, 14, pt 5, p.212.

16.73. Kay, A.F.: "Millimetre-Wave Antennas", Proc. IEEE, 1966, 54, p.641.

16.74. Brunning, R.C.E., and Macleod, A.J.M.: "Design and Development of a Cross-Polarized Feed for a 3-m Paraboloid", Marconi Rev., 1966, 29, p.79.

16.75. Das, R.: "Concentric Ring Array", Trans. IEEE, 1966, AP-14, p.398.

16.76. Wanselow, R.D., and Milligan, D.W.: "Broadband Slotted Cone Antenna", Trans. IEEE, 1966, AP-14, p.179.

16.77. Ludwig, A.C.: "Radiation Pattern Synthesis for Circular-Aperture Horn Antennas", Trans. IEEE, 1966, AP-14, p.434.

16.78. Mei, K.K., and Johnstone, D.: "Broadside Log-Periodic Antenna", Proc. IEEE, 1966, 54, p.889.

16.79. Rusch, W.V.T.: "Phase Error and Associated Cross-Polarization Effects in Cassegrainian-Fed Microwave Antennas", Trans. IEEE, 1966, AP-14, p.266.

16.80. Karjala, D.S., and Mittra, R.: "Radiation from a Modulated Corrugated Surface Excited by a Waveguide", Proc. IEE, 1966, 113, p.1143.

16.81. Kopp, E.H.: "Coupled-Waveguide Antennas", Trans. IEEE, 1966, AP-14, p.416.

16.82. Reitzig, R.: "Determination of Reflection Planes of TE_{01} and TM_{01} Modes Aperiodically Damped in a Conical Waveguide", Frequenz, 1966, 20, no.4, p.113.

16.83. Tang, C.H., and Wang, V.W.: "Directive Gain of Nonuniformly Spaced Arrays", Trans. IEEE, 1966, AP-14, p.505.

16.84. Poggio, A.J., Ziolkowski, F.P., and Mayes, P.E.: "Superdirective Array of Normal-Mode Helical Dipoles", Int. Conv. Rec. IEEE, 1966, 14, pt 5, p.143.

16.85. Lewis, B.L.: "Large-Diameter, Dielectric-Rod, End-Fire Antennas", Trans. IEEE, 1966, AP-14, p.239.

16.86. Stearns, C.O.: "Computed Performance of Moderate Size, Super-Gain, End-Fire Antenna Arrays", Trans. IEEE, 1966, AP-14, p.241.

16.87. Stephenson, D.T., and Mayes, P.E.: "Normal Mode Helices as Moderately Superdirective Antennas", Trans. IEEE, 1966, AP-14, p.108.

16.88. Wu, C.P., and Galindo, V.: "Properties of a Phased Array of Rectangular Waveguides with Thin Walls", Trans. IEEE, 1966, AP-14, p.163.

16.89. Galindo, V., and Wu, C.P.: "Numerical Solutions for an Infinite Phased Array of Rectangular Waveguides with Thick Walls", Trans. IEEE, 1966, AP-14, p.149.

16.90. Galindo, V., and Wu, C.P.: "Asymptotic Behaviour of the Coupling Coefficient for an Infinite Array of Thin-Walled Rectangular Waveguides", Trans. IEEE, 1966, AP-14, p.248.

16.91. Wrigley, W.B., Ray, N.R., and Paris, D.T.: "Effects of Dielectric Scaling on Radiation Characteristics of an Antenna", Trans. IEEE, 1966, AP-14, p.99.

16.92. Muller, K.E., and Strecker, S.: "Multiple Utilization of a Parabolic Horn Antenna", Tech. Mitt. RFZ, 1966, 10, p.97.

16.93. Feix, G.: "Investigation of Phase Centres of Extremely Long Pyramidal Horn Radiators", Nachr. Tech. Z., 1966, 19, p.441.

16.94. Yu, J.S., Rudduck, R.C., and Peters, L.: "Comprehensive Analysis for E-Plane Patterns of Horn Antennas by Edge Diffraction Theory", Trans. IEEE, 1966, AP-14, p.138.

16.95. Kitsuregawa, T., et al.: "Radiation Characteristics of the Conical Horn-Reflector Antenna Excited in Higher Modes", Int. Conv. Rec. IEEE, 1966, 14, pt 5, p.252.

16.96. Rumsey, V.H.: "Horn Antennas with Uniform Power Patterns Around their Axes", Trans. IEEE, 1966, AP-14, p.656.

16.97. Cornbleet, S.: "Pattern Synthesis from Zoned Circular Apertures", Trans. IEEE, 1966, AP-14, p.646.

16.98. Yevstropov, G.A., and Tsarapkin, S.A.: "Calculation of Slotted-Waveguide Antennas Taking into Account Interaction between Radiators", Radiotekh. Elektron., 1966, 11, p.822, and Radio Eng. Electron. Phys., 1966, 11, p.709.

16.99. Merkulov, V.V.: "Beam Pattern of Antenna Arrays with Randomly Located Radiators", Radiotekh. Elektron., 1966, 11, p.928, and Radio Eng. Electron. Phys., 1966, 11, p.801.

16.100. Manarini, A.M.M.: "Propagation of Waves in Sectoral Horns", Rend. Ist. Lomb., 1966, 100A, p.247.

16.101. Lader, L.J., and Winderman, J.B.: "Method of Reducing Sidelobes of Directive Antennas", Can. J. Phys., 1966, 44, p.2765.

16.102. Schlosser, W.: "Calculation and Optimization of Excitation Horns for Dielectric Waveguides", Arch. Elektr. Ubertrag., 1966, 20, p.451.

16.103. Zakzon, M.B., and Merkulov, V.V.: "Plane Antennas with Randomly Spaced Radiators", Radiotekh. Elektron., 1966, 11, p.1128, and Radio Eng. Electron. Phys., 1966, 11, p.985.

16.104. Karbowiak, A.E., and Chu, P.L.: "Radiation from Corners in Surface-Wave Lines", Electron. Lett., 1966, 2, p.465.

16.105. Itakura, K., Azakami, T., and Hamii, M.: "Focusing Lens for Electromagnetic Waves", Technol. Rep. Osaka Univ., 1966, 16, p.459.

16.106. Studd, A.C.: "Rear Feed for Paraboloidal Reflectors", Microwave J., 1966, 9, no.2, p.50.

16.107. Ludwig, A.C.: "Gain Computations from Pattern Integration", Trans. IEEE, 1967, AP-15, no.2, pp.309-11.

16.108. James, J.R.: "Theoretical Investigation of Cylindrical Dielectric-Rod Antennas", Proc. IEE, 1967, 114, no.3, pp.309-19,

16.109. Turrin, R.H.: "Dual-Mode Small-Aperture Antennas", Trans. IEEE, 1967, AP-15, no.2, pp.307-8.

16.110. Sharpe, C.B., and Crane, R.B.: "Effects of Dispersion in Linear Arrays", Trans. IEEE, 1967, AP-15, no.2, pp.295-6.

16.111. Kaiser, P.: "Inclined Log-Spiral Antenna", Trans. IEEE, 1967, AP-15, no.2, pp.304-5.

16.112. Parad, L.I.: "Input Admittance of a Slotted Array With or Without a Dielectric Sheet", Trans. IEEE, 1967, AP-15, no.2, pp.302-4.

16.113. Burrows, M.L., and Ricardi, L.J.: "Aperture Feed for a Spherical Reflector", Trans. IEEE, 1967, AP-15, no.2, pp.227-30.

16.114. Anastasovski, P., and Ristov, M.: "Microwave Lenses with Nonlinear Refractive Index", Electron. Lett., 1967, 3, no.3, pp.101-2.

16.115. Das, R.: "Broadbanding of Concentric Planar Ring Arrays by Space Tapering", Radio Electron. Eng., 1967, 33, no.4, pp.211-22.

16.116. Rudin, V.Yu., and Sazonov, V.V.: "Field Statistics of Linear Antenna Arrays Having Periodically Correlated Phase Errors", Radiotekhnika, 1967, 22, no.2, pp.43-8, and Telecommun. Radio Eng. Pt 2, 1967, 22, no.2.

16.117. Ostrovskii, D.B.: "Modulus Distribution of the Directivity Characteristic", Radiotekh. Elektron., 1967, 12, no.1, pp.128-31, and Radio Eng. Electron. Phys., 1967, 12, no.1.

16.118. Ivanov, V.N.: "Dispersion Characteristics of Complex Helices in the Region of Small Delays", Radiotekh. Elektron., 1967, 12, no.1, pp.142-3, and Radio Eng. Electron. Phys., 1967, 12, no.1.

16.119. Kopeikin, V.I.: "Directive Gain of Rectangular Apertures in the Fresnel Zone", Radiotekh. Elektron., 1967, 12, no.1, pp.132-6, and Radio Eng. Electron. Phys., 1967, 12, no.1.

16.120. Tseng, F.I., and Cheng, D.K.: "Gain Optimization for Arbitrary Antenna Arrays Subject to Random Fluctuations", Trans. IEEE, 1967, AP-15, no.3, pp.356-65.

16.121. Kmetzo, J.L.: "Analytical Approach to the Coverage of a Hemisphere by N Planar Phased Arrays", Trans. IEEE, 1967, AP-15, no.3, pp.367-71.

16.122. Angelakos, D.J., and Kajfez, D.: "(Mode Reduction) on Axial-Mode Helical Antenna", Proc. IEEE, 1967, 55, no.4, pp.558-9.

16.123. Carver, K.R.: "Helicon. Circularly Polarized Antenna with Low Sidelobe Level", Proc. IEEE, 1967, 55, no.4, p.559.

16.124. D'Auria, G., and Solimini, D.: "Fresnel Diffraction and Focusing Properties of Apertures Under Partially Coherent Illumination", Trans. IEEE, 1967, AP-15, no.3, pp.480-1.

16.125. Hamid, M.A.K.: "Mutual Coupling between Sectoral Horns Side by Side", Trans. IEEE, 1967, AP-15, no.3, pp.475-7.

16.126. Claydon, B.: "Study of the Performance of Cassegrain Aerials", Marconi Rev., 1967, 30, pp.98-115.

16.127. Allen, J.L.: "Simple Model for Mutual Coupling Effects on Patterns of Unequally Spaced Arrays", Trans. IEEE, 1967, AP-15, no.4, pp.530-3.

16.128. Hamid, M.A.K.: "Reflection Factor at a Horn-Waveguide Junction", Trans. IEEE, 1967, AP-15, no.4, pp.564-5.

16.129. Chute, F.S.: "Reaction Torque on an Axial Multipole Radiator", Trans. IEEE, 1967, AP-15, no.4, pp.585-6.

16.130. Fiser, K.: "Radiation of a Dipole in a Waveguide of General Cross Section", Cesk. Cas. Fis., 1967, 17A, no.1, pp.10-20.

16.131. Anderson, J.B.: "Radiation from Surface-Wave Antennas", Electron. Lett., 1967, 3, no.6, p.251.

16.132. McKenzie, J.F.: "Dipole Radiation in Moving Media", Proc. Phys. Soc., 1967, 91, pt 3, pp.537-51.

16.133. Loh, S.C., and Jacobsen, J.: "Radiation Characteristics of the Backfire Helical and Zigzag Antennas", Radio Electron. Eng., 1967, 35, no.5, pp.317-24.

16.134. Loh, S.C., and Jacobsen, J.: "Backfire Meander-Line Antenna", Electron. Eng., 1967, 39, no.9, pp.571-5.

16.135. Mironenko, I.G.: "Synthesis of Finite-Aperture Antenna Radiating Maximum Power into a Given Solid Angle", Radiotekhnika, 1967, 22, no.4, pp.43-9, and Telecommun. Radio Eng. Pt 2, 1967, 22, no.4.

16.136. Ricardi, L.J.: "Directivity of an Array of Slots on the Surface of a Cylinder", Electron. Eng., 1967, 39, no.9, pp.578-81.

16.137. Lewin, L.: "Field Representation in Cylindrical Coordinates", Electron. Lett., 1967, 3, no.7, pp.308-9.

16.138. Shubarin, Yu.V., and Chebotarev, V.I.: "Waveguide Slot Radiator with Controlled Polarization", Radiotekh. Elektron., 1967, 12, no.1, pp.140-2, and Radio Eng. Electron. Phys., 1967, 12, no.1, p.136.

16.139. Mentzoni, M.H., and Donohoe, J.: "Experimental Determination of the Mutual Admittance of Slot Antennas", Electron. Lett., 1967, 3, no.7.

16.140. Zich, R.: "Stepped Aperture Distribution for Horn Antenna Sidelobe Reduction", Alta Freq., 1967, 36, no.8, pp.736-44.

16.141. Komissorov, Ya.S., and Pavlyuk, V.A.: "Properties of Slanted Gratings", Radiotekh. Elektron., 1967, 12, no.1, pp.126-8, and Radio Eng. Electron. Phys., 1967, 12, no.1.

16.142. Akao, Y., and Miyazaki, Y.: "Beam Transformation Formulae for Dielectric Lenses", J. Inst. Electr. Commun. Eng. Jap., 1967, 50, no.5, pp.918-9.

16.143. Cain, F.L., and Johnson, R.C.: "Investigation of Choke Blinders for Feed Horns of Radar Antennas", Trans. IEEE, 1967, EMC-9, no.2, pp.65-72.

16.144. Potter, P.D.: "Application of Spherical Wave Theory to Cassegrainian-Fed Paraboloids", Trans. IEEE, 1967, AP-15, no.6, pp.727-36.

16.145. Charlton, G.G.: "Application of Grating-Lobe Series for Calculating Admittance Variation of a Phased-Array Antenna", Trans. IEEE, 1967, AP-15, no.6, pp.818-20.

16.146. Becker, K.D.: "Calculation of Radiation Field of a Dielectric Rod Using Laplace Transform and Method of Images", Arch. Elektr. Ubertrag., 1967, 21, no.9, pp.483-6.

16.147. Croswell, W.F., Rudduck, R.C., and Hatcher, D.M.: "Admittance of a Rectangular Waveguide Radiating into a Dielectric Slab", Trans. IEEE, 1967, AP-15, no.5, pp.627-33.

16.148. Bailey, M.C.: "Design of Dielectric-Covered Resonant Slots in a Rectangular Waveguide", Trans. IEEE, 1967, AP-15, no.5, pp.594-8.

16.149. Galindo, V., and Wu, C.P.: "Composite Planar Array of Parallel Plates", Trans. IEEE, 1967, AP-15, no.6, pp.826-8.

16.150. Lee, S.W.: "Radiation from an Infinite Aperiodic Array of Parallel-Plate Waveguides", Trans. IEEE, 1967, AP-15, no.5, pp.598-606.

16.151. Musil, J.: "Microwave Energy Concentration in the Fresnel Region", Slab. Obz., 1967, 28, no.8, pp.489-95.

16.152. Takeshima, T.: "X-Band Omnidirectional Double-Slot Array Antenna", Electron. Eng., 1967, 39, no.10, pp.617-21.

16.153. Fee, M.L.: "Loop-Slot Radiating Element for Obtaining Circular Polarization", Microwave J., 1967, 10, no.12, p.75.

16.154. Neumann, E.G.: "Field at the Free End of a Dielectric Rod. I and II", Z. Angew. Phys., 1967, 24, no.1, pp.1-11 and 12-18.

16.155. Shubarin, Yu.V.: "Effect of Mirrors on the Primary Source in a Cassegrain System", Radiotekh. Elektron., 1967, 12, no.9, pp.1666-8, and Radio Eng. Electron. Phys., 1967, 12, no.9.

16.156. Panchenko, B.A.: "Self and Mutual Conductances of Transverse Slots in a Cylinder", Radiotekhnika, 1967, 22, no.11, pp.61-6, and Telecommun. Radio Eng. Pt 2, 1967, 22, no.11.

16.157. Rao, B.V., and Reddy, K.K.: "Study of Microwave Horn and Lens Antennas", J. Inst. Telecommun. Eng. 1967, 13, no.10, pp.390-9.

16.158. Jull, E.V.: "Behaviour of Electromagnetic Horns", Proc. IEEE, 1968, 56, no.1, pp.106-8.

16.159. Sengupta, D.L., Smith, T.M., and Larson, R.W.: "Radiation Characteristics of a Spherical Array of Circularly Polarized Elements", Trans. IEEE, 1968, AP-16, no.1, pp.2-7.

16.160. Mott, H., and McQuiddy, D.N.: "Simple Waveguide System for Radiating Elliptically Polarized Waves", Trans. IEEE, 1968, AP-16, no.1, pp.134-5.

16.161. DuFort, E.C.: "Design of Corrugated Plates for Phased Array Matching", Trans. IEEE, 1968, AP-16, no.1, pp.37-46.

16.162. Love, A.W., and Gustincic, J.J.: "Line-Source Feed for Spherical Reflector", Trans. IEEE, 1968, AP-16, no.1, pp.132-4.

16.163. Koob, K.: "Area Efficiency of Rectangular Horn Radiators", Nachr. Tech. Z., 1968, 21, no.4, pp.204-5.

16.164. Wu, C.P., and Galindo, V.: "Surface-Wave Effects on Phased Arrays of Rectangular Waveguides Loaded with Dielectric Plugs", Trans. IEEE, 1968, AP-16, no.3, pp.358-60.

16.165. Munk, B.A., and Peters, L.: "Launcher for Helical Antenna", Trans. IEEE, 1968, AP-16, no.3, pp.362-3.

16.166. Barband, N.: "Waveguide-Fed Logperiodic Antennas", Trans. IEEE, 1968, AP-16, no.3, pp.357-8.

16.167. Kumar, A., and Chatterjee, R.: "Tapered Dielectric Rod Antennas", Proc. Natl. Inst. Sci. India, 1968, 34,no.1,p.1.

16.168. Cornbleet, S.: "Beam Shaping for Optimum Illumination Patterns", Electron. Lett., 1968, 4, p.502.

16.169. Geldmacher, E.: "Design Development of Large Parabolic Antennas", Tech. Mitt. Krupp Forsch., 1968, 26, p.69.

16.170. Trentini, G.V., Romeiser, K.P., and Reitzig, R.: "Possibilities of Change and Improvement in the Design of Large Cassegrain Antennas with Horn Feed", Frequenz, 1968, 22, p.216.

16.171. Walton, K.L., and Sundberg, C.: "Constant-Beamwidth Antenna Development", Trans. IEEE, 1968, AP-16, p.510.

16.172. Jones, J.E., et al.: "Admittance of a Parallel-Plate Waveguide Aperture Illuminating a Metal Sheet", Trans. IEEE, 1968, AP-16, p.528.

16.173. Scheffer, H.: "Radiation of Open Coaxial Line Excited by H Modes", Arch. Elektr. Ubertrag., 1968, 22, p.514.

16.174. Pagones, M.J.: "Gain Factor of an Offset-Feed Paraboloidal Reflector", Trans. IEEE, 1968, AP-16, p.536.

16.175. Chen, K.M., Nyquist, D.P., and Lin, J.L.: "Radiation Fields of the Short Backfire Antenna", Trans. IEEE, 1968, AP-16, p.596.

16.176. Carver, K.R.: "Polarization Stability of the Helical Beam Antenna", Trans. IEEE, 1968, AP-16, p.604.

16.177. Sanyal, G.S., and Singh, M.: "Fresnel Zone Plate Antenna", J. Inst. Telecommun. Eng., 1968, 14, p.265.

16.178. Lee, S.W., and Mittra, R.: "Radiation from Dielectric-Loaded Arrays of Parallel-Plate Waveguides", Trans. IEEE, 1968, AP-16, p.513.

16.179. Chatterjee, J.S., and Roy, M.N.: "Helical Logperiodic Array", Trans. IEEE, 1968, AP-16, p.592.

16.180. Holtzman, J.C.: "Dual (X and S)-Band Array", Trans. IEEE, 1968, AP-16,p.603.

16.181. Wegrowicz, L.: "Aperture Radiation and Superdirectivity", Bull. Acad. Polon. Sci., 1968, 16, p.381.

16.182. Voskresenskii, D.I., and Filippov, V.S.: "Gain of Convex-Type Pencil-Beam Antenna Arrays", Izv. VUZ Radioelektron., 1968, 11, p.413.

16.183. Ponomarev, L.I.: "Maximum Gain of Spherical and Conical Antennas", Izv. VUZ Radioelektron., 1968, 11, p.426.

16.184. Tsandoulas, G.N.: "Effect of Surface Impedance Variation Along a Rod Waveguide", Trans. IEEE, 1968, MTT-16, p.886.

16.185. Solosko, R.B., and Laxpati, S.R.: "Log-Periodic Antenna with Vertically Polarized Omnidirectional Radiation", Trans. IEEE, 1968, AP-16, p.752.

16.186. Roederer, A.G.: "Log-Periodic Cavity-Backed Slot Array", Trans. IEEE, 1968, AP-16, p.756.

16.187. Pogorzelski, S.: "Critical Section of a Cosecant-Squared Reflector", Pr. PIT, 1968, 18, no.61, p.21.

16.188. Voitovich, N.N., and Semionov, V.V.: "Shaping of Field with a Given Pattern", Radiotekh. Elektron., 1968, 13, p.1213, and Radio Eng. Electron. Phys., 1968, 13, no.7, pp.1054-61.

16.189. Kajfez, D.: "Dispersion Diagram of a Helical Antenna", Trans. IEEE, 1968, E-11, p.255.

16.190. Nair, K.G., Srivastava, G.P., and Singh, S.B.: "Effects of Metal Flanges on Radiation Patterns of H-Plane Sectoral Horns", J. Inst. Telecommun. Eng., 1968, 14, p.352, and Int. J. Electron., 1968, 25, p.153.

16.191. Knittel, G.H., Hessel, A., and Oliner, A.A.: "Element Pattern Nulls in Phased Arrays", Proc. IEEE, 1968, 56, p.1822.

16.192. Diamond, B.L.: "Generalized Approach to Analysis of Infinite Planar Arrays", Proc. IEEE, 1968, 56, p.1837.

16.193. Wheeler, H.A.: "Systematic Approach to Design of a Radiator Element for Phased Array", Proc. IEEE, 1968, 56, p.1940.

16.194. Shubarin, Yu.V.: "Effect of Waveguide End Reflection on Polarization Diagram of a Slotted Waveguide", Izv. VUZ Radioelektron., 1968, 11, p.643.

16.195. Breithaupt, R.W.: "Conductance Data for Offset Series Slots in Stripline", Trans. IEEE, 1968, MTT-16, p.969.

16.196. Kumar, A., and Chatterjee, R.: "Radiation from Tapered Dielectric Rods", J. Indian Inst. Sci., 1968, 50, p.374.

16.197. Yesepkina, N.A., and Braude, B.V.: "Feasibility of Adjusting Two-Mirror Antennas in the Near Field", Radiotekh. Elektron., 1968, 13, no.12, pp.1956-61.

16.198. Amitay, N., and Galindo, V.: "Analysis of Circular Waveguide Phased Arrays", Bell Syst. Tech. J., 1968, 47, p.1903.

16.199. Smits, V.C.: "Rear Gain Control of Dielectric Rod Antenna", Microwave J., 1968, 11, no.12, p.65.

16.200. Koshy, V.K., Nair, K.G., and Srivastava, G.P.: "Effect of Conducting Flanges on H-Plane Radiation Pattern of E-Plane Sectoral Horns", J. Inst. Telecommun. Eng., 1968, 14, p.519.

16.201. Pruzhanovskii, V.A.: "Radiation Pattern of a Coaxial Optical Waveguide", Zh. Tekh. Fiz., 1968, 38, p.1744, and Sov. Phys.-Tech. Phys., 1969, 13, p.1411.

16.202. Tocquec, Y.: "Adaptation of a Pair of Reflectors to a Primary Source", Onde Electr., 1969, 49, p.204.

16.203. Jones, K.E., and Mayes, P.E.: "Continuously Scaled Transmission Lines with Applications to Logperiodic Antennas", Trans. IEEE, 1969, AP-17, p.2.

16.204. Nair, K.G., Srivastava, G.P., and Hariharan, S.: "Sharpening of E-Plane Radiation Patterns from E-Plane Sectoral Horns by Metallic Grills", Trans. IEEE, 1969, AP-17, p.91.

16.205. Chow, Y.L., and Wu, S.C.: "Synthesis of Array Space Factors", Can. J. Phys., 1969, 47, p.291.

16.206. Thiele, G.A.: "Analysis of Yagi-Uda Antennas", Trans. IEEE,1969,AP-17,p.24.

16.207. Mailloux, R.J.: "Radiation and Near-Field Coupling between Two Collinear Open-Ended Waveguides", Trans. IEEE, 1969, AP-17, p.49.

16.208. Perini, J.: "Unusually Simple Technique for Sidelobe Reduction", Trans. IEEE, 1969, EMC-11, p.29.

16.209. Mikoshiba, K., and Kamimura, M.: "Leaky Waveguide. TE_{01} Circular Waveguide with Periodic Array of Circular Apertures", Trans. IEEE, 1969, MTT-17, p.15.

16.210. Arnbak, J.: "Leaky Waves on a Dielectric Rod", Electron. Lett., 1969, 5, p.41.

16.211. Takeshima, T., and Isogai, Y.: "Frequency Bandwidth of Slotted Array Antenna System", Electron. Eng., 1969, 41, p.201.

16.212. Bacon, J.M., and Medhurst, R.G.: "Superdirective Antenna Array Containing only One Fed Element", Proc. IEE, 1969, 116, p.365.

16.213. Shigesawa, H., and Takiyama, K.: "Approximate Pattern Calculation of a Leaky Waveguide", Trans. IEEE, 1969, AP-17, p.36.

16.214. Carberry, T.F.: "Analysis Theory for Shaped-Beam Doubly-Curved Reflector Antenna", Trans. IEEE, 1969, AP-17, p.131.

16.215. Prochazka, M.: "Periodic Receiving Broadband Antenna Arrays", Trans. IEEE, 1969, AP-17, p.221.

16.216. Chatterjee, R., Balasubramanian, V., and Chatterjee, S.K.: "Investigations on Dielectric-Rod Antenna Excited in Mixed HE_{11} and E_{01} Modes", J. Indian Inst. Sci., 1969, 51, p.77.

16.217. Kliger, G.A.: "Analysis of a Vertical Zigzag Logperiodic Antenna", Radiotekhnika, 1969, 24, no.1, p.57, and Telecommun. Radio Eng. Pt 2, 1969, 24, no.1.

16.218. Borgiotti, G.V., and Balzano, Q.: "Analysis of Periodic Phased Arrays of Circular Apertures", Trans. IEEE, 1969, AP-17, p.224.

16.219. Strait, B.T., and Hirasawa, K.: "Array Design for Specified Pattern by Matrix Methods", Trans. IEEE, 1969, AP-17, p.237.

16.220. Mikoshiba, K., and Nishida, S.: "Helix Leaky Waveguide", Trans. IEEE, 1969, MTT-17, p.66.

16.221. Brunner, A.: "Possible Design of Doubly Curved Reflectors for Search-Radar Antennas", Frequenz, 1969, 23, no.5, p.152.

16.222. Tokumaru, S.: "Double-Sheath Helices. Leaky-Wave Antenna", Trans. IEEE, 1969, AP-17, p.138.

16.223. Anders, R., and Wohlleben, R.: "Phase Velocity on a Conical Two-Armed Logarithmic Foil-Type Spiral Antenna", Trans. IEEE, 1969, AP-17, p.233.

16.224. Panicali, A.R., and Lo, Y.T.: "Probabilistic Approach to Large Circular and Spherical Arrays", Trans. IEEE, 1969, AP-17, p.514.

16.225. Jacobsen, J.: "Radiation from Log-Periodic Antennas", Forskning, 1969, 78, no.4, p.110.

16.226. Benenson, L.S.: "Mutual Impedance between Radiators in Arrays", Radiotekh. Elektron., 1969, 14, no.7, p.1202, and Radio Eng. Electron. Phys., 1969, 14, no.7, pp.1038-50.

16.227. James, J.R., and Longdon, L.W.: "Calculation of Radiation Patterns", Electron. Lett., 1969, 5, p.567.

16.228. Clarricoats, P.J.B., and Saha, P.K.: "Radiation from Wide-Flare-Angle Scalar Horns", Electron. Lett., 1969, 5, pp.376-378.

16.229. Yu, J.S., and Rudduck, R.C.: "H-Plane Pattern of a Pyramidal Horn", Trans. IEEE, 1969, AP-17, p.651.

16.230. Glaser, J.I.: "Design of Elliptical and Circular Aperture Pencil-Beam Antennas Using the Modified Lambda Functions", Trans. IEEE, 1969, AP-17, p.655.

16.231. Thomas, B.M.: "Bandwidth Properties of Corrugated Conical Horns", Electron. Lett., 1969, 5, p.561.

16.232. Bhartia, P., and Hamid, M.A.K.: "Eigenvalues for Higher-Order Conical-Horn Modes", Electron. Lett., 1969, 5, p.684.

16.233. Bao, V.T.: "Optimization of Efficiency of Deep Reflectors", Trans. IEEE, 1969, AP-17, p.811.

16.234. Wu, C.P.: "Resonances in Waveguide Antennas with Dielectric Plugs", Bell Syst. Tech. J., 1969, 48, p.2305.

16.235. Jeuken, M.E.J., and Lambrechtse, C.W.: "Small Corrugated Conical-Horn Antenna with Wide Flare Angle", Electron. Lett., 1969, 5, p.489.

16.236. Jeuken, M.E.J.: "Radiation Pattern of Corrugated Conical-Horn Antenna with Small Flare Angle", Electron. Lett., 1969, 5, p.485.

16.237. Balling, P.: "Periodically Modulated, Dielectrically Filled, Waveguide as a Microwave Antenna", Electron. Lett., 1969, 5, p.508.

16.238. Al-Hakkak, M.J.: "Experimental Investigation of Input-Impedance Characteristics of an Antenna in a Rectangular Waveguide", Electron. Lett., 1969, 5, p.513.

16.239. Sandrin, W.A., Glatt, C.R., and Hague, D.S.: "Design of Arrays with Unequal Spacing and Partially Uniform Amplitude Taper", Trans. IEEE, 1969, AP-17, p.642.

16.240. Tartakovskii, L.B., and Rubinshtein, A.I.: "Mutual Effects between Simple Plane Waveguide Type of Radiators with Open Ends", Radiotekh. Elektron., 1969, 14, no.8, p.1369, and Radio Eng. Electron. Phys., 1969, 14, no.8, pp.1187-93.

16.241. Swift, C.T.: "Admittance of a Waveguide-Fed Aperture Loaded with a Dielectric Plug", Trans. IEEE, 1969, AP-17, p.356.

16.242. Narasimhan, M.S., and Rao, B.V.: "Radiation from a Conical Horn", Int. J. Electron., 1969, 26, p.377.

16.243. Zav'yalov, A.S.: "Surface-Wave Antenna in the Form of a Directional Coupler", Izv. VUZ Radioelektron., 1969, 12, no.6, p.630.

16.244. Lyakhovetskii, G.Ya.: "Approximate Method for Calculating Parameters of Yagi Antennas on a Power-Criterion Basis", Elekrosvyaz, 1969, 23, no.1, p.28, and Telecommun. Radio Eng. Pt 1, 1969, 23, no.1.

16.245. Uslenghi, P.L.E.: "Electromagnetic and Optical Behaviour of Two Classes of Dielectric Lenses", Trans. IEEE, 1969, AP-17, p.235.

16.246. Chan, A.K., and Sigelmann, R.A.: "Experimental Investigation on Spherical Arrays", Trans. IEEE, 1969, AP-17, p.348.

16.247. Van Atta, L.C., Mailloux, R.J., and Levenson, M.M.: "Simple Antenna with Circular Polarization", Trans. IEEE, 1969, AP-17, p.360.

16.248. James, J.R.: "Leaky Waves on a Dielectric Rod", Electron. Lett., 1969, 5, p.252.

16.249. Ludwig, E., Reis, G., and Zocher, E.: "Improved Horn Feed with Two Modes for Shallow Paraboloids", Arch. Elektr. Ubertrag., 1969, 23, p.209.

16.250. Roy, M.N.: "Investigations on Normal-Mode Helices", Int. J. Electron., 1969, 26, p.573.

16.251. Agarwal, V.D., and Lo, Y.T.: "Distribution of Sidelobe Level in Random Arrays", Proc. IEEE, 1969, 57, p.1764.

16.252. Shigesawa, H., and Takiyama, K.: "Approximated Radiation Pattern of Leaky Waveguides with Periodic Loading Slots", Electron. Commun. Jap., 1969, 52, no.8, p.49.

16.253. Wu, C.P.: "Integral Equation Solutions for Radiation from a Waveguide through a Dielectric Slab", Trans. IEEE, 1969, AP-17, p.733.

16.254. Mal'tsev, V.P., et al.: "Leaky Waves in Waveguide with Two Separated Layers", Izv. VUZ Radiofiz., 1969, 12, p.1855.

16.255. Nefedov, Ye.I., and Popichenko, V.A.: "Broadening the Radiation Pattern from End of a Wide Waveguide", Izv. VUZ Radioelektron., 1969, 12, p.1430.

16.256. Roederer, A.G.: "Calculation of Field Radiated by a Logperiodic Dipole Antenna", Onde Electr., 1969, 49, p.223.

16.257. Narasimhan, M.S., and Rao, B.V.: "Hybrid Modes in Corrugated Conical Horns", Electron. Lett., 1970, 6, p.32.

16.258. Kyle, R.H.: "Mutual Coupling between Logperiodic Antennas", Trans. IEEE, 1970, AP-18, p.15.

16.259. Alton, E.D., and Groe, L.G.: "Computer Calculation of the Far-Field Radiation Pattern of an Antenna Array", Trans. IEEE, 1970, BC-16, p.8.

16.260. Booker, D.D., and McInnes, P.A.: "Computer-Predicted Performance of Corrugated Conical Feeds Using Experimental Primary-Radiation Patterns", Electron. Lett., 1970, 6, p.18.

16.261. Hamid, M.A.K., et al.: "Radiation Characteristics of Dielectric-Loaded Horn Antennas", Electron. Lett., 1970, 6, p.20.

16.262. Boswell, A.: "Logperiodic Dipole Arrays", Marconi Rev., 1970, 33, no.178, p.225.

16.263. Forman, B.J.: "Novel Directivity Expression for Planar Antenna Arrays", Radio Sci., 1970, 5, no.7, p.1077.

16.264. Deryugin, L.N., Marchuk, A.N., and Sotin, V.Ye.: "Radiation from a Plane Dielectric Waveguide", Izv. VUZ Radioelektron., 1970, 13, no.3, p.309.

16.265. Oh, L.L., Peng, S.Y., and Lunden, C.D.: "Effects of Dielectrics on Radiation Patterns of an Electromagnetic Horn", Trans. IEEE, 1970, AP-18, p.553.

16.266. Jull, E.V.: "Finite-Range Gain of Sectoral and Pyramidal Horns", Electron. Lett., 1970, 6, no.21, pp.680-1.

16.267. Nicolau, E.: "Synthesis of Linear Antenna Arrays", Stud. Cercet. Energ. Electrotech., 1970, 20, no.4, pp.919-35.

16.268. Kobrina, G.A., Pokras, A.M., and Fel'd, N.A.: "Amplitude and Phase Patterns of a Conical Horn at Distances Commensurable with its Aperture", Radiotekhnika, 1970, 25, no.10, pp.103-7, and Telecommun. Radio Eng. Pt 2, 1970, 25, no.10.

16.269. Panchenko, B.A.: "Effect of Interaction between Elements on the Radiation Pattern of a Linear Phased Array", Radiotekh. Elektron., 1970, 15, no.6, p.1294, and Radio Eng. Electron. Phys., 1970, 15, no.6, pp.1098-1100.

16.270. Rudge, A.W., and Withers, M.J.: "Design of Flared-Horn Primary Feeds for Parabolic Reflectors", Proc. IEE, 1970, 117, p.1741.

16.271. Arnbak, J.: "Quasi-Equispaced Slot Arrays with Extremely Low Cross Polarization", Electron. Lett., 1970, 6, p.585.

16.272. Cogdell, J.R., et al.: "High-Resolution Millimetre Reflector Antennas", Trans. IEEE, 1970, AP-18, p.515.

16.273. Lee, S.W.: "Ray Theory of Diffraction by Open-Ended Waveguides. I", J. Math. Phys., 1970, 11, no.9, pp.2830-50.

16.274. Lewin, L.: "Complementary Field Component of Radiation from a Cassegrain Subreflector", Electron. Lett., 1970, 6, no.22, pp.700-1.

16.275. Arnold, P.W.: "Circularly Polarized Octave-Bandwidth Unidirectional Antenna Using Conical Dipoles", Trans. IEEE, 1970, AP-18, no.5, pp.696-8.

16.276. Bantin, C.C., and Balmain, K.G.: "Study of Compressed Logperiodic Dipole Antennas", Trans. IEEE, 1970, AP-18, p.195.

16.277. Klein, C.F.: "Equivalent Circuit for Tee-Fed Slot Antennas", Trans. IEEE, 1970, AP-18, p.280.

16.278. Prasad, S.M., and Das, B.N.: "Circular Loop Antenna with TW Current Distribution", Trans. IEEE, 1970, AP-18, p.278.

16.279. Ivanov, I.I.: "Super-Wideband Antenna", Izv. VUZ Radioelektron., 1970, 13, p.74.

16.280. Van Koughnett, A.L.: "Mutual Coupling Effects in Linear Antenna Arrays", Can. J. Phys., 1970, 48, p.659.

16.281. Goebels, F.G., and Anderson, R.K.: "Dual-Band Slot Array Technique", Trans. IEEE, 1970, AP-18, p.282.

16.282. Bailey, M.C., Beck, F.B., and Croswell, W.R.: "Vertically Polarized Stacked Arrays of Omnidirectional Antennas", Trans. IEEE, 1970, AP-18, p.285.

16.283. Bates, C.P.: "Internal Reflections at the Open End of a Semi-Infinite Waveguide", Trans. IEEE, 1970, AP-18, p.230.

16.284. Marcuse, D.: "Radiation Losses of Tapered Dielectric Slab Waveguides", Bell Syst. Tech. J., 1970, 49, p.273.

16.285. Wu, C.P.: "Characteristics of Coupling between Parallel-Plate Waveguides With and Without Dielectric Plugs", Trans. IEEE, 1970, AP-18, p.188.

16.286. Wolter, J.: "Theory of Yagi Antennas", Nachr. Tech. Z., 1970, 23, p.180.

16.287. Baldwin, R., and McInnes, P.A.: "Surface-Wave Radiation From a Corrugated Horn", Electron. Lett., 1970, 6, p.259.

16.288. Bailey, M.C., et al.: "Waveguide Antennas Illuminating a Metal Sheet", Trans. IEEE, 1970, AP-18, p.396.

16.289. Zucker, F.J., and Strom, J.A.: "Experimental Resolution of Surface-Wave Antenna Radiation into Feed and Terminal Patterns", Trans. IEEE, 1970, AP-18, p.420.

16.290. Koshy, V.K., Nair, K.G., and Srivastava, G.P.: "Analysis of Radiation from a Flanged Aperture Antenna", Trans. IEEE, 1970, AP-18, p.407.

16.291. Rao, B.L.J., and Chen, S.N.C.: "Illumination Efficiency of a Shaped Cassegrain System", Trans. IEEE, 1970, AP-18, p.411.

16.292. Cornbleet, S.: "Superdirective Property of the Microwave Axicon", Proc. IEE, 1970, 177, p.869.

16.293. Wu, C.P.: "Analysis of Finite Parallel-Plate Waveguide Arrays", Trans. IEEE, 1970, AP-18, p.328.

16.294. Cheng, D.K., and Tseng, F.I.: "Pencil-Beam Synthesis for Large Circular Arrays", Proc. IEEE, 1970, 117, p.1232.

16.295. Narasimhan, M.S., and Rao, B.V.: "Diffraction by Wide-Flare-Angle Corrugated Conical Horns", Electron. Lett., 1970, 6, p.469.

16.296. Ap Rhys, T.L.: "Design of Radially Symmetric Lenses", Trans. IEEE, 1970, AP-18, p.497.

16.297. Neumann, E.G.: "Radiation Mechanism of Dielectric-Rod and Yagi Antennas", Electron. Lett., 1970, 6, p.528.

16.298. Roederer, A.G.: "Calculation of the Field Radiated by a Logperiodic Dipole Antenna", Philips Res. Rep., 1970, 25, no.3, p.175.

16.299. Coleman, H.P.: "Iterative Technique for Reducing Sidelobes of Circular Arrays", Trans. IEEE, 1970, AP-18, p.566.

16.300. Zucker, H.: "Computation of the Far Field of Open Cassegrain Antennas", Bell Syst. Tech. J., 1970, 49, no.7, pp.1595-602.

16.301. Gniss, H., and Ries, G.: "Remarks on the Concept of Equivalent Parabolas for Cassegrain Antennas", Electron. Lett., 1970, 6, no.23, pp.737-9.

16.302. Bailey, M.C.: "Impedance Properties of Dielectric-Covered Narrow Radiating Slots in the Broad Face of a Rectangular Waveguide", Trans. IEEE, 1970, AP-18, no.5, pp.596-603.

16.303. Crout, P.D.: "Determination of Radiation Patterns of n-Arm Antennas by Bicomplex Functions", Trans. IEEE, 1970, AP-18, no.5, pp.686-9.

16.304. Burnside, W.D., Delton, E.L., and Peters, L.: "Analysis of Finite Parallel-Plate Waveguide Arrays", Trans. IEEE, 1970, AP-18, no.5, pp.701-5.

16.305. Arora, R.K., and Vijayaraghavan, S.: "Scattering of a Shielded Surface Wave by a Wall-Impedance Discontinuity", Trans. IEEE, 1970, MTT-18, no.10, pp.734-6.

16.306. Casimir, H.B.G.: "Supergain Antennas", Philips Res. Rep., 1970, 25, no.4, pp.237-43.

16.307. Hong, M.H., Nyquist, D.P., and Chen, K.M.: "Radiation Fields of Open-Cavity Radiators and a Backfire Antenna", Trans. IEEE, 1970, AP-18, no.6, pp.813-15.

16.308. Au, H.K.: "Hybrid Modes in Corrugated Conical Horns with Narrow Flare Angle and Arbitrary Length", Electron. Lett., 1970, 6, no.24, pp.769-71.

16.309. Muehldorf, E.I.: "Phase Centre of Horn Antennas", Trans. IEEE, 1970, AP-18, no.6, pp.753-60.

16.310. Lee, S.H., and Mei, K.K.: "Analysis of Zigzag Antennas", Trans. IEEE, 1970, AP-18, no.6, pp.760-4.

16.311. Strecker, S.: "Wideband Radiation Behaviour of Quadratic Pyramidal Horn Radiators", Tech. Mitt. RFZ, 1970, 14, no.3, pp.138-42.

16.312. Chang, V.W.H.: "Experimental Study of Collinear and Planar Arrays", Trans. IEEE, 1970, AP-18, no.6, pp.791-5.

16.313. Cox, R.M., and Rupp, W.E.: "Circularly Polarized Phased-Array Antenna Element", Trans. IEEE, 1970, AP-18, no.6, pp.804-7.

16.314. Hansen, R.C.: "Directivity of Chebyshev Arrays", Trans. IEEE, 1970, AP-18, no.6, pp.815-8.

16.315. Itoh, K., Suzuki, M., and Matsumoto, T.: "Leaky Waves of a Slotted Rectangular Waveguide", Electron. Commun. Jap., 1970, 53, no.5, pp.12-3.

16.316. Garb, Kh.L., and Fridberg, P.Sh.: "Theory of Weakly Radiating Slots", Radiotekh. Elektron., 1970, 15, no.4, and Radio Eng. Electron. Phys., 1970, 15, no.4, pp.599-606.

16.317. Cottony, H.V.: "Wideband Array with Suppressed Grating Lobes", Trans. IEEE, 1970, AP-18, no.6, pp.774-9.

16.318. Tokatly, V.I.: "Theory of Horns with Small Flare Angle", Radiotekh. Elektron., 1970, 15, no.10, and Radio Eng. Electron. Phys., 1970, 15, no.10, pp.1815-23.

16.319. Tereshin, O.N., and Kuznetsov, L.N.: "Backward-Wave Antenna with Modulated Phase Velocity Using a Rod Slow-Wave Structure", Radiotekhnika, 1970, 25, no.12, pp.92-5, and Telecommun. Radio Eng. Pt 2, 1970, 25, no.12.

16.320. Gupta, K.C.: "Narrow-Beam Antennas Using an Artificial Dielectric Medium with Permittivity Less than Unity", Electron. Lett., 1971, 7, no.1, pp.16-8.

16.321. Satoh, T.: "Dielectric-Loaded Horn Antenna", Electron. Commun. Jap., 1971, 54, no.9, pp.57-63.

16.322. Yerokhin, G.A.: "Synthesis of a Horn Antenna", Radiotekhnika, 1971, 26, no.12, pp.74-81, and Telecommun. Radio Eng. Pt 2, 1971, 26, no.12, pp.104-10.

16.323. Narasimhan, M.S., and Rao, B.V.: "Modes in a Conical Horn", Proc. IEEE, 1971, 118, no.2, pp.287-92.

16.324. Hissink, A.J.: "Radar Antenna with Near-Field Cylindrical Obstruction", Proc. IEEE, 1971, 118, no.2, pp.293-300.

16.325. Vu, T.B.: "Bandwidth Characteristics of Corrugated Feed Horn", Int. J. Electron., 1971, 30, no.2, pp.189-92.

16.326. Miller, P.: "Yagi Antennas and Hansen-Woodyard Condition", Electron. Lett., 1971, 7, no.1, p.11.

16.327. Stutzman, W.L.: "Synthesis of Shaped-Beam Radiation Patterns Using the Iterative Sampling Method", Trans. IEEE, 1971, AP-19, no.1, pp.36-41.

16.328. Brunner, A.: "Dimensioning Doubly Curved Reflectors for Azimuth-Search Radar Antennas", Trans. IEEE, 1971, AP-19, no.1, pp.52-7.

16.329. Harrison, C.W., and Chang, D.C.: "Theory of the Annular Slot Antenna Based on Duality", Trans. IEEE, 1971, EMC-13, no.1, pp.8-14.

16.330.   Deschamps, G.A., and Dyson, J.D.: "Logarithmic Spiral in a Single-Aperture Multimode Antenna System", Trans. IEEE, 1971, AP-19, no.1, pp.90-6.

16.331.   Stewart, G.E., and Golden, K.E.: "Mutual Admittance for Axial Rectangular Slots in a Large Conducting Cylinder", Trans. IEEE, 1971, AP-19, no.1, pp.120-2.

16.332.   Cubley, H.D., and Hayre, H.S.: "Radiation Field of Spiral Antennas Employing Multimode Slow-Wave Techniques", Trans. IEEE, 1971, AP-19, no.1, pp.126-8.

16.333.   Hessel, A., and Sureau, J.C.: "(Theorem on) Realized Gain of Arrays", Trans. IEEE, 1971, AP-19, no.1, pp.122-4.

16.334.   Munger, A.D., and Provencher, J.H.: "Mutual Coupling on a Cylindrical Array of Waveguide Elements", Trans. IEEE, 1971, AP-19, no.1, pp.131-4.

16.335.   Bhatnagar, P.S., et al.: "Helical Infinite Yagi Structure", Proc. IEEE, 1971, 59, no.2, pp.289-91.

16.336.   Howard, A.Q.: "Truncation Error in the Analysis of Aperture Radiation", Electron. Lett., 1971, 7, no.5-6, pp.129-31.

16.337.   Winter, C.F.: "Dual-Vertical-Beam Properties of Doubly Curved Reflectors", Trans. IEEE, 1971, AP-19, no.2, pp.174-80.

16.338.   Wood, P.J.: "Field-Correlation Diffraction Theory of the Symmetrical Cassgrainian Antenna", Trans. IEEE, 1971, AP-19, no.2, pp.191-7.

16.339.   Zadvornov, V.S., and Timirev, N.P.: "Limits of the Active Range of Conical Equiangular Spiral Antennas", Izv. VUZ Radioelektron., 1971, 14, no.1, pp.40-8.

16.340.   Shen, L.C.: "Characteristics of Propagating Waves on Yagi-Uda Structures", Trans. IEEE, 1971, MTT-19, no.6, pp.536-42.

16.341.   Itoh, T., and Mittra, R.: "Method of Solution for Radiation from a Flanged Waveguide", Proc. IEEE, 1971, 59, no.7, pp.1131-3.

16.342.   Bakhrakh, L.D., and Karapetyan, K.E.: "Calculation of Two-Reflector Antenna", Izv. Akad. Nauk Arm. SSR Fiz., 1971, 6, no.1, pp.26-33.

16.343.   Kumar, A., and Murthy, P.K.: "Optimum Spacing for Chebyshev Arrays", Electron. Lett., 1971, 7, no.11, pp.292-3.

16.344.   Atia, A.E., and Mei, K.K.: "Analysis of Multiple-Arm Conical Log-Spiral Antennas", Trans. IEEE, 1971, AP-19, no.3, pp.320-31.

16.345.   Holland, R.: "Optimization Criterion for Illuminating Circular Antenna Apertures", Trans. IEEE, 1971, AP-19, no.3, pp.436-43.

16.346.   Magoulas, P.: "Gain of Standard Horns", Electron. Lett., 1971, 7, no.12, pp.325-7.

16.347.   Dubost, G., and Daniel, J.P.: "Study of Coupling between Dipoles and Application to a Thick Logperiodic Antenna", Ann. Telecommun., 1971, 26, no.3-4, pp.105-33.

16.348.   Lee, S.W.: "Aperture Matching for an Infinite Circularly Polarized Array of Rectangular Waveguides", Trans. IEEE, 1971, AP-19, no.3, pp.332-42.

16.349.   Lopez, A.R.: "Line-Source Excitation for Maximum Aperture Efficiency with Given Sidelobe Level", Trans. IEEE, 1971, AP-19, no.4, pp.530-2.

16.350.   Clarricoats, P.J.B., and Olver, A.D.: "Near-Field Radiation Characteristics of Corrugated Horns", Electron. Lett., 1971, 7, no.16, pp.446-8.

16.351.   Josefsson, L.G.: "Broadband Twist Reflector", Trans. IEEE, 1971, AP-19, no.4, pp.552-4.

16.352.   Clarricoats, P.J.B., and Salema, C.E.R.C.: "Propagation and Radiation Characteristics of Low-Permittivity Dielectric Cones", Electron. Lett., 1971, 7, no.17, pp.483-5.

16.353.   Clarricoats, P.J.B., and Saha, P.K.: "Propagation and Radiation Behaviour of Corrugated Feeds. I", Proc. IEE, 1971, 118, no.9, pp.1167-76.

16.354.   Borgiotti, G.V.: "Design of Circular Apertures for High Beam Efficiency and Low Sidelobes", Alta Freq., 1971, 40, no.8, pp.652-7.

16.355.   Pratt, T., and Claydon, B.: "Prediction of Polar Diagrams of Large Cassegrain Antennas", Marconi Rev., 1971, 34, no.18, pp.1-26.

16.356.   Kim, O.K., and Dyson, J.D.: "Log-Spiral Antenna with Selectable Polarization", Trans. IEEE, 1971, AP-19, no.5, pp.675-7.

16.357.   Narasimhan, M.S.: "Radiation from Conical Horns with Large Flare Angles", Trans. IEEE, 1971, AP-19, no.5, pp.678-81.

16.358.   Schiller, M., and Linhuber, E.: "Logperiodic Antenna for 1-15 GHz", News Rohde Schwarz, 1971, 11, no.49, pp.22-4.

16.359.   Bojsen, J.H., et al.: "Maximum Gain of Yagi-Uda Arrays", Electron. Lett., 1971, 7, no.18, pp.531-2.

16.360.   Cowles, P.R., and Parker, E.A.: "Helical Feeds at Millimetre Wavelengths", Electron. Lett., 1971, 7, no.18, pp.513-5.

16.361.   Ewell, G.W.: "Polarization-Transforming Antenna Feed Horns", Trans. IEEE, 1971, AP-19, no.5, pp.681-2.

16.362. Singh, M.D., Kosta, S.P., and Chaudhuri, M.: "Technique for Increasing the Gain of Logperiodic Dipole Antenna", Int. J. Electron., 1971, 31, no.6, pp.573-8.

16.363. Yur'yev, A.N.: "Minimization of Sidelobe Level of Antennas with Circular Aperture", Radiotekh. Elektron., 1971, 16, no.7, pp.1144-51, and Radio Eng. Electron. Phys., 1971, 16, no.7, pp.1106-12.

16.364. Papovkin, V.I., and Matorin, A.V.: "Synthesis of Antenna Array Consisting of Slot Radiators with Passive Elements", Radiotekh. Elektron., 1971, 16, no.7, pp.1133-44, and Radio Eng. Electron. Phys., 1971, 16, no.7, pp.1098-106.

16.365. Lee, S.W., and Jones, W.R.: "Surface Waves on Two-Dimensional Corrugated Structures", Radio Sci., 1971, 6, no.8-9, pp.811-8.

16.366. Kostelnicek, R.J., and Mittra, R.: "Radiation from a Parallel-Plate Waveguide into a Dielectric Layer", Radio Sci., 1971, 6, no.11, pp.981-90.

16.367. Kinber, B.Ye., and Yseytlin, V.B.: "Phase Centres of Parabolic Antennas", Radiotekh. Elektron., 1971, 16, no.2, pp.249-61, and Radio Eng. Electron. Phys., 1971, 16, no.2, pp.218-29.

16.368. Khlopov, G.I., Churilov, V.P., and Goroshko, A.I.: "Radiation from Open End of a Hollow Dielectric Guide", Radiotekh. (Kharkov), 1971, no.18, pp.3-9.

16.369. McInnes, P.A., Munro, E.W., and Whitaker, A.J.T.: "Radiation Patterns of Paraboloid with Logperiodic Dipole Feed", Electron. Lett., 1971, 7, no.22, pp.669-70.

16.370. Shen, L.C., Cubley, H.D., and Eggers, D.S.: "Measurement of Propagating Waves on Yagi-Uda Arrays", Trans. IEEE, 1971, AP-19, no.6, pp.776-9.

16.371. Thomas, D.T.: "Analysis and Design of Elementary Blinders for Large Horn-Reflector Antennas", Bell Syst. Tech. J., 1971, 50, no.9, pp.2979-95.

16.372. Koshy, V.K., Nair, K.G., and Srivastava, G.P.: "Gain Improvement of Horns by Flange-Angle Variation", Indian J. Pure Appl. Phys., 1971, 9, no.7,pp.476-8.

16.373. Sangster, A.J., and Hawkins, D.C.: "Arbitrarily Located Resonant Slots in Waveguide. Exact Condition for Circular Polarization", Electron. Lett., 1971, 7, no.25, pp.741-2.

16.374. Singh, M.D., Kosta, S.P., and Singh, A.: "Logperiodic Antenna with Loop Elements", Int. J. Electron. 1972, 32, no.1, pp.81-4.

16.375. Satoh, T.: "Dielectric-Loaded Horn Antennas", Trans. IEEE, 1972, AP-20, no.2, pp.199-201.

16.376. Amitay, N., and Zucker, H.: "Attenuation due to Ohmic Losses in Periodic Dipole and Slot Arrays", Trans. IEEE, 1972, MTT-20, no.2, pp.148-55.

16.377. Jull, E.V.: "Reflection from the Aperture of a Long E-Plane Sectoral Horn", Trans. IEEE, 1972, AP-20, no.1, pp.62-8.

16.378. Tsandoulas, G.N., and Fitzgerald, W.D.: "Aperture Efficiency Enhancement in Dielectrically Loaded Horns", Trans. IEEE, 1972, AP-20, no.1, pp.69-74.

16.379. Hamid, M.A.K., Towaij, S.J., and Martens, G.O.: "Dielectric-Loaded Circular-Waveguide Antenna", Trans. IEEE, 1972, AP-20, no.1, pp.96-7.

16.380. Louange, F., and Munier, J.: "Improved Angular Ambiguity and Accuracy by Aperture-Modulation Technique", Electron. Lett., 1972, 8, no.9, pp.233-5.

16.381. Yagjian, A.D., and Kornhauser, E.T.: "Modal Analysis of the Dielectric Rod Antenna Excited by HE_{11} Mode", Trans. IEEE, 1972, AP-20, no.2, pp.122-8.

16.382. Wrixon, G.T., and Welch, W.J.: "Gain Measurements of Standard Horns in the K Bands", Trans. IEEE, 1972, AP-20, no.2, pp.136-42.

16.383. Ludwig, A.C.: "Conical-Reflector Antennas", Trans. IEEE, 1972, AP-20, no.2, pp.146-52.

16.384. Lovis, D.: "Radiation from a Parallel-Plate Line with Diaphragm and with Finite Plate Thickness", Arch. Elektron. Ubertrag., 1972, 26, no.1, pp.49-54.

16.385. Howard, A.Q.: "Mathematical Theory of Radiation from Flanged Waveguides", J. Math. Phys., 1972, 13, no.4, pp.482-90.

16.386. Kauffman, J.F.: "Lens Antennas that Provide Both Amplitude and Phase Transformation", Appl. Opt., 1972, 11, no.2, pp.435-9.

16.387. Schejbal, V.: "Parabolic Horn Antenna", Slab. Obz., 1972, 33, no.2,pp.57-65.

16.388. Hristov, H.D., and Taylor, D.: "Rectangular Backfire Antenna with Dielectric Surface-Wave Structure", Electron. Lett., 1972, 8, no.6, pp.163-5.

16.389. Pridmore-Brown, C., and Stewart, G.E.: "Radiation from Slot Antennas on Cones", Trans. IEEE,1972,AP-20,no.1,pp.36-9.

16.390. Pridmore-Brown, D.C.: "Diffraction Coefficients for a Slot-Excited Conical Antenna", Trans. IEEE, 1972, AP-20, no.1, pp.40-9.

16.391. Lee, S.W.: "Ray Theory of Diffraction by Open-Ended Waveguides. II", J. Math. Phys., 1972, 13, no.5, pp.656-64.

16.392. Cha, A.G.: "Wave Propagation on Helical Antennas", Trans. IEEE, 1972, AP-20, no.5, pp.554-8.

16.393. Evans, B.G.: "Performance and Design of 2:1 Bandwidth Logperiodic Dipole Arrays", Radio Electron. Eng., 1972, 42, no.5, pp.225-35.

16.394. Dijk, J., Berends, J.M., and Maanders, E.J.: "Near Sidelobes in Cassegrain Antenna Systems", Arch. Elektron. Ubertrag., 1972, 26, no.7-8, pp.362-4.

16.395. Jansen, J.K.M., and Jeuken, M.E.J.: "Surface Waves in the Corrugated Conical Horn", Electron. Lett., 1972, 8, no.13, pp.342-4.

16.396. Jeuken, M.E.J.: "Corrugated Conical Horn Antennas with Small Flare Angles", Ingenieur, 1972, 84, no.34, p.ET88-94.

16.397. Uenakada, K., and Yasunaga, K.: "Omni-Directional Biconical Horn Antenna Excited by the TE_{11} Mode in a Circular Waveguide", NHK Tech. J., 1972, 24, no.3, pp.1-11.

16.398. Clarricoats, P.J.B., Salema, C.E.R.C., and Lim, S.H.: "Design of Cassegrain Antennas Employing Dielectric Cone Feeds", Electron. Lett., 1972, 8, no.15, pp.384-5.

16.399. Brauer, J.R.: "Rectangular Beam Waveguide Resonator and Antenna", Trans. IEEE, 1972, AP-20, no.5, pp.593-9.

16.400. Knop, C.M., and Wiesenfarth, H.J.: "Radiation from an Open-Ended Corrugated Pipe Carrying the HE_{11} Mode", Trans. IEEE, 1972, AP-20, no.5, pp.644-8.

16.401. Fitzgerald, W.D.: "Efficiency of Near-Field Cassegrainian Antennas", Trans. IEEE, 1972, AP-20, no.5, pp.648-50.

16.402. Dijk, J., Maanders, E.J., and Sniekers, J.P.F.: "Efficiency and Radiation Patterns of Mismatched Shaped Cassegrainian Antenna Systems", Trans. IEEE, 1972, AP-20, no.5, pp.653-5.

16.403. Narasimhan, M.S., and Rao, B.V.: "Radiation Properties of Conical Scalar Horns", Proc. IEE, 1972, 119, no.8, pp.1092-4.

16.404. Kosta, S.P.: "Theory of Helical Yagi Antenna", Nachr. Tech. Z., 1972, 25, no.7, pp.342-4.

16.405. Watson, P.A., and Ghobrial, S.I.: "Off-Axis Polarization Characteristics of Cassegrainian and Front-Fed Paraboloids", Trans. IEEE, 1972, AP-20, no.6, pp.691-8.

16.406. Fee, M.L.: "Trough-Waveguide Dual-Frequency Antenna", Trans. IEEE, 1972, AP-20, no.6, pp.781-4.

16.407. Pridmore-Brown, D.C.: "Radiation Patterns from Slot-Excited Cones", Trans. IEEE, 1972, AP-20, no.6, pp.815-6.

16.408. Yoshimura, Y.: "Microstrip Slot Antenna", Trans. IEEE, 1972, MTT-20, no.11, pp.760-2.

16.409. Rogers, A.: "Wideband Squintless Linear Arrays", Marconi Rev., 1972, 35, no.187, pp.221-43.

16.410. Lewis, L.R., Hessel, A., and Knittel, G.H.: "Performance of a Protruding-Dielectric Waveguide Element in a Phased Array", Trans. IEEE, 1972, AP-20, no.6, pp.712-22.

16.411. Shen, L.C.: "Directivity and Bandwidth of Single- and Double-Band Yagi Arrays", Trans. IEEE, 1972, AP-20, no.6, pp.778-80.

16.412. Hongo, K.: "Diffraction by a Flanged Parallel-Plate Waveguide", Radio Sci., 1972, 7, no.10, pp.955-63.

16.413. Kuchikyah, L.M.: "Interference Pattern at the Output End of a Fibre-Optic Wedge", Opt.-Mekh. Prom., 1972, 39, no.2, p.60, and Sov. J. Opt. Technol., 1972, 39, no.2, pp.114-5.

16.414. Massey, G.A.: "Beam-Diverging Lens System for High-Power Laser Transmitters", Appl. Opt., 1972, 11, no.12, p.2981.

16.415. Poulton, G.T.: "Class of Stepped-Reflector Antennas with Improved Frequency Response", Electron. Lett., 1972, 8, no.25, pp.605-6.

16.416. Poulton, G.T., Lim, S.H., and Masterman, P.H.: "Calculation of Input VSWR for a Reflector Antenna", Electron. Lett., 1972, 8, no.25, pp.610-1.

16.417. Shankara, K.N., and Chatterjee, S.K.: "Corrugated and Uniform Dielectric-Rod Antennas Excited in E_0 Mode", J. Indian Inst. Sci., 1972, 54, no.3, pp.146-81.

16.418. Banerjee, P.K., and Subrahmaniam, V.: "Logperiodic Structure with Asymmetric Dipole Elements", J. Inst. Telecommun. Eng., 1972, 18, no.11, pp.544-8.

16.419. Kuo, S.C.: "Size-Reduced Logperiodic Dipole Array Antenna", Microwave J., 1972, 15, no.12, pp.27-33.

16.420. Kornienko, L.G.: "Optimization of Antenna Parameters in Presence of Random Errors", Radiotekh. Elektron., 1972, 17, no.6, pp.1171-6, and Radio Eng. Electron. Phys., 1972, 17, no.6, pp.911-5.

16.421. Churilov, V.P., and Khlopov, G.I.: "Constructional Features of Millimetric Cassegrain Antennas", Radiotekhnika, 1972, 27, no.8, pp.104-5, and Telecommun. Radio Eng. Pt 2, 1972, 27, no.8, pp.132-4.

16.422. Goto, N.: "Directivities of Planar Arrays with Triangular Arrangement of Elements", Electron. Commun. Jap., 1972, 55, no.4, pp.66-71.

16.423. Kinber, B.Ye., and Popichenko, V.A.: "Radiation from a Sectoral Horn", Radiotekh. Elektron., 1972, 17, no.10, pp.2035-43, and Radio Eng. Electron. Phys., 1972, 17, no.10, pp.1621-7.

16.424. Fong, T.T.: "Radiation from an Open-Ended Waveguide with Extended Dielectric Loading", Radio Sci., 1972, 7, no.10, pp.965-72.

16.425. Neelakantaswamy, P.S., and Banerjee, D.K.: "Corrugated Spherical Antenna", Electron. Lett., 1972, 8, no.22, pp.534-6.

16.426. Williamson, A.G., and Otto, D.V.: "Cylindrical Antenna in a Rectangular Waveguide Driven from a Coaxial Line", Electron. Lett., 1972, 8, no.22, pp.545-7.

16.427. Jull, E.V.: "Errors in the Predicted Gain of Pyramidal Horns", Trans. IEEE, 1973, AP-21, no.1, pp.25-31.

16.428. Jull, E.V.: "Aperture Fields and Gain of Open-Ended Parallel-Plate Waveguides", Trans. IEEE, 1973, AP-21, no.1, pp.14-18.

16.429. Howard, A.Q.: "Comparison of Mode Match, Geometrical Theory of Diffraction, and Kirchhoff Radiation", Trans. IEEE, 1973, AP-21, no.1, pp.100-2.

16.430. Clarricoats, P.J.B., and Seng, L.M.: "Influence of Length on Radiation Pattern of Oblique-Flare-Angle Corrugated Horn", Electron. Lett., 1973,9,no.1,pp.15-6.

16.431. Towaij, S.J., Bhartia, P., and Hamid, M.A.K.: "Diffraction by a Sectoral Horn with Cylindrical Aperture", Int. J. Electron., 1973, 34, no.3, pp.381-3.

16.432. Holst, D.W.: "Radiation Patterns of Radial Waveguides with TM-Mode Excitation", Trans. IEEE, 1973, AP-21, no.2, pp.238-41.

16.433. Kosta, S.P.: "Phase Velocity of Propagation along an Infinite Zigzag Antenna", J. AEU, 1973, 5, no.4, pp.15-7.

16.434. Wood, P.J.: "Depolarization with Cassegrainian and Front-Fed Reflectors", Electron. Lett., 1973,9,no.8-9,pp.181-3.

16.435. Narasimhan, M.S., and Rao, V.V.: "Radiation Characteristics of Corrugated E-Plane Sectoral Horns", Trans. IEEE, 1973, AP-21, no.3, pp.320-7.

16.436. Wong, W.C.: "Equivalent-Parabola Technique to Predict the Performance Characteristics of a Cassegrainian System with Offset Feed", Trans. IEEE, 1973, AP-21, no.3, pp.335-9.

16.437. Chu, T.S., and Turrin, R.H.: "Depolarization Properties of Offset-Reflector Antennas", Trans. IEEE, 1973, AP-21, no.3, pp.339-45.

16.438. Baldwin, R.: "Radiation Patterns of Dielectric Loaded Rectangular Horns", Trans. IEEE, 1973, AP-21, no.3, pp.375-6.

16.439. Neelakantaswamy, P.S., and Banerjee, D.K.: "Radiation Characteristics of a Waveguide-Excited Dielectric Sphere Backed by a Metallic Hemisphere", Trans. IEEE, 1973, AP-21, no.3, pp.384-5.

16.440. Prasad, S.M., and Das, B.N.: "Studies on Waveguide-Fed Slot Antennas", Proc. IEE, 1973, 120, no.5, pp.539-40.

16.441. Chen, C.C.: "Broadband Impedance Matching of Rectangular-Waveguide Phased Arrays", Trans. IEEE, 1973, AP-21, no.3, pp.298-302.

16.442. De Vito, G.: "Design of Logperiodic Dipole Antennas", Trans. IEEE, 1973, AP-21, no.3, pp.303-8.

16.443. Young, L., Robinson, L.A., and Hacking, C.A.: "Meander-Line Polarizer", Trans. IEEE, 1973, AP-21, no.3, pp.376-8.

16.444. Collins, G.W.: "Shaping of Subreflectors in Cassegrainian Antennas for Maximum Aperture Efficiency", Trans. IEEE, 1973, AP-21, no.3, pp.309-13.

16.445. Chen, M.H.: "Radial-Mode Analysis of Propagation on Slotted Cylindrical Structures", Trans. IEEE, 1973, AP-2,no.3,pp.314-20.

16.446. Tuan, H.S.: "Radiation and Reflection of Surface Waves at a Discontinuity", Trans. IEEE, 1973, AP-21, no.3, pp.351-6.

16.447. Dey, K.K., and Khastgir, P.: "Study of the Characteristics of a Microwave Spherical Zone-Plate Antenna", Int. J. Electron., 1973, 35, no.1, pp.97-103.

16.448. Reisdorf, F., and Schminke, W.: "Application of a Swischenmedium to the Radiation from a Flanged Waveguide", Proc. IEE, 1973, 120, no.7, pp.739-40.

16.449. Clarricoats, P.J.B., and Salema, C.E.R.C.: "Antennas Employing Conical Dielectric Horns. I and II", Proc. IEE, 1973, 120, no.7, pp.741-9 and 750-6.

16.450. Righini, G.C., et al.: "Geodesic Lenses for Guided Optical Waves", Appl. Opt., 1973, 12, no.7, pp.1477-81.

16.451. Zeininger, S.: "Geometric Deformation of Spherical Dielectric-Lens Antennas", Arch. Elektron. Ubertrag., 1973, 27, no.7-8, pp.289-92.

16.452. Khac, T.Vu., and Carson, C.T.: "Impedance Properties of a Longitudinal Slot Antenna in the Broad Surface of a Rectangular Waveguide", Trans. IEEE, 1973, AP-21, no.5, pp.708-10.

16.453. Kerr, J.L.: "Short Axial Length Broadband Horns", Trans. IEEE, 1973, AP-21, no.5, pp.710-4.

16.454. Kajfez, D.: "Nonlinear Optimization Reduces Sidelobes of Yagi Antenna", Trans. IEEE, 1973, AP-21, no.5, pp.714-5.

16.455. Cheng, D.K., and Chen, C.A.: "Optimum Element Spacings for Yagi-Uda Arrays", Trans. IEEE, 1973, AP-21, no.5, pp.615-23.

16.456. Sangster, A.J.: "Circularly Polarized Linear Waveguide Array", Trans. IEEE, 1973, AP-21, no.5, pp.704-5.

16.457. Reitzig, R.: "Polarization Characteristics of Phased Arrays of Elliptically Polarized Elementary Radiators", Frequenz, 1973, 27, no.8, pp.205-13.

16.458. Banerjee, P.K., and Subrahmaniam, V.: "Input Impedance, Radiation Resistance, and Gain of Logperiodic Antenna with Asymmetric Dipole Elements", Indian J. Pure Appl. Phys., 1973, 11, no.3, pp.209-11.

16.459. Bar-Isaac, C., and Hardy, A.: "Diffraction of One-Dimensional Gaussian Beam by an Amplitude-Sinusoidal Grating", Opt. Acta, 1973, 20, no.1, pp.59-67.

16.460. Clarricoats, P.J.B., and Seng, L.M.: "Propagation and Radiation Characteristics of Corrugated Horns", Electron. Lett., 1973, 9, no.1, pp.7-9.

16.461. Teichman, M.: "Determination of Horn Antenna Phase Centres by Edge Diffraction Theory", Trans. IEEE, 1973, AES-9, no.6, pp.875-82.

16.462. Chatterjee, R., Govind, S., and Vedavathy, T.S.: "Dielectric Spherical Antennas Excited in the Unsymmetric Hybrid Mode", J. Indian Inst. Sci., 1973, 55, no.1, pp.5-15.

16.463. Inoue, T., and Hashiguchi, H.: "20-GHz Cassegrain Antenna Fundamental Characteristics", Rev. Electr. Commun. Lab., 1973, 21, no.9-10, pp.637-46.

16.464. Rao, T.C., and Chatterjee, R.: "Circular Cylindrical Dielectric-Coated Metal Rod Excited in Symmetric TM_{01} Mode. I", J. Indian Inst. Sci., 1973, 55, no.4, pp.199-231.

16.465. Chatterjee, R., and Rao, T.C.: "Circular Cylindrical Dielectric-Coated Metal Rod Excited in Symmetric TM_{01} Mode. II", J. Indian Inst. Sci., 1973, 55, no.4, pp.232-54.

16.466. Calla, O.P.N., et al.: "Dual Polarized Feed Systems for Communications Antennas", J. Inst. Electron. Telecommun. Eng., 1973, 19, no.8, pp.459-63.

16.467. Fel'd, Ya.N., et al.: "Diffraction by a System of Plane-Parallel Waveguides of Finite Length", Radiotekh. Elektron., 1973, 18, no.5, pp.897-908, and Radio Eng. Electron. Phys., 1973, 18, no.5, pp.655-63.

16.468. Goto, N.: "Pattern Synthesis for Multibeam Array Antennas", Electron. Commun. Jap., 1973, 56, no.5, pp.83-6.

16.469. Dorokhov, A.P., and Marchuk, V.S.: "Application of Higher-Order Modes in Backfire Antennas", Radiotekh. (Kharkov), 1973, no.27, pp.41-6.

16.470. Gorobets, N.N., and Lytov, Yu.V.: "Study of Waveguide Radiators with Bevelled Aperture Edges", Radiotekh. (Kharkov), 1973, no.27, pp.51-7.

16.471. Timofeeva, A.A.: "Wideband Horn Radiator with Axisymmetric Polar Diagram", Radiotekhnika, 1973, 28, no.9, pp.94-5, and Telecommun. Radio Eng. Pt 2, 1973, 28, no.9, pp.121-2.

16.472. Dey, K.K., and Khastgir, P.: "Comparitive Focusing Properties of Spherical and Plane Microwave Zone-Plate Antennas", Int. J. Electron., 1973, 35, no.4, pp.497-506.

16.473. Sangster, A.J., and Hawkins, D.C.: "Radiating Apertures in a Corrugated Rectangular Waveguide", Electon. Lett., 1973, 9, no.15, pp.329-31.

16.474. Mizusawa, M., and Kitsuregawa, T.: "Beam-Waveguide Feed having a Symmetric Beam for Cassegrain Antennas", Trans. IEEE, 1973, AP-21, no.6, pp.884-6.

16.475. Hogg, D.C., and Legg, W.E.: "Finline Radiator", Bell Syst. Tech. J., 1973, 52, no.7, pp.1249-53.

16.476. Lewin, L.: "Local Form of the Radiation Conditions. Application to Curved Dielectric Structures", Electron. Lett., 1973, 9, no.20, pp.468-9.

16.477. Watson, P.A., and Ghobrial, S.I.: "Cross Polarization in Cassegrain and Front-Fed Antennas", Electron. Lett., 1973, 9, no.14, pp.297-8.

16.478. Dijk, J., and Maanders, E.J.: "Errors in the Calculation of Radiation Patterns of Reflector Antennas Using Kirchhof Integration", Electron. Lett., 1973, 9, no.21, pp.510-2.

16.479. James, G.L., and Poulton, G.T.: "Modified Diffraction Coefficient for Focusing Reflectors", Electron. Lett., 1973, 9, no.23, pp.537-8.

16.480. Narashimhan, M.S., and Rao, V.V.: "Correction to the Available Radiation Formula for E-Plane Sectoral Horns", Trans. IEEE, 1973, AP-21, no.6, pp.878-9.

16.481. James, G.L., and Kerdemelidis, V.: "Selective Reduction in Back Radiation from Paraboloidal Reflectors", Trans. IEEE, 1973, AP-21, no.6, pp.886-7.

16.482. Wood, P.J.: "Crosspolarization with Cassegrainian and Front-Fed Reflectors", Electron. Lett., 1973, 9, no.25, pp.597-8.

16.483. Rudge, A.W.: "Offset-Reflector Antennas with Offset Feeds", Electron. Lett., 1973, 9, no.26, pp.611-3.

16.484. Mizusawa, M., Takeda, F., and Betsudan, S.: "Radiation Characteristics of a Corrugated Conical Horn", Electron. Commun. Jap., 1973, 56, no.1, pp.42-7.

16.485. Satoh, T.: "Dielectric-Loaded Horn Antenna and Crosspolarization Characteristics", KDD Tech. J. 1973, no.77, pp.1-12.

16.486. Wood, P.J.: "Near-Field Defocusing in the Pencil-Beam Cassegrain Antenna", Marconi Rev., 1973, 36, no.191, pp.201-23.

16.487. Kaloshin, V.A., and Shcherbenkov, V.Ya.: "Generalized Luneberg Problem for Anisotropic Medium", Radiotekh. Elektron., 1973, 18, no.1, pp.26-31, and Radio Eng. Electron. Phys., 1973, 18, no.1, pp.22-6.

16.488. Munson, R.E.: "Conformal Microstrip Antennas and Phased Arrays", Trans. IEEE, 1974, AP-22, no.1, pp.74-8.

16.489. Brooking, N., Clarricoats, P.J.B., and Olver, A.D.: "Radiation Patterns of Pyramidal Dielectric Waveguides", Electron. Lett., 1974, 10, no.3, pp.33-4.

16.490. Balzano, Q.: "Analysis of Periodic Arrays of Waveguide Apertures on Conducting Cylinders Covered by Dielectric", Trans. IEEE, 1974, AP-22, no.1, pp.25-34.

16.491. Munger, A.D., et al.: "Conical Array Studies", Trans. IEEE, 1974, AP-22, no.1, pp.35-43.

16.492. Narasimhan, M.S.: "Eigenvalues of Spherical Surface-Wave Modes in Corrugated Conical Horns", Trans. IEEE, 1974, AP-22, no.1, pp.122-3.

16.493. Bandukov, V.P., and Pokras, A.M.: "Study of Two-Mirror Antennas with Modified Mirror Surfaces", Radiotekhnika, 1974, 29, no.2, pp.38-45, and Telecommun. Radio Eng. Pt 2, 1974, 29, no.2, pp.86-91.

16.494. Mentzer, C.A., and Peters, L.: "Properties of Cutoff Corrugated Surfaces for Horn Design", Trans. IEEE, 1974, AP-22, no.2, pp.175-80.

16.495. Jull, E.V., and Allan, L.E.: "Gain of an E-Plane Sectoral Horn. Failure of the Kirchhoff Theory and a Proposal", Trans. IEEE, 1974, AP-22, no.2, pp.205-10.

16.496. Martin, A.G., and Oxtoby, A.J.A.: "Waveguide-Fed Spherical Dielectric Antennas", Trans. IEEE, 1974, AP-22, no.2, pp.322-3.

16.497. Raffoul, G.W., and Hilburn, J.L.: "Radiation Efficiency of an X-Band Waveguide Array", Trans. IEEE, 1974, AP-22, no.2, pp.339-40.

16.498. Dragone, C., and Hogg, D.C.: "Radiation Pattern and Impedance of Offset and Symmetrical Near-Field Cassegrainian and Gregorian Antennas", Trans. IEEE, 1974, AP-22, no.3, pp.472-5.

16.499. Lu, H.S.: "Computation of Radiation Pattern of a Zoned Waveguide Lens", Trans. IEEE, 1974, AP-22, no.3, pp.483-4.

16.500. Manwarren, T., and Farrar, A.: "Pattern Shaping with Hybrid Mode Corrugated Horns", Trans. IEEE, 1974, AP-22, no.3, pp.484-7.

16.501. Chen, C.C.: "Quadruple Ridge-Loaded Circular Waveguide Phased Arrays", Trans. IEEE, 1974, AP-22, no.3, pp.481-3.

16.502. Lewin, L.: "Radiation from Curved Dielectric Slabs and Fibres", Trans. IEEE, 1974, MTT-22, no.7, pp.718-27.

16.503. Neelakantaswamy, P.S., and Banerjee, D.K.: "Radiation Characteristics of a Circular Waveguide Aperture with Curved Disc", Arch. Elektron. Ubertrag., 1974, 28, no.6, pp.277-9.

16.504. Free, W.R., et al.: "High-Power Constant-Index Lens Antennas", Trans. IEEE, 1974, AP-22, no.4, pp.581-4.

16.505. Narasimhan, M.S., and Rao, V.V.: "Radiation from Wide-Flare Corrugated E-Plane Sectoral Horns", Trans. IEEE, 1974, AP-22, no.4, pp.603-8.

16.506. Ivanov, I.I.: "Impedance of a Symmetrical Dipole Antenna Using Coaxial Helices", Izv. VUZ Radioelektron., 1974, 17, no.5, pp.111-4.

16.507. Clavin, A., Huebner, D.A., and Kilburg, F.J.: "Improved Element for Use in Array Antennas", Trans. IEEE, 1974, AP-22, no.4, pp.521-6.

16.508. Lewis, L.R., Kaplan, L.J., and Hanfling, J.D.: "Synthesis of a Waveguide Phased Array Element", Trans. IEEE, 1974, AP-22, no.4, pp.536-40.

16.509. Chen, M.H., and Tsandoulas, G.N.: "Dual-Reflector Optical Feed for Wideband Phased Arrays", Trans. IEEE, 1974, AP-22, no.4, pp.541-5.

16.510. Yee, H.Y.: "Impedance of a Narrow Longitudinal Shunt Slot in a Slotted-Waveguide Array", Trans. IEEE, 1974, AP-22, no.4, pp.589-92.

16.511. Steyskal, H.: "Mutual Coupling Analysis of a Finite Planar Waveguide Array", Trans. IEEE, 1974, AP-22, no.4, pp.594-7.

16.512. Bailey, M.C.: "Mutual Coupling between Circular Waveguide-Fed Apertures in a Rectangular Ground Plane", Trans. IEEE, 1974, AP-22, no.4, pp.597-9.

16.513. Truman, W.M., and Balanis, C.A.: "Optimum Design of Horn Feeds for Reflector Antennas", Trans. IEEE, 1974, AP-22, no.4, pp.585-6.

16.514. Kajfez, D., Harrison, M.G., and Sterling, C.E.: "Electric Tripole Antenna for Circular Polarization", Trans. IEEE, 1974, AP-22, no.5, pp.647-50.

16.515. Narasimhan, M.S.: "Corrugated Conical Horn as Space Feed for Phased-Array Illumination", Trans. IEEE, 1974, AP-22, no.5, pp.720-2.

16.516. Cugiani, C., et al.: "Ridged Twist-Reflector for Controlled Pattern Application", Alta Freq., 1974, 43, no.5, pp.281-6.

16.517. Orefice, M.: "Increased Bandwidth Resonant Slot Array with Bidirectional Radiation Pattern", Electron. Lett., 1974, 10, no.19, pp.396-7.

16.518. Sinha, M., and Das, B.N.: "Utilization Factor of a Waveguide-Fed Slot Array", Electron. Lett., 1974, 10, no.19, pp.399-400.

16.519. de Vecchis, M., et al.: "Biconical Horn for Circular Polarization", Microwave J., 1974, 17, no.6, pp.20B-20D.

16.520. Das, B.N.: "Impedance Characteristics of Series Slots", Proc. IEE, 1974, 121, no.11, pp.1360-2.

16.521. Han, C.C., and Wickert, A.N.: "Multimode Rectangular Horn Antenna Generating a Circularly Polarized Elliptical Beam", Trans. IEEE, 1974, AP-22, no.6, pp.746-51.

16.522. Elmoazzen, Y.E., and Shafai, L.: "Mutual Coupling between Two Circular Waveguides", Trans. IEEE, 1974, AP-22, no.6, pp.751-60.

16.523. Tetenbaum, S.J.: "Experimental VSWR's and Radiation Patterns of an Axial Rectangular Slot on Conducting Cylinders of Varying Curvature", Trans. IEEE, 1974, AP-22, no.6, pp.835-7.

16.524. Basart, J.P.: "Short Design Method for Nonuniformly Spaced Antenna Arrays", Trans. IEEE, 1974, AP-22, no.6, pp.822-3.

16.525. Dragone, C.: "Improved Antenna for Microwave Radio Systems Consisting of Two Cylindrical Reflectors and a Corrugated Horn", Bell Syst. Tech. J., 1974, 53, no.7, pp.1351-77.

16.526. Gobert, J.F.: "Design of Wide- and Narrow-Angle Planar Bootlace Microwave Lenses", Frequenz, 1974, 28, no.8, pp.211-6.

16.527. Shcherbinin, V.Ya., et al.: "Partial Radiation Patterns of Phased Arrays", Izv. VUZ Radiofiz., 1974, 17, no.10, pp.1539-43.

16.528. Chuikov, V.D.: "Parabolic Phased Arrays", Izv. VUZ Radioelektron., 1974, 17, no.8, pp.95-8.

16.529. Neelakantaswamy, P.S., and Banerjee, D.K.: "Waveguide-Fed Dielectric Spherical Antennas", Electron. Lett., 1974, 10, no.25-26, pp.540-1.

16.530. Ojha, B., and Sharma, K.P.: "Radiation from a Metallic Disc over a Circular-Rod Surface Waveguide", Indian J. Radio Space Phys., 1974, 3, no.1, pp.17-20.

16.531. Zernike, F.: "Luneberg Lens for Optical Waveguide Use", Opt. Commun., 1974, 12, no.4, pp.379-81.

16.532. Galimov, G.K.: "Bifocal Two-Mirror Antennas", Radiotekhnika, 1974, 29, no.3, pp.64-70, and Telecommun. Radio Eng. Pt 2, 1974, 29, no.3, pp.105-10.

16.533. Zaikin, B.M., Konin, B.M., and Platonova, Zh.K.: "Method of Determining the Scattering Parameters of Phased Antenna Arrays", Kibern. Vychisl. Tekh., 1974, no.26, pp.53-8.

16.534. Stark, L.: "Microwave Theory of Phased-Array Antennas. Review", Proc. IEEE, 1974, 62, no.12, pp.1661-701.

16.535. Haneishi, M., Kobayashi, Y., and Tanaka, S.: "Characteristics of Dielectric-Plate Antenna", Sci. Eng. Rep. Saitama Univ., 1974, no.8, pp.21-6.

16.536. Usin, V.A.: "Statistics of Linear Antenna Arrays", Radiotekh. (Kharkov), 1974, no.30, pp.48-56.

16.537. Shifrin, Ya.S., and Kornienko, L.G.: "Limiting Levels for Sidelobes of Antenna Arrays Subject to Random Phase Errors", Radiotekh. (Kharkov), 1974, no.30, pp.75-84.

16.538. Rusch, W.V.T.: "Double-Aperture Blocking by Two Wavelength-Sized Feed-Support Struts", Electron. Lett., 1974, 10, no.15, pp.296-7.

17. MANUFACTURING TECHNIQUES

17.1. Sitaram, R.V.S., and Nagaraj, S.: "Electroforming Method to Produce Precision Microwave Components", J. Inst. Telecommun. Eng., 1965, 11, p.418.

17.2. Hyltin, T.M.: "Microstrip Transmission on Semiconductor Dielectrics", Trans. IEEE, 1965, MTT-13, p.777.

17.3. Caulton, M., Hughes, J.J., and Sobol, H.: "Properties of Microstrip Lines for MIC's", RCA Rev., 1966, 27, p.377.

17.4. Semplak, R.A., and Turrin, R.H.: "Pressure-Formed Parabolic Reflectors for Millimetre Wavelengths", Trans. IEEE, 1968, AP-16, p.762.

17.5. Wen, C.P.: "Integrated Circuit, Metallized-Plastic, Symmetrical Millimetre Trough Waveguide with Nonreciprocal Elements", RCA Rev., 1969, 30, p.724.

17.6. Meyeroff, R.W.: "Fabrication of Niobium RF Cavities", J. Appl. Phys., 1969, 40, p.2011.

17.7. Bohnke, R.D., and Zimmer, H.: "Technology of Manufacture of Superconducting Lead Resonators", J. Less-Common Metals, 1969, 17, no.2, p.235.

17.8. Miyakawa, T., Takizawa, A., and Hayashi, H.: "MIC Doubler", Fujitsu Sci. Tech. J., 1969, 5, no.1, p.89.

17.9. Tarowsky, N.: "Assembly of Hybrid MIC's", Microwaves, 1969, 8, no.8, p.52.

17.10. Letron, Y., and Guidevaux, J.: "Utilization of Ferrites and Garnets in Microwave Integrated Electronics", Microelectron. Rel., 1969, 8, no.4, p.319.

17.11. Hershenov, B., and Ernst, R.L.: "Miniature Microstrip Circulators Using High-Dielectric-Constant Substrates", RCA Rev., 1969, 30, p.541.

17.12. Huntt, R.L., Blankenship, A.G., and West, R.G.: "Ferrimagnetic Substrates for MIC's", Trans. IEEE, 1969, MAG-5,p.482.

17.13. Leppavuori, S., Lofgren, K.E., and Mannersalo, K.: "Investigation of MIC's Made by Thin-Film Technique", State Inst. Tech. Res. Rep. II, 1969, no.25, p.3.

17.14. Muto, T., and Sakurai, S.: "Thin-Film (Microwave) Technology", J. Inst. Electron. Commun. Eng. Jap., 1969, 52, no.11, p.1367.

17.15. Ota, T., Nakahara, S., and Horikiri, K.: "Substrates (for MIC's)", J. Inst. Electron. Commun. Eng. Jap., 1969, 52, no.11, p.1381.

17.16. Matsumoto, A., and Nagai, N.: "Passive Networks Made with Multiwire Stripline", J. Inst. Electron. Commun. Eng. Jap., 1969, 52, no.11, p.1433.

17.17. Schmitt, H.J., and Lemke, M.: "Miniaturized Strip-Conductor Components", Int. Elektron. Rdsch., 1969, 23,no.10,p.270.

17.18. Field, J.S.: "Noncontacting Air Gauges for Internal Measurements on Waveguides", J. Sci. Instrum., 1969, 2, p.195.

17.19. Almassy, G.: "Construction and Technology of Microwave Devices", Nachr. Tech., 1970, 20, p.111.

17.20. Kinter, M.L., Weissman, I., and Stein, W.W.: "Chemical Polish for Niobium (Superconducting) Microwave Structures", J. Appl. Phys., 1970, 41, p.828.

17.21. Goll, F.M.: "Microwave Integrated Circuits", Western Electr. Eng., 1970, 14, no.1, p.26.

17.22. Aho, O., and Leppavuori, S.: "Design and Manufacture of Microstrip Directional Couplers", Saehkoe,1970,43,no.3,p.87.

17.23. Peterson, N.C.: "Microwave Film Attenuators for Ferrite Devices", J. Vac. Sci. Technol., 1970, 7, p.493.

17.24. Odam, L.: "Integrated Circuits by Electron Beams", Elteknik, 1970, no.7-8, pp.36-7.

17.25. Leppavuori, S., and Mannersalo, K.: "Fabrication Technique of Microstrip Circuits", Elektronikka, 1970, 23, no.1, pp.13-15.

17.26. Edelman, H.C.M.: "Ferrite Technology Improves MIC's", Microelectron. Rel., 1970, 9, no.5, p.434.

17.27. Aschmoneit, E.K.: "Integrated Optical Circuits for Laser Telecommunications Links", Elektro-Tech. Z., 1970, 22B, no.21, pp.499-501.

17.28. Berman, R.: "Diamond Heat Sinks", Electron. Eng., 1970, 42, no.510, p.43.

17.29. Alley, G.D.: "Interdigital Capacitors and Application to Lumped-Element MIC's", Trans. IEEE, 1970, MTT-18, no.12, pp.1028-33.

17.30. Riches, E.E., et al.: "Ferrite Devices Using Field Effects for MIC's", Trans. IEEE, 1970, MAG-6, no.3, pp.670-3.

17.31. Goell, J.E., and Standley, R.D.: "Integrated Optical Circuits", Proc. IEEE, 1970, 58, no.10, p.1504.

17.32. Chiron, B.: "Design and Realization of Latching Ferrite Devices in Microwave Microelectronics", Onde Electr., 1970, 50, no.9, pp.779-85.

17.33. Caulton, M., and Sobol, H.: "MIC Technology. Survey", J. Solid-State Circuits IEEE, 1970, SC-5, no.6, pp.292-303.

17.34. Van Nie, A.G., and van Heuven, J.H.C.: "Integration of Microwave Circuits", Ned. Elektron. Radiogenoot., 1970, 35, no. no.11, pp.163-71.

17.35. Grabowska, K., et al.: "Technology of MIC Components on Ceramic and Ferrite Substrates", Elektronika, 1970, no.11, pp.449-55.

17.36. Krems, M., and Schmidt, W.: "Substrate Materials for MIC's", Frequenz, 1971, 25, no.1, pp.2-9.

17.37. Shubert, R., and Harris, J.H.: "Optical Guided-Wave Focusing and Diffraction", J. Opt. Soc. Am., 1971, 61, no.2, pp.154-61.

17.38. Pankratov, V.M. and Stepovoi, V.A.: "Technology of Fabrication of a Microwave Resonator Grating", Kvantovaya Elektron., 1971, no.6, pp.135-6.

17.39. Chang, W.S.C., and Loh, K.W.: "Experimental Observation of 10.6-micron Guided Waves in Ge Thin Films", Appl. Opt., 1971, 10, no.10, pp.2361-2.

17.40. Hasegawa, H., Furukawa, M., and Yanai, H.: "Slow-Wave Propagation along a Microstrip Line on $Si-SiO_2$ Systems", Proc. IEEE, 1971, 59, no.2, pp.297-9.

17.41. Caulton, M.: "Film Technology in MIC's", Proc. IEEE, 1971, 59, no.10, pp.1481-9.

17.42. Gutteberg, O., and Solbakken, K.: "(Integrated-Circuit) Techniques for Micro-Wave Systems", Elektrotek. Tidsskr., 1971, 84, no.4, pp.31-7.

17.43. Meszaros, L.: "Waveguides Realized in Multiple-Plane Printed-Circuit Boards", Feingerate Tech., 1971, 20, no.2, pp.67-8.

17.44. Heng, T.M.S.: "Trimming of Microstrip Circuits Utilizing Microcantilever Air Gaps", Trans. IEEE, 1971, MTT-19, no.7, pp.652-4.

17.45. Harrison, G.R., et al.: "Ferrimagnetic Parts for MIC's", Trans. IEEE, 1971, MTT-19, no.7, pp.577-88.

17.46. Caulton, M., et al.: "Status of Lumped Elements in MIC's. Present and Future", Trans. IEEE, 1971, MTT-19, no.7, pp.588-99.

17.47. Kindermann, W.J., and Knab, E.D.: "On-Site Bending of Rigid Circular Waveguide", Bell Lab. Rec., 1971, 49, no.6, pp.180-4.

17.48. Goodman, E.T.J.: "Manufacture of Waveguide Components by Electroforming", Post Office Electr. Eng. J., 1971, 64, pt 3, pp.160-3.

17.49. Boros, G.J.: "Production of Laser Mirrors", Finommech., 1971, 10, no.4, pp.97-102.

17.50. Diepers, H., et al.: "Method of Electropolishing Niobium (Superconducting Cavities)", Phys. Lett., 1971, 37A, no.2, pp.139-40.

17.51. Bisbee, D.L.: "Optical Fibre Joining Technique", Bell Syst. Tech. J., 1971, 50, no.10, pp.3153-8.

17.52. Miller, S.E.: "Survey of Integrated Optics", J. Quantum Electron. IEEE, 1972, QE-8, no.2, pp.199-205.

17.53. Goell, J.E., et al.: "Ion Bombardment Fabrication of Optical Waveguides Using Electron Resist Masks", Appl. Phys. Lett., 1972, 21, no.2, pp.72-3.

17.54. van Heuven, J.H.C., and Vlek, T.H.A.M.: "Anisotropy in Alumina Substrates for Microstrip Circuits", Trans. IEEE, 1972, MTT-20, no.11, pp.775-7.

17.55. Batchman, T.E., and Rashleigh, S.C.: "Mode-Selective Properties of a Metal-Clad/ Dielectric-Slab Waveguide for Integrated Optics", J. Quantum Electron. IEEE, 1972, QE-8, no.11, pp.848-50.

17.56. Gedeon, A.: "Spectral Properties of TE_0 and TM_0 Modes in Dielectric Thin-Film Waveguides Excited with a Prism Light-Coupler", Opt. Acta, 1972,19,no.9,pp.765-79.

17.57. Kasai, T., Noda, J., and Ida, I.: "Precision Machining of $LiTaO_3$ Crystal for Electrooptic Elements", Rev. Electr. Commun. Lab., 1972, 20, no.7-8, pp.739-46.

17.58. McFee, J.H., et al.: "Guided-Wave Propagation at 10.6 micron in AgBr Thin Films", Appl. Phys. Lett., 1972, 21, no.11, pp.534-6.

17.59. Garmire, E., et al.: "Optical Waveguiding in Proton-Implanted GaAs", Appl. Phys. Lett., 1972, 21, no.3, pp.87-8.

17.60. Taylor, H.F., et al.: "Fabrication of Single-Crystal Semiconductor Optical Waveguides by Solid-State Diffusion", Appl. Phys. Lett., 1972, 21, no.3, pp.95-8.

17.61. Ramaswamy, V.: "Epitaxial Electro-optic Mixed Crystal $(NH_4K)H_2PO_4$ Film Waveguide", Appl. Phys. Lett., 1972, 21, no.5, pp.183-5.

17.62. Tien, P.K., et al.: "Optical Waveguides of Single-Crystal Garnet Films", Appl. Phys. Lett., 1972, 21, no.5, pp.207-9.

17.63. Fujimori, M., Okamoto, A., and Nishimura, Y.: "Ta_2O_5 Thin Films for Light-Guide", Fujitsu Sci. Tech. J., 1972, 8, no.3, pp.177-89.

17.64. Izawa, T., and Nakagome, H.: "Optical Waveguide Formed by Electrically Induced Migration of Ions in Glass Plates", Appl. Phys. Lett., 1972, 21, no.12, pp.584-6.

17.65. Ince, W.J., et al.: "Method of Fabrication for Waveguide Nonreciprocal Toroidal Ferrite Phasers", Trans. IEEE, 1972, MTT-20, no.10, pp.705-7.

17.66. Weber, H.P., et al.: "Light-Guiding Structures of Photoresist Films", Appl. Phys. Lett., 1972, 20, no.3, pp.143-5.

17.67. Martin, W.E., and Hall, D.B.: "Optical Waveguide by Diffusion of II-VI Compounds", Appl. Phys. Lett., 1972, 21, no.7, pp.325-7.

17.68. Wessler, G.R., Rodel, K.H., and Friese, P.: "Light Waveguides in Thin Organic Films", Exp. Tech. Phys., 1972, 20, no.5, pp.435-8.

17.69. Budagyan, I.F., Dubrovin, V.F., and Mirovitskii, D.I.: "Method of Partial Waves in the Theory of Functional-Junction Integrated Optics", Opt. Spekt., 1972, 33, no.4, pp.736-41, and Opt. Spectrosc., 1972, 33, no.4, pp.405-8.

17.70. Rand, M.J., and Standley, R.D.: "Silicon-Oxynitride Films on Fused Silica for Optical Waveguides", Appl. Opt., 1972, 11, no.11, pp.2482-8.

17.71. Sosnowski, T.P., and Weber, H.P. "Thin Birefringent Polymer Film for Integrated Optics", Appl. Phys. Lett., 1972, 21, no.7, pp.310-1.

17.72. Hammer, J.M., et al.: "Low-Loss Epitaxial ZnO Optical Waveguides", Appl. Phys. Lett., 1972, 21, no.8, pp.358-60.

17.73. Ogilvie, G.J., and Esdaile, R.J.: "Transmission Loss of Tetrachloroethylene-Filled Liquid-Core-Fibre Light-Guide", Electron. Lett., 1972, 8, no.22, pp.533-4.

17.74. Yamamoto, S., Koyamada, Y., and Makimoto, T.: "Normal-Mode Analysis of Anisotropic and Gyrotropic Thin-Film Waveguides for Integrated Optics", J. Appl. Phys., 1972, 43, no.12, pp.5090-7.

17.75. Goell, J.E.: "Electron-Resist Fabrication of Bends and Couplers for Integrated Optical Circuits", Appl. Opt., 1973, 12, no.4, pp.729-36.

17.76. Dubois, J.C., et al.: "Fabrication of Optical Waveguides in Electron-Resist Films", Opt. Commun., 1973,7,no.3,pp.237-8.

17.77. Dalgoutte, D.G.: "High Efficiency Thin Grating Coupler for Integrated Optics", Opt. Commun., 1973, 8, no.2, pp.124-7.

17.78. Suematsu, Y., Furuya, K., and Kambayashi, T.: "Focusing Properties of Thin-Film Lenslike Light-Guide for Integrated Optics", Appl. Phys. Lett., 1973, 23, no.2, pp.78-9.

17.79. Takano, T., and Hamasaki, J.: "Propagating Modes of a Metal-Clad, Dielectric-Slab, Waveguide for Integrated Optics", Electron. Commun. Jap., 1973,55,no.4,pp.112-8.

17.80. Holmes, S., Klugman, A., and Kraatz, P.: "Copper Mirror Surfaces for High-Power IR Lasers", Appl. Opt., 1973, 12, no.8, pp.1743-5.

17.81. Wills, K.G.: "Electroforming Waveguides", Electroplat. Met. Finish., 1973, 12, no.8, pp.1743-5.

17.82. Anikin, V.I.: "Study of Thin-Film Micro-Waveguides for the Middle-IR Region", Izv. VUZ Radioelektron., 1973, 16, no.8, pp.5-11.

17.83. Wei, D.T.Y., Lee, W.W., and Bloom, L.R.: "Quartz Optical Waveguide by Ion Implantation", Appl. Phys. Lett., 1973, 22, no.1, pp.5-7.

17.84. Giallorenzi, T.G., et al.: "Optical Waveguides Formed by Thermal Migration of Ions in Glass", Appl. Opt., 1973, 12, no.6, pp.1240-5.

17.85. Shick, L.K., et al.: "Garnet Thin-Film Optical Waveguides", J. Electron. Mater., 1973, 2, no.4, pp.609-15.

17.86. Wessler, G.R., Rodel, K.H., and Friese, P.: "Light-Guides in Liquid Thin Films", Exp. Tech. Phys., 1973, 21, no.4, pp.343-8.

17.87. Goell, J.E.: "Barium-Silicate Films for Integrated Optical Circuits", Appl. Opt., 1973, 12, no.4, pp.737-42.

17.88. Kaminow, I.P.: "Optical Waveguiding Layers in $LiNbO_3$ and $LiTaO_3$", Appl. Phys. Lett., 1973, 22, no.7, pp.326-8.

17.89. Goizhevskii, V.A., et al.: "Effect of Tolerances on Parameters of Printed-Circuit Directional Couplers", Izv. VUZ Radioelektron., 1973, 16, no.3, pp.89-92.

17.90. Tangonan, G., Pastor, A.C., and Pastor, R.C.: "Single-Crystal KCl Fibres for 10.6-micron Integrated Optics", Appl. Opt., 1973, 12, no.6, pp.1110-1.

17.91. Mar'in, V.I., Meriakri, V.V., and Lagerev, L.I.: "Quality Control for Circular Waveguide Sections", Izv. VUZ Radiofiz., 1973, 16, no.5, pp.51-4.

17.92. Tracy, J.C., et al.: "Three-Dimensional Light-Guides in Single-Crystal GaAs-(AlGa)As", Appl. Phys. Lett., 1973, 22, no.10, pp.511-2.

17.93. Yajima, H.: "Dielectric Thin-Film Optical Branching Waveguide", Appl. Phys. Lett., 1973, 22, no.12, pp.647-9.

17.94. Takahashi, S., Hashimoto, S., and Ota, T.: "Hobbing Process for Magnetron Anodes", J. Jap. Soc. Prec. Eng., 1973, 39, no.6, pp.641-7.

17.95. Garmire, E.: "Optical Waveguides in Single Layers of (GaAl)As Grown on GaAs Substrates", Appl. Phys. Lett., 1973, 23, no.7, pp.403-4.

17.96. Pitt, C.W.: "Sputtered-Glass Optical Waveguides", Electron. Lett., 1973, 9, no.17, pp.401-3.

17.97. Laybourn, P.J.R., and Millar, C.A.: "Sandwich-Ribbon Optical Waveguides", Electron. Lett., 1974, 10, no.10, pp.175-6.

17.98. Deitch, R.H., et al.: "Sputtered Thin Films for Integrated Optics", Appl. Opt., 1974, 13, no.4, pp.712-5.

17.99. Sandor, J.M.: "Metallurgical Evaluation of Fabrication Technologies for High-Precision Microwave Filters for Space Application", COMSAT Tech. Rev., 1974, 4, no.1, pp.69-89.

17.100. Auracher, F.: "Photoresist Coupler for Optical Waveguides", Opt. Commun., 1974, 11, no.2, pp.196-200.

17.101. Passaret, M.: "Characteristic Properties of Materials for Optical Fibres", Ann. Telecommun., 1974, 29, no.5-6,pp.179-88.

17.102. Tracy, J.C., et al.: "Gratings for Integrated Optics by Electron Lithography", Appl. Opt., 1974, 13, no.7, pp.1695-702.

17.103. Blum, F.A., Shaw, D.W., and Holton, W.C.: "Optical Striplines for Integrated Optical Circuits in Epitaxial GaAs", Appl. Phys. Lett., 1974, 25,no.2,pp.116-8.

17.104. Wilson, C., and Stevens, P.J.: "Polishing Technique for Optical Waveguide Terminations", J. Phys. E, 1974, 7, no.8, pp.614-5.

17.105. Macchesney, J.B., O'Connor, P.B., and Presby, H.M.: "Technique for Preparation of Low-Loss and Graded-Index Optical Fibres", Proc. IEEE, 1974, 62, no.9, pp.1282-3.

17.106. Sheem, S.: "Drawn (Black-) Pitch Threads as Masks for Microfabrication", Appl. Opt., 1974, 13, no.8, pp.1757-9.

17.107. Hill, K.O.: "Aperiodic Distributed-Parameter Waveguides for Integrated Optics", Appl. Opt.,1974,13,no.8,pp.1853-6.

17.108. Cheng, Y.C., and Westwood, W.D.: "Technique of Making Graded-Refractive-Index Slab Waveguides", Electron. Lett., 1974, 10, no.15, pp.315-6.

17.109. Hamasaki, J., and Nosu, K.: "Partially Metal-Clad Dielectric-Slab Waveguide for Integrated Optics", J. Quantum Electron. IEEE, 1974, QE-10, no.10, pp.822-5.

17.110. Narula, R.C., and Thyagarajan, R.: "Simple Technique for Grinding Garnet Spheres", J. Phys. E, 1974, 7,no.9,pp.703-4.

17.111. Lee, Y.K., and Wang, S.: "Ta_2O_5 Light Guide on $LiTaO_3$", Appl. Phys. Lett., 1974, 25, no.3, pp.164-6.

17.112. Bennett, F.C., and Fynn, D.E.: "Automated Production of Microwave Circuit Masters", Marconi Rev., 1974, 37, no.194, pp.150-63.

17.113. Chrepta, M.M., and Jacobs, H.: "Millimetre-Wave Integrated Circuits", Microwave J., 1974, 17, no.11, pp.45-7.

17.114. Aumitter, G.D., et al.: "Nanometre-Resolution Replication of Relief Pattersn for Integrated Optics", J. Appl. Phys., 1974, 45, no.10, pp.4557-62.

17.115. Auracher, F., and Witte, H.H.: "Directional Coupler for Integrated Optics", J. Appl. Phys., 1974, 45, no.11, pp.4997-9.

17.116. Tynes, A.R.: "Partially Clad Triangular-Cored Glass Optical Fibres and Lasers", J. Opt. Soc. Am., 1974, 64, no.11, pp.1415-23.

17.117. Bryantsev, V.I., et al.: "Thin-Film Metallized Dielectric Waveguides", Kvantovaya Elektron., 1974, 1, no.8, pp.1747-51.

17.118. Meier, P.J.: "Low-Cost, High-Performance, Millimetre Integrated Circuits Constructed by Fin-Line Techniques", Microwave J., 1974, 17, no.11, pp.53-4.

17.119. Stanley, C.R., Duncan, W., and McMurray, J.A.: "IR Waveguides in Thin Films of CdS", Appl. Phys. Lett., 1974, 24, no.8, pp.380-2.

17.120. Sopori, B.L., Chang, W.S.C., and Vann, R.: "Low-Loss Two-Dimensional GaAs Epitaxial Waveguides at 10.6 micron", Trans. IEEE, 1974, MTT-22, no.7, pp.754-5.

17.121. Takada, S., et al.: "Optical Waveguides of Single-Crystal $LiNbO_3$ Film Deposited by RF Sputtering", Appl. Phys. Lett., 1974, 24, no.10, pp.490-2.

17.122. Charlton, D.A.: "Low-Cost Construction Technique for Garnet and Li-Ferrite Phaseshifters", Trans. IEEE, 1974, MTT-22, no.6, pp.614-7.

17.123. Singh, A.: "Developments in Microwave Devices and Technology. I. MIC's", J. Inst. Electron. Telecommun. Eng., 1974, 20, no.6, pp.293-304.

Part III

AMPLIFICATION AND GENERATION

18. ELECTRON-BEAM TUBES

18.1. Dunn, D.A., and Sackman, G.L.: "Interaction of Two Beams with a Cavity", Int. J. Electron., 1965, 18, p.181.

18.2. Snedkov, B.A.: "Influence of Space Charge in the Buncher of a Klystron", Zh. Tekh. Fiz., 1965, 35, p.282.

18.3. Mouthaan, K.: "Wave-Mechanical Approach to Nonlinear Theory of O-Type TWT's", Int. J. Electron., 1965, 18, p.301.

18.4. Sushkov, A.D., and Meas, V.A.: "Generation of Subnanosecond Pulses by Klystron Method. I and II", Zh. Tekh. Fiz., 1965, 35, pp.723 and 739, and Sov. Phys.-Tech. Phys., 1965, 10, no.4, pp.565-75 and 576-82.

18.5. Tsikin, B.G.: "Approximate Theory of Isochronous TWT's", Radiotekh. Elektron., 1965, 10, p.312, and Radio Eng. Electron. Phys., 1965, 10, p.259.

18.6. Ober, J.: "Large-Signal Behaviour of a TWT with an Attenuating Central Helix Section", Philips Res. Rep., 1965,20,p.357.

18.7. Sovetov, N.M., and Kazakov, G.T.: "Nonlinear Theory of a TWT with Relativistic Effects", Radiotekh. Elektron., 1965, 10, p.302, and Radio Eng. Electron. Phys., 1965, 10, p.251.

18.8. Hung-Kau, W., and Jih-Lung, H.: "Coupled-Mode Description of Multiple-Beam TWT's", Acta Electron. Sin., 1965,no.1,p.101.

18.9. Pohl, W.J.: "Design and Demonstration of a Wideband, Multiple-Beam, TW Klystron", Trans. IEEE, 1965, ED-12, p.351.

18.10. Loshakov, L.N., Ol'derogge, Ye.B., and Pchel'nikov, Yu.N.: "Possibility of Obtaining a Negative Coefficient of Depression in a TWT", Radiotekh. Elektron., 1965, 10, p.681, and Radio Eng. Electron. Phys., 1965, 10, no.4, pp.579-85.

18.11. Bobroff, D.L.: "Buildup of Oscillations in an Electron-Beam BWO", Trans. IEEE, 1965, ED-12, p.307.

18.12. Taranenko, V.P., and Shevchenko, V.I.: "Influence of Electron-Beam Ripples in TWT's on Power and Efficiency", Radiotekh. Elektron., 1965, 10, p.1269, and Radio Eng. Electron. Phys., 1965, 10, no.7,pp.1091-1101.

18.13. Kuklev, Yu.I.: "Determining the Region of Negative-Coefficient Depression in a TWT with Small Helix Parameters", Radiotekh. Electron., 1965, 10, p.1157, and Radio Eng. Electron. Phys., 1965, 10,no.6,pp.995-7.

18.14. Yung-Wan, K., and Chin-Sheng, T.: "Load Effects and Parasitic Resonance in Reflex Klystrons", Acta Electron. Sin., 1965, no.2-3, p.109.

18.15. Bahr, A.J.: "Coupled-Monotron Analysis of Band-Edge Oscillations in High-Power TWT's", Trans. IEEE, 1965, ED-12,p.547.

18.16. Petin, G.P.: "Synchronous Waves in an Electron Stream", Radiotekh. Elektron., 1965, 10, p.429, and Radio Eng. Electron. Phys., 1965, 10, p.368.

18.17. Curtice, W.R.: "Approximate Analytical Expression for Current Bunching in Three-Cavity Klystrons", Proc. IEEE, 1965, 53, p.1743.

18.18. Nishihara, H., Ura, K., and Terada, M.: "Initial Loss of a TWT", Technol. Rep. Osaka Univ., 1965, 16, p.389.

18.19. Igarashi, E., Miya, M., and Murakami, H.: "Experiments on Multicircular-Sector, Slow-Wave, Structure for Millimetre-Wave Tubes", Proc. IEEE, 1965, 53, p.1677.

18.20. Ayaki, K., and Goto, H.: "Suppression of Spurious-Mode Influence in the Helix-Type TWT", J. Inst. Electr. Commun. Eng. Jap., 1965, 48, p.1589.

18.21. Kornilov, S.A.: "Spectral Width of a Reflex Klystron", Zh. Tekh. Fiz., 1966, 36, p.163, and Sov. Phys.-Tech. Phys., 1966, 11, no.1, pp.117-24.

18.22. Hamilton, J.J., and Trench, R.J.: "Grid-Controlled-Reflector Reflex Klystron", Trans. IEEE, 1966, ED-13, p.377.

18.23. Day, W.R., and Noland, J.A.: "Millimetre-Wave Extended-Interaction Oscillator", Proc. IEEE, 1966, 54, p.539.

18.24. Chodorow, M., and Kulke, B.: "Extended-Interaction Klystron. Efficiency and Bandwidth", Trans. IEEE, 1966, ED-13, p.439.

18.25. Hamilton, J.J.: "Dual Reflex Klystron Cavity-Beam Interaction", Trans. IEEE, 1966, MTT-14, p.209.

117

18.26. Luebke, W., and Caryotakis, G.: "Development of a 1-MW CW Klystron", Microwave J., 1966, 9, no.8, p.43.

18.27. Dyott, R.B.: "Coupling Impedance for a Rippled Electron Beam in a Cylindrical Guide Carrying the E_{01} Mode", Electron. Lett., 1966, 2, p.70.

18.28. Dyott, R.B., and Davies, M.C.: "Interaction between an Electron Beam of Periodically Varying Diameter and Waves in a Cylindrical Guide", Trans. IEEE, 1966, ED-13, p.374.

18.29. Walder, J., and McIsaac, P.R.: "Experimental Analysis of Biased-Gap Klystron", Trans. IEEE, 1966, ED-13, no.12, pp.950-5.

18.30. Pond, N.H., and Twiggs, R.J.: "Improvement of TWT Efficiency through Period Tapering", Trans. IEEE, 1966, ED-13, no.12, pp.956-61.

18.31. Raue, G.E., and Ishii, T.K.: "Analysis of Observed Spectrum in Pulling of Millimetre-Wave Reflex Klystrons", Proc. IEEE, 1966, 54, no.12, pp.1942-3.

18.32. Khaikov, A.Z.: "Energy Relationships in a Klystron Amplifier with Dual Interaction in the Output Circuit", Radiotekhnika, 1966, 21, no.10, pp.37-44, and Telecommun. Radio Eng. Pt 2, 1966, 21, no.10, pp.95-100.

18.33. Malyshev, V.A.: "Nonlinear Theory of Amplification in O-Type TWT's with High Gain", Radiotekh. Elektron., 1966, 11, no.9, pp.1704-7, and Radio Eng. Electron. Phys., 1966, 11, no.9, pp.1497-5000.

18.34. Andrushkevich, V.S., and Toreev, A.I.: "Frequency Response of O-Type BWO's in Presence of Reflections", Radiotekh. Elektron., 1966, 11, no.9, pp.1634-44, and Radio Eng. Electron. Phys., 1966, 11, no.9, pp.1428-38.

18.35. Tateishi, M., et al.: "Medium-Power Millimetre-Wave Klystron. Laddertron", Mitsubishi Denki Lab. Rep., 1966, 7, no.3-4, pp.189-201.

18.36. Baya, G.P., Sant'agostino, M., and Zito, G.: "Gain/Frequency Characteristics of a TWT", Alta Freq., 1966, 35, no.12, pp.944-50.

18.37. Everett, W.W., and Mackenzie, L.A.: "Axially Symmetric Output Cavity for High-Power Klystron", Proc. IEEE, 1966, 54, no.12, pp.2021-2.

18.38. Solntsev, V.A., and Mchedlidze, G.G.: "Simplified Nonlinear TWT Equations for Finite Values of Gain Parameter", Radiotekh. Elektron., 1966, 11, no.1, pp.58-67, and Radio Eng. Electron. Phys., 1966, 11, no.1, pp.46-54.

18.39. Solntsev, V.A.: "Solution of the Characteristic TWT Equation for Large Space-Charge Parameter", Radiotekh. Elektron., 1966, 11, no.1, pp.68-74, and Radio Eng. Electron. Phys., 1966, 11, no.1, pp.54-60.

18.40. Anderson, L.B., et al.: "Determination of Microwave-Tube Transfer Functions", Proc. IEEE, 1967, 114, no.7, pp.873-8.

18.41. Mihran, T.G.: "Effect of Drift Length, Beam Radius, and Perveance, on Klystron Power Conversion", Trans. IEEE, 1967, ED-14, no.4, pp.201-6.

18.42. Galaktionov, S.V., and Filimonov, G.F.: "Suppression of Back Radiation at Input of a TWT Using the Output Reflection Factor", Radiotekh. Elektron., 1967, 12, no.8, pp.1501-3, and Radio Eng. Electron. Phys., 1967, 12, no.8, pp.1397-1400.

18.43. Vikulov, I.K., and Tager, A.S.: "Study of Buildup of Oscillations in O-Type BWO's", Radiotekh. Elektron., 1967, 12, no.12, pp.2146-55, and Radio Eng. Electron. Phys., 1967, 12, no.12.

18.44. Malyshev, V.A.: "Frequency Response of a Two-Stage Klystron Power Amplifier", Radiotekhnika, 1967, 22, no.10, pp.70-4, and Telecommun. Radio Eng. Pt 2, 1967, 22, no.10.

18.45. Eaton, J.L., and Whyte, D.J.: "Comparison between Measured and Theoretical Nonlinear Behaviour of Four-Cavity Klystrons", Electron. Lett., 1967, 3, no.2, pp.55-7.

18.46. Craig, E.J.: "Beam-Loading Admittance of Gridless Klystron Gaps", Trans. IEEE, 1967, ED-14, no.5, pp.273-8.

18.47. Curtice, W.R.: "Modulation Processes in Multicavity Klystron Amplifiers", Int. J. Electron., 1967, 22, no.2, pp.101-18.

18.48. Khaikov, A.Z.: "Output-Circuit Parameters of a Dual-Interaction Klystron Amplifier", Radiotekhnika, 1967, 22, no.5, and Telecommun. Radio Eng. Pt 2, 1967, 22, no.5.

18.49. Bessonov, V.I., and Zhelezovskii, B.Ye.: "Calculation of Gain in a Parametric TWT Amplifier by Method of Coupled Waves", Radiotekh. Elektron., 1967, 12, no.3, pp.545-9, and Radio Eng. Electron. Phys., 1967, 12, no.3.

18.50. Curtice, W.R.: "First-Order Nonlinear Effects in TWT's", Proc. IEE, 1967, 114, no.8, pp.1048-50.

18.51. Nio, K., and Tamai, S.: "Power Gain Variation in a Four-Cavity Klystron Caused by a Small Change of Beam Voltage", J. Inst. Electr. Commun. Eng. Jap., 1967, 50, no.1, pp.29-30.

18.52. Plantinga, G.H., and Westerhof, T.J.: "Experimental Reflex Klystron for 1.5-mm Wavelength", Philips Tech. Rev., 1967, 28, no.9, pp.284-5.

18.53. Sun, C., and Dalman, G.C.: "Large-Signal Behaviour of Distributed Klystrons", Trans. IEEE, 1968, ED-15, no.2, pp.60-9.

18.54. Curtice, W.R.: "Sources of Residual Frequency Modulation in X-Band Two-Cavity Klystron Oscillators", Trans. IEEE, 1968, ED-15, no.1, pp.6-16.

18.55. Lally, P., and Saharian, A.: "Self-Consistent Computer Solutions on a TWT Circuit", Trans. IEEE, 1968, ED-15, no.1, pp.44-5.

18.56. Niclas, K.B., and Gerchberg, R.W.: "Efficiency Improvement of TW Interaction Process", Trans. IEEE, 1968, ED-15, no.2, pp.49-50.

18.57. Belyavskii, Ye.D.: "Demodulation of Electron Streams in O-Type Devices", Radiotekh. Elektron., 1968, 13, p.1296, and Radio Eng. Electron. Phys., 1968, 13, no.7, pp.1124-30.

18.58. Galaktionov, S.V., and Filimonov, G.F.: "Effect of Backward Radiation on Self-Excitation Conditions of a TWT", Radiotekh. Elektron., 1968, 13, p.1425, and Radio Eng. Electron. Phys., 1968, 13, no.8, pp.1238-40.

18.59. Bava, E., Bava, G.P., and Zito, G.: "Performance of TWT Amplifiers with Negative Feedback", Alta Freq., 1968, 37, no.2, pp.147-52.

18.60. Solntsev, V.A.: "Forces Acting on an Electron Beam in a TWT", Zh. Tekh. Fiz., 1968, 38, no.1, pp.109-17, and Sov. Phys.-Tech. Phys., 1968, 13, no.1.

18.61. Vasil'ev, Ye.I., Kanavets, V.I., and Lopukhin, V.M.: "Effect of Constant Field Components on Resonator Efficiency", Radiotekh. Elektron., 1968, 13, no.1, pp.157-9, and Radio Eng. Electron. Phys., 1968, 13, no.1.

18.62. Shul'ga, V.G., and Konovalov, V.Ye.: "Linear Theory of Double-Beam O-Type TWT", Radiotekh. Elektron., 1968, 13, no.1, pp.69-74, and Radio Eng. Electron. Phys., 1968, 13, no.1.

18.63. Galwas, B.: "Design and Construction of Millimetre-Wave Reflex Klystrons", Przegl. Elektron., 1968, 9, no.4, pp.181-90.

18.64. Rusin, F.S., and Bogomolov, G.D.: "Orotron. Electron Device with Open Resonator and Reflecting Grating (for Millimetre Waves)", Izv. VUZ Radiofiz., 1968, no.5, pp.756-62.

18.65. Wallander, S.O.: "Large-Signal Computer Analysis of Klystron Waves", Int. J. Electron., 1968, 24, no.2, pp.185-96.

18.66. Galaktionov, S.V., and Filimonov, G.F.: "Effect of Backward Radiation on the Operation of a TWT in an Amplifying Mode", Radiotekh. Elektron., 1968, 13, no.4, pp.669-78, and Radio Eng. Electron. Phys., 1968, 13, no.4.

18.67. Giarola, A.J.: "Theoretical Description for Multiple-Signal Operation of a TWT", Trans. IEEE, 1968, ED-15, no.6, pp.381-95.

18.68. Machulka, G.A.: "Study of TWT Conditions for Oscillations at the Band Edges of the Delay Lines", Radiotekh. Elektron., 1968, 13, no.3, pp.560-2, and Radio Eng. Electron. Phys., 1968, 13, no.3.

18.69. Cohen, L.D.: "BWO's for the 50- to 300-GHz Range", Trans. IEEE, 1968, ED-15, no.6, pp.403-4.

18.70. Zhurakhovskii, V.A.: "Theory of Optimal Control Applied to Microwave Electronic Devices with Distributed Interaction", Radiotekh. Elektron., 1968, 13, no.5, pp.931-4, and Radio Eng. Electron. Phys., 1968, 13, no.5.

18.71. Wallander, S.O.: "Large-Signal Analytical Study of Bunching in Klystrons", Trans. IEEE, 1968, ED-15, no.8, pp.595-603.

18.72. Chandra, K., Parshad, R., and Kumar, R.C.: "Synchronous Operation of Reflex Klystrons in Parallel under Pulse, Square-Wave, and Frequency, Modulation", J. Inst. Telecommun. Eng., 1968, 14, no.3, pp.138-40.

18.73. Vlasov, A.D.: "Small-Signal Interaction of an Electron Beam with Microwave Cavities. Electronic Conductance", Radiotekh. Elektron., 1968, 13, no.5, pp.865-76, and Radio Eng. Electron. Phys., 1968, 13, no.5.

18.74. Belyavskii, Ye.D.: "Application of Variational Methods for Analysis of Interaction in Microwave Devices", Radiotekh. Elektron., 1968, 13, no.6, pp.1136-8, and Radio Eng. Electron. Phys., 1968, 13, no.6.

18.75. Belyavskii, Ye.D.: "Demodulation of Multivelocity Electron Streams in O-Type Devices", Radiotekh. Elektron., 1968, 13, no.10, p.1909, and Radio Eng. Electron. Phys., 1968, 13, no.10, pp.1679-80.

18.76. Kupchinov, N.F.: "Influence of Electron Deceleration in a TWT upon Deformation of Tape Beam", Izv. VUZ Radiofiz., 1968, 11, p.1606.

18.77. King, R.C.M.: "Symmetrical Rod-Coupled Circuit for High-Power TWT's", Nachr. Tech. Fachber., 1968, 35, p.67.

18.78. Nejib, U.R., and Elco, R.A.: "Growing Waves in Transverse-Electron-Beam Fast-Wave Interactions", Int. J. Electron., 1968, 25, p.557.

18.79. Boltz, G.: "3-mm BW Tube with Wide Tuning Range", Nachr. Tech. Fachber., 1968, 35, p.105.

18.80. Lien, E.L.: "Design of Coupled-Cavity Extended-Interaction Output Resonators for Klystron Amplifiers", Nachr. Tech. Fachber., 1968, 35, p.115.

18.81. Hechtel, J.R.: "Double-Gap Catcher Improving Efficiency of Klystrons", Nachr. Tech. Fachber., 1968, 35, p.121.

18.82. James, B.G., and Zitelli, L.T.: "Kilowatt CW Klystron Amplifiers at K Bands", Nachr. Tech. Fachber., 1968, 35, p.145.

18.83. Fourcade, D., Reignac, J., and Lacroix, G.: "Study of the Initiation of Reflex-Klystron Oscillations", C. R. Acad. Sci. (Paris), 1968, 267B, p.503.

18.84. Gupta, K.C.: "Problem of Coupling in Beam-Plasma Devices", J. Inst. Telecommun. Eng., 1968, 14, p.467.

18.85. Wallander, S.O.: "Large-Signal Theory of Amplitude and Phase Nonlinearities in TWT's", Nachr. Tech. Fachber., 1968, 35, p.11.

18.86. Nishida, S.: "Design and Performance of an X-Band CW TWT", Nachr. Tech. Fachber., 1968, 35, p.36.

18.87. Bryant, M.O.: "High-Power CW Ring-Bar TWT", Nachr. Tech. Fachber.,1968,35,p.42.

18.88. Tsikin, B.G., and Kornoukhov, G.M.: "Mechanism of Power Saturation in Isochronous TWT", Radiotekh. Elektron., 1968, 13, p.1608, and Radio Eng. Electron. Phys., 1968, 13, no.9, pp.1400-4.

18.89. Serebryanik, A.N.: "Problems in Theory of O-Type BWO's with Finite Gain", Radiotekh. Elektron., 1968, 13, p.1613, and Radio Eng. Electron. Phys., 1968, 13, no.9, pp.1405-13.

18.90. Nilsson, O., and Wallander, S.O.: "Distortion and Saturation in TWT's", Elteknik, 1968, 11, p.213.

18.91. Filimonov, G.F., et al.: "Influence of Lamination Effect on Operation of a TWT", Radiotekh. Elektron., 1968, 13, p.2000, and Radio Eng. Electron. Phys., 1968, 13, no.11, pp.1754-63.

18.92. Belov, N.Ye.: "Admittance Introduced into a Cavity by a Tubular Beam with Large Space-Charge Density", Radiotekh. Elektron., 1968, 13, p.2086, and Radio Eng. Electron. Phys., 1968, 13, no.11, pp.1830-6.

18.93. Craig, E.J.: "Relativistic Beam-Loading Admittance", Trans. IEEE, 1969, ED ED-16, p.139.

18.94. Day, W.R.: "Small-Signal Gain Calculations for Electrostatically Focused Klystrons", Trans. IEEE, 1969, ED-16,p.486.

18.95. Barybin, A.A., and Ter-Martirosyan, L.T.: "Electrodynamic Analysis of Interaction between an Electron Beam and a Slow-Wave System", Radiotekh. Elektron., 1969, 14, p.277, and Radio Eng. Electron. Phys., 1969, 14, no.2, pp.237-45.

18.96. King, R.C.M.: "Symmetrical Rod-Coupled Circuit for High-Power TWT's", Electron. Eng. 1969, 41, no.498, p.40.

18.97. Sushkov, A.D., and Fedyaev, V.K.: "Experimental Study of Current Harmonics in a Triode-Klystron", Izv. VUZ Radioelektron., 1969, 12, no.1, p.69.

18.98. Polishchuk, A.A.: "Effect of Space Charge on Starting Current and Reflector Potential in a Reflex Klystron", Izv. VUZ Radioelektron., 1969, 12, no.3, p.295.

18.99. Malyshev, V.A.: "Characteristics of Narrowband Microwave Oscillators with Extra Cavity", Izv. VUZ Radioelektron., 1969, 12, no.9, p.1041.

18.100. Belyavskii, Ye.D.: "Demodulation of Beams in O-Type Devices in Conditions of Energy Exchange with Field", Radiotekh. Elektron., 1969, 14, no.7, p.1342, and Radio Eng. Electron. Phys.,1969,14,no.7,pp.1167-9.

18.101. Saakov, E.O.: "Excitation of (Microwave) Open Resonator by an Electron Stream", Radiotekh. Elektron., 1969, 14, no.9, p.1703, and Radio Eng. Electron. Phys., 1969, 14, no.9, pp.1482-5.

18.102. Taranenko, V.P., and Shevchenko, V.I.: "Nonlinear Equations for a TWT. I", Izv. VUZ Radioelektron., 1969,12,no.9,p.1006.

18.103. Zhelezovskii, B.Ye., and Mashnikov, V.V.: "Theory of Two-Frequency Operating Range of O-Type TWT's", Izv. VUZ Radioelektron., 1969, 12, no.9, p.1094.

18.104. Burneika, K.P., and Kanavets, V.I.: "Parameter Optimization of a Two-Cavity Klystron", Izv. VUZ Radioelektron., 1969, 12, no.9, p.1018.

18.105. Zabalkanskii, E.S.: "Analysis of Convection Currents during Amplification of Complex-Shape Signals (in Multicavity Klystrons)", Izv. VUZ Radioelektron., 1969, 12, no.9, p.1023.

18.106. Lucken, J.A.: "Aspects of Circuit Power Dissipation in High-Power CW Helix TWT's. I and II", Trans. IEEE, 1969, ED-16, pp.813 and 821.

18.107. Gerchberg, R.W., and Niclas, K.B.: "Positively Tapered TWT", Trans. IEEE, 1969, ED-16, p.827.

18.108. Edgcombe, C.J., and Heppinstall, R.: "Studying Nonlinearity in Klystrons", Des. Electron., 1969, 7, no.1, p.46.

18.109. Fedyaev, V.K.: "Origin of Higher Harmonics of the Current in a Klystron", Radiotekh. Elektron., 1969, 14, no.6, p.1116, and Radio Eng. Electron. Phys., 1969, 14, no.6, pp.968-9.

18.110. Belyavskii, Ye.D.: "Effect of Transverse Motion in Nonlinear Demodulation Theory of Electron Streams in O-Type Devices", Radiotekh. Elektron., 1969, 14, no.7, p.1250, and Radio Eng. Electron. Phys., 1969, 14, no.7, pp.1080-6.

18.111. Tallerico, P.J., and Rowe, J.E.: "Relativistic Effects in the TW Amplifier", Trans. IEEE, 1970, ED-17, p.549.

18.112. Burneika, K.P., and Kanavets, V.I.: "Electron Bunching in a Klystron Using Nonlinear Wave Processes", Izv. VUZ Radioelektron., 1970, 13, no.3, p.370.

18.113. Wallander, S.O." Nonlinear Multi-Signal TWT Theory", Int. J. Electron., 1970, 29, p.201.

18.114. Kazakov, G.T., Kazakova, N.I., and Sovetov, N.M.: "Design of a TWT and a Twistron Using a Phase-Plane Method", Radiotekh. Elektron., 1970, 15, no.5, p.993, and Radio Eng. Electron. Phys., 1970, 15, no.5, pp.827-35.

18.115. Steele, C.W.: "Equivalent Generators for a Cavity with Bunched Charged Beam", Trans. IEEE, 1970, ED-17, p.160.

18.116. Nishihara, H., and Terada, M.: "Effects of Attenuator on Nonlinear Phase Distortion of a TWT", Trans. IEEE, 1970, ED-17, p.638.

18.117. Bitovets, V.V., et al.: "Excitation of a Klystron Output Resonator by Idealized Bunches", Vestn. Mosk. Univ., Fiz. Astron., 1970, no.6, pp.618-21.

18.118. Loshakov, L.N., and Ol'derogge, Ye.B.: "Calculation of Propagation Coefficients of Waves in a Helix TWT", Radiotekhnika, 1970, 25, no.8, and Telecommun. Radio Eng. Pt 2, 1970, 25, no.8, pp.131-5.

18.119. Jui-Pang, H.: "Calculation Method for Electron Admittance of Reflex Klystron", Electron. Commun. Jap., 1970, 53, no.1, pp.66-72.

18.120. Rikenglaz, L.E., and Yanovskii, S.A.: "Theory of Interaction between an Electron Beam and an External Harmonic Microwave Field", Radiotekh. Elektron., 1970, 15, no.9, and Radio Eng. Electron. Phys., 1970, 15, no.9, pp.1641-6.

18.121. Yajima, H., and Fujita, H.: "Large-Signal Electronic Admittance in a Long-Transit-Angle Cavity", Trans. IEEE, 1970, ED-17, p.76.

18.122. Nejib, U.R., and Elco, R.A.: "Distributed Circuit Model for Transverse Current-Wave Interaction Devices", Int. J. Electron., 1970, 28, p.349.

18.123. Jasper, L.J.: "Effects of a Gyromagnetic Material in a Double-Coupled Helix TWT", Trans. IEEE, 1970, ED-17, p.353.

18.124. Gvozdover, S.D.: "Variational Methods in the Analysis of a TWT with a Helical Lossless Delay Surface", Radiotekh. Elektron., 1970, 15, p.114, and Radio Eng. Electron. Phys., 1970, 15, no.1, pp.92-100.

18.125. van Iperen, B.B., and Tjassens, H.: "Experimental CW Klystron for Generating 0.4-mm Waves", Trans. IEEE, 1970, ED-17,p.380.

18.126. Verbitskii, I.L., and Buzik, L.M.: "Theory of Interaction of an Electron Current with Spatial Harmonics of a Comb System", Radiotekh. Elektron., 1970, 15, no.6, p.1200, and Radio Eng. Electron. Phys., 1970, 15, no.6, pp.1015-20.

18.127. Vasil'ev, Ye.I., Kanavets, V.I., and Lopukhin, V.M.: "Parameter Optimization of a Single-Gap-Klystron Output Cavity", Radiotekh. Elektron., 1970, 15, no.6, p.1189, and Radio Eng. Electron. Phys., 1970, 15, no.6, pp.1006-11.

18.128. Poizner, B.N.: "Experimental Study of a Klystron in a Multifrequency Mode with an Incident External Signal", Radiotekh. Elektron., 1970, 15, no.6, p.1196, and Radio Eng. Electron. Phys.,1970,15,no.6,pp.1012-5.

18.129. Hechtel, J.R.: "Effect of Potential Beam Energy on Performance of Linear Beam Devices", Trans. IEEE, 1970, ED-17, no.11, pp.999-1009.

18.130. Vorobeichikov, E.S., et al.: " "Forced Operation of a Multifrequency Reflex Klystron", Izv. VUZ Radioelektron., 1970, 13, no.8, pp.923-33.

18.131. Mizuno, K., Ono, S., and Shibata, Y.: "Electron Tube with Fabry-Perot Resonator for Generation of Submillimetre Waves", Rep. Res. Inst. Electr. Commun. Tohoku Univ., 1970, 21, no.3-4, pp.113-38.

18.132. Welch, G.V., and Ishii, T.K.: "Hybrid-T Coupled Oscillators", Electron. Lett., 1970, 6, no.22, pp.717-8.

18.133. Lebedev, I.V.: "Threshold Input Signals of Microwave Amplifiers", Izv. VUZ Radioelektron., 1970, 13, no.8, pp.909-15.

18.134. Grow, R.W., and Gunderson, D.R.: "Starting Conditions for BWO's with Large Loss and Large Space Charge", Trans. IEEE, 1970, ED-17, no.12, pp.1032-9.

18.135. Konovalov, V.Ye., and Shul'ga, V.G.: "Boundary Conditions and Gain of TWT's with Space-Distributed Currents", Izv. VUZ Radioelektron., 1970, 13, no.9, pp.1079-84.

18.136. Casimir, H.B.G.: "Microwave and Optical Generation and Amplification", Ingenieur, 1970, 82, no.52, pp.181-3.

18.137. Kaminski, K.: "Analysis of Reflex-Klystron Repeller Space", Electron. Commun. Jap., 1970, 53, no.7, pp.12-3.

18.138. Savel'ev, V.S., and Kushchenko, G.I.: "Experimental Investigation of a TWT with Radial Electron Stream", Radiotekh. Elektron., 1970, 15, no.12, and Radio Eng. Electron. Phys., 1970, 15, no.12, pp.2267-72.

18.139. Buzik, L.M.: "Effect of Nonsynchronous Spatial Harmonics on O-Type Interactions", Radiotekh. Elektron., 1970, 15, no.12, and Radio Eng. Electron. Phys., 1970, 15, no.12, pp.2324-6.

18.140. Burneika, K.P., and Kanavets, V.I.: "Electron Bunching in a Three-Cavity Klystron for Finite Perveance", Izv. VUZ Radioelektron., 1971, 14, no.2, pp.163-8.

18.141. Tellerico, P.J.: "Design Considerations for High-Power Multicavity Klystron", Trans. IEEE, 1971, ED-18, no.6, pp.374-82.

18.142. Bessonov, V.I., et al.: "Improving Gain of TWT's (by Parametric Effects)", Izv. VUZ Radioelektron., 1971, 14,no.5,pp.516-21.

18.143. Okamoto, T., et al.: "Millimetre-Wave High-Power TWT's", Toshiba Rev., 1971, no.57, pp.28-32.

18.144. Vasil'ev, Ye.I., et al.: "Effect of Space Charge on Efficiency of a Twin-Gap Output Cavity", Izv. VUZ Radioelektron., 1971, 14, no.5, pp.583-5.

18.145. Barakat, R.: "Solution of TWT Amplifier Equations", Int. J. Electron., 1971, 31, no.3, pp.297-9.

18.146. Englefield, C.G.: "Analysis of Dielectrically Loaded TWT", Trans. IEEE, 1971, ED-18, no.10, pp.892-7.

18.147. Mashnikov, V.V., Zhelezovskii, B.Ye., and Petrova, V.N.: "Study of Total Suppression of One Signal in a Two-Frequency TWT", Izv. VUZ Radioelektron., 1971, 14, no.9, pp.1027-31.

18.148. Levchenko, Ye.G., and Chaika, V.Ye.: "Spatially-Staggered Delay Structures Formed by Coupled Resonators", Izv. VUZ Radioelektron., 1971,14,no.9,pp.1105-11.

18.149. Gaiduk, V.I.: "Linear Theory of Microwave Devices with Curvilinear Electron Beams Assuming Space-Charge Conditions", Izv. VUZ Radioelektron., 1971, 14, no.1, pp.17-26.

18.150. Mihran, T.G., Branch, G.M., and Griffin, C.J.: "Design and Demonstration of a Klystron with 62% Efficiency", Trans. IEEE, 1971, ED-18, no.2, pp.124-33.

18.151. Kirichenko, A.Ya., Lysova, L.A., and Suvorov, A.N.: "Experimental Study of Ring-Plane Slow-Wave Systems", Izv. VUZ Radioelektron., 1971, 14, no.10, pp.1234-6.

18.152. Belyavskii, Ye.D.: "Operating Conditions of O-Type Devices with Capture of Electron Bunches by Electromagnetic Wave", Radiotekh. Elektron., 1971, 16, no.1, and Radio Eng. Electron. Phys., 1971, 16, no.1, pp.186-8.

18.153. Tereshchenko, A.I., and Grebenyuk, A.F.: "Design of Resonators for Radial Klystrons", Radiotekh. (Kharkov), 1971, no.17, pp.6-13.

18.154. Tereshchenko, A.I., et al.: "Parameters of Ring Resonators for Tubular-Beam Klystrons", Radiotekh. (Kharkov), 1971, no.18, pp.35-40.

18.155. Zboronskii, A.V., et al.: "Effect of Static-Field Irregularities on Starting Conditions of O-Type BWO's with Electrostatic Focusing", Izv. VUZ Radioelektron., 1971, 14, no.10, pp.1160-6.

18.156. Vorobeichikov, E.S., and Poizner, B.N.: "Signal Generation Processes in Multi-frequency Reflex Klystrons", Izv. VUZ Radioelektron., 1971, 14, no.12, pp.1519-20.

18.157. Burneika, K.P., et al.: "Bunching and Electronic Efficiency of a Four-Resonator Klystron", Radiotekh. Elektron., 1971, 16, no.4, and Radio Eng. Electron. Phys., 1971, 16, no.4, pp.641-4.

18.158. Tolmachev, M.M., and Tseitlin, M.B.: "Effect of Medium on Propagation of Space-Charge Waves in a Bounded Electron Beam", Radiotekh. Elektron., 1971, 16, no.3, and Radio Eng. Electron. Phys., 1971, 16, no.3, pp.507-11.

18.159. Zhelezovskii, B.Ye., Mashnikov, V.V., and Rusin, V.I.: "Analysis of the Combination Components of Higher-Order Currents in a TWT Frequency Converter", Radiotekh. Elektron., 1971, 16, no.3, and Radio Eng. Electron. Phys., 1971, 16, no.3, pp.565-7.

18.160. Algazinov, E.K., et al.: "Analysis of Properties of the Nonlinear Operation Mode of a TWT when Amplifying Multifrequency Signals", Radiotekh. Elektron., 1971, 16, no.6, pp.1028-32, and Radio Eng. Electron. Phys., 1971, 16, no.6, pp.1010-3.

18.161. Ezura, E., Kano, T., and Kamiryo, K.: "Effects of Local Attenuator on the Large-Signal Behaviour of TWT's", Rec. Electr. Commun. Eng. Conversaz. Tohoku Univ., 1971, 40, no.4, pp.43-7.

18.162. Galaktionov, S.V., et al.: "Effect of Localized Fields in a Cutoff Intermediate Section of the Slow-Wave Structure on a TWT", Radiotekh. Elektron., 1972, 16, no.9, pp.1673-9, and Radio Eng. Electron. Phys., 1972, 16, no.9, pp.1520-6.

18.163. Mihran, T.G., Branch, G.M., and Griffin, G.J.: "Electron Bunching and Output-Gap Interaction in Broadband Klystrons", Trans. IEEE, 1972, ED-19, no.9, pp.1011-7.

18.164. Denisov, A.I., and Rapoport, G.N.: "Approximate Analysis of Maximum Efficiency of an O-Type Generator with Resonant Delay Structure in the Presence of Losses", Izv. VUZ Radioelektron. 1972, 15, no.3,pp.296-301.

18.165. Brzhechko, L.V.: "Self-Consistent Theory of Laddertron Oscillators", Radiotekh. (Kharkov), 1972, no.20, pp.32-8.

18.166. Denisov, A.I., and Chaika, V.Ye.: "Nonlinear Analysis of Resonance O-Type Oscillators with Fixed Signal Amplitude in the Presence of Losses", Izv. VUZ Radioelektron., 1972, 15, no.8, pp.1022-6.

18.167. Shul'ga, V.G., and Bondarenko, B.N.: "Parametric Analysis of a Two-Beam TWT for Large Input Signals", Izv. VUZ Radioelektron., 1972, 15, no.8, pp.1027-32.

18.168. Bondarenko, B.N., et al.: "Experimental Study of a Two-Beam TWT", Izv. VUZ Radioelektron., 1972, 15, no.8, pp.1033-6.

18.169. Ezura, E., Kano, T., and Kamiryo, K.: "Large-Signal Analysis of Efficiency and Nonlinear Phase Distortion in a Phase-Velocity-Tapered TWT", Rec. Electr. Commun. Eng. Conversaz. Tohoku Univ., 1972, 41, no.2, pp.114-9.

18.170. Koroza, V.I.: "Theory of Waves in Slow-Wave Structures Loaded by an Electron Beam", Radiotekh. Elektron., 1972, 17, no.3, pp.577-86, and Radio Eng. Electron. Phys., 1972, 17, no.3, pp.449-57.

18.171. Koroza, V.I., and Sukhovskii, E.S.: "Interaction of Beams of Charged Particles with Slow-Wave Systems", Radiotekh. Elektron., 1972, 17, no.4, pp.794-804, and Radio Eng. Electron. Phys., 1972, 17, no.4, pp.619-28.

18.172. Yur'yev, V.I., et al.: "Experimental Investigation of Interaction between Electron-Beam Waves Synchronous with the Structure Travelling Wave", Radiotekh. Elektron., 1972, 17, no.4, pp.830-4, and Radio Eng. Electron. Phys., 1972, 17, no.4, pp.648-51.

18.173. Bobrovskii, Yu.L., and Ovchinnikov, K.D.: "Nonlinear Analysis of Reflex Klystrons with High Electronic Conductance", Radiotekh. Elektron., 1972, 17, no.4, pp.818-25, and Radio Eng. Electron. Phys., 1972, 17, no.4, pp.639-45.

18.174. Bessonov, B.I., Zhelezovskii, B.Ye., and Tuyrin, S.V.: "Starting Conditions for a BWO Employing Beam Premodulation", Izv. VUZ Radioelektron., 1972, 15, no.10, pp.1216-9.

18.175. Lind, B.I., and Askne, J.I.H.: "Energy Flow Relations in Multiwave Systems with Absorption and Dispersion", Trans. IEEE, 1972, ED-19, no.2, pp.239-45.

18.176. Wallander, S.O.: "Reflections and Gain Ripple in TWT's", Trans. IEEE, 1972, ED-19, no.5, pp.655-60.

18.177. Briggs, R.J., and Harris, N.W.: "TWT Delay Lines. I", Trans. IEEE, 1972, ED-19, no.5, pp.661-5.

18.178. Harris, N.W.: "TWT Delay Lines. II", Trans. IEEE, 1972, ED-19, no.5, pp.666-71.

18.179. Loshakov, L.N., and Ol'derogge, Ye.B.: "Distribution of Longitudinal Electric Field in Electron Stream of a TWT", Radiotekhnika, 1972, 27, no.6, pp.90-1, and Telecommun. Radio Eng. Pt 2, 1972, 17, no.6, pp.130-2.

18.180. Dudnik, R.A.: "Evaluation of Electronic Efficiency of a Transverse-Current Tube", Radiotekh. Elektron., 1972, 17, no.8, pp.1663-70, and Radio Eng. Electron. Phys., 1972, 17, no.8, pp.1311-7.

18.181. Mchedlidze, G.G., and Solntsev, V.A.: "Wave Method of Solving Nonlinear TWT Equations", Radiotekh. Elektron., 1972, 17, no.10, pp.2227-30, and Radio Eng. Electron. Phys., 1972, 17, no.10, pp.1789-92.

18.182. Vasil'ev, Ye.I., Kanavets, V.I., and Lapukhin, V.M.: "Electronic Admittance and Efficiency of Klystron Output Cavity", Izv. VUZ Radioelektron., 1972, 15, no.9, pp.1146-53.

18.183. Petrosyan, M.L.: "Contribution to Kinematic Theory of Multistage Bunching of Charged Particles", Radiotekh. Elektron., 1972, 17, no.4, pp.826-9, and Radio Eng. Electron. Phys., 1972, 17, no.4, pp.645-8.

18.184. Bava, E., et al.: "Reduction of Phase Nonlinearities in TWT's", Alta Freq., 1972, 41, no.11, pp.836-42.

18.185. Korutin, L.I., et al.: "Laddertron-Type (Electron-Beam) Generator", Izv. VUZ Radioelektron., 1973, 16, no.1, pp.91-6.

18.186. Smorgonskii, A.V.: "Nonlinear Theory of a Relativistic Monotron", Izv. VUZ Radiofiz., 1973, 16, no.1, pp.150-5.

18.187. Shein, A.G.: "Linear Theory of an O-Type TWT Operating in a Two-Frequency Mode", Radiotekh. (Kharkov), 1973, no.24, pp.104-12.

18.188. Kosmahl, H.G., and Branch, G.M.: "Generalized Representation of Electric Fields in Interaction Gaps of Klystrons and TWT's", Trans. IEEE, 1973, ED-20, no.7, pp.621-9.

18.189. Mizuno, K., Ono, S., and Shibata, Y.: "Two Different Mode Interactions in an Electron Tube with a Fabry-Perot Resonator. Ledatron", Trans. IEEE, 1973, ED-20, no.8, pp.749-52.

18.190. Ogura, K., et al.: "High-Efficiency High-Power Klystron", Toshiba Rev., 1973, no.81, pp.28-32.

18.191. Duving, V.G., and Chebotarev, G.N.: "Frequency Instability of a Self-Running Oscillator Using a Transit-Time, Multicavity, Klystron Amplifier", Izv. VUZ Radioelektron., 1973, 16, no.5, pp.111-4.

18.192. Lagranskii, L.M., et al.: "Effect of Absorber on Power Characteristics of a TWT", Izv. VUZ Radioelektron., 1973, 16, no.11, pp.29-35.

18.193. Nikolaenko, Yu.V., and Raevskii, S.B.: "Electron-Beam Excitation of an Open Resonator", Izv. VUZ Radioelektron., 1973, 16, no.9, pp.22-8.

18.194. Taranenko, V.P., Shevchenko, V.I., and Raichuk, B.F.: "Improving Efficiency of High-Power TWT's with Large Space-Charge Parameter", Izv. VUZ Radioelektron., 1973, 16, no.10, pp.57-67.

18.195. Borisenko, V.D., Ishchenko, A.I., and Perekupko, V.A.: "Design of Broadband X-Band TWT's", Izv. VUZ Radioelektron., 1973, 16, no.10, pp.68-75.

18.196. Pavlov, O.I., and Filimonov, G.F.: "Operating Characteristics of a Klystron with a High Space-Charge Density", Izv. VUZ Radioelektron., 1973, 16, no.10, pp.76-84.

18.197. Grebenyuk, A.F., Pospelov, L.A., and Tereshchenko, A.I.: "Feasibility of Reducing the Operating Wavelength of Klystrons", Izv. VUZ Radioelektron., 1973, 16, no.10, pp.85-9.

18.198. Khaikov, A.Z.: "Optimum Electron Bunching Modes in Multiresonator Klystrons", Radiotekhnika, 1973, 28, no.3, pp.77-84, and Telecommun. Radio Eng. Pt 2, 1973, 28, no.3, pp.114-9.

18.199. Pruslin, V.Z.: "Efficient Output-Power Utilization of a TWT", Radiotekhnika, 1973, 28, no.4, pp.50-9, and Telecommun. Radio Eng. Pt 2, 1973, 28, no.4, pp.83-5.

18.200. Pospelov, L.A.: "Theory of a Resonant Oscillator with Distributed Interaction", Radiotekh. (Kharkov), 1973, no.27, pp.64-74.

18.201. Nigmatullin, U.A., and Yur'yev, V.I.: "Gain Coefficient of a TWT for the Transverse Synchronous Wave of the Electron Beam", Radiotekh. Elektron., 1973, 18, no.8, pp.1747-9, and Radio Eng. Electron. Phys., 1973, 18, no.8, pp.1281-3.

18.202. Grebenyuk, A.F., Kanarik, G.G., and Vorotnikov, Ye.P.: "Analysis of Ring-Type Cavity Parameters for a Multiple-Beam Klystron", Izv. VUZ Radioelektron., 1974, 17, no.5, pp.10-6.

18.203. Cornet, G., and Raoult, G.: "Centimetre-Wave Generator Using a Periodic-Slit Cavity", Onde Electr., 1974, 54, no.7, pp.353-9.

18.204. Vyrovoi, S.I.: "Dispersion Equation for a Thin Pipe Polyscrew Electron Beam in a Circular Waveguide", Izv. VUZ Radiofiz., 1974, 17, no.9, pp.1378-91.

18.205. Maloney, E.D., and Faillon, G.: "High-Power Klystron for Industrial Processing Using Microwaves", J. Microwave Power, 1974, 9, no.3, pp.231-9.

18.206. Taranenko, V.P., and Mikhin, A.A.: "Methods of Suppression of Spurious Signals in O-Type TWT's", Izv. VUZ Radioelektron., 1974, 17, no.11, pp.5-17.

18.207. Shevchenko, V.I., and Zemlyak, A.M.: "Application of Differential Methods of Solution of Poisson's Equation for Electron-Beam Space-Charge Forces in an O-Type TWT", Izv. VUZ Radioelektron., 1974, 17, no.11, pp.18-26.

18.208. Algazinov, E.M., and Korobova, A.K.: "Analysis of Nonlinear Amplification of Multifrequency Signals in a TWT with External Feedback", Izv. VUZ Radioelektron., 1974, 17, no.11, pp.27-32.

18.209. Gassanov, L.G., et al.: "Theory of O-Type Devices Employing Nonidentical Coupled Cavity Chains", Izv. VUZ Radioelektron., 1974, 17, no.11, pp.33-42.

18.210. Amirov, R.Sh., et al.: "Study of Pre-Oscillatory Conditions in O-Type BW Tube", Izv. VUZ Radioelektron., 1974, 17, no.11, pp.52-8.

18.211. Molokovskii, S.I., and Tregubov, V.F.: "Motion of Modulated Beam in the Output Stage of a Klystron Employing Electrostatic Focusing", Izv. VUZ Radioelektron., 1974, 17, no.9, pp.20-4.

18.212. Khaikov, A.Z., and Podgrebel'naya, N.V.: "Effect of Higher Space-Charge Modes on Parameters of a Floating-Drift-Tube Klystron", Izv. VUZ Radioelektron., 1974, 17, no.11, pp.74-80.

18.213. Filimonov, G.F.: "Origin and Magnitude of Energy Flow in O-Type Devices", Izv. VUZ Radioelektron., 1974, 17, no.12, pp.3-9.

18.214. Sovetov, N.M., Sokolov, V.N., and Kazakov, G.T.: "Electrical Modelling of Twistrons and Klystrons with Distributed Interaction", Izv. VUZ Radioelektron., 1974, 17, no.12, pp.46-51.

18.215. Grebenyuk, A.F., and Vorotnikov, Ye.P.: "Oscillations in an Annular Resonator for Multibeam Klystrons", Radiotekh. (Kharkov), 1974, no.30, pp.128-33.

18.216. Zarnitsina, I.G., and Nusinovich, G.S.: "Stability of Single-Mode Oscillations in a Gyromonotron", Izv. VUZ Radiofiz., 1974, 17, no.12, pp.1858-67.

18.217. Vyrovoi, S.I., and Rapoport, G.N.: "Bandwidth of the Output Stage of a Gyrotwistron", Izv. VUZ Radioelektron., 1974, 17, no.9, pp.96-8.

28.218. Dement'ev, A.S., and Pavlenko, Yu.G.: "Microwave Maser With Nonrelativistic Electrons", Zh. Tekh. Fiz., 1974, 44, no.5, pp.1101-3, and Sov. Phys.-Tech. Phys., 1974, 19, no.5, pp.694-5.

19. CROSSED-FIELD AND HIGH-ENERGY GENERATORS

19.1. Lominadze, D.G.: "Cherenkov Radiation from a Ring Current in a Brillouin Cloud", Zh. Tekh. Fiz., 1965, 35, p.568, and Sov. Phys.-Tech. Phys., 1965, 10, no.3, pp.441-3.

19.2. Alybin, V.G., Guttsait, E.M., and Sokolova, L.I.: "Characteristics of Regenerative Magnetron Amplifiers", Izv. VUZ Radiotekh., 1965, 8, no.6, pp.647-51, and Sov. Radio Eng., 1965, 8, no.6, pp.482-4.

19.3. Dudnik, R.A.: "Analysis of Dispersion Equation of an M-Type Crossed-Field Tube", Izv. VUZ Radiofiz., 1965, 8, no.6, pp.1250-2.

19.4. Gazazyan, E.D., and Mergelyan, O.S.: "Radiation from a Point Charge in a Waveguide in the Presence of an External Magnetic Field", Zh. Tekh. Fiz., 1965, 35, p.158, and Sov. Phys.-Tech. Phys., 1965, 10, no.1, pp.125-6.

19.5. Kolpakov, O.A., Kotov, V.I., and Kha, O.S.: "Propagation of Slow Waves in a Waveguide Structure and Radiation from a Charge Moving Along the Axis", Zh. Tekh. Fiz., 1965, 35, p.26, and Sov. Phys.-Tech. Phys., 1965, 10, no.1, pp.18-23.

19.6. Trubetskov, D.I.: "Linear Theory of Magnetron-Type Beam Devices", Radiotekh. Elektron., 1965, 10, p.321, and Radio Eng. Electron. Phys., 1965, 10, p.267.

19.7. Okabe, T.: "Waveguide Magnetrons", Trans. IEEE, 1965, ED-12, p.421.

19.8. Gayduk, V.I., and Tseitlin, M.B.: "Theory of Waveguides with a Rotating Electron Stream", Radiotekh. Elektron., 1965, 10, p.284, and Radio Eng. Electron. Phys., 1965, 10, p.235.

19.9. Buck, D.C., and Symes, J.: "Stability of Magnetrons Operating Above the Cyclotron Frequency", Proc. IEEE, 1965, 53, p.540.

19.10. Raev, A., and Iliev, I.: "Pre-Oscillation in Heavily Cutoff, Small-Cathode, Magnetrons", Int. J. Electron., 1965, 19, p.31.

19.11. Kuraev, A.A., and Kuvshinov, Yu.N.: "Approximate Kinematic Analysis of Nonlinear Characteristics of the Helitron", Radiotekh. Elektron., 1965, 10, p.379, and Radio Eng. Electron. Phys., 1965, 10, p.322.

19.12. Gandhi, O.P., and Khandelwal, D.D.: "Improved Version of Split-Folded, Waveguide, Slow-Wave Structure", Int. J. Electron., 1965, 19, p.245.

19.13. Gandhi, O.P., and Khandelwal, D.D.: "Design Analysis of Split-Folded, Waveguide, Slow-Wave Structure", Int. J. Electron., 1965, 19, p.253.

19.14. Minich, V.V.: "Investigations of a Magnetron Exciting a H_{01} Wave in a Circular Waveguide", Radiotekh. Elektron., 1965, 10, p.1692, and Radio Eng. Electron. Phys., 1965, 10, p.1453.

19.15. Trifonov, Yu.M.: "Approximate Solution of the Steady-State Operation of a Multicavity Magnetron", Radiotekh. Elektron., 1965, 10, p.1453, and Radio Eng. Electron. Phys., 1965, 10, p.1252.

19.16. Groshkov, L.M., and Nechaev, V.E.: "Experimental Investigation of Electron Motion in an Oscillating Magnetron", Izv. VUZ Radiofiz., 1965, 8, p.413, and Sov. Radiophys., 1965, 8, p.297.

19.17. Ivanov, V.N.: "Effect of Side Cavities on the Frequency Spectrum of a Magnetron Anode Block", Radiotekh. Elektron., 1965, 10, p.1155, and Radio Eng. Electron. Phys., 1965, 10, p.993.

19.18. Sokolov, D.V., and Trubetskov, D.I.: "Effect of Non-Rectilinearities of Static Trajectories on the Operation of Magnetron-Type Electron-Beam Devices", Radiotekh. Elektron., 1965, 10, p.1540, and Radio Eng. Electron. Phys., 1965, 10, p.1330.

19.19. Megill, L.R.: "Excitation of Optical Radiation by High-Power-Density Radio Beams", J. Res. Natl. Bur. Stand., 1965, 69D, p.77.

19.20. Iliev, I., Nikolov, N., and Raev, A.: "Voltage-Tunable Oscillation in Small-Cathode Magnetrons", C. R. Acad. Bulg. Sci., 1966, 19, p.345.

19.21. Sasaki, A., and van Duzer, T.: "Coupled-Mode Analysis of Interactions in Crossed Fields", Trans. IEEE, 1966, ED-13, p.494.

19.22. Danilov, V.N.: "Theory of a Relativistic Magnetron", Radiotekh. Elektron., 1966, 11, no.12, pp.2160-75, and Radio Eng. Electron. Phys., 1966, 11, no.12, pp.1906-19.

19.23. Bayburin, V.B., and Sobolev, G.L.: "Analysis of Nonlinear Region of a Planar Multicavity Magnetron Taking Space Charge into Account", Radiotekh. Elektron., 1967, 12, no.3, pp.479-88, and Radio Eng. Electron. Phys., 1967, 12, no.3, pp.440-8.

19.24. Bayburin, V.B., and Sobolev, G.L.: "Design of Basic Electrical Parameters of Multicavity Magnetrons", Radiotekh. Elektron., 1967, 12, no.9, pp.1600-5, and Radio Eng. Electron. Phys., 1967, 12, no.9.

19.25. Tseitlin, M.B.: "Theory of Cylindrical M-Type Inverted Device", Radiotekh. Elektron., 1967, 12, no.12, pp.2250-1, and Radio Eng. Electron. Phys., 1967, 12, no.12.

19.26. Raev, G.Ya., et al.: "Electronics of a Single-Frequency Partly Cutoff Magnetron", Zh. Tekh. Fiz., 1967, 37, no.7, pp.1329-39, and Sov. Phys.-Tech. Phys., 1968, 13, no.7.

19.27. Filimonov, G.F.: "Adiabatic Method in Theory of Magnetron Devices", Zh. Tekh. Fiz., 1967, 37, no.7, pp.1320-8, and Sov. Phys.-Tech. Phys., 1968, 13, no.7.

19.28. Kulikov, M.N., and Stal'makhov, V.S.: "Experimental Study of the Diocotron Amplification in M-Type Electron Streams as a Function of Beam Thickness", Radiotekh. Elektron., 1968, 13, no.5, pp.935-7, and Radio Eng. Electron. Phys., 1968, 13, no.5.

19.29. Gourzo, V.V., Kulikov, M.N., and Stal'makhov, V.S.: "Nonresonance Coupling Elements in M-Type Electron Beams Realized on Rapid Cyclotron Waves", Izv. VUZ Radiofiz., 1968, no.8, pp.1236-48, and Sov. Radiophys., 1968, no.8.

19.30. Demassa, T.A., and Rowe, J.E.: "Space-Charge and Secondary-Emission Effects in Computer Simulation of Crossed-Field Distributed-Emission Amplifiers", Trans. IEEE, 1968, ED-15, no.2, pp.85-97.

19.31. Vural, B.: "Double-Stream Cyclotron-Wave Amplifier", Trans. IEEE, 1968, ED-15, no.1, pp.2-6.

19.32. Heathcote, V.A.: "Rugged Pulsed Magnetrons", GEC J. Sci. Technol., 1968, 35, no.1, pp.25-31.

19.33. Vaughan, J.R.M.: "Injection-Locked Voltage-Tunable Magnetron", Microwave J., 1968, 11, no.10, p.45.

19.34. Dvoryankin, V.V.: "Magnetron with Magnetic-Field Inclination", Ukr. Fiz. Zh., 1968, 13, p.1476.

19.35. Muller, Ya.N.: "Calculation of the Main Parameters of a Cascaded Amplifier of the Magnetron Type", Izv. VUZ Radioelektron., 1968, 11, p.922.

19.36. Borisov, D.G., and Gryzlov, A.I.: "Pulsed Magnetron Triggered by Trains of Sync Pulses", Prib. Tekh. Eksp., 1968, no.2, p.132, and Instrum. Exp. Tech., 1968, no.2, p.379.

19.37. Lewis, P.F.: "Experiment in the Parallel Operation of High-Power Magnetrons", Nachr. Tech. Fachber., 1968, 35, p.259.

19.38. Golombek, W., and Timmermans, F.: "Improvement of Stability of CW Magnetrons Against Load Reflections by Methods Outside the Tube", Nachr. Tech. Fachber., 1968, 35, p.264.

19.39. Kantorowicz, G.: "Measurements of a Controlled-Delay Crossed-Field Tube", Nachr. Tech. Fachber., 1968, 35, p.303.

19.40. Lebedev, I.V., et al.: "Magnetron Amplifier with Stepped Interaction Space", Izv. VUZ Radioelektron., 1968, 11, p.913.

19.41. Barsukov, K.A., and Svobodina, S.A.: "Cherenkov Radiation in a Helical Waveguide", Izv. VUZ Radiofiz., 1968, 11, p.1522.

19.42. Seshadri, S.R.: "Spectrum of Cherenkov Radiation in a Warm Anisotropic Plasma", Int. J. Electron., 1968, 25, p.401.

19.43. Bachheimer, J.P.: "Interaction (Smith-Purcell) Radiation of an Electron Beam with a Conducting Network", C. R. Acad. Sci. (Paris), 1969, 268B, p.599.

19.44. Neureuther, A.R., and Mittra, R.: "Self-Consistent Investigation of Smith-Purcell Radiation from a Narrow Tape Helix", Can. J. Phys., 1969, 15, p.435.

19.45. Betskii, O.V.: "Reduction of Starting Current in M-Type BWO's", Radiotekh. Elektron., 1969, 14, p.355, and Radio Eng. Electron. Phys., 1969, 14, no.2, pp.306-10.

19.46. Betskii, O.V., and Kokonchev, L.Z.: "Type-M BW Amplifier with Stepwise Interaction Space", Radiotekh. Elektron., 1969, 14, no.4, p.725, and Radio Eng. Electron. Phys., 1969, 14, no.4.

19.47. Sutherland, A.D.: "Tunable Cyclotron Oscillations in Smooth-Bore Magnetrons with Collision-Dominated Axial Transport", Proc. IEEE, 1969, 57, p.1325.

19.48. Roshal', A.S., Romanov, P.V., and Bavin, V.B.: "Statistical Theory of Transfer of Electrons in a Planar Magnetron with Continuous Anode", Radiotekh. Elektron., 1969, 14, no.3, p.552, and Radio Eng. Electron. Phys., 1969, 14, no.3, pp.478-81.

19.49. Kulikov, M.N., and Stal'makhov, V.S.: "Effect of a Beam of Finite Width on Starting Conditions of an M-Type Carcinotron", Izv. VUZ Radiofiz., 1969, 12, no.4, p.612.

19.50. Kulikov, M.N., and Stal'makhov, V.S.: "Spectrum of First Higher Kind of Signals in M-Type Carcinotrons", Radiotekh. Elektron., 1969, 14, no.6, p.1113, and Radio Eng. Electron. Phys., 1969, 14, no.6, pp.965-7.

19.51. Levin, G.Ya., et al.: "Single-Stage Electronically Tuned Magnetron", Radiotekh. Elektron., 1969, 14, no.7, p.1246, and Radio Eng. Electron. Phys., 1969, 14, no.7, pp.1077-80.

19.52. Barybin, A.A., and Gorin, Yu.N.: "Interaction of Cyclotron Waves in an Electron Beam with Travelling Waves", Radiotekh. Elektron., 1969, 14, no.7, p.1257, and Radio Eng. Electron. Phys., 1969, 14, no.7, pp.1087-92.

19.53. Lock, R.G.: "Non-Reentrant Crossed-Field Amplifier with Cycloidal Injected Beam", Trans. IEEE, 1969, ED-16, p.986.

19.54. Vyse, B.: "Observation of Pulsed-Magnetron Resonant Frequency in the Pre-oscillation Regime", Trans. IEEE, 1969, ED-16, p.1078.

19.55. Misra, V.K., and Wadhwa, R.P.: "Coupled-Mode Theory of Crossed-Field Instability", Proc. IEEE, 1969, 57, p.828.

19.56. Welsch, W.: "Nonlinear Phenomena in M-Type TWT's", Nachr. Tech. Z., 1969, 22, no.7, p.416.

19.57. Sutherland, A.D.: "Numerical Analysis of Smooth-Bore Magnetron with Mobility-Dominated Transport", Proc. IEEE, 1969, 57, p.1430.

19.58. Vigdorchik, V.I., and Kontorovich, V.M.: "Kinetic Theory of Static Regime of a Plane Magnetron", Izv. VUZ Radiofiz., 1969, 12, p.1882.

19.59. Tseitlin, M.B., and Tsitson, I.T.: "Nonlinear Theory of M-Type Beam Amplifier with Stepped Spatial Interaction", Izv. VUZ Radioelektron., 1969, 12, no.9, p.976.

19.60. Kotsarenko, N.Ya., et al.: "Amplification of Fast Waves by Rotating Electron Beam", Radiotekh. Elektron., 1969, 14, no.7, p.1277, and Radio Eng. Electron. Phys., 1969, 14, no.7, pp.1103-9.

19.61. Masunov, E.S.: "Radiation of a Charged Particle in a Waveguide with Dielectric Filling", Izv. VUZ Radiofiz., 1969, 12, no.10, p.1566.

19.62. Lepilov, V.A., and Neganov, V.A.: "Analysis of Starting Conditions in M-Carcinotrons with Negatively Charged Delay Line", Izv. VUZ Radioelektron.,1970,13,no.5,p.547.

19.63. Kato, Y., et al.: "Natural Cooling Technique of Magnetron", Natl. Tech. Rep., 1970, 16, no.1, p.97.

19.64. Vaughan, J.R.M.: "Magnetron Anode Erosion in Presence of Cathodes Containing Oxides", Trans. IEEE, 1970, ED-17, p.377.

19.65. Sobolev, G.L., and Vol'fson, A.O.: "Analysis of Dynamic Range of a Cylindrical Magnetron Taking Space Charge into Account", Radiotekh. Elektron., 1970, 15, p.130, and Radio Eng. Electron. Phys., 1970, 15, no.1, pp.105-10.

19.66. Bacal, M., Critescu, M., and Voci, C.: "Influence of Emission Velocities and Current Density on Cutoff Characteristic of a Plane Magnetron", Int. J. Electron., 1970, 28, p.149.

19.67. Dewan, E.M., and DeVito, P.A.: "Phase-Locked Magnetron and Lashinsky Spectrum", Proc. IEEE, 1970, 58, p.161.

19.68. Shukla, P.K., Singh, R.P., and Singh, R.N.: "Effect of Electron Collisions on Cherenkov Radiation from Magnetoplasma", Int. J. Electron., 1970, 28, p.421.

19.69. Kotsarenko, N.Ya., et al.: "Interaction of Half-Screw Electron Beam with Electromagnetic Waves Propagating Obliquely to the Magnetic Field", Izv. VUZ Radiofiz., 1970, 13, no.12, pp.1805-9.

19.70. Trubetskov, D.I., and Sharaevskii, Yu.P.: "Approximate Analysis of M-Type Tube in Large-Signal Regime", Radiotekh. Elektron., 1970, 15, no.11, and Radio Eng. Electron. Phys., 1970, 15, no.11, pp.2036-42.

19.71. Lebedev, I.V., and Meshkichev, V.N.: "Relationship between Saturation of Magnetron Oscillation and Threshold Power of M-Type Amplifiers", Radiotekh. Elektron., 1970, 15, no.12, and Radio Eng. Electron. Phys., 1970, 15, no.12, pp.2272-7.

19.72. Kawamura, Y.: "Analysis of Crossed-Field Tube Including Inhomogeneous Magnetic Field", Fujitsu Sci. Tech. J., 1970, 6, no.2, pp.45-74.

19.73. Bondarev, A.S., and Ovsepyan, Zh.M.: "Interaction of Transverse Electron-Beam Cyclotron Waves with a Resonant Bifilar Helical Coupler", Izv. VUZ Radioelektron., 1970, 13, no.8, pp.948-55.

19.74. Roshal', A.S., and Romanov, P.V.: "Statistical Model of Stationary States of a Planar Magnetron", Izv. VUZ Radioelektron., 1970, 13, no.9, pp.1092-8.

19.75. Petin, G.P., and El'bert, A.Ya.: "Theory of Magnetron-Type Devices", Izv. VUZ Radioelektron., 1970, 13, no.12, pp.1435-42.

19.76. Yulpatov, V.K.: "Excitation of Oscillations in the Hollow Resonator by a Relativistic Beam", Izv. VUZ Radiofiz., 1970, 13, no.12, pp.1784-8.

19.77. Kotsarenko, N.Ya., et al.: "Interaction of Cyclotron and Space-Charge Waves with Fast EM Waves", Radiotekh. Elektron., 1970, 15, no.12, and Radio Eng. Electron. Phys., 1970, 15, no.12, pp.2260-7.

19.78. Trubetskov, D.I., and Sharaevskii, Yu.P.: "Equation for Induced Field and Law of Energy Conservation in Nonlinear Theory of M-Type Beam Devices", Radiotekh. Elektron., 1971, 16, no.2, pp.442-3, and Radio Eng. Electron. Phys., 1971, 16, no.2, pp.377-8.

19.79. Kotsarenko, N.Ya., et al.: "Interaction of Polyhelical Beam of Electrons With Fast TM and TE Waves of a Cylindrical Guide", Radiotekh. Elektron., 1971, 16, no.2, pp.380-5, and Radio Eng. Electron. Phys., 1971, 16, no.2, pp.322-7.

19.80. Lepilov, V.A., and Neganov, V.A.: "Feasibility of Designing a Convection-Current-Signal Suppressor in M-Type Beam Devices Employing a Negatively-Charged Delay System", Izv. VUZ Radioelektron., 1971, 14, no.1, pp.27-33.

19.81. Nechaev, V.E., Rodionov, V.V., and Fuks, M.I.: "Linear Analysis of Waves in Cylindrical Magnetron Systems Oriented along the Axis", Izv. VUZ Radiofiz., 1971, 14, no.2, pp.317-22.

19.82. Tseitlin, M.B., et al.: "Study of Microwave Hybrid High-Power M-Type Oscillators", Izv. VUZ Radioelektron., 1971, 14, no.3, pp.235-42.

19.83. Azumi, A.: "Mode Instability of Magnetron", Fujitsu Sci. Tech. J., 1971, 7, no.1, pp.85-100.

19.84. Briggs, R.J.: "Nonlinear Behaviour of Transverse-Wave Amplifiers", J. Appl. Phys., 1971, 42, no.7, pp.2662-9.

19.85. Sharaevskii, Yu.P., and Aref'ev, Yu.A.: "Approximate Large-Signal Analysis of Output Characteristics of Sectionalized M-Type Tubes", Izv. VUZ Radioelektron., 1971, 14, no.5, pp.508-15.

19.86. Briggs, R.J., Paik, S.F., and Gottfried, A.H.: "Transverse-Wave Tubes as High-Efficiency Microwave Amplifiers", Trans. IEEE, 1971, ED-18, no.8, pp.511-20.

19.87. Rowe, J.F.: "Saturation Effects in Cyclotron Resonance Oscillators", Int. J. Electron., 1971, 31, no.1, pp.33-45.

19.88. Beck, A.H.W., and Mills, W.P.C.: "Millimetre-Wave Generator Working at Half the Cyclotron Resonance Frequency", Electron. Lett., 1971, 7, no.18, pp.533-4.

19.89. Adamchuk, A.S., and Naryshkina, L.G.: "Transient Radiation Theory of Resonator with Dielectric", Izv. VUZ Radiofiz., 1971, 14, no.8, pp.1260-7.

19.90. Ambrozhevich, V., and Raevski, L.: "Method of Designing the Amplitron Slow-Wave Structure", Pr. PIT, 1971, 21, no.72, pp.15-23.

19.91. Vilkova, L.P., Gaiduk, V.I., and Nefedov, Ye.I.: "Interaction of Nonaxisymmetric Tubular Helical Beam with Fast Waves in a Cylindrical Waveguide", Izv. VUZ Radioelektron., 1971, 14, no.9, pp.1009-20.

19.92. Filimonov, G.F.: "Methods of Nonlinear Analysis of M-Type Amplifiers", Izv. VUZ Radioelektron.,1971,14,no.9,pp.1042-54.

19.93. Bochkarev, V.V., and Sobolev, G.L.: "Analysis of Performance of Amplifiers with Distributed Cathodes", Izv. VUZ Radioelektron., 1971, 14, no.9, pp.1055-61.

19.94. Misra, V.K., and Wadhwa, R.P.: "Design Considerations for Improved Efficiency of M-Type BWO's", Trans. IEEE, 1971, ED-18, no.1, pp.35-42.

19.95. Okabe, T., Okamura, S., and Tanaka, I.: "Production of 1.3-mm Waves with Waveguide Magnetrons", Rep. Fac. Eng. Shizuoka Univ., 1971, no.22, pp.103-8, and Trans. IEEE, 1971, ED-18, no.1, pp.42-5.

19.96. Lebedev, I.V., and Kondratovskii, I.S.: "M-Type Amplifier Employing a Staggered Negative Electrode", Izv. VUZ Radioelektron., 1971, 14, no.9, pp.1062-6.

19.97. Tseitlin, M.B., Betskii, O.V., and Tsitson, I.T.: "Large-Amplitude Analysis of M-Type Two-Beam Staggered Amplifier", Izv. VUZ Radioelektron., 1971, 14, no.9,pp.1067-74.

19.98. Makarov, V.N.: "Threshold Input Signals of an M-Type Amplifier with Secondary-Emission Cathode", Izv. VUZ Radioelektron., 1971, 14, no.9, pp.1075-81.

19.99. Tereshchenko, A.I., and Shein, A.G.: "Comparitive Study of Frequency Stability of Magnetrons with Different Anode Configurations", Radiotekh. (Kharkov), 1971, no.17, pp.3-6.

19.100. Zhdanov, N.N., and Starostenko, V.V.: "Dispersion and Amplitude Spectrum of Spatial Harmonics in Rising-Sun Comb Structures", Radiotekh. (Kharkov), 1971, no.17, pp.22-7.

19.101. Vanke, V.A., and Timofeev, Yu.M.: "Saturation of Amplification in a Cyclotron Synchronous-Wave Amplifier", Izv. VUZ Radiofiz., 1971, 15, no.4, pp.615-24.

19.102. Kuznetsov, M.I., Zheleztsova, I.N., and Stepanov, S.V.: "Natural Oscillations of an Electron Cloud in a Cylindrical Magnetron", Izv. VUZ Radiofiz., 1972, 15, no.5, pp.787-91.

19.103. Berbasov, V.A., et al.: "Experimental Verification of Static Synchronous Electron Cloud in Pre-Oscillation Magnetron", Izv. VUZ Radiofiz., 1972, 15, no.6,pp.944-7.

19.104. Tsvyk, A.I., and Tsvyk, L.I.: "Excitation of Electromagnetic Waves by an Electron Beam Moving Sinusoidally above a Diffraction Grating", Radiotekh. (Kharkov), 1972, no.20, pp.24-32.

19.105. Balaklitskii, I.M., et al.: "Operational Features of (Electron-Beam) Diffraction Generators", Radiotekh. (Kharkov), 1972, no.20, pp.208-15.

19.106. Skrynnik, B.K.: "Study of Spectra of (Electron-Beam) Diffraction Generators", Radiotekh. (Kharkov), 1972, no.20, pp.216-22.

19.107. Muller, A.: "Measurements on an M-Type BW Amplifier Tube with Two Delay Lines", Arch. Elektron. Ubertrag., 1972, 26, no.10, pp.449-51.

19.108. Trubetskov, D.I., et al.: "Theory of Signal Suppression Effect in an M-Type TWT Using Beam Premodulation", Izv. VUZ Radioelektron., 1972, 15, no.8, pp.1007-14.

19.109. Buchegger, O., and Wolfram, R.: "Experimental Demonstration of Axial Propagation of Slow Waves in a Cylindrical Magnetron", Arch. Elektron. Ubertrag., 1972, 26, no.11, pp.493-500.

19.110. Gusarov, V.N., and Serebryakov, Yu.N.: "Radiation of a Charged Particle in a Semiinfinite Iris-Loaded Waveguide", Zh. Tekh. Fiz., 1972, 17, no.5, pp.991-4, and Soviet Phys.-Tech. Phys., 1972, 17, no.5, pp.785-7.

19.111. Shein, A.G.: "Calculating the Transient Characteristics of Magnetrons", Radiotekh. (Kharkov), 1972, no.22, pp.75-82.

19.112. Bayburin, V.B.: "Calculation of Electron Trajectories in a Multicavity Magnetron Taking Space Charge into Account", Radiotekh. Elektron., 1972, 17, no.3, pp.645-7, and Radio Eng. Electron. Phys., 1972, 17, no.3, pp.507-10.

19.113. Kietzmann, H.: "Investigation into Phase Response Characteristics of a Platinotron", Nachr. Tech. Z., 1972, 25, no.12, pp.560-3.

19.114. Prokof'yev, L.V., and Skobelkin, V.I.: "Stability of the Amplitron", Radiotekh. Elektron., 1972, 17, no.1, pp.119-26, and Radio Eng. Electron. Phys., 1972, 17, no.1, pp.89-95.

19.115. Kulke, B.: "Limitations on Millimetre-Wave Power Generation with Spiralling Electron Beams", Trans. IEEE, 1972, ED-19, no.1, pp.71-9.

19.116. Ohguchi, Y., and Yamamoto, K.: "Turbulent Diffusion in Smooth-Bore Magnetron", J. Appl. Phys., 1972, 43, no.3, pp.1307-8.

19.117. Kuznetsov, M.I., Berbasov, V.A., and Zheleztsova, I.N.: "Pre-Oscillation Charge Distribution in a Cylindrical Magnetron", Izv. VUZ Radiofiz., 1972, 15, no.2, pp.283-90.

19.118. Mergelyan, O.S.: "Radiation from a Charged Particle Moving Near a Dielectric Medium with Periodic Variations in Density", Izv. Akad. Nauk Arm. SSR Fiz., 1972, 7, no.5, pp.329-39.

19.119. Mergelyan, O.S.: "Field of a Charged Particle Moving Along a Channel in a Laminar Medium", Izv. Akad. Nauk Arm. SSR Fiz., 1972, 7, no.5, pp.324-8.

19.120. Levin, G.Ya., and Vigdorchik, V.I.: "Effect of Emission on Properties of a Planar Magnetron", Radiotekh. Elektron., 1972, 17, no.8, pp.1762-5, and Radio Eng. Electron. Phys., 1972, 17, no.8, pp.1400-4.

19.121. Korolev, V.I.: "Beat Regime of Generator Synchronization by an External Sinusoidal Force", Izv. VUZ Radiofiz., 1972, 15, no.10, pp.1527-37.

19.122. Farenik, V.I., et al.: "Experimental Investigation of Kinetic Instabilities in the Gabor Magnetron", Zh. Tekh. Fiz., 1972, 42, no.8, pp.1625-8, and Sov. Phys.-Tech. Phys., 1973, 17, no.8, pp.1298-300.

19.123. Kuteev, B.V., and Sominskii, G.G.: "Space-Charge Accumulation in Crossed-Field Devices", Zh. Tekh. Fiz., 1972, 42, no.8, pp.1761-3, and Sov. Phys.-Tech. Phys., 1973, 17, no.8, pp.1407-9.

19.124. Katyshev, E.G., et al.: "Emission from a Modulated Electron Beam", Zh. Tekh. Fiz., 1972, 42, no.11, p.2446, and Sov. Phys.-Tech. Phys., 1973, 17, no.11, p.1905.

19.125. Alferov, D.F., Bashmakov, Yu.A., and Bessnov, E.G.: "Radiation of Relativistic Particles in an Undulator", Zh. Tekh. Fiz., 1972, 42, no.9, pp.1921-6, and Sov. Phys.-Tech. Phys., 1973, 17, no.9, pp.1540-3.

19.126. Barsukov, K.A., and Kadantsev, V.N.: "Radiation of a Charged Particle Moving in a Radially Inhomogeneous Medium", Zh. Tekh. Fiz., 1972, 42, no.9, pp.1927-30, and Sov. Phys.-Tech. Phys., 1973, 17, no.9, pp.1544-6.

19.127. Lagranskii, L.M., et al.: "Amplitude and Phase Characteristics of an Injected Beam, Single-Stage, M-Type Amplifier", Izv. VUZ Radiofiz., 1973, 16, no.5, pp.63-70.

19.128. Gerard, W.A.: "Frequency-Agile Coaxial Magnetrons", Microwave J., 1973, 16, no.3, pp.29-33.

19.129. Lagranskii, L.M., et al.: "Amplitude and Phase Characteristics of an M-Type TWT Amplifier", Izv. VUZ Radioelektron., 1973, 16, no.5, pp.63-70.

19.130. Vaughan, J.R.M.: "Model for Calculation of Magnetron Performance", Trans. IEEE, 1973, ED-20, no.9, pp.818-26.

19.131. Nosko, B.F.: "Inertialess Fine Frequency Adjustment in Magnetrons", Radiotekh. (Kharkov), 1973, no.24, pp.162-6.

19.132. Eremka, V.D., Kirichenko, A.Ya., and Protsai, V.F.: "Electron Bunch Radiation above a Periodic Structure", Izv. VUZ Radiofiz., 1973, 16, no.10, pp.1599-1604.

19.133. Vanke, V.A., and Timofeev, Yu.M.: "Nonlinear Phenomena in Cyclotron Synchronous-Wave Amplifier", Izv. VUZ Radiofiz., 1973, 16, no.11, pp.1751-8.

19.134. Stevenson, J.R., Ellis, H., and Bartlett, R.: "Synchrotron Radiation as an IR Source", Appl. Opt., 1973, 12, no.12, pp.2884-9.

19.135. Kalinin, B.N., et al.: "Polarization of a Coherent Bremsstrahlung Beam from an Electron Synchrotron", Prib. Tekh. Eksp., 1973, 16, no.3, pp.24-7, and Instrum. Exp. Tech., 1973, 16, no.3, pp.684-6.

19.136. Levin, Yu.I., Trubetskov, D.I., and Shevchik, V.N.: "Approximate Linear Theory of Magnetron Beam Devices with a Nonuniform Magnetic Field", Izv. VUZ Radioelektron., 1973, 16, no.10, pp.106-15.

19.137. Il'in, S.N., Sobolev, G.L., and Rodina, L.P.: "Injection-Locked Voltage-Tunable Magnetron", Izv. VUZ Radioelektron., 1973, 16, no.10, pp.116-21.

19.138. Vyrovoi, S.I., and Rapoport, G.N.: "Study of Output Stage of a Gyrotwistron", Izv. VUZ Radioelektron., 1973, 16, no.10, pp.96-105.

19.139. van den Berg, P.M.: "Smith-Purcell Radiation from a Point Charge Moving Parallel with a Reflection Grating", J. Opt. Soc. Am., 1973, 63, no.12, pp.1588-97.

19.140. Muller, Ya.N., and Shlifer, E.D.: "Lower Limite of Generation in a Multicavity Magnetron", Izv. VUZ Radioelektron., 1973, 16, no.12, pp.23-9.

19.141. Vol'fson, A.O., and Sobolev, G.L.: "Averaged Electron Trajectories in a Magnetron for Large Space Charge", Radiotekh. Elektron., 1973, 18, no.8, pp.1283-4, and Radio Eng. Electron. Phys., 1973, 18, no.8, pp.1283-4.

19.142. Shmat'ko, Ye.I., and Zhdanov, N.N.: "Delay Structures for Crossed-Field TWT's", Radiotekh. (Kharkov), 1973, no.27, pp.128-32.

19.143. Shein, A.G.: "Parametric Interaction in Beam-Type Crossed-Field Devices", Radiotekh. (Kharkov), 1973, no.27, pp.81-8.

19.144. Winterberg, F.: "Possibility of Generating Powerful Bursts of Microwaves by Interaction of an Intense Relativistic Electron Beam with a Magnetized Plasma", Atomkernenergie, 1973, 22, no.3, pp.205-7.

19.145. Bondarev, A.S., and Kantyuk, S.P.: "Dynamic Desynchronization in Cyclotron Microwave Amplifiers Allowing for Space Charge", Izv. VUZ Radioelektron., 1973, 16, no.12, pp.30-8.

19.146. Beck, A.H.W.: "Millimetre-Wave Generator Using a Spiralling Electron Beam", Proc. IEE, 1973, 120, no.2, pp.197-205.

19.147. Lalor, E.: "Three-Dimensional Theory of Smith-Purcell Effect", Phys. Rev., 1973, 7A, no.2, pp.435-46.

19.148. Skowron, J.F.: "Continuous-Cathode (Sole) Crossed-Field Amplifier", Proc. IEEE, 1973, 61, no.3, pp.330-56.

19.149. Petelin, M.I., and Smorgonskii, A.V.: "Ubitron Nonlinear Theory", Izv. VUZ Radiofiz., 1973, 16, no.1, pp.294-304.

19.150. Shein, A.G., Sova, A.V., and Starostenko, V.V.: "Nonlinear Theory of a Three-Dimensional Crossed-Field TWT", Radiotekh. (Kharkov), 1973, no.27, pp.74-81.

19.151. Carmel, Y., and Nation, J.A.: "Application of Intense Relativistic Electron Beams to Microwave Generation", J. Appl. Phys., 1973, 44, no.12, pp.5268-74.

19.152. Kovalev, I.S., Kolosov, S.V., and Kuraev, A.A.: "Calculation of Transverse Electric Fields in Axisymmetric Gyroresonant Devices", Radiotekh. Elektron., 1973, 18, no.7, pp.1525-8, and Radio Eng. Electron. Phys., 1973, 18, no.7, pp.1128-31.

19.153. Friedman, M., and Herndon, M.: "Emission of Coherent Microwave Radiation from a Relativistic Electron Beam Propagating in a Spatially Modulated Field", Phys. Fluids, 1973, 16, no.11, pp.1982-95.

19.154. Rusbuildt, D., and Thimm, K.: "Synchrotron as Radiation Standard for the Vacuum UV", Nucl. Instrum. Methods, 1974, 116, no.1, pp.125-40.

19.155. Zhdanov, N.N., Mikhaleva, A.I., and Shein, A.G.: "Interference between Signal Modes in Magnetron Resonators", Izv. VUZ Radioelektron., 1974, 17, no.2, pp.101-4.

19.156. Sprangle, P.: "Fast-Wave Interaction with a Relativistic Electron Beam", J. Plasma Phys., 1974, 11, pt 2, pp.299-309.

19.157. Beck, A.H.W., and Caldwell-Nichols, C.J.: "Oscillator for 2-mm Waveband", Electron. Lett., 1974, 10, no.10, pp.171-2.

19.158. Granatstein, V.L., et al.: "Coherent Synchrotron Radiation from an Intense Relativistic Electron Beam", J. Quantum Electron. IEEE, 1974, QE-10, no.9, pp.651-4.

19.159. Schneider, S., and Spitzer, R.: "Interaction of Coherent Electromagnetic Waves with Relativistic Electrons in a Medium", Nature, 1974, 250, no.5468, pp.643-5.

19.160. Manheimer, W.M., and Ott, E.: "Theory of Microwave Generation by an Intense Relativistic Electron Beam in a Rippled Magnetic Field", Phys. Fluids, 1974, 17, no.2, pp.463-73.

19.161. Vorontsov, V.I., and Mausunbaev, S.S.: "Theory of Cyclotron-Resonance Maser", Poluprov. Tekh. Mickroelektron., 1974, no.6, pp.41-8.

19.162. Bykov, Yu.V., Gaponov, A.V., and Petelin, M.I.: "Theory of a TW, Transverse-Beam, Maser", Izv. VUZ Radiofiz., 1974, 17, no.8, pp.1219-23.

19.163. Bratman, V.I., and Tokarev, A.E.: "Linear Theory of Relativistic Cyclotron-Resonance Maser", Izv. VUZ Radiofiz., 1974, 17, no.8, pp.1224-8.

19.164. Carmel, Y., et al.: "Intense Coherent Cherenkov Radiation due to Interaction of a Relativistic Electron Beam with a Slow-Wave Structure", Phys. Rev. Lett., 1974, 33, no.21, pp.1278-82.

19.165. Roshal', A.S., and Romanov, P.V.: "Signal Generation Processes in Magnetron Oscillators", Izv. VUZ Radioelektron., 1974, 17, no.8, pp.63-8.

19.166. Kiselev, V.A.: "Conditions for Self-Excitation of Surface Waves in an Activated Waveguide with a Sinusoidally Corrugated Surface", Kvantovaya Elektron., 1974, 1, no.2, pp.441-4, and Sov. J. Quantum Electron., 1974, 4, no.2, pp.252-3.

19.167. Shein, A.G., and Gerasimov, V.P.: "Multifrequency Operation of Magnetron-Type TWT's", Izv. VUZ Radioelektron., 1974, 17, no.11, pp.64-8.

19.168. Lagranskii, L.M., and Usherovich, B.L.: "Synchronization of M-Type BWO's by External Harmonic Signals", Izv. VUZ Radioelektron., 1974, 17, no.11, pp.69-73.

19.169. Levin, Yu.I., and Nikulin, S.B.: "Application of Quate-Mullen Series to a Linear Theory of Magnetron Beam Devices", Izv. VUZ Radioelektron., 1974, 17, no.9, pp.15-9.

19.170. Pivovarova, A.G., Pospelov, L.A., and Saenko, V.I.: "Calculation of Displacement of an Electron in a Two-Section Magnetron", Radiotekhnika (Kharkov), 1974, no.28, pp.55-60.

19.171. Vyrovoi, S.I., and Rapoport, G.N.: "Bandwidth of the Output Stage of a Gyrotwistron", Izv. VUZ Radioelektron., 1974, 17, no.9, pp.96-8.

19.172. Muller, Ya.N., and Shlifer, E.D.: "Comparison of Operating Modes of Magnetron Oscillators", Izv. VUZ Radioelektron., 1974, 17, no.1, pp.39-48.

19.173. Dement'ev, A.S., and Pavlenko, Yu.G.: "Microwave Maser with Nonrelativistic Electrons", Zh. Tekh. Fiz., 1974, 44, no.5, pp.1101-3, and Sov. Phys.-Tech. Phys., 1974, 19, no.5, pp.694-5.

19.174. Granatstein, V.L., et al.: "Strong Submillimetre Radiation from Intense Relativistic Electron Beams", Trans. IEEE, 1974, MTT-22, no.12, pp.1000-5.

19.175. Kvochka, V.I., et al.: "Extraction (by Reflector) of Synchrotron Radiation", Zh. Tekh. Fiz., 1974, 44, no.6, pp.1210-7, and Sov. Phys.-Tech. Phys., 1974, 19, no.6, pp.762-6.

19.176. Zarnitsina, I.G., and Nusinovich, G.S.: "Stability of Single-Mode Oscillations in a Gyromonotron", Izv. VUZ Radiofiz., 1974, 17, no.12, pp.1858-67.

19.177. Alekseev, G.A.: "Synchronization of a Magnetron by an Electron Beam with Multiple Frequency Conversion", Radiotekh. (Kharkov), 1974, no.30, pp.159-65.

20. THEORY OF STIMULATED EMISSION

20.1. Greenstein, H.: "Theory of a Single-Mode Gas Laser", Phys. Rev., 1968, 175, p.438.

20.2. Bauer, A.: "Kirchhoff's and Lambert's Laws with Stimulated Emission", Lichttechnik, 1968, 20, no.12, p.146a.

20.3. Perel', V.I., and Rogova, I.V.: "Theory of Lasers with Additional Mirror. I", Opt. Spekt., 1968, 25, no.5, p.716, and Opt. Spectrosc., 1968, 25, no.5, p.401.

20.4. Rogova, I.V.: "Competition of Generation in Two Transitions with Same Upper State", Opt. Spekt., 1968, 25, p.401, and Opt. Spectrosc., 1968, 25, p.217.

20.5. Rubinov, A.N., and Mikhnov, S.A.: "Effect of Stationary Waves on Characteristics of a Laser with Plane Mirrors", Opt. Spekt., 1969, 25, p.903, and Opt. Spectrosc., 1969, 25, p.502.

20.6. Gnutzmann, U.: "Quantum-Mechanical Calculation of Multitime-Correlation Functions", Z. Phys., 1969, 225, no.5, p.416.

20.7. Hofelich-Abate, E., and Hofelich, F.: "Theory of Damped and Undamped Intensity Oscillations in Lasers", Z. Phys., 1969, 221, p.362.

20.8. Haken, H.: "Exact Stationary Solution of a Fokker-Planck Equation for Multimode Laser Action Including Phase Locking", Z. Phys., 1969, 219, p.246.

20.9. Kostanyan, R.B., and Pogosyan, P.S.: "Effect of Coherence on Process of Light Amplification", Radiotekh. Elektron., 1969, 14, no.4, p.730, and Radio Eng. Electron. Phys., 1969, 14, no.4, pp.630-3.

20.10. Schwarz, S.E., and Gordon, P.L.: "Hamilton's Principle and Maximum-Emission Coincidence", J. Appl. Phys., 1969, 40, p.4441.

20.11. Arutyunyan, V.M., and Melikyan, A.O.: "Analysis of Q-Switched Lasers", Radiotekh. Elektron., 1969, 14, no.10, p.1901, and Radio Eng. Electron. Phys., 1969, 14, no.10, pp.1645-7.

20.12. Karlov, N.V., and Konev, Yu.B.: "Theory of a Laser Pumped by Another Laser in Same Resonator", Radiotekh. Elektron., 1969, 14, no.10, p.1906, and Radio Eng. Electron. Phys., 1969, 14, no.10, pp.1650-2.

20.13. Stenholm, S., and Lamb, W.E.: "Semiclassical Theory of a High-Intensity Laser", Phys. Rev., 1969, 181, p.618.

20.14. Kovalenko, Ye.S., and Shangina, L.I.: "Theory of Laser with Inhomogeneous Pumping", Izv. VUZ Radiofiz., 1969, 12, no.6, p.846.

20.15. Il'inov, M.P., and Il'inova, T.M.: "Wave Processes in a Laser Amplifier Taking into Account the Transverse Distribution of Population Inversion", Radiotekh. Elektron., 1969, 14, no.3, p.478, and Radio Eng. Electron. Phys., 1969, 14, no.3, pp.413-9.

20.16. Tuchin, V.V., and Sedel'nikov, V.A.: "Random Spectral Width of Gas-Laser Radiation", Izv. VUZ Radioelektron., 1969, 12, no.3, p.221.

20.17. Armstrong, J.A., and Courtens, E.: "Pi-Pulse Propagation in the Presence of Host Dispersion", J. Quantum Electron. IEEE, 1969, QE-5, p.249.

20.18. Statz, H., and Bass, M.: "Locking in Multimode Solid-State Lasers", J. Appl. Phys., 1969, 40, p.377.

20.19. Perel', V.I., and Rogova, I.V.: "Theory of Lasers with Additional Mirror. II", Opt. Spekt., 1969, 25, no.6, p.943, and Opt. Spectrosc., 1969, 25, no.6, p.520.

20.20. Cubeddu, R.R., and Svelto, O.: "Theory of Laser Self-Locking in the Presence of Host Dispersion", J. Quantum Electron. IEEE, 1969, QE-5, p.495.

20.21. Kotomtseva, L.A., and Samson, A.M.: "Rise Period and Spectral Width of a Monopulse in a Laser with a Transparent Filter", Zh. Prikl. Spekt., 1969, 10, no.3, p.443.

20.22. Brunner, W., and Paul, H.: "Quantum-Mechanical Treatment of Lasers with Non-resonant Feedback", Ann. Phys. (Leipzig), 1969, 23, p.384, 23, p.152, and 24, p.38.

20.23. Allen, L.: "Comparison of Semi-Classical and Quantum Theories for a Single-Mode Laser", Proc. Phys. Soc., 1969, 2A, p.433.

20.24. Comly, J.C., Yariv, A., and Garmire, E.M.: "Stable, Chirped, Ultrashort Pulses in Lasers Using the Optical Kerr Effect", Appl. Phys. Lett., 1969, 15, p.148.

20.25. Lamb, G.L.: "Pi-Pulse Propagation in a Lossless Amplifier", Phys. Lett., 1969, 29A, p.507.

20.26. Riska, D.O., and Stenholm, S.: "Influence of Mode Structure on the Quantum Theory of the Laser", Phys. Lett., 1969, 30A, p.16.

20.27. Haake, F.: "Non-Markovian Effects in the Laser", Z. Phys., 1969, 227, no.2, p.179.

20.28. Perel', V.I., and Rogova, I.V.: "Types of Oscillation and Threshold Conditions in a Three-Mirror Resonator", Zh. Tekh. Fiz., 1969, 39, no.3, p.513, and Sov. Phys.-Tech. Phys., 1969, 14, no.3, pp.379-82.

20.29. Korzhenevich, I.M., and Ratner, A.M.: "Decay of Laser Relaxation Oscillations Caused by Diffusion of Excitations", Opt. Spekt., 1969, 26, no.1, p.108, and Opt. Spectrosc., 1969, 26, no.1, p.57.

20.30. Ratner, A.M., and Chernov, V.S.: "Kinetics of Radiation of a Q-Switched Laser. II", Zh. Tekh. Fiz., 1969, 39, no.5, p.885, and Sov. Phys.-Tech. Phys., 1969, 39, no.5, pp.662-8.

20.31. Smith, S.W.: "Effect of Host Dispersion on Mode-Locked Lasers", Appl. Phys. Lett., 1969, 15, p.194.

20.32. Ernst, V.: "Origin of Laser Coherence", Z. Phys., 1969, 229, no.3-5, p.432.

20.33. Beterov, I.M., and Chebotaev, V.P.: "Three-Level Gas Laser", Zh. Eksp. Teor. Fiz. Pis'ma, 1969, 9, no.4, p.216, and JETP Lett., 1969, 9, no.4, p.127.

20.34. Doronin, V.G., and Ostapchenko, E.P.: "Calculating the Radiation Intensity of a Pulsed Gas Laser", Zh. Prikl. Spekt., 1969, 11, p.805.

20.305. Matveev, V.I.: "Output Power of Gas Lasers", Izv. VUZ Fiz., 1969, no.3, p.149.

20.36. Bambini, A., and Burlamacchi, P.: "Analysis of Gain Interaction in a Multimode Gas Laser", J. Appl. Phys., 1969, 40, p.4424.

20.37. Magdich, L.N.: "Temporal Characteristics of Laser Radiation with Modulation Resonance Parameters", Zh. Tekh. Fiz., 1969, 39, no.3, p.578, and Sov. Phys.-Tech. Phys. 1969, 14, no.3, pp.383-90.

20.38. Odintsov, A.I., et al.: "Influence of Spatial Inhomogeneity of Laser Field on the Saturation of Amplification", Zh. Tekh. Fiz., 1969, 39, no.5, p.879, and Sov. Phys.-Tech. Phys., 1969, 14, no.5, pp.657-61.

20.39. Gurevich, G.L., and Otmakhov, Yu.A.: "Nonstationary Processes in TW Laser with Saturation Filter", Izv. VUZ Radiofiz., 1969, 12, no.2, p.208.

20.40. Bonifacio, R.: "Short-Pulse Propagation in a Laser Amplifier with Saturable Losses", Lett. Nuovo Cim., 1969, 1, no.14, p.671.

20.41. Bakeev, A.A., and Cheburkin, N.V.: "Natural Frequencies of a Three-Mirror Resonator", Radiotekh. Elektron., 1969, 14, no.7, p.1302, and Radio Eng. Electron. Phys., 1969, 14, no.7, pp.1125-30.

20.42. Sibert, E., and Tittel, F.: "Superfluorescence Limitations on the Inversion Process in Large-Aperture Laser Amplifiers", J. Appl. Phys., 1969, 40, p.4434.

20.43. Akanaev, B.A.: "Amplification of Ultrashort Optical Pulses in a Mixed Active Medium", Radiotekh. Elektron., 1969, 14, no.8, p.1521, and Radio Eng. Electron. Phys., 1969, 14, no.8, pp.1321-3.

20.44. Sanderson, R.L., and Streifer, W.: "Laser Resonators with Tilted Reflectors", Appl. Opt., 1969, 8, p.2241.

20.45. Booth, D.J., and Troup, G.J.: "Saturation-Induced Phase Distortion in Laser Amplifiers", J. Quantum Electron. IEEE, 1969, QE-5, p.547.

20.46. Schonnagel, H.: "Influence of Gain in the Radial Profile of a Laser with Aperture Coupling", Monatsber. Dtsch. Akad. Wiss. Berlin, 1969, 11, no.5-6, p.346.

20.47. Pokasov, V.V., and Khmelyevtsov, S.S.: "Relationship between Spatial Coherence of Laser Radiation and the Number of Generated Frequencies", Opt. Spekt., 1969, 25, p.945, and Opt. Spectrosc., 1969, 25, p.522.

20.48. Kabaev, N.I.: "Contour of Output Dip in a Gas Laser in the Presence of Elastic Collisions", Zh. Prikl. Spekt., 1969, 10, no.5, p.753.

20.49. Davidenko, D.F., Zubarev, T.N., and Tarasov, Yu.A.: "Pulsations of Radiation Power in Lasers", Opt. Spekt., 1969, 26, no.5, p.801, and Opt. Spectrosc., 1969, 26, no.5, p.437.

20.50. Valignat, S., and Erbeia, A.: "Spatial Coherence of a Laser Beam", Opt. Commun., 1969, 1, no.3, p.135.

20.51. Icsevgi, A., and Lamb, W.E.: "Propagation of Light Pulses in a Laser Amplifier", Phys. Rev., 1969, 185, p.517.

20.52. Grutter, A.A., Weber, H.P., and Dandliker, R.: "Imperfectly Mode-Locked Laser Emission and its Effects on Nonlinear Optics", Phys. Rev., 1969, 185, p.629.

20.53. Poizner, B.N.: "Aspects of Two-Mode Behaviour of Externally-Synchronized Gas Lasers", Radiotekh. Elektron., 1969, 14, no.12, p.2179, and Radio Eng. Electron. Phys., 1969, 14, no.12, pp.1886-91.

20.54. Chechenina, E.P., and Chekalinskaya, Yu.I.: "Frequency (Gain) Characteristics of a Linear Amplifier with Plain and Complex Resonators", Zh. Prikl. Spekt., 1969, 11, no.2, p.242.

20.55. Carruthers, J.A., and Bieber, T.: "Pulse Velocity in a Self-Locked Laser", J. Appl. Phys., 1969, 40, p.426.

20.56. Laussade, J.P., and Yariv, A.: "Analysis of Mode Locking and Ultrashort Laser Pulses with Nonlinear Refractive Index", J. Quantum Electron. IEEE, 1969, QE-5, p.435.

20.57. Mal'ishev, V.I., et al.: "Evaluation of Limit Duration of Ultrashort Pulses in Ruby and Nd^{3+}:Glass Lasers with Passive Shutter", Zh. Prikl. Spekt., 1969, 11, no.4, p.655.

20.58. Apanasevich, P.A., et al.: "Method of Normal Vibrations in Theory of Laser Mode Interaction", Zh. Prikl. Spekt., 1969, 11, no.4, p.644.

20.59. Basov, N.G., and Morozov, V.N.: "Theory of Dynamics of Injection Lasers", Zh. Eksp. Teor. Fiz., 1969, 57, no.2, p.617, and Sov. Phys.-JETP, 1970,30,no.2,pp.338-43.

20.60. Zel'dovich, B.Ya., Perelomov, A.M., and Popov, V.S.: "Relaxation of Lasers in Presence of an External Force", Zh. Eksp. Teor. Fiz., 1969, 57, no.1, p.196, and Sov. Phys.-JETP, 1970, 30, no.1, pp.111-6.

20.61. Kuznetsova, T.K.: "Statistics of Ultrashort Light Pulses in Laser with Bleachable Absorber", Zh. Eksp. Teor. Fiz., 1969, 57, no.5, p.1673, and Sov. Phys.-JETP, 1970, 30, no.5, pp.904-9.

20.62. Popov, V.S., and Perelomov, A.M.: "Parametric Excitation of a Laser. II", Zh. Eksp. Teor. Fiz., 1969, 57, no.5, p.1684, and Sov. Phys.-JETP, 1970, 30, no.5,pp.910-3.

20.63. Allen, L., and Peters, G.I.: "Superradiance, Coherence Brightening, and Amplified Spontaneous Emission", Phys. Lett., 1970, 31A, p.95.

20.64. Sayers, M.D., and Allen, L.: "Amplitude, Competition, Self-Locking, Beat Frequency, and Time Development, in a Three-Mode Gas Laser", Phys. Rev., 1970, 1A,p.1730.

20.65. Riska, D.O., and Stenholm, S.: "Quantum Theory of a Laser with Inhomogeneous Broadening", Proc. Phys. Soc.,1970,3A,p.189.

20.66. Diels, J.C.: "Light-Pulse Propagation in Homogeneously Broadened Amplifiers", Phys. Lett., 1970, 31A, p.26.

20.67. Gordon, J.P., and Aslaksen, E.W.: "Intensity Distribution for a Q-Switched Laser", J. Quantum Electron. IEEE, 1970, QE-6, p.428.

20.68. Schwartz, J., Weiler, W., and Chang, R.K.: "Laser-Pulse Shaping Using Absorption with Controlled Q-Switch Time Behaviour", J. Quantum Electron. IEEE, 1970, QE-6, p.442.

20.69. Chekalinskaya, Yu.I., and Chechenina, E.P.: "Amplifying a Wide-Spectrum Composition Signal with a Laser Amplifier", Zh. Prikl. Spekt., 1970, 12, no.4, p.657.

20.70. Kovalenko, E.G.: "Theory of Mode Synchronization in Solid-State Lasers", Izv. VUZ Fiz., 1970, no.1, p.65.

20.71. Grigor'yants, V.V.: "Analysis of Luminescence from the Active Substance of a Free-Running Laser", Zh. Eksp. Teor. Fiz., 1970, 58, no.5, p.1593, and Sov. Phys.-JETP, 1970, 31, no.5, pp.853-9.

20.72. Davis, L.W.: "Field of an AM Phase-Locked Laser", Am. J. Phys., 1970, 38, p.918.

20.73. Tsikunov, V.N.: "Resonance Pheno-mena During Forced Oscillations of Laser Intensity", Zh. Eksp. Teor. Fiz., 1970, 58, no.5, p.1646, and Sov. Phys. JETP, 1970, 31, no.5, pp.883-5.

20.74. Domelunksen, V.G., and Tolchinskaya, T.B.: "Laser Oscillations in a Three-Mirror Non-Coordinated Resonator", Opt. Spekt., 1970, 28, no.1, p.183, and Opt. Spectrosc., 1970, 28, no.1, p.95.

20.75. Svelto, O.: "Interpretation of Subpicosecond Structure and Frequency Chirp of Mode-Locked Lasers", Appl. Phys. Lett., 1970, 17, p.83.

20.76. Privalov, V.E.: "Mode Volume and Radiation Power of a Laser", Opt. Spekt., 1970, 28, no.3, p.524, and Opt. Spectrosc., 1970, 28, no.3, p.281.

20.77. Kochelap, V.A., and Pekar, S.I.: "Kinetics of Lasers Based on Light-Induced Chemical Reaction in Stationary Regime", Ukr. Fiz. Zh., 1970, 15, no.7, p.1057.

20.78. Graham, R., and Haken, H.: "Laser Light. First Example of a Second-Order Phase Transition Far from Thermal Equilibrium", Z. Phys., 1970, 237, no.1, p.31.

20.79. Zalesskii, V.Yu.: "Specific Energy Output of a Pulsed Laser", Zh. Prikl. Spekt., 1970, 12, no.3, p.441.

20.80. Gurevich, G.L.: "Theory of TW Lasers", Izv. VUZ Radiofiz., 1970, 13, no.7, p.1019.

20.81. Feldman, B.J., and Feld, M.S.: "Theory of High-Intensity Gas Laser", Phys. Rev., 1970, 1A, p.1375.

20.82. Marlow, W.C.: "Approximate Lasing Condition", J. Appl. Phys., 1970, 41, p.4019.

20.83. Riska, D.O., and Stenholm, S.: "Effects of Detuning on the Quantum Theory of an Inhomogeneously Broadened Laser", J. Phys. A, 1970, 3, p.536.

20.84. Hubner, H.: "Master and Fokker-Planck Equations for Coupled Modes of a Laser with Axial Magnetic Field", Z. Phys., 1970, 239, no.2, p.103.

20.85. Samson, A.M., and Ry'bakov, V.A.: "Autooscillation Regime of a Laser with a Light Filter", Zh. Prikl. Spekt., 1970, 12, no.6, p.997.

20.86. Poizner, B.N.: "Change of Spectrum of a Gas Laser Under Action of a Light Signal", Izv. VUZ Fiz., 1970, no.8, p.153.

20.87. Miyashita, T., and Ikenoue, J.: "Phase-Locking of the Internally Loss-Modulated Ring Laser", Jap. J. Appl. Phys., 1970, 9, no.6, p.720.

20.88. Casperson, L., and Yariv, A.: "Longitudinal Modes in a High-Gain Laser", Appl. Phys. Lett., 1970, 17, no.6, pp.259-61.

20.89. Faxvog, F.R., and Carruthers, J.A.: "Pulse Velocity and Mode Pulling in a Laser with Equally Spaced Modes", J. Appl. Phys., 1970, 41, p.2437.

20.90. Tursunov, A.T.: "Kinematic Modula-tion of Radiation Intensity from a Travelling-Medium Laser", Zh. Eksp. Teor. Fiz., 1970, 58, no.6, p.1919, and Sov. Phys.-JETP, 1970, 31, no.6.

20.91. Kafri, O., Speiser, S., and Kimel, S.: "Theory of Laser Q-Switching by Rotating Mirror", Phys. Lett., 1970, 33A, p.5.

20.92. Livshitz, B.L., and Tursunov, A.T.: "Ultrasonic Oscillations of Radiation Inten-sity of Travelling-Medium Lasers", Dokl. Akad. Nauk SSSR, 1970, 195, no.2, p.813, and Sov. Phys.-Dokl., 1970, 15, no.2, p.112.

20.93. Daree, K., and Puell, H.: "Temporal Development of Giant Pulses in Passively Q-Switched Lasers", Z. Naturforsch., 1970, 25a, no.6, p.909.

20.94. Dmitrevsky, S.: "Mechanism of Un-damped Output Pulsations in Maser Oscilla-tors", J. Appl. Phys., 1970, 41, p.1549.

20.95. Malyshev, V.A.: "Theory of a Laser with Instantaneous Q-Switching", Radiotekh. Elektron., 1970, 15, p.147, and Radio Eng. Electron. Phys., 1970, 15, no.1, pp.119-26.

20.96. Gnutzmann, U.: "Transient Solution of the Laser Master Equation for Several Switching Processes", Z. Phys., 1970, 233, p.380.

20.97. Ratner, A.M., and Chernov, V.S.: "Kinetics of Coupled Lasers", Ukr. Fiz. Zh., 1970, 15, p.331.

20.98. Anan'ev, Yu.A., et al.: "Telescopic Resonator Laser", Zh. Eksp. Teor. Fiz., 1970, 58, p.786, and Sov. Phys.-JETP, 1970, 31, no.3, pp.420-4.

20.99. Stenholm, S.: "Exact Solution of a Case of Semiclassical Laser Equations", Phys. Rev., 1970, 1B, p.15.

20.100. Scharf, G.: "Exact Analysis of a Quantum-Mechanical Maser Model", Phys. Lett., 1970, 31A, p.497.

20.101. Fleck, J.A.: "Ultra-Short Pulse Generation by Q-Switched Lasers", Phys. Rev., 1970, 1B, p.84.

20.102. Dandliker, R., Grutter, A.A., and Weber, H.P.: "Statistical Amplitude and Phase Variations in Mode-Locked Lasers", J. Quantum Electron. IEEE, 1970, QE-6, no.11, p.687-93.

20.103. Siegman, A.E.: "Modulator Frequency Detuning Effects in FM Mode-Locked Laser", J. Quantum Electron. IEEE, 1970, QE-6, no.12, p.803-8.

20.104. Bonifacio, R., and Schwendimann, P.: "Superradiant (Pi-Pulse) Laser", Lett. Nuovo Cim., 1970, 3, no.15, p.509-11.

20.105. Bonifacio, R., and Schwendimann, P.: "Master Equation Approach to (Pi-Pulse) Superradiance", Lett. Nuovo Cim., 1970, 3, no.15, p.512-14.

20.106. McCall, S.L., and Hahn, E.L.: "Pulse-Area/ Pulse-Energy Description of a TW Laser Amplifier", Phys. Rev., 1970, 2A, no.3, p.861-70.

20.107. Kuizenga, D.J., and Siegman, A.E.: "FM and AM Mode Locking of the Homogeneous Laser. I", J. Quantum Electron. IEEE, 1970, QE-6, no.11, p.694-708.

20.108. Kulish, N.P., and Lisitsa, M.P.: "Effect of Resonator Spacing on Parameters of Laser with Passive Q-Switch", Opt. Spekt., 1970, 29, no.2, pp.360-4, and Opt. Spectrosc., 1970, 29, no.2, pp.191-3.

20.109. Vorobeichikov, E.S., and Poizner, B.N.: "Comparison of Optical and Microwave Multi-Frequency Oscillators", Radiotekh. Elektron., 1970, 15, no.9, and Radio Eng. Electron. Phys., 1970, 15, no.9, pp.1732-4.

20.110. Tanno, N., and Inaba, H.: "Theoretical Analysis of Transient Features of Laser Oscillation", Electron. Commun. Jap., 1970, 53, no.1, pp.82-9.

20.111. Weber, H.P., and Grutter, A.A.: "Influence of Spatial Hole-Burning in Single-Mode Q-Switched Operation", Opto-Electron., 1970, 2, no.2, pp.63-7.

20.112. Egorov, Yu.P., and Petrov, A.S.: "Synchronization between Types of Oscillations of a Gas Laser During Permittivity Modulation", Izv. VUZ Fiz., 1970, no.8, pp.17-22.

20.113. Sargent, M., Scully, M.O., and Lamb, W.E.: "Buildup of Laser Oscillations from Quantum Noise", Appl. Opt., 1970, 9, no.11, pp.2423-7.

20.114. Hordvik, A.: "Pulse Stretching by Two-Photon Induced Light Absorption", J. Quantum Electron. IEEE, 1970, QE-6, p.199.

20.115. Hope, L.L., and Vassell, M.O.: "Pulse Amplification by Stimulated Two-Photon Emission", Phys. Lett., 1970, 31A, p.256.

20.116. Kazantsev, A.P., and Surdutovich, G.I.: "Quantum Model of a Laser with Nonlinear Absorption", Zh. Eksp. Teor. Fiz., 1970, 58, no.1, p.245, and Sov. Phys.-JETP, 1970, 31, no.1, pp.133-7.

20.117. Roess, D.: "Spectral Transients in Laser Oscillators", Opto-Electron., 1970, 2, no.4, pp.183-92.

20.118. Tiunova, T.I., and Ratner, A.M.: "Region of Regular Self-Oscillations of a Ruby Laser", Zh. Prikl. Spekt., 1970, 13, no.5, pp.915-7.

20.119. Kafri, O., Speiser, S., and Kimel, S.: "Doppler-Effect Mechanism for Laser Q-Switching with Rotating Mirror", J. Quantum Electron. IEEE, 1971, QE-7, no.3, pp.122-6.

20.120. Meltzer, D., and Mandel, L.: "Dynamics of a Q-Switched Laser near Threshold", Phys. Rev., 1971, 3A, no.5, pp.1763-72.

20.121. Milovskii, N.D.: "Stability of Single-Frequency Stationary Regime of a TW Laser on a Uniformly Broadened Active Material", Izv. VUZ Radiofiz.,1971,14,no.1,pp.93-9.

20.122. Stenholm, S.: "Theory of Atomic Scattering in a He-Ne Laser", Phys. Rev. Lett., 1971, 26, no.4, pp.159-61.

20.123. Keller, M., et al.: "(Theory of) Spectral Behaviour of Q-Switched Lasers", Z. Angew. Math. Phys., 1971, 22, no.1, pp.43-53.

20.124. Grigor'yants, V.V., et al.: "Effective Cross Section of Stimulated Radiation in a Nonuniformly Broadened Active Medium", Zh. Prikl. Spekt., 1971, 14, no.1, pp.154-7.

20.125. Kugushev, A.M., Golubeva, N.S., and Rozhdestvin, V.N.: "Synchronization of Nine Solid-State Laser Oscillators", Izv. VUZ Radioelektron., 1971, 14, no.4, pp.363-70.

20.126. Khanin, Ya.I.: "Mode Locking in Lasers", Zh. Eksp. Teor. Fiz., 1971, 60, no.4, pp.1282-90.

20.127. Mukhtarov, Ch.K.: "Effect of Internal Dielectric Boundaries of Optical Resonators on Generation of Stimulated Radiation", Zh. Eksp. Teor. Fiz., 1971, 60, no.3, pp.929-42, and Sov. Phys.-JETP, 1971, 33, no.3, pp.502-9.

20.128. Powell, H.T., and Wolga, G.J.: "Repetitive Passive Q-Switching of Single-Frequency Lasers", J. Quantum Electron. IEEE, 1971, QE-7, no.6, pp.213-9.

20.129. Grutter, A.A., and Weber, H.P.: "Pulse-Shaping Effects in Q-Switched Laser Oscillators", Opto-Electron., 1971, 3, no.1, pp.13-8.

20.130. Ingarden, R.S.: "Information Approach to the Theory of Lasers", Bull. Acad. Pol. Sci., 1971, 19, no.1, pp.77-82.

20.131. Danielmeyer, H.G.: "Effects of Drift and Diffusion of Excited States on Spatial Hole Burning and Laser Oscillation", J. Appl. Phys., 1971, 42, no.8, pp.3125-32.

20.132. Falk, J., Yarborough, J.M., and Ammann, E.O.: "Internal Optical Parametric Oscillation", J. Quantum Electron. IEEE, 1971, QE-7, no.7, pp.359-69.

20.133. Vlasov, S.N., et al.: "Calculation of Spatial-Temporal Characteristics of Laser Radiation", Izv. VUZ Radiofiz., 1971, 14, no.6, pp.828-39.

20.134. Grutter, A.A., and Weber, H.P.: "Non-Ideal Giant Pulses", Z. Angew. Math. Phys., 1971, 22, no.3, pp.465-72.

20.135. Mathieu, E., and Weber, H.P.: "Model for the Evolution of Short Light Pulses from Noise", Z. Angew. Math. Phys., 1971, 22, no.3, pp.458-65.

20.136. Fesquet, J., and Roig, J.: "Establishment of a Pseudosteady State in a Ruby Laser with Plane Mirrors", C. R. Acad. Sci. (Paris), 1971, 272B, no.25, pp.1414-17.

20.137. Creighton, J.R., and Jackson, J.L.: "Simplified Theory of Picosecond Pulses in Lasers", J. Appl. Phys., 1971, 42, no.9, pp.3409-14.

20.138. Kushch, G.G., and Kovalenko, Ye.S.: "Self-Synchronization of Transverse Modes in the Multimode-Generation Regime", Izv. VUZ Fiz., 1971, no.7, pp.68-72.

20.139. Haug, C., and Dabby, F.W.: "Transverse-Mode Locking Using Cylindrically Symmetric Modes", J. Quantum Electron. IEEE, 1971, QE-7, no.10, pp.489-91.

20.140. Barjot, G.: "Influence of Rotating Prism in Q-Switching of a Laser", J. Appl. Phys., 1971, 42, no.9, pp.3641-2.

20.141. Bambini, A., and Vallauri, R.: "Conditions for Self-Pulsing in a Laser Oscillator", Opt. Commun., 1971, 3, no.4, pp.209-12.

20.142. Uegami, S., and Tanaka, S.: "Temporal Coherence of Q-Switched Ruby Laser", Jap. J. Appl. Phys., 1971, 10, no.9, pp.1234-7.

20.143. Sherstobitov, V.E., Shorokhov, O.A., and Anan'ev, Yu.A.: "Calculation of the Efficiency of a Laser Exhibiting Large Radiation Losses", Kvantovaya Elektron., 1971, no.1, pp.91-5.

20.144. Kalestynski, A., and Zardecki, A.: "Diffraction Investigation of Higher-Order Laser Modes", Opt. Commun., 1971, 4, no.1, pp.5-8.

20.145. Cho, Y., and Geffers, H.: "Nonlinear Resonance Effects in Loss-Modulated Lasers", Opt. Commun., 1971, 4, no.1, pp.29-32.

20.146. Boyd, J.T.: "Threshold for Spontaneous Mode-Locking in Q-Switched Lasers", J. Quantum Electron. IEEE, 1971, QE-7, no.12, pp.575-6.

20.147. Locchi, F.: "Oscillating Modes in Open Resonators with Saturable-Gain Medium", Atti. Fond. Giorgio Ronchi, 1971, 26, no.5, pp.641-59.

20.148. Klinkov, V.K., and Mukhtarov, Ch.K.: "Effect of Redistribution of Energy Density in a Cavity on Laser Generation", Zh. Eksp. Teor. Fiz., 1971, 61, no.6, pp.2248-58 and Sov. Phys.-JETP, 1971, 34, no.6, pp.1203-9.

20.149. Dytynko, V.M., et al.: "Pulsed Laser with Delay Line in Resonator", Radiotekh. Elektron., 1971, 16, no.3, and Radio Eng. Electron. Phys.,1971,16,no.3,pp.563-4.

20.150. Landa, P.S., and Kovalev, A.S.: "Influence of Spatial Modulation of Populations on Dynamics and Flucturation Characteristics of a Laser", Kvantovaya Elektron., 1971, no.4, pp.67-76.

20.151. Ratner, A.M.: "Statistics of Self-Mode-Locking of a Solid-State Laser with Slowly Relaxing Filter", Radiotekh. Elektron., 1971, 16, no.5, pp.813-24, and Radio Eng. Electron. Phys., 1971, 16, no.5, pp.838-47.

20.152. Vinokurov, G.N.; "Theory of Synchronization of Laser Spikes by a Periodic Action", Opt. Spekt., 1971, 31, no.3, pp.472-5, and Opt. Spectrosc., 1971, 31, no.3, pp.251-2.

20.153. Melekhin, G.V.: "Competition of Laser Oscillations in a System with a Common Upper Level", Opt. Spekt., 1971, 31, no.4, pp.628-36, and Opt. Spectrosc., 1971, 31, no.4, pp.330-5.

20.154. Kuznetsova, T.I.: "Broadening of Laser Radiation Spectrum as a Result of an Increase in Transmission of a Nonlinear Filter", Kvantovaya Elektron., 1971, no.3, pp.102-5.

20.155. Newstein, M.C.: "Spontaneous Emission from a Driven Doppler Line", J. Quantum Electron. IEEE, 1972, QE-8, no.2, pp.38-46.

20.156. Nelson, T.J.: "Coupled Mode Analysis of Mode Locking in Homogeneously Broadened Lasers", J. Quantum Electron. IEEE, 1972, QE-8, no.2, pp.29-33.

20.157. Barone, S.R.: "Stability of a Steady State Pi-Pulse", J. Quantum Electron. IEEE, 1972, QE-8, no.2, pp.89-90.

20.158. Ripper, J.E., and Paoli, T.L.: "Mode Configurations in Second-Order Mode-Locked Lasers", J. Quantum Electron. IEEE, 1972, QE-8, no.2, pp.74-9.

20.159. Rozanov, N.N.: "Theory of Phase-Mode Laser Interaction", Izv. VUZ Radiofiz., 1972, 15, no.1, pp.43-50.

20.160. Golubev, Yu.M., and Shatsev, A.N.: "Interaction of External Radiation with Substance Inside a Resonant Cavity. II", Opt. Spekt., 1972, 32, no.4, pp.792-4, and Opt. Spectrosc., 1972, 32, no.4, pp.417-8.

20.161. New, G.H.C.: "Mode Locking of Quasi-CW Lasers", Opt. Commun., 1972, 6, no.2, pp.188-92.

20.162. Balashov, I.F., Berezin, B.G., and Ermakov, B.A.: "Monoimpulsive Laser with Non-Instantaneous Q-Switching", Zh. Tekh. Fiz., 1972, 42, no.2, pp.385-90, and Sov. Phys.-Tech. Phys., 1972, 17, no.2, pp.306-10.

20.163. Kogelnik, H., and Shank, C.V.: "Coupled-Wave Theory of Distributed-Feedback Lasers", J. Appl. Phys., 1972, 43, no.5, pp.2327-35.

20.164. Gurevich, G.L., and Paskhin, V.M.: "Steady-State Regime of Mode Locking in a Laser with Saturable Absorber", Izv. VUZ Radiofiz., 1972, 15, no.2, pp.221-6.

20.165. Kornienko, L.S., Kravtsov, N.V., and Naumkin, N.I.: "Dynamics of Ruby Laser with Optical Delay Line in Resonator", Dokl. Akad. Nauk SSSR, 1972, 199, no.6, pp.1284-5, and Sov. Phys.-Dokl.,1972,16,no.8,pp.659-60.

20.166. Nesterenko, T.M., and Khapalyuk, A.P.: "Spectral Width of a Laser", Zh. Prikl. Spekt., 1972, 17, no.4, pp.623-32.

20.167. Emanuel, G.: "Gain Saturation in a Laser Amplifier Driven by a CW Oscillator", J. Quant. Spectrosc.,1972,12,no.5,pp.913-24.

20.168. Davis, L.W.: "Theory of a Wave Packet in a Laser Cavity", Phys. Rev., 1972, 5A, no.6, pp.2594-7.

20.169. Chesler, R.B.: "Optimized TEM_{00} Output from a Uniformly Pumped Four-Level Laser", J. Quantum Electron. IEEE, 1972, QE-8, no.6, pp.493-6.

20.170. Bambini, A., and Vallauri, R.: "Energy and Duration of Self-Pulses in a Laser", Opt. Commun., 1972, 5, no.1, pp.23-6.

20.171. Kim, D.M., Tittel, F.K., and Rabson, T.A.: "Effect of Saturated Frequency Chirping on Mode-Locked Laser Pulses", J. Appl. Phys., 1972, 43, no.6, pp.2899-901.

20.172. Voitovich, A.P., and Smirnov, A.Ya.: "Stability of Oscillation in Gas Lasers with Nonlinear Selective Losses", Zh. Prikl. Spekt., 1972, 16,no.4,pp.633-7.

20.173. Lax, M., Louisell, W.H., and McKnight, W.B.: "Mode Competition in High-Power Homogeneously Broadened Lasers", J. Appl. Phys., 1972, 43, no.7, pp.3136-41.

20.174. Le Floch, A., and Stephan, G.: "Study of Frequency-Locking Region of a Monomode Anisotropic Zeeman Laser", Phys. Rev., 1972, 6A, no.2, pp.845-7.

20.175. Fontana, J.R.: "Theory of Spontaneous Mode Locking in Lasers Using a Circuit Model", J. Quantum Electron. IEEE, 1972, QE-8, no.8, pp.699-703.

20.176. Weber, H., and Wilbrandt, R.: "Buildup of Multimode Laser Oscillations", Opt. Commun., 1972, 5, no.3, pp.215-7.

20.177. Richter, P.H., and Grossmann, S.: "Spectral Linewidths of Two-Mode Lasers", Z. Phys., 1972, 255, no.1, pp.59-75.

20.178. Mirels, H., and Batdorf, S.B.: "Centreline Laser Radiation Intensity in an Unstable Cavity", Appl. Opt., 1972, 11, no.10, pp.2384-6.

20.179. Halford, C.E.: "Improved Model for a Self-Pulsing Ring Laser", Phys. Lett., 1972, 41A, no.3, pp.253-4.

20.180. Vinokurov, G.N.: "Periodic Motions of a System of Weakly Interacting Solid-State Laser Modes", Opt. Spekt., 1972, 32, no.2, pp.418-21, and Opt. Spectrosc., 1972, 32, no.2, pp.218-9.

20.181. Kryukov, P.G., and Letokhov, V.S.: "Fluctuation Mechanism of Ultrashort Pulse Generation by Laser with Saturable Absorber", J. Quantum Electron. IEEE, 1972, QE-8, no.10, pp.766-82.

20.182. Swain, S.: "Exact Analysis of Spontaneous Emission by a Single Two-Level Atom in the Rotating-Wave Approximation. I and II", J. Phys. A, 1972, 5, no.11, pp.1587-600 and 1601-18.

20.183. Dohm, V.: "Nonequilibrium Phase Transition in Laser-Active Media", Solid-State Commun., 1972, 11, no.9, pp.1273-6.

20.184. Aronowitz, F.: "Effects of Radiation Trapping on Mode Competition and Dispersion in the Ring Laser", Appl. Opt., 1972, 11, no.10, pp.2146-52.

20.185. Sochor, V.: "Dependence of Characteristics of a Gas Laser on the Parameters of an Intracavity Absorber", Czech. J. Phys., 1972, 22B, no.11, pp.1095-101.

20.186. Bulthuis, K.: "Time History of Laser Power Pulses from Molecular Lasers", Phys. Lett., 1972, 42A, no.1, pp.38-40.

20.187. Zel'dovich, B.Ya., and Kuznetsova, T.I.: "Generation of Ultrashort Light Pulses by Means of Lasers", Usp. Fiz. Nauk, 1972, 15, no.1, pp.48-84, and Sov. Phys.-Usp., 1972, 15, no.1, pp.25-44.

20.188. Kozel, S.M., and Kuznetsov, Ye.P.: "Nonlinear Interaction of Spontaneous Radiation Field with Active Medium of High-Gain Lasers", Izv. VUZ Radiofiz., 1972, 15, no.10, pp.1486-92.

20.189. Korolev, V.I.: "Beat Regime of Generator Synchronization by an External Sinusoidal Force", Izv. VUZ Radiofiz., 1972, 15, no.10, pp.1527-37.

20.190. Walls, D.F.: "Higher-Order Effects in the Single-Atom/ Field-Mode Interaction", Phys. Lett., 1972, 42A, no.3, pp.217-8.

20.191. Arecchi, F.T., et al.: "Atomic Coherent States in Quantum Optics", Phys. Rev., 1972, 6, no.6, pp.2211-37.

20.192. Lama, W.L., and Mandel, L.: "Source-Field Approach to the Two-Level Atom in a Closed Cavity", Phys. Rev., 1972, 6A, no.6, pp.2247-51.

20.193. Dement'ev, V.A., and Zubarev, T.N.: "Spiking Operation in a Single-Mode Laser", Dokl. Akad. Nauk SSSR, 1972, 201, no.5, pp.66-9, and Sov. Phys.-Dokl., 1972, 17, no.5, pp.450-2.

20.194. Rivlin, L.A.: "Coherent Inter-action of a Beam of Particles with Matter", Kvantovaya Elektron., 1972, no.3, pp.90-2, and Sov. J. Quantum Electron., 1972, 2, no.3, pp.274-5.

20.195. Rivlin, L.A.: "Optics of Moving Media Exhibiting Negative Absorption", Kvantovaya Elektron., 1972, no.3, pp.92-3, and Sov. J. Quantum Electron., 1972, 2, no.3, pp.276-7.

20.196. Farist, A., et al.: "Coherent Radiation in Interaction with a Two-Level System", Helv. Phys. Acta, 1972, 45, no.6, pp.956-9.

20.197. Mashkevich, V.S.: "Spectral Theory of Laser Radiation. I", Kvantovaya Elektron., 1972, no.6, pp.171-205, and Sov. J. Quantum Electron., 1972, 2, no.6.

20.198. Rivlin, L.A.: "Spatial Locking of Laser Modes", Kvantovaya Elektron., 1972, no.5, pp.46-52, and Sov. J. Quantum Electron., 1972, 2, no.5, pp.427-31.

20.199. Kryukov, P.G., et al.: "Formation of Ultrashort Laser Pulses with the Aid of a Two-Component Medium", Zh. Eksp. Teor. Fiz. Pis'ma, 1972, 16, no.3, pp.117-9, and JETP Lett., 1972, 16, no.3, pp.81-3.

20.200. Godomskii, O.N., et al.: "Theory of Radiation from a System of Weakly Inter-acting Particles", Zh. Eksp. Teor. Fiz., 1972, 63, no.3, pp.813-9, and Sov. Phys.-JETP, 1973, 36, no.3, pp.426-9.

20.201. Bykov, V.P.: "Excited Molecules in a Medium with Negative Permittivity", Zh. Eksp. Teor. Fiz., 1972, 63, no.4, pp.1226-34, and Sov. Phys.-JETP,1973,36,no.4,pp.646-50.

20.202. Barnard, T.W.: "2N-pi Ultrashort Light Pulses", Phys. Rev., 1973, 7A, no.1, pp.373-6.

20.203. Ingarden, R.S.: "Generalized Irreversible Thermodynamics and Application to Lasers. I", Acta Phys. Pol., 1973, 43A, no.1, pp.3-14.

20.204. Mandel, P.: "Nonstationary Beha-viour of a Monomode Laser", Physica, 1973, 63, no.3, pp.553-69.

20.205. Dekker, H.: "Corrections to Third-Order Lamb Theory and Application to Mode-Locking Phenomena", Phys. Lett., 1973, 42A, no.6, pp.410-2.

20.206. Wang, Y.K., and Hioe, F.T.: "Phase Transition in the Dicke Model of Superradi-ance", Phys. Rev., 1973, 7A, no.3,pp.831-6.

20.207. Oliver, G., and Tallet, A.: "Quantum Fluctuations in Cooperative Emission of Radiation", Phys. Rev., 1973, 7A, no.3, pp.1061-4.

20.208. Glas, P.: "Investigation of the Kinetics of a Laser without External Feed-back", Exp. Tech. Phys.,1973, 21, no.3, pp.209-20.

20.209. Kovalenko, Ye.S., and Kushch, G.G.: "Structure of the Radiation Field under Synchronization of Transverse Modes in a Laser", Izv. VUZ Fiz., 1973, no.2, pp.156-8.

20.210. Kim, D.M., Marathe, S., and Rabson, T.A.: "Eigenfunction Analysis of Mode-Locking Process", J. Appl. Phys., 1973, 44, no.4, pp.1673-5.

20.211. Vlasov, S.N., Makarov, A.I., and Khizhnyak, A.I.: "Influence of External Self-Focusing on Laser Amplifier Operation", Izv. VUZ Radiofiz., 1973, 16, no.1,pp.217-21.

20.212. Watanabe, A., Hill, K.O., and MacDonald, R.I.: "Amplification of Light in an Optical Waveguide with Evanescent Pump-ing", Can. J. Phys., 1973, 51,no.7,pp.761-71.

20.213. Garside, B.K., and Lim, T.K.: "Laser Mode Locking Using Saturable Absor-bers", J. Appl. Phys., 1973, 44, no.5, pp.2335-42.

20.214. Dmitriev, L.S.: "Output Oscilla-tion Amplitude Stability of the Linear Solid-State Laser Amplifier", Ukr. Fiz. Zh., 1973, 18, no.4, pp.544-8.

20.215. Schell, A., and Barakat, R.: "Approach to Equilibrium of Single-Mode Radiation in a Cavity", J. Phys. A, 1973, 6, no.6, pp.826-36.

20.216. Kobayashi, K.: "Analysis of Pul-sations in Coupled-Cavity Structure Semi-conductor Lasers", J. Quantum Electron. IEEE, 1973, QE-9, no.4, pp.449-58.

20.217. Tarantovich, T.M., and Khon'kin, V.M.: "Dynamics of Two-Frequency Self-Oscil-lator with Fixed Excitation", Izv. VUZ Radio-fiz., 1973, 16, no.2, pp.222-6.

20.218. Malakhov, A.N., and Sandler, M.S.: "Influence of Field Inhomogeneity on Natural Width of Laser Spectral Line", Izv. VUZ Radiofiz., 1973, 16, no.2, pp.308-10.

20.219. Bershtein, I.L.: "Influence of Reflected Signal on Laser Operation", Izv. VUZ Radiofiz., 1973, 16, no.4, pp.526-30.

20.220. Chinn, S.R.: "Effects of Mirror Reflectivity in a Distributed-Feedback Laser", J. Quantum Electron. IEEE, 1973, QE-9, no.6, pp.574-80.

20.221. Sasada, T.: "Non-Markovian Effect on Superradiance from Harmonic Oscillators", J. Phys. Soc. Jap., 1973, 35, no.1, pp.33-9.

20.222. Armstrong, L., and Feneuille, S.: "Influence of Photon Statistics on the Bloch-Siegert Shift", J. Phys. B, 1973, 6, no.7, pp.L182-7.

20.223. McGeoch, M.W.: "Computational Theory of Short-Pulse Generation", Appl. Phys., 1973, 1, no.6, pp.293-9.

20.224. Casperson, L.W.: "Spectral Narrow-ing in Double-Pass Superradiant Lasers", Opt. Commun., 1973, 8, no.1, pp.85-7.

20.225. Efimenko, L.V., and Mashkevich, V.S.: "Theory of Two-Channel Laser Generation in Spectral-Inhomogeneous Media. I", Ukr. Fiz. Zh., 1973, 18, no.5, pp.756-71.

20.226. Boystov, V.F., and Slusarev, S.G.: "Coherent States Derived from the Excited States of Lasers", Vestn. Leningr. Univ. Fiz. Khim., 1973, no.4, pp.161-2.

20.227. DeWames, R.E., and Hall, W.F.: "Conditions for Laser Oscillations in Distributed-Feedback Waveguides", Appl. Phys. Lett., 1973, 23, no.1, pp.28-30.

20.228. Zakharov, S.M., and Manykin, E.A.: "Effects of the Optical Nutation Type in Multiple-Photon Echo", Zh. Eksp. Teor. Fiz. Pis'ma, 1973, 17, no.8, pp.431-4, and JETP Lett., 1973, 17, no.8, pp.308-11.

20.229. Lu, E.Y.C., and Wood, L.E.: "Generation of Up- or Down-Converted Photon Echoes in Three-Level Resonant Medium", Phys. Lett., 1973, 44A, no.1, pp.68-70.

20.230. Maystre, P., et al.: "Coherent Radiation in Interaction with a Two-Level System", Helv. Phys. Acta, 1973, 46, no.1, p.43.

20.231. Smithers, M.E., and Lu, E.Y.C.: "Exact Closed Form Solutions and Photon Statistics of Coherent Spontaneous Emission in a Cavity", Phys. Lett., 1973, 44A, no.5, pp.355-6.

20.232. Drosdziok, S.: "Harmonic Oscillator. Example of Non-Thermal Phase Transitions", Z. Phys., 1973, 261, no.5, pp.431-5.

20.233. Lamb, G.L.: "Phase Variation in Coherent-Optical-Pulse Propagation", Phys. Rev. Lett., 1973, 31, no.4, pp.196-9.

20.234. Shagidullin, A.G., and Samartsev, V.V.: "Influence of Nonresonance States on Optical Induction and Echo", Fiz. Tver. Tela, 1973, 15, no.1, pp.330-2, and Sov. Phys.-Solid State, 1973, 15, no.1, pp.246-7.

20.235. Saunders, R., and Bullough, R.K.: "Perturbation Theory of Superradiance. I and II", J. Phys. A, 1973, 6, no.9, pp.1348-59 and 1360-74.

20.236. Stenholm, S.: "Quantum Theory of RF Resonances. Semiclassical Limit", J. Phys. B, 1973, 6, no.8, pp.1656-69.

20.237. Apanasevich, P.A., Fainberg, B.D., Kiselev, B.A.: "Theory of Susceptibility of Quantum Systems with Degenerate Levels", Opt. Spekt., 1973, 34, no.2, pp.209-13, and Opt. Spectrosc., 1973, 34, no.2, pp.117-9.

20.238. Ernst, G.J., and Witteman, W.J.: "Mode Structure of Active Resonators", J. Quantum Electron. IEEE, 1973, QE-9, no.9, pp.911-8.

20.239. Chang, C.S., and Stehle, P.: "Quantum Theory of Electron-Cavity Interaction", Phys. Rev., 1973, 8, no.1, pp.318-27.

20.240. Alekseeva, A.N., Ostapchenko, E.P., and Teselkin, V.V.: "Analysis of an Active Optical Resonator with Opening in the Mirror", Zh. Prikl. Spekt., 1973, 18, no.5, pp.816-20.

20.241. Nesterenko, T.M., and Khapatyuk, A.P.: "Generation in a Complex Resonator with Thin Lenses", Zh. Prikl. Spekt., 1973, 18, no.5, pp.821-8.

20.242. Carmichael, H.J., and Walls, D.F.: "Master Equation for Strongly Interacting Systems", J. Phys. A, 1973, 6, no.10, pp.1552-64.

20.243. Kopvillem, U.Kh., and Nagibarov, V.R.: "Theory of Coherent Luminescence of Inverted Systems", J. Lumin., 1973, 6, no.5, pp.420-4.

20.244. Aihara, M., and Inaba, H.: "Doubly Resonant Photon Echoes in a Three-Level System", Opt. Commun., 1973, 8, no.4, pp.280-4.

20.245. Diels, J.C., and Hahn, E.L.: "Carrier-Frequency Distance Dependence of a Pulse Propagating in a Two-Level System", Phys. Rev., 1973, 8A, no.2, pp.1084-10.

20.246. Chester, A.N.: "Gain Thresholds for Diffuse (-Reflected) Parasitic Laser Modes", Appl. Opt., 1973, 12, no.9, pp.2139-46.

20.247. Wang, Y.K., and Lamb, W.E.: "Quantum Theory of a Laser. VI. Transient Behaviour", Phys. Rev., 1973, 8A, no.2, pp.866-73.

20.248. Terletskii, A.Ya.: "Calculation of Mirror Nonideality in Laser Theory", Vestn. Mosk. Univ. Fiz. Astron., 1973, 14, no.3, pp.373-4.

20.249. Chirkin, A.S.: "Excitation Efficiency at Different Frequencies of Multimode Emission", Zh. Prikl. Spekt., 1973, 19, no.1, pp.56-60.

20.250. Chekalinskaya, Yu.I., and Ledneva, G.P.: "Calculation of Power of Polarized Laser Emission with an Anisotropic Resonator", Zh. Prikl. Spekt., 1973, 19, no.1, pp.147-50.

20.251. Milinkevich, A.V., and Smason, A.M.: "Problem of Limiting Uses of Equilibrium Equations", Zh. Prikl. Spekt., 1973, 19, no.1, pp.61-7.

20.252. Narducci, L.M., Bowden, C.M., and Coulter, C.A.: "Atomic Coherent-State Representation of Superradiance", Lett. Nuovo Cim., 1973, 8, no.1, pp.57-62.

20.253. Sharma, R.D.: "Possibility of Obtaining Stimulated Emission without Inversion", Phys. Lett., 1973, 45A, no.2, pp.153-4.

20.254. Davies, E.B.: "Exact Dynamics of an Infinite-Atom Dicke Maser Model", Commun. Math. Phys., 1973, 33, no.3, pp.187-205.

20.255. Cherpak, N.T., and Shamfarov, Ya.L.: "Effect of Pump-Field Distribution in the Active Substance on Maser Characteristics", Izv. VUZ Radioelektron., 1973, 16, no.8, pp.53-6.

20.256. Aihara, M., and Inaba, H.: "Theoretical Study of Photon Echoes Associated with the Coherent Nonlinear Optical Effect in a Resonant Three-Level System", J. Phys. A, 1973, 6, no.11, pp.1709-24.

20.257. Aihara, M., and Inaba, H.: "Photon Echoes in a Resonant Three-Level System with Arbitrary Level Degeneracy", J. Phys. A, 1973, 6, no.11, pp.1725-42.

20.258. Swain, S.: "Quantum and Semi-classical Theories of a Two-Level Atom Interacting with a Single-Field Mode", J. Phys. A, 1973, 6, no.12, pp.L169-73.

20.259. Haken, H.: "Laser Light Emission from Interacting Chaotic Fields", Z. Phys., 1973, 265, no.2, pp.105-18.

20.260. Ratner, A.M., and Fisher, A.M.: "Applicability of Phenomenological Equations of a Multi-Mode Laser", Izv. VUZ Radiofiz., 1973, 16, no.10, pp.1510-21.

20.261. Alekseev, A.I., and Evseev, I.V.: "Properties of Light Echo in Gases", Izv. Akad. Nauk SSSR, Ser. Fiz., 1973, 37, no.10, pp.2046-55.

20.262. Gadomskii, O.N., Nagibarov, V.R., and Solovarov, N.K.: "Concentration Dependence of the Intensity of Light Echo", Izv. Akad. Nauk SSSR, Ser. Fiz., 1973, 37, no.10, pp.2125-8.

20.263. Lu, E.Y.C., and Wood, L.E.: "Single-Peaked Self-Induced Transparency Pulses in a Degenerate Resonant Medium", Phys. Lett., 1973, 45A, no.5, pp.373-4.

20.264. Kuroda, K., and Ogura, I.: "Convergence and Asymptotic Behaviour of the Solution for a High-Intensity Single-Mode Gas Laser in the Form of Continued Fractions", Jap. J. Appl. Phys., 1973, 12, no.11, pp.1758-65.

20.265. Troup, G.J., and Bambini, A.: "Use of Modified Kramers-Kronig Relation in the Rate-Equation Approach of Laser Theory", Phys. Lett., 1973, 45A, no.5, pp.393-4.

20.266. Demchenko, V.V., and El-Siragy, N.M.: "Semi-Classical Treatment of Many-Level Molecular Systems", Acta Phys. Acad. Sci. Hung., 1973, 34, no.2-3, pp.125-38.

20.267. Il'inova, T.M., Klyukach, I.L., and Khokhlov, V.V.: "Excitation of Superluminescence of Three-Level Systems by Short Laser Pulses", Izv. Akad. Nauk SSSR, Ser. Fiz., 1973, 37, no.10, pp.2203-6.

20.268. Salomaa, M., and Salomaa, R.: "Theory of High-Intensity TW Laser", Phys. Fenn., 1973, 8, no.4, pp.289-333.

20.269. Carmichael, H.J., Gardiner, C.W., and Walls, D.F.: "Higher-Order Corrections to the Dicke Superradiant Phase Transition", Phys. Lett., 1973, 46A, no.1, pp.47-8.

20.270. Goreslavskii, S.P., and Yakovlev, V.P.: "Two-Level System in a Resonating Field of Variable Amplitude", Izv. Akad. Nauk SSSR, Ser. Fiz., 1973, 37, no.10, pp.2211-3.

20.271. Lariontsev, E.G., and Serkin, V.N.: "Interaction of Opposite Super-Short Light Pulses in the Laser with Absorbing Filter", Izv. VUZ Radiofiz., 1973, 16, no.11, pp.1671-5.

20.272. Boitsov, V.F., and Slyusarev, S.G.: "Statistical Theory of Induced Brillouin Scattering", Opt. Spekt., 1973, 35, no.1, pp.175-8, and Opt. Spectrosc., 1973, 35, no.1, pp.101-3.

20.273. Picard, R.H., and Willis, C.R.: "Coupled Superradiance Master Equation", Phys. Rev., 1973, 8A, no.3, pp.1536-61.

20.274. Kornienko, L.S., et al.: "Injection of a Short Light Pulse into a Laser with a Long Resonator", Dokl. Akad. Nauk SSSR, 1973, 209, no.4-6, pp.826-8, and Sov. Phys.-Dokl., 1973, 18, no.4, pp.234-5.

20.275. Kozhevnikov, N.M.: "Perturbation Method in the Analysis of Polarization of Anisotropic Laser Resonators", Zh. Tekh. Fiz., 1973, 43, no.4, pp.878-80, and Sov. Phys.-Tech. Phys., 1973, 18, no.4, pp.557-9.

20.276. Idiatulin, V.S., and Uspenskii, A.V.: "Influence of the Gain Profile on Laser Dynamics", Kvantovaya Elektron., 1973, no.3, pp.51-6, and Sov. J. Quantum Electron., 1973, 3, no.3, pp.208-10.

20.277. Rubin, P.L.: "Interaction of Longitudinal Oscillation Modes Emitted by a Gas Laser with Rapidly Decaying Lower Level", Kvantovaya Elektron., 1973, no.3, pp.27-34, and Sov. J. Quantum Electron., 1973, 3, no.3, pp.193-7.

20.278. Mak, A.A., Ustyugov, V.I., and Fromzel, V.A.: "Gain of a Laser Amplifier in Regenerative and Superregenerative Modes", Opt. Spekt., 1973, 35, no.5, pp.911-8, and Opt. Spectrosc., 1973, 35, no.5, pp.528-31.

20.279. Strizhevskii, V.I., and Klimenko, V.M.: "Theory of Stimulated Raman Emission Accompanied by the Excitation of Polaritons in an Optical Resonator", Kvantovaya Elektron., 1973, no.3, pp.79-87, and Sov. J. Quantum Electron., 1973, 3, no.3, pp.224-8.

20.280. Weiss, R.: "Analysis of Nonresonator Maser Model", Helv. Phys. Acta, 1973, 46, no.4, pp.546-71.

20.281. Mashkevich, V.S.: "Spectral Theory of Laser Radiation. II", Kvantovaya Elektron., 1973, no.7, pp.77-122, and Sov. J. Quantum Electron., 1973, 3, no.7.

20.282. Potapov, S.K., et al.: "Influence of Stark Modulation of Molecular Vibrations on Stimulated Raman Scattering", Kvantovaya Elektron., 1973, no.2, pp.112-4, and Sov. J. Quantum Electron., 1973, 3, no.2, pp.170-1.

20.283. O'Bryan, C.L., and Sargent, M.: "Theory of Multimode Laser Operation", Phys. Rev., 1973, 8A, no.6, pp.3071-92.

20.284. Melekhin, G.V., and Melekhina, G.P.: "Effect of Radiation Polarization on Interaction of Modes of Two Lasing Channels Coupled by a Common Level", Opt. Spekt., 1973, 35, no.4, pp.724-35, and Opt. Spectrosc., 1973, 35, no.4, pp.420-5.

20.285. Levedev, S.A., Volkov, V.M., and Kogan, B.Ya.: "Gain for Light Internally Reflected from a Medium with Inverted Population", Opt. Spekt., 1973, 35, no.5, pp pp.976-7, and Opt. Spectrosc., 1973, 35, no.5, pp.565-6.

20.286. Markov, V.B., et al.: "Frequency Tunable Generation of Giant Pulses in Spectrally Inhomogeneous Media", Zh. Eksp. Teor. Fiz., 1973, 64, no.5, pp.1538-48, and Sov. Phys.-JETP, 1973, 37, no.5, pp.778-83.

20.287. Letokhov, V.S.; "Problem of Nuclear-Transition Gamma-Laser", Zh. Eksp. Teor. Fiz., 1973, 64, no.5, pp.1555-67, and Sov. Phys.-JETP, 1973, 37, no.5, pp.787-93.

20.288. Grasyuk, A.Z., et al.: "Dynamics of Emission and Amplification of Light in Stimulated Raman Scattering", Kvantovaya Elektron., 1973, no.5, pp.27-35, and Sov. J. Quantum Electron., 1973, 3, no.5, pp.380-4.

20.289. Uchiyama, T., and Fujioka, T.: "Theory of AM Transverse Mode Locking", Electron. Commun. Jap., 1973, 56, no.6, pp.115-21.

20.290. Uchiyama, T., and Fujioka, T.: "Theory of Laser Oscillators Considering Transverse Modes", Electron. Commun. Jap., 1973, 56, no.7, pp.97-102.

20.291. Ernst, G.J., and Witteman, W.J.: "Effect of Radial Radiation Transport on Intensity Characteristics and Oscillation Frequency of Homogeneously Broadened Lasers", J. Quantum Electron. IEEE, 1974, QE-10, no.1, pp.37-44.

20.292. Selden, A.C.: "Threshold for Laser Generation via Backward Scattering", Opt. Commun., 1974, 10, no.1, pp.1-3.

20.293. Dekker, H.: "Stationary Momentum-Space Solution of the Fokker-Planck Equation for a Simple Model of a Laser Oscillator Exhibiting Spatial Dispersion", Opt. Commun., 1974, 10, no.2, pp.114-9.

20.294. Chinn, S.R., and Kelley, P.L.: "Analysis of the Transmission, Reflection, and Noise, Properties of Distributed Feedback Laser Amplifiers", Opt. Commun., 1974, 10, no.2, pp.123-6.

20.295. Kaufman, Y.J., and Oppenheim, U.P.: "Rate Equations of High-Gain Lasers and Determination of Laser Parameters", Appl. Opt., 1974, 13, no.2, pp.374-8.

20.296. Seybold, K., and Risken, H.: "Theory of a Detuned Single-Mode Laser near Threshold", Z. Phys.,1974,267,no.4,pp.323-30.

20.297. Mandel, P.: "Kinetic Equation for a High-Intensity Monomode Laser", Phys. Lett., 1974, 47A, no.4, pp.307-8.

20.298. Wang, S.: "Principles of Distributed Feedback and Distributed Bragg-Reflection Lasers", J. Quantum Electron. IEEE, 1974, QE-10, no.4, pp.413-27.

20.299. Gibberd, R.W.: "Equivalence of the Dicke Maser Model and the Ising Model at Equilibrium", Aust. J. Phys., 1974, 27, no.2, pp.241-7.

20.300. Welsch, D.: "Linewidths in a Solid-State anti-Stokes Raman Oscillator", Phys. Lett., 1974, 47A, no.6, pp.487-8.

20.301. Vitrishchak, I.B., Soms, L.N., and Tarasov, A.A.: "Polarization of a Cavity with a Thermally Deformed Active Element", Zh. Tekh. Fiz., 1974, 44, no.5, pp.1055-62, and Sov. Phys.-Tech. Phys.,1974,19,no.5,pp.664-8.

20.302. Barnes, N.P.: "Mode-Locking Dynamics of Homogeneously Broadened Lasers", J. Appl. Phys., 1974, 45, no.3, pp.1291-7.

20.303. Goldstein, J.C., and Hopf, F.A.: "Laser Amplifier Design", Opt. Commun., 1974, 11, no.2, pp.118-22.

20.304. Cordero, R.F., and Wang, S.: "Threshold Condition for Thin-Film Distributed-Feedback Lasers", Appl. Phys. Lett., 1974, 24, no.10, pp.474-6.

20.305. Kennedy, C.J., and Barry, J.D.: "Stability of an Intracavity Frequency-Doubled Nd^{3+}:YAG Laser", J. Quantum Electron. IEEE, 1974, QE-10, no.8, pp.596-9.

20.306. Ivanov, V.A., Kovalev, A.A., and Lebedev, V.I.: "Optimal Working Conditions for a Passive Shutter in a Laser", Zh. Prikl. Spekt., 1974, 20, no.4, pp.597-60.

20.307. Narducci, L.M., Coulter, C.A., and Bowden, C.M.: "Exact Diffusion Equation for a Model for Superradiant Emission", Phys. Rev., 1974, 9A, no.2, pp.829-45.

20.308. Dekker, H.: "Boundary Conditions in Laser Oscillators with Spatial Dispersion", Appl. Phys., 1974, 4, no.2, pp.105-7.

20.309. Walgraef, D.: "Quantum Statistics of a Monomode-Laser Model", Physica, 1974, 72, no.2, pp.578-96.

20.310. Efimenko, L.V., and Mashkevich, V.S.: "Theory of Two-Channel Laser Generation in Spectrally Heterogeneous Media. II and III", Ukr. Fiz. Zh., 1974, 19, no.7, pp.1185-92 and pp.1193-202.

20.311. Pasmanik, G.A., and Sandler, M.S.: "Linewidth of a TW Raman Laser Excited by Nonmonochromatic Radiation", Zh. Eksp. Teor. Fiz., 1974, 66, no.1, pp.74-80, and Sov. Phys.-JETP, 1974, 39, no.1.

20.312. Anikeev, B.V.: "Dynamics of Active Mode Phasing in a Pulsed Laser", Zh. Eksp. Teor. Fiz. Pis'ma, 1974, 19, no.1, pp.34-8, and JETP Lett., 1974, 19, no.1.

20.313. Casperson, L.W.: "Mode Stability of Lasers and Periodic Optical Systems", J. Quantum Electron. IEEE, 1974, QE-10, no.9, pp.629-34.

20.314. Belanger, P.A.: "Analysis of the Emerging Beam of a Multimode Laser", Can. J. Phys., 1974, 52, no.13, pp.1189-94.

20.315. Ahmad, F.: "Frequency Shifts in a Single-Mode Laser", J. Phys. A., 1974, 7, no.15, pp.1934-43.

20.316. Texter, J.A., and Bergmann, E.E.: "Stability Boundaries for Two-Mode Lasers in the Strong-Coupling Limit", Phys. Rev., 1974, 9, no.6, pp.2649-51.

20.317. Pekar, V.S.: "Theory of Spontaneous and Stimulated Emission in One-Dimensionally Inhomogeneous Media and Resonators", Zh. Eksp. Teor. Fiz., 1974, 67, no.2, pp.471-80, and Sov. Phys.-JETP,1974,40,no.2.

20.318. Yuen, S.Y., and Lax, B., and Wolff, P.A.: "Theory of TW Electronic Raman Lasers. Stokes/ anti-Stokes Coupling in the Steady State", Phys. Rev., 1974, 10, no.1, pp.416-34.

20.319. Dekker, H.: "Theory of Self-Locking Phenomena in the Pressure-Broadened, Three-Mode, He-Ne Laser", Appl. Phys., 1974, 4, no.3, pp.257-63.

20.320. Carter, W.H.: "Stable and Unstable Cylindrical-Laser-Resonator Theory in an Angular-Spectrum Representation", J. Opt. Soc. Am., 1974, 64, no.8, pp.1100-6.

20.321. Lu, E.Y.C., and Wood, L.E.: "Distortionless Propagation of Zero-Degree Pulses in Degenerate Resonant Medium", Appl. Phys. Lett., 1974, 24, no.8, pp.382-4.

20.322. Cha, H.H.: "Theory of a Giant-Pulse Laser Modulated Actively and Passively Simultaneously", Sci. Sin., 1974, 17, no.3, pp.363-91.

20.323. Hafele, H.G.: "Spin-Flip Raman Laser", Appl. Phys., 1974, 5, no.2, pp.97-108.

20.324. Byrne, J., Peters, G.I., and Allen, L.: "Stimulated Emission from Nuclei", Appl. Opt., 1974, 13, no.11, pp.2499-504.

20.325. Nagel, D.J.: "Approaches to High-Photon-Energy Lasers", Phys. Fenn., 1974, 9, pp.381-8.

20.326. Wang, S., and Tsang, W.T.: "Analysis of Ring Distributed-Feedback Lasers", J. Appl. Phys., 1974, 45, no.9, pp.3978-80.

20.327. Scharf, G., and Weiss, R.: "Photon Correlations in the Dicke Maser Model", Helv. Phys. Acta, 1974, 47, no.4, pp.505-16.

20.328. Keszthelyi, C.P.: "Chemistry in Lasers. VII", Spectrosc. Lett., 1974, 7, no.11, pp.537-46.

20.329. Siegman, A.E.: "Mode Calculations in Unstable Resonators with Flowing Saturable Gain", Appl. Opt., 1974, 13, no.12, pp.2775-92.

20.330. Birman, A.Ya., and Savushkin, A.F.: "Theory of Diffraction Phenomena in a Ring Laser", Opt. Spekt., 1974, 37, no.2, pp.317-21, and Opt. Spectrosc., 1974, 37, no.2, pp.181-3.

20.331. Anan'ev, Yu.A., et al.: "Calculation of Efficiency of Lasers with Unstable Resonators", Kvantovaya Elektron., 1974, 1, no.5, pp.1201-11, and Sov. J. Quantum Electron., 1974, 4, no.5, pp.659-64.

20.332. Mashkevich, V.S.: "Spectral Theory of Laser Emission", Kvantovaya Elektron. (Kiev), 1974, no.8, pp.12-35.

20.333. Mandel, P.: "Kinetic Equation for a High Intensity Monomode Laser", Physica, 1974, 77, no.1, pp.174-91.

20.334. Vlasov, A.G., and Sklyarov, O.P.: "Theory of Open Cavity Partially Filled with Active Dielectric", Opt.-Mekh. Prom., 1974, 41, no.2, pp.9-11, and Sov. J. Opt. Technol., 1974, 41, no.2, pp.72-4.

20.335. Anghelov, D.A., Kircheva, P.P., and Misheva, M.A.: "Time Development of Stimulated Emission from Organic Dyes With Inhomogeneously Broadened Electron Levels", Bulg. J. Phys., 1974, 1, no.1, pp.5-12.

20.336. Andreichev, V.A., et al.: "Dependence of Ruby-Laser Efficiency With a Passive Shutter on Type of Illuminator", Zh. Prikl. Spekt., 1974, 21, no.5, pp.807-10.

20.337. Draganescu, V., et al.: "Theoretical Description of Gas-Transport Electrically Excited He-N_2-CO_2 Laser", Rev. Roum. Phys., 1974, 19, no.9, pp.899-904.

20.338. Reshetnyak, S.A., and Shelepin, L.A.: "Kinetics of Physical Processes in Pinch-Discharge Lasers", Zh. Prikl. Mekh. Tekh. Fiz., 1974, no.4, pp.14-21.

20.339. Letokhov, V.S.: "Pumping of Nuclear Levels by X-Ray Radiation of a Laser Plasma", Kvantovaya Elektron., 1974, 3, no.4, pp.125-7, and Sov. J. Quantum Electron., 1974, 3, no.4, pp.360-1.

21. PARAMAGNETIC-MATERIAL LASERS

21.1. Kamiryo, K., Kano, T., and Matsuzawa, H.: "Optimum Design of Elliptic-Cylinder Chambers for Pumping Solid-State Lasers", Rep. Res. Inst. Electr. Commun. Tohoku Univ., 1968, 20, p.67.

21.2. Berzing, E.G., et al.: "Q-Switching a Laser Based on Triplet-Triplet Transitions of Organic Molecules", Opt. Spekt., 1968, 25, p.421, and Opt. Spectrosc.,1968,25,p.227.

21.3. Deserno, U., Roess, D., and Zeidler, G.: "Quasi-CW Giant-Pulse Emission of $^4F-^4I$ Transitions at 1320nm in Nd^{3+}:YAG", Phys. Lett., 1968, 28A, p.422.

21.4. Clark, J.C., and Grow, R.W.: "Optically Pumped Millimetre-Wave Maser", Nachr. Tech. Fachber., 1968, 35, p.745.

21.5. Yamanaka, C., et al.: "Tandem Amplifier System of Glass and $SeOCl_2$-Liquid Lasers Doped with Nd^{3+}", Nachr. Tech. Fachber., 1968, 35, p.791.

21.6. Yamaguchi, G., et al.: "Organic Dye Solution Laser", Technol. Rep. Osaka Univ., 1968, 18, p.425.

21.7. Varga, P., et al.: "IR Generation with Nd^{3+}:Glass Laser Using a Polymethine Dye", Zh. Eksp. Teor. Fiz. Pis'ma, 1968, 8, p.501, and JETP Lett., 1968, 8, p.307.

21.8. Morgenshtern, Z.L., and Neustruev, V.B.: "Influence of Colour Centres on Lasing Threshold of Ruby", Izv. Akad. Nauk SSSR, Ser. Fiz., 1968, 32, no.1, p.1, and Bull. Acad. Sci. USSR, Phys. Ser., 1968, 32, no.1, p.7.

21.9. Anan'ev, Yu.A., et al.: "Characteristics of Sm^{2+}:CaF_2 Laser", Opt.-Mekh. Prom., 1968, 35, no.5, p.30, and Sov. J. Opt. Technol., 1968, 35, no.5, p.313.

21.10. Kaminskii, A.A., et al.: "CW Laser Based on Nd^{3+}:$LaNa(MoO_4)_2$ Crystals Operating at $300^\circ K$", Zh. Prikl. Spect., 1968, 9, p.884.

21.11. Derkacheva, L.D., et al.: "Stimulated Luminescence of Dyes in the 720-920 nm Region", Opt. Spekt., 1968, 25, no.5, p.723, and Opt. Spectrosc., 1968, 25, no.5, p.404.

21.12. Kotsubanov, V.D., et al.: "Laser Action in Solutions of Organic Luminophors in the 400-650 nm Range", Opt. Spekt., 1968, 25, no.5, p.727, and Opt. Spectrosc., 1968, 25, no.5, p.406.

21.13. Stepanov, B.I., and Rubinov, A.N.: "Lasers Based on Solutions of Organic Dyes", Usp. Fiz. Nauk, 1968, 95, no.1, p.45, and Sov. Phys.-Usp., 1968, 11, no.3, p.304.

21.14. Kaminskii, A.A.: "Laser with Combined Active Medium", Dokl. Akad. Nauk SSSR, 1968, 180, p.59, and Sov. Phys.-Dokl., 1968, 13, p.413.

21.15. Kotsubanov, V.D., et al.: "Xanthene-Dye Laser Excited by Second Harmonic of Nd^{3+}:Glass Laser", Zh. Tekh. Fiz., 1968, 38, no.7, p.1114, and Sov. Phys.-Tech. Phys., 1969, 13, no.7, pp.923-4.

21.16. Stepanov, B.I.: "Sources of Losses in Organic Dye Lasers", Dokl. Akad. Nauk SSSR, 1968, 182, p.545, and Sov. Phys.-Dokl., 1969, 13, p.933.

21.17. Kovalenko, Ye.S., and Tikhomirov, A.A.: "Possibility of Obtaining Population Inversion between Sublevels of a Laser Ground State", Izv. VUZ Fiz., 1969, no.1, p.125.

21.18. Sheherbakov, A.A., and Berezhnaya, V.P.: "Problem of Optimal Matching of Pumping Source with Active Laser Medium", Zh. Prikl. Spekt., 1969, 11, no.2, p.260.

21.19. Rubinov, A.N.: "Influence of Pumping-Pulse Leading Edge Upon Generation Threshold of Organic Dyes", Zh. Prikl. Spekt., 1969, 11, no.3, p.436.

21.20. Rubinov, A.N., and Korda, I.M.: "Lasing in a Dye Solution by Pumping Liquid Through the Active Zone", Prib. Tekh. Eksp., 1969, no.6, pp.174-6, and Instrum. Exp. Tech., 1969, no.6, pp.1569-70.

21.21. Weber, M.J., and Bass, M.: "Frequency- and Time-Dependent Gain Characteristics of Dye Lasers", J. Quantum Electron. IEEE, 1969, QE-5, p.175.

21.22. Snavely, B.B., and Schafer, F.P.: "Feasibility of Operation of Dye-Lasers", Phys. Lett., 1969, 28A, p.728.

21.23. Segre, J.P.: "High-PRF Operation of Nd^{3+}:Glass Slab Laser", J. Quantum Electron. IEEE, 1969, QE-5, p.342.

21.24. Bruce, R.E., et al.: "Erbium Laser. Material and Propagation Aspects", J. Quantum Electron. IEEE, 1969, QE-5, p.479.

21.25. Deutsch, T.F.: "Laser-Pumped Dye Lasers Near 400 nm", J. Quantum Electron. IEEE, 1969, QE-5, p.260.

21.26. Birnbaum, M., and Fincher, C.L.: "Self-Q-Switched Nd^{3+}:YAG and Ruby Lasers", Proc. IEEE, 1969, 57, p.804.

21.27. Johnson, L.F., and Ballman, A.A.: "Coherent Emission from Rare-Earth Ions in Electrooptic Crystals", J. Appl. Phys., 1969, 40, p.297.

21.28. Soffer, B.H., and Evtuhov, V.: "Quasi-CW Dye Laser", J. Quantum Electron. IEEE, 1969, QE-5, p.386.

21.29. Peterson, O.G., McColgin, W.C., and Eberly, J.H.: "Triplet-State Effects in Dye Lasers at Threshold", Phys. Lett., 1969, 29A, p.399.

21.30. Varga, P., et al.: "Emission of Polymethine Dye Used in Nd^{3+}:Glass Lasers", Opt. Spekt., 1969, 26, no.6, p.1006, and Opt. Spectrosc., 1969, 26, no.6, p.545.

21.31. Voron'ko, Yu.K., et al.: "Liquid Laser with Active Substance Based on Nd^{3+}:$POCl_3$", Zh. Prikl. Spekt., 1969, 10, no.2, p.233.

21.32. Olver, J.R., and Barnes, F.S.: "Comparison of Rare-Gas Flashlamps", J. Quantum Electron. IEEE, 1969, QE-5, p.232.

21.33. Olver, J.R., and Barnes, F.S.: "Rare-Gas Pumping Efficiencies for Nd^{3+} Lasers", J. Quantum Electron. IEEE, 1969, QE-5, p.225.

21.34. Macfarlane, W.A.R., Hughes, J.L., and Winokuroff, A.: "Increasing Excitation Efficiency of Rod Laser", J. Phys. E, 1969, 2, p.442.

21.35. Belostotskii, B.R.: "Temperature Regime of Pulsed Hollow Laser Rod", Zh. Prikl. Spekt., 1969, 10, p.49.

21.36. Laderman, A.J., Byron, S.R., and Lawrence, W.W.: "Shock-Tube Pumping of Laser Crystals", Appl. Opt., 1969, 8, p.1743.

21.37. Furumoto, H.W., and Ceccon, H.L.: "Optical Pumps for Organic Dye Lasers", Appl. Opt., 1969, 8, p.1613.

21.38. Ragul'skii, V.V.: "Nd^{3+}:$LiYF_4$ Crystal Laser", Opt. Spekt., 1969, 27, p.859, and Opt. Spectrosc., 1969, 27, p.469.

21.39. Snavely, B.B.: "Flashlamp-Excited Organic Dye Lasers", Proc. IEEE, 1969, 57, p.1374.

21.40. Buzhinskii, I.M., Toimetov, D.N., and Koryagina, E.I.: "Efficiency of Nd^{3+}: Glass Laser as a Function of Active and Non-active Absorption", Zh. Prikl. Spekt., 1969, 11, no.1, p.67.

21.41. Aleshkevich, V.A., et al.: "Nd^{3+} Laser with Regulated Pulse Duration", Zh. Eksp. Teor. Fiz. Pis'ma, 1969, 9, no.4, p.209, and JETP Lett., 1969, 9, no.4, p.123.

21.42. Bied-Charreton, P., et al.: "Spectral Condensation of Energy Emitted by Dye Lasers", C. R. Acad. Sci. (Paris), 1969, 268B, p.1377.

21.43. Gibson, A.J.: "Flashlamp-Pumped Dye Laser for Resonance Scattering Studies of Upper Atmosphere", J. Sci. Instrum., 1969, 2E, p.802.

21.44. Ahmed, S.A., and Yu, C.: "Threshold Effects and Liquid-Laser Wavelengths", Proc. IEEE, 1969, 57, p.1686.

21.45. Chartier, G.: "Operation of Ruby Laser with Different Polarizations", C. R. Acad. Sci. (Paris), 1969, 269B, p.759.

21.46. Mikaelyan, A.L., et al.: "Q-Switched Unidirectional Ruby Ring Lasers", J. Quantum Electron. IEEE, 1969, QE-5, p.617.

21.47. Gardash'yan, V.M., et al.: "Pulsed Ruby Laser Oscillators Using Mercury Pump Sources", Radiotekh. Elektron., 1969, 14, no.6, p.1069, and Radio Eng. Electron. Phys., 1969, 14, no.6, pp.922-3.

21.48. Mack, M.E.: "Superradiant TW Dye Laser", Appl. Phys. Lett., 1969, 15, p.166.

21.49. Furumoto, H.W., and Ceccon, H.L.: "Flashlamp-Pumped Organic Scintillator Lasers", J. Appl. Phys., 1969, 40, p.4204.

21.50. Bradley, D.J., and O'Neill, F.: "Passive Mode-Locking of Flashlamp-Pumped Rhodamine Dye Lasers", Opto-Electron., 1969, 1, no.2, p.69.

21.51. Wetzels, W., and Alfs, A.: "Double-Giant-Pulse Ruby Laser with Extended Coherence Length", Rev. Sci. Instrum., 1969, 40, p.1642.

21.52. Kohlmannsperger, J.: "Organic Laser. Coronene in Methylcyclohexane/Isopentane at $100^\circ K$", Z. Naturforsch., 1969, 24a, no.10, p.1547.

21.53. Samson, A.M., Ry'bakov, V.A., and Stashkevich, N.K.: "Transient Generation of a Ruby Laser with a Filter", Zh. Prikl. Spekt., 1969, 10, no.2, p.244.

21.54. Huth, B.G., Farmer, G.I., and Kagan, M.R.: "Temperature-Dependent Measurements of a Flashlamp-Pumped Dye Laser", J. Appl. Phys., 1969, 40, p.5145.

21.55. Allen, R.B., and Scalise, S.J.: "CW Operation of Nd^{3+}:YAG Laser by Injection Luminescent Pumping", Appl. Phys. Lett., 1969, 14, p.188.

21.56. Remski, R.L., et al.: "Pulsed Laser Action in $Er^{3+}Ho^{3+}$:$LiYF_4$ at $77^\circ K$", J. Quantum Electron. IEEE, 1969, QE-5, p.214.

21.57. Artamonov, S.A., Yermakov, B.A., and Lukin, A.V.: "Giant-Pulse, Periodic-Operation, Ruby Laser", Opt.-Mekh. Prom., 1969, 36, no.5, pp.33-4, and Sov. J. Opt. Technol., 1969, 36, no.5, pp.685-6.

21.58. Vanyukov, M.P., et al.: "Giant-Pulse Nd^{3+}:Glass Laser with Diffraction Divergence", Opt.-Mekh. Prom., 1969, 36, no.5, pp.79-80, and Sov. J. Opt. Technol., 1969, 36, no.5, pp.725-6.

21.59. Galant, Ye.I., et al.: "Activated Glasses for Lasers", Opt.-Mekh. Prom., 1969, 36, no.6, pp.48-65, and Sov. J. Opt. Technol., 1969, 36, no.6, pp.770-88.

21.60. Skinner, D.R.: "Effect of Laser-Rod Properties on Efficiency of Pumping Chambers Using Helical Flash Lamps", Appl. Opt., 1969, 8, p.1467.

21.61. Goldstein, A., and Dacol, F.H.: "Reliable Flashlamp-Pumped Tunable Organic Dye Laser", Rev. Sci. Instrum., 1969, 40, p.1597.

21.62. Srinivasan, R.: "Materials for Flash-Pumped Organic Lasers", J. Quantum Electron. IEEE, 1969, QE-5, p.552.

21.63. Kravchenko, V.I., et al.: "Tunable Lasers with Dispersion Resonators Containing Organic Dyes", Zh. Prikl. Spekt., 1969, 11, p.796.

21.64. Derkachyova, L.D., et al.: "Mode-Locking in Polymethine Dye Lasers", Opt. Spekt., 1969, 26, p.1051, and Opt. Spectrosc., 1969, 26, p.572.

21.65. Aristov, A.V., et al.: "Influence of Certain Quenchers on Laser Threshold in Organic Phosphors", Opt. Spekt., 1969, 27, p.1009, and Opt. Spectrosc., 1969, 27, p.548.

21.66. Bagdasarov, Kh.S., and Kaninskii, A.A.: "YAlO$_3$ with Trivalent Rare Earths as Active Laser Medium", Zh. Eksp. Teor. Fiz. Pis'ma, 1969, 9, no.9, p.501, and JETP Lett., 1969, 9, no.9, p.303.

21.67. Kaminskii, A.A., Bodretsova, A.I., and Levikov, S.I.: "Quasi-CW Laser with Pyrotechnic Excitation", Zh. Tekh. Fiz., 1969, 39, no.3, p.535, and Sov. Phys.-Tech. Phys., 1969, 14, no.3.

21.68. Liberman, I., and Grassel, R.L.: "Comparison of Lamps for Use in High-CW-Power Nd^{3+}:YAG Lasers", Appl. Opt., 1969, 8, p.1875.

21.69. Fawcett, B.C.: "Experimental Comparison of Tuned Organic Lasers", J. Quantum Electron. IEEE, 1970, QE-6, p.473.

21.70. von Gutfeld, R.J., Webber, B., and Tynan, E.E.: "Increased Laser Tunability by Acidification of Organic Dyes", J. Quantum Electron. IEEE, 1970, QE-6, p.528.

21.71. Bagdasorov, Kh. S., et al.: "CW Laser Based on Nd^{3+}:Ca(NbO$_3$)$_2$", Kristallogr., 1970, 15, no.2, p.380, and Sov. Phys.-Cryst., 1970, 15, no.2, pp.323-4.

21.72. Dzyubenko, M.I., et al.: "Green-Light Laser with Organic Dye Excited by Pulse Lamp", Ukr. Fiz. Zh., 1970, 15,no.2,p.342.

21.73. Derr, V.E., et al.: "Effect of Pulsed Ionizing Radiation on Laser Action in Cr^{3+}Nd^{3+}:YAG", Proc. IEEE, 1970,58,p.846.

21.74. Birnbaum, M., and Fincher, C.L.: "Laser Characteristics of Nd^{3+}-Doped Lithium-Germanate Glass", J. Appl. Phys., 1970, 41, p.2470.

21.75. Ribotta, R.: "Maser Oscillating on Lateral Bands", C. R. Acad. Sci. (Paris), 1970, 270B, p.696.

21.76. Ziermann, A.: "Pump Energy Density in a Solid-State Laser", Frequenz, 1970, 24, no.4, p.98.

21.77. Walther, H., and Hall, J.L.: "Tunable Dye Laser with Narrow Spectral Output", Appl. Phys. Lett., 1970, 17, no.6, pp.239-42.

21.78. Peterson, O.G., Tuccio, S.A., and Snavely, B.B.: "CW Operation of an Organic Dye-Solution Laser", Appl. Phys. Lett., 1970, 17, no.6, pp.245-7.

21.79. Aristov, A.V., et al.: "Structure of Tetrakis Benzoylacetonate Compound of Eu^{3+} for Lasers", Teor. Eksp. Khim., 1970, 6, no.1, pp.61-6, and Theor. Exp. Chem., 1970, 6, no.1, pp.53-7.

21.80. Myer, J.A., et al.: "Dye-Laser Excitation with a Pulsed N$_2$ Laser at 337.1 nm", Appl. Phys. Lett., 1970, 16, p.3.

21.81. Myer, J.A., Itzkan, I., and Kierstead, E.: "Dye Lasers in the UV", Nature, 1970, 225, p.544.

21.82. Lidholt, L.R.: "Dye Laser Pumped by UV N$_2$ Laser", J. Quantum Electron. IEEE, 1970, QE-6, p.162.

21.83. Rubinov, A.N.: "Duration of Generation in Organic Dye Solutions", Zh. Prikl. Spekt., 1970, 12, p.57.

21.84. Ziermann, A.: "Calculation of Output Power of Ruby Lasers", Z. Angew. Phys., 1970, 19, p.22.

21.85. Pappalardo, R., Samelson, H., and Lempicki, A.: "Long-Pulse Laser Emission from Rhodamine 6G in Cyclooctatetraene", Appl. Phys. Lett., 1970, 16, p.267.

21.86. Samelson, H., et al.: "Characteristics of Lasers Based on Nd^{3+} in Aprotic Solvents", J. Appl. Phys., 1970, 41, p.2459.

21.87. Folin, K.G., et al.: "Dynamics of a Ruby Laser in Free-Oscillation Conditions", Zh. Eksp. Teor. Fiz., 1970, 58, p.1146, and Sov. Phys.-JETP, 1970, 31, no.4, pp.613-9.

21.88. Shank, C.V., et al.: "Near-UV to Yellow Tunable Laser Emission from an Organic Dye", Appl. Phys. Lett., 1970, 16, p.405.

21.89. Furumoto, H.W., and Ceccon, H.L.: "UV Organic-Liquid Lasers", J. Quantum Electron. IEEE, 1970, QE-6, p.262.

21.90. Gregg, D.W., et al.: "Wavelength Tunability of Flashlamp-Pumped Laser Dyes", J. Quantum Electron. IEEE, 1970, QE-6,p.270.

21.91. Keller, R.A.: "Effect of Quenching of Molecular Triplet States in Organic Dye Lasers", J. Quantum Electron. IEEE, 1970, QE-6, p.411.

21.92. Stepanov, B.I.: "Transient Generation of Organic Dyes", Zh. Prikl. Spekt., 1970, 12, no.4, p.627.

21.93. Rubinov, A.N., Mostovnikov, V.A., and Loiko, M.M.: "Energy Characteristics of Polymethine Dye Generation in Longitudinal and Transverse Variants of Excitation", Zh. Prikl. Spekt., 1970, 12, no.4, p.634.

21.94. Clark, J.C., and Davies T.J.: "Stimulated Emission from Organic Dye Solutions Pumped by a Small Coaxial N$_2$ Laser", Appl. Opt., 1970, 9, p.1725.

21.95. Wortman, D.E.: "Ground-State Levels and Possible Effect on Laser Action for Er^{3+}: $CaWO_4$", J. Opt. Soc. Am., 1970, 60, p.1143.

21.96. Lidholt, L.R.: "Scintillator and Dye Lasers in the Range 300-600 nm Pumped by a 200-kW N_2 Laser", Opto-Electron., 1970, 2, no.1, p.21.

21.97. Aristov, A.V., et al.: "Relation between Lasing Threshold and Quantum Yield of Fluorescence of Organic Phosphors", Opt. Spekt., 1970, 28, no.3, p.293, and Opt. Spectrosc., 1970, 28, no.3, p.546.

21.98. Schubert, J.: "Organic Dye Lasers", Bull. Assoc. Suisse Electr., 1970, 61, no.15, p.657.

21.99. Banse, K., Gassmann, M.H., and Seelig, W.H.: "Continuous Pumping of a Dye Laser", Phys. Lett., 1970, 32A, p.544.

21.100. Dmitriev, V.G., et al.: "Tunable Organic-Dye Laser", Zh. Prikl. Spekt., 1970, 12, no.6, p.1023.

21.101. Borisevich, N.A., et al.: "Organic Molecule Solution Laser in Near-UV Region", Zh. Prikl. Spekt., 1970, 12, no.6, p.1111.

21.102. Forrest, M.J., and Magyar, G.: "Short Dye-Switched Giant (Ruby) Pulse", Opto-Electron., 1970, 2, no.1, p.48.

21.103. Rubinov, A.N.: "Dye-Laser Characteristics with Pulsed Lamp Excitation", Zh. Prikl. Spekt., 1970, 12, no.5, p.837.

21.104. Carruthers, J.A., and Coutts, G.W.: "Spontaneous Giant Pulsing in a Ruby Laser with One Output Beam Reflected Back into Cavity", Appl. Phys. Lett.,1970,17,p.36.

21.105. Marling, J.B., Gregg, D.W., and Thomas, S.J.: "Effect of Oxygen on Flashlamp-Pumped Dye Lasers", J. Quantum Electron. IEEE, 1970, QE-6, p.570.

21.106. Fricke, W.C.: "Fundamental-Mode Nd^{3+}:YAG Laser Analysis", Appl. Opt., 1970, 9, p.2045.

21.107. Chesler, R.B.: "Stabilizing Sleeve for the Nd^{3+}:YAG Laser", Appl. Opt., 1970, 9, p.2190.

21.108. Massey, G.A.: "Criterion for Selection of CW-Laser Host Materials to Increase Available Power", Appl. Phys. Lett., 1970, 17, p.213.

21.109. Clobes, A.R., and Brienza, M.J.: "Pulsed Nd^{3+}:YAG Laser Utilizing a Flowing Dye Cell", J. Quantum Electron. IEEE, 1970, QE-6, p.651.

21.110. Weber, M.J., et al.: "Stimulated Emission at 1662 nm from Er^{3+}:$YAlO_3$", J. Quantum Electron. IEEE, 1970, QE-6, p.654.

21.111. Danielmeyer, H.G.: "Low-Frequency Dynamics of Homogeneous Four-Level CW Lasers", J. Appl. Phys., 1970, 41, p.4014.

21.112. Faries, D.W., and Shen, Y.R.: "Simultaneous Q-Switching of Two Independent Ruby Lasers", Rev. Sci. Instrum., 1970, 41, p.216.

21.113. Mashkevich, V.S.: "Theory of Laser Generation with Two Crystals in a Resonator", Ukr. Fiz. Zh., 1970, 15, p.396.

21.114. Remski, R.L., and Smith, D.J.: "Temperature Dependence of Pulsed Laser Threshold in $Er^{3+}Tm^{3+}Ho^{3+}$:YAG", J. Quantum Electron. IEEE, 1970, QE-6, no.11, pp.750-1.

21.115. Whittaker, B.: "Low-Threshold Laser Action of a Rare-Earth Chelate in Liquid and Solid Hosts", Nature, 1970, 228, no.5267, pp.157-9.

21.116. Findlay, D.: "Experimental Analysis of a Continuously Pumped, Repetitively Q-Switched, Nd^{3+}:YAG Laser. I", Opto-Electron., 1970, 2, no.2, pp.51-8.

21.117. Aplin, C., et al.: "High-Coherence Nd^{3+}:Glass Laser", Acta Cient., 1970, 3, no.2, pp.69-74.

21.118. Escudero, R., and Magar, R.: "Giant-Pulse Ruby Laser", Rev. Mex. Fis., 1970, 19, pp.91-114.

21.119. Bayha, W.T., Creedon, J.E., and Schneider, S.: "Alkali-Vapour Light Sources as Optical Pumps for Nd^{3+}:YAG Lasers", Trans. IEEE, 1970, ED-17, no.8, pp.612-16.

21.120. Fountain, W.D., Osterink, L.M., and Foster, J.D.: "Comparison of Kr and Xe Flashlamps for Nd^{3+}:YAG Lasers", J. Quantum Electron. IEEE, 1970, QE-6, no.11, pp.684-7.

21.121. Lotsch, H.K.V., and Matovich, E.: "High-Power CW Solid-State Laser with Increased Efficiency through Pump-Light Recirculation", Optik, 1970, 32, no.2, pp.95-115.

21.122. Kalinin, Yu.A., and Mak, A.A.: "Solid-State Laser Optical-Pumping Systems", Opt.-Mekh. Prom., 1970, 37, no.7, pp.129-39, and Sov. J. Opt. Technol., 1970, 37, no.7, pp.61-71.

21.123. Kalinin, Yu.A.: "Design of Pumping System with Hollow Lamp for Solid-State Lasers", Opt.-Mekh. Prom., 1970, 37, no.1, pp.3-6, and Sov. J. Opt. Technol., 1970, 37, no.1, pp.1-4.

21.124. Calatroni, J., et al.: "Study on Size of Elliptical Chamber of Solid-State Laser for Maximum Efficiency", Acta Cient., 1970, 3, no.2, pp.75-9.

21.125. Shank, C.V., Dienes, A., and Silfvast, W.T.: "Single-Pass Gain of Exciplex 4-MU and Rhodamine-6G Dye-Laser Amplifiers", Appl. Phys. Lett., 1970, 17, no.7, pp.307-9.

21.126. Crozet, P., and Meyer, Y.: "Rhodamine-6G Solution Laser", C. R. Acad. Sci. (Paris), 1970, 271B, no.14, pp.718-21.

21.127. Pappalardo, R., Samelson, H., and Lempicki, A.: "Long-Pulse Laser Emission from Rhodamine 6G", J. Quantum Electron. IEEE, 1970, QE-6, no.11, pp.716-25.

21.128. Fill, E.E.: "Subnanosecond Pulses from a Nd^{3+} Liquid Laser", J. Appl. Phys., 1970, 41, no.11, pp.4749-50.

21.129. Lang, R.S.: "Production of Giant Pulses by Means of Inorganic Nd^{3+} Liquid Lasers", Z. Naturforsch., 1970, 25A, no.8-9, pp.1354-55.

21.130. Douglass, H.S., and Campbell, C.K.: "Gain Characteristics of Nd^{3+}:Glass Pulsed Laser Amplifier", Can. J. Phys., 1970, 48, no.22, pp.2714-9.

21.131. Chesler, R.B., Karr, M.A., and Guesic, J.E.: "Experimental and Theoretical Study of High-PRF, Q-Switched, Nd^{3+}:YAG Lasers", Proc. IEEE, 1970, 58, no.12, pp.1899-914.

21.132. Streifer, W., and Whinnery, J.R.: "Analysis of a Dye Laser Tuned by Acousto-optic Filter", Appl. Phys. Lett., 1970, 17, no.8, pp.335-7.

21.133. Brecher, C., et al.: "Transmission Losses in Aprotic Liquid Lasers", J. Appl. Phys., 1970, 41, no.11, pp.4578-81.

21.134. Ketskemety, I., et al.: "Wavelength and Energy of Pulses from Liquid Lasers with Flashlamp Excitation", Z. Naturforsch., 1970, 25a, no.10, pp.1512-3.

21.135. Bass, M., and Weber, M.J.: "$Nd^{3+}Cr^{3+}$:YAlO$_3$ Laser Tailored for High-Energy Q-Switched Operation", Appl. Phys. Lett., 1970, 17, no.9, pp.395-8.

21.136. Kovalenko, Ye.S., et al.: "CW Generation in a Ruby Laser at Room Temperature", Izv. VUZ Fiz., 1970, no.10, pp.156-7.

21.137. Nikashchin, V.A., et al.: "Amplifier of Single-Frequency Laser Radiation", Prib. Tekh. Eksp., 1970, no.1, pp.194-5, and Instrum. Exp. Tech., 1970, no.1, pp.223-4.

21.138. Naboikin, Yu.V., et al.: "Spectral and Energy Characteristics of Organic Molecule Lasers in Polymers and Toluene", Opt. Spekt., 1970, 28, no.5, pp.974-85, and Opt. Spectrosc., 1970, 28, no.5, pp.528-32.

21.139. Kaminskii, A.A., Klevtsov, P.V., and Pavlyuk, A.A.: "Stimulated Emission from Nd^{3+}:KY(MoO$_4$)$_2$", Phys. Status Solidi, 1970, 1, no.3, pp.K91-4.

21.140. Koechner, W.: "Multihundred-Watt Nd^{3+}:YAG CW Laser", Rev. Sci. Instrum., 1970, 41, no.12, pp.1699-706.

21.141. Capelle, G., and Phillips, D.: "Tuned N$_2$-Laser-Pumped Dye Laser", Appl. Opt., 1970, 9, no.12, pp.2742-45.

21.142. Miyazoe, Y., and Maeda, M.: "Polymethine Dye Lasers", Opto-Electron., 1970, 2, no.4, pp.227-33.

21.143. Abakumov, G.A., et al.: "Laser Action in Single and Binary Organic Scintillator Solutions", Opto-Electron., 1970, 2, no.4, pp.235-7.

21.144. Kupper, F.P., Mastop, W.J., and Sterk, A.B.: "Double-Light Pulses from a Q-Switched Ruby Laser", Opto-Electron., 1970, 2, no.3, pp.153-4.

21.145. Aplin, C., Fleuret, J., and Gaggioli, N.G.: "Highly Coherent Nd^{3+}:Glass Laser", Rev. Phys. Appl., 1970, 5, no.6, pp.841-4.

21.146. Kornienko, L.S., et al.: "Properties of a Solid-State Laser with Long Resonator", Dokl. Akad. Nauk SSSR, 1970, 193, no.4-6, pp.1280-2, and Sov. Phys.-Dokl., 1971, 15, no.8, pp.764-5.

21.147. Kravchenko, V.I., Smirnov, A.A., and Soskin, M.S.: "Rhodamine-6Zh Laser with Increased Spectral Intensity and Retunable Frequency", Dokl. Akad. Nauk SSSR, 1970, 193, no.1, pp.69-71, and Sov. Phys.-Dokl., 1971, 15, no.7, pp.664-5.

21.148. Devor, D.P., Soffer, B.H., and Robinson, M.: "Stimulated Emission from Ho^{3+} at 2 micron in HoF$_3$", Appl. Phys. Lett., 1971, 18, no.4, pp.122-4.

21.149. Wallace, R.W.: "Oscillation of the 1.833-micron Line in Nd^{3+}:YAG", J. Quantum Electron. IEEE, 1971, QE-7, no.5, pp.203-4.

21.150. Baldwin, G.D.: "Output-Power Calculations for a Continuously Pumped Q-Switched Nd^{3+}:YAG Laser", J. Quantum Electron. IEEE, 1971, QE-7, no.6, pp.220-4.

21.151. Kimura, T., Otsuka, K., and Saruwatari, M.: "Spatial Hole-Burning Effects in a Nd^{3+}:YAG Laser", J. Quantum Electron. IEEE, 1971, QE-7, no.6, pp.225-30.

21.152. Hansch, T.W., Pernier, M., and Schawlow, A.L.: "Laser Action of Dyes in Gelatine", J. Quantum Electron. IEEE, 1971, QE-7, no.1, pp.45-6.

21.153. Turek, C.A., and Yardley, J.T.: "N$_2$-Pumped UV Dye Lasers", J. Quantum Electron. IEEE, 1971, QE-7, no.2, p.102.

21.154. Schafer, F.P., and Muller, H.: "Tunable Dye Ring Laser", Opt. Commun., 1971, 2, no.8, pp.407-9.

21.155. Zeidler, G.: "Optical-Waveguide Technique with Organic-Dye Lasers", J. Appl. Phys., 1971, 42, no.2, pp.884-5.

21.156. Ostermayer, F.W.: "Ga(AsP)-Diode Pumped Nd^{3+}:YAG Lasers", Appl. Phys. Lett., 1971, 18, no.3, pp.93-6.

21.157. Chesler, R.B., and Maydan, D.: "Calculation of Nd^{3+}:YAG Cavity Dumping", J. Appl. Phys., 1971, 42, no.3, pp.1028-30.

21.158. Osche, G.R.: "Cavity-Loss Dependence of Er^{3+}:Glass Laser Lines", J. Quantum Electron. IEEE, 1971, QE-7, no.6, pp.252-3.

21.159. Mak, A.A., et al.: "Effect of Induced Anisotropy of Active Material on Operation of a Nd^{3+}:Glass Laser", Opt. Spekt., 1971, 30, no.6, pp.1081-7, and Opt. Spectrosc., 1971, 30, no.6, pp.579-82.

21.160. Hopkins, R.H., et al.: "Silicate Oxyapatites. High-Energy-Storage Laser Hosts for Nd^{3+}", J. Electrochem. Soc., 1971, 118, no.4, pp.637-9.

21.161. Moeller, C.E., Verber, C.M., and Adelman, A.H.: "Laser Pumping by Excitation Transfer in Dye Mixtures", Appl. Phys. Lett., 1971, 18, no.7, pp.278-80.

21.162. Peterson, O.G., Webb, J.P., and McColgin, W.C.: "Organic Dye Laser Threshold", J. Appl. Phys., 1971, 42, no.5, pp.1917-28.

21.163. Hammond, P.R., and Hughes, R.S.: "Search for Wide-Range-Tunable Dye-Laser Systems", Nature, 1971, 231, no.20, pp.59-60.

21.164. Weber, J.: "Photochemical Instability of Rhodamine-6G Dye Laser", Z. Angew. Phys., 1971, 31, no.1, pp.7-9.

21.165. Borisevich, N.A., et al.: "Generation and Retuning of Emission Bands in a Laser with Solutions of Organic Compounds", Zh. Prikl. Spekt., 1971, 14, no.1, pp.41-4.

21.166. DePratti, N.P.: "Pumping Technique for 100-Hz Operation of Nd^{3+}:Glass Laser", J. Phys. E, 1971, 4, no.3, pp.253-6.

21.167. Wildey, C.G.: "Efficiency of Twin Circular Flashtube Reflectors for Laser Pumping", J. Phys. E, 1971, 4, no.3, pp.257-9.

21.168. Mahlein, H.F., and Zeidler, G.: "Pump-Light Distribution in Laser Rod Pumped Exfocally in a Rotational Ellipsoid", Appl. Opt., 1971, 10, no.4, pp.872-9.

21.169. Borisevich, N.A., et al.: "Generation by Coumarin Solutions Excited by Flash Lamp", Zh. Prikl. Spekt., 1971, 14, no.1, pp.148-50.

21.170. Antonov, I.V., et al.: "Generation by Rhodamine 6Zh Pumped by z-Pinch Discharge", Zh. Prikl. Spekt., 1971, 14, no.1, pp.151-3.

21.171. Vlasov, S.N., Talanov, V.I., and Khizhnyak, A.I.: "Influence of Saturable Absorber on Transverse Radiation Structure of Pulsed Solid-State Laser", Izv. VUZ Radiofiz., 1971, 14, no.4, pp.570-4.

21.172. Barabanova, V.N., et al.: "Performance of Nd^{3+}:Glass Lasers Using Pump Lamps of Different Noble Gases", Zh. Prikl. Spekt., 1971, 14, no.2, pp.222-5.

21.173. Strome, F.C., and Webb, J.P.: "Flashtube-Pumped Dye Laser with Multiple-Prism Tuning", Appl. Opt., 1971, 10, no.6, pp.1348-53.

21.174. Shank, C.V., Bjorkholm, J.E., and Kogelnik, H.: "Tunable Distributed-Feedback Dye Laser", Appl. Phys. Lett., 1971, 18, no.9, pp.395-6.

21.175. Alves, R.V., et al.: "Nd^{3+}:La_2O_2S. High-Gain Laser Material", J. Appl. Phys., 1971, 42, no.8, pp.3043-8.

21.176. Kaminow, I.P., Weber, H.P., and Chandross, E.A.: "Perspex (Lucite) Dye Laser with Internal Diffraction Grating", Appl. Phys. Lett., 1971, 18, no.11, pp.497-9.

21.177. Collier, F., Michon, M., and Le Sergent, C.: "Laser Parameters of Nd^{3+} Liquid Host $POCl_3$-$SnCl_4$ Compared with YAG and Glass", C. R. Acad. Sci. (Paris), 1971, 272B, no.16, pp.945-7.

21.178. Hirono, M., Uchino, O., and Makino, Y.: "Characteristics of Tunable Narrowband and High-Power Polymethine Dye Lasers", Jap. J. Appl. Phys., 1971, 10, no.7, pp.960-1.

21.179. Carboni, G., et al.: "Flashlamp-Excited Dye Laser in the Near IR", Lett. Nuovo Cim., 1971, 1, no.23, pp.979-82.

21.180. Kohn, R.L.: "Intercavity-Pumped CW Dye Laser", Opt. Commun., 1971, 3, no.3, pp.177-80.

21.181. Danielmeyer, H.G., and Barro, J.M.: "Laser Pump Cavity with Conical Geometry", Appl. Opt., 1971, 10, no.8, pp.1983-4.

21.182. Yoshikawa, S., Iwamoto, K., and Washio, K.: "Efficient Arc Lamps for Optical Pumping of Nd^{3+} Lasers", Appl. Opt., 1971, 10, no.7, pp.1620-3.

21.183. Massey, G.A., and Yarborough, J.M.: "High-Average-Power Operation and Nonlinear Optical Generation with the Nd^{3+}:$YAlO_3$ Laser", Appl. Phys. Lett., 1971, 18, no.12, pp.576-9.

21.184. Reinberg, A.R., et al.: "GaAs:Si-Diode Pumped Yb^{3+}:YAG Laser", Appl. Phys. Lett., 1971, 9, no.1, pp.11-13.

21.185. Weber, H.P., and Ulrich, R.: "Thin-Film (Dye-Plastic) Ring Laser", Appl. Phys. Lett., 1971, 19, no.2, pp.38-40.

21.186. Johnson, L.F., and Guggenheim, H.J.: "IR-Pumped (BaY_2F_8) Visible Laser", Appl. Phys. Lett., 1971, 19, no.2, pp.44-7.

21.187. Hopkins, R.H., et al.: "Crystal Growth and Properties of $CaY_4(SiO_4)_3O$. Laser Host for Ho^{3+}", J. Cryst. Growth, 1971, 10, no.3, pp.218-22.

21.188. Hultzsch, R.: "Correlations between Active and Passive Parameters of Solid-State Laser Media. I", Phys. Status Solidi, 1971, 5A, no.3, pp.539-58.

21.189. Zharkov, A.P., et al.: "KDP-Switched Giant-Pulse Ruby Laser", Opt.-Mekh. Prom., 1971, 37, no.9, pp.30-4, and Sov. J. Opt. Technol., 1971, 37, no.9, pp.593-6.

21.190. Kortenski, T., et al.: "Conditions for Producing a Liquid Organic Laser from Hydrocarbon Components of Oil", C. R. Acad. Bulgar. Sci., 1971, 24, no.4, pp.439-42.

21.191. Goncharov, V.A., et al.: "Effect of Triplet Levels on Energy Characteristics of Laser–Pumped Xanthene Dye Lasers", Opt. Spekt., 1971, 30, no.1, pp.151–3, and Opt. Spectrosc., 1971, 30, no.1, pp.78–9.

21.192. Chicklis, E.P., et al.: "High-Efficiency Room–Temperature 2.06-micron Laser Using Sensitized Ho^{3+}:YLF", Appl. Phys. Lett., 1971, 19, no.4, p.119.

21.193. Weber, M.J., et al.: "Ho^{3+} Laser Action in $YAlO_3$ at 2.119 micron", J. Quantum Electron. IEEE, 1971, QE–7, no.10, pp.497–8.

21.194. Ketskemety, I., et al.: "Method for Frequency Tuning of Organic Dye Lasers", Acta Phys. Chem. Szeged.,1971, 17, no.1-2, pp.9–13.

21.195. Mack, M.E.: "0.2-W Repetitively Pulsed Flashlamp–Pumped Dye Laser", Appl. Phys. Lett., 1971, 19, no.4, pp.108–10, and 1972, 20, no.2, p.71.

21.196. Marling, J.B., Wood, L.L., and Gregg, D.W.: "Long–Pulse Dye–Laser Emission Across the Visible Spectrum", J. Quantum Electron. IEEE, 1971, QE–7, no.10, pp.498–9.

21.197. Yanait, Yu.A., et al.: "UV Generation with Flashlamp–Excited Paraterphenyl Solution", Zh. Eksp. Teor. Fiz. Pis'ma, 1971, 13, no.11, pp.616–9, and JETP Lett., 1971, 13, no.11, pp.438–40.

21.198. Chin, S.L., and Bedard, G.: "High-Efficiency Superradiant TW Dye Laser", Opt. Commun., 1971, 3, no.5, pp.299–300.

21.199. Konno, Y., et al.: "Oscillation Characteristics of N_2–Laser–Pumped Dye Lasers", Rec. Electr. Commun. Conversaz. Tohoku Univ., 1971, 40, no.1, pp.40–5.

21.200. Ketskemety, I., et al.: "Study of Possible Adjustments of Liquid Lasers by Changes of Concentration", Zh. Prikl. Spektr., 1971, 14, no.6, pp.1000–3.

21.201. Varsanyi, F.: "Surface (Thin) Lasers", Appl. Phys. Lett., 1971, 19, no.6, pp.169–71.

21.202. Hercher, M., and Pike, H.A.: "CW Dye Laser Emission at 522–657 nm", J. Quantum Electron. IEEE, 1971, QE–7, no.9, p.473.

21.203. Ostermayer, F.W., Allen, R.B., and Dierschke, E.G.: "Room–Temperature CW Operation of a Ga(AsP)–Diode–Pumped Nd^{3+}:YAG Laser", Appl. Phys. Lett., 1971, 19, no.8, pp.289–92.

21.204. Popovichev, V.I., Ragul'skii, V.V., and Faizullov, F.S.: "Generation of 1-MW Pulses by a Ruby Laser under Free Conditions", Kvantovaya Elektron., 1971, no.1, pp.135–6.

21.205. Malyshev, B.N., et al.: "Spatial and Energy Characteristics of a Liquid Nd^{3+}:$POCl_3$.$SnCl_4$ Circulation-Type Laser", Kvantovaya Elektron.,1971,no.1,pp.139–40.

21.206. Strome, F.C., and Tuccio, S.A.: "Triplet Quenching and CW Laser Action in Three Fluorescein Dyes", Opt. Commun., 1971, 4, no.1, pp.58–9.

21.207. Rapp, W.: "Frequency/Time Behaviour of an Organic Dye Laser", Z. Naturforsch., 1971, 26a, no.9, pp.1519–23.

21.208. Chin, S.L., and Bedard, G.: "Wavelength Tuning of a Superradiant Dye Laser", Opt. Commun., 1971, 4, no.2, pp.148–9.

21.209. Bradley, D.J., Caughey, W.G.I., and Vukusic, J.I.: "High-Efficiency Interferometric Tuning of Flashlamp–Pumped Dye Lasers", Opt. Commun., 1971, 4, no.2,pp.150–3.

21.210. Zalewski, E.F., and Keller, R.A.: "Tunable Multiple–Wavelength Organic-Dye Laser", Appl. Opt., 1971, 10, no.12,pp.2773–5.

21.211. Sonsa, J.A., and Roach, J.F.: "Tuning Method for Dye Lasers", Rev. Sci. Instrum., 1971, 42, no.11, pp.1736–7.

21.212. Bodretsova, A.I., et al.: "Crystal Lasers with Pyrolamp Illuminators", Prib. Tekh. Eksp., 1971, 14, no.3, pp.180–1, and Instrum. Exp. Tech., 1971, 14, no.3,pp.866–7.

21.213. Volynkin, V.M., Mikhailov, Yu.N., and Pogodaev, A.K.: "UV Filtering of Pump Radiation of Nd^{3+}:Glass Lasers", Kvantovaya Elektron., 1971, no.3, pp.117–8.

21.214. Kaminskii, A.A., et al.: "Stimulated Emission from Nd^{3+}:$KY(WO_4)_2$", Phys. Status Solidi, 1971, 5A, no.2, pp.K79–81, and Kvantovaya Elektron., 1971,no.4,pp.113–6.

21.215. Vanyukov, M.P.: "Parallel–Element High–Power Nd^{3+}:Glass Lasers", Kvantovaya Elektron., 1971, no.4, pp.117–20.

21.216. Grasyuk, A.Z., Zubarev, I.G., and Mulikov, V.F.: "Generation of Narrow Spectral Lines with Nd^{3+}:Glass Laser", Zh. Prikl. Spekt., 1971, 15, no.5, pp.806–9.

21.217. Bushik, B.A., Mikhnov, S.A., and Rubinov, A.N.: "Construction of Wideband Dye Laser Using a Double–Impulse Pump Lamp", Zh. Prikl. Spekt., 1971, 15, no.4, pp.732–4.

21.218. Huth, B.G., and Agan, M.R.: "Dynamics of a Flashlamp–Pumped Rhodamine–6G Laser", IBM J. Res. Dev., 1971, 15, no.4, pp.278–92.

21.219. Runge, P.K.: "CW Mode–Locked Dye Laser Pumped in the Red", Opt. Commun., 1971, 4, no.3, pp.195–8.

21.220. Gassmann, M.H., and Weber, H.P.: "Superradiance from High–Gain Flashlamp–Pumped Dye Laser", Z. Angew. Math. Phys., 1971, 22, no.5, pp.975–8.

21.221. Zhitkova, M.B., et al.: "Vortex Discharge as Pump Source for CW Lasers", Kvantovaya Elektron., 1971, no.3, pp.48–53.

21.222. Veduta, A.P., and Kirsanov, B.P.: "Four-Photon Parametric Frequency Selection in Wide Stimulated-Emission Lines", Kvantovaya Elektron., 1971, no.3, pp.73-81.

21.223. Podgaetskii, V.M., et al.: "Influence of Lamp Filling on Pump Conditions in Pulsed Nd^{3+}:YAG Laser", Kvantovaya Elektron., 1971, no.3, pp.110-3.

21.224. Gassmann, M.H., and Weber, H.: "Flashlamp-Pumped High-Gain Laser Dye Amplifiers", Opto-Electron., 1971, 3, no.4, pp.177-84.

21.225. Smol'skaya, T.I., and Rubinov, A.N.: "Effect of Different Quenchers on Energy Characteristics of Dye Lasers", Opt. Spekt., 1971, 31, no.3, pp.440-7, and Opt. Spectrosc., 1971, 31, no.3, pp.235-8.

21.226. Hongyo, M., et al.: "Characteristics of Nd^{3+}:$POCl_3$-$ZrCl_4$ Liquid Lasers", Technol. Rep. Osaka Univ., 1971, 21, no.995-1026, pp.529-35.

21.227. Mikhnov, S.A., Zybin, M.I., and Strizhnev, V.S.: "Dye Laser Based on Rhodamine 6J with 0.75% Efficiency", Zh. Prikl. Spekt., 1971, 15, no.5, pp.947-8.

21.228. Kalosha, I.I., Maslennikova, V.P., and Tsukerman, S.V.: "Laser Generation by Solutions of Dipyrazolinylbenzene Derivatives", Zh. Prikl. Spekt., 1971, 15, no.5, pp.960-1.

21.229. Mishin, V.I.: "Ruby Laser for Generating Microsecond Light Pulses having a Narrow Spectrum", Prib. Tekh. Eksp., 1971, 14, no.4, pp.181-2, and Instrum. Exp. Tech., 1971, 14, no.4, pp.1163-4.

21.230. Venkin, G.V., et al.: "Single-Mode Laser with Continuously Variable Pulse Duration", Kvantovaya Elektron., 1971, no.6, pp.97-100.

21.231. Kalinin, Yu.A., et al.: "Influence of Gas Pressure in Arc Lamps on Pumping Efficiency of CW Garnet Lasers", Kvantovaya Elektron., 1971, no.6, pp.102-4.

21.232. Abakumov, G.A., et al.: "Absorption of Pump Radiation by Excited Molecules and Efficiency of Organic-Solution Lasers", Kvantovaya Elektron., 1971, no.5,pp.116-20.

21.233. Alekseev, V.A., et al.: "Energy Characteristics of Laser Action in Rhodamine 6G Pumped by a Pinched Discharge", Kvantovaya Elektron., 1971, no.6, pp.100-2.

21.234. Hayashi, R., and Igarashi, T.: "Theoretical Analysis and Experiments on Characteristics of Linear-Flashlamp-Pumped Dye Laser", Rev. Radio Res. Lab., 1971, 17, no.92, pp.416-36.

21.235. Bereza, V.N., et al.: "Study of Optimal Conditions of Laser Generation in Solutions of Organic Compounds for 710-1100 nm", Zh. Prikl. Spekt., 1971, 15, no.4, pp.630-5.

21.236. Kruzhalov, S.V.: "TW Laser Using Faraday Effect in a Longitudinal Field", Zh. Tekh. Fiz., 1971, 41, no.12, pp.2621-2, and Sov. Phys.-Tech. Phys., 1972, 16, no.12, pp.2081-82.

21.237. Planner, A., and Szymanski, M.: "Nd^{3+}:Glass Laser Permitting Giant Pulses", Acta Phys. Pol., 1970, 41A, no.2, pp.241-4.

21.238. Milovskii, N.D., and Popova, L.L.: "Stability of a Single-Frequency Laser on Inhomogeneously Broadened Active Material", Izv. VUZ Radiofiz., 1972, 15, no.1, pp.19-26.

21.239. Koechner, W., et al.: "Characteristics and Performance of High-Power CW Kr Arc Lamps for Nd^{3+}:YAG Laser Pumping", J. Quantum Electron. IEEE, 1972, QE-8, no.3, pp.310-6.

21.240. Kepros. J.G., and Eyring, E.M.: "Dye Amplifiers Pumped by One- and Two-Photon Processes", Appl. Phys. Lett., 1972, 20,n no.4, pp.160-2.

21.241. Chandra, S., Takeuchi, N., and Hartmann, S.R.: "Prism Dye Laser", Appl. Phys. Lett., 1972, 21, no.4, pp.144-6.

21.242. Pappalardo, R., Samelson, H., and Lempicki, A.: "Calculated Efficiency of Dye Lasers as a Function of Pump Parameters and Triplet Lifetime", J. Appl. Phys., 1972, 43, no.9, pp.3776-87.

21.243. Stokes, E.D., et al.: "High-Efficiency Dye Laser Tunable from UV to IR", Opt. Commun., 1972, 5, no.4, pp.267-70.

21.244. Runge, P.K.: "Wavelength Tuning of an Intracavity-Pumped CW Mode-Locked Dye Laser", Opt. Commun., 1972,5,no.5,pp.311-4.

21.245. Gacoin, P., and Flamant, P.: "High-Efficiency Cresyl-Violet Laser", Opt. Commun., 1972, 5, no.5, pp.351-3.

21.246. Goncharov, V.A., Zverev, G.M., and Martynov, A.D.: "Effect of Triplet Levels on the Energy Characteristics of Xanthene Dye Lasers when Pumped with a Mode-Locked Laser", Opt. Spekt., 1972, 32, no.1, pp.218-9, and Opt. Spectrosc., 1972, 32, no.1, pp.112-3.

21.247. Hecht, D.L., et al.: "Dye Lasers with Ultrafast Transverse Flow", J. Quantum Electron. IEEE, 1972, QE-8, no.1, pp.15-9.

21.248. Drake, J.M., Tam, E.M., and Morse, R.I.: "Use of Light Converters to Increase Power of Flashlamp-Pumped Dye Lasers", J. Quantum Electron. IEEE, 1972, QE-8, no.2, pp.92-4.

21.249. Itzkan, I., and Cunningham, F.W.: "Oscillator/Amplifier Dye-Laser System Using N_2-Laser Pumping", J. Quantum Electron. IEEE, 1972, QE-8, no.2, pp.101-5.

21.250. Andreou, D., et al.: "Output Characteristics of Q-Switched Liquid-Laser System Nd^{3+}:$POCl_3$-$ZrCl_4$", J. Phys. D, 1972, 5, no.1, pp.59-63.

21.251. Hook, W.R., Dishington, R.H., and Hilberg, R.P.: "Xe Flashlamp Triggering for Laser Applications", Trans. IEEE, 1972, ED-19, no.3, pp.308-14.

21.252. Grigor'yants, V.V., et al.: "Lasing and Spectral-Line Characteristics in Phosphate Glasses and Inorganic Liquids with Nd^{3+}", J. Quantum Electron. IEEE, 1972, QE-8, no.2, pp.196-8.

21.253. Chicklis, E.P., et al.: "Stimulated Emission in Multiply Doped Ho^{3+}:YLF and :YAG. Comparison", J. Quantum Electron. IEEE, 1972, QE-8, no.2, pp.225-30.

21.254. Devor, D.P., and Soffer, B.H.: "2.1-micron Laser of 20-W Output and 4% Efficienciy from Ho^{3+}:YAG", J. Quantum Electron. IEEE, 1972, QE-8, no.2, pp.231-4.

21.255. Draegert, D.A.: "Efficient Single-Longitudinal-Mode Nd^{3+}:YAG Laser", J. Quantum Electron. IEEE,1972,QE-8,no.2,pp.235-9.

21.256. Barton, I.J., and Goodwin, D.W.: "CW Laser Action in $YAlO_3$", J. Phys. D, 1972, 5, no.2, pp.228-34.

21.257. Huchital, D.A., and Steinberg, G.N.: "RF-Excited Kr Arc Lamps for Nd^{3+}:YAG Lasers", Proc. IEEE, 1972, 60, no.2, p.233.

21.258. Tuccio, S.A., and Strome, F.C.: "Design and Operation of a Tunable CW Dye Laser", Appl. Opt., 1972,11,no.1,pp.64-73.

21.259. Hongyo, M., et al.: "High-Power Nd^{3+}:$POCl_3$ Liquid-Laser System", J. Quantum Electron. IEEE, 1972, QE-8, no.2, pp.192-6.

21.260. Loth, C., Astier, R., and Meyer, Y.H.: "Small Repetitive Dye Laser", J. Phys. E, 1972, 5, no.2, pp.169-70.

21.261. Dienes, A., Ippen, E.P., and Shank, C.V.: "High-Efficiency Tunable CW Dye Laser", J. Quantum Electron. IEEE, 1972, QE-8, no.3, p.388.

21.262. Schmidt, W., Appt, W., and Wittekindt, N.: "Characteristics of a Cresyl-Violet Laser", Z. Naturforsch., 1972, 27a, no.1, pp.37-41.

21.263. Fill, E.E.: "Nd^{3+}:$POCl_3$ Laser Amplifier", Z. Angew, Phys., 1972, 32, no.5-6, pp.356-8.

21.264. Pohl, D.: "Operation of Ruby Laser in Purely TE_{01} Mode", Appl. Phys. Lett., 1972, 20, no.7, pp.266-7.

21.265. Bedilov, M.R., and Khaidarov, K.: "Properties of Ruby Laser Irradiated with ^{60}Co Gamma Rays", Zh. Tekh. Fiz., 1972, 42, no.2, pp.391-4, and Sov. Phys.-Tech. Phys., 1972, 17, no.2, pp.311-3.

21.266. Hansch, T.W.: "Repetitively Pulsed Tunable Dye Laser for High-Resolution Spectroscopy", Appl. Opt., 1972, 11, no.4, pp.895-8.

21.267. Kaminskii, A.A., et al.: "CW Crystal Lasers", Zh. Eksp. Teor. Fiz. Pis'ma, 1972, 16, no.10, pp.548-51, and JETP Lett., 1972, 16, no.10, pp.387-9.

21.268. Devor, D.P.: "Sensitizers for Lamp-Pumped Rods", Laser Focus, 1972, 18, no.11, pp.38-42.

21.269. Chin, S.L., Leclerc, L., and Bedard, G.: "Emission Band from a Superradiant TW Rhodamine-6G Dye Laser", Opt. Commun., 1972, 6, no.3, pp.264-6.

21.270. Leonov, G.S., et al.: "Efficient Pumping of CW YAG Laser by Water-Cooled/Metal-Halide Lamps", Kvantovaya Elektron., 1972, no.2, pp.112-5, and Sov. J. Quantum Electron., 1972, 2, no.2, pp.190-2.

21.271. Kaminskii, A.A., et al.: "Laser 4_F-4_I Transitions in Nd^{3+}:$KY(WO_4)_2$", J. Quantum Electron. IEEE, 1972, QE-8, no.5, pp.457-8.

21.272. Chang, M.S., et al.: "Light Amplification in a Thin Film", Appl. Phys. Lett., 1972, 20, no.8, pp.313-4.

21.273. Birnbaum, M., and Gelbwachs, J.A.: "Stimulated-Emission Cross Section of Nd^{3+} at 1060 nm in $POCl_3$, YAG, $CaWO_4$, E-D2 Glass, and LG55 Glass", J. Appl. Phys., 1972, 43, pp.2325-8.

21.274. Maeda, M., and Miyazoe, Y.: "Flashlamp-Excited Organic Liquid Laser in Range 342-889 nm", Jap. J. Appl. Phys., 1972, 11, no.5, pp.692-8.

21.275. Bokut', B.V., et al.: "Intense Generation with Variable Spectrum in the Range 280-385 nm", Zh. Eksp. Teor. Fiz. Pis'ma, 1972, 15, no.1, pp.26-30, and JETP Lett., 1972, 15, no.1, pp.18-20.

21.276. Bergman, A., David, R., and Jortner, J.: "Powerful Broadband-Tunable Dye Laser", Opt. Commun., 1972, 4, no.5,pp.431-3.

21.277. Kuhl, J., et al.: "Simple Reliable Dye Laser for Spectroscopy", Z. Naturforsch., 1972, 27a, no.4, pp.601-4.

21.278. Marowsky, G., Ringwelski, L., and Schafer, F.P.: "Characteristics of a Tunable TW Dye Ring Laser", Z. Naturforsch., 1972, 27a, no.4, pp.711-3.

21.279. Mikhnov, S.A., and Strizhnev, V.S.: "Effect of Excitation on the Energy Parameters of Rhodamine 6J", Zh. Prikl. Spekt., 1972, 16, no.2, pp.262-9.

21.280. Kovalenko, Ye.S., Shangina, L.I., and Babchenko, T.N.: "Azimuthal Heterogeneity of Distribution of Pumping in Elliptical Chambers", Zh. Prikl. Spekt., 1972, 16, no.2, pp.274-8.

21.281. Steinbruegge, K.B., et al.: "Laser Properties of Nd^{3+}- and Ho^{3+}-Doped Crystals with Apatite Structure", Appl. Opt., 1972, 11, no.5, pp.999-1012.

21.282. Birnbaum, M., Tucker, A.W., and Pomphrey, P.J.: "Nd^{3+}:YAG Laser $^4F-^4I$ Transition", J. Quantum Electron. IEEE, 1972, QE-8, no.6, p.501.

21.283. Tseng, D.Y.: "High-PRF Continuously Pumped Ruby Laser", J. Quantum Electron. IEEE, 1972, QE-8, no.7, pp.675-6.

21.284. Magyar, G., and Schneider-Muntau, H.J.: "Dye Laser Forced Oscillator", Appl. Phys. Lett., 1972, 20, no.10, pp.406-8.

21.285. Ferrar, C.M.: "Vortex-Confined Pumping Discharge in Dye-Laser Solution", Appl. Phys. Lett., 1972, 20, no.11, pp.419-20.

21.286. Gronau, B., Lippert, E., and Rapp, W.: "Dye Laser Properties and Orientational Relaxation Processes", Ber. Bunsenges. Phys. Chem., 1972, 76, no.5, pp.432-7.

21.287. Brun, P., and Caro, P.: "Laser Effect in a $POCl_3-D_2O-Nd_2O_3$ Solution", C. R. Acad. Sci. (Paris), 1972, 274B, no.18, pp.1072-4.

21.288. Ferguson, J., and Mau, A.V.H.: "Laser Emission from Meso-Substituted Anthracenes at Low Temperatures", Chem. Phys. Lett., 1972, 14, no.2, pp.245-8.

21.289. Shah, J.: "Output Coupling Scheme for CW Dye Lasers", Appl. Phys. Lett., 1972, 20, no.12, pp.479-80.

21.290. Kato, K.: "Wavelength-Tunable UV Dye Laser Pumped by Fourth Harmonic of Nd^{3+}:YAG Laser", Jap. J. Appl. Phys., 1972, 11, no.6, pp.912-3.

21.291. Klein, M.B.: "Locking of a CW Dye Laser to Inverted Atomic Transitions", Opt. Commun., 1972, 5, no.2, pp.114-6.

21.292. Anliker, P., Gassmann, M.H., and Weber, H.: "12-J Rhodamine-6G Laser", Opt. Commun., 1972, 5, no.2, pp.137-8.

21.293. Yamanaka, C., Tanaka, H., and Yamaguchi, G.: "Properties of Dye-Laser Amplifiers", Technol. Rep. Osaka Univ., 1972, 22, no.1027-1052, pp.187-91.

21.294. Farmer, G.I., and Woodall, J.M.: "Solid-State Pumping Source for Nd^{3+}:YAG Lasers with Integrated Focusing", IBM Tech. Disclosure Bull., 1972, 15, no.1, pp.149-50.

21.295. Mercurio, P., and Milburn, R.H.: "Hybrid YAG-YAlO$_3$ Laser Operation at 1064 nm", Appl. Opt., 1972, 11, no.9, pp.2097-100.

21.296. Baues, P., Hundelshausen, U.V., and Mockel, P.: "Concept for Generation of Reproducible and Controllable Giant Pulses", Appl. Phys. Lett., 1972, 21, no.4, pp.135-7.

21.297. Kaminskii, A.A., et al.: "Stimulated Emission from Nd^{3+}:LiGd(MoO$_4$)$_2$", Phys. Status Solidi, 1972, 12A, no.2, pp.K73-5.

21.298. Mikhnov, S.A., and Strizhnev, V.S.: "Characteristics of Rhodamine 6J Excited by Flashlamps of Different Types", Zh. Prikl. Spekt., 1972, 17, no.1, pp.38-42.

21.299. Krupke, W.F.: "Assessment of a Promethium:YAG Laser", J. Quantum Electron. IEEE, 1972, QE-8, no.8, pp.725-6.

21.300. Kepros, J.G., Eyring, E.M., and Cagle, F.W.: "Experimental Evidence on an X-Ray Laser", Proc. Natl. Acad. Sci., 1972, 69, no.7, pp.1744-5.

21.301. Andreou, D., Selden, A.C., and Little, V.I.: "Amplification of Mode-Locked Trains with Nd^{3+}:POCl$_3$-ZrCl$_4$", J. Phys. D, 1972, 5, no.8, pp.1405-7.

21.302. Neumann, G., and Wieder, I.: "Longitudinal Excitation of Short-Cavity Tunable Dye Laser by N$_2$ Laser", Opt. Commun., 1972, 5, no.3, pp.197-9.

21.303. Siegman, A.E.: "Dispersive Explanation of Spectral Behaviour of Runge's Mode-Locked Dye Laser", Opt. Commun., 1972, 5, no.3, pp.200-1.

21.304. Bloom, A.L.: "CW-Pumped Dye Lasers", Opt. Eng., 1972, 11, no.1, pp.1-8.

21.305. de Witte, O., Signore, R., and Gauss, J.: "Principle and Design of a Fluorescent Dye Laser", Rev. Tech. Thomson-CSF, 1972, 4, no.2, pp.283-332.

21.306. Dzyubenko, M.I., et al.: "Coaxial Flashlamp for Pumping of Solutions of Organic Dyes", Prib. Tekh. Eksp., 1972, 15, no.1, pp.171-3, and Instrum. Exp. Tech., 1972, 15, no.1, pp.198-200.

21.307. Morey, W.W.: "Active (Flashlamp) Filtering for Nd^{3+} Lasers", J. Quantum Electron. IEEE, 1972, QE-8, no.10, pp.818-9.

21.308. Gibson, A.J.: "Flashlamp-Pumped Dye Lasers for Investigations of the Upper Atmosphere", J. Phys. E, 1972, 5, no.10, pp.971-3.

21.309. Rapp, C.F., and Chrysachoos, J.: "Nd^{3+}:Glass-Ceramic Laser Material", J. Mater. Sci., 1972, 7, no.9, pp.1090-2.

21.310. Arifov, U.A., et al.: "Properties of Ruby Laser Acted on by ^{60}Co Radiation", Dokl. Akad. Nauk SSSR, 1972, 203, no.1, pp.68-70, and Sov. Phys.-Dokl., 1972, 17, no.3, pp.222-4.

21.311. Blaszczak, Z., Patkowski, A., and Dobek, A.: "Setup for Obtaining Giant Pulses from a Nd^{3+}:Glass Laser", Acta Phys. Pol., 1972, 42A, no.3, pp.349-50.

21.312. Hansch, T.W., Schawlow, A.L., and Toschek, P.: "Ultrasensitive Response of a CW Dye Laser to Selective Extinction", J. Quantum Electron. IEEE, 1972, QE-8, no.10, pp.802-4.

21.313. Kindt, T., Lippert, E., and Rapp, W.: "Influence of Excitation Intensity on Laser Spectra of 4-Methylumbelliferone", Z. Naturforsch., 1972, 27a, no.8-9, pp.1370-2.

21.314. Ippen, E.P., and Shank, C.V.: "Evanescent-Field-Pumped Dye Laser", Appl. Phys. Lett., 1972, 21, no.7, pp.301-2.

21.315. Pilloff, H.S.: "Simultaneous Two-Wavelength Selection in N_2-Laser-Pumped Dye Laser", Appl. Phys. Lett., 1972, 21, no.8, pp.339-40.

21.316. Siegman, A.E., Phillion, D.W., and Kuizenga, D.J.: "Rotational Relaxation and Triplet-State Effects in the CW Dye Laser", Appl. Phys. Lett., 1972, 21, no.8, pp.345-8.

21.317. Bunkenburg, J.: "11-MW 6.8-J Flashlamp-Pumped Coaxial Liquid Dye Laser", Rev. Sci. Instrum., 1972, 43, no.11, pp.1611-2.

21.318. Lax, B.: "Quantitative Aspects of a Soft-X-Ray Laser", Appl. Phys. Lett., 1972, 21, no.8, pp.361-3.

21.319. Runge, P.K., and Rosenberg, R.: "Unconfined Flowing Films for CW Dye Lasers", J. Quantum Electron. IEEE, 1972, QE-8, no.12, pp.910-1.

21.320. Britt, A.D., and Moniz, W.B.: "Effect of pH on Photobleaching of Organic Laser Dyes", J. Quantum Electron. IEEE, 1972, QE-8, no.12, pp.913-4.

21.321. Caristi, R.F., Myer, J.A., and Itzkan, I.: "Design Problems in Wide-Tuning-Range Lasers", Image Technol., 1972, 14, no.3, pp.19-22.

21.322. Rosenkrantz, L.J.: "GaAs-Diode-Pumped Nd^{3+}:YAG Laser", J. Appl. Phys., 1972, 43, no.11, pp.4603-5.

21.323. van der Ziel, J.P., et al.: "Coherent Emission from Ho^{3+} Ions in Epitaxial Thin Al-Garnet Films", Phys. Lett., 1972, 42A, no.1, pp.105-6.

21.324. Yajima, H., Kawase, S., and Sekimoto, Y.: "Amplification at 1060 nm by a Nd^{3+}:Glass Thin-Film Waveguide", Appl. Phys. Lett., 1972, 21, no.9, pp.407-9.

21.325. Zeidler, G.: "Measurements on Thin-Film (Dye) Lasers", Arch. Elektron. Ubertrag., 1972, 26, no.12, pp.533-6.

21.326. Aristov, A.V., et al.: "Dye Laser with Spiral Pump Lamp", Prib. Tekh. Eksp., 1972, 15, no.2, pp.169-70, and Instrum. Exp. Tech., 1972, 15, no.2, pp.504-5.

21.327. Friedman, H.A., and Bell, J.T.: "Investigation of Am^{3+} in Liquid $POCl_3$", J. Inorg. Nucl. Chem., 1972, 34, no.12, pp.3928-30.

21.328. Dunning, F.B., and Stokes, E.D.: "Tunable IR Generation by N_2-Laser-Pumped Dye Laser", Opt. Commun., 1972, 6, no.2, pp.160-2.

21.329. Mory, S., Leupold, D., and Konig, R.: "Mechanism of Rhodamine-6G Fluorescence Quenching", Opt. Commun., 1972, 6, no.4, pp.394-8.

21.330. Rulliere, C., et al.: "Laser Action in Perylene at 473 nm", Opt. Commun., 1972, 6, no.4, pp.407-9.

21.331. Aristov, A.V., and Maslyukov, Yu.S.: "Influence of Anthracene Compounds on Stimulated Emission of Organic Luminophors", Opt. Spekt., 1972, 32, no.6, pp.1244-5, and Opt. Spectrosc., 1972, 32, no.6, p.678.

21.332. Tikhonov, E.A., and Shpak, M.T.: "Experimental Investigation of Pulsed Lasers Employing Solutions of Organic Dyes for 710-1100 nm", Kvantovaya Elektron., 1972, no.6, pp.48-71.

21.333. Dzybenko, M.I., et al.: "Investigation of Characteristics of Organic-Compound Lasers with Dispersion Resonators", Kvantovaya Elektron., 1972, no.6, pp.109-19.

21.334. Kaminskii, A.A., et al.: "Spectroscopic and Generation Study of Laser Crystal. Nd^{3+}:$KY(WO_4)_2$", Izv. Akad Nauk SSSR Neorg. Mater., 1972, 8, no.12, pp.2153-63, and Inorg. Mater., 1972, 8, no.12, pp.1896-904.

21.335. Bodretsova, A.I., et al.: "High-Power Nd^{3+}:YAG Laser with Explosion-Type Lamp", Kvantovaya Elektron., 1972, no.2, pp.107-8, and Sov. J. Quantum Electron., 1972, 2, no.2, pp.183-4.

21.336. Tomin, V.I., Bushuk, B.A., and Rubinov, A.N.: "Study of Gain Curve and Triplet-Triplet Absorption in a Rhodamine-6G Laser", Opt. Spekt., 1972, 32, no.5, pp.983-8, and Opt. Spectrosc., 1972, 32, no.5, pp.527-9.

21.337. Alekseev, B.A., et al.: "Characteristics of Rhodamine-6G Laser with Pinched-Discharge Pump", Zh. Prikl. Spekt., 1972, 17, no.2, pp.212-7.

21.338. Tomin, B.I., and Bushuk, B.A.: "Spectral Control of Dye-Laser Radiation", Zh. Prikl. Spekt., 1972, 17, no.2, pp.218-22.

21.339. Ketskemety, I., and Kozma, L.: "Threshold Pump Intensity of Flashlamp-Pumped Liquid Lasers", Z. Naturforsch., 1972, 27a, no.11, pp.1685-6.

21.340. Bhawalkar, D.D., and Pandit, L.: "Improving Pumping Efficiency of a Nd^{3+}: Glass Laser Using Dyes", J. Quantum Electron. IEEE, 1972, QE-9, no.1, pp.43-6.

21.341. Karl, N.: "Laser Emission from an Organic Molecular Crystal", Phys. Status Solidi, 1972, 13A, no.2, pp.651-5.

21.342. Kopvillem, U.Kh., et al.: "Photon Echo in Ruby", Fiz. Tver. Tela, 1972, 14, no.6, pp.1794-5, and Sov. Phys.-Solid State, 1972, 14, no.6, pp.1544-5.

21.343. Baltakov, F.N., et al.: "Pulsed 110-J Laser Using Rhodamine 6G in Ethyl Alcohol", Zh. Tekh. Fiz., 1972, 42, no.7, pp.1459-61, and Sov. Phys.-Tech. Phys., 1972, 17, no.7, pp.1161-3.

21.344. Loth, C., and Meyer, Y.H.: "Study of a 1-W Repetitive Dye Laser", Appl. Opt., 1973, 12, no.1, pp.123-5.

21.345. Hirth, A., et al.: "Near-IR Coupled-Mode Dye Laser Excited by a Shock Tube", C. R. Acad. Sci. (Paris), 1973, 276B, no.4, pp.153-5.

21.346. Strome, F.C., and Tuccio, S.A.: "Loss Analysis and Design Improvement for a CW Dye Laser", J. Quantum Electron. IEEE, 1973, QE-9, no.2, pp.230-5.

21.347. Marowsky, G.: "Tunable Flashlamp-Pumped Dye Ring Laser with Extremely Narrow Line", J. Quantum Electron. IEEE, 1973, QE-9, no.2, pp.245-6.

21.348. Bethea, C.G.: "Megawatt Power at 1318 nm in Nd^{3+}:YAG with Simultaneous Oscillation at 1060 nm", J. Quantum Electron. IEEE, 1973, QE-9, no.2, p.254.

21.349. Fahlen, T.S.: "High-Average-Power Q-Switched Liquid Laser", J. Quantum Electron. IEEE, 1973, QE-9, no.4, pp.493-6.

21.350. Wu, C.Y., and Lombardi, J.R.: "Effect of Electric Field on Active Medium in a Dye Laser", J. Quantum Electron. IEEE, 1973, QE-9, no.1, pp.26-9.

21.351. Birnbaum, M., and Tucker, A.W.: "Nd^{3+}:$YAlO_3$ Oscillation at 930 nm at 300°K", J. Quantum Electron. IEEE, 1973, QE-9, no.1, p.46.

21.352. Elton, R.C., et al.: "Further Evidence of Collimated X-Ray Emission from $CuSO_4$-Doped Gelatine", Appl. Opt., 1973, 12, no.1, p.155.

21.353. Jacobs, R.R., Samelson, H., and Lempicki, A.: "Losses in CW Dye Lasers", J. Appl. Phys., 1973, 44, no.1, pp.263-72.

21.354. Flamant, P., and Meyer, Y.H.: "Steady-State Gain Equation for Flashpumped Dye Amplifier", Opt. Commun., 1973, 7, no.2, pp.146-9.

21.355. Decker, C.D., and Tittel, F.K.: "Broadly Tunable, Narrow-Line, Dye-Laser Emission in the Near IR", Opt. Commun., 1973, 7, no.2, pp.155-7.

21.356. Weber, J.: "Effect of Concentration on Laser Threshold of Organic-Dye Laser", Z. Phys., 1973, 258, no.3, pp.277-83.

21.357. Barnes, N.P.: "Diode-Pumped Solid-State Lasers", J. Appl. Phys., 1973, 44, no.1, pp.230-7.

21.358. Boster, T.A.: "Questions on the Evidence of Laser X-Ray Emission from $CuSO_4$-Doped Gelatine", Appl. Opt., 1973, 12, no.2, pp.433-4.

21.359. Pavlopoulos, T.G.: "Prediction of Laser Action Properties of Organic Dyes from their Structure and the Polarization Characteristics of their Electronic Transitions", J. Quantum Electron. IEEE, 1973, QE-9, no.5, pp.510-6.

21.360. Gale, G.M.: "Single-Mode Flashlamp-Pumped Dye Laser", Opt. Commun., 1973, 7, no.1, pp.86-8.

21.361. Hansch, T.W., Schawlow, A.L., and Toschek, P.: "Single Dye Laser Repetitively Pumped by a Xe-Ion Laser", J. Quantum Electron. IEEE, 1973, QE-9, no.5, pp.553-4.

21.362. Rulliere, C., and Denariez-Roberge, M.M.: "Laser Action in Organic Compounds at 370-500 nm", Opt. Commun., 1973, 7, no.2, pp.166-8.

21.363. Tuccio, S.A., Drexhage, K.H., and Reynolds, G.A.: "CW Laser Emission from Coumarin Dyes in the Blue and Green", Opt. Commun., 1973, 7, no.3, pp.248-52.

21.364. Shank, C.V., et al.: "Evidence for Diffusion Independent Triplet Quenching in the Rhodamine-6G, Ethylene-Glycol, CW Laser", Opt. Commun., 1973, 7, no.3, pp.176-7.

21.365. Wu, C.Y., and Lombardi, J.R.: "Simultaneous Two-Frequency Oscillation in a Dye Laser", Opt. Commun., 1973, 7, no.3, pp.233-6.

21.366. Sources, J.M., Goldman, L.M., and Lubin, M.J.: "Spatial Distribution of Inversion in Face-Pumped Nd^{3+}:Glass Laser Slabs", Appl. Opt., 1973, 12, no.5, pp.927-8.

21.367. Sevast'yanov, B.K., et al.: "Lasing on Cr^{3+} Ions in YAG Crystals", Zh. Eksp. Teor. Fiz. Pis'ma, 1973, 17, no.2, pp.69-71, and JETP Lett., 1973, 17, no.2, pp.47-8.

21.368. Konjevic, R., and Konjevic, N.: "Coaxial Glass Flashlamp for Organic Dye Laser", Fizika, 1973, 5, no.1, pp.49-51.

21.369. Ewanizky, T.F., Wright, R.H., and Theissing, H.H.: "Shock-Wave Termination of Laser Action in Coaxial-Flash-Lamp Dye Lasers", Appl. Phys. Lett., 1973, 22, no.10, pp.520-1.

21.370. Streifer, W., and Saltz, P.: "Transient Analysis of an Electronically Tunable Dye Laser. II", J. Quantum Electron. IEEE, 1973, QE-9, no.6, pp.563-9.

21.371. Schimitschek, E.J., et al.: "Improved Laser Dye for the Blue-Green Spectral Region", J. Quantum Electron. IEEE, 1973, QE-9, no.7, pp.781-2.

21.372. Jacobs, R.R., Lempicki, A., and Samelson, H.: "Efficient and Damage-Resistant Tunable CW Dye Laser", J. Appl. Phys., 1973, 44, no.6, pp.2775-80.

21.373. Hirth, A., et al.: "Flashlamp-Excited Dye Lasers in the Near IR", Opt. Commun., 1973, 7, no.4, pp.339-42.

21.374. Green, J.M., Hohimer, J.P., and Tittel, F.K.: "TW Operation of a Tunable CW Dye Laser", Opt. Commun., 1973, 7, no.4, pp.349-50.

21.375. Schafer, F.P., and Ringwelski, L.: "Triplet Quenching by Oxygen in a Rhodamine 6G Laser", Z. Naturforsch., 1973, 28a, no.5, pp.792-3.

21.376. Weber, H.P., et al.: "Nd^{3+}-Ultra-phosphate Laser", Appl. Phys. Lett., 1973, 22, no.10, pp.534-6.

21.377. Birnbaum, M., and Klein, C.F.: "Stimulated Emission Cross Section at 1061 nm in Nd^{3+}:YAG", J. Appl. Phys., 1973, 44, no.6, pp.2928-30.

21.378. Bina, M.J., and Jones, C.R.: "Characteristics of Nd^{3+} Lasers Operating in the 1350-nm Range", Opt. Commun., 1973, 7, no.4, pp.400-1.

21.379. Kaminskii, A.A., et al.: "Investigation of Stimulated Emission in the $^4F-^4I$ Transition of Nd^{3+} Ions in Crystals", Phys. Status Solidi, 1973, 17A, no.1, pp.K75-7.

21.380. Aussenegg, F., and Gilly, H.: "Simple Organic Dye Laser", Acta Phys. Austriaca, 1973, 37, no.3, pp.254-8.

21.381. Dienes, A., Jain, R.K., and Lin, C.: "Formation Mechanisms in an Excited-State-Reaction Dye Laser", Appl. Phys. Lett., 1973, 22, no.12, pp.632-4.

21.382. Dezauzier, P., Eranian, A., and de Witte, O.: "Amplification Competition in a Double-Cavity Flash-Pumped Dye Laser", Appl. Phys. Lett., 1973, 22, no.12, pp.664-6.

21.383. Dienes, A., Shank, C.V., and Kohn, R.L.: "Characteristics of the 4-Methylumbelliferone Laser Dye", J. Quantum Electron. IEEE, 1973, QE-9, no.8, pp.833-43.

21.384. van der Ziel, J.P., et al.: "Laser Oscillation from Ho^{3+} and Nd^{3+} Ions in Epitaxially Grown Thin Aluminium Garnet Films", Appl. Phys. Lett., 1973, 22, no.12, pp.656-7.

21.385. Danielmeyer, H.G., Blatte, M., and Balmer, P.: "Fluorescence Quenching in Nd^{3+}:YAG", Appl. Phys., 1973, 1, no.5, pp.269-74.

21.386. Blatte, M., Danielmeyer, H.G., and Ulrich, R.: "Energy Transfer and the Complete Level System of Nd^{3+}-Ultraphosphate", Appl. Phys., 1973, 1, no.5, pp.275-8.

21.387. Kaminskii, A.A., et al.: "Investigation of Stimulated Emission from Nd^{3+}:$LiLa(MoO_4)_2$", Phys. Status Solidi, 1973, 17A, no.2, pp.K115-7.

21.388. Peters, R.L.St., and Taylor, D.J.: "Face-Pumped High-Average Power Low-Distortion Dye Laser", Appl. Phys. Lett., 1973, 23, no.2, pp.90-1.

21.389. Johnson, L.F., and Guggenheim, H.J.: "Laser Emission at 3 micron from Dy^{3+}:BaY_2F_8", Appl. Phys. Lett., 1973, 23, no.2, pp.96-8.

21.390. Weaver, E.A., Stewart, D.R., and Neilson, G.F.: "Lasing in a Phase-Separated Glass", J. Am. Ceram. Soc., 1973, 56, no.2, pp.68-72.

21.391. Okada, M., Shimizu, S., and Ieri, S.: "Dye Laser Pumped by Third Harmonic of Nd^{3+}:YAG Laser", Jap. J. Appl. Phys., 1973, 12, no.8, pp.1284-5.

21.392. Wortman, D.E., and Morrison, C.A.: "Laser Considerations for Promethium in $LiYF_4$", J. Quantum Electron. IEEE, 1973, QE-9, no.9, pp.956-8.

21.393. Zabiyakin, Yu.E., Smirnov, V.S., and Bakhshiev, N.G.: "Continuous Frequency Tuning of a Laser with Mixed Dye Solution", Opt. Spekt., 1973, 34, no.1, pp.148-50, and Opt. Spectrosc., 1973, 34, no.1, pp.81-2.

21.394. Rubinov, A.N., Batyrev, V.A., and Efendiev, T.Sh.: "Problem of Generation Spectra of Organic Dye Solutions", Zh. Prikl. Spekt., 1973, 18, no.5, pp.806-12.

21.395. Riseberg, L.A., Brown, R.M., and Holton, W.C.: "Class of Intermediate-Gain Laser Materials. Mixed Garnets", Appl. Phys. Lett., 1973, 23, no.3, pp.127-9.

21.396. Gavrilov, O.D., et al.: "Special Features of Generation at 920 nm with Nd^{3+}: Glass", Opt. Spekt., 1973, 34, no.1, pp.141-7, and Opt. Spectrosc., 1973, 34, no.1, pp.77-80.

21.397. Kaminskii, A.A., et al.: "Investigation of Stimulated Emission from $Lu_3Al_5O_{12}$ with Ho^{3+}, Er^{3+}, and Tm^{3+} Ions", Phys. Status Solidi, 1973, 18A, no.1, pp.K31-4.

21.398. Dzhibladze, M.I., and Murina, T.M.: "Concentration Dependence of Laser Generation Parameters in Dy^{2+}:CaF_2", Dokl. Akad. Nauk SSSR, 1973, 208, no.4-6, pp.1318-20, and Sov. Phys.-Dokl., 1973, 18, no.2, pp.134-5.

21.399. Friesem, A.A., Ganiel, U., and Neumann, G.: "Simultaneous Multiple-Wavelength Operation of a Tunable Dye Laser", Appl. Phys. Lett., 1973, 23, no.5, pp.249-51.

21.400. Letouzey, J.P., and Sari, S.O.: "Continuous Pulse Train Dye Laser Using an Open Flowing Passive Absorber", Appl. Phys. Lett., 1973, 23, no.6, pp.311-3.

21.401. Gillard, P.G., Foltz, N.D., and Cho, C.W.: "Gain in DTTC-Methanol Solutions as a Function of Pump Power, Wavelength, and Time", Appl. Phys. Lett., 1973, 23, no.6, pp.325-7.

21.402. Alekseev, N.E., et al.: "Effect of Thionyl Chloride on the Laser Characteristics of Nd^{3+}:$POCl_3-SnCl_4$", Izv. Akad. Nauk SSSR, Neorg. Mater., 1973, 9, no.2, pp.239-42, and Inorg. Mater., 1973, 9, no.2, pp.215-7.

21.403. Burlamacchi, P., Pratesi, R., and Salimbeni, R.: "Waveguide Dye-Laser Amplifier", J. Appl. Phys., 1973, 44, no.9, pp.4248-50.

21.404. Weber, J.: "Waveguide Dye Lasers Prepared by Diffusion", Opt. Commun., 1973, 8, no.5, pp.316-7.

21.405. Hirth, A., Faure, J., and Lougnot, D.: "Quenching Effects in Flashlamp-Excited Polymethine Dye Lasers", Opt. Commun., 1973, 8, no.4, pp.318-22.

21.406. Gerlach, H., and Scharf, H.: "Dye Laser with a Transverse-Flow Cuvette", Opt. Laser Technol., 1973, 5, no.4, pp.162-5.

21.407. Sieganthaler, K.E., et al.: "Further Observations Relating to X-Ray Laser Emission from $CuSO_4$-Doped Gelatine", Appl. Opt., 1973, 12, no.9, pp.2005-6.

21.408. Brown, D.C.: "Parasitic Oscillations in Large-Aperture Nd^{3+}:Glass Amplifiers", Appl. Opt., 1973, 12, no.10, pp.2215-7.

21.409. Chesler, R.B., and Draegert, D.A.: "Miniature Diode-Pumped Nd^{3+}:YAG Lasers", Appl. Phys. Lett., 1973, 23, no.5, pp.235-6.

21.410. Wilbrandt, R., and Weber, H.: "Overshoot Inversion of a Nd^{3+} Laser System", Opt. Commun., 1973, 8, no.4, pp.307-9.

21.411. Sevast'yanov, B.K., et al.: "Laser Based on Cr^{3+}:YAG", Kristallogr., 1973, 18, no.2, pp.308-10, and Sov. Phys.-Crystallogr., 1973, 18, no.2, pp.191-2.

21.412. Leontovich, A.M., and Mozharovskii, A.M.: "Self-Termination of Free Oscillations in Ruby at Low Temperatures", Kvantovaya Elektron., 1973, no.6, pp.69-73, and Sov. J. Quantum Electron., 1973, 2, no.6, pp.544-6.

21.413. Gol'danskii, V.I., Kagan, Yu., and Namiot, V.A.: "Two-Stage Excitation of Nuclei to Obtain Stimulated Gamma-Radiation", Zh. Eksp. Teor. Fiz. Pis'ma, 1973, 18, no.1, pp.61-3, and JETP Lett., 1973, 18, no.1, pp.34-5.

21.414. Ahmed, S.A., Infante, D.A., and Gergely, J.S.: "Simultaneous Three-Colour Output from a Three-Dye-Mixture Laser", J. Opt. Soc. Am., 1973, 63, no.10, p.1321.

21.415. Nakashima, M., Clapp, R.C., and Sousa, J.A.: "Benzocoumarins. Family of Laser Dyes", Nature, 1973, 245, no.147, pp.124-6.

21.416. Burlamacchi, P., Pratesi, R., and Vanni, U.: "Refractive Index Gradient Effects in a Superradiant Slab Dye Laser", Opt. Commun., 1973, 9, no.1, pp.31-4.

21.417. Pinter, F., et al.: "Effect of Photodecomposition on Generation of a Dye Laser with Pump Lamp", Zh. Prikl. Spekt., 1973, 19, no.2, pp.246-9.

21.418. Aristov, A.V., et al.: "Effect of Structure and Acidity of Rhodamine Dyes on their Laser Characteristics", Zh. Prikl. Spekt., 1973, 19, no.2, pp.250-3.

21.419. Ivanov, I.G., and Sem, M.F.: "Generation Lines in Thalline", Zh. Prikl. Spekt., 1973, 19, no.2, pp.358-60.

21.420. Stone, J., and Burrus, C.A.: "Nd^{3+}:Silica Lasers in End-Pumped Fibre Geometry", Appl. Phys. Lett., 1973, 23, no.7, pp.388-9.

21.421. Weber, M.J., et al.: "Laser Action from Ho^{3+}, Er^{3+}, and Tm^{3+} in $YAlO_3$", J. Quantum Electron. IEEE, 1973, QE-9, no.11, pp.1079-86.

21.422. Scully, M.O., and Louisell, W.H.: "Soft-X-Ray Laser Utilizing Charge Exchange", Opt. Commun., 1973, 9, no.3, pp.246-8.

21.423. Berlman, I.B., Rokni, M., and Goldschmidt, C.R.: "Lasing in Aromatic Couples by Energy Transfer", Chem. Phys. Lett., 1973, 22, no.3, pp.458-60.

21.424. Teschke, O., and Dienes, A.: "Solvent Effects on Triplet-State Population in Jet-Stream CW Dye Lasers", Opt. Commun., 1973, 9, no.2, pp.128-31.

21.425. Hirth, A., Vollrath, K., and Fouassier, J.P.: "Optimizing the Emission Characteristics of a Rhodamine 6G Dye Laser", Opt. Commun., 1973, 9, no.2, pp.139-45.

21.426. Meyer, Y.H., and Flamant, P.: "Study of Stimulated Emission in Attenuating and Amplifying Dye Solutions", Opt. Commun., 1973, 9, no.3, pp.227-30.

21.427. Draegert, D.A.: "Single-Diode/End-Pumped Nd^{3+}:YAG Laser", J. Quantum Electron. IEEE, 1973, QE-9, no.12, pp.1146-9.

21.428. Bedilov, M.R., and Khaidarov, K.: "Emission of a Ruby Laser under ^{60}Co Gamma Radiation", Opt. Spektr., 1973, 34, no.4, pp.765-7, and Opt. Spectrosc., 1973, 34, no.4, pp.441-2.

21.429. Yanait, Yu.A., Anisimov, Yu.M., and Pchelkin, V.I.: "Lamp Pumping System for Lasers Based on Solutions of Organic Compounds", Prib. Tekh. Eksp., 1973, 16, no.2, pp.181-3, and Instrum. Exp. Tech., 1973, 16, no.2, pp.546-8.

21.430. Grove, R.E., et al.: "Jet-Stream CW Dye Laser for High-Resolution Spectroscopy", Appl. Phys. Lett., 1973, 23, no.8, pp.442-4.

21.431. Burlamacchi, P., and Pratesi, R.: "High-Efficiency Coaxial Waveguide Dye Laser with External Excitation", Appl. Phys. Lett., 1973, 23, no.8, pp.475-6.

21.432. Ferrar, C.M.: "Spark-Pumped Dye Laser with High PRF and Low Threshold", Appl. Phys. Lett., 1973, 23, no.10, pp.548-9.

21.433. Baczynski, A., et al.: "Influence of Molecular Parameters on Laser Properties of Dye Solutions", Acta Phys. Pol., 1973, 44A, no.6, pp.805-12.

21.434. Kruhler, W.W., Jeser, J.P., and Danielmeyer, H.G.: "Properties and Laser Oscillation of Nd^{3+}:YP_5O_{14}", Appl. Phys., 1973, 2, no.6, pp.329-33.

21.435. Danielmeyer, H.G., et al.: "CW Oscillation of a Nd^{3+}:ScP_5O_{14} Laser with 4-mW Pump Threshold", Appl. Phys., 1973, 2, no.6, pp.335-8.

21.436. Andreev, R.B., and Volosov, V.D.: "Two-Frequency Nd^{3+}:Glass Laser", Opt. Spekt., 1973, 34, no.4, pp.810-2, and Opt. Spectrosc., 1973, 34, no.4, pp.468-9.

21.437. Sahar, E., and Wieder, I.: "Excited-Singlet-State Absorption Spectrum with Tunable Dye Lasers", Chem. Phys. Lett., 1973, 23, no.4, pp.518-21.

21.438. Korobov, A.M., et al.: "Two-Band Generation of a Rhodamine-6G Laser", Opt. Commun., 1973, 9, no.4, pp.336-7.

21.439. Lotem, H.: "Tunable Dye Laser in the Range 1000-1145 nm", Opt. Commun., 1973, 9, no.4, pp.346-7.

21.440. Kaporskii, L.N., and Kalabushkin, O.I.: "Lasing in a Nd^{3+}:$POCl_3$-$SnCl_4$ Liquid Laser", Zh. Tekh. Fiz., 1973, 43, no.5, pp.1097-9, and Sov. Phys.-Tech. Phys., 1973, 18, no.5, pp.700-1.

21.441. Greskovich, C., and Chernoch, J.P.: "Polycrystalline Nd^{3+}:Ceramic Lasers", J. Appl. Phys., 1973, 44, no.10, pp.4599-606.

21.442. Mori, K.: "Effect of Induced Transient Absorption in Nd^{3+}:YAG on Laser Threshold and Efficiency", NEC Res. Dev., 1973, no.30, pp.22-6.

21.443. Gapontsev, V.P., et al.: "Effective Stimulated Emission at 1054-1540 nm", Zh. Eksp. Teor. Fiz. Pis'ma, 1973, 18, no.7, pp.428-31, and JETP Lett., 1973, 18, no.7, pp.251-3.

21.444. Burmasov, V.S., et al.: "Keramidonines. Class of Organic Compounds with Low Generation Thresholds", Zh. Prikl. Spekt., 1973, 19, no.3, pp.545-9.

21.445. Damen, T.C., Weber, H.P., and Tofield, B.C.: "Nd^{3+}:LaP_5O_{14} Laser Performance", Appl. Phys. Lett., 1973, 23, no.9, pp.519-20.

21.446. Makogon, M.M., Ponomarev, Yu.N., and Serduykov, V.I.: "Self-Q-Switched Nd^{3+} Laser", Kvantovaya Elektron., 1973, no.2, pp.59-61, and Sov. J. Quantum Electron., 1973, 3, no.2, pp.121-2.

21.447. Kryukov, P.G., et al.: "Mechanism of Power Limitation in Amplification of Ultrashort Pulses in Nd^{3+}:Glass Lasers", Kvantovaya Elektron., 1973, no.2, pp.102-5, and Sov. J. Quantum Electron., 1973, 3, no.2, pp.161-2.

21.448. Wilbrandt, R., Weber, H., and Keller, M.: "Passive Q-Switch Conditions for Nd^{3+}:Glass Lasers", Z. Angew. Math. Phys., 1973, 24, no.3, p.452.

21.449. Zabiyakin, Yu.E., Smirnov, V.S., and Babhshiev, N.G.: "Experimental Study of Lasing Characteristics of Solutions of Phthalimide Substitutes with Lamp Pumping", Opt. Spekt., 1973, 35, no.5, pp.958-9, and Opt. Spectrosc., 1973, 35, no.5, pp.553-4.

21.450. Aristov, A.V., and Maslyukov, Yu.S.: "Amplification and Induced Absorption in Solutions of Organic Luminophors", Opt. Spekt., 1973, 35, no.6, pp.1138-41, and Opt. Spectrosc., 1973, 35, no.6, pp.660-1.

21.451. Zabiyakin, Yu.E., Smirnov, V.S., and Bakhshiev, N.G.: "Lasing of Pyrilium-Perchlorate Solutions by Lamp Pumping", Opt. Spekt., 1973, 35, no.6, pp.1167-8, and Opt. Spectrosc., 1973, 35, no.6, pp.675-6.

21.452. Venkin, G.V., et al.: "(Dye) Laser with Waveguide-Type Ring Resonator", Kvantovaya Elektron., 1973, no.7, pp.108-9, and Sov. J. Quantum Electron., 1973, 3, no.1, p.60.

21.453. Gol'danskii, V.I., and Kagan, Yu.: "Possibility of Creating a Nuclear Gamma Laser", Zh. Eksp. Teor. Fiz., 1973, 64, no.1, pp.90-7, and Sov. Phys.-JETP, 1973, 37, no.1, pp.49-52.

21.454. Imagawa, T., et al.: "CW Oscillation of a Dye Laser", Oyo Buturi, 1973, 42, no.7, pp.752-5.

21.455. Sharp. E.J., Horowitz, D.J., and Miller, J.E.: "High-Efficiency Nd^{3+}:$LiYF_4$", J. Appl. Phys., 1973, 44, no.12, pp.5399-401.

21.456. Chesler, R.B., and Singh, S.: "Performance Model for End-Pumped Miniature Nd^{3+}:YAG Lasers", J. Appl. Phys., 1973, 44, no.12, pp.5441-3.

21.457. Maeda, M., Noda, Y., and Miyazoe, Y.: "Dye Laser Pumped by a Flashlamp with Fast Rise Period", Technol. Rep. Kyushu Univ., 1973, 46, no.6, pp.708-13.

21.458. Ketskemety, I., and Kozma, L.: "Calculation of Frequency of Dye Lasers in Quasistationary Regime", Acta Phys. Chem. Szeged., 1973, 19, no.3, pp.217-20.

21.459. Neporent, B.S., Shilov, V.B., and Lukomskii, G.V.: "Spectral Kinetics of Laser Action in Solution of Various Organic Substances", Opt. Spekt., 1973, 35, no.3, pp.535-9, and Opt. Spectrosc., 1973, 35, no.3, pp.312-4.

21.460. Derkacheva, L.D., and Petukhov, V.A.: "Luminescence and Stimulated Emission from Photoprolytic Forms of Dyes", Kvantovaya Elektron., 1973, no.2, pp.89-93, and Sov. J. Quantum Electron., 1973, 3, no.2, pp.149-51.

21.462. Galant, Ye.I., et al.: "Stimulated Emission of Nd^{3+} Ions in Quartz Glass", Zh. Eksp. Teor. Fiz. Pis'ma, 1973, 18, no.10, pp.635-7, and JETP Lett., 1973, 18, no.10, pp.372-3.

21.463. Bonch-Bruevich, A.M., Kaporskii, L.N., and Kalabushkin, O.I.: "Inorganic Liquid Lasers", Opt.-Mekh. Prom., 1973, 40, no.12, pp.49-58, and Sov. J. Opt. Technol., 1973, 40, no.12, pp.770-81.

21.464. Balashov, I.F., Berenberg, V.A., and Ermakov, B.A.: "Controlled-Duration Microsecond Pulses in a Ruby Laser", Zh. Tekh. Fiz., 1973, 43, no.7, pp.1523-9, and Sov. Phys.-Tech. Phys., 1974, 18, no.7, pp.964-7.

21.465. Krupke, W.F.: "Induced-Emission Cross Sections in Nd^{3+} Laser Glasses", J. Quantum Electron. IEEE, 1974, QE-10, no.4, pp.450-7.

21.466. Stemme, R., Herziger, G., and Weber, H.: "Power and Halfwidth of First Laser Spike", Opt. Commun., 1974, 10, no.3, pp.221-6.

21.467. Elliott, C.J.: "Gain Saturation and Self-Focusing Considerations in Design of Optical Amplifiers", Appl. Phys. Lett., 1974, 24, no.2, pp.91-3.

21.468. Rowley, P.D., and Billman, K.W.: "Experimental Attempts to Confirm X-Ray Lasing from $CuSO_4$", Appl. Opt., 1974, 13, no.3, pp.453-5.

21.469. Dinev, S.G., and Tomov, I.V.: "X-Ray Collimated Emission from $CuSO_4$-Doped Gelatine", Opto-Electron., 1974, 6, no.2, pp.197-8.

21.470. Ammann, E.O., Decker, C.D., and Falk, J.: "High-Peak-Power 532-nm-Pumped Dye Laser", J. Quantum Electron. IEEE, QE-10, no.4, pp.463-5.

21.471. Ketskemety, I., and Kozma, L.: "Remarks Concerning the Theory of Quasistationary Dye Lasers", Acta Phys. Acad. Sci. Hung., 1974, 35, no.1-4, pp.63-71.

21.472. Fielding, S.J.: "Intermediate-Laser/ Dye-Laser System", J. Phys. E, 1974, 7, no.4, pp.250-2.

21.473. Maeda, M., and Miyazoe, Y.: "Compact Lamp-Pumped Dye Laser with Fast Flash Rise Period", Jap. J. Appl. Phys., 1974, 13, no.2, pp.369-70.

21.474. Takakusa, M., and Itoh, U.: "Extended Tunability of a Dye Laser Based on 4MU in an Ethanol/Water Solution", Opt. Commun., 1974, 10, no.1, pp.8-10.

21.475. Kato, K., and Fujisawa, A.: " "Longitudinally Pumped High-Power Dye Laser in the Blue", Opt. Commun., 1974, 10, no.1, pp.21-2.

21.476. Wang, G.: "IR Dye Laser Excited by a Diode Laser", Opt. Commun., 1974, 10, no.2, pp.149-53.

21.477. Basting, D., Schafer, F.P., and Steyer, B.: "Laser Dyes", Appl. Phys., 1974, 3, no.1, pp.81-8.

21.478. Loth, C., and Megie, G.: "High Spectral Luminance Dye Amplifier", J. Phys. E, 1974, 7, no.2, pp.80-2.

21.479. Magde, D., Bushaw, B.A., and Windsor, M.W.: "Q-Switching and Mode-Locking Nd^{3+}:Glass Laser with Nickel Dithienes", J. Quantum Electron. IEEE, 1974, QE-10, no.3, p.394.

21.480. Mollenauer, L.F., and Olson, D.H.: "Broadly Tunable CW Laser Using Colour Centres", Appl. Phys. Lett., 1974, 24, no.8, pp.386-8.

21.481. Weber, H.P., Liao, P.F., and Tofield, B.C.: "Emission Cross Section and Fluorescence Efficiency of Nd^{3+}:Pentaphosphate", J. Quantum Electron. IEEE, 1974, QE-10, no.7, pp.563-7.

21.482. Watts, R.K., and Holton, W.C.: "Intermediate-Gain Laser Materials. Nd^{3+}:$Y_3(AlGa)_5O_{12}$", J. Appl. Phys., 1974, 45, no.2, pp.873-81.

21.483. Singh, S., et al.: "Laser Emission at 1065 nm from Nd^{3+}:$CeCl_3$ at Room Temperature", Appl. Phys. Lett., 1974, 24, no.1, pp.10-3.

21.484. Srinivasan, R., et al.: "Anomalous Fluorescence and Laser Emission from 7-Alkylamino Coumarins in Acid Solutions", Chem. Phys. Lett., 1974, 25, no.4, pp.537-40.

21.485. Liberman, S., and Pinard, J.: "Single-Mode CW Dye Laser with Large Frequency-Range Tunability", Appl. Phys. Lett., 1974, 24, no.3, pp.142-4.

21.486. Andreou, D.: "High-Power Liquid Laser Amplifier", J. Phys. D, 1974, 17, no.8, pp.1073-7.

21.487. Edwards, J.G., and Sandoe, J.N.: "Theoretical Study of Nd:Yb:Er:Glass Laser", J. Phys. D, 1974, 7, no.8, pp.1078-95.

21.488. Bokut', B.V., et al.: "Organic Dye Laser Resonator with Longitudinal Linear Pumping", Zh. Prikl. Spekt., 1974, 20, no.1, pp.38-41.

21.489. Kepros, J.G.: "Theoretical Model Explaining Aspects of the Utah X-Ray Laser Experiments", Appl. Opt., 1974, 13, no.4, pp.695-6.

21.490. Measures, R.M.: "Prospects for Developing a Laser Based on Electrochemiluminescence", Appl. Opt., 1974, 13, no.5, pp.1121-33.

21.491. Baczynski, A., et al.: "Numerical Calculations of Flash-Pumped Dye Laser", Acta Phys. Pol., 1974, 45A, no.5, pp.793-801.

21.492. Wellegehausen, B., Welling, H., and Beigang, R.: "Narrowband Jet-Stream Dye Laser", Appl. Phys., 1974, 3, no.5, pp.387-91.

21.493. Maeda, M., and Miyazoe, Y.: "Efficient UV Organic Liquid Laser Pumped by a High-Power N_2 Laser", Jap. J. Appl. Phys., 1974, 13, no.5, pp.827-34.

21.494. Hildebrand, O.: "N_2-Laser Excitation of Polymethine Dyes for Wavelengths up to 950 nm", Opt. Commun., 1974, 10, no.4, pp.310-2.

21.495. Owyoung, A.: "Efficient, Ruby-Laser-Pumped, Diffraction-Limited Dye Laser", Opt. Commun., 1974, 11, no.1, pp.14-7.

21.496. Aoyagi, Y., and Namba, S.: "(Dye) Laser Oscillation in Simple Corrugated Optical Waveguide", Appl. Phys. Lett., 1974, 24, no.11, pp.537-9.

21.497. Aoyagi, Y., and Namba, S.: "Wavelength Control of Thin-Film Dye Laser by Changing Thickness", Jap. J. Appl. Phys., 1974, 13, no.6, pp.1031-2.

21.498. Brinkschulte, H., Perchermeier, J., and Schimitschek, E.J.: "Repetitively Pulsed, Q-Switched, Inorganic Liquid Laser", J. Phys. D, 1974, 7, no.10, pp.1361-8.

21.499. Burlamacchi, P., Pratesi, R., and Salimbeni, R.: "High-Energy Planar Self-Guiding Dye Laser", Opt. Commun., 1974, 11, no.2, pp.109-11.

21.500. Dunning, F.B., and Stebbings, R.F.: "Efficient Generation of Tunable Near-UV Radiation Using a N_2-Laser-Pumped Dye", Opt. Commun., 1974, 11, no.2, pp.112-4.

21.501. Stone, J., and Burrus, C.A.: "Nd^{3+}:Glass Fibre Lasers. Room Temperature CW Operation with an Injection-Laser Pump", Appl. Opt., 1974, 13, no.6, pp.1256-8.

21.502. Zaraga, F.: "Single-Mode N_2-Laser-Pumped Tunable Dye Ring Laser", Appl. Phys., 1974, 4, no.1, pp.87-8.

21.503. Johnson, L.F., and Guggenheim, H.J.: "Electronic and Phonon-Terminated Laser Emission from Ho^{3+}:BaY_2F_8", J. Quantum Electron. IEEE,1974,QE-10,no.4,pp.442-9.

21.504. De Martini, F., Grassano, U.M., and Simoni, F.: "Stimulated Emission from F-Centres in KCl", Opt. Commun., 1974, 11, no.1, pp.8-10.

21.505. Yarborough, J.M.: "CW Dye Laser Emission Spanning the Visible Spectrum", Appl. Phys. Lett., 1974,24,no.12,pp.629-30.

21.506. Saruwatari, M., and Izawa, T.: "Nd^{3+}:Glass Laser with Three-Dimensional Optical Waveguide", Appl. Phys. Lett., 1974, 24, no.12, pp.603-5.

21.507. Yariu, A.: "Analytical Considerations of Bragg Coupling Coefficients and Distributed-Feedback X-Ray Lasers in Single Crystals", Appl. Phys. Lett., 1974, 25, no.2, pp.105-7.

21.508. Eckardt, R.C., DeRosa, J.L., and Letellier, J.P.: "Characteristics of a Nd^{3+}:CaLaSOAP Mode-Locked Oscillator", J. Quantum Electron. IEEE, 1974, QE-10, no.8, pp.620-2.

21.509. Fisher, R.A.: "Possibility of a Distributed-Feedback X-Ray Laser", Appl. Phys. Lett., 1974, 24, no.12, pp.598-9.

21.510. Schimitschek, E.J., et al.: "Laser Performance and Stability of Fluorinated Coumarin Dyes", Opt. Commun., 1974, 11, no.4, pp.352-5.

21.511. Nakashima, M., and Sousa, J.A.: "Polychromatic Pulsed Dye Laser", Spectrosc. Lett., 1974, 7, no.1, pp.15-7.

21.512. Zhuravleva, L.N., Kromskii, G.I., and Shcherbakov, A.A.: "Optical Efficiencies of Coaxial Systems of Pumping Liquid Lasers", Zh. Prikl. Spekt., 1974, 20, no.6, pp.981-6.

21.513. Baltakov, F.N., Barikhin, B.A., and Sukhanov, L.V.: "Laser Using Rhodamine 6G in Ethyl Alcohol with Output Pulses of 400J", Zh. Eksp. Teor. Fiz. Pis!ma, 1974, 19, no.5, pp.300-2, and JETP Lett.,1974, 19, no.5.

21.514. Bogdanov, V.L., and Klochkov, V.P.: "Near-Threshold Ruby Laser under Passive Q-Switching", Opt. Spekt., 1974, 36, no.2, pp.405-9, and Opt. Spectrosc., 1974, 36, no.2, pp.233-5.

21.515. Leontovich, A.M., Baeva, E.D., and Mozharovskii, A.M.: "Low-Temperature Emission from a Ruby Laser with Passive Q-Switch", Kvantovaya Elektron., 1974, no.4, pp.106-8, and Sov. J. Quantum Electron., 1974, 3, no.4, pp.343-4.

21.516. Dmitriev, V.F., and Shuryak, E.V.: "Feasibility of the Gamma-Ray Laser", Zh. Eksp. Teor. Fiz., 1974, 67, no.2, pp.494-502, and Sov. Phys.-JETP, 1974, 40, no.2.

21.517. Ornstein, M.H., and Derr, V.E.: "Prepulse Enhancement of Flashlamp-Pumped Dye Laser", Appl. Opt., 1974, 13,no.9,pp.2100-14.

21.518. Evans, D.E., Purie, J., and Yeoman, M.L.: "Quenching of Laser Action in Cresyl Violet by 694.3-nm Radiation", Appl. Phys. Lett., 1974, 25, no.3, pp.151-2.

21.519. Jethwa, J., and Schafer, F.P.: "Reliable High-Average-Power Dye Laser", Appl. Phys., 1974, 4, no.4, pp.299-302.

21.520. Ganiel, U., and Neumann, G.: "Power Output Coupling in a Dye Laser Pumped by a N_2 Laser", Opt. Commun., 1974, 12, no.1, pp.5-7.

21.521. Drake, J.M., and Morse, R.I.: "Operating Characteristics of a Linear-Flashlamp-Pumped Laser Using a Coaxial Dye Cell", Opt. Commun., 1974, 12, no.2, pp.132-5.

21.522. Antonov, E.N., et al.: "Influence of the Solvent on the Output Parameters of a CW Dye Laser", Kvantovaya Elektron., 1974, 4, no.1, pp.204-6, and Sov. J. Quantum Electron., 1974, 4, no.1, pp.126-7.

21.523. Adrain, R.S., et al.: "Amplification of Picosecond Dye-Laser Pulses", Opt. Commun., 1974, 12, no.2, pp.140-2.

21.524. Brughera, S., Polloni, R., and Scattorin, M.: "Simple Single-Transverse-Mode, Flashlamp-Pumped, Dye Laser", Alta Freq., 1974, 43, no.7, pp.456-8.

21.525. Cherpak, N.T., Shamfarov, Ya.L., and Smirnova, T.A.: "Anomalous Increase in Population Inversion in Ruby at Low Temperature", Fiz. Tver. Tela, 1974, 16, no.2, pp.587-8, and Sov. Phys.-Solid State, 1974, 16, no.2, pp.380-1.

21.526. Eremenko, A.S., et al.: "Enclosure with Diffusely Reflecting Layer in a Pumping System for a Solid-State Laser", Kvantovaya Elektron., 1973, no.5, pp.124-6, and Sov. J. Quantum Electron., 1974, 3, no.5, pp.442-3.

21.527. Marling, J.B., et al.: "Lasing Characteristics of Seventeen Visible-Wavelength Dyes Using a Coaxial-Flashlamp-Pumped Laser", Appl. Opt., 1974, 13, no.10, pp.2317-20.

21.528. Kozlov, N.P., and Protasov, Yu.S.: "Use of Plasma Accelerator for Optical Pumping of Lasers", Zh. Tekh. Fiz., 1974, 44, no.3, pp.575-9, and Sov. Phys.-Tech. Phys., 1974, 19, no.3, pp.358-60.

21.529. Efendiev, T.Sh., and Rubinov, A.N.: "Tunable Dye Laser with Distributed Feedback and Narrow Emission Line", Zh. Prikl. Spekt., 1974, 21, no.3, pp.526-8.

21.530. Conant, L.C., and Reno, C.W.: "GaAs-Laser-Diode Pumped Nd^{3+}:YAG Laser", Appl. Opt., 1974, 13, no.11, pp.2457-8.

21.531. Farkas, G., and Horvath, Z.G.: "Metal Monocrystals as Possible Distributed-Feedback X-Ray-Laser Materials in Pre-Plasma State", Phys. Lett., 1974,50A,no.1,pp.45-6.

21.532. Burlamacchi, P.: "Superradiant Guided Modes in Flashlamp Pumped Capillary Dye Lasers", Opto-Electron., 1974, 6, no.6, pp.465-72.

21.533. Andreichev, V.A., et al.: "Dependence of Ruby-Laser Efficiency with a Passive Shutter on Type of Illuminator", Zh. Prikl. Spekt., 1974, 21, no.5, pp.807-10.

21.534. Everev, G.M., et al.: "Stimulated Emission of Er^{3+}:YAG", Zh. Prikl. Spekt., 1974, 21, no.5, pp.820-3.

21.535. Winters, B.H., Mandelberg, H.I., and Mohr, W.B.: "Photochemical Products in Coumarin Laser Dyes", Appl. Phys. Lett., 1974, 25, no.12, pp.723-5.

21.536. Izumi, T.: "Design and Characteristics of a Repetitively Pulsed and Tunable Dye Laser", Anritsu Tech. Bull., 1974, no.31, pp.52-60.

21.537. Chis, I.D., Julea, Th.N., and Popescu, I.M.: "Flashlamp-Pumped Rhodamine-6G Dye Laser", Rev. Roum. Phys., 1974, 19, no.9, pp.1005-7.

21.538. Greskovich, C., and Chernoch, J.P.: "Improved Polycrystalline Ceramic Lasers", J. Appl. Phys., 1974, 45, no.10, pp.4495-502.

21.539. Nikolaev, F.A., et al.: "Optical Pumping of Nd^{3+}:Glass by a High-Current Discharge", Kvantovaya Elektron., 1974, 1, no.4, pp.858-62, and Sov. J. Quantum Electron., 1974, 4, no.4, pp.471-3.

21.540. Anghelov, D.A., Kircheva, P.P., and Misheva, M.A.: "Time Development of Stimulated Emission from Organic Dyes with Inhomogeneously Broadened Electron Levels", Bulg. J. Phys., 1974, 1, no.1, pp.5-12.

21.541. Okada, M., Shimizu, S., and Ieiri, S.: "Tunable Dye Lasers Excited by Harmonics of Nd^{3+}:YAG Laser", NHK Tech. J., 1974, 26, no.4, pp.161-71.

21.542. Stepanov, B.I., and Batyrev, V.A.: "Tuning of Organic Lasers", Opt. Spekt., 1974, 37, no.1, pp.166-70, and Opt. Spectrosc., 1974, 37, no.1, pp.92-4.

21.543. Zaretskii, A.I., et al.: "Characteristics of Nd^{3+}:$POCl_3$-$SnCl_4$ Inorganic Liquid Laser", Kvantovaya Elektron., 1974, 1, no.5, pp.1180-4, and Sov. J. Quantum Electron., 1974, 4, no.5, pp.646-8.

21.544. Alekseev, V.A., Protasov, Yu.S., and Rubinov, A.N.: "High-Power UV Laser with Paraterphenyl Solution Excited by Magneto-plasma Compressor", Zh. Eksp. Teor. Fiz. Pis'ma, 1974, 20, no.11, pp.716-8, and JETP Lett., 1974, 20, no.11.

21.545. Galaktionova, N.M., et al.: "CW Nd^{3+}:Glass Lasers", Opt. Spekt., 1974, 37, no.1, pp.162-5, and Opt. Spectrosc., 1974, 37, no.1, pp.90-1.

21.546. Aristov, A.V., and Shevandin, V.S.: "Nd^{3+}:Glass Laser Pumped by a Flash-Excited Dye Laser", Opt. Spekt., 1974, 37, no.3, pp.596-8, and Opt. Spectrosc., 1974, 37, no.3, pp.336-7.

21.547. Bagdasarov, Kh.S., et al.: "Stimulated Radiation of Nd^{3+}:$Gd_3Ga_5O_{12}$", Dokl. Akad. Nauk SSSR, 1974, 216, no.4-6, pp.1018-21, and Sov. Phys.-Dokl., 1974, 19, no.6, pp.353-5.

21.548. Bagdasarov, Kh.S., et al.: "Stimulated Emission of Yb^{3+} in Al Garnets", Dokl. Akad. Nauk SSSR, 1974, 216, no.4-6, pp.1247-9, and Sov. Phys.-Dokl., 1974, 19, no.6, pp.358-9.

21.549. Ketskemety, I., et al.: "Investigation of Dye Lasers Pumped with N_2 Laser", Acta Phys. Chem. Szeged., 1974, 20, no.3, pp.191-7.

21.550. Basov, Yu.G., and Vorob'ev, M.Yu.: "Pulsed Xe Lamp with Spiral Discharge Channel", Prib. Tekh. Eksp., 1974, 17, no.3, pp.171-2, and Instrum. Exp. Tech., 1974, 17, no.3, pp.825-7.

21.551. Kalinin, V.N., et al.: "Lasing Properties of $Yb^{3+}Er^{3+}$:Glass Pumped by a Laser", Zh. Tekh. Fiz., 1974, 44, no.6, pp.1328-31, and Sov. Phys.-Tech. Phys., 1974, 19, no.6, pp.835-6.

21.552. Rubinov, A.N., et al.: "Applicability of Empirical Formulae of Absorption and Fluorescence Bands of Dyes to Calculation of Laser Parameters", Acta Phys. Chem. Szeged., 1974, 20, no.3, pp.295-8.

21.553. Dzyubenko, M.I., et al.: "Features of Generation Spectra of a Rhodamine Laser", Kvantovaya Elektron. (Kiev), 1974, no.8, pp.126-33.

21.554. Zharikov, E.V., et al.: "Stimulated Emission from Er^{3+}:YAG at 2.94 micron", Kvantovayo Elektron., 1974, 1, no.8, pp.1867-9.

21.555. Dzyubenko, M.I., Matveev, A.Ya., and Naumenko, I.G.: "Increase in Efficiency of Dye Lasers", Opt. Spekt., 1974, 37, no.4, pp.745-9, and Opt. Spectrosc., 1974, 37, no.4, pp.423-5.

22. GASEOUS-PHASE LASERS

22.1. Noon, J.H., Blaszuk, P.R., and Holt, E.H.: "Electron Radiation-Temperature Measurements in a CO_2-Laser Amplifier", J. Appl. Phys., 1968, 39, p.5518.

22.2. Fendley, J.R.: "CW UV (Gas) Lasers", J. Quantum Electron. IEEE, 1968, QE-4, p.627.

22.3. Klement, E., and Rosenberger, D.: "Increasing Power of a Sealed CO_2 Laser by Rotation of the Filling Gas", Nachr. Tech. Fachber., 1968, 35, p.703.

22.4. Kneubuhl, F.K.: "Theory and Construction of Gas Lasers for 0.10-0.77 mm", Nachr. Tech. Fachber., 1968, 35, p.754.

22.5. Preobrazhenskii, N.G., and Shaparev, N.Ya.: "Quenching of Laser Action in an Ion Laser", Opt. Spekt., 1968, 25, p.317, and Opt. Spectrosc., 1968, 25, p.172.

22.6. Babaev, I.K., et al.: "Thermodynamic Approach to Design of Molecular Lasers", Radiotekh. Elektron., 1968, 13, p.2262, and Radio Eng. Electron. Phys., 1968, 13, no.12, pp.1987-9.

22.7. Orlov, L.N., and Rubanov, V.S.: "Method of Calculating Temperature Dependence of He-Ne Laser Gain", Zh. Prikl. Spekt., 1968, 9, no.5, p.820.

22.8. Tunitskii, L.N., and Cherkasov, E.M.: "Pure-Oxygen Gas Laser", Zh. Prikl. Spekt., 1968, 9, no.5, p.812, Zh. Tekh. Fiz., 1968, 38, no.7, p.1200, and Sov. Phys.-Tech. Phys., 1969, 13, no.7, pp.993-4.

22.9. Smirnov, B.M.: "Molecular-Ion Gas Laser", Dokl. Akad. Nauk SSSR, 1968, 183, no.3, p.554, and Sov. Phys.-Dokl., 1969, 13, no.11, p.1148.

22.10. Herziger, G., Makosch, G., and Weber, J.: "Gain, Bandwidth, and Radiance of 633-nm He-Ne Amplifier", Z. Phys., 1969, 228, no.1, p.89.

22.11. Dyubko, S.F., Svich, V.A., and Balitov, R.A.: "Submillimetre CW Laser Working with HCN and H_2O Vapours", Zh. Tekh. Fiz., 1969, 39, no.6, p.1135, and Sov. Phys.-Tech. Phys., 1969, 14, no.6, pp.855-8.

22.12. Basov, N.G., and Letokhov, V.S.: "Two-Level Gas Laser with Coherent Optical Pumping", Zh. Eksp. Teor. Fiz. Pis'ma, 1969, 9, no.12, p.660, and JETP Lett., 1969, 9, no.12, p.409.

22.13. Kabashnikov, V.P.: "Competition of Rotational Transitions in a CO_2 Laser", Zh. Prikl. Spekt., 1969, 11, p.805.

22.14. Murai, A.: "Life of a Pulsed-Discharge, Sealed-Off, HCN Laser", Jap. J. Appl. Phys., 1969, 8, p.250.

22.15. Hassler, J.C., and Coleman, P.D.: "Far-IR Lasing in H_2S, OCS, and SO_2", Appl. Phys. Lett., 1969, 14, p.135.

22.16. Pogorelyi, P.A., and Tibilov, A.S.: "Mechanism of Laser Action in H_2-Na Mixture", Opt. Spekt., 1969, 25, p.542, and Opt. Spectrosc., 1969, 25, p.301.

22.17. Giuliano, C.R., and Hess, L.D.: "Reversible Photodissociative Laser System", J. Appl. Phys., 1969, 40, p.2428.

22.18. Karlov, N.V., et al.: "Inversion in a CO_2 Laser Using Pulsed Pumping", Radiotekh. Elektron., 1969, 14, p.320, and Radio Eng. Electron. Phys., 1969, 14, no.2, pp.273-7.

22.19. Frayne, P.G.: "Repetitive Q-Modulated HCN Gas Laser", J. Phys. B, 1969, 2, p.247.

22.20. Garavaglia, M., Gallardo, M., and Massone, C.A.: "Interaction between 1^+ and 2^+ Laser Systems in N_2", Phys. Lett., 1969, 28A, p.787.

22.21. Bergmann, K., and Demtroder, W.: "Cascade Laser Transition in He-Ne Mixture", Phys. Lett., 1969, 29A, p.94.

22.22. Crafer, R.C., et al.: "Time-Dependent Processes in CO_2 Laser Amplifiers", J. Phys. D, 1969, 2, p.183.

22.23. Crafter, R.C., Gibson, A.F., and Kimmitt, M.F.: "Rotational Relaxation and Gain Saturation in CO_2 Laser Amplifiers", J. Phys. D, 1969, 2, p.1135.

22.24. Akitt, D.P., and Jeffers, W.Q.: "Correlation Effects in H_2O and D_2O Laser Transitions", J. Appl. Phys.,1969,40,p.429.

22.25. Orlov, L.N.: "Effect of Reabsorption on Gas-Laser Output", Zh. Prikl. Spekt., 1969, 10, no.1, p.146.

22.26. Osgood, R.M., Nichols, E.R., and Eppers, W.C.: "Q-Switching of CO Laser", Appl. Phys. Lett., 1969, 15, p.69.

22.27. Carbone, R.J., and Witteman, W.J.: "Vibrational Energy Transfer in CO_2 Under Laser Conditions", J. Quantum Electron. IEEE, 1969, QE-5, p.442.

22.28. Goldsborough, J.P., and Bloom, A.L.: "Near-IR Laser", J. Quantum Electron. IEEE, 1969, QE-5, p.459.

22.29. Bridges, W.B., and Mercer, G.N.: "CW Operation of High Ionization States in a Xe Laser", J. Quantum Electron. IEEE, 1969, QE-5, p.476.

22.30. Korolev, F.A., and Baikov, S.S.: "TW Generation in He-Ne Laser at 3.39 micron", Zh. Prikl. Spekt., 1969,10,no.3,p.441.

22.31. Batovskii, O.M., et al.: "Chemical Laser Operating on Branched Chain Reaction of F_2 with H_2", Zh. Eksp. Teor. Fiz. Pis'ma, 1969, 9, no.6, p.341, and JETP Lett., 1969, 9, no.6, p.200.

22.32. Markova, S.V., et al.: "Current Changes in CO_2-N_2-He Discharge due to Generation", Zh. Prikl. Spekt.,1969,10,no.3,p.421.

22.33. Silfvast, W.T.: "CW Metal-Vapour Laser Transitions in Cd, Sn, and Zn", Appl. Phys. Lett., 1969, 15, p.23.

22.34. Deutsch, T.F., Horrigan, F.A., and Rudko, R.I.: "CW Operation of High-Pressure Flowing CO_2 Lasers", Appl. Phys. Lett., 1969, 15, p.88.

22.35. Rosenberger, D.: "Optimum Transmission of Folded High-Gain Lasers", Z. Naturforsch. 1969, 24a, no.5, p.867.

22.36. Ali, A.W.: "Study of N_2-Laser Power Density and Design Considerations", Appl. Opt., 1969, 8, p.993.

22.37. McFarlane, R.A., and Fretz, L.H.: "High-Power Operation of Pulsed H_2O Laser and Precision-Wavelength Measurement", Appl. Phys. Lett., 1969, 14, p.385.

22.38. Morantz, D.J., and Regan, A.J.: "Impurity Studies in a He-Ne Laser", J. Phys. D, 1969, 2, p.1474.

22.39. Henry, A., Arditi, I., and Henry, L.: "Passive Q-Switching of HCl Chemical Laser", C. R. Acad. Sci. (Paris), 1969, 268B, p.1245.

22.40. Tyte, D.C.: "Mean Electron Energy in a CO_2 Laser Plasma", Electron. Lett., 1969, 5, p.447.

22.41. Alimpiev, S.S., et al.: "Influence of Dissociation of Inversion on a CO_2 Laser with Pulsed Pumping", Zh. Eksp. Teor. Fiz. Pis'ma, 1969, 9, no.7, p.377, and JETP Lett., 1969, 9, no.7, p.223.

22.42. Jones, R.G., et al.: "Transient Phenomena in the 337-micron Laser", Appl. Opt., 1969, 8, p.701.

22.43. Chang, T.Y., Wang, C.H., and Cheo, P.K.: "Passive Q-Switching of CO_2 Laser by CH_3F and PF_3 Gases", Appl. Phys. Lett., 1969, 15, p.157.

22.44. Goldsborough, J.P.: "Stable Long-Life CW Excitation of He-Cd Lasers by DC Cataphoresis", Appl. Phys. Lett., 1969, 15, p.159.

22.45. Marcus, S.: "Two Saturable Absorbers for Extending the Wavelength Limits of a Passively Q-Switched CO_2 Laser", Appl. Phys. Lett., 1969, 15, p.217.

22.46. Taieb, G., and Legay, F.: "Effect of Certain Sulphurized Compounds on Operation of a CO-N_2 Laser", C. R. Acad. Sci. (Paris), 1969, 269B, p.371.

22.47. Bennett, W.R.: "Metal-Vapour Lasers", Comments At. Mol. Phys., 1969, 1, no.1, p.15.

22.48. McClung, F.J., and Close, D.H.: "Operation of a Transverse Mode-Controlled Hydrogen Stokes Laser", J. Appl. Phys., 1969, 40, p.3978.

22.49. Berry, M.J., and Pimental, G.C.: "HF-Elimination Chemical Laser", J. Chem. Phys., 1969, 49, p.5190.

22.50. Hattori, S., and Goto, T.: "Anomalous Saturation of Argon-Ion Lasers and Influence of Pinch Effects", Jap. J. Appl. Phys., 1969, 8, no.9, p.1159.

22.51. Belousova, I.M., et al.: "Investigation of Gas Temperature Effects on Power of a He-Ne Laser at 633 nm", Opt. Spekt., 1969, 26, no.1, p.87, and Opt. Spectrosc., 1969, 26, no.1, p.44.

22.52. Tunitskii, L.N., and Cherkasov, E.M.: "Role of Collision in Gas Lasers", Opt. Spectrosc., 1969, 26, no.2, p.146.

22.53. Donin, V.I.: "CW Ar[II] Laser Action at High Current Densities", Opt. Spekt., 1969, 26, no.2, p.298, and Opt. Spectrosc., 1969, 26, no.2, p.160.

22.54. Brunet, H., and Lavarini, B.: "Thermal Effects in Molecular Lasers", Phys. Lett., 1969, 30A, p.181.

22.55. Jensen, R.C., Collins, G.J., and Bennett, W.R.: "Charge-Exchange Excitation and CW Oscillation in the Zn^{II} Laser", Phys. Rev. Lett., 1969, 23, p.363.

22.56. Stefanov, V.J., and Petrova, M.: "Experimental Investigation of Resonator Power Distribution of a CO_2 Laser", Izv. Fiz. Inst. ANEB, 1969, 18, p.233.

22.57. O'Brien, D.E., and Bowen, J.R.: "Kinetic Model for Iodine Photodissociation Laser", J. Appl. Phys., 1969, 40, p.4767.

22.58. Ohtsuka, Y., and Yoshinaga, H.: "Q-Switched CO_2 Laser with Absorbing Gas in Cavity", Jap. J. Appl. Phys., 1969, 8, p.1319.

22.59. Tunitskii, L.N., and Cherkasov, E.M.: "Pulse Generation in an $Ar-O_2$ Laser", Opt. Spekt., 1969, 26, no.4, p.630.

22.60. Ahmed, S.A., and Campillo, A.J.: "He-Ne-Cd Laser with Two-Colour Output", Proc. IEEE, 1969, 57, p.2084.

22.61. Bugaev, V.A.: "Generation Mechanism in a Submillimetric H_2O Laser", Radiotekh. Elektron., 1969, 14, no.6, p.1126, and Radio Eng. Electron. Phys.,1969,14,no.6,pp.979-80.

22.62. Krupnov, A.F., et al.: "Effect of Gas Addition on Operation of Submillimetre H_2O and D_2O Lasers", Radiotekh. Elektron. 1969, 14, no.7, p.1345, and Radio Eng. Electron. Phys., 1969, 14, no.7, pp.1170-1.

22.63. Pollack, M.A.: "Far-IR-Laser Gain Resulting from Rotational Perturbations", J. Quantum Electron. IEEE, 1969,QE-5,p.558.

22.64. Dumanchin, R., and Rocca-Serra, J.: "Increase in Volume Density of Energy and Power in a Pulsed CO_2 Laser", C. R. Acad. Sci. (Paris), 1969, 269B, p.916.

22.65. Herceg, J.E., and Miley, G.H.: "Characteristics of a Low-Voltage He-Ne Laser", J. Appl. Phys., 1969, 40, p.465.

22.66. Bokhan, P.A., and Talankina, G.I.: "Possibility of Optical Pumping of Gases by their Own Radiation", Opt. Spekt., 1969, 25, p.536, and Opt. Spectrosc., 1969,25,p.298.

22.67. Maitland, A.: "Plasma Jet as Cathode for an Argon Laser", J. Phys. D, 1969, 2, p.535.

22.68. Ahmed, S.A., Campillo, A.J., and Cody, R.P.: "RF Inductively Excited Ion Laser with Aperture Magnetic Confinement", J. Quantum Electron. IEEE, 1969, QE-5, p.267.

22.69. Cool, T.A., and Stephens, R.R.: "Chemical Laser by Fluid Mixing", J. Chem. Phys., 1969, 51, p.5175.

22.70. Fein, M.E., Verdeyen, J.T., and Cherrington, B.E.: "Thermally Pumped CO_2 Laser", Appl. Phys. Lett.,1969,14,p.337.

22.71. Bokhan, P.A.: "Optical Pumping of a Molecular Laser by Blackbody Radiation", Opt. Spekt., 1969, 26, no.5, p.773, and Opt. Spectrosc., 1969, 26, p.423.

22.72. Pixton, R.M., and Fowles, G.R.: "Laser Oscillation in H_2 at 752.5 nm", J. Quantum Electron. IEEE, 1969, QE-5, p.478.

22.73. McKnight, W.B.: "Excitation Mechanisms in Pulsed CO_2 Lasers", J. Appl. Phys., 1969, 40, p.2810.

22.74. Bazarov, Ye.N., et al.: "Laser Using Chemically Derived CO_2 in an Electric Discharge", Radiotekh. Elektron., 1969, 14, no.4, p.733, and Radio Eng. Elektron. Phys., 1969, 14, no.4, pp.634-5.

22.75. Troitskii, Yu.V., and Khyuppenen, V.P.: "Characteristics of Competing Transitions in Single-Mode He-Ne Laser", Opt. Spekt., 1969, 27, p.87, and Opt. Spectrosc., 1969, 27, p.172.

22.76. Zakharenko, Yu.G., and Privalov, V.E.: "Plasma Oscillations and Radiated Power of He-Ne Laser", Opt. Spekt., 1969, 27, p.821, and Opt. Spectrosc., 1969, 27, p.447.

22.77. Berry, M.J., and Pimental, G.C.: "Hydrogen-Halide Photoelimination Chemical Lasers", J. Chem. Phys., 1969, 51, p.2274.

22.78. Goto, T., Nakamura, K., and Hattori, S.: "Characteristics of Output Powers of Pulsed Ar^{II} Lasers with Gas Mixtures", Mem. Fac. Eng. Nagoya Univ., 1969, 21, no.2,p.294.

22.79. Hattori, S., and Goto, T.: "Excitation Mechanism and Plasma Parameters in Pulsed Argon-Ion Lasers", J. Quantum Electron. IEEE, 1969, QE-5, p.531.

22.80. Korolev, F.A., et al.: "Population of Working Levels in an Argon Laser", Zh. Prikl. Spekt., 1969, 11, no.2, p.351.

22.81. Lotkova, E.N., and Ochkin, V.N.: "IR Radiation of a Discharge Applied in a CO_2 Laser", Zh. Prikl. Spekt., 1969, 11, no.4, p.739.

22.82. Kobayashi, S., Kurikawa, K., and Kamiyama, M.: "Studies of Pulsed Noble-Gas Ion Lasers", Annu. Rep. Eng. Res. Inst. Univ. Tokyo, 1969, 28, pp.146-53.

22.83. Murai, A., and Okajima, S.: "Far-IR Laser with H_2+O_2 and D_2+O_2 in Large Discharge Tube", Mem. Fac. Eng. Osaka Univ., 1969, 11, pp.137-52.

22.84. Gilson, V.A.: "Gas Additives for Stable Pulsed Argon-Ion Lasers", Rev. Sci. Instrum., 1969, 40, p.448.

22.85. Bazarov, Ye.N., et al.: "Gas Laser with Atmospheric Air", Zh. Prikl. Spekt., 1969, 10, no.2, p.324.

22.86. Brunet, H., and Voignier, F.: "Passive Q-Switching of N_2O Laser Using Ethylene", Appl. Phys. Lett., 1969, 15,p.423.

22.87. Pekar, S.I.: "High-Pressure Chemical Lasers and Chemical Reactions Stimulated by Light", Dokl. Akad. Nauk SSSR, 1969, 187, no.3, p.555, and Sov. Phys.-Dokl., 1970, 14, no.7, p.691.

22.88. Baranov, M.D., Kaslin, V.M., and Petrash, G.G.: "Pulsed Laser Action in CO Electronic Transitions at Cooling of the Gas", Zh. Eksp. Teor. Fiz., 1969, 57, no.2, p.375, and Sov. Phys.-JETP, 1970, 30, no.2, pp.205-12.

22.89. Papayoanou, A., and Gumeiner, I.M.: "High-Power Xe Laser Action in Pinched Discharges", Appl. Phys. Lett., 1970, 16, p.5.

22.90. Csillag, L., et al.: "Investigation on a CW, 441.6-nm, Cd^{II} Laser", J. Phys. D, 1970, 3, p.64.

22.91. Carbone, R.J., and Witteman, W.J.: "Tantalum Hollow Cathode in Argon-Ion Laser", Rev. Sci. Instrum., 1970, 41, p.689.

22.92. Batanov, V.A., et al.: "Gasdynamic Molecular Laser with Optical Pumping", Dokl. Akad. Nauk SSSR, 1970, 191, no.4-6, pp.1267-9, and Sov. Phys.-Dokl., 1970, 15, no.4, pp.408-9.

22.93. Bronfin, B.R., Boedeker, L.R., and Cheyer, J.P.: "Thermal Laser Excitation by Mixing in a Highly Convective Flow", Appl. Phys. Lett., 1970, 16, p.214.

22.94. Gasilevich, E.S., et al.: "Investigation of Composition of Gas-Discharge Plasma in a CO_2 Laser", Zh. Prikl. Spekt., 1970, 13, no.4, pp.712-4.

22.95. Sugawara, Y., et al.: "CW Laser Oscillations in He-Cd and He-Zn Hollow-Cathode Lasers", Jap. J. Appl. Phys., 1970, 9, no.12, p.1537.

22.96. Franz, F.A.: "Population Inversion between Hyperfine Levels of Rb and Cs", Appl. Phys. Lett., 1970, 16, p.391.

22.97. Papakin, V.F., and Sem, M.F.: "Application of Isotopes in Cd and Zn Vapour Lasers", Izv. VUZ Fiz., 1970, no.2, p.117.

22.98. Privalov, V.E., and Fridrikhov, S.A.: "Optimum Mixture in a He-Ne Laser Generating Simulta eously at 633 nm and 3.39 micron", Zh. Prikl. Spekt., 1970, 12, no.3, p.446.

22.99. Morris, G.E., et al.: "Role of CO_2 in Reducing Current Threshold in HCN Laser", Proc. IEEE, 1970, 58, p.477.

22.100. Ronn, A.M.: "Effect of Substituting $^{15}N_2$ for $^{14}N_2$ in the CO_2-N_2 and N_2O-N_2 Lasers", J. Appl. Phys., 1970, 41, p.2246.

22.101. Airey, J.R.: "Cl+HBr Pulsed Chemical Laser", J. Chem. Phys., 1970, 52, p.156.

22.102. Nagai, H., et al.: "Simultaneous Oscillation of He-Ne Laser at 633 nm and 3.39 micron", Jap. J. Appl. Phys., 1970, 9, no.1, p.109.

22.103. Fendley, J.R., et al.: "Characteristics of Sealed-Off $^3He-^{114}Cd$ Laser", J. Quantum Electron. IEEE, 1970, QE-6, p.8.

22.104. Weigand, W.J., Fowler, M.C., and Benda, J.A.: "CO Formation in CO_2 Lasers", Appl. Phys. Lett., 1970, 16, p.237.

22.105. Jung, C.K., Ronn, A.M., and LaTourrette, J.T.: "Passive Q-Switching of a CO_2-N_2-He Laser with CH_3Br Gas", Chem. Phys. Lett., 1970, 5, no.2, p.67.

22.106. Lebedeva, V.V., Mashtakov, D.M., and Odintsov, A.I.: "Role of Multistage Excitation in the Argon Laser", Opt. Spekt., 1970, 28, no.2, p.350, and Opt. Spectrosc., 1970, 28, no.2, p.187.

22.107. Losev, V.V., Papulovskii, V.F., and Fedina, T.A.: "Q-Switching of a Molecular Gas Laser by Saturable Filters", Opt. Spekt., 1970, 28, no.2, p.420, and Opt. Spectrosc., 1970, 28, no.2, p.226.

22.108. Heer, C.V.: "Broadband UV H_2 Laser", Phys. Lett., 1970, 31A, p.160.

22.109. Boersch, H., et al.: "Saturation of CW UV Ion Laser with Large-Bore Tubes", Phys. Lett., 1970, 31A, p.188.

22.110. Kuehn, D.M., and Monson, D.J.: "Experiments with a CO_2 Gasdynamic Laser", Appl. Phys. Lett., 1970, 16, p.48.

22.111. Malacara, D.: "Dependence of Power Output of He-Ne Lasers on the Cavity Configuration", Am. J. Phys., 1970, 38, p.327.

22.112. Buczek, C.J., et al.: "Magnetically Stabilized Cross-Field CO_2 Laser", Appl. Phys. Lett., 1970, 16, p.321.

22.113. Fourcin, F., Menard, J., and Henry, L.: "Study of the Power of HCl Gas Chemical Lasers", C. R. Acad. Sci. (Paris), 1970, 270B, p.494.

22.114. Osgood, R.M., Eppers, W.C., and Nichols, E.R.: "Investigation of High-Power CO_2 Lasers", J. Quantum Electron. IEEE, 1970, QE-6, p.145.

22.115. Polman, J.: "Electron-Radiation Temperature Measurements in a Sealed-Off CO_2-Laser System", J. Quantum Electron. IEEE, 1970, QE-6, p.154.

22.116. Goto, T., Nakamura, K., and Hattori, S.: "Properties of Argon-Ion Lasers with He Mixture", J. Quantum Electron. IEEE, 1970, QE-6, p.159.

22.117. Bloom, A.L., and Goldsborough, J.P.: "CW Laser Transitions in Cd and Zn", J. Quantum Electron. IEEE, 1970, QE-6, p.164.

22.118. Cool, T.A., and Stephens, R.R.: "HBr-CO_2 CW Chemical Laser", J. Chem. Phys., 1970, 52, p.3304.

22.119. Becklake, E.J.S., and Smith, M.A.: "Cyanide Gas Lasers for Submillimetric Wavelengths", Radio Electron. Eng., 1970, 39, p.161.

22.120. Kasymdzhanov, M.A.: "Properties of Stimulated Emission of Pulsed UV Crossed-Field N₂ Laser", Vestn. Mosk. Univ. Fiz. Astron., 1970, 11, p.83.

22.121. D'yakonov, M.I., and Perel', V.I.: "Effect of Resonant Radiation Capture on the Characteristics of a Gas Laser", Zh. Eksp. Teor. Fiz., 1970, 58, p.1090, and Sov. Phys.-JETP, 1970, 31, no.3, pp.585-8.

22.122. Eletskii, A.V., Levinson, G.R., and Sviridov, A.N.: "Thermal Regime of a Pulsed CO₂ Laser", Zh. Prikl. Spekt., 1970, 12, no.3, p.543.

22.123. Spencer, D.J., Kwok, M.A., et al.: "Comparison of HF and DF CW Chemical Lasers. I and II", Appl. Phys. Lett., 1970, 16, pp.384 and 386.

22.124. Hill, A.E.: "Role of Thermal Effects and Fast-Flow Power Scaling Techniques in CO₂-N₂-He Lasers", Appl. Phys. Lett., 1970, 16, p.423.

22.125. Mann, M.M., Bhaumik, M.L., and Lacina, W.B.: "Room-Temperature CO Laser", Appl. Phys. Lett., 1970, 16, p.430.

22.126. Schuebel, W.K.: "CW Cd-Vapour Laser Transitions in a Hollow-Cathode Structure", Appl. Phys. Lett., 1970, 16, p.470.

22.127. Beaulieu, A.J.: "TEA CO₂ Lasers", Appl. Phys. Lett., 1970, 16, p.504.

22.128. Witteman, W.J., and Carbone, R.J.: "Rotational Transition Competition in a Single-Mode CO₂ Laser", J. Quantum Electron. IEEE", 1970, QE-6, p.462.

22.129. Simmons, W.W., and Witte, R.S.: "High-Power Pulsed Xe-Ion Lasers", J. Quantum Electron. IEEE, 1970, QE-6, p.466.

22.130. Willet, C.S.: "Near-IR Operating Characteristics of the Hg^{II} Laser", J. Quantum Electron. IEEE, 1970, QE-6, p.469.

22.131. Sarjeant, W.J., Kucerovsky, Z., and Brannen, E.: "Enhancement of Laser Action in Water Vapour by Addition of Foreign Gases", J. Quantum Electron. IEEE, 1970, QE-6, p.270.

22.132. Ohtsuka, Y.: "Proposal for Far-IR Laser Using Molecular Rotations and Q-Switching Technique", Jap. J. Appl. Phys., 1970, 9, no.4, p.408.

22.133. Dolgov-Savel'ev, G.G., et al.: "Possibility of Designing a CO₂ Laser with Electron-Beam Pumping", Zh. Prikl. Spekt., 1970, 12, no.4, p.737.

22.134. Alekseev, E.I., and Bazarov, Ye.N.: "Theory of Rb Maser Using Optical Pumping", Radiotekh. Elektron., 1970, 15, no.5, p.1044, and Radio Eng. Electron. Phys., 1970, 15, no.5, pp.851-7.

22.135. Leonard, E.T., Yaffee, M.A., and Billman, K.W.: "White-Light Laser", Appl. Opt., 1970, 9, p.1209.

22.136. Jeffers, W.Q., and Wiswall, C.E.: "Transverse-Flow CO Chemical Laser", Appl. Phys. Lett., 1970, 17, p.67.

22.137. Brunne, M., et al.: "Interaction of Monochromatic Radiation with a Flowing Active Medium. I-IV", Bull. Acad. Polon. Sci., 1970, 18, no.3, pp.25 and 33, and no.6, pp.57-65 and 67-72.

22.138. Miroshnichenko, V.I.: "Effects of Saturation and Dispersion in a Gas Laser", Ukr. Fiz. Zh., 1970, 15, p.672.

22.139. Dobrov, W.I., and Washwell, E.R.: "CO₂-Laser Pumping by a DC/ Tesla-Coil Combination", Appl. Opt., 1970, 9, p.1485.

22.140. Hodges, D.T.: "He-Cd Laser Parameters", Appl. Phys. Lett., 1970, 17, p.11.

22.141. Naegeli, D.W., and Ultee, C.J.: "CW HCl Chemical Laser", Chem. Phys. Lett., 1970, 6, no.2, p.121.

22.142. Day, G.W., Gaddy, O.L., and Jungling, K.C.: "Electrooptic Q-Switching of CO₂ Laser", J. Quantum Electron. IEEE, 1970, QE-6, p.553.

22.143. McCoy, J.H.: "Passively Q-Switched N₂O Laser", J. Quantum Electron. IEEE, 1970, QE-6, p.567.

22.144. Dauger, A.B., and Stafsudd, O.M.: "Observation of CW Laser Action in Cl₂, Ar, and He Gas Mixtures", J. Quantum Electron. IEEE, 1970, QE-6, p.572.

22.145. Bhaumik, M.L., Lacina, W.B., and Mann, M.M.: "Enhancement of CO-Laser Efficiency by Addition of Xe", J. Quantum Electron. IEEE, 1970, QE-6, p.575.

22.146. Young, R.T., Willet, C.S., and Maupin, R.T.: "Effect of Helium on Population Inversion in the He-Ne Laser", J. Appl. Phys., 1970, 41, p.2936.

22.147. Wittig, C., Hassler, J.C., and Coleman, P.D: "CW Oscillation in a CO Chemical Laser", Nature, 1970, 226, p.845.

22.148. Hansch, Th., and Toschek, P.: "Theory of a Three-Level Gas-Laser Amplifier", Z. Phys., 1970, 236, p.213.

22.149. Dolgol-Savel'ev, G.G., et al.: "Pulsed Generation in Inert Gases with Pressure Up to 1 atm and with Fast-Electron-Beam Pumping", Zh. Prikl. Spekt., 1970, 12, no.5, p.930.

22.150. Lebedeva, V.V., Mashtakov, D.M., and Odintsov, A.I.: "Inversion Saturation in A^{II} Laser with High Current Density", Zh. Prikl. Spekt., 1970, 12, no.5, p.934.

22.151. Willet, C.S., and Janney, G.M.: "Amplification at 10.6 micron in the Negative Glow of a Hollow-Cathode Discharge in a CO₂-He Mixture", J. Quantum Electron. IEEE, 1970, QE-6, p.568.

22.152. Bhaumik, M.L.: "High-Efficiency CO Laser at Room Temperature", Appl. Phys. Lett., 1970, 17, p.188.

22.153. Graham, W.J., et al.: "Pulsed Behaviour and Inversion Mechanism in the CO Laser", Appl. Phys. Lett., 1970, 17, p.194.

22.154. Legay, F., Taieb, G., et al.: "Mechanism of a CO-N_2 Laser. I and II", Can. J. Phys., 1970, 48, pp.1949 and 1956.

22.155. Ultee, C.J.: "Pulsed HF Lasers", J. Quantum Electron. IEEE, 1970, QE-6, p.647.

22.156. Schuebel, W.K.: "CW-Laser Transitions in Cd^{II} and Zn^{II}", J. Quantum Electron. IEEE, 1970, QE-6, p.654.

22.157. Offenberger, A.A., and Rose, D.J.: "Roles of He and N_2 in CO_2 Laser Excitation", J. Appl. Phys., 1970, 41, p.3908.

22.158. Cool, T.A., Stephens, R.R., and Shirley, J.A.: "HCl, HF, and DF, Partially Inverted CW Chemical Lasers", J. Appl. Phys., 1970, 41, p.4038.

22.159. Jung, C.K., Ronn, A.M., and LaTourrette, J.T.: "Passive Q-Switching and Mode-Locking of a CO_2 Laser with CH_3Br, PF_5, or SF_6", J. Appl. Phys., 1970, 41, p.4240.

22.160. Yano, K., Dote, T., Ichimiya, T.: "Experiments on Current Change with Interruption of Laser Beam", Jap. J. Appl. Phys., 1970, 9, p.1002.

22.161. Takeuchi, N.: "Study of Multi-Wavelength Oscillations in H_2O Laser by Rate Equations", Jap. J. Appl. Phys., 1970, 9, p.1119.

22.162. Dronov, A.P., et al.: "Gasdynamic CO_2 Laser with Escape of Shock-Tube-Heated Mixture through a Slit", Zh. Eksp. Teor. Fiz. Pis'ma, 1970, 11, no.11, p.353, and JETP Lett., 1970, 11, no.11, p.516.

22.163. Eletskii, A.V., and Smirnov, B.M.: "Pulsed CO_2 Laser", Dokl. Akad. Nauk SSSR, 1970, 187, no.2, p.809, and Sov. Phys.-Dokl., 1970, 15, no.2, p.109.

22.164. Orlov, L.N.: "Effect of ^{3}He Upon Generation in He-Ne Laser", Zh. Prikl. Spekt., 1970, 12, no.6, p.994.

22.165. Hodgson, R.T.: "Vacuum-UV Laser Action Observed in the Lyman Bands of H_2", Phys. Rev. Lett., 1970, 25, p.494.

22.166. Holt, H.K.: "Theory of Gas Lasers and Application to an Experiment", Phys. Rev., 1970, 2A, p.233.

22.167. Gensel, P., Kompa, K.L., and Warmer, J.: "IF_5-H_2 HF Chemical Laser Involving a Chain Reaction", Chem. Phys. Lett., 1970, 5, no.3, p.179.

22.168. Markovic, B., Persin, A., and Persin, M.: "Amplification Factor of a He-Ne Mixture in a Detuned Resonator", Fizika, 1970, 2, no.1, p.1.

22.169. Tulip, J.: "Gain Saturation of CO_2 Laser", J. Quantum Electron. IEEE, 1970, QE-6, p.206.

22.170. Akitt, D.P., and Yardley, J.T.: "Far-IR Laser Emission in Gas Discharges Containing Boron Trihalides", J. Quantum Electron. IEEE, 1970, QE-6, p.113.

22.171. Green, W.H., and Whitney, W.T.: "Competition Effects in a $^{12}C^{16}O_2$-$^{13}C^{16}O_2$ Laser", J. Appl. Phys., 1970, 41, p.437.

22.172. Chang, T.Y., Bridges, T.J., and Burkhardt, E.G.: "CW Submillimetre Laser Action in Optically Pumped CH_3F, CH_3OH, and Vinyl-Chloride, Gases", Appl. Phys. Lett., 1970, 17, no.6, pp.249-51.

22.173. Schwarz, S.E., and DeTemple, T.A.: "High-Pressure Pulsed Xe Laser", Appl. Phys. Lett., 1970, 17, no.7, pp.305-6.

22.174. Raffo, C.A., Lavarini, B., and Michon, M.: "Influence of Decomposition of the Gas Mixture on Output at 10.6 micron of a CO_2-N_2-He Amplifier", C. R. Acad. Sci. (Paris), 1970, 270B, pp.788-91.

22.175. Lin, M.C.: "HCl Chemical Laser from $H+Cl_2O$", Chem. Phys. Lett., 1970, 7, no.2, pp.209-10.

22.176. Jonathan, N., Melliar-Smith, C.M., and Slater, D.H.: "IR Emission from the Reaction of Atomic Fluorine with HCl", Chem. Phys. Lett., 1970, 7, no.2, pp.257-9.

22.177. Gallardo, M.: "Unidentified Ionized-Xe Laser Lines", J. Quantum Electron. IEEE, 1970, QE-6, no.11, p.745-7.

22.178. Wittig, C., Hassler, J.C., and Coleman, P.D.: "Collisionally Induced Lasing in CO_2 and N_2O Using Vibrationally Excited CO", J. Quantum Electron. IEEE, 1970, QE-6, no.11, pp.754-6.

22.179. Hoffmann, V., and Toschek, P.: "Laser Emission from Ionized Xe", J. Quantum Electron. IEEE, 1970, QE-6, no.11, p.757.

22.180. Hodges, D.T., and Tang, O.L.: "CW Ion Laser Transitions in Ar, Kr, and Xe", J. Quantum Electron. IEEE, 1970, QE-6, no.11, pp.757-9.

22.181. Menzies, R.T., George, N., and Bhaumik, M.L.: "Spectral Coincidences between Emission Lines of the CO Laser and Absorption Lines of Nitrogen Oxides", J. Quantum Electron. IEEE, 1970, QE-6, no.12, pp.800-2.

22.182. Hernqvist, K.G.: "Long-Pulse Operation of Argon Lasers", Appl. Opt., 1970, 9, no.10, pp.2247-9.

22.183. Lee, S.M., Gamss, L.A., and Ronn, A.M.: "Passive Q-Switching of a CO_2-N_2-He Laser with Ethylene", Chem. Phys. Lett., 1970, 7, no.4, pp.463-4.

22.184. Sedgwick, G., and Seguin, H.J.: "Low-Voltage CO_2-Laser Excitation", Appl. Opt., 1970, 9, no.12, pp.2737-41.

22.185. Lin, M.C., and Green, W.H.: "Pulsed-Discharge-Initiated Chemical Lasers. I", J. Chem. Phys., 1970, 53,no.8,pp.3383-4.

22.186. Chester, A.N.: "Vibrational Inversion in Chemically Pumped Molecular Lasers", J. Chem. Phys., 1970, 53, no.9, pp.3595-8.

22.187. Chang, T.Y.: "Accurate Frequencies and Wavelengths of CO_2 Laser Lines", Opt. Commun., 1970, 2, no.2, pp.77-80.

22.188. Odintsov, A.I.: "Calculation of Output Power of a Gas Laser Taking into Account Energy Exchange", Vestn. Mosk. Univ. Fiz. Astron., 1970, no.4, pp.391-9.

22.189. Stefanov, V.J.: "Use of Current Changes for Detection of Laser Action", J. Phys. E, 1970, 3, no.12, pp.1027-8.

22.190. Stenholm, S.: "Pressure Effects in a High-Intensity Gas Laser", Phys. Rev., 1970, 2A, no.5, pp.2089-107.

22.191. Jacobson, T.V., and Kimbell, G.H.: "Transversely Spark-Initiated Chemical Laser with High Pulse Energies", J. Appl. Phys., 1970, 41, no.13, p.5210-2.

22.192. Waynant, R.W., et al.: "Vacuum-UV Laser Emission from Molecular Hydrogen", Appl. Phys. Lett., 1970, 17, no.9, pp.383-4.

22.193. Hinchen, J.J., and Banas, C.M.: "CW HF Electric-Discharge Mixing Laser", Appl. Phys. Lett., 1970, 17, no.9, pp.386-8.

22.194. Brown, C.O.: "High-Power CO_2 Electric-Discharge Mixing Laser", Appl. Phys. Lett., 1970, 17, no.9, pp.388-91.

22.195. Silfvast, W.T., and Klein, M.B.: "CW Laser Action of 24 Visible Lines in Se^{II}", Appl. Phys. Lett., 1970, 17, no.9, pp.400-3.

22.196. Konyukhov, V.K., et al.: "Gasdynamic CW Laser Using Mixture of CO_2, N_2, and H_2O", Zh. Eksp. Teor. Fiz. Pis'ma, 1970, 12, no.10, pp.461-4, and JETP Lett., 1970, 12, no.10, pp.321-3.

22.197. Krupnov, A.F.: "Investigation of a Submillimetre $D_2O + D_2$ Laser", Opt. Spekt., 1970, 29, no.2, pp.409-10, and Opt. Spectrosc., 1970, 29, no.2, p.217.

22.198. Mazan'ko, I.P., and Sviridov, M.V.: "He-Ne Laser Cavity Amplifier at 3.39 micron", Opt.-Mekh. Prom., 1970, 37, no.7, pp.18-9, and Sov. J. Opt. Technol., 1970, 37, no.7, pp.433-4.

22.199. Dyatlov, M.K., et al.: "Generation Mechanism of Cd-Ion Laser", Opt. Spekt., 1970, 29, no.5, pp.1014-5, and Opt. Spectrosc., 1970, 29, no.5, pp.539-40.

22.200. Dyubko, S.F., et al.: "Submillimetre-Wave Laser", Prib. Tekh. Eksp. 1970, no.1, pp.187-9, and Instrum. Exp. Tech., 1970, no.1, pp.215-7.

22.201. Lecayer, A., and Legay-Sommaire, N.: "Study of a Chemically Formed CO Laser", C. R. Acad. Sci. (Paris), 1970, 271B, no.25, pp.1212-5.

22.202. Gerry, E.T.: "Gasdynamic Lasers", Laser Focus, 1970, 6, no.12, pp.27-31.

22.203. Spencer, D.J., Mirels, H., and Jacobs, T.A.: "Initial Performance of CW Chemical Laser", Opto-Electron., 1970, 2, no.3, pp.155-60.

22.204. Karlov, N.V., and Konev, Yu.B.: "Dependence of Parameters of CO_2 Lasers on Pulse Rate for Continuous Pumping", Radiotekh. Elektron., 1970, 15, no.8, and Radio Eng. Electron. Phys., 1970, 15, no.8, pp.1435-8.

22.205. Sugawara, Y., and Iijima, T.: "Mixing Effects of Ne, Ar, and Kr, on Oscillation Characteristics of He-Zn Hollow-Cathode Laser", Technol. Rep. Seikei Univ., 1970, no.10, pp.885-6.

22.206. Alimpiev, S.S., et al.: "Effect of Dissociation on Inversion in a Pulsed CO_2 Laser", Radiotekh. Elektron., 1970, 15, no.11, pp.2072-7, and Radio Eng. Electron. Phys., 1970, 15, no.11.

22.207. Yabumoto, T., Arai, T., and Takeno, A.: "Electron Temperature and Density in He-Ne Laser", Res. Rep. Fac. Eng. Meiji Univ., 1970, no.25, pp.245-51.

22.208. Sugawara, Y., and Tokiwa, Y.: "CW Hollow-Cathode Laser Action in Zn^{II} and Cd^{II}", Technol. Rep. Seikei Univ., 1970, no.9, pp.759-60.

22.209. Jeffers, W.Q., and Wiswall, C.E.: "Laser Action in Atomic Fluorine Based on Collisional Dissociation of HF", Appl. Phys. Lett., 1970, 17, no.10, pp.444-7.

22.210. McKenzie, R.L.: "Laser Power at 5 micron from Supersonic Expansion of CO", Appl. Phys. Lett., 1970, 17, no.10, pp.462-4.

22.211. Booth, D.J., and Gibbs, W.E.K.: "Interferometric Measurement of Refractive-Index and Temperature Profiles in a Pulsed CO_2 Laser Amplifier", J. Quantum Electron. IEEE, 1971, QE-7, no.1, pp.17-22.

22.212. Riseberg, L.A., and Schearer, L.D.: "Excitation Mechanism of He-Zn Laser", J. Quantum Electron. IEEE, 1971, QE-7, no.1, pp.40-1.

22.213. Green, W.H., and Lin, M.C.: "Pulsed Discharge Initiated Chemical Lasers. II", J. Quantum Electron. IEEE, 1971, QE-7, no.1, pp.98-101.

22.214. Gensel, P., Hohla, K., and Kompa, K.L.: "Energy Storage of CF_3I Photodissociation Laser", Appl. Phys. Lett., 1971, 18, no.2, pp.48-50.

22.215. Jensen, R.C., Collins, G.J., and Bennett, W.R.: "Low-Noise CW Hollow-Cathode Zn[II] Laser", Appl. Phys. Lett., 1971, 18, no.2, pp.50-1.

22.216. Pilloff, H.S., et al.: "Effect of Molecular Chlorine on CO_2 Laser", J. Quantum Electron. IEEE, 1971, QE-7, no.3, pp.134-5.

22.217. Powell, F.X., and Djeu, N.I.: "CW Atomic Oxygen Laser at 4.56 micron", J. Quantum Electron. IEEE, 1971, QE-7, no.4, pp.176-7.

22.218. Lin, M.C.: "Chemical HF Lasers from NF_3-H_2 and NF_3-C_2H_6 Systems", J. Phys. Chem., 1971, 75, no.2, pp.284-6.

22.219. Deutsch, T.F.: "Parameter Studies of Pulsed HF Lasers", J. Quantum Electron. IEEE, 1971, QE-7, no.4, pp.174-5.

22.220. Eckbreth, A.C., Davis, J.W., and Pinsley, E.A.: "Investigation of a CO_2 Laser Pulse Amplifier", Appl. Phys. Lett., 1971, 18, no.3, pp.73-5.

22.221. Wood, O.R., et al.: "High-Pressure Laser Action in 13 Gases with Transverse Excitation", Appl. Phys. Lett., 1971, 18, no.4, pp.112-5.

22.222. Fortin, R.: "Preliminary Measurements of a TEA CO_2 Laser", Can. J. Phys., 1971, 49, no.2, pp.257-65.

22.223. Brunet, H., and Mabru, M.: "Study of a DF-CO_2 Laser Using a Transverse Gas Flow", C. R. Acad. Sci. (Paris), 1971, 272B, no.3, pp.232-5.

22.224. Lavarini, B., et al.: "(CO_2) Laser with Electrical Excitation and Adiabatic Relaxation", C. R. Acad. Sci. (Paris), 1971, 272B, no.5, pp.335-8.

22.225. Jacobson, T.V., and Kimbell, G.H.: "Transversely-Pulse-Initiated Chemical Laser", Chem. Phys. Lett., 1971, 8, no.3, pp.309-11.

22.226. Lamberton, H.M., and Pearson, P.R.: "Improved Excitation Techniques for Atmospheric-Pressure CO_2 Lasers", Electron Lett., 1971, 7, no.5-6, pp.141-2.

22.227. Singh, G., DiLavore, P., and Alley, C.O.: "GaAs-Laser-Induced Population Inversion in the Ground-State Hyperfine Levels of [133]Cs", J. Quantum Electron. IEEE, 1971, QE-7, no.5, pp.196-8.

22.228. Robinson, A.M.: "Effect of Inductance on Small-Signal Gain of a TEA Discharge in CO_2", J. Quantum Electron. IEEE, 1971, QE-7, no.5, pp.199-200.

22.229. Waynant, R.W., et al.: "Laser Emission in the Vacuum UV from H_2", Proc. IEEE, 1971, 59, no.4, pp.679-84.

22.230. Csillag, L., et al.: "Laser Oscillation at 441.6 nm in a Ne-Cd Discharge", Phys. Lett., 1971, 34A, pp.110-1.

22.231. Collins, G.J., Jensen, R.C., and Bennett, W.R.: "Excitation Mechanisms in the Zn-Ion Laser", Appl. Phys. Lett., 1971, 18, no.7, pp.282-4.

22.232. Osgood, R.M., Goldhar, J., and McNair, R.: "High-Pressure Transverse-Discharge CO Laser", J. Quantum Electron. IEEE, 1971, QE-7, no.6, pp.253-5.

22.233. Sorokin, P.P., and Lankard, J.R.: "IR Lasers Resulting from Giant-Pulse Laser Excitation of Alkali-Metal Molecules", J. Chem. Phys., 1971, 54, no.5, pp.2184-90.

22.234. Green, W.H., and Lin, M.C.: "Pulsed-Discharge-Initiated Chemical Lasers. III", J. Chem. Phys., 1971, 54, no.7, pp.3222-3.

22.235. Pichamuthu, J.P., and Coleman, P.D.: "Regenerative Amplification at 28 micron", Proc. IEEE, 1971, 59, no.6, pp.1028-9.

22.236. Barry, J.D., Boney, W.E., and Brandelik, J.E.: "Laser Emission from He-Air-CH_4 and He-Air-C_3H_8 Mixtures", Appl. Phys. Lett., 1971, 18, no.1, pp.15-6.

22.237. Eletskii, A.V., et al.: "Optically Pumped CO_2 Laser", Dokl. Akad. Nauk SSSR, 1971, 194, no.1-3, pp.298-301, and Sov. Phys.-Dokl., 1971, 15, no.9, pp.843-5.

22.238. Bienert, H., Schafer, G., and Theiss, F.J.: "Investigation of Superradiance in Neon at 614.3 nm", Z. Angew. Phys., 1971, 31, no.1, pp.9-11.

22.239. Brechignac, Ph., and Legay, F.: "Continuous Self-Mode-Locking in a CO-He-O_2 Laser", Appl. Phys. Lett., 1971, 18, no.10, pp.424-6.

22.240. Rosenwaks, S., and Yatsiv, S.: "CO Chemical Laser by Flash Photolysis of CS_2+NO_2", Chem. Phys. Lett., 1971, 9, no.3, pp.266-8.

22.241. Crooker, A., and Lamberton, H.M.: "Output Pulse Characteristics and Self-Mode-Locking of Atmospheric-Pressure CO_2 Lasers", Electron. Lett., 1971, 7, no.10, pp.272-3.

22.242. Stenholm, S.: "Theory of Atomic Scattering in a He-Ne Laser", Phys. Rev. Lett., 1971, 26, no.4, pp.159-61.

22.243. Shupe. D.M., and Beaubien, M.W.: "Operation of TEA CO_2 Laser at 9.6 micron", Phys. Lett., 1971, 35A, no.1, pp.13-4.

22.244. Gorog, I.: "Comparitive Analytical Study of the Performance of Argon-Laser Amplifiers and Oscillators", RCA Rev., 1971, 32, no.1, pp.88-114.

22.245. Mash, L.D., Rabkin, B.M., and Rybakov, B.V.: "Investigation of the Cataphoresis Effect in a Cd-Vapour Laser", Zh. Eksp. Teor. Fiz. Pis'ma, 1971, 13, no.5, pp.240-3, and JETP Lett., 1971, 13, no.5, pp.169-71.

22.246. Hodges, D.T.: "CW Laser Oscillation in Mg^{II}", Appl. Phys. Lett., 1971, 18, no.10, pp.454-6.

22.247. Freed, C.: "Sealed-Off Operation of Stable CO Lasers", Appl. Phys. Lett., 1971, 18, no.10, pp.458-61.

22.248. Klein, M.B., and Silfvast, W.T.: "CW Laser Transitions in Se^{II}", Appl. Phys. Lett., 1971, 18, no.11, pp.482-5.

22.249. Watt, W.S.: "CO Gasdynamic Laser", Appl. Phys. Lett., 1971, 18, no.11, pp.487-9.

22.250. Hubner, G., et al.: "Assignments of the Far-IR SO_2 Laser Lines", Appl. Phys. Lett., 1971, 18, no.11, pp.511-3.

22.251. Suchard, S.N., and Pimentel, G.C.: "DF Vibrational Overtone Chemical Laser", Appl. Phys. Lett., 1971, 18, no.12, pp.530-1.

22.252. Ganley, T., Verdeyen, J.T., and Miley, G.H.: "Enhancement of CO_2 Laser Power and Efficiency by Neutron Irradiation", Appl. Phys. Lett., 1971, 18, no.12, pp.568-9.

22.253. Denes, L.J., and Farish, O.: "High-Pressure Pulsed CO_2 Laser with Uniform Excitation", Electron. Lett., 1971, 7, no.12, pp.337-8.

22.254. Havey, M.E., and Barry, J.D.: "Small-Signal Gain in the CO_2-Air-He Laser", J. Quantum Electron. IEEE, 1971, QE-7, no.7, pp.370-2.

22.255. Biriukov, A.S., et al.: "Gasdynamic CO_2-He-N_2 Laser Investigations", J. Quantum Electron. IEEE, 1971, QE-7, no.8, pp.388-91.

22.256. Lotkova, E.N., Ochkin, V.N., and Sobolev, N.N.: "Dissociation of CO_2 and Inversion in CO_2 Laser", J. Quantum Electron. IEEE, 1971, QE-7, no.8, pp.396-400.

22.257. Jeffers, W.Q., and Wiswall, C.E.: "Excitation and Relaxation in a High-Pressure CO Laser", J. Quantum Electron. IEEE, 1971, QE-7, no.8, pp.407-12.

22.258. Wheeler, J.P.: "Xe Laser Line Observed", J. Quantum Electron. IEEE, 1971, QE-7, no.8, p.429.

22.259. Rich, J.W.: "Kinetic Modelling of High-Power CO Laser", J. Appl. Phys., 1971, 42, no.7, pp.2719-30.

22.260. Vonyukhov, V.K., and Prokhorov, A.M.: "Feasibility of an Adsorption Gasdynamic Laser", Zh. Eksp. Teor. Fiz. Pis'ma, 1971, 13, no.4, pp.216-8, and JETP Lett., 1971, 13, no.4, pp.153-5.

22.261. Blaszczak, Z., and Dymaczewski, H.: "He-Cd CW Gas Laser", Postepy Fiz., 1971, 22, no.1, pp.123-6.

22.262. Yardley, J.T.: "Population Inversion and Energy Transfer in CO Lasers", Appl. Opt., 1971, 10, no.8, pp.1760-7.

22.263. Kulakov, B.P., and Nurmukhametov, V.K.: "Amplitude Characteristic of a He-Xe Laser", Radiotekh. Elektron., 1971, 16, no.1, and Radio Eng. Electron. Phys., 1971, 16, no.1, pp.178-81.

22.264. Mishakov, V.G., et al.: "Laser Action in H_2-Na and H_2-K Mixtures with Pulsed Injection of Metal Vapour", Opt. Spekt., 1971, 31, no.2, pp.324-5, and Opt. Spectrosc., 1971, 31, no.2, pp.176-7.

22.265. Hess, L.D.: "Chain-Reaction Chemical Laser Using H_2-F_2-He Mixtures", Appl. Phys. Lett., 1971, 19, no.1, pp.1-3.

22.266. Rich, J.W., and Thompson, H.M.: "IR Sidelight Studies in the High-Power CO Laser", Appl. Phys. Lett., 1971, 19, no.1, pp.3-5.

22.267. Pilloff, H.S., Searles, S.K., and Djeu, N.: "CW CO Laser from CS_2-O_2 Flame", Appl. Phys. Lett., 1971, 19, no.1, pp.9-11.

22.268. Fortin, R., and Gravel, M.: "Helical Transversely Excited CO_2 Lasers", Can. J. Phys., 1971, 49, no.13, pp.1783-93.

22.269. Basov, N.G., Gromov, V.V., et al.: "CW Chemical Laser Using DF-CO_2", Zh. Eksp. Teor. Fiz. Pis'ma, 1971, 13, no.9, pp.496-8, and JETP Lett., 1971, 13, no.9, pp.352-4.

22.270. Ultee, C.J.: "Compact Pulsed HF Lasers", Rev. Sci. Instrum., 1971, 42, no.8, pp.1174-6.

22.271. Yatsiv, S., et al.: "Pulsed CO_2 Gasdynamic Laser", Appl. Phys. Lett., 1971, 19, no.3, pp.65-8.

22.272. Eckbreth, A.C., and Davis, J.W.: "Cross-Beam Electric-Discharge Convection Laser", Appl. Phys. Lett., 1971, 19, no.4, pp.101-3.

22.273. Chang, T.Y., and McGee, J.D.: "Submillimetre Laser Action in Symmetric-Top Molecules Optically Pumped", Appl. Phys. Lett., 1971, 19, no.4, pp.103-5.

22.274. Brunne, M., et al.: "Two-Fluid Model of Active Medium in Gasdynamic Lasers", Bull. Acad. Pol. Sci. Tech., 1971, 19, no.3, pp.17-24.

22.275. Ernst, G.J., and Witteman, W.J.: "Transition Selection with Adjustable Outcoupling for a CO_2 Laser", J. Quantum Electron. IEEE, 1971, QE-7, no.10, pp.484-8.

22.276. Marcus, S., and Carbone, R.J.: "Performance of a Transversely Excited Pulsed HF Laser", J. Quantum Electron. IEEE, 1971, QE-7, no.10, pp.493-4.

22.277. Giallorenzi, T.G., and Ahmed, S.A.: "Saturation and Discharge Studies in He-Cd Laser", J. Quantum Electron. IEEE, 1971, QE-7, no.1, pp.11-7.

22.278. Schuebel, W.K.: "CW Visible and Near-IR Laser Action in Hg^{II}", J. Quantum Electron. IEEE, 1971, QE-7, no.1,pp.39-40.

22.279. Barry, J.D., and Boney, W.E.: "CO Laser Emission Below 5 micron", J. Quantum Electron. IEEE, 1971, QE-7, no.2, p.101.

22.280. Callear, A.B., and Van den Burgh, H.E.: "Hydroxyl-Radical IR Laser", Chem. Phys. Lett., 1971, 8, no.1, pp.17-8.

22.281. Boscher, J., Kindt, T., and Schafer, G.: "Saturation of Laser Power and Optimum Electron Temperature of Argon-Ion Laser", Z. Phys., 1971, 241, no.3, pp.280-90.

22.282. Voignier, F., Brunet, H., and Mabru, M.: "Pulsed Chemical Laser Initiated by Transverse Electric Discharge", C. R. Acad. Sci. (Paris),1971,273B,no.22,pp.972-4.

22.283. Searles, S.K., and Djeu, N.: "Characteristics of CW CO Laser Resulting from a CS_2-O_2 Additive Flame", Chem. Phys. Lett., 1971, 12, no.1, pp.53-6.

22.284. Dezenberg, G.J., and Willett, C.S.: "Unidentified High-Gain Oscillation at 486.1 nm and 434.0 nm in Neon", J. Quantum Electron. IEEE,1971,QE-7,no.10,pp.491-3.

22.285. Franzen, D.L.: "Laser Action in a Flowing Mixture of Formic Acid, N_2, and He", J. Quantum Electron. IEEE, 1971, QE-7, no.10, pp.494-5.

22.286. Tulip, J., and Seguin, H.J.: "Gasdynamic $CO_2^{'}$ Laser Pumped by Combustion of Hydrocarbons", J. Appl. Phys., 1971, 42, no.9, pp.3393-3401.

22.287. Waynant, R.W., Ali, A.W., and Julienne, P.S.: "Experimental Observations and Calculated Band Strengths for the D_2 Lyman-Band Laser", J. Appl. Phys., 1971, 42, no.9, pp.3406-8.

22.288. Knyazev, I.N., and Letokhov, V.S.: "Excitation of Far-Vacuum-UV Lasers by Fast Heating of Plasma Electrons in Ultrashort Pulsed Optical Fields", Opt. Commun., 1971, 3, no.5, pp.332-4.

22.289. Davis, C.C., and King, T.A.: "Time-Resolved Gain Measurements and Excitation Mechanisms of the Pulsed Argon-Ion Laser", Phys. Lett., 1971,36A,no.3,pp.169-70.

22.290. Collins, G.J., Jensen, R.C., and Bennett, W.R.: "Charge-Exchange Excitation in the He-Cd Laser", Appl. Phys. Lett., 1971, 19, no.5, pp.125-8.

22.291. Glaze, J.A.: "Gain and Spectral Characteristics of a CW HF/DF Chemical Laser", Appl. Phys. Lett., 1971, 19, no.5, pp.135-6.

22.292. Brandelik, J.E., Barry, J.D., and Boney, W.E.: "Laser Emission from $(He-Air-C_2N_2)$-CO", Appl. Phys. Lett., 1971, 19, no.5, pp.141-2.

22.293. Skribanowitz, N., et al.: "Possibility of a Unidirectional Laser Amplifier Produced by Monochromatic Optical Pumping of a Coupled Doppler-Broadened Transition", Appl. Phys. Lett., 1971, 19, no.5,pp.161-4.

22.294. Stone, F.T., and Bresman, J.M.: "Variable-Gain Laser Amplifier at 3.39 micron", Appl. Phys. Lett., 1971, 19, no.6, pp.190-1.

22.295. Pan, Y.L., Turner, C.E., and Pettipiece, K.J.: "Characteristics of an Electron-Beam Initiated Pulsed Chemical Laser", Chem. Phys. Lett., 1971, 10, no.5, pp.577-9.

22.296. Glas, P.: "Experimental Results with a He-Cd Laser", Exp. Tech. Phys., 1971, 19, no.3, pp.207-11.

22.297. Barry, J.D., Boney, W.E., and Brandelik, J.E.: "CO_{2-1} Laser Transitions from He-Air-CH_4", J. Quantum Electron. IEEE, 1971, QE-7, no.9, pp.461-2.

22.298. Bridges, W.B., and Chester, A.N.: "Identification of Xenon-Ion Laser Lines", J. Quantum Electron. IEEE, 1971, QE-7, no.9, pp.471-2.

22.299. Florin, A.E., and Jensen, R.J.: "Pulsed Laser Oscillation at 731.1 nm from F Atoms", J. Quantum Electron. IEEE, 1971, QE-7, no.9, p.472.

22.300. Florin, A.E., and Jensen, R.J.: "$F+H_2O$ Chemical Laser", J. Quantum Electron. IEEE, 1971, QE-7, no.9, pp.472-3.

22.301. Jacobson, T.V., and Kimbell, G.H.: "Transversely Pulse Initiated Chemical Lasers", J. Appl. Phys., 1971, 42, no.9, pp.3402-5.

22.302. Collins, G.J.: "Properties of the He-Ne-Zn Laser", J. Appl. Phys., 1971, 42, no.10, pp.3812-5.

22.303. Goto, T., et al.: "Electron Temperature and Density in Positive-Column He-Cd^{II} Lasers", J. Appl. Phys., 1971, 42, no.10, pp.3816-8.

22.304. Turner, R., and Poehler, T.O.: "Characteristics of HCN Laser Radiation at High Excitation Currents", J. Appl. Phys., 1971, 42, no.10, pp.3819-26.

22.305. Emanuel, G.: "Analytical Model for a CW Chemical Laser", J. Quant. Spectrosc., 1971, 11, no.10, pp.1481-520.

22.306. Dzhidzhoev, M.S., et al.: "Detonation Gasdynamic Laser", Zh. Eksp. Teor. Fiz. Pis'ma, 1971, 14, no.2, pp.73-6, and JETP Lett., 1971, 14, no.2, pp.47-9.

22.307. Cason, C., and Horton, T.E.: "Estimating Output Power from Gasdynamic Lasers", Phys. Lett., 1971, 36A, no.4, pp.303-4.

22.308. Agarbiceanu, I.I., et al.: "CW He-[113]Cd Laser", Rev. Roum. Phys., 1971, 16, no.6, pp.607-12.

22.309. Vasiliu, V.: "Study of Stimulated Emission in a He-Ne Mixture", Stud. Cercet. Fiz., 1971, 23, no.6, pp.637-69.

22.310. Goto, T., Kawahara, A., and Hattori, S.: "Quantitative Level Population Mechanism in Gas Discharges for Pulsed Argon-Ion Lasers", J. Quantum Electron. IEEE, 1971, QE-7, no.12, pp.555-60.

22.311. Brus, L.E., and Lin, M.C.: "Chemical HF Lasers from Flash Photolysis of Various N_2F_4+RH Systems", J. Chem. Phys., 1971, 75, no.17, pp.2546-50.

22.312. Nakatsuka, M., et al.: "Optimum Condition of Mixing Ratio on TEA-Type CO_2-N_2 Laser", Jap. J. Appl. Phys., 1971, 10 no.10, pp.1480-1.

22.313. Stephens, R.R., and Cool, T.A.: "CW Chemical Laser for Laser-Induced Fluorescence Studies", Rev. Sci. Instrum., 1971, 42, no.10, p.1489-94.

22.314. Raff, G.J.: "Electron Density Measurements in CW Argon-Ion Lasers", Q. Progr. Rep. Res. Lab. Electron. MIT, 1971, 101, pp.109-14.

22.315. Mirels, H., and Spencer, D.J.: "Power and Efficiency of a CW HF Chemical Laser", J. Quantum Electron. IEEE, 1971, QE-7, no.11, pp.501-7.

22.316. Basov, N.G., Danilychev, V.A., and Popov, Yu.M.: "Stimulated Emission in the Vacuum UV", Kvantovaya Elektron, 1971, no.1, pp.29-34.

22.317. Garnsworthy, R.K., Mathias, L.E.S., and Carmichael, C.H.H.: "Atmospheric-Pressure Pulsed CO_2 Laser Utilizing Preionization by High-Energy Electrons", Appl. Phys. Lett., 1971, 19, no.12, p.506.

22.318. Pichamuthu, J.P., Hassler, J.C., and Coleman, P.D.: "Excitation Mechanism of the Water-Vapour Laser", Appl. Phys. Lett., 1971, 19, no.12, pp.510-2.

22.319. Ultee, C.J.: "Premixed CW Electric-Discharge CO Chemical Lasers", Appl. Phys. Lett., 1971, 19, no.12, pp.535-7.

22.320. Targ, R., and Sasnett, M.W.: "He-Xe Laser at High Pressure and High PRF", Appl. Phys. Lett., 1971,19,no.12, pp.537-9.

22.321. Grimblatov, V.M., et al.: "Stable Single-Colour Emission of Argon-Ion Laser under Line Competition", Kvantovaya Elektron., 1971, no.4, pp.88-91.

22.322. Karatsu, O., Katoh, M., and Ogura, I.: "Competition Effects between the 647.1 nm and 568.2 nm CW Lines in Kr[II] Laser", Oyo Buturi, 1971,40,no.9,pp.987-90.

22.323. Sugawara, Y., Tokiwa, Y., and Iijima, T.: "Oscillation Characteristics and Excitation Mechanisms in He-Cd and He-Zn Hollow-Cathode Lasers", Technol. Rep. Seikei Univ., 1971, no.12, pp.961-74.

22.324. Chekalinskaya, Yu.I., and Chechenina, E.P.: "Calculation of Power Output of Gas Lasers", Zh. Prikl. Spekt., 1971, 15, no.5, pp.925-6.

22.325. Mikaberidze, A.A., and Ochkin, V.N.: "Vibrational Temperatures in CO_2 Lasers", Kvantovaya Elektron., 1971, no.3, pp.96-9.

22.326. Tessier, M., and Vanier, J.: "Theory of the [87]Rb Maser", Can. J. Phys., 1971, 49, no.21, pp.2680-9.

22.327. Dauger, A.B., and Stafsudd, O.M.: "Characteristics of the CW Neutral Argon Laser", Appl. Opt., 1971,10,no.12,pp.2690-7.

22.328. Center, R.E., and Caledonia, G.E.: "Theoretical Description of the Electrical CO Laser", Appl. Phys. Lett., 1971, 19, no.7, pp.211-3.

22.329. Wittig, C., Hassler, J.C., and Coleman, P.D.: "CW CO Chemical Laser", J. Chem. Phys., 1971, 55, no.12, pp.5523-32.

22.330. Fradkin, E.E., and Khayutin, L.M.: "Effect of Radiation Trapping on Mode Competition in a Gas Laser", Opt. Spektr., 1971, 30, no.5, pp.978-9, and Opt. Spectrosc., 1971, 30, no.5, pp.521-2.

22.331. Gibbs, W.E.K., and McLeary, R.: "Uniform Discharges in Flowing CO_2 Laser Mixtures at Atmospheric Pressure", Phys. Lett., 1971, 37A, no.3, pp.229-30.

22.332. Basov, N.G., Belenov, E.M., et al.: "Gas Lasers at High Pressures", Zh. Eksp. Teor. Fiz. Pis'ma, 1971, 14, no.7, pp.421-6, and JETP Lett., 1971, 14, no.7, pp.285-8.

22.333. DeTemple, T., and Nurmikko, A.: "Dynamics of Single-Mode Operation of High-Pressure CO_2 Laser by Saturable Absorbers", Opt. Commun., 1971, 4, no.3, pp.231-3.

22.334. Glaze, J.A., Finzi, J., and Krupke, W.F.: "Transverse Flow CW HCl Chemical Laser", Appl. Phys. Lett., 1971, 18, no.5, pp.173-5.

22.335. Fowler, M.C.: "Analysis of the Dependence of CO_2-Laser Performance on Electric-Discharge Properties", Appl. Phys. Lett., 1971, 18, no.5, pp.175-8.

22.336. Gilbert, J., and Lachambre, J.L.: "Self-Locking of Axial Modes in a CO_2 TEA Laser", Appl. Phys. Lett., 1971, 18, no.5, pp.187-9.

22.337. Rich, J.W., et al.: "Electrically Excited Gasdynamic CO Laser", Appl. Phys. Lett., 1971, 19, no.7, pp.230-2.

22.338. Hill, A.E.: "Uniform Electrical Excitation of Large-Volume High-Pressure Near-Sonic He-N_2-CO_2 Flowstream", Appl. Phys. Lett., 1971, 18, no.5, pp.194-7.

22.339. Novgorodov, M.Z., et al.: "Electron Energy Distribution in CO_2 Laser Discharges", J. Quantum Electron. IEEE, 1971, QE-7, no.11, pp.508-12.

22.340. Laurie, K.A., and Hale, M.M.: "Pin-Electrode Atmospheric-Pressure CO_2 Laser", J. Quantum Electron. IEEE, 1971, QE-7, no.11, pp.530-1.

22.341. Djeu, N.I., and Powell, F.X.: "More Laser Transitions in Atomic Iodine", J. Quantum Electron. IEEE, 1971, QE-7, no.11, pp.537-8.

22.342. Mullaney, G.J., Ahlstrom, H.G., and Christiansen, W.H.: "Pulsed N_2O Molecular Laser Studies", J. Quantum Electron. IEEE, 1971, QE-7, no.12, pp.551-5.

22.343. Kaslin, V.M., Knyazev, I.N., and Petrash, G.G.: "Pulse Emission in the 1^+ System of N_2 Bands under Cooling Conditions", Kvantovaya Elektron., 1971, 1, no.5, pp.44-52.

22.344. Bashkin, A.S., Oraevskii, A.N., and Yuryshev, N.N.: "Possible Utilization of Photorecombination of Radicals and Atoms in CW Lasers", Kvantovaya Elektron., 1971, 1, no.6, pp.89-91.

22.345. Sugawara, Y., Tokiwa, Y., and Iijima, T.: "CW Oscillation of a Zn^{II}-Cd^{II} Laser by Means of Hollow Cathode Discharge", Oyo Buturi, 1971, 40, no.2, pp.211-6.

22.346. Lyon, D.L.: "Analysis of TEA CO_2 Laser", Q. Progr. Rep. Res. Lab. Electron. MIT, 1971, no.103, pp.51-9.

22.347. Elkind, M.S., and Hoff, P.W.: "Gain and Relaxation Studies of a TEA CO_2 Laser", Q. Progr. Rep. Res. Lab. Electron. MIT, 1971, no.103, pp.59-68.

22.348. Donnerhacke, K.H., et al.: "Line Selection in Q-Switched CO_2 Laser", Exp. Tech. Phys., 1971, 19, no.5, pp.345-51.

22.349. Beterov, I.M., Lisitsyn, V.N., and Chebotaev, V.P.: "Saturation Effects and Mode Selection in He-Ne Lasers. II", Opt. Spekt., 1971, 30, no.6, pp.1108-17, and Opt. Spectrosc., 1971, 30, no.6, pp.592-7.

22.350. Basov, N.G., Igoshin, V.I., et al.: "Dynamics of Chemical Lasers (Review)", Kvantovaya Elektron., 1971, no.2, pp.3-24.

22.351. Goncharuk, I.N., et al.: "He-Ne Laser with Mercury Cathode", Prib. Tekh. Eksp., 1971, 14, no.3, pp.182-3, and Instrum. Exp. Tech., 1971, 14, no.3, pp.868-9.

22.352. Basov, N.G., Galochkin, V.T., et al.: "Chemical Lasers Utilizing Mixtures of Fluorine or Nitrogen Fluorides with Deuterium", Kvantovaya Elektron., 1971, no.4, pp.50-7.

22.353. Gur'ev, T.T., Kyun, V.V., and Shevchenko, Yu.N.: "Optimum Conditions for Emission at 520 nm, 568 nm, and 647 nm in Kr^{II} Laser", Opt. Spekt., 1971, 31, no.5, pp.763-5, and Opt. Spectrosc., 1971, 31, no.5, pp.410-1.

22.354. Hernqvist, K.G.: "He-Cd Lasers Using Recirculation Geometry", J. Quantum Electron. IEEE, 1971, QE-8, no.9, pp.740-3.

22.355. Davis, C.C., and King, T.A.: "Laser Action on Unclassified Xe Transitions in Highly Ionized Plasma", J. Quantum Electron. IEEE, 1971, QE-8, no.9, pp.755-7.

22.356. Murai, A., et al.: "Output Variation of a Pulse Excited H_2O Far-IR Laser due to Superposition of Microwave Power", Mem. Fac. Eng. Osaka City Univ., 1971, 12, Dec., pp.257-63.

22.357. Raffo, C.A.: "Thermal Effects and Molecular Dissociation in CO_2 Lasers of High CW Power", Cienc. Tec., 1971, 137, no.695, pp.35-46.

22.358. Glas, P.: "Amplification of a N_2-He-CO_2 Laser", Monatsber. Dtsch. Akad. Wiss. Berlin, 1971, 13, no.10-12, pp.786-92.

22.359. Knyazev, I.N.: "Dynamic Processes in a N_2 Laser", Zh. Eksp. Teor. Fiz., 1971, 61, no.1, pp.72-90, and Sov. Phys.-JETP, 1972, 34, no.1, pp.38-47.

22.360. Zalesskii, V.Yu.: "Kinetics of CF_3I Photodissociation Laser", Zh. Eksp. Teor. Fiz., 1971, 61, no.3, pp.892-905, and Sov. Phys.-JETP, 1972, 34, no.3, pp.474-80.

22.361. Margulis, V.M., and Margolin, A.D.: "Gain of a Diffusion Molecular Laser", Zh. Tekh. Fiz., 1971, 41, no.12, pp.2590-3, and Sov. Phys.-Tech. Phys., 1972, 16, no.12, pp.2056-9.

22.362. Waynant, R.W.: "Observations of Gain by Stimulated Emission in the Werner Band of H_2", Phys. Rev. Lett., 1972, 28, no.9, pp.533-5.

22.363. Hodgson, R.T., and Dreyfus, R.W.: "Vacuum-UV Laser Action Observed in H_2 Werner Bands at 116.1-124.0 nm", Phys. Rev. Lett., 1972, 28, no.9, pp.536-9.

22.364. Girardeau-Montaut, J.P., et al.: "Laser Amplifier giving 50 MW at 337 nm in N_2", C. R. Acad. Sci. (Paris), 1972, 274B, no.9, pp.607-10.

22.365. Smith, R.C.: "Computer Secondary-Electron and Electric-Field Distributions in an Electron-Beam-Controlled Gas-Discharge Laser", Appl. Phys. Lett., 1972, 21, no.8, pp.352-5.

22.366. Seguin, H.J., and Tulip, J.: "Photoinitiated and Photosustained Laser", Appl. Phys. Lett., 1972, 21, no.9, pp.414-5.

22.367. Brunet, H., and Mabru, M.: "Electrical CO-Mixing Gasdynamic Laser", Appl. Phys. Lett., 1972, 21, no.9, pp.432-3.

22.368. Fujimoto, T.: "Collisional-Radiative Calculation of Population Inversion in the Argon-Ion Laser", Jap. J. Appl. Phys., 1972, 11, no.10, pp.1501-8.

22.369. Wenzel, R.G., and Arnold, G.P.: "Double-Discharge-Initiated HF Laser", J. Quantum Electron. IEEE, 1972, QE-8, no.1, pp.26-7.

22.370. Hoffman, A.L., and Vlases, G.C.: "Simplified Model for Predicting Gain, Saturation, and Pulse Duration for Gasdynamic Lasers", J. Quantum Electron. IEEE, 1972, QE-8, no.2, pp.46-53.

22.371. Dezenberg, G.J., Roy, E.L., and McKnight, W.B.: "Performance of High-Voltage Axially Pulsed CO_2 Lasers", J. Quantum Electron. IEEE, 1972, QE-8, no.2, pp.58-65.

22.372. Bhaumik, K.L., Lacina, W.B., and Mann, M.M.: "Characteristics of a CO Laser", J. Quantum Electron. IEEE, 1972, QE-8, no.2, pp.150-60.

22.373. Avivi, P., et al.: "Role of CO in CO_2 Lasers", Phys. Lett., 1972, 42A, no.1, pp.22-4.

22.374. Efremenkova, L.Ya., and Smirnov, B.M.: "Vacuum-UV Laser Using Lyman Transition", Dokl. Akad. Nauk SSSR, 1972, 203, no.4, pp.779-82, and Sov. Phys.-Dokl., 1972, 17, no.4, pp.336-8.

22.375. Ali, A.W., and Kepple, P.C.: "H_2 Lyman and Werner Bands Laser Theory", Appl. Opt., 1972, 11, no.11, pp.2591-6.

22.376. Skribanowitz, N., Herman, I.P., and Feld, M.S.: "Laser Oscillation and Anisotropic Gain in the 1-0 Vibrational Band of Optically Pumped HF", Appl. Phys. Lett., 1972, 21, no.10, pp.466-70.

22.377. Brandenberg, W.M., Bailey, M.P., and Texeira, P.D.: "Supersonic TE Laser", J. Quantum Electron. IEEE, 1972, QE-8, no.4, pp.414-8.

22.378. Spencer, D.J., Mirels, H., and Durran, D.A.: "Performance of CW HF Chemical Laser with N_2 or He Diluent", J. Appl. Phys., 1972, 43, no.3, pp.1151-6.

22.379. Suart, R.D., Dawson, P.H., and Kimbell, G.H.: "CS_2-O_2 Chemical Laser Characteristics", J. Appl. Phys., 1972, 43, no.3, pp.1022-32.

22.380. Fenstermacher, C.A., et al.: "Electron-Beam-Controlled Electrical Discharge as Method of Pumping Large Volumes of CO_2 Laser Media at High Pressure", Appl. Phys. Lett., 1972, 20, no.2, pp.56-60.

22.381. Wilson, J., and Stephenson, J.S.: "Atmospheric-Pressure Pulsed Chemical Laser", Appl. Phys. Lett., 1972, 20, no.2, pp.64-6.

22.382. Wood, O.R., and Chang, T.Y.: "Transverse-Discharge Hydrogen-Halide Lasers", Appl. Phys. Lett., 1972, 20, no.2, pp.77-9.

22.383. Hernqvist, K.G., and Pultorak, D.C.: "Study of He-Se^{II} Laser Performance", Rev. Sci. Instrum., 1972, 43, no.2, pp.290-2.

22.384. Sarjeant, W.J., Kucerovsky, Z., and Brannen, E.: "Excitation Processes and Relaxation Rates in Pulse Water-Vapour Laser", Appl. Opt., 1972, 11, no.4, pp.735-41.

22.385. Fetterman, H.R., Schlossberg, H.R., and Waldman, J.: "Submillimetre Lasers Optically Pumped Off Resonance", Opt. Commun., 1972, 6, no.2, pp.156-9.

22.386. Burkhardt, E.G., Bridges, T.J., and Smith, P.W.: "BeO-Capillary CO_2 Waveguide Laser", Opt. Commun., 1972, 6, no.2, pp.193-5.

22.387. Nighan, W.L.: "Electron Kinetic Processes in CO Lasers", Appl. Phys. Lett., 1972, 20, no.2, pp.96-9.

22.388. Kan, T., Stregack, J.A., and Watt, W.S.: "Electric-Discharge Gasdynamic Laser", Appl. Phys. Lett., 1972, 20, no.3, pp.137-9.

22.389. Spencer, D.J., Durran, D.A., and Bixler, H.A.: "CW Chemical-Laser Cavity Studies", Appl. Phys. Lett., 1972, 20, no.4, pp.164-7.

22.390. Padrick, T.D., and Pimental, G.C.: "Addition-Elimination HF Chemical Laser", Appl. Phys. Lett., 1972, 20, no.4, pp.167-8.

22.391. Dreyfus, R.W., and Hodgson, R.T.: "Electron-Beam Excitation of N_2 Laser", Appl. Phys. Lett., 1972, 20, no.5, pp.195-7.

22.392. Hess, L.D.: "Use of MoF_6 to Increase Reaction Rates in H_2-F_2 Mixtures", J. Appl. Phys., 1972, 43, no.3, pp.1157-60.

22.393. Burak, I., et al.: "TEA Chemical Lasers from H_2+Cl_2 and H_2+Br_2", Chem. Phys. Lett., 1972, 13, no.3, pp.322-4.

22.394. Hartwick, T.S., and Walder, J.: "Effect of O_2 on CO-Laser Performance", J. Quantum Electron. IEEE, 1972, QE-8, no.5, pp.455-6.

22.395. Watanabe, S., Chihara, M., and Ogura, I.: "CW Oscillation at 570.8 nm in He-Te^{II} Laser", Jap. J. Appl. Phys., 1972, 11, no.4, p.600.

22.396. Reilly, J.P.: "Pulser/Sustainer Electric-Discharge Laser", J. Appl. Phys., 1972, 43, no.8, pp.3411-6.

22.397. Fowler, M.C.: "Influence of Plasma Kinetic Processes on Electrically Excited CO_2-Laser Performance", J. Appl. Phys., 1972, 43, no.8, pp.3480-7.

22.398. Basting, D., Schafer, F.P., and Steyer, B.: "Simple, High-Power, N_2 Laser", Opto-Electron., 1972, 4, no.1, pp.43-9.

22.399. Rao, B.S.S., et al.: "Helium Permeation in He-Ne Lasers at Room Temperature", Am. J. Phys., 1972, 40, no.6, pp.916-7.

22.400. Janossy, M., Itagi, V.V., and Csillag, L.: "Excitation Mechanism and Operative Parameters of 441.6-nm He-Cd Laser", Acta Phys. Acad. Sci. Hung., 1972, 32, no.1-4, pp.149-63.

22.401. Seguin, H.J., and Sedgwick, G.: "Low-Voltage Gas-Transport TE CO_2 Laser", Appl. Opt., 1972, 11, no.4, pp.745-8.

22.402. Lotkova, E.N., Mercer, G.N., and Sobolev, N.N.: "Vibration Population, Gain, and Excitation Mechanism of the CO Laser", Appl. Phys. Lett., 1972, 20, no.8, pp.309-11.

22.403. Deutsch, T.F.: "Effect of H_2 on CO_2 TEA Lasers", Appl. Phys. Lett., 1972, 20, no.8, pp.315-6.

22.404. Fortin, R., Laflamme, A.K., and Rheault, F.: "Double-Discharge TEA Laser Beams", Can. J. Phys., 1972, 50, no.6, pp.583-9.

22.405. Vlases, G.C., and Moeny, W.M.: "Numerical Modelling of Pulsed Electric CO_2 Lasers", J. Appl. Phys., 1972, 43, no.4, pp.1840-4.

22.406. Basov, N.G., Zavorotnyi, S.I., et al.: "Pulsed Chemical High-Pressure Laser Using $D_2+F_2+CO_2$", Zh. Eksp. Teor. Fiz. Pis'ma, 1972, 15, no.3, pp.135-7, and JETP Lett., 1972, 15, no.3, pp.93-4.

22.407. Kerber, R.L., Emanuel, G., and Whittier, J.S.: "Computer Modelling and Parametric Study for a Pulsed H_2+F_2 Laser", Appl. Opt., 1972, 11, no.5, pp.1112-23.

22.408. Massone, C.A., et al.: "Investigation of a Pulsed N_2 Laser at Low Temperature", Appl. Opt., 1972, 11, no.6, pp.1317-28.

22.409. Pummer, H., and Kompa, K.L.: "Investigation of a 1-J Pulsed Discharge-Initiated HF Laser", Appl. Phys. Lett., 1972, 20, no.9, pp.356-7.

22.410. Vallach, E., et al.: "Transverse Excitation Pulsed Laser in Gasdynamically Cooled Mixtures", Appl. Phys. Lett., 1972, 20, no.10, pp.395-7.

22.411. Bridges, T.J., Burkhardt, E.G., and Smith, P.W.: "CO_2 Waveguide Lasers", Appl. Phys. Lett., 1972, 20, no.10, pp.403-5.

22.412. Byer, R.L., et al.: "Optically Pumped I_2 Vapour-Phase Laser", Appl. Phys. Lett., 1972, 20, no.11, pp.463-6.

22.413. Seals, R.K., et al.: "Theory and Experiment of Electric-Discharge CO_2 Convection Lasers", AIAA J., 1972, 10, no.4, pp.369-70.

22.414. Pettipiece, K.J.: "TEM_{00} Short-Pulse HF Oscillator", Chem. Phys. Lett., 1972, 14, no.2, pp.261-3.

22.415. Sabotinov, N.V.: "Investigation of a He-Cd Laser at 441.6 nm", Elektro Prom. Prib., 1972, 7, no.1, pp.21-3.

22.416. Linford, G.J.: "High-Gain Neutral Laser Lines in Pulsed Noble-Gas Discharges", J. Quantum Electron. IEEE, 1972, QE-8, no.6, pp.477-82.

22.417. Brown, F., et al.: "Ten-Watt CH_3F Laser at 496 micron", J. Quantum Electron. IEEE, 1972, QE-8, no.6, pp.499-500.

22.418. Cahuzac, P.; "IR Laser Lines in Mg Vapour", J. Quantum Electron. IEEE, 1972, QE-8, no.6, p.500.

22.419. Wynne, J.J., and Shimizu, F.: "Passive Q-Switching of a CO_2 Laser near 9.2 micron Using CF_2Cl_2", J. Quantum Electron. IEEE, 1972, QE-8, no.7, pp.676-7.

22.420. Okajima, A., and Murai, A.: "Far-IR Laser Emission from H_2CO in a Large Gas Tube", J. Quantum Electron. IEEE, 1972, QE-8, no.7, pp.677-9.

22.421. Weisbach, M.F., and Chackerian, C.: "CW Operation in CO Lines below 5 micron", J. Quantum Electron. IEEE, 1972, QE-8, no.7, p.679.

22.422. Collins, G.J., et al.: "CW Laser Oscillation at 612.7 nm in Ionized Iodine", J. Quantum Electron. IEEE, 1972, QE-8, no.7, pp.679-80.

22.423. Ahlborn, B., Gensel, P., and Kompa, K.L.: "Transverse-Flow Transverse-Pulsed Chemical CO Laser", J. Appl. Phys., 1972, 43, no.5, pp.2487-9.

22.424. Howgate, D.W., Roberts, T.G., and Barr, T.A.: "Numerical Calculation of Arc-Driven Supersonic Laser Operating in the Gasdynamic Mode", J. Appl. Phys., 1972, 43, no.6, pp.2799-804.

22.425. Sadie, F.G., Buger, P.A., and Malan, O.G.: "CW Overtone Bands in a CS_2-O_2 Chemical Laser", J. Appl. Phys., 1972, 43, no.6, pp.2906-7.

22.426. Tanaka, A., et al.: "Study of a Semi-Tunable N_2O Laser Using the $^{14}NH_3$ Absorption Line", Jap. J. Appl. Phys., 1972, 11, no.5, pp.768-9.

22.427. Ciura, A.I., et al.: "CO_2 TEA Laser with High Output Pulses", Rev. Roum. Phys., 1972, 17, no.3, pp.399-400.

22.428. Draganescu, V., et al.: "High-Power CO_2 Lasers", Stud. Cercet. Fiz., 1972, 24, no.4, pp.389-400.

22.429. Rusbuildt, D., and Hartwig, H.: "Comparison between Different Excitation Methods for TEA CO_2 Lasers", Atomkernenergie, 1972, 19, no.3, pp.211-6.

22.430. Seguin, H.J., Tulip, J., and White, B.: "Sealed Room-Temperature $CO-CO_2$ Laser Operating at 5 or 10 micron", Appl. Phys. Lett., 1972, 20, no.11, pp.436-8.

22.431. Kan, T., and Whitney, W.T.: "Forced-Convective-Flow CO Laser", Appl. Phys. Lett., 1972, 21, no.5, pp.213-5.

22.432. Marcus, S.: "Excitation of a Long-Pulse CO_2 Laser with a Short-Pulse Longitudinal Electron Beam", Appl. Phys. Lett., 1972, 21, no.1, pp.18-9.

22.433. Chang, T.Y., and Wood, O.R.: "Optically Pumped Atmospheric-Pressure CO_2 Laser", Appl. Phys. Lett., 1972, 21, no.1, pp.19-21.

22.434. Eckbreth, A.C., and Davis, J.W.: "RF Augmentation in CO_2 Closed-Cycle DC Electric-Discharge Convection Lasers", Appl. Phys. Lett., 1972, 21, no.1, pp.25-7.

22.435. Hohla, K., and Kompa, K.L.: "Energy Transfer in a Photochemical Iodine Laser", Chem. Phys. Lett., 1972, 14, no.4, pp.445-8.

22.436. Katsurai, M., and Sekiguchi, T.: "Microwave-Excited Ion Laser with External Magnetic Field", Electron. Commun. Jap., 1972, 54, no.1, pp.61-8.

22.437. Popescu, I.M., Preda, A.M., and Enache, A.: "Influence of Alpha Particles on Performance of a CO_2 Laser", Rev. Roum. Phys., 1972, 17, no.2, pp.121-3.

22.438. Hohla, K., and Kompa, K.L.: "Kinetic Processes in a Photochemical Iodine Laser", Z. Naturforsch., 1972, 27a, no.6, pp.938-47.

22.439. Bristow, T.C., et al.: "High-Intensity X-Ray Spectra and Stimulated Emission from Laser Plasmas", Opt. Commun., 1972, 5, no.5, pp.315-8.

22.440. Emanuel, G., and Whittier, J.S.: "Closed-Form Solution to Rate Equations for $F+H_2$ Laser", Appl. Opt., 1972, 11, no.9, pp.2047-56.

22.441. Piper, J.A., Collins, G.J., and Webb, C.E.: "CW Laser Oscillation in Ionized Iodine", Appl. Phys. Lett., 1972, 21, no.5, pp.203-5.

22.442. Poehler, T.O., Shandor, M., and Walker, R.E.: "High-Pressure Pulsed CO_2 Chemical-Transfer Laser", Appl. Phys. Lett., 1972, 20, no.12, pp.497-9.

22.443. Lacina, W.B., and Mann, M.M.: "Transient Oscillator Analysis of a High-Pressure Electrically Excited CO Laser", Appl. Phys. Lett., 1972, 21, no.5, pp.224-6.

22.444. Andriyakhin, V.M., et al.: "High-Pressure Gas Laser Preionized by a Reactor", Zh. Eksper. Teor. Fiz. Pis'ma, 1972, 15, no.12, pp.637-9, and JETP Lett., 1972, 15, no.12, pp.451-3.

22.445. Zaroslov, D.Yu., et al.: "Plasma Jet CO_2 Laser", Zh. Eksp. Teor. Fiz. Pis'ma, 1972, 15, no.12, pp.665-8, and JETP Lett., 1972, 15, no.2, pp.470-2.

22.446. Wetherall, A.T., and Sharp. L.E.: "High-Power Pulsed HCN Laser", Appl. Opt., 1972, 11, no.8, pp.1737-41.

22.447. Alekseeva, A.N., and Pyatkova, L.M.: "Single-Colour Laser Action in CO_2", Opt. Spekt., 1972, 32, no.1, pp.163-7, and Opt. Spectrosc., 1972, 32, no.1, pp.82-4.

22.448. Voitovich, A.P.: "Hysteresis in a Gas Laser by Conversion from Single- to Double-Frequency Generation", Zh. Prikl. Spekt., 1972, 17, no.1, pp.43-50.

22.449. Kabashnikov, V.P.: "Spectrum of Stationary Generation of a CO_2 Laser at Reduced Pressure", Zh. Prikl. Spekt., 1972, 17, no.1, pp.51-8.

22.450. Dyer, P.E., James, D.J., and Ramsden, S.A.: "Single-Transverse-Mode Operation of a Pulsed Volume-Excited, Atmospheric-Pressure, CO_2 Laser with Unstable Resonator", Opt. Commun., 1972, 5, no.4, pp.236-8.

22.451. Kuehn, D.M.: "Importance of Nozzle Geometry to High-Pressure Gasdynamic Lasers", Appl. Phys. Lett., 1972, 21, no.3, pp.112-4.

22.452. Milewski, J., et al.: "CW Gasdynamic Thermally Excited and Selectively Pumped CO_2-N_2 Laser", Bull. Acad. Pol. Sci., 1972, 20, no.4, pp.313-9.

22.453. Smith, P.W.: "Effect of Cross Relaxation on the Behaviour of Gas Lasers", J. Quantum Electron. IEEE, 1972, QE-8, no.8, pp.704-10.

22.454. McLeary, R.: "Calculations of Gain and Power Output for a Gasdynamic Laser", J. Quantum Electron. IEEE, 1972, QE-8, no.8, pp.716-8.

22.455. Nichols, D.B., and Brandenberg, W.M.: "Radio-Frequency Preionization in a Supersonic Transverse-Discharge Laser", J. Quantum Electron. IEEE, 1972, QE-8, no.8, pp.718-9.

22.456. Targ, R.: "N_2 Laser at High PRF", J. Quantum Electron. IEEE, 1972, QE-8, no.8, pp.726-8.

22.457. Ramsay, I.A.: "Elimination of Unwanted Lasing at 640.1 nm in a 633-nm He-Ne Laser", Appl. Opt., 1970,11,no.10,pp.2386-7.

22.458. Boedeker, L.R., Shirley, J.A., and Bronfin, B.R.: "Arc-Excited Flowing CO Chemical Laser", Appl. Phys. Lett., 1972, 21, no.6, pp.247-9.

22.459. Jeffers, W.Q.: "R-Branch Emission from a CW CO Chemical Laser", Appl. Phys. Lett., 1972, 21, no.6, pp.267-9.

22.460. Suchard, S.N., Ching, A., and Whittier, J.S.: "Efficient Pulsed Chemical Laser", Appl. Phys. Lett., 1972, 21, no.6, pp.274-5.

22.461. Patterson, E.L., Gerardo, J.B., and Johnson, A.W.: "Intense-Electron-Beam Excitation of the 337.1-nm N_2 Laser System", Appl. Phys. Lett., 1972, 21, no.6, pp.293-5.

22.462. Brunne, M., et al.: "Approximate Theory of the CW Gasdynamic Laser with an Unstable Resonator", Bull. Acad. Pol. Sci., 1972, 20, no.5, pp.395-405.

22.463. Brunne, M., et al.: "Multisectional CW Gasdynamic Laser", Bull. Acad. Pol. Sci., 1972, 20, no.5, pp.407-15.

22.464. Brunne, M., et al.: "Elements of a Theory of CW Gasdynamic Lasers", Bull. Acad. Pol. Sci., 1972, 20, no.6, pp.477-87.

22.465. Beattie, W.H., Arnold, G.P., and Wenzel, R.G.: "Chemical Efficiency in a Pulsed HF Laser", Chem. Phys. Lett., 1972, 16, no.1, pp.164-8.

22.466. English, J.R., et al.: "HCl Chemical Lasers with $SOCl_2$, SO_2Cl_2, Cl_2CNCl, and ClCN", Chem. Phys. Lett., 1972, 16, no.1, pp.180-2.

22.467. Ivanov, I.G., et al.: "Plasma Parameters and Pumping Mechanism in He-Cd Laser", Izv. VUZ Fiz., 1972, no.8, pp.85-90.

22.468. Ultee, C.J.: "Compact Pulsed DF Laser", J. Quantum Electron. IEEE, 1972, QE-8, no.10, p.820.

22.469. Javan, A., and Levine, J.S.: "Feasibility of Producing Laser Plasmas via Photoionization", J. Quantum Electron. IEEE, 1972, QE-8, no.11, pp.827-32.

22.470. English, J.R., Gardner, H.C., and Merritt, J.A.: "Pulsed Stimulated Emission from N, C, Cl, and F Atoms", J. Quantum Electron. IEEE, 1972, QE-8, no.11, pp.843-4.

22.471. Palmer, A.J., and McGowan, J.W.: "Laser Excitation Processes in the Cathode Region of a Glow Discharge through Metal-Vapour/ Rare-Gas Mixtures", J. Appl. Phys., 1972, 43, no.10, pp.4084-8.

22.472. Sadie, F.G., Buger, P.A., and Malan, O.G.: "Investigations on the CS_2-O_2 Laser", Z. Naturforsch., 1972, 27a, no.8-9, pp.1260-3.

22.473. Cohn, D.B.: "CO TEA Laser at $77°K$", Appl. Phys. Lett., 1972, 21, no.8, pp.343-5.

22.474. Ahistrom, H.G., et al.: "Cold-Cathode Electron-Beam-Controlled CO_2 Laser Amplifier", Appl. Phys. Lett., 1972, 21, no.10, pp.492-4.

22.475. Gundel, H.: "Kinetics of the Pulsed CO_2 Gas Laser", Beitr. Plasma Phys., 1972, 12, no.3, pp.159-77.

22.476. Robinson, A.M.: "Gain Distribution in a CO_2 TEA Laser", Can. J. Phys., 1972, 50, no.20, pp.2471-4.

22.477. Bailly, D., et al.: "Stationary State in CO_2-N_2, CO_2-He, and CO_2-N_2-He, Plasmas Produced by Electric Discharges", Can. J. Phys., 1972, 50, no.21, pp.2605-13.

22.478. Greiner, N.R.: "Submicrosecond Pulses from a HF Laser with High Energy Density and Quantum Efficiency", J. Quantum Electron. IEEE, 1972, QE-8, no.12, pp.872-6.

22.479. Dauger, A.B., and Stafsudd, O.M.: "Line Competition in the Ar^I Laser", J. Quantum Electron. IEEE, 1972, QE-8, no.12, pp.912-3.

22.480. DePoorter, G.L., and Balog, G.: "IR Laser Line in OCS and Method for C-Atom Lasing", J. Quantum Electron. IEEE, 1972, QE-8, no.12, pp.917-8.

22.481. Abraham, G., and Fisher, E.R.: "Modelling of a Pulsed $CO-N_2$ Molecular Laser System", J. Appl. Phys., 1972, 43, no.11, pp.4621-31.

22.482. Suchard, S.N., et al.: "Effect of H_2 Pressure on Pulsed H_2-F_2 Laser", J. Chem. Phys., 1972, 57, no.12, pp.5065-75.

22.483. Sakurai, T.: "Discharge Current Dependence of Saturation Parameter of a He-Ne Laser", Jap. J. Appl. Phys., 1972, 11, no.12, pp.1832-6.

22.484. Zharov, V.F., et al.: "Effectiveness of Excitation of Lasing in a H_2-F_2 Mixture by a Beam of Relativistic Electrons", Zh. Eksp. Teor. Fiz. Pis'ma, 1972, 16, no.4, pp.219-22, and JETP Lett., 1972, 16, no.4, pp.154-6.

22.485. Bokhan, P.A.: "Experiment on Optical Pumping of a CO_2 Laser", Opt. Spekt., 1972, 32, no.4, pp.826-7, and Opt. Spectrosc., 1972, 32, no.4, pp.435-6.

22.486. Vetter, R.: "Isotopic Shifts in the 3.99-micron Laser Transition in Xe", Phys. Lett., 1972, 42A, no.3, pp.231-2.

22.487. McKenzie, R.L.: "Diatomic Gasdynamic Lasers", Phys. Fluids, 1972, 15, no.12, pp.2163-73.

22.488. Brandi, H.S.: "Excitation of Laser States in a CW A^{II} Laser", Phys. Rev. Lett., 1972, 29, no.23, pp.1539-41.

22.489. Healy, J.J., and Morse, T.F.: "Cavity Detuning and Multimode Operation of an Optically Pumped Gas Laser", Phys. Rev., 1972, 6, no.6, pp.2457-69.

22.490. Golubev, S.A., et al.: "Effect of Proton Beam on a CO_2 Gas Laser", Zh. Eksp. Teor. Fiz., 1972, 62, no.2, pp.458-65, and Sov. Phys.-JETP, 1972, 35, no.2, pp.244-7.

22.491. Bradley, C.C.: "Gain and Frequency Characteristics of a 20-mW CW Water-Vapour Laser at 118.6 micron", Infrared Phys., 1972, 12, no.4, pp.287-99.

22.492. Russell, G.R., Nerheim, N.M., and Pivorotto, T.J.: "Supersonic Electrical-Discharge Copper-Vapour Laser", Appl. Phys. Lett., 1972, 21, no.12, pp.565-7.

22.493. Manes, K.R., and Seguin, H.J.: "Analysis of the CO_2 TEA Laser", J. Appl. Phys., 1972, 43, no.12, pp.5073-8.

22.494. Feld, M.S., et al.: "Selective Reabsorption Leading to Multiple Oscillations in the 844.6-nm Atomic-Oxygen Laser", Phys. Rev., 1972, 7A, no.1, pp.257-62.

22.495. Domash, L.H., Feldman, B.J., and Feld, M.S.: "Interactions among Multiple Lines in the 844.6-nm Atomic-Oxygen Laser", Phys. Rev., 1972, 7A, no.1, pp.262-9.

22.496. Baklanov, Ye.V., and Chebotaev, V.P.: "Theory of Interaction between a Standing-Wave Field and Gas", Zh. Eksp. Teor. Fiz., 1972, 62, no.2, pp.541-50, and Sov. Phys.-JETP, 1972, 35, no.2, pp.287-91.

22.497. Volk, R.: "Operating Conditions of a CW Cyanide Laser with Mixtures of CH_4, N_2, and He", Phys. Lett., 1972, 42A, no.4, pp.321-2.

22.498. Sadie, F.G., Buger, P.A., and Malan, O.G.: "Reaction Mechanisms of the CS_2-O_2 Chemical Laser", J. Appl. Phys., 1972, 43, no.12, pp.5141-2.

22.499. Croshko, V.N., Soloukhin, R.I., and Wolanski, P.: "Population Inversion by Mixing in a Shock-Tube Flow", Opt. Commun., 1972, 6, no.3, pp.275-7.

22.500. Knyazev, I.N., and Letokhov, V.S.: "Stimulated Emission in the Far Vacuum UV by Rapid Heating of Plasma Electrons with Ultrashort Light Pulses", Opt. Spekt., 1972, 33, no.1, pp.110-4, and Opt. Spectrosc., 1972, 33, no.1, pp.59-61.

22.501. Gembarzhevskii, G.V., et al.: "Amplification Factor of CO_2-N_2-He Mixture Expanding in a Supersonic Jet", Zh. Eksp. Teor. Fiz., 1972, 62, no.3, pp.844-7, and Sov. Phys.-JETP, 1972, 35, no.3, pp.447-8.

22.502. Dyubko, S.F., Svich, V.A., and Fesenko, L.D.: "Submillimetre Gas Laser Pumped by a CO_2 Laser", Zh. Eksp. Teor. Fiz. Pis'ma, 1972, 16, no.11, pp.592-4, and JETP Lett., 1972, 16, no.11, pp.418-9.

22.503. Knyazev, I.N., Letokhov, V.S., and Movshev, V.G.: "TEA N_2 UV Laser with Reduced Spectrum", Opt. Commun., 1972, 6, no.3, pp.250-2.

22.504. Dubrovin, A.N., Tibilov, A.S., and Shevtsov, M.K.: "Lasing on Cd, Zn, and Mg Lines", Opt. Spekt., 1972, 32, no.6, pp.1252-3, and Opt. Spectrosc., 1972, 32, no.6, p.685.

22.505. Okajima, S., and Murai, A.: "Far-IR Laser Emission from Formaldehyde in Large Gas Tube", Mem. Fac. Eng. Osaka City Univ., 1972, 13, pp.153-62.

22.506. Tarasenko, V.F., Kurbatov, Yu.A., and Bychkov, Yu.I.: "Pulsed N_2 Laser at 337.1 nm", Kvantovaya Elektron., 1972, no.2, pp.84-5, and Sov. J. Quantum Electron., 1972, 2, no.2, pp.155-6.

22.507. Nakano, T., and Yamanaka, C.: "Properties of SF_6-H_2 Chemical Laser", Technol. Rep. Osaka Univ., 1972, 22, no.1053-89, pp.555-61.

22.508. Moskalenko, V.P., Ostapchenko, E.P., and Chernikov, V.A.: "Mechanism of Pulsed Laser Action in a He-Xe Mixture", Opt. Spekt., 1972, 33, no.2, pp.308-13, and Opt. Spectrosc., 1972, 33, no.2, pp.163-6.

22.509. Mazan'ko, I.P., and Sviridov, M.V.: "Effect of Spontaneous Emission on the Operation of a TW Gas-Laser Amplifier", Opt. Spekt., 1972, 33, no.2, pp.314-20, and Opt. Spectrosc., 1972, 33, no.2, pp.167-70.

22.510. Bezukh, B.A., and Khodyko, Yu.V.: "Electrodeless Induction RF Discharge as Source of Inversion in a CO_2-N_2 Mixture", Opt. Spekt., 1972, 33, no.2, pp.360-1, and Opt. Spectrosc., 1972, 33, no.2, pp.192-3.

22.511. Smirnov, V.S., and Zhelnov, B.L.: "Quantum Theory of a Gas Laser under High-Energy Conditions", Opt. Spekt., 1972, 33, no.3, pp.505-12, and Opt. Spectrosc., 1972, 33, no.3, pp.272-6.

22.512. Demin, A.I., et al.: "Gasdynamic Laser with High Water-Vapour Content", Kvantovaya Elektron., 1972, no.3, pp.72-3, and Sov. J. Quantum Electron., 1972, 2, no.3, pp.251-3.

22.513. Hofland, R., and Mirels, H.: "Flame-Sheet Analysis of CW Diffusion-Type Chemical Lasers. II", AIAA J., 1972, 10, no.10, pp.1271-80.

22.514. Balczewski, L.E.: "Optical Coherence of the 633-nm Laser Transition in Ne[I]", Acta Phys. Pol., 1972, 42A, no.6, pp.749-51.

22.515. Crane, R.A., and Waksberg, A.L.: "Gain Correlation with Sidelight and Plasma Impedance Properties of a CO_2 Laser Discharge", Can. J. Phys., 1972, 50, no.23, pp.3067-9.

22.516. Schiffner, G.: "Calculation of Accurate CO_2 Laser Transition Frequencies", Opto-Electron., 1972, 4, no.3, pp.215-23.

22.517. Ishchenko, V.N., Lisitsyn, V.N., and Chapovskii, P.I.: "Lines of Pulsed Oscillation in the (2,0) Band of the 1[+] System of N_2", Opt. Spekt., 1972, 33, no.2, pp.366-7, and Opt. Spectrosc., 1972, 33, no.2, pp.196-7.

22.518. Tkach, Yu.V., et al.: "Plasma-Beam Discharge Laser", Zh. Eksp. Teor. Fiz., 1972, 62, no.5, pp.1702-16, and Sov. Phys.-JETP, 1972, 35, no.5, pp.886-92.

22.519. King, W.S., and Mirels, H.: "Numerical Study of Diffusion-Type Chemical Laser", AIAA J., 1972, 10, no.12, pp.1647-54.

22.520. Greenberg, R.A., et al.: "Rapid Expansion Nozzles for Gasdynamic Lasers", AIAA J. 1972, 10, no.11, pp.1494-8.

22.521. Kato, I., and Shimizu, T.: "Micro-Wave-Pulse Excited Argon-Ion Laser", Electron. Commun. Jap., 1972, 55, no.7,pp.108-15.

22.522. Rogova, I.V.: "Nonlinear Polarizibility of Gas in Laser Transition with Resonance Radiation Trapping", Opt. Spekt., 1972, 33, no.4, pp.720-4, and Opt. Spec-rosc. trosc., 1972, 33, no.4, pp.397-9.

22.523. Gudzenko, L.I., and Yakovlenko, S.I.: "Electron Transitions of Molecules in a Plasma Laser", Dokl. Akad. Nauk SSSR, 1972, 207, no.4-6, pp.1085-7, and Sov. Phys.-Dokl., 1972, 16, no.12, pp.1172-3.

22.524. Averin, V.G., Karchevskii, A.I., and Yurkin, G.V.: "Stimulated Emission by Pumping with a Pulsed Electron Beam", Zh. Eksp. Teor. Fiz., 1972, 63, no.1, pp.85-91, and Sov. Phys.-JETP, 1973, 36, no.1,pp.44-7.

22.525. Matsuoka, T., and Furuse, T.: "TEM_{10}-Mode Gain of He-Ne Laser", Electron. Commun. Jap., 1972, 55, no.8, pp.75-81, and NEC Res. Dev., 1973, no.30, pp.32-9.

22.526. Gordon, E.B., et al.: "Kinetics of a Pulsed Chemical CO Laser with Photo-initiated Oxidation of CS_2", Zh. Eksp. Teor. Fiz., 1972, 63, no.4, pp.1159-72, and Sov. Phys.-JETP, 1973, 36, no.4, pp.611-8.

22.527. Dolgov-Savel'ev, G.G., and Podminogin, A.A.: "Pulsed Laser Utilizing a H_2+F_2 Mixture", Kvantovaya Elektron., 1972, no.4, pp.69-76, and Sov. J. Quantum Electron., 1973, 2, no.4, pp.348-52.

22.528. Afanas'ev, Yu.V., et al.: "Distribution of Molecules between Vibrational Levels under Time-Dependent External Pumping Conditions", Kvantovaya Elektron., 1972, no.4, pp.97-9, and Sov. J. Quantum Electron., 1973, 2, no.4, pp.372-3.

22.529. Dolgov-Savel'ev, G.G., and Chumak, G.M.: "ClF_3 Chemical Laser", Kvantovaya Elektron., 1972, no.4, pp.108-10, and Sov. J. Quantum Electron., 1973, 2, no.4, pp.383-4.

22.530. Isaev, A.A., Kazaryan, M.A., and Petrash, G.G.: "Pulsed Lead-Vapour Laser with High Peak and Average Powers", Kvantovaya Elektron., 1972, no.5, p.100, and Sov. J. Quantum Electron., 1973, 2, no.5, p.470.

22.531. Vinogradov, A.V., and Sobel'man, I.I.: "Problem of Laser Radiation in Far-UV and X-Ray Regions", Zh. Eksp. Teor. Fiz., 1972, 63, no.6, pp.2113-20, and Sov. Phys.-JETP, 1973, 36, no.6, pp.1115-9.

22.532. Kaslin, V.M., Zun'kova, Z.E., and Petrash, G.G.: "Emission of IR H_2 Lines from a Cooled-Gas Laser", Kvantovaya Elektron., 1972, no.5, pp.101-3, and Sov. J. Quantum Electron., 1973, 2, no.5, pp.471-3.

22.533. Bugaev, V.A., and Kukhta, A.V.: "Correlation between Output Power and Composition of Discharge Products in a Water-Vapour Laser", Kvantovaya Elektron., 1972, no.5, pp.111-4, and Sov. J. Quantum Electron., 1973, 2, no.5, pp.482-4.

22.534. Bashkin, A.S., and Yuryshev, N.N.: "Output Parameters of a Chemical CS_2+O_2 Laser", Kvantovaya Elektron., 1972, no.5, pp.129-31, and Sov. J. Quantum Electron., 1973, 2, no.5, pp.499-500.

22.535. Poehler, T.O., Pirkle, J.C., and Walker, R.E.: "High-Pressure Pulsed CO_2 Chemical Transfer Laser", J. Quantum Electron. IEEE, 1973, QE-9, no.1, pp.83-93.

22.536. Searles, S.K., and Djeu, N.: "Gain Measurements on CO P-Branch Transitions in a $C_2H_2-O_2$ Flame", J. Quantum Electron. IEEE, 1973, QE-9, no.1, pp.116-20.

22.537. Rockwood, S.D., et al.: "Time-Dependent Calculations of CO Laser Kinetics", J. Quantum Electron. IEEE, 1973, QE-9, no.1, pp.120-9.

22.538. Lyon, D.L.: "Comparison of Theory and Experiment for a TE High-Pressure CO_2 Laser", J. Quantum Electron. IEEE, 1973, QE-9, no.1, pp.139-53.

22.539. Stratton, U.F., et al.: "Electron-Beam-Controlled CO_2 Laser Amplifiers", J. Quantum Electron. IEEE, 1973, QE-9, no.1, pp.157-63.

22.540. Kerber, R.L., Cohen, N., and Emanuel, G.: "Chemical Transfer Laser", J. Quantum Electron. IEEE, 1973, QE-9, no.1, pp.94-113.

22.541. Hirose, Y., Hassler, J.C., and Coleman, P.D.: "CW CO Chemical Laser from Reaction of Active N_2 with O_2+CS_2", J. Quantum Electron. IEEE, 1973, QE-9,no.1,pp.114-6.

22.542. Schenck, P., and Metcalf, H.: "Low-Cost N_2-Laser Design for Dye-Laser Pumping", Appl. Opt., 1973, 12, no.2,pp.183-6.

22.543. Levine, J.S., and Javan, A.: "Observation of Oscillation in a 1-atm CO_2-N_2-He Laser Pumped by an Electrically Heated Plasma", Appl. Phys. Lett., 1973, 20, no.2, pp.55-7.

22.544. Casperson, L.W.: "Saturation and Power in a High-Gain Gas Laser", J. Quantum Electron. IEEE, 1973, QE-9, no.2, pp.250-2.

22.545. Hassler, J.C., Hubner, G., and Coleman, P.D.: "Excitation Mechanism of the Far-IR SO_2 Molecular Laser", J. Appl. Phys., 1973, 44, no.2, pp.795-801.

22.546. Aldridge, F.T.: "High-Pressure Iodine Photodissociation Laser", Appl. Phys. Lett., 1973, 22, no.4, pp.180-2.

22.547. Rice, W.W., and Jensen, R.J.: "Aluminium-Fluoride Exploding-Wire Laser", Appl. Phys. Lett., 1973, 22,no.2,pp.67-8.

22.548. Padrick, T.D., and Gusinow, M.A.: "Energy and Threshold Characteristics of Chemical Lasers", Appl. Phys. Lett., 1973, 22, no.4, pp.183-5.

22.549. Hohla, K., and Kompa, K.L.: "Gigawatt Photochemical Iodine Laser", Appl. Phys. Lett., 1973, 22, no.2, pp.77-8.

22.550. Nelson, L.Y., Mullaney, G.J., and Byron, S.R.: "Superfluorescence in N_2 and H_2 Electron-Beam Stabilized Discharges", Appl. Phys. Lett., 1973, 22, no.2, pp.79-80.

22.551. Chang, T.Y., and Wood, O.R.: "Optically Pumped N_2O Laser", Appl. Phys. Lett., 1973, 22, no.3, pp.93-4.

22.552. Judd, O.P.: "Efficient Electrical CO_2 Laser Using Preionization by UV Radiation", Appl. Phys. Lett., 1973, 22, no.3, pp.95-6.

22.553. Pichamuthu, J.P., et al.: "Role of He in the H_2O Laser", J. Quantum Electron. IEEE, 1973, QE-9, no.2, pp.244-5.

22.554. Boney, W.E., Barry, J.D., and Brandelik, J.E.: "CO and CO_2 Laser Action by Organic-Molecule Oxidation", J. Quantum Electron. IEEE, 1973, QE-9, no.2, pp.246-7.

25.555. Davit, J., and Charles, C.: "Performance of an Unstable Repetitively Pulsed CO_2 Laser", Appl. Phys. Lett., 1973, 22, no.5, pp.248-50.

22.556. Rice, W.W., and Beattie, W.H.: "Metal-Atom Oxidation Lasers", Chem. Phys. Lett., 1973, 19, no.1, pp.82-5.

22.557. Arnold, G.P., and Wenzel, R.G.: "Improved Performance of an Electrically Initiated HF Laser", J. Quantum Electron. IEEE, 1973, QE-9, no.4, pp.491-3.

22.558. Jacobson, T.V., Kimbell, G.H., and Snelling, D.R.: "High-PRF Chemical HF Laser", J. Quantum Electron. IEEE, 1973, QE-9, no.4, pp.496-7.

22.559. Cubeddu, R., and Curry, S.M.: "Simple High-Power Pulsed N_2 Laser", J. Quantum Electron. IEEE, 1973, QE-9, no.4, pp.499-500.

22.560. Ultee, C.J.: "IR Laser Emission from Discharges through Gaseous Sulphur Compounds", J. Appl. Phys., 1973, 44, no.3, p.1406.

22.561. Turner, E.B., Adams, W.D., and Emanuel, G.: "Numerical Formulation for Constant-Gain Chemical Laser Calculations", J. Comput. Phys., 1973, 11, no.1, pp.15-27.

22.562. Healy, J.J., and Morse, T.F.: "Theory of an Optically Pumped Gas Laser", J. Quant. Spectrosc., 1973, 13, no.3, pp.235-54.

22.563. Skribanowitz, N., et al.: "Observation of Dicke Superradiance in Optically Pumped HF Gas", Phys. Rev. Lett., 1973, 30, no.8, pp.309-12.

22.564. Suzuki, K., et al.: "Triggering Characteristics of TEA CO_2 Laser", Jap. J. Appl. Phys., 1973, 12, no.3, pp.483-4.

22.565. Mkrtchyan, M.M., and Platonenko, V.T.: "Feasibility of High-Pressure Noble-Gas Lasers", Zh. Eksp. Teor. Fiz. Pis'ma, 1973, 17, no.1, pp.28-31, and JETP Lett., 1973, 17, no.1, pp.19-21.

22.566. Baird, K.M., Smith, D.S., and Berger, W.E.: "Wavelength of CH_4 Line at 3.39 micron", Opt. Commun., 1973, 7, no.2, pp.107-9.

22.567. Jassby, D.L., Marhic, M.E., and Regan, D.R.: "High-Power Pulsed HCN Laser with Auxiliary DC Discharge", Appl. Opt., 1973, 12, no.7, pp.1403-4.

22.568. Searles, S.K., and Airey, J.R.: "Chemically Pumped CO_2 and N_2O Lasers", Appl. Phys. Lett., 1973, 22, no.10, pp.513-4.

22.569. Hill, A.E.: "Continuous Uniform Excitation of Medium-Pressure CO_2 Laser Plasmas by Controlled Avalanche Ionization", Appl. Phys. Lett., 1973, 22, no.12, pp.670-3.

22.570. Davies, T.J., and Nelson, M.A.: "Pulsed Neon Laser with 350-ps Pulse Duration and Subnanosecond Jitter at 614.3 nm", Appl. Opt., 1973, 12, no.4, pp.880-1.

22.571. Poehler, T.O., and Walker, R.E.: "Transverse-Discharge Pulsed CO_2 Chemical-Transfer Laser", Appl. Phys. Lett., 1973, 22, no.6, pp.282-3.

22.572. Linevsky, M.J., and Carabetta, R.A.: "CW Laser Power from CS_2 Flames", Appl. Phys. Lett., 1973, 22, no.6, pp.288-91.

22.573. Lis, L.: "Characteristics of $3s_2$-$3p_1$ (4.218 micron) Laser Action in Ne", Acta Phys. Pol., 1973, A43, no.3, pp.453-9.

22.574. Barry, J.D., Boney, W.E., and Brandelick, J.E.: "Near-IR Lasers from HCHO", J. Appl. Phys., 1973, 44, no.4, pp.1915-6.

22.575. Kochelap, V.A., and Kukibnyi, Yu.A.: "Kinetics of Chemical Lasers of High Pressure", Ukr. Fiz. Zh., 1973, 18, no.3, pp.378-88.

22.576. Klement'ev, V.M., and Solov'ev, M.V.: "Characteristics of a Mercury-Vapour Laser", Zh. Prikl. Spekt., 1973, 18, no.1, pp.41-5.

22.577. Munjee, S.A., and Christiansen, W.H.: "Mixed Mode Contributions to Absorption in CO_2 at 10.6 micron", Appl. Opt., 1973, 12, no.5, pp.993-6.

22.578. Piper, J.A., and Webb, C.E.: "CW Laser Oscillation in As^{II}", J. Phys. B, 1973, 6, no.5, pp.L116-20.

22.579. Pummer, H., et al.: "Parameter Study of a 10-J HF Laser", Appl. Phys. Lett., 1973, 22, no.7, pp.319-20.

22.580. Chubb, D.L., and Rose, J.R.: "Population Inversion Calculations Using Near-Resonant Charge Exchange as a Pumping Mechanism", Appl. Phys. Lett., 1973, 22, no.8, pp.417-8.

22.581. Waynant, R.W.: "Vacuum-UV Laser Emission from C^{IV}", Appl. Phys. Lett., 1973, 22, no.8, pp.419-20.

22.582. Lee, J.H.S., Bui, T.D., and Knystautus, H.: "Population Inversion in Blast Waves", Appl. Phys. Lett., 1973, 22, no.9, pp.434-6.

22.583. Parker, J.V., and Stephens, R.R.: "Pulsed HF Chemical Laser with High Electrical Efficiency", Appl. Phys. Lett., 1973, 22, no.9, pp.450-2.

22.584. Dawson, P.H.: "Evolution of the CO Vibrational Energy Distribution in a Transverse Flow Laser", Can. J. Phys., 1973, 51, no.9, pp.1026-9.

22.585. Yermachenko, V.: "Stability Condition for an Intense Two-Mode Region of a Gas Laser", C. R. Acad. Sci. (Paris), 1973, 276B, no.14, pp.611-4.

22.586. Avivi, P., et al.: "Influence of CO on the Population Inversion in CO_2 Lasers", J. Appl. Phys., 1973, 44, no.4, pp.17-21.

22.587. Basov, N.G., Danilychev, V.A., et al.: "Population Inversion in the Active Medium of an Electro-Ionization CO_2 Laser at a Pressure of 20 bar", Zh. Eksp. Teor. Fiz. Pis'ma, 1973, 17, no.3, pp.147-50, and JETP Lett., 1973, 17, no.3, pp.102-4.

22.588. Bokhan, P.A., and Yegorova, Ye.S.: "Feasibility of Increasing Output Power from a CO_2-N_2-He Mixture", Laser Unconv. Opt. J., 1973, no.43, pp.17-20.

22.589. Anh, T.D., and Dietel, W.: "Homogeneous Broadening in a 633 nm Single-Mode He-Ne Laser", Opto-Electron., 1973, 5, no.3, pp.243-8.

22.590. Glaze, J.A., and Linford, G.J.: "Design and Performance Characteristics of a Small Subsonic-Flow HF Chemical Laser", Rev. Sci. Instrum., 1973, 44, no.5, pp.600-4.

22.591. Finzel, R., and Schafer, G.: "Mutual Influence of the A^{II}-Laser Transitions at 488 nm and 514.5 nm", Z. Phys., 1973, 259, no.4, pp.355-64.

22.592. Fuhs, A.E.: "Density Inhomogeneity in a Laser Cavity due to Energy Release", AIAA J., 1973, 11, no.3, pp.374-5.

22.593. Klimek, D.E., and Berry, M.J.: "Formyl Fluoride Photochemical Laser", Chem. Phys. Lett., 1973, 20, no.1, pp.141-5.

22.594. Il'yushko, V.G., Papakin, V.F., and Sem, M.F.: "Influence of Isotopic Splitting on Characteristics of Cd- and Zn-Vapour Lasers", Izv. VUZ Fiz., 1973, no.3, pp.138-9.

22.595. Papayoanou, A., Buser, R.G., and Gumeiner, I.M.: "Parameters in a Dynamically Compressed Xenon Plasma Laser", J. Quantum Electron. IEEE, 1973, QE-9, no.6, pp.580-5.

22.596. Lacina, W.B., Mann, M.M., and McAllister, G.L.: "Transient Oscillator Analysis of a High-Pressure Electrically Excited CO Laser", J. Quantum Electron. IEEE, 1973, QE-9, no.6, pp.588-93.

22.597. Kerber, R.L., et al.: "Efficiently Initiated Pulsed H_2+F_2 Laser", J. Quantum Electron. IEEE, 1973, QE-9, no.6, pp.607-9.

22.598. Fahlen, T.S., and Targ, R.: "High-Average-Power Xe Laser", J. Quantum Electron. IEEE, 1973, QE-9, no.6, p.609.

22.599. Linford, G.J.: "Pulsed Laser Lines in Kr", J. Quantum Electron. IEEE, 1973, QE-9, no.6, pp.610-1.

22.600. Linford, G.J.: "Pulsed and CW Laser Lines in Heavy Noble Gases", J. Quantum Electron. IEEE, 1973, QE-9, no.6, pp.611-2.

22.601. Pearson, R.K., et al.: "Pressure Dependency of the NF_3-H_2 Transverse-Discharge Pulse Initiated HF Chemical Laser", J. Quantum Electron. IEEE, 1973, QE-9, no.7, pp.723-30.

22.602. Gerardo, J.B., and Johnson, A.W.: "High-Pressure Xe Laser at 173 nm", J. Quantum Electron. IEEE, 1973, QE-9, no.7, pp.748-55.

22.603. Kano, H., Goto, T., and Hattori, S.: "Electron Temperature and Density in the He-CdI$_2$ Positive Column Used for an I^{II} Laser", J. Quantum Electron. IEEE, 1973, QE-9, no.7, pp.776-8.

22.604. Barry, J.D., et al.: "Simultaneous CO and CO_2 Laser", J. Quantum Electron. IEEE, 1973, QE-9, no.7, pp.779-80.

22.605. Osgerby, I.T.: "Perfectly Stirred Reactor. Concept for CW Chemical Laser", J. Appl. Phys., 1973, 44, no.6, pp.2627-30.

22.606. Suhre, D.R., Coleman, P.D., and DeTemple, T.A.: "Electron Excitation Efficiency of CO_2 Lasers", J. Appl. Phys., 1973, 44, no.6, pp.2923-4.

22.607. Browne, P.G., and Dunn, M.H.: "Metastable Densities and Excitation Processes in the He-Cd Laser Discharge", J. Phys. B, 1973, 6, no.6, pp.1103-17.

22.608. Maitland, A., and Dunn, M.H.: "Flow Graphs for Solution of Linear Rate Equations of Gas Discharges and Lasers", J. Phys. D, 1973, 6, no.10, pp.1266-73.

22.609. Ovchinnikov, A.A.: "Electron-Vibrational Inversion in the Oxidation of CS_2", Zh. Eksp. Teor. Fiz. Pis'ma, 1973, 17, no.5, pp.259-62, and JETP Lett., 1973, 17, no.5, pp.185-8.

22.610. Ferrario, A.: "Excitation Mechanism in Hg^{II} Laser", Opt. Commun., 1973, 7, no.4, pp.376-8.

22.611. Piltch, M.: "Multiline Pulsed CO_2 Laser", Opt. Commun., 1973, 7, no.4, pp.397-9.

22.612. Byszewski, W.W.: "High-Pressure CO_2-N_2 Laser Excited by Electrical Discharge Controlled by an Electron Beam", Bull. Acad. Pol. Sci., 1973, 21, no.2, pp.151-5.

22.613. Persin, A., and Vukicevic, D.: "Current Dependence of the Ne $^3s_2$-Level Population in a 633-nm Laser", Fizika, 1973, 5, no.2, pp.77-82.

22.614. Schotzau, H.J., Kneubuhl, F.K., and Veprek, S.: "Formation of Laser Active Molecules in a CW HCN Laser", Helv. Phys. Acta, 1973, 46, no.1, p.33.

22.615. McLeary, R., and Gibbs, W.E.K.: "CW CO_2 Laser at Atmospheric Pressure", J. Quantum Electron. IEEE, 1973, QE-9, no.8, pp.828-33.

22.616. Mosburg, E.R.: "Study of the CW 28-micron Water-Vapour Laser", J. Quantum Electron. IEEE, 1973, QE-9, no.8, p.843-51.

22.617. Lam, M.F., Jassby, D.L., and Casperson, L.W.: "Transverse-Excitation Pulsed HCN Laser", J. Quantum Electron. IEEE, 1973, QE-9, no.8, pp.851-2.

22.618. Bergmann, E.E., and Eberhardt, N.: "Short High-Power TE Nitrogen Laser", J. Quantum Electron. IEEE, 1973, QE-9, no.8, pp.853-4.

22.619. Ferrar, C.M.: "Copper-Vapour Laser with Closed-Cycle Transverse Vapour Flow", J. Quantum Electron. IEEE, 1973, QE-9, no.8, pp.856-7.

22.620. Sabotinov, N.V., and Telbizov, P.K.: "He-Cd-Se Gas Laser", J. Quantum Electron. IEEE, 1973, QE-9, no.8, pp.837-9.

22.621. Bulthuis, K.: "Laser Power and Vibrational Energy Transfer in CO_2 Lasers", J. Chem. Phys., 1973, 58, no.12, pp.5786-94.

22.622. Bykovskii, Yu.A., et al.: "Optical Pumping of ^{133}Cs Vapour by Injection Laser", Zh. Eksp. Teor. Fiz. Pis'ma, 1973, 17, no.6, pp.302-5, and JETP Lett., 1973, 17, no.6, pp.216-9.

22.623. Ducloy, M., Gorza, M.P., and Decomps, B.: "Higher-Order Nonlinear Effects in a Gas Laser", Opt. Commun., 1973, 8, no.1, pp.21-5.

22.624. Wagner, R.J., Zelano, A.J., and Ngai, L.H.: "Submillimetre Laser Lines in Optically Pumped Gas Molecules", Opt. Commun., 1973, 8, no.1, pp.46-7.

22.625. Isaev, A.A., Kazaryan, M.A., and Petrash, G.G.: "Pulsed, High-PRF, Lasers Based on Pb, Mn, Cu, and Au Vapours", Zh. Prikl. Spekt., 1973, 18, no.3, pp.483-4.

22.626. Kerber, R.L.: "Simple Model of a Line Selected, Long Chain, Pulsed DF-CO_2 Chemical Transfer Laser", Appl. Opt., 1973, 12, no.6, pp.1157-64.

22.627. Nighan, W.L., Weigand, W.J., and Haas, R.A.: "Ionization Instability in CO_2 Laser Discharges", Appl. Phys. Lett., 1973, 22, no.11, pp.579-82.

22.628. Jeffers, W.Q., et al.: "CW CO Chemical Laser Directly Fuelled by CS", Appl. Phys. Lett., 1973, 22, no.11, pp.587-9.

22.629. Mirels, H., Hofland, R., and King, W.S.: "Simplified Model of CW Diffusion-Type Chemical Laser", AIAA J., 1973, 11, no.2, pp.156-64.

22.630. Rheault, F., et al.: "Saturation Properties of TEA CO_2 Amplifiers in the Nano-second-Pulse Regime", Opt. Commun., 1973, 8, no.2, pp.132-5.

22.631. Takemura, M., Kobayasi, T., and Inaba, H.: "TEA UV N_2 Laser", Rec. Electr. Commun. Eng. Conversaz. Tohoku Univ., 1973, 42, no.1, pp.27-34.

22.632. Rutkovskii, F.K.: "Probability of Radiation Absorption in a Gas Laser with Pumping from Internal Light Sources", Zh. Prikl. Spekt., 1973, 18, no.4, pp.614-20.

22.633. Colliex, C., and Mourier, G.: "Theoretical Study of Inversion Mechanism and Output of a CW Argon-Ion Laser", Rev. Tech. Thomson-CSF, 1973, 5, no.1, pp.81-112.

22.634. Suchard, S.N.: "Lasing from the Upper Vibrational Levels of Flash-Initiated H_2-F_2 Laser", Appl. Phys. Lett., 1973, 23, no.2, pp.68-70.

22.635. Abrosimov, G.V., Andreev, N.G., and Odintsov, A.I.: "Study of Pulsed Super-radiance in Thallium Vapours", Vestn. Mosk. Univ. Fiz. Astron., 1973, 14, no.3, pp.207-91.

22.636. Gembarzhevskii, G.V.: "Approximate Determination of Population Inversion and Gas Multiplication in Adiabatic Expansion in a Nozzle", Zh. Prikl. Mekh. Tekh. Fiz., 1973, no.3, pp.34-40.

22.637. Collins, G.J.: "Excitation Mechanisms in He-Cd and He-Zn Ion Lasers", J. Appl. Phys., 1973, 44, no.10, pp.4633-52.

22.638. Shirahata, H., and Fujisawa, A.: "Aerodynamically Mixed Electric-Discharge CO_2 Laser", Appl. Phys. Lett., 1973, 23, no.2, pp.80-1.

22.639. Benard, D.J., Benson, R.C., and Walker, R.E.: "N_2O Pure Chemical CW Flame Laser", Appl. Phys. Lett., 1973, 23, no.2, pp.82-4.

22.640. Liu, C.S., Sucov, E.W., and Weaver, L.A.: "Copper Superradiant Emission from Pulsed Discharges in Copper-Iodide Vapour", Appl. Phys. Lett., 1973, 23, no.2, pp.92-3.

22.641. Ebert, W.: "Investigations of CW Inert-Gas Ion Lasers", Beitr. Plasma Phys., 1973, 12, no.5, pp.227-44.

22.642. Cason, C., Dezenberg, G.J., and Huff, R.J.: "Operation of a Cold-Cathode Electron-Beam-Controlled CO_2 Laser at 1-3 atm", Appl. Phys. Lett., 1973, 23, no.2, pp.110-1.

22.643. Walker, R.E., et al.: "Vibrational Disequilibrium in a Low-Pressure Na-Catalysed $CO-N_2O$ Flame", Chem. Phys. Lett., 1973, 20, no.6, pp.528-33.

22.644. Patterson, E.L.: "Superradiant Laser Action in N_2 Excited by High-Energy Electron Beam", J. Appl. Phys., 1973, 44, no.7, pp.3193-7.

22.645. Antonov, V.S., et al.: "Hydrogen Laser in Vacuum UV at Atmospheric Pressure", Zh. Eksp. Teor. Fiz. Pis'ma, 1973, 17, no.10, pp.545-8, and JETP Lett., 1973, 17, no.10, pp.393-5.

22.646. Provorov, A.S., and Chebotaev, V.P.: "CW Generation in CO_2-N_2-He Mixtures at Atmospheric Pressure", Dokl. Akad. Nauk SSSR, 1973, 208, no.1, pp.318-20, and Sov. Phys.-Dokl. 1973, 18, no.1, pp.56-7.

22.647. Nachshon, Y., and Oppenheim, U.P.: "Gain Saturation in the CO_2 Laser", Appl. Opt., 1973, 12, no.8, pp.1934-9.

22.648. George, E.V., and Rhodes, C.K.: "Kinetic Model of UV Inversions in High-Pressure Rare-Gas (Laser) Plasmas", Appl. Phys. Lett., 1973, 23, no.3, pp.139-41.

22.649. Girard, A., Pepin, H., and Vallee, J.G.: "Parametric Study of a Helical TEA CO_2 Laser", Can. J. Phys., 1973, 51, no.16, pp.1705-8.

22.650. Atanasov, P.A.: "Effect of Temperature and Composition of Gas Mixture on Population Inversion in Pulsed CO_2 Laser", C. R. Acad. Bulg. Sci. 1973, 26, no.3, pp.327-30.

22.651. Dudkin, V.A.: "Anomalous Effect of Additives on Emission of CS_2-O_2 Flame Laser", Fiz. Goreniya Vzryva, 1973, 9, no.3, pp.458-9.

22.652. Belous, V.V., and Kostin, V.N.: "Investigation of Influence of Inhomogeneous RF Electric Field on He-Ne Laser", Izv. VUZ Fiz., 1973, no.7, pp.154-6.

22.653. Pearson, R.K., et al.: "Relative Performance of a Variety of NF_2+Hydrogen-Donor Transverse-Discharge HF Chemical-Laser Systems", J. Quantum Electron. IEEE, 1973, QE-9, no.9, pp.879-89.

22.654. Fournier, G., and Pigache, D.: "Oversaturation of High-Gain CO_2 Lasers", J. Quantum Electron. IEEE, 1973, QE-9, no.10, pp.1030-1.

22.655. Ault, E.R., and Bhaumik, M.L.: "Xenon Molecular Laser in Vacuum UV", J. Quantum Electron. IEEE, 1973, QE-9, no.10, pp.1031-2.

22.656. Burnett, N.H., and Offenberger, A.A.: "Simple Electrode Configuration for UV-Initiated High-Power TEA Laser Discharges", J. Appl. Phys., 1973, 44, no.8, pp.3617-8.

22.657. Sato, K., and Sekiguchi, T.: "Effect of Water Vapour on Output Power of CO_2 Gasdynamic Laser", J. Phys. Soc. Jap., 1973, 35, no.1, p.315.

22.658. Soldatov, A.N., Evtushenko, G.S., and Murav'ev, I.I.: "Excitation of Neon in a Glow Discharge", Opt. Spekt., 1973, 34, no.1, pp.13-8, and Opt. Spectrosc., 1973, 34, no.1, pp.6-8.

22.659. Kolosovskaya, L.A.: "Calculation of Population Relaxation for Rotational Levels", Opt. Spekt., 1973, 34, no.1, pp.184-5, and Opt. Spectrosc., 1973, 34, no.1, pp.101-2.

22.660. Kochelap, V.A., and Kukibnyi, Yu.A.: "Theory of a High-Pressure IR Chemical Laser", Opt. Spekt., 1973, 34, no.2, pp.328-36, and Opt. Spectrosc., 1973, 34, no.2, pp.186-90.

22.661. Buger, P.A., Sadie, F.G., and Malan, O.G.: "Investigations on the $C_2H_2-O_2$ Chemical Laser", Z. Naturforsch., 1973, 28a, no.7, pp.1221-2.

22.662. Pugnin, V.I., Rudelov, S.A., and Stepanov, A.F.: "Generation on Ionic Transitions in Iodine", Zh. Prikl. Spekt., 1973, 18, no.5, pp.912-3.

22.663. Hodges, D.T., and Hartwick, T.S.: "Waveguide (Methyl-Alcohol) Laser for the Far IR Pumped by a CO_2 Laser", Appl. Phys. Lett., 1973, 23, no.5, pp.252-3.

22.664. Fill, E., and Schmid, W.: "Amplification of Short Pulses in CO_2 Laser Amplifiers", Phys. Lett., 1973, 45A, no.2, pp.145-6.

22.665. Pihlman, P., and Stenholm, S.: "Correlation between Phase- and Velocity-Changing Collisions in a Gas Laser", Phys. Fenn., 1973, 8, no.1, pp.13-31.

22.666. Ali, A.W.: "N_2^{II} Meinel and O_2^{II} Second-Negative Bands. Laser Theory", Appl. Opt., 1973, 12, no.10, pp.2243-5.

22.667. Harris, S.E., et al.: "Stimulated Emission in Multiple-Phonon-Pumped Xe and Ar Excimers", Appl. Phys. Lett., 1973, 23, no.5, pp.232-4.

22.668. Hoff, P.W., Swingle, J.C., and Rhodes, C.K.: "Observations of Stimulated Emission from High-Pressure Kr and Ar-Xe Mixtures", Appl. Phys. Lett., 1973, 23, no.5, pp.245-6.

22.669. McArthur, D.A., Miller, G.H., and Tollefsrud, P.B.: "Pumping of High-Pressure CO_2 Laser Media via a Fast-Burst Reactor and Electrical Sustainer", Appl. Phys. Lett., 1973, 23, no.6, pp.303-5.

22.670. Djeu, N.: "CW Single-Line CO Laser on the v_{1-0} Band", Appl. Phys. Lett., 1973, 23, no.6, pp.309-10.

22.671. Stark, E.E., et al.: "Comparison of Theory and Experiment for Nanosecond-Pulse Amplification in High-Gain CO_2 Systems", Appl. Phys. Lett., 1973, 23, no.6, pp.322-4.

22.672. Wang, S.C., and Siegman, A.E.: "Hollow-Cathode Transverse-Discharge He-Ne and He-Cd^{II} Lasers", Appl. Phys., 1973, 2, no.3, pp.143-50.

22.673. Nygaard, K.J.: "Effect of Caesium in Photoionization Laser Plasmas", J. Quantum Electron. IEEE, 1973, QE-9, no.10, pp.1020-3.

22.674. Tarasenko, V.F., and Kurbatov, Yu.A.: "Nitrogen Laser with Longitudinal Discharge and High Specific Power", Prib. Tekh. Eksp., 1973, 16, no.1, pp.182-3, and Instrum. Exp. Tech., 1973, 16, no.1, pp.219-20.

22.675. Tarasenko, V.F., and Bychkov, Yu.I.: "Nitrogen Laser with Transverse Discharge", Prib. Tekh. Eksp., 1973, 16, no.1, pp.183-4, and Instrum. Exp. Tech., 1973, 16, no.1, pp.221-2.

22.676. Buimistrov, V.M., and Trahtenberg, L.T.: "Free-Free Electron Transitions and Light Amplification", Opt. Commun., 1973, 8, no.4, pp.289-90.

22.677. Danilychev, V.A., Kerimov, O.M., and Kovsh, I.V.: "Electroionization Pulsed CO_2 Laser", Prib. Tekh. Eksp., 1973, 16, no.1, pp.184-5, and Instrum. Exp. Tech., 1973, 16, no.1, pp.223-4.

22.678. Gerardo, J.B., and Johnson, A.W.: "173-nm Radiation Dominated by Stimulated Emission from High-Pressure Xenon", J. Appl. Phys., 1973, 44, no.9, pp.4120-4.

22.679. Chen, C.J.: "Manganese Laser", J. Appl. Phys., 1973, 44, no.9, pp.4246-7.

22.680. Rabideau, S.W.: "Atom-Molecule Kinetics. $H+ClF_3$ Reaction", J. Chem. Phys., 1973, 59, no.3, pp.1533-4.

22.681. Andriyakhin, V.M., et al.: "Quasi-stationary Atmospheric-Pressure CO_2 Laser with Non-Maintaining Discharge Controlled by a Neutron Flux", Zh. Eksp. Teor. Fiz. Pis'ma, 1973, 18, no.1, pp.15-9, and JETP Lett., 1973, 18, no.1, pp.7-9.

22.682. Gorza, M.P., Decomps, B., and Ducloy, M.: "Higher-Order Nonlinear Effects in a Gas Laser", Opt. Commun., 1973, 8, no.4, pp.323-8.

22.683. Hopf, F.A., and Rhodes, C.K.: "Influence of Vibrational, Rotational, and Reorientational Relaxation on Pulse Amplification in Molecular Amplifiers", Phys. Rev., 1973, 8, no.2, pp.912-29.

22.684. Gullberg, K., Hartmann, B., and Kleman, B.: "Submillimetre Emission from Optically Pumped $^{14}NH_3$", Phys. Scr., 1973, 8, no.5, pp.177-82.

22.685. Tarasenko, V.F., and Savin, V.V.: "High-Pressure CO_2 Laser with a Transverse Discharge", Zh. Tekh. Fiz., 1973, 43, no.2, pp.353-4, and Sov. Phys.-Tech. Phys., 1973, 18, no.2, pp.227-8.

22.686. Biryukov, A.S., and Shelepin, L.A.: "Kinetics of Electrical Gasdynamic Lasers", Zh. Tekh. Fiz., 1973, 43, no.2, pp.355-60, and Sov. Phys.-Tech. Phys., 1973, 18, no.2, pp.229-31.

22.687. Doronin, V.G., and Ostapchenko, E.P.: "Normal Competitive Mode Generated in Channels with a Common Upper Level", Zh. Prikl. Spekt., 1973, 19, no.1, pp.50-5.

22.688. Herziger, G., Luthi, H.R., and Seelig, W.H.: "Temperatures of Ion Laser Plasmas", Z. Phys., 1973, 264, no.1, pp.61-71.

22.689. Voitovich, A.P., and Smirnov, A.Ya.: "Effect of Double Optical Resonance on Interactions and Frequency Selection in Gas Lasers", Zh. Prikl. Spekt., 1973, 19, no.1, pp.27-32.

22.690. Isaev, A.A., Kazaryan, M.A., and Petrash, G.G.: "Pulsed Gas Lasers Based on Vapours of Nonvolatile Substances", Prib. Tekh. Eksp., 1973, 16, no.1, pp.188-9, and Instrum. Exp. Tech., 1973, 16, no.1, pp.228-9.

22.691. Avizonis, P.V., Dean, D.R., and Grotbeck, R.: "Determination of Vibrational and Translational Temperatures in Gasdynamic Lasers", Appl. Phys. Lett., 1973, 23, no.7, pp.375-8.

22.692. Hughes, W.M., et al.: "High-Power UV Laser Radiation from Molecular Xenon", Appl. Phys. Lett., 1973, 23, no.7, pp.385-7.

22.693. Spencer, D.J., and Varwig, R.L.: "Experimental CW Chemical Laser Studies", AIAA J., 1973, 11, no.7, pp.1000-5.

22.694. Chung, P.M.: "Performance of Turbulent Chemical Laser", AIAA J., 1973, 11, no.7, pp.1040-2.

22.695. Feldman, B.J.: "Short-Pulse Multi-line and Multiband Energy Extraction in High-Pressure CO_2-Laser Amplifiers", J. Quantum Electron. IEEE, 1973, QE-9, no.11, pp.1070-8.

22.696. Greiner, N.R.: "Laser Action from Atmospheric-Pressure H_2-F_2 Mixture Made at 300°K", J. Quantum Electron. IEEE, 1973, QE-9, no.11, pp.1123-4.

22.697. Sanders, J.H., and Thomson, J.E.: "High-Gain Transitions in Neon", J. Phys. B, 1973, 6, no.10, pp.2177-83.

22.698. Brown, F., et al.: "Characteristics of a 30-kW-Peak, 496-micron, CH_3F Laser", Opt. Commun., 1973, 9, no.1, pp.28-30.

22.699. Kozlov, G.I., Ivanov, V.N., and Korablev, A.S.: "Gasdynamic Laser Using Hydrocarbon-Air Combustion Products", Zh. Eksp. Teor. Fiz. Pis'ma, 1973, 17, no.12, pp.651-4, and JETP Lett., 1973,17, no.12.

22.700. Theiss, F.J.: "Dependence of Peak Power and Pulse Duration for UV N_2 Superradiant Pulses on Operating Conditions in a Coaxial Laser Device", Opt. Commun., 1973, 9, no.1, pp.25-7.

22.701. Willgoss, R.A., and Thomas, G.C.: "Role of Striations on the Inversion Mechanism in the 441.6 nm He-Cd Cataphoretic Laser", J. Phys. D, 1973, 6,no.16,pp.L121-3.

22.702. Hopf, F.A.: "Short Pulse Energy Extraction in CO Amplifiers", Opt. Commun., 1973, 9, no.1, pp.38-41.

22.703. Tagliaferri, A.A., et al.: "UV Stimulated Emission from N_2 and NO", Phys. Lett., 1973, 45A, no.3, pp.211-2.

22.704. Janossy, M.: "Laser Oscillation in a Pulsed Ar-Cd Discharge", Phys. Lett., 1973, 45A, no.3, p.243.

22.705. Jolly, J.: "Argon-Ion Laser Stabilized by High-Temperature Walls", Rev. Phys. Appl., 1973, 8, no.3, pp.271-7.

22.706. Tunitskii, L.N.: "Study of Pulsed Argon-Ion Lasers", Zh. Prikl. Spekt., 1973, 19, no.2, pp.233-40.

22.707. Leont'ev, V.G., and Ostapchenko, E.P.: "Linear Heterogeneity of Amplification in He-Ne Laser Excited by Constant Current", Zh. Prikl. Spekt., 1973, 19, no.2, pp.241-5.

22.708. Steinvall, O.: "Design and Coherence Properties of a Simple N_2 Laser", J. Phys. E, 1973, 6, no.11, pp.1125-8.

22.709. Chang, T.V., and Wood, O.R.: "Optically Pumped 33-atm. CO_2 Laser", Appl. Phys. Lett., 1973, 23, no.7, pp.370-2.

22.710. Gordon, R.J., and Lin, M.C.: "Chemical HF Laser Emission from $CHF+O_2$ Reaction", Chem. Phys. Lett., 1973, 22, no.1, pp.107-12.

22.711. Patel, B.S., and Joshi, J.C.: "Transversely Excited, High-Pressure, CO_2 Laser", Indian J. Pure Appl. Phys., 1973, 11, no.4, pp.271-3.

22.712. Hodges, D.T., Reel, R.D., and Barker, D.H.: "Low-Threshold CW Submillimetre Laser Action in CO_2-Laser-Pumped $C_2H_4F_2$, $C_2H_2F_2$, and CH_3OH", J. Quantum Electron. IEEE, 1973, QE-9, no.12, pp.1159-60.

22.713. Patel, B.S.: "Role of Helium in TEA CO_2 Lasers", J. Quantum Electron. IEEE, 1973, QE-9, no.12, pp.1160-1.

22.714. Howgate, D.W., and Barr, T.A.: "Dynamics of the CS_2-O_2 Flame", J. Chem. Phys., 1973, 59, no.6, pp.2815-29.

22.715. Nagata, I., and Kimura, Y.: "Compact High-Power Nitrogen Laser", J. Phys. E, 1973, 6, no.12, pp.1193-5.

22.716. Rosenwaks, S., and Smith, I.W.M.: "Laser Emission from CO Formed in the Flash-Initiated Reactions of $O(^3P)$ Atoms with CS and CSe", J. Chem. Soc. Faraday Trans. II, 1973, 69, pt 9, pp.1416-24.

22.717. Belland, P., and Veron, D.: "Compact CW HCN Laser with High Stability and Power Output", Opt. Commun., 1973, 9, no.2, pp.146-8.

22.718. Adam, B., Schotsau, H.J., and Kneubuhl, F.K.: "Standard Transverse Excitation of the HCN Laser 337-micron Emission", Phys. Lett., 1973, 45A, no.5, pp.365-6.

22.719. Kokorin, V.V., and Los, V.F.: "Optical Pumping of an X-Ray (Plasma) Laser", Phys. Lett., 1973, 45A, no.6, pp.487-8.

22.720. Sudhanshuns, J.: "Stimulated Emission of X-Rays from Plasmas Generated by Short-Pulse Laser Heating of Solid Targets", Pramana, 1973, 1, no.2, pp.88-97.

22.721. Cristescu, C.P., Popescu, I.M., and Preda, A.M.: "Excitation Mechanisms in the He-Cd Plasma of a Hollow Cathode Discharge", Rev. Roum. Phys., 1973, 18, no.7, pp.859-65.

22.722. Biryukov, A.S., and Shelepin, L.A.: "Role of Nonresonant Transfer Processes in Gasdynamic Lasers", Zh. Prikl. Mekh. Tekh. Fiz., 1973, no.4, pp.25-32.

22.723. Boulanger, P., et al.: "Double-Discharge, High-Power, TEA CO_2 Laser", Z. Angew. Math. Phys., 1973,24,no.3,pp.439-42.

22.724. Barry, J.D., and Brandelik, J.E.: "CO-Laser Gas Temperature", Appl. Opt., 1973, 12, no.12, pp.2809-10.

22.725. Mayer, S.W., Taylor, D., and Kwok, M.A.: "HF Chemical Lasing at Higher Vibrational Levels", Appl. Phys. Lett., 1973, 23, no.8, pp.434-6.

22.726. Gross, R.W.F., and Wesner, F.: "Electron-Beam-Initiated Chemical Laser in SF_6-H_2 Mixtures", Appl. Phys. Lett., 1973, 23, no.10, pp.559-61.

22.727. Alcock, A.J., Leopold, K., and Richardson, M.C.: "Continuously Tunable High-Pressure CO_2 Laser with UV Photo-Preionization", Appl. Phys. Lett., 1973, 23, no.10, pp.562-4.

22.728. Byron, S.R., Nelson, L.Y., and Mullaney, G.J.: "HF and DF Lasers by Direct Electrical Discharge Excitation", Appl. Phys. Lett., 1973, 23, no.10, pp.565-7.

22.729. Curry, B.P., Zwick, S.A., and Aliprantis, C.D.: "Closed-Form Solution for Excited-State Populations in Temporally and Spatially Varying Lasing Media", Appl. Phys. Lett., 1973, 23, no.10, pp.574-5.

22.730. Asamo, Y., Fujii, K., and Takahashi, T.: "Hollow-Cathode He-Cd[II] Laser as a Source of White Light", J. Inst. Telev. Eng. Jap., 1973, 27, no.8, pp.617-23.

22.731. Emanuel, G., Cohen, N., and Jacobs, T.A.: "Theoretical Performance of HF Chemical CW Laser", J. Quant. Spectrosc., 1973, 13, no.12, pp.1365-93.

22.732. Bokhan, P.A., Klimkin, V.M., and Prokop'ev, V.E.: "Gas Laser Using Ionized Europium", Zh. Eksp. Teor. Fiz. Pis'ma, 1973, 18, no.2, pp.80-2, and JETP Lett., 1973, 18, no.2, pp.44-5.

22.733. Bagratashvili, V.N., et al.: "Electrochemical High-Pressure HF Laser", Zh. Eksp. Teor. Fiz. Pis'ma, 1973, 18, no.2, pp.110-3, and JETP Lett., 1973, 18, no.2, pp.62-4.

22.734. Grin, Yu.I., Polyakov, V.M., and Testov, V.G.: "Experimental Study of Gas-dynamic N_2O-N_2-He Amplification", Zh. Eksp. Teor. Fiz. Pis'ma, 1973, 18, no.4, pp.260-3, and JETP Lett., 1973, 18, no.4, pp.155-7.

22.735. Dufresne, D., et al.: "Experimental Study of a TEA CO_2 Laser Preionized by Electron Beam from a Cold-Cathode Gun", C. R. Acad. Sci. (Paris) 1973, 277B, no.22, pp.667-9.

22.736. Christiansen, W.H., and Greenfield, E.: "Analysis of a Collisionally Induced Dipole Laser", Appl. Phys. Lett., 1973, 23, no.11, pp.623-5.

22.737. Velculescu, V.G., and Udrea, M.V.: "Dynamics of CO_2 TEA Laser", Rev. Roum. Phys., 1973, 18, no.10, pp.1177-83.

22.738. Mikaberidze, A.A., Ochkin, V.N., and Sobolev, N.N.: "Population of the Lower Active Level in a CO_2 Laser", Kvantovaya Elektron., 1973, no.7, pp.41-6, and Sov. J. Quantum Electron., 1973, 3, no.1, pp.21-3.

22.739. Keidan, V.F., et al.: "Laser Action in Se[II]", Kvantovaya Elektron., 1973, no.7, pp.75-8, and Sov. J. Quantum Electron., 1973, 3, no.1, pp.40-2.

22.740. Lotkova, E.N., et al.: "Stimulated Emission of 5-micron Radiation from a CO_2-N_2-He Mixture", Kvantovaya Elektron., 1973, no.7, pp.137-9, and J. Quantum IEEE, 1973, 3, no.1, pp.87-8.

22.741. Chen, C.J., Nerheim, N.M., and Russell, G.R.: "Double-Discharge Copper-Vapour Laser with the Chloride as Lasant", Appl. Phys. Lett., 1973, 23, no.9, pp.514-5.

22.742. Lowke, J.J., Phelps. A.V., and Irwin, B.W.: "Predicted Electron Transport Coefficients and Operating Characteristics of CO_2-N_2-He Laser Mixtures", J. Appl. Phys., 1973, 44, no.10, pp.4664-71.

22.743. Seguin, H.J., Tulip, J., and McKen, D.: "Enhancement of Photoelectron Density in TEA Lasers Using Additives", Appl. Phys. Lett., 1973, 23, no.9, pp.527-9.

22.744. Wallace, S.C., Hodgson, R.T., and Dreyfus, R.W.: "Short-Pulse Excitation of Xenon Molecular Dissociation Laser at 172.9 nm by Relativistic Electrons", Appl. Phys. Lett., 1973, 23, no.12, pp.672-4.

22.745. Fetterman, H.R., Parker, C.D., and Schlossberg, H.R.: "CW Submillimetre Laser by Optically Pumped Stark-Tuned NH_3", Appl. Phys. Lett., 1973, 23, no.12, pp.684-6.

22.746. Fisher, E.R.: "Modelling of a Pulsed CO_2-N_2 Molecular Laser System. II", J. Appl. Phys., 1973, 44, no.11, pp.5031-4.

22.747. Isaev, A.A., Kazaryan, M.A., and Petrash, G.G.: "Copper-Vapour Pulsed Laser with PRF of 10-kHz", Opt. Spekt., 1973, 35, no.3, pp.528-30, and Opt. Spectrosc., 1973, 35, no.3, pp.307-8.

22.748. Yukov, E.A.: "Elementary Processes in the Active Medium of an Iodine Photodissociation Laser", Kvantovaya Elektron., 1973, no.2, pp.53-8, and Sov. J. Quantum Electron., 1973, 3, no.2, pp.117-20.

22.749. Gadetskii, N.P., et al.: "Stimulated Emission of Visible Light due to Transitions in Cl[II] and I[II]", Kvantovaya Elektron., 1973, no.2, pp.110-2, and Sov. J. Quantum Electron., 1973, 3, no.2, pp.168-9.

22.750. Just, T., and Roth, P.: "Calculation of Inversion and Laser Power in Expanding Combustible Gas. CO_2-N_2-He", Z. Flugwiss., 1973, 21, no.7, pp.242-8.

22.751. Margulis, V.M., Margolin, A.D., and Kaganova, Z.I.: "Field Concentration of Excited Molecules in Diffusion CO_2-N_2 Laser", Fiz. Goreniya Vzryva, 1973, 9, no.6, pp.818-22.

22.752. Patel, B.S.: "Parametric Study of TE CO_2-N_2 Laser", Indian J. Pure Appl. Phys., 1973, 11, no.8, pp.591-3.

22.753. Wilson, J., et al.: "Electron-Beam Dissociation of Fluorine", J. Appl. Phys., 1973, 44, no.12, pp.5447-54.

22.754. Boyer, K., Henderson, D.B., and Morse, R.L.: "Spatial Distribution of Ionization in Electron-Beam-Controlled Discharge Lasers", J. Appl. Phys., 1973, 44, no.12, pp.5511-2.

22.755. Henderson, D.B.: "Electron Transport in Gas Discharge Lasers", J. Appl. Phys., 1973, 44, no.12, pp.5513-6.

22.756. Paulson, R.F.: "Effect of Organic Gas Additive on a Pulsed HF Chemical Laser", J. Appl. Phys., 1973, 44, no.12, pp.5633-4.

22.757. Lin, M.C.: "Mechanism of CO Laser Emission from CH+NO Reaction", J. Phys. Chem., 1973, 77, no.3, pp.2726-9.

22.758. Basov, N.G., Belenov, E.M., et al.: "Electric Ionization Lasers", Zh. Eksp. Teor. Fiz., 1973, 64, no.1, pp.108-21, and Sov. Phys.-JETP, 1973, 37, no.1, pp.58-64.

22.759. Zakharenko, Yu.G., and Privalov, V.E.: "Oscillations in Discharge Gap of He-Ne Laser and their Effect on Emission Parameters", Opt. Spekt., 1973, 35, no.4, pp.750-8, and Opt. Spectrosc., 1973, 35, no.4, pp.434-9.

22.760. Sadie, F.G., Buger, P.A., and Malan, O.G.: "CW CO Chemical Laser from Oxidation of Acetylene", Z. Naturforsch., 1973, 28a, no.2, pp.309-10.

22.761. Padminogin, A.A.: "HF (or DF) Pulsed Laser", Kvantovaya Elektron., 1973, no.3, pp.88-90, and Sov. J. Quantum Electron., 1973, 3, no.3, pp.229-31.

22.762. Kulke, D.: "Collisions of Excited Ne Atoms in a He-Ne 633-nm Laser", Kvantovaya Elektron., 1973, no.3, pp.97-9, and Sov. J. Quantum Electron., 1973, 3, no.3, pp.238-9.

22.763. Tarasenko, V.F., et al.: "Characteristics of a High-Power N_2 Laser", Kvantovaya Elektron., 1973, no.3, pp.103-4, and Sov. J. Quantum Electron., 1973, 3, no.3, pp.244-5.

22.764. Gavrikov, V.F., et al.: "Nonequilibrium Flow in a Gasdynamic Laser", Kvantovaya Elektron., 1973, no.3, pp.109-12, and Sov. J. Quantum Electron., 1973, 3, no.3, pp.250-2.

22.765. Konovalov, I.N., et al.: "Threshold Pumping Energy of a High-Pressure Pulsed CO_2 Laser", Kvantovaya Elektron., 1973, no.3, pp.112-5, and Sov. J. Quantum Electron., 1973, 3, no.3, pp.253-4.

22.766. Bychkov, Yu.I., Osipov, V.V., and Tarasenko, V.F.: "CO_2 Pulsed Laser Excited by Double-Discharge Method", Kvantovaya Elektron., 1973, no.3, pp.122-4, and Sov. J. Quantum Electron., 1973, 3, no.3, pp.262-3.

22.767. Latush, E.L., and Sem, M.F.: "Stimulated Emission due to Transitions in Alkaline-Earth Metal Ions", Kvantovaya Elektron., 1973, no.3, pp.66-71, and Sov. J. Quantum Electron., 1973, 3, no.3, pp.216-9.

22.768. Ross, W., and Seliger, K.: "Vacuum-UV Laser Emission from H_2 in a Short-Pulse Electric Discharge", Exp. Tech. Phys., 1973, 21, no.6, pp.483-7.

22.769. Fujii, K.: "Characteristics and Design Method of Negative-Glow He-Cd^{II} Laser", Electr. Eng. Jap., 1973, 93, no.1, pp.106-14.

22.770. Yudin, V.I.: "Investigation of He-Ne Laser with High-Frequency Excitation", Kvantovaya Elektron., 1973, no.3, pp.134-6, and Sov. J. Quantum Electron., 1973, 3, no.3, pp.274-5.

22.771. Biryukov, A.S., et al.: "Shock-Tube Studies on a Gasdynamic Laser", High Temp.Pressures, 1973, 5, no.4, pp.389-92.

22.772. Baranov, V.Yu., Borisov, V.M., and Strel'tsov, A.P.: "Pulsed CO_2 Laser Having Increased Pressure and Preionization on the Cathode", Prib. Tekh. Eksp., 1973, 16, no.5, pp.188-90, and Instrum. Exp. Tech., 1973, 16, no.5, pp.1522-4.

22.773. Alekseev, V.A., Andreeva, T.L., and Sobel'man, I.I.: "Theory of Nonlinear Power Resonances in Gas Lasers", Zh. Eksp. Teor. Fiz., 1973, 64, no.3, pp.813-24, and Sov. Phys.-JETP, 1973, 37, no.3, pp.413-8.

22.774. Leshenyuk, N.S., and Orlov, L.N.: "Temperature Field in CO Laser", Zh. Tekh. Fiz., 1973, 43, no.11, pp.2382-7, and Sov. Phys.-Tech. Phys., 1974, 18,no.11,pp.1504-7.

22.775. Sabotinov, N.V., Telbizov, P.K., and Kalchev, S.D.: "He-Cd-Se Laser", Zh. Tekh. Fiz., 1973, 43, no.12, pp.2621-4, and Sov. Phys.-Tech. Phys., 1974, 18, no.12, pp.1646-7.

22.776. Dyubko, S.F., Svich, V.A., and Fesenko, L.D.: "Submillimetre CH_3OH and CH_3OD Lasers with Optical Pumping", Zh. Tekh. Fiz., 1973, 43, no.8, pp.1772-3, and Sov. Phys.-Tech. Phys., 1974,18,no.8,p.1121.

22.777. Kroshko, V.N., and Soloukhin, R.I.: "Optical Inversion Modes during Thermal Excitation by Mixing in a Supersonic Stream", Dokl. Akad. Nauk SSSR, 1973, 211, no.4-6, pp.829-32, and Sov. Phys.-Dokl., 1974, 18, no.8, pp.554-5.

22.778. Golger, A.L., and Letokhov, V.S.: "Population Inversion of Vibrational Levels in a High-Pressure Molecular Gas Pumped by Laser Radiation", Kvantovaya Elektron., 1973, no.5, pp.106-15, and Sov. J. Quantum Electron., 1974, 3, no.5, pp.428-32.

22.779. Gavrikov, V.F., et al.: "Gasdynamic N_2O Laser", Kvantovaya Elektron., 1973, no.5, pp.119-20, and Sov. J. Quantum Electron., 1974, 3, no.5, pp.437-9.

22.780. Dyubko, S.F., Svich, V.A., and Fesenko, L.D.: "Submillimetre Laser Using Formic-Acid Vapour Pumped by a CO_2 Laser", Kvantovaya Elektron., 1973, no.5, pp.128-9, and Sov. J. Quantum Electron., 1974, 3, no.5, p.446.

22.781. Anokhin, A.V., Markova, S.V., and Petrash, G.G.: "Pulsed Stimulated Emission due to Vibrational Transitions in CO in Presence of He and Ar", Kvantovaya Elektron., 1973, no.5, pp.100-5, and Sov. J. Quantum Electron., 1974, 3, no.5, pp.424-7.

22.782. Judd, O.P., and Wada, J.Y.: "Investigation of a UV Preionized Electrical Discharge and CO_2 Laser", J. Quantum Electron. IEEE, 1974, QE-10, no.1, pp.12-20.

22.783. Bin-Nun, E., and Rokni, M.: "CW
10.6-micron CO_2 Laser Pumped by Chemical
Action of NO with Nitrogen Atoms", J. Quantum Electron. IEEE, 1974, QE-10, no.1,
pp.89-91.

22.784. Nishihara, H., and Kronast, B.:
"TE Duolaser Providing Simultaneous Lasing
on CO_2 and N_2O Transitions", J. Quantum
Electron. IEEE, 1974, QE-10, no.1, pp.91-2.

22.785. Tucker, A.W., and Birnbaum, M.:
"Pulsed-Ion Laser Performance in N_2, O_2, Kr,
Xe and Ar", J. Quantum Electron. IEEE, 1974,
QE-10, no.1, pp.99-100.

22.786. Ultee, C.J., and Bonczyk, P.A.:
"Performance and Characteristics of a Chemical CO Laser", J. Quantum Electron IEEE,
1974, QE-10, no.2, pp.105-10.

22.787. Friedland, L., et al.: "Semi-
Empirical Approach to CO_2 Laser Kinetics",
J. Phys. D, 1974, 7, no.2, pp.303-13.

22.788. Smith, A.L.S., and Austin, J.M.:.
"Dissociation Mechanism in Pulsed and CW
CO_2 Lasers", J. Phys. D, 1974, 7, no.2,
pp.314-22.

22.789. Piper, J.A.: "Increased Efficiency
and CW Transitions in the He-I_2 Laser System", J. Phys. D, 1974, 7, no.2, pp.323-8.

22.790. McArthur, D.A., and Poukey, J.W.:
"Theory of Plasma Electron Contribution to
the Electron-Beam-Excited N_2 Laser", Phys.
Rev. Lett., 1974, 32, no.3, pp.89-92.

22.791. Gerber, R.A., and Patterson,
E.L.: "Intense Electron-Beam Initiation of
a High-Energy HF Laser", J. Quantum Electron. IEEE, 1974, QE-10, no.3, pp.333-7.

22.792. Chang, N.C., and Tavis, M.T.:
"Gain (Analysis) of High-Pressure CO_2
Lasers", J. Quantum Electron. IEEE, 1974,
QE-10, no.3, pp.372-5.

22.793. Taylor, D., Mayer, S.W., and
Suchard, S.N.: "HCl Chemical Lasing with HI
and Chloro Compounds", J. Quantum Electron.
IEEE, 1974, QE-10, no.3, pp.389-90.

22.794. Fowles, G.R., Zuryk, J.A., and
Jensen, R.C.: "IR Laser Lines in P^I", J.
Quantum Electron. IEEE, 1974, QE-10, no.3,
pp.394-5.

22.795. Sato, K., and Sekiguchi, T.:
"Experimental Studies of CO_2 Gasdynamic
Laser by Shock Tube", J. Phys. Soc. Jap.,
1974, 36, no.3, pp.808-14.

22.796. McKenzie, A.L.: "Radial Profiles
of Upper-Laser-Level Emission in a He-Se
Discharge", J. Phys. B, 1974, 7, no.4,
pp.L141-5.

22.797. Koda, T., Twai, T., and Nakashima,
Y.: "Time-Resolved Behaviours of Singly-
and Doubly-Ionized Argon Laser Lines", Jap.
J. Appl. Phys., 1974, 13, no.1, pp.170-6.

22.798. Kitazima, I.: "Efect of He and
N_2 on Pulsed Operation of CO_2 Lasers", Opt.
Commun., 1974, 10, no.2, pp.141-4.

22.799. Hocker, L.O.: "Identification and
Prediction of Optically Pumped Laser Lines
in Ozone", Opt. Commun., 1974, 10, no.2,
pp.157-9.

22.800. Wood, O.R.: "High-Pressure Pulsed
Molecular Lasers", Proc. IEEE, 1974, 62,
no.3, pp.355-97.

22.801. Janossy, M., et al.: "CW Laser
Oscillation in a Hollow-Cathode He-Kr Discharge", Phys. Lett., 1974, 46A, no.6,
pp.379-80.

22.802. Austin, J.M., Smith, A.L.S., and
Browne, P.G.: "Dissociation in TEA CO_2
Lasers", Phys. Lett.,1974,46,no.6,pp.427-8.

22.803. Freund, I.: "Optically Stimulated
X-Ray Laser", Appl. Phys. Lett., 1974, 24,
no.1, pp.13-5.

22.804. Bohm, W.L.: "Possible Population
Inversions for Vacuum-UV and Soft-X-Ray
Transitions in Hydrogen-Like Ions", Appl.
Phys. Lett., 1974, 24, no.1, pp.15-7.

22.805. Masek, K., and Vokaty, E.: "Solution of Boltzmann's Equation for Electrons
in the Discharges in Helium/ Metal-Vapour
Mixtures", Czech. J. Phys., 1974, 24B, no.3,
pp.267-83.

22.806. Henry, A., and Thomas, J.M.: "Increase in Gain and Power of CO_2 Lasers",
C. R. Acad. Sci. (Paris), 1974, 278B, no.1,
pp.5-6.

22.807. Cohn, D.B.: "Photoinitiated TEA
CO_2 Laser at Low Temperatures", J. Quantum
Electron. IEEE, 1974, QE-10, no.4,pp.459-61.

22.808. Peterson, A.B.: "C^{II} Laser at
678.3 nm, 657.8 nm, and 514.5 nm", J. Quantum Electron. IEEE, 1974, QE-10,no.4,p.468.

22.809. Menzies, R.T.: "Effect of Centrifugal Distortion on CO_2-Laser Frequencies",
J. Quantum Electron. IEEE, 1974, QE-10,
no.5, pp.486-9.

22.810. Ahmed, S.A., and Keeffe, W.M.:
"Parametric and Discharge Studies of Three-
Colour Gas-Mix Ion Lasers", J. Appl. Phys.,
1974, 45, no.1, pp.182-6.

22.811. Arecchi, F.T., Banfi, G.P., and
Malvezzi, A.M.: "Threshold Evaluations for
an X-Ray Laser", Opt. Commun., 1974, 10,
no.3, pp.214-8.

22.812. Steyer, B., and Schafer, F.P.:
"Vapour-Phase Dye Laser", Opt. Commun.,
1974, 10, no.3, pp.219-20.

22.813. Mahr, H., and Roeder, U.: "Use of
Metastable Ions for a Soft-X-Ray Laser",
Opt. Commun., 1974, 10, no.3, pp.227-8.

22.814. Bikmukhametov, K.A., and Klement'ev,
V.M.: "Mercury Vapour Laser", Avtometriya,
1974, no.1, pp.101-3.

22.815. Fuhs, A.E., et al.: "Experimental Verification of Density Inhomogeneity due to Lasing in a Gasdynamic Laser", Appl. Phys. Lett., 1974, 24, no.3, pp.132-4.

22.816. Murray, E.R., Kruger, C., and Mitchner, M.: "Measurement of 9.6-micron CO_2 Laser Transition Probability and Optical Broadening Cross Section", Appl. Phys. Lett., 1974, 24, no.4, pp.180-1.

22.817. Chang, T.Y., and Wood, O.R.: "Optical Transfer 42-atm. N_2O Laser", Appl. Phys. Lett., 1974, 24, no.4, pp.182-3.

22.818. Hon, J.F., and Novak, J.R.: "CW HF Chemical Laser from the Reaction of F Atoms with C_3H_8 and C_4H_{10}", Appl. Phys. Lett., 1974, 24, no.4, pp.202-4.

22.819. Aprahamian, R., et al.: "Pulsed Electron-Beam-Initiated Chemical Laser Operating on the H_2-F_2 Chain Reaction", Appl. Phys. Lett., 1974, 24, no.5, pp.239-42.

22.820. Brown, R.T., and Smith, D.C.: "Optically Pumped Electric-Discharge UV Laser", Appl. Phys. Lett., 1974, 24, no.5, pp.236-8.

22.821. Collins, C.B., et al.: "Stimulated Emission from the Recombining Afterglow of an Electron-Beam Discharge in Several Atmospheres of Helium", Appl. Phys. Lett., 1974, 24, no.5, pp.245-7.

22.822. Oodate, H., Obara, M., and Fujioka, T.: "Enhanced HBr Chemical Laser Output with Addition of SF_6", Appl. Phys. Lett., 1974, 24, no.6, pp.272-4.

22.823. Lin, M.C.: "Photoexcitation and Photodissociation Lasers. I", J. Quantum Electron. IEEE, 1974, QE-10, no.6, pp.516-21.

22.824. Girard, A., and Beaulieu, A.J.: "TEA CO_2 Laser with Output Pulse Duration Adjustable from 50ns to over 0.05ms", J. Quantum Electron. IEEE, 1974, QE-10, no.6, pp.521-4.

22.825. Gallardo, M., Massone, C.A., and Garavaglia, M.: "Inversion Mechanism of H_2 Vacuum-UV Laser", J. Quantum Electron. IEEE, 1974, QE-10, no.6, pp.525-6.

22.826. Hattori, S., et al.: "CW Iodine-Ion Laser in a Positive-Column Discharge", J. Quantum Electron. IEEE, 1974, QE-10, no.6, pp.530-1.

22.827. Gaidish, V.A., et al.: "Iodine Laser with Output of 20J and Pulse Duration of 3ns", Zh. Eksp. Teor. Fiz. Pis'ma, 1974, 20, no.4, pp.243-6, and JETP Lett., 1974, 20, no.4.

22.828. Smith, A.L.S., and Austin, J.M.: "Atomic Oxygen Recombination in CO_2 Laser Gases", J. Phys. B, 1974, 7, no.6, pp.L191-4.

22.829. Goujon, P., and Clerc, M.: "TE Lasers in the UV", J. Chin. Phys., 1974, 71, no.2, pp.206-10.

22.830. Yamabe, C., et al.: "Effect of Adding Xylene in TEA Double-Discharge CO_2 Laser", Jap. J. Appl. Phys., 1974, 13, no.3, pp.569-70.

22.831. Kiselevskii, L.I., Skutov, D.K., and Sokolov, S.A.: "Electrodischarge Laser with Mixed N_2 and CO_2 Gases and Without Helium", Zh. Prikl. Spekt., 1974, 20, no.1, pp.35-7.

22.832. Broadwell, J.E.: "Effect of Mixing Rate on HF Chemical Laser Performance", Appl. Opt., 1974, 13, no.4, pp.962-7.

22.833. Aprahamian, R., et al.: "Repetitively Pulsed Electron-Beam-Initiated Chemical Lasers from SF_6-H_2 and SF_6-D_2 Mixtures", Appl. Phys. Lett., 1974, 24, no.8, pp.384-6.

22.834. Puzewicz, Z., and Trzesowski, Z.: "Law of Energy Conservation in Flow Lasers", Bull. Acad. Pol. Sci., Ser. Sci. Tech., 1974, 22, no.2, pp.141-7.

22.835. Mizeraczyk, J., and Konieczyk, J.: "Two-Anode He-Cd^{II} Laser", Bull. Acad. Pol. Sci., 1974, 22, no.3, pp.237-8.

22.836. Dreyfus, R.W., Hodgson, R.T., and Grischkowsky, D.R.: "Matching the Electron-Beam and Light-Beam Velocities in Longitudinally Pumped Gas Lasers", IBM Tech. Disclosure Bull., 1974, 16, no.9, pp.3112-3.

22.837. Johnson, A.W., and Gerardo, J.B.: "Diluent Cooling of a Vacuum-UV High-Pressure Xe Laser", J. Appl. Phys., 1974, 45, no.2, pp.867-72.

22.838. Collins, C.B., et al.: "Stimulated Emission from Charge-Transfer Reactions in the Afterglow of an Electron-Beam Discharge in High-Pressure He-N_2 Mixtures", Appl. Phys. Lett., 1974, 24, no.10, pp.477-8.

22.839. Hughes, W.M., Shannon, J., and Hunter, R.: "126.1-nm Molecular Argon Laser", Appl. Phys. Lett., 1974, 24, no.10, pp.488-90.

22.840. Chen, C.J.: "Manganese Laser Using the Chloride as Lasant", Appl. Phys. Lett., 1974, 24, no.10, pp.499-500.

22.841. Belous, V.V., and Kostin, V.N.: "Influence of an Inhomogeneous RF Electric Field on a CO_2-Laser Plasma", Izv. VUZ Fiz., 1974, no.4, pp.67-71.

22.842. Hall, R.J., and Eckbreth, A.C.: "Kinetic Modelling of CW CO Electric-Discharge Lasers", J. Quantum Electron. IEEE, 1974, QE-10, no.8, pp.580-90.

22.843. Palmer, R.E., and Gusinow, M.A.: "Gain v. Time in the CF_3I Iodine Photodissociation Laser", J. Quantum Electron. IEEE, 1974, QE-10, no.8, pp.615-6.

22.844. Ault, E.R., Bhaumik, M.L., and Olson, N.T.: "High-Power Ar-N_2 Transfer Laser at 357.7 nm", J. Quantum Electron. IEEE, 1974, QE-10, no.8, pp.624-6.

22.845. Blandchard, M., et al.: "Super-atmospheric Double-Discharge CO_2 Laser", J. Appl. Phys., 1974, 45, no.3, pp.1311-4.

22.846. Mills, C.B.: "CO_2 Gas Laser Excitation and Gain", J. Appl. Phys., 1974, 45, no.3, pp.1336-41.

22.847. Hoffman, J.M., Bingham, F.W., and Moreno, J.B.: "Parametric Study of a Cold-Cathode Electron-Beam-Controlled CO_2 Laser", J. Appl. Phys., 1974, 45, no.4, pp.1798-805.

22.848. Hinchen, J.J.: "Operation of a Small Single-Mode Stable CW HF Laser", J. Appl. Phys., 1974, 45, no.4, pp.1818-21.

22.849. Patterson, E.L., Gerber, R.A., and Blair, L.S.: "HF Laser Initiated by 300-keV Electron Beam", J. Appl. Phys., 1974, 45, no.4, pp.1822-5.

22.850. Palmer, R.E., and Gusinow, M.A.: "Late-Time Gain of CF_3I Iodine Photodissociation Laser", J. Appl. Phys., 1974, 45, no.5, pp.2174-9.

22.851. Obara, M., and Fujioka, T.: "Parametric Study of TE HF Chemical Lasers", Jap. J. Appl. Phys., 1974, 13, no.6, pp.995-1000.

22.852. Borgstrom, S.A.: "Simple Low-Divergence H_2 Laser at 160 nm", Opt. Commun., 1974, 11, no.2, pp.105-8.

22.853. Wauchop. T.S., Schiff, H.I., and Welge, K.H.: "Pulsed-Discharge IR OH Laser", Rev. Sci. Instrum., 1974, 45, no.5, pp.633-5.

22.854. Dyubko, S.F., Svich, V.A., and Fesenko, L.D.: "Stimulated Emission of Submillimetre Waves by Hydrazine Molecules Excited by CO_2 Laser", Zh. Prikl. Spekt., 1974, 20, no.4, pp.718-9.

22.855. Golden, D.E., and Ormonde, S.: "Negative-Ion Lasers", Appl. Phys. Lett., 1974, 24, no.12, pp.618-29.

22.856. Guenoche, H., Lee, J.H.S., and Sedes, C.: "Population Inversion in Blast Waves Propagating in H_2-F_2-He Mixtures", Combust. Flame, 1974, 22, no.2, pp.237-41.

22.857. Patel, B.S.: "Effect of Rotational Relaxation on Sequentially Q-Switched CO_2 Laser Transitions", Indian J. Pure Appl. Phys., 1974, 12, no.2, pp.129-32.

22.858. Greiner, N.R., Blair, L.S., and Bird, P.F.: "0.2-GW Pulsed H_2-F_2 Chemical Laser Initiated by an Electron Beam", J. Quantum Electron. IEEE, 1974, QE-10, no.9, pp.646-7.

22.859. Ferrar, C.M.: "Buffer Gas Effects in a Rapidly Pulsed Copper-Vapour Laser", J. Quantum Electron. IEEE, 1974, QE-10, no.9, pp.655-7.

22.860. Sakurai, T.: "Electronic Collisions. Induced Transition between Upper Levels of a He-Cd Laser", J. Appl. Phys., 1974, 45, no.6, pp.2666-7.

22.861. De Koker, J.G., and Rice, W.W.: "MgF_2, Metal-Atom, Oxidation Laser from Explosively Generated Vapour", J. Appl. Phys., 1974, 45, no.6, pp.2770-2.

22.862. Bihl, S., Fouassier, J.P., and Joeckle, R.: "Calculation of IR Emission of a CO_2 Laser Medium", J. Quant. Spectrosc., 1974, 14, no.9, pp.819-27.

22.863. Hashino, Y., Katsuyama, Y., and Fukuda, K.: "UV Laser Oscillations of Multiply Ionized Ions in z-Pinch Discharge", Jap. J. Appl. Phys., 1974, 13, no.7, pp.1134-44.

22.864. Yoshihiro, K., and Yamanouchi, C.: "Stable CW HCN Laser", Rev. Sci. Instrum., 1974, 45, no.6, pp.767-8.

22.865. Struk, I.I., et al.: "Iodine Photodissociation Laser Using Group V Compounds", Zh. Eksp. Teor. Fiz. Pis'ma, 1974, 19, no.1, pp.44-7, and JETP Lett., 1974, 19, no.1.

22.866. Letokhov, V.S.: "Pumping of Nuclear Levels by X-Ray Radiation of a Laser Plasma", Kvantovaya Elektron., 1974, 3, no.4, pp.125-7, and Sov. J. Quantum Electron., 1974, 3, no.4, pp.360-1.

22.867. Gundel, H., and Ross, W.: "Excitation and Emission in a Pulsed N_2 Gas Laser", Ann. Phys. (Leipzig), 1974, 31, no.3, pp.263-76.

22.868. Fischer, H., Girnus, R., and Ruhl, F.: "Low Threshold Coaxial N_2 Laser with Resonator", Appl. Opt., 1974, 13, no.8, pp.1759-60.

22.869. Searles, S.K., and Hart, G.A.: "Laser Emission at 357.7 nm and 380.5 nm in Electron-Beam-Pumped Ar-N_2 Mixtures", Appl. Phys. Lett., 1974, 25, no.1, pp.79-82.

22.870. Hughes, W.M., Shannon, J., and Hunter, R.: "Efficient High-Energy-Density Molecular Xenon Laser", Appl. Phys. Lett., 1974, 25, no.1, pp.85-7.

22.871. Cummings, J.C., Dube, C.M., and Witte, A.B.: "Comparison of H_2 and HBr as Cavity Fuels in a CW HF Laser", Appl. Phys. Lett., 1974, 25, no.1, pp.89-92.

22.872. Peterson, A.B., and Wittig, C.: "Chemically Pumped CO_2 Laser from CS_2-O_2 Reaction", Chem. Phys. Lett., 1974, 27, no.3, pp.442-4.

22.873. Brechignac, Ph., Martin, J.P., and Taieb, G.: "Small-Signal Gain Measurements and Vibrational Distribution in CO", J. Quantum Electron. IEEE, 1974, QE-10, no.10, pp.797-802.

22.874. Kitaeva, V.F., et al.: "Probe Measurements of Ar[II]-Laser Plasma Parameters", J. Quantum Electron. IEEE, 1974, QE-10, no.10, pp.803-9.

22.875. Inoue, G., and Tsuchiya, S.: "Mechanism of Transversely Excited Chemical HBr Laser", Jap. J. Appl. Phys., 1974, 13, no.9, pp.1421-8.

22.876. Dreyfus, R.W., and Hodgson, R.T.: "Molecular-Hydrogen Laser. 109.8-161.3 nm", Phys. Rev., 1974, 9A, no.6, pp.2635-48.

22.877. Isaev, A.A., Kazaryan, M.A., and Petrash, G.G.: "Laser Pulses due to Transitions from a Resonance to a Metastable Level in Ba Vapour", Kvantovaya Elektron., 1974, no.4, pp.123-5, and Sov. J. Quantum Electron., 1974, 3, no.4, pp.358-9.

22.878. Bergstedt, K., and Hulsmann, H.G.: "Experimental Investigation of Emission Properties of a Coaxial Nitrogen Laser", Z. Phys., 1974, 269, no.3, pp.195-303.

22.879. Ishchenko, V.N., et al.: "Superradiance in the 2^+ and 1^- Bands of Nitrogen in a Discharge at 10 atm.", Zh. Eksp. Teor. Fiz. Pis'ma, 1974, 19, no.7, pp.429-33, and JETP Lett., 1974, 19, no.7.

22.880. Balykin, V.I., et al.: "CO_2 Laser Operating at 10.6 micron with Optical Pumping at 9.6 micron", Zh. Eksp. Teor. Fiz. Pis'ma, 1974, 19, no.7, pp.482-5, and JETP Lett., 1974, 19, no.7.

22.881. Smith, P.W., et al.: "Optically Excited Organic-Dye Vapour Laser", Appl. Phys. Lett., 1974, 25, no.3, pp.144-6.

22.882. Werner, C.W., et al.: "Dynamic Model of High-Pressure Rare-Gas Excimer Lasers", Appl. Phys. Lett., 1974, 25, no.4, pp.235-8.

22.883. Singer, S.: "Observations of Anomalous Gain Coefficients in TEA Double-Discharge CO_2 Lasers", J. Quantum Electron. IEEE, 1974, QE-10, no.11, pp.829-31.

22.884. Whittier, J.S., and Kerber, R.L.: "Performance of an HF Chain-Reaction Laser with High Initiation Efficiency", J. Quantum Electron. IEEE, 1974, QE-10,no.11,pp.844-7.

22.885. Kitazima, I.: "Effects of Foreign Gases on Pulsed Operation of a CO_2 Laser", J. Appl. Phys., 1974, 45, no.7,pp.2997-3004.

22.886. Gudzenko, L.I., Yakovlenko, S.I., and Yevstigneev, V.V.: "Application of Principle of a Plasma Layer to Amplification in the X-Ray Region", Phys. Lett., 1974, 48A, no.6, pp.419-20.

22.887. Anokhin, A.V., Markova, S.V., and Petrash, G.G.: "Pulsed Stimulated Emission due to Vibrational Transitions in the CO Molecule in Mixtures with Xe", Kvantovaya Elektron., 1974, 1, no.1, pp.96-101, and Sov. J. Quantum Electron., 1974, 4, no.1, pp.51-4.

22.888. Lee, G.: "Quasi-One-Dimensional Solution for the Power of CO_2 Gasdynamic Lasers", Phys. Fluids,1974,17,no.3,pp.644-9.

22.889. Vasilik, N.Ya., Margolin, A.D., and Margulis, V.M.: "Theory of the Gasdynamic Laser Based on a Binary Mixture", Zh. Prikl. Mekh. Tekh. Fiz., 1974, no.3, pp.23-30.

22.890. Akinfiev, N.N., et al.: "Chemical Laser Triggered by IR Radiation", Zh. Eksp. Teor. Fiz. Pis'ma, 1974, 19, no.12, pp.745-7, and JETP Lett., 1974, 19, no.12.

22.891. Sabotinov, N.V., and Telbizov, P.K.: "Effect of Ne Admixtures on Generation Conditions of He-Se Laser", C. R. Acad. Bulg. Sci., 1974, 27, no.6, pp.763-6.

22.892. Pappalardo, R.: "Observation of Afterglow Character and High Gain in the Laser Lines of O^{III}", J. Appl. Phys., 1974, 45, no.8, pp.3547-53.

22.893. Girardeau-Montaut, J.P., and Girardeau-Montaut, C.: "Temporal Variation of Laser Parameters in N_2", Nouv. Rev. Opt., 1974, 5, no.3, pp.179-83.

22.894. Kasuya, K., Kumusaka, T., and Murasaki, T.: "Output-Power Characteristics of a CW Gasdynamic Laser with Supersonic Gas Mixing", Oyo Buturi, 1974, 43, no.5,pp.477-9.

22.895. Zalesskii, V.Yu.: "Gas Discharge Laser Operating on the 1315-nm Iodine Transition", Zh. Eksp. Teor. Fiz., 1974, 67, no.1, pp.30-7, and Sov. Phys.-JETP, 1975, 40,no.1.

22.896. Gavrikov, V.F., et al.: "Gasdynamic CO Laser", Kvantovaya Elektron., 1974, 1, no.1, pp.183-5, and Sov. J. Quantum Electron., 1974, 4, no.1, pp.108-9.

22.897. Tkach, Yu.V., et al.: "High-Power UV Laser Pumped by a Transverse Discharge", Kvantovaya Elektron., 1974, 1, no.1, pp.189-91, and Sov. J. Quantum Electron., 1974, 4, no.1, pp.114-5.

22.898. Downey, G.D., and Robinson, D.W.: "Single-Line Far-IR Water (-Vapour) Laser", Chem. Phys. Lett., 1974,24,no.1,pp.108-10.

22.899. Andersson, H.E.B., and Tobin, R.C.: "Electrical Breakdown and Pumping in an Axial-Field N_2 Laser", Phys. Scr., 1974, 9, no.1, pp.7-14.

22.900. My, L.T., Peyron, M., and Puget, P.: "Kinetic Study of CW HF Laser by Reaction of Fluorine with Hydrogen and Alkanes", J. Chim. Phys., 1974, 71, no.3, pp.377-82.

22.901. Emanuel, G., and Cohen, N.: " "Theoretical Performance of HCl Chemical CW Laser", J. Quant. Spectrosc., 1974, 14, no.7, pp.613-36.

22.902. Prokhorov, A.M., et al.: "Atmospheric-Pressure CO_2 Laser with Semi-Self-Maintained Discharge Regulated by UV Radiation", Zh. Eksp. Teor. Fiz. Pis'ma, 1974, 20, no.2, pp.108-11, and JETP Lett., 1974, 20, no.2.

22.903. Kerimov, O.M., et al.: "High-Pressure UV Laser Operating on a Mixture of Ar and N_2", Zh. Eksp. Teor. Fiz. Pis'ma, 1974, 20, no.2, pp.124-8, and JETP Lett., 1974, 20, no.2.

22.904. Liberman, I., et al.: "High-PRF Copper-Iodide Laser", Appl. Phys. Lett., 1974, 25, no.6, pp.334-5.

22.905. Collins, C.B., Cunningham, A.J., and Stockton, M.: "Nitrogen Ion Laser Pumped by Charge Transfer", Appl. Phys. Lett., 1974, 25, no.6, pp.344-5.

22.906. Djeu, N., and Burnham, R.: "Optically Pumped CW Hg Laser at 546.1 nm", Appl. Phys. Lett., 1974, 25, no.6, pp.350-1.

22.907. McGee, T.J., and Powell, F.X.: "CO_2-Laser Pumping via Transfer from N_2 Species Produced from the Reaction of NO with Discharged Nitrogen", J. Quantum Electron. IEEE, 1974, QE-10, no.12,pp.853-60.

22.908. Pappalardo, R.: "Observations on Multiply Ionized Xenon Laser Lines", J. Quantum Electron. IEEE, 1974, QE-10, no.12, pp.897-8.

22.909. Shirley, J.A., Hall, R.J., and Bronfin, B.R.: "Stimulated Emission from CO Transitions below 5 micron Excited in Supersonic Electric Discharge", J. Appl. Phys., 1974, 45, no.9, pp.3934-6.

22.910. LaCour, B., and Michon, M.: "Use of Picosecond Laser to Produce Coherent X-Rays", Onde Electr., 1974, 54, no.9, pp.474-8.

22.911. Izatt, J.R., Caudle, G.F., and Bean, B.L.: "Q-Switching and Mode-Locking of CO_2 Laser with Aromatic Halogenated Hydrocarbons", Appl. Phys. Lett., 1974, 25, no.8, pp.446-8.

22.912. Fahlen, T.S.: "Hollow-Cathode Copper-Vapour Laser", J. Appl. Phys., 1974, 45, no.9, pp.4132-3.

22.913. Csillag, L., et al.: "Near-IR CW Laser Oscillation in Cu^{II}", Phys. Lett., 1974, 50A, no.1, pp.13-4.

22.914. Beletskii, L.K., et al.: "Gasdynamic Laser Using Products of Combustion of Nitro-Compounds", Kvantovaya Elektron., 1974, 1, no.2, pp.439-41, and Sov. J. Quantum Electron., 1974, 4, no.2, pp.250-1.

22.915. Reshetnyak, S.A., and Shelepin, L.A.: "Kinetics of Physical Processes in Pinch-Discharge Lasers", Zh. Prikl. Mekh. Tekh. Fiz., 1974, no.4, pp.14-21.

22.916. Brown, F., Kronheim, S., and Silver, E.: "Tunable Far-IR CH_3F Laser Using Transverse Optical Pumping", Appl. Phys. Lett., 1974, 25, no.7, pp.394-6.

22.917. Chen, C.J.: "Lead Laser Using the Chloride as Lasant", J. Appl. Phys., 1974, 45, no.10, pp.4663-4.

22.918. Gudzenko, L.I., Shelepin, L.A., and Yakovlenko, S.I.: "Amplification in a Recombining Plasma", Usp. Fiz. Nauk, 1974, 114, no.3, pp.457-85.

22.919. Wallace, S.C., and Dreyfus, R.W.: "Continuously Tunable Xe Laser at 172 nm", Appl. Phys. Lett., 1974, 25, no.4,pp.498-500.

22.920. Nelson, L.Y., Fisher, C.H., and Byron, S.R.: "High-Pressure CS_2 Electric-Discharge Laser", Appl. Phys. Lett., 1974, 25, no.9, pp.517-20.

22.921. Silfvast, W.T., Szeto, L.H., and Wood, O.R.: "C_3F_7I Photodissociation Laser Initiated by a CO_2-Laser-Produced Plasma", Appl. Phys. Lett., 1974, 25, no.10,pp.593-5.

22.922. Powell, H.T., Murray, J.R., and Rhodes, C.K.: "Laser Oscillation on the Green Bands of XeO and KrO", Appl. Phys. Lett., 1974, 25, no.12, pp.730-2.

22.923. Searles, S.K.: "Superfluorescent Laser Emission from Electron-Beam-Pumped Ar-N_2 Mixtures", Appl. Phys. Lett., 1974, 25, no.12, pp.735-7.

22.924. Judd, O.P.: "Effect of Gas Mixture on Electron Kinetics in the Electrical CO_2 Gas Laser", J. Appl. Phys., 1974, 45, no.10, pp.4572-5.

22.925. Iijima, T., and Sugawara, Y.: "CW Laser Oscillations in He-Zn Hollow-Cathode Laser", J. Appl. Phys., 1974, 45, no.11, pp.5091-2.

22.926. Draganescu, V., et al.: "Theoretical Description of Gas-Transport Electrically Excited He-N_2-CO_2 Laser", Rev. Roum. Phys., 1974, 19, no.9, pp.899-904.

22.927. Rozsa, K., et al.: "Investigation of He-Cd Laser with Hollow Cathode", Kvantovaya Elektron., 1974, 1, no.4, pp.953-5, and Sov. J. Quantum Electron., 1974, 4, no.4, pp.523-4.

22.928. Batovskii, O.M., and Gur'ev, V.I.: "Photostimulated Pulsed HF Laser", Kvantovaya Elektron., 1974, 1, no.6, pp.1446-51, and Sov. J. Quantum Electron., 1974, 4, no.6,pp.801-4.

22.929. Pirkle, J.C., and Sigillito, V.G.: "Analysis of Optically Pumped CO_2 Laser", Appl. Opt., 1974, 13, no.12, pp.2799-807.

22.930. Baumann, W., et al.: "Computer Simulation of CW DF-CO_2 Chemical Transfer Laser", Appl. Opt., 1974,13,no.12,pp.2823-34.

22.931. Gagne, J.M., Mah, S.Q., and Conturie, Y.: "Transverse Flow Quasi-CW HF Chemical Laser", Appl. Opt., 1974, 13, no.12, pp.2835-9.

22.932. Smith, N.S., Hassan, H.A., and McInville, R.M.: "Small Signal Gain Calculations for High-Flow CO Discharge Lasers", AIAA J., 1974, 12, no.12, pp.1619-20.

22.933. Stefanov, V.I., and Atanasov, P.A.: "Dependence of Generation Duration on Discharge Current in a Pulsed CO_2 Laser", Bulg. J. Phys., 1974, 1, no.3, pp.311-5.

22.934. Ross, W., and Gundel, H.: "Excitation and Emission in the Pulsed N_2 Laser", Exp. Tech. Phys., 1974, 22, no.6, pp.541-8.

22.935. Donnerhacke, K.H., Rentsch, M., and Rose, J.: "Stable CO_2 TEA Laser with Double Discharge System", Exp. Tech. Phys., 1974, 22, no.6, pp.549-56.

22.936. Chang, T.Y.: "Optically Pumped Submillimetre-Wave Sources", Trans. IEEE, 1974, MTT-22, no.12, pp.983-8.

22.937. Plant, T.K., et al.: "High-Power Optically Pumped Far-IR Lasers", Trans IEEE, 1974, MTT-22, no.12, pp.988-90.

22.938. Kamenev, Yu.E., et al.: "IR Laser Based on Iodine Vapour", Prib. Tekh. Eksp., 1974, 17, no.3, pp.167-8, and Instrum. Exp. Tech., 1974, 17, no.3, pp.820-1.

22.939. Akchurin, G.G., et al.: "Dispersion Characteristics of He-Ne Laser at 633 nm", Opt. Spekt., 1974, 37, no.1, pp.157-61, and Opt. Spectrosc., 1974, 37, no.1, pp.87-9.

22.940. Dyubko, S.F., Svich, V.A., and Fesenko, L.D.: "Submillimetre Laser with CH_3I Excited by CO_2 Laser", Opt. Spekt., 1974, 37, no.1, p.208, and Opt. Spectrosc., 1974, 37, no.1, p.118.

22.941. Tarasenko, V.F., and Bychkov, Yu.I.: "IR N_2 Laser Using a Transverse Discharge", Zh. Tekh. Fiz., 1974, 44, no.5, pp.1100-1, and Sov. Phys.-Tech. Phys., 1974, 19, no.5, p.693.

22.942. Basov, N.G., Zavorotnyi, S.I., et al.: "Pulsed $D_2+F_2+CO_2+He$ Chemical Laser", Kvantovaya Elektron., 1974, 1, no.3, pp.560-4, and Sov. J. Quantum Electron., 1974, 4, no.3, pp.311-3.

22.943. Domnin, Yu.S., Tatarenkov, V.M., and Shumyatskii, P.S.: "Emission Lines of a CH_3OH Laser Pumped by CO_2 Laser", Kvantovaya Elektron., 1974, 1, no.3, pp.703-6, and Sov. J. Quantum Electron., 1974, 4, no.3, pp.401-2.

22.944. Ishchenko, V.N., Lisitsyn, V.N., and Starinskii, V.N.: "Pulsed UV N_2 Laser", Opt.-Mekh. Prom., 1974, 41, no.3, pp.32-4, and Sov. J. Opt. Technol., 1974, 41, no.3, pp.155-7.

22.945. Kudryavtsev, N.N., Novikov, S.S., and Svetlichnyi, I.B.: "Amplification of CO_2 Laser Emission in the Products of the $CO+N_2O$ Reaction", Zh. Prikl. Mekh. Tekh. Fiz., 1974, no.5, pp.9-15.

22.946. Churakov, V.V.: "Calculation of Gain and Inversion in a Molecular Laser at High Active-Medium Pressures", Zh. Prikl. Spekt., 1974, 21, no.6, pp.997-1004.

23. SEMICONDUCTOR LASERS

23.1. Chakravarti, A.N.: "Dielectric-Waveguide Modes in Laser Diodes in the Presence of a Transverse Magnetic Field", Int. J. Electron., 1968, 25, p.393.

23.2. Kruzhilin, Yu.I., and Sheveikin, V.I.: "Maximum Efficiency of an Injection Laser", Radiotekh. Elektron., 1968, 13, p.1628, and Radio Eng. Electron. Phys., 1968, 13, no.9, pp.1418-21.

23.3. Kuryl'ev, V.V., et al.: "Super-radiance Spectra of Injection Lasers and Distribution of Inhomogeneities Along a p-n Junction", Zh. Eksp. Teor. Fiz. Pis'ma, 1968, p.317, and JETP Lett., 1968,8,p.194.

23.4. Bykovskii, Yu.A., Goncharov, I.G., and Maslov, V.A.: "Observation of Absorption by Nonequilibrium Carriers in a GaAs Diode under Laser Conditions", Fiz. Tekh. Poluprov., 1969, 3, no.2, p.264, and Sov. Phys.-Semicond., 1969, 3, no.2, pp.217-8.

23.5. Rode, D.L.: "Theory of Wideband (Microwave) Laser Action in GaAs", J. Appl. Phys., 1969, 40, p.4123.

23.6. Hayashi, I., Panish, M.B., and Foy, P.W.: "Low-Threshold, Room-Temperature, Injection Laser", J. Quantum Electron. IEEE, 1969, QE-5, p.211.

23.7. Keune, D.L., et al.: "Time Behaviour of Exciton Formation and Laser Emission in Cd(SSe) Platelets", Appl. Phys. Lett., 1969, 14, p.99.

23.8. Kressel, H., and Nelson, H.: "Close-Confinement GaAs p-n Junction Lasers with Reduced Optical Loss at Room Temperature", RCA Rev., 1969, 30, p.106.

23.9. Chakravarti, A.N., and Rakshit, S.: "Dependence of Total Stimulated Power and Gain Factor in GaAs Junction Lasers on Length", Int. J. Electron., 1969, 26, no.2, p.191.

23.10. Paoli, T.L., Ripper, J.E., and Zachos, T.H.: "Resonant Modes of GaAs Junction Lasers. II", J. Quantum Electron. IEEE, 1969, QE-5, p.271.

23.11. Lavine, J.M., Mozzi, R.L., and Adams, A.: "Nonuniform Emission Characteristics of Electron-Beam-Pumped, Dislocation-Free, GaAs Lasers", J. Quantum Electron. IEEE, 1969, QE-5, p.421.

23.12. Chakravarti, A.N., Biswas, S.N., and Rakshit, S.: "Doping Dependence of Mode Spectrum in GaAs Junction Lasers Operated at Constant Current", Int. J. Electron., 1969, 26, p.499.

23.13. Byer, N.E.: "Role of Optical Flux and Current Density in Gradual Degradation of GaAs Injection Lasers", J. Quantum Electron. IEEE, 1969, QE-5, p.242.

23.14. Rossi, J.A., Holonyak, N., and Dapkus, P.D.: "Time Behaviour of Laser Modes in GaAs Platelet Lasers", J. Appl. Phys., 1969, 40, p.1934.

23.15. Kressel, H., and Byer, N.E.: "Physical Basis of Noncatastrophic Degradation in GaAs Injection Lasers", Proc. IEEE, 1969, 57, p.25.

23.16. Golulbev, G.P., et al.: "Investigation of the Characteristics of a CdTe Laser Excited by Electron Bombardment", Fiz. Tekh. Poluprov., 1969, 3, p.287, and Sov. Phys.-Semicond. 1969, 3, no.2, pp.240-1.

23.17. Adams, M.J.: "Simple Approximation for High-Temperature Properties of Injection Laser", J. Phys. D, 1969, 2, p.1565.

23.18. Rossi, J.A., et al.: "Laser Recombination Transition in p-GaAs", Appl. Phys. Lett., 1969, 15, p.109.

23.19. Bogdankevich, O.V., Letokhov, V.S., and Suchkov, A.F.: "Theory of the Effects of Excitation Inhomogeneity on Semiconductor Laser Pumped With an Electron Beam", Fiz. Tekh. Poluprov., 1969, 3, no.5, p.665, and Sov. Phys.-Semicond., 1969, 3, no.5, pp.566-70.

23.20. Popov, Yu.M., Strakhovskii, G.M., and Shuikin, N.N.: "Influence of Decrease of Lifetime With Increasing Current on Mode Excitation in an Injection Laser", Fiz. Tekh. Poluprov. 1969, 3, no.6, p.803, and Sov. Phys.-Semicond., 1969, 3, no.6, pp.685-9.

23.21. Chakravarti, A.N., Rakshit, S., and Biswas, S.N.: "Dependence of Total Stimulated Power in GaAs Junction Lasers on Reflectivity", Int. J. Electron., 1969, 26, p.399.

23.22. Burrell, G.J., Moss, T.S., and Hetherington, A.: "Distribution of Energy States at Band Edges in GaAs Laser Diodes", Solid-State Electron., 1969, 12, no.10, p.787.

23.23. Adams, M.J.: "Theoretical Effects of Exponential Band Tails on the Properties of the Injection Laser", Solid-State Electron., 1969, 12, no.8, p.661.

23.24. Pogorelova, E.V.: "Nonlinear Theory of Semiconductor Lasers With Optical Excitation", Vestn. Mosk. Univ. Fiz. Astron., 1969, no.4, p.100.

23.25. Kruzhilin, Yu.I., et al.: "Effect of Thermoelastic Potentials in Injection Lasers", Izv. VUZ Radioelektron., 1969, 12, no.7, p.692.

23.26. Holonyak, N., et al.: "Laser Operation and Spectroscopy of Epitaxial GaAs Gunn-Oscillator Wafers", J. Appl. Phys., 1969, 40, p.4998.

23.27. Paoli, T.L., and Ripper, J.E.: "Coupled Longitudinal-Mode Pulsing in Semiconductor Lasers", Phys. Rev. Lett., 1969, 22, p.1085.

23.28. Bille, J., et al.: "Continuously Tunable Laser Emission by Cd(SSe) Graded-Bandgap Crystals", Phys. Status Solidi, 1969, 36, p.771.

23.29. Bille, J.: "Exciton Diffusion and Multimode Behaviour in Electron-Beam-Pumped CdS Lasers", Phys. Status Solidi, 1969, 36, p.775.

23.30. Kuryl'ev, V.V., and Senatorev, K.Ya.: "Space-Spectral Structure of Irradiation of Generation Channels in Injection Lasers", Vestn. Mosk. Univ. Fiz. Astron., 1969, no.6, p.118.

23.31. Chakravarti, A.N., Rakshit, S., and Biswas, S.N.: "Dependence of Differential External Quantum Efficiency and Gain in GaAs Laser Diodes on Laser Reflectivity", Int. J. Electron., 1969, 26, no.2, p.179.

23.32. Chakravarti, A.N., Rakshit, S., and Biswas, S.N.: "Dependence of Differential External Quantum Efficiency and Gain in GaAs Laser Diodes on Laser Length", Int. J. Electron., 1969, 26, no.2, p.185.

23.33. Nelson, H., and Kressel, H.: "Improved Red and IR Emitting (GaAl)As Laser Diodes Using Close-Confinement Structure", Appl. Phys. Lett., 1969, 15, p.7.

23.34. Chakravarti, A.N., Rakshit, S., and Biswas, S.N.: "Dependence of Maximum Duration of Output Pulse and Adiabatic Heating in GaAs Junction Lasers on Length", Int. J. Electron., 1969, 27, p.387.

23.35. Chakravarti, A.N., Rakshit, S., and Biswas, S.N.: "Dependence of Maximum Duration of Output Pulse and Adiabatic Heating in GaAs Junction Lasers on Reflectivity", Int. J. Electron., 1969, 27, p.393.

23.36. Iwai, S., and Namba, S.: "GaAs Laser Under Electron-Beam Excitation", Sci. Pap. Inst. Phys. Chem. Res., 1969, 63, no.3, p.61.

23.37. Aritome, H., et al.: "Electron-Beam-Pumped CdS Laser", Sci. Pap. Inst. Phys. Chem. Res., 1969, 63, no.3, p.66.

23.38. Kononenko, V.K.: "Consideration of Absorption by Free Carriers in Active Zone of a Semiconductor Laser", Zh. Prikl. Spekt., 1969, 11, no.6, p.1012.

23.39. Broom, R.F.: "Analysis of a Diode Laser Having Two Orthogonal Fabry-Perot Cavities", J. Quantum Electron. IEEE, 1969, QE-5, p.539.

23.40. Chashchin, S.P., et al.: "Dependence of Stimulated-Emission Threshold of PbS Laser Diodes on Resonator Length", Fiz. Tekh. Poluprov., 1969, 3, no.10, p.1572, and Sov. Phys.-Semicond., 1970, 3, no.10, pp.1317-8.

23.41. Alferov, Zh.I., et al.: "GaAs-AlAs Heterojunction Injection Laser With Low Room-Temperature Threshold", Fiz. Tekh. Poluprov., 1969, 3, no.9, p.1328, and Sov. Phys.-Semicond., 1970, 3, no.9, pp.1107-10.

23.42. Molchanov, A.G., and Popov, Yu.M.: "Excitation of Semiconductor Lasers With X- and Gamma-Rays", Fiz. Tver. Tela, 1969, 11, no.7, p.1965, and Sov. Phys.-Solid State, 1970, 11, no.7, pp.1580-1.

23.43. Sharapov, B.N.: "Semiconductor Injection Laser With Large Emitting Area", Fiz. Tekh. Poluprov., 1969, 3, no.10, p.1566, and Sov. Phys.-Semicond., 1970, 3, no.10, pp.1311-2.

23.44. Popov, Yu.M., Strakhovskii, G.M., and Shuikin, N.N.: "Influence of Spatial Inhomogeneities on Excitation of Semiconductor Laser Modes", Fiz. Tekh. Poluprov., 1969, 3, no.8, p.1113, and Sov. Phys.-Semicond., 1970, 3, no.8, pp.943-7.

23.45. Vlasov, A.N., et al.: "Stimulated Radiation of Mixed Compounds (ZnCd)Te Excited With Fast Electrons", Fiz. Tekh. Poluprov., 1969, 3, no.9, p.1428, and Sov. Phys.-Semicond., 1970, 3, no.9, p.1198.

23.46. Karpenko, V.A., and Goncharenko, A.M.: "Electromagnetic Theory of Injection Lasers", Zh. Prikl. Spekt., 1970, 13, no.1, p.158.

23.47. Burnham, R.D.: "GaAs Junction Lasers Containing Amphoteric Dopants Ge and Si", Solid-State Electron., 1970, 13, no.2, p.199.

23.48. Hayashi, I., et al.: "Low-Threshold Heterostructure Injection Laser", J. Quantum Electron. IEEE, 1970, QE-6, p.4.

23.49. Shewchun, J., Kawasaki, B.S., and Garside, B.K.: "Characteristics of Stimulated Emission from CdS by Electron-Beam Pumping", J. Quantum Electron. IEEE, 1970, QE-6, p.133.

23.50. Kozina, G.S., et al.: "Semiconductor Laser Using Electron Excitation", Radiotekh. Elektron., 1970, 15, no.2, p.365, and Radio Eng. Electron. Phys., 1970, 15, no.2, pp.316-8.

23.51. Hayashi, I., and Panish, M.B.: "(GaAl)As-GaAs Heterostructure Injection Lasers which Exhibit Low Thresholds at Room Temperature", J. Appl. Phys., 1970, 41, p.150.

23.52. Brodin, M.S., et al.: "Effect of Temperature on Stimulated Radiation of (ZnCd)S Crystals With Two-Photon Excitation", Fiz. Tekh. Poluprov., 1970, 4, p.522, and Sov. Phys.-Semicond., 1970, 4, no.3, pp.435-8.

23.53. Keune, D.L., et al.: "Thin-Semiconductor-Laser to Thin-Platelet Optical Coupler", Appl. Phys. Lett., 1970, 16, p.18.

23.54. Bishop, S.G., Moore, W.J., and Swiggard, E.M.: "Optically Pumped Cd_3P_2 Laser", Appl. Phys. Lett., 1970, 16, p.459.

23.55. Nicoll, F.H.: "Low-Threshold Electron-Beam-Pumped CdS Lasers with End-Pumped Configuration", Appl. Phys. Lett., 1970, 16, p.501.

23.56. Herzog, D.G., and Kressel, H.: "Thermoelectrically Cooled (GaAl)As Laser Illuminator", Appl. Opt., 1970, 9, p.2249.

23.57. Antcliffe, G.A., and Wrobel, J.S.: "Spontaneous and Laser Emission from (PbSn)Te Diodes", Appl. Phys. Lett., 1970, 17, no.7, pp.290-2.

23.58. Chakravarti, A.N., Biswas, S.N., and Rakshit, S.: "Dependence of Junction Temperature Rise on Laser Length in GaAs Diodes Operated CW at 77°K", Int. J. Electron., 1970, 28, p.185.

23.59. Popov, Yu.M., and Shuikin, N.N.: "Compound Resonator for Semiconductor Lasers", Fiz. Tekh. Poluprov., 1970, 4, no.1, p.45, and Sov. Phys.-Semicond., 1970, 4, no.1, pp.34-8.

23.60. Shotov, A.P., and Muminov, R.A.: "Properties of Injection Lasers Based on Electron-Hole Plasma in InSb", Fiz. Tekh. Poluprov., 1970, 4, no.1, p.145, and Sov. Phys.-Semicond., 1970, 4, no.1, pp.115-8.

23.61. Allakhverdyan, R.G., et al.: "Influence of Waveguide Properties of a p-n Junction on Coherent Emission of GaAs", Fiz. Tekh. Poluprov., 1970, 4, p.341, and Sov. Phys.-Semicond., 1970, 4, no.2, pp.277-81.

23.62. Yonezu, H., Sakuma, I., Nannichi, Y.: "Gain Coefficient and Loss in a (GaAl)As-GaAs Laser Diode", Jap. J. Appl. Phys., 1970, 9, p.231.

23.63. Gribkovskii, V.P., and Kononenko, V.K.: "Radiative Generation of Junctions With Gaussian Impurity Zones", Zh. Prikl. Spekt., 1970, 12, p.45.

23.64. Panish, M.B., Hayashi, I., and Sumski, S.: "Double-Heterostructure Injection Lasers with Room-Temperature Thresholds as Low as 2.3 kA/cm^2", Appl. Phys. Lett., 1970, 16, p.326.

23.65. Gubanov, A.I., and Sharapov, B.N.: "Threshold Current of an Injection Laser with p-n Heterojunction", Fiz. Tekh. Poluprov., 1970, 4, no.3, p.433, and Sov. Phys.-Semicond., 1970, 4, no.3, pp.367-71.

23.66. Dobkin, A.S., et al.: "Semiconductor Laser with Local Mirrors", Fiz. Tekh. Poluprov., 1970, 4, p.613, and Sov. Phys.-Semicond., 1970, 4, no.3, pp.515-6.

23.67. Iwai, S., and Namba, S.: "ZnO Laser by Electron-Beam Excitation", Appl. Phys. Lett., 1970, 16, p.354.

23.68. Kressel, H., Nelson, H., and Hawrylo, F.Z.: "Control of Optical Losses in p-n Junction Lasers by Use of a Heterojunction", J. Appl. Phys., 1970, 41,p.2019.

23.69. Ulbrich, R., and Pilkuhn, M.H.: "Longitudinal Photon Flux Distribution in Low-Q Semiconductor Lasers", Appl. Phys. Lett., 1970, 16, p.516.

23.70. Kressel, H., Lockwood, H.F., and Nelson, H.: "Low-Threshold (GaAl)As Visible- and IR-Emitting Diode Lasers", J. Quantum Electron. IEEE, 1970, QE-6, p.278.

23.71. Goodwin, A.R., and Selway, P.R.: "Gain and Loss Processes in (GaAl)As-GaAs Heterostructure Lasers", J. Quantum Electron. IEEE, 1970, QE-6, p.285.

23.72. Ulmer, E.A., and Hayashi, I.: "Internal Q-Switching in (GaAl)As-GaAs Heterostructure Lasers", J. Quantum Electron. IEEE, 1970, QE-6, p.297.

23.73. Goodwin, A.R., and Thompson, G.H.B.: "Superlinear Dependence of Gain on Current Density in GaAs Injection Lasers", J. Quantum Electron. IEEE, 1970, QE-6,p.311.

23.74. Ulbrich, R., and Pilkuhn, M.H.: "Influence of Reflectivity on External Quantum Efficiency of GaAs Injection Lasers", J. Quantum Electron. IEEE, 1970, QE-6,p.314.

23.75. Carran, J.H., et al.: "GaAs Lasers Utilizing Light Propagation Along Curved Junctions", J. Quantum Electron. IEEE, 1970, QE-6, p.367.

23.76. Patel, N., and Yariv, A.: "Electrical and Optical Characteristics of InAs Junction Lasers", J. Quantum Electron. IEEE, 1970, QE-6, p.383.

23.77. Bogdankevich, O.V., et al.: "Influence of Electron Energy on the Characteristics of Beam-Pumped Semiconductor Lasers", J. Quantum Electron. IEEE, 1970, QE-6,p.389.

23.78. Magee, C.J., and Haug, H.: "Optically Excited Bulk Semiconductor Lasers", J. Quantum Electron. IEEE, 1970,QE-6,p.392.

23.79. Keune, D.L.: "Stimulated Emission in Lossy Semiconductor Laser Modes", J. Appl. Phys., 1970, 41, p.2725.

23.80. Popov, Yu.M., and Shuikin, N.N.: "Dependence of Multiple-Mode Excitation in Semiconductor Lasers on Nonlinearity of Absorption", Zh. Eksp. Teor. Fiz., 1970, 58, no.5, p.1727, and Sov. Phys.-JETP, 1970, 31, no.5.

23.81. Sharapov, B.N.: "Threshold Currents in Injection Lasers With One or Two Heterojunctions", Fiz. Tekh. Poluprov., 1970, 4, no.6, pp.1121, and Sov. Phys.-Semicond., 1970, 4, no.6, pp.948-53.

23.82. Chashchin, S.P., et al.: "Properties of PbSe Laser Diodes at 77°K", Fiz. Tekh. Poluprov., 1970, 4, no.6, p.1170, and Sov. Phys.-Semiconduct., 1970, 4, no.6.

23.83. Novikov, A.A., and Sukhov, Ye.G.: "Calculation of Dynamic Range of a Semiconductor Laser Under Steady-State Operating Conditions", Radiotekh. Elektron., 1970, 15, no.10, and Radio Eng. Electron. Phys., 1970, 15, no.10, pp.1924-7.

23.84. Dapkus, P.D., et al.: "Near-Bandgap, Narrow-Spectrum, Low-Loss Volume-Excited GaAs Laser (at 77°K)", J. Appl. Phys., 1970, 41, no.13, pp.5215-7.

23.85. Rivlin, L.A., and Yakubovich, S.D.: "Extended Semiconductor Laser With Radiating Lattice", Zh. Eksp. Teor. Fiz. Pis'ma, 1970, 12, no.6, pp.282-6, and JETP Lett., 1970, 12, no.6, pp.190-3.

23.86. Adams, M.J.: "Theory of Pulsations in Output of GaAs Junction Lasers", Phys. Status Solidi, 1970, 1A, no.1, pp.143-52.

23.87. Butler, J.K., Sommers, H.S., and Kressel, H.: "High-Order Transverse Cavity Modes in Heterojunction Diode Lasers", Appl. Phys. Lett., 1970, 17, no.9, pp.403-6.

23.88. Lockwood, H.F., et al.: "Efficient Large-Cavity Injection Laser", Appl. Phys. Lett., 1970, 17, no.11, pp.499-502.

23.89. Hwang, C.J.: "Properties of Spontaneous and Stimulated Emission in GaAs Junction Lasers. I and II", Phys. Rev., 1970, 2, no.10, pp.4117-25 and 4126-34.

23.90. Kawabe, M., et al.: "Abnormal Laser Emission from Electron-Beam-Excited GaAs", Jap. J. Appl. Phys., 1970,9,no.7,p.850.

23.91. Yonezu, H.: "Analysis of Mode Confinement in (GaAl)As-GaAs Laser Diode", Jap. J. Appl. Phys., 1970, 9, no.8, p.1013.

23.92. Unno, Y., Yamamoto, M., and Iida, A.: "Internal Q-Switching in (GaAl)As-GaAs Heterojunction Lasers", Jap. J. Appl. Phys., 1970, 9, no.9, p.1181.

23.93. Gill, R.B.: "Room-Temperature Close-Confinement GaAs Laser With External Quantum Efficiency of 40%", Proc. IEEE, 1970, 58, p.949.

23.94. Keune, D.L., et al.: "Spontaneous-
and Stimulated-Carrier Lifetime and Spectral
Output of CdSe at 77°K", Appl. Phys. Lett.,
1970, 17, p.42.

23.95. Vlasov, G.K., and Mashkevich, V.S.:
"Theory of Laser Emission Using Indirect
Magnetooptic Transitions with Free-Carrier
Participation", Fiz. Tekh. Poluprov., 1970,
4, no.4, p.663, and Sov. Phys.-Semicond.,
1970, 4, no.4, pp.562-5.

23.96. Abdullaev, G.B., et al.: "Laser
Emission by GaSe Under Two-Photon Optical
Excitation", Fiz. Tekh. Poluprov., 1970,
4, no.7, p.1395, and Sov. Phys.-Semicond.,
1971, 4, no.7, pp.1189-90.

23.97. Kurbatov, L.N.: "Investigation of
Superluminescence of a GaAs Diode", Fiz.
Tekh. Poluprov., 1970, 4, no.11, pp.2025-31,
and Sov. Phys.-Semicond., 1971, 4, no.11,
pp.1739-44.

23.98. Bogdankevich, O.V., et al.: "In-
fluence of Irradiation on a GaAs Laser
Excited by an Electron Beam", Fiz. Tekh.
Poluprov., 1970, 4, no.7, p.1209, and Sov.
Phys.-Semicond., 1971, 4, no.7, pp.1027-32.

23.99. Grasyuk, A.Z., et al.: "Spectral
and Threshold Characteristics of Radiation-
Excited GaAs Lasers", Fiz. Tekh. Poluprov.,
1970, 4, no.7, p.1411, and Sov. Phys.-Semi-
cond., 1971, 4, no.7, pp.1207-9.

23.100. Aritome, H., Masuda, K., and
Namba, S.: "Time Delay in CdS Laser Pumped
by an Electron Beam", Opto-Electron., 1970,
2, no.4, pp.239-44, and J. Quantum Electron. IEEE, 1971, QE-7, no.3, pp.118-22.

23.101. Chashchin, S.P., et al.: "Properties of PbS Laser Diodes at 77°K", Fiz.
Tekh. Poluprov., 1970, 4, no.8, pp.1546-8,
and Sov. Phys.-Semicond., 1971, 4, no.8,
pp.1320-2.

23.102. Khalfin, V.B.: "Threshold Current
of a Heterojunction Laser", Fiz. Tekh.
Poluprov., 1970, 4, no.9, pp.1644-9, and
Sov. Phys.-Semicond., 1971, 4, no.9, pp.1414-8.

23.103. Alferov, Zh.I., et al.: "Influence of AlAs-GaAs Heterostructure Parameters on Laser Threshold Current", Fiz.
Tekh. Poluprov., 1970, 4, no.9, pp.1826-9,
and Sov. Phys.-Semicond., 1971, 4, no.9,
pp.1573-5.

23.104. Nikitina, V.Yu., and Poluektov,
I.A.: "Saturation of Gain in Semiconductor
Lasers and Amplifiers", Fiz. Tekh. Poluprov.,
1970, 4, no.9, p.1834, and Sov. Phys.-Semi-
cond., 1971, 4, no.9, p.1578.

23.105. Garside, B.K., Shewchun, J., and
Kawasaki, B.S.: "Wavelength Tuning of GaAs
and CdSe Electron-Beam Pumped Lasers", J.
Quantum Electron. IEEE, 1971, QE-7, no.2,
pp.88-93.

23.106. Hodgson, R.T., von Gutfeld, R.J.,
and Woodall, J.M.: "High-Power, Optically
Pumped, Multilayer Semiconductor Laser", IBM
Tech. Disclosure Bull., 1971, 14, no.7,
pp.2246-7.

23.107. Kurbatov, L.N., et al.: "Study of
a Multibeam GaAs Injection-Laser Amplifier",
Radiotekh. Elektron., 1971, 16, no.4,
pp.639-41, and Radio Eng. Electron. Phys.,
1971, 16, no.4, pp.702-3.

23.108. Aleksanyan, A.G., Poluektov, I.A.,
and Papov, Yu.M.: "Influence of Impurity Concentration on Threshold Characteristics of
Semiconductor Lasers", Kvantovaya Elektron.,
1971, no.3, pp.15-22.

23.109. Bogdankevich, O.V., et al.:
"Characteristics of a GaAs Laser Pumped by
High-Energy Electron Beam", Kvantovaya Elek-
tron., 1971, no.3, pp.29-33.

23.110. Rivlin, L.A.: "Infinitely Long
Semiconductor Injection Laser With Distributed
Radiative Losses", Kvantovaya Elektron., 1971,
no.3, pp.34-41.

23.111. Kressel, H., Lockwood, H.F., and
Hawrylo, F.Z.: "Low-Threshold, Large-Cavity,
GaAs Injection Lasers", Appl. Phys. Lett.,
1971, 18, no.2, pp.43-5.

23.112. Hwang, C.J.: "Excitation and Doping
Dependencies of Electron Diffusion Length in
GaAs Junction Lasers", J. Appl. Phys., 1971,
42, no.2, pp.757-61.

23.113. Scifres, D.R., et al.: "GaAs-
Pumped GaAs Lasers", J. Appl. Phys., 1971,
42, no.2, pp.896-7.

23.114. Chakravarti, A.N., Biswas, S.N.,
and Rakshit, S.: "Reflectivity Dependence of
Junction Temperature Rise and Stimulated
Power in GaAs CW Lasers at 77°K", Int. J.
Electron., 1971, 30, no.3, pp.291-7.

23.115. Alferov, Zh.I., et al.: "Efficient
Generation in Injection Heterolasers", Fiz.
Tekh. Poluprov., 1971, 5, no.5, pp.972-3,
and Sov. Phys.-Semicond., 1971, 5, no.5,
pp.858-9.

23.116. Bykovskii, Yu.A., et al.: "Spectral Characteristics of a Semiconductor
Laser With Nonuniform Electron-Beam Pumping",
Fiz. Tekh. Poluprov., 1971, 5, no.5,
pp.1005-7, and Sov. Phys.-Semicond., 1971,
5, no.5, pp.893-4.

23.117. Era, K., and Langer, D.W.: "Sti-
mulated Emission Spectra of CdS Platelets
Under Various Excitation Levels", J. Appl.
Phys., 1971, 42, no.3, pp.1021-7.

23.118. Sakuma, I., et al.: "CW Operation
of Junction Lasers at Room Temperature",
Jap. J. Appl. Phys., 1971, 10, no.2, pp.282-3.

23.119. Burnham, R.D., et al.: "Spectral
Behaviour, Carrier Lifetime, and Pulsed and
CW Laser (77°K) Operation of (InGa)P", Appl.
Phys. Lett., 1971, 18, no.4, pp.160-2.

23.120. Bykovskii, Yu.A., et al.: "Electron-Beam-Pumped GaAs Laser With a Waveguide Structure", Fiz. Tekh. Poluprov., 1971, 5, no.1, pp.187-8, and Sov. Phys.-Semicond., 1971, 5, no.1, pp.164-5.

23.121. Hayashi, I., Panish, M.B., and Reinhart, F.K.: "(GaAl)As-GaAs Double Heterostructure Injection Lasers", J. Appl. Phys., 1971, 42, no.5, pp.1929-41.

23.122. Craford, M.G., Groves, W.O., and Fox, M.J.: "GaAs-Ga(AsP) Heterostructure Injection Lasers", J. Electrochem. Soc., 1971, 118, no.2, pp.355-8.

23.123. Miller, B.I., et al.: "Semiconductor Lasers Operating CW in the Visible at Room Temperature", Appl. Phys. Lett., 1971, 18, no.9, pp.403-5.

23.124. Chakravarti, A.N., and Parui, D.P.: "Time Dependence of Shift of Peak Emission in GaAs Junction Lasers Operated With a Flat-Topped Current Pulse at 77°K", Int. J. Electron., 1971, 30, no.6, pp.589-93.

23.125. Bille, J., Kramer, B.M., and Ruppel, W.: "Optical Losses and Efficiency of Electron-Beam-Pumped CdS Lasers", Phys. Status Solidi, 1971, 4A, no.3, pp.731-5.

23.126. Johnston, W.D.: "Characteristics of Optically Pumped Platelet Lasers of ZnO, CdS, CdSe, and Cd(SSe), at 80°-300°K", J. Appl. Phys., 1971, 42, no.7, pp.2731-40.

23.127. Nicoll, F.H.: "Decrease in Spontaneous Emission at Onset of Lasing in Semiconductors", J. Appl. Phys., 1971, 42, no.7, pp.2743-6.

23.128. Ettenberg, M., et al.: "Control of Facet Damage in GaAs Laser Modes", Appl. Phys. Lett., 1971, 18, no.12, pp.571-3.

23.129. Dingle, R., et al.: "Stimulated Emission and Laser Action in GaN", Appl. Phys. Lett., 1971, 19, no.1, pp.5-7.

23.130. Burnham, R.D., et al.: "Double-Heterojunction (GaAl)(AsP) Quarternary Lasers", Appl. Phys. Lett., 1971, 19, no.2, pp.25-8.

23.131. Biswas, S.N., Rakshit, S., and Chakravarti, A.N.: "Doping Dependence of Internal Quantum Efficiency as a Function of Beam Voltage in Electron-Beam-Excited n-GaAs at 300°K", Opt. Commun., 1971, 3, no.5, pp.324-7.

23.132. Kressel, H., et al.: "Mode Guiding in Symmetrical (GaAl)As-GaAs Heterojunction Lasers With Very Narrow Active Regions", RCA Rev., 1971, 32, no.3, pp.393-401.

23.133. Arora, B.M., Ahlburn, B.T., and Compton, W.D.: "Radiation-Induced Laser Action in CdSe", Trans. IEEE, 1971, NS-18, no.6, pp.40-4.

23.134. Marinace, J.C.: "Experimental Fabrication of One-Dimensional GaAs Laser Arrays", IBM J. Res. Dev., 1971, 15, no.4, pp.258-64.

23.135. Sprokel, G.J.: "Fabrication and Properties of Monolithic Laser Diode Arrays", IBM J. Res. Dev., 1971, 15, no.4, pp.265-71.

23.136. Molochev, V.I., Nikitin, V.V., and Samoilov, V.D.: "Optical Interaction of Inhomogeneously Excited Injection Lasers", Kvantovaya Elektron., 1971, no.2, pp.25-31.

23.137. Bogdankevich, O.V., et al.: "Electron-Beam-Pumped High-Power Semiconductor Laser", Kvantovaya Elektron., 1971, no.2, pp.92-3.

23.138. Osvenskii, V.B., Proshko, G.P., and Sizov, S.M.: "Parameters of Injection Lasers Made of GaAs With Different Dislocation Densities", Kvantovaya Elektron., 1971, no.2, pp.94-6.

23.139. Eliseev, P.G.: "Optimal Thickness of Active Layer in a Heterojunction Laser", Kvantovaya Elektron., 1971, no.3, pp.120-1.

23.140. Akimov, Yu.A., et al.: "Generation of Controllable Light Pulses in Electron-Beam-Pumped Laser", Kvantovaya Elektron., 1971, no.6, pp.105-6.

23.141. Semenov, A.T.: "Injection Laser Under Self-Q-Switching Conditions", Kvantovaya Elektron., 1971, no.6, pp.107-10.

23.142. Chakravarti, A.N., Rakshit, S., and Biswas, S.N.: "Dependence of Total Power from CW GaAs Junction Lasers at 77°K on Current and Length", Indian J. Pure Appl. Phys., 1971, 9, no.4, pp.243-5.

23.143. Ahn, B.H., Trussell, C.W., and Shurtz, R.R.: "Amphoterically Si-Doped GaAs Laser", Appl. Phys. Lett., 1971, 19, no.10, pp.408-10.

23.144. Miller, B.I., et al.: "Highly Uniform (GaAl)As-GaAs Double-Heterostructure Lasers at Room Temperature", Appl. Phys. Lett., 1971, 19, no.9, pp.340-3.

23.145. Smiley, V.N., Taylor, H.F., and Lewis, A.L.: "Laser Emission in Thin Dielectric-Coated CdSe Lasers", J. Appl. Phys., 1971, 42, no.13, pp.5859-61.

23.146. Kawasaki, B.S., Shewchun, J., and Garside, B.K.: "Relationship Between Wavelength Tuning and Lasing Threshold in Electron-Beam-Pumped GaAs Lasers", J. Appl. Phys., 1971, 42, no.13, pp.5877-9.

23.147. Nahory, R.E., Shaklee, K.L., and Leheny, R.F.: "Indirect-Bandgap Superradiant Laser in GaP Containing Isoelectronic Traps", Phys. Rev. Lett., 1971, 27, no.24, pp.1647-50.

23.148. Pak, G.T., et al.: "Internal Parameters of Injection Lasers at 300°K", Kvantovaya Elektron., 1971, no.5, pp.99-101.

23.149. Gorbylev, V.A., et al.: "Dependence of Threshold of Injection Lasers on Duration of Pumping-Current Pulses", Kvantovaya Elektron., 1971, 1, no.5, pp.97-9.

23.150. Bogdankevich, O.V., et al.: "Multielement Semiconductor Laser of the Emitting-Mirror Type", Kvantovaya Elektron., 1971, no.5, pp.95-6.

23.151. Allakhverdyan, R.G., et al.: "Influence of Refractive-Index Nonlinearity on Dynamics of Semiconductor Lasers", Kvantovaya Elektron., 1971, no.6, pp.53-9.

23.152. Eliseev, P.G., and Sukhov, Ye.G.: "Analysis of CW Lasing Conditions of a Semiconductor at Room Temperature", Radiotekh. Elektron., 1971, 16, no.6, pp.1005-9, and Radio Eng. Electron. Phys., 1971, 16, no.6, pp.991-4.

23.153. Singh, V.B.: "Laser Action in Thin Platelets of Cd(SSe) Solid Solutions", Labdev J. Sci. Technol., 1971, 9A, no.2, pp.85-92.

23.154. Butler, J.K.: "Theory of Transverse-Cavity Mode Selection in Homojunction and Heterojunction Semiconductor Lasers", J. Appl. Phys., 1971, 42, no.11, pp.4447-57.

23.155. Reinhart, F.K., Hayashi, I., and Panish, M.B.: "Mode Reflectivity and Waveguide Properties of Double-Heterostructure Injection Lasers", J. Appl. Phys., 1971, 42, no.11, pp.4466-79.

23.156. Aritome, H., et al.: "Internal Q-Switching in a CdS Laser Pumped by an Electron Beam", Jap. J. Appl. Phys., 1971, 10, no.11, p.1655.

23.157. Elesin, V.F.: "Semiconductor Laser Theory", Opt. Spekt., 1971, 30, no.3, pp.569-70, and Opt. Spectrosc., 1971, 30, no.3, pp.308-9.

23.158. Unger, H.G.: "Minimum Threshold Current of Double-Heterojunction Lasers", Arch. Elektron. Ubertrag., 1971, 25, no.11, pp.539-40.

23.159. Stern, F., and Woodall, J.M.: "Injection Laser for Ambient Temperature Operation", IBM Tech. Disclosure Bull., 1971, 14, no.2, p.354.

23.160. Mestvirishvili, A.N., et al.: "Investigation of Powerful Semiconductor Lasers With Electronic Excitation", Prib. Tekh. Eksp., 1971, 14, no.1, pp.199-202, and Instrum. Exp. Tech., 1971, 14, no.1, pp.234-7.

23.161. Adams, M.J., and Cross, M.: "Theory of Heterostructure Injection Lasers", Solid-State Electron., 1971, 14, no.9, pp.865-83.

23.162. Macksey, H.M., et al.: "(InGa)P p-n Junction Lasers", Appl. Phys. Lett., 1971, 19, no.8, pp.271-3.

23.163. Chashchin, S.P., et al.: "Epitaxial p-n Heterojunction in the (PbSn)Se-PbSe System", Fiz. Tekh. Poluprov., 1971, 5, no.8, p.1632, and Sov. Phys.-Semicond., 1972, 5, no.8, p.1427.

23.164. Kononenko, V.K., and Gribkovskii, V.P.: "Influence of Radiation Noise on Threshold and Output Power of an Injection Laser", Fiz. Tekh. Poluprov., 1971, 5, no.10, pp.1875-81, and Sov. Phys.-Semicond., 1972, 5, no.10, pp.1631-5.

23.165. Bykovskii, Yu.A., et al.: "Influence of a Waveguide on Characteristics of a Semiconductor Laser Excited by Electron Bombardment", Fiz. Tekh. Poluprov., 1971, 5, no.8, pp.1666-9, and Sov. Phys.-Semicond., 1972, 5, no.8, pp.1462-4.

23.166. Gribkovskii, V.P., et al.: "Output Power and Efficiency of an Injection Laser", Fiz. Tekh. Poluprov., 1971, 5, no.8, pp.1606-8, and Sov. Phys.-Semicond., 1972, 5, no.8, pp.1400-1.

23.167. Scifres, D.R., et al.: "Optically Pumped, Volume-Excited, CW Room Temperature (InGa)P Platelet Lasers", Appl. Phys. Lett., 1972, 20, no.5, pp.184-6.

23.168. Scifres, D.R., et al.: "Optically Pumped (InGa)P Platelet Lasers from the IR to Yellow (Regions)", J. Appl. Phys., 1972, 43, no.3, pp.1019-21.

23.169. Ettenberg, M., and Kressel, H.: "Dependence of Threshold Current Density and Efficiency on Cavity Parameters. Heterojunction Laser Diodes", J. Appl. Phys., 1972, 43, no.3, pp.1204-10.

23.170. Pinkas, E., et al.: "(GaAl)As-GaAs Double-Heterostructure Lasers. Effect of Doping", J. Appl. Phys., 1972, 43, no.6, pp.2827-35.

23.171. Dyment, J.C., et al.: "Proton-Bombardment Formation of Stripe-Geometry Heterostructure Lasers for 300°K CW Operation", Proc. IEEE, 1972, 60, no.6, pp.726-8.

23.172. Ludeke, R., and Harris, E.P.: "Tunable GaAs Laser in an External Dispersive Cavity", Appl. Phys. Lett., 1972, 20, no.12, pp.499-500.

23.173. Holonyak, N., et al.: "N:(InGa)P Laser Operation on the A-Line Transition", Appl. Phys. Lett., 1972, 20, no.1, pp.11-4.

23.174. Rossi, J.A., Chinn, S.R., and Mooradian, A.: "Optically Pumped Room-Temperature (InGa)As Lasers", Appl. Phys. Lett., 1972, 20, no.2, pp.84-6.

23.175. Kressel, H., Lockwood, H.F., and Hawrylo, F.Z.: "Large-Cavity (GaAl)As-GaAs Heterojunction Laser", J. Appl. Phys., 1972, 43, no.2, pp.561-7.

23.176. Holonyak, N., et al.: "Laser Operation Associated With Deep Isoelectronic Traps in Indirect Semiconductors", Phys. Rev. Lett., 1972, 28, no.4, pp.230-3.

23.177. Kressel, H., and Lockwood, H.F.: "Lasing Transitions in p^+-n-n^+ (GaAl)As-GaAs Heterojunctions", Appl. Phys. Lett., 1972, 20, no.4, pp.175-7.

23.178. Godenko, L.P., and Mashkevich, V.S.: "Threshold Theory for Injection Laser With Transitions Between Exponential Tails", Kvantovaya Elektron., 1972, no.6, pp.141-65.

23.179. Nicoll, F.H.: "Low-Voltage Electron-Beam-Pumped Lasing of CdS", Trans. IEEE, 1972, ED-19, no.6, pp.838-9.

23.180. Holonyak, N., et al.: "Long-Wavelength Shift in the Operation of Lightly Doped Semiconductor Lasers", J. Appl. Phys., 1972, 43, no.5, pp.2302-6.

23.181. Scifres, D.R., et al.: "Stimulated Emission of Direct and Indirect Ga(AsP) on Nitrogen Isoelectronic Trap Transitions", J. Appl. Phys., 1972, 43, no.5,pp.2368-75.

23.182. Selway, P.R., and Goodwin, A.R.: "Properties of Double Heterostructure Lasers With Very Narrow Active Regions", J. Phys. D, 1972, 5, no.5, pp.904-14.

23.183. Tsukada, T., et al.: "Very-Low-Current Operation of Mesa-Stripe-Geometry Double-Heterostructure Injection Lasers", Appl. Phys. Lett., 1972, 20,no.9,pp.344-5.

23.184. Ikegami, T.: "Oscillation Mode in Double-Heterostructure Injection Lasers", J. Quantum Electron. IEEE, 1972, QE-8, no.6, pp.470-6.

23.185. Chou, P.T., and Ballantyne, J.M.: "Photon Loss and Gain in CdS Electron-Beam-Pumped Lasers", J. Quantum Electron. IEEE, 1972, QE-8, no.6, pp.483-6.

23.186. Phillip-Rutz, E.M.: "High-Radiance Room-Temperature GaAs Laser With Single Transverse Mode", J. Quantum Electron. IEEE, 1972, QE-8, no.7, pp.632-41.

23.187. Gribkovskii, V.P.: "Absorption Saturation in Semiconducting Lasers With Optical Pumping", Zh. Prikl. Spekt., 1972, 16, no.4, pp.627-32.

23.188. Holloway, H., et al.: "Injection Luminescence and Laser Action in Epitaxial PbTe Diodes", Appl. Phys. Lett., 1972, 21, no.1, pp.5-6.

23.189. Bogdankevich, O.V., et al.: "Waveguide Resonator Structure for Semiconductor Lasers With Electron-Beam Pumping", Dokl. Akad. Nauk SSSR, 1972, 201, no.6, pp.1316-8, and Sov. Phys.-Dokl., 1972, 16, no.12, pp.1071-3.

23.190. Byrne, F.T., and Culver, W.H.: "Increasing Radiance from Injection Lasers", IBM Tech. Disclosure Bull.,1972,15,no.1,pp pp.79-80.

23.191. Paoli, T.L.: "Reduction in the Rate of Increase of Spontaneous Emission from Double-Heterostructure Injection Lasers at Threshold", Appl. Phys. Lett., 1972, 21, no.3, pp.101-2.

23.192. Ripper, J.E., Patel, N.B., and Brosson, P.: "Behaviour of Spontaneous Emission Across Threshold in GaAs Junction Lasers", Appl. Phys. Lett., 1972, 21, no.3, pp.98-100.

23.193. Dupuis, R.D., et al.: "Mode-Coupling Effects in Thin-Platelet Semiconductor Lasers", J. Appl. Phys., 1972, 43, no no.9, pp.3801-3.

23.194. Bille, J., et al.: "Tunable Laser Emission of Cd(SSe) Graded-Bandgap Crystals by Two-Photon Excitation", Phys. Status Solidi, 1972, 12A, no.2, pp.K91-3.

23.195. Chakravarti, A.N., and Parui, D.P.: "Method for Determining the Coefficient of Superlinear Dependence of Gain on Current Density in GaAs Junction Lasers", Indian J. Pure Appl. Phys., 1972, 10, no.2, pp.173-4.

23.196. Butler, J.K., and Kressel, H.: "Transverse-Mode Selection in Injection Lasers With Widely Spaced Heterojunctions", J. Appl. Phys., 1972, 43, no.8, pp.3403-11.

23.197. Dolocan, V.: "Effect of Impurity Gradient on Time Delays and Q-Switching in Junction Lasers", Phys. Status Solidi, 1972, 12A, no.1, pp.81-7.

23.198. Burnham, R.D., et al.: "(GaAl)(AsP) Solid Solutions and Use in Double-Heterojunction Injection Lasers", Fiz. Tekh. Poluprov., 1972, 6, no.1, pp.97-102, and Sov. Phys.-Semicond., 1972, 6, no.1,pp.77-81.

23.199. Pleshkov, A.A., Prozorov, O.N., and Trukhan, V.G.: "Self-Q-Switched Semiconductor Laser With a Nonlinear Photovoltaic Cell", Fiz. Tekh. Poluprov., 1972, 6, no.1, pp.163-6, and Sov. Phys.-Semicond., 1972, 6, no.1, pp.132-4.

23.200. Eliseev, P.G., et al.: "Influence of Composition of (GaAl)As Solid Solutions on Optical Channelling in Heterostructures", Fiz. Tekh. Poluprov., 1972, 6, no.1, pp.177-9, and Sov. Phys.-Semicond., 1972, 6, no.1, pp.145-7.

23.201. Cross, M., and Adams, M.J.: "Waveguiding Properties of Stripe-Geometry Double-Heterostructure Injection Lasers", Solid-State Electron., 1972, 15, no.8, pp.919-21.

23.202. Rossi, J.A., and Hsieh, J.J.: "Double-Heterostructure GaAs:Si Diode Lasers", Appl. Phys. Lett., 1972, 21, no.6,pp.287-9.

23.203. Hammermann, B.: "Investigation of Dynamic Behaviour of Semiconductor Lasers by Equations of Balance", Nachr. Tech., 1972, 22, no.10, pp.327-8.

23.204. Alferov, Zh.I., et al.: "Efficient Heterojunction Injection Lasers Operating in the Range 740-900 nm", Fiz. Tekh. Poluprov., 1972, 6, no.3, pp.568-9, and Sov. Phys.-Semicond., 1972, 6, no.3, pp.495-6.

23.205. Kawasaki, B.S., Garside, B.K., and Shewchun, J.: "Laser Dynamics and Wavelength Tuning in Electron-Beam-Pumped GaAs", Appl. Phys. Lett., 1972, 21, no.10, pp.477-9.

23.206. Antcliffe, G.A., Parker, S.G., and Bate, R.T.: "CW Tunable Operation Using Diode Laser of (PbGe)Te", Appl. Phys. Lett., 1972, 21, no.10, pp.505-7.

23.207. Matthews, M.R., Dyott, R.B., and Carling, W.P.: "Filaments as Optical Waveguides in GaAs Lasers", Electron. Lett., 1972, 8, no.23, pp.570-2.

23.208. Rossi, J.A., and Chinn, S.R.: "Efficient Optically Pumped InP and (InGa)As Lasers", J. Appl. Phys., 1972, 43, no.11, pp.4806-7.

23.209. Chakravarti, A.N., and Parui, D.P.: "Condition for Population Inversion in Semiconductors at Relatively High Temperatures", Phys. Status Solidi, 1972, 14A, no.2, pp.K139-42.

23.210. Stolyarov, S.N.: "Influence of Waveguide Properties of Heterojunction Layers on the Principal Characteristics of Injection Lasers", Kvantovaya Elektron., 1972, no.2, pp.69-76, and Sov. J. Quantum Electron., 1972, 2, no.2, pp.144-9.

23.211. Aleksanyan, A.G., Poluektov, I.A., and Popov, Yu.M.: "Optical Gain of Heavily Doped Semiconductors", Kvantovaya Elektron., 1972, no.2, pp.77-83, and Sov. J. Quantum Electron., 1972, 2, no.2, pp.150-4.

23.212. Bogdankevich, O.V., et al.: "Waveguide Resonator Structure of an Electron-Beam-Pumped Semiconductor Laser", Kvantovaya Elektron., 1972, no.2, pp.61-8, and Sov. J. Quantum Electron., 1972, 2, no.2, pp.138-43.

23.213. Eliseev, P.G., and Strakhov, V.P.: "Single-Mode Generation in Injection Lasers", Zh. Eksp. Teor. Fiz. Pis'ma, 1972, 16, no.11, pp.606-8, and JETP Lett., 1972, 16, no.11, pp.428-9.

23.214. Gladkii, B.I., and Potykevich, I.V.: "Absorption Coefficient and Gain of a GaAs Injection Laser", Opt. Spekt., 1972, 32, no.6, pp.1163-6, and Opt. Spectrosc., 1972, 32, no.6, pp.630-2.

23.215. Kurbatov, L.N., et al.: "Characteristics of a Miniature Pulsed Laser With Electron Excitation", Radiotekh. Elektron., 1972, 17, no.6, pp.964-8.

23.216. Ettenberg, M., Lockwood, H.F., and Sommers, H.S.: "Radiation Trapping in Laser Diodes", J. Appl. Phys., 1972, 43, no.12, pp.5047-51.

23.217. Gomenyuk, A.S., and Ratner, Ye.S.: "Spectral Characteristics of InAs Semiconductor Laser", Opt.-Mekh. Prom., 1972, 39, no.4, pp.16-7, and Sov. J. Opt. Technol., 1972, 39, no.4, pp.201-2.

23.218. Chakravarti, A.N., and Parui, D.P.: "Current Limitations and Maximum Output as Functions of Length in CW GaAs Junction Lasers at 77°K", Indian J. Pure Appl. Phys., 1972, 10, no.8, pp.610-3.

23.219. Kazarinov, R.F., and Suris, R.A.: "Injection Heterojunction Laser With Diffraction Grating on Contact Surface", Fiz. Tekh. Poluprov., 1972, 6, no.7, pp.1359-65, and Sov. Phys.-Semicond., 1973, 6, no.7, pp.1184-9.

23.220. Shakhidzhanov, S.S.: "Nonlinear Theory of Amplification in p-n Junctions", Fiz. Tekh. Poluprov., 1972, 6, no.8, pp.1424-31, and Sov. Phys.-Semicond., 1973, 6, no.8, pp.1241-6.

23.221. Kastal'skii, A.A.: "Proposal for Semiconductor Laser Utilizing Landau Levels", Fiz. Tekh. Poluprov., 1972, 6, no.8, pp.1576-81, and Sov. Phys.-Semicond., 1973, 6, no.8, pp.1359-63.

23.222. Eliseev, P.G.: "Kinetics of Aging of Injection Lasers", Fiz. Tekh. Poluprov., 1972, 6, no.9, pp.1655-61, and Sov. Phys.-Semicond., 1973, 6, no.9, pp.1431-6.

23.223. Rivlin, L.A.: "Asymptotic Nature of Threshold Conditions and Multimode Laser Emission", Kvantovaya Elektron., 1972, no.5, pp.94-7, and Sov. J. Quantum Electron., 1973, 2, no.5, pp.464-6.

23.224. Biswas, S.N., and Kumar, N.: "Reflectivity Dependence of Photon Lifetime in GaAs Junction Lasers", Indian J. Pure Appl. Phys., 1973, 11, no.7, pp.538-9.

23.225. Galitskii, V.M., and Elesin, V.F.: "Electron Kinetics and Stationary Generation in Semiconductor Lasers", Zh. Eksp. Teor. Fiz., 1973, 64, no.2, pp.691-702, and Sov. Phys.-JETP, 1973, 37, no.2, pp.351-6.

23.226. Biswas, S.N., and Kumar, N.: "Dependence of Photon Lifetime in GaAs Junction Lasers on Length", Indian J. Pure Appl. Phys., 1973, 11, no.11, pp.855-6.

23.227. Biswas, S.N., Rakshit, S., and Chakravarti, A.N.: "Doping Dependence of Photon Yield as Function of Excitation Energy in Optically Excited n-GaAs at 300°K", Int. J. Electron., 1973, 34, no.1, pp.135-8.

23.228. Doerbeck, F.H., Blacknall, D.M., and Carroll, R.L.: "(GaAl)As-GaAs Heterostructure Lasers Amphoterically Si-Doped", J. Appl. Phys., 1973, 44, no.1, pp.529-30.

23.229. Hvam, J.M.: "Exciton-Exciton Interaction and Laser Emission in High-Purity ZnO", Solid-State Commun., 1973, 12, no.2, pp.95-7.

23.230. Paoli, T.L., Hakki, B.W., and Miller, B.I.: "Zero-Order Transverse-Mode Operation of GaAs Double Heterostructure Lasers With Thick Waveguides", J. Appl. Phys., 1973, 44, no.3, pp.1276-80.

23.231. Ahn, B.H.: "Dependence of Threshold Current Density and Beamspread on Potential-Barrier Height in Close-Confinement Lasers", J. Appl. Phys., 1973, 44, no.3, pp.1411-2.

23.232. Cross, M.: "Selection Mechanisms of Transverse Modes in Semiconductor Injection Lasers", J. Quantum Electron. IEEE, 1973, QE-9, no.5, pp.517-22.

23.233. Adams, M.J.: "Rate Equations and Transient Phenomena in Semiconductor Lasers", Opto-Electron., 1973, 5, no.2, pp.201-15.

23.234. Cross, M.: "Theory of Transient Evolution and Self-Focusing Behaviour of Lasing Filaments in Injection Lasers", Phys. Status Solidi, 1973, 16A, no.1, pp.167-79.

23.235. Gribkovskii, V.P., et al.: "Injection Laser With Nonplanar p-n Transition", Zh. Prikl. Spekt., 1973, 18,no.1,pp.140-1.

23.236. Nicoll, F.H.: "Semiconductor Lasers Pumped by Pulsed Electric Discharge in Vacuum", Appl. Phys. Lett., 1973, 22, no.8, pp.363-4.

23.237. Dupuis, R.D., et al.: "Laser Operation of N:Ga(AsP) on Photopumped NN Pair Transitions", Appl. Phys. Lett., 1973, 22, no.8, pp.369-71.

23.238. Manlief, S.K., and Palik, E.D.: "Two-Photon-Excited Stimulated Recombination Emission in InSb", Appl. Phys. Lett., 1973, 22, no.9, pp.443-5.

23.239. Nakamura, M., et al.: "Optically Pumped GaAs Surface Laser With Corrugation Feedback", Appl. Phys. Lett., 1973, 22, no.10, pp.515-6.

23.240. Panish, M.B., et al.: "Reduction of Threshold Current Density in (GaAl)As-GaAs Heterostructure Lasers by Separate Optical and Carrier Confinement", Appl. Phys. Lett., 1973, 22, no.11, pp.590-1.

23.241. Nakashima, H., et al.: "Effects of Composition Profile on Characteristics of (GaAl)As-GaAs Double-Heterostructure Lasers", J. Appl. Phys., 1973, 44, no.6, pp.2688-9.

23.242. Walpole, J.N., et al.: "High Power Output in (PbSn)Te Diode Lasers With Improved Mirror Quality", J. Appl. Phys., 1973, 44, no.6, pp.2905-7.

23.243. Godenko, L.P., and Mashkevich, V.S.: "Threshold Theory of Laser Generation in p-n Junctions With Exponential Band Tails", Phys. Status Solidi, 1973, 17A, no.1, pp.125-39.

23.244. Kurbatov, L.N., et al.: "Stimulated Emission from Solid Solutions of Tin and Lead Chalcogenides Near 10 micron", Kvantovaya Elektron., 1973, no.3, pp.97-9, and Sov. J. Quantum Electron., 1973, 2, no.3, pp.281-3.

23.245. Akimov, Yu.A., et al.: "Electron-Beam-Pumped CdS Laser", Kvantovaya Elektron., 1973, no.3, pp.99-101, and Sov. J. Quantum Electron., 1973, 2, no.3, pp.284-5.

23.246. Kirkby, P.A., and Thompson, G.H.B.: "High Peak Power from (GaAl)As-GaAs Double-Heterostructure Injection Lasers", Appl. Phys. Lett., 1973, 22, no.12, pp.638-40.

23.247. Rossi, J.A., Chinn, S.R., and Heckscher, H.: "High-Power Narrow-Line Operation of GaAs Diode Lasers", Appl. Phys. Lett., 1973, 23, no.1, pp.25-7.

23.248. Grundorfer, S., and Adams, M.J.: "Theoretical Considerations of Time Delays in Semiconductor Lasers", J. Quantum Electron. IEEE, 1973, QE-9, no.8, pp.814-9.

23.249. Brinkman, W.F., and Lee, P.A.: "Coulomb Effects on Gain Spectrum of Semiconductors", Phys. Rev. Lett., 1973, 31, no.4, pp.237-40.

23.250. Verdeyen, J.T., et al.: "Electron-Beam-Pumped Semiconductor Laser Using Gas-Plasma Gun", Appl. Phys. Lett., 1973, 23, no.2, pp.102-3.

23.251. Hwang, C.J., and Dyment, J.C.: "Dependence of Threshold and Electron Lifetime on Acceptor Concentration in (GaAl)As-GaAs Lasers", J. Appl. Phys., 1973, 44, no.7, pp.3240-4.

23.252. Catalano, I.M., Cingolani, A., and Minafra, A.: "Spontaneous and Stimulated Luminescence in CdS and ZnS Excited by Multiphoton Optical Pumping", Phys. Rev., 1973, 8B, no.4, pp.1488-92.

23.253. Bogdankevich, O.V., et al.: "Dynamics of an Electron-Beam-Excited Semiconductor Laser of the Radiating-Mirror Type", Fiz. Tekh. Poluprov., 1973, 7, no.2, pp.242-5, and Sov. Phys.-Semicond., 1973, 7, no.2, pp.175-7.

23.254. Hartman, R.L., and Hartman, A.R.: "Strain-Induced Degradation of GaAs Injection Lasers", Appl. Phys. Lett., 1973, 23, no.3, pp.147-9.

23.255. Reinhart, F.K., and Logan, R.A.: "Interface Stress of (GaAl)As-GaAs Layer Structures", J. Appl. Phys., 1973, 44, no.7, pp.3171-5.

23.256. Hartman, R.L., et al.: "CW Operation of (GaAl)As-GaAs Double-Heterostructure Lasers With 30°C Half-Lives Exceeding 100h", Appl. Phys. Lett., 1973, 23, no.4, pp.181-3.

23.257. Sommers, H.S.: "Experimental Properties of Injection Lasers. IV", J. Appl. Phys., 1973, 44, no.8, pp.3601-8.

23.258. Rossi, J.A., et al.: "Time Delays in External-Cavity-Controlled (GaAl)As-GaAs Single-Heterostructure Diode Lasers", Appl. Phys. Lett., 1973, 23, no.5, pp.254-6.

23.257. Johnston, W.D., and Miller, B.I.: "Degradation Characteristics of CW Optically Pumped (GaAl)As Heterostructure Lasers", Appl. Phys. Lett., 1973, 23, no.4, pp.192-4.

23.260. Halak, A.: "Determination of Threshold Current, Quantum Efficiency, and Losses, of Single-Heterostructure (GaAl)As-GaAs Injection Lasers", Phys. Status Solidi, 1973, 18A, no.1, pp.K39-43.

23.261. Nicoll, F.H.: "Room-Temperature Lasing of CdS Crystals in a Glow Discharge", Trans. IEEE, 1973, ED-20, no.10, pp.905-6.

23.262. Nakamura, M., et al.: "Laser Oscillation in Epitaxial GaAs Waveguides With Corrugation Feedback", Appl. Phys. Lett., 1973, 23, no.5, pp.224-5.

23.263. Rossi, J.A., Heckscher, H., and Chinn, S.R.: "Threshold, Spectral, and Output-Power Characteristics of (GaAl)As-GaAs Single-Heterostructure Diode Lasers", Appl. Phys. Lett., 1973, 23, no.5, pp.257-9.

23.264. Grossman, B., et al.: "Physical Basis for Negative Resistance in Double Heterostructure Injection Lasers", Appl. Phys., 1973, 2, no.4, pp.173-6.

23.265. Kressel, H., and Hawrylo, F.Z.: "Red-Light Emitting (GaAl)As Heterojunction Laser Diodes", J. Appl. Phys., 1973, 44, no.9, pp.4222-3.

23.266. Kobayashi, T., and Sugiyama, K.: "Effect of Uniaxial Stress on Double Heterostructure Lasers", Jap. J. Appl. Phys., 1973, 12, no.9, pp.1388-92.

23.267. Yonezu, H., Kobayashi, K., and Sakuma, I.: "Threshold Current Density and Lasing Transverse Mode in a (GaAl)As-GaAs Double-Heterostructure Laser", Jap. J. Appl. Phys., 1973, 12, no.10, pp.1593-9.

23.268. Borisov, N.A., et al.: "Influence of Mechanical Treatment of the Resonator on the Parameters of an Electron-Beam-Pumped CdS Laser", Kvantovaya Elektron., 1973, no.6, pp.115-6, and Sov. J. Quantum Electron., 1973, 2, no.6, p.574.

23.269. Thompson, G.H.B., and Kirkby, P.A.: "Low Threshold-Current Density in Five-Layer-Heterostructure (GaAl)As-GaAs Localized-Gain-Region Injection Lasers", Electron. Lett., 1973, 9, no.13, pp.295-6.

23.270. Dumke, W.P.: "Current Thresholds in Stripe-Contact Injection Lasers", Solid-State Electron., 1973, 16, no.11, pp.1279-81.

23.271. Nicoll, F.H.: "Room-Temperature Lasing in CdS Excited by High-Voltage RF Current Pulses", Appl. Phys. Lett., 1973, 23, no.8, pp.465-6.

23.272. Petroff, P., and Hartman, R.L.: "Defect Structure Introduced During Operation of Heterojunction GaAs Laser", Appl. Phys. Lett., 1973, 23, no.8, pp.469-71.

23.273. Weber, W.H., and Yeung, K.F.: "Waveguide and Luminescent Properties of Thin-Film Pb-Salt Injection Lasers", J. Appl. Phys., 1973, 44, no.11, pp.4991-5000.

23.274. Kazirinov, R.F.: "Maximum Reduction of Threshold Current Density in Double-Heterojunction Injection Lasers", Fiz. Tekh. Poluprov., 1973, 7, no.4, pp.763-74, and Sov. Phys.-Semicond., 1973, 7, no.4, pp.525-31.

23.275. Holonyak, N., et al.: "Pumping of N:Ga(AsP) by an Electron Beam from a Gas Plasma", J. Appl. Phys., 1973, 44, no.12, pp.5517-21.

23.276. Martinez, G.: "Band Inversion in (PbSn)Se Alloys Under Hydrostatic Pressure. III", Phys. Rev., 1973, 8B, no.10, pp.4693-707.

23.277. Bogdankevich, O.V., et al.: "Electron-Beam-Excited Semiconductor Laser Made of GaAs Doped With Group IV Elements", Fiz. Tekh. Poluprov., 1973, 7, no.7, pp.1263-9, and Sov. Phys.-Semicond., 1974, 7, no.7, pp.849-52.

23.278. Kuznetsova, E.M.: "Generation of Coherent Light in a Semiconductor", Izv. VUZ Fiz., 1973, no.11, pp.131-4.

23.279. Walpole, J.H., et al.: "Single-Heterojunction (PbSn)Te Diode Lasers", Appl. Phys. Lett., 1973, 23, no.11, pp.620-2.

23.280. Nash, F.R.: "Mode Guidance Parallel With the Junction Plane of Double-Heterostructure GaAs Lasers", J. Appl. Phys., 1973, 44, no.10, pp.4696-707.

23.281. Salathe, R., Voumard, C., and Weber, H.: "Optical Coupling of Two Diode Lasers", Phys. Status Solidi, 1973, 20A, no.2, pp.527-34.

23.282. Biswas, S.N., and Kumar, N.: "Reflectivity Dependence of Lasing Photon Energy in GaAs Junction Lasers at 77°K", Int. J. Electron., 1974, 36, no.3, pp.427-9.

23.283. Aleksanyan, A.G., Poluektov, I.A., and Popov, Yu.M.: "Theory of Optical Gain and Threshold Properties of Semiconductor Lasers", J. Quantum Electron. IEEE, 1974, QE-10, no.3, pp.297-305.

23.284. Nishimura, Y.: "Electron Scattering Periods in GaAs Injection Lasers", Jap. J. Appl. Phys., 1974, 13, no.1, pp.109-17.

23.285. Nuese, C.J., Enstrom, R.W., and Ettenberg, M.: "Room-Temperature Laser Operation of (InGa)As p-n Junctions", Appl. Phys. Lett., 1974, 24, no.2, pp.83-5.

23.286. Guerra, J.M., Escudero, J.L., and Sancho, J.: "Configurational Coordinates Model in the Alkali-Halide Exciton Laser", J. Quantum Electron. IEEE, 1974, QE-10, no.1, pp.1-5.

23.287. Itoh, K.: "619-nm Emission at 77°K of (GaAl)As Double-Heterostructure Lasers", Appl. Phys. Lett., 1974, 24, no.3, pp.127-9.

23.288. Nuese, C.J., et al.: "CW Laser Diodes and High-Power Arrays of (InGa)As for 1.06-micron", Appl. Phys. Lett., 1974, 24, no.5, pp.224-7.

23.289. Voges, E.: "Losses and Transverse Mode Selection in Stripe-Geometry Double-Heterojunction Lasers", Arch. Elektron. Ubertrag., 1974, 28, no.4, pp.183-6.

23.290. Stern, F., and Woodall, J.M.: "Photon Recycled Injection Laser", IBM Tech. Disclosure Bull., 1974, 16, no.9, p.3076.

23.291. Hakki, B.W.: "Mode Gains and Junction Current in GaAs Under Lasing Conditions", J. Appl. Phys., 1974, 45, no.1, pp.288-94.

23.292. Schlosser, W.O., et al.: "(GaAl)As-GaAs Heterostructure Laser With Separate Optical and Carrier Confinement", J. Appl. Phys., 1974, 45, no.1, pp.322-33.

23.293. Rakshit, S., Biswas, S.N., and Chakravarti, A.N.: "Effect of Wave Confinement on Threshold Current in Heterostructure Injection Lasers", Int. J. Electron., 1974, 36, no.5, pp.593-9.

23.294. Lockwood, H.P., et al.: "GaAs p-n-p-n Laser Diode", J. Quantum Electron. IEEE, 1974, QE-10, no.7, pp.567-9.

23.295. Zory, P.S.: "Zero-Mode Diode Laser", IBM Tech. Disclosure Bull., 1974, 16, no.9, p.3187.

23.296. Ushiku, K., et al.: "Stimulated Emission from CdS Thin Films Excited by N_2 Laser", Jap. J. Appl. Phys., 1974, 13, no.5, pp.909-10.

23.297. Nakamura, M., and Yariv, A.: "Analysis of the Threshold of Double Heterojunction (GaAl)As-GaAs Lasers With Corrugated Interface", Opt. Commun., 1974, 11, no.1, pp.18-20.

23.298. Blum, F.A., et al.: "Optically Pumped (Epitaxially Grown)GaAs Mesa Surface Laser", Appl. Phys. Lett., 1974, 24, no.9, pp.430-2.

23.299. Dyment, J.C., et al.: "Threshold Reduction by Addition of Phosphorous to the Ternary Layers of Double-Heterostructure GaAs Lasers", Appl. Phys. Lett., 1974, 24, no.10, pp.481-4.

23.300. Namizaki, H., et al.: "Current Dependence of Spontaneous Carrier Lifetimes in (GaAl)As-GaAs Double-Heterostructure Lasers", Appl. Phys. Lett., 1974, 24, no.10, pp.486-7.

23.301. Gobel, G.: "Recombination Without k-Selection Rules in Dense Electron-Hole Plasmas in High-Purity GaAs Lasers", Appl. Phys. Lett., 1974, 24, no.10, pp.492-4.

23.302. Lee, M.H., et al.: "Behaviour of Above-Gap NN Pair States in Radiative Recombination in N:Ga(AsP)", J. Appl. Phys., 1974, 45, no.4, pp.1775-8.

23.303. Sommers, H.S., and North, D.O.: "Spontaneous Power and Coherent State of Injection Lasers", J. Appl. Phys., 1974, 45, no.4, pp.1787-93.

23.304. Hakki, B.W., and Hwang, C.J.: "Mode Control in GaAs Large-Cavity Double-Heterostructure Lasers", J. Appl. Phys., 1974, 45, no.5, pp.2168-73.

23.305. Lang, R.: "Saturation Behaviour of Optical Gain in GaAs Injection Lasers", J. Quantum Electron. IEEE, 1974, QE-10, no.10, pp.825-6.

23.306. Stoll, H.M., and Seib, D.H.: "Distributed Feedback GaAs Homojunction Injection Laser", Appl. Opt., 1974, 13, no.9, pp.1981-2.

23.307. Shank, C.V., Schmidt, R.V., and Miller, B.I.: "Double-Heterostructure GaAs Distributed-Feedback Laser", Appl. Phys. Lett., 1974, 25, no.4, pp.200-1.

23.308. Scifres, D.R., Burnham, R.D., and Streifer, W.: "Distributed-Feedback Single-Heterojunction GaAs Diode Laser", Appl. Phys. Lett., 1974, 25, no.4, pp.203-6.

23.309. Grundorfer, S., Adams, M.J., and Thomas, B.: "Theory of Internal Q-Switching in Semiconductor Lasers", Electron. Lett., 1974, 10, no.17, pp.354-6.

23.310. Birzhis, S.V., Godenko, L.P., and Mashkevich, V.S.: "Theory of Laser Threshold of a Homogeneous Heavily Doped Semiconductor Subjected to a Quantizing Magnetic Field", Kvantovaya Elektron., 1974, 1, no.1, pp.69-77, and Sov. J. Quantum Electron., 1974, 4, no.1, pp.36-40.

23.311. Kryukova, I.V., et al.: "Electron-Beam-Pumped GaSb Laser", Kvantovaya Elektron., 1974, 1, no.1, pp.114-8, and Sov. J. Quantum Electron., 1974, 4, no.1, pp.62-4.

23.312. Namizaki, H., et al.: "Characteristics of the Junction-Stripe-Geometry (GaAl)As-GaAs Double-Heterostructure Lasers", Jap. J. Appl. Phys., 1974, 13, no.10, pp.1618-23.

23.313. Kurnosov, V.D., and Semenov, A.T.: "Injection Laser With Two Strongly Coupled Resonators", Kvantovaya Elektron., 1974, 1, no.1, pp.54-61, and Sov. J. Quantum Electron., 1974, 4, no.1, pp.28-31.

23.314. Borodulin, V.I., and Shveikin, V.I.: "Excitation of TE$_n$ Waves with High Transverse Index in Laser Diodes", Kvantovaya Elektron., 1974, 1, no.1, pp.54-61, and Sov. J. Quantum Electron., 1974, 4, no.1, pp.28-31.

23.315. Aleksanyan, A.G., Poluektov, I.A., and Popov, Yu.M.: "Theory of the Gain of Semiconductor Lasers", Kvantovaya Elektron., 1974, 1, no.1, pp.62-8, and Sov. J. Quantum Electron., 1974, 4, no.1, pp.32-5.

23.316. Bykovskii, Yu.A., et al.: "Electron-Beam-Pumped Heterostructure Lasers", Kvantovaya Elektron., 1974, 1, no.1, pp.141-3, and Sov. J. Quantum Electron., 1974, 4, no.1, pp.78-9.

23.317. Dolginov, L.M., et al.: "Parameters of Electron-Beam-Pumped (GaAl)As Visible-Band Lasers", Kvantovaya Elektron., 1974, 1, no.1, pp.178-80, and Sov. J. Quantum Electron., 1974, 4, no.1, pp.104-5.

23.318. Groves, S.H., Nill, K.W., and Strauss, A.J.: "Double Heterostructure (PbSn)Te-PbTe Lasers with CW Operation at 77°K", Appl. Phys. Lett., 1974, 25, no.6, pp.331-3.

23.319. Tomasetta, L.R., and Fonstad, C.G.: "Threshold Reduction in (PbSn)Te Laser Diodes Through Use of Double Heterojunctions", Appl. Phys. Lett., 1974, 25, no.8, pp.440-2.

23.320. Thompson, G.H.B., et al.: "Role of Optical Guiding in Critical-Temperature Behaviour, Delays, and Q-Switching, in Single-Heterostructure (GaAl)As-GaAs Lasers", Electron. Lett., 1974, 10, no.22, pp.456-7.

23.321. Salathe, R., Voumard, C., and Weber, H.: "Rate-Equation Approach for Diode Lasers. I and II", Opto-Electron., 1974, 6, no.6, pp.451-6 and 457-63.

23.322. Bogdankevich, O.V., and Vlasyuk, V.N.: "Influence of Excitation Inhomogeneity on Threshold of Electron-Beam-Pumped (Semiconductor) Lasers", Kvantovaya Elektron., 1974, 1, no.2, pp.357-64, and Sov. J. Quantum Electron., 1974, 4, no.2, pp.198-202.

23.323. Nakamura, M., et al.: "(GaAl)As-GaAs Double-Heterostructure Distributed-Feedback Diode Lasers", Appl. Phys. Lett., 1974, 25, no.9, pp.487-8.

23.324. Nuese, C.J., Ettenberg, M., and Olsen, G.H.: "Room-Temperature Heterojunction Laser Diodes from Vapour-Grown (InGa)P-GaAs", Appl. Phys. Lett., 1974, 25, no.10, pp.612-4.

23.325. Paoli, T.L., and Ripper, J.E.: "Single-Longitudinal-Mode Operation of CW Junction Lasers by Frequency-Selective Optical Feedback", Appl. Phys. Lett., 1974, 25, no.12, pp.744-6.

23.326. Minden, H.T., and Premo, R.: "High-Temperature GaAs Single Heterojunction Laser Diodes", J. Appl. Phys., 1974, 45, no.10, pp.4520-7.

23.327. Tsukada, T.: "(GaAl)As-GaAs Buried-Heterostructure Injection Lasers", J. Appl. Phys., 1974, 45, no.11, pp.4899-906.

23.328. Sleger, K.J., et al.: "Single-Heterostructure Pb(SSe) Diode Lasers", J. Appl. Phys., 1974, 45, no.11, pp.5069-71.

23.329. Rossi, J.A., et al.: "Comparison of Optical to Injection Excitation in GaAs Heterostructure Lasers", J. Appl. Phys., 1974, 45, no.12, pp.5383-8.

23.330. Shur, M.S.: "Possibility of Stimulated Emission of Light from a TRAPATT Diode", Fiz. Tekh. Poluprov., 1974, 8, no.5, pp.857-60, and Sov. Phys.-Semicond., 1974, 8, no.5, pp.554-5.

23.331. Anderson, D.B., August, R.R., and Coker, J.E.: "Distributed-Feedback Double-Heterostructure GaAs Injection Laser with Fundamental Grating", Appl. Opt., 1974, 13, no.12, pp.2742-4.

23.332. Nannichi, Y., and Hayashi, I.: "Degradation (Improvement) of (GaAl)As Double-Heterostructure Diode Lasers", J. Cryst. Growth, 1974, 27, no.1, pp.126-32.

23.333. Takuma, H.: "Tunable Semiconductor Laser", Oyo Buturi, 1974, 43, no.10, pp.1035-7.

23.334. Morozov, V.N.: "Theory of Multimode Coherent Emission from Semiconductor Lasers", Kvantovaya Elektron., 1974, 1, no.3, pp.634-44, and Sov. J. Quantum Electron., 1974, 4, no.3, pp.354-9.

23.335. Borovich, L.N., et al.: "Gain of CdS and CdSe Under Electron-Beam Excitation", Kvantovaya Elektron., 1974, 1, no.3, pp.653-9, and Sov. J. Quantum Electron., 1974, 4, no.3, pp.365-8.

23.336. Eliseev, P.G., and Ts'ai, C.M.: "Investigation of Relations Governing Multimode Excitation in Injection Lasers", Kvantovaya Elektron., 1974, 1, no.5, pp.1138-44, and Sov. J. Quantum Electron., 1974, 4, no.5, pp.622-5.

23.337. Borodulin, V.I., et al.: "Investigation of Injection Lasers With Wide Active Region", Kvantovaya Elektron., 1974, 1, no.5, pp.1220-3, and Sov. J. Quantum Electron., 1974, 4, no.5, pp.670-1.

23.338. Kressel, H., et al.: "Heterojunction Laser Diodes for Room-Temperature Operation", Opt. Eng., 1974, 13, no.5, pp.416-22.

23.339. Blum, F.A., et al.: "Monolithic GaAs Injection Mesa Lasers With Grown Optical Facets", Appl. Phys. Lett., 1974, 25, no.10, pp.620-1.

24. CONSTRUCTION AND OPERATION OF LASERS

24.1.   Orlov, L.N., and Rubanov, V.S.: "Effect of Pumping Intensity Upon Relation between Gain and Temperature of He-Ne Laser", Zh. Prikl. Spekt., 1968, 9, p.947.

24.2.   Gruzinskii, V.V., and Levchik, V.L.: "Application of Interference to Obtaining Nonpolarized Single-Frequency Radiation from a Gas Laser", Zh. Prikl. Spekt., 1968, 9, p.959.

24.3.   Latimer, I.D.: "High-Power Quasi-CW UV Ion Laser", Appl. Phys. Lett., 1968, 13, p.333.

24.4.   Rohlicek, F.: "Resonator of a Ruby Laser Ended by a Rectangular Prism", Jemna Mech., Opt., 1968, no.12, p.383.

24.5.   Veduta, A.P.: "Distribution of Temperature and Population of Ruby-Rod Lasers During Pumping", Zh. Prikl. Spekt., 1968, 9, p.964.

24.6.   Kulybin, V.M., and Rinkevichyus, B.S.: "Influence of a Transverse Magnetic Field on the Output and Polarization of a He-Ne Laser at 3.39 micron", Izv. VUZ Fiz., 1968, no.10, p.130.

24.7.   Bokhan, P.A.: "Possibility of Using Noncontracted Discharge to Enhance the Power of a CO_2-N_2-He Laser", Opt. Spekt., 1968, 25, p.417, and Opt. Spectrosc., 1968, 25, p.225.

24.8.   Belostotskii, B.R., et al.: "Vortex-Flow Cooled Laser", Opt.-Mekh. Prom., 1968, 35, no.4, p.35, and Sov. J. Opt. Technol., 1968, 35, no.4, p.450.

24.9.   Mikaelyan, A.L., Savel'ev, V.G., and Turkov, Yu.G.: "Transmission of Highly Coherent Ruby-Laser Emission Through an Amplifier", Radiotekh. Elektron., 1968, 13, no.10, p.1819, and Radio Eng. Electron. Phys., 1968, 13, no.10, pp.1592-5.

24.10.   Morozov, V.V.: "Generation of Ruby-Laser Giant Pulse in Narrow Spectral Interval", Prib. Tekh. Eksp., 1968, no.2, 11, p.179, and Instrum. Exp. Tech., 1968, no.2, p.430.

24.11.   Doyle, W.M., and White, M.B.: "Experimental Studies of Weakly Anisotropic Gas Laser", Nachr. Tech. Fachber., 1968, 35, p.683.

24.12.   Nicoll, F.H.: "Mirrors for Semiconductor Injection Lasers", RCA Tech. Notes, 1968, no.21, p.1.

24.13.   Balashov, I.F., et al.: "Heat Removal from Laser Rod by Metallic Conductor", Opt.-Mekh. Prom., 1968, 35, no.4, p.5, and Sov. J. Opt. Technol., 1968, 35, no.2, p.221.

24.14.   Kon'kov, I.D., et al.: "Effect of Longitudinal Alternating Magnetic Field on Oscillations in Ionized Argon", Radiotekh. Elektron., 1968, 13, no.12, p.2280, and Radio Eng. Electron. Phys., 1968, 13, no.12, pp.2008-9.

24.15.   Rundle, W.J.: "Ruby Laser Modified for Pulse-Transmission-Mode Cavity Dumping", J. Appl. Phys., 1968, 39, p.5338.

24.16.   Yamanaka, C., et al.: "High-Power ($SeOCl_2$) Nonlinear Laser Amplifier", Technol. Rep. Osaka Univ., 1968, 18, p.169.

24.17.   Privalov, V.E., and Fridrikhov, S.A.: "Dependence of Radiation Power of He-Ne Laser on Cross Section of Discharge", Zh. Tekh. Fiz., 1968, 38, no.12, p.2080, and Sov. Phys.-Tech. Phys., 1969, 13, no.12, pp.1667-70.

24.18.   Rogova, I.V.: "Coupled Lasers", Zh. Tekh. Fiz., 1968, 38, no.11, p.1897, and Sov. Phys.-Tech. Phys., 1969, 13, no.11, p.1525.

24.19.   Corcoran, V.J., Smith, W.T., and Gallagher, J.J.: "CW Gain Characteristics of the 890-GHz HCN Laser Line", J. Quantum Electron. IEEE, 1969, QE-5, p.292.

24.20.   Sayers, M.D.: "Single-Pass Gain as a Function of Discharge Current for the 488-nm Argon-Ion Laser", Phys. Lett., 1969, 29A, p.591.

24.21.   Gruzinskii, V.V., and Matusevich, L.A.: "Gas Laser With Window Mirrors", Zh. Prikl. Spekt., 1969, 10, no.3, p.418.

24.22.   Levinson, G.R., et al.: "Investigating a CO_2 Laser in Pulsed Regime", Zh. Prikl. Spekt., 1969, 10, no.3, p.425.

24.23.   Kuryl'ev, V.V., et al.: "GaAs Injection Laser Amplifier", Radiotekh. Elektron., 1969, 14, no.6, p.1072, and Radio Eng. Electron. Phys., 1969, 14, no.6, pp.924-6.

24.24.   Michelangeli, G.B., and Martellucci, S.: "Optical Distortion in Nd^{3+}:Glass for Repetitive Pumping", Appl. Opt., 1969, 8, p.1447.

24.25.   Hagen, W.F.: "Diffraction-Limited High-Radiance Nd^{3+}:Glass Laser System", J. Appl. Phys., 1969, 40, p.511.

24.26.   Balashov, I.F., et al.: "Laser Amplifier with Double Passage of Signal", Zh. Tekh. Fiz., 1969, 39, no.5, p.926, and Sov. Phys.-Tech. Phys., 1969, 14, no.5, pp.692-3.

24.27.   Treacy, E.B.: "Diffractive Coupling from a CO_2 Laser", Appl. Opt., 1969, 8, no.6, p.1107.

24.28.   Cool, T.A.: "Power and Gain Characteristics of High-Flow Lasers", J. Appl. Phys., 1969, 40, p.3563.

24.29.   Kikuchi, B., et al.: "Power Dependence on Repetition Period of Q-Switched CO_2 Laser", Jap. J. Appl. Phys., 1969, 8, no.8, p.1060.

24.30. Nasini, M.: "Ceramic Plasma Container for Ion Lasers", Rev. Sci. Instrum., 1969, 40, p.1473.

24.31. Anan'ev, Yu.A., et al.: "High-Efficiency Nd^{3+}:Glass Laser", Opt.-Mekh. Prom., 1969, 33, no.9, p.26, and Sov. J. Opt. Technol., 1969, 33, no.5, p.561.

24.32. Hernqvist, K.G.: "Effects of Transverse Magnetic Field for He-Cd Laser", J. Appl. Phys., 1969, 40, p.5399.

24.33. Diels, J.C., and Trum, H.M.G.J.: "Gain Measurements at 10.6 micron in Pulse-Generated CO_2 Plasma at High Pressures", Phys. Lett., 1969, 29A, p.697.

24.34. Peele, J.R., and Whitney, W.T.: "Flexible Adhesive/Sealants for CO_2 Laser Components", Rev. Sci. Instrum., 1969, 40, p.1114.

24.35. Giedt, R.R., and Gross, R.W.F.: "Mirror Mount for a Shock-Tube Laser Cavity", Rev. Sci. Instrum., 1969, 40, p.1238.

24.36. Currie, G.D.: "High-Power, Single-Mode, He-Ne Laser", Appl. Opt., 1969, 8, p.1068.

24.37. Swain, J.E., et al.: "Large-Aperture Glass-Disc Laser System", J. Appl. Phys., 1969, 40, p.3973.

24.38. Korolev, F.A., and Baikov, S.S.: "Investigation of Gain of a TW He-Ne Amplifier at 3.39 micron", Zh. Prikl. Spekt., 1969, 10, no.2, p.260.

24.39. Kokhanenko, P.N., and Antipov, A.B.: "Possibility of Determining the Wavelength of an Active Ruby from its Temperature. I and II", Izv. VUZ Fiz., 1969, no.5, pp.33 and 37.

24.40. Roess, D., and Zeidler, G.: "Time Dependence and Compensation of Thermal Resonator Bending of Solid-State Lasers", Z. Naturforsch., 1969, 24a, p.2027.

24.41. Wolinski, W.: "Studies of Excitation and Operation Conditions of He-Ne Lasers", Electron Technol.,1969,2,no.1,p.83.

24.42. Adamowicz, T., Kesik, J., and Wolinski, W.: "Measurement of 633-nm Gain Coefficient in He-Ne Mixture", Electron Technol., 1969, 2, no.1, p.169.

24.43. Molchanov, M.I., and Saushkin, A.F.: "Gain Measurements in He-Ne Mixture", Radiotekh. Elektron., 1969, 14, no.11, p.2020, and Radio Eng. Electron. Phys., 1969, 14, no.11.

24.44. Barchukov, A.I., and Terin, V.S.: "Output Mirror for CO_2 Laser", Radiotekh. Elektron., 1969, 14, no.11, p.2072, and Radio Eng. Electron. Phys., 1969, 14, no.11, pp.1796-7.

24.45. Horrigan, F.A., et al.: "Windows for High-Power Lasers", Microwaves, 1969, 8, no.1, p.68.

24.46. Crocker, A., and Wills, M.S.: "CO_2 Laser With High Power per Unit Length", Electron. Lett., 1969, 5, p.63.

24.47. Bridges, T.J., and Cheo, P.K.: "Spontaneous Self-Pulsing and Cavity Dumping in a CO_2 Laser With Electrooptic Q-Switching", Appl. Phys. Lett., 1969, 14, p.262.

24.48. Powell, T.: "Simple Method for Rapid Assembly of CO_2 Laser Tubes", J. Phys. E, 1969, 2, p.542.

24.49. Crafer, R.C., Gibson, A.F., and Kimmitt, M.F.: "High Peak Power from CO_2 Laser", Radio Electron. Eng., 1969, 38, no.6, p.354.

24.50. Tomlinson, W.J., and Fork, R.L.: "Anisotropy Effects in a Zeeman Laser", Phys. Rev., 1969, 180, p.628.

24.51. Swain, J.E., and Rainer, F.: "Many-Pass Resonant Laser Amplifier", J. Quantum Electron. IEEE, 1969, QE-5, p.385.

24.52. Edmonds, H.D.: "External Resonator Turn-On Time-Delay Effects With GaAs Injection Lasers at Room Temperature", Proc. IEEE, 1969, 57, p.1307.

24.53. Calawa, A.R., et al.: "Magnetic-Field Dependence of Laser Emission in (PbSn)Se Diodes", Phys. Rev. Lett., 1969, 23, no.1, p.7.

24.54. Stafsudd, O.M., and Yeh, Y.C.: "CW Gain Characteristics of Several Gas Mixtures at 337 micron", J. Quantum Electron. IEEE, 1969, QE-5, p.377.

24.55. McKnight, W.B.: "Considerations on High Peak Powers from CO_2 Lasers", J. Quantum Electron. IEEE, 1969, QE-5, p.420.

24.56. Day, G.W., Gaddy, O.L., and Jungling, K.C.: "Investigation of a Q-Switched Pulsed-Discharge CO_2 Laser", J. Quantum Electron. IEEE, 1969, QE-5, p.423.

24.57. Fridrikhov, S.A., Terekhin, D.K., and Lapshin, G.M.: "Energy Characteristics of a He-Ne Laser in a Transverse Magnetic Field", Zh. Prikl. Spekt., 1969,10,no.1,p.38.

24.58. Evtuhov, V., and Neeland, J.K.: "Continuously Pumped, Repetitively-Q-Switched, Ruby Laser", J. Quantum Electron. IEEE, 1969, QE-5, p.207.

24.59. Cordover, R.H., and Bonczyk, P.A.: "Effects of Collisions on the Saturation Behaviour of the 633-nm Transition of Ne Studied with He-Ne Laser", Phys. Rev., 1969, 188, p.696.

24.60. Afonnikov, N.A., et al.: "Argon-Ion Laser With Output of Several Watts", Zh. Prikl. Spekt., 1969, 11, p.886.

24.61. Donon, V.I.: "Saturation in a CW Argon-Ion Laser with High-Density Discharge Current", Zh. Prikl. Spekt., 1969, 11, p.889.

24.62. Virdi, S.S.: "Polishing the Ends of a Laser Tube at Brewster's Angle", Indian J. Pure Appl. Phys., 1969, 7, p.206.

24.63. Dyment, J.C., Ripper, J.E., and Zachos, T.H.: "Optimum Stripe Width for CW Operation of GaAs Junction Lasers", J. Appl. Phys., 1969, 40, p.1802.

24.64. Dienes, A.: "Polarization-Dependent Gain Saturation and Nonlinearly Induced Anisotropy in 3.39-micron He-Ne Laser Amplifier", J. Quantum Electron. IEEE, 1969, QE-5, p.162.

24.65. Buzhinskii, I.M., and Mamonov, S.K.: "Experimental Solid-State-Laser Head", Opt.-Mekh. Prom., 1969, 36, no.1, p.74, and Sov. J. Opt. Technol., 1969,36,no.1,p.81.

24.66. Gorban', I.S., and Kononchuck, G.L.: "Determining the Internal Losses of a Ruby Laser", Zh. Prikl. Spekt., 1969, 11, p.450.

24.67. Tiffany, W.B., Targ, R., and Foster, J.D.: "Kilowatt CO_2 Gas-Transport Laser", Appl. Phys. Lett., 1969, 15, p.91.

24.68. Smith, D.C.: "Q-Switched CO_2 Laser", J. Quantum Electron. IEEE, 1969, QE-5, p.291.

24.69. Harman, T.C., et al.: "Temperature and Compositional Dependence of Laser Emission in (PbSn)Se", Appl. Phys. Lett., 1969, 14, p.333.

24.70. Le Floch, A., and Brun, P.: "Nonlinear Effects in Output Intensity of He-Ne Laser Located in a Magnetic Field", C. R. Acad. Sci. (Paris), 1969, 269B, p.23.

24.71. Christensen, C.P., Freed, C., and Haus, H.A.: "Gain Saturation and Diffusion in CO_2 Lasers", J. Quantum Electron. IEEE, 1969, QE-5, p.276.

24.72. Magyar, G., and Selden, A.C.: "Lasers With Random Stack Mirrors", Appl. Opt., 1970, 9, no.9, p.2040.

24.73. Dezenberg, G.J., et al.: "Properties of High-Energy Electrically Pulsed CO_2 Lasers", J. Quantum Electron. IEEE, 1970, QE-6, p.652.

24.74. Tyte, D.C.: "Compact 20-W CW CO_2 Laser", J. Phys. E, 1970, 3, p.734.

24.75. Belousova, I.M., and Pantilov, V.G.: "Experimental Investigation of Pulsed Operation of a He-Ne/ Nd^{3+}:Glass Laser", Zh. Prikl. Spekt., 1970, 12, no.6, p.1012.

24.76. Rossi, J.A., et al.: "Threshold Requirements and Carrier Interaction Effects in GaAs Platelet Lasers", J. Appl. Phys., 1970, 41, p.312.

24.77. Nester, J.F.: "Dynamic Optical Properties of CW Nd^{3+}:YAG Lasers", J. Quantum Electron. IEEE, 1970, QE-6, p.97.

24.78. Dezenberg, G.J., Roy, E.L., and Merritt, J.A.: "Properties of a 150-mm-diam. Multipath CO_2 Laser", Appl. Opt.,1970,9,p.516.

24.79. Freiberg, R.J., and Clark, P.O.: "CO_2 Transverse-Discharge Lasers", J. Quantum Electron. IEEE, 1970, QE-6, p.105.

24.80. Regan, A.J.: "(Design of) Sealed CO_2-N_2-He Laser", J. Phys. E, 1970, 3, p.95.

24.81. Eliseev, P.G.: "Temperature Dependence of Gain in Injection Lasers. I", Fiz. Tekh. Poluprov., 1970, 4, p.51, and Sov. Phys.-Semicond., 1970, 4, no.1, pp.39-42.

24.82. Eichler, H., et al.: "Investigation of Spatial Hole-Burning in a Ruby Laser by Light Diffraction", Z. Angew. Phys., 1970, 28, p.303.

24.83. Delcroix, J.L., et al.: "Delay Between Discharge Current Impulse and Output of an Argon-Ion Laser", C. R. Acad. Sci. (Paris), 1970, 270B, p.347.

24.84. McGeoch, M.W.: "Prism Reflectors for High-Power Lasers", Opto-Electron., 1970, 2, no.2, pp.85-9.

24.85. Ivanov, V.A., and Lebedev, V.I.: "Determination of Thermal Parameters of a Ruby Laser Operating With Repeated Pulses", Zh. Prikl. Spekt., 1970, 13, no.1, pp.40-5.

24.86. Cool, T.A., Shirley, J.A., and Stephens, R.R.: "Operating Characteristics of a Transverse-Flow DF-CO_2 Purely Chemical Laser", Appl. Phys. Lett., 1970, 17, no.7, pp.278-81.

24.87. Franzen, D.L., and Collins, R.J.: "High-Gain Small-Bore Cooled CO_2 Amplifier", J. Quantum Electron. IEEE, 1970, QE-6, p.163.

24.88. Jennings, W.C., Noon, J.H., and Holt, E.H.: "Comparison of Hollow-Cathode and Conventional Argon-Ion Lasers", Rev. Sci. Instrum., 1970, 41, p.322.

24.89. Arkadeev, D.I., et al.: "Giant-Pulse Laser Using Ruby and Nd^{3+}:Glass", Radiotekh. Elektron., 1970, 15, p.523, and Radio Eng. Electron. Phys., 1970, 15, no.3, pp.445-50.

24.90. Chisler, E.V.: "Mercury (-Pool) Argon-Ion Laser", Opt. Spektr., 1970, 29, no.2, p.208, and Opt. Spectrosc., 1970, 29, no.2, pp.109-10.

24.91. Mak, A.A., et al.: "Effect of Cavity-Mirror Transmittance on Characteristics of a Single-Pulse Laser", Opt.-Mekh. Prom., 1970, 37, no.8, pp.7-9, and Sov. J. Opt. Technol., 1970, 37, no.8, pp.499-501.

24.92. Petru, F., and Vesula, Z.: "Output Power of He-Ne Lasers", Jemna Mech. Opt., 1970, 15, no.4, pp.89-96.

24.93. Petru, F., and Vesula, Z.: "Experimental Values of Output Power of He-Ne Lasers at 633 nm", Jemna Mech. Opt., 1970, 15, no.5, pp.122-6.

24.94. Keyes, R.W.: "Thermal Problems of the Pulsed Injection Laser", IBM J. Res. Dev., 1970, 14, p.158.

24.95. Lee, T.P., and Roldan, R.H.: "Repetitively Q-Switched Light Pulses from GaAs Injection Lasers With Tandem Double-Section Stripe Geometry", J. Quantum Electron. IEEE, 1970, QE-6, p.339.

24.96. Ripper, J.E.: "Reliability of GaAs Stripe-Geometry Junction Lasers", J. Quantum Electron. IEEE, 1970, QE-6, p.372.

24.97. Ahearn, W.E., and Crowe, J.W.: "High-Peak-Power Room-Temperature GaAs Laser Arrays", J. Quantum Electron. IEEE, 1970, QE-6, p.377.

24.98. Laurie, K.A., and Hale, M.M.: "Folded-Path Atmospheric-Pressure CO_2 Laser", J. Quantum Electron. IEEE, 1970, QE-6,p.526.

24.99. Hernqvist, K.G., and Pultorak, D.C.: "Simplified Construction and Processing of a He-Cd Laser", Rev. Sci. Instrum., 1970, 41, p.696.

24.100. Herziger, G., and Theiss, F.J.: "Coaxial Gas Laser for Generation of Short and Intensive Laser Pulses", Z. Angew. Phys., 1970, 29, no.3, p.157.

24.101. Fotiadi, A.E., and Fridrikhov, S.A.: "Effect of Transverse Magnetic Field Upon Operation of CW Argon-Ion Laser", Zh. Prikl. Spekt., 1970, 12, no.4, p.743.

24.102. Freudenthal, J.: "Deposits in a Sealed-Off CO_2 Laser-Type Discharge", J. Quantum Electron. IEEE, 1970, QE-6,p.503.

24.103. Csillag, L., Janossy, M., and Salamon, T.: "Time Delay of Laser Oscillations on the Green Transitions of a Pulsed He-Cd Laser", Phys. Lett., 1970, 31A, p.532.

24.104. Belousova, I.M., et al.: "Investigation of Optical Inhomogeneities of the Active Substance in a CF_3I Photodissociation Laser", Zh. Eksp. Teor. Fiz., 1970, 58, no.5, p.1481, and Sov. Phys.-JETP, 1970, 31, no.5, pp.791-3.

24.105. Hayashi, I., et al.: "Junction Lasers which Operate CW at Room Temperature", Appl. Phys. Lett., 1970, 17, p.109.

24.106. Aritome, H., Masuda, K., and Namba, S.: "Large Time Delay of CdS Laser Pumped by Electron Beam", Jap. J. Appl. Phys., 1970, 9, no.5, p.579.

24.107. Shaw, D.A., and Thornton, P.R.: "Catastrophic Degradation in GaAs Laser Diodes", Solid-State Electron., 1970, 13, no.7, p.919.

24.108. Mahlein, H.F., and Schollmeier, G.: "Giant-Pulse, 25-mm, Ruby Laser With PRF of 50 Hz", Z. Naturforsch., 1970, 26a, no.5, p.768.

24.109. McQuillan, A.K., and Carswell, A.I.: "Spatially Resolved Gain Measurements in a Flowing CO_2 Amplifier", Appl. Phys. Lett., 1970, 17, p.158.

24.110. Boersch, H., and Stahl, H.: "Bandwidth in the Threshold Region of the 633-nm He-Ne Laser", Z. Phys., 1970, 237,no.1,p.58.

24.111. Schafer, G., and Seelig, W.H.: "Ion-Laser Tube of Anodically Oxidised Aluminium Segments", Z. Angew, Phys., 1970, 29, no.4, p.246.

24.112. Koechner, W.: "Absorbed Pump Power, Thermal Profile, and Stresses, in a CW-Pumped Nd^{3+}:YAG Crystal", Appl. Opt., 1970, 9, p.1429.

24.113. Koechner, W., and Rice, D.K.: "Effect of Birefringence on Performance of Linearly Polarized Nd^{3+}:YAG Lasers", J. Quantum Electron. IEEE, 1970, QE-6, p.557.

24.114. Galaktionova, N.M., et al.: "Investigation of a CW Nd^{3+}:YAG Laser", Opt. Spekt., 1970, 28, no.1, p.96, and Opt. Spectrosc., 1970, 28, no.1, p.49.

24.115. Schuebel, W.K.: "Transverse-Discharge Slotted Hollow-Cathode Laser", J. Quantum Electron. IEEE, 1970, QE-6, p.574.

24.116. Lapshin, G.M., et al.: "Interferometric Investigation of Transverse-Magnetic Field Effect Upon He-Ne Laser Radiation Spectrum", Zh. Prikl. Spekt., 1970, 12, no.5, p.824.

24.117. Privalov, V.E., and Fridrikhov, S.A.: "He-Ne Laser With Conical-Cross-Section Discharge Tube", Zh. Prikl. Spekt., 1970, 12, no.5, p.937.

24.118. Vinogin, Yu.P., et al.: "Narrow-Line Ruby Laser", Opt. Spekt., 1970, 28, no.1, p.168, and Opt. Spectrosc., 1970, 28, no.1, p.85.

24.119. Gans, F., Troyanowsky, C., and Valat, P.: "Optically Homogeneous Nd^{3+}:Glass Laser", C. R. Acad. Sci. (Paris), 1970, 270B, p.1343.

24.120. Anan'ev, Yu.A., and Grishmanova, N.I.: "Possible Dynamic Compensation of Thermal Deformation of a Laser Resonator", Zh. Prikl. Spekt., 1970, 12, no.6, p.1109.

24.121. Hieslmair, H., Bickart, C.J., and Fulton, J.N.: "Small-Signal Gain of a CO_2-Laser Amplifier Utilizing a White Optical-Reflector Design", J. Quantum Electron. IEEE, 1970, QE-6, p.86.

24.122. Menzies, R.T., Dienes, A., and George, N.: "Axial-Magnetic-Field Effects on a Saturated He-Ne Laser Amplifier", J. Quantum Electron. IEEE, 1970, QE-6, p.117.

24.123. Brannen, E., and Sarjeant, W.J.: "Far-IR Laser Action Using Compound Grating Resonators", J. Quantum Electron. IEEE, 1970, QE-6, p.138.

24.124. Condor, P.C., and Foster, H.: "Sealed-Off, Beryllia-Tube, Argon-Ion Laser", Radio Electron. Eng., 1970, 39, no.2, p.97.

24.125. Moran, J.M.: "Coupling of Power from a Circular Confocal Laser With Output Aperture", J. Quantum Electron. IEEE, 1970, QE-6, p.93.

24.126. Goodwin, F.E., Trimble, F.C., and Nussmeier, T.A.: "One-Year Operation of Sealed-Off CO_2 Laser", J. Quantum Electron. IEEE, 1970, QE-6, no.11, pp.756-7.

24.127. Silver, M., Hartwick, T.S., and Posakony, M.J.: "Gain Measurements in CO_2 Isotope Lasers", J. Appl. Phys., 1970, 41, no.11, pp.4566-8.

24.128. Peters, W.N., and Stein, E.K.: "Helium-Permeation Compensation Techniques for Long-Life Gas Lasers", J. Phys. E, 1970, 3, no.9, pp.719-21.

24.129. Avivi, P., and Dothan-Deutsch, F.: "Metal-Tube Gas Laser", J. Phys. E, 1970, 3, no.9, p.750.

24.130. De La Forest-Divonne, A., and Bensimon, J.: "CW Ionic Lasers, Particularly Working Under Pulsed Excitation", Opt. Commun., 1970, 2, no.2, pp.55-8.

24.131. Gibbs, W.E.K., and Booth, D.J.: "Gain Measurements at 10.6 micron on Pulsed High-Voltage Discharges in CO_2-N_2-He Mixtures", Phys. Lett., 1970, 33A, no.4, pp.261-2.

24.132. Ehlers, K.W., and Brown, I.G.: "Rejuvenation of He-Ne Lasers", Rev. Sci. Instrum., 1970, 41, no.10, pp.1505-6.

24.133. Wittman, H.R., and Smith, J.L.: "Bulk Changes during Catastrophic Degradation of GaAs Laser Diodes", Phys. Status Solidi, 1970, 1A, no.2, pp.279-82.

24.134. Ter-Pogosyan, A.S.: "Laser Thermal Regime at High Pumping PRF", Zh. Prikl. Spekt., 1970, 13, no.3, pp.418-24.

24.135. Koechner, W.: "Thermal Lensing in a Nd^{3+}:YAG Laser Rod", Appl. Opt., 1970, 9, no.11, pp.2548-53.

24.136. Dianov, E.M., and Prokhorov, A.M.: "Thermal Distortions of Laser Resonators With Rectangular-Plate Active Rods", Dokl. Akad. Nauk SSSR, 1970, 192, no.1-3, pp.531-3, and Sov. Phys.-Dokl., 1970, 15, no.5, pp.481-2.

24.137. Smith, D.C., and DeMaria, A.J.: "Parametric Behaviour of Atmospheric-Pressure Pulsed CO_2 Laser", J. Appl. Phys., 1970, 41, no.13, pp.5212-14.

24.138. Ling, H., Colombo, J., and Fisher, C.L.: "Wide-Bore, Helical-Inductively-Pumped, Argon Laser", Rev. Sci. Instrum., 1970, 41, no.10, pp.1436-7.

24.139. Laflamme, A.K.: "Double Discharge Excitation for Atmospheric-Pressure CO_2 Lasers", Rev. Sci. Instrum., 1970, 41, no.11, pp.1578-81.

24.140. Redaelli, G.: "Single Sealed-Off TEM_{00} Ion-Laser Ceramic Tube", Appl. Opt., 1970, 9, no.11, pp.2593-4.

24.141. Farkas, G.: "Roof-Prism Output Coupler for Gigawatt Laser Pulses", Opt. Laser Technol., 1970, 2, no.4, pp.204-5.

24.142. Ripper, J.E., and Paoli, I.L.: "Optical Coupling of Adjacent Stripe-Geometry Junction Lasers", Appl. Phys. Lett., 1970, 17, no.9, pp.371-3.

24.143. Leheny, R.F., et al.: "Model for Temperature-Dependent CdS Laser", Appl. Phys. Lett., 1970, 17, no.11, pp.494-7.

24.144. Vanyukov, M.P., et al.: "Nd^{3+}: Glass Laser With High Radiation Density", Opt. Spekt., 1970, 28, no.5, pp.1008-12, and Opt. Spectrosc., 1970, 28, no.5, pp.544-7.

24.145. Patlach, A.M.: "(GaAs) Laser Packaging", IBM Tech. Disclosure Bull., 1970, 13, no.2, pp.337-8.

24.146. Wolinski, W., et al.: "He-Ne Laser With Non-Tuned Internal Resonator", Elektronika, 1970, no.7-8, pp.312-3.

24.147. Anan'ev, Yu.A., et al.: "Energy Characteristics of a Nd^{3+}:Glass Laser With Polarized Radiation", Zh. Prikl. Spekt., 1970, 13, no.2, pp.227-31.

24.148. Brunet, H., Lavarini, B., and Voignier, F.: "Gas Cooling Processes in Fast-Flow CO_2 Laser", Phys. Lett., 1970, 33A, no.8, pp.497-8.

24.149. Kuznetsov, V.V., and Orishich, A.M.: "Pulsed Generation in CO_2 at High Pressure", Zh. Prikl. Spekt., 1970, 13, no.4, pp.599-601.

24.150. Damle, R.V.: "Construction of CW CO_2 Laser", Indian J. Technol., 1970, 8, no.10, pp.387-8.

24.151. Gromov, Yu.N., Tychinskii, V.P., and Khaikin, N.Sh.: "Study of CO_2 Laser With Ge Brewster Windows", Prib. Tekh. Eksp., 1970, no.5, pp.182-3, and Instrum. Exp. Tech., 1970, no.5, pp.1455-6.

24.152. Belostotskii, B.R.: "Analysis of Temperature Stresses in Active Elements of a Laser", Inzh.-Fiz. Zh., 1970, 19, no.2, pp.272-5, and J. Eng. Phys., 1970, 19, no.2, pp.985-7.

24.153. Kertesz, I., et al.: "Effect of Pump-Induced Birefringence in Nd^{3+}:Glass on Laser Operation", Zh. Eksp. Teor. Fiz., 1970, 59, no.4, pp.1115-24, and Sov. Phys.-JETP, 1971, 32, no.4, pp.606-11.

24.154. Chakravarti, A.N., and Parui, D.P.: "Effect of a Longitudinal Magnetic Field on Diffusion Coefficient of Minority Carriers in InSb Junction Lasers", Int. J. Electron., 1971, 30, no.3, pp.275-9.

24.155. Evenson, K.M., et al.: "Variable-Output Coupling Far-IR Michelson Laser", J. Appl. Phys., 1971, 42, no.3, pp.1233-4.

24.156. Witte, H.H.: "Influence of Thermal Lens Effect on the Resonator Geometry of Gas Lasers", Opt. Laser Technol., 1971, 3, no.1, pp.31-5.

24.157. Scheuermann, W., and Ritter, G.J.: "300-mW Folded He-Ne Laser", Opt. Laser. Technol., 1971, 3, no.1, pp.45-6.

24.158. Beaulieu, A.J.: "High-Peak-Power (CO_2) Gas Lasers", Proc. IEEE, 1971, 59, no.4, pp.667-74.

24.159. Johnson, D.C.: "Excitation of an Atmospheric-Pressure CO_2-N_2-He Laser by Capacitor Discharges", J. Quantum Electron. IEEE, 1971, QE-7, no.5, pp.185-9.

24.160. Chun, M.K.: "Thermal Transient Effects in Optically Pumped Repetitively Pulsed Lasers", J. Quantum Electron. IEEE, 1971, QE-7, no.5, pp.200-2.

24.161. Kogelnik, H., and Shank, C.V.: "Stimulated Emission in a Periodic Structure", Appl. Phys. Lett., 1971, 18, no.4, pp.152-4.

24.162. Ripper, J.E., et al.: "Stripe-Geometry Double-Heterostructure Junction Lasers", Appl. Phys. Lett., 1971, 18, no.4, pp.155-7.

24.163. Barnes, C.E.: "Neutron Damage in Epitaxial GaAs Laser Diodes", J. Appl. Phys., 1971, 42, no.5, pp.1941-9.

24.164. Corcoran, V.J., et al.: "CW (Gain) Measurements of the CO Molecular Laser", J. Quantum Electron. IEEE, 1971, QE-7, no.6, pp.246-8.

24.165. Lee, G., and Gowen, F.E.: "Gain of CO_2 Gasdynamic Lasers", Appl. Phys. Lett., 1971, 18, no.6, pp.237-9.

24.166. Levine, F.A.: "TEM_{00} Enhancement in CW Nd^{3+}:YAG by Thermal-Lensing Compensation", J. Quantum Electron. IEEE, 1971, QE-7, no.4, pp.170-2.

24.167. Vallese, L.M.: "Temperature Dependence of Semiconductor-Laser Characteristics", Solid-State Technol., 1971, 14, no.1, pp.38-9.

24.168. Mansell, D.M., Love, J.A., and Snell, W.L.: "Investigations of Unstable Confocal Resonator on a Chemical-Laser System", J. Quantum Electron. IEEE, 1971, QE-7, no.4, p.177.

24.169. Flamant, P., and Meyer, Y.H.: "Absolute Gain Measurements in a Multistage Dye Amplifier", Appl. Phys. Lett., 1971, 19, no.11, pp.491-3.

24.170. Dmitriev, V.G., et al.: "Thermal Stresses in Active Elements Under Continuous Pumping", Kvantovaya Elektron., 1971, no.2, pp.80-6.

24.171. Vinevich, B.S.: "Measurement of Focal Length of Gaseous Lens in Active Medium of CO_2 Laser", Opt. Spekt., 1971, 30, no.6, pp.1146-7, and Opt. Spectrosc., 1971, 30, no.6, pp.611-2.

24.172. Bykov, V.P., et al.: "Influence of Resonator Matching on Output Power of Solid-State Laser", Kvantovaya Elektron., 1971, no.2, pp.53-6.

24.173. Zheltov, G.I., Pubanov, A.S., and Chalei, A.V.: "Thermal Deformation of the Active Elements at Laser Frequencies", Zh. Prikl. Spekt., 1971, 14, no.2, pp.226-30.

24.174. Koechner, W., and Rice, D.K.: "Birefringence of Nd^{3+}:YAG Laser Rods as a Function of Growth Direction", J. Opt. Soc. Am., 1971, 61, no.6, pp.758-66.

24.175. August, H., and Weber, H.: "Temperature Dependence of the Single-Pass Gain and Superradiance of a Ruby-Laser Amplifier", Z. Angew. Phys., 1971, 31, no.2, pp.111-6.

24.176. Valyavko, V.V., and Boiko, B.B.: "Effect of a Magnetic Field on the Output of a Ruby Laser", Zh. Prikl. Spekt., 1971, 14, no.2, pp.325-7.

24.177. Weigand, W.J., Fowler, M.C., and Benda, J.A.: "Influence of Discharge Properties on CO_2 Laser Gain", Appl. Phys. Lett., 1971, 18, no.9, pp.365-7.

24.178. Gibson, A.F., Kimmitt, M.F., and Patel, B.S.: "Gain of Transversely Excited CO_2 Lasers at Pressures up to 500 torr", J. Phys. D, 1971, 4, no.7, pp.882-7.

24.179. Robinson, A.M.: "High-Gain Pulsed CO_2 Transverse-Discharge Amplifier", Phys. Lett., 1971, 35A, no.1, pp.47-8.

24.180. Chakravarti, A.N., and Parui, D.P.: "Effect of a Transverse Magnetic Field on the Maximum Duration of the Output Pulse in InAs Junction Lasers", Int. J. Electron., 1971, 31, no.1, pp.95-9.

24.181. Kimura, T., and Otsuka, K.: "Thermal Effects of a Continuously Pumped Nd^{3+}:YAG Laser", J. Quantum Electron. IEEE, 1971, QE-7, no.8, pp.403-7.

24.182. Soderholm, L.G.: "Portable CO_2 Laser Uses Sealed, Temperature-Controlled, System", Des. News, 1971, 26, no.5, p.33.

24.183. Buczek, C.J., Freiberg, R.J., and Skolnick, M.L.: "CO_2 Regenerative Ring Power Amplifiers", J. Appl. Phys., 1971, 42, no.8, pp.3133-7.

24.184. Matsuda, M., et al.: "Multi-Module Different-Phase Discharge CO_2 Laser", Jap. J. Appl. Phys., 1971, 10, no.7, pp.958-9.

24.185. Maitland, A.: "Theory of Segmented Metal Discharge Tubes for Argon Lasers", J. Phys. D, 1971, 4, no.7, pp.907-15.

24.186. Lomnes, R.K., and Taylor, J.C.W.: "Cold-Cathode (Low-Cost) Pulsed Gas Laser", Rev. Sci. Instrum., 1971, 42, no.6, pp.766-9.

24.187. Laderman, A.J., and Byron, S.R.: "Temperature Rise and Radial Profiles in CO_2 Lasers", J. Appl. Phys., 1971, 42, no.8, pp.3138-44.

24.188. Malacara, D., Morales, A., and Rizo, I.: "Losses in Brewster's-Angle Windows of He-Ne Lasers Due to Mechanical Stresses", Appl. Opt., 1971, 10, no.8, pp.1984-5.

24.189. Inoue, A., et al.: "Design and Use of UV N_2 Laser", Rep. Inst. Phys. Chem. Res., 1971, 47, no.2, pp.40-5.

24.190. Gruzinskii, V.V., and Stratskevich, L.K.: "Effect of Linear Magnetic Field on Generating Power of a He-Ne Laser", Zh. Prikl. Spekt., 1971, 14, no.5, pp.804-8.

24.191. Mills, R.W., Peggs, I.D., and Schulte, E.H.: "Determination of the Radial Energy Distribution in a Pulsed (Ruby-) Laser Beam", J. Phys. E, 1971, 4, no.9, pp.700-2.

24.192. Turgeon, M.F.: "High-PRF (Transverse) CO_2 Laser", J. Quantum Electron. IEEE, 1971, QE-7, no.10, pp.495-7.

24.193. Ben-Yosef, N., et al.: "Electrode Configuration and Power Output for a Transverse-Flow CO_2 Laser", J. Phys. E, 1971, 4, no.9, pp.708-9.

24.194. Ducloy, M.: "Nonlinear Effects in a Multimode Gas Laser. Saturation of a J_{1-0} Transition in a Magnetic Field", Opt. Commun., 1971, 3, no.4, pp.205-8.

24.195. Samid, I., and Shimony, U.: "Quenching of GaAs Laser Diodes Due to Heating", Israel J. Technol., 1971, 9, no.3, pp.233-8.

24.196. Grisar, R., et al.: "Resonantly Pumped Tunable Stimulated Recombination Radiation in InSb", Opt. Commun., 1971, 3, no.6, pp.415-7.

24.197. Salathe, R., and Mohn, E.: "Influence of Doping Gradient on Temperature Dependence of Threshold Current Density of GaAs Injection Lasers", Solid-State Electron., 1971, 14, no.9, pp.843-7.

24.198. Smith, P.W.: "(Hollow-Dielectric) Waveguide Gas Laser", Appl. Phys. Lett., 1971, 19, no.5, pp.132-4.

24.199. Smith, D.C.: "Velocity Dependence of Gain of a CO_2 Laser", J. Quantum Electron. IEE, 1971, QE-7, no.9, pp.459-60.

24.200. Malysh, A.G., et al.: "Observation of Birefringence in Solid-State Lasers Due to Pumping", Prib. Tekh. Eksp., 1971, 14, no.2, pp.207-9, and Instrum. Exp. Tech., 1971, 14, no.2, pp.569-71.

24.201. Fotiadi, A.E., Fridrikhov, S.A., and Elagin, V.V.: "Experimental Study of Intensity of Argon-Ion Laser in a Magnetic Field", Zh. Prikl. Spekt., 1971, 15, no.4, pp.735-6.

24.202. Bjorkholm, J.E., Damen, T.C., and Shah, J.: "Improved Use of Gratings in Tunable Lasers", Opt. Commun., 1971, 4, no.4, pp.283-4.

24.203. Barchukov, A.I., and Konev, Yu.B.: "Amplitude-Phase Distortions in CO_2 Power Amplifier With Periodic Correction", Radiotekh. Elektron., 1971, 16, no.4, pp.549-53, and Radio Eng. Electron. Phys., 1971, 16, no.4, pp.633-6.

24.204. Karpushko, F.V.: "Dynamic Cavity Resonator for Pulsed Lasers", Prib. Tekh. Eksp., 1971, 14, no.3, pp.186-9, and Instrum. Exp. Tech., 1971, 14, no.3, pp.874-7.

24.205. Kamenskii, E.I., and Kozlov, V.V.: "Lasers With Polyhedral Energy Guides", Kvantovaya Elektron., 1971, no.4, pp.77-86.

24.206. Vlasenko, V.Ye., et al.: "Design Features of an Injection Laser Operating at Room Temperature with a High PRF", Radiotekh. Elektron., 1971, 16, no.2, pp.437-8, and Radio Eng. Electron. Phys., 1971, 16, no.2, pp.372-4.

24.207. Basov, N.G., Belenov, E.M., et al.: "High-Pressure Pulsed CO_2 Laser", Kvantovaya Elektron., 1971, no.3, pp.121-2.

24.208. Kitaeva, V.F., Ostrovskaya, L.Ya., and Sobolev, N.N.: "Dependences of the Populations of A^{II} Levels in a CW Laser on the Discharge-Tube Diameter and Magnetic Field", Kvantovaya Elektron., 1971, no.4, pp.41-9.

24.209. Blagodarov, Yu.A., et al.: "Temperature Effects in He-Ne Laser", Zh. Prikl. Spekt., 1971, 15, no.6, pp.993-6.

24.210. Senatskii, Yu.V.: "Active Elements for High-Power Nd^{3+} Lasers", Kvantovaya Elektron., 1971, no.5, pp.109-12.

24.211. Karube, N.: "Production of High Power from a CO_2 Laser", Oyo Buturi, 1971, 40, no.2, pp.201-5.

24.212. Privalov, V.E.: "Effect of a Transverse Magnetic Field on Power Radiated by a He-Ne Laser", Opt. Spektr., 1971, 31, no.6, pp.970-5, and Opt. Spectrosc., 1971, 31, no.6, pp.524-6.

24.213. de Lang, H., Polder, D., and van Haeringen, W.: "Optical Polarization Effects in a Gas Laser", Philips Tech. Rev., 1971, 32, no.6-8, pp.190-204.

24.214. Korneev, N.E., and Folomeev, A.V.: "Laser With Plano-Convex Cavity and Mirror Transmission that Varies Over the Cross Section", Radiotekh. Elektron., 1971, 16, no.11, pp.2230-2, and Radio Eng. Electron. Phys., 1971, 16, no.11, pp.1976-8.

24.215. Freiberg, R.J., and Buczek, C.J.: "Saturated Gain-Bandwidth of CO_2 Regenerative Ring Amplifiers", Opt. Commun., 1971, 4, no.2, pp.139-43.

24.216. Mozan'ko, I.P., et al.: "Gain Distribution Measurement in He-Ne Lasers", Opt. Spekt., 1971, 30, no.5, pp.927-31, and Opt. Spectrosc., 1971, 30, no.5, pp.495-7.

24.217. Klement'ev, V.M.: "He-Ne Laser in a Strong Magnetic Field", Zh. Prikl. Spekt., 1971, 15, no.3, pp.421-5.

24.218. Sommers, H.S.: "Current Dependence of the Intensity of Spontaneous Emission of GaAs Injection Lasers", Appl. Phys. Lett., 1971, 19, no.10, pp.424-6.

24.219. Kucerovsky, Z., Sarjeant, W.J., and Brannen, E.: "Analysis of Gain Measurements of a Pulsed Two-Level Laser System", Appl. Opt., 1971, 10, no.9, pp.2070-3.

24.220. Merchant, V., and Irwin, J.C.: "Atmospheric-Pressure Pulsed CO_2 Laser With Metal Electrodes", Rev. Sci. Instrum., 1971, 42, no.10, pp.1437-9.

24.221. Redaelli, G.: "Criteria for Design and Construction of a Compact High-Gain TEM_{00} He-Ne Laser", Vacuum, 1971, 21, no.6, pp.207-10.

24.222. Klinkov, V.K.: "Generation by a Ruby Laser in a Magnetic Field", Dokl. Akad. Nauk SSSR, 1971, 194, no.1-3, pp.565-7, and Sov. Phys.-Dokl., 1971, 16, no.5, pp.384-7.

24.223. Yamanaka, M., Yamauchi, T., and Yoshinaga, H.: "Time Behaviour of Pulsed Gain of HCN Laser at 337 micron", Jap. J. Appl. Phys., 1971, 10, no.11, pp.1601-3.

24.224. Taylor, D.J., Harris, S.E., and Nieh, S.T.K.: "Electronic Tuning of a Dye Laser Using an Acoustooptic Filter", Appl. Phys. Lett., 1971, 19, no.8, pp.269-71.

24.225. Tulip, J., and Seguin, H.J.: "Explosion-Pumped Gasdynamic CO_2 Laser", Appl. Phys. Lett., 1971, 19, no.8, pp.263-5.

24.226. Sparks, M.: "Optical Distortion by Heated Windows in High-Power Laser Systems", J. Appl. Phys., 1971, 42, no.12, pp.5029-46.

24.227. Hvam, J.M.: "Temperature-Induced Wavelength Shift of Electron-Beam-Pumped Lasers from CdSe, CdS, and ZnO", Phys. Rev., 1971, 4B, no.12, pp.4459-64.

24.228. Artamonov, S.A., et al.: "6-Hz Giant-Pulse Ruby Laser", Opt.-Mekh. Prom., 1971, 38, no.1, pp.23-6, and Sov. J. Opt. Technol., 1971, 38, no.1, pp.19-21.

24.229. Silfvast, W.T., and Szeto, L.H.: "Simplified Low-Noise He-Cd Laser With Segmented Bore", Appl. Phys. Lett., 1971, 19, no.10, pp.445-7.

24.230. Seguin, H.J., Tulip, J., and White, B.: "Sealed CO Laser at Room Temperature", Can. J. Phys., 1971, 49, no.21, pp.2731-2.

24.231. Keyes, R.W.: "Thermal Problems of CW Injection Laser", IBM J. Res. Dev., 1971, 15, no.5, pp.401-4.

24.232. Yang, E.S.: "Injected-Carrier Lifetime and Ageing of GaAs Injection Lasers", J. Appl. Phys., 1971, 42, no.13, pp.5635-9.

24.233. Bachert, H., et al.: "Energy Dependence of Internal Losses in Injection Lasers", Phys. Status Solidi, 1971, 8A, no.2, pp.477-82.

24.234. Deryagin, V.N., and Marasin, L.E.: "Distribution of Oscillation Delay Periods Over Emitting Surface of a Semiconductor Laser", Fiz. Tekh. Poluprov., 1971, 5, no.10, pp.1981-3, and Sov. Phys.-Semicond., 1972, 5, no.10, pp.1716-7.

24.235. Lapitskaya, G.A., Pleshkov, A.A., and Trukhan, V.G.: "Influence of Temperature on Radiation from Inhomogeneously Excited Semiconductor Laser", Fiz. Tekh. Poluprov., 1971, 5, no.11, pp.2226-8, and Sov. Phys.-Semicond., 1972, 5, no.11.

24.236. Eckbreth, A.C., and Davis, J.W.: "Cross-Beam Electric Discharge Convection Laser", J. Quantum Electron. IEEE, 1972, QE-8, no.2, pp.139-44.

24.237. Pearson, P.R., and Lamberton, H.M.: "Atmospheric-Pressure CO_2 Lasers Giving High Output Energy per Unit Volume", J. Quantum Electron. IEEE, 1972, QE-8, no.2, pp.145-9.

24.238. Yatsiv, S., et al.: "Experiments With a Pulsed CO_2 Gasdynamic Laser", J. Quantum Electron. IEEE, 1972, QE-8, no.2, pp.161-3.

24.239. Dyment, J.C., Ripper, J.E., and Lee, T.P.: "Interpretation of Long Spontaneous Lifetimes in Double Heterostructure Lasers", J. Appl. Phys., 1972, 43, no.2, pp.452-7.

24.240. Steffen, J., et al.: "(Thermal Lensing of) Fundamental-Mode Radiation With Solid-State Laser", J. Quantum Electron. IEEE, 1972, QE-8, no.2, pp.239-45.

24.241. Dumanchin, R., et al.: "Extension of TEA-CO_2 Laser Capabilities", J. Quantum Electron. IEEE, 1972, QE-8, no.2, pp.163-5.

24.242. Targ, R., and Sasnett, M.W.: "High-PRF Xe Laser With Transverse Excitation", J. Quantum Electron. IEEE, 1972, QE-8, no.2, pp.166-9.

24.243. Strome, F.C.: "Transient Gain Measurements on Laser Dyes", J. Quantum Electron. IEEE, 1972, QE-8, no.2, pp.98-101.

24.244. Bliss, E.S.: "Importance of Self-Focusing in a Laser Amplifier With Large Beam Diameter", J. Quantum Electron. IEEE, 1972, QE-8, no.2, pp.273-6.

24.245. Franzen, D.L., and Jennings, D.A.: "Gain Saturation Measurements in CO_2 TEA Amplifiers", J. Appl. Phys., 1972, 43, no.2, pp.729-30.

24.246. Newman, D.H., Ritchie, S., and O'Hara, S.: "Experimental Tests of Proposed Mechanisms for Gradual Degradation of GaAs Double-Heterostructure Injection Lasers", J. Quantum Electron. IEEE, 1972, QE-8, no.3, pp.379-82.

24.247. Tan, K.P., Makios, V., and Morrison, R.W.: "TEA CO_2 Laser Driven by a 200-kV Marx Generator", Phys. Lett., 1972, 38A, no.4, pp.225-6.

24.248. Hebner, R.E., and Nygaard, K.J.: "Method for Studying Model Behaviour and Temperature Tuning of GaAs Lasers", Physica, 1972, 58, no.2, pp.225-8.

24.249. Franzen, D.L., and Collins, R.J.: "Radial Gain Profiles in CO_2 Laser Discharges", J. Quantum Electron. IEEE, 1972, QE-8, no.4, pp.400-4.

24.250. Chester, A.N., and Hess, L.D.: "Study of HF Chemical Laser by Pulse-Delay Measurements", J. Quantum Electron. IEEE, 1972, QE-8, no.1, pp.1-13.

24.251. Fill, E.E., and Finckenstein, K.G.V.: "Comparison of Performance of Different Laser-Amplifier Media", J. Quantum Electron. IEEE, 1972, QE-8, no.1, pp.24-6.

24.252. Lee, G., Gowen, F.E., and Hage, J.R.: "Gain and Power of CO_2 Gasdynamic Lasers", AIAA J., 1972, 10, no.1, pp.65-71.

24.253. Wang, C.P., and Lin, S.C.: "Experimental Study of Argon-Ion Laser Discharge at High Current", J. Appl. Phys., 1972, 43, no.12, pp.5068-73.

24.254. Belostotskii, B.R., et al.: "Non-linear Aspects of Cooling of a Strongly Anisotropic Optical Element of a Laser", Kvantovaya Elektron., 1972, no.2, pp.23-9, and Sov. J. Quantum Electron., 1972, 2, no.2, pp.106-10.

24.255. Andreev, R.B., et al.: "Powerful Single-Pulsed Laser With Stable Spectrum and Radiation Alignment", Zh. Prikl. Spekt., 1972, 17, no.2, pp.355-7.

24.256. McQuillan, A.K., Carswell, A.I., and Jammu, K.S.: "Spatially Resolved Gain Measurements in a CO_2 Amplifier", Can. J. Phys., 1972, 50, no.8, pp.769-77.

24.257. Garside, B.K., Ballik, E.A., and Reid, J.: "Pulse Delays in TEA CO_2 Lasers", J. Appl. Phys., 1972, 43, no.5, pp.2387-90.

24.258. Rink, J.P.: "TEA-Laser Power for Atmospheric and Higher Pressures", J. Appl. Phys., 1972, 43, no.5, p.2441.

24.259. Garnsworthy, R.K.: "Pressure-Dependent Performance of a TE Pulse CO_2 Laser", J. Phys. D, 1972, 5, no.5, pp.901-3.

24.260. Powell, T.: "Design and Operation of a Modular CO_2 Laser System", Opt. Laser Technol., 1972, 4, no.1, pp.19-23.

24.261. Pan, Y.L., Bernhardt, A.F., and Simpson, J.R.: "Construction and Operation of a Double-Discharge TEA CO_2 Laser", Rev. Sci. Instrum., 1972, 43, no.4, pp.662-6.

24.262. Vitale, G.F.: "Thermal Properties of Semiconductor Lasers", Alta Frequenza, 1972, 41, no.2, pp.90-3.

24.263. Massey, G.A.: "Measurements of Device Parameters for Nd^{3+}:$YAlO_3$ Lasers", J. Quantum Electron. IEEE, 1972, QE-8, no.7, pp.669-74.

24.264. Deutsch, T.F., and Rudko, R.I.: "Spatial and Temporal Dependence of the Gain of a Transversely Excited Pulsed CO_2 Laser", Appl. Phys. Lett., 1972, 20, no.11, pp.423-5.

24.265. Marcus, S., and Carbone, R.J.: "Gain and Relaxation Studies in Transversely Excited HF Lasers", J. Quantum Electron. IEEE, 1972, QE-8, no.7, pp.651-5.

24.266. Marcuse, D.: "Hollow Dielectric Waveguide for Distributed-Feedback Lasers", J. Quantum Electron. IEEE, 1972, QE-8, no.7, pp.661-9.

24.267. Jensen, R.E., and Tobin, M.S.: "CO_2 Waveguide Gas Laser", Appl. Phys. Lett., 1972, 20, no.12, pp.508-10.

24.268. Smol'skaya, T.I., and Rubinov, A.N.: "Effect of Transverse Pumping Distribution on Kinetics and Thermooptical Distortion Profile of Solutions of Rhodamine 6J", Zh. Prikl. Spekt., 1972, 16, no.4, pp.618-26.

24.269. Gurski, T.R.: "Optical Contact Approach to Laser Rod Support", Appl. Optics, 1972, 11, no.9, pp.2105-6.

24.270. Ripper, J.E., Patel, N.B., and Brosson, P.: "Effect of Uniaxial Pressure on the Threshold Current of Double-Heterostructure GaAs Lasers", Appl. Phys. Lett., 1972, 21, no.4, pp.124-5.

24.271. Zory, P., and Woodall, J.: "Superlattice Surface Laser", IBM Tech. Disclosure Bull., 1972, 15, no.1, p.163.

24.272. Mason, J.F.: "Power Raised in Zigzag Laser by Linking it to Two Smaller Lasers", Electron. Des., 1972, 20, no.13, pp.32-3.

24.273. Vanyukov, M.P., et al.: "High-Power, Giant-Pulse, Laser With Unstable Cavity", Opt.-Mekh. Prom., 1972, 39, no.1, p.58, and Sov. J. Opt. Technol., 1972, 39, no.1, pp.51-2.

24.274. Berger, P.J., and Smith, D.C.: "Gas Breakdown in the Laser as the Limitation of Pulsed High-Pressure CO_2 Lasers", Appl. Phys. Lett., 1972, 21, no.4, pp.167-70.

24.275. Morrison, R.W., and Swail, C.: "Method of Exciting Uniform Discharges for High-Pressure Lasers", Phys. Lett., 1972, 40A, no.5, pp.375-7.

24.276. Hidson, D.J., Makios, V., and Morrison, R.W.: "Transverse CO_2 Laser Action at Several Atmospheres", Phys. Lett., 1972, 40A, no.5, pp.413-4.

24.277. Rubinov, A.N., and Anufrik, S.S.: "Possible Dynamic Compensation of Thermooptical Distortion in a Liquid Laser Resonator", Zh. Prikl. Spekt., 1972, 17, no.1, pp.33-7.

24.278. Chakravarti, A.N., Biswas, S.N., and Rakshit, S.: "Dependence of Power and Efficiency of CW (at $77^{\circ}K$) GaAs Junction Lasers on Current and Length", Indian J. Pure Appl. Phys., 1972, 10, no.2, pp.128-30.

24.279. Bogatov, A.P., et al.: "Pulsed and CW Output Characteristics of Stripe-Geometry Symmetrical Heterojunction Lasers at $300^{\circ}K$", Fiz. Tekh. Poluprov., 1972, 6, no.1, pp.43-8, and Sov. Phys.-Semicond., 1972, 6, no.1, pp.34-8.

24.280. Anderson, J.D.: "Effect of Na on N_2-CO_2 Gasdynamic Laser Gain", J. Appl. Phys., 1972, 43, no.2, pp.534-6.

24.281. Ninnis, R.M., and Rieckhoff, K.E.: "Effects of Thermal Lensing in Glass Lasers", Can. J. Phys., 1972, 50, no.14, pp.1656-8.

24.282. Yamabe, C., et al.: "Gain Measurements of Matrix-Type TEA CO_2 Laser", Jap. J. Appl. Phys., 1972, 11, no.8, pp.1227-8.

24.283. Petru, F., and Vesela, Z.: "Output Power of 633-nm He-Ne Lasers", Opto-Electron., 1972, 4, no.1, pp.1-20.

24.284. Petru, F., and Vesela, Z.: "Experimental Output Powers of 633-nm He-Ne Lasers", Opto-Electron., 1972, 4, no.1, pp.21-30.

24.285. Ferrario, A., and Sironi, A.: "Metal-Tube Plasma Container for Argon-Ion Lasers", Rev. Sci. Instrum., 1972, 43, no.8, pp.1216-8.

24.286. Gruzinskii, V.V., and Stratskevich, L.K.: "Effect of Impulse Magnetic Field on Power of a He-Ne Laser at 1150 nm", Zh. Prikl. Spekt., 1972, 16, no.6, pp.978-84.

24.287. Robinson, A.M.: "Effect of Capacitance on Gain in a Transversely Pulsed CO_2 Discharge", Can. J. Phys., 1972, 50, no.18, pp.2138-48.

24.288. Ames, H.S.: "Effects of Axial Magnetic Field on Output Power of a Pulsed Xenon Laser", J. Quantum Electron. IEEE, 1972, QE-8, no.10, pp.808-9.

24.289. Maitland, A., and Cornish, J.C.L.: "Electron Temperatures in a Segmented-Metal Argon-Ion Laser", J. Phys. D, 1972, 5, no.10, pp.1807-14.

24.290. Goncharuk, I.N., et al.: "250-mW He-Ne Laser With a Mercury Cathode", Opt. Spekt., 1972, 32, no.2, pp.427-8, and Opt. Spectrosc., 1972, 32, no.2, pp.223-4.

24.291. Shimauchi, A., et al.: "Effect of Axial Magnetic Field on Mode Spectrum and Output Power of a He-Ne Laser", Sci. Light, 1972, 21, no.1, pp.22-43.

24.292. Bjorkholm, J.E., and Shank, C.V.: "Distributed-Feedback Lasers in Thin-Film Optical Waveguides", J. Quantum Electron. IEEE, 1972, QE-8, no.11, pp.833-8.

24.293. Abrams, R.L.: "Coupling Losses in Hollow Waveguide Laser Resonators", J. Quantum Electron. IEEE, 1972, QE-8, no.11, pp.838-43.

24.294. Zeidler, G.: "Thermal Resonator Effects in Organic Dye Lasers", Z. Naturforsch., 1972, 27a, no.8-9, pp.1272-7.

24.295. Kogan, B.Ya., et al.: "Superluminescence and Stimulated Radiation Under Internal-Reflection Conditions", Zh. Eksp. Teor. Fiz. Pis'ma, 1972, 16, no.3, pp.144-7, and JETP Lett., 1972, 16, no.3, pp.100-1.

24.296. Rampton, D.T., and Gandhi, O.P.: "Performance Characteristics of a Helical TEA CO_2 Laser", Appl. Phys. Lett., 1972, 21, no.10, pp.457-60.

24.297. Brown, C.O., and Davis, J.W.: "Closed-Cycle Performance of a High-Power Electric-Discharge Laser", Appl. Phys. Lett., 1972, 21, no.10, pp.480-1.

24.298. Dyer, P.E., James, D.J., and Ramsden, S.A.: "Operational Characteristics of a Volume-Excited TEA CO_2 Laser", J. Phys. E, 1972, 5, no.12, pp.1162-4.

24.299. Sakurai, T.: "Dependence of He-Ne Laser Output Power on Discharge Current, Gas Pressure, and Tube Radius", Jap. J. Appl. Phys., 1972, 11, no.12, pp.1826-31.

24.300. Leont'ev, V.G., Ostapchenko, E.P., and Sedov, G.S.: "Optimum Conditions of He-Ne Laser in the TEM_{00} Axial Mode", Opt. Spekt., 1972, 32, no.4, pp.795-7, and Opt. Spectrosc., 1972, 32, no.4, pp.418-9.

24.301. Leont'ev, V.G., Ostapchenko, E.P., and Sedov, G.S.: "He-Ne Laser With Metallic Internal Wall Surface", Opt. Spekt., 1972, 32, no.4, pp.798-801, and Opt. Spectrosc., 1972, 32, no.4, pp.420-1.

24.302. Agofitei, A., et al.: "(Rotating-Prism) Q-Switched Nd^{3+}:Glass Laser", Rev. Roum. Phys., 1972, 17, no.8, pp.1001-4.

24.303. Bykovskii, Yu.A., et al.: "Method of Increasing Radiation Coherence of a Pulsed Semiconductor Laser", Opt. Spekt., 1972, 32, no.3, pp.621-4, and Opt. Spectrosc., 1972, 32, no.3, p.331.

24.304. Basting, D., and Steyer, B.: "Gain Measurements in a N_2 Laser Amplifier", Z. Naturforsch., 1972, 27a, no.10, pp.1517-8.

24.305. Vasil'ev, A.M., and Loginov, A.V.: "Laboratory Argon-Ion Laser With BeO Discharge Tube", Avtometriya, 1972, no.5, pp.125-6.

24.306. Belanger, P.A., et al.: "Atmospheric Pressure CO_2 Pulsed Laser with Semiconducting Plastic Electrodes", Can. J. Phys., 1972, 50, no.22, pp.2753-6.

24.307. Bykovskii, Yu.A., et al.: "Heating of Semiconductor Injection Pulse Lasers", Prib. Tekh. Eksp., 1972, 15, no.3,pp.208-10.

24.308. Loiko, M.M., and Rubinov, A.N.: "Optical Elements of Lasers Based on Dyes with Laser Pumping", Prib. Tekh. Eksp., 1972, 15, no.3, pp.201-2, and Instrum. Exp. Tech., 1972, 15, no.3, pp.854-6.

24.309. Donin, V.I.: "Output-Power Saturation With Discharge Current in Powerful CW Argon Lasers", Zh. Eksp. Teor. Fiz., 1972, 62, no.5, pp.1648-60, and Sov. Phys.-JETP, 1972, 35, no.5, pp.858-64.

24.310. Juyal, D.P.: "Design of RF-Excited He-Ne Gas Laser", Def. Sci. J., 1972, 22, no.4, pp.245-8.

24.311. Chakravarti, A.N., and Parui, D.P.: "Effect of Longitudinal Magnetic Field on InSb Lasers at Low Temperatures", Indian J. Pure Appl. Phys., 1972, 10, no.9,pp.686-9.

24.312. Dukhovnyi, A.M., et al.: "Controllable Single-Pulse Nd^{3+}:Glass Laser", Opt. Spekt., 1972, 33, no.4, pp.783-5, and Opt. Spectrosc., 1972, 33, no.4, pp.403-5.

24.313. Zhitkova, M.B., et al.: "CW Laser With Vortex-Stabilized Lamp", Kvantovaya Elektron., 1972, no.3, pp.24-9, and Sov. J. Quantum Electron., 1972, 2, no.3, pp.215-8.

24.314. Kuzin, V.E., and Suchkov, A.F.: "Operation of a Laser With a Planar Resonator at High Pumping Levels", Kvantovaya Elektron., 1972, no.3, pp.53-8, and Sov. J. Quantum Electron., 1972, 2, no.3, pp.236-9.

24.315. Gibadullin, N.S., Kulakov, B.P., and Nurmukhametov, V.K.: "Analysis of Limiting Amplification Characteristics of a Superregenerative Gas Laser", Radiotekh. Elektron., 1972, 17, no.7, pp.1439-44, and Radio Eng. Electron. Phys., 1972, 17, no.7, pp.1127-31.

24.316. Anan'ev, Yu.A., et al.: "Laser With Hollow Pumping Lamp", Opt.-Mekh. Prom., 1972, 39, no.9, p.35, and Sov. J. Opt. Technol., 1972, 39, no.9, pp.546-7.

24.317. Bokhan, P.A.: "Sealed CO_2 Laser With Long Life", Prib. Tekh. Eksp., 1972, 15, no.6, pp.166-7, and Instrum. Exp. Eksp., 1972, 15, no.6, pp.1807-8.

24.318. Arai, T., Iwase, T., and Yabumoto, T.: "Effect of Transverse Magnetic Field on Electron Characteristics and Output Power of He-Ne Laser", Res. Rep. Fac. Eng. Meiji Univ., 1972, no.26-27, pp.289-96.

24.319. Eliseev, P.G., Morozov, D.N., and Fedorov, Yu.F.: "Statistical Distribution of Failure of Injection Lasers", Kvantovaya Elektron., 1972, no.3, p.107, and Sov. J. Quantum Electron., 1972, 2, no.3, pp.292-3.

24.320. Borodulin, V.I., et al.: "Aspects of Degradation of Heterojunction Lasers", Kvantovaya Elektron., 1972, no.3, pp.108-10, and Sov. J. Quantum Electron., 1972, 2, no.3, pp.294-6.

24.321. Tychinskaya, M.P., et al.: "Thermal Deformation of Injection-Laser Crystal During Pulse Pumping", Kvantovaya Elektron., 1972, no.4, pp.101-3, and Sov. J. Quantum Electron., 1973, 2, no.4, pp.376-7.

24.322. Goncharuk, I.N., Davydov, V.Yu., and Chisler, E.V.: "High-Current Gas Lasers With Mercury Cathode", Prib. Tekh. Eksp., 1972, 15, no.5, pp.189-91, and Instrum. Exp. Tech., 1972, 15, no.5, pp.1512-4.

24.323. Thomas-Andraud, M., et al.: "High-Pressure Gasdynamic Laser Powered by a Slow-Compression Heater", Rech. Aerosp., 1972, no.6, pp.325-32.

24.324. Luthi, H.R., and Seelig, W.H.: "Radial-Gain Profile of 488-nm A^{II} Laser and Distribution of Charge Carriers in a Wall-Stabilized Low-Pressure Arc", Z. Angew. Math. Phys., 1972, 23, no.4, pp.665-72.

24.325. Rubinov, A.N., and Asimov, M.M.: "Time Dependence of Gain of an Optically Pumped Solution of Rhodamine 6G", Kvantovaya Elektron., 1972, no.2, pp.108-10, and Sov. J. Quantum Electron., 1972, 2, no.2, pp.185-7.

24.326. Isaev, S.K., et al.: "Investigation of Delay of Emission from Dy^{2+}:CaF_2 Laser Relative to Pumping Pulses", Kvantovaya Elektron., 1972, no.4, pp.48-54, and Sov. J. Quantum Electron., 1973, 2, no.4, pp.335-8.

24.327. Lachambre, J.L., et al.: "Performance Characteristics of a TEA Double-Discharge Grid Amplifier", J. Quantum Electron. IEEE, 1973, QE-9, no.4, pp.459-68.

24.328. Gasperson, L.W., and Romero, C.: "Properties of a Radial-Mode CO_2 Laser", J. Quantum Electron. IEEE, 1973, QE-9, no.4, pp.484-8.

24.329. Degnan, J.J., et al.: "Gain and Saturation Intensity Measurements in a Waveguide CO_2 Laser", J. Quantum Electron. IEEE, 1973, QE-9, no.4, pp.489-91.

24.330. Richardson, M.C., et al.: "300-J Multigigawatt CO_2 Laser", J. Quantum Electron IEEE, 1973, QE-9, no.2, pp.236-43.

24.331. Lubin, M.J., Sources, J.M., and Goldman, L.M.: "Large-Aperture Nd^{3+}:Glass Laser Amplifier for High-Peak-Power Application", J. Appl. Phys., 1973, 44, no.1, pp.347-50.

24.332. Uchida, T., et al.: "CW Oscillation and Amplification in Light-Focusing Glass Lasers", Jap. J. Appl. Phys., 1973, 12, no.1, pp.126-34.

24.333. Figueira, J.F., et al.: "Nanosecond Pulse Amplification in Electron-Beam-Pumped CO_2 Lasers", Appl. Phys. Lett., 1973, 22, no.5, pp.216-8.

24.334. Rothhardt, L., Rentsch, M., and Jahn, G.: "Pulsed Noble-Gas Ion Laser", Exp. Tech. Phys., 1973, 21, no.1, pp.15-20.

24.335. Hill, K.O., and Watanabe, A.: "Passive-Core Corrugated-Waveguide Laser", Appl. Opt., 1973, 12, no.2, pp.430-3.

24.336. Bjorkholm, J.E., Sosnowski, T.P., and Shank, C.V.: "Distributed-Feedback Lasers in Optical Waveguides Deposited on Anisotropic Substrates", Appl. Phys. Lett., 1973, 22, no.4, pp.132-4.

24.337. Kogelnik, H., Shank, C.V., and Bjorkholm, J.E.: "Hybrid Scattering in Periodic Waveguides (for Lasers)", Appl. Phys. Lett., 1973, 22, no.4, pp.135-7.

24.338. Nagai, H., and Taniguchi, I.: "Simple, Single-Frequency, He-Ne Laser", Jap. J. Appl. Phys., 1973,12,no.3,pp.434-8.

24.339. Bina, M.J., and Jones, C.R.: "Temperature-Induced Wavelength Tuning and Anomalous Behaviour of a 1.35-micron Nd^{3+}: $YAlO_3$ Laser", J. Opt. Soc. Am., 1973, 63, no.4, pp.463-4.

24.340. Matyushin, G.A., Byalko, N.G., and Tolkachev, B.V.: "Influence of the Spectral Characteristics of Liquid Filters on Thermal Properties of a Nd^{3+} Laser", Zh. Prikl. Spekt., 1973, 18, no.1, pp.142-4.

24.341. Smith, B.: "Laser Brewster's-Window Sealing Techniques", J. Quantum Electron. IEEE, 1973, QE-9, no.5, pp.546-8.

24.342. Chang, T.Y.: "Improved Uniform-Field Electrode Profiles for TEA Laser", Rev. Sci. Instrum., 1973, 44, no.4,pp.405-7.

24.343. Chodsko, R.A., Spencer, D.J., and Mirels, H.: "Zero-Power Gain Measurements in a CW HF Laser Using a Pulse-Probe Laser", J. Quantum Electron. IEEE, 1973, QE-9, no.5, pp.550-3.

24.344. Oettinger, P.E.: "Parametric Studies on CO_2 TEA Lasers With Extra- and Intra-Cavity Electrodes", Appl. Phys. Lett., 1973, 22, no.9, pp.465-6.

24.345. Kast, S.J., and Cason, C.: "Performance Comparison of Pulsed-Discharge and Electron-Beam-Controlled CO_2 Lasers", J. Appl. Phys., 1973, 44, no.4, pp.1631-7.

24.346. Dyer, P.E., James, D.J., and Ramsden, S.A.: "Time-Resolved Gain of a Volume-Excited TEA CO_2 Laser Amplifier", J. Appl. Phys., 1973, 44, no.5, pp.2408-10.

24.347. Burlamacchi, P.: "Waveguide Super-radiant Dye Laser", Appl. Phys. Lett., 1973, 22, no.7, pp.334-5.

24.348. Aoyagi, Y., and Namba, S.: "Temperature Tuning of 4-Methylumbelliferone Dye Laser", Jap. J. Appl. Phys., 1973, 12, no.4, pp.624-5.

24.349. Wang, S., and Sheem, S.: "Two-Dimensional Distributed-Feedback Lasers and Applications", Appl. Phys. Lett., 1973, 22, no.9, pp.460-2.

24.350. Milam, D., and Schlossberg, H.: "Emission Characteristics of a Tube-Shaped Laser Oscillator", J. Appl. Phys., 1973, 44, no.5, pp.2297-9.

24.351. Fesquet, J., et al.: "Excitation Period of Ruby Laser with Plane Mirrors", C. R. Acad. Sci. (Paris), 1973, 276B, no.14, pp.595-8.

24.352. Siegrist, M., Adam, B., and Kneubuhl, F.K.: "Pulse Delays of TEA-CO_2-Laser Emissions", Helv. Phys. Acta, 1973, 46, no.1, pp.33-4.

24.353. Krylov, K.I., Baloshin, Yu.A., and Aver'yanov, N.E.: "Effect of Coupling Aperture (Diameter) of a CO_2 Gas Laser on Output Power", Izv. VUZ Prib., 1973, 16, no.3, pp.122-7.

24.354. Landry, M.J.: "Pulsed Ruby Laser with Individually Q-Switched Multiple Cavities", J. Quantum Electron. IEEE, 1973, QE-9, no.6, pp.604-6.

24.355. Jones, C.R.: "Gain and Energy Measurements on an HF/DF Electrically Pulsed Chemical Laser", Appl. Phys. Lett., 1973, 22, no.12, pp.653-5.

24.356. Kasuya, K., Kumusaka, T., and Murasaki, T.: "Output-Power Characteristics of a CW Gasdynamic Laser", Jap. J. Appl. Phys., 1973, 12, no.5, pp.771-2.

24.357. Girard, A., and Pepin, H.: "Performance Characteristics of a TEM_{00} Mode TEA CO_2 Laser", Opt. Commun., 1973, 8, no.1, pp.68-72.

24.358. Carlsten, J.L., and McIlrath, T.J.: "Dye Laser. Tunable High Powers Without Grating Damage", Opt. Commun., 1973, 8, no.1, pp.52-5.

24.359. Johns, T.W., and Nation, J.A.: "Resistive Electrode, High Energy, Transverse Laser Discharge", Rev. Sci. Instrum., 1973, 44, no.2, pp.169-71.

24.360. Zory, P.: "Laser Oscillation in Leaky Corrugated Waveguides", Appl. Phys. Lett., 1973, 22, no.4, pp.125-8.

24.361. Lee, T.P., Burrus, C.A., and Miller, B.I.: "Stripe-Geometry Double-Heterostructure (Superradiant)Diode", J. Quantum Electron. IEEE, 1973, QE-9, no.8, pp.820-8.

24.362. De Loach, B.C., et al.: "Degradation of CW GaAs Double-Heterojunction Lasers at 300°K", Proc. IEEE, 1973, 61, no.7, pp.1042-4.

24.363. Breton, J.: "Parameters of a Pulsed He-Kr^{II} Laser", J. Quantum Electron. IEEE, 1973, QE-9, no.8, pp.854-5.

24.364. Woodward, B.W., Ehlers, V.J., and Lineberger, W.C.: "Reliable Repetitively Pulsed High-Power N_2 Laser", Rev. Sci. Instrum., 1973, 44, no.7, pp.882-7.

24.365. Yang, L.C., and Menichelli, V.J.: "High-Efficiency, Small, Solid-State Laser for Pyrotechnic Ignition", JPL Q. Tech. Rev., 1973, 2, no.4, pp.29-37.

24.366. Champagne, L.F.: "Cryogenically Cooled He-CO TEA Laser", Appl. Phys. Lett., 1973, 23, no.3, pp.158-60.

24.367. Richardson, M.C., Leopold, K., and Alcock, A.J.: "Large-Aperture CO_2 Laser Discharges", J. Quantum Electron. IEEE, 1973, QE-9, no.9, pp.934-9.

24.368. Abrams, R.L., and Bridges, W.B.: "Characteristics of Sealed-Off Waveguide CO_2 Lasers", J. Quantum Electron. IEEE, 1973, QE-9, no.9, pp.940-6.

24.369. Razmadze, N.A., Chkuaseli, Z.D., and Butov, I.Ya.: "Argon-Ion Laser With Gas-Discharge Tube of Comparitively Large Diameter", Zh. Prikl. Spekt., 1973, 18, no.5, pp.793-7.

24.370. Hotz, R.F.: "Thermal Transient Effects in Repetitively Pulsed Flashlamp-Pumped Nd^{3+}:YAG and $Nd^{3+}Lu^{3+}$:YAG Laser Material", Appl. Opt. 1973, 12, no.8, pp.1834-8.

24.371. McMahon, J.M., et al.: "Glass-Disc Laser Amplifier", J. Quantum Electron. IEEE, 1973, QE-9, no.10, pp.992-9.

24.372. Koechner, W.: "Transient Thermal Profile in Optically Pumped Laser Rods", J. Appl. Phys., 1973, 44, no.7, pp.3162-70.

24.373. Cornish, J.C.L., and Maitland, A.: "Demountable Argon-Ion Laser of All-Metal Construction", J. Phys. E, 1973, 6, no.9, pp.80-4.

24.374. White, M.S., and Dangor, A.E.: "Parametric Study of Performance of a TEA CO_2 Laser", J. Phys. E, 1973, 6, no.9, pp.891-4.

24.375. Fahlen, T.S.: "CO_2 Laser Design Procedure", Appl. Opt., 1973, 12, no.10, pp.2381-90.

24.376. Brown, R.T.: "High-PRF Effects in TEA Lasers", J. Quantum Electron. IEEE, 1973, QE-9, no.11, pp.1120-2.

24.377. Dote, T., Yamaguchi, N., and Nakamura, T.: "Effects of RF Electric Field on He-Ne Laser Output", Phys. Lett., 1973, 45A, no.1, pp.29-30.

24.378. Antcliffe, G.A., and Parker, S.G.: "Characteristics of Tunable (PbSn)Te Junction Lasers in the 8-12 micron Region", J. Appl. Phys., 1973, 44, no.9, pp.4145-60.

24.379. Patel, B.S., Mallik, A., and Charan, S.: "Brewster's-Window Losses in 633-nm He-Ne Lasers", J. Phys. E, 1973, 6, no.10, pp.1014-6.

24.380. Milovskii, N.D.: "Frequency-Locking Band of a TW Laser", Kvantovaya Elektron., 1973, no.6, pp.96-102, and Sov. J. Quantum Electron., 1973, 2, no.6, pp.559-62.

24.381. Sparks, M.: "Stress and Temperature Analysis for Surface Cooling or Heating of Laser Window Materials", J. Appl. Phys., 1973, 44, no.9, pp.4137-44.

24.382. Denes, L.J., and Weaver, L.A.: "Laser Gain Characteristics of Near-Atmospheric CO_2-N_2-He Glows in a Planar Electrode Geometry", J. Appl. Phys., 1973, 44, no.9, pp.4125-36.

24.383. Hakki, B.W., and Paoli, T.L.: "CW Degradation at 300°K of GaAs Double-Heterostructure Junction Lasers. II", J. Appl. Phys., 1973, 44, no.9, pp.4113-9.

24.384. Sato, K., and Sekiguchi, T.: "Dependence of Output Power of CO_2 Gasdynamic Laser on the Distance from Nozzle Throat", J. Phys. Soc. Jap., 1973, 35, no.2, p.630.

24.385. Glaros, S., Manes, K.R., and Nilson, J.: "Characterization of a Large-Aperture CO_2 Amplifier", J. Quantum Electron. IEEE, 1973, QE-9, no.11, pp.1122-3.

24.386. Gundersen, M., and Harper, C.D.: "High-Power Pulsed Xenon-Ion Laser", J. Quantum Electron. IEEE, 1973, QE-9, no.12, p.1160.

24.387. Marcus, S., et al.: "Variable-Pulse-Duration Electron-Beam CO_2 Laser", J. Appl. Phys., 1973, 44, no.9, pp.4232-3.

24.388. Yonezu, H., et al.: "(GaAl)As-GaAs Double-Heterostructure Planar Stripe Laser", Jap. J. Appl. Phys., 1973, 12, no.10, pp.1485-92.

24.389. Parmentier, E.M., and Greenberg, R.A.: "Supersonic Flow Aerodynamic Windows for High-Power Lasers", AIAA J., 1973, 11, no.7, pp.943-9.

24.390. Charan, S., Patel, B.S., and Swarup, P.: "Construction of a Pulsed Argon-Ion Laser", Indian J. Pure Appl. Phys., 1973, 11, no.4, p.300.

24.391. Leshenyuk, N.S., and Orlov, L.N.: "Increasing the Service Life of a CO_2 Laser", Prib. Tekh. Eksp., 1973, 16, no.2, pp.175-6, and Instrum. Exp. Tech., 1973, 16, no.2, p.539.

24.392. Berezovskii, V.V.: "Characteristics of Ge and Te Mirrors for the Cavity of a CO_2 Laser", Prib. Tekh. Eksp., 1973, 16, no.2, pp.176-7, and Instrum. Exp. Tech., 1973, 16, no.2, p.540.

24.393. Thompson, A.G.: "GaAs for Laser-Window Applications", J. Electron. Mater., 1973, 2, no.1, pp.47-70.

24.394. Takusagawa, M., et al.: "Internally Striped Planar Laser with 3-micron Stripe Width Oscillating in Transverse Single Mode", Proc. IEEE, 1973, 61, no.12, pp.1758-9.

24.395. Gembarzhevskii, G.V., Generalov, N.A., and Kozlov, G.I.: "Experimental Investigation of Gain in $CO_2+N_2+He(H_2O)$ Expanding in a Supersonic Nozzle", Zh. Prikl. Mekh. Tekh. Fiz., 1973, no.4, pp.18-24.

24.396. Tsarkov, V.A., and Molchanov, M.I.: "Measurement of Gain Distribution in the Tube of a He-Ne, 633-nm, Laser Under RF Excitation", Opt. Spekt., 1973, 35, no.2, pp.328-9, and Opt. Spectrosc., 1973, 35, no.2, p.191.

24.397. Hernqvist, K.G.: "Vented-Bore He-Cd Lasers", RCA Rev., 1973, 34, no.3, pp.401-7.

24.398. Andriyakhin, V.M., et al.: "Nuclear Pumping in Molecular Gas Lasers", Zh. Eksp. Teor. Fiz., 1973, 63, no.5, pp.1635-44, and Sov. Phys.-JETP, 1973, 36, no.5, pp.865-9.

24.399. Matoba, M., Sasaki, M., and Yamanaka, C.: "High-Circulating-Flow CW CO_2 Laser", Technol. Rep. Osaka Univ., 1973, 23, no.1121-54, pp.443-53.

24.400. Terletskii, A.Ya.: "Effect of External Magnetic Field on a Kr^{II} Laser", Vestn. Mosk. Univ. Fiz. Astron., 1973, 14, no.4, pp.496-8.

24.401. Zuev, M.G., Shalyapin, A.L., and Gravrilov, F.F.: "Measuring Technique of Cavity Losses in Solid-State Lasers", Opt. Spekt., 1973, 34, no.4, pp.797-8, and Opt. Spectrosc., 1973, 34, no.4, pp.459-60.

24.402. Wang, C.P., and Lin, S.C.: "Performance of Large-Bore High-Power Argon-Ion Laser", J. Appl. Phys., 1973, 44, no.10, pp.4681-2.

24.403. Mazan'ko, I.P., Ogurok, N.D.D., and Sviridov, M.V.: "Measurement of Saturation Parameter of a He-Ne Mixture at 3.39 micron", Opt. Spekt., 1973, 35, no.3, pp.563-4, and Opt. Spectrosc., 1973, 35, no.3, pp.327-8.

24.404. Apostol, I., et al.: "High-Reliability, High-Efficiency, TEA CO_2 Laser With Xylene Vapour Addition", Rev. Roum. Phys., 1973, 18, no.10, pp.1185-8.

24.405. Ciura, A.I., and Popescu, I.M.: "Temperature and Electron-Density Measurements in He-Cd Laser with Cataphoretic Excitation", Stud. Cercet. Fiz., 1973, 25, no.8, pp.897-900.

24.406. Gleason, T.J., Kruger, J.S., and Curnutt, R.M.: "Thermally Induced Focusing in Nd^{3+}:YAG Laser Rod at Low Input Powers", Appl. Opt., 1973, 12, no.12, pp.2942-6.

24.407. Smith, P.W., Maloney, P.J., and Wood, O.R.: "Waveguide TEA (CO_2) Laser", Appl. Phys. Lett., 1973, 23, no.9, pp.524-6.

24.408. Lebedeva, L.M., et al.: "Effect of Temperature on Power of He-Ne Laser", Zh. Prikl. Spekt., 1973, 19, no.5, pp.911-3.

24.409. Varnado, S.G.: "Degradation in Long-Pulse Dye Laser Emission Under Fast-Flow Conditions", J. Appl. Phys., 1973, 44, no.11, pp.5067-8.

24.410. Gorban', I.S., and Kononchuk, G.L.: "Electronic Component of Change in Refractive Index of Ruby during Pumping", Kvantovaya Elektron., 1973, no.7, pp.123-9.

24.411. Stussi, E., Weber, H., and Keller, H.: "Measurement of Lens Action in Optically Pumped Nd^{3+}:Glass Rods", Z. Angew. Math. Phys., 1973, 24, no.3, p.451.

24.412. Chinn, S.R., Rossi, J.A., and Wolfe, C.M.: "Temperature Dependence of the Lasing Transition in High-Purity GaAs", Appl. Phys. Lett., 1973, 23, no.12, pp.699-701.

24.413. Hakki, B.W.: "Carrier and Gain Spatial Profiles in Stripe-Geometry Lasers", J. Appl. Phys., 1973, 44, no.11, pp.5021-8.

24.414. Zverev, V.A., et al.: "Power and Aberration Characteristics of Multiple-Pass Laser Amplifiers", Opt.-Mekh. Prom., 1973, 40, no.6, pp.25-7, and Sov. J. Opt. Technol., 1973, 40, no.6, pp.359-61.

24.415. Boiko, B.B., et al.: "Use of Adjustable Telescopic System in Laser Resonators (to Overcome Lens Effect)", Zh. Prikl. Spekt., 1973, 19, no.5, pp.808-11.

24.416. Freiberg, R.J., Chenausky, P.P., and Buczek, C.J.: "Unidirectional Unstable Ring Lasers", Appl. Opt., 1973, 12, no.6, pp.1140-4.

24.417. Patel, B.S.: "Power Coupling from CO_2 Laser by Rotatable Reflector", Appl. Opt., 1973, 12, no.5, pp.943-5.

24.418. Susaki, W., et al.: "(Stripe-) Geometry Double Heterostructure Injection Laser for Room-Temperature CW Operation", J. Appl. Phys., 1973, 44, no.6, pp.2893-4.

24.419. Gromov, Yu.N., Khaikin, N.Sh., and Yurist, B.V.: "Dependence of Power from Encapsulated CO_2 Laser on Wall Temperature", Prib. Tekh. Eksp., 1973, 16, no.4, pp.204-6, and Instrum. Exp. Tech., 1973, 16, no.4, pp.1224-6.

24.420. Gorokhov, Yu.A., and Kompanets, O.N.: "TW Cooled CO_2 Amplifier Having Large Gain", Prib. Tekh. Eksp., 1973, 16, no.4, pp.207-8, and Instrum. Exp. Tech., 1973, 16, no.4, pp.1227-8.

24.421. Koechner, W.: "Transient Thermal Profile in Optically Pumped Laser Rods", J. Vac. Sci. Technol., 1973, 10, no.6, p.1130.

24.422. Alferov, G.N., Donin, V.I., and Yurshin, B.Ya.: "CW Argon Laser With 0.5-kW Output Power", Zh. Eksp. Teor. Fiz. Pis'ma, 1973, 18, no.10, pp.629-31, and JETP Lett., 1973, 18, no.10, pp.369-70.

24.423. Basov, N.G., Belenov, E.M., et al.: "Gain of a CO_2 ($-N_2$) Electric Ionization Laser", Kvantovaya Elektron., 1973, no.3, pp.46-50, and Sov. J. Quantum Electron., 1973, 3, no.3, pp.205-7.

24.424. Zaroslov, D.Yu., et al.: "Shape and Energy of Pulses Emitted by a CO_2 Laser With Helical Distribution of Electrodes", Kvantovaya Elektron., 1973, no.3, pp.116-9, and Sov. J. Quantum Electron., 1973, 3, pp.257-9.

24.425. Masda, I.I., et al.: "High-Power Nitrogen and Neon Pulsed Gas Lasers", Kvantovaya Elektron., 1973, no.3, pp.119-22, and Sov. J. Quantum Electron., 1973, 3, no.3, pp.260-1.

24.426. Johnson, L.F., and Ingersoll, K.A.: "Elimination of Degradation in the Laser Output from Ho^{3+}:YAG (by Filter)", J. Appl. Phys., 1973, 44, no.12, pp.5444-6.

24.427. Parui, D.P., and Chakravarti, A.N.: "Temperature Dependence of Threshold Current in Electron-Beam-Pumped Lasers Using n-GaAs", Indian J. Pure Appl. Phys., 1973, 11, no.8, pp.588-90.

24.428. Kornienko, L.S., Kravtsov, N.V., and Shelaev, A.N.: "Characteristics of a CW Solid-State Ring Laser", Opt. Spekt., 1973, 35, no.4, pp.775-6, and Opt. Spectrosc., 1973, 35, no.4, p.449.

24.429. Fromm, D., and Schamberger, H.: "Compact He-Ne Laser", Tech.-Wiss. Abh. Osram-Ges., 1973, 11, pp.206-14.

24.430. Fraas, L.M.: "Injection-Laser-Pumped Dye Laser Tunable in the IR", Rev. Bras. Fis., 1973, 3, no.3, pp.425-30.

24.431. Chis, I., et al.: "Parametric Measurements on a CO_2 TEA Laser With Rogowsky Electrodes", Stud. Cercet. Fiz., 1973, 25, no.7, pp.875-8.

24.432. Orishich, A.M., et al.: "Power Characteristics and Stability of Double-Transverse Discharge when Pumping a CO_2 Laser", Dokl. Akad. Nauk SSSR, 1973, 212, no.4-6, pp.1099-102, and Sov. Phys.-Dokl., 1974, 18, no.10, pp.671-2.

24.433. Baranova, N.B., et al.: "Possibility of Construction of High-Power Laser With Funnel-Shaped Active Media", Kvantovaya Elektron., 1973, no.5, pp.57-67, and Sov. J. Quantum Electron., 1974, 3, no.5, pp.398-404.

24.434. Mak, A.A., and Shcherbakov, A.A.: "Method for Thermodynamic Design of Laser Pumping Systems", Kvantovaya Elektron., 1973, no.5, pp.68-76, and Sov. J. Quantum Electron., 1974, 3, no.5, pp.405-9.

24.435. Kovsh, I.V., et al.: "Electrically Driven 200-J CO_2 Laser", Zh. Tekh. Fiz., 1973, 43, no.11, pp.2357-63, and Sov. Phys.-Tech. Phys., 1974, 18, no.11, pp.1488-91.

24.436. Privalov, V.E., and Khadovoi, V.A.: "He-Ne Laser With Discharge Gap of Rectangular Cross Section", Opt. Spekt., 1974, 37, no.4, pp.797-9, and Opt. Spectrosc., 1974, 37, no.4, pp.455-7.

24.437. Stein, A.: "Thermooptically Perturbed Resonators", J. Quantum Electron. IEEE, 1974, QE-10, no.4, pp.427-34.

24.438. Shubert, R.: "Theory of Optical Waveguide Distributed Lasers With Nonuniform Gain and Coupling", J. Appl. Phys., 1974, 45, no.1, pp.209-15.

24.439. Yonezu, H., et al.: "Degradation Mechanism of (GaAl)As Double-Heterostructure Laser Diodes", Appl. Phys. Lett., 1974, 24, no.1, pp.18-9.

24.440. Draegert, D.A.: "Selection and Stabilization of Longitudinal Modes by Spacing of Laser Cavity Components", J. Quantum Electron. IEEE, 1974, QE-10, no.5, pp.476-9.

24.441. Vanderleeden, J.C.: "Resonant Cavities With Mirrors made from Graded-Index Rods", J. Appl. Phys., 1974, 45, no.1, pp.201-8.

24.442. Belousov, P.Ya., Koronkevich, V.P., and Nagornyi, V.N.: "Small-Scale Isotropic He-Ne 633-nm Laser with External Mirrors", Avtometriya, 1974, no.1, pp.80-3.

24.443. Lokhmatov, A.I., Sergeeva, N.S., and Shatalov, V.A.: "Small-Scale He-Ne Laser With Life Exceeding 5000h", Avtometriya, 1974, no.1, pp.99-101.

24.444. Asawa, C.K.: "Compact 1.1-W Sealed-Off Waveguide CO Laser", Appl. Phys. Lett., 1974, 24, no.3, pp.121-3.

24.445. Degnan, J.L.: "Phenomenological Approach to the Design of Highly Tunable Pressure-Broadened Gas Lasers", J. Appl. Phys., 1974, 45, no.1, pp.257-62.

24.446. Fukuda, S., and Miya, M.: "Metal-Ceramic He-Cd^{II} Laser With Sectioned Hollow Cathode Giving Simultaneous Oscillations", Jap. J. Appl. Phys., 1974, 13, no.4, pp.667-74.

24.447. Pan, Y.L., et al.: "Generation of Multigigawatt Nanosecond CO_2 Pulse", J. Quantum Electron. IEEE, 1974, QE-10, no.1, pp.44-8.

24.448. Ahlstrom, H.G., et al.: "Results of Time-Dependent Gain with High-Current Electron-Beam-Sustained Discharge in CO_2", J. Quantum Electron. IEEE, 1974, QE-10, no.1, pp.26-9.

24.449. Deutsch, T.F.: "Gain Measurements on Uniformly Excited HF/DF TEA Lasers", J. Quantum Electron. IEEE, 1974, QE-10, no.1, pp.84-6.

24.450. Brown, D.C., and Swift, T.M.: "Investigation of Effect of Axial Magnetic Field on He-Cd^{II} Laser", J. Quantum Electron. IEEE, 1974, QE-10, no.1, pp.94-5.

24.451. Otis, G., and Tremblay, R.: "Lensing Effect in Helical TEA Lasers", Can. J. Phys., 1974, 52, no.3, pp.257-64.

24.452. Afonin, Yu.V., et al.: "Gain and Power Characteristics of Electron-Beam-Controlled Discharge TEA CO_2 Laser", Opt. Commun., 1974, 10, no.1, pp.11-3.

24.453. Danielmeyer, H.G., and Leibolt, W.N.: "Stable Tunable Single-Frequency Nd^{3+}:YAG Laser", Appl. Phys., 1974, 3, no.3, pp.193-8.

24.454. Richter, K.R., and Koechner, W.: "Electrical Analogy of Transient Heat Flow in Laser Rods", Appl. Phys., 1974, 3, no.3, pp.205-12.

24.455. Patel, B.S.: "Estimation of Output Power and Optimum Transmittance through a Coupling-Out Hole for a CW CO_2 Laser", Appl. Opt., 1974, 13, no.1, pp.19-21.

24.456. Rudko, R.I.: "Pulsed Waveguide Laser", J. Quantum Electron. IEEE, 1974, QE-10, no.5, pp.497-8.

24.457. Belomestnov, P.I., et al.: "Use of Induced-Flow, Glowing-Gas, Discharge in a CO_2 Laser in a Closed Loop with Convection Cooling", Zh. Prikl. Mekh. Tekh. Fiz., 1974, no.1, pp.4-12.

24.458. Apollonov, V.V., Barchukov, A.I., and Prokhorov, A.M.: "Optical Distortion of Heated Mirrors in CO_2-Laser Systems", J. Quantum Electron. IEEE, 1974, QE-10, no.6, pp.505-8.

24.459. Witte, H.H.: "Designing Unstable Optical Resonators with High Power Output and Satisfactory Mode Discrimination", Siemens Forsch. Entwick., 1974, 3,no.2,pp.65-9.

24.460. Mikhnov, S.A., and Matyushkov, V.E.: "Distortion of a Ruby Laser Resonator", Zh. Prikl. Spekt., 1974, 20, no.2,pp.195-8.

24.461. Sato, K., and Sekiguchi, T.: "Output Power Enhancement by Multireflection Cavity Scheme in CO_2 Gasdynamic Laser Using a Shock Tube", J. Phys. Soc. Jap., 1974, 36, no.2, p.621.

24.462. Vanderleeden, J.C.: "Two-Dimensional Arrays of Miniature Nd^{3+}:YAG Lasers", Opto-Electron., 1974, 6, no.3, pp.245-7.

24.463. Katsaros, V., and Petalas, P.: "Method for Pumping a Solid-State Laser by Optical Fibres", Opt. Acta, 1974, 21, no.6, pp.445-52.

24.464. Yonezu, H., et al.: "Lasing Characteristics in a Degraded GaAs-(AlGa)As Double-Heterostructure Laser", Jap. J. Appl. Phys., 1974, 13, no.5, pp.835-42.

24.465. Dube, G., and Boling, N.L.: "Liquid Cladding for Face-Pumped Nd^{3+}:Glass Lasers", Appl. Opt., 1974, 13,no.4,pp.699-700.

24.466. Campillo, A.J., et al.: "Soft Apertures for Reducing Damage to High-Power Laser-Amplifier Systems", Opt. Commun., 1974, 10, no.4, pp.313-5.

24.467. Brinkschulte, H., and Lang, R.: "Gain Measurements in a TEA CO_2 Laser", Phys. Lett., 1974, 47A, no.6, pp.455-6.

24.468. Ebert, W., and Redlich, L.: "High-Power CW Noble-Gas Ion Laser", Exp. Tech. Phys., 1974, 22, no.3, pp.197-207.

24.469. Harper, C.D., and Gundersen, M.: "Construction of a High-Power Xenon-Ion Laser", Rev. Sci. Instrum., 1974, 45, no.3, pp.400-2.

24.470. Fesquet, J., Irla, H., and Roig, J.: "Radial Distribution of Refractive Index and Gain in a Ruby Laser With Plane Mirrors", C. R. Acad. Sci. (Paris), 1974, 278B, no.21, pp.935-8.

24.471. Corcoran, V.J., McMillan, R.W., and Barnoski, S.K.: "Flashlamp-Pumped Nd^{3+}: YAG Laser Action at Kilohertz Rates", J. Quantum Electron. IEEE, 1974, QE-10, no.8, pp.618-20.

24.472. Farmer, G.I., and Kiang, Y.C.: "Low-Current-Density LED-Pumped Nd^{3+}:YAG Laser Using a Solid Cylindrical Reflector", J. Appl. Phys., 1974, 45, no.3, pp.1356-71.

24.473. Ralston, R.W., et al.: "High CW Output Power in Stripe Geometry PbS Diode Lasers", J. Appl. Phys., 1974, 45, pp.1323-5.

24.474. Andreeva, E.Yu., Gulyaev, S.N., and Terekhin, D.K.: "(Practical) Study of a 3.39-micron Laser Amplifier", Zh. Prikl. Spekt., 1974, 20, no.4, pp.713-5.

24.475. Banse, K., Luthi, H.R., and Seelig, W.H.: "Influence of Axial Magnetic Fields on High-Power A^{II} Lasers", Appl. Phys., 1974, 4, no.2, pp.141-5.

24.476. Figueira, J.F.: "Parametric Study of a UV Preionized CO_2 Laser", Opt. Commun., 1974, 11, no.3, pp.220-4.

24.477. Namizaki, H., et al.: "Transverse-Junction, Stripe-Geometry, Double-Heterostructure Lasers With Low Threshold Current", J. Appl. Phys., 1974, 45, no.6, pp.2785-6.

24.478. Osinski, M.: "Electromagnetic Theory of Injection Laser Radiation", Rozpr. Electrotech., 1974, 20, no.2, pp.331-58.

24.479. Richardson, M.C.: "Multiline Mode-Locked UV-Preionized CO_2 Laser", Appl. Phys. Lett., 1974, 25, no.1, pp.31-3.

24.480. Fotiadi, A.E., and Fridrikhov, S.A.: "LF Beats in the Emission of a CW Ar[II] Laser in a Magnetic Field", Opt. Spekt., 1974, 36, no.2, pp.430-2, and Opt. Spectrosc., 1974, 36, no.2, pp.247-8.

24.481. Baranov, V.Yu., et al.: "Pulsed CO_2 Laser With a Radiation Energy of 150J", Zh. Eksp. Teor. Fiz. Pis'ma, 1974, 19, no.4, pp.15-21, and JETP Lett., 1974, 19, no.4.

24.482. Velikhov, E.P., et al.: "Quasi-stationary Atmospheric Pressure CO_2 Laser With a Semi-Self-Maintained Discharge Controlled by an Electron Beam", Zh. Eksp. Teor. Fiz. Pis'ma, 1974, 19, no.6, pp.364-8, and JETP Lett., 1974, 19, no.6.

24.483. Akhmanov, S.A., and Lyakhov, G.A.: "Distributed Feedback in Lasers Using Non-stationary Pumping", Zh. Eksp. Teor. Fiz. Pis'ma, 1974, 19, no.7, pp.470-4, and JETP Lett., 1974, 19, no.7.

24.484. Rao, T.A.P., and Seetharaman, N.: "Mechanical Tuning of a Superradiant Dye Laser", Jap. J. Appl. Phys., 1974, 13, no.8, pp.1329-30.

24.485. Mikaelyan, A.L., et al.: "Determination of Laser-Radiation Parameters of a YAG Crystal", Kvantovaya Elektron., 1974, no.4, pp.13-9, and Sov. J. Quantum Electron., 1974, 3, no.4, pp.284-7.

24.486. Bubnov, M.M., et al.: "Enhancement of the Brightness of Nd^{3+}:Glass Lasers by Selection of the Active-Element Matrix", Kvantovaya Elektron., 1974, no.4, pp.113-5, and Sov. J. Quantum Electron., 1974, 3, no.4, pp.349-50.

24.487. Borisov, V.I., and Lebedev, V.I.: "Growth of Power in a Ruby Laser with Self-Q-Switching", Zh. Prikl. Spekt., 1974, 20, no.6, pp.987-9.

24.488. Belomestnov, P.I., et al.: "Tunable Resonator with Mirror of Variable Curvature", Kvantovaya Elektron., 1974, no.4, pp.110-3, and Sov. J. Quantum Electron., 1974, 3, no.4, pp.347-8.

24.489. Harris, N.W., O'Neill, F., and Whitney, W.T.: "Operation of a 15-atm. Electron-Beam-Controlled CO_2 Laser", Appl. Phys. Lett., 1974, 25, no.3, pp.148-51.

24.490. Lind, R.C., et al.: "Long-Pulse High-Energy CO_2 Laser Pumped by UV-Sustained Electric Discharge", J. Quantum Electron. IEEE, 1974, QE-10, no.10, pp.818-21.

24.491. Bradley, D.J., et al.: "Megawatt Vacuum-UV Xenon Laser Employing Coaxial Electron-Beam Excitation", Opt. Commun., 1974, 11, no.4, pp.335-8.

24.492. Nakada, O., et al.: "Continuous Operation Over 2500h of Double-Heterostructure Laser Diodes With Output Powers more than 80mW", Jap. J. Appl. Phys., 1974, 13, no.9, pp.1485-6.

24.493. Tanaka, A., et al.: "Optically Pumped Far-IR Waveguide Lasers", Jap. J. Appl. Phys., 1974, 13, no.9, pp.1491-2.

24.494. Hoag, E., et al.: "Performance Characteristics of a 10-kW Industrial CO_2 Laser System", Appl. Opt., 1974, 13, no.8, pp.1959-64.

24.495. Gerber, R.A., et al.: "Multikilo-joule HF Laser Using Intense-Electron-Beam Initiation of H_2-F_2 Mixtures", Appl. Phys. Lett., 1974, 25, no.5, pp.281-3.

24.496. Donnerhacke, K.H.: "Investigations of Sealed-Off CO_2 Lasers", Exp. Tech. Phys., 1974, 22, no.5, pp.415-24.

24.497. Idiatulin, V.S., and Uspenskii, A.V.: "Non-Uniform Population Inversion Effects in the Dynamics of a Single-Mode Solid-State Laser", Opt. Acta, 1974, 21, no.10, pp.773-82.

24.498. Forrester, P.A., Alexander, V.J., and Evans, H.W.: "Pocket-Size Nd^{3+}:YAG Pulsed Laser", Opt. Laser Technol., 1974, 6, no.4, pp.174-6.

24.499. Mikhalevich, V.G., and Shipulo, G.P.: "Characteristics of a High-Power Nd^{3+}:YAG CW Laser", Kvantovaya Elektron., 1974, 1, no.1, pp.129-33, and Sov. J. Quantum Electron., 1974, 4, no.1, pp.71-3.

24.500. Borodulin, V.I., et al.: "Injection Laser With an Average Output Power of 200mW", Kvantovaya Elektron., 1974, 1, no.1, pp.163-4, and Sov. J. Quantum Electron., 1974, 4, no.1, p.94.

24.501. Tarasenko, V.F., Fedorov, A.I., and Losev, V.F.: "Influence of Cavity Geometry on Power of a N_2 Laser", Kvantovaya Elektron., 1974, 1, no.1, pp.200-3, and Sov. J. Quantum Electron., 1974, 4, no.1, pp.123-4.

24.502. Litvinov, I.I., and Poduval'tsev, V.V.: "Design of a Selectively Reflecting Pump Source for a Laser", Kvantovaya Elektron., 1974, 1, no.1, pp.211-5, and Sov. J. Quantum Electron., 1974, 4, no.1, pp.131-3.

24.503. Mockel, P., Oberbacher, R., and Rauscher, W.: "Flashlamp-Pumped Nd^{3+}:YAG Waveguide Laser", J. Appl. Phys., 1974, 45, no.8, pp.3460-2.

24.504. Barnes, C.E.: "Increased (Neutron) Radiation Hardness of GaAs Laser Diodes at High Current Densities", J. Appl. Phys., 1974, 45, no.8, pp.3485-9.

24.505. Petroff, P., and Hartman, R.L.: "Rapid Degradation Phenomenon in Heterojunction (GaAl)As-GaAs Lasers", J. Appl. Phys., 1974, 45, no.9, pp.3899-903.

24.506. Stern, F., and Woodall, J.M.: "Photon Recycling in Semiconductor Lasers", J. Appl. Phys., 1974, 45, no.9, pp.3904-6.

24.507. Hakki, B.W., and Nash, F.R.: "Catastrophic Failure in GaAs Double-Heterostructure Injection Lasers", J. Appl. Phys., 1974, 45, no.9, pp.3907-12.

24.508. Abrams, R.L.: "Gigahertz Tunable Waveguide CO_2 Laser", Appl. Phys. Lett., 1974, 25, no.5, pp.304-6.

24.509. Nakamura, O., and Kikuchi, B.: "Stabilized High-Power Ar^{II} Laser", Fujitsu Sci. Tech. J., 1974, 10, no.3, pp.147-71.

24.510. Antonov, V.S., Khyazev, I.N., and Movshev, V.G.: "Output Radiation of UV N_2 Laser Excited Transversely in an Open-Air Cell", Kvantovaya Elektron., 1974, 1, no.2, pp.433-5, and Sov. J. Quantum Electron., 1974, 4, no.2, pp.246-7.

24.511. Shokin, A.A.: "Dependence of Evolution of Heat in Nd^{3+}:YAG on the Output", Kvantovaya Elektron., 1974, 1, no.2, pp.423-5, and Sov. J. Quantum Electron., 1974, 4, no.2, pp.238-9.

24.512. Chodzko, R.A.: "Multiple-Selected-Line Unstable Resonator", Appl. Opt., 1974, 13, no.10, pp.2321-5.

24.513. Rensch, D.B.: "Three-Dimensional Unstable Resonator Calculations With Laser Medium", Appl. Opt., 1974, 13, no.11, pp.2546-61.

24.514. Ewanizky, T.F.: "Unstable-Resonator Flashlamp-Pumped Dye Laser", Appl. Phys. Lett., 1974, 25, no.5, pp.295-7.

24.515. Brueckner, K.A., Jorna, S., and Moncur, K.: "Double-Pass Laser Amplifiers", Appl. Opt., 1974, 13, no.10, pp.2183-5.

24.516. Clark, W.M., and Lind, R.C.: "Space- and Time-Resolved Gain Measurements of a UV-Sustained CO_2 Laser", Appl. Phys. Lett., 1974, 25, no.5, pp.284-6.

24.517. Browne, P.G., and Smith, A.L.S.: "Long-Lived CO_2 Lasers With Distributed Heterogeneous Catalysis", J. Phys. D, 1974, 7, no.18, pp.2464-70.

24.518. Wang, S., Cordero, R.F., and Tseng, C.C.: "Analysis of Distributed-Feedback and Distributed-Bragg-Reflector Laser Structures", J. Appl. Phys., 1974, 45, no.9, pp.3975-7.

24.519. Saito, T.T., and Simmons, L.B.: "Performance Characteristics of Single-Point-Diamond Machined Metal Mirrors for IR Laser Applications", Appl. Opt., 1974, 13, no.11, pp.2647-50.

24.520. Levatter, J.I., and Lin, S.C.: "High-Power Generation from a Parallel-Plates-Driven Pulse Nitrogen Laser", Appl. Phys. Lett., 1974, 25, no.12, pp.703-5.

24.521. McMullen, J.D., Anderson, D.B., and Davis, R.L.: "Gain Measurements on a CW Transverse CO_2 Planar-Waveguide Laser", J. Appl. Phys., 1974, 45, no.11, pp.5084-7.

24.522. Tanaka, A., et al.: "Characteristics of 10.8-micron N_2O Lasers in CW and Q-Switched Operations", Jap. J. Appl. Phys., 1974, 13, no.12, pp.2009-13.

24.523. Kiselevskii, L.I., Skutov, D.K., and Sokolov, S.A.: "Use of High-Frequency Induction Discharge for Obtaining CW Laser Generation", Zh. Prikl. Spekt., 1974, 21, no.5, pp.951-5.

24.524. Komarov, V.S., Seleznev, V.G., and Korobogatov, G.A.S.: "Method for Determining Absolute Characteristics of CF_3I Laser Emission", Zh. Tekh. Fiz., 1974, 44, no.4, pp.875-8, and Sov. Phys.-Tech. Phys., 1974, 19, no.4, pp.558-60.

24.525. Babaev, I.K., et al.: "Electron-Beam-Controlled CO_2 Laser Operating Under Pulse-Repetition Conditions", Kvantovaya Elektron., 1974, 1, no.6, pp.1407-10, and Sov. J. Quantum Electron., 1974, 4, no.6, pp.777-9.

24.526. Bakos, J.S., et al.: "Distributed Feedback Dye Laser of Wavelength Tunable Over 747-840 nm", Phys. Lett., 1974, 50A, no.3, pp.227-8.

24.527. Novokreshchenov, V.K., and Shkunov, N.V.: "Determination of Lasing Losses in Solid-State CW Lasers", Prib. Tekh. Eksp., 1974, 17, no.2, pp.207-8, and Instrum. Exp. Tech., 1974, 17, no.2, pp.545-6.

24.528. Isbasescu, M., and Stratan, A.: "High-Energy Nd^{3+}:Glass Laser Amplifier", Rev. Roum. Phys., 1974, 19, no.10, pp.1107-9.

24.529. Mikaelyan, A.L., and D'yachenko, V.V.: "Lasers With Waveguide Resonators", Kvantovaya Elektron., 1974, 1, no.4, pp.937-49, and Sov. J. Quantum Electron., 1974, 4, no.4, pp.514-20.

24.530. Adamson, N.A., Bronnikova, L.K., and Fomin, Yu.A.: "Effect of Resonator Distortion on Zeeman He-Ne Laser Parameters", Zh. Prikl. Spekt., 1974, 21, no.5, pp.920-2.

24.531. Hasson, V., et al.: "Transverse Double-Discharge High-Pressure Glow Excitation of UV N_2 Laser", Appl. Phys. Lett., 1974, 25, no.11, pp.654-6.

24.532. Mas, G., Blancher, H., and Roig, J.: "Light Intensity and Polarization of a (Magnetized) He-Ne Gas Laser Without Brewster Windows", Appl. Opt., 1974, 13, no.12, pp.2771-3.

24.533. Brown, F., and Cohn, D.R.: "Mega-watt-Level Far-IR Lasers", Trans. IEEE, 1974, MTT-22, no.12, pp.1112-3.

24.534. Bychkov, Yu.I., et al.: "Electro-ionization CO_2 Laser With Radiation Energy of 2.5J in a Pulse", Prib. Tekh. Eksp., 1974, 17, no.3, pp.165-7, and Instrum. Exp. Tech., 1974, 17, no.3, pp.817-9.

24.535. Biryukov, A.S., and Shelepin, L.A.: "Kinetics of Gasdynamic Lasers. Nozzle Geometry", Zh. Tekh. Fiz., 1974, 44, no.6, pp.1232-43, and Sov. Phys.-Tech. Phys., 1974, 19, no.6, pp.775-81.

24.536. Demin, A.I., et al.: "Maximum Permissible Water-Vapour Content in CO_2-H_2O-N_2 Laser", Kvantovaya Elektron., 1974, 1, no.3, pp.528-33, and Sov. J. Quantum Electron., 1974, 4, no.3, pp.293-5.

24.537. Lotkova, E.N., Goncharova, S.G., and Pisarenko, V.V.: "Influence of Gas Composition of a (Water-Cooled) CO Laser on Output Power", Kvantovaya Elektron., 1974, 1, no.3, pp.542-6, and Sov. J. Quantum Electron., 1974, 4, no.3, pp.301-3.

24.538. Demin, A.I., et al.: "Influence of Water-Vapour Condensation on a CO_2 Gasdynamic Laser", Kvantovaya Elektron., 1974, 1, no.3, pp.403-4.

24.539. Ivanov, I.G., Il'yushko, V.G., and Sem, M.F.: "Dependences of the Gain of Cataphoretic Lasers on He Pressure and Tube Diameter", Kvantovaya Elektron., 1974, 1, no.5, pp.1081-8, and Sov. J. Quantum Electron., 1974, 4, no.5, pp.589-93.

24.540. Mikaelyan, A.L., et al.: "Transverse-Flow CO_2 Laser Operated as an Amplifier", Kvantovaya Elektron., 1974, 1, no.5, pp.1175-9, and Sov. J. Quantum Electron., 1974, 4, no.5, pp.643-5.

24.541. Tatasenko, V.F., Fedorov, A.I., and Bychkov, Yu.I.: "High-Power N_2 Laser", Kvantovaya Elektron., 1974, 1, no.5, pp.1226-7, and Sov. J. Quantum Electron., 1974, 4, no.5, p.674.

24.542. Bychkov, Yu.I., Kudryashov, V.P., and Osipov, V.V.: "Pulsed CO_2 Laser With Output of 15J", Kvantovaya Elektron., 1974, 1, no.5, pp.1256-8, and Sov. J. Quantum Electron., 1974, 4, no.5, pp.695-6.

24.543. Gudzenko, L.I., Iakoba, I.S., and Yakovlenko, S.I.: "Pulsed Laser Using Molecules in an Expanding Plasma", Kvantovaya Elektron., 1974, 1, no.5, pp.1273-5, and Sov. J. Quantum Electron., 1974, 4, no.5, pp.709-10.

24.544. Glaze, J.A., Guch, S., and Trenholme, J.B.: "Parasitic Suppression in Large Aperture Nd^{3+}:Glass Disc Laser Amplifiers", Appl. Opt., 1974, 13, no.12, pp.2808-11.

24.545. Cheremiskin, I.V., and Chekhlova, T.K.: "Thin-Film Waveguide Laser With Distributed Feedback and Gain Modulation", Kvantovaya Elektron., 1974, 1, no.3, pp.686-8, and Sov. J. Quantum Electron., 1974, 4, no.3, pp.387-8.

24.546. Eliseev, P.G., Pinsker, I.Z., and Fedorov, Yu.E.: "Degradation of Injection Lasers Under the Influence of Fast Particles", Kvantovaya Elektron., 1974, 1, no.5, pp.1271-3, and Sov. J. Quantum Electron., 1974, 4, no.5, pp.707-8.

24.547. Akkerman, D., et al.: "Injection Laser With Diffraction Grating in Resonator", Kvantovaya Elektron., 1974, 1, no.5, pp.1145-9, and Sov. J. Quantum Electron., 1974, 4, no.5, pp.626-8.

24.548. Sabotinov, N.V.: "Effect of Self-Cleaning of Brewster Windows in a He-Se Laser", Prib. Tekh. Eksp., 1974, 17, no.3, p.170, and Instrum. Exp. Tech., 1974, 17, no.3, p.824.

24.549. Zharkov, V.D., Lapushonok, L.Yu., and Chebykhin, N.N.: "Optimization of Parameters of a CO_2 Gasdynamic Laser", Zh. Prikl. Mekh. Tekh. Fiz., 1974, no.5, pp.3-8.

24.550. Brunne, M.: "Estimation of Influence of Turbulence-Induced Scattering on Gasdynamic-Laser Characteristics", Phys. Fluids, 1974, 18, no.4, pp.458-63.

24.551. Seelig, W.H.: "Gain of a He-Ne Laser With Optically Thin Discharge Tube", Z. Angew. Math. Phys., 1974, 25, no.6, pp.727-33.

24.552. Ivanov, P., et al.: "Criteria for Optimization of a He-Ne Laser", Elektro Prom. Prib., 1974, 9, no.9, pp.310-2.

24.553. Zemskov, K.I., et al.: "Use of Unstable Resonators in Achieving Diffraction-Limited Radiation from High-Gain Pulsed Gas Lasers", Kvantovaya Elektron., 1974, 1, no.4, pp.863-9, and Sov. J. Quantum Electron., 1974, 4, no.4, pp.474-7.

24.554. Mak, A.A., et al.: "Unidirectional CW Oscillation in a Solid-State Ring Laser With Return Mirror", Zh. Tekh. Fiz., 1974, 44, no.4, pp.868-70, and Sov. Phys.-Tekh. Phys., 1974, 19, no.4, pp.552-3.

25. SOLID-STATE SOURCES

25.1. Suzuki, K.: "Generation of Micro-wave Radiation from InSb", Jap. J. Appl. Phys., 1965, 4, p.42.

25.2. Buchsbaum, S.J., Chynoweth, A.G., and Feldmann, W.L.: "Microwave Emission from InSb", Appl. Phys. Lett., 1965, 6, p.67.

25.3. Sterzer, F.: "Analysis of GaAs Tunnel-Diode Oscillators", Trans. IEEE, 1965, ED-12, p.242.

25.4. Allen, J.W., et al.: "Microwave Oscillations in Ga(AsP) Alloys", Appl. Phys. Lett., 1965, 7, p.78.

25.5. Heeks, J.S., Woode, A.D., and Sandbach, C.P.: "Coherent High-Field Oscillations in Long Samples of GaAs", Proc. IEEE, 1965, 53, p.554.

25.6. Rivier, E.: "Broadband Tunable Tunnel-Diode Oscillator for Millimetre Waves", Proc. IEEE, 1965, 53, p.1675.

25.7. Haydl, W.H., and Quate, C.F.: "Microwave Emission from n-CdS", Appl. Phys. Lett., 1965, 7, p.45.

25.8. Johnston, R.L., De Loach, B.C., and Cohen, B.G.: "Si-Diode Microwave Oscillator", Bell Syst. Tech. J., 1965, 44, p.369.

25.9. Hakki, B.W., and Irvin, J.C.: "CW Microwave Oscillations in GaAs", Proc. IEEE, 1965, 53, p.80.

25.10. Hakki, B.W., and Knight, S.: "Phenomenological Aspects of CW Microwave Oscillations in GaAs", Solid-State Commun., 1965, 3, no.5, p.89.

25.11. Hilsum, C., et al.: "CW X- and K-Band Radiation from GaAs Epitaxial Layers", Electron. Lett., 1965, 1, p.178.

25.12. Swift, J.: "Microwave Tunnel-Diode Oscillators", Electron. Eng., 1965, 37, p.508.

25.13. Burrus, C.A.: "Millimetre-Wave Oscillations from Avalanching p-n Junctions in Si", Proc. IEEE, 1965, 53, p.1256.

25.14. Brand, F.A., et al.: "Microwave Generation from Avalanching Varactor Diodes", Proc. IEEE, 1965, 53, p.1276.

25.15. Quist, T.M., and Foyt, A.G.: "S-Band GaAs Gunn-Effect Oscillators", Proc. IEEE, 1965, 53, p.303.

25.16. Thomson, G.W.: "Microwave Oscillators Using Series and Series-Parallel Combinations of Tunnel Diodes", Proc. IEEE, 1965, 53, p.535.

25.17. Vorontsov, Yu.I., and Polyakov, I.V.: "Investigation of Oscillation Processes in Circuits With Several Series Tunnel Diodes", Radiotekh. Elektron., 1965, 10, p.890, and Radio Eng. Electron. Phys., 1965, 10.

25.18. Ferry, D.K., and Dougal, A.A.: "Microwave Emission from Bulk n-InAs", Appl. Phys. Lett., 1965, 7, p.318.

25.19. Hornbostel, D.H.: "5-10 GHz YIG-Tuned Tunnel-Diode Oscillator", Proc. IEEE, 1965, 53, p.1731.

25.20. Straub, W.D., Ayer, J.A., and Roth, H.: "CW X-Band GaAs Microwave Generators", Solid-State Electron., 1966, 9, p.281.

25.21. Muja, M., and Terai, M.: "Microwave Emission from CdS", Jap. J. Appl. Phys., 1966, 5, p.186.

25.22. Day, G.F.: "Microwave Oscillations in High-Resistivity GaAs", Trans. IEEE, 1966, ED-13, p.88.

25.23. Hakki, B.W., and Knight, S.: "Microwave Phenomena in Bulk GaAs", Trans. IEEE, 1966, ED-13, p.94.

25.24. Thim, H.W., and Barber, M.R.: "Microwave Amplification in GaAs Bulk Semiconductor", Trans. IEEE, 1966, ED-13, p.110.

25.25. Hasty, T.E., Cunningham, P.A., and Wisseman, W.R.: "Microwave Oscillations in Epitaxial Layers of GaAs", Trans. IEEE, 1966, ED-13, p.114.

25.26. Allen, J.W., Shyam, M., and Pearson, G.L.: "Gunn Oscillations in InAs", Appl. Phys. Lett., 1966, 9, p.39.

25.27. Das, P., and Staecker, P.W.: "Upper Frequency Limit for Gunn Oscillators Imposed by Carrier Energy-Relaxation Period", Electron. Lett., 1966, 2, p.258.

25.28. Dow, D.G., Mosher, C.H., and Vane, A.B.: "High-Peak-Power GaAs Oscillators", Trans. IEEE, 1966, ED-13, p.105.

25.29. Musha, T., Lindvall, F., and Hagglund, J.: "Microwave Emission from InSb for Low Electric Fields", Appl. Phys. Lett., 1966, 8, p.157.

25.30. Yamashita, E., and Baird, J.R.: "Theory of Tunnel-Diode Oscillator in Microwave Structure", Proc. IEEE, 1966, 54, p.606.

25.31. Kataoka, S., and Naito, H.: "Generation of Microwave Power by Magneto-Resistance Effect in Semiconductors", Electron. Lett., 1966, 2, p.29.

25.32. De Loach, B.C., and Johnston, R.L.: "Avalanche Transit-Time Microwave Oscillators and Amplifiers", Trans. IEEE, 1966, ED-13, p.181.

25.33. Irvin, J.C.: "GaAs Avalanche Microwave Oscillators", Trans. IEEE, 1966, ED-13, p.208.

25.34. Johnston, R.L., and Josenhans, J.G.: "Improved Performance of Microwave Read Diodes", Proc. IEEE, 1966, 54, p.412.

25.35. Miyai, Y., Nakashima, S., and Mizuno, H.: "Microwave Oscillations from p-n Junctions in Ge", J. Phys. Soc. Jap., 1966, 21, p.563.

25.36. Carroll, J.E.: "Oscillations Covering 4-31 GHz from a Single Gunn Diode", Electron. Lett., 1966, 2, p.141.

25.37. Cawsey, D.: "Design of Wide-Range Varactor-Tuned Microwave Tunnel-Diode Oscillators", Proc. IEE, 1966, 113, p.943.

25.38. Vincent, B.T., Cooke, H.E., and Anderson, A.J.: "Microwave Power Generation and Amplification Using Transistors", Microwave J., 1966, 4, no.7, p.63.

25.39. Hobson, G.S.: "Small-Signal Admittance of a Gunn-Effect Device", Electron. Lett., 1966, 2, p.207.

25.40. Socci, R.J., and Harrison, R.I.: "Multidiode Avalanche Oscillation for Increased CW Microwave Power Output", Proc. IEEE, 1966, 54, p.1006.

25.41. Lee, H.C.: "Microwave Power Generation Using Overlay Transistors", RCA Rev., 1966, 27, p.199.

25.42. Kennedy, W.K.: "Power Generation in GaAs at Frequencies Far in Excess of the Intrinsic Gunn Frequency", Proc. IEEE, 1966, 54, p.710.

25.43. Nakashima, S., Takeshima, M., and Miyai, Y.: "CW Microwave Oscillation from Ge Diodes", Jap. J. Appl. Phys., 1966, 5, p.639.

25.44. Thim, H.W.: "Linear Negative-Conductance Amplification With Gunn Oscillators", Proc. IEEE, 1967, 55, no.3, pp.446-7.

25.45. Kennedy, W.F., and Eastman, L.F.: "High-Power Pulsed Microwave Generation in GaAs", Proc. IEEE, 1967, 55, no.3, pp.434-5.

25.46. Saunders, T.E., and Stark, P.D.: "Integrated 4-GHz Balanced Transistor Amplifier", J. Solid-State Circuits IEEE, 1967, SC-7, no.1, pp.4-10.

25.47. Yamashita, E., and Baird, J.R.: "Tunnel-Diode Oscillator in an Open Structure", Trans. IEEE, 1967, MTT-15, no.7, pp.415-21.

25.48. Liu, S.G.: "GaAs Avalanche Oscillator With 1-W Output", Proc. IEEE, 1967, 55, no.5, pp.689-90.

25.49. Gelnovatch, V.G.: "Design of Distributed Transistor Amplifiers for Microwaves", Microwave J., 1967, 10, no.2, pp.41-7.

25.50. Thim, H.W., and Lemner, M.M.: "Linear Millimetre-Wave Amplifier With GaAs Wafers", Proc. IEEE, 1967, 55, no.5, pp.718-9.

25.51. Dawson, R.W.: "Planar 5-W X-Band Avalanche Oscillator", Proc. IEEE, 1967, 55, no.6, pp.1114-5.

25.52. Fleming, P.L.: "Observations Above and Below Twice Gunn Threshold", Proc. IEEE, 1967, 55, no.8, pp.1538-9.

25.53. Gilden, M., and Moroney, W.: "High-Power Pulsed Avalanche-Diode Oscillator for Microwaves", Proc. IEEE, 1967, 55, no.7, pp.227-8.

25.54. Vautey, B.: "Design and Construction of Tunnel-Diode Microwave Amplifiers", Onde Electr., 1967, 47, pp.105-10.

25.55. Sewell, K.G., and Boatner, L.A.: "Multimode Operation in Gunn Oscillator Induced by Cooling and Illumination", Proc. IEEE, 1967, 55, no.7, pp.1228-9.

25.56. Kriksunov, V.G., and Boychuk, B.A.: "Frequency-Tunable Tunnel-Diode Oscillator", Radiotekhnika, 1967, 22, no.1, pp.41-6, and Telecommun. Radio Eng. Pt 2, 1967, 22, no.1, pp.97-100.

25.57. Rulison, R.L., Gibbons, G., and Josenhans, J.G.: "Improved Performance of IMPATT Ge Diodes", Proc. IEEE, 1967, 55, no.2, pp.223-4.

25.58. Carroll, J.E.: "Series Operation of Gunn Diodes for High RF Power", Electron. Lett., 1967, 3, no.10, pp.455-6.

25.59. Holmstrom, F.R.: "Small Signal Behaviour of Gunn Diodes", Trans. IEEE, 1967, ED-14, no.9, pp.464-8.

25.60. Misawa, T.: "Microwave Si Avalanche Diode With Abrupt Junction", Trans. IEEE, 1967, ED-14, no.9, pp.580-4.

25.61. Copeland, J.A.: "Doping Nonuniformity and Geometry of LSA Oscillator Diodes", Trans. IEEE, 1967, ED-14, no.9, pp.497-500.

25.62. Gibbons, G., and Sze, S.M.: "Avalanche Breakdown in Read and PIN Diodes", Solid-State Electron., 1968, 11, no.2, pp.225-32.

25.63. Kataoka, S., et al.: "Microwave Oscillation and Amplification in a Long Bulk GaAs Sample with $BaTiO_3$ Sheets on the Surface", Nachr. Tech. Fachber., 1968, 35, p.454.

25.64. Wasse, M.P., Pearson, A., and King, G.: "Microstrip Circuit Module for Gunn Oscillator", Nachr. Tech. Fachber. 1968, 35, p.470.

25.65. Magarshack, J., and Spitalnik, R.: "Magnetically Tunable Gunn Oscillators Using a YIG Sphere", Nachr. Tech. Fachber., 1968, 35, p.475.

25.66. Toussaint, H.N.: "Transistorized Microwave Power Oscillators With Electronic Wideband Tuning", Nach. Tech. Fachber., 1968, 35, p.493.

25.67. Kataoka, S., and Tacano, M.: "Microwave Emission from InSb Avalanche Plasma Under a Magnetic Field", Nach. Tech. Fachber., 1968, 35, p.612.

25.68. Schenk, J.F., and Midford, T.A.: "Failure Mechanisms of Solid-State Microwave Sources", Des. Electron., 1968, 6, no.3, p.3.

25.69. Wagner, R.J., Gray, W.W., and Cooper, P.V.: "X-Band IMPATT Microstrip Power Sources", J. Solid-State Circuits IEEE, 1968, SC-3, p.221.

25.70. Blum, S.C.: "10-W S-Band Solid-State Amplifier", J. Solid-State Circuits IEEE, 1968, SC-3, p.233.

25.71. Midford, T.A., and Bowers, H.C.: "Two-Port IMPATT-Diode TW Amplifier", Proc. IEEE, 1968, 56, p.1724.

25.72. Harth, W., and Jaenicke, R.: "Evaluation of Microwave Emission from n-InSb", Nachr. Tech. Fachber., 1968, 35, p.628.

25.73. Aitchison, C.S., et al.: "Stable 100-mW Q-Band Solid-State Source", Nachr. Tech. Fachber., 1968, 35, p.637.

25.74. Fuchs, J.A.: "Transient Phase Response of Solid-State Pulsed X-Band Source", Proc. IEEE, 1968, 56, p.2196.

25.75. Ayaki, K., and Kajiwara, Y.: "4-GHz Transistor Amplifier", NEC Res. Dev., 1968, no.12, p.16.

25.76. Gorshkov, A.S., and Lavrova, O.G.: "Model of a BW Parametric Oscillator", Radiotekh. Elektron., 1968, 13, p.1515, and Radio Eng. Electron. Phys., 1968, 13, no.8, pp.1319-21.

25.77. Olsson, K.O.I.: "LSA-Diode Theory for Long Samples", Trans. IEEE, 1969, ED-16, p.202.

25.78. Kvasil, B.: "Linear Method of Exciting a Resonator by a Gunn Diode", Acta Tech. CSAV, 1969, 14, no.5, p.515.

25.79. Yeh, C., Liu, S.G., and Hawrylo, F.Z.: "Microwave Oscillations in (GaAl)As Avalanche Diodes", Proc. IEEE, 1969, 57, p.1785.

25.80. Glance, B.: "Microstrip IMPATT Oscillator with High Locking Figure of Merit", Proc. IEEE, 1969, 57, p.2052.

25.81. Petrov, V.A., and Prokhorov, E.D.: "Operational Features of Gunn Diode in a Circuit Containing Inductance", Radiotekh. Elektron., 1969, 14, no.9, p.1713.

25.82. Berson, B.E., Enstrom, R.E., and Reynolds, J.F.: "High-Efficiency Transferred-Electron Oscillators", Proc. IEEE, 1969, 57, no.9, p.1692.

25.83. Furukawa, S., and Ohmi, T.: "High-Efficiency Operation of Gunn and IMPATT Diodes", J. Inst. Electron. Commun. Eng. Jap., 1969, 52, no.11, p.1384.

25.84. Frey, W.: "Admittance of Phase-Locked Gunn Oscillator at High Input Powers", Electron. Lett., 1969, 5, no.26, p.672.

25.85. Taylor, B.C., and Howes, M.J.: "LSA Operation of GaAs Layers in Large-Scale Tunable Microwave Circuits", Trans. IEEE, 1969, ED-16, p.928.

25.86. Schneider, H.M., and Kennedy, W.K.: "Cavity Design for Millimetre-Wavelength GaAs Devices", Proc. IEEE, 1969, 57, p.1213.

25.87. Baughan, K.M., and Myers, F.A.: "Multiple Series Operation of Gunn-Effect Oscillators", Electron. Lett., 1969,5,p.371.

25.88. Shackle, P.W.: "High-Efficiency Self-Pulsed IMPATT Oscillator Circuit", Electron. Lett., 1969, 5, p.395.

25.89. Jones, S.: "Frequency-Stable Microstrip X-Band Gunn Oscillator", Proc. IEEE, 1969, 57, p.364.

25.90. Fank, F.B., and Day, G.F.: "High-CW-Power K-Band Gunn Oscillators", Proc. IEEE, 1969, 57, p.339.

25.91. Mantena, N.R., and Wright, M.L.: "Circuit Model Simulation of Gunn-Effect Devices", Trans. IEEE, 1969, MTT-17, p.363.

25.92. Sekido, K., et al.: "CW Oscillations in GaAs Planar-Type Bulk Diodes", Proc. IEEE, 1969, 57, p.815.

25.93. Vorob'ev, V.N., and Etkin, V.S.: "Study of Gunn Effect in Magnetic Field", Radiotekh. Elektron., 1969, 14, p.358, and Radio Eng. Electron. Phys., 1969, 14.

25.94. Rode, D.L.: "Self-Resonant LSA Diode", Proc. IEEE, 1969, 57, p.1216.

25.95. Camp, W.O.: "Computer Simulation of Multifrequency LSA Oscillations in GaAs", Proc. IEEE, 1969, 57, p.220.

25.96. Koike, T.: "Mode of Low-Field Microwave Emission from n-InSb", Jap. J. Appl. Phys., 1969, 8, p.403.

25.97. Tsai, W.C., and Rosenbaum, F.J.: "Bias-Circuit Oscillations in Gunn Devices", Trans. IEEE, 1969, ED-16, p.196.

25.98. Kulke, B., and Wilmarth, R.W.: "Small-Signal and Saturation Characteristics of an X-Band Cyclotron-Resonance Oscillator", Proc. IEEE, 1969, 57, p.219.

25.99. Vorob'ev, V.N., and Etkin, V.S.: "Effect of Magnetic Field on Operation of Avalanche Transit Diode Oscillators", Radiotekh. Elektron., 1969, 14, p.357, and Radio Eng. Electron. Phys., 1969, 14.

25.100. Steele, M.C., Califano, F.P., and Larrabee, R.D.: "High-Efficiency Series Operation of Gunn Devices", Electron. Lett., 1969, 5, p.81.

25.101. Suematsu, Y., et al.: "Small-Signal Characteristics of Gunn-Effect Devices Higher than Transit-Period Frequency", Electron. Commun. Jap., 1969, 52, p.118.

25.102. van Iperen, B.B., Tjassens, H., and Goedbloed, J.J.: "Relation Between Microwave Parameters of IMPATT Diodes", Proc. IEEE, 1969, 57, p.1341.

25.103. Nagano, S., et al.: "Millimetre-Wave CW Gunn Oscillators", NEC Res. Dev., 1969, no.15, p.62.

25.104. Rokushima, K., et al.: "Characteristics of Waveguide-Mounted Tunnel-Diode Oscillators", Electron. Commun. Jap., 1969, 52, no.7, p.36.

25.105. Fawcett, W., Hilsum, C., and Rees, H.D.: "Optimum Semiconductor for Microwave Devices", Electron. Lett., 1969, 5, p.313.

25.106. Noisten, J.: "K-Band Power Amplifier With Avalanche Diodes", Int. Elektron. Rdsch., 1970, 24, no.3, p.67.

25.107. Brayne, T., et al.: "Problems of Design and Fabrication of a CW Si IMPATT Diode", Microelectron. Reliabil., 1970, 9, p.157.

25.108. Groll, H.: "Impedance Behaviour of Gunn Oscillators With External Synchronization", Nachr. Tech., 1970, 20, p.151.

25.109. Cawsey, D.: "Wide-Range Tuning of Solid-State Microwave Oscillators", J. Solid-State Circuits IEEE, 1970, SC-5, p.82.

25.110. Hilsum, C., and Rees, H.D.: "Three-Level Oscillator. Form of Transferred-Electron Device", Electron. Lett., 1970, 6, p.277.

25.111. Colliver, D.J., Gibbs, S.E., and Taylor, B.C.: "Material Selection for Efficient Transferred-Electron Devices at Q Band", Electron. Lett., 1970, 6, p.353.

25.112. Holliday, H.R.: "Effect of Operating Parameters on the Phase Angle of a Locked X-Band Gunn Oscillator", Trans. IEEE, 1970, 17, p.527.

25.113. Bowman, D.R.: "S-Band Solid-State Swept-Frequency Oscillators", Electron. Eng., 1970, 42, no.7, p.37.

25.114. Hilsum, C., et al.: "Instabilities of InP Three-Level Transferred-Electron Oscillators", Electron. Lett., 1970, 6, p.307.

25.115. Shyam, M.: "CW Operation of LSA Oscillators in R Band", Electron. Lett., 1970, 6, p.315.

25.116. Colliver, D.J., et al.: "Microwave Generation by InP Three-Level Transferred-Electron Oscillators", Electron. Lett., 1970, 6, p.436.

25.117. Seidel, T.E., and Scharfetter, D.L.: "High-Power Millimetre-Wave IMPATT Oscillators", Proc. IEEE, 1970, 58, p.1135.

25.118. Romanyuk, V.A.: "Effect of Domain Capacity on Oscillation Frequency of the Gunn Oscillator", Radiotekhnika, 1970, 25, no.3, p.50, and Telecommun. Radio Eng. Pt 2, 1970, 25, no.3.

25.119. Kulikov, S.M., and Polyakov, I.V.: "TW Amplifier Using a Nonuniform Line with Tunnel Diodes", Radiotekh. Elektron., 1970, 15, p.505, and Radio Eng. Electron. Phys., 1970, 15, no.3, pp.432-8.

25.120. Starosel'skii, V.I.: "Threshold Power in the Gunn Effect", Fiz. Tekh. Poluprov., 1970, 4, p.565, and Sov. Phys.-Semicond., 1970, 4, no.3, pp.468-9.

25.121. Blouke, M.M., Tolar, N.J., and Leedy, H.M.: "Efficient Read-Type Diodes", Proc. IEEE, 1970, 58, p.805.

25.122. Khandelwal, D.D., and Curtice, W.R.: "Study of Single-Frequency Quenched-Domain Mode Gunn Oscillator", Trans. IEEE, 1970, MTT-18, p.178.

25.123. Cottam, M.G.: "High-Efficiency Oscillations in GaAs Avalanche Diodes", Electron. Lett., 1970, 6, p.71.

25.124. Becker, R., Bosch, B.G., and Engelmann, R.W.H.: "Domains and Guided Waves in GaAs Striplines", Electron. Lett., 1970, 6, p.604.

25.125. White, M.H., and Thurston, M.O.: "Charactierization of Microwave Transistors", Solid-State Electron., 1970, 13, p.523.

25.126. Ohmori, M., et al.: "CW Oscillations at Millimetre Wavelengths with Si p-n Junction Avalanche Diodes", Rev. Electr. Commun. Lab., 1970, 18, no.9-10, pp.663-9.

25.127. Ohmi, T., and Matsudaira, M.: "Submillimetre Oscillations by Ideal Crystals With Periodic Energy-Band Structure", Electron. Commun. Jap., 1970, 53, no.1, pp.107-14.

25.128. Arendar', V.N., et al.: "Hahn Microwave Diodes from GaAs", Prib. Tekh. Eksp., 1970, no.5, pp.230-1, and Instrum. Exp. Tech., 1970, no.5, pp.1512-3.

25.129. Rode, D.L.: "Dielectric-Loaded Self-Resonant LSA Diode", Trans. IEEE, 1970, ED-17, p.47.

25.130. Bott, I.B., and Holliday, H.R.: "Effects of Changes in Operating Conditions on the Phase of a Frequency-Locked Gunn Oscillator in X Band", Electron. Lett., 1970, 6, p.206.

25.131. Rode, D.L.: "Axial Modes and Self-Resonance in LSA Diodes", J. Appl. Phys., 1970, 41, p.2402.

25.132. Cottam, M.G.: "Theory for High-Efficiency Oscillations in GaAs Avalanche Diodes", J. Phys. D, 1970, 3, p.1033.

25.133. Baynham, A.C.: "Emission of TEM Waves Generated Within an n-Ge Cavity", Electron. Lett., 1970, 6, p.306.

25.134. Liu, S.G., Risko, J.J., and Chang, K.K.N.: "High-Power K-Band Si Avalanche-Diode Oscillators", Proc. IEEE, 1970, 58, p.919.

25.135. Owens, J.M.: "GaAs-on-Sapphire Gunn-Effect Devices", Proc. IEEE, 1970, 58, p.930.

25.136. Ying, R.S., and Kramer, N.B.: "X-Band Si TRAPATT Diodes", Proc. IEEE, 1970, 58, p.1285.

25.137. de Biasi, R.S., and Lee, S.S.: "Bulk Germanium LSA-Mode Oscillator", Proc. IEEE, 1970, 58, p.1301.

25.138. Abdel-Fattakh, Kh.A., and Rzhevkin, K.S.: "Amplification of Microwaves in Bulk GaAs During Generation of Domains", Radiotekh. Elektron., 1970, 15, no.6, p.1247, and Radio Eng. Electron. Phys., 1970, 15, no.6, pp.1056-9.

25.139. Jones, E.L., et al.: "Velocity of Gunn Domains", J. Appl. Phys., 1970, 41, p.3498.

25.140. Higashisaka, A.: "Temperature Dependence of Gunn Effect in GaAs Over the Range 77-545°K", Jap. J. Appl. Phys., 1970, 9, no.5, p.583.

25.141. Quine, J.P.: "LSA-Mode Transferred-Electron-Diode Oscillator for Microstrip", Proc. IEEE, 1970, 58, p.1291.

25.142. Law, H.C., and Kao, K.C.: "Gunn Effect in Two-Valley Semiconductors With Traps", Solid-State Electron. 1970, 13, no.8, p.1119.

25.143. Baynham, A.C., and Colliver, D.J.: "Mode of Microwave Emission from GaAs", Electron. Lett., 1970, 6, p.498.

25.144. Narayan, S.Y., Enstrom, R.E., and Gobat, A.R.: "High-Power CW Transferred-Electron Oscillators", Electron. Lett., 1970, 6, p.17.

25.145. Bhattacharya, T.K.: "Single Analysis of Tapered Gunn Oscillators", Phys. Status Solidi, 1970, 1A, no.4, pp.757-64.

25.146. Malyshev, V.A., and Alekseev, Yu.I.: "Study of Current in a Kinematic Model of a Gunn Diode", Izv. VUZ Radioelektron., 1970, 13, no.8, pp.1027-30.

25.147. Glendinning, W.B., Mark, A., and Harmatz, M.: "Si Etch-Refill p$^+$-p-n$^+$ Epitaxial X-Band IMPATT Device", Proc. IEEE, 1970, 58, no.11, pp.1867-8.

25.148. Yanai, H., et al.: "Experimental Analysis for Large-Amplitude, High-Efficiency, Mode of Oscillation With Si Avalanche Diodes", Trans. IEEE, 1970, ED-17, no.12, pp.1067-76.

25.149. Reynolds, J.F., Berson, B.E., and Enstrom, R.E.: "Microwave Circuits for High-Efficiency Operation of Transferred-Electron Oscillators", Trans. IEEE, 1970, MTT-18, no.11, pp.827-34.

25.150. Snider, D.M.: "One-Watt CW High-Efficiency X-Band Avalanche-Diode Amplifier", Trans. IEEE, 1970, MTT-18, no.11, pp.963-7.

25.151. Spiwak, R.R.: "Step-Iris Resonator for LSA Operation", Trans. IEEE, 1970, MTT-18, no.11, pp.973-5.

25.152. Kondo, M., et al.: "Effect of Junction Area of Si IMPATT Diodes on Output Power", NEC Res. Dev., 1970, no.18,pp.36-42.

25.153. Migitaka, M., Miyazaki, M., and Saito, K.: "High-Power Gunn-Oscillator Diodes on Type-IIa-Diamond Heat Sinks", Trans. IEEE, 1970, MTT-18, no.11, pp.1004-5.

25.154. Yu, S.P., and Young, J.D.: "Measurement of Interaction Impedance of Microwave Circuits for Solid-State Devices", Trans. IEEE, 1970, MTT-18, no.11, pp.999-1001.

25.155. Fray, S.J., and Taylor, B.C.: "Frequency-Tuning Characteristics of Waveguide-Mounted Transferred-Electron Oscillators", Electron. Lett., 1970, 6, no.22, pp.708-10.

25.156. Owens, R.P., and Cawsey, D.: "Microwave Equivalent-Circuit Parameters of Gunn-Effect Device Packages", Trans. IEEE, 1970, MTT-18, no.11, pp.790-8.

25.157. Clouser, P.L., and Risser, V.V.: "C-Band FET Amplifiers", J. Solid-State Circuits IEEE, 1970, SC-5, no.6, pp.323-7.

25.158. D'Aiello, R.V., and Assour, J.M.: "Characteristics of Si Avalanche Diodes as Oscillators and Power Amplifiers in S Band", J. Solid-State Circuits IEEE, 1970, SC-5, no.6, pp.358-61.

25.159. Taylor, B.C., Fray, S.J., and Gibbs, S.E.: "Frequency-Saturation Effects in Transferred-Electron Oscillators", Trans. IEEE, 1970, MTT-18, no.11, pp.799-807.

25.160. Tsai, W.C.: "Circuit Analysis of Waveguide-Cavity Gunn-Effect Oscillator", Trans. IEEE, 1970, MTT-18, no.11, pp.808-17.

25.161. Pollmann, H., et al.: "Load Dependence of Gunn-Oscillator Performance", Trans. IEEE, 1970, MTT-18, no.11, pp.817-27.

25.162. Kumabe, K., and Kanbe, H.: "Mechanism of Coupling between Space-Charge and EM Waves in a Bulk GaAs TW Amplifier", Rev. Electr. Commun. Lab., 1970, 18, no.11-12, pp.913-20.

25.163. Romanyuk, V.A.: "Design of Gunn Oscillators in the Quenched-Domain Mode", Radiotekhnika, 1970, 25, no.7, and Telecommun. Radio Eng. Pt 2, 1970, 25, no.7, pp.100-6.

25.164. Andreev, V.S.: "Quasi-Linear Analysis of Tunnel-Diode Microwave Oscillators", Radiotekhnika, 1970, 25, no.7, and Telecommun. Radio Eng. Pt 2, 1970, 25, no.7, pp.107-14.

25.165. Grasl, L.M., and Zimmerl, O.F.: "Frequency Behaviour of Space-Charge-Wave Amplifiers in GaAs", Phys. Status Solidi, 1970, 2A, no.2, pp.391-405.

25.166. Nii, R., et al.: "GaAs Sandwich-Type Gunn Oscillators", Rev. Electr. Commun. Lab., 1970, 18, no.9-10, pp.654-62.

25.167. Ikoma, T., et al.: "Characteristics of Transferred-Electron Devices", J. Fac. Eng. Univ. Tokyo, 1970, 30B, no.4, pp.347-94.

25.168. Arendar', V.N., Prokhorov, E.D., and Bagrov, G.V.: "Investigation of Characteristics of Microwave Gunn Diodes", Radiotekh. Elektron., 1970, 15, no.10, and Radio Eng. Electron. Phys., 1970, 15, no.10, pp.1873-8.

25.169. Fikart, J.: "Semiconductor Generator for 23-GHz Band", Slab. Obz., 1970, 31, no.9, pp.409-12.

25.170. Kimura, T., Yanai, H., and Kamiyama, M.: "Optical Triggering of a Gunn-Effect Device", Annu. Rep. Eng. Res. Inst. Univ. Tokyo, 1970, 29, pp.105-10.

25.171. Kondo, M., et al.: "CW Performance of Si IMPATT Diodes", NEC Res. Dev., 1970, no.18, pp.7-15.

25.172. Okamura, S., et al.: "Microwave (Gunn) Oscillator Using GaAs", Annu. Rep. Eng. Res. Inst. Univ. Tokyo, 1970, 29, pp.119-24.

25.173. Kennedy, W.K., and Rossiter, E.L.: "LSA Microstrip Phase-Locked Power Amplifier", Electron. Lett., 1970, 6, no.26, pp.852-3.

25.174. Bechteler, M.: "Buildup Behaviour of IMPATT Oscillators in Linear and Nonlinear Ranges", Electron. Lett., 1970, 6, no.26, pp.856-8.

25.175. Kastral'skii, A.A., Leonov, E.I., and Shur, M.S.: "Gunn Devices With Varying Energy Gap", Fiz. Tekh. Poluprov., 1970, 4, no.8, pp.1609-11, and Sov. Phys.-Semicond., 1971, 4, no.8, pp.1384-6.

25.176. Rodgers, J.M., and Pomeroy, R.C.: "Q-Band GaAs Schottky-Barrier IMPATT", Electron. Lett., 1971, 7, no.1, p.21.

25.177. Narayan, S.Y., Huang, H.C., and Gobat, A.R.: "Operation of Transferred-Electron Oscillators in Ridged-Waveguide Circuit", Electron. Lett., 1971, 7, no.2, pp.31-2.

25.178. Bybokas, J., and Farrell, B.: "Gunn Flange. Building Block for Low-Cost Microwave Oscillators", Electron., 1971, 41, no.5, pp.47-50.

25.179. Perlman, B.S., Upadhyayula, C.L., and Siekanowicz, W.W.: "Microwave Properties and Applications of Negative-Conductance Transferred-Electron Devices", Proc. IEEE, 1971, 59, no.8, pp.1229-37.

25.180. Novak, S.: "Double-Cavity Tuning of Gunn Oscillators at Millimetre Wavelengths", Proc. IEEE, 1971, 59, no.6, pp.1026-7.

25.181. Nakamura, M., et al.: "High-Speed Pulse Response of Planar-Type Gunn Diodes", Proc. IEEE, 1971, 59, no.6, pp.1039-40.

25.182. Paxman, D.H., and Tree, R.J.: "Characteristics of Bulk InP Microwave Oscillators", Electron. Lett., 1971, 7, no.10, pp.240-1.

25.183. Coleman, D.J., and Sze, S.M.: "Low-Noise Metal-Semiconductor-Metal Microwave Oscillator", Bell Syst. Tech. J., 1971, 50, no.5, pp.1695-9.

25.184. Jethwa, C.P., and Gunshor, R.L.: "Circuit Characterization of Waveguide-Mounted Gunn-Effect Oscillators", Electron. Lett., 1971, 7, no.15, pp.433-6.

25.185. Tucker, R.S.: "Synthesis of Broadband Microwave Transistor Amplifiers", Electron. Lett., 1971, 7, no.16, pp.455-6.

25.186. Wright, G.T.: "Transit-Time Oscillator With Velocity-Limited Injection", Electron. Lett., 1971, 7, no.16, pp.449-51.

25.187. Eddolls, D.V., Ward, F.S., and Whitehead, A.J.: "High-Power High-Efficiency CW Gunn Oscillators in X Band", Electron. Lett., 1971, 7, no.16, pp.472-3.

25.188. Braddock, P.W.: "Experimental Characteristics of Avalanche-Diode Reflection Amplifiers and Locked Oscillators", Electron. Lett., 1971, 7, no.2, pp.42-4.

25.189. Colliver, D.J., Taylor, B.C., and Morgan, J.R.: "Performance of InP Three-Level Oscillators at 18-40 GHz", Electron. Lett., 1971, 7, no.2, pp.50-1.

25.190. Monroe, J.W., and Camp, W.O.: "LSA Operation of a GaAs Device in Microstrip", Trans. IEEE, 1971, ED-18, no.1, pp.69-70.

25.191. Kurokawa, K., and Magalhaes, F.M.: "X-Band 10-W Multiple-IMPATT Oscillator", Proc. IEEE, 1971, 59, no.1, pp.102-3.

25.192. Curtice, W.R., and Khandelwal, D.D.: "Multifrequency Operation of Quenched-Domain-Mode Gunn-Effect Device", Proc. IEEE, 1971, 59, no.3, pp.416-7.

25.193. Lazarus, M.J., Novak, S., and Bullimore, E.D.: "Use of Voltage-Controlled Cap Resonance to Obtain Higher Power and Frequencies from Millimetre-Wave Gunn Oscillators", Proc. IEEE, 1971, 59, no.4, pp.716-7.

25.194. Gibbons, G., and White, P.M.: "InP Pulsed and CW Millimetre-Wave Oscillators", Electron. Lett., 1971, 7, no.7, pp.150-1.

25.195. Kawamoto, H., and Liu, S.G.: "Anti-Parallel Pair of High-Efficiency Avalanche Diodes", Proc. IEEE, 1971, 59, no.3, pp.427-8.

25.196. Ozasa, M., and Inoue, T.: "Reflection-Type Broadband Esaki-Diode Amplifier", Mem. Res. Inst. Sci. Eng. Ritumeikan Univ., 1971, no.21, pp.27-32.

25.197. Camp, W.O.: "High-Efficiency GaAs Transferred-Electron-Device Operation and Circuit Design", Trans. IEEE, 1971, ED-18, no.12, pp.1175-84.

25.198. Taylor, B.C., and Colliver, D.J.: "InP Microwave Oscillators", Trans. IEEE, 1971, ED-18, no.10, pp.835-40.

25.199. Rosen, A., Reynolds, J.F., and Assour, J.M.: "Broadband High-Power TRAPATT Diode Amplifier at S Band", Electron. Lett., 1971, 7, no.26, pp.778-9.

25.200. Hamilton, C.H.: "Electronically Tuned Sources for Microwave Sweepers", Marconi Instrum., 1971, 13, no.2, pp.35-7.

25.201. Hartmann, K., Kotyczka, W., and Strutt, M.J.O.: "Equivalent Networks for Three Different Microwave Bipolar Transistor Packages in the 2-10 GHz Range", Electron. Lett., 1971, 7, no.18, pp.510-1.

25.202. Kocabiyikoglu, Z.U., Hobson, G.S., and De Sa, B.A.E.: "Relationship of Starting Delay Time and Frequency/Temperature Characteristics of X-Band Gunn Oscillators", Electron. Lett., 1971, 7, no.18, pp.550-2.

25.203. Kawamoto, H., and Allen, E.L.: "Wideband Microwave Amplifier With Two Antiparallel High-Efficiency Diode Pairs", Electron. Lett., 1971, 7, no.20, pp.602-3.

25.204. Baechtold, W.: "O-Band GaAs FET Amplifier and Oscillator", Electron. Lett., 1971, 7, no.10, pp.275-6.

25.205. Cowley, A.M., and Patterson, R.C.: "High-Power Parallel-Array IMPATT Diodes", Electron. Lett., 1971, 7, no.11, pp.301-3.

25.206. Dobrowolski, J.: "Properties of Gunn-Diode Oscillator With Coaxial Cavity", Elektronika, 1971, no.3, pp.97-102.

25.207. Jeppsson, B.I.: "LSA Relaxation Oscillations in a Waveguide Iris Circuit", Trans. IEEE, 1971, ED-18, no.7, pp.432-9.

25.208. Wilson, P.G., and Minakovic, B.: "Development of an FM Pulsed Gunn Oscillator at X Band", Trans. IEEE, 1971, ED-18, no.7, p.450.

25.209. Perlman, B.S.: "Microwave Amplification Using Transferred-Electron Devices in Prototype Filter Equalization Networks", RCA Rev., 1971, 32, no.1, p.3-23.

25.210. Seidel, T.E., Davis, R.E., and Iglesias, D.E.: "Double-Drift-Region Ion-Implanted Millimetre-Wave IMPATT", Proc. IEEE, 1971, 59, no.8, pp.1222-8.

25.211. Schroeder, W.E., and Haddad, G.I.: "Effect of Temperature on Operation of an IMPATT Diode", Proc. IEEE, 1971, 59, no.8, pp.1242-4.

25.212. Camp, W.O.: "Experimental Observations of Relaxation Waveforms in GaAs", Proc. IEEE, 1971, 59, no.8, pp.1248-50.

25.213. Driver, M.C., Kim, H.B., and Barret, D.L.: "GaAs Self-Aligned-Gate FET", Proc. IEEE, 1971, 59, no.8, pp.1244-5.

25.214. Chaffin, R.J.: "Temperature Dependence of TRAPATT Oscillators", Proc. IEEE, 1971, 59, no.8, pp.1270-1.

25.215. Chang, N.S., and Matsuo, Y.: "High-Efficiency Operation of Transferred-Electron Oscillators", Mem. Inst. Sci. Ind. Res. Osaka Univ., 1971, 28, pp.29-43.

25.216. Kobyzev, V.N., and Tager, A.S.: "Coherent Microwave Radiation of n-InSb", Zh. Eksp. Teor. Fiz. Pis'ma, 1971, 13, no.11, pp.607-11, and JETP Lett., 1971, 13, no.11, pp.433-5.

25.217. Parkes, E.P., Taylor, B.C., and Colliver, D.J.: "Performance of Planar Gunn Oscillators in X Band", Trans. IEEE, 1971, ED-18, no.10, pp.840-3.

25.218. Barrera, J.S.: "GaAs LSA V-Band Oscillators", Trans. IEEE, 1971, ED-18, no.10, pp.866-72.

25.219. Kurokawa, K.: "Single-Cavity Multi-Device (IMPATT) Oscillator", Trans. IEEE, 1971, MTT-19, no.10, pp.793-801.

25.220. Mircea, A.E., Farrayre, A., and Kramer, B.: "X-Band GaAs Diffused IMPATT Diodes for High Efficiency", Proc. IEEE, 1971, 59, no.9, pp.1376-7.

25.221. Rothbauer, M., and Tomiak, Z.: "Gunn Diode for Continuous Operation in the 3-cm Band", Sdelovaci Tech., 1971, 19, no.7, pp.208-10.

25.222. Gandhi, O.P., and Grow, R.W.: "Microwave Emission from Plasmas in InSb With and Without Magnetic Fields", Trans. IEEE, 1971, ED-18, no.10, pp.853-65.

25.223. Scharfetter, D.L.: "Power-Impedance-Frequency Limitations of IMPATT Oscillators Calculated from a Scaling Approximation", Trans. IEEE, 1971, ED-18, no.8, pp.536-43.

25.224. Gupta, M.S., and Lomax, R.J.: "Self-Consistent Large-Signal Analysis of a Read-Type IMPATT Diode Oscillator", Trans. IEEE, 1971, ED-18, no.8, pp.544-5.

25.225. Clorfeine, A.S.: "Guidelines for Design of High-Efficiency Mode Avalanche-Diode Oscillators", Trans. IEEE, 1971, ED-18, no.8, pp.550-6.

25.226. De Sa, B.A.E., and Hobson, G.S.: "Thermal Effects in the Bias-Circuit Frequency Modulation of Gunn Oscillators", Trans. IEEE, 1971, ED-18, no.8, pp.557-62.

25.227. Thust, H.: "Avalanche-Diode Oscillator for X Band", Nachr. Tech., 1971, 21, no.6, pp.215-9.

25.228. Sze, S.M., and Ryder, R.M.: "Microwave Avalanche Diodes", Proc. IEEE, 1971, 59, no.8, pp.1140-54.

25.229. Sterzer, F.: "Transferred-Electron Amplifiers and Oscillators for Microwave Applications", Proc. IEEE, 1971, 59, no.8, pp.1155-63.

25.230. Turner, J.A., et al.: "Dual-Gate GaAs Microwave FET", Electron. Lett., 1971, 7, no.22, pp.661-2.

25.231. Majewski, M.: "Avalanche Transit-Time Diode Microwave Oscillator", Arch. Elektrotech., 1971, 20, no.3, pp.811-5.

25.232. Aitchison, C.S., et al.: "Lumped-Circuit Elements at Microwave Frequencies", Trans. IEEE, 1971, MTT-19, no.12, pp.928-37.

25.233. Yu, S.P., and Glover, G.H.: "Heat-Sink Temperature Distribution of Annular Solid-State Microwave Diode", Electron. Lett., 1971, 7, no.8, pp.182-3.

25.234. Culshaw, B.: "Time-Domain Model of the Device-Circuit Characteristics of Trapped-Plasma-Mode, Avalanche-Diode, Oscillator", Electron. Lett., 1971, 7, no.12, pp.339-40.

25.235. Downing, B.J., and Myers, F.A.: "Broadband Varactor-Tuned X-Band Gunn Oscillator", Electron. Lett., 1971, 7, no.14, pp.407-9.

25.236. Teszner, T.L.: "Tunable Gunn Oscillator by Semiconductor Surface Loading", Electron. Lett., 1971, 7, no.7, pp.146-8.

25.237. Dobrowolski, J.: "Influence of Loaded Q-Factor of a Resonator on Gunn-Diode Oscillator Power", Elektronika, 1971, no.8, pp.320-4.

25.238. Ito, Y., et al.: "Experimental and Computer Simulation Analysis of a Gunn Diode", Trans. IEEE, 1971, MTT-19, no.12, pp.900-5.

25.239. Kawamoto, H.: "High-Power Microwave Amplifier Using an Antiparallel Avalanche-Diode Pair", Trans. IEEE, 1971, MTT-19, no.12, pp.911-5.

25.240. Fallmann, W.F., and Hartnagel, H.L.: "Aspects of Planar Gunn Diodes for High CW Output Power", Solid-State Electron., 1971, 14, no.10, pp.909-12.

25.241. Sene, A., and Rosenbaum, F.J.: "Wideband Gunn-Effect CW Waveguide Amplifier", Trans. IEEE. 1972, MTT-20, no.10, pp.645-50.

25.242. Glance, B.: "Low-Q Microstrip IMPATT Oscillator at 30 GHz", Proc. IEEE, 1972, 60, no.9, pp.1105-6.

25.243. Ollivier, P.M.: "Microwave YIG-Tuned Transistor Oscillator Design for C Band", J. Solid-State Circuits IEEE, 1972, SC-7, no.1, pp.54-60.

25.244. Wasse, M.P., Mun, J., and Heeks, J.S.: "Optimum Loading for Relaxation LSA Diode", Electron. Lett., 1972, 8, no.14, pp.364-6.

25.245. Kaneda, S., and Abe, M.: "Analysis of Microwave Characteristics of Hot Electrons in n-GaAs", Electron. Commun. Jap., 1972, 53, no.12, pp.112-20.

25.246. Eisenhart, R.L., and Khan, P.J.: "Tuning Characteristics and Oscillation Conditions of a Waveguide Mounted Transferred-Electron Diode Oscillator", Trans. IEEE, 1972, ED-19, no.9, pp.1050-5.

25.247. Bains, A.S.: "High-Power Microwave Amplifier Using IMPATT Diodes", Electron. Lett., 1972, 8, no.16, pp.427-8.

25.248. Jethwa, C.P., and Gunshor, R.L.: "Analytical Equivalent Circuit Representation for Waveguide-Mounted Gunn Oscillators", Trans. IEEE, 1972, MTT-20, no.9, pp.565-72.

25.249. Slaymaker, N.A., and Carroll, J.E J.E.: "Verification of Simple TRAPATT-Oscillator Model", Proc. IEE, 1972, 119, no.8, pp.1113-8.

25.250. Paik, S.F.: "Iterated Solid-State Microwave Power Amplifier", Trans. IEEE, 1972, MTT-20, no.3, pp.202-9.

25.251. Carroll, J.E., and Crede, R.H.: "Computer Simulation of TRAPATT Circuits", Int. J. Electron., 1972, 32, no.3, pp.273-96.

25.252. White, J.F.: "Simplified Theory for Post Coupling Gunn Diodes to Waveguide", Trans. IEEE, 1972, MTT-20, no.6, pp.372-8.

25.253. Dalman, G.C., Zappert, F.G., and Lee, C.A.: "Relaxing-Avalanche-Mode Reflection Amplifier", Electron. Lett., 1972, 8, no.9, pp.243-4.

25.254. Coleman, D.J.: "Transit-Time Oscillations in BARITT Diodes", J. Appl. Phys., 1972, 43, no.4, pp.1812-8.

25.255. Mircea, A.E.: "Computer-Optimized Design of Pulsed Gunn Oscillators", Trans. IEEE, 1972, ED-19, no.1, pp.21-6.

25.256. Colliver, D.J., Gray, K.W., and Joyce, B.D.: "High-Efficiency Microwave Generation in InP", Electron. Lett., 1972, 8, no.1, pp.11-2.

25.257. Snapp, C.P.: "Experiments Concerning the Nature of Trapped-Plasma-Mode Harmonic Extraction from p^+-n-n^+ Avalanche Diodes", Trans. IEEE, 1972, ED-19, no.2, pp.172-81.

25.258. Okean, H.C., Sard, E.W., and Pflieger, R.H.: "Microwave Integrated Oscillators for Broadband High-Performance Receivers", Trans. IEEE, 1972, MTT-20, no.2, pp.155-64.

25.259. Miyai, Y., et al.: "Ge Microwave Oscillators", Natl. Tech. Rep., 1972, 18, no.3, pp.296-302.

25.260. Kantelberg, G.: "Behaviour of Gunn Diodes in Coaxial Resonator", Nachr. Tech. 1972, 22, no.9, pp.309-10.

25.261. Sultan, N.B., and Wright, G.T.: "Electronic Tuning of Punch-Through Injection Transit-Time Microwave Oscillator", Trans. IEEE, 1972, MTT-20, no.11, pp.773-5.

25.262. Guha, S.: "Studies on Gunn Diodes in Resistive and Reactive Circuits", J. AEU, 1972, 5, no.2, pp.50-5.

25.263. Takayama, Y.: "Dynamic Behaviour of Nonlinear Power Amplifiers in Stable and Injection-Locked Modes", Trans. IEEE, 1972, MTT-20, no.9, pp.591-5.

25.264. Baynham, A.C.: "Microwave Emission Spectrum from Large Samples of GaAs in which TEM Waves Experience Round-Trip Gain", Electron. Lett., 1972, 8, no.25, pp.606-8.

25.265. Bullimore, E.D., et al.: "CW 0.25-W, O-Band, Gunn Oscillator", Electron. Lett., 1972, 8, no.26, pp.629-30.

25.266. White, P.M., and Gibbons, G.: "High-Efficiency CW Operation of Anomalous InP Microwave Oscillators", Electron. Lett., 1972, 8, no.6, pp.166-8.

25.267. Kondo, A., and Ishii, T.: "Performance of Read Diodes at K Band", Trans. IEEE, 1972, ED-19, no.5, p.695.

25.268. Dalman, G.C., and Lee, C.A.: "Synchronous Detuned Microwave-Oscillator Power Combiners", Electron. Lett., 1972, 8, no.5, pp.125-7.

25.269. Cowley, A.M., and Hamilton, S.: "Using IMPATT Diodes at Microwaves. II", Electron. Aust., 1972, 33, no.11, pp.52-55.

25.270. Baskaran, S., and Robson, P.N.: "Gain and Noise Factor of GaAs Transferred-Electron Amplifiers at 34 GHz", Electron. Lett., 1972, 8, no.5, pp.109-10.

25.271. Tomizawa, K., and Kataoka, S.: "Dependence of Transverse Spreading Velocity of a High-Field Domain in GaAs on the Bias Electric Field", Electron. Lett., 1972, 8, no.5, pp.130-2.

25.272. Habovsik, P.: "Problems of Wideband Retuning of Gunn Diodes", Elektrotech. Cas., 1972, 23, no.9, pp.590-5.

25.273. Braddock, P.W., Owen, N.C., and Genner, R.: "High-Power Avalanche IMPATT Reflection Amplifier Using the Rucker Combining Circuit", Electron. Lett., 1972, 8, no.23, pp.562-4.

25.274. Jones, D., and Rees, H.D.: "Accumulation Transit Mode in Transferred-Electron Oscillators", Electron. Lett., 1972, 8, no.23, pp.566-7.

25.275. Ito, Y., et al.: "K-Band High-Power Single-Tuned IMPATT Oscillator Stabilized by Hybrid-Coupled Cavities", Trans. IEEE, 1972, MTT-20, no.12, pp.799-805.

25.276. Ruttan, T.G.: "42-GHz Push-Pull Gunn Oscillator", Proc. IEEE, 1972, 61, no.11, pp.1441-2.

25.277. Rosen, A., et al.: "Wideband Class-C TRAPATT Amplifiers", RCA Rev., 1972, 33, no.4, pp.729-36.

25.278. Eliseev, N.I., and Kats, S.I.: "Determination of Power Dissipated in Individual Sections of a Series of Microwave Four-Poles", Radiotekhnika, 1972, 27, no.9, pp.100-1, and Telecommun. Radio Eng. Pt 2, 1972, 27, no.9, pp.134-5.

25.279. Rakshit, P.C., Paria, H., and Nag, B.R.: "Continuously Tunable Gunn Oscillator Using a Coaxial-Line Resonant Circuit", Indian J. Pure Appl. Phys., 1972, 10, no.5, pp.348-50.

25.280. Davydova, N.S., et al.: "Linear Theory of an IMPATT-Diode Distributed Microwave Amplifier", Radiotekhnika, 1972, 27, no.8, pp.77-81, and Telecommun. Radio Eng. Pt 2, 1972, 27, no.8, pp.112-5.

25.281. Narayan, S.Y., and Paczkowski, J.P.: "Integral-Heat-Sink Transferred-Electron Oscillators", RCA Rev., 1972, 33, no.4, pp.752-65.

25.282. Eldridge, A.L.: "Frequency Stability of Gunn Oscillators With Variation of Ambient Temperature", Solid-State Electron., 1972, 15, no.11, pp.1187-96.

25.283. Sinha, J.K., Mohan, R., and Kataria, B.K.: "Development of Cavities for Microwave Solid-State Sources", J. Inst. Telecommun. Eng., 1972, 18, no.12, pp.595-9.

25.284. Gibbons, G., et al.: "50-GHz GaAs IMPATT Oscillator", Electron. Lett., 1972, 8, no.21, pp.513-4.

25.285. Borodovskii, P.A., Bul'dygin, A.E., and Utkin, K.K.: "Series Operation of Gunn Diodes in a Coaxial Resonator", Izv. VUZ Radioelektron., 1972, 15, no.8, pp.954-8.

25.286. Mackintosh, I.W.: "Circuit Model for TRAPATT-Oscillator Cavities", Proc. IEE, 1972, 119, no.11, pp.1529-37.

25.287. Chigogidze, Z.N., et al.: "Mechanism of Failure of Gunn Diodes", Fiz. Tekh. Poluprov., 1972, 6, no.9, pp.1670-6, and Sov. Phys.-Semicond., 1973, 6, no.9, pp.1443-7.

25.288. Kim, C.: "High-Power, High-Efficiency, Operation of Read-Type IMPATT Oscillators", Electron. Lett., 1973, 9, no.8-9, pp.173-4.

25.289. Allen, S.G.: "Simple TRAPATT Circuit Model", Electron. Lett., 1973, 9, no.8-9, pp.178-80.

25.290. Watts, B.E., Howard, A.M., and Gibbons, G.: "Double-Drift Millimetre-Wave IMPATT Diodes Prepared by Epitaxial Growth", Electron. Lett., 1973, 9, no.8-9, pp.183-4.

25.291. Frey, W.: "Optimum RF-Power Transport in nd-Limited GaAs TW Amplifiers", Electron. Lett., 1973, 9, no.1, pp.12-4.

25.292. Hanson, D.C., and Heinz, W.W.: "Integrated Electrically Tuned X-Band Power Amplifier Utilizing Gunn and IMPATT Diodes", J. Solid-State Circuits IEEE, 1973, SC-8, no.1, pp.3-14.

25.293. Sweet, A.A., Collinet, J.C.R., and Wallace, R.N.: "Multistage Gunn Amplifiers for FM CW Systems", J. Solid-State Circuits IEEE, 1973, SC-8, no.1, pp.20-8.

25.294. Mackintosh, I.W.: "Circuit Mode Chart for High-Efficiency Avalanche Diode Oscillators", J. Solid-State Circuits IEEE, 1973, SC-8, no.1, pp.44-53.

25.295. Baechtold, W.: "X- and K-Band Amplifiers with GaAs Schottky-Barrier FET's", J. Solid-State Circuits IEEE, 1973, SC-8, no.1, pp.54-8.

25.296. Niehaus, W.C., Seidel, T.E., and Iglesias, D.E.: "Double-Drift IMPATT Diodes Near 100 GHz", Trans. IEEE, 1973, ED-20, no.9, pp.765-71.

25.297. Parkes, E.P., and Taylor, B.C.: "Effects of Temperature on Epitaxial InP Transferred-Electron Devices", Trans. IEEE, 1973, ED-20, no.10, pp.852-5.

25.298. Joshi, J.S., and Cornick, J.A.F.: "Characteristics of Electromechanically Tuned Gunn Oscillators", Trans. IEEE, 1973, MTT-21, no.9, pp.582-6.

25.299. Tomizawa, K., Hariu, T., and Hartnagel, H.: "Analysis of Microwave Circuit for Characterization of Negative-Conductance Devices by Transients", Trans. IEEE, MTT-21, no.9, pp.596-7.

25.300. Wilson, W.L.: "Precise Frequency and Phase Control of LSA Oscillators", Trans. IEEE, 1973, MTT-21, no.3, pp.146-9.

25.301. Salmer, G., et al.: "Theoretical and Experimental Study of GaAs IMPATT Oscillator Efficiency", J. Appl. Phys., 1973, 44, no.1, pp.314-24.

25.302. Nigrin, J., Goud, P.A., and Rybczynski, A.M.: "Pulse Avalanche Diode Oscillators With Injected CW Signal", Proc. IEEE, 1973, 61, no.3, pp.397-8.

25.303. Baynham, A.C., et al.: "Wave Propagation in Multilayered Drifted Solid-State Plasmas", Trans. IEEE, 1973, MTT-21, no.2, pp.111-3.

25.304. Shimizu, N., Kumabe, K., and Kanbe, H.: "Characteristics of a GaAs TW Amplifier With Schottky-Barrier Contacts", Electron. Lett., 1973, 9, no.2, pp.29-30.

25.305. Lekholm, A., and Mayr, J.: "Computer Optimisation of Double-Drift-Region IMPATT Diodes", Electron. Lett., 1973, 9, no.3, pp.64-6.

25.306. Dean, M., and Howes, M.J.: "J-Band Transferred-Electron Oscillators", Trans. IEEE, 1973, MTT-21, no.3, pp.121-7.

25.307. Huang, H.C.: "Modified GaAs IMPATT Structure for High-Efficiency Operation", Trans. IEEE, 1973, ED-20, no.5, pp.482-6.

25.308. Tantraporn, W., and Yu, S.P.: "Efficiencies of Schottky-Barrier GaAs and Both Complementary Structures of Si IMPATT Diodes", Trans. IEEE, 1973, ED-20, no.5, pp.492-6.

25.309. Tan, H.S.: "Computer Study of Phase-Locking Characteristics of LSA-Mode Diode Oscillator", Trans. IEEE, 1973, ED-20, no.5, pp.510-2.

25.310. Su, S., and Sze, S.M.: "Design Considerations of High-Efficiency GaAs IMPATT Diodes", Trans. IEEE, 1973, ED-20, no.6, pp.541-3.

25.311. Nakamura, M., Kordera, H., and Migitaka, M.: "Computer Study on GaAs Schottky-Barrier IMPATT Diodes", Solid-State Electron. 1973, 16, no.6, pp.163-7.

25.312. Grierson, J.R., and O'Hara, S.: "Comparison of Si and GaAs Large-Signal IMPATT-Diode Behaviour at 10-100 GHz", Solid-State Electron. 1973, 16, no.6, pp.719-41.

25.313. Curtice, W.R.: "Quenched-Domain Mode Oscillation in Waveguide Circuits", Trans. IEEE, 1973, MTT-21, no.6, pp.369-74.

25.314. Talwar, A.K., and Curtice, W.R.: "Experimental Study of Stabilized Transferred-Electron Amplifiers", Trans. IEEE, 1973, MTT-21, no.7, pp.477-81.

25.315. Nagano, S., and Ohnaka, S.: "Highly Stabilized IMPATT Oscillators at Millimetre Wavelengths", Trans. IEEE, 1973, MTT-21, no.7, pp.491-2.

25.316. Wen, C.P., Chiang, Y.S., and Young, A.F.: "High-Frequency Silicon-on-Sapphire IMPATT Oscillator", Proc. IEEE, 1973, 61, no.6, pp.794-5.

25.317. Chao, C., and Haddad, G.I.: "Non-linear Behaviour and Bias Modulation of an IMPATT-Diode Oscillator", Trans. IEEE, 1973, MTT-21, no.10, pp.619-30.

25.318. Lefeuvre, S., and Hanna, V.F.: "Coupled-Mode Analysis for Solid-State TW Amplifiers", Int. J. Electron., 1973, 35, no.2, pp.145-62.

25.319. Mahapatra, S., and Joshi, J.S.: "Ge Tunnel-Diode Oscillators and Amplifiers. Large-Signal Analysis", Int. J. Electron., 1973, 35, no.2, pp.169-76.

25.320. Hambleton, K.G., and Robson, P.N.: "Design Considerations for Resonant TW IMPATT Oscillators", Int. J. Electron., 1973, 35, no.2, pp.225-44.

25.321. Conn, D.R., and Mitchell, R.H.: "Design of Microstrip Transistor Oscillators", Int. J. Electron., 1973, 35, no.3, pp.385-95.

25.322. Kurokawa, K.: "Injection Locking of Microwave Solid-State Oscillators", Proc. IEEE, 1973, 61, no.10, pp.1386-410.

25.323. Gough, R.A., and Newton, B.H.: "Integrated Wideband Varactor-Tuned Gunn Oscillator", Trans. IEEE, 1973, ED-20, no.10, pp.863-5.

25.324. Aishima, A., Uchlike, H., and Fukushima, Y.: "Estimation of Saturation Power of Gunn-Diode Amplifiers", Electron. Commun. Jap,, 1973, 56, no.10, pp.93-9.

25.325. Baxter, G.K.: "Thermal Response of Microwave Transistors under Pulsed-Power Operation", Trans. IEEE, 1973, PHP-9, no.3, pp.185-93.

25.326. Goldwasser, R.E.: "35-GHz Transferred Electron Amplifiers", Proc. IEEE, 1973, 61, no.10, pp.1502-4.

25.327. Cox, N.W., et al.: "X-Band CW TRAPATT Oscillators Using Ring Diodes on Diamond Heat Spreaders", Electron. Lett., 1973, 9, no.12, pp.269-70.

25.328. Pitzalis, O., and Gilson, R.A.: "Broadband Microwave Class-C Transistor Amplifiers", Trans. IEEE, 1973, MTT-21, no.11, pp.660-8.

25.329. Laton, R.W., and Haddad, G.I.: "Characteristics of IMPATT-Diode Reflection Amplifiers", Trans. IEEE, 1973, MTT-21, no.11, pp.668-80.

25.330. Peterson, D.F.: "Device Characterization and Circuit Design Procedure for Realizing High-Power Millimetre-Wave IMPATT-Diode Amplifiers", Trans. IEEE, 1973, MTT-21, no.11, pp.681-9.

25.331. Gupta, M.S.: "Large-Signal Equivalent Circuit for IMPATT-Diode Characterization", Trans. IEEE, 1973, MTT-21, no.11, pp.689-94.

25.332. Kuno, H.J.: "Analysis of Nonlinear Characteristics and Transient Response of IMPATT Amplifiers", Trans. IEEE, 1973, MTT-21, no.11, pp.694-702.

25.333. Kuno, H.J., and English, D.L.: "Nonlinear and Large-Signal Characteristics of Millimetre-Wave IMPATT Amplifiers", Trans. IEEE, 1973, MTT-21, no.11, pp.703-6.

25.334. Willing, H.A.: "Two-Stage IMPATT-Diode Amplifier", Trans. IEEE, 1973, MTT-21, no.11, pp.707-16.

25.335. Komizo, H., et al.: "Improvement of Nonlinear Distortion in an IMPATT Stable Amplifier", Trans. IEEE, 1973, MTT-21, no.11, pp.721-8.

25.336. Yarrington, L.I., and Hawkins, P.W.: "Analysis of Phase Characteristics as a Function of Ambient Temperature of IMPATT Amplifiers", Trans. IEEE, 1973, MTT-21, no.11, pp.728-30.

25.337. Monroe, J.W.: "Effects of Package Parasitics on the Stability of Microwave Negative-Resistance Devices", Trans. IEEE, 1973, MTT-21, no.11, pp.731-5.

25.338. Lidgey, F.J., and Foulds, K.W.H.: "X-Band LSA Amplifier", Trans. IEEE, 1973, MTT-21, no.11, pp.736-8.

25.339. Ivanek, F.: "Increasing the Locking Bandwidth of a Waveguide-Cavity Oscillator Through Use of a Double-Tuned Circuit", Electron. Lett., 1973, 9, no.19, pp.444-5.

25.340. Pfund, G., Podell, A., and Tarakci, U.: "Pulsed Si Double-Drift IMPATT's for Millimetre-Wave Applications", Electron. Lett., 1973, 9, no.22, pp.518-20.

25.341. Hobson, G.S.: "Ambient-Temperature Effects in Transferred-Electron Amplifiers", Electron. Lett., 1973, 9, no.24, pp.559-61.

25.342. Hansson, B.: "Hysteresis Effects in Microwave Amplifiers and Phase-Locked Oscillators Caused by Amplitude-Dependent Susceptance", Trans. IEEE, 1973, MTT-21, no.11, pp.739-41.

25.343. Yoon, C.Y.: "Study on Impedance Matching of Microwave IMPATT-Diode Oscillator", J. Korean Inst. Electron. Eng., 1973, 10, no.4, pp.74-9.

25.344. Morgan, G.B.: "Linear Microwave Amplification Using IMPATT Diodes", Radio Electron. Eng., 1973, 43, no.10, pp.625-30.

25.345. De Sa, B.A.E., and Hobson, G.S.: "Design Criteria for CW Gunn Oscillators With Good Frequency/Temperature Stability", Solid-State Electron., 1973, 16, no.11, pp.1261-6.

25.346. Manzelev, I.A., and Popov, V.I.: "Push-Pull Gunn-Diode Oscillators", Radiotekhnika, 1973, 28, no.1, pp.99-101, and Telecommun. Radio Eng. Pt 2, 1973, 28, no.1, pp.126-8.

25.347. Mark, I.I.: "Analogue Models of Synchronized Tunnel- and IMPATT-Diode Microwave Generators", Radiotekhnika, 1973, 28, no.7, pp.93-6, and Telecommun. Radio Eng. Pt 2, 1973, 28, no.7, pp.121-3.

25.348. Arendar', V.N., and Prokhorov, E.D.: "Frequency Jump in Gunn Oscillators", Radiotekh. Elektron., 1973, 18, no.6, pp.1320-1, and Radio Eng. Electron. Phys., 1973, 18, no.6, pp.977-8.

25.349. Ohmori, M., and Ino, M.: "Characteristics of Millimetre-Wave Si IMPATT Diodes With Abrupt Junctions", Electron. Commun. Jap., 1973, 56, no.5, pp.108-14.

25.350. Matino, H.: "Microwave GaAs Schottky-Gate FET", Electron. Commun. Jap., 1973, 56, no.7, pp.90-6.

25.351. Porsev, V.I., and Orlov, O.E.: "Frequency Stability of a Gunn Oscillator", Elektrosvyaz, 1973, 27, no.12, pp.60-3, and Telecommun. Radio Eng. Pt 1, 1973, 27, no.12, pp.51-3.

25.352. Risko, J.J., et al.: "8-10 GHz TRAPATT Diode Sources", Electron. Lett., 1973, 9, no.24, pp.572-3.

25.353. Kenyon, N.D.: "Single-Tuned Solid-State Microwave Oscillators", Int. J. Circuit Theory Appl., 1973, 1, no.4, pp.387-93.

25.354. Bosch, R., and Thim, H.W.: "Computer Simulation of Transferred-Electron Devices Using Displaced Maxwellian Approach", Trans. IEEE, 1974, ED-21, no.1, pp.16-25.

25.355. Shackle, P.W.: "Experimental Study of Distributed Effects in a Microwave Bipolar Transistor", Trans. IEEE, 1974, ED-21, no.1, pp.32-9.

25.356. Wahl, A.J.: "Distributed Theory for Microwave Bipolar Transistors", Trans. IEEE, 1974, ED-21, no.1, pp.40-9.

25.357. Swartz, G.A., et al.: "Performance of p-Type Epitaxial Si Millimetre-Wave IMPATT Diodes", Trans. IEEE, 1974, ED-21, no.2, pp.165-71.

25.358. Migitaka, M., et al: "20-GHz High-Power GaAs DDR-IMPATT Diodes With a p^+-p-n-n^+ Structure", Proc. IEEE, 1974, 62, no.1, pp.141-2.

25.359. Watanabe, T., Kodera, H., and Migitaka, M.: "GaAs 50-GHz Schottky-Barrier IMPATT Diodes", Electron. Lett., 1974, 10, no.1, pp.7-8.

25.360. Culshaw, B., and Giblin, R.A.: "High-Efficiency Microwave Oscillations Using Four-Layer Structure", Electron. Lett., 1974, 10, no.3, pp.27-9.

25.361. Ruttan, T.G.: "High-Frequency Gunn Oscillators", Trans. IEEE, 1974, MTT-22, no.2, pp.142-4.

25.362. Carroll, J.E.: "Effects of Series Loss on Circuit-Controlled TRAPATT Diodes", Int. J. Electron., 1974, 36, no.2, pp.145-61.

25.363. Salmon, J.: "MIC Phase-Locked-Loop Avalanche Oscillator in X Band", Trans. IEEE, 1974, MTT-22, no.4, pp.464-6.

25.364. Knerr, R.H., and Murray, J.H.: "Microwave Amplifier Using Several IMPATT Diodes in Parallel", Trans. IEEE, 1974, MTT-22, no.5, pp.569-72.

25.365. Corbey, C.D.: "Characteristics of IMPATT Diodes in Relation to Wideband Varactor Tuned Oscillators", Acta Electron., 1974, 17, no.2, pp.1-6.

25.366. Berenz, J.J., Ying, R.S., and Lee, D.H.: "CW Operation of Ion-Implanted GaAs Read-Type IMPATT Diodes", Electron. Lett., 1974, 10, no.9, pp.157-8.

25.367. Colliver, D.J., et al.: "High-Efficiency InP Transferred-Electron Oscillators", Electron. Lett., 1974, 10, no.11, pp.221-2.

25.368. Bullimore, E.D., Downing, B.J., and Myers, F.A.: "Frequency-Stable Pulsed Gunn Oscillators", Electron. Lett., 1974, 10, no.11, pp.220-1.

25.369. Iwai, F., Hayashi, H., and Fujita, T.: "Millimetre-Wave IMPATT Oscillator", Fujitsu Sci. Tech. J., 1974, 10, no.1, pp.55-74.

25.370. El-Sayed, O.L.: "Impedance Characterization of a Two-Post Mounting Structure for Varactor-Tuned Gunn Oscillators", Trans. IEEE, 1974, MTT-22, no.8, pp.769-76.

25.371. Bastida, E.M., and Conciauro, G.: "Influence of Harmonics on Power Generated by Waveguide-Tunable Gunn Oscillators", Trans. IEEE, 1974, MTT-22, no.8, pp.796-8.

25.372. Tozer, R.C., Charlton, R., and Hobson, G.S.: "Characteristics of Microwave Circuits Using an IMPATT Diode Biased Below Breakdown", Trans. IEEE, 1974, MTT-22, no.8, pp.806-8.

25.373. Chan, C.K., and Cole, R.S.: "Stable Microwave Integrated-Circuit X-Band Gunn Oscillator", Trans. IEEE, 1974, MTT-22, no.8, p.815.

25.374. Al-Charchafchi, S.H.: "Microwave Tunnel-Diode Amplifier", Int. J. Electron., 1974, 36, no.6, pp.753-66.

25.375. Cripps, S.C., Orton, R.S., and Carroll, J.E.: "Combined Theoretical and Experimental Studies of a Push-Pull TRAPATT Circuit", Int. J. Electron., 1974, 37, no.1, pp.1-21.

25.376. Das, A., Rakshit, P.C., and Paria, H.: "Circuit Dependence of Fundamental Power in a Multifrequency Gunn Oscillator", Int. J. Electron., 1974, 37, no.1, pp.33-40.

25.377. Ho, P.T., and Curtice, W.R.: "Microstrip TRAPATT Amplifier for X-Band Operation", Proc. IEEE, 1974, 62, no.7, pp.1029-30.

25.378. Zalud, V.: "Useful Semiconductor Microwave Circuits", Sdelovaci Tech., 1974, 22, no.5, pp.175-9.

25.379. Willing, H.A.: "Two-Stage IMPATT Diode Amplifier", Telecommun. J., 1974, 41, no.7, pp.439-45.

25.380. Sawayama, Y., et al.: "Stabilized Double-Resonator-Coupled Gunn Oscillator", Toshiba Rev., 1974, no.90, pp.31-6.

25.381. Jeremy, M.L., and Howes, M.J.: "Large-Signal Circuit Characterization of Solid-State Microwave Oscillator Devices", Trans. IEEE, 1974, ED-21, no.8, pp.488-99.

25.382. Dean, M., and Howes, M.J.: "Electronic Tuning of Stable Transferred-Electron Oscillators", Trans. IEEE, 1974, ED-21, no.9, pp.563-9.

25.383. Izgagin, L.N., et al.: "Computer-Aided Design of Very Wideband Transistor Microwave Amplifiers", Izv. VUZ Radioelektron., 1974, 17, no.6, pp.112-7.

25.384. Geleji, V., and Henk, T.: "Test of Gunn Oscillators With Domain Operation", Hiradastechnika, 1974, 25, no.5, pp.134-41.

25.385. Grubin, H.L.: "Bias-Dependent Oscillations in 10 micron-Long Transferred-Electron Oscillators", Electron. Lett., 1974, 10, no.18, pp.371-2.

25.386. Sechi, F.N.: "Stable Frequency/Temperature Characteristics for TEO's", Microwave J., 1974, 17, no.7, pp.33-45.

25.387. Hammershaimb, E., Jeppesen, P., and Schjaer-Jacobsen, H.: "Computer-Aided Design of Broadband Reflection-Type (Transferred-Electron) Amplifiers", Int. J. Circuit Theory Appl., 1974, 2, no.3, pp.261-8.

25.388. Bastida, E.M., Conciauro, G., and Pierini, G.: "General Programme for Simulation of Gunn-Diode Microwave Oscillators", Alta Freq., 1974, 43, no.7, pp.407-12.

25.389. Obah, C.O.G., et al.: "Single-Diode 0.5-kW TRAPATT Oscillator", Electron. Lett., 1974, 10, no.21, pp.430-1.

25.390. Blakey, P.A., Culshaw, B., and Giblin, R.A.: "Efficiency Enhancement in Avalanche Diodes by Depletion-Region-Width Modulation", Electron. Lett., 1974, 10, no.21, pp.435-6.

25.391. Rzhevkin, K.S., and Snigirev, O.V.: "Synchronization Characteristic of Gunn-Effect Oscillators", Izv. VUZ Radioelektron., 1974, 17, no.11, pp.81-6.

25.392. Cullen, A.L., and Forrest, J.R.: "Analytic Theory of the IMPATT Diode and Application to Calculations of Oscillator Locking Characteristics", Proc. IEE, 1974, 121, no.12, pp.1467-74.

25.393. Stevens, R., and Myers, F.A.: "Temperature Compensation of Gunn Oscillators", Electron. Lett., 1974, 10, no.22, pp.463-4.

25.394. Heng, T.M.S., and Nathanson, H.C.: "Vertical MOS Transistor Geometry for Power Amplification at Gigahertz Frequencies", Electron. Lett., 1974, 10 no.23, pp.490-2.

25.395. Stevens, R., Tarrant, D., and Myers, F.A.: "Pulsed Gunn-Diode Oscillator. 40W at 16 GHz", Electron. Lett., 1974, 10, no.25-26, pp.531-3.

25.396. Dobrowolski, J.: "Experimental Analysis of Behaviour of X-Band Gunn-Diode Oscillator", Elektronika, 1974, 15, no.9, pp.397-400.

25.397. Kondo, A., Ishii, T., and Shirahata, K.: "Simple Stabilizing Method for Solid-State Microwave Oscillators", Trans. IEEE, 1974, MTT-22, no.11, pp.970-2.

25.398. Seddik, M.M., and Haddad, G.I.: "Properties of Millimetre-Wave IMPATT Diodes", Trans. IEEE, 1974, ED-21, no.12, pp.809-11.

25.399. Nagano, S., and Ohnaka, S.: "Low-Noise 80-GHz Si IMPATT Oscillator Highly Stabilized with a Transmission Cavity", Trans. IEEE, 1974, MTT-22, no.12, pp.1152-9.

25.400. Trew, R.J., Masnari, N.A., and Haddad, G.I.: "Optimization of S-Band TRAPATT Oscillators", Trans. IEEE, 1974, MTT-22, no.12, pp.1166-70.

25.401. Glance, B., and Schneider, M.V.: "Millimetre-Wave Microstrip Oscillators", Trans. IEEE, 1974, MTT-22, no.12, pp.1281-3.

25.402. Gover, A., Burrell, K.H., and Yariv, A.: "Solid-State TW Amplification in the Collisionless Regime", J. Appl. Phys., 1974, 45, no.11, pp.4847-51.

25.403. Dannecker, R.F., and Gunn, M.W.: "Design for Waveguide Gunn-Effect Oscillator", Proc. IREE Aust., 1974, 35, no.7, pp.210-2.

25.404. Tomizawa, K., and Hartnagel, H.: "Characterization Methods of Gunn Oscillators", Radio Electron. Eng., 1974, 44, no.12, pp.667-72.

25.405. Rosloniec, S.: "Power Frequency Characteristic of IMPATT Oscillator", Elektronika, 1974, 15, no.12, pp.547-8.

25.406. Nagao, H., and Katayama, S.: "Si IMPATT Device Incorporating Diamond Heat Sink", NEC Res. Dev., 1974, no.35, pp.67-76.

25.407. O'Reilly, G.T., and Neidert, R.E.: "Designing Microstrip Matching Networks for Microwave-Transistor Power Amplifiers", Trans. IEEE, 1974, MTT-22, no.12, pp.1323-5.

25.408. Cox, N.W., et al.: "X-Band TRAPATT Amplifier", Trans. IEEE, 1974, MTT-22, no.12, pp.1325-8.

25.409. Lee, C.K.: "Tunnel-Diode Oscillator as Self-Excited Mixer for Moving Targets", J. Korean Inst. Electron. Eng., 1974, 11, no.1, pp.40-6.

25.410. Noesen, P., Winzeler, H.R., and Bodmer, F.: "Semiconductor Oscillator for 11-GHz Beamed Links of Large Channel Capacity", Bull. Assoc. Suisse Electr., 1974, 65, no.11, pp.813-20.

26. INDIRECT OPERATION

26.1.   Demidov, B.A., et al.: "Anomalous Resistance and Microwave Radiation from a Plasma in a Strong Electric Field", Zh. Eksp. Teor. Fiz., 1965, 48, p.454, and Sov. Phys.-JETP, 1965, 21.

26.2.   Idehara, T., Abe, H., and Kubo, H.: "Microwave Radiation from a Bounded Plasma", J. Phys. Soc. Jap., 1965, 20, p.2298.

26.3.   Chiyoda, K.: "Theory of Harmonic Generation in a Gaseous Plasma", J. Phys. Soc. Jap., 1965, 20, p.290.

26.4.   Kaufman, I., and Oltman, G.: "Harmonic Generation by Electron-Beam Pattern Motion. Bermutron", Trans. IEEE, 1965, ED-12, p.31.

26.5.   Scanlan, J.O., and Laybourn, P.J.R.: "Large-Signal Analysis of Varactor Harmonic Generators Without Idlers", Proc. IEE, 1965, 112, p.1515.

26.6.   Moriyana, M., and Sumi, M.: "Microwave Generation of Harmonics in a Static Magnetic Field", J. Phys. Soc. Jap., 1965, 20, p.138.

26.7.   Tanaka, S., Kubo, H., and Mitani, K.: "Microwave Radiation Near Electron Cyclotron Harmonics", J. Phys. Soc. Jap., 1965, 20, p.462.

26.8.   Matsuo, G., Takeyama, M., and Mitani, K.: "Microwave Resonance in Afterglow Plasma", J. Phys. Soc. Jap., 1965, 20, p.288.

26.9.   van Iperen, B.B., and Kuypers, W.: "Experimental CW Klystron Multiplier for Submillimetre Waves", Philips Res. Rep., 1965, 20, p.462.

26.10.   Katoh, H., et al.: "Microwave Amplification by Interaction of an Electron Beam with a Caesium Plasma", J. Inst. Electr. Commun. Eng. Jap., 1965, 48, p.21.

26.11.   Collin, R.E.: "Electromagnetic Potentials and Field Expansion for Plasma Radiation in Waveguides", Trans. IEEE, 1965, MTT-13, p.413.

26.12.   Murphy, B.: "Harmonic Generation in a Microwave Discharge", Phys. Fluids, 1965, 8, p.1534.

26.13.   Shih-gu, T.: "Experimental Results on Harmonic Generation Using a Varactor", Acta Phys. Sin., 1965, 21, p.1581.

26.14.   Terumichi, Y., et al.: "Negative Absorption of Microwave Radiation in a Weakly Ionized Xe Plasma in a Magnetic Field", J. Phys. Soc. Jap., 1965, 20, p.1705.

26.15.   Waniek, R.W., Swanson, D.G., and Grannan, R.T.: "Intense Microwave Radiation from Collective Effects in a Plasma", Phys. Rev. Lett., 1965, 15, p.444.

26.16.   Pavlichenko, O.S., et al.: "Instability of Plasma Produced by an Oscillating Electron Discharge. I", Zh. Tekh. Fiz., 1965, 3, p.1394, and Sov. Phys.-Tech. Phys., 1966, 10, no.8, pp.1082-7.

26.17.   Golant, V.E., et al.: "Microwave Radiation from a Beam-Plasma System in a Magnetic Field", Zh. Tekh. Fiz., 1965, 35, p.2034, and Sov. Phys.-Tech. Phys., 1966, 10, no.11, pp.1559-65.

26.18.   Morisaki, H., and Inuishi, Y.: "Harmonic Generation in Microwave Emission from InSb", Jap. J. Appl. Phys., 1966, 5, p.343.

26.19.   Shaw, H.J., et al.: "Microwave Generation in Pulsed Ferrites", J. Appl. Phys., 1966, 37, p.1060.

26.20.   Kita, S., and Seki, S.: "Millimetre-Wave Pulse Generator by Multiplier", Proc. IEEE, 1966, 54, p.71.

26.21.   Deryugin, I.A., et al.: "Frequency Doubling in Ferrites", Radiotekh. Elektron., 1966, 11, p.150, and Radio Eng. Electron. Phys., 1966, 11, p.124.

26.22.   Idehara, T., et al.: "Analysis of Negative Absorption in Plasma-Cavity Systems", J. Phys. Soc. Jap., 1966, 21, p.778.

26.23.   Sodha, M.S., and Srivastava, H.K.: "Microwave Third-Harmonic Generation in Semiconductors at Low Temperatures", Proc. Phys. Soc., 1967, 90, p.435.

26.24.   Brodin, M.S., and Volovik, N.V.: "Optical SHG in GaSe Crystals", Ukr. Fiz. Zh., 1967, 12, p.1503.

26.25.   Glenn, W.H.: "Parametric Amplification of Ultrashort Laser Pulses", Appl. Phys. Lett., 1967, 11, p.250.

26.26.   Bass, M., and Andringa, K.: "Reproducible Optical SHG Using a Mode-Locked Laser", J. Quantum Electron. IEEE, 1967, QE-3, p.621.

26.27.   Grasyuk, A.Z., et al.: "Laser Based on Raman Scattering in Liquid Nitrogen", Zh. Eksp. Teor. Fiz. Pis'ma, 1968, 8, no.9, p.474, and JETP Lett., 1968, 8, no.9, p.291.

26.28.   Corbey, C.D., Davies, R., and Newton, B.H.: "Varactor Harmonic-Generator Chain as Pump Source for Parametric Amplifier", Electron. Lett., 1968, 4, p.397.

26.29.   Vanyukov, M.P., Volosov, V.D., and Rashchektaeva, M.I.: "Effect of Radiation Parameters of a Laser on Efficiency of Frequency Doubling in a KDP Crystal", Opt. Spekt., 1968, 25, no.5, p.735, and Opt. Spectrosc., 1968, 25, no.5, p.410.

26.30.   Sachdev, D.K.: "Microwave Sources With Frequency Multipliers", J. Inst. Telecommun. Eng., 1968, 14, p.324.

26.31. Hansen, N.P.R., and McIntosh, R.E.: "Microwave Emission from a Plasma-Filled Capacitor", Int. J. Electron., 1968,25,p.597.

26.32. Vanyukov, M.P., et al.: "Pulsed Nd^{3+}:Glass Laser with Selected Second-Harmonic Radiation", Opt.-Mekh. Prom., 1968, 35, no.4, p.26, and Sov. J. Opt. Technol., 1968, 3, no.2, p.240.

26.33. van Iperen, B.B., Kuypers, W., and Tjassens, H.: "Klystron Frequency Multiplier for Submillimetre Waves Generating its own Driving Power", Nachr. Tech. Fachber., 1968, 35, p.165.

26.34. Dudnik, R.L., and Sheremet'ev, E.V.: "Parametric Oscillator With Semiconductor-Capacitance Frequency Adjustment", Izmer. Tekh., 1969, no.5, p.60, and Meas. Tech., 1969, no.5, p.679.

26.35. Schmid, Ch.: "Coherence and Saturation Properties of SHG With Nonlinear Crystal Inside the Laser Cavity. I", Z. Phys., 1969, 222, p.314.

26.36. Yajima, T., and Inoue, K.: "Submillimetre-Wave Generation by Difference-Frequency Mixing of Ruby-Laser Lines in ZnTe", J. Quantum Electron. IEEE, 1969, QE-5, p.140.

26.37. Hagen, W.F., and Magnante, P.C.: "Efficient SHG With Diffraction-Limited Nd^{3+}:Glass Lasers", J. Appl. Phys., 1969, 40, p.219.

26.38. Barkhudarova, T.M., et al.: "Dependence of Second-Harmonic Conversion Coefficient Upon Space-Time Distribution of Laser Power", Zh. Prikl. Spekt., 1969, 10, no.1, p.33.

26.39. Bjorkholm, J.E.: "Analysis of Doubly Resonant Optical Parametric Oscillator Without Power-Dependent Reflections", J. Quantum Electron. IEEE, 1969, QE-5,p.293.

26.40. Dmitriev, V.G., Eremeeva, R.A., and Epshov, A.G.: "Observation of Parametric Generation in a Nonlinear Resonator Without Reflecting Dielectric Coatings", Zh. Prikl. Spekt., 1969, 10, no.4, p.658.

26.41. Yarborough, J.M., et al.: "Efficient, Tunable, Optical Emission from LiNbO$_3$ Without a Resonator", Appl. Phys. Lett., 1969, 15, p.102.

26.42. Sodha, M.S., and Gupta, G.P.: "Nonlinear Second-Harmonic and Sum-and-Difference Generation in an Inhomogeneous Semiconductor at Low Temperatures", Indian J. Pure Appl. Phys., 1969, 7, p.289.

26.43. Belohoubek, E.F., and Rosen, A.: "Coupled Microstrip Varactor Doublers", Trans. IEEE, 1969, MTT-17, p.286.

26.44. Harris, S.E.: "Tunable Optical Parametric Oscillators", Proc. IEEE, 1969, 57, p.2096.

26.45. Tamaru, T., Chiyoda, K., and Maejima, Y.: "Harmonic Generation in a Millimetre-Wave Gas Discharge at 55 GHz", Proc. IEEE, 1969, 57, p.851.

26.46. Bloembergen, N., Burns, W.K., and Matsuoka, M.: "Reflected Third Harmonic Generated by Picosecond Laser Pulses", Opt. Commun., 1969, 1, no.4, p.195.

26.47. Akmanov, A.G., Kovrigin, A.I., and Podsotskaya, N.K.: "Frequency Differentiation in Optical Harmonic Generators", Radiotekh. Elektron., 1969, 14, no.8, p.1516, and Radio Eng. Electron. Phys., 1969, 14, no.8, pp.1315-7.

26.48. Davidis, K.Ye., Minakova, I.I., and Semenova, T.A.: "Three-Photon Frequency Multiplier Using Ammonia", Radiotekh. Elektron., 1969, 14, no.9, p.1661, and Radio Eng. Electron. Phys., 1969, 14, no.9, pp.1438-42.

26.49. Orlov, R.Yu., Usmanov, T., and Chirkin, A.S.: "Doubling of Laser Radiation Frequency Under Nonstationary (Mode-Locked) Conditions", Zh. Eksp. Teor. Fiz., 1969, 57, no.4, p.1069, and Sov. Phys.-JETP, 1970, 30, no.4, pp.584-9.

26.50. Deserno, U.: "Stable High-Power Giant-Pulse Laser at 530 nm Using LiIO$_3$", Phys. Lett., 1969, 30A, p.483.

26.51. Asmussen, J., and Beyer, J.B.: "Microwave Harmonic Generation in a Plasma Capacitor", Trans. IEEE, 1969, ED-16, p.18.

26.52. Weinberg, D.L.: "Tunable Optical Parametric Amplifiers and Generators", Laser Focus, 1969, 5, no.4, p.35.

26.53. Machi, Y.: "Microwave-Frequency Multiplication Using Hot Electrons in Semiconductors", Trans. IEEE, 1969,MTT-17,p.333.

26.54. Glenn, W.H.: "SHG by Picosecond Optical Pulses", J. Quantum Electron. IEEE, 1969, QE-5, p.284.

26.55. Lopasov, V.P., and Makogon, M.M.: "SHG by a Nd^{3+} Laser with Lenses Inside the Resonator", Opt. Spektr., 1969, 26, no.6, p.1035, and Opt. Spectrosc., 1969, 26, no.6, p.561.

26.56. Tamaru, T., Chiyoda, K., and Maejima, Y.: "Harmonic Radiation in a Microwave Gas Discharge at 34 GHz", Jap. J. Appl. Phys., 1969, 8, p.104.

26.57. Belyaev, Yu.N., Kiselev, A.M., and Freidman, G.I.: "Investigation of Parametric Generator With Feedback in Only One of the Waves", Zh. Eksp. Teor. Fiz. Pis'ma, 1969, 9, no.8, p.441, and JETP Lett., 1969, 9, no.8, p.263.

26.58. Sodha, M.S., and Gupta, G.P.: "Generation of Sum and Difference Frequencies in Plasmas by Nonuniform Microwave Fields", Indian J. Pure Appl. Phys.,1969,7,no.1,p.54.

26.59. Inaba, H., and Hidaka, T.: "Dif-ference-Frequency Generation Due to Multi-ple-Photon Process during Laser Action", Nature, 1969, 324, p.57.

26.60. Kovrigin, A.I., et al.: "Study of the Angular Structure of the Second Optical Harmonic. II", Opt. Spekt., 1969, 26, no.3, p.393, and Opt. Spectrosc., 1969, 26, no.3, p.215.

26.61. Mednikov, O.I.: "Theory of a Two-Stage Parametric Oscillator", Radiotekh. Elektron., 1969, 14, no.6, p.1120, and Radio Eng. Electron. Phys.,1969,14,no.6,pp.973-6.

26.62. Cooper, B.F.C., and Wells, G.A.: "Six-Times Multiplier With 2-W Output at 2.7 GHz", Proc. IREE Aust., 1969, 30, no.10, p.340.

26.63. Frey, W., et al.: "Influence of Second-Harmonic-Frequency Termination on Gunn-Oscillator Performance", Electron. Lett., 1969, 5, p.691.

26.64. Wang, C.C., and Baardsen, E.L.: "Optical Third-Harmonic Generation Using Mode-Locked and Non-Mode-Locked Lasers", Appl. Phys. Lett., 1969, 15, p.396.

26.65. Volosov, V.D., and Andreev, R.B.: "Generation of Second Harmonic of Nonmono-chromatic Laser Radiation in a Nonlinear Crystal", Opt. Spekt., 1969, 26, no.5, p.809, and Opt. Spectrosc., 1969, 26, no.5, p.441.

26.66. Dathe, G., et al.: "Microwave Frequency Doubling in Ruby", J. Quantum Electron. IEEE, 1969, QE-5, p.169.

26.67. Kirsanov, B.P., et al.: "Laser with Smooth Change of Generation Frequency. IV", Opt. Spekt., 1969, 26, no.5, p.780, and Opt. Spectrosc., 1969, 26, no.5, p.426.

26.68. Chattopadhyay, D., and Nag, B.R.: "Harmonic Generation Using the Nonlinearity in Hot-Carrier Characteristics of GaAs", Int. J. Electron., 1969, 27, p.443.

26.69. Volosov, V.D., et al.: "Efficient Generation of Second-Harmonic Ruby-Laser Radiation", Opt.-Mekh. Prom., 1969, 36, no.5, pp.3-4, and Sov. J. Opt. Technol., 1969, 36, no.5, pp.656-7.

26.70. Oshimoto, A., Kikuchi, K., and Yoshinari, H.: "Millimetre-Wave Frequency Multiplication Using Ohmic Hot-Carrier Diode", Mem. Def. Acad., 1969,8,no.3,p.635.

26.71. Byer, R.L., Kovrigin, A., and Young, J.F.: "CW Ring-Cavity Parametric Oscillator", Appl. Phys. Lett., 1969, 15, p.136.

26.72. Bass, M., et al.: "Optical SHG in Crystals of Organic Dyes", Appl. Phys. Lett., 1969, 15, p.393.

26.73. Lopasov, V.P., and Makogon, M.M.: "SHG by Ruby Laser with Diffuse Reflector", Izv. VUZ Fiz., 1969, no.5, p.133, Opt. Spekt., 1969, 27, p.83, and Opt. Spectrosc., 1969, 27, p.165.

26.74. Bessenov, V.I., et al.: "Applica-tion of Power Series in Analysis of Three-Frequency Parametric Amplifier", Izv. VUZ Radioelektron., 1969, 12, no.9, p.1097.

26.75. Gelbwachs, J.A., et al.: "Tunable Stimulated Raman Oscillator", Appl. Phys. Lett., 1969, 14, p.258.

26.76. Hansch, Th., and Toschek, P.: "Parametric Generation of Multiple Radiation in a Gas Laser", Z. Phys., 1970, 236, no.4, p.373.

26.77. Aslaksen, E.W.: "Threshold of the Optical BW Parametric Oscillator", J. Quan-tum Electron. IEEE, 1970, QE-6, p.612.

26.78. Levitsky, S.M., and Filonenko, E.G.: "Microwave Amplification in an Arc Discharge with Hot Cathode", Ukr. Fiz. Zh., 1970, 15, no.2, p.269.

26.79. Smith, R.G.: "Theory of Intracavity Optical SHG", J. Quantum Electron. IEEE, 1970, QE-6, p.215.

26.80. McGeoch, M.W., and Smith, R.C.: "Optimum SHG in $LiNbO_3$", J. Quantum Electron. IEEE, 1970, QE-6, p.203.

26.81. Suematsu, Y.: "Tunable Parametric Oscillator Using a Guided-Wave Structure", Jap. J. Appl. Phys., 1970, 9, p.798.

26.82. Volosov, V.D., and Rashchektaeva, M.I.: "Efficient SHG from Nd^{3+}:Glass Laser Radiation", Opt. Spekt., 1970, 28, no.1, p.105, and Opt. Spectrosc., 1970, 28, no.1, p.53.

26.83. Goldberg, L.S.: "Optical Paramet-ric Oscillation in $LiIO_3$", Appl. Phys. Lett., 1970, 17, no.11, pp.489-91.

26.84. Wallace, R.W.: "Stable, Efficient, Optical Parametric Oscillators Pumped With Doubled Nd^{3+}:YAG (Emission)", Appl. Phys. Lett., 1970, 17, no.11, pp.497-9.

26.85. Watanabe, T., and Takita, S.: "High-Power Frequency Multiplier Using Var-actor Diodes", Rev. Electr. Commun. Lab., 1970, 18, no.7-8, pp.475-92.

26.86. Bjorkholm, J.E., Ashkin, A., and Smith, R.G.: "Improvement of Optical Para-metric Oscillators by Nonresonant Pump Ref-lection", J. Quantum Electron. IEEE, 1970, QE-6, no.12, pp.797-9.

26.87. Dionne, N.: "Harmonic Generation in Octave-Bandwidth TWT's", Trans. IEEE, 1970, ED-17, p.365.

26.88. Ammann, E.O.: "Efficient Internal Optical Parametric Oscillation", Appl. Phys. Lett., 1970, 16, p.309.

26.89.　Corcoran, V.J., et al.: "Nonlinear Optical Effects Using a CO_2 Laser and a Klystron", Appl. Phys. Lett., 1970, 16, p.316.

26.90.　Bokut', B.V.: "Special-Geometry Nonlinear Frequency Converter", Zh. Prikl. Spekt., 1970, 12, p.223.

26.91.　Kruzhilin, Yu.I., and Koloskov, Yu.I.: "Second Harmonic in Injection Laser Radiation", Zh. Prikl. Spekt., 1970, 12, p.334.

26.92.　Chesler, R.B., Karr, M.A., and Geusic, J.E.: "Repetitively Q-Switched, Nd^{3+}:YAG and $LiIO_3$, 530-nm Harmonic Source", J. Appl. Phys., 1970, 41, p.4125.

26.93.　Chi, C.: "Second-Harmonic Generation of a GaAs Laser", J. Appl. Phys., 1970, 41, p.3184.

26.94.　Gandrud, W.B., and Abrams, R.L.: "Reduction in SHG Efficiency in Te by Photoionized Carriers", Appl. Phys. Lett., 1970, 17, no.7, pp.302-5.

26.95.　Bhawalkar, D.D., Colles, M.J., and Smith, R.C.: "Mode Conversion in Optical SHG", Opto-Electron., 1970, 2, no.2, pp.90-7.

26.96.　Aslaksen, E.W.: "Optical BW Parametric Oscillator Above Threshold", Opt. Commun., 1970, 2, no.2, pp.69-72.

26.97.　Ammann, E.O., and Yarborough, J.M.: "Optical Parametric Oscillation in Proustite", Appl. Phys. Lett., 1970, 17, no.6, pp.233-5.

26.98.　Nath, G., and Mehmanesch, H.: "Efficient Conversion of a Ruby Laser to 347 nm in Low-Loss $LiIO_3$", Appl. Phys. Lett., 1970, 17, no.7, pp.286-8.

26.99.　Ammann, E.O., and Montgomery, P.C.: "Threshold Calculations for an Optical Parametric Oscillator Employing a Hemispherical Resonator", J. Appl. Phys., 1970, 41, no.13, pp.5270-4.

26.100.　White, D.R., Dawes, E.L., and Marburger, J.H.: "Theory of SHG with High-Conversion Efficiency", J. Quantum Electron. IEEE, 1970, QE-6, no.12, pp.793-6.

26.101.　Sushchik, M.M., and Freidman, G.I.: "Influence of Homogeneity of Pump Radiation on Spatial Locking of Parametrically Amplified Waves", Izv. VUZ Radiofiz., 1970, 13, no.9, pp.1354-60.

26.102.　Turkin, A.A.: "SHG in Two-Circuit Parametric Oscillator", Radiotekh. Elektron., 1970, 15, no.7, and Radio Eng. Electron. Phys., 1970, 15, no.7, pp.1324-7.

26.103.　Edmonds, H.D., and Smith, A.W.: "Second-Harmonic Generation with GaAs Laser", J. Quantum Electron. IEEE, 1970, QE-6, p.356.

26.104.　Ippen, E.P.: "Low-Power Quasi-CW Raman Oscillator", Appl. Phys. Lett., 1970, 16, p.303.

26.105.　Derkacheva, L.D., Krymova, A.I., and Sopina, N.P.: "Obtaining Second Harmonic of a Nd^{3+}:Glass Laser in Dye Powders", Zh. Eksp. Teor. Fiz. Pis'ma, 1970, 11, no.10, p.319, and Sov. Phys.-JETP, 1970, 11, no.10, p.469.

26.106.　Smith, R.G., and Parker, J.V.: "Experimental Observation of and Comments on Optical Parametric Oscillation Internal to the Laser Cavity", J. Appl. Phys., 1970, 41, p.3401.

26.107.　Furukawa, S., and Nakagami, T.: "Design Theory of Self-Excited Frequency-Multiplying Oscillator", J. Solid-State Circuits IEEE, 1970, SC-5, p.87.

26.108.　Izrailenko, A.I., Kovrigin, A.I., and Nikles, P.V.: "Parametric Generation of Light in High-Efficiency Nonlinear $LiIO_3$ and HIO_3 Crystals", Zh. Eksp. Teor. Fiz. Pis'ma, 1970, 12, no.10, pp.475-8, and JETP Lett., 1970, 12, no.10, pp.331-3.

26.109.　Shinohara, S.: "Millimetre-Wave Frequency Multiplier Circuit Using a Semiconductor Diode", Rep. Fac. Eng. Shizuoka Univ., 1970, no.21, pp.67-77.

26.110.　Mushta, A.I., and Novozhilov, O.P.: "Maximum Convertible Power of a Threefold (Varactor) Parametric Frequency Converter", Radiotekh. Elektron., 1970, 15, no.9, and Radio Eng. Electron. Phys., 1970, 15, no.9, pp.1743-6.

26.111.　Golovin, E.S., and Kravchenko, G.I.: "Mechanisms for Formation of Deterministic Phases in Unbalanced Parametric Oscillations", Radiotekh. Elektron., 1970, 15, no.12, and Radio Eng. Electron. Phys., 1970, 15, no.12, pp.2278-80.

26.112.　Grasyuk, A.Z., et al.: "Increase of Radiation Brightness in a Brillouin Laser", Zh. Eksp. Teor. Fiz. Pis'ma, 1970, 12, no.6, pp.286-9, and JETP Lett., 1970, 12, no.6, pp.193-5.

26.113.　Kielich, S.: "Optical Harmonic Generation and Laser Light Frequency Mixing Processes in Nonlinear Media", Opto-Electron., 1970, 2, no.3, pp.125-51.

26.114.　Akahori, H.: "Effects of an Electric Field on Phase Matching in Optical SHG", Bull. Electrotech. Lab., 1970, 34, no.8, pp.658-72.

26.115.　Teich, M.C.: "Photon-Correlation Enhancement of SHG at 10.6 micron", Opt. Commun., 1970, 2, no.5, pp.206-8.

26.116.　Sushchik, M.M., Fortus, V.M., and Freidman, G.I.: "Resonatorless Parametric Light Oscillator", Izv. VUZ Radiofiz., 1971, 14, no.2, pp.263-8.

26.117.　Moshkina, T.V., and Tarantovich, A.S.: "Theory of a Resistive Parametric Oscillator", Izv. VUZ Radiofiz., 1971, 14, no.3, pp.459-67.

26.118. Lacina, W.B., et al.: "CO_2-Laser Frequency Doubling Using Te Reflector", Appl. Opt., 1971, 10, no.1, pp.221-2.

26.119. Bardati, F., Fossati, M., and Gerosa, G.: "Second-Harmonic Generation in Ferrites", Alta Freq., 1971, 40, no.2, pp.147-51.

26.120. Yarborough, J.M., and Ammann, E.O.: "Simultaneous Optical Parametric Oscillation, SHG, and Difference-Frequency Generation", Appl. Phys. Lett., 1971, 18, no.4, pp.145-7.

26.121. Falk, J.: "Instabilities in the Doubly Resonant Parametric Oscillator. Theoretical Analysis", J. Quantum Electron. IEEE, 1971, QE-7, no.6, pp.230-5.

26.122. Henningsen, T., Feichtner, J.D., and Melamed, N.T.: "Frequency Doubling of 2.06-micron Ho^{3+}:Apatite Laser Output in Ag_3AsS_3", J. Quantum Electron. IEEE, 1971, QE-7, no.6, pp.248-50.

26.123. McDonald, D.G., et al.: "Harmonic Mixing of Microwave and Far-IR Laser Radiation Using a Josephson Junction", Appl. Phys. Lett., 1971, 18, no.4, pp.162-4.

26.124. Laurence, C., and Tittel, F.: "Visible CW Parametric Oscillator Using $Ba_2NaNb_5O_{15}$", J. Appl. Phys., 1971, 42, no.5, pp.2137-8.

26.125. Borodovskii, P.A., and Bul'dygin, A.F.: "(Parametric) Amplification in a Gunn Diode with External Signal", Fiz. Tekh. Poluprov., 1971, 5, no.2, pp.247-50, and Sov. Phys.-Semicond., 1971, 5,no.2,pp.211-4.

26.126. Hanna, D.C., Smith, R.C., and Stanley, C.R.: "Generation of Tunable Medium-IR Radiation by Optical Mixing in Proustite", Opt. Commun., 1971, 4, no.4, pp.300-3.

26.127. Hamadani, S.M., and Magyar, G.: "Tunable Near-UV Emission by Second Harmonic of a Dye Laser", Opt. Commun., 1971, 4, no.4, pp.310-2.

26.128. Wallace, R.W.: "Generation of Tunable Radiation Over 261-315 nm", Opt. Commun., 1971, 4, no.4, pp.316-8.

26.129. Smith, R.G.: "Effect of Competing Laser Transitions on Intracavity Optical SHG", J. Quantum Electron. IEEE, 1971, QE-7, no.4, pp.150-2.

26.130. Bjorkholm, J.E.: "Effects of Spatially Nonuniform Pumping in Pulsed Optical Parametric Amplifiers", J. Quantum Electron. IEEE, 1971, QE-7, no.3, pp.109-18.

26.131. Snapp. C.P., and Baukus, J.P.: "High-Harmonic Content in Trapped-Plasma-Mode Oscillators", Electron. Lett., 1971, 7, no.15, pp.426-8.

26.132. Krivoshchekov, G.V., Nikulin, N.G., and Sokolovskii, R.I.: "SHG Using Pulse-Modulated Light Waves", Avtometriya, 1971, no.1, pp.89-101.

26.133. Riley, J.R.: "Factors Influencing Conversion Efficiency in Gaseous-Plasma Harmonic Generators at Microwave Frequencies", Proc. IEE, 1971, 118, no.10, pp.1339-44.

26.134. Liu, S.G.: "Harmonic Extraction from High-Efficiency Avalanche Diodes", Proc. IEEE, 1971, 59, no.8, pp.1216-21.

26.135. Norinskii, L.V., and Kolosov, V.A.: "Effective Generation of Powerful Coherent UV Radiation", Zh. Eksp. Teor. Fiz. Pis'ma, 1971, 13, no.4, pp.189-92, and JETP Lett., 1971, 13, no.4, pp.133-5.

26.136. Fischer, R.: "Calculation of Efficiency of an Optical Parametric Oscillator With Resonance Only with One of the Generated Waves", Ann. Phys. (Leipzig), 1971, 27, no.1, pp.101-6.

26.137. Campillo, A.J., and Tang, C.L.: "Extending the Range of Tunable Oscillators by Upconversion", Appl. Phys. Lett., 1971, 19, no.2, pp.36-8.

26.138. Anderson, D.B., and Boyd, J.T.: "Wideband CO_2-Laser SHG Phase Matched in GaAs Thin-Film Waveguides", Appl. Phys. Lett., 1971, 19, no.8, pp.266-8.

26.139. Dmitriev, V.G., et al.: "Generation of Optical Harmonics Under High-PRF Conditions", Kvantovaya Elektron., 1971, no.1, pp.116-9.

26.140. Brunner, W., Paul, H., and Bandilla, A.: "Optical Parametric Oscillators. I and II", Ann. Phys. (Leipzig), 1971, 27, no.1, pp.69-81 and 82-90.

26.141. Boyd, G.D., and Nash, F.R.: "Parametric Oscillation Thresholds in Symmetric and Half-Symmetric Configurations in the Absence of Double Refraction", J. Appl. Phys., 1971, 42, no.7, pp.2815-7.

26.142. Okada, M., and Ieiri, S.: "Efficiency in Optical Mixing Between Waves at 1060 nm and 530 nm", Jap. J. Appl. Phys., 1971, 10, no.6, p.808.

26.143. Dewey, C.F., and Hocker, L.O.: "IR Difference-Frequency Generation Using a Tunable Dye Laser", Appl. Phys. Lett., 1971, 18, no.2, pp.58-60.

26.144. Young, J.F., et al.: "Pump Line-width Requirement for Optical Parametric Oscillators", J. Appl. Phys., 1971, 42, no.1, pp.497-8.

26.145. Boissel, P.: "Generation of Coherent Second Harmonic in Macroscopically Isotropic Medium", Opt. Commun., 1971, 2, no.8, p.410.

26.146. Bey, P.P., Giuliani, J.F., and Rabin, H.: "Enhanced Optical Third-Harmonic Generation by Coupled Nonlinear Absorption", J. Quantum Electron. IEEE, 1971, QE-7, no.2, pp.86-8.

26.147. Bradley, D.J., Nicholas, J.V., and Shaw, J.R.D.: "Megawatt Tunable Second-Harmonic and Sum-Frequency Generation at 280 nm from a Dye Laser", Appl. Phys. Lett., 1971, 19, no.6, pp.172-3.

26.148. Petrov, A.S., and Tabarin, V.A.: "Investigation of a Parametric Amplifier Using a Single Crystal of Ferrite", Radiotekh. Elektron., 1971, 16, no.4, pp.544-8, and Radio Eng. Electron. Phys., 1971, 16, no.4, pp.630-3.

26.149. Voronina, L.A., et al.: "Frequency Doubler Employing Diodes With Metal-Semiconductor Barrier", Radiotekh. Elektron., 1971, 16, no.5, pp.880-2, and Radio Eng. Electron. Phys., 1971, 16, no.5, pp.899-900.

26.150. Kravchenko, V.I., Smirnov, A.A., and Soskin, M.S.: "Frequency Tuning and High-Efficiency Extraction of SHG from Prism Dispersion Resonator of a Nd^{3+} Laser", Kvantovaya Elektron., 1971, no.5, pp.131-3.

26.151. Dmitriev, V.G., et al.: "Efficient Generation of Second Harmonic of CW Nd^{3+}:YAG Laser Radiation in $LiNbO_3$", Kvantovaya Elektron., 1971, no.5, pp.133-6.

26.152. Kaneda, S., and Abe, M.: "Microwave Harmonic Generator Using Nonlinearity of Negative Resistance in n-GaAs", Electron. Commun. Jap., 1971, 54, no.10, pp.82-7.

26.153. Kovrigin, A.I., and Nikles, P.V.: "Resonatorless Parametric Light Generator Using HIO_3", Zh. Eksp. Teor. Fiz. Pis'ma, 1971, 13, no.8, pp.313-5, and JETP Lett., 1971, 13, no.8, pp.440-3.

26.154. Genkin, R.O., et al.: "SHG in a Resonator", Opt. Spekt., 1971, 30, no.1, pp.137-9, and Opt. Spectrosc., 1971, 30, no.1, pp.72-3.

26.155. Chemla, D.S., and Kupecek, Ph.J.: "Analysis of SHG Experiments", Rev. Phys. Appl., 1971, 6, no.1, pp.31-50.

26.156. Tomishima, K., Okada, K., and Ito, S.: "SHG With Giant-Pulse Laser and KDP", Mitsubishi Denki Giho, 1971, 45, no.2, pp.250-5.

26.157. Vakulenko, V.M., et al.: "Single-Pulse Laser With Cascaded Multipliers", Prib. Tekh. Eksp., 1971, 14, no.5, pp.197-200, and Instrum. Exp. Tech., 1971, 14, no.5, pp.1496-8.

26.158. Kielich, S., and Zawodny, R.: "DC-Magnetic-Field Induced SHG of Laser Beam", Opt. Commun., 1971, 4, no.2, pp.132-4.

26.159. Gol'din, Yu.A., Dmitriev, V.G., and Lisovskii, L.P.: "Method of Shortening Pulse for SHG in Nonlinear Resonator", Izv. VUZ Radiofiz., 1971, 14, no.12, pp.1801-4.

26.160. Sushchik, M.M., and Freidman, G.I.: "Optimal Focusing of Pumping for Parametrically Coupled Oscillations in Resonators", Izv. VUZ Radiofiz., 1971, 14, no.8, pp.1176-81.

26.161. Rice, R.R., and Burkhart, G.H.: "Efficient Mode-Locked Frequency-Doubled Operation of a Nd^{3+}:$YAlO_3$ Laser", Appl. Phys. Lett., 1971, 19, no.7, pp.225-7.

26.162. Stafsudd, O.M., and Alexander, D.H.: "CW SHG in Single-Crystal CdTe", Appl. Opt., 1971, 10, no.11, pp.2566-7.

26.163. Karr, M.A.: "Nd^{3+}:YAG/ $Ba_2NaNb_5O_{15}$ 660 nm Harmonic Source", J. Appl. Phys., 1971, 42, no.11, pp.4517-9.

26.164. Kozlov, V.A., Navrotskii, V.I., and Vizel', A.A.: "Study of Varactor Frequency-Doubler Operation at 77°K", Radiotekh. Elektron., 1971, 16, no.3, and Radio Eng. Electron. Phys., 1971, 16, no.3, pp.554-6.

26.165. Belyaev, Yu.N., Kiselev, A.M., and Freidman, G.I.: "Optical Parametric Oscillator With Two Interacting Regions", Izv. VUZ Radiofiz., 1971, 14, no.8, pp.1182-8.

26.166. Brueck, S.R.J., and Mooradian, A.: "Efficient, Single-Mode, CW Tunable Spin-Flip Raman Laser", Appl. Phys. Lett., 1971, 18, no.6, pp.229-30.

26.167. Aggarwal, R.L., et al.: "High-Intensity Tunable InSb Spin-Flip Raman Laser", Appl. Phys. Lett., 1971, 18, no.9, pp.383-5.

26.168. Shaw, E.D.: "Spin-Flip Laser", Phys. Bull., 1971, 22, pp.389-90.

26.169. Grasyuk, A.Z., et al.: "Increase of Intensity of Emission by a Brillouin Laser", Kvantovaya Elektron., 1971, no.1, pp.70-8.

26.170. Pidgeon, C.R., et al.: "Tunable Coherent (Spin-Flip plus CO_2) Source in the 5-micron Region", Appl. Phys. Lett., 1971, 19, no.9, pp.333-5.

26.171. Kovalev, V.I., et al.: "Single-Frequency Brillouin Laser Using Methane", Zh. Eksp. Teor. Fiz. Pis'ma, 1971, 14, no.9, pp.503-7, and JETP Lett., 1971, 14, no.9, pp.344-7.

26.172. Patel, C.K.N.: "Tunable Spin-Flip Raman Laser at Fields as Low as 400G", Appl. Phys. Lett., 1971, 19, no.10, pp.400-3.

26.173. Burkley, C.J., Phelan, P.T., and Sexton, M.C.: "Bias Characteristics of an X- to Q-Band GaAs Varactor Tripler", Int. J. Electron. 1971, 30, no.5, pp.473-80.

26.174. Hitz, C.B., and Osterink, L.M.:
"Simultaneous Intracavity Frequency Doubling
and Mode Locking in a Nd^{3+}:YAG Laser",
Appl. Phys. Lett., 1971, 18, no.9,pp.378-80.

26.175. Laurence, C., and Tittel, F.:
"Prediction of the Tuning Characteristics
of an Optical Oscillator Using Parametric
Fluorescence", Opto-Electron., 1971, 3,
no.2, pp.1-4.

26.176. Crosignani, B., et al.: "Effect
of Pump Coherence on Frequency Conversion
and Parametric Amplification", J. Quantum
Electron. IEEE, 1971, QE-8, no.9, pp.731-9.

26.177. Meltzer, D.W., and Goldberg,
L.S.: "Tunable IR Difference-Frequency
Generation in $LiIO_3$", Opt. Commun., 1972,
5, no.3, pp.209-11.

26.178. Dunning, F.B., Stokes, E.D., and
Stebbings, R.F.: "Efficient Generation of
Coherent Radiation Tunable Over 250-325 nm",
Opt. Commun., 1972, 6, no.1, pp.63-6.

26.179. Hanna, D.C., et al.: "Reliable
Operation of a Proustite Parametric Oscillator", Appl. Phys. Lett., 1972, 20, no.1,
pp.34-6.

26.180. Kozlov, V.A., and Piskarev, V.I.:
"Frequency Multiplication in Millimetre
Range With n-InSb", Izv. VUZ Radiofiz.,
1972, 15, no.2, pp.300-4.

26.181. Zemon, S., et al.: "High-Power
Effects in Nonlinear Optical Waveguides",
Appl. Phys. Lett., 1972, 21, no.7,pp.327-9.

26.182. Johnston, R.H., and Kiddle, E.R.:
"Harmonic Tuning for High-Efficiency GaAs
Oscillators", Proc. IEEE, 1972, 61, no.11,
pp.1449-51.

26.183. Herbst, R.L., and Byer, R.L.:
"Singly Resonant CdSe IR Parametric Oscillator", Appl. Phys. Lett., 1972, 21, no.5,
pp.189-91.

26.184. Okada, M., and Ieiri, S.: "Generation of UV Coherent Radiation by Nonlinear
Optical Processes", NHK Tech. J., 1972, 24,
no.2, pp.49-57.

26.185. Dinev, S.G., Stamenov, K.V., and
Tomov, I.V.: "Generation of Tunable Radiation in the Range 216-234 nm", Opt. Commun.,
1972, 5, no.5, pp.419-21.

26.186. Isyanova, E.D., and Ovchinnikov,
V.M.: "Generation of Single Pulses of Sec-
ond-Harmonic Radiation in a Resonator",
Opt. Spekt., 1972, 32, no.1, pp.168-73, and
Opt. Spectrosc., 1972, 32, no.1, pp.85-7.

26.187. Kuhl, J., and Spitschan, H.:
"Efficient Second-Harmonic and Sum-Frequency
Generation from a Flashlamp-Pumped Dye
Laser", Opt. Commun., 1972, 5, no.5,pp.382-8.

26.188. Thomas, J.M.R., and Taran, J.P.E.:
"Pulse Distortions in Mismatched SHG", Opt.
Commun., 1972, 4, no.5, pp.329-34.

26.189. Bridges, T.J., and Strnad, A.R.:
"Submillimetre-Wave Generation by Difference-
Frequency Mixing in GaAs", Appl. Phys. Lett.,
1972, 20, no.10, pp.382-4.

26.190. Wallace, R.W.: "Rapidly Tunable
Dye-Laser-Pumped Parametric Oscillator", J.
Quantum Electron. IEEE, 1972, QE-8, no.10,
pp.819-20.

26.191. Davis, J.I., and Krupke, W.F.:
"Theory of Pulsed Internal Optical Parametric
Oscillators", J. Appl. Phys., 1972, 43, no.10,
pp.4171-83.

26.192. Panyakeow, S., et al.: "Temperature Dependence of SHG in Te With CO_2 Laser",
J. Appl. Phys., 1972, 43, no.10, pp.4268-9.

26.193. Tolomanenko, A.F.: "Transfer Coefficient of a Transistor Parametric Microwave Multiplier", Izv. VUZ Radioelektron.,
1972, 15, no.3, pp.391-3.

26.194. Andreev, R.B., and Volosov, V.D.:
"Properties of SHG of a Twin-Frequency
Laser", Z. Naturforsch., 1972, 27a, no.2,
pp.363-4.

26.195. Rabson, T.A., et al.: "Efficient
SHG of Picosecond Laser Pulses", Appl. Phys.
Lett., 1972, 20, no.8, pp.282-4.

26.196. Tseitlin, M.B., et al.: "Frequency-Multiplying Properties of a Magnetron
Amplifier", Radiotekh. Elektron., 1972, 16,
no.9, pp.1666-72, and Radio Eng. Electron.
Phys., 1972, 16, no.9, pp.1515-20.

26.197. Zhdanov, B.V., Kovrigin, A.I.,
and Pershin, S.M.: "Stable Oscillator Producing Harmonics of a Nd^{3+}:Glass Laser",
Prib. Tekh. Eksp., 1972, 15, no.3, pp.206-8,
and Instrum. Exp. Tech., 1972, 15, no.3,
pp.861-3.

26.198. Brunner, W.: "Optical Parametric
Oscillator", Fortschr. Phys., 1972, 20,
no.11, pp.629-99.

26.199. Bhar, G.C., et al.: "Tunable Downconversion from an Optical Parametric Oscillator", Opt. Commun., 1972, 6,no.4,pp.323-6.

26.200. Volosov, V.D., et al.: "Second-
Harmonic Conversion of Nd^{3+}:Glass Laser
Radiation", Kvantovaya Elektron., 1972, no.2,
pp.101-2, and Sov. J. Quantum Electron.,
1972, 2, no.2, pp.175-6.

26.201. Vizel', A.A., et al.: "Frequency
Multipliers for Millimetre Waves Employing
GaAs Diodes", Radiotekh. Elektron., 1972,
17, no.6, pp.1335-6, and Radio Eng. Electron. Phys., 1972, 17, no.6, pp.1045-7.

26.202. Burneika, K.P., et al.: "Parametric Generation of Ultrashort Pulses of
Frequency-Tunable Radiation", Zh. Eksp.
Teor. Fiz. Pis'ma, 1972, 16, no.7, pp.365-7,
and JETP Lett., 1972, 16, no.7, pp.257-8.

26.203. Akhmanov, S.A., et al.: "Formation of Subpicosecond UV Pulses by Multiple Nonlinear Conversion", Zh. Eksp. Teor. Fiz. Pis'ma, 1972, 16, no.8, pp.471-5, and JETP Lett., 1972, 16, no.8, pp.335-7.

26.204. Kosolobov, S.N., et al.: "Conversion of Broad IR Spectra Into the Visible Band under Critical Vector Phase-Matching Conditions", Zh. Eksp. Teor. Fiz. Pis'ma, 1972, 16, no.8, pp.475-9, and JETP Lett., 1972, 16, no.8, pp.338-40.

26.205. Dmitriev, V.G., et al.: "Optimization of the Parameters of a Quasi-CW Nd^{3+}:YAG Laser With Nonlinear Element in Resonator", Kvantovaya Elektron., 1972, no.2, pp.111-2, and Sov. J. Quantum Electron., 1972, 2, no.2, pp.188-9..

26.206. Fischer, R., and Frahm, J.: "Efficiency of an Optical Parametric Four-Photon Oscillator", Exp. Tech. Phys., 1972, 20, no.6, pp.533-8.

26.207. Campillo, A.J.: "Internal Upconversion and Doubling of an Optical Parametric Oscillator to Extend the Tuning Range", J. Quantum Electron. IEEE, 1972, QE-8, no.12, pp.914-6.

26.208. Weller, J.F., Giallorenzi, T.G., and Andrews, R.A.: "Time-Resolved Spectral Output of a Doubly Resonant CW Optical Parametric Oscillator", J. Appl. Phys., 1972, 43, no.11, pp.4650-2.

26.209. Stroganov, V.I., Tarasov, V.M., and Samarin, V.I.: "Interaction of Light Rays in a Highly Focused Beam", Opt. Spekt., 1972, 32, no.4, pp.834-6, and Opt. Spectrosc., 1972, 32, no.4, pp.441-2.

26.210. Giallorenzi, T.G., and Reilly, M.H.: "Theory of Internal Upconversion and Frequency Doubling in Optical Parametric Oscillators", J. Quantum Electron. IEEE, 1972, QE-8, no.3, pp.302-10.

26.211. Bey, P.P., and Tang, C.L.: "Plane-Wave Theory of Parametric Oscillator and Coupled Oscillator-Upconverter", J. Quantum Electron. IEEE, 1972, QE-8, no.3, pp.361-9.

26.212. Pearson, J.E., Yariv, A., and Ganiel, U.: "Parametric-Oscillator Tuning Curve from Observations of Total Parametric Fluorescence", J. Quantum Electron. IEEE, 1972, QE-8, no.3, pp.383-5.

26.213. Campillo, A.J.: "Properties of a Pulsed $LiIO_3$, Doubly Resonant, Parametric Oscillator", J. Quantum Electron. IEEE, 1972, QE-8, no.10, pp.809-11.

26.214. Gable, C., and Hercher, M.: "Continuously Tunable Source of Coherent UV Radiation", J. Quantum Electron. IEEE, 1972, QE-8, no.11, pp.850-1.

26.215. Hindin, H.J.: "First Parametric Generator With Full Visible Spectrum", Microwaves, 1972, 11, no.1, p.9.

26.216. Kaneda, S., and Horima, H.: "Microwave Second-Harmonic Generator of High Conversion Efficiency Using Nonlinearity of Negative Resistance in n-GaAs", Electron. Lett., 1972, 8, no.6, pp.141-2.

26.217. Boyd, G.D., et al.: "Phase-Matched Submillimetre-Wave Generation by Difference-Frequency Mixing in $ZnGeP_2$", Appl. Phys. Lett., 1972, 21, no.11, pp.553-5.

26.218. Kuizenga, D.J.: "Optimum Focusing Conditions for Parametric Gain in Crystals with Double Refraction", Appl. Phys. Lett., 1972, 21, no.12, pp.570-2.

26.219. Gorokhov, Yu.A., et al.: "SHG in an Argon-Ion Laser", Vestn. Mosk. Univ. Fiz. Astron., 1972, 13, no.2, pp.252-4.

26.220. Suematsu, Y.: "Focusing Effect on Optical Parametric Oscillations for Three- and Four-Wave Interactions", Jap. J. Appl. Phys., 1972, 11, no.3, pp.387-95.

26.221. Ernst, G.J., and Witteman, W.J.: "SHG in Proustite with CW CO_2 Laser", J. Quantum Electron. IEEE, 1972, QE-8, no.3, pp.382-3.

26.222. Meadors, J.G., and Poirier, M.A.: "Generation of IR Radiation by Raman Scattering and Difference-Frequency Mixing with Nd^{3+}:YAG Pump", J. Quantum Electron. IEEE, 1972, QE-8, no.4, pp.427-8.

26.223. Schinke, D.P.: "Generation of UV Light Using the Nd^{3+}:YAG Laser", J. Quantum Electron. IEEE, 1972, QE-8, no.2, pp.86-7.

26.224. Kabyalka, V.I., et al.: "Optical Parametric Oscillator With Long Nonlinear Interaction and a Low Feedback", Litov. Fiz. Sb., 1972, 12, no.5, pp.863-8.

26.225. Kovrigin, A.I., et al.: "Effect of Crystal Quality on SHG in $LiNbO_3$", Opt. Spekt., 1972, 33, no.4, pp.752-6, and Opt. Spectrosc., 1972, 33, no.4, pp.413-6.

26.226. Voronenko, V.P., et al.: "Experimental Investigation of Millimetre Frequency Converter in n-InSb at $4.2^{\circ}K$", Radiotekh. Elektron., 1972, 17, no.8, pp.1632-8, and Radio Eng. Electron. Phys., 1972, 17, no.8, pp.1287-92.

26.227. Gorokhov, Yu.A., and Krindach, D.P.: "Extraction of Second-Harmonic Radiation from a Laser Resonator", Prib. Tekh. Eksp., 1972, 15, no.6, pp.170-2, and Instrum. Exp. Tech., 1972, 15, no.6, pp.1811-3.

26.228. Basu, R., and Steier, W.H.: "Bandwidth and Threshold Calculations for Angle-Tuned Parametric Oscillators", J. Quantum Electron. IEEE, 1972, QE-8, no.8, pp.693-9.

26.229. Allwood, R.L., et al.: "Tunable Spin-Flip Magneto-Raman IR Laser", Radio Electron. Eng., 1972, 42, no.5, pp.243-6.

26.230. Lidiard, A.B., and Smith, S.D.: "Stimulated Raman Emission and Tunable IR Sources", Comments Solid-State Phys., 1972, 4, no.2, pp.52-61.

26.231. Vinogradov, A.V., and Yukov, E.A.: "(SRS) Method of Increasing Frequency of High-Power Pulsed Lasers", Zh. Eksp. Teor. Fiz. Pis'ma, 1972, 16, no.11, pp.631-4, and JETP Lett., 1972, 16, no.11, pp.447-8.

26.232. Boyd, J.T.: "Theory of Parametric Oscillation Phase Matched in GaAs Thin-Film Waveguides", J. Quantum Electron. IEEE, 1972, QE-8, no.10, pp.788-96.

26.233. Maruyama, M., Kano, T., and Kamiryo, K.: "Frequency Multiplication With a TWT. I and II", Rec. Electr. Commun. Eng. Conversaz. Tohoku Univ., 1972, 41, no.2, pp.124-7.

26.234. Umegaki, S., Yabumoto, S., and Tanaka, S.: "Noncollinearly Phase-Matched SHG in $LiIO_3$", Appl. Phys. Lett., 1972, 21, no.8, pp.400-2.

26.235. Panina, T.A., and Utkin, G.M.: "Lumped Two-Frequency Parametric Oscillators", Izv. VUZ Radiofiz., 1972, 15, no.10, pp.1509-16.

26.236. Postnikov, L.V., and Mel'nikova, V.A.: "Amplification on the Principle of Three-Frequency Interaction", Izv. VUZ Radiofiz., 1972, 15, no.10, pp.1517-26.

26.237. Kitaeva, V.F., et al.: "Spontaneous Parametric Emission by Polaritons in alpha-HIO_3", Dokl. Akad. Nauk SSSR, 1972, 207, no.4-6, pp.1322-3, and Sov. Phys.-Dokl., 1973, 17, no.12, pp.1189-91.

26.238. Ivleva, L.I., et al.: "SHG in $LiNbO_3$", Fiz. Tver. Tela, 1972, 14, no.11, pp.3137-42, and Sov. Phys.-Solid State, 1973, 14, no.11, pp.2686-9.

26.239. Krivoshchekov, G.V., Nikulin, N.G., and Sokolovskii, R.I.: "Excitation of Second Harmonic by a Periodic Sequence of Ultrashort Light Pulses", Kvantovaya Elektron., 1972, no.4, pp.63-8, and Sov. J. Quantum Electron., 1973, 2, no.4, pp.344-7.

26.240. D'yakov, Yu.E., and Kovrigin, A.I.: "Theory of Shape of Pulses Produced by Transient Parametric Generation of Light", Kvantovaya Elektron., 1972, no.4, pp.86-9, and Sov. J. Quantum Electron., 1973, 3, no.4, pp.362-4.

26.241. Dmitriev, V.G., et al.: "Engineering Design and Optimization of Parameters of Frequency Doublers for the Visible Range", Kvantovaya Elektron., 1972, no.5, pp.72-9, and Sov. J. Quantum Electron., 1973, 2, no.5, pp.445-9.

26.242. Tunkin, V.G., Usmanov, T., and Shakirov, V.A.: "Generation of Fifth Picosecond Laser Harmonic", Kvantovaya Elektron., 1972, no.5, pp.117-8, and Sov. J. Quantum Electron., 1973, 2, no.5, pp.487-8.

26.243. Vinogradov, A.V., and Pustovalov, V.V.: "SHG in an Inhomogeneous Laser Plasma", Zh. Eksp. Teor. Fiz., 1972, 63, no.3, pp.940-50, and Sov. Phys.-JETP, 1973, 36, no.3, pp.492-7.

26.244. Borodovskii, P.A., et al.: "Parametric Excitation of Microwave Oscillations by a Travelling Domain in a Gunn Diode", Fiz. Tekh. Poluprov., 1972, 6, no.9, pp.1703-8, and Sov. Phys.-Semicond., 1973, 6, no.9, pp.1471-5.

26.245. Matsumoto, N., and Yajima, T.: "Far-IR Generation by Self-Beating of Dye-Laser Light", Jap. J. Appl. Phys., 1973, 12, no.1, pp.90-7.

26.246. Kinnemann, G.: "(Diode) Parametric Amplification Combined With Frequency Multiplication", Nachr. Tech. Elektron., 1973, 23, no.1, pp.22-4.

26.247. Smith, R.G.: "Study of Factors Affecting the Performance of a Continuously Pumped Doubly-Resonant Optical Parametric Oscillator", J. Quantum Electron. IEEE, 1973, QE-9, no.5, pp.530-41.

26.248. Hsu, H., and Yu, C.: "Photon Conversion Efficiency of BW Parametric Amplification and Oscillation", Opt. Commun., 1973, 7, no.1, pp.80-2.

26.249. Dunning, F.B., Tittel, F.K., and Stebbings, R.F.: "Generation of Tunable Coherent Radiation at 230-300 nm Using Lithium Formate", Opt. Commun., 1973, 7, no.3, pp.181-3.

26.250. Kung, A.H., Young, J.F., and Harris, S.E.: "Generation of 118.2-nm Radiation in Phase-Matched Mixtures of Inert Gases", Appl. Phys. Lett., 1973, 22, no.6, pp.301-2.

26.251. Jahnig, L.: "Parametric Amplification With Focused Beams in $Ba_2NaNb_5O_{15}$", Arch. Elektron. Ubertrag., 1973, 27, no.4, pp.195-6.

26.252. Aggarwal, R.L., Lax, B., and Favrot, G.: "Noncollinear Phase Matching in GaAs", Appl. Phys. Lett., 1973, 22, no.7, pp.329-30.

26.253. Sorokin, P.P., Wynne, J.J., and Lankard, J.R.: "Tunable Coherent IR Source Based Upon Four-Wave Parametric Conversion in Alkali Metal Vapours", Appl. Phys. Lett., 1973, 22, no.7, pp.342-4.

26.254. Decker, C.D., and Tittel, F.K.: "High-Power, Broadly Tunable, Difference-Frequency Generation in Proustite", Appl. Phys. Lett., 1973, 22, no.8, pp.411-3.

26.255. Hanna, D.C., Luther-Davies, B., and Smith, R.C.: "Singly Resonant Proustite Parametric Oscillator Tunable Over 1.22-8.5 micron", Appl. Phys. Lett., 1973, 22, no.9, pp.440-2.

26.256. Sato, T.: "Continuously Tunable Radiation Source around 253.5 nm", J. Appl. Phys., 1973, 44, no.5, pp.2257-9.

26.257. Fischer, R.: "Comparison of Properties of Double-Resonant Optical Parametric Oscillators", Exp. Tech. Phys., 1973, 21, no.1, pp.21-33.

26.258. Bates, H.E.: "Burst-Mode Frequency-Doubled Nd^{3+}:YAG Laser for Time-Sequenced High-Speed Photography and Holography", Appl. Opt., 1973, 12, no.6, pp.1172-8.

26.259. Bethea, C.G.: "Tunable Difference-Frequency Mixing in the 1.5-1.7 micron Region at High PRF's", Appl. Opt., 1973, 12, no.6, pp.1104-5.

26.260. Pearson, J.E., Yariv, A., and Ganiel, U.: "Observations of Parametric Fluorescence and Oscillation in the IR (Region)", Appl. Opt., 1973, 12, no.6, pp.1165-71.

26.261. Okada, M., and Ieiri, S.: "Influences of Self-Induced Thermal Effects on Optical SHG", NHK Tech. J., 1973, 25, no.2, pp.72-80.

26.262. Auston, D.H., Glass, A.M., and LeFur, P.: "Tunable Far-IR Generation by Difference-Frequency Mixing of Dye Lasers in Reduced (Black) $LiNbO_3$", Appl. Phys. Lett., 1973, 23, no.1, pp.47-8.

26.263. Uematsu, Y., and Fukuda, T.: "Characteristics and Performance of Nd^{3+}:YAG/ $KNbO_3$ Intracavity SHG", Jap. J. Appl. Phys., 1973, 12, no.6, pp.841-4.

26.264. Hanna, D.C., Rampal, V.V., and Smith, R.C.: "Tunable IR Downconversion in Silver Thiogallate", Opt. Commun., 1973, 8, no.2, pp.151-3.

26.265. Bridges, T.J., and Nguyen, Van T.: "Generation of Tunable Far-IR Radiation by Difference-Frequency Mixing Using Conduction Electron Spin Nonlinearity in InSb", Appl. Phys. Lett., 1973, 23, no.2, pp.107-9.

26.266. Decker, C.D., and Tittel, F.K.: "Difference-Frequency Generation by Optical Mixing of Two Dye Lasers in Proustite", Opt. Commun., 1973, 8, no.3, pp.244-7.

26.267. Harris, S.E.: "Generation of Vacuum-UV and Soft-X-Ray Radiation Using High-Order Nonlinear Optical Polarizabilities", Phys. Rev. Lett., 1973, 31, no.6, pp.341-4.

26.268. Schumacher, F.: "High-Power Frequency Multiplier Using MIS Varactors", Trans. IEEE, 1973, MTT-21, no.10, pp.648-9.

26.269. Bernecker, O.: "Limitations for Mode-Locking Enhancement of Internal SHG in a Laser", J. Quantum Electron. IEEE, 1973, QE-9, no.9, pp.897-900.

26.270. Vanherzeele, H.: "Efficiency of Phasematched SHG Process", Rev. HF, 1973, 9, no.2, pp.27-40.

26.271. Suematsu, Y., Sasaki, Y., and Shibata, K.: "SHG Due to a Guided-Wave Structure of Quartz Coated with a Glass Film", Appl. Phys. Lett., 1973, 23, no.3, pp.137-8.

26.272. Conwell, E.M.: "Theory of SHG in Optical Waveguides", J. Quantum Electron. IEEE, 1973, QE-9, no.9, pp.867-79.

26.273. Yacoby, Y., Aggarwal, R.L., and Lax, B.: "Phase Matching by Periodic Variation of Nonlinear Coefficients", J. Appl. Phys., 1973, 44, no.7, pp.3180-1.

26.274. Basov, N.G., Zritskii, A.R., et al.: "Generation of High-Power Light Pulses at 1060 nm and 530 nm", Kvantovaya Elektron., 1973, no.6, pp.50-5, and Sov. J. Quantum Electron., 1973, 2, no.6, pp.533-5.

26.275. Miles, R.B., and Harris, S.E.: "Optical Third-Harmonic Generation in Alkali Metal Vapours", J. Quantum Electron. IEEE, 1973, QE-9, no.4, pp.470-84.

26.276. Tang, C.L., and Bey, P.P.: "Phase Matching in SHG Using Artificial Periodic Structures", J. Quantum Electron. IEEE, 1973, QE-9, no.1, pp.9-17.

26.277. Gonzalez, D.G., Nieh, S.T.K., and Steier, W.H.: "Two-Pass Internal SHG Using a Prism Coupler", J. Quantum Electron. IEEE, 1973, QE-9, no.1, pp.23-6.

26.278. Koster, A., Vossoughi, A., and Chartier, G.: "Production of Coherent Submillimetre Radiation by Heterodyne Method", C. R. Acad. Sci. (Paris), 1973, 276B, no.1, pp.43-6.

26.279. Nath, G., and Pauli, G.: "Efficient Pulsed Optical Parametric Oscillator With Tuning Range 400-2100 nm", Appl. Phys. Lett., 1973, 22, no.2, pp.75-6.

26.280. Fery, H., and Hermann, F.: "Noncollinear SHG in KDP", Opt. Commun., 1973, 8, no.4, pp.291-6.

26.281. Kaneda, S., and Takabe, S.: "Microwave Harmonic Generation Using Nonlinearity of Negative Resistance in n-GaAs Under Small-Signal Excitation", Int. J. Electron., 1973, 35, no.5, pp.577-90.

26.282. Hsu, H., and Yu, C.: "Complete Photon Conversion in BW Parametric Amplification and Oscillation", Electron. Lett., 1973, 9, no.19, pp.442-4.

26.283. Sushchik, M.M., and Freidman, G.I.: "Optimal Focusing of the Pump in Single-Resonator Parametric Light Generators", Izv. VUZ Radiofiz., 1973, 16, no.6, pp.898-902.

26.284. Davydov, A.A., et al.: "Tunable IR Parametric Oscillator in a CdSe Crystal", Opt. Commun., 1973, 9, no.3, pp.234-6.

26.285. Sushchik, M.M., and Fortus, V.M.: "Efficiency of Parametric Conversion in High-Quality Optical Resonators", Izv. VUZ Radiofiz., 1973, 16, no.10, pp.1522-9.

26.286. Frahm, J.: "Theory of Parametric Four-Photon Processes. IV", Ann. Phys. (Leipzig), 1973, 30, no.2, pp.173-84.

26.287. Volosov, V.D., and Krylov, V.N.: "High-Efficiency Scheme of Intracavity Generation of Harmonics and Parametric Oscillation", Opt. Spekt., 1973, 35, no.1, pp.120-4, and Opt. Spectrosc., 1973, 35, no.1, pp.69-71.

26.288. Brunner, W., Fischer, R., and Paul, H.: "Singly Resonant Optical Parametric Oscillator", Ann. Phys. (Leipzig), 1973, 30, no.3-4, pp.299-308.

26.289. Bokut', B.V., et al.: "Generation of Various Frequencies by Mixing Ruby and Nd^{3+} Laser Emissions in a $LiNbO_3$ Crystal", Zh. Prikl. Spekt., 1973, 19, no.4, pp.712-5.

26.290. Melkov, G.A., and Lutsenko, A.L.: "Frequency Doubling by Means of Ferrites", Radiotekh. Elektron., 1973, 18, no.2, pp.350-4, and Radio Eng. Electron. Phys., 1973, 18, no.2, pp.250-3.

26.291. Kirichenko, A.Ya., and Lysova, L.A.: "Millimetre-Wave Frequency Multiplier Using a Ring-Plane Periodic System", Izv. VUZ Radioelektron., 1973, 16, no.10, pp.90-5.

26.292. Yang, K.H., et al.: "Phase-Matched Far-IR Generation by Optical Mixing of Dye-Laser Beams", Appl. Phys. Lett., 1973, 23, no.12, pp.669-71.

26.293. Shen, Y.R.: "Theory of Far-IR Generation by Optical Mixing and Stimulated Raman Scattering via Spin-Flip Transitions in InSb", Appl. Phys. Lett., 1973, 23, no.9, pp.516-8.

26.294. Hori, K., Maekawa, S., and Furuya, T.: "High-Power SHG Using (BSN in a) Nd^{3+}: YAG Laser", Fujitsu Sci. Tech. J., 1973, 9, no.4, pp.105-22.

26.295. Persico, F., and Vetri, G.: "Theory of SHG at Microwave Frequencies by Paramagnetic Materials", Phys. Rev., 1973, 8, no.8, pp.3512-36.

26.296. Burns, W.K., and Andrews, R.A.: "Index Tuning of Phase-Matched SHG in Optical Waveguides", Appl. Opt., 1973, 12, no.10, pp.2249-50.

26.297. Rao, B.R.: "Nonlinear Power Amplification at Second Harmonic in TRAPATT Diodes", Electron. Lett., 1973, 9, no.22, pp.526-7.

26.298. Deryugin, I.A., Kurashov, V.N., and Mar'enko, V.V.: "Direct IF Generation in a Maser", Izv. VUZ Radioelektron., 1973, 16, no.5, pp.109-11.

26.299. Shkalikov, V.N., and Lutin, E.A.: "Phase Characteristics of Varactor Frequency Multipliers", Radiotekhnika, 1973, 28, no.10, pp.60-5, and Telecommun. Radio Eng. Pt 2, 1973, 28, no.10, pp.103-7.

26.300. Shul'ga, V.G., et al.: "Harmonic Components in a Double-Beam TWT", Radiotekh. (Kharkov), 1973, no.27, pp.92-5.

26.301. Kazak, N.S., and Mashchenko, A.G.: "Regeneration in the 241-nm Region", Zh. Prikl. Spekt., 1973, 19, no.5, pp.914-5.

26.302. Apanasevich, P.A., Afanas'ev, A.A., and Orlovich, V.A.: "Amplifier Based on Stimulated Raman Effect With Lateral Pumping", Zh. Prikl. Spekt., 1973, 18, no.3, pp.406-9.

26.303. Berezovskii, V.V.: "Dependences of the Spectra of Second Harmonics Generated in Te and Proustite on the Phase-Matching Band", Kvantovaya Elektron., 1973, no.2, pp.96-9, and Sov. J. Quantum Electron., 1973, 3, no.2, pp.155-7.

26.304. Lax, B., Aggarwal, R.L., and Favrot, G.: "Far-IR Step-Tunable Coherent Radiation (by Mixing of CO_2 Lasers)", Appl. Phys. Lett., 1973, 23, no.12, pp.679-81.

26.305. Klyshko, D.N., and Nazarova, N.I.: "Transient Parametric Superluminescence", Zh. Tekh. Fiz., 1973, 43, no.10, pp.2158-62, and Sov. Phys.-Tech. Phys., 1974, 18, no.10.

26.306. Genkin, V.N., et al.: "Frequency Tripling in the Millimetre Range in p-InSb", Fiz. Tekh. Poluprov., 1973, 7, no.7, pp.1408-11, and Sov. Phys.-Semicond., 1974, 7, no.7, pp.941-2.

26.307. Haldemann, H.: "Influence of the Sum Frequency on Optically Pumped Parametric Amplifiers", Appl. Phys., 1974, 3, no.4, pp.291-8.

26.308. Hanna, D.C., Rampal, V.V., and Smith, R.C.: "Tunable Medium-IR Generation in $AgGaS_2$ by Downconversion in Flash-Pumped Dye-Laser Radiation", J. Quantum Electron. IEEE, 1974, QE-10, no.4, pp.461-2.

26.309. Dahele, J.S., and Hill, D.R.: "Slimguide Frequency Multipliers for Microwave Applications", Electr. Commun., 1974, 49, no.1, pp.65-71.

26.310. Hamilton, C.H.: "Transistor Harmonic Oscillators with X-Band Output", Nachr. Tech. Z., 1974, 27, no.5, pp.196-7.

26.311. Burns, W.K., and Lee, A.B.: "Observation of Non-Critically Phase-Matched SHG in an Optical Waveguide", Appl. Phys. Lett., 1974, 24, no.5, pp.222-4.

26.312. Frahm, J., and Fischer, R.:
"Theory of the Parametric Four-Photon Oscillator. III", Ann. Phys. (Leipzig), 1974,
31, no.2, pp.143-70.

26.313. Hordvik, A., and Sackett, P.B.:
"Characteristics of the Optical Parametric
Oscillator", Appl. Opt., 1974, 13, no.5,
pp.1060-4.

26.314. Massey, G.A.: "Efficient Upconversion of Long-Wavelength UV Light Into
the 200-235 nm Band", Appl. Phys. Lett.,
1974, 24, no.8, pp.371-3.

26.315. Kildal, H., and Mikkelsen, J.C.:
"Efficient Doubling and CW Difference-Frequency Mixing in the IR Using the Chalcopyrite $CdGeAs_2$", Opt. Commun., 1974, 10, no.4,
pp.306-9.

26.316. Rabus, W.: "Harmonic-Frequency
Generation along Schottky-Contact Microstrip Lines", Arch. Elektron. Ubertrag.,
1974, 28, no.1, pp.1-11.

26.317. Hodgson, R.T., Sorokin, P.P.,
and Wynne, J.J.: "Tunable Coherent Vacuum-
UV Generation in Atomic Vapours", Phys. Rev.
Lett., 1974, 32, no.7, pp.343-6.

26.318. Yuen, S.Y., and Lax, B.: "Output
Behaviour of Electronic Spin-Flip Raman
Lasers", Opt. Commun., 1974, 10, no.1, pp.4-7.

26.319. Colles, M.J., et al.: "Study of
Factors Affecting the CW InSb Spin-Flip
Raman Laser", Opt. Commun., 1974, 10, no.2,
pp.145-8.

26.320. Gupta, S.C., and Srivastava,
G.P.: "Spatial and Temporal Profile Effects
on SHG", Opt. Acta, 1974, 21, no.3, pp.221-41.

26.321. Kato, K., Alcock, A.J., and
Richardson, M.C.: "Conversion of High-Power
Ruby Laser Radiation to the UV in RDP", Opt.
Commun., 1974, 11, no.1, pp.5-7.

26.322. Hwang, D.M., and Salin, S.A.:
"Mixing of Ar^{II} and CO_2 Laser Radiations
in Uniaxial $RbClO_3$", Phys. Rev., 1974, 9B,
no.4, pp.1884-96.

26.323. Stolen, R.H., Bjorkholm, J.E.,
and Ashkin, A.: "Phase-Matched Three-Wave
Mixing in Silica-Fibre Waveguides", Appl.
Phys. Lett., 1974, 24, no.7, pp.308-10.

26.324. Weiss, J.A., and Goldberg, L.S.:
"Singly Resonant CdSe Parametric Oscillator
Pumped by an HF Laser", Appl. Phys. Lett.,
1974, 24, no.8, pp.389-91.

26.325. Lee, P., et al.: "Harmonic Generation and Frequency Mixing in Laser-Produced Plasmas", Appl. Phys. Lett., 1974, 24,
no.9, pp.406-8.

26.326. Bloom, D.M., et al.: "IR Upconversion with Resonantly Two-Photon-Pumped
Metal Vapours", Appl. Phys. Lett., 1974,
24, no.9, pp.427-8.

26.327. Gray, G., Jahnig, L., and Sauta,
E.: "CW Optical Parametric Oscillation in
$Ba_2NaNb_5O_{15}$", Arch. Elektron. Ubertrag.,
1974, 28, no.7-8, pp.340-2.

26.328. Yuen, S.Y., Lax, B., and Wolff,
P.A.: "Stress-Tuned Stimulated Light Scatter-
ing in p-Type Semiconductors", Solid-State
Commun., 1974, 14, no.11, pp.1079-82.

26.329. Brueck, S.R.J., and Mooradian, A.:
"Frequency Stabilization and Fine-Tuning
Characteristics of a CW InSb Spin-Flip
Laser", J. Quantum Electron. IEEE, 1974,
QE-10, no.9, pp.634-42.

26.330. Bel'dyugin, I.M., et al.: "Possible Method for Stabilization of a Raman
Laser at the First Stokes Frequency", Kvantovaya Elektron., 1974, no.4, pp.118-20, and
Sov. J. Quantum Electron., 1974, 3, no.4,
pp.353-4.

26.331. Laubereau, A., Greiter, L., and
Kaiser, W.: "Intense Tunable Picosecond
Pulses in the IR", Appl. Phys. Lett., 1974,
25, no.1, pp.87-9.

26.332. Ghoul, F., Besser, L., and Hsieh,
C.: "Design of a High-Power S-Band Doubler",
Microwaves, 1974, 13, no.6, pp.58-65.

26.333. Hanna, D.C., et al.: "CdSe Downconverter Tuned over 9.5-24 micron", Appl.
Phys. Lett., 1974, 25, no.3, pp.142-4.

26.334. van der Ziel, J.P., et al.:
"Phase-Matched SHG in GaAs Optical Waveguides
by Focused Laser Beams", Appl. Phys. Lett.,
1974, 25, no.4, pp.238-40.

26.335. Brignall, N., et al.: "Tunable
Far-IR Generation by Difference-Frequency
Mixing in InSb", Opt. Commun., 1974, 12,
no.1, pp.17-20.

26.336. Abdullin, U.A., and Chirkin, A.S.:
"Quasi-CW Generation of a Tunable Submilli-
metric Difference Frequency", Vestn. Mosk.
Univ. Fiz. Astron., 1974, 15, no.3, pp.340-6.

26.337. Adonts, G.G., Kocharyan, L.M.,
and Shakhazaryan, N.V.: "Four-Photon Parametric Interaction in Three-Level Resonance
Medium", Izv. Akad. Nauk Arm. SSS Fiz., 1974,
9, no.4, pp.338-41.

26.338. Frey, R., and Pradere, F.: "Powerful Tunable IR Generation by Stimulated Raman
Scattering", Opt. Commun., 1974, 12, no.1,
pp.98-101.

26.339. Tashiro, H., and Yajima, T.:
"Tunable IR Difference-Frequency Generation
with a High-PRF Dye Laser", Opt. Commun.,
1974, 12, no.2, pp.129-31.

26.340. Zarshchikov, V.A., Lobov, G.D.,
and Shtykov, V.V.: "Conversion of Coherent
IR Radiation into Millimetric Radiation at
the Contact of Two Metals", Radiotekh. Elektron., 1974, 18, no.7, pp.1545-6, and Radio
Eng. Electron. Phys., 1974, 18, no.7, pp.1143-4.

26.341. Akanaev, B.A., Bel'dyugin, I.M., and Gerasimov, V.B.: "Influence of the Width of a Spectral Line on Excitation of Stimulated Scattering in Resonators", Kvantovaya Elektron., 1974, 1, no.1, pp.164-7, and Sov. J. Quantum Electron., 1974,4,no.1,pp.95-6.

26.342. Bel'dyugin, I.M., et al.: "Raman Laser with Plane-Parallel Resonator Pumped by a Converging Beam", Kvantovaya Elektron., 1974, 1, no.1, pp.16-26, and Sov. J. Quantum Electron., 1974, 4, no.1, pp.6-11.

26.343. Brunner, W., Fischer, R., and Paul, H.: "Calculation of the Rise Period in a Double-Resonant Optical Parametric Oscillator", Ann. Phys. (Leipzig), 1974, 31, no.4, pp.343-51.

26.344. Massey, G.A., and Elliott, R.A.: "Tunable IR Parametric Generation in CDA", J. Quantum Electron. IEEE, 1974, QE-10, no.12, pp.899-900.

26.345. Atanesyan, V.G., Karmenyan, K.V., and Sarkisyan, S.A.: "Wideband-Tunable Radiation Source Using SHG in $LiIO_3$", Zh. Eksp. Teor. Fiz. Pis'ma, 1974, 20, no.8, pp.537-40, and JETP Lett., 1974, 20, no.8.

26.346. Gyuzalyan, R.N.: "Frequency Tunable Raman Laser Using Inclined Polaritons", Zh. Eksp. Teor. Fiz. Pis'ma, 1974, 20, no.1, pp.48-51, and JETP Lett., 1974, 20, no.1.

26.347. Shul'ga, V.G., et al.: "Study of Second-Harmonic Excitation in a Double-Beam TWT", Izv. VUZ Radioelektron., 1974, 17, no.11, pp.102-4.

26.348. Pospelov, L.A.: "Nonlinear Theory of the Multiplier Magnetron. I", Radiotekh. (Kharkov), 1974, no.28, pp.46-55.

26.349. Aggarwal, R.L., et al.: "CW Generation of Tunable Narrowband Far-IR Radiation", J. Appl. Phys., 1974, 45, no.9, pp.3972-4.

26.350. Becker, M.F., et al.: "Analytic Expressions for Ultrashort Pulse Generation in Mode-Locked Optical Parametric Oscillators", J. Appl. Phys., 1974, 45, no.9, pp.3996-4005.

26.351. Gerlach, H.: "Difference-Frequency Generation in $LiIO_3$ Using Two Tunable Dye Lasers", Opt. Commun., 1974, 12, no.4, pp.405-8.

26.352. Ito, H., Vesugi, N., and Inaba, H.: "Phase-Matched, Guided-Optical, SHG in Oriented ZnS Polycrystalline Thin-Film Waveguides", Appl. Phys. Lett., 1974, 25, no.7, pp.385-7.

26.353. Glushak, N.S., et al.: "Operational Efficiency of Active Components in a Parametric TW Transmission Line", Izv. VUZ Radioelektron., 1974,17,no.10,pp.108-10.

26.354. Bonch-Bruevich, A.M., Tkachuk, A.M., and Fedorov, A.A.: "Frequency-Doubling Single-Mode Laser for Picosecond Pulses", Zh. Tekh. Fiz., 1974, 44, no.4, pp.864-7, and Sov. Phys.-Tech. Phys., 1974, 19, no.4, pp.549-51.

26.355. Ganley, J.T., Harrison, F.B., and Leland, W.T.: "Spin-Flip Raman Laser as a Tunable IR Source", J. Appl. Phys., 1974, 45, no.11, pp.4980-1.

26.356. Volosov, V.D., and Kalintsev, A.G.: "Design of Optical Frequency Doublers", Kvantovaya Elektron., 1974, 1, no.4, pp.825-9, and Sov. J. Quantum Electron., 1974, 4, no.4, pp.451-3.

26.357. Sattler, J.P., Weber, B.A., and Nemarich, J.: "Tunable Spin-Flip Raman Scattering in (HgCd)Te", Appl. Phys. Lett., 1974, 25, no.9, pp.491-3.

26.358. Chen, B.V., Tang, C.L., and Telle, J.M.: "CW Harmonic Generation in the UV Using a Thin-Film Waveguide on a Nonlinear Substrate", Appl. Phys. Lett., 1974, 25, no.9, pp.495-8.

26.359. Herbert, R.L., Flemming, R.N., and Byer, R.L.: "1.4-4 micron High-Energy Angle-Tuned $LiNbO_3$ Parametric Oscillator", Appl. Phys. Lett., 1974, 25, no.9, pp.520-2.

26.360. Kung, A.H.: "Generation of Tunable Picosecond Vacuum-UV Radiation", Appl. Phys. Lett., 1974, 25, no.11, pp.653-4.

26.361. Lax, B., and Aggarwal, R.L.: "Tunable Radiation Sources in the Submillimetre Region", Microwave J., 1974, 17, no.11, pp.31-3.

26.362. Stroganov, V.I., and Samarin, V.I.: "Interference Effects during Excitation of Optical Harmonics", Opt. Spekt., 1974, 37, no.2, pp.300-2, and Opt. Spectrosc., 1974, 37, no.2, pp.172-3.

26.363. Bogdanova, M.V., Sukhorukov, A.P., and Sukhorukova, A.K.: "Optical Parametric Oscillators With Phase-Modulated Pumping", Kvantovaya Elektron., 1974, 1, no.4, pp.840-7, and Sov. J. Quantum Electron., 1974, 4, no.4, pp.457-9.

26.364. Obukhovskii, V.V., and Strizhevskii, V.L.: "Theory of Polariton Parametric Oscillator", Kvantovaya Elektron., 1974, 1, no.6, pp.1395-406, and Sov. J. Quantum Electron., 1974, 4, no.6, pp.770-6.

26.365. Srivastava, G.P., and Gupta, S.C.: "SHG from Naphthalene Raman Laser", Opt. Acta, 1974, 21, no.2, pp.157-67.

26.366. Takuma, H.: "Raman Lasers", Oyo Buturi, 1974, 43, no.10, pp.1029-34.

26.367. Grasyak, A.Z.: "Raman Lasers", Kvantovaya Elektron., 1974, 1, no.3, pp.485-509, and Sov. J. Quantum Electron., 1974, 4, no.3, pp.269-82.

26.368. Zubarev, I.G., and Mikhailov, S.I.: "Influence of Excitation Linewidth on Coherent Emission Due to SRS", Kvantovaya Elektron., 1974, 1, no.3, pp.629-33, and Sov. J. Quantum Electron., 1974, 4, no.3, pp.351-3.

26.369. Boikova, R.E., Zanadvorov, P.N., and Furlinska, M.G.: "Optical SHG in a Fabry-Perot Resonator", Kvantovaya Elektron., 1974, 1, no.3, pp.1150-5, and Sov. J. Quantum Electron., 1974, 4, no.3, pp.629-31.

26.370. Scholz, M., Konig, R., and Leupold, D.: "Generation of Tunable Coherent UV Radiation by Dye Lasers", Exp. Tech. Phys., 1974, 22, no.6, pp.557-63.

26.371. Thompson, D.E., and Coleman, P.D.: "Step-Tunable Far-IR Radiation by Phase-Matched Mixing in Planar-Dielectric Waveguides", Trans. IEEE, 1974, MTT-22, no.12, pp.995-1000.

26.372. Medvedev, B.A., and Parshov, O.M.: "Amplification of IR Radiation by SRS in Excited Vibrational Levels of Molecules", Opt. Spekt., 1974, 37, no.3, pp.476-81, and Opt. Spectrosc., 1974, 37, no.3, pp.270-2.

26.373. Gorokhov, Yu.A., et al.: "Influence of Thermal Self-Interactions on CW SHG", Kvantovaya Elektron., 1974, 1, no.3, pp.679-83, and Sov. J. Quantum Electron., 1974, 4, no.3, pp.382-4.

Part IV

TRANSMISSION AND RECEPTION

27. RADIATION PROPERTIES OF GENERATORS

27.1. Stewart, J.A.C.: "Magnetic Tuning of Coaxial IMPATT Oscillators", Electron. Lett., 1968, 4, p.526.

27.2. Rosenbaum, F.J., and Tsai, W.C.: "Gunn-Effect Swept-Frequency Oscillator", Proc. IEEE, 1968, 56, p.2164.

27.3. Iida, S.: "Time Variation of Spectral Output of GaAs Diode Laser", Jap. J. Appl. Phys., 1968, 7, no.11, p.1414.

27.4. Zharkov, A.P., et al.: "Laser With Electrooptic Quartz Q-Switch", Opt.-Mekh. Prom., 1968, 35, no.3, p.3, and Sov. J. Opt., Technol., 1968, 35, no.2, p.175.

27.5. Johnston, W.D., and Smith, P.W.: "Competition and Stimulated Switching of Transverse Laser Modes", J. Quantum Electron. IEEE, 1968, QE-4, p.372.

27.6. Tikhonov, E.A., and Shpak, M.T.: "Passive Q-Switching in Solid-State Lasers Based on Stimulated Brillouin Scattering", Zh. Eksp. Teor. Fiz. Pis'ma, 1968, 8, p.282, and JETP Lett., 1968, 8, p.173.

27.7. Freiberg, R.J., and Halsted, A.S.: "Transverse Modes in Gas Lasers", Laser Focus, 1968, 4, no.23, p.21.

27.8. Balashov, I.F., et al.: "Single-Pulse Ruby Lasers With Optomechanical Shutters", Opt.-Mekh. Prom., 1968, 35, no.2, p.1, and Sov. J. Opt. Technol., 1968, 35, no.2, p.145.

27.9. Zharkov, A.P., et al.: "Optical Alignment of Lasers With Electrooptic Q-Switching", Opt.-Mekh. Prom., 1968, 35, no.2, p.5, and Sov. J. Opt. Technol., 1968, 35, no.2, p.148.

27.10. Kalinin, Yu.A., and Stepanov, A.I.: "Q-Switching of Lasers by Shutters Consisting of Exploding Metal Films", Opt.-Mekh. Prom., 1968, 35, no.4, p.59, and Sov. J. Opt. Technol., 1968, 35, no.4, p.474.

27.11. Davis, L.W.: "Effects of Transverse-Mode Selection on Wavefront of a Ruby Laser", J. Appl. Phys., 1968, 39, p.5331.

27.12. Rezvov, A.V.: "Measurement of Gain (and Optical Nonuniformity) of a Liquid Laser", Opt. Spekt., 1968, 25, p.429, and Opt. Spectrosc., 1968, 25, p.231.

27.13. Bykovskii, Yu.A., et al.: "Characteristic Features of Radiation of p-n Junction GaAs Laser With Nonlinear Passive Element in the Resonator", Fiz. Tekh. Poluprov., 1968, 2, p.1831, and Sov. Phys.-Semicond., 1969, 2, no.12, pp.1522-3.

27.14. Mak, A.A., and Sedov, B.M.: "Stabilization Effect of Frequency of Solid-State Laser", Zh. Tekh. Fiz., 1968, 38, no.12, p.2119, and Sov. Phys.-Tech. Phys., 1969, 13, no.12, pp.1704-6.

27.15. Kalinin, Yu.A., and Mak, A.A.: "Influence of Inverse Population Distribution on Transverse Modes of a Laser", Zh. Tekh. Fiz., 1968, 38, no.7, p.1108, and Sov. Phys.-Tech. Phys., 1969, 13, no.7.

27.16. Troitskii, Yu.V.: "Method of Emission-Line Selection in Lasers", Opt. Spekt., 1969, 26, no.5, p.858, and Opt. Spectrosc., 1969, 26, no.5, p.468.

27.17. Dzhibladze, M.I., et al.: "Time Characteristics of a Single-Mode $Dy^{2+}:CaF_2$ Pulsed Laser", Opt. Spekt., 1969, 27, p.464, and Opt. Spectrosc., 1969, 27, p.249.

27.18. Kerdiles, J.: "Modulation of a Saturable-Absorber Q-Switched Laser by a Train of Picosecond Pulses", Opto-Electron., 1969, 1, no.4, p.193.

27.19. Chartier, G.: "Polarization of Light Output from a Weakly Birefringent Solid-State Laser", Phys. Lett., 1969, 30A, p.526.

27.20. Cleverley, M.E., and Norbury, J.R.: "Technique for Improving the Frequency/Temperature Characteristics of 3cm-Band Gunn Oscillators", Electron. Lett., 1969,5,p.499.

27.21. Shakhov, V.O.: "Experimental Investigation of Transient Processes in a Molecular Generator", Izv. VUZ Radiofiz., 1969, 12, no.10, p.1472.

27.22. Day, G.W., Gaddy, O.L., and Jungling, K.C.: "Wavelength Selection in a Q-Switched CO_2 Laser Using Selective Absorption in Gases", Proc. IEEE, 1969, 57, p.2060.

27.23. Mak, A.A., et al.: "Width of Emission Spectrum of Solid-State Lasers", Opt. Spekt., 1969, 26, no.2, p.276, and Opt. Spectrosc., 1969, 26, no.2, p.149.

27.24. Bebchuk, A.S., et al.: "Q-Switching a Ruby Laser With LiNbO$_3$ Electrooptic Shutter", Radiotekh. Elektron., 1969, 14, no.6, p.1065, and Radio Eng. Electron. Phys., 1969, 14, no.6, pp.919-21.

27.25. Yoshino, K., Kawabe, K., and Inuishi, Y.: "Q-Switching of Ruby Laser by Chlorophyll", Technol. Rep. Osaka Univ., 1969, 19, no.853-875, p.131.

27.26. McGroddy, J.C., et al.: "Effects of Hydrostatic Pressure on Hot-Electron Phenomena in n-InSb", IBM J. Res. Dev., 1969, 13, p.580.

27.27. Lee, T.P., and Standley, R.D.: "Frequency Modulation of a Millimetre-Wave IMPATT-Diode Oscillator and Related Harmonic Generation Effects", Bell Syst. Tech. J., 1969, 48, p.143.

27.28. Tajime, T., Cho, Y., and Matsuo, Y.: "Composite-Cavity Tunable Laser", Appl. Opt., 1969, 8, p.1509.

27.29. Antsiferov, V.V., et al.: "Effect of Spatial Nonuniformity of Radiation Field on the Dynamics of a Ruby Laser", Zh. Eksp. Teor. Fiz., 1969, 56, no.2, p.526, and Sov. Phys.-JETP, 1969, 29, no.2.

27.30. Zakharov, Yu.P., et al.: "Investigation of Pulsed Operation of a GaAs Injection Laser", Fiz. Tekh. Poluprov., 1969, 3, no.6, p.864, and Sov. Phys.-Semicond., 1969, 3, no.6, pp.729-32.

27.31. Dyment, J.C., Ripper, J.E., and Roldan, R.H.: "Spiking in Light Pulses from GaAs Q-Switched Junction Lasers", J. Quantum Electron. IEEE, 1969, QE-5, p.415.

27.32. Gerber, E., and Ahlstrom, E.R.: "Solid-State Laser With Vibrating Reflector", J. Quantum Electron. IEEE, 1969, QE-5, p.403.

27.33. Prishivalko, A.P., Zhukovskii, V.V., and Astaf'eva, L.G.: "Investigation of Stability Regions of Resonators With Focusing Elements", Zh. Prikl. Spekt., 1969, 10, no.2, p.321.

27.34. Mackintosh, I.W.: "Double-Etalon Q-Switching of a Continuously Pumped Nd^{3+}: YAG Laser", Appl. Opt., 1969, 8, p.1991.

27.35. Broom, R.F.: "Self-Modulation at Gigahertz Frequencies of a Diode Laser Coupled to an External Cavity", Electron. Lett., 1969, 5, p.571.

27.36. Arakelyan, V.S., and Karlov, N.V.: "Translucent (BCl$_3$) Filter in CO$_2$ Laser Using Active Q-Switching", Radiotekh. Elektron., 1969, 14, no.3, p.561, and Radio Eng. Electron. Phys., 1969,14,no.3,pp.488-9.

27.37. Kan, T., Powell, H.T., and Wolga, G.J.: "Observation of Central Tuning Dip in N$_2$O and CO$_2$ Molecular Lasers", J. Quantum Electron. IEEE, 1969, QE-5, p.299.

27.38. Hanst, P.L., and Morreal, J.A.: "Wavelength-Selective, Repetitively Pulsed, CO$_2$ Laser", Appl. Opt., 1969, 8, p.109.

27.39. Kasuya, T., et al.: "Undulation in the Output of a Pulsed Water-Vapour Laser", Jap. J. Appl. Phys., 1969, 8, p.478.

27.40. Dienes, A.: "Photon Echoes and Combination Tones. Polarization Dependence", J. Quantum Electron. IEEE, 1969, QE-5,p.246.

27.41. Gordeev, D.V., et al.: "Coherence of He-Ne Laser Radiation With Pure Transversal Oscillations", Zh. Prikl. Spekt., 1969, 11, p.1034.

27.42. Willem, A.A., and Van Beeck, W.P.: "Mode Structure of a Laser", Rev. HF, 1969, 7, p.313.

27.43. Berezin, B.G., and Ermakov, B.A.: "Production of Polarized Radiation in a Nd^{3+}:Glass Laser", Opt. Spekt., 1969, 27, p.310, and Opt. Spectrosc., 1969, 27, no.2, p.163.

27.44. Isyanova, E.D., et al.: "Induced Polarization of Radiation in a Nd^{3+}:Glass Laser", Opt. Spekt., 1969, 27, p.686, and Opt. Spectrosc., 1969, 27, no.4, p.371.

27.45. Rivlin, L.A.: "Dynamic Instabilities of Semiconductor-Laser Radiation", Izv. VUZ Radiofiz., 1969, 12, p.1796.

27.46. Andreeva, E.Yu., et al.: "Polarization of a Single-Frequency He-Ne Laser", Opt. Spekt., 1969, 27, p.809, and Opt. Spectrosc., 1969, 27, p.441.

27.47. Gonchukov, S.A., et al.: "Competition of Longitudinal Modes in a He-Ne Laser at 633 nm", Opt. Spekt., 1969, 27, p.813, and Opt. Spectrosc., 1969, 27, p.443.

27.48. Paoli, T.L., and Ripper, J.E.: "Optical Pulses from CW GaAs Injection Lasers", Appl. Phys. Lett., 1969,15,p.105.

27.49. Chakravarti, A.N., Biswas, S.N., and Rakshit, S.: "Temperature Dependence of Stimulated Light Power from Semiconductor Junction Lasers", Int. J. Electron., 1969, 26, p.95.

27.50. Malyshev, V.I., Markin, A.S., and Sychev, A.A.: "Kinetic Spectrum of Free Laser Generation in Ruby", Zh. Tekh. Fiz., 1969, 39, no.2, p.334, and Sov. Phys.-Tech. Phys., 1969, 14, no.2, pp.241-5.

27.51. Meyer, C., et al.: "Stabilization and Power of a Monochromatic Single-Mode CO$_2$ Laser", Can. J. Chem., 1969,47,p.2565.

27.52. Smol'skaya, T.I., and Pubinov, A.N.: "Spectral, Temporal, and Phase Characteristics of Composite Ruby-Laser Radiation", Zh. Prikl. Spekt., 1969, 10, no.3, p.433.

27.53. Bondarenko, A.N., et al.: "Single-Frequency Ruby Laser With Active Q-Switch", Zh. Prikl. Spekt., 1969, 9, no.2, p.57.

27.54. McMahon, J.M.: "Laser Mode Selection with Slowly Opened Q-Switches", J. Quantum Electron. IEEE, 1969, QE-5, p.489.

27.55. Ku, R.T., Verdeyen, J.T., and Cherrington, B.E.: "Frequency Shift at 3.39 micron Due to Competition by 633-nm Laser Radiation", J. Appl. Phys., 1969, 40, p.3860.

27.56. Szabo, A., and Erickson, L.E.: "Study of Saturable Absorber Switching Efficiencies", J. Appl. Phys., 1969, 40, p.3574.

27.57. Mohn, E.: "Regular Pulsing of GaAs-Diode Laser", Electron. Lett., 1969, 5, p.261.

27.58. Rohr, H.: "Spectral Behaviour of Flashlamp-Pumped Dye (Rhodamine-B) Laser", Z. Phys., 1969, 228, no.5, p.465.

27.59. van Haeringen, W.: "Role of Linear Phase Anisotropy in a Zeeman Laser", Phys. Rev., 1969, 180, p.624.

27.60. Valakh, M.Ya.: "Q-Switching of a Laser by Variation of the Plasma Reflectance of a Semiconductor Under Carrier-Heating Conditions", Fiz. Tekh. Poluprov., 1969, 3, p.426, and Sov. Phys.-Semicond., 1969, 3, no.3, pp.360-1.

27.61. Oraevskii, I.N., et al.: "Spectral Characteristics of Injection-Diode Amplifiers", Phys. Status Solidi, 1969, 32, p.55.

27.62. Briquet, G., Jego, J.M., and Terneaud, A.: "Generation of a Monomode Giant-Pulse Laser by Coherent Pumping", Appl. Phys. Lett., 1969, 14, p.282.

27.63. Aberbakh, V.S., and Vlasov, S.N.: "Experimental Study of He-Ne Gas Laser with Non-Spherical Mirrors", Radiotekh. Elektron., 1969, 14, no.9, p.1709, and Radio Eng. Electron. Phys., 1969, 14, no.9, pp.1489-91.

27.64. Lapshin, G.M., et al.: "Frequency Characteristics of He-Ne Laser in a Transverse Magnetic Field", Zh. Prikl. Spekt., 1969, 10, no.2, p.256.

27.65. Bardo, W.S., and Laine, D.C.: "Induced Spiking in an NH_3-Maser Oscillator", Electron. Lett., 1969, 5, p.688.

27.66. Eichler, H., and Wiesemann, W.: "Frequency Reaction of an Optical Resonator on a Laser", Z. Angew. Phys., 1969, 28, no.3, p.125.

27.67. Eichler, H., and Wiesemann, W.: "Frequency Reaction of an Optical Resonator on a Laser in the Case of Mismatch", Z. Angew. Phys., 1969, 28, no.3, p.127.

27.68. Callen, W.R., Pantell, R.H., and Warszawski, J.: "Pulse Stretching of Q-Switched Lasers", Opto-Electron., 1969, 1, no.3, p.123.

27.69. Vinokurov, G.N., et al.: "Nd^{3+}:YAG Laser With Spectrum Width Less than 28Hz", Zh. Eksp. Teor. Fiz. Pis'ma, 1969, 10, no.8, p.232, and JETP Lett., 1969, 10, no.8.

27.70. Daneu, V.: "Use of a Low-Order Resonator for IR Laser Line Selection", Appl. Opt., 1969, 8, p.1745.

27.71. Malyutin, A.A., et al.: "Investigation of Temporal Structure of Nd^{3+}:Glass Laser Emission in Self-Mode-Locking Regime", Zh. Eksp. Teor. Fiz. Pis'ma, 1969, 9, no.8, p.445, and JETP Lett., 1969, 9, no.8, p.266.

27.72. Nikitin, V.V., Semenov, A.S., and Strakhov, V.P.: "Investigation of the Radiation Pulsations of a CW GaAs Injection Laser", Zh. Eksp. Teor. Fiz. Pis'ma, 1969, 9, no.9, p.516, and JETP Lett., 1969, 9, no.9, p.313.

27.73. Belostotskii, B.R., et al.: "Monoblock Q-Switch With Temperature Compensation for Misalignment", Zh. Prikl. Spekt., 1969, 11, no.2, p.257.

27.74. Rubinov, A.N.: "Transient Losses of Single-Pulse Laser Radiation", Zh. Prikl. Spekt., 1969, 11, no.6, p.1137.

27.75. Soskin, M.S., and Kravchenko, V.I.: "Retuning, Stabilization, and Loss-Factor Determination in Solid-State Lasers by Use of Dispersive Resonators", Electron Technol., 1969, 2, no.2-3, p.169.

27.76. Terekhin, D.K., Fridrikhov, S.A., and Antonov, G.G.: "Polarization of He-Ne Laser Radiation at 633 nm in a Transverse Magnetic Field", Opt. Spekt., 1969, 26, no.4, p.653, and Opt. Spectrosc., 1969, 26, no.4, p.358.

27.77. Gonchukov, S.A., et al.: "Line Shapes and Gap Widths in the Amplification Curve of the He-Ne Laser at 633 nm", Zh. Tekh. Fiz., 1969, 39, no.3, p.528, and Sov. Phys.-Tech. Phys., 1969, 14, no.3, pp.391-5.

27.78. Khvostenko, G., and Choika, M.: "Linewidth of Spontaneous Emission in a Gas Laser", Opt. Spekt., 1969, 26, no.3, p.482, and Opt. Spectrosc., 1969, 26, no.3, p.268.

27.79. Gorlanov, A.V., Lyubimov, V.V., and Petrov, V.F.: "Single-Pulse Generation With Controllable Liquid-Film Shutter", Prib. Tekh. Eksp., 1969, no.6, pp.176-7, and Instrum. Exp. Tech., 1969, no.6, pp.1571-2.

27.80. Hoeksema, M., Sarjeant, W.J., and Brannen, E.: "Far-IR Gas Lasers as Sources of Linearly Polarized Radiation", J. Quantum Electron. IEEE, 1969, QE-5, p.477.

27.81. Pugovkin, A.V.: "Effect of Crystal Orientation Upon Generation Spectrum of Ruby Laser", Zh. Prikl. Spekt., 1969, 10, no.1, p.149.

27.82. Kimura, T., and Otsuka, K.: "Effects of Perturbations on Output of a CW Nd^{3+}:YAG Laser", J. Appl. Phys., 1969, 40, p.5399.

27.83. Newbery, A.R.: "Method for Producing Controllable Double Pulses from a Q-Switched Laser", Opto-Electron., 1969, 1, no.3, p.134.

27.84. Livshitz, B.L.: "Travelling-Medium Laser", Usp. Fiz. Nauk, 1969, 98, p.393, and Sov. Phys.-Usp., 1969, 12, p.430.

27.85. Mak, A.A., et al.: "Angular Divergence and Transverse Modes in a Nd^{3+}:Glass Laser With Spherical Resonator", Opt. Spekt., 1969, 26, no.5, p.793, and Opt. Spectrosc., 1969, 26, no.5, p.433.

27.86. Freiberg, R.J., and Halsted, A.S.: "Properties of Low-Order Transverse Modes in Argon-Ion Lasers", Appl. Opt., 1969, 8, p.355.

27.87. Mikaelyan, A.L., et al.: "Construction of a Directional Ruby Ring Laser", Zh. Eksp. Teor. Fiz., 1969, 57, no.1, p.38, and Sov. Phys.-JETP, 1970, 30, no.1.

27.88. Anan'ev, Yu.A., et al.: "Features of a Monopulse Laser With Monostable Resonator", Zh. Tekh. Fiz., 1969, 39, no.7, p.1325, and Sov. Phys.-Tech. Phys., 1970, 14, no.7, pp.997-9.

27.89. McAllister, G.L., Mann, M.M., and DeShazer, L.G.: "Transverse-Mode Distortions in Giant-Pulse Lasers", J. Quantum Electron. IEEE, 1970, QE-6, p.44.

27.90. Broom, R.F., et al.: "Microwave Self-Modulation of a Diode Laser Coupled to an External Cavity", J. Quantum Electron. IEEE, 1970, QE-6, p.328.

27.91. Lopasov, V.P., and Makogon, M.M.: "Operation of a Ruby Laser With Surface Scatterers", Izv. VUZ Fiz., 1970, no.3, p.127.

27.92. Chang, I.C., Lean, E.G.H., and Powell, C.G.: "Dynamics of Feedback-Controlled Nd^{3+}:YAG Laser", J. Quantum Electron. IEEE, 1970, QE-6, p.436.

27.93. Mahlein, H.F., and Schollmeier, G.: "Periodic Multiplate Resonant Reflector for a Nd^{3+}:YAG Laser at 1318 nm", J. Quantum Electron. IEEE, 1970, QE-6, p.525.

27.94. Murray, J.E., and Harris, S.E.: "Pulse Lengthening via Overcoupled (Q-Switched) Internal SHG", J. Appl. Phys., 1970, 41, p.609.

27.95. Paoli, T.L., and Ripper, J.E.: "Self-Stabilization and Narrowing of Optical Pulses from GaAs Junction Lasers by Injection-Current Feedback", Appl. Phys. Lett., 1970, 16, p.96.

27.96. Hillman, G., Tulip, J., and Seguin, H.J.: "PRF Control and Stabilization in a Passively Q-Switched CO_2 Laser", Appl. Opt., 1970, 9, p.515.

27.97. Bogdankevich, O.V., et al.: "Spectral Characteristics and Directionality of a Semiconductor Laser With External Mirror", Fiz. Tekh. Poluprov., 1970, 4, p.29, and Sov. Phys.-Semicond., 1970, 4, no.1, pp.22-5.

27.98. Korolev, F.A., et al.: "Fine Structure of Superradiance Spectrum in a Pulsed Neon Laser", Opt. Spekt., 1970, 28, no.3, p.290, and Opt. Spectrosc., 1970, 28, no.3, p.540.

27.99. Bennett, W.R.: "Hole-Burning Effects in Gas Lasers With Saturable Absorbers", Comments At. Mol. Phys., 1970, 2, no.1, p.10.

27.100. Smirnov, V.S., and Tumaikin, A.M.: "Polarization of Radiation from a Gas Laser", Zh. Eksp. Teor. Fiz., 1970, 58, no.6, p.2023, and Sov. Phys.-JETP, 1970, 32, no.6.

27.101. Isyanova, E.D., Marugin, A.M., and Ovchinnikov, V.M.: "Single-Frequency Ruby Laser With Electrooptic Q-Switching", Zh. Prikl. Spekt., 1970, 12, no.5, p.834.

27.102. Wang, S.C., Byer, R.L., and Siegman, A.E.: "Observation of an Enhanced Lamb Dip with a Pure Xe Gain Cell Inside a 3.51-micron He-Xe Laser", Appl. Phys. Lett., 1970, 17, p.120.

27.103. Malyshev, V.I., Masalov, A.V., and Sychev, A.A.: "Structure of Spectra of Solid-State Lasers in Free-Generation Regime", Zh. Eksp. Teor. Fiz. Pis'ma, 1970, 11, no.7, p.324, and JETP Lett., 1970, 11, no.7, p.215.

27.104. Smiley, V.N., Lewis, A.L., and Taylor, H.F.: "Mode Hopping and Tuning of Thin CdSe Platelet Lasers", J. Opt. Soc. Am., 1970, 60, p.1321.

27.105. Weber, H.P., and Gloge, D.: "Short-Term Mode Behaviour of GaAs Lasers", Appl. Phys. Lett., 1970, 17, p.231.

27.106. Anokhov, S.P., Kravchenko, V.I., and Soskin, M.S.: "Spectral and Kinetic Characteristics of Nd^{3+}:Glass Laser in Frequency-Scanning Regime", Ukr. Fiz. Zh., 1970, 15, no.8, p.1342.

27.107. Little, V.I., and Rowley, D.M.: "Use of a Rough-Stack Mirror in a Mode-Locked Ruby Laser", J. Phys. E, 1970, 3, p.469.

27.108. Mikaelyan, A.L., et al.: "Method for Generating a Giant Pulse in Lasers", Zh. Eksp. Teor. Fiz. Pis'ma, 1970, 11, no.5, p.244, and JETP Lett., 1970, 11, no.5, p.155.

27.109. Salzmann, H.: "Mode-Locking of Giant-Pulse Ruby Lasers", Phys. Lett., 1970, 32A, p.40.

27.110. Livshitz, B.L., and Tursunov, A.T.: "Spectral Kinetics of Radiation from a Nd^{3+}:Glass Travelling-Medium Laser", Zh. Eksp. Teor. Fiz., 1970, 58, no.5, p.1518, and Sov. Phys.-JETP, 1970, 31, no.5, pp.812-4.

27.111. Soskin, M.S., et al.: "Application of Nonlinear Absorption for Correction of Solid-State-Laser Radiation Wavefront", Zh. Prikl. Spekt., 1970, 12, no.4, p.740.

27.112. Boiko, B.B., et al.: "Monopulse Ruby Laser With Phototropic Ruby Shutter", Zh. Prikl. Spekt., 1970, 12, no.4, p.1757.

27.113. Belousova, I.M., Danilov, O.B., and Zapryagaev, A.F.: "Experimental Investigation of the Memory Effect of He-Ne Lasers", Zh. Eksp. Teor. Fiz. Pis'ma, 1970, 11, no.2, p.59, and JETP Lett., 1970, 11, no.2, p.97.

27.114. Kogelnik, H., et al.: "Hologram Wavelength Selector for Dye Lasers", Appl. Phys. Lett., 1970, 16, p.499.

27.115. Bennett, W.R.: "Theory of Cavity-Mode-Mixing Effects in Internally Scanned Lasers", Phys. Rev., 1970, 2A, p.458.

27.116. Hard, T.M.: "Laser Wavelength Selection and Output Coupling by a Grating", Appl. Opt., 1970, 9, p.1825.

27.117. Lopasov, V.P., and Makogon, M.M.: "Laser Frequency Control by Birefringent Crystals", Opt. Spekt., 1970, 28, no.3, p.291, and Opt. Spectrosc., 1970, 28, no.3, p.543.

27.118. Jensen, R.E., and Tobin, M.S.: "Passive Q-Switching of CO_2 Laser by C_2H_4 Gas and CH_3OH Vapour", J. Quantum Electron. IEEE, 1970, QE-6, p.477.

27.119. Paoli, T.L., and Ripper, J.E.: "Frequency Stabilization and Narrowing of Optical Pulses from CW GaAs Injection Lasers", J. Quantum Electron. IEEE, 1970, QE-6, p.335.

27.120. Kurnosov, V.D., et al.: "Instability of a Semiconductor Laser With Inhomogeneous Injection", Zh. Eksp. Teor. Fiz. Pis'ma, 1970, 11, no.8, p.385, and JETP Lett., 1970, 11, no.8, p.258.

27.121. Lindner, F.W., and Fischer, H.: "Properties of Nanosecond-Pulsed GaAs Laser Emission", Proc. IEEE, 1970, 58, p.925.

27.122. Scotland, R.M.: "Mode-Controlled Q-Switched Tunable Ruby Laser", Appl. Opt., 1970, 9, p.1211.

27.123. Stamenov, K., et al.: "Rotating-Prism Q-Switching for Lasers", Annu. Univ. Sofia Fac. Phys., 1970, 64-5, pp.289-93.

27.124. Michon, M., Auffret, R., and Dumanchin, R.: "Selection and Multiple-Pass Amplification of a Single Mode-Locked Optical Pulse", J. Appl. Phys., 1970, 41, p.2739.

27.125. Zasavitskii, I.I., et al.: "Tuning the Coherent Radiation Frequency of InSb by a Magnetic Field", Fiz. Tech. Poluprov. 1970, 4, p.337, and Sov. Phys.-Semicond., 1970, 4, no.2, pp.274-6.

27.126. Korobkin, V.V., Malyutin, A.A., and Shchelev, M.Ya.: "Dynamics of Radiation and Spectrum Changes of a Nd^{3+} Laser With Self-Mode-Locking", Zh. Eksp. Tekh. Fiz. Pis'ma, 1970, 11, p.103, and JETP Lett., 1970, 11, p.168.

27.127. Cawsey, D.: "Varactor-Tuned Gunn-Effect Oscillators", Electron. Lett., 1970, 6, p.246.

27.128. Izatt, J.R.: "Spectral Anomalies Due to Inhomogeneous Optical Pumping in the Ruby Laser", J. Appl. Phys., 1970, 41, no.11, pp.4569-77.

27.129. Key, M.H., and Preston, D.A.: "Pulse-Transmission-Mode Laser", J. Phys. E, 1970, 3, no.11, pp.932-3.

27.130. Pakhomycheva, L.A., et al.: "Line Structure of Lasers With Inhomogeneous Broadening", Zh. Eksp. Teor. Fiz. Pis'ma, 1970, 12, no.2, pp.60-3, and JETP Lett., 1970, 12, no.2, pp.43-5.

27.131. Malota, F.: "Regular Relaxation Pulses in a Ruby Laser", Opto-Electron., 1970, 2, no.2, pp.99-101.

27.132. Mikaelyan, A.L.: "Coherent Combination of Fields of Single-Frequency Ruby-Laser Radiation", Dokl. Akad. Nauk SSSR, 1970, 195, no.3, pp.565-7, and Sov. Phys.-Dokl., 1970, 15, no.3, pp.300-1.

27.133. Marcus, S., and McCoy, J.H.: "Self-Mode-Locking and Saturation Pulse Sharpening in a Rotating-Mirror Q-Switched CO_2 Laser", Appl. Phys. Lett., 1970, 16, p.11.

27.134. Ambartsumyan, R.V., et al.: "Investigation of Emission Spectrum of a He-Xe Laser With Nonresonant Feedback", Zh. Eksp. Teor. Fiz., 1970, 58, p.441, and Sov. Phys.-JETP, 1970, 31, no.2, pp.234-41.

27.135. Dolgopyatov, R.M., et al.: "Effect of Laser Field Structure on the Spectrum of Modulated Emission", Radiotekhnika, 1970, 25, no.2, p.48, and Telecommun. Radio Eng. Pt 2, 1970, 25, no.2.

27.136. Genkin, R.O., et al.: "Radiation Spectrum and Selection of Longitudinal Modes of a Laser With Electrooptic Q-Switching", Zh. Prikl. Spekt., 1970, 12, p.227.

27.137. Evans, D.E., and Forrest, M.J.: "Two-Pulse Q-Switched High-Power Ruby Laser", J. Phys. D, 1970, 3, no.12, pp.1931-4.

27.138. Korobkin, V.V., et al.: "Phase Self-Modulation and Self-Focusing of Nd^{3+}-Laser Radiation with Mode-Locking", Zh. Eksp. Teor. Fiz. Pis'ma, 1970, 12, no.5, pp.216-20, and JETP Lett., 1970, 12, no.5, pp.150-2.

27.139. Royce, G.A., and Sargent, M.: "Isotope Dependence of Gas-Laser Intensity Profiles", Appl. Opt., 1970, 9, no.11, pp.2428-34.

27.140. Alfano, R.R., and Shapiro, S.L.: "Picosecond Pulse Emission from a Mode-Locked Nd^{3+}:$POCl_3$ Liquid Laser", Opt. Commun., 1970, 2, no.2, pp.90-2.

27.141. Everett, P.N.: "Design of Nd^{3+}: Glass Mode-Locked Laser With Frequency Doubling and Pulse Selection", Rev. Sci. Instrum., 1970, 41, no.10, pp.1495-500.

27.142. Korzhenevich, I.M., et al.: "Optimal Conditions for Narrowing the Spectrum of a Laser With Highly Degenerate Oscillations", Radiotekh. Elektron., 1970, 15, no.4, and Radio Eng. Electron. Phys., 1970, 15, no.4, pp.693-5.

27.143. Vinevich, B.S.: "Divergence of the Beam from a Laser Resonator With a Coupling Hole", Opt. Spekt., 1970, 28, no.5, pp.853-5, and Opt. Spectrosc., 1970, 28, no.5, pp.463-4.

27.144. Kondilenko, I.I., Korotkov, P.A., and Koshel, O.N.: "Polarization Characteristics of Two Series-Mounted Ruby Lasers", Zh. Prikl. Spekt., 1970, 13, no.2,pp.220-2.

27.145. Salje, E., Welling, H., and Pfeiffer, B.: "Polarization Effects of Nd^{3+}:YAG Laser Radiation", Z. Angew. Phys., 1970, 30, no.4, pp.286-7.

27.146. Danielmeyer, H.G., and Turner, E.H.: "Electrooptic Elimination of Spatial Hole Burning in Lasers", Appl. Phys. Lett., 1970, 17, no.12, pp.519-21.

27.147. Virnik, Ya.Z., Kovalev, S., and Lariontsev, E.G.: "Inherent Frequency Fluctuations in Laser with Locked Modes", Izv. VUZ Radiofiz., 1970, 13, no.12, pp.1769-74.

27.148. Paoli, T.L., and Ripper. J.E.: "Observation of Intrinsic Quantum Fluctuations in Semiconductor Lasers", Phys. Rev., 1970, 2A, no.6, pp.2551-5.

27.149. Goldina, N.D., Kirin, Yu.M., and Troitskii, Yu.V.: "Narrowing of Ruby-Laser Spectrum by Diffraction Selector", Opt. Spekt., 1970, 28, no.5, pp.1005-7, and Opt. Spectrosc., 1970, 28,no.5,pp.543-4.

27.150. Arakelyan, V.S., et al.: "Fine Structure of a Giant Pulse in a CO_2 Laser with Transverse Modes", Radiotekh. Elektron., 1970, 15, no.4, pp.885-6, and Radio Eng. Electron. Phys., 1970,15,no.4,pp.725-6.

27.151. Buchman, W., Koechner, W., and Rice, D.K.: "Vibrating Mirror as a Repetitive Q-Switch", J. Quantum Electron. IEEE, 1970, QE-6, no.11, pp.747-9.

27.152. Carman, R.L., Reuntjes, J., and Furumoto, H.: "Time Synchronization of a Single Mode-Locked, Nd^{3+}:Glass-Laser, Pulse With a Q-Switched Ruby-Laser Pulse", J. Quantum Electron. IEEE, 1970, QE-6, no.11, pp.751-2.

27.153. Pardue, A.L., and McDuff, O.P.: "Reactive Q-Switching in CO_2, Multimode-Multipass, Laser", J. Quantum Electron. IEEE, 1970, QE-6, no.11, pp.753-4.

27.154. Willenbring, G.R., and Carruthers, J.A.: "Pulse Behaviour in a Self-Locked 441.6-nm He-^{114}Cd Laser", J. Appl. Phys., 1970, 41, no.12, pp.5040-1.

27.155. Gudkov, Yu.P.: "Effect of Collisions on Interaction of Axial Modes in a Gas Laser", Opt. Spekt., 1970, 29, no.2, pp.128-37, and Opt. Spectrosc., 1970, 29, no.2, pp.68-73.

27.156. Kliot-Dashinskii, M.I.: "Transference of a Modulated Signal by Coupled Transitions in Gas Lasers", Opt. Spekt., 1970, 29, no.2, pp.141-7, and Opt. Spectrosc., 1970, 29, no.2, pp.75-8.

27.157. Kulish, N.P., and Lisitsa, M.P.: "Effect of Resonator Spacing on Parameters of a Laser with Passive Q-Switch", Opt. Spekt., 1970, 29, no.2, pp.360-4, and Opt. Spectrosc., 1970, 29, no.2, pp.191-3.

27.158. Zakharenko, Yu.G., and Privalov, V.E.: "Effects of Regular Oscillations in the Discharge on Output Power of He-Ne Laser", Opt. Spekt., 1970, 29, no.2, pp.236-42, and Opt. Spectrosc., 1970, 29, no.2, pp.124-8.

27.159. Valuev, A.D., et al.: "Diffraction Splitting of Frequencies in 3.39-micron Laser", Opt. Spekt., 1970, 29, no.2, pp.410-2, and Opt. Spectrosc., 1970, 29,no.2,pp.217-8.

27.160. Glas, P.: "Coherence and Emission Properties of a Laser Without Resonator", Monatsber. Dtsch. Akad. Wiss. Berlin, 1970, 12, no.8, pp.593-9.

27.161. Kovrigin, A.I., et al.: "Spectral Characteristics of Parametric Light Generator with Single-Mode Pumping", Vestn. Mosk. Univ. Fiz. Astron., 1970, no.5, pp.535-9.

27.162. Demidov, Yu.P., et al.: "Radiation Directivity Pattern of a Semiconducting Multilayer Structure", Zh. Eksp. Teor. Fiz. Pis'ma, 1970, 12, no.4, pp.169-72, and JETP Lett., 1970, 12, no.4, pp.117-9.

27.163. Zeidler, G.: "Stabilization of Giant-Pulse Emission With Nonlinear Resonator Damping", Z. Naturforsch., 1970, 25a, no.10, pp.1511-2.

27.164. Shimizu, F.: "Q-Switching of N_2O and CO_2 Lasers by Stark Effect of Ammonia", Appl. Phys. Lett., 1970, 16, p.368.

27.165. Shimauchi, A., et al.: "Variation of Mode Intensity Spectrum of a He-Ne Laser in an Axial Magnetic Field", Sci. Light, 1970, 19, no.1, p.14.

27.166. Veiko, V.P., and Suslov, G.P.: "Stability of Energy in Pulsed Solid-State Lasers", Zh. Prikl. Spekt., 1970, 12, p.41.

27.167. Goster, V.I., et al.: "Contact Electrooptic Shutters for Q-Switching in a Ruby Laser", Opt.-Mekh. Prom., 1970, 37, no.5, pp.30-5, and Sov. J. Opt. Technol., 1970, 37, no.5, pp.302-6.

27.168. Vargashkin, A.I., et al.: "Spontaneous Transient Processes in Emission Spectrum of a Semiconductor Laser", Zh. Eksp. Teor. Fiz. Pis'ma, 1970, 12, no.1, pp.3-6, and JETP Lett., 1970, 12, no.1, pp.1-4.

27.169. Kalinin, N.A., Latysheva, E.I., and Efremov, Yu.P.: "Frequency Shift of Direct-Current Lasers", Opt. Spekt., 1970, 29, no.5, pp.1020-1, and Opt. Spectrosc., 1970, 29, no.5, p.543.

27.170. Zhekov, V.I., et al.: "High-Repetition-Rate Giant-Pulsed Dy^{2+}:CaF_2 Laser With $LiNbO_3$ Electrooptic Shutter", Radiotekh. Elektron., 1970, 15, no.10, and Radio Eng. Electron. Phys., 1970, 15, no.10, pp.1857-60.

27.171. Bhawalkar, D.D., Chatterjee, U.K., and Pant, H.C.: "Temporal Behaviour of Light Emission from GaAs Injection Lasers", Indian J. Pure Appl. Phys.,1970,8,no.12,pp.849-50.

27.172. Vlasov, A.N., et al.: "Spectrochronographic Study of a Semiconductor Laser with Electron-Beam Excitation", Radiotekh. Elektron., 1970, 15, no.9, and Radio Eng. Electron. Phys., 1970, 15,no.9,pp.1735-7.

27.173. Ikegami, T., et al.: "Transient Behaviour of Semiconductor Injection Lasers", Electron. Commun. Jap.,1970,53,no.5,pp.15-6.

27.174. Bespalov, V.I., et al.: "Modular Electrooptic Ruby-Laser Q-Switches", Opt.-Mekh. Prom., 1970, 37, no.6, pp.11-3, and Sov. J. Opt. Technol., 1970, 37, no.6, pp.350-2.

27.175. Gaponov, S.V., et al.: "Giant Pulse in a Solid-State Laser With Organic Solvents", Zh. Eksp. Teor. Fiz. Pis'ma, 1970, 11, no.8, p.370, and JETP Lett., 1970, 11, no.8, p.248.

27.176. Sharapov, B.N.: "Possible Polarization of a Semiconductor p-n Heterojunction Laser", Fiz. Tekh. Poluprov., 1970, 4, no.12, pp.2389-91, and Sov. Phys.-Semicond., 1971, 4, no.12, pp.2059-60.

27.177. Leontovich, A.M., and Churkin, V.L.: "Kinetics of a Ruby Laser", Zh. Eksp. Teor. Fiz., 1970, 59, no.1, p.7, and Sov. Phys.-JETP, 1971, 32, no.1, pp.4-10.

27.178. Malyshev, V.I., Masalov, A.V., and Sychev, A.A.: "Spectral-Line Technique for Investigation of Partial Mode Locking in Ruby and Nd^{3+}:Glass Lasers", Zh. Eksp. Teor. Fiz., 1970, 59, no.1, p.48, and Sov. Phys.-JETP, 1971, 32, no.1, pp.27-30.

27.179. Alferov, Zh.I., et al.: "Spatial Distribution of Emission by GaAs-AlAs Heterojunction Lasers at Room Temperature", Fiz. Tekh. Poluprov., 1970, 4, no.9, pp.1697-703, and Sov. Phys.-Semicond., 1971, 4, no.9, pp.1457-62.

27.180. Dzhibladze, M.I., et al.: "Time Characteristics of a Dy^{2+}:CaF_2 Laser Under Monochromatic Pumping", Dokl. Akad. Nauk SSSR, 1970, 195, no.4-6, pp.1078-81, and Sov. Phys.-Dokl., 1971, 15, no.12,pp.1146-8.

27.181. Gloge, D., and Lee, T.P.: "Signal Structure of Self-Pulsing CW GaAs Lasers", J. Quantum Electron. IEEE, 1971, QE-7, no.1, pp.43-5.

27.182. Abrosimov, G.V.: "Spatial and Temporal Coherence of Pulsed Ne and Tl Gas Lasers", Opt. Spekt., 1971, 31, no.1, pp.106-10, and Opt. Spectrosc., 1971, 31, no.1, pp.54-6.

27.183. Fujiwara, H.: "Spatial Coherence of Laser Light With Multimode Oscillations", Oyo Buturi, 1971, 40, no.3, pp.281-6.

27.184. Vvedenskii, B.S., et al.: "Anisotropy of Active Field and Polarization of GaAs Injection Laser", Vestn. Mosk. Univ. Fiz. Astron., 1971, 12, no.6, pp.743-5.

27.185. Pardue, A.L., et al.: "Mode Competition and Self-Locking in a Multipass CO_2 Laser", J. Quantum Electron. IEEE, 1971, QE-7, no.1, pp.22-4.

27.186. Waksberg, A.L., Boag, J.C., and Sizgorie, S.: "Signature Variations With Mirror Separation for Small Sealed CO_2 Lasers", J. Quantum Electron. IEEE, 1971, QE-7, no.1, pp.29-35.

27.187. Miyazoe, Y., and Maeda, M.: "Spiking Phenomenon in Organic Dye Lasers", J. Quantum Electron. IEEE, 1971, QE-7, no.1, pp.36-7.

27.188. Smith, A.W., and Edmonds, H.D.: "Picosecond Structure in Self-Pulsing GaAs Injection Lasers", J. Quantum Electron. IEEE, 1971, QE-7, no.2, pp.94-5.

27.189. Dumanchin, R., et al.: "Analysis of Giant-Pulse Amplification in Nd^{3+}:Glass", J. Quantum Electron. IEEE, 1971, QE-7, no.2, pp.53-8.

27.190. Burak, I., et al.: "Mechanism of Passive Q-Switching in CO_2 Lasers", J. Quantum Electron. IEEE,1971,QE-7,no.2,pp.73-82.

27.191. Kan, T., and Wolga, G.J.: "Influence of Collisions on Radioactive Saturation and Lamb-Dip Formation in CO_2 Molecular Lasers", J. Quantum Electron. IEEE, 1971, QE-7, no.4, pp.141-50.

27.192. Dube, G.: "Background Energy Content of Mode-Locked Laser Pulses", Appl. Phys. Lett., 1971, 18, no.3, pp.69-70.

27.193. Bogdankevich, O.V., et al.: "Space-Time Characteristics of Radiation from an Electron-Beam-Pumped Semiconductor Laser", Zh. Eksp. Teor. Fiz., 1971, 60, no.1, pp.132-5, and Sov. Phys.-JETP, 1971, 33, no.1, pp.74-6.

27.194. Kondilenko, I.I., et al.: "Peculiarities of Many-Element Ruby-Laser Kinetics", Phys. Status Solidi, 1971, 4A, no.1, pp.43-52.

27.195. Harris, E.P.: "Spiking in Current-Modulated CW GaAs External Cavity Lasers", J. Appl. Phys., 1971, 42, no.2, pp.892-3.

27.196. Hovel, W.: "Phase Coupling Between Modes of Two (Common-Medium) He-Ne Lasers", Nachr. Tech. Z., 1971, 24, no.3, pp.133-6.

27.197. Nurmikko, A., DeTemple, T.A., and Schwarz, S.E.: "Single-Mode Operation and Mode-Locking of High-Pressure CO_2 Lasers by Saturable Absorbers", Appl. Phys. Lett., 1971, 18, no.4, pp.130-2.

27.198. Ohtsuka, Y.: "Spatial-Coherence Properties of a CO_2 Laser", Phys. Lett., 1971, 34A, no.5, pp.279-80.

27.199. Dobryden', V.A., and Tsarenko, V.T.: "Design of Microwave Power Stabilizer With Nonlinear Delayed Feedback", Izv. VUZ Radioelektron., 1971, 14, no.4, pp.424-9.

27.200. Bykovskii, Yu.A., et al.: "Investigation of Kinetics of Spectrum of a Semiconductor Laser Under Nonuniform Injection Conditions", Fiz. Tekh. Poluprov., 1971, 5, no.5, pp.939-42, and Sov. Phys.-Semicond., 1971, 5, no.5, pp.825-8.

27.201. Wieder, H.: "Sequential Excitation of Longitudinal Modes in GaAs Lasers", Appl. Phys. Lett., 1971, 18, no.6, pp.223-4.

27.202. Karr, M.A.: "Nd^{3+}:YAG Laser Cavity Loss Due to an Internal Brewster Polarizer", Appl. Opt., 1971, 10, no.4, pp.893-5.

27.203. Scott, W.C., and de Wit, M.: "Birefringence Compensation and TEM_{00}-Mode Enchancement in a Nd^{3+}:YAG Laser", Appl. Phys. Lett., 1971, 18, no.1, pp.3-4.

27.204. Young, J.F., et al.: "Q-Switched Laser With Controllable Pulse Duration", Appl. Phys. Lett., 1971,18,no.4,pp.129-30.

27.205. Vinokurov, G.N., et al.: "Spike Structure of Solid-State Lasers", Zh. Eksp. Teor. Fiz., 1971, 60, no.2, pp.489-99.

27.206. Eguchi, R.G., et al.: "Simultaneous Mode Locking and Pulse Coupling of the CO_2 Laser", Appl. Phys. Lett., 1971, 18, no.9, pp.406-8.

27.207. Klochkov, V.P., et al.: "Q-Switching of Ruby Laser by Dye Vapours", Zh. Eksp. Teor. Fiz. Pis'ma, 1971, 13, no.1, pp.47-8.

27.208. Svelto, O.: "Subpicosecond Structure and Frequency Sweep of Mode-Locked Lasers", Opt. Commun., 1971, 3, no.2,pp.105-6.

27.209. Livshitz, B.L.: "Nature of Spike Generation of Lasers", Dokl. Akad. Nauk SSSR, 1971, 199, no.10, pp.1298-300, and Sov. Phys.-Dokl., 1971, 16, no.10, pp.949-51.

27.210. Erickson, L.E., and Szabo, A.: "Spectral Narrowing of Dye-Laser Output by Injection of Monochromatic Radiation into the Laser Cavity", Appl. Phys. Lett., 1971, 18, no.10, pp.433-5.

27.211. Hall, D.R., and Pao, Y.H.: "High-Efficiency Active Q-Switching of CO_2 Laser Using the Stark Effect", J. Quantum Electron. IEEE, 1971, QE-7, no.8, pp.427-9.

27.212. Yajima, T., and Takeuchi, N.: "Spectral Properties and Tunability of Far-IR Difference-Frequency Radiation Produced by Picosecond Laser Pulses", Jap. J. Appl. Phys., 1971, 10, no.7, pp.907-15.

27.213. Ripper, J.E., and Paoli, T.L.: "Self-Pulsing of Junction Lasers Operating CW at Room Temperature", Appl. Phys. Lett., 1971, 18, no.10, pp.466-8.

27.214. Vrehen, Q.H.F.: "Spectral Distribution of the Stimulated Emission of a Rhodamine-B Dye Laser", Opt. Commun., 1971, 3, no.3, pp.144-6.

27.215. Vyshlov, S.S., et al.: "Transverse Mode Locking in an Injection Laser", Zh. Eksp. Teor. Fiz. Pis'ma, 1971, 13, no.3, pp.131-3, and JETP Lett., 1971, 13, no.3, pp.90-2.

27.216. Aritome, H., Masuda, K., and Namba, S.: "Time Behaviour of Electron-Beam-Pumped CdS Laser", Sci. Pap. Inst. Phys. Chem. Res., 1971, 65, no.1, pp.10-8.

27.217. Kitaeva, V.F., Osipov, Yu.I., and Sobolev, N.N.: "Widening of Spectral Lines in an Ar^{II} Laser", J. Quantum Electron. IEEE, 1971, QE-7, no.8, pp.391-6.

27.218. Ladygin, M.V., and Tsarkov, B.A.: "Effect of a Longitudinal Magnetic Field on the Characteristics of a He-Ne Laser Operating at 3.39 micron", Opt. Spekt., 1971, 30, no.1, pp.133-6, and Opt. Spectrosc., 1971, 30, no.1, pp.69-71.

27.219. Ohtsuka, Y.: "Spatial Coherence in CO_2 Laser Beams out of Diffraction-Coupled Resonators", Phys. Lett., 1971, 36A, no.2, pp.151-2.

27.220. Malyshev, V.I., Sychev, A.A., and Babenko, V.A.: "Investigation of a Nd^{3+}: Glass (Mode-Locked) Laser with Passive Shutter Having a Finite Relaxation Period", Zh. Eksp. Teor. Fiz. Pis'ma, 1971, 13, no.11, pp.588-92, and JETP Lett., 1971, 13, no.11, pp.419-22.

27.221. de Lang, H.: "Flow Lines on the Poincare Sphere as Aid to Study of Mode Polarization in Lasers", J. Quantum Electron. IEEE, 1971, QE-7, no.9, pp.441-4.

27.222. Adams, M.J., and Cross, M.: "Polarization of Radiation from Double-Heterostructure Injection Lasers", Electron. Lett., 1971, 17, no.19, pp.569-70.

27.223. Wieder, H.: "Mode Perturbations and Filamentary Coupling in GaAs Lasers", J. Appl. Phys., 1971, 42, no.10, pp.3839-43.

27.224. Masuyama, A., et al.: "Internal Q-Switching in n-GaAs Lasers Under Electron-Beam Excitation", Jap. J. Appl. Phys., 1971, 10, no.9, p.1281.

27.225. Wang, C.C., and Davis, L.I.: "Transients Effects in Mode Locking of Nd^{3+}:Glass Lasers", Appl. Phys. Lett., 1971, 19, no.6, pp.167-9.

27.226. Rice, R.R.: "Pulse Shape of a Mode-Locked Frequency-Doubled Nd^{3+}:YAG Laser", J. Appl. Phys., 1971, 42, no.10, pp.4109-11.

27.227. White, M.B., and Gerber, W.D.: "Mode-Locked Dual-Polarization Operation of a CO_2 Laser", J. Quantum Electron. IEEE, 1971, QE-7, no.9, pp.445-50.

27.228. Yamanaka, M., et al.: "Polarization Flip in the Far-IR HCN Laser", J. Quantum Electron. IEEE, 1971, QE-7, no.9, pp.457-9.

27.229. Ivanov, V.D., and Leontovich, A.M.: "Spatial Coherence of Radiation from a Q-Switched Ruby Laser", Kvantovaya Elektron., 1971, no.1, pp.96-101.

27.230. Aussenegg, F., and Mockel, P.: "Instabilities in the Emission Distribution of Dye-Switched Ruby Lasers", Z. Naturforsch., 1971, 26a, no.7, pp.1201-5.

27.231. Moskalenko, V.F., et al.: "Radiation of a Pulsed Ion Laser", Opt. Spekt., 1971, 30, no.2, pp.369-71, and Opt. Spectrosc., 1971, 30, no.2, pp.201-2.

27.232. Mikaelyan, A.L., et al.: "Investigation of Emission by a Ruby Laser under Self-Q-Switched Conditions", Kvantovaya Elektron., 1971, no.1, pp.102-9.

27.233. Eckardt, R.C., Lee, C.H., and Bradford, J.N.: "Temporal and Spectral Development of Mode Locking in a Ring-Cavity Nd^{3+}:Glass Laser", Appl. Phys. Lett., 1971, 19, no.10, pp.420-3.

27.234. Anan'ev, Yu.A.: "Angular Divergence of Radiation of Solid-State Lasers", Usp. Fiz. Nauk, 1971, 14, no.2, pp.705-38, and Sov. Phys.-Usp., 1971, 4, no.2, pp.197-215.

27.235. Runge, P.K.: "Mode-Locking of He-Ne Lasers With Saturable Organic Dyes", Opt. Commun., 1971, 3, no.6, pp.434-6.

27.236. Otis, G., and Tremblay, R.: "Time-Dependent Lensing Effects in Transversely Excited Atmospheric CO_2 Lasers", Opt. Commun., 1971, 3, no.6, pp.418-20.

27.237. Yoshida, K., et al.: "Generation and Amplification of Laser Pulse With a Variable Duration Using PTM Method", Jap. J. Appl. Phys., 1971, 10, no.11, pp.1643-4.

27.238. Vanyukov, M.P., et al.: "Formation of High-Power Pulses With Steep Leading Edges in a Laser System With Passive Nonlinear Elements", Kvantovaya Elektron., 1971, no.1, pp.35-41.

27.239. Suchard, S.N., Gross, R.W.F., and Whittier, J.S.: "Time-Resolved Spectroscopy of a Flash-Initiated HF Laser", Appl. Phys. Lett., 1971, 19, no.10, pp.411-3.

27.240. Bagratashvili, V.N., Knyazev, I.N., and Letokhov, V.S.: "Tunable IR Gas Lasers", Opt. Commun., 1971, 4, no.2, pp.154-6.

27.241. Myers, S.A.: "Improved Line-Narrowing Technique for a Dye Laser Excited by a N_2 Laser", Opt. Commun., 1971, 4, no.2, pp.187-9.

27.242. Arthurs, E.G., Bradley, D.J., and Roddie, A.G.: "Frequency-Tunable Transform-Limited Picosecond Dye-Laser Pulses", Appl. Phys. Lett., 1971, 19, no.11, pp.480-2.

27.243. Ohtsuka, Y.: "Effects of Output-Coupling Apertures on Spatial Coherence of a CO_2 Laser", Opt. Acta, 1971, 18, no.12, pp.879-90.

27.244. Kravtsov, N.V., and Yatsenko, Yu.P.: "Time Characteristics of Nd^{3+}:Glass Lasers With Long Resonators", Vestn. Mosk. Univ. Fiz. Astron., 1971, 12, no.6, pp.734-6.

27.245. Magdich, L.N.: "Pulsed Q-Switching of a CW Laser Cavity", Opt. Spekt., 1971, 31, no.2, pp.301-3, and Opt. Spectrosc., 1971, 31, no.2, pp.161-2.

27.246. Tsetsegova, E.I.: "He-Ne Dispersion Characteristics at 3.39 micron", Opt. Spekt., 1971, 31, no.2, pp.319-20, and Opt. Spectrosc., 1971, 31, no.2, pp.172-3.

27.247. Robinson, L.C., and Whitbourn, L.B.: "Experimental Investigation of a Pulsed 337-micron HCN Laser", Proc. IREE Aust., 1971, 32, no.10, pp.355-60.

27.248. Zakharov, Yu.P., et al.: "Influence of Output Intensity Pulsations on Emission Spectrum of an Injection Laser", Kvantovaya Elektron., 1971, no.4, pp.99-103.

27.249. Andreev, A.G., et al.: "Fluctuations of the Output Energy of a Ruby Laser", Kvantovaya Elektron., 1971, no.4, pp.120-1.

27.250. Hanna, D.C., et al.: "Two-Step Q-Switching Technique for Producing High Power in a Single Longitudinal Mode", Opto-Electron., 1971, 3, no.4, pp.163-9.

27.251. Tikhomirov, A.A., and Shandarov, S.M.: "Selection of Spectral Lines in a Ruby Laser at 77°K", Zh. Prikl. Spekt., 1971, 15, no.5, pp.803-5.

27.252. Gubin, M.A., Popov, A.I., and Protsenko, E.D.: "Investigation of Competition Between Two Axial Modes in a Laser With a Homogeneously Broadened Line", Kvantovaya Elektron., 1971, no.4, pp.34-40.

27.253. Bazarov, Ye.N., and Gerasimov, G.A.: "Passive Q-Switching of a CO_2 Laser by a Saturable Filter Containing OsO_4 Vapour", Kvantovaya Elektron., 1971, no.4, pp.87-8.

27.254. Zargar'yants, M.N., et al.: "Adding the Radiation Power of Injection Lasers", Radiotekh. Elektron., 1971, 16, no.3, and Radio Eng. Electron. Phys., 1971, 16, no.3, pp.560-3.

27.255. Anan'ev, Yu.A., Chernov, V.N., and Sherstobitov, V.E.: "Solid-State Laser With High Spatial Coherence of Radiation", Kvantovaya Elektron., 1971, no.4, pp.112-3.

27.256. Gromov, Yu.N., et al.: "Internal Frequency Selection of a CO_2 Laser Using a Fabry-Perot Interferometer", Prib. Tekh. Eksp., 1971, 14, no.3, pp.183-6, and Instrum. Exp. Tech., 1971, 14, no.3, pp.870-3.

27.257. Bogatov, A.P., et al.: "Comparison of Instantaneous and Averaged Emission Spectra of an Injection Laser Operating Under Spiking Conditions", Kvantovaya Elektron., 1971, no.5, pp.93-5.

27.258. Allakhverdyan, R.G., et al.: "Influence of Refractive-Index Nonlinearity on the Dynamics of Emission from Semiconductor Lasers", Kvantovaya Elektron., 1971, no.6, pp.53-9.

27.259. Dzhibladze, M.I., et al.: "Regular Pulsations of Intensity of a Nd^{3+}:Glass Fibre Laser", Kvantovaya Elektron., 1971, no.5, pp.120-2.

27.260. Sedel'nikov, V.A., Sinichkin, Yu.P., and Tuchin, V.V.: "Properties of Emission Spectrum of Argon-Ion Lasers", Opt. Spekt., 1971, 31, no.5, pp.761-2, and Opt. Spectrosc., 1971, 31, no.5, pp.408-9.

27.261. Kurbatov, L.N., et al.: "Time Characteristics of Heterojunction Injection Lasers", Kvantovaya Elektron., 1971, no.6, pp.110-2.

27.262. Goloyadova, V.I., et al.: "Shape of a Giant Pulse in a Cavity With Nearly Transparent Reflector", Radiotekh. Elektron., 1971, 16, no.10, pp.1839-45, and Radio Eng. Electron. Phys., 1971, 16, no.10, pp.1653-7.

27.263. Grigor'yants, V.V.: "Effective Amplification and Shape of Spectral Dip in Laser Materials With Inhomogeneous Broadening", Radiotekh. Elektron., 1971, 16, no.10, pp.1865-72, and Radio Eng. Electron. Phys., 1971, 16, no.10, pp.1672-8.

27.264. Poizner, B.N.: "Spectrum of a Multimode Gas Laser with Injected Light Signal", Radiotekh. Elektron., 1971, 16, no.10, pp.1852-8, and Radio Eng. Electron. Phys., 1971, 16, no.10, pp.1662-7.

27.265. Yermakov, B.A., and Lukin, A.V.: "Passive Shutter for Periodically Operating Laser", Opt.-Mekh. Prom., 1971, 38, no.9, pp.39-42, and Sov. J. Opt. Technol., 1971, 38, no.9, pp.551-3.

27.266. de Lang, H., Polder, D., and van Haeringen, W.: "Optical Polarization Effects in a Gas Laser", Philips Tech. Rev., 1971, 32, no.6-8, pp.190-204.

27.267. Korneev, N.E.: "Problem of Q-Switching a Convex Optical Cavity", Radiotekh. Elektron., 1971, 16, no.12, p.2325, and Radio Eng. Electron. Phys., 1971, 16, no.12, p.2133.

27.268. Mikaelyan, A.L., et al.: "High-Coherence Ruby Laser With Slow Q-Switching", Kvantovaya Elektron., 1971, no.2, pp.96-9.

27.269. Klockhov, V.P., Bogdanov, V.L., and Neporent, B.S.: "Q-Switching of a Ruby Laser by Phthalocyanine Vapour", Opt. Spekt., 1971, 30, no.6, pp.1088-91, and Opt. Spectrosc., 1971, 30, no.6, pp.583-4.

27.270. Smirnov, A.G., Staselko, D.I., and Terent'ev, V.E.: "Coherent Characteristics of a Ruby Laser With Controlled Lasing Period", Opt. Spekt., 1971, 31, no.1, pp.103-5, and Opt. Spectrosc., 1971, 31, no.1, pp.52-3.

27.271. Landry, M.J.: "Variably Spaced Giant Pulses from Multiple Laser Cavities in a Single Lasing Medium", Appl. Phys. Lett., 1971, 18, no.11, pp.494-6.

27.272. Shilov, V.B., and Neporent, B.S.: "Spectral Shifts in Laser Emission of Laser-Pumped Dye Solutions", Opt. Spekt., 1971, 30, no.6, pp.1074-80, and Opt. Spectrosc., 1971, 30, no.6, pp.576-9.

27.273. Kobayashi, K., et al.: "Observation of Pulsation from Double-Heterostructure Injection Laser Due to Lateral Optical Coupling", Appl. Phys. Lett., 1971, 19, no.9, pp.323-4.

27.274. Antsiferov, V.V., et al.: "Free Laser Generation in Ruby with Spherical Resonator With Electrooptic Spatial Smoothing", Zh. Tekh. Fiz., 1971, 41, no.12, pp.2594-9, and Sov. Phys.-Tech. Phys., 1972, 16, no.12, pp.2060-5.

27.275. Brzhazovskii, Yu.V., et al.: "Investigation of Radiation Pulsation from a CO_2 Laser with Nonlinear Absorbing Cell", Zh. Eksp. Teor. Fiz., 1971, 61, no.2, pp.500-10, and Sov. Phys.-JETP, 1972, 34, no.2, pp.265-70.

27.276. Casperson, L.W., and Yariv, A.: "Time Behaviour and Spectra of Relaxation Oscillations in a High-Gain Laser", J. Quantum Electron. IEEE, 1972, QE-8,no.2,pp.69-73.

27.277. Nakatsuka, M., et al.: "Self-Mode Locking of TEA CO_2 Laser With Helical Pumping", Jap. J. Appl. Phys., 1972, 11, no.1, pp.114-5.

27.278. Brackett, C.A.: "Second-Order Dispersion in Oscillating GaAs Junction Lasers", J. Quantum Electron. IEEE, 1972, QE-8, no.2,pp.66-9.

27.279. Casperson, L.W., and Yariv, A.: "Spectral Narrowing in High-Gain Lasers", J. Quantum Electron. IEEE, 1972, QE-8, no.2, pp.80-5.

27.280. Tobin, M.S., and Jensen, R.E.: "Q-Switching of CO_2 Laser by Molecular Gases Exhibiting the Stark Effect", J. Quantum Electron. IEEE, 1972, QE-8,no.1,pp.21-2.

27.281. Arthurs, E.G., Bradley, D.J., and Roddie, A.G.: "Passive Mode Locking of Flashlamp-Pumped Dye Lasers Tunable Over 580-700 nm", Appl. Phys. Lett., 1972, 20, no.3, pp.125-7.

27.282. Ono, K., et al.: "Correlation Between 28- and 119-micron Lines and Pressure Effect in CW H_2O Laser", Jap. J. Appl. Phys., 1972, 11, no.2, pp.221-3.

27.283. Nishizawa, J., et al.: "Conservation of Polarization in GaAs Junction Laser", Jap. J. Appl. Phys., 1972, 11, no.3, pp.419-20.

27.284. Yoshino, T.: "Polarization Properties of Internal-Mirror He-Ne Lasers at 633 nm", Jap. J. Appl. Phys., 1972, 11, no.2, pp.263-5.

27.285. Whitbourn, L.B., Robinson, L.C., and Tait, G.D.: "Origin of Spiking Pulses in HCN Gas Lasers", Phys. Lett., 1972, 38A, no.5, pp.315-7.

27.286. Houston, P.L., Sutton, D.G., and Steinfeld, J.I.: "Behaviour of Pulsed-Discharge Laser With Intracavity Absorber", J. Appl. Phys., 1972, 43, no.4,pp.2014-5.

27.287. Kuehn, D.M.: "Artificial Laser-Intensity Oscillations Associated With Beam Splitters", Appl. Opt., 1972, 11, no.6, pp.1431-3.

27.288. Otsuka, K.: "Separation of Mode-Locked States in the Intracavity Phase-Modulated Nd^{3+}:YAG Laser", J. Quantum Electron. IEEE, 1972, QE-8, no.6, pp.496-7.

27.289. Koechner, W.: "Output Fluctuations of Continuously Pumped Nd^{3+}:YAG Lasers", J. Quantum Electron. IEEE, 1972, QE-8, no.7, pp.656-61.

27.290. Nishihara, H., and Kronast, B.: "Investigation of the Line Structure of a Helical TEA CO_2 Laser", Opt. Commun., 1972, 5, no.1, pp.65-7.

27.291. Orlov, L.N., and Rubanov, V.S.: "Gas Laser Using External Mirrors Generating Nonpolarized Emission", Zh. Prikl. Spekt., 1972, 16, no.4, pp.744-5.

27.292. Selden, A.C.: "Transverse Modes in a Liquid Laser", Opt. Commun., 1972, 5, no.1, pp.62-4.

27.293. Firester, A.H., Gayeski, T.E., and Heller, M.E.: "Efficient Generation of Laser Beams With an Elliptic Cross Section", Appl. Opt., 1972, 11, no.7, pp.1648-9.

27.294. Danielmeyer, H.G., and Ostermayer, F.W.: "Diode-Pumped-Modulated Nd^{3+}:YAG Laser", J. Appl. Phys., 1972, 43, no.6, pp.2911-3.

27.295. Lucht, R.A., Allario, F., and Jarrett, O.: "Simple Technique for Sequential Q-Switching of Molecular Lasers", Appl. Opt., 1972, 11, no.7, pp.1568-71.

27.296. Wentz, J.L.: "Electrooptic Q-Switching for Randomly Polarized Lasers", Proc. IEEE, 1972, 60, no.3, pp.343-4.

27.297. Scott, W.C., and de Wit, M.: "Efficient Variable Threshold Acoustooptic Q-Switching of Flash-Pumped Nd^{3+}:YAG", Appl. Phys. Lett., 1972, 20, no.3, pp.141-3.

27.298. Ohtsuka, Y.: "Competition Effects on Spatial Coherence in a CO_2 Laser", Opt. Commun., 1972, 5, no.2, pp.106-10.

27.299. Dahlstrom, L.: "Passive Mode-Locking and Q-Switching of High-Power Lasers by Optical Kerr Effect", Opt. Commun., 1972, 5, no.3, pp.157-62.

27.300. Mushiake, Y., et al.: "Generation of Radially Polarized Optical Beam Mode by Laser", Proc. IEEE, 1972, 60,no.9,pp.1107-9.

27.301. Bespalov, V.I., et al.: "Single-Crystal Electrooptic Shutter for Q-Switching Lasers With Unpolarized Radiation", Opt.-Mekh. Prom., 1972, 38, no.12, pp.30-2, and Sov. J. Opt. Technol., 1972, 38, no.12, pp.739-41.

27.302. Balashov, I.F., Berezin, B.G., and Ermakov, B.A.: "Single-Pulse Lasing With Noninstantaneous Q-Switching", Zh. Tekh. Fiz., 1972, 17, no.2, pp.385-90, and Sov. Phys.-Tech. Phys., 1972, 17, no.2, pp.306-10.

27.303. Puschert, W.: "Automatic Mode Locking of CW Lasers", Opt. Commun., 1972, 5, no.5, pp.380-1.

27.304. Beterov, I.M., Kuz'min, Yu.E., and Chebotaev, V.P.: "Polarization of a Three-Level Gas Laser", Opt. Spekt., 1972, 32, no.1, pp.220-2, and Opt. Spectrosc., 1972, 32, no.1, pp.114-5.

27.305. Burak, I., et al.: "Q-Switched CO_2 Lasers With Variable Pulse Delay", Rev. Sci. Instrum., 1972, 43, no.9, pp.1390-2.

27.306. Akhmanov, S.A., Golyaev, Yu.D., and Dmitriev, V.G.: "Spectral Characteristics and Oscillation Dynamics in Nd^{3+}:YAG Quasi-CW Lasers", Zh. Eksp. Teor. Fiz., 1972, 62, no.1, pp.133-43, and Sov. Phys.-JETP, 1972, 35, no.1, pp.70-5.

27.307. Freiberg, R.J., Chenausky, P.P., and Buczek, C.J.: "Experimental Study of Unstable Confocal CO_2(-Laser) Resonators", J. Quantum Electron. IEEE, 1972, QE-8, no.12, pp.882-92.

27.308. Zakharenko, Yu.G.: "Effect of Gas Pressure on the Range of Existence of Relaxation Oscillations in a Discharge", Opt. Spekt., 1972, 32, no.3, pp.455-7, and Opt. Spectrosc., 1972, 32, no.3, pp.238-9.

27.309. Mukhtarov, Ch.K.: "Spectrum of Stimulated Radiation in a Flat-Mirror Resonator", Dokl. Akad. Nauk SSSR, 1972, 203, no.5, pp.70-3, and Sov. Phys.-Dokl., 1972, 17, no.5, pp.453-6.

27.310. Hanna, D.C., Luther-Davies, B., and Smith, R.C.: "Active Q-Switching Technique for Producing High Laser Power in a Single Longitudinal Mode", Electron. Lett., 1972, 8, no.15, pp.369-70.

27.311. Becker, M.F., Kuizenga, D.J., and Siegman, A.E.: "Harmonic Mode Locking of Nd^{3+}:YAG Laser", J. Quantum Electron. IEEE, 1972, QE-8, no.8, pp.687-93.

27.312. Chun, M.K., and Bischoff, J.T.: "Multipulsing Behaviour of Electrooptically Q-Switched Lasers", J. Quantum Electron. IEEE, 1972, QE-8, no.8, pp.715-6.

27.313. Sizgoric, S., and Waksberg, A.L.: "Experimental Analysis of Vibrational-Rotational Line Content of a Q-Switched CO_2 Laser", Can. J. Phys., 1972, 50, no.13, pp.1465-70.

27.314. Chang, T.Y., and Wood, O.R.: "Simple Self-Mode-Locked Atmospheric-Pressure CO_2 Laser", J. Quantum Electron. IEEE, 1972, QE-8, no.8, pp.721-3.

27.315. Devir, A.D.: "Doppler Q-Switching in a Single-Mode CO_2 Laser by a Rotating Mirror", J. Appl. Phys., 1972, 43, no.8, pp.3397-8.

27.316. Kato, I., Seki, N., and Shimizu, T.: "Ring-Shaped Beam Formation in Microwave-Pulse-Excited He-Kr^{II} Laser", Jap. J. Appl. Phys., 1972, 11, no.8, pp.1236-7.

27.317. Fountain, W.D.: "Far-Field Brightness of Amplifier Laser Systems", Appl. Opt., 1972, 11, no.10, pp.2383-4.

27.318. Sorokin, P.P., and Lankard, J.R.: "Alkali-Metal-Vapour Q-Switches for Synchronizing Mode-Locked-Laser Pulse Trains With External Events", J. Quantum Electron. IEEE, 1972, QE-8, no.10, pp.813-4.

27.319. Sal'kova, E.N., and Sukhoverkhova, L.G.: "Investigation of Optical Scattering in a Plane-Parallel Resonator", Ukr. Fiz. Zh., 1972, 17, no.9, pp.1552-4.

27.320. Kopvillem, U.Kh., et al.: "Resonance-Frequency Shift of Intense Laser Pulses", Ukr. Fiz. Zh., 1972, 17, no.9, pp.1557-8.

27.321. Garside, B.K., Shewchun, J., and Kawasaki, B.S.: "Time Variations in the Far-Field Patterns of Electron-Beam-Pumped Semiconductor Lasers", J. Appl. Phys., 1972, 43, no.10, pp.4050-60.

27.322. Gleason, T.J., et al.: "Self-Mode-Locking of a TE N_2 Laser in the First Positive System at 1048 nm", Appl. Phys. Lett., 1972, 21, no.6, pp.276-7.

27.323. Davis, D.T., Smith, D.L., and Koval, J.S.: "Generation of Single 1-ns Pulses at 10.6 micron", J. Quantum Electron. IEEE, 1972, QE-8, no.11, pp.846-8.

27.324. Ippen, E.P., Shank, C.V., and Dienes, A.: "Passive Mode Locking of the CW Dye Laser", Appl. Phys. Lett., 1972, 21, no.8, pp.348-50.

27.325. Alekseev, V.A., et al.: "Time Dependence of Divergence of Rhodamine Laser Pumped by a Pinched Discharge", Kvantovaya Elektron., 1972, no.7, pp.64-7, and Sov. J. Quantum Electron., 1972, 2, no.1, pp.52-5.

27.326. Terent'ev, V.E., and Chertkov, A.A.: "Periodic Control of Emission from a Ruby Laser by a Q-Switch Using Transverse Electrooptic Effect", Kvantovaya Elektron., 1972, no.7, pp.88-9, and Sov. J. Quantum Electron., 1972, 2, no.1, pp.81-2.

27.327. Sommers, H.S.: "Experimental Studies of Injection Lasers. Spontaneous Spectrum at Room Temperature", J. Appl. Phys., 1972, 43, no.10, pp.4067-74.

27.328. Bykovskii, Yu.A., et al.: "Coherence of a Pulsed Single-Mode Injection Laser", Dokl. Akad. Nauk SSSR, 1972, 203, no.5, pp.1027-9, and Sov. Phys.-Dokl., 1972, 17, no.4, pp.359-61.

27.329. Rysakov, V.M., and Akatov, L.L.: "Dynamics of the Emission Spectrum of a GaAs Laser", Fiz. Tekh. Poluprov., 1972, 6, no.4, pp.728-30, and Sov. Phys.-Semicond., 1972, 6, no.4, pp.627-8.

27.330. Antsiferov, V.V., et al.: "Dynamics of Generation in Solid-State Lasers", Avtometriya, 1972, no.5, pp.98-105.

27.331. Arsen'ev, V.V., Dneprovskii, V.S., and Klyshko, D.N.: "Control of Laser Pulse Duration by Nonlinear Absorption in Semiconductors", Kvantovaya Elektron., 1972, no.7, pp.33-7, and Sov. J. Quantum Electron. 1972, 2, no.1, pp.23-6.

27.332. Makarov, A.K.: "Investigation of Dynamics of Pulse Phase-Locked Lasers", Izv. VUZ Radiofiz., 1972, 15, no.10, pp.1538-46.

27.333. Isaev, A.A., Kazaryan, M.A., and Petrash, G.G.: "Shape and Duration of Super-radiance Pulses Corresponding to Ne Lines", Kvantovaya Elektron., 1972, no.7, pp.62-4, and Sov. J. Quantum Electron., 1972, 2, no.1, pp.49-51.

27.334. Dandawate, V.D., Thomas, G.C., and Zemrod, A.: "Time Behaviour of a TEA Xe Laser", J. Quantum Electron. IEEE, 1972, QE-8, no.12, pp.918-9.

27.335. Goodwin, A.R., Lovelace, D.H., and Selway, P.R.: "Near- and Far-Field Double-Heterostructure Lasers", Opto-Electron., 1972, 4, no.3, pp.311-21.

27.336. Kirkby, P.A., and Thompson, G.H.B.: "Effect of Double-Heterojunction Waveguide Parameters on the Far-Field Emission Patterns of Lasers", Opto-Electron., 1972, 4, no.3, pp.323-34.

27.337. Andreou, D., and Little, V.I.: "Effect of Frequency Shifts on the Power Gain of a Laser Amplifier", Opt. Commun., 1970, 6, no.2, pp.180-4.

27.338. Mikhnenko, G.A., Protsenko, E.D., and Sedoi, E.A.: "Investigation of 633-nm Line Shift in a He-^{20}Ne Laser With an Absorption Cell", Opt. Spekt., 1972, 32, no.4, pp.809-13, and Opt. Spectrosc., 1972, 32, no.4, pp.425-7.

27.339. Antsiferov, V.V., et al.: "Selection and Change of Frequency in a Q-Switched Ruby Laser", Avtometriya, 1972,no.5,pp.94-7.

27.340. Desbois, J., and Tournois, P.: "Procedure for Optical Coupling of the Cavities of Two Lasers Allowing the Stable Generation of Ultrashort Light Pulses", C. R. Acad. Sci. (Paris), 1972, 275B, no.12, pp.415-8.

27.341. Bykovskii, Yu.A., Goncharov, I.G., and Uzkii, A.F.: "Use of Compound Resonator to Improve Monochromaticity of an Injection Laser", Opt. Spekt., 1972, 33, no.1, pp.135-7, and Opt. Spectrosc., 1972, 33, no.1, pp.72-3.

27.342. Skoborodko, P.A., and Yakobi, Yu.A.: "Population Inversion and Radiation Density in a CO_2 Laser With Varying Q-Factor", Zh. Prikl. Mekh. Tekh. Fiz., 1972, no.6, pp.18-23.

27.343. O'Neill, F.: "Picosecond Pulses from a Passively Mode-Locked CW Dye Laser", Opt. Commun., 1972, 6, no.4, pp.360-3.

27.344. Magyar, G.: "Frequency Shift in a Mode-Selected Dye Laser", Opt. Commun., 1972, 6, no.4, pp.388-90.

27.345. Ruiz, H.J., Turner, J.J., and Rabson, T.A.: "Interpulse Coherence of Picosecond Pulses", Opt. Commun., 1972, 6, no.4, pp.435-7.

27.346. Kozlov, V.V.: "Emission Spectrum of a Polyhedral-Resonator Laser Pumped by Long Pulses", Kvantovaya Elektron., 1972, 2, no.2, pp.30-6, and Sov. J. Quantum Electron., 1972, 2, no.2, pp.111-6.

27.347. Artushenko, K.A., et al.: "Synchronization of Radiation from Q-Switched Lasers", Prib. Tekh. Eksp., 1972, 15, no.3, pp.210-1, and Instrum. Exp. Tech., 1972, 15, no.3, pp.867-8.

27.348. Venkin, G.V., et al.: "Nd^{3+}:Glass Laser With Adjustable Pulse Duration", Vestn. Mosk. Univ. Fiz. Astron., 1972, 13, no.6, pp.734-5.

27.349. Boersch, H., and Theiss, F.J.: "Properties of Coaxial N_2 Superradiant Source at Low PRF", Z. Naturforsch., 1972, 27a, no.8-9, pp.1264-71.

27.350. Soloukhin, R.I., and Yakobi, Yu.A.: "Q-Switching CO_2 Laser With Active Gas Cell", Zh. Prikl. Mekh. Tekh. Fiz., 1972, no.4, pp.171-3.

27.351. Kornienko, L.S., Kravtsov, N.V., and Naumkin, N.I.: "Structure of Laser Pulse With Delay Line Inside Cavity", Radiotekh. Elektron., 1972, 17, no.8, pp.1760-2, and Radio Eng. Electron. Phys., 1972, 17, no.8, pp.1399-400.

27.352. Kryukov, P.G., et al.: "Investigation of Shape of Pulses Emitted by a Self-Mode-Locked Laser", Zh. Eksp. Teor. Fiz., 1972, 52, no.6, pp.2036-43, and Sov. Phys.-JETP, 1972, 35, no.6, pp.1062-6.

27.353. Eremenko, A.S., et al.: "Investigation of Pulsed Laser Utilizing Exploding-Film Q-Switch", Kvantovaya Elektron., 1972, no.3, pp.30-5, and Sov. J. Quantum Electron., 1972, 2, no.3, pp.219-23.

27.354. Zubarev, I.G., and Mulikov, V.F.: "Single-Frequency Nd^{3+}:Glass Lasers Under Nonspiking Free-Oscillation and Q-Switched Conditions", Kvantovaya Elektron., 1972, no.3, pp.13-7, and Sov. J. Quantum Electron., 1972, 2, no.3, pp.207-10.

27.355. Lyubimov, V.V., Orlova, I.B., and Fromzel, V.A.: "Influence of Inversion Inhomogeneity on the Transverse Structure of Oscillations in Solid-State Lasers", Kvantovaya Elektron., 1972, no.3, pp.94-6, and Sov. J. Quantum Electron., 1972, 2, no.3, pp.278-80.

27.356. Korolev, F.A., Abrosimov, G.V., and Odintsov, A.I.: "Effect of Excitation Conditions on Coherence Properties of Pulsed Superradiance of Neon", Opt. Spekt., 1972, 33, no.4, pp.725-8, and Opt. Spectrosc., 1972, 33, no.4, pp.399-401.

27.357. Corti, M.: "Pulsed Neon Laser at 540.1 nm with Subnanosecond Emission", Opt. Commun., 1972, 4, no.5, pp.373-6.

27.358. Wieder, H.: "Uniformly Polarized GaAs-Laser Array", Appl. Phys. Lett., 1972, 20, no.8, pp.311-2.

27.359. Brinkschulte, H., Fill, E., and Lang, R.: "Spectral Output Properties of $Nd^{3+}:POCl_3$ Inorganic Liquid Laser", J. Appl. Phys., 1972, 43, no.4, pp.1807-11.

27.360. Vrehen, Q.H.F., and Breimer, A.J.: "Spectral Properties of a Pulsed Dye Laser With Monochromatic Injection", Opt. Commun., 1972, 4, no.5, pp.416-20.

27.361. Bergman, A., David, R., and Jortner, J.: "Powerful Broadband Tunable Dye Laser", Opt. Commun., 1972, 4, no.5, pp.431-3.

27.362. Korolev, F.A., et al.: "Effect of Laser Field on Contours of Lines of Neighbouring Transitions in Ar^{II} Laser", Zh. Prikl. Spekt., 1972, 17, no.6, pp.980-3.

27.363. Belousova, I.M., Kiselev, V.M., and Kurzenkov, V.N.: "Linewidth for Induced Emission due to ^2P Transition of Atomic Iodine", Opt. Spekt., 1972, 33, no.2, pp.210-3, and Opt. Spectrosc., 1972, 33, no.2, pp.115-6.

27.364. Vize, L., Pinter, F., and Gati, L.: "First-Order Coherence of Radiation from a Dye Laser", Acta Phys. Chem. Szeged., 1972, 18, no.3-4, pp.107-14.

27.365. Borisova, M.S.: "Self-Mode-Locking in an Argon-Ion Laser With Nonlinear Absorber", Opt. Spekt., 1972, 33, no.6, pp.1134-8, and Opt. Spectrosc., 1972, 33, no.6, pp.620-2.

27.366. Belousov, N.D., et al.: "Spectra of Nd^{3+} Laser with $CaWO_4/$ $LaNa(WO_4)_2$ Composite Medium", Opt. Spekt., 1972, 33, no.5, pp.1002-3, and Opt. Spectrosc., 1972, 33, no.5, pp.550-1.

27.367. Vanyukov, M.P., et al.: "High-Power Multi-Laser $Nd^{3+}:$Glass System Emitting Picosecond Pulses", Opt.-Mekh. Prom., 1972, 39, no.12, pp.31-2, and Sov. J. Opt. Technol., 1972, 39, no.12, pp.751-2.

27.368. Koval'chuk, L.V., and Sventsitskaya, N.A.: "Methods for Alignment of Lasers With Unstable Resonators", Kvantovaya Elektron., 1972, no.5, pp.80-5, and Sov. J. Quantum Electron., 1973, 2, no.5, pp.450-3.

27.369. Anton'yants, V.Ya., et al.: "Ruby Laser With Wide Emission Spectrum", Kvantovaya Elektron., 1972, no.5, pp.106-8, and Sov. J. Quantum Electron., 1973, 2, no.5, pp.477-8.

27.370. Makhnev, V.P., and Telegin, G.G.: "Fluctuations of Radiation Rise Period in a Gas Laser With Nonlinear Resonant Absorption", Zh. Eksp. Teor. Fiz., 1972, 63, no.4, pp.1212-20, and Sov. Phys.-JETP, 1973, 36, no.4, pp.638-42.

27.371. Gonchukov, S.A., et al.: "Amplitude Characteristics of He-Ne Laser at 633 nm in Region of Strong Interaction Between Two Modes", Kvantovaya Elektron., 1972, no.4, pp.113-5, and Sov. J. Quantum Electron., 1973, 2, no.4, pp.387-9.

27.372. Gurevich, G.L., and Ingel', L.Kh.: "Influence of Self-Focusing on Stability of Steady-State Laser Emission", Kvantovaya Elektron., 1972, no.4, pp.95-7, and Sov. J. Quantum Electron., 1973, 2, no.4, pp.370-1.

27.373. Klinkov, V.K., and Mukhtarov, Ch.K.: "Generation in Ruby Laser With Moving Mirror and Selector in the Resonator", Dokl. Akad. Nauk SSSR, 1972, 207, no.4-6, pp.817-20, and Sov. Phys.-Dokl., 1973, 17, no.12, pp.1163-5.

27.374. Ivanov, L.P., et al.: "Self-Switching in Single-Heterojunction Injection Lasers", Kvantovaya Elektron., 1972, no.5, pp.92-4, and Sov. J. Quantum Electron., 1973, 2, no.5, pp.461-3.

27.375. Andreev, V.M., et al.: "Spatial Distribution of Radiation of Heterojunction Laser", Fiz. Tekh. Poluprov., 1972, 6, no.9, pp.1739-48, and Sov. Phys.-Semicond., 1973, 6, no.9, pp.1500-6.

27.376. Rozanov, N.N.: "Pulsations from a Laser With Frequency Dispersion", Zh. Eksp. Teor. Fiz., 1972, 63, no.6, pp.2033-42, and Sov. Phys.-JETP, 1973, 36, no.6, pp.1074-9.

27.377. Golubev, Yu.M., and Fradkin, E.E.: "Effect of Combination Coupling on Spectral and Statistical Properties of Multimode Fluctuations in a TW Laser", Zh. Eksp. Teor. Fiz., 1972, 63, no.6, pp.2082-93, and Sov. Phys.-JETP, 1973, 36, no.6, pp.1099-104.

27.378. Chernov, V.A.: "Effects of Cavity Configuration and Position of the Active Medium on the Temporal Characteristics of a Solid-State Laser", Zh. Tekh. Fiz., 1973, 43, no.5, pp.1095-7, and Sov. Phys.-Tech. Phys., 1973, 18, no.5, pp.698-9.

27.379. Gonda, S.I., Matsushima, Y., and Makita, Y.: "Possibility of Light-Pulse Generation from Junction Lasers by (External) Magnetic Field", Bull. Electrotech. Lab., 1973, 37, no.7, pp.659-88.

27.380. Brunner, W., Klose, E., and Paul, H.: "Formation of Picosecond Impulses by Two-Photon Absorption", Ann. Phys. (Leipzig), 1973, 30, no.3-4, pp.279-90.

27.381. Stemme, R., and Weber, H.: "Power and Pulse Duration in Spiking Region of Pulsed Lasers", Z. Angew. Math. Phys., 1973, 24, no.3, p.452.

27.382. Solamaa, R., and Stenholm, S.: "Gas Laser with Saturable Absorber. I and II", Phys. Rev., 1973, 8A, no.5, pp.2695-711.

27.383. Mal'tsev, A.A.: "Natural Linewidth of Three Modes of an Equally Spaced Self-Mode-Locked Spectrum", Radiotekh. Elektron., 1973, 18, no.3, pp.572-80, and Radio Eng. Electron. Phys., 1973, 18, no.3, pp.415-22.

27.384. Fortin, R., et al.: "Powerful Nanosecond Pulses by Stable Passive Mode-Locking of TEA CO_2 Lasers", Can. J. Phys., 1973, 51, no.4, pp.414-7.

27.385. Fesquet, J., Irla, H., and Dejardin, J.L.: "Characteristics of a Ruby Laser With Plane Mirrors", C. R. Acad. Sci. (Paris), 1973, 276B, no.11, pp.429-31.

27.386. Ivanov, V.A., and Lebedev, V.I.: "Effect of Thermal Distortion of Passive Shutter on Operation of a Ruby Laser", Zh. Prikl. Spekt., 1973, 18, no.3, pp.400-5.

27.387. Sommers, H.S., and Lockwood, H.F.: "Experimental Properties of Injection Lasers. III", J. Appl. Phys., 1973, 44, no.4, pp.1902-4.

27.388. Chodzko, R.A., et al.: "Application of a Single-Frequency Unstable Cavity to a CW HF Laser", J. Quantum Electron. IEEE, 1973, QE-9, no.5, pp.523-30.

27.389. Lim, T.K., and Garside, B.K.: "Generalized Model for a Self-Pulsing Ring Laser", Phys. Lett., 1973, 43A, no.3, pp.251-2.

27.390. Dahlstrom, L.: "Mode-Locking of High-Power Lasers by a Combination of Intensity- and Time-Dependent Q-Switching", Opt. Commun., 1973, 7, no.1, pp.89-92.

27.391. Falk, J., and Hitz, C.B.: "Stabilized Two-(Mode-Locked)-Pulse Operation of a Frequency-Doubled (Nd^{3+}:YAG) Laser", Opt. Commun., 1973, 7, no.3, pp.277-9.

27.392. Kleiman, H., and Marcus, S.: "CO_2 Laser Pulse Shaping With Saturable Absorbers", J. Appl. Phys., 1973, 44, no.4, pp.1646-8.

27.393. Ross, I.N., and Gates, J.W.C.: "Small Ruby Laser With Simple Rotating-Mirror Q-Switch", J. Phys. E, 1973, 6, no.2, pp.125-7.

27.394. Freed, C.: "Lamb Dip in CO Lasers", J. Quantum Electron. IEEE, 1973, QE-9, no.2, pp.219-26.

27.395. Muuss, H., and Scholz, V.: "Investigation of Coherence of Pulsed Argon-Ion Laser", Optik, 1973, 37, no.1, pp.26-30.

27.396. Hill, K.O., and Campbell, C.K.: "Studies of Porro-Prism Q-Switched Nd^{3+}: Glass Laser", Can. J. Phys., 1973, 51, no.1, pp.20-4.

27.397. Baker, H.J., and King, T.A.: "Self-Mode-Locking in He-Se Laser", J. Phys. D, 1973, 6, no.4, pp.395-9.

27.398. Piloff, G.S.: "Dye Laser With Coherent Output at Two Frequencies", Izv. Akad. Nauk SSSR, 1973, 37, no.2, pp.387-90.

27.399. Andreou, D., and Little, V.I.: "Spiking Behaviour of Laser Having Combined Liquid and Glass Active Media", J. Phys. D, 1973, 6, no.4, pp.390-4.

27.400. Maeda, H., and Yariv, A.: "Narrowing and Rebroadening of Amplified Spontaneous Emission in High-Gain Laser Media", Phys. Lett., 1973, 43A, no.4, pp.383-5.

27.401. Krivoshchekov, G.V., et al.: "Formation of an UV Light Pulse in Ruby Laser With Resonance Loss Modulation", Izv. VUZ Radiofiz., 1973, 16, no.3, pp.369-74.

27.402. Motorin, I.I., and Khanin, Ya.I.: "Spectral-Width Dependence of (Multimode) Laser Generation in an Inhomogeneously Broadened Solid", Izv. VUZ Radiofiz., 1973, 16, no.3, pp.386-92.

27.403. Bykov, V.P.: "Spike Operation and Self-Q-Switching in a Solid-State Laser", Dokl. Akad. Nauk SSSR, 1973, 206, no.5, pp.1078-81, and Sov. Phys.-Dokl., 1973, 17, no.10, pp.987-9.

27.404. Zargar'yants, M.N., et al.: "Directivity and Spectral Content of Radiation from Multi-Injection-Diode Arrays", Opt.-Mekh. Prom., 1973, 39, no.7, pp.18-20, and Sov. J. Opt. Technol., 1973, 39, no.7, pp.395-7.

27.405. McFarland, B.B.: "Effect of Rotational Level Coupling on Pulse Sharpening in CO_2 Amplifiers", J. Quantum Electron. IEEE, 1973, QE-9, no.7, pp.731-6.

27.406. Kohn, R.L., Shank, C.V., and Dienes, A.: "Observation of Inhomogeneity in the Gain Spectrum of a Coumarin Laser Dye", Opt. Commun., 1973, 7, no.4, pp.309-12.

27.407. Cheng, D.: "Instability of Cavity-Dumped YAG Laser Due to Time-Varying Reflections", J. Quantum Electron. IEEE, 1973, QE-9, no.6, pp.585-8.

27.408. Lotem, H., and Koren, G.: "Problems Associated With Backward Reflections Using a Giant-Pulse Laser", J. Phys. E, 1973, 6, no.7, p.672.

27.409. Atkinson, J.B., and Pace, P.W.: "Spectral Linewidth of a Flashlamp-Pumped Dye Laser", J. Quantum Electron. IEEE, 1973, QE-9, no.6, pp.569-74.

27.410. Breitschwerdt, K.G.: "Frequency Shift of Dye Laser Emission in the Linear Approximation", Phys. Lett., 1973, 43A, no.6, pp.615-6.

27.411. Gordon, E.I.: "Mode Selection in GaAs Injection Lasers Resulting from Fresnel Reflection", J. Quantum Electron. IEEE, 1973, QE-9, no.7, pp.772-6.

27.412. Teng, T.C., Gerlach, R., and Pao, Y.H.: "Mode-Locking of a 611.8-nm He-Ne Laser by Use of a Ne Discharge Cell", J. Quantum Electron. IEEE, 1973, QE-9, no.7, p.783.

27.413. Signore, R.: "Possibility of Obtaining Locking of Transverse Modes of a Laser", Rev. Phys. Appl., 1973, 8, no.1, pp.33-48.

27.414. Glas, P.: "Investigation of Properties of Radiation from a Laser With External Feedback", Ann. Phys. (Leipzig), 1973, 29, no.2, pp.121-36.

27.415. Stiehl, W.A., and Hoff, P.W.: "Measurement of Spectrum of a Helical TEA CO_2 Laser", Appl. Phys. Lett., 1973, 22, no.12, pp.680-2.

27.416. Menzies, R.T., and Shumate, M.S.: "Beat-Frequency Measurements Between $^{12}C^{16}O_2$ and $^{12}C^{18}O_2$ Lasers", J. Quantum Electron. IEEE, 1973, QE-9, no.8, p.862.

27.417. Hoff, P.W., Swingle, J.C., and Rhodes, C.K.: "Temporal Coherence, Spatial Coherence, and Threshold Effects, in the Molecular Xenon Laser", Opt. Commun., 1973, 8, no.2, pp.128-31.

27.418. Ciura, A.I., and Popescu, I.M.: "Features of the Self-Locking Mode in He-Cd Lasers", Rev. Roum. Phys., 1973, 18, no.6, pp.775-8.

27.419. Antropov, E.T.: "Spectral Characteristics of a Selective CO_2 Laser With Diffraction Grating", Zh. Prikl. Spekt., 1973, 18, no.4, pp.621-4.

27.420. Schroder, H.W., Welling, H., and Wellegehausen, B.: "Narrowband Single-Mode Dye Laser", Appl. Phys., 1973, 1, no.6, pp.343-8.

27.421. von der Linde, D., and Rodgers, K.F.: "Suppression of Spectral Narrowing Effect in Lasers Mode-Locked by Saturable Absorbers", Opt. Commun., 1973, 8, no.1, pp.91-3.

27.422. Dabu, R., and Isbasescu, M.: "Electrooptic Q-Switched Nd^{3+}:Glass Laser", Rev. Roum. Phys., 1973, 18, no.5, pp.655-60.

27.423. Zheltov, G.I., and Rubanov, A.S.: "Emission Polarization for Glass Lasers", Zh. Prikl. Spekt., 1973, 18, no.4, pp.625-8.

27.424. Rowley, W.R.C., and Wallard, A.J.: "Wavelength Values of the 633-nm Laser Stabilized With $^{127}I_2$ Saturated Absorption", J. Phys. E, 1973, 6, no.7, pp.647-52.

27.425. Ershov, B.V., et al.: "Improving the Angular Divergence of a Nd^{3+}:Glass, High-Energy, Pulsed Laser", Dokl. Akad. Nauk SSSR, 1973, 208, no.1, pp.70-2, and Sov. Phys.-Dokl., 1973, 18, no.1, pp.43-4.

27.426. Zubarev, T.N., and Dement'ev, V.A.: "Stability of Monochromatic Generation Mode in Multimode Solid-State Lasers", Dokl. Akad. Nauk SSSR, 1973, 208, no.1, pp.310-3, and Sov. Phys.-Dokl., 1973, 18, no.1, pp.50-2.

27.427. Krivoshchekov, G.V., and Smirnov, V.A.: "Excitation of Ultrashort Light Pulses With Stable Parameters in a Laser with Active Modulation", Zh. Prikl. Mekh. Tekh. Fiz., 1973, no.2, pp.163-4.

27.428. Glas, V.P.: "Investigation of Shape and Alignment of Laser Radiation Without External Feedback", Exp. Tech. Phys., 1973, 21, no.4, pp.299-315.

27.429. Patel, B.S.: "Effect of Rotational Relaxation on Duration of Q-Switched CO_2-Laser Pulses", Indian J. Pure Appl. Phys., 1973, 11, no.3, pp.194-8.

27.430. Blaney, T.G., et al.: "Absolute-Frequency Measurement of R-12 Transition of CO_2 at 9.3 micron", Nature, 1973, 244, no.5417, p.504.

27.431. Schniffner, G.: "Improved Determination of Accurate CO_2-Laser-Transition Frequencies and their Standard Deviation", Opto-Electron., 1973, 5, no.5, pp.411-3.

27.432. Simmons, J.D., and Keller, R.A.: "Interferometric Effects on the Output of Organic Dye Lasers", Appl. Opt., 1973, 12, no.9, p.2033.

27.433. Weichel, H.: "Improving the Reliability of Nd^{3+}:Glass Mode-Locked Lasers", J. Appl. Phys., 1973, 44, no.8, pp.3635-7.

27.434. Ito, R., Nakashima, H., and Nakada, O.: "Simple Method for Evaluating Luminescent Pattern of Double-Heterostructure Crystals", Jap. J. Appl. Phys., 1973, 12, no.8, pp.1272-4.

27.435. Isyanova, E.D., Kyznetsov, B.V., and Ovchinnikov, V.M.: "Instability of Growth Periods for an Emission Pulse", Zh. Prikl. Spekt., 1973, 18, no.6, pp.998-1002.

27.436. Iwaoka, H., and Fujioka, T.: "Anomalous Transient Behaviour of Sealed CO_2 Laser", Jap. J. Appl. Phys., 1973, 12, no.7, pp.1047-9.

27.437. Patel, B.S., and Swarup. P.: "Rotational-Line Overlap in CO_2 Laser Transitions", J. Phys. D, 1973,6,no.14,pp.1670-3.

27.438. Glaze, J.A.: "Linewidth Parameters from the Lamb Dip in a CW HF Chemical Laser", Appl. Phys. Lett., 1973, 23, no.6, pp.300-2.

27.439. Schappert, G.T.: "Rotational Relaxation Effects in Short-Pulse CO_2 Amplifiers", Appl. Phys. Lett., 1973, 23, no.6, pp.319-21.

27.440. Bonczyk, P.A.: "Passive Q-Switching of a CO Laser by Nitric Oxide", J. Appl. Phys., 1973, 44, no.9, p.4251.

27.441. Patel, B.S.: "Collision Broadening of High-Pressure CO and CO_2 Laser Transitions", Phys. Lett.,1973,45A,no.2,pp.137-8.

27.442. Vdovin, Yu.A., et al.: "Mode Competition in the $3s_2$-$3p_4$ Transition in a Neon Laser with a Methane Absorbing Cell", Kvantovaya Elektron., 1973, no.6, pp.105-7, and Sov. J. Quantum Electron., 1973, 2, no.6, pp.565-6.

27.443. Liao, P.R., and Hartmann, S.R.: "Magnetic-Field- and Concentration-Dependent Photon Echo Relaxation in Ruby With Simple Exponential Decay", Opt. Commun., 1973, 8, no.4, pp.310-1.

27.444. Morgun, Yu.F., Muravitskii, M.A., and Lavrovskii, L.A.: "Emission Spectrum of a Ruby Laser and its Dependence on Etalon Thickness", Zh. Prikl. Spekt., 1973, 19, no.1, pp.33-8.

27.445. Sommers, H.S.: "Experimental Properties of Injection Lasers", J. Appl. Phys., 1973, 44, no.3, pp.1263-75.

27.446. Eranian, A., Dezauzier, P., and de Wilte, O.: "Two-Nanosecond Pulses from Double-Cavity Dye Laser", Opt. Commun., 1973, 7, no.2, pp.150-4.

27.447. Carman, R.L., Johnson, B.C., and Steinmetz, L.L.: "Self-Driven Laser Oscillator for Directly Producing Bandwidth-Limited Pulses of about 1ns", Opt. Commun., 1973, 7, no.2, pp.169-71.

27.448. Kryzhanovskii, V.I., Serebryakov, V.A., and Starikov, A.D.: "Depolarization of Radiation of Giant-Pulse Nd^{3+}:Glass Lasers", Opt.-Mekh. Prom., 1973, 40, no.8, pp.14-5, and Sov. J. Opt. Technol., 1973, 40, no.8, pp.478-9.

27.449. Paoli, T.L., and Hakki, B.W.: "CW Degradation at $300^{\circ}K$ of GaAs Double-Heterostructure Junction Lasers. I", J. Appl. Phys., 1973, 44, no.9, pp.4108-12.

27.450. Baues, P., von Hundelshausen, V., and Mockel, P.: "Generation of Reproducible Giant Pulses With Optically Regenerative Q-Switch", J. Appl. Phys., 1973, 44, no.9, pp.4067-71.

27.451. Laine, D.C., and Lefrere, P.R.: "Nonlinear Effects on a Molecular-Beam Zeeman Maser Oscillator", Phys. Lett., 1973, 45A, no.4, pp.279-80.

27.452. Cody, R.J., and Pilloff, H.S.: "Simultaneous Two-Wavelength HF Lasers", J. Quantum Electron. IEEE, 1973, QE-9, no.12, pp.1152-4.

27.453. Shimoda, K.: "Frequency Shifts in Methane-Stabilized Lasers", Jap. J. Appl. Phys., 1973, 12, no.9, pp.1393-1402.

27.454. Tanaka, A., et al.: "Self-Mode-Locking in N_2O Laser With Rotating-Mirror Q-Switching", Jap. J. Appl. Phys., 1973, 12, no.10, pp.1650-1.

27.455. Ross, J.N.: "Transient Phenomena in the Anode Region of a Metal-Clad Argon Laser", J. Phys. D, 1973, 6, no.16,pp.1917-28.

27.456. Marowsky, G.: "Spectral Narrowing in a Dye Laser With Nonresonant Feedback", Appl. Phys., 1973, 2, no.4, pp.213-7.

27.457. Hackett, C.E., and Dewey, C.F.: "Improved Temporal Stability of Polymethine Laser Dyes in Aqueous Solutions", J. Quantum Electron. IEEE, 1973, QE-9,no.11,pp.1119-20.

27.458. Nagata, I., and Nakaya, T.: "Polarization of Dye-Laser Light", J. Phys. D, 1973, 6, no.16, pp.1870-82.

27.459. Bradley, D.J., and Sibbett, W.: "Streak-Camera Studies of Picosecond Pulses from a Mode-Locked Nd^{3+}:Glass Laser", Opt. Commun., 1973, 9, no.1, pp.17-20.

27.460. Iwai, S., and Namba, S.: "Thermal Shift of Laser Wavelength in CdS During Excitation Pulse of Electron Beam", Jap. J. Appl. Phys., 1973, 12, no.9, pp.1382-7.

27.461. Bogdankevich, O.V., et al.: "Divergence of Output Radiation of Electron-Beam-Pumped Radiating-Mirror Lasers", Kvantovaya Elektron., 1973, no.6, pp.110-1, and Sov. J. Quantum Electron., 1973, 2, no.6, pp.569-70.

27.462. Caruso, A., and Gratton, R.: "Mode-Locked Ring-Laser and -Amplifier Characteristics", J. Quantum Electron. IEEE, 1973, QE-9, no.11, pp.1039-43.

27.463. Zubarev, T.N., Martynov, V.M., and Tarasov, Yu.A.: "LF Modulation of Solid-State Lasers in Mode-Locking Conditions", Opt. Spekt., 1973, 34, no.4, pp.752-4, and Opt. Spectrosc., 1973, 34, no.4, pp.433-4.

27.464. Kats, M.L., et al.: "Dispersion Characteristic of Three-Mode Gas Laser at Modulation of Relative Excitation", Izv. VUZ Radiofiz., 1973, 16, no.6, pp.892-7.

27.465. Hill, G.A., James, D.J., and Ramsden, S.A.: "Generation of Fast-Rise, Variable-Duration, Pulses of 10.6-micron TEA CO_2 Laser Radiation", Opt. Commun., 1973, 9, no.3, pp.237-9.

27.466. Saprykin, E.G., Yudin, R.N., and Atutov, S.N.: "Self-Mode Selection of a 633-nm Laser With Increased Pressure in the Discharge Tube", Opt. Spekt., 1973, 34, no.4, pp.755-61, and Opt. Spectrosc., 1973, 34, no.4, pp.435-8.

27.467. Cristescu, C.P., Popescu, I.M., and Preda, A.M.: "Method for Study of Gain and Oscillating Modes of a CO_2 Laser", Rev. Roum. Phys., 1973, 18, no.8, pp.1001-6.

27.468. Friesem, A.A., et al.: "Tunable Dye Laser With a Composite Holographic Wavelength Selector", Opt. Commun., 1973, 9, no.2, pp.149-51.

27.469. Tohma, K.: "Analysis of Frequency Locking of Organic Dye Lasers by Faraday Effect", Rev. Radio Res. Lab., 1973, 19, no.101, pp.57-68.

27.470. Hockham, G.A.: "Radiation (Pattern) from a Solid-State Laser", Electron. Lett., 1973, 9, no.17, pp.389-91.

27.471. Kuizenga, D.J., et al.: "Simultaneous Q-Switching and Mode-Locking in the CW Nd^{3+}:YAG Laser", Opt. Commun., 1973, 9, no.3, pp.221-6.

27.472. Makogon, M.M., and Ponomarev, Yu.N.: "Self-Modulation Mechanism of a Ruby Laser", Opt. Spekt., 1973, 34, no.4, pp.762-4, and Opt. Spectrosc., 1973, 34, no.4, pp.439-40.

27.473. Gonda, S.I., Matsushima, Y., and Makita, Y.: "Possibility of Light-Pulse Generation from Junction Lasers by an External Magnetic Field", J. Quantum Electron. IEEE, 1973, QE-9, no.12, pp.1154-5.

27.474. Troitskii, Yu.V.: "Oscillographic Recording of the Appearance of Off-Axial Modes in a Gas Laser", Prib. Tekh. Eksp., 1973, 16, no.2, pp.179-80, and Instrum. Exp. Tech., 1973, 16, no.2, pp.544-5.

27.475. Wilbrandt, R., and Weber, H.: "Effect of Thermal Blooming in a Liquid Filter on Emission of a Nd^{3+}:Glass Laser", Opt. Commun., 1973, 9, no.3, pp.231-3.

27.476. Kostanyan, R.B., and Pogasyan, P.S.: "Effect of Coherence on Spectral Composition of Radiation Resulting from Interfering Waves in Ruby", Izv. Akad. Nauk SSSR, Ser. Fiz., 1973, 37, no.10, pp.2111-4.

27.477. Voitovich, A.P., and Smirnov, A.Ya.: "Generation of Two Frequencies in a Gas Laser With Nonlinear Selective Losses", Opt. Spekt., 1973, 34, no.5, pp.925-30, and Opt. Spectrosc., 1973, 34, no.5, pp.533-6.

27.478. Bikmukhametov, K.A., Klement'ev, V.M., and Chebotaev, V.P.: "Collision Broadening in Hg-He-Ne of the 1.53-micron Line of Hg", Opt. Spekt., 1973, 34, no.6, pp.1062-5, and Opt. Spectrosc., 1973, 34, no.6, pp.616-7.

27.479. Zborovskii, V.A., et al.: "Measurement of the Natural Linewidth of a TW He-Ne Laser in the 633-nm Region", Opt. Spekt., 1973, 34, no.6, pp.1213-4, and Opt. Spectrosc., 1973, 34, no.6, pp.704-5.

27.480. Polishchuk, V.A., et al.: "Longitudinal Alignment of the Lasing Levels of a He-Ne Laser", Opt. Spekt., 1973, 34, no.6, pp.1220-2, and Opt. Spectrosc., 1973, 34, no.6, pp.709-11.

27.481. Bazhenov, S.V., Brazovskii, V.E., and Telegin, G.G.: "Statistical Phenomena in the Transient of a He-Ne Laser With Given Initial Photon Distribution", Opt. Spekt., 1973, 35, no.1, pp.106-16, and Opt. Spectrosc., 1973, 35, no.1, pp.62-6.

27.482. Chernov, V.A.: "Phase Relations Between Longitudinal Modes of a Free-Running Laser", Zh. Tekh. Fiz., 1973, 43, no.4, pp.884-6, and Sov. Phys.-Tech. Phys., 1973, 18, no.4, pp.562-3.

27.483. Allakhverdyan, R.G., et al.: "Dynamics of a Semiconductor Laser With Intensity-Dependent Index of Refraction", Zh. Tekh. Fiz., 1973, 43, no.5, pp.1024-8, and Sov. Phys.-Tech. Phys., 1973, 18, no.5, pp.647-9.

27.484. Anan'ev, Yu.A., and Sherstobitov, V.E.: "Unstable Resonators With a Central Coupling Aperture in Laser Oscillators and Amplifiers", Zh. Tekh. Fiz., 1973, 43, no.5, pp.1013-23, and Sov. Phys.-Tech. Phys., 1973, 18, no.5, pp.640-6.

27.485. Winer, I.M.: "Production of Mode-Locked Laser Pulses in Nd^{3+}:YAG Using a Glass Filter Absorber", Appl. Opt., 1973, 12, no.12, p.2809.

27.486. Pashchenko, V.Z., and Rubin, L.B.: "Control of Laser Pulse Shape With an Organic Dye Shutter", Zh. Tekh. Fiz., 1973, 43, no.5, pp.1004-8, and Sov. Phys.-Tech. Phys., 1973, 18, no.5, pp.635-7.

27.487. Belokrinitskii, N.S., et al.: "Study of Spatial Coherence of Gas-Laser Radiation", Ukr. Fiz. Zh., 1973, 18, no.11, pp.1809-13.

27.488. Anokhin, A.V., Markova, S.V., and Petrash, G.G.: "Study of Lasing Spectra in the Vibrational Transitions of the CO Molecule", Opt. Spekt., 1973, 35, no.1, pp.166-7, and Opt. Spectrosc., 1973, 35, no.1, pp.94-5.

27.489. Berezovskii, V.V., et al.: "Four-Frequency Structure of the CO_2 Laser Pulse", Opt. Spekt., 1973, 35, no.1, pp.171-3, and Opt. Spectrosc., 1973, 35, no.1, pp.98-100.

27.490. Bikmukhametov, K.A., and Klement'ev, V.M.: "Simultaneous Lasing on Hg and Ne Transitions in the Composite Mixture Hg-He-Ne", Opt. Spekt., 1973, 35, no.1, p.181, and Opt. Spectrosc., 1973, 35, no.1, p.105.

27.491. Dindarov, V.E., and Kotov, O.I.:
"633-nm Ne Laser With Nonlinear Absorption",
Zh. Tekh. Fiz., 1973, 43, no.5, pp.1009-12,
and Sov. Phys.-Tech. Phys., 1973, 18, no.5,
pp.638-9.

27.492. Chernov, V.A.: "Transient Processes in Multimode Lasers With Inhomogeneous
Broadening", Zh. Tekh. Fiz., 1973, 43, no.4,
pp.797-802, and Sov. Phys.-Tech. Phys.,
1973, 18, no.4, pp.500-3.

27.493. Lambert, L.Q., Abella, I.D., and
Compaan, A.: "Short-Time-Interval Behaviour
of Photon Echoes in Ruby Near Level Crossings", Phys. Rev., 1973, 8A, no.3, pp.1641-3.

27.494. Mak, A.A., and Ustyugov, V.I.:
"Spontaneous Single-Frequency Generation of
a Solid-State Ring Laser", Zh. Eksp. Teor.
Fiz. Pis'ma, 1973, 18, no.4, pp.253-5, and
JETP Lett., 1973, 18, no.4, pp.151-2.

27.495. Sakane, T., and Kashiwagi, H.:
"Forced Mode-Locking of a Transversely Exci-
ted High-Pressure CO_2 Laser", Jap. J. Appl.
Phys., 1973, 12, no.12, pp.1950-1.

27.496. Tait, G.D., Whitbourn, L.B., and
Robinson, L.C.: "Short Pulses from a Plasma-
Controlled HCN Gas Laser", Phys. Lett.,
1973, 46A, no.4, pp.239-40.

27.497. Korolev, F.A., Salimov, V.M.,
and Odintsov, A.I.: "Frequency Spectrum
and Self-Mode-Locking in an Argon-Ion
Laser", Radiotekh. Elektron., 1973, 18,
no.1, pp.209-11, and Radio Eng. Electron.
Phys., 1973, 18, no.1, pp.159-61.

27.498. Zeiger, S.G.: "Instability of a
Three-Mode Gas Laser With Symmetrical Frequency Configuration", Zh. Tekh. Fiz.,
1973, 43, no.6, pp.1308-10, and Sov. Phys.-
Tech. Phys-Tech. Phys., 1973, 43, no.6,
pp.832-3.

27.499. Ermachenko, V.M.: "Determination
of Spectral Width of Laser Transition from
Lamb Dip", Kvantovaya Elektron., 1973,
no.7, pp.134-5, and Sov. J. Quantum Electron., 1973, 3, no.1, pp.83-4.

27.500. Gandel'man, I.L., and Tikhonov,
E.A.: "Automatic Synchronization of Modes
in Dye Laser with Nanosecond Pulsing", Ukr.
Fiz. Zh., 1973, 18, no.10, pp.1730-2.

27.501. Stemme, R., et al.: "Undamped
Resonance Spiking by Internal Modulation
of Pulsed Solid-State Lasers", Opt. Commun.,
1973, 9, no.4, pp.338-41.

27.502. Dolotko, V.I., Krichevskii, V.I.,
and Shevchenko, V.V.: "Investigation of
Spatial Coherence of Batch-Manufactured Gas
Lasers", Prib. Tekh. Eksp., 1973, 16, no.4,
pp.211-3, and Instrum. Exp. Tech., 1973,
16, no.4, pp.1232-4.

27.503. Nestrizhenko, Yu.A.: "Two-Frequency
Laser With Controlled Polarization", Prib.
Tekh. Eksp., 1973, 16, no.4, pp.209-11, and
Instrum. Exp. Tech., 1973, 16, no.4, pp.1229-31.

27.504. Casey, H.C., Panish, M.B., and
Merz, J.L.: "Beam Divergence of Emission from
Double-Heterostructure Injection Lasers", J.
Appl. Phys., 1973, 44, no.12, pp.5470-5.

27.505. Makhorin, V.I., Popov, A.I., and
Protsenko, E.D.: "Tuning (by Methane Cell)
of He-Ne Laser Over Range 3.3912-3.3922 micron", Kvantovaya Elektron., 1973, no.7,
pp.47-55, and Sov. J. Quantum Electron.,
1973, 3, no.1, pp.24-8.

27.506. Dyubko, S.F., and Topkov, A.N.:
"Improvement (by Adding CO) of Monochromaticity of HCN Laser", Kvantovaya Elektron.,
1973, no.7, pp.103-5, and Sov. J. Quantum
Electron., 1973, 3, no.1, pp.56-7.

27.507. Berezovskii, V.V., Bykovskii,
Yu.A., and Remizov, A.N.: "Parameters of a
Four-Frequency CO_2 Laser With Transverse
Discharge", Kvantovaya Elektron., 1973,
no.2, pp.75-7, and Sov. J. Quantum Electron.,
1973, 3, no.2, pp.136-7.

27.508. Vas'kov, V.A., et al.: "Frequency
Characteristics of a Double-Mode Gas Laser
With Internal Absorbing Cell", Kvantovaya
Elektron., 1973, no.2, pp.107-10, and Sov.
J. Quantum Electron., 1973, 3, no.2, pp.165-7.

27.509. Klinkov, V.K., and Mukhtarov,
Ch.K.: "Modulation of Ruby-Laser Action With
a Moving Selector", Kvantovaya Elektron.,
1973, no.2, pp.61-4, and Sov. J. Quantum
Electron., 1973, 3, no.2, pp.123-5.

27.510. Gaprindashivili, Kh.I., et al.:
"Threshold, Temporal, and Spectral, Characteristics of a Fibre Laser", Kvantovaya Elektron., 1973, no.2, pp.25-30, and Sov. J.
Quantum Electron., 1973, 3, no.2, pp.100-3.

27.511. Box, S.J., and John, P.K.: "Q-
Switching a CO_2 Laser by a (Shock-Tube)
Plasma", J. Appl. Phys., 1973, 44, no.11,
pp.5167-8.

27.512. Salathe, R., and Voumard, C.:
"Optical Coupling of Two Laser Diodes", Z.
Angew. Math. Phys., 1973, 24, no.3, p.448.

27.513. Kalchev, S.D., Pocheva, Y.H., and
Sabotinov, N.V.: "Study of Oscillation Spec-
trum of He-Se Gas Laser", C. R. Acad. Bulg.
Sci., 1973, 26, no.10, pp.1323-6.

27.514. Neporent, B.S., et al.: "Spectral
Kinetics of Organic Dye Lasers", Acta Phys.
Chem. Szeged., 1973, 19, no.1-2, pp.3-9.

27.515. Rubinov, A.N., and Efendiev,
T.Sh.: "Generation of Stimulated Radiation
in a Dye Solution on Reflection from a Non-
linear Self-Induced Mirror", Kvantovaya
Elektron., 1973, no.3, pp.129-30, and Sov.
J. Quantum Electron., 1973, 3, no.3, pp.268-9.

27.516. Pol'skii, Yu.E., and Yakutenkov, A.A.: "Experimental Investigation of Generation Kinetics of a Ruby Laser With Nonstationary Resonator", Zh. Eksp. Teor. Fiz., 1973, 64, no.2, pp.438-45, and Sov. Phys.-JETP, 1973, 37, no.2, pp.223-6.

27.517. Melekhin, G.V.: "Effect of Resonator Lengths of Coupled Lasing Channels on the Interaction of Standing Waves", Opt. Spekt., 1973, 35, no.5, pp.984-5, and Opt. Spectrosc., 1973, 35, no.5, pp.571-2.

27.518. Anan'ev, Yu.A., et al.: "Stabilization of Direction of Emission from Unstable Prism Resonators", Kvantovaya Elektron., 1973, no.3, pp.115-6, and Sov. J. Quantum Electron., 1973, 3, no.3, pp.255-6.

27.519. Bobrik, V.I., Kolomnikov, Yu.D., and Chebotaev, V.P.: "He-Ne Laser With Nonlinear Absorbing Gas as Delay Line", Opt. Spekt., 1973, 35, no.6, pp.1179-80, and Opt. Spectrosc., 1973, 35, no.6, pp.682-3.

27.520. Antsiferov, V.V., et al.: "Spike Structure of Emission from Solid-State Lasers", Kvantovaya Elektron., 1973, no.3, pp.57-65, and Sov. J. Quantum Electron., 1973, 3, no.3, pp.211-5.

27.521. Krivoshchekov, G.V., et al.: "Spectral and Kinetic Features of Emission from a Ruby Laser Excited by External Light Pulses", Kvantovaya Elektron., 1973, no.3, pp.105-6, and Sov. J. Quantum Electron., 1973, 3, no.3, pp.246-7.

27.522. Linnik, V.P., et al.: "Cophasing of Radiation Emitted from Multichannel Solid-State Laser", Kvantovaya Elektron., 1973, no.3, pp.131-3, and Sov. J. Quantum Electron., 1973, 3, no.3, pp.270-1.

27.523. Kornienko, L.S., et al.: "Features of Solid-State Ring Lasers Associated With Diffraction Elements", Vestn. Mosk. Univ. Fiz. Astron., 1973,14,no.6,pp.719-21.

27.524. Boiko, B.B., et al.: "Giant-Pulse Laser With Double Resonator", Zh. Prikl. Spekt., 1973, 19, no.6, pp.1010-3.

27.525. Klochkov, V.P., and Bogdanov, V.L.: "Q-Switching of Ruby Laser by Cu-Phthalocyanine Vapours", Zh. Prikl. Spekt., 1973, 19, no.6, pp.1014-6.

27.526. Wolinski, W., et al.: "Selecting Lines Emitted by Argon-Ion Laser Using a Littrow Prism", Electron Technol., 1973, 6, no.1-2, pp.195-200.

27.527. Fotiadi, A.E., and Fridrikhov, S.A.: "Polarization of CW Argon-Ion Laser Emission in a Transverse Magnetic Field", Opt. Spekt., 1973, 35, no.5, pp.961-3, and Opt. Spectrosc., 1973, 35, no.5, pp.556-7.

27.528. Troitskii, Yu.V.: "Investigation of Stimulated-Emission Spectrum of He-Ne Laser", Kvantovaya Elektron., 1973, no.3, pp.35-9, and Sov. J. Quantum Electron., 1973, 3, no.3, pp.198-200.

27.529. Vize, L., Pinter, F., and Gati, L.: "First-Order Coherence of Radiation of a Dye Laser", Acta Phys. Chem. Szeged., 1973, 19, no.4, pp.417-21.

27.530. Ivanov, V.A., et al.: "Removal of Mode Degeneracy in CO_2 Laser", Radiotekh. Elektron., 1973, 18, no.5, pp.1080-2, and Radio Eng. Electron. Phys., 1973, 18, no.5, pp.800-2.

27.531. Burmatov, I.F., Vasilenko, L.S., and Shishaev, A.V.: "Measurement of Angular Divergence and Transverse Intensity Distribution of Radiation from a CO_2 Laser", Prib. Tekh. Eksp., 1973, 16, no.6, pp.141-2, and Instrum. Exp. Tech., 1973, 16, no.6, pp.1779-81.

27.532. Neporent, B.S., et al.: "Spectral-Line Characteristics of Dye Laser Pumped by a Train of Ultrashort Pulses", Opt. Spekt., 1973, 35, no.3, pp.531-4, and Opt. Spectrosc., 1973, 35, no.3. pp.309-11.

27.533. Chekalinskaya, Yu.I., and Chechenina, E.P.: "Effect of Anisotropic Resonator on Polarization of a Gas-Laser Amplifier", Zh. Prikl. Spekt., 1973, 19, no.5, pp.812-20.

27.534. Veduta, A.P., et al.: "Use of Nd^{3+} Niobium-Phosphate Glasses in Lasers With Passive Mode Locking", Kvantovaya Elektron., 1973, no.5, pp.36-40, and Sov. J. Quantum Electron., 1973, 3, no.5, pp.385-7.

27.535. Alferov, Zh.I., et al.: "Polarization of Injection-Type Heterojunction Lasers", Fiz. Tekh. Poluprov., 1973, 7, no.8, pp.1638-41, and Sov. Phys.-Semicond., 1974, 7, no.8, pp.1095-6.

27.536. Anan'ev, Yu.A., Grishmanova, N.I., and Sventsitskaya, N.A.: "Laser With Grid Mirror", Zh. Tekh. Fiz., 1973, 43, no.7, pp.1530-6, and Sov. Phys.-Tech. Phys., 1974, 18, no.7, pp.968-71.

27.537. Kobak, I.A., et al.: "Temporal Characteristics of GaAs Semiconductor Lasers at Near-Threshold Currents", Fiz. Tekh. Poluprov., 1973, 7, no.11, pp.2070-3, and Sov. Phys.-Semicond.,1974,7,no.11,pp.1384-5.

27.538. Zaitsev, Yu.I.: "Intensity Fluctuations from a Two-Mode He-Ne Laser", Kvantovaya Elektron., 1973, no.5, pp.77-86, and Sov. J. Quantum Electron., 1974, 3, no.5, pp.410-5.

27.539. Troitskii, Yu.V.: "Investigation of Saturation in a He-Ne Laser", Kvantovaya Elektron., 1973, no.5, pp.87-94, and Sov. J. Quantum Electron., 1974, 3, no.5,pp.416-20.

27.540. Isaev, S.K., Kornienko, L.S., and Lariontsev, E.G.: "Kinetics of the Stimulated Emission of a Dy^{2+}:CaF_2 Laser with Hemispherical Resonator", Kvantovaya Elektron., 1973, no.5, pp.41-6, and Sov. J. Quantum Electron., 1974, 3, no.5, pp.388-91.

27.541. Bogatov, A.P., et al.: "Kinetics of Emission of an Injection Laser and Collapse Single-Mode Emission", Kvantovaya Elektron., 1973, no.5, pp.14-20, and Sov. J. Quantum Electron., 1974, 3, no.5, pp.372-5.

27.542. Klochan, E.L., et al.: "Characteristics of a Laser With Delay Line and Passive Q-Switch", Kvantovaya Elektron., 1973, no.5, pp.47-51, and Sov. J. Quantum Electron., 1974, 3, no.5, pp.392-4.

27.543. Steinvall, O.: "Studies of Time-Resolved Spectra from Pulsed GaAs Lasers", Phys. Scr., 1974, 10, no.4, pp.186-90.

27.544. Young, M., and Hicks, A.: "Holographic Ruby Laser With Long Coherence and Precise Timing", Appl. Opt., 1974, 13, no.11, pp.2486-8.

27.545. Veduta, A.P., Fedotov, N.B., and Furzikov, N.P.: "Generation of Single Ultrashort Light Pulses With Aid of a Combustible Mirror", Kvantovaya Elektron., 1974, 1, no.2, pp.408-10, and Sov. J. Quantum Electron., 1974, 4, no.2, pp.228-9.

27.546. Odintsov, A.I., and Yakunin, V.P.: "Observation of Coherent Interaction Effects During Amplification of Short Light Pulses in Neon", Zh. Eksp. Teor. Fiz. Pis'ma, 1974, 20, no.4, pp.233-5, and JETP Lett., 1974, 20, no.4.

27.547. Garside, B.K., and Lim, T.K.: "Passive Mode-Locking in Flashlamp-Pumped Dye Lasers", Opt. Commun., 1974, 12, no.3, pp.240-5.

27.548. Isobe, K., and Tanaka, S.: "Temporal Coherence of Nd^{3+}:Glass Laser", Jap. J. Appl. Phys., 1974, 13, no.11, pp.1811-6.

27.549. Miyashita, T., and Ikenoue, J.: "High Peak Power and Short Pulses from a Ring Laser Mode-Locked by Internal Modulation", J. Quantum Electron. IEEE, 1974, QE-10, no.3, pp.387-9.

27.550. Landry, M.J.: "Laser Q-Switching by a PLZT Shutter", J. Quantum Electron. IEEE, 1974, QE-10, no.3, pp.356-8.

27.551. Drexhage, K.H., and Reynolds, G.A.: "Dye Solutions for Mode Locking IR Lasers", Opt. Commun., 1974, 10, no.1, pp.18-20.

27.552. Babenko, V.A., et al.: "Spectrum of a Giant Laser Pulse Under Frequency Self-Modulation Conditions", Kvantovaya Elektron., 1974, no.2, pp.19-24, and Sov. J. Quantum Electron., 1974, 3, no.2, pp.97-9.

27.553. Hultzsch, R.: "Passive Q-Switching of Ruby Laser With Additively Coloured La: CaF_2 Crystals", Phys. Status Solidi, 1974, 23A, no.2, pp.K117-21.

27.554. Zimmerman, J., and Gaddy, O.L.: "Self-Pulsing by Discharge Relaxation Oscillation in CO_2 Waveguide Laser", J. Quantum Electron. IEEE, 1974, QE-10, no.1, pp.92-3.

27.555. Eckardt, R.C., Lee, C.H., and Bradford, J.N.: "Effect of Self-Phase-Modulation on Evolution of Picosecond Pulses in Nd^{3+}:Glass Laser", Opto-Electron., 1974, 6, no.1, pp.67-85.

27.556. Somekh, S.: "Transverse Mode Control in a Distributed-Feedback Semiconductor Laser", Proc. IEEE, 1974, 62, no.2, pp.277-8.

27.557. Siegman, A.E., and Kuizenga, D.J.: "Active Mode-Coupling Phenomena in Pulsed and CW Lasers", Opto-Electron., 1974, 6, no.1, pp.43-66.

27.558. Penzkofer, A.: "Generation of Picosecond Light Pulses With Saturable Absorbers", Opto-Electron., 1974, 6, no.1, pp.87-98.

27.559. Schotzau, H.J., and Kneubuhl, F.K.: "Chemical Excitation and Suppression of HCN-Laser Modes", Phys. Lett., 1974, 46A, no.6, pp.415-6.

27.560. Scavennec, A., and Nahman, N.S.: "Simple Passively Mode-Locked CW Dye Laser", J. Quantum Electron. IEEE, 1974, QE-10, no.1, pp.95-6.

27.561. New, G.H.C.: "Pulse Evolution in Mode-Locked Quasi-CW Lasers", J. Quantum Electron. IEEE, 1974, QE-10, no.2, pp.115-24.

27.562. Wynne, J.J.: "Generation of Rotationally Symmetric TE_{01} and TM_{01} Modes from a Wavelength-Tunable Laser", J. Quantum Electron. IEEE, 1974, QE-10, no.2, pp.125-7.

27.563. Eckardt, R.C.: "Self-Focusing in Mode-Locked Nd^{3+}:Glass Lasers", J. Quantum Electron. IEEE, 1974, QE-10, no.1, pp.48-56.

27.564. Whitford, B.G., Siemsen, K.J., and Riccius, H.D.: "Absolute Frequency Measurements of $^{12}C^{16}O$ and $^{13}C^{16}O$ Laser Transitions", Opt. Commun., 1974, 10, no.3, pp.288-9.

27.565. Sabotinov, N.V., and Telbizov, P.K.: "Mixed Gas Laser of the Three Basic Colours", Opto-Electron., 1974, 6, no.2, pp.185-7.

27.566. Sawatari, T., and Shupe, D.M.: "Application of an Organic Dye for Wavefront Multiplication", Appl. Phys. Lett., 1974, 24, no.2, pp.95-7.

27.567. Thomas, B., Mistry, D., and Davies, C.F.L.: "Temperature Dependence of Spectral Emission from Semiconductor Lasers Showing Long Time Delays", J. Quantum Electron. IEEE, 1974, QE-10, no.4, pp.401-5.

27.568. Wieder, H.: "High-Efficiency Polarized GaAs Laser Array", IBM Tech. Disclosure Bull., 1974, 16, no.8, pp.2465-6.

27.569. Sommers, H.S.: "Experimental Properties of Injection Lasers. V", J. Appl. Phys., 1974, 45, no.1, pp.237-42.

27.570. Linford, G.J., et al.: "Very Long (30 km) Lasers", Appl. Opt., 1974, 13, no.2, pp.379-90.

27.571. Poizner, B.N., Portnova, T.S., and Tsidulko, I.M.: "Natural Width of Spectrum of Two-Frequency Gas-Laser Modes", Izv. VUZ Radiofiz., 1974, 17, no.3, pp.350-3.

27.572. Wang, C.P., and Lin, S.C.: "Mode Structure and Beam Divergence of a Large-Bore High-Power Argon-Ion Laser", J. Appl. Phys., 1974, 45, no.1, pp.350-6.

27.573. Obara, M., and Fujioka, T.: "Time-Resolved Spectroscopic Studies on the TE HF Chemical Lasers Using H_2-SF_6 and CH_4-SF_6 Mixtures", Jap. J. Appl. Phys., 1974, 13, no.4, pp.675-83.

27.574. Vas'kov, V.A., et al.: "Study of Stabilization Method for Two-Mode Generation in a He-Ne Laser", Zh. Prikl. Spekt., 1974, 20, no.2, pp.192-4.

27.575. Lariontsev, E.G., and Skuibina, I.P.: "(Nonlinear) Interaction of Radiation With (Solid-State) Active Medium in Multimode Regime", Izv. VUZ Radiofiz., 1974, 17, no.3, pp.354-9.

27.576. Kennedy, C.J.: "Pulse Chirping in a Nd^{3+}:YAG Laser", J. Quantum Electron. IEEE, 1974, QE-10, no.6, pp.528-30.

27.577. Chinone, N., and Ito, R.: "Spectral Behaviour of Self-Pulsing Double-Heterostructure Injection Lasers", Jap. J. Appl. Phys., 1974, 13, no.3, pp.575-6.

27.578. Rohr, H., and Kellerer, L.: "Resonance Reflector for Suppressing Undesired Rotational Lines from TEA CO_2 Lasers", Appl. Phys. Lett., 1974, 24, no.3, pp.124-5.

27.579. Lewin, L.: "Obliquity Factor for Radiation from Solid-State Laser", Electron. Lett., 1974, 10, no.8, pp.134-5.

27.580. Isbasescu, M.: "(Dye-) Mode-Locked Nd^{3+}:Glass Laser", Rev. Roum. Phys., 1974, 19, no.3, pp.363-4.

27.581. Gibson, A.F., Kimmitt, M.F., and Norris, B.: "Generation of Bandwidth-Limited Pulses from a TEA CO_2 Laser Using p-Ge", Appl. Phys. Lett., 1974, 24, no.7, pp.306-7.

27.582. Yamanaka, M., et al.: "Transverse Mode in an Optically Pumped Far-IR NH_3 Laser", Jap. J. Appl. Phys., 1974, 13, no.5, pp.843-50.

27.583. Pike, C.T.: "Spatial Hole Burning in CW Dye Lasers", Opt. Commun., 1974, 10, no.1, pp.14-7.

27.584. Andersson, H.E.B., and Borgstrom, S.A.: "Time-Resolved Analysis of a TE N_2 Laser", Opto-Electron., 1974, 6, no.3, pp.225-34.

27.585. Baumhacker, H., and Lang, R.S.: "Actively Mode-Locked TEA CO_2 Laser With High-Power Multiband Output", Phys. Lett., 1974, 47A, no.6, pp.429-30.

27.586. Andreeva, E.Yu., Terekhin, D.K., and Fridrikhov, S.A.: "Effect of Resonator Anisotropy on Emission Polarization of a He-Ne Laser in a Magnetic Field", Zh. Prikl. Spekt., 1974, 20, no.3, pp.389-92.

27.587. Maeda, M., and Miyazoe, Y.: "Saturable Absorbers for Rhodamine-6G Dye Laser to Generate Picosecond Pulses", Jap. J. Appl. Phys., 1974, 13, no.1, pp.193-4.

27.588. Belland, P., Ciura, A.I., and Whitbourn, L.B.: "Gain Saturation and Oscillation Linewidth of a CW 337-micron HCN Laser", Opt. Commun., 1974, 11, no.1, pp.21-6.

27.589. Haug, C., and Whinnery, J.R.: "Experiments on Transverse Mode Locking of Circular Cylindrical and Certain Flat Beams", J. Quantum Electron. IEEE, 1974, QE-10, no.4, pp.406-8.

27.590. Teng, T.C., and Pao, Y.H.: "Hysteresis Effect in Mode Locking of He-Ne Lasers by Ne Discharge Cell", J. Quantum Electron. IEEE, 1974, QE-10, no.5, pp.494-6.

27.591. Andreeva, E.Yu., Terekhin, D.K., and Fridrikhov, S.A.: "Nonlinear Properties of He-Ne Amplifier at 3.39 micron", Zh. Prikl. Spekt., 1974, 20, no.3, pp.513-5.

27.592. Shank, C.V., and Ippen, E.P.: "Sub-Picosecond Kilowatt Pulses from Mode-Locked CW Dye Laser", Appl. Phys. Lett., 1974, 24, no.8, pp.373-5.

27.593. Sanchez, F., and Lecompte, C.: "Q-Switched Nd^{3+}:Glass Laser of Variable Temporal Coherence", Appl. Opt., 1974, 13, no.5, pp.1071-6.

27.594. Boiko, B.B., Mikhnov, S.A., and Matyushkov, V.E.: "Effect of Optical Discharges of a Passive Shutter on Energy Parameters of a Monoimpulse Ruby Laser", Zh. Prikl. Spekt., 1974, 20, no.3, pp.507-9.

27.595. Yang, E.S., et al.: "Degradation-Induced Microwave Oscillations in Double-Heterostructure Injection Lasers", Appl. Phys. Lett., 1974, 24, no.7, pp.324-7.

27.596. Gurevich, G.L., and Ingel', L.Kh.: "Mode Locking of Laser at Resonance Modulation of Resonator Parameters", Izv. VUZ Radiofiz., 1974, 17, no.2, pp.219-27.

27.597. Yu, W., and Alfano, R.R.: "Satellite Picosecond Pulses Investigated With an Optical Kerr Gate", Opto-Electron., 1974, 6, no.3, pp.243-4.

27.598. Oppenheim, U.P., and Kaufman, Y.J.: "Molecular Saturation and Criterion for Passive Q-Switching", J. Quantum Electron. IEEE, 1974, QE-10, no.7, pp.533-40.

27.599. Graubner, F., Hermann, G., and Scharmann, A.: "Effect of Collisions on Mode Crossing Signals of a He-Ne Laser", Z. Phys., 1974, 269, no.1, pp.79-82.

27.600. Leshenyuk, N.S., and Orlov, L.N.: "Study of Spectral Composition of Q-Switched CO_2-Laser Emission", Zh. Prikl. Spekt., 1974, 20, no.4, pp.601-5.

27.601. Voitovich, A.P., and Shkadarevich, A.P.: "Energy Characteristics of Competing Transitions in a He-Ne Laser with Magnetic Field", Zh. Prikl. Spekt., 1974, 20, no.4, pp.606-11.

27.602. Linford, G.J., and Hill, L.W.: "Nd^{3+}:YAG Long (to 6.3 km) Lasers", Appl. Opt., 1974, 13, no.6, pp.1387-94.

27.603. Fisher, R.A., and Bischel, W.K.: "Pulse Compression for More Efficient Operation of Solid-State Amplifier Chains", Appl. Phys. Lett., 1974, 24, no.10, pp.468-70.

27.604. Muller, A., and Willenbring, G.R.: "Reduction of Time Jitter in a Passively Q-Switched and Mode-Locked Ruby Laser by a Double-Pulse Technique", Appl. Phys., 1974, 4, no.1, pp.47-50.

27.605. Konishi, S., Kobayashi, T., and Sueta, T.: "FM Mode-Locking of a He-Ne, 3.39-micron, Laser Using Large Locking Signal", Jap. J. Appl. Phys., 1974, 13, no.7, pp.1189-90.

27.606. Verreault, M., Otis, G., and Tremblay, R.: "Active Mode Selection in CO_2 TEA Lasers With Helicoidal Configuration of Electrodes", Opt. Commun., 1974, 11, no.3, pp.227-30.

27.607. Fan, B., et al.: "High-PRF Nd^{3+}: Glass Laser Mode-Locked by a Saturable Absorber", J. Quantum Electron. IEEE, 1974, QE-10, no.9, pp.654-5.

27.608. Jackson, J.E., and Rice, R.R.: "Output Fluctuations of High-Frequency Pulse-Pumped Nd^{3+}:YAG Laser", J. Appl. Phys., 1974, 45, no.5, pp.2353-5.

27.609. Drexhage, K.H., and Elsenthal, K.B.: "Observation of Compressed Picosecond Pulses of High Repetition Rate from a Nd^{3+}: Glass Laser", J. Appl. Phys., 1974, 45, no.6, pp.2614-5.

27.610. Chow, B.S.K.: "Measurement of Gas Laser Cavity Stability", Opt. Commun., 1974, 11, no.3, pp.231-4.

27.611. Czerlinski, G., and Bracokova, V.: "Coaxial Alignment of Laser Beams for Perturbation Experiments", Appl. Opt., 1974, 13, no.7, pp.1639-45.

27.612. Bates, H.E., Phillips, R.W., and Sasser, T.: "Test of a Semiempirical Equation Describing Laser Beam Divergence", J. Appl. Phys., 1974, 45, no.6, pp.2808-10.

27.613. Prakash, H., and Chandra, N.: "Conversion of Random-Phase Light Into Phase-Coherent Light", Phys. Rev., 1974, 9, no.5, pp.2167-9.

27.614. Maunders, E.A., McAllister, G.L., and Steier, W.H.: "Experiments on Improved Unstable Mode Profiles by Aperture Shaping", J. Quantum Electron. IEEE, 1974, QE-10, no.10, pp.821-2.

27.615. Neuhauser, W., and Toschek, P.E.: "Coupled Laser Mode Locking by Two-Quantum Interaction", Opt. Commun., 1974, 11, no.4, pp.331-4.

27.616. Klose, E., Dahne, S., and Durr, H.: "Dyes for Passive Switches in Ruby Lasers", Kvantovaya Elektron., 1974, no.4, pp.5-12.

27.617. Henshall, G.D., and Whiteaway, J.E.A.: "Far-Field Emission Patterns of Single Heterostructure GaAs Lasers", Electron. Lett., 1974, 10, no.15, pp.326-7.

27.618. Butler, J.K., and Zoroofchi, J.: "Radiation Fields of GaAs-(AlGa)As Injection Lasers", J. Quantum Electron. IEEE, 1974, QE-10, no.10, pp.809-15.

27.619. Iida, S., and Watanabe, Y.: "Spectral Characteristics and Inhomogeneities Near Active Regions of (GaAl)As-GaAs Lasers", Jap. J. Appl. Phys., 1974,13,no.8,pp.1249-58.

27.620. Akandev, B.A., and Bel'dyugin, I.M.: "Locking of Transverse Modes in a Raman Laser", Kvantovaya Elektron., 1974, no.4, pp.69-75, and Sov. J. Quantum Electron., 1974, 3, no.4, pp.317-20.

27.621. Vinokurov, G.N., and Terent'ev, V.E.: "Radiation of Lasers With Controllable Phototropic Shutters", Opt. Spekt., 1974, 36, no.2, pp.398-404, and Opt. Spectrosc., 1974, 36, no.2, pp.229-32.

27.622. Nestrizhenko, Yu.A., et al.: "Efficient Polarization of Laser Radiation by a Prism With Small Birefringence", Opt. Spekt., 1974, 36, no.3, pp.557-60, and Opt. Spectrosc., 1974, 36, no.3, pp.321-3.

27.623. Rubinov, A.N., and Korda, I.M.: "Nonlinear Total Internal Reflection and its Utilization in Ruby Laser Mode Locking", Kvantovaya Elektron., 1974, no.4, pp.96-7, and Sov. J. Quantum Electron., 1974, 3, no.4, p.334.

27.624. Bicanic, D.D., and Dymanus, A.: "Experiments With a 4-m CW HCN Laser", Infrared Phys., 1974, 14, no.3, pp.153-63.

27.625. Girard, A.: "Effects of Insertion of a CW, Low-Pressure, CO_2 Laser Into a TEA CO_2 Laser Cavity", Opt. Commun., 1974, 11, no.4, pp.346-51.

27.626. Bagaev, S.N., Dmitriev, A.K., and Chebotaev, V.P.: "Sharp Resonances in a Two-Frequency Gas Laser", Opt. Spekt., 1974, 36, no.3, pp.531-8, and Opt. Spectrosc., 1974, 36, no.3, pp.307-10.

27.627. Vdovin, Yu.A., et al.: "Interaction of Modes With Orthogonal and Parallel Polarizations in a Gas Laser", Kvantovaya Elektron., 1974, no.4, pp.35-42, and Sov. J. Quantum Electron., 1974, 3, no.4,pp.297-pp.297-301.

27.628. Avtonomov, V.P., et al.: "Selection of Vibration-Rotation CO_2 Laser Lines by a Diffraction Grating in the Resonator", Kvantovaya Elektron., 1974, no.4, pp.108-10, and Sov. J. Quantum Electron., 1974, 3, no.4, pp.345-6.

27.629. Marowsky, G., and Zaraga, F.: "Dual-Wavelength Operation of Two-Coupled Dye Lasers", Opt. Commun., 1974, 11, no.4, pp.343-5.

27.630. Andreeva, E.Y., Gulyaev, S.N., and Terekhin, D.K.: "Effect of Gas Pressure on LF Beats in a He-Ne Laser", Opt. Spekt., 1974, 36, no.2, pp.379-81, and Opt. Spectrosc., 1974, 36, no.2, pp.219-20.

27.631. Melekhin, G.V.: "Lasing Line Shapes in the $3s_2-2p_4$ and $2s_2-2p_4$ Transitions of He-Ne Mixture", Opt. Spekt., 1974, 36, no.2, pp.382-5, and Opt. Spectrosc., 1974, 36, no.2, pp.221-2.

27.632. Clark, W.M.: "Optical Homogeneity of a UV-Preionized CO_2 Laser Discharge", Appl. Opt., 1974, 13, no.9, pp.1995-7.

27.633. Dreizin, Yu.A., and Dykhne, A.M.: "Free-Oscillating Emission Instability of Fast Flowing Lasers Employing Unstable Resonators", Zh. Eksp. Teor. Fiz. Pis'ma, 1974, 19, no.12, pp.718-22, and JETP Lett., 1974, 19, no.12.

27.634. Flach, R., Shahin, I.S., and Yen, W.M.: "Application of Pressure Scanning to the Tuning of a High-Resolution Dye Laser", Appl. Opt., 1974, 13, no.9, pp.2095-9.

27.635. Lim, T.K., and Garside, B.K.: "Saturable Absorber Mode-Locking and Laser Pulse Compression", Opt. Commun., 1974, 12, no.1, pp.8-13.

27.636. Schaefer, R.B., and Willis, C.R.: "Effects of Triplet State Losses on Coherent Properties of Organic-Dye Lasers", Phys. Lett., 1974, 48A, no.6, pp.465-6.

27.637. Bocharov, V.V., and Zubarev, I.G.: "Frequency Locking of a Solid-State Laser With Externally Controlled Spectral Gain Lines", Izv. VUZ Radiofiz., 1974, 17, no.7, pp.964-9.

27.638. Bakhorin, V.A., et al.: "Single-Frequency TW Laser Using Nd^{3+}:Glass and Active Q-Switching", Zh. Eksp. Teor. Fiz. Pis'ma, 1974, 19, no.12, pp.758-61, and JETP Lett., 1974, 19, no.12.

27.639. Hasuo, S., and Ohmi, T.: "Spatial Distribution of Light Intensity in Injection Lasers", Jap. J. Appl. Phys., 1974, 13, no.9, pp.1429-34.

27.640. Allakhverdyan, R.G., Morozov, V.N., and Suchkov, A.F.: "Influence of Intensity Pulsations on Stimulated Emission Spectrum of a Semiconductor Laser", Kvantovaya Elektron., 1974, 1, no.1, pp.102-7, and Sov. J. Quantum Electron., 1974, 4, no.1, pp.55-7.

27.641. Litvinov, V.F., et al.: "Dynamic Instability of Semiconductor Laser at Low Temperatures", Zh. Eksp. Teor. Fiz. Pis'ma, 1974, 19, no.12, pp.747-50, and JETP Lett., 1974, 19, no.12.

27.642. Troitskii, Yu.V.: "Inhomogeneous Extraction of the Energy of Higher Transverse Modes", Kvantovaya Elektron., 1974, 1, no.1, pp.124-8, and Sov. J. Quantum Electron., 1974, 4, no.1, pp.68-70.

27.643. Arakelyan, S.M., et al.: "Maximum Coherence of TEM_{00} Modes of CW Laser Radiation", Kvantovaya Elektron., 1974, 1, no.1, pp.215-7, and Sov. J. Quantum Electron, 1974, 4, no.1, pp.134-5.

27.644. Zav'yalov, V.V., and Bogomolov, G.D.: "Frequency Beats Between Orthogonal Polarizations in a Water-Vapour Laser", Zh. Eksp. Teor. Fiz. Pis'ma, 1974, 20, no.6, pp.393-5, and JETP Lett., 1974, 20, no.6.

27.645. Arthurs, E.G., Bradley, D.J., and Glynn, T.J.: "Effect of Saturable Absorber Lifetime in Picosecond Pulse Generation. I. Ruby Laser", Opt. Commun., 1974, 12, no.2, pp.136-9.

27.646. Ershov, B.V., and Federov, V.B.: "Energy Characteristics of Millisecond Stimulated Emission from Nd^{3+}:Glass Lasers", Kvantovaya Elektron., 1974, 1, no.1, pp.108-13, and Sov. J. Quantum Electron., 1974, 4, no.1, pp.58-61.

27.647. Arapov, A.P., Muratov, V.R., and Sidorenko, Yu.K.: "Nd^{3+}:Glass Laser With High-Frequency Q-Switching", Kvantovaya Elektron., 1974, 1, no.1, pp.134-7, and Sov. J. Quantum Electron., 1974, 4, no.1,pp.74-5.

27.648. Kozlov, V.V.: "Laser With Polyhedral Active Element Emitting Long Nanosecond Pulses", Kvantovaya Elektron., 1974, 1, no.1, pp.197-200, and Sov. J. Quantum Electron., 1974, 4, no.1, pp.121-2.

27.649. Bogdankevich, O.V., et al.: "Investigation of Dynamics of Emission from a Radiating-Mirror Semiconductor Laser With External Resonator", Kvantovaya Elektron., 1974, 1, no.1, pp.149-51, and Sov. J. Quantum Electron., 1974, 4, no.1, pp.84-5.

27.650. Lisitsa, M.P., Mozol, P.E. and Fekeshgazi, I.V.: "Laser Pulse Lengthening With Nonlinear Absorbing Semiconductors CdP_2 and ZnP_2", Kvantovaya Elektron., 1974, 1, no.1, pp.185-7, and Sov. J. Quantum Electron., 1974, 4, no.1, pp.110-1.

27.651. Girard, G., et al.: "Experiments and Theory of Nanosecond Pulse Extraction in an Atmospheric-Pressure CO_2 Amplifier", J. Quantum Electron. IEEE, 1974, QE-10, no.12, pp.901-2.

27.652. Craig, A.D., and Perkin, R.M.: "Premature Gain-Switch Initiation by Amplifier Interaction in a TEA CO_2 Oscillator-Amplifier Laser System", Opt. Commun., 1974, 12, no.3, pp.256-9.

27.653. Brazovskii, V.E., and Telegin, G.G.: "Transient Processes and Statistical Phenomena in a He-Ne Laser With Slowly Varying Pump Parameter", Opt. Spektr., 1974, 36, no.4, pp.739-42, and Opt. Spectrosc., 1974, 36, no.4, pp.429-30.

27.654. Eliseev, P.G., et al.: "Influence of Microwave Modulation on the Emission Spectrum of an Injection Laser", Kvantovaya Elektron., 1974, 1, no.1, pp.151-4, and Sov. J. Quantum Electron., 1974, 4, no.1, pp.86-7.

27.655. Senatorov, K.Ya., et al.: "(Self-) Modulation of an Injection Laser With Double-Sided Heterostructure", Kvantovaya Elektron., 1974, 4, no.1, pp.92-3.

27.656. Prozorov, O.N., et al.: "Investigation of a Multibeam Semiconductor Laser With an Emitting Array", Kvantovaya Elektron., 1974, 1, no.1, pp.169-72, and Sov. J. Quantum Electron., 1974, 4, no.1, pp.98-9.

27.657. Kurbatov, L.N., et al.: "Time Characteristics of Double-Heterojunction Lasers", Kvantovaya Elektron., 1974, 1, no.2, pp.191-3, and Sov. J. Quantum Electron., 1974, 4, no.1, pp.116-7.

27.658. Eliseev, P.G., et al.: "Control of Polarization of Heterolaser Radiation by Uniaxial Compression", Kvantovaya Elektron., 1974, 1, no.1, pp.196-7, and Sov. J. Quantum Electron., 1974, 4, no.1, p.120.

27.659. Kryukov, P.G., Matveets, Yu.A., and Shatberashvili, O.B.: "Broadening of the Emission Spectrum of a Self-Mode-Locked Laser", Kvantovaya Elektron., 1974, 1, no.2, pp.450-2, and Sov. J. Quantum Electron., 1974, 4, no.2, pp.258-9.

27.660. Milinkevich, A.V., et al.: "High-Frequency Automodulation of Giant Pulse", Zh. Prikl. Spekt., 1974, 21, no.4, pp.604-12.

27.661. Selway, P.R., et al.: "Measurement of Effect of Injected Carriers on the p-n Refractive-Index Step in Single Heterostructure Diode Lasers", Electron. Lett., 1974, 10, no.22, pp.453-5.

27.662. Pugh, E.R., et al.: "Optical Quality of Pulsed Electron-Beam Sustained Lasers", Appl. Opt., 1974, 13, no.11, pp.2512-7.

27.663. Gromova, V.G., Gromov, Yu.N., and Khaikin, N.Sh.: "Selection of Vibrational-Rotational Transitions of a CO_2 Laser by Internal Fabry-Perot Interferometers", Prib. Tekh. Eksp., 1974, 17, no.1, pp.175-6, and Instrum. Exp. Tech., 1974, 17, no.1, pp.200-2.

27.664. Johnson, R.H.: "Ar-Ion, Mode-Locked, Cavity-Dumped Laser", Opt. Spectra, 1974, 9, no.9, pp.33-5.

27.665. Ohta, T., Seriu, J., and Ogawa, T.: "Isotope Effects in the Two-Mode Operation of a He-Ne Laser", Opto-Electron., 1974, 6, no.6, pp.433-42.

27.666. Owens, D.K., and Weiss, R.: "Measurement of the Phase Fluctuations in a He-Ne Zeeman Laser", Rev. Sci. Instrum., 1974, 45, no.9, pp.1060-2.

27.667. Ageikin, V.A., et al.: "Emission Spectrum of an Electric-Ionization High-Pressure CO_2 Laser", Kvantovaya Elektron., 1974, 1, no.2, pp.334-40, and Sov. J. Quantum Electron., 1974, 4, no.2, pp.185-8.

27.668. Lisitsyn, V.N., Chapovskii, P.I., and Chernenko, A.A.: "Monochromatization and Tuning of Frequency of IR Radiation of a N_2 Laser", Kvantovaya Elektron., 1974, 1, no.2, pp.341-5, and Sov. J. Quantum Electron., 1974, 4, no.2, pp.189-91.

27.669. Chan, C.K., and Sari, S.O.: "Tunable Dye Laser Pulse Converter for Production of Picosecond Pulses", Appl. Phys. Lett., 1974, 25, no.7, pp.403-6.

27.670. Osinski, M.: "Dynamical Properties of Injection Laser Radiation", Postepy Fiz., 1974, 25, no.5, pp.517-36.

27.671. Zasavitskii, I.I., et al.: "Influence of Hydrostatic Pressure on Emission Spectra of (PbSn)Se", Fiz. Tekh. Poluprov., 1974, 8, no.4, pp.732-6, and Sov. Phys.-Semicond., 1974, 8, no.4, pp.467-70.

27.672. Alferov, Zh.I., et al.: "Semiconductor Laser with Extremely Low Divergence of Radiation", Fiz. Tekh. Poluprov., 1974, 8, no.4, pp.832-3, and Sov. Phys.-Semicond., 1974, 8, no.4, pp.541-2.

27.673. Vanderleeden, J.C.: "Proposal for Wavelength Tuning and Stabilization of GaAs Lasers With Graded-Index Fibre Segment in a Dispersive Cavity", Opto-Electron., 1974, 6, no.6, pp.443-9.

27.674. Anan'ev, Yu.A., et al.: "Angular Characteristics of Radiation Emitted by a Laser With Resonator of Large Effective Length", Kvantovaya Elektron., 1974, 1, no.2, pp.296-301, and Sov. J. Quantum Electron., 1974, 4, no.2, pp.164-7.

27.675. Bonczyk, P.A.: "Observation of Lamb Dip for HF Chemical Laser", J. Appl. Phys., 1974, 45, no.11, pp.5077-8.

27.676. Mizuno, K., et al.: "Correlation Between the 337- and 311-micron Lines in a CW HCN Laser", J. Appl. Phys., 1974, 45, no.12, pp.5464-5.

27.677. Belousova, I.M., Kiselev, V.M., and Kurzenkov, V.N.: "Characteristics of Stimulated Emission from Iodine Atoms in Pulsed Magnetic Fields", Kvantovaya Elektron., 1974, 1, no.6, pp.1389-94, and Sov. J. Quantum Electron., 1974, 4, no.6, pp.767-9.

27.678. Datskevich, N.P., et al.: "Space-Time Characteristics of Pulses Emitted from a Double-Discharge CO_2 Laser", Kvantovaya Elektron., 1974, 1, no.6, pp.1416-9, and Sov. J. Quantum Electron., 1974, 4, no.6, pp.783-5.

27.679. Royt, T.R., et al.: "Temporarily Coincident Ultrashort Pulses from Synchronously Pumped Tunable Dye Lasers", Appl. Phys. Lett., 1974, 25, no.9, pp.514-6.

27.680. Hilborn, R.C., and Brayman, H.C.: "Simultaneous Two-Wavelength Output from Multiple-Dye Pulsed Tunable Lasers", J. Appl. Phys., 1974, 45, no.11, pp.4912-4.

27.681. Sugiura, Y., Matsunaga, Y., and Fujioka, T.: "Wavelength Measurements of Simultaneously Excited Superradiance and Laser Emission from Cryptocyanine Dye", J. Appl. Phys., 1974, 45, no.11, pp.4969-70.

27.682. Baltakov, F.N., Bariklin, B.A., and Sukhanov, L.V.: "Spatial and Temporal Characteristics of a 100-J Laser Using Rhodamine 6G in Ethanol", Kvantovaya Elektron., 1974, 1, no.4, pp.973-7, and Sov. J. Quantum Electron., 1974, 4, no.4, pp.537-9.

27.683. Antsiferov, V.V., and Folin, K.G.: "Time Behaviour of the Spectrum of a Ruby Laser With Spherical Mirrors in a Quasistationary Operating Mode", Avtometriya, 1974, no.6, pp.103-4.

27.684. Eichler, H.J., and Eichler, J.: "Narrowing of the Emission Spectrum of Ruby Laser by Inhomogeneous Pumping", J. Appl. Phys., 1974, 45, no.11, pp.4950-3.

27.685. Sharp. E.J., et al.: "Optical Spectra and Laser Action in Nd^{3+}: $CaY_2Mg_2Ge_3O_{12}$", J. Appl. Phys., 1974, 45, no.11, pp.4974-9.

27.686. Antsiferov, V.V., et al.: "High-Power Single-Frequency Ruby Ring Laser With Smooth Emission Frequency Adjustment", Avtometriya, 1974, no.6, pp.97-9.

27.687. Alekseev, V.N., Anan'ev, Yu.A., and Dauengauer, E.F.: "Powerful Solid-State Laser With Telescopic Resonator", Dokl. Akad. Nauk SSSR, 1974, 214, no.1-3, pp.535-8, and Sov. Phys.-Dokl., 1974, 19, no.1, pp.23-5.

27.688. Galaktionova, N.M., Mak, A.A., and Khyuppenen, A.P.: "Parasitic Amplitude Modulation of a Stabilized Nd^{3+}:YAG Laser", Zh. Tekh. Fiz., 1974, 44, no.4, pp.770-7, and Sov. Phys.-Tech. Phys., 1974, 19, no.4, pp.486-90.

27.689. Kuzovkova, T.A., et al.: "Pulsed Nd^{3+}:Glass Laser With Periodic Q-Switching", Zh. Tekh. Fiz., 1974, 44, no.4, pp.797-802, and Sov. Phys.-Tech. Phys., 1974, 19, no.4, pp.503-5.

27.690. Mironov, A.B., and Shatberashvili, O.B.: "Spectrum of a Single Ultrashort Pulse Emitted by a Nd^{3+}:Glass Laser", Kvantovaya Elektron., 1974, 1, no.6, pp.1452-4, and Sov. J. Quantum Electron., 1974, 4, no.6, pp.805-6.

27.691. Antsiferov, V.V., Derzhi, N.M., and Folin, K.G.: "Dynamics of a Nd^{3+}:Glass Laser With Spherical Mirrors and Elimination of Inhomogeneous Inversion", Zh. Prikl. Spekt., 1974, 21, no.5, pp.917-9.

27.692. Alferov, Zh.I., et al.: "Temporal Characteristics of Near Field of Injection Laser", Zh. Tekh. Fiz., 1974, 44, no.4, pp.863-4, and Sov. Phys.-Tech. Phys., 1974, 19, no.4, p.548.

27.693. Arifov, U.A., Bedilov, M.R., and Egamov, U.: "Effect of Ionization of Cr^{3+} Ions on Radiation Kinetics of a Ruby Laser", Dokl. Akad. Nauk SSSR, 1974, 214, no.1-3, pp.305-7, and Sov. Phys.-Dokl., 1974, 19, no.1, pp.21-2.

27.694. Klochan, E.L., et al.: "Unidirectional Oscillation in a Solid-State Ring Laser", Dokl. Akad. Nauk SSSR, 1974, 215, no.1-3, pp.313-6, and Sov. Phys.-Dokl., 1974, 19, no.3, pp.129-30.

27.695. Mak, A.A., et al.: "Unidirectional CW Oscillation in a Solid-State Ring Laser With Return Mirror", Zh. Tekh. Fiz., 1974, 44, no.4, pp.868-70, and Sov. Phys.-Tech. Phys., 1974, 19, no.4, pp.552-3.

27.696. Zhiryakov, B.M., Popov, N.I., and Fannibo, A.K.: "Dynamics of Laser Pulses Generated With High Spatial Homogeneity of Pumping", Kvantovaya Elektron., 1974, 1, no.4, pp.835-9, and Sov. J. Quantum Electron., 1974, 4, no.4, pp.457-9.

27.697. Zemskov, K.I., et al.: "Use of Unstable Resonators in Achieving Diffraction-Limited Radiation from High-Gain Pulsed Gas Lasers", Kvantovaya Elektron., 1974, 1, no.4, pp.863-9, and Sov. J. Quantum Electron., 1974, 4, no.4, pp.474-7.

27.698. Krivoshchekov, G.V., and Smirnov, V.A.: "Generation of Ultrashort Pulses in a Laser With Combined Active and Passive Modulation", Avtometriya, 1974, no.6, pp.91-6.

27.699. Zheriknin, A.N., et al.: "Origin of Temporal Structure of Ultrashort Laser Pulses", Kvantovaya Elektron., 1974, 1, no.4, pp.956-9, and Sov. J. Quantum Electron., 1974, 4, no.4, pp.525-6.

27.700. Andreev, R.B., and Volosov, V.D.: "Temporal and Spatial Coherence of Second Harmonic of Radiation from a Multimode Nd^{3+}: Glass Laser", Kvantovaya Elektron., 1974, 1, no.6, pp.1355-9, and Sov. J. Quantum Electron., 1974, 4, no.6, pp.746-8.

27.701. Rice, D.K.: "Spectral-Line Selection of CO Lasers", Appl. Opt., 1974, 13, no.12, pp.2812-5.

27.702. Kasuya, T.: "Broadband-Tunable Gas Laser With Superconducting Solenoid", Oyo Buturi, 1974, 43, no.10, pp.1050-4.

27.703. Batovskii, O.M., and Gur'ev, V.I.: "Emission Spectra of HF and DF Lasers", Kvantovaya Elektron., 1974, 1, no.3, pp.676-9, and Sov. J. Quantum Electron., 1974, 4, no.3, pp.380-1.

27.704. Ivanov, I.G., Il'yushko, V.G., and Sem, M.F.: "Broadening of Emission Spectrum of Cataphoretic Lasers", Kvantovaya Elektron., 1974, 1, no.3, pp.716-9, and Sov. J. Quantum Electron., 1974, 4, no.3, pp.411-2.

27.705. Pankratov, V.A.: "Rb Laser with Zero Frequency Shift Caused by a Buffer Gas", Kvantovaya Elektron., 1974, 1, no.3, pp.720-1, and Sov. J. Quantum Electron., 1974, 4, no.3, p.413.

27.706. Galaktionova, N.M., Gershun, V.V., and Mak, A.A.: "Single-Mode CW Nd^{3+}: YAG Lasers", Opt. Spekt., 1974, 37, no.2, pp.322-5, and Opt. Spectrosc., 1974, 37, no.2, pp.184-5.

27.707. Danileiko, Yu.K., Manenkov, A.A., and Nechitailo, V.S.: "Single-Frequency Ruby Laser With Spatially Homogeneous Radiation Field and Variable Duration of Nanosecond Pulses", Kvantovaya Elektron., 1974, 1, no.3, pp.604-8, and Sov. J. Quantum Electron., 1974, 4, no.3, pp.336-8.

27.708. Zubarev, I.G., and Mikhailov, S.I.: "External-Signal Control of Emission Spectrum of a Q-Switched Nd^{3+} Laser", Kvantovaya Elektron., 1974, 1, no.3, pp.625-8, and Sov. J. Quantum Electron., 1974, 4, no.3, pp.348-50.

27.709. Kytina, I.G., and Nesterenko, V.M.: "Laser With Improved Homogeneity of Radiation Field", Kvantovaya Elektron., 1974, 1, no.3, pp.721-3, and Sov. J. Quantum Electron., 1974, 4, no.3, pp.414-5.

27.710. Anan'ev, Yu.A., et al.: "Comparison of Divergence of Emission by a Multistage System and by a Laser with Telescopic Resonator", Kvantovaya Elektron., 1974, 1, no.5, pp.1247-50, and Sov. J. Quantum Electron., 1974, 4, no.5, pp.689-90.

27.711. Vvedenskii, B.S., Logginov, A.S., and Senatorov, K.S.: "Temporal Coherence of Injection Lasers", Kvantovaya Elektron., 1974, 1, no.5, pp.1232-4, and Sov. J. Quantum Electron., 1974, 4, no.5, pp.678-9.

27.712. Bogdankevich, O.V., et al.: "Dynamics of a Semiconductor Laser Excited Longitudinally by an Electron Beam", Kvantovaya Elektron., 1974, 4, no.5, pp.702-3.

27.713. Boitsov, V.F., and Murina, T.A.: "Angular Divergence of Field and Minimum Light Spots in Ring Cavity With Gaussian Diaphragm", Opt. Spekt., 1974, 37, no.1, pp.152-6, and Opt. Spectrosc., 1974, 37, no.1, pp.84-6.

27.714. Alekseev, V.A., et al.: "Reproducibility of Frequency of a Ring Gas Laser With Nonlinear Absorption Cell", Kvantovaya Elektron., 1974, 1, no.5, pp.1089-98, and Sov. J. Quantum Electron., 1974, 4, no.5, pp.594-9.

27.715. Kovalev, A.A., et al.: "Dynamics of Giant Pulses in a Laser With Different Types of Q-Switch", Kvantovaya Elektron., 1974, 1, no.5, pp.1191-4, and Sov. J. Quantum Electron., 1974, 4, no.5, pp.653-5.

27.716. Szmolszhaja, T.I., et al.: "First-Order Coherence of Radiation of a Pulsed Dye Laser", Acta Phys. Chem. Szeged., 1974, 20, no.3, pp.305-13.

27.717. Hori, K., Bamba, Y., and Furuya, T.: "Coherence Study of a CW Nd^{3+}:YAG Laser and Second Harmonic", Fujitsu Sci. Tech. J., 1974, 10, no.4, pp.173-91.

27.718. Berzing, E.G., et al.: "Effect of a Ruby Laser Spectrum", Kvantovaya Elektron. (Kiev), 1974, no.8, pp.90-7.

27.719. Nestrizhenko, Yu.A., and Pyatikop, A.P.: "Laser With Variable Spectral Interval Generation Lines", Kvantovaya Elektron. (Kiev), 1974, no.8, pp.124-6.

27.720. Korda, I.M., and Rubinov, A.N.: "Utilization of Nonlinear Total Internal Reflection in Two Q-Switching Methods for a Ruby Laser", Kvantovaya Elektron., 1974, 1, no.8, pp.1877-80.

27.721. Troitskii, Yu.V.: "Uniform Illumination Using a Gas Laser", Opt. Spekt., 1974, 37, no.5, pp.973-8, and Opt. Spectrosc., 1974, 37, no.5, pp.554-7.

27.722. Dzyubenko, M.I., et al.: "Features of Generation Spectra of a Rhodamine Laser", Kvantovaya Elektron. (Kiev), 1974, no.8, pp.126-33.

28. FREQUENCY FILTERS

28.1.   Seo, W.Y.: "Discriminator for Above 10 GHz", Proc. IEEE, 1965, 53, p.179.

28.2.   Bava, G.P.: "Waveguide Cavities Terminated by Irises of Arbitrary Thickness", Alta Freq., 1965, 34, p.24.

28.3.   Levinson, I.B., and Fridberg, P.Sh.: "Coupling Between Two Cavities via a Narrow Slot", Radiotekh. Elektron., 1965, 10, p.260, and Radio Eng. Electron. Phys., 1965, 10.

28.4.   Flannery, W.T., et al.: "Microwave Rejection Networks", Trans. IEEE, 1965, EMC-7, no.1, p.25.

28.5.   Craven, G.: "Channel-Separating Filters for Long-Distance Communication by Waveguide", Proc. IREE Aust., 1965,26,p.11.

28.6.   Stone, R.H.: "Open Periodic Filters", Microwave J., 1965, 8, no.10, p.31.

28.7.   Haas, W., and Godtmann, H.D.: "Microwave Bandpass Filters Using Open Resonators", Arch. Elektr. Ubertrag., 1965, 19, p.551.

28.8.   Horton, M.C., and Wenzel, R.J.: "General Theory and Design of Optimum Quarter-Wave TEM Filters", Trans. IEEE, 1965, MTT-13, p.316.

28.9.   Taub, J.J., and Sleven, R.L.: "Design of Bandstop Filters in the Presence of Dissipation", Trans. IEEE, 1965, MTT-13, p.589.

28.10.   Schiffman, B.M., Matthaei, G.L., and Young, L.: "Rectangular-Waveguide Filter Using Trapped-Mode Resonators", Trans. IEEE, 1965, MTT-13, p.575.

28.11.   Matthaei, G.L., and Weller, D.B.: "Circular TE_{011}-Mode, Trapped-Mode, Band-pass Filters", Trans. IEEE, 1965, MTT-13, p.581.

28.12.   Wenzel, R.J.: "Exact Theory of Interdigital Bandpass Filters and Related Coupled Structures", Trans. IEEE, 1965, MTT-13, p.559.

28.13.   Mumford, W.W.: "Tables of Stub Admittances for Maximally Flat Filters Using Short-Circuited Quarter-Wave Stubs", Trans. IEEE, 1965, MTT-13, p.695.

28.14.   Dishal, M.: "Simple Design Procedure for Small-Percentage-Bandwidth Round-Rod Interdigital Filters", Trans. IEEE, 1965, MTT-13, p.696.

28.15.   Caputo, J., and Bell, F.: "Waffle-Iron Harmonic Suppression Filter", Trans. IEEE, 1965, MTT-13, p.701.

28.16.   Carter, P.S.: "Sidewall-Coupled, Stripline, Magnetically Tunable Filters", Trans. IEEE, 1965, MTT-13, p.306.

28.17.   Riblet, H.J.: "General Design Procedure for Quarter-Wavelength Inhomogeneous Impedance Transformers Having Approximately Equal-Ripple Performance", Trans. IEEE, 1965, MTT-13, p.622.

28.18.   Cumming, R.C., and Howell, D.W.: "YIG Filters as Envelope Limiters", Trans. IEEE, 1965, MTT-13, p.616.

28.19.   Model', A.M.: "Amplitude/Frequency Characteristic of a Waveguide Bandpass Filter at Frequencies Well Off Resonance", Elektrosvyaz,1965, no.1, and Telecommun. Radio Eng. Pt 1, 1965, no.1, p.45.

28.20.   Skeie, H.: "Application of Ferrite Monocrystals for Nonreciprocal Microwave Bandpass and Bandstop Filters", Elektrotek. Tidsskr., 1965, 78, no.9, p.145.

28.21.   Model', A.M.: "Series Resonant Circuit in a Waveguide", Radiotekhnika, 1965, 20, no.8, p.23, and Telecommun. Radio Eng. Pt 2, 1965, 20, no.8, p.92.

28.22.   Schiffman, B.M.: "Capacitively-Coupled Stub Filter", Trans. IEEE, 1965, MTT-13, p.253.

28.23.   Craven, G., Stopp, D.W., and Thomas, R.R.: "Resonant-Slot Hybrid Junctions and Channel-Dropping Filters", Proc. IEE, 1965, 112, p.669.

28.24.   Chen, T.S.: "Theory and Performance of Coupled-Slot Waveguide Gain Equalizers", Int. J. Electron., 1965, 18, p.369.

28.25.   Schuegraf, E.: "Frequency-Selective H_{10}-H_{01} Mode Changer and Development as a Channel Separating Filter", Frequenz, 1965, 19, p.341.

28.26.   Pfitzenmaier, G.: "Continuously Tunable Delay Equalizer for Microwaves", Frequenz, 1965, 19, p.338.

28.27.   Cristal, E.G.: "Method for Design of Nonreflecting High-Power Microwave Bandpass Filters", Microwave J., 1966, 9, no.6, p.69.

28.28.   Kurzrok, R.M.: "Design of Comb-Line Bandpass Filters", Trans. IEEE, 1966, MTT-14, p.351.

28.29.   Nicholson, B.F.: "Resonant Frequency of Interdigital Filter Elements", Trans. IEEE, 1966, MTT-14, p.250.

28.30.   Schiffman, B.M.: "Two Nomograms for Coupled-Line Sections for Bandstop Filters", Trans. IEEE, 1966, MTT-14, p.297.

28.31.   Ricardi, L.J.: "Diplexer Using Hybrid Junctions", Trans. IEEE, 1966, MTT-14, p.364.

28.32.   Cristal, E.G.: "Bandpass Spurline Resonators", Trans. IEEE, 1966, MTT-14,p.296.

28.33. Korenev, Yu.V., and Ratbil', E.L.: "Design of Microwave Bandstop Radial Filters", Elektrosvyaz,1966, no.4, p.1, and Telecommun. Radio Eng. Pt 1, 1966,no.4,p.60.

28.34. Kurzrok, R.M.: "General Four-Resonator Filters at Microwave Frequencies", Trans. IEEE, 1966, MTT-14, p.296.

28.35. Gunston, M.A.R., and Nicholson, B.F.: "Interdigital and Comb-Line Filters", Marconi Rev., 1966, 29, p.133.

28.36. Craven, G.: "Waveguide Bandpass Filters Using Evanescent Modes", Electron. Lett., 1966, 2, p.251.

28.37. Craven, G.: "Tuning Techniques for Multisection Waveguide Bandpass Filters Using Evanescent Modes", Electron. Lett., 1966, 2, p.419.

28.38. Cohen, J., and Taub, J.: "Confocal Resonator Bandpass Filters", Trans. IEEE, 1966, MTT-14, no.12, pp.698-9.

28.39. Abele, T.A.: "High-Quality Waveguide Directional Filter", Bell Syst. Tech. J., 1967, 46, p.81.

28.40. Karp. A., and Young, L.: "Control of Resonant Frequency of YIG-Disc Filter by Doublet Tuning", Trans. IEEE, 1967, MTT-15, p.193.

28.41. Reed, J.: "Precision Design of Direct-Coupled Filters", Trans. IEEE, 1967, MTT-15, no.2, pp.134-6.

28.42. Abele, T.A., and Wang, H.C.: "Adjustable Narrowband Microwave Delay Equalizer", Trans. IEEE, 1967, MTT-15, no.10, pp.566-74.

28.43. Scanlan, S.O., and Rhodes, J.D.: "Microwave Networks With Constant Delay", Trans. IEEE, 1967, CT-14, no.3, pp.290-7.

28.44. Bava, G.P., Scaglia, C., and Bava, E.: "Tentative Approach to the Design of Waveguide-Below-Cutoff Bandpass Filters", Electron. Lett., 1967, 3, no.8, pp.361-2.

28.45. Skedd, R.F., and Craven, G.F.: "Type of Magnetically Tunable Multisection Bandpass Filter in Ferrite-Loaded Evanescent Waveguide", Trans. IEEE, 1967, MAG-3, no.3, pp.397-400.

28.46. Tokheim, R.E., et al.: "Nonreciprocal Filters", Trans. IEEE, 1967, MAG-3, no.3, pp.383-91.

28.47. Matthaei, G.L., and Leedom, D.A.: "Lowpass, Quasi-Optical, Filters Using Dielectric With Metal-Strip Inclusions", Proc. IEEE, 1967, 55, no.11, pp.2056-7.

28.48. Wenzel, R.J.: "Wideband High-Selectivity Diplexers Utilizing Digital Elliptic Filters", Trans. IEEE, 1967, MTT-15, no.12, pp.669-80.

28.49. Standley, R.D.: "Millimetre-Wave, Two-Pole, Circular-Electric-Mode, Channel-Dropping Filter Structure", Bell Syst. Tech. J., 1967, 46, no.10, pp.2261-76.

28.50. Standley, R.D.: "Millimetre-Wavelength Diplexing Filters Utilizing Circular TE_{011}-Mode Resonators", Trans. IEEE, 1967, MTT-16, no.1, pp.50-1.

28.51. Wang, H.C.: "Improved Design of Waveguide Band-Rejection Filters", Bell Syst. Tech. J., 1968, 47, no.1, pp.1-15.

28.52. Scanlan, S.O., and Rhodes, J.D.: "Microwave Allpass Networks. I and II", Trans. IEEE, 1968, MTT-16, no.2, pp.62-72 and 72-9.

28.53. Schussler, H.: "Null Position n-Stage Minimum Phase Chebyshev Filter With Various Couplings", Arch. Elektr. Ubertrag., 1968, 22, no.3, pp.161-2.

28.54. Craven, G.: "Relationship Between Direct-Coupled Waveguide Filters and Evanescent-Mode Filters", Electron. Lett., 1968, 4, no.3, pp.44-6.

28.55. Wenzel, R.J.: "Printed-Circuit Complementary Filters for Narrowband Multiplexers", Trans. IEEE, 1968, MTT-16, no.3, pp.147-52.

28.56. Muller, H.M.: "Dielectric Resonators and Application as Microwave Filters", Z. Angew. Phys., 1968, 24, no.3, pp.142-7.

28.57. Taub, J.: "Minimum Volume of an Equal-Element Bandpass Filter", Trans. IEEE, 1968, MTT-16, no.4, pp.264-5.

28.58. Harrison, W.H.: "Miniature High-Q Bandpass Filter Employing Dielectric Resonators", Trans. IEEE, 1968, MTT-16, no.4, pp.210-8.

28.59. Conning, S.W.: "Direct-Coupled Waveguide Filters With Post Doublets", Electr. Commun., 1968, 43, p.233.

28.60. Easter, B.: "Direct-Coupled Resonator Filters With Improved Selectivity", Electron. Lett., 1968, 4, p.415.

28.61. Kurzrok, R.M.: "Frequency Behaviour of Post-Coupled TEM Comb-Line Resonators", Trans. IEEE, 1968, MTT-16, p.888.

28.62. Ito, Y., and Meguro, T.: "Bandpass and Bandstop Microwave Filter Using Quarter-Wave Rod Resonators", Fujitsu Sci. Tech. J., 1968, 4, no.3, p.29.

28.63. Grossbach, R.: "Improved Tunable Ferrimagnetic Filter Using a Hybrid-T", Proc. IEEE, 1968, 56, p.2077.

28.64. Hupert, J.J., and Vigil, J.: "Evanescent-Mode Resonator of Pure TE Mode", Electron. Lett., 1968, 4, p.569.

28.65. Fjallbrant, T.T.: "Non-Minimum-Phase Microwave Filters", Trans. IEEE, 1968, MTT-16, p.990.

28.66. Garault, Y., and Fonchy, R.: "Microwave Selective Filter Using a Resonant Cavity With One Coupling Element", C. R. Acad. Sci. (Paris), 1968, 267B, p.576.

28.67. Matthaei, G.L., and Leedom, D.A.: "Lowpass Quasi-Optical Filters for Oversized or Focused-Beam Waveguide Applications", Trans. IEEE, 1968, MTT-16, p.1038.

28.68. Rupke, H.D.: "Magnetically Tunable Microwave Filters with High Q-Factors", Z. Angew. Phys., 1968, 26, p.32.

28.69. Vollerner, I.F., et al.: "Design of a Set of Filters Using Delay Lines", Izv. VUZ Radioelektron., 1968, 11, p.1105.

28.70. Krul, L.: "Stripline Filters in Microwave Amplifiers", Ingenieur, 1969, 81, no.8, p.23.

28.71. Saha, A.L., and Barlow, H.E.M.: "Channel-Dropping Filter for Millimetre Waves in Circular Waveguide", Proc. IEE, 1969, 116, p.941.

28.72. Rhodes, J.D.: "Stepped Digital Elliptic Filter", Trans. IEEE, 1969, MTT-17, p.178.

28.73. Rhodes, J.D.: "Design and Synthesis of a Class of Microwave Bandpass Linear-Phase Filters", Trans. IEEE, 1969, MTT-17, p.189.

28.74. Gerdine, M.A.: "Frequency-Stabilized Microwave Bandreject Filter Using High-Permittivity Resonators", Trans. IEEE, 1969, MTT-17, p.354.

28.75. Harris, S.E., Nieh, S.T.K., and Winslow, D.K.: "Electronically Tunable Acoustooptic Filter", Appl. Phys. Lett., 1969, 15, p.325.

28.76. Mashkovtsev, B.M., and Tkachenko, K.A.: "Wave Method for Synthesizing Stripline Single-Loop Directional Filters", Elektrosvyaz, 1969, 23, no.6, p.15, and Telecommun. Radio Eng. Pt 1, 1969, 23, no.6.

28.77. Carlin, H.J., and Gupta, O.P.: "Computer Design of Filters With Lumped-Distributed Elements of Frequency-Variable Terminations", Trans. IEEE, 1969, MTT-17, p.598.

28.78. Tu, P.J.: "Computer-Aided Design of a Microwave Delay Equalizer", Trans. IEEE, 1969, MTT-17, p.626.

28.79. Kislyakovskii, A.V.: "Waveguide Bandpass Filters", Izv. VUZ Radioelektron., 1969, 12, no.6, p.627.

28.80. Weir, W.B., and Adams, D.K.: "Wideband Multiplexers Using Directional Filters", Microwaves, 1969, 8, no.5, p.94.

28.81. Braeckelmann, W.: "Bandstop Filter for H_{10} Mode in a Rectangular Waveguide", Electron. Lett., 1969, 5, p.500.

28.82. Gaglione, S., and Dydyk, M.: "External Q of Ellipsoidal Ferrimagnetic Resonator Coupled to Stripline", Electron. Lett., 1969, 5, p.640.

28.83. Honicke, H.: "Transmission Characteristics of Narrowband Microwave Bandpass Filters", Frequenz, 1969, 23, p.81.

28.84. Ohtomo, I., and Shimada, S.: "Channel Dropping Filter Using Ring Resonators for Millimetre-Wave Communication", Electron. Commun. Jap., 1969, 52, p.57.

28.85. Lewin, L.: "Equivalent Circuit of Evanescent-Mode Inductive Element", Electron. Lett., 1969, 5, p.415.

28.86. Stuber, F.M., and Davis, L.E.: "Resonance-Frequency Behaviour of a Cavity With Evanescent Section", Electron. Lett., 1969, 5, p.418.

28.87. Roy, S.C.D.: "Highpass Transmission-Line Directional Couplers", Trans. IEEE, 1969, MTT-17, p.440.

28.88. Davis, R.M.: "Three-Quarter-Wavelength Parallel-Staggered Microwave Filters", Trans. IEEE, 1969, MTT-17, p.404.

28.89. Suzuki, N., and Shimada, S.: "Cosine-Type Cutoff Filter for Millimetre Waves", Electron. Commun. Jap., 1969, 52, no.12, pp.116-24.

28.90. Bernardi, P.: "Tunable Absorbing Bandstop Filter by Field Rotation", Trans. IEEE, 1969, MTT-17, p.62.

28.91. Rupke, H.D.: "Magnetodynamic Modes in Ferrite Spheres for Microwave Filters", Trans. IEEE, 1969, MAG-5, p.481, and 1970, MAG-6, p.80.

28.92. Rhodes, J.D.: "Generalized Direct-Coupled Cavity Linear-Phase Filter", Trans. IEEE, 1970, MTT-18, p.308.

28.93. Gupta, O.P., and Wenzel, R.J.: "Design Tables for a Class of Optimum Microwave Bandstop Filters", Trans. IEEE, 1970, MTT-18, p.402.

28.94. Pregla, R.: "Microwave Filters of Coupled Lines and Lumped Capacitances", Trans. IEEE, 1970, MTT-18, p.278.

28.95. Archer, J.L., Bongianni, W.L., and Collins, J.H.: "Magnetically Tunable Microwave Bandstop Filters Using Epitaxial YIG Film Resonators", J. Appl. Phys., 1970, 41, p.1359.

28.96. Rhodes, J.D.: "Lowpass Prototype Network for Microwave Linear-Phase Filters", Trans. IEEE, 1970, MTT-18, p.290.

28.97. Varanasi, P.: "Filter for 3.39-micron He-Ne Laser", Appl. Opt., 1970, 9, p.2191.

28.98. Mishra, S.R., and Wadhwa, R.P.: "Development of X-Band Waveguide Frequency Discriminator", Trans. IEEE, 1970, MTT-18, p.660.

28.99. Rakovich, B.D., and Jovanovich, A.D.: "Transmission Factors of Microwave Filters With Prescribed Attenuation and Group Delay", Radio Electron. Eng., 1970, 40, no.3, p.121.

28.100. Halim, M.A., and Hamid, M.A.K.: "Fringing Capacitance in Thick-Strip Transmission-Line Filters", Electron. Lett., 1970, 6, p.413.

28.101. Wardrop, B.: "Stripline Microwave Group-Delay Equalizers", Marconi Rev., 1970, 33, no.177, p.150.

28.102. Vol'man, V.I., and Sarkis'yants, A.G.: "Improvement of Resonant Characteristics of Filters", Radiotekhnika, 1970, 25, no.5, p.104, and Telecommun. Radio Eng. Pt 2, 1970, 25, no.5.

28.103. Shimada, S., et al.: "Millimetre-Wave Directional Filter", Trans. IEEE, 1970, MTT-18, p.61.

28.104. Snyder, R.V., and Bozarth, D.L.: "Analysis and Design of a Microwave Transistor Active Filter", Trans. IEEE, 1970, MTT-18, p.2.

28.105. Halim, M.A., and Hamid, M.A.K.: "End Effect in Parallel-Coupled Transmission-Line Filters", Electron. Lett., 1970, 6, p.130.

28.106. Cuilwik, A.W., and Mulligan, J.H.: "Synthesis of a Class of Lossy Distributed Networks", Trans. IEEE, 1970, CT-17, p.19.

28.107. Anderson, N., and Arthurs, A.M.: "Bounds for Capacities in Microwave Filter Problems", Int. J. Electron., 1970,28,p.259.

28.108. Leedom, D.A., and Matthaei, G.L.: "Bandpass and Pseudo-Highpass Quasi-Optical Filters", Trans. IEEE, 1970, MTT-18, p.253.

28.109. Edson, W.A., and Wakabayashi, J.: "Input Manifolds for Microwave Channelizing Filters", Trans. IEEE, 1970, MTT-18, p.270.

28.110. Bogdanov, G.B., and Malyakin, A.K.: "Ferrite Microwave Filters Using a TWT", Radiotekh. Elektron., 1970, 15, no.2, p.405, and Radio Eng. Electron. Phys., 1970, 15, no.2, pp.354-7.

28.111. Fjerstad, R.L.: "Design Considerations and Realizations of Iris-Coupled YIG-Tuned Filters in the 12-40 GHz Region", Trans. IEEE, 1970, MTT-18, p.205.

28.112. Sugahara, H., Nara, T., and Yamada, K.: "Small-Sized Channel Dropping Filter for Microwaves", Rev. Electr. Commun. Lab., 1970, 18, no.7-8, pp.493-511.

28.113. Hamilton, C.H.: "Electronically Tuned Filter Using PIN Diodes", Electron. Lett., 1970, 6, no.22, pp.697-8.

28.114. Scanlan, S.O., and Pantzaris, T.P.: "Microwave Networks With Equiripple Delay Characteristics", Trans. IEEE, 1970, MTT-18, p.15.

28.115. Frigyes, I.: "Design of a Class of Parallel-Coupled Quasi-Stripline Filters by Conformal Mapping", Trans. IEEE, 1970, MTT-18, p.170.

28.116. Agafonov, V.M.: "Polynomial Microwave Filters", Radiotekh. Elektron., 1970, 15, no.10, and Radio Eng. Electron. Phys., 1970, 15, no.10, pp.1917-9.

28.117. Nunotani, T., and Nakagawa, I.: "Three-Guide Type of Channel-Dropping Filter for Millimetre Waves", Electron. Commun. Jap., 1970, 53, no.6, pp.65-72.

28.118. Dorsi, D., and Panigata, M.: "Thermal Compensation of Microwave Filters", Rend. Riun. Ass. Elettrotec. Ital., 1970, 45, no.2.03, pp.1-9.

28.119. Bertusi, F., and D'Oro, E.C.: "Branching Filter for High-Power Microwave Transmitters", Rend. Riun. Ass. Elettrotec. Ital., 1970, 45, no.2.04, pp.1-5.

28.120. Kurzl, A., Maurer, P., and Ottremba, K.: "Hybrid Microwave Circuits Realized by Quasi-Concentrated Components", Rend. Riun. Ass. Elettrotec. Ital., 1970, 45, no.2.06, pp.1-6.

28.121. Maltese, U., Gamberini, G., and Lupano, S.: "Microwave Matching Circuits and Filters Using Microstrip with Suspended Dielectric", Rend. Riun. Ass. Elettrotec. Ital., 1970, 45, no.2.07, pp.1-4.

28.122. Dorsi, D., and Giavarini, A.: "Considerations in the Design and Thermal Stability of Interdigital and Comb-Type Filters", Rend. Riun. Ass. Elettrotec. Ital., 1970, 45, no.2.08, pp.1-7.

28.123. Swain, G.R.: "Calculation of Transient Response of Long Chains of Slightly Lossy Coupled Resonators", Trans. IEEE, 1970, NS-17, no.4, pp.10-13.

28.124. Hagele, W.: "Microwave Filter With Non-Adjacent Coupling for Producing Attenuation Poles", Frequenz, 1970, 24, no.11, pp.340-1.

28.125. Kudsia, C.M.: "Synthesis of Optimum Reflection-Type Microwave Equalizers", RCA Rev., 1970, 31, no.3, pp.571-95.

28.126. Horna, J.: "Microwave Filters With Low Temperature Dependence", Tesla Electron., 1970, 3, no.4, pp.99-103.

28.127. Mariani, E.A., and Agrios, J.P.: "Slot-Line Filters and Couplers", Trans. IEEE, 1970, MTT-18, no.12, pp.1089-95.

28.128. Williams, A.E.: "Four-Cavity Elliptic Waveguide Filter", Trans. IEEE, 1970, MTT-18, no.12, pp.1109-14.

28.129. Wenzel, R.J.: "Small Elliptic-Function Lowpass Filters", Trans. IEEE, 1970, MTT-18, no.12, pp.1150-8.

28.130. Craven, G.F., and Mok, C.K.: "Design of Evanescent-Mode Waveguide Bandpass Filters for a Prescribed Insertion-Loss Characteristic", Trans. IEEE, 1971, MTT-19, no.3, pp.295-308.

28.131. Kurilin, B.I.: "Synthesis of Microwave Oscillatory Structures Using Sections of Nonuniform Transmission Line", Izv. VUZ Radioelektron., 1971, 14, no.4, pp.450-3.

28.132. Cristal, E.G.: "Design Equations for a Class of Microwave Filters", Trans. IEEE, 1971, MTT-19, no.5, pp.486-90.

28.133. Standley, R.D.: "Time-Delay Equalizer (TE_{01}-Circular) Using Directional Filter Cascades", Trans. IEEE, 1971, MTT-19, no.5, pp.497-8.

28.134. Halim, M.A., and Hamid, M.A.K.: "Fringing Capacitance in Printed Stripline Filters", Tijdschr. Ned. Elektron. Radiogenoot., 1971, 36, no.1, pp.8-14.

28.135. Ohtomo, I., Shimada, S., and Suzuki, N.: "Two-Cavity, Ring-Type, Channel-Dropping Filters for Millimetre Communication Guide", Rev. Electr. Commun. Lab., 1971, 19, no.1-2, pp.87-98, and Trans. IEEE, 1971, MTT-19, no.5, pp.481-4.

28.136. Kuhn, E., and Pregla, R.: "Commutating Microwave Channel-Separating Filters With Improved Transfer Characteristics", Arch. Elektron. Ubertrag., 1971, 25, no.1, pp.54-5.

28.137. Unrau, U.: "Periodic Band-Multiplexer for a TE_{01}-Guide Communication System", Arch. Elektron. Ubertrag., 1971, 25, no.1, pp.56-7.

28.138. Mayer, K.: "Synthesis of Optimum Multisection Quarter-Wave Transformers With Stubs", Arch. Elektron. Ubertrag., 1971, 25, no.2, pp.61-8.

28.139. Roberts, R.J.: "Effect of Tolerances on Performance of Microstrip Parallel-Coupled Bandpass Filters", Electron. Lett., 1971, 7, no.10, pp.255-7.

28.140. Iveland, T.D.: "Dielectric Resonator Filters for Application in MIC's", Trans. IEEE, 1971, MTT-19, no.7, pp.643-52.

28.141. Wenzel, R.J.: "Synthesis of Combine and Capacitively Loaded Interdigital Bandpass Filters of Arbitrary Bandwidth", Trans. IEEE, 1971, MTT-19, no.8, pp.678-86.

28.142. Iveland, T.D.: "Dielectric Bandpass Filters for Microstrip Circuits", Elektrotek. Tidsskr., 1971, 84, no.11, pp.30-34.

28.143. Scanlan, S.O., and Pantzaris, T.P.: "Class of Minimum-Phase Microwave Filters With Simultaneous Conditions on Amplitude and Decay", Trans. IEEE, 1971, MTT-19, no.9, pp.749-59.

28.144. Sobol, H.: "Microwave Hybrid With Impedance Transforming Properties", Trans. IEEE, 1971, MTT-19, no.9, pp.774-6.

28.145. Ravenscroft, I.A., and Kidd, G.P.: "Improved Commutating Channel Separator for Millimetric-Waveguide Systems", Electron. Lett., 1971, 7, no.16, pp.466-8.

28.146. Gulczynski, J.: "S-Band Comb-Line Filter", Prace PIT, 1971, 21, no.71, pp.49-53.

28.147. Snell, W.W.: "Low-Loss Microstrip Filters Developed by Frequency Scaling", Bell Syst. Tech. J., 1971, 50, no.6, pp.1919-31.

28.148. Hagele, W., and Hirsch, G.: "Drive System With Mechanical Tracking Compensation for Multistage Microwave Filters", Tech. Mitt. AEG-Telefunken, 1971, 61, no.6, pp.341-4.

28.149. Tae, P.K., Arc. L.J., and Chan, W.K.: "Magnetically Tunable Narrowband Stop and/or Pass Directional YIG Filter", J. Korean Inst. Electron. Eng., 1971, 8, no.4, pp.176-81.

28.150. Bachinina, Ye.L., et al.: "Microwave-Filter Loss and Miniaturization Problems", Radiotekhnika, 1971, 26, no.10, pp.46-52, and Telecommun. Radio Eng. Pt 2, 1971, 26, no.10, pp.87-92.

28.151. Baynham, A.C., and Dunsmore, M.R.B.: "High-Power Magnetically Tunable Microwave Filter", Electron. Lett., 1971, 7, no.4, pp.90-2.

28.152. Shelamov, G.N.: "Microwave Bandstop Filters Using Dielectric Resonators", Izv. VUZ Radioelectron., 1971, 14, no.12, pp.1460-4.

28.153. Mashkovtsev, B.M., and Lipatov, A.A.: "Synthesis of Microwave Filters from Nonresonant Sections With Concentrated Couplings", Radiotekh. Elektron., 1971, 16, no.3, and Radio Eng. Electron. Phys., 1971, 16, no.3, pp.417-22.

28.154. Atia, A.E., and Williams, A.E.: "Types of Waveguide Bandpass Filters for Satellite Transponders", COMSAT Tech. Rev., 1971, 1, no.1, pp.21-43.

28.155. Shakhov, V.V.: "Symmetrical Stripline for Shaping Powerful Microsecond Pulses", Prib. Tekh. Eksp., 1971, 14, no.1, pp.114-7, and Instrum. Exp. Tech., 1971, 14, no.1, pp.127-9.

28.156. Wanselow, R.D.: "Direct-Coupled Waveguide Resonator Equalizer Networks", J. Franklin Inst., 1971, 292, no.3, pp.179-92.

28.157. Kizlyakovskii, A.V., et al.: "Multi-Purpose Functional Waveguide (Ferrite) Device", Izv. VUZ Radioelektron., 1971, 14, no.10, pp.1130-6.

28.158. Kashyap, S.C., and Hamid, M.A.K.: "Frequency Response of Waveguide Filters With Thick Diaphragms", Int. J. Electron., 1972, 32, no.2, pp.169-80.

28.159. Van Dijk, M.H.H., Hagstrom, C.E., and Killberg, E.L.: "Properties of Stripline Filters on Dielectric Substrates", Electron. Lett., 1972, 8, no.3, pp.70-1.

28.160. Rhodes, J.D., and Ismail, M.Z.: "Cascade Synthesis of Selective Linear-Phase Filters", Trans. IEEE, 1972, CT-19, no.2, pp.183-9.

28.161. Jachimovits, L.: "Design and Measurement of a Single-Cavity Radial Filter and the Optimal Design of Multicavity Filters", Hiradastechnika, 1972, 23, no.1, pp.10-6.

28.162. Atia, A.E., and Williams, A.E.: "Narrow-Bandpass Waveguide Filters", Trans. IEEE, 1972, MTT-20, no.4, pp.258-65.

28.163. Irish, R.T.: "Annular Resonant Structures and Uses as Microwave Filters", Radio Electron. Eng., 1972, 42, no.2, pp.85-90.

28.164. Mok, C.K., Stopp, D.W., and Craven, G.: "Susceptance-Loaded Evanescent-Mode Waveguide Filters", Proc. IEE, 1972, 119, no.4, pp.416-20.

28.165. Hoffman, G.: "Design of Stripline Directional Filters with One Ring Resonator", Rev. HF, 1972, 8, no.9, pp.209-14.

28.166. Wolf, I.: "Microstrip Bandpass Filter Using Degenerate Modes of a Ring Resonator", Electron. Lett., 1972, 8, no.12, pp.302-3.

28.167. Masse, D.J., and Pucel, R.A.: "Temperature-Stable Bandpass Filter Using Dielectric Resonators", Proc. IEE, 1972, 119, no.6, pp.730-1.

28.168. Reeder, T.M.: "Broadband Coupling to High-Q Resonant Loads", Trans. IEEE, 1972, MTT-20, no.7, pp.453-8.

28.169. Kihlen, R.: "Stripline Triplexer for Use in Narrowband Multichannel Filters", Trans. IEEE, 1972, MTT-20, no.7, pp.486-8.

28.170. Hoffman, G., and Vanooteghem, H.: "Computer Programme for Optimum Synthesis of TEM Transmission-Line Filters", Trans. IEEE, 1972, MTT-20, no.10,pp.709-10.

28.171. Bergmann, R., and Bommas, G.: "Calculation Method for Minimum-Phase-Shift Microwave Filters Using Transmission-Line Elements", Nachr. Tech. Z., 1972, 25, no.8, pp.363-8.

28.172. Suzuki, N., Ohtomo, I., and Shimada, S.: "Branching Filters for 75-100 GHz Band", Electr. Commun. Lab. Tech. J., 1972, 21, no.3, pp.475-503, and Rev. Electr. Commun. Lab., 1972, 20, no.11-12,pp.1002-20.

28.173. Shimada, S., Ohtomo, I., and Suzuki, N.: "Millimetre-Wave Branching Filter System for 4-Phase PSK Transmission", Electr. Commun. Lab. Tech. J., 1972, 21, no.5, pp.873-900.

28.174. Ohtomo, I., Shimada, S., and Yamada, K.: "Three-Cavity Ring-Type Filter", Rev. Electr. Commun. Lab., 1972, 20, no.9-10, pp.837-46.

28.175. Nakagami, T., and Takenaka, S.: "Precision Design of Millimetre-Wave Bandpass Filter", Fujitsu Sci. Tech. J., 1972, 8, no.4, pp.91-104.

28.176. Bogdanov, G.B., et al.: "Experimental Characteristics of Semi-Open, Multi-Circuit, Microwave Ferrite Filters", Radiotekh. Elektron., 1972, 17, no.10, pp.2043-7, and Radio Eng. Electron. Phys., 1972, 17, no.10, pp.1627-30.

28.177. Bogdanov, G.B.: "Theory of Half-Open, Ferrite, Microwave Filters", Radiotekh. Elektron., 1972, 17, no.10, pp.2169-75, and Radio Eng. Electron. Phys., 1972, 17, no.10, pp.1731-7.

28.178. Ishida, N.: "4-GHz Equalizer Using a Meander Line", Electr. Commun. Lab. Tech. J., 1972, 21, no.6, pp.1001-23.

28.179. Fox, A.: "Temperature-Stable Low-Loss Microwave Filters Using Dielectric Resonators", Electron. Lett., 1972, 8, no.23, pp.582-3.

28.180. Ren, C.L.: "Design of a Channel Diplexer for Millimetre Waves", Trans. IEEE, 1972, MTT-20, no.12, pp.820-7.

28.181. Igarashi, M., and Naito, Y.: "Properties of a Four-Port Nonreciprocal Circuit Utilizing YIG on Stripline", Trans. IEEE, 1972, MTT-20, no.12, pp.828-33.

28.182. Rhodes, J.D.: "Waveguide Bandstop Elliptic-Function Filters", Trans. IEEE, 1972, MTT-20, no.11, pp.715-8.

28.183. Cristal, E.G., and Frankel, S.: "Hairpin-Line and Hybrid Hairpin-Line/ Half-Wave Parallel-Coupled-Line Filters", Trans. IEEE, 1972, MTT-20, no.11, pp.719-28.

28.184. Helszajn, J.: "Synthesis of Quarter-Wave-Coupled Circulators With Chebyshev Characteristics", Trans. IEEE, 1972, MTT-20, no.11, pp.764-9.

28.185. Park, K.S.: "Characteristics of Linearly Tapered Coupled Stripline Filters", J. Korean Inst. Electron. Eng., 1972, 9, no.2, pp.59-74.

28.186. Kireev, V.S., and Gudkov, K.G.: "Calculation of a Single-Circuit Ferrite Filter", Izv. VUZ Radiofiz., 1972, 15, no.10, pp.1567-71.

28.187. Lind, L.F., and Massar, R.E.: "Generalized Interdigital Helical Filter", Electron. Lett., 1972, 8, no.21, pp.525-6.

28.188. Levdikova, T.L., and Detinko, V.N.: "Multisection Circuits Using Below-Cutoff Waveguide", Izv. VUZ Radioelektron., 1972, 15, no.7, pp.899-905.

28.189. Craven, G.: "Slimguide Microwave Components", Electr. Commun., 1972, 47, no.4, pp.245-58.

28.190. Lebedov, V.K.: "Branch-Arm (Bandpass) Filter", Izv. VUZ Radioelektron., 1972, 15, no.10, pp.1296-8.

28.191. Blunden, D.F.: "Analysis of Corrugated Waveguide Filters", Marconi Rev., 1972, 35, no.187, pp.244-59.

28.192. Powell, I.: "Design and Manufacture of Delay Equalized Comb-Line Filters", Marconi Rev., 1972, 35, no.187, pp.272-94.

28.193. Ohtomo, I., Shimada, S., and Ohi, K.: "Ring-Type Channel-Dropping Filters for Millimetre Waveguide", Microwave J., 1972, 15, no.11, pp.35-40.

28.194. Ohtomo, I., Yamada, K., and Shimada, S.: "Channel-Dropping Filters for a 20-GHz PCM Radio Relay", Rev. Electr. Commun. Lab., 1972, 20, no.11-12, pp.1021-40.

28.195. Model', A.M., Stuzhin, V.A., and Grossman, V.B.: "Broadband Waveguide Filter Having Constant Input Impedance", Elektrosvyaz, 1972, 26, no.9, pp.13-5, and Telecommun. Radio Eng. Pt 1, 1972, 26, no.9, pp.11-3.

28.196. Prinsen, P.J.A.: "Class of Highpass Digital MTI Filters With Nonunifrom PRF", Proc. IEEE, 1973, 61, no.8, pp.1147-8.

28.197. Rudeshko, G.A.: "Dimensioning of Hybrid Integrated-Circuit Bandpass Filters at Microwaves", Izv. VUZ Radioelektron., 1973, 16, no.5, pp.20-4.

28.198. Vobian, J.: "Calculation and Construction of Filters in the Submillimetre Band", Fernmelde-Ing., 1973, 27, no.6, pp.1-32.

28.199. Levy, R.: "Tapered Corrugated Waveguide Lowpass Filters", Trans. IEEE, 1973, MTT-21, no.8, pp.526-32.

28.200. Helszajn, J.: "Frequency Response of Quarter-Wave-Coupled Reciprocal Stripline Junctions", Trans. IEEE, 1973, MTT-21, no.8, pp.533-7.

28.201. Kowalski, G.: "Delay Equalization by Tapered Meander Lines", Arch. Elektron. Ubertrag., 1973, 27, no.2, pp.65-9.

28.202. Easter, B., and Richings, J.G.: "Microstrip Bandpass Filters With Reduced Radiation Effects", Electron. Lett., 1973, 9, no.4, pp.93-4.

28.203. Moore, W.S.: "Design, Analysis, and Performance, of Resonant and Nonresonant Microwave Transmission Devices With Theoretically Infinite Rejection", Rev. Sci. Instrum., 1973, 44, no.2, pp.158-64.

28.204. Kimura, T., Saruwatari, M., and Otsuka, K.: "Birefringent Branching Filters for Wideband Optical FDM Communications", Appl. Opt., 1973, 12, no.2, pp.373-9.

28.205. Suzuki, N.: "Millimetre-Wave Broadband Michelson-Interferometer Bandsplitting Filters", Electron. Commun. Jap., 1973, 56, no.12, pp.51-8.

28.206. Steiner, K.H., Treheux, M., and Bouillie, R.: "Multimode Waveguide. Lowpass Filter", Appl. Opt., 1973, 12, no.11, pp.2732-5.

28.207. Matheau, J.G., and Seoane, A.R.: "Programming Synthesis of Waveguide Bandpass Filters", Electron. Fis. Apl., 1973, 16, no.2, pp.194-6.

28.208. Jackson, L.A., and Hadi, Z.M.A.: "Channel-Dropping Coupled-Line Filter", Electron. Lett., 1973, 9, no.19, pp.462-4.

28.209. Hung, P.C.: "Design of Microstrip Bandpass Filter", Hiradastechnika, 1973, 24, no.8, pp.240-6.

28.210. Konishi, Y., et al.: "Microwave Filter With Mounted Planar Circuit in Waveguide", NHK Lab. Note, 1973, no.162, pp.1-12.

28.211. Loele, H., and Hotho, K.: "Magnetically Tunable Filters With Ferrimagnetic Materials in Microstrip", Nachr. Tech. Elektron., 1973, 23, no.10, pp.382-6.

28.212. Bodonyi, J.: "Improvement of Channel Distribution in Millimetric Multiplexers by a Phase-Correcting Network", Marconi Rev., 1973, 36, no.191, pp.224-36.

28.213. Mashkovtsev, B.M., and Tkachenko, K.A.: "Frequency Characteristics of Stripline Loop Directional Filters", Radiotekhnika, 1973, 28, no.1, pp.37-41, and Telecommun. Radio Eng. Pt 2, 1973, 28, no.1, pp.85-8.

28.214. Krausse, N.: "Investigations of Waveguide Filters by a Computer Programme", Wiss. Ber. AEG-Telefunken, 1973, 46, no.2, pp.40-4.

28.215. Carli, E., Corzani, T., and Stracca, G.B.: "Design and Characteristics of Waveguide Interferential Branching Filters", Alta Freq., 1973, 42, no.10, pp.519-28.

28.216. Mok, C.K.: "Design of Evanescent-Mode Waveguide Diplexers", Trans. IEEE, 1973, MTT-21, no.1, pp.43-8.

28.217. Roschmann, P.: "Compact YIG Band-pass Filter with Finite-Pole Frequencies for Applications in Microwave Integrated Circuits", Trans. IEEE, 1973, MTT-21, no.1, pp.52-7.

28.218. Unrau, U.: "Bandsplitting Filters in Oversized Rectangular Waveguide", Electron. Lett., 1973, 9, no.2, pp.30-1.

28.219. Bergaud, G.: "Design of TE_{011}-Mode Filters for Satellite Output Multiplexers", Rev. Tech. Thomson-CSF, 1973, 5, no.3, pp.625-39.

28.220. Shimada, S., Ohtomo, I., and Suzuki, N.: "Experimental Channel Multiplexing and Demultiplexing Network for 4-phase PCM Millimetre Waveguide System", Rev. Electr. Commun. Lab., 1973, 21, no.7-8, pp.507-20.

28.221. Levitan, G.I.: "Narrowband Filters With Chebyshev Group-Delay Time Characteristic", Elektrosvyaz, 1973, 27, no.4, pp.65-9, and Telecommun. Radio Eng. Pt 1, 1973, 27, no.4, pp.54-8.

28.222. Osipenkov, V.M., Bachinina, Ye.L., and Fel'dshtein, A.L.: "Aspects of Lossy Microwave Filter Design", Radiotekhnika, 1973, 28, no.4, pp.25-30, and Telecommun. Radio Eng. Pt 2, 1973, 28, no.4, pp.70-3.

28.223. Saha, A.L., and Singh, R.K.: "Channel-Dropping Filter for Millimetre Waves in Circular Waveguide", Indian J. Pure Appl. Phys., 1973, 11, no.10, pp.750-4.

28.224. Schaefer, D.: "Investigations on Tunable Microwave Bandpass Filters With TE_{011}-Mode Resonators", Robotron. Tech. Commun., 1973, 15, no.10, pp.49-61.

28.225. Rodnikov, V.V., and Kravchenko, A.T.: "Microwave Bandpass Filters With Electronic Tuning", Radiotekhnika, 1973, 28, no.8, pp.34-8, and Telecommun. Radio Eng. Pt 2, 1973, 28, no.8, pp.86-9.

28.226. Iwakura, H., and Arakawa, T.: "Synthesis Method for Interdigital and Comb-Line Bandpass Filters Using Circular Inner Conductors", Rep. Univ. Electro-Commun., 1973, 24, no.1, pp.143-8.

28.227. Bessalov, A.V.: "Design of TEM Filters Using Distributed and Lumped-Constant Parameters", Izv. VUZ Radioelektron., 1973, 16, no.11, pp.120-3.

28.228. Carli, E., Corzani, T., and Stracca, G.B.: "Characteristics of an Interferential Branching Filter for a Millimetre Waveguide Communication System", Alta Freq., 1973, 42, no.10, pp.529-34.

28.229. Gruner, K.: "Method of Synthesizing Nonuniform Waveguides", Trans. IEEE, 1974, MTT-22, no.3, pp.317-22.

28.230. Rhodes, J.D., and Ismail, M.Z.: "In-Line Waveguide Selective Linear-Phase Filters", Trans. IEEE, 1974, MTT-22, no.1, pp.1-5.

28.231. Wanselow, R.D., and Taggart, D.A.: "Circularly Polarized Equalizer Networks", Trans. IEEE, 1974, MTT-22, no.1, pp.63-6.

28.232. Lee, Y.S.: "Mode Compensation Applied to Parallel-Coupled Microstrip Directional-Filter Design", Trans. IEEE, 1974, MTT-22, no.1, pp.66-9.

28.233. Rhodes, J.D.: "Waveguide Sandwich Filters", Trans. IEEE, 1974, MTT-22, no.4, pp.394-9.

28.234. Atia, A.E., and Williams, A.E.: "Nonminimum-Phase Optimum-Amplitude Bandpass Waveguide Filters", Trans. IEEE, 1974, MTT-22, no.4, pp.425-31.

28.235. Cristal, E.G., and Gysel, U.H.: "Compact Channel-Dropping Filter for Stripline and MIC's", Trans. IEEE, 1974, MTT-22, no.5, pp.499-504.

28.236. Gysel, U.H.: "Theory and Design for Hairpin-Line Filters", Trans. IEEE, 1974, MTT-22, no.5, pp.523-31.

28.237. Suzuki, N.: "4-120 GHz Michelson-Interferometer-Type Band-Splitting Filter", Trans. IEEE, 1974, MTT-22, no.5, pp.565-6.

28.238. Arnaud, J.A., Saleh, A.A.M., and Ruscio, J.T.: "Walk-Off Effects in Fabry-Perot (Submillimetre) Diplexers", Trans. IEEE, 1974, MTT-22, no.5, pp.486-93.

28.239. Yano, T., and Watanabe, A.: "Non-collinear Acoustooptic Tunable Filter Using Birefringence in p-Tellurite", Appl. Phys. Lett., 1974, 24, no.6, pp.256-8.

28.240. Flanders, D.C., et al.: "Grating Filters for Thin-Film Optical Waveguides", Appl. Phys. Lett., 1974, 24, no.4, pp.194-6.

28.241. Allen, J.L., and Barnes, W.J.: "Unequal Coupled Mode-Velocity C-Section Filters", Electron. Lett., 1974, 10, no.8, pp.128-9.

28.242. Il'chenko, M.Ye., and Mirskikh, G.A.: "Design Principles for High-Selectivity Cascade Filters", Izv. VUZ Radioelektron., 1974, 17, no.2, pp.126-8.

28.243. Jelenski, A., and Twarowski, J.: "Miniature Filter With (High-Permittivity) Dielectric Resonator", Nachr. Tech. Elektron., 1974, 24, no.2, pp.59-61.

28.244. Shimada, S., et al.: "Channel Multiplexing Network for Millimetre-Waveguide Transmission System", Trans. IEEE, 1974, COM-22, no.5, pp.714-21.

28.245. Zaki, K.A., and Newcomb, R.W.: "Scattering Matrix Synthesis Using Multiport Cavities", Int. J. Electron., 1974, 36, no.5, pp.607-21.

28.246. Craven, G.: "Waveguide Filters Using High-Impedance Mode Conversion/Reconversion", Electron. Lett., 1974, 10, no.11, pp.216-7.

28.247. Saleh, A.A.M.: "Adjustable Quasi-Optical Bandpass Filter. I and II", Trans. IEEE, 1974, MTT-22, no.7, pp.728-34 and 734-9.

28.248. Rao, B.R.: "Effect of Loss and Frequency Dispersion on Performance of Microstrip Directional Couplers and Coupled-Line Filters", Trans. IEEE, 1974, MTT-22, no.7, pp.747-50.

28.249. Kowalski, G.: "Coplanar Printed Meander Lines", Arch. Elektron. Ubertrag., 1974, 28, no.6, pp.257-62.

28.250. Bell, H.C., and Sugundo, El.: "Canonical Lowpass Prototype Network for Symmetric Coupled-Resonator Bandpass Filters", Electron. Lett., 1974, 10, no.13, pp.265-6.

28.251. Ishida, N., and Yokoyama, H.: "1.7-GHz Band-Delay Equalizer for Millimetre-Wave Communication System", Electr. Commun. Lab. Tech. J., 1974, 23, no.1, pp.137-49.

28.252. Koyama, K., and Shimada, S.: "Multiplexer Operating in 4-26 GHz Bands for Experimental Earth Station", Rev. Electr. Commun. Lab., 1974, 22, no.5-6, pp.513-21.

28.253. Schmidt, R.V., et al.: "Narrowband Grating Filters for Thin-Film Optical Waveguides", Appl. Phys. Lett., 1974, 25, no.11, pp.651-2.

28.254. Vincze, A.D.: "Practical Design Approach to Microstrip Comb-Line Filters", Trans. IEEE, 1974, MTT-22, no.12, pp.1171-81.

28.255. Allen, J.L.: "Inhomogeneous Coupled-Line Filters With Large Mode-Velocity Ratios", Trans. IEEE, 1974, MTT-22, no.12, pp.1182-6.

28.256. Ren, C.L., and Wang, H.C.: "Class of Waveguide Filters for Over-Moded Applications", Trans. IEEE, 1974, MTT-22, no.12, pp.1202-9.

28.257. Fleischmann, U.: "Calculation and Construction of Bandpass Filters in Stripline. I", Funk-Tech., 1974, no.2, pp.51-4.

28.258. Konishi, Y., and Uenakada, K.: "Design of a Bandpass Filter with Inductive Strip. Planar Circuit Mounted in Waveguide", Trans. IEEE, 1974, MTT-22, no.10, pp.869-73.

28.259. Ohm, E.A.: "Lowloss Branching Filter for Broad, Widely Spaced, Bands", Trans. IEEE, 1974, MTT-22, no.10, pp.891-4.

28.260. Ishida, N.: "4-GHz-Band Delay Equalizer Using a Meander Line", Rev. Electr. Commun. Lab., 1974, 22, no.1-2, pp.101-13.

28.261. Pegov, A.A., Zav'yalov, A.S., and Lavrov, V.I.: "Bandpass Filters Based on Laminated Structures", Izv. VUZ Fiz., 1974, no.8, pp.36-9.

28.262. Suzuki, N., and Shimada, S.: "Semicircular, 43-87 GHz, Waveguide Band-splitting Filter", Rev. Electr. Commun. Lab., 1974, 22, no.7-8, pp.728-40.

28.263. Dupuis, P.A., and Cristal, E.G.: "Folded-Line and Hybrid-Folded-Line Bandstop Filters", Trans. IEEE, 1974, MTT-22, no.12, pp.1312-6.

28.264. Irish, R.T.: "Narrowband Microwave Filter Using Exponential Lines", Electron. Lett., 1974, 10, no.15, pp.302-4.

28.265. Tajima, Y., and Sawayama, Y.: "Design and Analysis of a Waveguide-Sandwich Microwave Filter", Trans. IEEE, 1974, MTT-22, no.9, pp.839-41.

28.266. Bodonyi, J.: "Development of Channelling Filters for Millimetre-Wave Multiplexing System", Marconi Rev., 1974, 37, no.193, pp.69-72.

28.267. Garault, Y., Jarry, P., and Clapeau, M.: "Microwave Filters With Flat Time Delay Both in Passband and Stopband", Trans. IEEE, 1975, CAS-22, no.5, pp.423-7.

28.268. Matsuhara, M., and Hill, K.O.: "Optical Waveguide Bandreject Filters", Appl. Opt., 1974, 13, no.12, pp.2886-8.

28.269. Snyder, R.V.: "Realization of Dual-Mode Filter", Microwave J., 1974, 17, no.12, pp.31-3.

28.270. Suzuki, N.: "Rectangular-Waveguide-Type Diplexers for Millimetre Waves", Electr. Commun. Lab. Tech. J., 1974, 23, no.12, pp.2837-54.

28.271. Lynes, G.D., et al.: "Design of Broadband 4-Bit Loaded Switched-Line Phase-Shifter", Trans. IEEE, 1974, MTT-22, no.6, pp.693-7.

28.272. Jacobs, H., and Chrepta, M.M.: "Electronic Phaseshifter for Millimetre-Wave Semiconductor Dielectric Integrated Circuits", Trans. IEEE, 1974, MTT-22, no.4, pp.411-7.

28.273. Hord, W.E., and Rosenbaum, F.J.: "Coupled-Mode Analysis of Longitudinally Magnetized Ferrite Phaseshifters", Trans. IEEE, 1974, MTT-22, no.2, pp.135-8.

28.274. Fong, T.T., and Lee, S.W.: "Modal Analysis of a Planar Dielectric Strip Waveguide for Millimetre-Wave Integrated Circuits", Trans. IEEE, 1974, MTT-22, no.8, pp.776-83.

29. IMAGE-FORMING ANTENNAS

29.1. Davies, D.E.N., and McCartney, B.S.: "Cylindrical Arrays With Electronic Beam Scanning", Proc. IEE, 1965, 112, p.497.

29.2. Tanaka, S.: "Scanning Antennas", J. Inst. Electr. Commun. Eng. Jap., 1965, 48, p.592.

29.3. Fujimoto, K.: "Active Antennas. Tunnel-Diode Loaded Dipoles", Proc. IEEE, 1965, 53, p.174.

29.4. Pratt, H.J., and Da Macogno, N.G.: "Parametric Upconverter Receiving Array", Microwave J., 1965, 8, no.12, p.35.

29.5. Thiele, G.A., and Rudduck, R.C.: "Geodesic Lens Antenna for Low-Angle Radiation", Trans. IEEE, 1965, AP-13, p.514.

29.6. Klement'ev, Y.M.: "Directional Radiation from a Fabry-Perot Resonator", Radiotekh. Elektron., 1965, 10, p.367, and Radio Eng. Electron. Phys., 1965,10,p.310.

29.7. Feix, G.: "Significance of Geometric Image Defects for Beam Rotation With Parabolic Antennas", Optek, 1965,22,p.507.

29.8. Rupp, W.E.: "Impedance Variation in Scanning Arrays", Microwave J., 1965, 8, no.12, p.52.

29.9. Knittel, G.H.: "Choosing the Number of Faces of a Phased-Array Antenna for Hemisphere Scan Coverage", Trans. IEEE, 1965, AP-13, p.878.

29.10. Deryugin, L.N., and Kuznetsov, M.G.: "Frequency Sensitivity of Antenna Arrays at Various Angles and Relation to the Feed Waveguide", Radiotekh. Elektron., 1965, 10, p.2119, and Radio Eng. Electron. Phys., 1965, 10, p.1809.

29.11. Whicker, L.R., and Jones, R.R.: "Digital Current-Controlled Latching Ferrite Phaseshifter", Int. Conv. Rec. IEEE, 1965, 13, pt 5, p.217.

29.12. Chien, T.M., and Unz, H.: "Radiation from Rectangular Waveguide With Ferrite Slabs", Trans. IEEE, 1965,MTT-13,p.137.

29.13. Shubarin, Yu.V., and Magda, A.N.: "Slot Radiator With Ferrite-Controlled Polarization", Izv. VUZ Radiotekh., 1965, 8, p.119, and Sov. Radio Eng. 1965,8,p.87.

29.14. Delany, W.D., and Kyle, R.F.: "Antenna for Rapid-Scan Decorrelation Radar", Radio Electron. Eng., 1966, 32, p.156.

29.15. Nishimura, S., et al.: "Frequency-Scanned Antenna Using Dielectric-Loaded Waveguide", J. Inst. Electr. Commun. Eng. Jap., 1966, 48, p.742.

29.16. Magill, E.G., and Wheeler, H.A.: "Wide-Angle Impedance Matching of a Planar Array Antenna by a Dielectric Sheet", Trans. IEEE, 1966, AP-14, p.49.

29.17. Ruben, R.W.: "Electronically Scanned Antenna Using Fresnel-Zone Techniques", Int. Conv. Rec. IEEE, 1966, 14, pt 5, p.244.

29.18. Russo, V., and Scheggi, A.M.: "Scanning Performance of a Multifrequency Diffraction Reflector", Alta Freq., 1966, 35, p.379.

29.19. Cooper, D.C., and Said, R.A.K.: "Phasing for a Beam-Steering Array by Use of Optical Techniques", Electron. Lett., 1966, 2, no.10, pp.377-8.

29.20. Assaly, R.N., and Ricardi, L.J.: "Theoretical Study of a Multi-Element Scanning Feed System for a Parabolic Cylinder", Trans. IEEE, 1966, AP-14, no.5, pp.601-5.

29.21. Aronov, F.A.: "Phasing a Multi-element Matrix Antenna Using Discrete Shifters", Radiotekh. Elektron., 1966, 11, no.7, pp.1181-8, and Radio Eng. Electron. Phys., 1966, 11, no.7, pp.1035-40.

29.22. Hannan, P.W.: "Proof that a Phased-Array Antenna is Impedance Matchable for All Scan Angles", Radio Sci., 1967, 2, no.3, pp.361-9.

29.23. Davies, D.E.N.: "Independent Angular Steering of Each Zero of the Directional Pattern for a Linear Array", Trans. IEEE, 1967, AP-15, no.2, pp.296-8.

29.24. Patton, W.T.: "Determinants of Electronically Steerable Antenna Arrays", RCA Rev., 1967, 28, no.1, pp.3-37.

29.25. Fradin, A.Z.: "Parameters of Beam-Scanning Antennas", Radiotekhnika, 1967, 22, no.1, pp.23-6, and Telecommun. Radio Eng. Pt 2, 1967, 22, no.1, pp.82-5.

29.26. Kmetzo, J.L.: "Analytical Approach to the Coverage of a Hemisphere by N Planar Phased Arrays", Trans. IEEE, 1967, AP-15, no.3, pp.367-71.

29.27. Payne, J.B.: "Wideband IF-Time-Delay Steering of Array Antennas", Trans. IEEE, 1967, AP-15, no.3, pp.474-5.

29.28. Katsenelenbaum, B.Z., and Semenov, V.V.: "Synthesis of Phase Correctors Forming a Given Field", Radiotekh. Elektron., 1967, 12, no.2, pp.244-52, and Radio Eng. Electron. Phys., 1967, 12, no.2.

29.29. Grineva, K.I., and Gostyukhin, V.L.: "Sectionalized Beam-Scanning Antennas", Radiotekhnika, 1967, 22, no.5, pp.27-34, and Telecommun. Radio Eng. Pt 2, 1967, 22, no.5.

29.30. Kaidanovskii, N.L.: "Control of Directivity Pattern of an Array Using the Phase Difference Between Local Oscillator and Signal", Radiotekh. Elektron., 1967, 12, no.3, pp.515-6, and Radio Eng. Electron. Phys., 1967, 12, no.3.

29.31. Magarshack, J.: "Gunn Oscillator Used as a Phased-Array Aerial Element", Electron. Lett., 1967, 3, no.12, pp.556-7.

29.32. Goebels, F.J., Foreman, B.J., and Nonemaker, C.H.: "Electronic Scanning of Linear Slot Arrays", Trans. IEEE, 1968, AP-16, no.1, pp.8-14.

29.33. Amitay, N., Butzien, P.E., and Heidt, R.C.: "Match Optimization of a Two-Port Phased Array Antenna Element", Trans. IEEE, 1968, AP-16, no.1, pp.47-57.

29.34. Kelly, A.J.: "Electronically Variable Time-Delay Network for Broadband Phased-Array Steering", Trans. IEEE, 1968, AES-4, p.837.

29.35. Winter, C.F.: "Phase-Scanning Experiments With Two-Reflector Antenna Systems", Proc. IEEE, 1968, 56, p.1984.

29.36. Tanaka, S., et al.: "Ceramic Rod Array Scanned with Ferrite Phaseshifters", Proc. IEEE, 1968, 56, p.2000.

29.37. Louapre, M.E., and Fujioka, J.K.: "High-Efficiency, Electronically Scanned, K-Band Phased Array for Spaceborne Radiometric Applications", Proc. IEEE, 1968, 56, p.2010.

29.38. White, J.F.: "Review of Semiconductor Microwave Phaseshifters", Proc. IEEE, 1968, 56, p.1924.

29.39. Schell, A.C., et al.: "Electronic Scanning", Electro-Technol., 1968, 82, no.5, p.29.

29.40. Pattan, B.: "K-Band, Mixer-Steering, Electronic Scanning Phased-Array System", Proc. IEEE, 1968, 56, p.1976.

29.41. Stark, L., et al.: "Microwave Components for Wideband Phased Arrays", Proc. IEEE, 1968, 56, p.1908.

29.42. Staiman, D., Breese, M., and Patton, W.T.: "Technique for Combining Solid-State Sources", J. Solid-State Circuits IEEE, 1968, SC-3, p.238.

29.43. Kinber, B.Ye., Safonova, S.S., and Tseitlin, V.B.: "Methods for Calculating Collimating Lenses", Radiotekh. Elektron., 1968, 13, no.12, p.2162, and Radio Eng. Electron. Phys., 1968, 13, no.12,pp.1897-904.

29.44. Glynn, T.: "Beam Steering of a Phased Array by Conjugate Element Phase Shifting", Trans. IEEE, 1968, AP-16, p.597.

29.45. Ardab'evskii, A.I., et al.: "Principle of Construction of Frequency-Independent Antennas with Electrical Scanning", Izv. VUZ Radioelektron., 1968, 11, p.465.

29.46. Warner, J.: "Spatial-Resolution Measurements in Upconversion from 10.6 micron to the Visible Region", Appl. Phys. Lett., 1968, 13, p.360.

29.47. Takeshima, T.: "Beam Scanning of Parabolic Antenna by Defocusing", Electron. Eng., 1969, 41, no.491, p.70.

29.48. Albaugh, N., and Wesseling, K.H.: "Novel Way of Beam Switching, Particularly Suitable at Millimetre Wavelengths", Trans. IEEE, 1969, AP-17, p.98.

29.49. McMahon, D.H.: "Relative Efficiency of Optical Bragg Diffraction as a Function of Interaction Geometry", Trans. IEEE, 1969, SU-6, p.41.

29.50. Arkhipov, V.K., et al.: "Deflection System for a Beam of Light Based on Kerr Effect", Radiotekh. Elektron., 1969, 14, no.12, p.2278, and Radio Eng. Electron. Phys., 1969, 14, no.12, pp.1974-7.

29.51. Deryugin, I.A., Kotov, V.V., and Oboznenko, Yu.L.: "Two-Dimensional Scanning of Optical Beams Using Acoustic Waves", Radiotekh. Elektron., 1969, 14, no.12, p.2185, and Radio Eng. Electron. Phys., 1969, 14, no.12, p.1891-4.

29.52. Kazel, S.: "High-Sensitivity Real-Time Microwave Holography and Imaging", Proc. IEEE, 1969, 57, p.1222.

29.53. Feix, G.: "Focus Broadening by Astigmatism of Large Microwave Parabolic Antennas", Appl. Opt., 1969, 8, p.1631.

29.54. Uslenghi, P.L.E.: "Generalized Luneberg Lenses", Trans. IEEE, 1969, AP-17, p.644.

29.55. Shelton, J.P.: "Reduced Sidelobes for Butler-Matrix-Fed Linear Arrays", Trans. IEEE, 1969, AP-17, p.645.

29.56. Wellman, W.J., and Shapiro, S.S.: "Beam Pointing Direction of TW Arrays" Microwaves, 1969, 8, no.6, p.76.

29.57. Clarricoats, P.J.B., and Lim, S.H.: "Proposed High-Efficiency Spherical-Reflector Antenna", Electron. Lett., 1969,5,p.709.

29.58. Torguet, R., and Dieulesaint, E.: "Continuous Deflection of Light With Longitudinal Acoustic Waves", Electron. Lett., 1969, 5, p.632.

29.59. Warner, J.: "Phase Matching for Optical Upconversion With Maximum Angular Aperture", Opto-Electron., 1969, 1, p.25.

29.60. Epstein, J., and Woodward, O.M.: "Step-Scanned Ring Array", RCA Rev., 1969, 30, p.114.

29.61. Andrews, R.A.: "Wide Angular Aperture Image Upconversion", J. Quantum Electron. IEEE, 1969, QE-5, p.548.

29.62. Singlar, O.P., et al.: "Electro-optic Deflection at 1 GHz", J. Quantum Electron. IEEE, 1969, QE-5, p.521.

29.63. Salmon, M., and Trigon, R.: "Array Optimized to Scanning Angle", Onde Electr., 1969, 49, p.209.

29.64. Golovanevskii, E.I., et al.: "Wollaston-Prism Device for Discrete Deflection of Laser Beam", Opt. Spekt., 1969, 26, no.2, p.289, and Opt. Spectrosc., 1969, 26, no.2, p.155.

29.65. Rudge, A.W., and Withers, M.J.: "Beam-Scanning Primary Feed for Parabolic Reflectors", Electron. Lett., 1969, 5, p.39.

29.66. Daniele, V., Uslenghi, P.L.E., and Zich, R.: "Note on the Generalized Luneberg Lens", Alta Freq., 1970, 39, p.128e.

29.67. Hunt, R.P., Magyary, K., and Dickey, B.C.: "Optical Scanning for a Magnetooptic Memory", J. Appl. Phys., 1970, 41, p.1399.

29.68. Goto, N., Sugie, M., and Fukumoto, K.: "Radiation Pattern of a Scanning Antenna Using Digital Phaseshifters", NEC Res. Dev., 1970, no.16, p.73.

29.69. Rumsey, V.H.: "Design and Performance of Feeds for Correcting Spherical Aberration", Trans. IEEE, 1970, AP-18, p.343.

29.70. Phillips, C.J.E., and Clarricoats, P.J.B.: "Optimum Design of a Gregorian Corrected Spherical-Reflector Antenna", Proc. IEE, 1970, 117, p.718.

29.71. Hoff, F.: "Acoustooptical Method of Scanning a Coherent-Light Beam", Slab. Obz., 1970, 31, no.1, p.25.

29.72. Malissin, R.: "Multibeam Millimetre Antenna", Rev. Tech. Thomson-CSF, 1970, 2, no.2, pp.217-28.

29.73. Okada, M., and Ieiri, S.: "Electrooptic Light Scanner", Jap. J. Appl. Phys., 1970, 9, p.153.

29.74. Uchida, N., and Ohmachi, Y.: "Acoustooptic Light Deflector Using TeO_2 Single Crystal", Jap. J. Appl. Phys., 1970, 9, no.1, p.155.

29.75. Yamada, Y., and Nomura, S.: "Laser Display", J. Inst. Electron. Commun. Eng. Jap., 1970, 53, no.4, p.472.

29.76. Ancona, C.: "Focusing Nonparabolic Mirrors by Using Phase-Corrected Multibeam Sources", Radio Sci., 1970, 5, no.4, p.707.

29.77. Hockham, G.A.: "Investigation of Modulated Corrugated Antenna for Frequency Scanning", Proc. IEE, 1970, 117, p.1729.

29.78. Goto, N., and Cheng, D.K.: "Sidelobe-Reduction Techniques for Phased Arrays Using Digital Phaseshifters", Trans. IEEE, 1970, AP-18, no.6, pp.769-73.

29.79. Inagaki, N.: "Relationship Between Element Density and Limit of Scan Angle in Planar Antennas", Bull. Nagoya Inst. Technol., 1970, no.22, pp.261-6.

29.80. Boyns, J.E., et al.: "Step-Scanned Circular-Array Antenna", Trans. IEEE, 1970, AP-18, no.5, pp.590-5.

29.81. Nagai, N.: "Electronic Scanning Antennas", J. Inst. Electr. Commun. Eng. Jap., 1970, 53, no.8, pp.1091-9.

29.82. Weller, J.F., and Andrews, R.A.: "Resolution Measurements in Parametric Upconversion of Images", Opto-Electron., 1970, 2, no.3, pp.171-6.

29.83. Katyl, R.H.: "Extended Range Surface-Wave Scannable Laser", IBM Tech. Disclosure Bull., 1970, 13, no.7, pp.1825-8.

29.84. Goto, N., Sugie, M., and Fukumoto, K.: "Sidelobe Reduction of Phased Array Using Digital Phaseshifters", Electron. Commun. Jap., 1970, 53, no.7, pp.10-1.

29.85. Tang, R., Burns, R.W., and Wong, N.S.: "Phased Array Antenna for Airborne Application", Microwave J., 1971, 14, no.1, pp.31-40.

29.86. Panicali, A.R., Lo, Y.T., and Deschamps, G.A.: "Reflector Antenna Corrected for Spherical, Coma, and Chromatic, Aberrations", Proc. IEEE, 1971, 59, no.2, pp.311-2.

29.87. Okada, M., and Ieiri, S.: "Synthesis of an Electrooptic Light Scanner With Arbitrary Beam Shape", Appl. Opt., 1971, 10, no.4, pp.845-57.

29.88. Farhat, N.H.: "Constant Doppler Wide-Angle Laser Beam Scanning", Appl. Opt., 1971, 10, no.5, pp.1180-1.

29.89. Hammer, J.M.: "Digital Electrooptic Grating Deflector and Modulator", Appl. Phys. Lett., 1971, 18, no.4, pp.147-9.

29.90. Hansch, T.W., Varsanyi, F., and Schawlow, A.L.: "Image Amplification by Dye Lasers", Appl. Phys. Lett., 1971, 18, no.4, pp.108-10.

29.91. Mikaelyan, A.L., et al.: "Discrete System of Beam Deflection Using Crystals of $LiNbO_3$", Radiotekh. Elektron., 1971, 15, no.8, and Radio Eng. Electron. Phys., 1971, 15, no.8, pp.1528-31.

29.92. Vaccaro, F.E., et al.: "Product Design of a High-Power S-Band Integrated Module for Phased Arrays", Trans. IEEE, 1971, MTT-19, no.7, pp.609-16.

29.93. Wasse, M.P.: "Array of Pulsed X-Band Microstrip Gunn-Diode Transmitters With Temperature Stabilization", Trans. IEEE, 1971, MTT-19, no.7, pp.616-22.

29.94. Latta, M.R.: "Laser Raster Scanner", IBM Tech. Disclosure Bull., 1971, 13, no.12, pp.3879-80.

29.95. Brenner, H., and Kraus, H.: "Computer-Aided Design for an S-Band Phaseshifter Using PIN Diodes for Antenna Arrays", Frequenz, 1971, 25, no.5, pp.138-45.

29.96. Rudge, A.W., and Withers, M.J.: "Technique for Beam Steering With Fixed-Parabolic Reflectors", Pric. IEE, 1971, 118, no.7, pp.857-63.

29.97.　Beasley, J.D.: "Electrooptic Laser Scanner for TV Projection Display", Appl. Opt., 1971, 10, no.8, pp.1934-6.

29.98.　Kikuchi, Y., Chubachi, N., and Yasuda, S.: "Experiments on Bragg Diffraction of He-Ne Laser Beams by Ultrasonic Waves", Rec. Electr. Commun. Eng. Conversaz. Tohoku Univ., 1971, 40, no.1, pp.60-4.

29.99.　Ekinge, R., and Josefsson, L.: "Electronically Steered Antenna Arrays", Elteknik, 1971, 14, no.7-8, pp.20-4.

29.100.　Joachim, R.J., and Pitasi, M.J.: "IR and Thermal Evaluation of (Active) Phased-Array Antenna", Mater. Eval., 1971, 29, no.9, pp.193-8.

29.101.　Okada, M., and Ieiri, S.: "Electrooptic Light Scanner", NHK Tech. J., 1971, 23, no.4, pp.8-24.

29.102.　Mikaelyan, A.L., Koblova, M.M., and Zasovin, E.A.: "Investigation of a System for Beam Deflection With $LiNbO_3$ Crystals", Kvantovaya Elektron, 1971, no.1, pp.120-4.

29.103.　Beiser, L.: "Intracavity Scanning of a CO_2 Laser by Electron-Beam-Trigger Excitation", Appl. Phys. Lett., 1971, 19, no.8, pp.251-3.

29.104.　Poulton, G.: "Efficiency of a Stepped Reflector Fed from Off-Axis Source", Electron. Lett., 1971, 7, no.22, pp.666-7.

29.105.　Poulton, G.: "Image-Region Fields for a Stepped Reflector", Electron. Lett., 1971, 7, no.27, pp.650-1.

29.106.　Uchida, N.: "Laser Beam Deflectors and Scanners", Solid-State Phys., 1971, 6, no.10, pp.28-36.

29.107.　Balakshii, V.I., et al.: "Acoustic Scanning of Light in Anisotropic Medium", Radiotekh. Elektron., 1971, 16, no.11, pp.2226-30, and Radio Eng. Electron. Phys., 1971, 16, no.11, pp.1973-6.

29.108.　Packard, J.R., Tait, W.C., and Dierssen, G.H.: "Two-Dimensionally Scannable Electron-Beam-Pumped Laser", Appl. Phys. Lett., 1971, 19, no.9, pp.338-40.

29.109.　Dickson, L.D.: "Optical Considerations for an Acoustooptic Deflector", Appl. Opt., 1972, 11, no.10, pp.2196-202.

29.110.　Dreyfus, R.W., and Hodgson, R.T.: "Electron-Beam-Pumped Solid-State Scan Laser", IBM Tech. Disclosure Bull., 1972, 15, no.2, pp.527-8.

29.111.　Vul', V.A.: "Noise Characteristics of a Digital System of Deflection of a Beam of Light", Izv. VUZ Prib., 1972, 15, no.7, pp.105-10.

29.112.　Tada, K., et al.: "Electrooptic Light-Beam Deflection With $(SrBa)Nb_2O_6$ Prism", Jap. J. Appl. Phys., 1972, 11, no.11, pp.1622-7.

29.113.　Gfeller, F.R., and Pitt, C.W.: "Collinear Acoustooptic Deflection in Thin Films", Electron. Lett., 1972, 8, no.22, pp.549-51.

29.114.　Warner, A.W., White, D.L., and Bonner, W.A.: "Acoustooptic Light Deflectors Using Optical Activity in Tellurite", J. Appl. Phys., 1972, 43, no.11, pp.4489-95.

29.115.　Amitay, N., and Zucker, H.: "Compensation of Spherical-Reflector Aberrations by Planar-Array Feeds", Trans. IEEE, 1972, AP-20, no.1, pp.49-56.

29.116.　Meyer, R.A.: "Optical Beam Steering Using a Multichannel $LiTaO_3$ Crystal", Appl. Opt., 1972, 11, no.3, pp.613-6.

29.117.　Gunderson, L.C.: "Analysis of a Cylindrical Homogeneous Lens", Trans. IEEE, 1972, AP-20, no.4, pp.476-9.

29.118.　Ore, F.R.: "Millimetre-Wave Antenna With Omnidirectional or Directional-Scannable Azimuth Pattern and a Directional Vertical Pattern", Trans. IEEE, 1972, AP-20, no.4, pp.481-2.

29.119.　Hilburn, J.L., et al.: "Frequency-Scanned X-Band Waveguide Array", Trans. IEEE, 1972, AP-20, no.4, pp.506-9.

29.120.　Catuneanu, V.M., Sterian, E.P., and Vlad, V.I.: "Acoustooptical System of Laser Beam Deflection", Posta Telecommun., 1972, 2, no.5, pp.217-20.

29.121.　Torguet, R., Bauza, J.M., and Carles, C.: "Applications of Light Deflection by Elastic Waves", J. Phys. (Paris), 1972, 33, no.11-12, sup.C6, pp.235-8.

29.122.　Akimov, Yu.A., et al.: "Electron-Beam Tube With Semiconductor Target. Scanning Laser", Kvantovaya Elektron., 1972, no.3, pp.110-2, and Sov. J. Quantum Electron., 1972, 2, no.3, pp.297-9.

29.123.　Arkhipov, V.K., Ershov, E.I., and Tarasov, R.P.: "Electrooptic Deflection of a Laser Beam With Multiple Utilization of the Deflecting Element", Radiotekh. Elektron., 1972, 17, no.1, pp.94-102, and Radio Eng. Electron. Phys., 1972, 17, no.1, pp.74-81.

29.124.　Baues, P.: "Beam Deflection from the Interior of Optical Resonators", Opt. Commun., 1972, 6, no.3, pp.226-9.

29.125.　Zverev, V.A., and Shagal, A.M.: "Analysis and Calculation of Focusing Lenses With Aspherical Surface", Opt.-Mekh. Prom., 1972, 39, no.5, pp.14-6, and Sov. J. Opt. Technol., 1972, 39, no.5, pp.260-2.

29.126.　Hess, H.: "Concentric Double-Spherical-Reflector System as a Simple Scanning Antenna With Limited Range of Angle", Tech. Z., 1973, 26, no.2, pp.54-6.

29.127.　Cogdell, J.R., and Davis, J.H.: "Astigmatism in Reflector Antennas", Trans. IEEE, 1973, AP-21, no.4, pp.565-7.

29.128. Smith, R.G.: "Use of Acoustooptic Light Deflector as an Optical Isolator", J. Quantum Electron. IEEE, 1973, QE-9, no.5, pp.545-6.

29.129. Anderson, A.P., and Dawoud, M.M.: "Performance of Transistor-Fed Monopoles in Active Antennas", Trans. IEEE, 1973, AP-21, no.3, pp.371-4.

29.130. Schmidt, U.J., Schroder, E., and Thust, W.: "Optimization Procedures for Digital Light-Beam Deflectors", Appl. Opt., 1973, 12, no.3, pp.460-6.

29.131. Ninomiya, Y.: "Ultrahigh Resolving Electrooptic Prism-Array Light Deflectors", J. Quantum Electron. IEEE, 1973, QE-9, no.8, pp.791-5.

29.132. Mailloux, R.J., and Forbes, G.R.: "Array Technique with Grating-Lobe Suppression for Limited-Scan Applications", Trans. IEEE, 1973, AP-21, no.5, pp.597-602.

29.133. Brandt, G.B., Gottlieb, M., and Conroy, J.J.: "Bulk-Acoustic-Wave Interaction With Guided Optical Waves", Appl. Phys. Lett., 1973, 23, no.2, pp.53-4.

29.134. Cohen, R.W., and Gorog, I.: "Frequency Response of Laser Scanners and its Optimization Through Apodization", J. Opt. Soc. Am., 1973, 63, no.9, pp.1071-9.

29.135. Shih, H., et al.: "Interacting (Scanning) Laser Beams in a Resonant Medium", Appl. Opt. 1973, 12, no.9, pp.2198-202.

29.136. Ishimaru, A., Sreenivasiah, I., and Wong, V.K.: "Double Spherical Cassegrain Reflector Antennas", Trans. IEEE, 1973, AP-21, no.6, pp.774-80.

29.137. McFee, J.H., et al.: "Beam Deflection of 10.6-micron Guided Waves by Free-Carrier Injection in (GaAl)As-GaAs Heterostructures", Appl. Phys. Lett., 1973, 23, no.10, pp.571-3.

29.138. Scibor-Rylski, M.T.V.: "Total-Internal-Reflection Electrooptic Diffraction Deflector/Modulator", Electron. Lett., 1973, 9, no.14, pp.309-10.

29.139. Tatuoka, S., Motoki, T., and Ninomiya, Y.: "Phase-Grating-Type Light Modulator and Deflector", NHK Lab. Note, 1973, no.166, pp.1-12.

29.140. Deryugin, L.N., Osovitskii, A.N., and Sotin, V.Ye.: "Possibility of Use of Combination-Frequency Radiation from Nonlinear Waveguides for Antenna Beam Scanning", Radiotekh. Elektron., 1973, 18, no.6, pp.1145-51, and Radio Eng. Electron. Phys., 1973, 18, no.6, pp.836-40.

29.141. Osipov, Yu.V.: "Multichannel Splitting of Laser Beam by Fly's-Eye Lens", Opt.-Mekh. Prom., 1973, 40, no.11, pp.12-4, and Sov. J. Opt. Technol., 1973, 40, no.11, pp.671-3.

29.142. Ninomiya, Y.: "Electrooptic Prism-Array Light Deflector", Electron. Commun. Jap., 1973, 56, no.6, pp.108-14.

29.143. Giallorenzi, T.G.: "Acoustooptical Deflection in Thin-Film Waveguides", J. Appl. Phys., 1973, 44, no.1, pp.242-53.

29.144. Goto, N.: "Directivity of a Scanning Antenna With Cylindrical Reflector", Electron. Commun. Jap., 1973,56,no.8,pp.60-4.

29.145. Steinberg, B.D.: "Design Approach for a High-Resolution Microwave Imaging Radio Camera", J. Franklin Inst., 1973, 296, no.6, pp.415-32.

29.146. Reitzig, R., and Dieges, W.: "Planar Array Antenna Featuring Electronic Beam Scanning", Siemens Forsch. Entwick., 1973, 2, no.6, pp.350-4.

29.147. Torguet, R., and Bauza, J.M.: "Light Deflection and Modulation by Acoustooptic Effect", Onde Electr., 1973, 53, no.10, pp.384-92.

29.148. Gregorwich, W.S.: "Electronically Despun Array Flush-Mounted on a Cylindrical Spacecraft", Trans. IEEE, 1974, AP-22, no.1, pp.71-4.

29.149. Kruger, J., Schulten, G., and Jasmer, W.: "Theoretical and Experimental Investigations of Scanlaser Resonators", Optik, 1974, 39, no.4, pp.327-50.

29.150. Ninomiya, Y.: "High Contrast Electrooptic Prism-Array Light Deflectors", J. Quantum Electron. IEEE, 1974, QE-10, no.3, pp.358-62.

29.151. Schroder, E.: "Compensation for Chromatic Aberrations of Birefringent Prisms in Digital Laser Beam Deflectors", Optik, 1974, 39, no.5, pp.488-98.

29.152. Enns, R.H., and Rangnekar, S.S.: "Diffraction by a Laser-Induced Thermal Phase Grating. I and II", Can. J. Phys., 1974, 52, no.2, pp.99-109 and no.6,pp.562-7.

29.153. Joshi, N.K., and Verma, J.S.: "Electronically Scannable Narrow-Beam Plasma Antenna System", Nachr. Tech. Z., 1974, 27, no.3, pp.118-22.

29.154. Hilburn, J.L., and Prestwood, F.H.: "K-Band Frequency-Scanned Waveguide Array", Trans. IEEE, 1974, AP-22, no.2, pp.324-6.

29.155. Sinclair, D.C.: "Lens Design for Laser Systems", J. Opt. Soc. Am., 1974, 64, no.3, pp.314-6.

29.156. Chen, C.C.: "Wideband Wide-Angle Impedance Matching and Polarization Characteristics of Circular-Waveguide Phased Arrays", Trans. IEEE, 1974, AP-22, no.3, pp.414-8.

29.157. Boyd, J.T.: "Integrated Electrooptic Laser Deflector", Appl. Opt., 1974, 13, no.5, pp.1041-4.

29.158. Al-Ani, A.H., Cullen, A.L., and Forrest, J.R.: "Phase-Locking Method for Beam Steering in Active Array Antennas", Trans. IEEE, 1974, MTT-22,no.6,pp.698-703.

29.159. Wille, D.A., and Hamilton, M.C.: "Acoustooptic Deflection in Ta_2O_5 Waveguides", Appl. Phys. Lett., 1974, 24, no.4,pp.159-60.

29.160. Terrio, F.G., Stockton, R.J., and Sato, W.D.: "Low-Cost PIN-Diode Phase-shifter for Airborne Phased-Array Antennas", Trans. IEEE, 1974, MTT-22, no.6, pp.688-92.

29.161. Das, S.N.: "Scanning Ferroelectric Apertures", Radio Electron. Eng., 1974, 44, no.5, pp.263-8.

29.162. Grib, B.N., et al.: "Electrooptical Gradient Light Deflectors With Distributed Parameters", Ukr. Fiz. Zh., 1974, 19, no.6, pp.951-8.

29.163. Kratirov, I.A., and Pavlov, V.M.: "Resolution of a (Scanning) Laser Beam", Tekh. Kino Telev., 1974, no.4, pp.63-7.

29.164. Rao, B.L.J.: "Bifocal Dual-Reflector Antenna", Trans. IEEE, 1974, AP-22, no.5, pp.711-4.

29.165. DeBenedictis, L.C., and Lucero, J.A.: "Optical Polarization Sensitivity of Lead Molybdate", Appl. Phys. Lett., 1974, 25, no.1, pp.62-4.

29.166. Spaulding, R.A.: "Three-Colour Modulation (Scanning) Divided by Three", Electro-Opt. Syst. Des., 1974, 6, no.6, pp.36-40.

29.167. Isobe, T., Umeno, M., and Kozuka, K.: "Light-Beam Deflector Using Gigahertz Spin Waves in YIG Crystals", Oyo Buturi, 1974, 43, pp.164-7.

29.168. Imbriale, W.A., Ingerson, P.G., and Wong, W.C.: "Large Lateral Feed Displacements in a Parabolic Reflector", Trans. IEEE, 1974, AP-22, no.6, pp.742-5.

29.169. Spaulding, R.A.: "Three-Colour Acoustooptic Modulator", J. Soc. Motion Pict. Telev. Eng., 1974, 83, no.10, p.843.

29.170. Ninomiya, Y.: "Light Deflectors and Modulators Using Electrooptic Prism Array", NHK Tech. J., 1974, 26, no.3, pp.1-17.

29.171. Mozhaiskii, V.N.: "Bragg Scanning of Light Using Lead Molybdate", Prib. Tekh. Eksp., 1974, 17, no.2, pp.200-3, and Instrum. Exp. Tech., 1974, 17, no.2, pp.537-40.

29.172. Tien, P.K., Sanseverino, S.R., and Ballman, A.A.: "Light Beam Scanning and Deflection in Epitaxial $LiNbO_3$ Electrooptic Waveguides", Appl. Phys. Lett., 1974, 25, no.10, pp.563-5.

29.173. Sheridan, J.P., and Giallorenzi, T.G.: "Electrooptically Induced Deflection in Liquid-Crystal Waveguides", J. Appl. Phys., 1974, 45, no.12, pp.5160-3.

29.174. Mozhaiskii, V.N., and Sonin, A.S.: "Theory of Acoustic Modulating and Deflecting Devices", Opt. Spekt., 1974, 37, no.2, pp.337-44, and Opt. Spectrosc., 1974, 37, no.2, pp.192-5.

29.175. Opran, M.E., et al.: "Laser Beam Digital Electrooptic Deflection", Stud. Cercet. Fiz., 1974, 26, no.7, pp.791-800.

29.176. Arumov, G.P., et al.: "Bragg Diffraction of 10.6-micron Radiation by Acoustic Waves in Proustite", Kvantovaya Elektron., 1974, 1, no.3, pp.699-701, and Sov. J. Quantum Electron., 1974, 4, no.3, pp.397-8.

29.177. Berezin, P.D., et al.: "Liquid-Crystal Deflector", Kvantovaya Elektron., 1974, 1, no.5, pp.1253-5, and Sov. J. Quantum Electron., 1974, 4, no.5, pp.693-4.

29.178. Corcoran, V.J., and Crabbe, I.A.: "Electronically Scanned Waveguide Laser Arrays", Appl. Opt., 1974, 13, no.8,pp.1755-7.

29.179. Dorokhov, A.P., and Protopopov, N.I.: "Stripline (Radiation-) Pattern Generating Devices", Radiotekh. (Kharkov), 1974, no.30, pp.43-7.

29.180. Stark, L.: "Microwave Theory of Phased-Array Antennas. Review", Proc. IEEE, 1974, 62, no.12, pp.1661-701.

30. MODULATION TECHNIQUES

30.1. Heinz, W.W., and Okwit, S.: "Microwave Modulator Utilizing Impact Ionization in Bulk Ge at $4.2^\circ K$", Proc. IEEE, 1965, 53, p.1263.

30.2. Seo, W.Y.: "Stripline Modulator", Proc. IEEE, 1965, 53, p.326.

30.3. Garver, R.V.: "Broadband Binary 180° Diode Phase Modulators", Trans. IEEE, 1965, MTT-13, p.32.

30.4. Jacobs, H., Benjamin, R.W., and Holmes, D.A.: "Semiconductor Reflection-Type Modulator", Solid-State Electron., 1965, 8, p.699.

30.5. Jaffe, J.S., and Mackey, R.C.: "Microwave Frequency Translator", Trans. IEEE, 1965, MTT-13, p.371.

30.6. Helszajn, J.: "Coupled-Wave Description of Absorption-Type Ferrite Modulators", Radio Electron. Eng., 1965,29,p.129.

30.7.   Deryugin, I.A., and Kurashov, V.N.: "Modulation of Microwave Fields Using Ferromagnetic Resonance", Radiotekh. Elektron., 1965, 10, no.10, p.1797, and Radio Eng. Electron. Phys., 1965, 10, p.1541-6.

30.8.   Hinton, L.J.T., and Burry, L.F.: "PIN-Diode Modulators for K and Q Bands", Radio Electron. Eng., 1966, 31, p.22.

30.9.   Rzepecka, M., Stuchly, S., and Kraszewski, A.: "Microwave Ferrite Modulators", Przegl. Elektron., 1966, 7, p.40.

30.10.   Stuchly, S., Kraszewski, A., and Rzepecka, M.: "Microwave Amplitude Modulator", Rozpr. Electrotech., 1966, 12, p.403.

30.11.   Aspesi, F.: "Klystron-Varactor Wideband Frequency Deviator", Alta Freq., 1966, 35, no.8, pp.639-44.

30.12.   Reggia, F.: "Amplitude and Phase Modulators in Rectangular Waveguides for 5-7 GHz", Trans. IEEE, 1966, MTT-14, no.3, pp.154-7.

30.13.   Vidallon, C.: "Ultrarapid Modulation of Millimetric Waves", C. R. Acad. Sci. (Paris), 1967, 264B, no.2, pp.112-5.

30.14.   Webb, D.C., and Moore, R.A.: "Solid-State YIG Serrodyne", Trans. IEEE, 1967, MTT-15, no.7, pp.421-9.

30.15.   Deryugin, I.A., Kuts, P.S., and Strizhevskii, V.L.: "Ferromagnetic Resonance Microwave Modulators", Radiotekhnika, 1967, 22, no.6, pp.81-8, and Relecommun. Radio Eng. Ot 2, 1967, 22, no.6.

30.16.   Malherbe, C.W.: "Frequency Changes in Reflex Klystrons During Square-Wave Modulation", Proc. IEEE, 1967, 55, no.5, pp.706-7.

30.17.   Liebe, H.J., and Senitzky, B.: "258-GHz Modulator With Stark Waveguide", Rev. Sci. Instrum., 1967, 38, no.11, pp.1678-80.

30.18.   Klein, G., and Dubrowsky, L.: "Digilator. Broadband Microwave Frequency Translator", Trans. IEEE, 1967, MTT-15, p.172.

30.19.   Murata, S., Nio, K., and Yamaguchi, K.: "AM to PhM Conversion in a Klystron Amplifier", Tech. J. Jap. Broadcast. Corp., 1968, 20, no.1, pp.9-17.

30.20.   Hannan, W.J., and Bordogna, J.: "Comparison of Electrooptic Modulation Methods", Trans. IEEE, 1968, AES-4, p.874.

30.21.   Hill, B., and Wencker, G.: "Optimal Dimensioned Resonance Modulators in the Microwave Region", Nach. Tech. Fachber., 1968, 35, p.718.

30.22.   Tan, B.T.G.: "Bulk-Effect Ge Microwave Modulation", Proc. IEEE, 1968, 56, p.1750.

30.23.   Rozhanskii, V.A., and Skomorovskii, Yu.A.: "Nonlinear Signal Distortions of Light Modulated by Pockels' Effect", Radiotekh. Elektron., 1968, 13, p.2095, and Radio Eng. Electron. Phys., 1968, 13, no.11, pp.1840-2.

30.24.   Majewski, S.: "Microwave Light Modulator and Measurement Techniques", Pr. PIE, 1968, 9, no.3-4, p.173.

30.25.   Petzinger, K.G.: "Gunn-Effect Light Modulator", RCA Tech. Notes, 1968, no.21, p.1.

30.26.   Lee, R.W.: "Linear Electrooptic Effect in Cubic Crystals. Wide-Angle Modulation", Appl. Opt., 1969, 8, p.1385.

30.27.   Laubereau, A.: "External Frequency Modulation and Compression of Picosecond Pulses", Phys. Lett., 1969, 29A, p.539.

30.28.   Riesz, R.P., and Biazzo, M.R.: "Gigahertz Optical Modulation", Appl. Opt. 1969, 8, p.1393.

30.29.   Kendall, A.J., and Pattenden, J.H.: "Computation of Electric-Field Configurations in Anisotropic Electrooptic Crystals", Electron. Lett., 1969, 5, p.446.

30.30.   Chow, K.K., Comstock, R.L., and Leonard, W.B.: "1.5-GHz Bandwidth Light Modulator", J. Quantum Electron. IEEE, 1969, QE-5, p.618.

30.31.   Auth, D.C.: "Half-Octave-Bandwidth TW X-Band Optical Phase Modulator", J. Quantum Electron. IEEE, 1969, QE-5, p.622.

30.32.   Yokoo, K., and Shibata, Y.: "Electronic Tunable Gunn-Diode Oscillator", Trans. IEEE, 1969, ED-16, p.494.

30.33.   Zalesskii, I.E., and Rutkovskii, I.Z.: "Modulating the Radiation of GaAs Injection Lasers at 1 GHz", Zh. Prikl. Spekt., 1969, 10, no.1, p.162.

30.34.   Broom, R.F.: "Self-Modulation at Gigahertz Frequencies of a Diode Laser Coupled to an External Cavity", Electron. Lett., 1969, 5, p.571.

30.35.   Califano, F.P.: "Frequency Modulation of Three-Terminal Gunn Devices by Optical Means", Trans. IEEE, 1969, ED-16, p.149.

30.36.   Fortus, V.M., and Freidman, G.I.: "Mode Locking in Optical Parametric Oscillator", Izv. VUZ Radiofiz., 1969, 12, p.1788.

30.37.   Cubeddu, R., and Svelto, O.: "Effect of Dispersion on Laser Self-Locking", Phys. Lett., 1969, 29A, p.78.

30.38.   Clobes, A.R., and Brienza, M.J.: "Passive Mode-Locking of a Pulsed Nd^{3+}:YAG Laser", Appl. Phys. Lett., 1969, 14, p.287.

30.39.   Nikolaev, I.V., Zasovin, E.A., and Koblova, M.M.: "Use of GaAs Crystals for Modulation", Radiotekh. Elektron., 1969, 14, no.9, p.1711, and Radio Eng. Electron. Phys., 1969, 14, no.9, pp.1491-3.

30.40.   Fay, H.: "Electrooptic Modulation of Light Propagating Near the Optic Axis in $LiNbO_3$", J. Opt. Soc. Am., 1969, 59, p.1399.

30.41.   Sugimoto, S.: "Ultra-High-Speed Diode Switch for 50 GHz Utilizing Avalanche Breakdown in Varactor Diodes", NEC Res. Dev., 1969, no.14, p.40.

30.42. Tron'ko, V.D.: "Faraday-Type Light Modulator With Inclined Magnetooptic Substance", Radiotekh. Elektron., 1969, 14, no.10, p.1848, and Radio Eng. Electron. Phys., 1969, 14, no.10, pp.1595-601.

30.43. Gires, F., and Tournois, P.: "Active Interferometer for Frequency Translation or Modulation of Laser Pulses", C. R. Acad. Sci. (Paris), 1969, 268B, p.313.

30.44. Siegman, A.E., and Kuizenga, D.J.: "Simple Analytic Expressions for AM and FM Mode-Locked Pulses in Homogeneous Lasers", Appl. Phys. Lett., 1969,14,p.181.

30.45. Mansell, J.R.: "Video-Frequency Light Modulator Having Wide Angular Aperture", Proc. IEE, 1969, 116, p.691.

30.46. Beterov, I.M., Lisitsyn, V.N., and Chebotaev, V.P.: "Selection and Self-Synchronization of Oscillatory Modes in a Laser With Nonlinear Absorption", Radiotekh. Elektron., 1969, 14, no.6, p.1127, and Radio Eng. Electron. Phys., 1969, 14, no.6, pp.981-2.

30.47. Kazaryan, R.A., and Sidorova, S.P.: "Microwave Modulation of He-Ne Lasers", Radiotekh. Elektron., 1969, 14, no.10, p.1899, and Radio Eng. Electron. Phys., 1969, 14, no.10, pp.1644-5.

30.48. Adrianova, I.I., et al.: "Broadband Internal Microwave Frequency Modulation of a He-Ne Laser", Opt. Spekt., 1969, 26, no.3, p.402, and Opt. Spectrosc., 1969, 26, no.3, p.220.

30.49. Bamberskii, M.V., and Nikolaev, V.S.: "Correction for the Controlling Signal in Microwave Ferrite Modulators", Radiotekhnika, 1969, 24, no.3, p.63, and Telecommun. Radio Eng. Pt 2, 1969,24,no.3.

30.50. Adrianova, I.I., et al.: "Performance of Wideband, Continuous, Optical Microwave Modulator", Opt.-Mekh. Prom., 1969, 36, no.2, p.59, and Sov. J. Opt. Technol., 1969, 36, no.2, p.149.

30.51. Trevelyan, B.: "Practical Design of a Laser Modulator Using 45°-Cut ADP Crystals", J. Phys. E, 1969, 2, p.425.

30.52. Hill, B., and Wencker, G.: "Electrooptic Resonance Modulation in the Microwave Region", Nachr. Tech. Z., 1969, 22, p.309.

30.53. Mak, A.A., and Fromzel, V.A.: "Self-Synchronization of Transverse Modes in a Solid-State Laser", Zh. Eksp. Teor. Fiz. Pis'ma, 1969, 10, no.7, p.313, and JETP Lett., 1969, 10, no.7, p.199.

30.54. Allen, L.B., Rice, R.R., and Mathews, R.F.: "Two-Cavity Mode-Locking of He-Ne Laser", Appl. Phys. Lett., 1969, 15, p.416.

30.55. Magnante, P.C.: "Mode-Locked and Bandwidth-Narrowed Nd^{3+}:Glass Laser", J. Appl. Phys., 1969, 40, p.4437.

30.56. Arakelyan, V.S., Karlov, N.V., and Prokhorov, A.M.: "Self-Synchronization of Transverse Modes of a CO_2 Laser", Zh. Eksp. Teor. Fiz. Pis'ma, 1969, 10, no.6, p.279, and JETP Lett., 1969, 10, no.6, p.178.

30.57. Hong, G.W., and Whinnery, J.R.: "Switching of Phase-Locked States in the Intracavity Phase-Modulated He-Ne Laser", J. Quantum Electron. IEEE, 1969, QE-5, p.367.

30.58. Belova, G.N., Kazantsev, V.F.: "Ultrasonic Modulation of a Laser", Akust. Zh., 1969, 15, no.1, p.5, and Sov. Phys.-Acoustics, 1969, 15, no.1, p.4.

30.59. Honda, T., and Sakurai, K.: "Microwave Modulation of Laser Beams. II", Bull. Electrotech. Lab., 1969, 33, no.12, p.19.

30.60. DeVito, P.A., Kearns, W.J., and Seavey, M.H.: "Phase-Pattern Control of Injection-Locked Pulsed Magnetrons", Proc. IEEE, 1969, 57, p.1436.

30.61. de Cremoux, B.: "Free-Carrier Absorption for 10-micron Modulation", Proc. IEEE, 1969, 57, p.1674.

30.62. Mustel', Ye.R., Parygin, V.N., and Solomatin, V.S.: "Internal Modulation of Gas Laser in Nonsynchronous Range", Radiotekh. Elektron., 1969, 14, no.6, p.1029, and Radio Eng. Electron. Phys., 1969, 14, no.6,pp.891-5.

30.63. Okada, M., and Ieiri, S.: "Light Modulators Using 45°-z-Cut KDP-Type Crystals", Tech. J. Jap. Broadcast. Corp., 1969, 21, no.6, p.21.

30.64. Adrianova, I.I., and Volkonskii, V.B.: "In-Cavity Microwave Modulation of Laser Emission", Opt.-Mekh. Prom., 1969, 36, no.2, p.1, and Sov. J. Opt. Technol., 1969, 36, no.2, pp.157-60.

30.65. Lee, T.P., and Roldan, R.H.: "Subnanosecond Light Pulses from GaAs Injection Lasers", J. Quantum Electron. IEEE, 1969, QE-5, p.551.

30.66. Ferrar, C.M.: "Mode-Locked Flashlamp-Pumped Coumarin-Dye Laser at 460 nm", J. Quantum Electron. IEEE, 1969, QE-5, p.550.

30.67. Kazaryan, R.A., et al.: "Adjustable Electrooptic UHF Light Modulator", Prib. Tekh. Eksp., 1969, no.6, p.178, and Instrum. Exp. Tech., 1969, no.6, pp.1573-4.

30.68. Garver, R.V.: "360° Varactor Linear Phase Modulator", Trans. IEEE, 1969, MTT-17, p.137.

30.69. Yamashita, E., and Atsuki, K.: "Proposed Microwave Structure and Design Method for TW Modulation of Light", Trans. IEEE, 1969, MTT-17, p.118.

30.70. Melngailis, I., and Tannenwald,
P.E.: "Far-IR and Submillimetre Impact
Ionization Modulator", Proc. IEEE, 1969,
57, p.806.

30.71. Carlson, D.G.: "Generation of Vari-
able-Width Subnanosecond Optical Pulse",
Proc. IEEE, 1969, 57, p.807.

30.72. Frova, A.: "Pulse Delay Effects
in He-Ne Lasers Mode-Locked by Ne Absorption
Cell", J. Appl. Phys., 1969, 40, p.3969.

30.73. Ito, H., and Inaba, H.: "Self-Mode-
Locking of Transverse Modes in CO_2 Laser
Oscillation", Opt. Commun., 1969, 1, no.2,
p.61.

30.74. Cubbeddu, R., et al.: "Picosecond
Pulses with TEM_{00} Mode-Locked Ruby Laser",
J. Quantum Electron. IEEE, 1969, QE-5, p.470.

30.75. Inaba, H., and Ito, H.: "Self-
Locking Operation of CO_2 Laser", Laser
Focus, 1969, 5, no.1, p.25.

30.76. Malyshev, V.I., et al.: "Mode-
Locking in Ruby and Nd^{3+} Lasers Operating
Under Free-Oscillation Conditions", Zh.
Eksp. Teor. Fiz., 1969, 57, no.3, p.827, and
Sov. Phys.-JETP, 1970, 30, no.3.

30.77. Smith, P.W.: "Mode Locking of
Lasers", Proc. IEEE, 1970, 58, p.1342.

30.78. Whitney, C.G., and Pratt, G.W.:
"Resolution of Sidebands in Semiconductor
Laser Frequency Modulated by Ultrasonic
Waves", J. Quantum Electron. IEEE, 1970,
QE-6, p.352.

30.79. Gambling, W.A., et al.: "Accurate
Tuning of Laser Mode-Locking Device", Electron. Lett., 1970, 6, p.570.

30.80. Faxvog, F.R., Willenbring, G.R.,
and Carruthers, J.A.: "Self-Pulsing in He-
Cd Laser", Appl. Phys. Lett., 1970, 16, p.8.

30.81. Lyon, D.L., and Kinsel, T.S.:
"Transitions Between Mode-Locked States in
Intracavity Phase-Modulated Lasers", Appl.
Phys. Lett., 1970, 16, p.89.

30.82. Hall, D., Yariv, A., and Garmire,
E.: "Optical Guiding and Electrooptic Modu-
lation in GaAs Epitaxial Layers", Opt.
Commun., 1970, 1, no.9, p.403.

30.83. Cooper, R.W., and Page, J.L.:
"Magnetooptic Light Modulators", Radio Electron. Eng., 1970, 39, p.302.

30.84. Grauling, C.H., and Geller, B.D.:
"Broadband Frequency Translator With 30-dB
Suppression of Spurious Sidebands", Trans.
IEEE, 1970, MTT-18, p.651.

30.85. Stadnik, B.: "Temperature Instability of Electrooptic Laser Modulators
With ADP, KDP, and DKDP", Acta Tech. CSAV,
1970, 15, p.65.

30.86. Bracale, M.: "Design of Broadband
Light Modulators", Radio Electron. Eng.,
1970, 39, p.185.

30.87. Shamburov, V.A.: "Electrooptic
Shutter for Generating Laser Pulses", Radio-
tekh. Elektron., 1970, 15, p.512, and Radio
Eng. Electron. Phys., 1970, 15, no.3, pp.438-45.

30.88. Zaslavskaya, V.R.: "Electrooptic
Modulation of He-Ne Laser With Three-Mirror
Resonator", Opt. Spektr., 1970, 28, no.1,
p.93, and Opt. Spectrosc., 1970, 28, no.1,
p.47.

30.89. Miyashita, T., and Ikenoue, J.:
"Phase-Locking of Gas Laser Loss-Modulated
With Double Mode Spacing", Jap. J. Appl.
Phys., 1970, 9, no.6, p.717.

30.90. Luk'yanov, D.P.: "Interaction
Between an Arbitrarily Polarized Light Wave
and a Circular Modulating Field in Cubic
Crystals", Radiotekh. Elektron., 1970, 15,
p.1052, and Radio Eng. Electron. Phys.,
1970, 15, no.5, pp.857-62.

30.91. Carra, A.A.: "Simple Arrangement
for Internal Audio Modulation of a He-Ne
Laser", Am. J. Phys., 1970, 38, p.926.

30.92. White, M.B., Gerber, W.D., and
Doyle, W.M.: "Intracavity Polarization
Modulation of a Nearly Isotropic CO_2 Laser",
J. Quantum Electron. IEEE, 1970, QE-6, p.457.

30.93. Landman, A.: "Efficient CO_2 Laser
Modulation Internal to the Cavity by Methyl-
Chloride Gas", J. Quantum Electron. IEEE,
1970, QE-6, p.472.

30.94. Kobayashi, T., et al.: "Mode-
Locking of Lasers by Feedback of Self-Beat
Between Cavity Modes", Jap. J. Appl. Phys.,
1970, 9, p.318.

30.95. Maydan, D.: "Fast Modulator for
Extraction of Internal Laser Power", J.
Appl. Phys., 1970, 41, p.1552.

30.96. Martin, B., and Hobson, G.S.:
"High-Speed Phase and Amplitude Modulation
of Gunn Oscillators", Electron. Lett., 1970,
6, p.244.

30.97. Cawsey, D.: "Varactor-Tuned Gunn-
Effect Oscillators", Electron. Lett., 1970,
6, p.246.

30.98. Marriott, S.P.A.: "Tunable Gunn-
Diode Oscillators", Des. Electron., 1970,
8, no.2, pp.72-4.

30.99. Christmas, T.M., and Wildey, C.G.:
"Precise PTM Control of a Ruby Laser", Electron. Lett., 1970, 6, no.22, pp.696-7.

30.100. Silfvast, W.T., and Smith, P.W.:
"Mode Locking of He-Cd Laser at 441.6 nm
and 325.0 nm", Appl. Phys. Lett., 1970, 17,
p.70.

30.101. Danilov, V.V., et al.: "Faraday-
Effect Light Modulator With Closed Magnetic
Circuit", Radiotekh. Elektron., 1970, 15,
no.2, p.362, and Radio Eng. Electron. Phys.,
1970, 15, no.2, pp.314-5.

30.102. Maydan, D.: "Acoustooptical Pulse Modulators", J. Quantum Electron. IEEE, 1970, QE-6, p.15.

30.103. Ley, J.M., Christmas, T.M., and Wildey, C.G.: "Solid-State Subnanosecond Light Switch", Proc. IEE, 1970,117,p.1057.

30.104. Solbakken, K.: "Digital Microwave Modulators With Special Reference to Phase Modulators", Telektronikk, 1970,no.1-2,p.1.

30.105. Hookabe, K., and Matsuo, Y.: "Novel Type of Cut for KDP Crystals for Low-Voltage Light Modulation", Electron. Lett., 1970, 6, p.550.

30.106. Page, P., and Pursey, H.: "Tunable Single-Sideband Electrooptic Ring Modulator", Opto-Electron. 1970, 2, no.1, p.1.

30.107. Kamach, Yu.E., et al.: "Monolithic LiNbO$_3$ Shutter Reflectors", Radiotekh. Elektron., 1970, 15, no.6, p.1323, and Radio Eng. Electron. Phys., 1970, 15, no.6, pp.1134-6.

30.108. John, P.I., and Sarkar, D.C.: " "Amplitude Modulation of Microwaves Propagated Through a Periodically Varying Plasma", Radio Sci., 1970, 5, no.1, p.101.

30.109. Rowe, S.H.: "Efficiency Enhancement of Light Modulators", Appl. Opt., 1970, 9, p.1222.

30.110. Fralick, R.D., and Zipf, E.C.: "Versatile Electrical Shutter for Optical Pumping Experiments", Rev. Sci. Instrum., 1970, 41, p.47.

30.111. Kaminow, I.P., Bridges, T.J., and Pollack, M.A.: "964-GHz TW Electrooptic Light Modulator", Appl. Phys. Lett., 1970, 16, p.416.

30.112. Okada, M., and Ieiri, S.: "Extinction Ratio of Electrooptic Light Modulator", J. Quantum Electron. IEEE, 1970, 6, p.522.

30.113. Ripper, J.E.: "Analysis of Frequency Modulation of Junction Lasers by Ultrasonic Waves", J. Quantum Electron. IEEE, 1970, QE-6, p.129.

30.114. Vane, A.B., and Dunn, V.E.: "Digitally Tuned Gunn-Effect Microstrip Oscillator", Proc. IEEE, 1970, 58, p.171.

30.115. Tsvirko, Yu.A.: "Tuning of Gunn-Effect Oscillator in a Resonant Circuit", Radiotekh. Elektron., 1970, 15, no.2, p.401, and Radio Eng. Electron. Phys., 1970, 15, no.2, pp.350-2.

30.116. Chow, K.K., and Leonard, W.B.: "Efficient Octave-Bandwidth Microwave Light Modulators", J. Quantum Electron. IEEE, 1970, QE-6, no.12, pp.789-93.

30.117. Chattopadhyay, D., and Nag, B.R.: "IR Modulation Using Hot-Electron Faraday Effect in n-InSb", Phys. Status Solidi, 1970, 40, no.2, pp.701-6.

30.118. Kalymnios, D., Ley, J.M., and Rashidi, K.: "Optimization of 45°-y-Cut Modulators", Electron. Lett., 1970, 6, no.4, pp.771-3.

30.119. Francois, G.E., Librecht, F.M., and Engelen, J.J.: "Optimum Cut in XDP Crystals for Transverse Light Modulation", Electron. Lett., 1970, 6, no.24, pp.778-9.

30.120. Von der Linde, D., Bernecker, O., and Laubereau, A.: "Fast Electrooptic Shutter for Selection of Single Picosecond Laser Pulses", Opt. Commun., 1970, 2, no.5,pp.215-8.

30.121. Henderson, D.M., and Abrams, R.L.: "Comparison of Acoustooptic and Electrooptic Modulators at 10.6 micron", Opt. Commun., 1970, 2, no.5, pp.223-6.

30.122. Kalymnios, D.: "Optimization of Electrooptic Crystal Modulators", Electron. Lett., 1970, 6, no.25, pp.804-5.

30.123. Sirieix, M.: "Large-Depth Wideband Modulation of a 10.6-micron Laser Beam", Onde Electr., 1970, 50, no.10, pp.864-8.

30.124. Boiko, B.B., et al.: "Electrooptical Shutter for a Ruby Laser With Low Control Voltage", Zh. Prikl. Spektr., 1970, 13, no.5, pp.812-15.

30.125. Goto, K., et al.: "Low-Voltage Light Modulator Using Polarization-Rotated Reflector", Opto-Electron., 1970, 2, no.3, pp.168-9.

30.126. Stern, F.: "Standing-Wave Acoustic Modulation of Semiconductor Lasers", IBM Tech. Disclosure Bull., 1970, 13, no.6, pp.1646-7.

30.127. Mann, M.M., et al.: "Mode Locking of CO$_2$ Laser by Intracavity Phase Modulation", Appl. Phys. Lett., 1970, 17, no.9, pp.393-5.

30.128. Lyon, D.L., George, E.V., and Haus, H.A.: "Observation of Spontaneous Mode Locking in a High-Pressure CO$_2$ Laser", Appl. Phys. Lett., 1970, 17, no.11, pp.474-6.

30.129. Smith, P.W.: "Phase Relationship Between Oscillating Modes in a Mode-Locked 633-nm He-Ne Laser", Opt. Commun., 1970, 2, no.6, pp.292-4.

30.130. Signore, R., and Becrelle, J.: "Self-Locking in He-Ne Lasers", Rev. Tech. Thomson-CSF, 1970, 2, no.4, pp.681-707.

30.131. Malota, F.: "Mode-Locking Effects in a Ruby Laser With Regularly Moving Crystal", Opto-Electron., 1970, 2, no.2,pp.101-3.

30.132. Kim, H.H., and Marantz, H.: "Continuous Self-Mode-Locking of IR Gas Lasers", J. Quantum Electron. IEEE, 1970, QE-6, no.11, pp.749-50.

30.133. Cohen, L.D.: "Direct Frequency Modulation of a Microwave Source", Rev. Sci. Instrum., 1970, 41, no.9, pp.1365-6.

30.134. Kuizenga, D.J., and Siegman, A.E.: "FM-Laser Operation of the Nd^{3+}:YAG Laser", J. Quantum Electron. IEEE, 1970, QE-6, no.11, pp.673-7.

30.135. Kuizenga, D.J., and Siegman, A.E.: "FM and AM Mode Locking of the Homogeneous Laser. II", J. Quantum Electron. IEEE, 1970, QE-6, no.11, pp.709-15.

30.136. Gribkov, V.A., et al.: "Kerr Shutter for High-Frequency Modulation of Laser Radiation", Prib. Tekh. Eksp., 1970, no.4, pp.213-6, and Instrum. Exp. Tech., 1970, no.4, pp.1206-9.

30.137. Wood, O.R., Abrams, R.L., and Bridges, T.J.: "Mode-Locking of a TEA CO_2 Laser", Appl. Phys. Lett., 1970, 17, no.9, pp.376-8.

30.138. Savel'ev, V.Ya.: "Electronic Stabilization of Operating Point of an Electrooptic Light Modulator", Radiotekh. Elektron., 1970, 15, no.9, and Radio Eng. Electron. Phys., 1970, 15, no.9, pp.1659-62.

30.139. Voronov, V.I., and Pol'skii, Yu.E.: "Multilayer Dielectric Mirror with Piezoelectric Support for Laser Modulation", Prib. Tekh. Eksp., 1970, no.6, pp.1748-50.

30.140. Ostapchenko, E.P., et al.: "Modulation of He-Ne Laser via Excitation Source", Radiotekh. Elektron., 1970, 15, no.11, and Radio Eng. Electron. Phys., 1970, 15, no.11, pp.2063-5.

30.141. Sueta, T., Goto, K., and Makimoto, T.: "TW Light-Intensity Modulators Using Polarization-Rotated Reflection", Electron. Commun. Jap., 1970, 53, no.8, pp.115-22.

30.142. Ikegami, T., and Suematsu, Y.: "Large-Signal Characteristics of Directly Modulated Semiconductor Injection Lasers", Electron. Commun. Jap., 1970, 53, no.9, pp.69-75.

30.143. Califano, F.P., and Vitale, G.F.: "Study of a Variable-Frequency Gunn Oscillator", Rend. Riun. Assoc. Elettrotec. Ital., 1970, 45, no.2.04, pp.1-4.

30.144. Glover, G.H., and Champlin, K.S.: "Investigation of Twist-Mode Propagation in InSb at 70 GHz", Trans. IEEE, 1970, MTT-18, p.570.

30.145. Benoit, J.: "CO_2 Laser Modulation by Hole Injection in n-InSb", Appl. Phys. Lett., 1970, 16, p.482.

30.146. Solbakken, K.: "Design of Digital Amplitude and Phase Modulators for 12 GHz", Telektronikk, 1970, no.1-2, p.10.

30.147. Okada, M., and Ieiri, S.: "Light Modulators Using 45°-x-Cut and 45°-y-Cut ADP Crystals", NHK Tech. J., 1970, 22, no.6, pp.50-60.

30.148. Dolgopyatov, R.M., et al.: "Gigahertz Optical Modulator Using the Quadratic Electrooptic Effect", Radiotekh. Elektron., 1970, 15, no.7, and Radio Eng. Electron. Phys., 1970, 15, no.7, pp.1335-8.

30.149. Hovel, W.: "Phase Coupling Between Modes of Two 633-nm He-Ne Lasers Through Nonlinear Gain Characteristics of the Inverted Population", J. Quantum Electron. IEEE, 1970, QE-6, p.82.

30.150. Bogdankevich, O.V., et al.: "Mode Locking in a Semiconductor Laser with Electronic Excitation of an Internal Nonlinear Medium", Zh. Eksp. Teor. Fiz. Pis'ma, 1970, 12, no.4, pp.184-5, and JETP Lett., 1970, 12, no.4, pp.128-9.

30.151. Dukhovnyi, A.M., Mak, A.A., and Fromzel, V.A.: "Transverse Mode Locking in a Solid-State Laser", Zh. Eksp. Teor. Fiz., 1970, 59, no.4, pp.1165-76, and Sov. Phys.-JETP, 1971, 32, no.4, pp.636-42.

30.152. Henoch, B.T., and Tamm, P.: "Reflection-Type Varactor Diode Phase Modulator", Trans. IEEE, 1971, MTT-19, no.1, pp.103-5.

30.153. Rashidi, K.: "Pulse Response of z-Cut ADP Modulators", Electron. Lett., 1971, 7, no.4, pp.114-5.

30.154. Preier, H., and Kuzmany, H.: "Light Modulation by Thin CdS Platelet Oscillators", Appl. Phys. Lett., 1971, 18, no.1, pp.19-22.

30.155. de Barros, M.A.R.P., and Wilson, M.G.F.: "Nanosecond Baseband Optical-Diffraction Modulator", Electron. Lett., 1971, 7, no.10, pp.267-9.

30.156. Johnston, A.R., and Melville, R.D.S.: "Stark-Effect Modulation of a CO_2 Laser by NH_2D", Appl. Phys. Lett., 1971, 19, no.12, pp.503-6.

30.157. Nikolaev, N.V., and Koblova, M.M.: "(Electrooptic) Modulation of 10.6-micron Radiation", Kvantovaya Elektron., 1971, no.2, pp.57-64.

30.158. Listvin, V.N., and Potapov, V.T.: "Semiconductor Modulator for Submillimetre Radiation", Radiotekh. Elektron., 1971, 16, no.7, pp.122-4, and Radio Eng. Electron. Phys., 1971, 16, no.7, pp.1168-70.

30.159. Kazovskii, L.G.: "Optimization of the Frequency Characteristic of a Broadband Electrooptic TW Modulator", Radiotekh. Elektron., 1971, 15, no.8, and Radio Eng. Electron. Phys., 1971, 15, no.8, pp.1427-30.

30.160. Hulme, K.F.: "Oblique-Cut Longitudinal Electrooptic Modulators", J. Quantum Electron. IEEE, 1971, QE-7, no.6, pp.236-9.

30.161. Biazzo, M.R.: "Fabrication of a $LiTaO_3$ Temperature-Stabilized Optical Modulator", Appl. Opt., 1971, 10, no.5, pp.1016-21.

30.162. Standley, R.D., and Mandeville, G.D.: "Performance of an 11-GHz Optical Modulator Using LiTaO₃", Appl. Opt., 1971, 10, no.5, pp.1022-3.

30.163. Hook, W.R., and Hilberg, R.P.: "Lossless DKDP Pockels' Cell for High-Power Q-Switching", Appl. Opt., 1971, 10, no.5, pp.1179-80.

30.164. Heising, S.J., and Jarrett, S.M.: "Mode Locking of Argon-Ion Laser in the UV", Appl. Phys. Lett., 1971, 18, no.11, pp.516-8.

30.165. Gibson, A.F., Kimmitt, M.F., and Rosito, C.A.: "Passive Mode Locking of a High-Pressure CO₂ Laser With a CO₂ Saturable Absorber", Appl. Phys. Lett., 1971, 18, no.12, pp.546-8.

30.166. Ferrario, A., and Querzola, B.: "He-Se Mode-Locked Laser With Internal Phase Modulation", Opt. Commun., 1971, 3, no.3, pp.161-4.

30.167. Librecht, F.M., and Francois, G.E.: "Optimum Cut for a LiNbO₃ Transverse Light Modulator", J. Quantum Electron. IEEE, 1971, QE-7, no.7, pp.374-6.

30.168. Ross, J.N.: "Phase-Modulated Axial Modes in a Passive Cavity", J. Phys. D, 1971, 4, no.8, pp.1092-9.

30.169. von Gutfeld, R.J.: "Picosecond Dye-Laser Pulses Using N₂-Laser Pumping", Appl. Phys. Lett., 1971, 81, no.11, pp.481-2.

30.170. Belousova, I.M., Danilov, O.B., and Zapryagaev, A.F.: "Radiation Spectrum of He-Ne Laser on Amplitude Modulation by a Movable External Mirror", Zh. Tekh. Fiz., 1971, 41, no.5, pp.1028-33, and Sov. Phys.-Tech. Phys., 1971, 16, no.5.

30.171. Feldtkeller, E.: "Magnetooptics and Application to Electrotechnology", Elektro-Anz., 1971, 24, no.12, pp.265-9.

30.172. Hall, D.R., Pao, Y.H., and Claspy, P.C.: "Stark-Effect Modulation of a Passively Q-Switched CO₂ Laser", Appl. Opt., 1971, 10, no.7, pp.1688-9.

30.173. White, M.B., and Gerber, W.D.: "Frequency Modulation of a Self-Mode-Locked Dual-Polarization CO₂ Laser", J. Quantum Electron. IEEE, 1971, QE-7, no.12, pp.577-8.

30.174. Alekseeva, L.L., et al.: "Frequency Modulation of Gas Laser via Excitation", Izv. VUZ Radiofiz., 1971, 14, no.9, pp.1336-42.

30.175. Gandrud, W.B.: "Reduced Modulator Drive-Power Requirements for 10.6-micron Guided Waves", J. Quantum Electron. IEEE, 1971, QE-7, no.12, pp.580-1.

30.176. Warner, J.: "Simulation of a Double-45°, z-Cut DKDP, Electrooptic Q-Switch by Desk-Top Computer", Opt. Laser Technol., 1971, 2, no.4, pp.215-7.

30.177. Scavennec, A.: "Various States of Production of Light Pulses by an Internally Modulated Argon Laser", C. R. Acad. Sci. (Paris), 1971, 272B, no.23, pp.1303-6.

30.178. Dianova, V.A., and Parygin, V.N.: "LiNbO₃ Modulator of Light", Izv. VUZ Radioelektron., 1971, 14, no.6, pp.606-12.

30.179. Hookabe, K., and Matsuo, Y.: "Novel Cut (45°z-45°y) of KDP-Type Crystals", Mem. Inst. Sci. Ind. Res. Osaka Univ., 1971, 28, pp.23-7.

30.180. Voitsekhovskii, A.V., Glasnov, M.A., and Petrov, A.S.: "Creation of an Electrooptic Modulator Based on GaAs", Izv. VUZ Fiz., 1971, no.7, pp.121-3.

30.181. Morgan, P.D., and Peacock, N.J.: "Nanosecond Laser Pulse Generation Using an Electrooptic Shutter External to the Oscillator Cavity", J. Phys. E, 1971, 4, no.9, pp.677-80.

30.182. Brooker, P.G., and Beynon, J.D.E.: "10-GHz Single-Sideband Modulator With 1-kHz Frequency Shift", Trans. IEEE, 1971, MTT-19, no.10, pp.829-34.

30.183. Ahmed, M.S., and Schroter, H.: "Four-Pole Transducer Analogues of Electrooptic Modulators", Opt. Laser Technol., 1971, 3, no.3, pp.140-3.

30.184. Seider, K.H., and Koppatz, P.: "TW Modulation of Laser Light at 4 GHz", Exp. Tech. Phys., 1971, 19, no.2, pp.113-8.

30.185. Campbell, J.P., and Steier, W.H.: "Rotating-Waveplate Optical-Frequency Shifting in LiNbO₃", J. Quantum Electron. IEEE, 1971, QE-7, no.9, pp.450-7.

30.186. Cuchy, Z., and Landovsky, J.: "Electrooptic Light Modulators", Kristall. Tech., 1971, 6, no.4, pp.563-72.

30.187. Henaff, J., and Feldmann, M.: "Acoustooptic Extraction of Internal Laser Power", Onde Electr., 1971, 51, no.9, pp.805-15.

30.188. Claspy, P.C., and Pao, Y.H.: "Basic Characteristics of High-Frequency Stark-Effect Modulation of CO₂ Lasers", J. Quantum Electron. IEEE, 1971, QE-7, no.11, pp.512-9.

30.189. Aitchison, C.S., and Newton, B.H.: "Varactor-Tuned X-Band Gunn Oscillator Using Lumped Thin-Film Circuits", Electron. Lett., 1971, 7, no.4, pp.93-4.

30.190. Norton, P., et al.: "Extended Wavelength Tuning of (PbSn)Te Lasers", Appl. Phys. Lett., 1971, 18, no.4, pp.158-9.

30.191. Gilbert, J., Lachambre, J.L., and Rheault, F.: "Active Mode Locking of a High-Pressure Pulsed N₂O Laser", J. Quantum Electron. IEEE, 1971, QE-7, no.9, pp.462-3.

30.192. Dienes, A., Ippen, E.P., and Shank, C.V.: "Mode-Locked CW Dye Laser", Appl. Phys. Lett., 1971, 19, no.8, pp.258-60.

30.193. Abrams, R.L., and Wood, O.R.: "Characteristics of a Mode-Locked TEA CO_2 Laser", Appl. Phys. Lett., 1971, 19, no.12, pp.518-20.

30.194. Baglikov, V.B., and Parygin, V.N.: "Unlocked Internal Modulation of Multifrequency Gas Laser", Radiotekh. Elektron., 1971, 16, no.3, and Radio Eng. Electron. Phys., 1971, 16, no.3, pp.490-5.

30.195. Baglikov, V.B., and Parygin, V.N.: "Coupling Modulation of a High-Gain Gas Laser", Radiotekh. Elektron., 1971, 16, no.11, pp.2144-52, and Radio Eng. Electron. Phys., 1971, 16, no.11, pp.1899-906.

30.196. Hoffmann, G.W., and Jovin, T.M.: "Nanosecond-Risetime Mechanical Chopper for Laser Light", Appl. Opt., 1971, 10, no.1, pp.218-9.

30.197. Pardue, A.L., and Dezenburg, G.J.: "CO_2-Laser Mode-Locking Produced by Sinusoidal Cavity-Length Modulation", J. Quantum Electron. IEEE, 1971, QE-7, no.2, pp.95-7.

30.198. Smith, D.C., and Berger, P.J.: "Mode-Locking of a TEA Pulsed CO_2 Laser", J. Quantum Electron. IEEE, 1971, QE-7, no.4, pp.172-4.

30.199. Heising, S.J., and Jarrett, S.M.: "Forced and Self Mode-Locking of Krypton-Ion Laser", J. Quantum Electron. IEEE, 1971, QE-7, no.5, pp.205-7.

30.200. Belogol'skii, V.A., and Kubarev, A.V.: "Passive Power Stabilization for Gas Lasers", Izmer. Tekh., 1971, 14, no.3, pp.89-90, and Meas. Tech., 1971, 14, no.3, pp.504-5.

30.201. White, G.: "(Electrooptic) Modulation at High Information Rates", Bell. Syst. Tech. J., 1971, 50, no.8, pp.2607-45.

30.202. Zemlyachev, Ye.Z., and Parygin, V.N.: "Optical Cavity Light Modulator", Radiotekh. Elektron., 1971, 16, no.6, pp.1010-6, and Radio Eng. Electron. Phys., 1971, 16, no.6, pp.995-1000.

30.203. Bekhtin, Yu.I., and Shabel'nikov, A.V.: "Diffraction-Doppler Light Modulator", Prib. Tekh. Eksp., 1971, 14, no.4, pp.207-8, and Instrum. Exp. Tech., 1971, 14, no.4, pp.1195-6.

30.204. Voronin, E.S., et al.: "Reduction of Output-Power Fluctuations of a He-Ne Gas Laser", Prib. Tekh. Eksp., 1971, 14, no.5, pp.200-1, and Instrum. Exp. Tech., 1971, 14, no.5, pp.1499-1500.

30.205. Adrianova, I.I., Popov, Yu.V., and Terent'ev, V.E.: "Intracavity Modulation of Ruby Laser With Frequency Close to Axial-Mode Difference", Opt. Spekt., 1971, 31, no.6, pp.976-80, and Opt. Spectrosc., 1971, 31, no.6, pp.527-9.

30.206. Udelson, B.J., and Hines, R.E.: "Frequency Modulation of CW Avalanche Oscillator by Injected RF Signal. II", Microwave J., 1971, 14, no.12, pp.43-6.

30.207. Baglikov, V.B., and Parygin, V.N.: "Asynchronous Modulation of Gas Laser Coupling", Radiotekh. Elektron., 1971, 16, no.8, pp.1411-7, and Radio Eng. Electron. Phys., 1971, 16, no.8, pp.1320-5.

30.208. Zubkov, V.I., and Shcheglov, V.I.: "Frequency Modulation in a Ferrite Delay Line", Radiotekh. Elektron., 1971, 16, no.7, pp.1261-2, and Radio Eng. Electron. Phys., 1971, 16, no.7, pp.1198-9.

30.209. Baranov, L.I., et al.: "Design of Microwave Semiconductor Waveguide Modulators", Radiotekh. Elektron., 1971, 16, no.8, pp.1437-42, and Radio Eng. Electron. Phys., 1971, 16, no.8, pp.1339-42.

30.210. Russo, D.P.G., and Harris, J.H.: "Electrooptic Modulation in a Thin-Film Waveguide", Appl. Opt., 1971, 10, no.12, pp.2786-8.

30.211. Katagiri, Y., Toge, T., and Ihaya, A.: "Mode-Locked Laser (Using KDP)", Fujitsu Sci. Tech. J., 1971, 7, no.3, pp.65-85.

30.212. Chiba, T., Motoki, T., and Sugiura, Y.: "Optical Communication System Using (Electrooptic) Laser Double-Modulation Method", J. Inst. Telev. Eng. Jap., 1971, 25, no.8, pp.635-40.

30.213. Kenyon, N.D., et al.: "Experimental Millimetric (PIN-Diode) Path-Length Modulator", Bell Syst. Tech. J., 1971, 50, no.9, pp.2917-45.

30.214. Kuliev, T.A., Mustel', Ye.R., and Parygin, V.N.: "Analysis of an Electrooptic Light Modulator Using a KDP Crystal Near the Curie Temperature", Vestn. Mosk. Univ. Fiz. Astron., 1971, no.5, pp.547-52.

30.215. Bykovskii, Yu.A., et al.: "Frequency Modulation of a Semiconductor Laser by Injection Current", Kvantovaya Elektron., 1971, no.3, pp.90-2.

30.216. Bogdankevich, O.V., et al.: "Pumping of a Semiconductor Laser by an Electron Beam Modulated at a Microwave Frequency", Kvantovaya Elektron., 1971, no.4, pp.97-9.

30.217. Brodovich, N.A.: "Mode-Locking Using Single-Sideband Modulation", Opt. Spekt., 1971, 31, no.3, pp.428-32, and Opt. Spectrosc., 1971, 31, no.3, pp.228-30.

30.218. Belova, G.N.: "Modulation of Laser Radiation by a Vibrating Mirror", Akust. Zh., 1971, 17, no.3, pp.365-70, and Sov. Phys.-Acoust., 1972, 17, no.3, pp.309-12.

30.219. Francois, G.E., and Librecht, F.M.: "ADP 45°-x-Cut Four-Crystal Light Modulator", Appl. Opt., 1972, 11, no.2, pp.472-3.

30.220. Muller, K.H., Nimtz, G., and Selders, M.: "Fast CO_2-Laser Modulation by Hot Carriers", Appl. Phys. Lett., 1972, 20, no.8, pp.322-3.

30.221. Hasegawa, T., and Sato, H.: "Polarization Reflection-Type Light Modulator Using Ferroelectrics", Ferroelectrics, 1972, 3, no.2-4, pp.183-90.

30.222. Crisp, M.D.: "Frequency Modulation and Transient Effects in the Resonant Propagation of Coherent Light Pulses", Appl. Opt., 1972, 11, no.5, pp.1124-32.

30.223. Chang, W.S.C., and Loh, K.W.: "Theoretical Design of Guided-Wave Structure for Electrooptic Modulation at 10.6 micron", J. Quantum Electron. IEEE, 1972, QE-8, no.6, pp.463-70.

30.224. Reinhart, F.K., and Miller, B.I.: "Efficient (GaAl)As-GaAs Double-Heterostructure Light Modulators", Appl. Phys. Lett., 1972, 20, no.1, pp.36-8.

30.225. Kuizenga, D.J., and Becker, M.F.: "Controlled Bistable Operation of a FM Mode-Locked Nd^{3+}:YAG Laser", J. Quantum Electron. IEEE, 1972, QE-8, no.3, pp.385-6.

30.226. Polloni, R.: "Single-Transverse-Mode Mode-Locked Ruby Laser", J. Quantum Electron. IEEE, 1972, QE-8, no.4, pp.428-9.

30.227. McAvoy, N., Osmundson, J., and Schiffner, G.: "Broadband CO_2 Laser Coupling Modulation", Appl. Opt., 1972, 11, no.2, pp.473-4.

30.228. Otsuka, K., and Kimura, T.: "Carrier-Frequency Controlled Gigabit Pulse Generation With a Nd^{3+}:YAG Laser", J. Quantum Electron. IEEE, 1972, QE-8, no.1, pp.23-4.

30.229. Patel, B.S.: "Cavity Dumping of a Tranversely Excited CO_2 Laser", J. Appl. Phys., 1972, 43, no.7, pp.3215-6.

30.230. Smith, P.W., et al.: "Mode-Locked High-Pressure Waveguide CO_2 Laser", Appl. Phys. Lett., 1972, 21, no.10, pp.470-2.

30.231. Rheault, F., et al.: "Selection of a Single Pulse from a Mode-Locked TEA CO_2 Laser", Can. J. Phys., 1972, 50, no.16, pp.1876-80.

30.232. Jensen, R.E., and Tobin, M.S.: "Investigation of Gases for Stark Modulating a CO_2 Laser", J. Quantum Electron. IEEE, 1972, QE-8, no.2, pp.34-8.

30.233. El-Shandwily, M.E., and El-Dinary, M.: "TW Coherent-Light Phase Modulator", Trans. IEEE, 1972, MTT-20, no.2, pp.132-7.

30.234. Kiefer, J.E., Nussmeier, T.A., and Goodwin, F.E.: "Intracavity CdTe Modulators for CO_2 Lasers", J. Quantum Electron. IEEE, 1972, QE-8, no.2, pp.173-9.

30.235. Henderson, D.M.: "Effects of Mode Conversion in Acoustooptical Modulation", J. Quantum Electron. IEEE, 1972, QE-8, no.2, pp.184-91.

30.236. Arapov, A.P., et al.: "Laser Working With High-Frequency Modulation", Zh. Prikl. Spekt., 1972, 16, no.4, pp.638-41.

30.237. Imazu, S., Hirasawa, S., and Yoshida, N.: "Dependence in a He-Ne Laser of the Internally Modulated Light Beam on Longitudinal Electric Field", Jap. J. Appl. Phys., 1972, 11, no.6, pp.920-1.

30.238. Kato, D., and Sato, T.: "Tuning a Dye Laser by Rotatory Dispersive Elements", Opt. Commun., 1972, 5, no.2, pp.134-6.

30.239. Zusman, M.I., Maneshin, N.K., and Parygin, V.N.: "Modulation of 10-micron Radiation by Ultrasonic Waves", Vestn. Mosk. Univ. Fiz. Astron., 1972, 13, no.2, pp.190-4.

30.240. Clemetson, W.J., et al.: "100-GHz-Band Pathlength Modulator", Proc. IEEE, 1972, 60, no.7, pp.912-3.

30.241. Yamamoto, T.: "Asymmetrical Sidebands in a TW Light Modulator", Electron. Commun. Jap. 1972, 53, no.12, pp.128-35.

30.242. de Barros, M.A.R.P., and Wilson, M.G.F.: "High-Speed Electrooptic Diffraction Modulator for Baseband Operation", Proc. IEE, 1972, 119, no.7, pp.807-14.

30.243. Gaponov, S.V., and Paramonov, L.V.: "Laser Pulse Modulation by Ultrasound and an Auxiliary Passive Resonator", Izv. VUZ Radiofiz., 1972, 15, no.8, pp.1262-4.

30.244. Ninomiya, Y., and Motoki, T.: "$LiNbO_3$ Light Modulator", NHK Tech. J., 1972, 24, no.2, pp.35-42.

30.245. Marugin, A.M., and Ovchinnikov, V.M.: "Effect of Electrode Arrangement and z-Cut KDP Crystal Configuration on the Control Voltage for Switching", Opt.-Mekh. Prom., 1972, 39, no.1, pp.6-7, and Sov. J. Opt. Technol., 1972, 39, no.1, pp.4-5.

30.246. Waksberg, A., and Wood, J.: "Automatic Optical Bias Control for Laser Modulators", Rev. Sci. Instrum., 1972, 43, no.9, pp.1271-3.

30.247. White, G., and Chin, G.M.: "TW Electrooptic Modulators", Opt. Commun., 1972, 5, no.5, pp.374-9.

30.248. Ninomaya, Y.: "Thermally Self-Induced Decline of Extinction Ratio of Light Modulators", J. Quantum Electron. IEEE, 1972, QE-8, no.8, pp.710-4.

30.249. Karatsu, O., and Ogura, I.: "Cross-Modulation Effects in the Kr^{II} Laser", Jap. J. Appl. Phys., 1972, 11, no.8, pp.1165-72.

30.250. Kislyakovskii, A.V., Kushch, S.N., and Kabakov, L.T.: "Selective Waveguide Modulator", Izv. VUZ Radioelektron., 1972, 15, no.9, pp.1179-82.

30.251. Sugiura, T.: "4-GHz, 400-Mbit/s Quadraphase Modulators", NEC Res. Dev., 1972, no.27, pp.41-9.

30.252. Piskarev, V.I., and Shchelokov, A.N.: "Microwave Modulation of Light by a Travelling Wave (in Circular Guide)", Radiotekh. Elektron., 1972, 17, no.5, pp.1010-7, and Radio Eng. Electron. Phys., 1972, 17, no.5, pp.786-92.

30.253. Hookabe, K., and Matsuo, Y.: "Optimum Cut of KDP-Type Crystals for Light Modulation", Electron. Commun. Jap., 1972, 54, no.5, pp.77-84.

30.254. Bertolotti, M., et al.: "Acoustic Modulation of Light by Nematic Liquid Crystals", Appl. Phys. Lett., 1972, 21, no.2, pp.74-5.

30.255. Hammer, J.M., and Channin, D.J.: "Simple Acoustic Grating Modulators", Appl. Opt., 1972, 11, no.10, pp.2203-9.

30.256. Ito, T.: "Performance of Light Modulators for Optical Multiplexing and Demultiplexing", Electr. Commun. Lab. Tech. J., 1972, 21, no.5, pp.737-55.

30.257. Sokolovskii, I.I., and Kostylev, S.A.: "Frequency Control of Gunn-Effect Oscillators by Magnetic Field", Izv. VUZ Radioelektron. 1972, 15, no.8, pp.945-8.

30.258. Goto, K., Sueta, T., and Makimoto, T.: "TW Light-Intensity Modulators Using Method of Polarization-Rotated Reflection", J. Quantum Electron. IEEE, 1972, QE-8, no.6, pp.486-93.

30.259. Mazan'ko, I.P., and Petrashko, G.A.: "Influence of Parasitic Generation at 3.39 micron on Radiation Fluctuations of He-Ne Laser at 633 nm", Zh. Eksp. Teor. Fiz. Pis'ma, 1972, 15, no.5, pp.263-5, and JETP Lett., 1972, 15, no.5, pp.183-5.

30.260. Sklizhov, G.V., and Fedotov, S.I.: "Electrooptic Shutter with Synchronization by Laser Radiation", Prib. Tekh. Eksp., 1972, 15, no.2, pp.176-8, and Instrum. Exp. Tech., 1972, 15, no.2, pp.512-4.

30.261. Luk'yanov, D.P., and Pupov, A.D.: "Experimental Investigation of Interaction of Light and Circular Modulation Waves in a Quadratic Electrooptic Medium", Radiotekh. Elektron., 1972, 17, no.3, pp.565-8, and Radio Eng. Electron. Phys., 1972, 17, no.3, pp.439-42.

30.262. Abramov, V.S., et al.: "Use of Semiconductor Laser Diode as Modulator for Gas-Laser Radiation", Kvantovaya Elektron., 1972, no.2, pp.96-8, and Sov. J. Quantum Electron., 1972, 2, no.2, pp.170-1.

30.263. Kobayashi, T., et al.: "High-PRF Optical Generator Using a Fabry-Perot Electrooptic Modulator", Appl. Phys. Lett., 1972, 21, no.8, pp.341-3.

30.264. Sharma, B.L.: "Intercavity Electrooptic Frequency Modulation of CW Gas Lasers", Indian J. Pure Appl. Phys., 1972, 10, no.6, pp.424-6.

30.265. Tien, P.K., and Martin, R.J.: "Switching and Modulation of Light in Magneto-optic Waveguides of Garnet Films", Appl. Phys. Lett., 1972, 21, no.8, pp.394-6.

30.266. Adrianova, I.I., et al.: "Microwave Modulation of Light Based on Electro-optic Effect in ZnSe Crystal", Kvantovaya Elektron., 1972, no.7, pp.81-2, and Sov. Quantum Electron., 1972, 2, no.1, pp.72-3.

30.267. Lee, T.P., and Standley, R.D.: "Frequency Modulation of a Millimetre-Wave IMPATT-Diode Oscillator", Polytech. Tijdschr. Elektrotech. Elektron., 1972, 27, no.3, pp.86-94.

30.268. Helsztynski, J.: "Light Beam Modulation", Elektronika, 1972, no.10, pp.405-8.

30.269. Aubrecht, L.: "Contribution to the Theory of Single-Mode Laser Modulation", Czech. J. Phys., 1972, 22B, no.12, pp.1211-9.

30.270. Hutcheson, L.D., and Hughes, R.S.: "Rapid Transverse Acoustooptic Tuning (of Dye Lasers)", Appl. Opt., 1972, 11, no.12, pp.2981-3.

30.271. Saltz, P., and Streifer, W.: "Transient Analysis of an Electronically Tunable Dye Laser. I", J. Quantum Electron. IEEE, 1972, QE-8, no.12, pp.893-9.

30.272. Giles, M.K., and Hughes, R.S.: "Electromechanical Techniques for Rapid Frequency Tuning of Lasers", J. Phys. E, 1972, 5, no.12, pp.1216-22.

30.273. Cooper, R.W.: "IR Modulators Using Faraday Rotation in YIG", Electron. Appl. Bull., 1972, 31, no.4, pp.244-57.

30.274. Otsuka, K., and Kimura, T.: "Higher-Order Mode-Locking and Separation of Mode-Locked States in a Nd^{3+}:YAG Laser", Rev. Electr. Commun. Lab., 1972, 20, no.7-8, pp.682-9.

30.275. Arapov, A.P., et al.: "Acoustooptic Modulator for Lasers", Prib. Tekh. Eksp., 1972, 15, no.2, pp.173-5, and Instrum. Exp. Tech., 1972, 15, no.2, pp.509-11.

30.276. Knyaz'kov, B.N., and Yanovskii, M.S.: "Single-Sideband Modulation in Quasi-Optical Channel", Radiotekhnika, 1972, 27, no.9, pp.7-11, and Telecommun. Radio Eng. Pt 2, 1972, 27, no.9, pp.63-6.

30.277. Slutskaya, V.V.: "Electrically Controlled Absorption Modulators for Microwave Frequencies", Elekrosvyaz, 1972, 26, no.9, p.16, and Telecommun. Radio Eng. Pt 1, 1972, 26, no.9, pp.13-4.

30.278. Sokolovskii, I.I., and Kostylev, S.A.: "Modulation Sensitivity of Gunn Oscillators", Radiotekh. Elektron., 1972, 17, no.8, pp.1713-7, and Radio Eng. Electron. Phys., 1972, 17, no.8, pp.1350-3.

30.279. Feygel's, V.I.: "Extinction Ratio of Electrooptic Light Modulators Employing Class-42m Crystals", Opt.-Mekh. Prom., 1972, 39, no.10, pp.8-10, and Sov. J. Opt. Tech-ol nol., 1972, 39, no.10, pp.594-5.

30.280. Ninomiya, Y.: "Deterioration of Extinction Ratio of Light Modulators Due to Absorption Heating", Electron. Commun. Jap., 1972, 55, no.9, pp.90-7.

30.281. Petrov, A.S., Popov, L.N., and Fromin, V.D.: "Calculation of Deviation for a FM Gas Laser", Radiotekh. Elektron., 1972, 17, no.8, pp.1758-60, and Radio Eng. Electron. Phys., 1972, 17, no.8, pp.1396-9.

30.282. Zubarev, T.N., Sokolov, A.K., and Yudin, L.I.: "(Electrooptic) Light Modulator for Laser With External Spherical Mirrors", Prib. Tekh. Eksp., 1972, 16, no.6, p.224, and Instrum. Exp. Tech., 1972, 16, no.6, p.1882.

30.283. Yoshida, K., et al.: "Investigation of Laser Pulse Forming by Spark Gaps and Pockels' Cells", Electr. Eng. Jap., 1972, 92, no.6, pp.117-22.

30.284. Nishiwaki, A., and Mikami, T.: "Characteristics of Total-Reflection-Type Light Modulator With Interdigital Electrodes", Mem. Chubu Inst. Technol., 1972, 8A, pp.51-5.

30.285. Lee, J.A.: "Frequency Variation of IMPATT Diode Coupled With YIG", J. Korean Inst. Electron. Eng., 1972, 9, no.6, pp.270-5.

30.286. Goto, K., and Sueta, T.: "Phase-Compensated TW Light Modulators With Alternately Overtuned Crystals", Electron. Commun. Jap., 1972, 55, no.6, pp.119-26.

30.287. Kamach, Yu.E., et al.: "Contrast in a Class-42m-Crystal Electrooptic Shutter", Opt.-Mekh. Prom., 1972, 39, no.8, pp.14-6, and Sov. J. Opt. Technol., 1972, 39, no.8, pp.456-9.

30.288. Kazaryan, R.K., et al.: "Modulator for Millimetre Wavelengths", Prib. Tekh. Eksp., 1972, 15, no.5, p.136, and Instrum. Exp. Tech., 1972, 15, no.5, pp.1444-5.

30.289. Aksenov, E.T., et al.: "High-Frequency Ultrasonic Bragg Modulators", Zh. Tekh. Fiz., 1972, 42, no.11, pp.2432-4, and Sov. Phys.-Tech. Phys., 1973, 17, no.11, pp.1891-2.

30.290. Zasavitskii, I.I., et al.: "Tuning of PbSe Injection Lasers by Hydrostatic Pressure", Fiz. Tekh. Poluprov., 1972, 6, no.11, pp.2206-10, and Sov. Phys.-Semicond., 1973, 6, no.11, pp.1859-62.

30.291. Anani, O.: "Internally FM Gas Laser", J. Quantum Electron. IEEE, 1973, QE-9, no.4, pp.497-8.

30.292. Maksimenkov, P.P., and Mikhailov, B.M.: "Millimetre-Wave Magnetoelastic Amplitude Modulator Employing an Antiferromagnetic (Hematite)", Radiotekh. Elektron., 1973, 17, no.6, pp.1256-9, and Radio Eng. Electron. Phys., 1973, 17, no.6, pp.975-8.

30.293. Giles, M.K., and Hughes, R.S.: "Angular Dispersion of an Acoustooptic Bragg Cell Used in Wavelength Tuning of an Organic-Dye Laser", Appl. Opt., 1973, 12, no.2, pp.420-1.

30.294. Giles, M.K., Hughes, R.S., and Thompson, J.L.: "Angular Dispersion of Diffraction Gratings Used for Tuning Organic-Dye Lasers", Appl. Opt., 1973, 12, no.2, pp.421-2.

30.295. Milam, D.: "Brewster's-Angle Pockels'-Cell Design", Appl. Opt., 1973, 12, no.3, pp.602-6.

30.296. Cheo, P.K.: "Pulse Amplitude Modulation of a CO_2 Laser in an Electrooptic Thin-Film Waveguide", Appl. Phys. Lett., 1973, 22, no.5, pp.241-4.

30.297. Rao, B.S.S., Subrahmanyam, A., and Swarup, P.: "Technique of Modulating Pulsed Semiconductor Lasers", Trans. IEEE, 1973, COM-21, no.4, pp.284-9.

30.298. Roschmann, P., and Tolksdorf, W.: "SHF Oscillators with YIG-Tuned Gunn Elements", Int. Elektron. Rundsch., 1973, 27, no.3, p.73.

30.299. Sokolovskii, I.I., and Kostylev, S.A.: "Use of a YIG Sphere for Controlling a Gunn-Diode Oscillator", Izv. VUZ Radio-elektron., 1973, 16, no.2, pp.118-20.

30.300. Maes, J.P., and Moore, M.: "Intensity Stabilizing Circuit for a CW Laser", J. Phys. E, 1973, 6, no.1, pp.15-7.

30.301. Joshi, J.S.: "Wideband Varactor-Tuned X-Band Gunn Oscillators in Full-Height Waveguide Cavity", Trans. IEEE, 1973, MTT-21, no.3, pp.137-9.

30.302. Chown, M., et al.: "Direct Modulation of Double-Heterostructure Lasers at Rates Up to 1 Gbit/s", Electron. Lett., 1973, 9, no.2, pp.34-6.

30.303. Johnson, R.H.: "Characteristics of Acoustooptic Cavity Dumping in a Mode-Locked Laser", J. Quantum Electron. IEEE, 1973, QE-9, no.2, pp.255-7.

30.304. Spiewick, F.: "Tuning of Ion Lasers (by Tilted Etalon)", Messtechnik, 1973, 81, no.1, pp.31-2.

30.305. Reno, C.W.: "High-Data-Rate YAG-Laser (Modulation) Techniques", Appl. Opt., 1973, 12, no.4, pp.883-5.

30.306. Karpushko, F.V., et al.: "Programmed Control of Spectral Emission of Ruby Laser", Zh. Prikl. Spekt., 1973, 18, no.1, pp.23-7.

30.307. Antsiferov, V.V., et al.: "Monofrequency Ruby Laser With Electrooptic Q-Switching", Zh. Prikl. Spekt., 1973, 18, no.1, pp.142-4.

30.308. Carroll, J.E., and Farrington, J.G.: "Short-Pulse Modulation of GaAs Lasers With TRAPATT Diodes", Electron. Lett., 1973, 9, no.7, pp.166-7.

30.309. Tewari, D.P., et al.: "Effects of Self-Focusing on Self-Distortion of Amplitude Modulated Microwaves in Nonparobolic Semiconductors", Opto-Electron., 1973, 5, no.2, pp.131-5.

30.310. Nakamura, T.: "FM Light Modulator Using GaAs LED's", J. Inst. Telev. Eng. Jap., 1973, 27, no.2, pp.97-102.

3030.311. Dean, M., and Howes, M.J.: "Electronically Tuned Gunn-Oscillator Circuit", Trans. IEEE, 1973, ED-20, no.6, pp.597-8.

30.312. Channin, D.J.: "Optical Waveguide Modulation Using Nematic Liquid Crystals", Appl. Phys. Lett., 1973, 22, no.8, pp.365-6.

30.313. Weiss, J.A., and Schnur, J.M.: "Heterodyne Detection of Frequency Sweeping in the Output of TE CO_2 Lasers", Appl. Phys. Lett., 1973, 22, no.9, pp.453-4.

30.314. Curry, S.M., Cubeddu, R., and Hansch, T.W.: "Intensity Stabilization of Dye-Laser Radiation by Saturated Amplification", Appl. Phys., 1973,1,no.3,pp.153-9.

30.315. Russet, P., and Schulz, S.: "Direct Modulation of a Double Heterostructure Laser at 2.3 Gbit/s", Arch. Elektron. Ubertrag., 1973, 27, no.4, pp.193-5.

30.316. Cirkovic, L.J., and Jovicic, J.: "Tuning of Dye Lasers by Use of Christiansen Filters", Fizika, 1973, 5, no.1,pp.53-5.

30.317. Motegi, M., and Ihaya, A.: "High-Speed Electrooptical Modulator", Fujitsu Sci. Tech. J., 1973, 9, no.1, pp.91-114.

30.318. Lotspeich, J.F.: "Characteristics of Fabry-Perot Electrooptic Modulators With Internal Loss", Appl. Opt., 1973, 12, no.6, pp.1109-10.

30.319. Kaminow, I.P., et al.: "Thin-Film $LiNbO_3$ Electrooptic Light Modulator", Appl. Phys. Lett., 1973, 22, no.10, pp.540-2.

30.320. Yariv, A., Nussmeier, T.A., and Kiefer, J.E.: "Frequency Response of Intracavity Laser Coupling Modulation", J. Quantum Electron. IEEE, 1973,QE-9,no.6,pp.594-7.

30.321. Fleck, J.A., and Carman, R.L.: "Laser Pulse Shaping Due to Self-Phase Modulation in Amplifying Media", Appl. Phys. Lett., 1973, 22, no.10, pp.546-8.

30.322. Sheridan, J.P., Schnur, J.M., and Giallorenzi, T.G.: "Electrooptic Switching in Lowloss Liquid-Crystal Waveguides", Appl. Phys. Lett., 1973,22,no.11,pp.560-1.

30.323. Moore, I.J.H.S., and Stewart, J.A.C.: "State-Space Analysis of a Magnetically Tuned IMPATT-Oscillator Lumped Model", Trans. IEEE, 1973, MTT-21, no.6, pp.422-5.

30.324. Glance, B.: "Magnetically Tunable Microstrip IMPATT Oscillator", Trans. IEEE, 1973, MTT-21, no.6, pp.425-6.

30.325. Steinmetz, L.L., Pouliot, T.W., and Johnson, B.C.: "Cylindrical, Ring-Electrode, KDDP Electrooptic Modulator", Appl. Opt., 1973, 12, no.7, pp.1468-71.

30.326. Motoki, T.: "Low-Voltage Optical Modulator Using Electrooptically Induced Phase Gratings", Appl. Opt., 1973, 12, no.7, pp.1472-6.

30.327. Ittu, M., Dabu, R., and Nicolau, I.F.: "KDP Electrooptic Light Modulator", Rev. Roum. Phys., 1973, 18, no.6,pp.781-3.

30.328. Gerlach, H.: "Rapid Tuning of a Dye Laser", Opt. Commun., 1973, 8, no.1, pp.41-5.

30.329. Flachenecker, G., and de la Fuente, P.: "180°-PSK Modulator for Broadband Microwave Application", Nachr. Tech. Z., 1973, 26, no.4, pp.250-3.

30.330. Shilov, A.F., Lavrukovich, V.I., and Shas', A.V.: "Electrooptic Modulator Based on a Coaxial Stepped Resonator", Izv. VUZ Prib., 1973, 16, no.4, pp.112-6.

30.331. Shah, M.L.: "Fast Acoustooptic Waveguide Modulators", Appl. Phys. Lett., 1973, 23, no.2, pp.75-7.

30.332. Kateev, I.G., and Basov, A.A.: "Production of Picosecond Light Pulses Tightly Synchronized to a Reference Signal", Radiotekh. Elektron., 1973, 18, no.2, pp.355-61, and Radio Eng. Electron. Phys., 1973, 18, no.2, pp.254-8.

30.333. Dudnik, O.F., Kopylov, Yu.L., and Kravechnko, V.B.: "Modulation of Light With $(BaSr)Nb_2O_6$", Zh. Eksp. Teor. Fiz. Pis'ma, 1973, 18, no.7, pp.407-9, and JETP Lett., 1973, 18, no.7, pp.239-40.

30.334. Levinshtein, M.E.: "TRAPATT Oscillation in High-Frequency Modulation of IR Radiation", Fiz. Tekh. Poluprov., 1973, 7, no.6, pp.1236-7, and Sov. Phys.-Semicond., 1973, 7, no.6, p.832.

30.335. Davies, D.E.N.: "(Resonant-Ring) Technique for Producing Short Microwave Pulses", Radio Electron. Eng., 1973, 43, no.7, pp.418-20.

30.336. Taylor, H.F.: "Optical Switching and Modulation in Parallel Dielectric Waveguides", J. Appl. Phys., 1973, 44, no.7, pp.3257-62.

30.337. Martin, W.E.: "Waveguide Electrooptic Modulation in II-VI Compounds", J. Appl. Phys., 1973, 44, no.8, pp.3703-7.

30.338. Gyunashyan, K.S., et al.: "Cavity Light Modulator With KDP Crystal for an Electrooptical Rangefinder", Izv. VUZ Radio-elektron., 1973, 16, no.7, pp.92-4.

30.339. Vorob'ev, K.I., et al.: "Electro-optical Light Modulator Based on $BaTiO_3$", Izv. VUZ Fiz., 1973, no.7, pp.35-8.

30.340. Vorobeichikov, E.S., et al.: "Frequency Modulation of a Gas Laser", Izv. VUZ Fiz., 1973, no.6, pp.111-5.

30.341. Eliseev, P.G., and Khaidarov, A.V.: "Electrooptic and Piezoelectric Tuning of a Compound Cavity of a Semiconductor Laser", Opt. Spekt., 1973, 34, no.2, pp.343-6, and Opt. Spectrosc., 1973, 34, no.2, pp.194-5.

30.342. Schicketanz, D.: "Modulation of GaAs Laser Diodes (by Current)", Siemens Forsch. Entwick., 1973, 2, no.4,pp.218-21.

30.343. Hammer, J.M., Channin, D.J., and Duffy, M.T.: "Fast Electrooptic Waveguide Deflector Modulator", Appl. Phys. Lett., 1973, 23, no.4, pp.176-7.

30.344. Russo, D.P.G., and Kumar, C.S.: "Sputtered Ferroelectric Thin-Film Electro-optic Modulator", Appl. Phys. Lett., 1973, 23, no.5, pp.229-31.

30.345. Magdich, L.N., Safronov, O.I., and Sasov, V.N.: "Electrooptic Modulation of IR Radiation", Kvantovaya Elektron., 1973, no.6, pp.111-2, and Sov. J. Quantum Electron., 1973, 2, no.6, p.571.

30.346. Mustel', Ye.R., et al.: "Three-Dimensional Electron-Beam Light Modulator", Kvantovaya Elektron., 1973, no.6, pp.113-5, and Sov. J. Quantum Electron., 1973, 2, no.6, pp.572-3.

30.347. Ley, J.M.: "Light Modulators and Applications", Phys. Bull., 1973, 24, pp.441-3.

30.348. Odulov, S.G.: "Possible Optical Modulation Mechanism With Ruby Resonator Technique", Ukr. Fiz. Zh., 1973, 18, no.7, pp.1215-8.

30.349. Uehara, S.: "Focusing-Type Optical Modulator", J. Quantum Electron. IEEE, 1973, QE-9, no.10, pp.984-6.

30.350. Downing, B.J., and Myers, F.A.: "38-GHz Varactor-Tuned Gunn Oscillators", Electron. Lett., 1973, 9, no.11, pp.244-5.

30.351. Waksberg, A.: "Comparitive Evaluation Between Internal and External Polarization Modulation of Lasers", J. Quantum Electron. IEEE, 1973, QE-9, no.11, pp.1086-97.

30.352. Bespalova, M.P., Pikhtelev, A.I., and Timofeev, Yu.V.: "Modulation Method for [87]Rb Laser Tuning", Izv. VUZ Radiofiz., 1973, 16, no.6, pp.956-7.

30.353. Warner, A.W., and Pinnow, D.A.: "Miniature Acoustooptic Modulators for Optical Communications", J. Quantum Electron. IEEE, 1973, QE-9, no.12, pp.1155-7.

30.354. Lee, M.K., and Shin, C.C.: "TW Laser Phase Modulator With Partially Loaded Rectangular Waveguide", J. Korean Inst. Electron. Eng., 1973, 10, no.2, pp.1-6.

30.355. Pavlidis, D., and Hartnagel, H.: "Microwave PSK Pulse Modulator", Arch. Elektron. Ubertrag., 1973, 27, no.10, pp.450-1.

30.356. Harth, W.: "Large-Signal Direct Modulation of Injection Lasers", Electron. Lett., 1973, 9, no.22, pp.532-3.

30.357. Zemlyachev, Ye.Z., and Parygin, V.N.: "Interference SHF Light Modulator", Vestn. Mosk. Univ. Fiz. Astron., 1973, 14, no.4, pp.426-30.

30.358. Sen, A., and Kaw, P.: "Resonant-Absorption Method of Laser Modulation", J. Phys. D, 1973, 6, no.17, pp.2091-7.

30.359. Motoki, T.: "Low-Voltage Optical Modulator Using Electrooptically Induced Phase Grating", NHK Tech. J., 1973, no.8, pp.21-8.

30.360. Kasuya, T.: "Broadband Frequency Tuning of a He-Xe Laser With a Superconducting Solenoid", Appl. Phys., 1973, 2, no.6, pp.339-43.

30.361. Schmidt, R.V., Kaminow, I.P., and Carruthers, I.R.: "Acoustooptic Diffraction (Modulation) of Guided Optical Waves in $LiNbO_3$", Appl. Phys. Lett., 1973, 23, no.8, pp.417-9.

30.362. Valov, V.A., Mazov, L.S., and Piskarev, V.I.: "Millimetre-Wave Modulator Based on Interband Flashover in p-InSb", Izv. VUZ Radiofiz., 1973, 16,no.7,pp.1122-5.

30.363. Shah, M.L.: "Fast Acoustic Diffraction-Type Optical Waveguide Modulation", Appl. Phys. Lett., 1973, 23, no.10,pp.556-8.

30.364. Adrianova, I.I., et al.: "Operating Temperatures of GaAs Electrooptic Modulators", Opt.-Mekh. Prom., 1973, 40, no.3, pp.11-4, and Sov. J. Opt. Technol., 1973, 40, no.3, pp.153-5.

30.365. Washio, K., et al.: "Frequency-Tunable Mode-Locked CW Nd^{3+}:Glass Laser", J. Quantum Electron. IEEE, 1973, QE-9, no.8, pp.807-13.

30.366. Garside, B.K., and Lim, T.K.: "Ultrashort Pulses from Mode-Locked CW Dye Lasers", Opt. Commun., 1973, 8, no.4, pp.297-301.

30.367. Bagratashvili, V.N., et al.: "Frequency Tuning of an Electron-Beam-Pre-Ionized High-Pressure CO_2 Laser", Opt. Commun., 1973, 9, no.2, pp.135-8.

30.368. Lopez, J.C., Daude, N., and Greco, Ch.: "Laser Power Stabilization by Pockels' Effect", Rev. Phys. Appl., 1973, 8, no.4, pp.387-8.

30.369. Figueira, J.F., Reichelt, W.H., and Singer, S.: "Single-Nanosecond-Pulse Generation at 10.6 micron Using a Brewster's-Angle Modulator", Rev. Sci. Instrum., 1973, 44, no.10, pp.1481-4.

30.370. Booth, A.D.: "High-Efficiency, Medium-Speed, Modulator for Optical Data Transmission", Eng. J., 1973, 56, no.11, pp.55-6.

30.371. Tuchin, V.V.: "Characteristics of Frequency Modulation of He-Ne Lasers", Opt. Spekt., 1973, 35, no.4, pp.746-9, and Opt. Spectrosc., 1973, 35, no.4, pp.432-3.

30.372. Bogomolov, G.D., et al.: "Current-Stabilizer for Submillimetre Gas Laser", Prib. Tekh. Eksp., 1973, 16, no.4, pp.187-9, and Instrum. Exp. Tech., 1973, 16, no.4, pp.1206-7.

30.373. Sukhorukov, A.P., and Shchednova, A.K.: "Effect of Simultaneous Spatial and Temporal Modulation of a Laser Wave on Parametric Amplification", Opt. Spekt., 1973, 35, no.5, pp.929-33, and Opt. Spectrosc., 1973, 35, no.5, pp.537-9.

30.374. Zusman, M.I., et al.: "Acousto-optic Light Modulator", Radiotekh. Elektron., 1973, 18, no.6, pp.1203-7, and Radio Eng. Electron. Phys., 1973, 18, no.6, pp.876-9.

30.375. Baranov, L.I., et al.: "Design of Semiconductor Modulators Employing Distributed p-n-n$^+$ Structures", Radiotekh. Elektron., 1973, 18, no.6, pp.1307-10, and Radio Eng. Electron. Phys., 1973, 18, no.6, pp.964-7.

30.376. Ovvyan, P.P.: "Electrooptic Waveguide Modulator", Zh. Tekh. Fiz., 1973, 43, no.11, pp.2402-6, and Sov. Phys.-Tech. Phys., 1974, 18, no.11, pp.1516-8.

30.377. Kompanets, O.N., Kukudzhanov, A.R., and Mikhailov, E.L.: "Stabilization of Output Power of a CO_2 Laser", Kvantovaya Elektron., 1973, no.5, pp.122-4, and Sov. J. Quantum Electron., 1974, 3, no.5, pp.440-1.

30.378. Zakharov, S.D., et al.: "Evolution of Emission of Ultrashort Pulses from Nd^{3+}:Glass Laser", Kvantovaya Elektron., 1973, no.5, pp.52-6, and Sov. J. Quantum Electron., 1974, 3, no.5, pp.395-7.

30.379. Forno, C., and Jones, O.C.: "Hexamine Electrooptic Light Modulators", J. Phys. E, 1974, 7, no.2, pp.101-4.

30.380. Wemple, S.H.: "Materials for Magnetooptic Modulators", J. Electron. Mater., 1974, 3, no.1, pp.243-63.

30.381. Sosnowski, T.P., and Boyd, G.D.: "Efficiency of Thin-Film Optical-Waveguide Modulators Using Electrooptic Films or Substrates", J. Quantum Electron. IEEE, 1974, QE-10, no.3, pp.306-11.

30.382. Templin, A.S., and Gunshor, R.L.: "Analytic Model for Varactor-Tuned Waveguide Gunn Oscillators", Trans. IEEE, 1974, MTT-22, no.5, pp.554-6.

30.383. Telle, J.M., and Tang, C.L.: "Method for Electrooptical Tuning of Tunable Lasers", Appl. Phys. Lett., 1974, 24, no.2, pp.85-7.

30.384. Hutcheson, L.D., and Hughes, R.S.: "Electronic Tuning of a Dye Laser With Simultaneous Multiple-Wavelength Output", J. Quantum Electron. IEEE, 1974, QE-10, no.4, pp.462-3.

30.385. Scholtz, A.L., and Schiffner, G.: "CO_2-Laser TW Intracavity GaAs Coupling Modulator for Nanosecond Pulses", J. Quantum Electron. IEEE, 1974, QE-10, no.4, pp.457-9.

30.386. Botineau, J., Gires, F., and Vanneste, C.: "Modulation of IR Wave Guided by Free Carriers", C. R. Acad. Sci. (Paris), 1974, 278B, no.5, pp.171-2.

30.387. Uehara, S., Yamauchi, Y., and Izawa, T.: "Optical Intensity Modulator With Waveguide Structure", Appl. Phys. Lett., 1974, 24, no.1, pp.19-21.

30.388. Yamashita, E., Atsuki, K., and Akamatsu, T.: "Application of Microstrip Analysis to the Design of a Broadband Electro-optical Modulator", Trans. IEEE, 1974, MTT-22, no.4, pp.462-4.

30.389. Arkhipkin, V.G., Popov, L.N., and Fomin, V.D.: "Frequency Modulation of Gas Lasers With Multimirror Resonators", Izv. VUZ Fiz., 1974, no.2, pp.145-6.

30.390. Lopasov, V.P., Makogon, M.M., and Tytyshkin, I.S.: "Laser With (Electro-optic) Frequency Scanning", Izv. VUZ Fiz., 1974, no.2, pp.123-5.

30.391. Dutu, D., and Klemet, E.: "Investigation of an Automatic Frequency Stabilized CO_2 Laser", Rev. Roum. Phys., 1974, 19, no.1, pp.3-15.

30.392. Kalymnios, D., and Scibor-Rylski, M.T.V.: "$LiNbO_3$ TIR Modulator as Q-Switch", J. Phys. D, 1974, 7, no.8, pp.L79-82.

30.393. Ippen, E.P., Shank, C.V., and Gustafson, T.K.: "Self-Phase Modulation of Picosecond Pulses in Optical Fibres", Appl. Phys. Lett., 1974, 24, no.4, pp.190-2.

30.394. Nowicki, T.: "Acoustooptic and Electrooptic Modulators. Basics and Comparisons", Electro-Opt. Syst. Des., 1974, 6, no.2, pp.23-8.

30.395. Harthun, N.: "Laser Modulator With Pockels' Cell", Nachr. Tech. Z., 1974, 27, no.5, p.167.

30.396. Figueira, J.F.: "Extinction Ratios of GaAs and CdTe Electrooptical Modulators", J. Quantum Electron. IEEE, 1974, QE-10, no.7, pp.572-3.

30.397. Richardson, M.C.: "Production of Single Subnanosecond Multiline 10.6-micron Laser Pulses by Fast Optical Gating", Opt. Commun., 1974, 10, no.4, pp.302-5.

30.398. Champagne, L.F., O'Neill, F., and Whitney, W.T.: "Reliable Half-Wave Operation of a GaAs Pockels' Cell", Opt. Commun., 1974, 11, no.1, pp.11-3.

30.399. Bortfeld, D.P.: "Analysis of Heterojunction Optical Waveguides With a Modulated Region Smaller than the Guide Width", J. Quantum Electron. IEEE, 1974, QE-10, no.7, pp.551-6.

30.400. Morawski, T., and Bogdanowicz, A.: "Study of a Microwave Varactor Phase Modulator", Elektronika, 1974, 15, no.1, pp.33-5.

30.401. Fukunishi, S., et al.: "Electro-optic Modulation of Optical Guided Waves in $LiNbO_3$ Thin Film", Appl. Phys. Lett., 1974, 24, no.9, pp.424-6.

30.402. Kenan, R.P., Verber, C.M., and Wood, V.E.: "Wide-Angle Electrooptic Switch", Appl. Phys. Lett., 1974, 24, no.9, pp.428-30.

30.403. Chen, F.S., and Benson, W.W.: "$LiNbO_3$ Light Modulator for Optical-Fibre Communications", Proc. IEEE, 1974, 62, no.1, pp.133-4.

30.404. Milam, D., et al.: "Production of Intense Subnanosecond Pulses by Cavity Dumping", J. Quantum Electron. IEEE, 1974, QE-10, no.1, pp.20-5.

30.405. Bonek, E., et al.: "Proposed CO_2-Laser Standing-Wave Intracavity Coupling Modulation for a 53-GHz CW Signal", J. Quantum Electron. IEEE, 1974, QE-10, no.2, pp.128-30.

30.406. Wilson, L.O., and Reinhart, F.K.: "Phase-Modulation Nonlinearity of Double-Heterostructure p-n Junction Diode Light Modulators", J. Appl. Phys., 1974, 45, no.5, pp.2219-28.

30.407. Weil, R., and Halido, D.: "Resonant-Piezoelectrooptic Light Modulation", J. Appl. Phys., 1974, 45, no.5, pp.2258-65.

30.408. Gribble, R.F., and Kristal, R.: "Multifringe Electrooptic Phase Modulator", Rev. Sci. Instrum., 1974, 45, no.4, pp.520-2.

30.409. Hutcheson, L.D., and Hughes, R.S.: "Rapid Acoustooptic Tuning of a Dye Laser", Appl. Opt., 1974, 13, no.6, pp.1395-8.

30.410. Allain, M., and Roblin, G.: "Operation and Optimization of an Optical Delay Modulator", Opt. Commun., 1974, 11, no.2, pp.196-200.

30.411. Kaminow, I.P., Ramaswamy, V., and Schmidt, R.V.: "$LiNbO_3$ Ridge-Waveguide Modulator", Appl. Phys. Lett., 1974, 24, no.12, pp.622-4.

30.412. Nomura, S., et al.: "Low-Voltage Light Modulation With $(BaSr)Nb_2O_6$ at Video Frequency", Jap. J. Appl. Phys., 1974, 13, no.7, pp.1185-6.

30.413. Mozhaiskii, V.N., Pankratov, V.M., and Petrova, T.V.: "$LiNbO_3$ Crystal for Light Modulator", Kvantovaya Elektron., 1974, no.4, pp.99-101, and Sov. J. Quantum Electron., 1974, 3, no.4, pp.337-8.

30.414. Landstorfer, F., and Heller, M.: "Modulation Characteristics of an IMPATT-Diode Amplifier", Nachr. Tech. Z., 1974, 27, no.9, pp.356-8.

30.415. Burgov, V.A., and Remizov, V.V.: "Possibilities of Designing Light Modulators for Photographic Sound Recording", Tekh. Kino Telev., 1974, no.6, pp.37-42.

30.416. Hamadani, S.M., et al.: "Coherent Optical Pulse Reshaping in a Resonant Molecular Absorber", Appl. Phys. Lett., 1974, 25, no.3, pp.160-3.

30.417. Noda, J., Uchida, N., and Saku, T.: "Electrooptic Diffraction Modulator Using Out-Diffused Waveguiding Layer in $LiNbO_3$", Appl. Phys. Lett., 1974, 25, no.3, pp.131-3.

30.418. Ito, T.: "Performance of Light Modulators for Optical Time-Division Multiplexing", Rev. Electr. Commun. Lab., 1974, 22, no.1-2, pp.92-100.

30.419. Harth, W., and Siemsen, D.: "Subharmonic Resonance in Direct Modulation of Injection Lasers", Arch. Elektron. Ubertrag., 1974, 28, no.9, pp.391-2.

30.420. Cheo, P.K., and Gilden, M.: "Microwave Modulation of CO_2 Lasers in GaAs Optical Waveguides", Appl. Phys. Lett., 1974, 25, no.5, pp.272-4.

30.421. Vakulenko, A.M., et al.: "High-Speed Electrooptic DKDP-Crystal Switch", Kvantovaya Elektron., 1974, 1, no.1, pp.138-41, and Sov. J. Quantum Electron., 1974, 4, no.1, pp.76-7.

30.422. Vanderleeden, J.C.: "Miniature Intracavity Modulated Solid-State Lasers", Opto-Electron., 1974, 6, no.5, pp.393-400.

30.423. Keller, H.: "Limiting the Peak Intensity of a Mode-Locked Nd^{3+}:Glass Laser by Electronic Feedback", Opto-Electron., 1974, 6, no.5, pp.419-20.

30.424. Menyuk, N., et al.: "Effects of Pressure on Optically Pumped GaSb, InAs, and InSb Lasers", J. Appl. Phys., 1974, 45, no.8, pp.3477-84.

30.425. Aitchison, C.S.: "Gunn-Oscillator Electronic Tuning Range and Reactance Compensation", Electron. Lett., 1974, 10, no.23, pp.488-9.

30.426. Aitchison, C.S.: "Figure of Merit for Varactor-Tuned Microwave Oscillator", Int. J. Electron., 1974, 37, no.5, pp.705-8.

30.427. Alcock, A.J., and Walker, A.C.: "Generation and Detection of 150ps Mode-Locked Pulses from a Multi-Atmosphere CO_2 Laser", Appl. Phys. Lett., 1974, 25, no.5, pp.299-301.

30.428. Kasuya, T.: "Development of a Zeeman-Tuned Atomic Gas Laser", Laser Elektro-Opt., 1974, 6, no.2, pp.36-8.

30.429. Berezovskii, V.V., et al.: "Variation of Duration of Pulses Emitted by a Mode-Locked CO_2 Laser", Kvantovaya Elektron., 1974, 1, no.2, pp.447-9, and Sov. J. Quantum Electron., 1974, 4, no.2, pp.256-7.

30.430. Grimblatov, V.M., Teselkin, V.V., and Chulyaeva, E.G.: "Tuning an Ar-Ion Laser by a Magnetic Field", Zh. Prikl. Spekt., 1974, 21, no.3, pp.537-9.

30.431. Degnan, J.J.: "Minimization of the Prime Power Consumption of a Coupling-Modulated Gas-Laser Transmitter", Appl. Opt., 1974, 13, no.11, pp.2489-98.

30.432. Bonek, E., and Magerl, G.: "Propagation Characteristics of Dielectrically Loaded Rectangular Waveguides for Laser-Beam Modulators", Arch. Elektron. Ubertrag., 1974, 28, no.12, pp.499-506.

30.433. Harrington, D., and Malmstadt, H.V.: "Digital-Scanning Tunable Dye Laser for Spectroanalytical Methods", Am. Lab., 1974, 6, no.3, pp.33-40.

30.434. Gavalda, J.: "Three-Channel Light Modulator", Rev. Esp. Electron., 1974, 21, no.241, pp.42-3.

30.435. Leclert, A., et al.: "Propagation and Modulation of Light at 1060 nm in Magnetic Garnets", Opt. Commun., 1974, 12, no.4, pp.414-5.

30.436. Shcherbov, V.A., Goroshko, A.I., and Savchenko, V.N.: "Quasi-Optical Frequency Shifter", Izv. VUZ Radioelektron., 1974, 17, no.7, pp.72-7.

30.437. Lotspeich, J.F.: "Single-Crystal Electrooptic Thin-Film Waveguide Modulators for IR Laser Systems", Appl. Opt., 1974, 13, no.11, pp.2529-39.

30.438. Tada, K., and Hirose, K.: "Light Modulator Using Perturbation of Synchronism Between Two Coupled Guides", Appl. Phys. Lett., 1974, 25, no.10, pp.561-2.

30.439. Figueira, J.F., and Sutphin, H.D.: "Generation of Multiband 1-ns Pulses in CO_2 Lasers", Appl. Phys. Lett., 1974, 25, no.11, pp.661-3.

30.440. Bonek, E., and Korecky, H.: "Intracavity Millimetre-Wave Coupling Modulation of a CO_2 Laser", Appl. Phys. Lett., 1974, 25, no.12, pp.740-1.

30.441. Zakharov, M.I., et al.: "Possible Developments of Angle-Controlled Extraction of Radiation from a Laser", Avtometriya, 1974, no.6, pp.40-5.

30.442. Patel, B.S.: "Modulation of CO_2 Lasers by GaAs", Indian J. Technol., 1974, 12, no.6, pp.266-8.

30.443. Basov, N.G., Kryukov, P.G., et al.: "Generation of High-Power Nanosecond Pulses in a Nd^{3+}:Glass Laser System", Kvantovaya Elektron., 1974, 1, no.6, pp.1428-34, and Sov. J. Quantum Electron., 1974, 4, no.6, pp.791-4.

30.444. Sasaki, H., Kushibiki, J., and Chubachi, N.: "Efficient Acoustooptic TE-TM Mode Conversion in ZnO Films", Appl. Phys. Lett., 1974, 25, no.9, pp.476-7.

30.445. Goncharov, V.N.: "Matching of a Broadband Microwave Modulator of Light Beams", Izv. VUZ Radioelektron., 1974, 17, no.10, pp.119-22.

30.446. Stefanov, V.J., and Apostolov, K.V.: "Increase of Efficiency of Kerr Cell Electrooptic Modulator", Bulg. J. Phys., 1974, 1, no.2, pp.189-98.

30.447. Kamenskii, E.I., and Kozlov, V.V.: "Spontaneous Mode Locking in a Laser With Multifaceted Active Element", Zh. Tekh. Fiz., 1974, 44, no.6, pp.1323-6, and Sov. Phys.-Tech. Phys., 1974, 19, no.6, pp.831-2.

30.448. Bogdankevich, O.V., et al.: "Forced Locking of Longitudinal Modes in an Electron-Beam-Pumped Semiconductor Laser", Kvantovaya Elektron., 1974, 1, no.5, pp.1264-5, and Sov. J. Quantum Electron., 1974, 4, no.5, p.701.

30.449. Perel', V.I., and Tendler, M.B.: "Effect of Movement of an Additional Mirror on the Intensity of a Gas Laser", Opt. Spekt., 1974, 37, no.3, pp.569-73, and Opt. Spectrosc., 1974, 37, no.3, pp.322-4.

30.450. Govor, I.N., and Nesterenko, V.M.: "Stabilization of Power of a Laser", Prib. Tekh. Eksp., 1974, 17, no.3, pp.168-9, and Instrum. Exp. Tech., 1974, 17, no.3, pp.822-3.

30.451. Toyoda, K., Aoyagi, Y., and Namba, S.: "Light Modulation in ZnS Thin-Film Waveguides by Free-Carrier Injection", Sci. Pap. Inst. Phys. Chem. Res., 1974, 68, no.4, pp.124-34.

31. RADIATION DETECTORS

31.1. Walsh, D., and Pearson, K.: "P-N Junction Diodes as Millimetre-Wave Detectors", Electron. Lett., 1965, 1, p.240.

31.2. Sobol, G.A.: "Detection of Centimetre Waves in a Gas-Discharge Plasma", Izv. VUZ Radiofiz., 1965, 8, p.420, and Sov. Radiophys., 1965, 8, p.303.

31.3. Harrison, R.I., and Zucker, J.: "Hot-Carrier Microwave Detector", Proc. IEEE, 1966, 54, p.588.

31.4. Genzabella, C., and Howell, C.M.: "Integrated S-Band Mixers", Int. Conv. Rec. IEEE, 1966, 14, pt 5, p.113.

31.5. Severin, P.J.W., and Van Nie, A.G.: "Simple and Rugged, Wideband, Gas-Discharge Detector for Millimetre Waves", Trans. IEEE, 1966, MTT-14, p.431.

31.6. Arams, F., et al.: "Millimetre Mixing and Detection in Bulk InSb", Proc. IEEE, 1966, 54, p.612.

31.7. Pashin, Yu.N., Khilov, Yu.K., and Etkin, V.S.: "Noise in Semiconductor Diodes Under Pulse Conditions", Radiotekh. Elektron., 1966, 11, no.8, pp.1528-30, and Radio Eng. Electron. Phys., 1966, 11, no.8.

31.8. Eng, S.T.: "Recent Results on Low 1/f-Noise Mixer Diodes", Proc. IEEE, 1966, 54, no.12, pp.1968-70.

31.9. Dickens, L.E.: "Spreading Resistance as a Function of Frequency", Trans. IEEE, 1967, MTT-15, no.2, pp.101-9.

31.10. Denker, S.P., and Scarmgella, D.: "Competing Detection Mechanisms in Hot-Carrier Microwave Diodes", Solid-State Electron., 1967, 10, no.8, pp.777-84.

31.11. Anand, Y., and Doherty, W.E.: "Reciprocity in Si Schottky Diodes", Electron. Lett., 1967, 3, no.6, p.236.

31.12. Fayne, P.G.: "Video-Crystal Detection of 337-micron Laser Radiation", Electron. Lett., 1967, 3, no.7, pp.338-9.

31.13. Vystavkin, A.N., et al.: "Investigation of the Influence of Impurities and a Magnetic Field on the Detection Properties of n-InSb in the Submillimetre Range", Fiz. Tekh. Poluprov., 1967, 1, no.6, pp.844-54, and Sov. Phys.-Semicond., 1967, 1, no.6.

31.14. Lippman, R.: "(Diode) Mixer for 60 GHz", Frequenz, 1967, 21, no.11,pp.360-2.

31.15. Barber, M.R.: "Noise Factor and Conversion Loss of Schottky Mixer Diode", Trans. IEEE, 1967, MTT-15, no.11,pp.629-35.

31.16. Deficis, A., Delfour, A., and Vidallon, C.: "Recovery Phenomena in Point Diodes", C. R. Acad. Sci. (Paris), 1967, 265B, no.17, pp.873-6.

31.17. McColl, M., et al.: "Improved 94-GHz GaAs Mixer Diodes Using Au-Cu Alloy Whiskers", Proc. IEEE, 1967, 55, no.12, pp.2169-70.

31.18. Kodali, V.P.: "Skin Effect in Microwave Semiconductor Diodes", Electron. Lett., 1968, 4, no.4, pp.67-8.

31.19. Radic, Z.: "Microwave Tunnel-Diode Mixer for J-Band", Elektrotehnika, 1968, 11, no.1, pp.32-42.

31.20. Oxley, T.H.: "Ge Microwave Diodes for Broadband Mixer and Low-Level Detector Applications", Mullard Tech. Commun., 1968, 10, pp.122-32.

31.21. Rudin, V.L.: "Investigation of an Opticoacoustic Receiver of Centimetric Waves", Opt. Spekt., 1968, 24, p.272, and Opt. Spectrosc., 1968, 24, p.139.

31.22. Roulston, D.J.: "Low-Noise Photoparametric Upconverter", J. Solid-State Circuits IEEE, 1968, SC-3, p.431.

31.23. Rzewuski, M.: "Photodetector With Frequency Conversion", Arch. Elektrotech., 1968, 17, p.681.

31.24. Kampa, J., and Sielanko, W.: "TW Phototube for Demodulation of Microwave-Modulated Laser Light", Pr. PIE, 1968, 9, p.159.

31.25. Vilisov, A.A., Vyatkin, A.P., and Dement'ev, V.A.: "GaAs Microwave Diodes", Izv. VUZ Radioelektron., 1968, 11, p.1080.

31.26. Anand, Y., and Howell, C.M.: "Burnout Criterion for Schottky Mixer Diodes", Proc. IEEE, 1968, 56, p.2098.

31.27. van Iperen, B.B.: "Impedance Relations in a Diode Waveguide Mount", Nachr. Tech. Fachber., 1968, 35, p.483.

31.28. Okabe, T., et al.: "Improvement of 50-GHz Si Point-Contact Mixer Diode by Ion Bombardment", Rep. Fac. Eng. Shizuoka Univ., 1968, no.19, p.25.

31.29. Lichtenberger, G.L.: "Simple Microwave Detector for 120-GHz Application", Rev. Sci. Instrum., 1969, 40, p.807.

31.30. Malyutenko, V.K., et al.: "Detector of IR Radiation Based on n-CdSb", Ukr. Fiz. Zh., 1969, 14, p.1570.

31.31. Svetlichnyi, V.M.: "Response of Hot-Carrier Microwave Detectors", Izv. VUZ Radioelektron., 1969, 12, p.1353.

31.32. Nesmelova, I.M., et al.: "Improved Sensitivity of InSb and InAs Detectors", Prib. Tekh. Eksp., 1969, no.3, p.185, and Instrum. Exp. Tech., 1969, no.3, p.739.

31.33. Vystavkin, A.N., et al.: "Measurement of Submillimetre Radiation With n-InSb Detectors", Electron Technol., 1969, 2, no.2-3, p.247.

31.34. Maher, A.T., Streetman, B.G., and Holonyak, N.: "IR Detection Properties of Zn-Doped Si PIN Diodes", Trans. IEEE, 1969, ED-16, p.963.

31.35. Stelzer, E.L., Schmit, J.L., and Tufte, O.N.: "(HgCd)Te as IR Detector Material", Trans. IEEE, 1969, ED-16, p.880.

31.36. Vlasov, V.G., and Lazneva, E.V.: "Threshold Sensitivity of Heterodyne Method for Detection of AM Radiation With a Photodiode", Opt.-Mekh. Prom., 1969, 36, no.3, p.19, and Sov. J. Opt. Technol., 1969, 36, no.3, p.395.

31.37. Berman, L.V., et al.: "Low-Temperature Detectors for Long-Wavelength IR Spectral Instruments", Prib. Tekh. Eksp., 1969, no.6, p.218, and Instrum. Exp. Tech., 1969, no.6, pp.1630-1.

31.38. Akaike, M., and Okamura, S.: "Semiconductor-Diode Mixer for Millimetre-Wave Region", Electron. Commun. Jap., 1969, 52, no.10, pp.84-93.

31.39. du Chatenier, F.J., and van den Broek, J.: "Electrical Properties of Vapour-Deposited Layers of PbO", Philips Res. Rep., 1969, 24, p.392.

31.40. Bassi, L., and Muzii, L.: "Au:Ge Photodetector for a CO_2 Laser", Note Recens. Not., 1969, 18, no.2, p.246.

31.41. Topp. J.A., et al.: "Improvement of Signal-to-Noise Ratio of Photomultipliers for Very Weak Signals", Rev. Sci. Instrum., 1969, 40, p.1164.

31.42. Rabinovich, A.I.: "Spectrum of Photocurrent Fluctuations and Sensitivity Threshold of a Photodiode", Fiz. Tekh. Poluprov. 1969, 3, p.424, and Sov. Phys.-Semicond., 1969, 3, no.3, pp.357-9.

31.43. Beerman, H.P.: "Pyroelectric Detector of IR Radiation", Trans. IEEE, 1969, ED-16, p.554.

31.44. Alday, J.R.: "Millimetre-Wave Detectors Using the Pyroelectric Effect", Trans. IEEE, 1969, ED-16, p.598.

31.45. Guoga, V.I., and Pozhela, Yu.K.: "Sensitivity of a Hot-Carrier Microwave Detector", Radiotekh. Elektron., 1969, 14, no.3, p.565, and Radio Eng. Electron. Phys., 1969, 14, no.3, pp.492-3.

31.46. Toloczko, M.: "Sensitivity of Heterodyne Microwave Receiver", Przegl. Telekomun., 1969, 42, no.4, p.97.

31.47. Aarons, B.D., and Manuel, J.: "Enhancement of Photomultiplier Sensitivity by an Optical Method", J. Phys. E, 1969, 2, p.734.

31.48. Yamamoto, J., Yoshinaga, H., and Kon, S.: "Far-IR Resonant Photoconductive Properties of n-InSb", Jap. J. Appl. Phys., 1969, 8, p.242.

31.49. Lindley, W.T., et al.: "GaAs Schottky-Barrier Avalanche Photodiodes", Appl. Phys. Lett., 1969, 14, p.197.

31.50. Becklake, E.J.S., Payne, C.D., and Prewer, B.E.: "Polarity Reversal in the Detected Output of a Point-Contact Diode at Submillimetre Wavelengths", Electron. Lett., 1969, 5, p.544.

31.51. Lauerenz, D.O., and Gaddy, O.L.: "Measurement of Duration of Sampling Function in the Dynamic Crossed-Field Photomultiplier", Proc. IEEE, 1969, 57, p.2153.

31.52. Abrams, R.L.: "Photomixing at 10.6 micron With (SrBa)Nb$_2$O$_6$ Pyroelectric Detectors", Appl. Phys. Lett., 1969,15,p.251.

31.53. Kohn, A.N., and Schlickman, J.J.: "1-2 micron (HgCd)Te Photodetectors", Trans. IEEE, 1969, ED-16, p.885.

31.54. Yonezu, H., and Kawaji, A.: "Computer-Aided Design of a Si Avalanche Photodiode", Trans. IEEE, 1969, ED-16, p.923.

31.55. McIntyre, R.J.: "Comparison of Photomultipliers and Avalanche Photodiodes for Laser Applications", Trans. IEEE, 1970, ED-17, p.347.

31.56. Rzewuski, M.: "Photodetectors With Microwave Polarization", Elektronika, 1970, no.3, p.103.

31.57. Nishida, K., et al.: "Avalanche Photodiode With a Tapered Light-Focusing Fibre Guide", Proc. IEEE, 1970, 58, p.790.

31.58. Albrecht, P., and Bechteler, M.: "Noise Factor and Conversion Loss of Self-Excited Gunn-Diode Mixers", Electron. Lett., 1970, 6, p.321.

31.59. Becklake, E.J.S., Payne, C.D., and Prewer, B.E.: "Submillimetre Performance of Diode Detectors Using Ge, Si, and GaAs", J. Phys. D, 1970, 3, p.473.

31.60. Sodha, M.S., Dubey, P.K., and Sirohi, R.S.: "Optimum Nonlinear Optical Mixing in n-InSb", Solid-State Commun., 1970, 8, no.2, p.117.

31.61. Davis, Q.V., and Kulczyk, W.K.: "Optical and Electronic Mixing in an Avalanche Photodiode", Electron. Lett., 1970, 6, p.25.

31.62. Gunter, W.D., Grant, G.R., and Shaw, S.A.: "Optical Devices to Increase Photocathode Quantum Efficiency", Appl. Opt., 1970, 9, p.251.

31.63. Day, H.M., Gleason, K.R., and Macpherson, A.C.: "Role of Edge Capacitance in Design of Microwave Schottky Detector Diodes", Solid-State Electron. 1970, 13, no.7, p.1111.

31.64. Shepherd, F.D., Yang, A.C., and Taylor, R.W.: "1-2 micron Si Avalanche Photodiode", Proc. IEEE, 1970, 58, p.1160.

31.65. Rolls, W.H., and Eddington, R.J.: "(PbSn)Te Photovoltaic Detectors for the 8-14 micron Band", Infrared Phys., 1970, 10, no.1, p.71.

31.66. Lakes, R., and Poultney, S.K.: "Photocathode Quantum Efficiency Enhancement of RCA Photomultiplier at 633 nm", Appl. Opt., 1970, 9, p.2192.

31.67. Abrams, R.L., and Gandrud, W.B.: "Heterodyne Detection of 10.6-micron Radiation by Metal-to-Metal Point-Contact Diodes", Appl. Phys. Lett., 1970, 17, p.150.

31.68. Boutot, J.P., and Pietri, G.: "Ultra-High-Speed Microchannel Photomultiplier", Trans. IEEE, 1970, ED-17, p.493.

31.69. Jennings, R.J., Gunter, W.D., and Grant, G.R.: "Quantum Efficiencies Greater than 50% from Commercially Available Photomultipliers", J. Appl. Phys., 1970, 41, p.2266.

31.70. Hopkins, J.B.: "Microwave Backward Diodes in InAs", Solid-State Electron., 1970, 13, p.697.

31.71. Voronkov, V.P., et al.: "Investigation of Receiver Based on Avalanche Photodiode", Izv. VUZ Fiz., 1970, no.5, p.116.

31.72. Nishida, K.: "Optimization of Multiplication Factor of Avalanche Photodiode", Jap. J. Appl. Phys., 1970, 9, no.5, p.481.

31.73. Burrus, C.A., and Sharpless, W.M.: "Planar p-n Junction Ge Photodiodes for Microwave Modulation Frequencies", Solid-State Electron., 1970, 13, no.9, p.1283.

31.74. Rolls, W.H., Lee, R., and Eddington, R.J.: "Preparation and Properties of (PbSn)Te Photodiodes", Solid-State Electron., 1970, 13, no.1, p.75.

31.75. Igo, T., Ohwada, K., and Noguchi, Y.: "Regenerative Light Pulse Detection Using the Gunn Effect", Jap. J. Appl. Phys., 1970, 9, no.10, pp.1283-5.

31.76. Van Nie, A.G.: "Low-Level Properties of Microwave Crystal Detectors", Philips Electron. Meas. Microwave Not., 1970, no.1, pp.1-9.

31.77. Contreras, B., and Gaddy, O.L.: "Nanosecond-Response Room-Temperature IR Detection With Thin-Film Bolometers", Appl. Phys. Lett., 1970, 17, no.10, pp.450-3.

31.78. Sun, C., and Walsh, T.E.: "Packaged System for a Solid-State Microwave-Biased Photoconductive Detector for 10.6 micron", Proc. IEEE, 1970, 58, no.10, pp.1732-6.

31.79. Iwasaki, H.: "Characteristics of BaTiO₃ Pyroelectric Detector for Optical Radiation", J. Radio Res. Lab., 1970, 17, no.90, pp.147-51.

31.80. Peyton, B.J., et al.: "Photodiode 10.6-micron Heterodyne Detection", Proc. IEEE, 1970, 58, no.10, pp.1769-70.

31.81. Igo, T., and Toyoshima, Y.: "Measured Noise in GaAs Avalanche Photodiodes", Jap. J. Appl. Phys., 1970, 9, no.10, pp.1286-7.

31.82. Raines, J.A., Parker, J.T., and Monahan, B.C.: "Stabilization of the Operating Point of an Avalanche Photodetector", J. Phys. E, 1970, 3, no.8, pp.621-3.

31.83. Jervis, M.H., and Morten, F.D.: "(HgCd)Te Detectors at 5 micron and Normal Ambient Temperature", Mullard Tech. Commun., 1970, 11, no.108, pp.182-4.

31.84. Gardiner, J.G., and Banerjee, A.R.: "Conversion-Loss Stability and Gain Comparison in Schottky-Diode Mixers", Electron. Lett., 1970, 6, no.26, pp.829-30.

31.85. Plaksii, V.T., and Svetlichnyi, V.M.: "Voltage/Power Sensitivity of BiSb Point-Contact Microwave Detectors", Izv. VUZ Radioelektron., 1970, 13, no.11, pp.1383-5.

31.86. Voronenko, V.P., and Kukushkin, V.V.: "Minimum Noise Factor of a Ge Tunnel Diode Autodyne Microwave Converter", Radiotekh. Elektron., 1970, 15, no.4, pp.740-3, and Radio Eng. Electron. Phys., 1970, 15, no.4, pp.740-3.

31.87. Godzinski, Z., et al.: "Equivalent Resistance of Microwave TW Phototube", Arch. Elektrotech., 1970, 19, no.4, pp.787-99.

31.88. Hallford, B.R.: "Low-Noise Microstrip Mixer on Plastic Substrate", Trans. IEEE, 1970, MTT-18, no.12, pp.1178-81.

31.89. Gulakov, I.R., et al.: "Single-Electron Distribution of Pulses in Photomultiplier With Large Electron Transit Coefficient", Prib. Tekh. Eksp., 1970, no.3, pp.209-11, and Instrum. Exp. Tech., 1970, no.3, pp.877-9.

31.90. Zanadvorov, P.N., and Moldavskaya, V.M.: "Optical Detection in a Matched Cavity", Radiotekh. Elektron., 1970, 15, no.9, and Radio Eng. Electron. Phys., 1970, 15, no.9, pp.1737-8.

31.91. Gertsenshtein, M.Ye., et al.: "Noise Factor of Microwave Receivers", Radiotekhnika, 1970, 25, no.1, p.70, and Telecommun. Radio Eng. Pt 2, 1970, 25, no.1.

31.92. Ballik, E.A.: "Area and Wavelength Sensitivity of a Photomultiplier", Appl. Opt., 1971, 10, no.3, pp.689-91.

31.93. Land, P.L.: "Discussion of the Region of Linear Operation of Photomultipliers", Rev. Sci. Instrum., 1971, 42, no.4, pp.420-5.

31.94. Kamenskii, N.N., et al.: "Dependence of Sensitivity and Response of Photodiodes on Size and Position of Light Spot", Izv. VUZ Radioelektron., 1971, 14, no.1, pp.72-5.

31.95. Satyukov, A.I., and Svetlichnyi, V.M.: "Voltage/Current Characteristics of Thermoelectric (Hot-Carrier) Microwave Detectors", Izv. VUZ Radioelektron., 1971, 14, no.1, pp.94-8.

31.96. Contreras, B., and Gaddy, O.L.: "Heterodyne Detection at 10.6 micron With Thin-Film Bolometers", Appl. Phys. Lett., 1971, 18, no.7, pp.277-8.

31.97. Armencha, N.N., and Tarkhin, D.V.: "Mechanisms Governing the Frequency of Current Pulses in Si Avalanche Photodiodes", Fiz. Tekh. Poluprov., 1971, 5, no.2, pp.273-9, and Sov. Phys.-Semicond., 1971, 5, no.2, pp.235-40.

31.98. Lobov, G.D.: "Metal-Dielectric-Metal Tunnel Junction for Signal Detection at Microwaves", Izv. VUZ Radioelektron., 1971, 14, no.3, pp.298-307.

31.99. Lazarus, M.J., Bullimore, E.D., and Novak, S.: "Sensitive Millimetre-Wave Self-Oscillating Gunn-Diode Mixer", Proc. IEEE, 1971, 59, no.5, pp.812-4.

31.100. Personick, S.D.: "Image-Band Interpretation of Optical Heterodyne Noise", Bell Syst. Tech. J., 1971, 50, no.1, pp.213-6.

31.101. Schaeffer, E.M., and Gaddy, O.L.: "Noise Properties of the Dynamic Crossed-Field Photomultiplier", Trans. IEEE, 1971, ED-18, no.6, pp.388-90.

31.102. Robben, F.: "Noise in the Measurement of Light With Photomultipliers", Appl. Opt., 1971, 10, no.4, pp.776-96.

31.103. Katoh, M., and Akaiwa, Y.: "4-GHz Integrated-Circuit Mixer", Trans. IEEE, 1971, MTT-19, no.7, pp.634-7.

31.104. Vilisov, A.A., et al.: "Sensitivity of GaAs Detector Diodes", Izv. VUZ Radioelektron., 1971, 14, no.5, pp.585-7.

31.105. Plaksii, V.I., Svetlichnyi, V.M., and Peresun'ko, O.A.: "Response Period of BiSb Point-Contact Microwave Detectors", Izv. VUZ Radioelektron., 1971, 14, no.5, pp.588-9.

31.106. Anand, Y., and Moroney, W.J.: "Microwave Mixer and Detector Diodes", Proc. IEEE, 1971, 59, no.8, pp.1182-90.

31.107. Haecker, W., Groezinger, O., and Pikuhn, M.H.: "IR Photon Counting by Ge Avalanche Diodes", Appl. Phys. Lett., 1971, 19, no.4, pp.113-5.

31.108. Stanley, C.R.: "Detectors for the Range 1.5-30 micron", Opt. Laser Technol., 1971, 3, no.3, pp.144-9.

3131.109. Krumm, C.F., and Haddard, G.I.: "Submillimetre-Wave Detection by Paramagnetic Materials", J. Quantum Electron. IEEE, 1971, QE-7, no.10, pp.475-84.

31.110. Bar-David, I.: "Extension of the Discrete Model for Optical Detection", Proc. IEEE, 1971, 59, no.11, p.1612.

31.111. Tseng, H.F., and Li, S.S.: "Photodetection Using Photoelectromagnetic and Dember Effects in Au:Si", Proc. IEEE, 1971, 59, no.12, pp.1719-20.

31.112. Genzow, D.: "IR Photovoltaic Detector With Te Single Crystals", Phys. Status Solidi, 1971, 7A, no.2, pp.K77-8.

31.113. Logothetis, E.M., et al.: "IR Detection by Schottky Barriers in Epitaxial PbTe", Appl. Phys. Lett., 1971, 19, no.9, pp.318-20.

31.114. Phelan, R.J., Mahler, R.J., and Cook, A.R.: "Sensitive Pyroelectric Polyvinylfluoride Detectors", Appl. Phys. Lett., 1971, 19, no.9, pp.337-8.

31.115. Moss, T.S.: "Photon Pressure Effects in Semiconductors", Phys. Status Solidi, 1971, 8A, no.1, pp.223-32.

31.116. Sharma, B.L., Mukerjee, S.N., and Modi, J.K.: "Detectivity Calculations for n-p Heterojunction Detectors", Infrared Phys., 1971, 11, no.4, pp.207-12.

31.117. Yamaka, E., Hayashi, T., and Matsumoto, M.: "PbTiO3 Pyroelectric IR Detector", Infrared Phys., 1971, 11, no.4, pp.247-8.

31.118. Afinogenov, V.M., and Trifonov, V.I.: "Low-Inertia Detector of Millimetric Radiation Based on n-InSb", Prib. Tekh. Eksp., 1971, no.6, pp.114-6, and Instrum. Exp. Tech., 1971, no.6.

31.119. Araki, T., and Hirayama, M.: "Quasi-Millimetric Integrated Mixer", Electr. Commun. Lab. Tech. J., 1971, 20, no.8, pp.1719-26.

31.120. Pirogov, Yu.A.: "Mechanism of Microwave Detection by Reflex Klystrons", Vestn. Mosk. Univ. Fiz. Astron., 1971, 12, no.6, pp.661-7.

31.121. Vasil'ev, A.M.: "Signal Noise in Photodiodes with a Wide Depletion Region", Radiotekh. Elektron., 1971, 16, no.3, and Radio Eng. Electron. Phys., 1971, 16, no.3, pp.530-7.

31.122. Dunaev, A.S.: "Noise Probability Density at the Output of a Photodetector for a Two-Level Background Model", Opt.-Mekh. Prom., 1971, 38, no.5, pp.15-7, and Sov. J. Opt. Technol., 1971, 38, no.5, pp.268-71.

31.123. Krumpholz, O.: "Signal/Noise Ratio of Avalanche Photodiodes", Wiss. Ber. AEG-Telefunken, 1971, 44, no.2, pp.80-4.

31.124. Kraikin, N.Sh., and Yurist, B.V.: "Analysis of Sensitivity of an Optical Heterodyne Receiver", Radiotekh. Elektron., 1971, 16, no.3, and Radio Eng. Electron. Phys., 1971, 16, no.3, pp.496-500.

31.125. Dem'yanova, E.A., Konochuk, L.V., and Markov, V.I.: "Investigation of Threshold Capabilities of Photoelectron Multipliers at 530 nm", Prib. Tekh. Eksp. 1971, 14, no.4, pp.179-81, and Instrum. Exp. Tech., 1971, 14, no.4, pp.1160-2.

31.126. Kukushkin, V.V., and Shinkarenko, V.G.: "Optimal Tunnel Microwave Mixers", Radiotekh. Elektron., 1971, 16, no.5, pp.882-5, and Radio Eng. Electron. Phys., 1971, 16, no.5, pp.900-2.

31.127. Kukushkin, V.V., and Shinkarenko, V.G.: "Microwave Mixer Employing Two Series-Connected Tunnel Diodes", Radiotekh. Elektron., 1971, 16, no.5, pp.885-8, and Radio Eng. Electron. Phys., 1971, 16, no.5, pp.902-5.

31.128. Hadni, A., Strimer, P., and Thomas, R.: "Study of Ga:Ge Bolometer for IR Detection", Nouv. Rev. Opt. Appl., 1971, 2, no.6, pp.379-88.

31.129. Zhuravle, V.A., Zelikson, D.L., and Petrov, G.D.: "Detection of Pulsed Laser Radiation by a Free Flame", Opt. Spekt., 1971, 31, no.5, pp.830-1, and Opt. Spectrosc., 1971, 31, no.5, p.447.

31.130. Gershenzon, Ye.M., et al.: "Ge Hot-Electron Narrowband Detector", Radiotekh. Elektron., 1971, 16, no.8, pp.1447-55, and Radio Eng. Electron. Phys., 1971, 16, no.8, pp.1346-53.

31.131. Arnold, R.M., and Rosenbaum, F.J.: "Transversely Magnetized Microwave (-Response) TW Photodetector", Trans. IEEE, 1971, ED-18, no.2, pp.134-5.

31.132. Watson, D.C., Grow, R.W., and Johnson, C.C.: "Cyclotron-Wave Rectifier for S and X Bands", Trans. IEEE, 1971, ED-18, no.1, pp.3-10.

31.133. Gitel'son, A.A., and Kovbasa, A.P.: "Phase Relations in Ferrite Mixers", Radiotekhnika, 1971, 26, no.10, pp.70-1, and Telecommun. Radio Eng. Pt 2, 1971, 26, no.10, pp.105-6.

31.134. Georgiyeskaya, Ye.A., et al.: "High-Frequency Si Photodiodes with PIN-Junction Structure", Radiotekh. Elektron., 1971, 16, no.11, pp.2232-4, and Radio Eng. Electron. Phys., 1971, 16, no.11, pp.1978-80.

31.135. Ohta, T., and Nakano, H.: "Characteristics of Reflected-Type Schottky-Diode Mixer", Electron. Commun. Jap., 1971, 54, no.11, pp.101-7.

31.136. Radhakrishnan, V., and Newhouse, V.L.: "Noise Analysis for Amplifiers With Superconducting Input", J. Appl. Phys., 1971, 42, no.1, pp.129-32.

31.137. Liu, S.G., and Risko, J.J.: "Low-Noise Punch-Through Semiconductor-Metal Microwave Diodes", RCA Rev., 1971, 32, no.4, pp.636-44.

31.138. Okamura, S., and Minorikawa, K.: "Semiconductor-Diode Mixer at Millimetre Waves", Ann. Rep. Eng. Res. Inst. Univ. Tokyo, 1971, 30, pp.137-43.

31.139. Okamura, S., and Okabe, Y.: "Equivalent Circuit of a Diode in the Millimetre Region", Ann. Rep. Eng. Res. Inst. Univ. Tokyo, 1971, 30, pp.131-6.

31.140. Eldumiati, I.I.: "Microwave-Biased Submillimetre-Wave Detector Using InSb", Trans. IEEE, 1972, ED-19, no.2, pp.257-67.

31.141. Emmons, S.P., and Ashley, K.L.: "Minority-Carrier Sweepout in 0.09-eV (HgCd)Te", Appl. Phys. Lett., 1972, 20, no.4, pp.162-3.

31.142. Kimmitt, M.F., Tyte, D.C., and Wright, M.J.: "Photon-Drag Radiation Monitors for Use with Pulsed CO_2 Lasers", J. Phys. E, 1972, 5, no.3, pp.239-40.

31.143. Allen, D.H., Isaacs, P.O., and Wright, J.J.: "Logarithmic Detector for Pulsed Lasers", Appl. Opt., 1972, 11, no.2, pp.476-7.

31.144. Verie, C., and Sirieix, M.: "Gigahertz-Cutoff-Frequency Capabilities of (HgCd)Te Photovoltaic Detectors at 10.6 micron", J. Quantum Electron. IEEE, 1972, QE-8, no.2, pp.180-4.

31.145. Tousek, J., and Kuzel, R.: "Characteristics of CdTe Photodiodes", Sdelovaci Tech., 1972, 20, no.1, p.7.

31.146. McIntyre, R.J.: "Distribution of Gains in Uniformly Multiplying Avalanche Photodiodes. Theory", Trans. IEEE, 1972, ED-19, no.6, pp.702-13.

31.147. Conradi, J.: "Distribution of Gains in Uniformly Multiplying Avalanche Photodiodes. Experimental", Trans. IEEE, 1972, ED-19, no.6, pp.713-8.

31.148. McDonald, D.G., et al.: "Heterodyne Detection of Far-IR Radiation Using Josephson Junctions", J. Quantum Electron. IEEE, 1972, QE-8, no.6, p.558.

31.149. Yamaka, E., et al.: "Pyroelectric IR Detectors", Natl. Tech. Rep., 1972, 18, no.2, pp.141-52.

31.150. Shigiyama, K., et al.: "IR Thermometer With Pyroelectric Detector", Natl. Tech. Rep., 1972, 18, no.2, pp.153-64.

31.151. Eldumiati, I.I., and Haddad, G.I.: "Effect of Magnetic Field on Performance of Millimetre-Wave Detectors Using Bulk InSb", Trans. IEEE, 1972, ED-19, no.9, pp.1061-3.

31.152. Stafsudd, O.M., and Pines, M.Y.: "Characteristics of KTN Pyroelectric Detectors", J. Opt. Soc. Am., 1972, 62, no.10, pp.1153-5.

31.153. Buchs, J.D.: "Noise of Microwave Schottky Diodes at 70 MHz", Frequenz, 1972, 26, no.8, pp.218-23.

31.154. Nishida, N., and Nakajima, M.: "Temperature Dependence and Stabilization of Avalanche Photodiodes", Rev. Sci. Instrum., 1972, 43, no.9, pp.1345-50.

31.155. Andrews, A.M., et al.: "High-Speed (PbSn)Te Photodiodes", Appl. Phys. Lett., 1972, 21, no.6, pp.285-7.

31.156. Kulczyk, W.K., and Davis, Q.V.: "Avalanche Photodiode as Electronic Mixer in an Optical Receiver", Trans. IEEE, 1972, ED-19, no.11, pp.1181-90.

31.157. Kudryashov, V.A., et al.: "Laws Governing the Noise Distribution of Optical Receivers Using Avalanche Photodiodes", Prib. Tekh. Eksp., 1972, 15, no.1, pp.177-8, and Instrum. Exp. Tech., 1972, 15, no.1, pp.204-5.

31.158. Afinogenov, V.M., et al.: "Sensitive Submillimetre Receiver Using n-GaAs", Izv. VUZ Radiofiz., 1972,15,no.10,pp.1572-9.

31.159. Panyakeow, S., et al.: "High-Performance Photon Drag Detector for a CO_2 Laser Using p-Te", Appl. Phys. Lett., 1972, 21, no.7, pp.314-6.

31.160. Soderman, D.A., and Pinkston, W.H.: "(HgCd)Te Photodiode Laser Receivers for the 1.3-micron Region", Appl. Opt., 1972, 11, no.10, pp.2162-8.

31.161. Dragone, C.: "Performance and Stability of Schottky-Diode Mixers", Bell Syst. Tech. J., 1972, 51, no.10,pp.216-95.

31.162. Roundy, C.B., and Byer, R.L.: "Subnanosecond Pyroelectric Detector", Appl. Phys. Lett., 1972, 21, no.10, pp.512-5.

31.163. Pykacz, H.: "Photon Detectors in the Range 3-25 micron", Postepy Fiz., 1972, 23, no.4, pp.369-98.

31.164. Deryugin, I.A., Kurashov, V.N., and Mar'yenko, V.V.: "Signal Detection in a Maser", Radiotekh. Elektron., 1972, 17, no.2, pp.351-3, and Radio Eng. Electron. Phys., 1972, 17, no.2, pp.267-9.

31.165. Deloron, M.A.: "Ultrahigh-Speed Phototube", Acta Electron., 1972, 15, no.4, pp.265-70.

31.166. Orazguleev, B.: "Si Photomagnetic IR-Radiation Receiver", Prib. Tekh. Eksp., 1972, 15, no.5, pp.193-4, and Instrum. Exp. Tech., 1973, 15, no.5, pp.1518-9.

31.167. Knyazev, A.A., Kozlov, G.A., and Lerner, N.B.: "Investigation of Phase-Sensitive TWT With Decreased Collector Potential (as a Detector)", Radiotekh. Elektron., 1973, 18, no.6, pp.1215-20, and Radio Eng. Electron. Phys., 1973, 18, no.6, pp.886-90.

31.168. Herczfeld, P.R., and Hanlon, L.R.: "DC- and Microwave-Biased Photoconductive Response in CdS Crystals", Trans. IEEE, 1973, MTT-21, no.2, pp.109-11.

31.169. Foss, N.A., and Ward, S.A.: "Large-Area MOS Avalanche Photodiodes", J. Appl. Phys., 1973, 44, no.2, pp.728-31.

31.170. Gross, C., Mattauch, R.J., and Viola, T.J.: "Characteristics of IR Photodetectors Produced by Radiation Doping", J. Appl. Phys., 1973, 44, no.2, pp.735-9.

31.171. Schonert, B., and Schmidt, H.: "Application Possibilities of III-V Semiconductor Compounds in Photomultipliers", Radio Fernsehen Elektron., 1973, 22,no.1,pp.33-5.

31.172. Wang, C.T., and Li, S.S.: "Grating-Type Au n-Si Schottky-Barrier Photodiode", Trans. IEEE, 1973, ED-20, no.6, pp.522-6.

31.173. Ostrowsky, D.B., et al.: "Integrated-Optics Photodetector", Appl. Phys. Lett., 1973, 22, no.9, pp.463-4.

31.174. Belyantsev, A.M., and Klishin, E.V.: "Far-IR Detector Using Superconductor Electron Heating", Izv. VUZ Radiofiz., 1973, 16, no.3, pp.479-81.

31.175. Allen, G.A.: "Calculations on the Performance of GaAs Photocathodes", Acta Electron., 1973, 16, no.3, pp.229-36.

31.176. Skibarko, A.P., Davydov, Yu.T., and Merkishin, G.V.: "Noise Characteristics of a Si Avalanche Photodiode", Izv. VUZ Radiofiz., 1973, 16, no.5, pp.99-100.

31.177. Faris, S.M., Gustafson, T.K., and Wiesner, J.C.: "Detection of Optical Radiation With DC-Biased Electron-Tunnelling Metal-Barrier-Metal Diodes", J. Quantum Electron. IEEE, 1973, QE-9, no.7, pp.737-45.

31.178. Patel, B.S.: "Photon-Drag Ge Detectors for CO_2 Lasers", Proc. IEEE, 1973, 61, no.6, pp.795-6.

31.179. Rolls, W.H., and Eddolls, D.V.: "High-Detectivity (PbSn)Te Photovoltaic Diodes", Infrared Phys., 1973, 13, no.2, pp.143-7.

31.180. Logan, R.M.: "Calculation of Temperature Distribution and Temperature Noise in a Pyroelectric Detector. II", Infrared Phys., 1973, 13, no.2, pp.91-8.

31.181. Bishop, P.J., Gibson, A.F., and Kimmitt, M.F.: "Performance of Photon-Drag Detectors at High Laser Intensities", J. Quantum Electron. IEEE, 1973, QE-9, no.10, pp.1007-11.

31.182. Block, W.H., and Gaddy, O.L.: "Thin-Metal-Film, Room Temperature, IR Bolometers for Nanosecond Response Period", J. Quantum Electron. IEEE, 1973, QE-9, no.11, pp.1044-53.

31.183. Kaneda, T., and Takanashi, H.: "Frequency Response of Ge Avalanche Photodiodes", Jap. J. Appl. Phys., 1973, 12, no.10, pp.1652-3.

31.184. Li, S.S., and Lindholm, F.A.: "Quantum Yield of PIN Photodiodes", Phys. Status Solidi, 1973, 15, no.1, pp.237-45.

31.185. Held, H.: "Mixing Properties of Gunn-Diode Oscillators", Nachr. Tech. Z., 1973, 26, no.12, pp.542-4.

31.186. Stoll, H., et al.: "Proton-Implanted Optical Waveguide Detectors in GaAs", Appl. Phys. Lett., 1973, 23,no.12,pp.664-5.

31.187. Donnelly, J.P., and Holloway, H.: "Photodiodes Fabricated in Epitaxial PbTe by Sb^+ Implantation", Appl. Phys. Lett., 1973, 23, no.12, pp.682-3.

31.188. Andreeva, L.I., Stepanov, B.M., and Talanov, Yu.V.: "Study of Fatigue of High-Speed Coaxial Photocells", Izmer. Tekh., 1973, 16, no.6, pp.33-4, and Meas. Tech., 1973, 16, no.6, pp.844-5.

31.189. Yurchenko, I.I., et al.: "Series-Connected Josephson Junctions for Detecting Microwave Radiation", Zh. Tekh. Fiz., 1973, 43, no.10, pp.2174-7, and Sov. Phys.-Tech. Phys., 1974, 18, no.10, pp.1368-9.

31.190. Gerber, K.: "Lightguide-to-Photomultiplier Coupler", Appl. Opt., 1974, 13, no.5, pp.1006-7.

31.191. Takanashi, H., Kaneda, T., and Sei, H.: "Ge Avalanche Photodetectors", Fujitsu Sci. Tech. J., 1974, 10, no.1, pp.119-34.

31.192. Gustafson, T.K., Schmidt, R.V., and Perucca, J.R.: "Optical Detection in Thin-Film Metal-Oxide-Metal Diodes", Appl. Phys. Lett., 1974, 24, no.12, pp.620-2.

31.193. Hohnke, D.K., and Holloway, H.: "Epitaxial PbSe Schottky-Barrier Diodes for IR Detection", Appl. Phys. Lett., 1974, 24, no.12, pp.633-5.

31.194. Mania, L., and Stracca, G.B.: "Effects of Diode Junction Capacitance on the Conversion Loss of Microwave Mixers", Trans. IEEE, 1974, COM-22, no.9,pp.1428-35.

31.195. Tae, P.K., Arc, L.J., and Ho, L.T.: "Study on the Self-Excited Mixing Effect of IMPATT Diodes", J. Korean Inst. Electron. Eng., 1974, 11, no.2, pp.5-11.

31.196. Patel, B.S.: "Photon Drag Ge Detectors for CW CO_2 Lasers", Proc. IEEE, 1974, 62, no.1, pp.143-4.

31.197. Meier, P.J.: "Low-Noise Mixer in Oversized Microstrip for 5-mm Band", Trans. IEEE, 1974, MTT-22, no.4,pp.450-1.

31.198. Felterman, H.R., et al.: "Submillimetre Detection and Mixing Using Schottky Diodes", Appl. Phys. Lett., 1974, 24, no.2, pp.70-2.

31.199. Le Borgne, A., Perichon, R., and Constant, E.: "Noise Measure and Temperature Effect in Small-Signal GaAs IMPATT Amplifiers", Proc. IEEE, 1974, 62, no.4,pp.535-7.

31.200. Vasil'ev, Yu.S., et al.: "Heterodyne Reception of Optical Signals at 10.6 micron", Kvantovaya Elektron., 1974, no.4, pp.86-91, and Sov. J. Quantum Electron., 1974, 3, no.4, pp.327-30.

31.201. Kono, A., and Hattori, S.: "Rapidly Gated Crossed-Field Photomultiplier", Appl. Opt., 1974, 13, no.9, pp.2002-3.

31.202. Piqueras, J., and Munoz, E.: "Novel GaAs Avalanche Photodiode", Appl. Phys. Lett., 1974, 25, no.4, pp.214-5.

31.203. Peterson, R.L., et al.: "Analysis of Response of Pyroelectric Optical Detectors", J. Appl. Phys., 1974, 45, no.8, pp.3296-3303.

31.204. Agafonov, V.G., et al.: "Detectors Based on Photon Drag in Semiconductors", Fiz. Tekh. Poluprov., 1974, 7, no.12, pp.2316-25, and Sov. Phys.-Semicond., 1974, 7, no.12, pp.1540-5.

31.205. Goranson, R.W., and Skipper, J.D.: "Signal/Noise Ratio in Optical Photodiode Receivers", Proc. IEEE, 1974, 62, no.10, pp.1404-6.

31.206. Pyee, M., Clairon, A., and Auvray, J.: "Using a Schottky Diode in Detecting and Mixing at 30 THz", Electron. LETT., 1974, 10, no.23, pp.500-1.

31.207. Koehler, T.: "(HgCd)Te Photodiodes in CO_2 Laser Systems", Electro-Opt. Syst. Des., 1974, 6, no.10, pp.24-8.

31.208. Walker, A.C., and Alcock, A.J.: "10-micron Detection System With 40-ps Resolution", Opt. Commun., 1974, 12, no.4, pp.430-2.

31.209. Borodovskii, P.A., Bul'dygin, A.F., and Utkin, K.K.: "Gunn-Diode Autodyne Mixer", Izv. VUZ Radioelektron., 1974, 17, no.12, pp.82-4.

31.210. Tamarchak, D.Ya., and Khotuntsev, Yu.L.: "Stabilization of (Microwave) Detector Characteristics Over a Temperature Range", Izv. VUZ Radioelektron., 1974, 17, no.12, pp.25-30.

31.211. Webb, P.P., McIntyre, R.J., and Conradi, J.: "Properties of Avalanche Photodiodes", RCA Rev., 1974, 35, no.2, pp.234-78.

31.212. Okamura, S., and Ijichi, K.: "Property of Metal-Metal Point-Contact Diode in IR Region", Annu. Rep. Eng. Res. Inst. Fac. Eng. Univ. Tokyo, 1974, 33, pp.121-6.

31.213. Beerman, H.P.: "Characteristics of Fully Deuterated Triglycine Sulphate", Ferroelectrics, 1974, 8, no.3-4, pp.653-6.

31.214. Judy, J.H., Park, H.D., and van der Ziel, A.: "Molecular Field Model Applied to Pyroelectric Detectors", Ferroelectrics, 1974, 8, no.3-4, pp.685-7.

31.215. Vetokhin, S.S., Gulakov, I.R., and Reznikov, I.V.: "Investigation of One-Electron Photomultipliers for UV Range", Prib. Tekh. Eksp., 1974, 17, no.4, pp.150-2, and Instrum. Exp. Tech., 1974, 17, no.4, pp.1120-2.

31.216. Fetterman, H.R., et al.: "Sub-millimetre Heterodyne Detection and Harmonic Mixing Using Schottky Diodes", Trans. IEEE, 1974, MTT-22, no.12, pp.1013-5.

31.217. Taur, Y., Claassen, J.H., and Richards, P.L.: "Josephson Junctions as Heterodyne Detectors", Trans. IEEE, 1974, MTT-22, no.12, pp.1005-9.

32. RECEIVER TECHNIQUES

32.1. Juillerat, R.: "Microwave Oscillator Noise Measurements", Onde Electr., 1965, 45, p.101.

32.2. Orang, H.S., and Shearman, E.D.R.: "Self-Balancing Microwave Comparison Radiometer", Electron. Lett., 1965, 1, p.222.

32.3. Klipper, H.: "Sensitivity of Crystal-Video Receivers With RF Preamplification", Microwave J., 1965, 8, no.8, p.85.

32.4. Kuckes, A.F., and Wong, A.Y.: "Microwave Emission from a Fully Ionized Plasma", Phys. Fluids, 1965, 8, p.1161.

32.5. Jordan, P.R., and Pascalar, K.G.: "Radiometric Performance of a Solid-State 35-GHz Receiver", Proc. IEEE, 1965, 53, p.1655.

32.6. O'Neill, H.J.: "Image Power in a Microwave Crystal Mixer", Electron. Lett., 1965, 1, p.17.

32.7. Elberson, J.E., and Everett, W.W.: "Increased Sensitivity of Direct-Detection Microwave Receivers for Compatibility Measurements", Proc. IEEE, 1965, 53, p.532.

32.8. Draysey, D.W., Court, W.P.N., and Bott, I.B.: "Noise Performance of Gunn Microwave Generators in X and J Bands", Electron. Lett., 1966, 2, p.125.

32.9. Kornilov, S.A.: "Spectral Width of a Reflex Klystron", Zh. Tekh. Fiz., 1966, 36, p.163, and Sov. Phys.-Tech. Phys., 1966, 11, no.1, pp.117-24.

32.10. Johnson, S.L., Smith, B.H., and Calder, D.A.: "Noise Spectrum Characteristics of Low-Noise Microwave Tubes and Solid-State Devices", Proc. IEEE, 1966, 54, p.258.

32.11. Healey, D.J.: "Noise Characteristics of an X-Band Triode Oscillator", Proc. IEEE, 1966, 54, p.304.

32.12. Sheremet'ev, A.G., and Kochetkov, R.M.: "Signal Detection by Photoreceiver with Quantum Input Amplifier", Probl. Peredachi. Inform., 1967, 3, no.3, and Problems Inform. Transm., 1967, 3, no.3, pp.23-7.

32.13. Faulkner, E.A., and Meade, M.L.: "FM Noise in Gunn Oscillators", Electron. Lett., 1967, 3, no.9, p.419.

32.14. Wadhwa, R.P.: "Noise Transport in an Injected-Beam Crossed-Field Device", Int. J. Electron., 1967, 23, no.2, pp.123-34.

32.15. Wadhwa, R.P., and van Duzer, T.: "Low-Noise Measurements on an Injected-Beam Medium-Power Crossed-Field Amplifier", Int. J. Electron., 1967, 23, no.2, pp.135-52.

32.16. Curtice, W.R.: "Sources of Residual Frequency Modulation in X-Band Two-Cavity Klystron Oscillators", Trans. IEEE, 1968, ED-15, no.1, pp.6-16.

32.17. Povarov, A.A.: "Effect of Local-Oscillator Amplitude Fluctuations on Frequency Conversion in a Microwave Crystal Mixer", Radiotekhnika, 1968, 23, p.35, and Telecommun. Radio Eng. Pt 2, 1968, 23.

32.18. Ashley, J.R., and Searles, C.B.: "Microwave-Oscillator Noise Reduction by Transmission Stabilizing Cavity", Trans. IEEE, 1968, MTT-16, p.741.

32.19. Peppiatt, H.J., Hall, J.A., and McDaniel, A.V.: "Low-Noise Class-C Oscillator Using a Directional Coupler", Trans. IEEE, 1968, MTT-16, p.746.

32.20. Troup, G.J.: "Statistics of Laser Light", Phys. Lett., 1968, 28A, p.252.

32.21. Rao, B.V., and Gambling, W.A.: "Noise Spectra of Beam-Type Microwave Oscillators", Radio Electron. Eng., 1968, 35, no.3, pp.165-73.

32.22. Kodali, V.P.: "AM- and FM-Noise Characteristics of Solid-State Microwave Oscillators", Electron. Lett., 1968, 4, no.8, pp.147-8.

32.23. Okabe, Y., and Okamura, S.: "Analysis of Stability and Noise of Solid-State Oscillators", Electron. Commun. Jap., 1969, 52, no.12, pp.102-10.

32.24. Sugiyama, M., and Sugimoto, S.: "Experimental Considerations on Millimetre-Wave Wideband Downconverter", Electron. Commun. Jap., 1969, 52, no.11, pp.117-24.

32.25. Rufer, R.P.: "Electrooptical Automatic Gain Control", Trans. IEEE, 1969, AED-5, p.463.

32.26. Dryagin, Yu.A., and Fedoseev, L.I.: "Millimetre Detector Radiometers", Izv. VUZ Radiofix., 1969, 12, no.6, p.813.

32.27. Al'bats, N.Ye., and Sidorenko, N.G.: "Noise in Cooled Microwave Devices", Radiotekhnika, 1969, 24, no.7, p.101, and Telecommun. Radio Eng. Pt 2, 1969, 24, no.7.

32.28. Kuznetsov, M.I.: "Mechanism of Fluctuations in a Magnetron", Izv. VUZ Radiofiz., 1969, 12, p.1873.

32.29. Bershtein, I.L., and Zaitsev, Yu. Yu.I.: "Radiation Fluctuations of Gas Lasers", Electron Technol., 1969, 2, no.2-3, p.105.

32.30. Gilson, V.A., and Stoll, H.M.: "Argon-Laser FM Detection", Appl. Opt. 1969, 8, p.717.

32.31. Karavaev, V.V.: "Theoretical Noise Limit of Linear Amplifiers", Radiotekh. Elektron., 1969, 14, no.10, p.1890, and Radio Eng. Electron. Phys., 1969, 14, no.10, pp.1633-5.

32.32. Karp, S.: "Comments on Optical Heterodyne Detection", Trans. IEEE, 1969, COM-17, p.574.

32.33. Vlasov, V.G., and Lazneva, E.V.: "Heterodyne Photodiode Detection of Radiation Amplitude Modulated at 5-50 MHz", Opt.-Mekh. Prom., 1969, 35, no.10, p.5, and Sov. J. Opt. Technol., 1969, 35, no.5, p.608.

32.34. Lee, S.J., and van der Ziel, A.: "Noise in Optical Mixing", Physica, 1969, 45, no.3, p.379.

32.35. Kuriksha, A.A., and Kurushin, A.D.: "Theory of Optical Receivers With Gain and Frequency Conversion", Radiotekh. Elektron., 1969, 14, no.11, p.1987, and Radio Eng. Electron. Phys., 1969, 14, no.11, pp.1719-27.

32.36. Hernqvist, K.G.: "Low-Noise He-Ne Laser", RCA Rev., 1969, 30, p.429.

32.37. Hongo, S., Kubo, U., and Inuishi, Y.: "Noise in He-Ne Laser Oscillators", Technol. Rep. Osaka Univ., 1969, 19, no.853-875, p.145.

32.38. Young, R.T., and Maupin, R.T.: "Spontaneous-Emission Noise (in DC-Excited Gas Lasers)", J. Appl. Phys., 1969, 40, p.3881.

32.39. Suzuki, T.: "Noise in Superradiation from Ruby", Jap. J. Appl. Phys., 1969, 8, p.1059.

32.40. Bandilla, A., and Paul, H.: "Laser Amplifier and Phase Uncertainty", Ann. Phys. (Leipzig), 1969, 23, no.7-8, p.323.

32.41. Sakuraba, I., and Tsubo, T.: "Directional Characteristics and Forms of the Detector Surface in Optical Heterodyne Detection Processes", Mem. Fac. Eng. Hokkaido Univ., 1969, 12, no.3, p.295.

32.42. Baklanov, Ye.V., et al.: "Fluctuations of Radiation Emission Buildup in Gas Lasers", Zh. Eksp. Teor. Fiz., 1969, 56, p.1120, and Sov. Phys.-JETP, 1969, 29.

32.43. van der Ziel, A.: "Radiation Noise in Optical Mixers for Laser Light Reception", Physica, 1970, 49, no.4, pp.613-14.

32.44. Ross, A.H.M.: "Optical Heterodyne Mixing Efficiency Invariance", Proc. IEEE, 1970, 58, no.10, pp.1766-7.

32.45. Leistiko, O., and Guldbrandsen, T.: "Simple Two-Phase Lock-In Detection System for a Microwave Interferometer", J. Phys. E, 1970, 3, p.224.

32.46. Eng. S.T., and Gudmundsen, R.A.: "Theory of Optical Heterodyne Detection Using the Pyroelectric Effect", Appl. Opt., 1970, 9, p.161.

32.47. Gelikonov, V.M., and Zaitsev, Yu.I.: "Natural He-Ne Laser Intensity Fluctuations at 1150 nm", Izv. VUZ Radiofiz., 1970, 13, no.6, p.904.

32.48. Kuhn, P.: "Noise in Gunn Oscillators Depending on Surface of Diode", Electron. Lett., 1970, 6, no.26, pp.845-7.

32.49. Guekos, G., and Strutt, M.J.O.: "Compensation of Injection-Laser Excess Noise by an External Modulator", Electron. Lett., 1970, 6, p.250.

32.50. Sweet, A.S., and Mackenzie, L.A.: "FM Noise of a CW Gunn Oscillator", Proc. IEEE, 1970, 58, p.822.

32.51. Suzuki, T.: "Discharge Current Noise in He-Ne Laser and Suppression", Jap. J. Appl. Phys., 1970, 9, p.309.

32.52. Saito, S., and Uehara, S.: "Modulation Noise of a 633-nm He-Ne Laser", Proc. IEEE, 1970, 58, p.598.

32.53. Ostapchenko, E.P., et al.: "Modulation of He-Ne Laser Emission via the Excitation Source", Radiotekh. Elektron., 1970, 15, no.1, p.143, and Radio Eng. Electron. Phys., 1970, 15, no.1, pp.116-9.

32.54. Johnson, W.A., Mori, T.T., and Shimabukuro, F.I.: "Design, Development, and Initial Measurements of a 1.4-mm Radiometric System", Trans. IEEE, 1970, AP-18, p.512.

32.55. Jauho, P., et al.: "Approximate Photon Distributions in Lasers", Phys. Lett., 1970, 31A, p.422.

32.56. Guekos, G., and Strutt, M.J.O.: "Reduction of Excess Noise of GaAs-Diode Lasers by Optoelectronic Feedback", Proc. IEEE, 1970, 58, p.949.

32.57. Guekos, G., and Strutt, M.J.O.: "Laser-Light Noise and Current Noise of GaAs CW Laser Diodes", J. Quantum Electron. IEEE, 1970, QE-6, p.423.

32.58. Suzuki, T.: "Dynamic Mode Competition in He-Ne Laser Noise", J. Appl. Phys., 1970, 41, p.3904.

32.59. Golubev, Yu.M.: "Quantum Noise in Laser Systems. II and III", Opt. Spekt., 1970, 28, no.3, p.283 and 29, no.2, pp.149-55, and Opt. Spectrosc., 1970, 28, no.3, p.528 and 29, no.2, pp.79-82.

32.60. Bensimon, J., and Divonne, F.: "Stabilization of the Power of an Argon-Ion Laser with LF Oscillations", J. Phys. D, 1970, 3, p.158.

32.61. Kuznetsov, M.I.: "Mechanism of Fluctuations in a Magnetron Diode", Izv. VUZ Radiofiz., 1970, 13, p.292.

32.62. Chandra, N., and Prakash, H.: "Anticorrelation in Two-Photon Attenuated Laser Beam", Phys. Rev., 1970, 1A, p.1696.

32.63. Vystavkin, A.N., Listvin, V.N., and Morenkov, A.D.: "Radiometer for Submillimetre Waves Using n-InSb Detector", Prib. Tekh. Eksp., 1970, no.3, pp.183-4, and Instrum. Exp. Tech., 1970, no.3, pp.846-7.

32.64. Rozsa, K., and Salamon, T.: "Low-Noise He-Ne Laser", Opt. Technol., 1970, 2, no.3, p.151.

32.65. Gertsenshtein, M.Ye., et al.: "Laser as Square-Law Radiation Detector", Radiotekh. Elektron., 1970, 15, no.10, and Radio Eng. Electron. Phys., 1970, 15, no.10, pp.1922-3.

32.66. Yamauchi, K., and Nakamura, M.: "Effect of Frequency Fluctuations in FM Radiometers", Rep. Himeji Inst. Technol., 1970, 23A, pp.35-40.

32.67. Gus'kov, N.A.: "Optical (-Resonator) Frequency Discriminator", Radiotekh. Elektron., 1970, 15, no.11, pp.2140-2, and Radio Eng. Electron. Phys., 1970,15,no.11.

32.68. Uehara, S., and Fujii, Y.: "Gas-Discharge Modulation and Noise of He-Ne Gas Laser", Electron. Commun. Jap., 1970, 53, no.8, pp.103-9.

32.69. Vlaardingerbroek, M.T., and Goedbloed, J.J.: "Theory of Noise and Injection Phase Locking of IMPATT Diode Oscillators", Philips Res. Rep., 1970, 25, no.6, pp.452-71.

32.70. Kurpis, G.P., and Taub, J.J.: "Wideband X-Band Microstrip Image-Rejection Balanced Mixer", Trans. IEEE, 1970, MTT-18, no.12, pp.1181-2.

32.71. Klimontovich, Yu.L.: "Natural Fluctuations in Solid-State Lasers", Zh. Eksp. Teor. Fiz., 1970, 59, no.2, pp.464-70, and Sov. Phys.-JETP, 1971, 32, no.2.

32.72. Walsh, W.M., and Rupp, L.W.: "Self-Detecting Microwave Marginal Oscillator", Rev. Sci. Instrum., 1971, 42, no.4,pp.468-70.

32.73. Kazantsev, A.P.: "Quantum Fluctuations of the Radiation from a Gas Laser", Zh. Eksp. Teor. Fiz., 1971, 60, no.2, pp.500-8, and Sov. Phys.-JETP, 1971, 33, no.2, pp.269-73.

32.74. Haken, H., and Vollmer, H.D.: "Fokker-Planck Equation of a Laser With Many Modes and Multilevel Atoms", Z. Phys., 1971, 242, no.5, pp.416-31.

32.75. Grossmann, S., and Richter, P.H.: "Laser Threshold and Nonlinear Landau Fluctuation Theory of Phase Transitions", Z. Phys., 1971, 242, no.5, pp.458-75.

32.76. Galehouse, D.C., et al.: "Investigation of Coherent Fluctuations in an Argon-Ion Laser", Appl. Phys. Lett., 1971, 18, no.1, pp.13-5.

32.77. Shofner, F.M., et al.: "Plasma-Induced Fluctuations in CO_2 Lasers", J. Quantum Electron. IEEE, 1971, QE-7, no.6, pp.245-6.

32.78. Nowicki, R.: "Output-Power Stabilization of a He-Ne Laser", J. Phys. E, 1971, 4, no.4, pp.274-6.

32.79. Magun, A., and Kunzi, K.: "Influence of Statistical Gain Fluctuations of the High-Frequency Amplifier on Sensitivity of a Dicke Radiometer", Z. Angew. Math. Phys., 1971, 22, no.3, pp.392-403.

32.80. Suzuki, T., and Ieiri, S.: "LF Noise in Unlocked Multimode He-Ne Lasers", Jap. J. Appl. Phys., 1971,10,no.9,pp.1238-43.

32.81. Haus, H.A., Statz, H., and Pucel, R.A.: "Optimum Noise Measure of IMPATT Diodes", Trans. IEEE, 1971, MTT-19, no.10, pp.801-3.

32.82. Alaman, A., Herbet, H., and Harth, W.: "Output Power and FM Noise of Multiple-Contact Gunn Oscillators", Arch. Elektron. Ubertrag., 1971, 25, no.8, pp.397-8.

32.83. Goedbloed, J.J.: "FM Noise of Low-Level-Operating IMPATT-Diode Oscillators", Electron. Lett., 1971, 7, no.16, pp.445-6.

32.84. Meade, M.L.: "Relationship Between FM Noise and Current Noise in a Cavity-Controlled Gunn-Effect Oscillator", Radio Electron. Eng. 1971, 41, no.3, pp.126-32.

32.85. Brookner, E.: "System Combining Best Features of Heterodyne and Direct-Detection Receivers", Appl. Opt. 1971, 10, no.5, pp.1009-11.

32.86. Allen, L., and Sayers, M.D.: "Mode Competition and Noise in the He-Ne Laser", J. Appl. Phys., 1971, 42, no.6, pp.2569-70.

32.87. Araki, T., and Hirayama, M.: "20-GHz Integrated Balanced Mixer", Trans. IEEE, 1971, MTT-19, no.7, pp.638-43.

32.88. Buchs, J.D.: "Stripline Mixer With Subharmonic Local Oscillator Using Two Schottky Diodes in an Antiparallel Connection", Arch. Elektron. Ubertrag., 1971, 25, no.1, pp.52-3.

32.89. Baldwin, G.D., and Basil, I.T.: "Parasitic Noise on the Output of a CW Nd^{3+}: YAG Laser", J. Quantum Electron. IEEE, 1971, QE-7, no.4, pp.179-81.

32.90. Malakhov, A.N., and Sandler, M.S.: "Investigation of Inherent Laser Fluctuations With Allowance for the Lumped Parameters", Izv. VUZ Radiofiz., 1971, 14, no.6, pp.845-53.

32.91. Brunner, W., Paul, H., and Bandillo, A.: "Fluctuations in an Optical Parametric Oscillator", Ann. Phys. (Leipzig), 1971, 27, no.1, pp.91-3.

32.92. Suzuki, T.: "Vibration Noise in Lasers", Jap. J. Appl. Phys., 1971, 10, no.7, p.949.

32.93. Back, I.: "FM Noise in an Injection-Locked Oscillator When Reverse Locking Exists", Electron. Lett., 1971, 7, no.12, pp.346-8.

32.94. Hashiguchi, S., and Okoshi, T.: "Determination of Equivalent-Circuit Parameters Describing Noise from a Gunn Oscillator", Trans. IEEE, 1971, MTT-19, no.8, pp.686-91.

32.95. Thaler, H.J., Ulrich, G., and Weidmann, G.: "Noise in IMPATT-Diode Amplifiers and Oscillators", Trans. IEEE, 1971, MTT-19, no.8, pp.692-705.

32.96. Kovalev, A.S.: "Amplitude Fluctuations in Two- and Four-Level Laser Models", Izv. VUZ Radiofiz., 1971, 14, no.6, pp.823-7.

32.97. Hongo, S., et al.: "Dynamic Behaviour and Noise of He-Ne Lasers", Electr. Eng. Jap., 1971, 90, no.4, pp.44-51.

32.98. Cantrell, C.D., and Smith, A.W.: "Calculation of Third-Order Intensity Correlation in a Laser Near Threshold", Phys. Lett., 1971, 37A, no.2, pp.167-8.

32.99. Suzuki, T.: "Discharge-Current Noise in DC-Excited Ar^{II} Lasers", Jap. J. Appl. Phys., 1971, 10, no.10, pp.1419-24.

32.100. Nichols, S.T., and Dennis, L.P.: "Influence of a Transmission Line on the Noise Spectra of Cavity-Stabilized Oscillators", Electron. Lett., 1971, 7, no.22, pp.659-61.

32.101. Mouthaan, K., and Rispert, H.P.M.: "Nonlinearity and Noise in the Avalanche Transit-Time Oscillator", Philips Res. Rep., 1971, 26, no.5, pp.391-413.

32.102. Evans, W.J.: "Noise Spectra of a CW Si TRAPATT Oscillator", Proc. IEEE, 1971, 59, no.12, p.1730.

32.103. Saito, T., Takagi, T., and Mano, K.: "Noise Effect in Oscillators Using Multiple Active Devices Connected in Series or Parallel", Proc. IEEE, 1971, 59, no.12, pp.1730-1.

32.104. Grossmann, S.: "Fluctuation Theory of Detuned Single-Mode Laser", Z. Phys., 1971, 248, no.3, pp.244-53.

32.105. Grossmann, S., and Richter, P.H.: "Intensity, Correlations, and Linewidths, of Two-Mode Lasers With Intensity Coupling Near Threshold", Z. Phys., 1971, 249, no.1, pp.43-57.

32.106. Kazaryan, R.A.: "Heterodyne Detector of Polarization-Modulated Light", Radiotekh. Elektron., 1971, 16, no.7, pp.1296-7, and Radio Eng. Electron. Phys., 1971, 16, no.7, pp.1231-2.

32.107. Araki, T., and Hirayama, M.: "20-GHz Integrated Balanced Mixer", Electron. Commun. Jap., 1971, 54, no.10, pp.56-60.

32.108. Suzuki, T.: "Mechanical-Vibration Noise in He-Ne Lasers", Oyo Buturi, 1971, 40, no.10, pp.1086-92.

32.109. Fedoseev, L.I., and Kulikov, Yu.Ya.: "Heterodyne Radiometers for Submillimetre Bands", Radiotekh. Elektron., 1971, 16, no.4, pp.554-60, and Radio Eng. Electron. Phys., 1971, 16, no.4, pp.637-41.

32.110. Personick, S.D.: "Statistics of a General Class of Avalanche Detectors With Applications to Optical Communication", Bell Syst. Tech. J., 1971, 50, no.10, pp.3075-95.

32.111. Evans, W.J.: "Noise Spectra of a CW Si TRAPATT Oscillator", Proc. IEEE, 1972, 60, no.1, pp.125-6.

32.112. Ruzicka, J.: "Determining the Noise Spectrum in a BW Oscillator", Slab. Obz., 1972, 33, no.1, pp.9-13.

32.113. Chopra, S., and Mandel, L.: "Intensity Correlation of a Laser Beam Near Threshold", J. Quantum Electron. IEEE, 1972, QE-8, no.3, pp.324-7.

32.114. Teich, M.C., and Yen, R.Y.: "Signal/Noise Ratio for Optical Heterodyne Detection", J. Appl. Phys., 1972, 43, no.5, pp.2480-1.

32.115. Kestenbaum, A., and Mishkin, E.A.: "Photon Statistics of Phase-Locked Coupled Light Beams", Phys. Rev., 1972, 5A, no.5, pp.2267-75.

32.116. De Cacqueray, A., Blasquez, G., and Graffeuil, J.: "Experimental Study of the Correlation of FM and LF Noise in Gunn Oscillators", Electron. Lett., 1972, 8, no.9, pp.217-8.

32.117. Coates, P.B.: "Photomultiplier Noise Statistics", J. Phys. D, 1972, 5, no.5, pp.915-30.

32.118. Teich, M.C.: "Photocounting Receiver Performance for Detection of Multimode Laser or Scattered Radiation", Trans. IEEE, 1972, AES-8, no.1, pp.13-8.

32.119. Chwalow, M.L.E.: "Dynamic Radiometer", IBM Tech. Disclosure Bull., 1972, 14, no.10, pp.2942-3.

32.120. Cierpisz, S.: "Dynamic Properties of Self-Adaptive Radiometer", Pomiary Autom. Kontr., 1972, 18, no.4, pp.162-4.

32.121. Morgan, D.J., and Adams, M.J.: "Quantum Noise in Semiconductor Lasers", Phys. Status Solidi, 1972, 11A, no.1, pp.243-53.

32.122. Lazarus, M.J., and Novak, S.: "Millimetre-Wave Gunn Mixer With Picowatt Sensitivity", Proc. IEE, 1972, 119, no.6, pp.747-8.

32.123. Yokoshima, I.: "Generalized Performance Analysis of Microwave Radiometers", Electron. Commun. Jap., 1972, 54, no.3, pp.75-82.

32.124. Schunemann, K., and Schiek, B.: "Synchronization and Noise Performance of Mutually Coupled Oscillators", Arch. Elektron. Ubertrag. 1972, 26, no.7-8, pp.310-8.

32.125. Ohtomo, M.: "Experimental Evaluation of Noise Parameters in Gunn and Avalanche Oscillators", Trans. IEEE, 1972, MTT-20, no.7, pp.425-37.

32.126. Schiek, B.: "Noise of Negative-Resistance Oscillators at High Modulation Frequencies", Trans. IEEE, 1972, MTT-20, no.10, pp.635-41.

32.127. Sjolund, A.: "Analysis of Large-Signal Noise in Read Oscillators", Solid-State Electron., 1972, 15, no.9, pp.971-8.

32.128. Gerhardt, H., Welling, H., and Guttner, A.: "Observation of Quantum-Phase and -Amplitude Noise for a Laser Below and Above Threshold", Phys. Lett., 1972, 40A, no.3, pp.191-3.

32.129. Bergou, J., et al.: "Probability Distribution of Optical Field Emission Counts", Acta Phys. Acad. Sci. Hung., 1972, 32, no.1-4, pp.319-22.

32.130. Klimontovich, Yu.L., Kovalev, A.S., and Landa, P.S.: "Natural Fluctuations in Lasers", Usp. Fiz. Nauk, 1972, 15, no.1, pp.279-313, and Sov. Phys.-Usp., 1972, 15, no.1, pp.95-113.

32.131. Ohtsuka, Y.: "Optical Heterodyning With Noncritical Angular Alignment Using a Pair of Ultrasonic Light Modulators", Jap. J. Appl. Phys., 1972, 11, no.10, pp.1446-54.

32.132. Schunemann, K., and Schiek, B.: "Noise in Cavity-Stabilized Microwave Oscillators", Philips Res. Rep., 1972, 27, no.5, pp.486-507.

32.133. Golubentsev, A.F., and Terzhov, I.I.: "Minimum Noise Factor of M-Type Microwave Beam Amplifier", Radiotekh. Elektron., 1972, 17, no.3, pp.569-76, and Radio Eng. Electron. Phys., 1972, 17, no.3, pp.443-9.

32.134. Borisov, V.S., and Koridalin, V.E.: "Investigation of Intensity Fluctuations of a Gas Laser at Low and IR Frequencies", Radiotekh. Elektron., 1972, 17, no.2, p.425, and Radio Eng. Electron. Phys., 1972, 17, no.2, pp.331-2.

32.135. Yurist, B.V., and Khaikin, N.Sh.: "Study of the Conversion Coefficient of the Mixer of an Optical Heterodyne Receiver", Radiotekh. Elektron., 1972, 17, no.1, pp.103-18, and Radio Eng. Electron. Phys., 1972, 17, no.1, pp.82-8.

32.136. Arsen'ev, V.V., et al.: "Simple Semiconductor Correlator for Picosecond Light Pulses", Kvantovaya Elektron., 1972, no.7, pp.82-4, and Sov. J. Quantum Electron., 1972, 2, no.1, pp.74-5.

32.137. Penin, N.A., Khaikin, N.Sh., and Yurist, B.V.: "Investigation of the Noise Factor of an Optical Heterodyne Receiver with Extrinsic Photodetector", Radiotekh. Elektron., 1972, 17, no.5, pp.1018-23, and Radio Eng. Electron. Phys., 1972, 17, no.5, pp.792-7.

32.138. Sobolev, G.L., Vol'fson, A.O., and Il'in, U.K.: "Fluctuations in Synchronized and Stabilized Voltage-Controlled Magnetrons", Radiotekh. Elektron., 1972, 17, no.5, pp.1039-45, and Radio Eng. Electron. Phys., 1972, 17, no.5, pp.808-13.

32.139. Wieder, I.: "Ruby Quantum Converter", Opt. Commun., 1972, 6, no.4, pp.309-13.

32.140. Asmaryan, E.A.: "Influence of Elastic Collisions on the Intensity of Natural Fluctuations in a He-Ne Laser", Vestn. Mosk. Univ. Fiz. Astron., 1972, 13, no.6, pp.728-31.

32.141. Deryugin, I.A., et al.: "Investigation of Statistical Properties of Multimode Laser Radiation by Photon Counting Method", Radiotekh. Elektron., 1972, 17, no.8, pp.1622-7, and Radio Eng. Electron. Phys., 1972, 17, no.8, pp.1279-83.

32.142. Adrianova, I.I., and Zaslavskaya, V.R.: "Detection of Phase-Modulated Light by Heterodyne Method", Opt.-Mekh. Prom., 1972, 39, no.10, p.11, and Sov. J. Opt. Technol., 1972, 39, no.10, pp.596-7.

32.143. Nagai, H.: "Laser Frequency Fluctuations Due to Mechanical Vibrations", Mitsubishi Denki Lab. Rep., 1972, 13, no.1-4, pp.1-18.

32.144.   Abrams, R.L., and White, R.C.: "Three-Frequency Heterodyne Detection of 10.6-micron Laser Signals", J. Quantum Electron. IEEE, 1972, QE-8, no.1, pp.13-5.

32.145.   Saito, S., Fujii, Y., and Iwamoto, A.: "Analysis and Measurement of Shot-Noise Reduction Factor", Electron. Commun. Jap., 1972, 55, no.10, pp.69-76.

32.146.   Deryugin, I.A., Kurashov, V.N., and Mirzaev, A.T.: "Study of Time Correlation of Multifrequency Laser Radiation by Photon-Coincidence Method", Opt. Spekt., 1972, 33, no.6, pp.1129-33, and Opt. Spectrosc., 1972, 33, no.6, pp.618-20.

32.147.   Kurbasov, V.V., and Manzhura, V.Ya.: "Amplifier and Coincidence Circuit for Photon Counter With Large Time Resolution", Prib. Eksp. Tekh., 1972, 15, no.6, pp.134-6, and Instrum. Exp. Tech., 1972, 15, no.6, pp.1771-3.

32.148.   Hubbard, W.M.: "Comparitive Performance of Twin- and Single-Channel Optical Receivers", Trans. IEEE, 1972, COM-20, no.6, pp.1079-86.

32.149.   Vorobev, F.A., and Sokolovskii, R.I.: "Variation of Statistical Properties of Radiation on Interaction With a Resonant Medium", Opt. Spekt., 1972, 32, no.4, pp.842-3, and Opt. Spectrosc., 1972, 32, no.4, pp.446-7.

32.150.   Nagai, H.: "Laser Frequency Fluctuations Due to Mechanical Vibrations", J. Quantum Electron. IEEE, 1972, QE-8, no.12, pp.857-65.

32.151.   Landa, P.S., and Slin'ko, E.F.: "Natural Fluctuations in Linear Quantum Amplifiers", Kvantovaya Elektron., 1972, no.5, pp.114-6, and Sov. J. Quantum Electron., 1973, 2, no.5, pp.485-6.

32.152.   Arsen'ev, V.V., et al.: "Investigation of Statistical Properties of Ultrashort Light Pulses in Two-Photon Absorption in Semiconductors", Zh. Eksp. Teor. Fiz., 1972, 63, no.3, pp.776-83, and Sov. Phys.-JETP, 1973, 36, no.3, pp.407-10.

32.153.   Ataman, A., and Harth, W.: "Intrinsic FM Noise of Gunn Oscillators", Trans. IEEE, 1973, ED-20, no.1, pp.12-4.

32.154.   Sjolund, A.: "Small-Signal Noise Analysis of BARITT Diodes", Electron. Lett., 1973, 9, no.1, pp.2-4.

32.155.   Herbst, H., and Ataman, A.: "Q-Dependence of Gunn Oscillator FM Noise", Trans. IEEE, 1973, MTT-21, no.2, pp.114-5.

32.156.   Chopra, S., and Mandel, L.: "Higher-Order Correlation Properties of a Laser Beam", Phys. Rev. Lett., 1973, 30, no.2, pp.60-3.

32.157.   Sjolund, A.: "Noise at RF Amplitudes in IMPATT Oscillators", Int. J. Electron., 1973, 34, no.4, pp.551-64.

32.158.   Lee, D.H.: "LF Noise of Ion-Implanted Double-Drift IMPATT Diodes", Proc. IEEE, 1973, 61, no.5, pp.666-70.

32.159.   Gutmann, R.J., and Kucharewiski, N.: "LF Noise Characteristics of Commercial Si and GaAs IMPATT Diodes", Proc. IEEE, 1973, 61, no.5, pp.676-8.

32.160.   Graham, R., and Smith, W.A.: "Intensity Fluctuations of Lasers in Intensity-Coupled Two-Mode Operation", Opt. Commun., 1973, 7, no.3, pp.289-91.

32.161.   Weller, K.P.: "Study of Millimetre-Wave GaAs IMPATT Oscillator", Trans. IEEE, 1973, ED-20, no.6, pp.517-21.

32.162.   Fikart, J.L., and Goud, P.A.: "Theory of Oscillator Noise and Application to IMPATT Diodes", J. Appl. Phys., 1973, 44, no.5, pp.2284-96.

32.163.   Zardecki, A., and Delisle, C.: "Distribution of the Sum of Photocounts in Gaussian Light of Any State of Coherence and Polarization", Can. J. Phys., 1973, 51, no.9, pp.1017-25.

32.164.   Bendjaballah, C.: "Analysis of Clipped Photocount Autocorrelation Formulae for non-Gaussian Light", J. Phys. A, 1973, 6, no.6, pp.837-43.

32.165.   Cantrell, C.D., and Fields, J.R.: "Effect of Spatial Coherence on the Photoelectric Counting Statistics of Gaussian Light", Phys. Rev., 1973, 7A, no.6, pp.2063-9.

32.166.   Pawluczyk, R., and Mroz, E.: "Unidirectional Optical Coherent Noise Elimination by the Time Averaging Method", Opt. Acta, 1973, 20, no.5, pp.379-86.

32.167.   Grossmann, S., Kummel, H., and Richter, P.H.: "Intensities and Correlations in Two-Mode Lasers Near Threshold", Appl. Phys., 1973, 1, no.5, pp.257-61.

32.168.   Blake, J., and Barakat, R.: "Two-Fold Photoelectron Counting Statistics. Clipped Correlation Function", J. Phys. A, 1973, 6, no.8, pp.1196-210.

32.169.   Kitazima, I.: "Quasi-Periodic Noises in a He-Ne Laser", Opt. Commun., 1973, 8, no.3, pp.257-9.

32.170.   Mista, L.: "Counting Distribution for M-Mode Superposition of Chaotic and Modulated Coherent Fields", Czech. J. Phys., 1973, 23B, no.7, pp.715-8.

32.171.   Zardecki, A.: "Brownian Motion Model for Superradiant Fluctuations", Phys. Lett., 1973, 44A, no.5, pp.363-4.

32.172.   Hughes, A.J., et al.: "Photon Correlation Spectroscopy", J. Phys. A, 1973, 6, no.9, pp.1327-36.

32.173. Tornau, N., and Echtermeyer, B.: "Photon Counting Statistics of Thermal Light Consisting of Two Spectral Lines", Ann. Phys. (Leipzig), 1973, 29, no.3, pp.233-45, and no.4, pp.289-301.

32.174. Smithers, M.E., and Lu, E.Y.C.: "Photon Statistics and Coherence of Stimulated Superradiant Pulses", Lett. Nuovo Cim., 1973, 7, no.15, pp.750-2.

32.175. Swofford, R.L., and McClain, W.M.: "Peak Height Measurement System for Pulsed Laser Experiments", Rev. Sci. Instrum., 1973, 44, no.8, pp.978-81.

32.176. Perina, J., et al.: "Accuracy of Formulae for Photon Counting Statistics of Mixed Coherent and Chaotic Radiation", Czech. J. Phys., 1973, B23, no.10, pp.1008-13.

32.177. Wonneberger, W., and Lempert, J.: "Photon-Counting Distributions for Interacting Laser Modes", Opt. Commun., 1973, 9, no.1, pp.4-7.

32.178. Voronov, V.N., and Kirakosyan, R.: "Radiometer for 3-4 mm with Input Switch", Izv. VUZ Radiofiz., 1973, 16, no.9, pp.1439-41.

32.179. Nicholls, N.S., Damon, L., and Scoffield, B.P.: "Reduction of Noise in High-Power Crossed-Field Amplifiers", Electron. Lett., 1973, 9, no.17, pp.398-9.

32.180. Cantrell, C.D., Lax, M., and Smith, W.A.: "Third- and Higher-Order Intensity Correlation in Laser Light", Phys. Rev., 1973, 7A, no.1, pp.175-81.

32.181. Lax, M., and Zwanziger, M.: "Exact Photocount Statistics. Lasers Near Threshold", Phys. Rev., 1973, 7A, no.2, pp.750-71.

32.182. Corti, M., Degiorgio, V., and Arecchi, F.T.: "Measurements of Fine Structure of Laser Intensity Correlations Near Threshold", Opt. Commun., 1973, 8, no.4, pp.329-32.

32.183. Wang, Y.K., and Lamb, W.E.: "Theory of Laser Noise Effects", Phys. Rev., 1973, 8A, no.2, pp.873-80.

32.184. Zavorotnyi, V.U., and Tabarskii, V.I.: "Quantum Fluctuations of Photon Flux in Space and in the Diffraction Pattern", Zh. Eksp. Teor. Fiz., 1973, 64, no.2, pp.453-62, and Sov. Phys.-JETP, 1973, 37, no.2, pp.231-5.

32.185. Klimontovich, Yu.L., Kovalev, A.S., and Landa, P.S.: "Spontaneous Emission in an Intense Field and Relationship to Noise in Lasers", Opt. Spekt., 1973, 35, no.6, pp.1013-8, and Opt. Spectrosc., 1973, 35, no.6, pp.590-2.

32.186. Voronov, V.N., et al.: "Results of the Application of Schottky Diodes in Mixers for Radiometers in the Range 0.9-3.8 mm", Izv. VUZ Radiofiz., 1973, 16, no.7, pp.1125-6.

32.187. Saleh, B.E.A., and Cardoso, M.F.: "Effect of Channel Correlation on the Accuracy of Photon Counting Digital Autocorrelators", J. Phys. A, 1973, 6, no.12, pp.1897-909.

32.188. Srinivasan, S.K., Sukavanam, S., and Sudarshan, E.C.G.: "Many-Time Photcount Distribution", J. Phys. A, 1973, 6, no.12, pp.1910-8.

32.189. Mista, L., Perina, J., and Braunerova, Z.: "Photon-Counting Distribution of Superposition of Linearly Polarized Coherent and Partially Polarized Chaotic Radiation", Opt. Commun., 1973, 9, no.2, pp.113-8.

32.190. Deryugin, I.A., et al.: "Statistical Properties of Multi-Frequency Laser Radiation", Izv. Akad. Nauk SSSR, Ser. Fiz., 1973, 37, no.10, pp.2115-20.

32.191. Haken, H., and Wohrstein, H.G.: "Atom-Field Correlation, Conservation Laws, and Phase Transition, of the Laser", Opt. Commun., 1973, 9, no.2, pp.123-7.

32.192. Swartz, G.A., et al.: "(Results of) FM-Noise Measurements on p- and n-Si IMPATT Oscillators", Electron. Lett., 1973, 9, no.25, pp.578-80.

32.193. Gupta, S., and Mehta, C.L.: "Determination of the Statistical Properties of Light by Photodetectors With a Finite Dead Time", Phys. Rev., 1973, 8A, no.3, pp.1630-2.

32.194. Spears, D.L., and Freed, C.: "$(HgCd)Te$ Varactor Photodiode Detection of CW CO_2 Laser Beats Beyond 60 GHz", Appl. Phys. Lett., 1973, 23, no.8, pp.445-7.

32.195. Dubkov, V.I., and Kiselev, B.A.: "Photomixing of Main and Reference Laser Beams in Analysis of Frequency Composition by Heterodyning", Opt. Spekt., 1973, 35, no.2, pp.325-7, and Opt. Spectrosc., 1973, 35, no.2.

32.196. Riedl, G.: "Noise Measurements on Pulsed High-Power Microwave Amplifiers", Siemens Forsch. Entwick., 1973, 2, no.6, pp.327-32.

32.197. Geneux, E., et al.: "Destruction of Coherence by Scattering of Radiation on Atoms. II", Helv. Phys. Acta, 1973, 46, no.4, pp.457-60.

32.198. Luk'yanchikova, N.B., Garbar, N.P., and Zargar'yants, M.N.: "Light and Voltage Noise of Lasers Based on $(GaAl)As$-GaAs Heterojunctions", Phys. Status Solidi, 1973, 20, no.2, pp.637-45.

32.199. Lang, R., and Scully, M.O.: "Fluctuations in Mode-Locked Single-Mode Laser Oscillation", Opt. Commun., 1973, 9, no.4, pp.331-5.

32.200. Voronin, V.I., Dunaev, A.S., and Mukhamedyarov, R.D.: "Model of a Photodetector as a Nonlinear Section of a Receiving Channel", Opt.-Mekh. Prom., 1973, 40, no.2, pp.9-10, and Sov. J. Opt. Technol., 1973, 40, no.2, pp.76-7.

32.201. Bendjaballah, C., and Perrot, F.: "Statistical Properties of Intensity-Modulated Coherent Radiation", J. Appl. Phys., 1973, 44, no.11, pp.5130-41.

32.202. Arutyunyan, A.G., et al.: "Spatial Field and Intensity Correlation Functions of Laser Radiation", Zh. Eksp. Teor. Fiz., 1973, 64, no.5, pp.1511-26, and Sov. Phys.-JETP, 1973, 37, no.5, pp.764-71.

32.203. Akaike, M., Kato, H., and Fukatsu, Y.: "Millimetre-Wave IMPATT-Diode Oscillator and Noise Characteristics", Electron. Commun. Jap., 1973, 56, no.5, pp.55-61.

32.204. Georgiyevskaya, Ye.A., Mansurov, A.N., and Pelezneva, I.A.: "Frequency and Noise Properties of Si p-n Avalanche and p-i-n Photodiodes", Radiotekh. Elektron., 1973, 18, no.7, pp.1482-8, and Radio Eng. Electron. Phys., 1973, 18, no.7, pp.1093-7.

32.205. Kimble, H.J., and Mandel, L.: "Spatial Coherence of Laser Output Far Below Threshold", J. Opt. Soc. Am., 1973, 63, no.12, pp.1550-2.

32.206. Deryugin, I.A., Kurashov, V.N., and Mashchenko, A.I.: "Effect of Pump-Field Instability on Signal Statistical Properties in Quantum Parametric Amplifiers", Kvantovaya Elektron., 1973, no.7, pp.135-42.

32.207. Kitazima, I.: "Photon-Counting Distribution Under Square-Wave Intensity Modulation", Opt. Commun., 1974, 10, no.2, pp.137-40.

32.208. Fikart, J.L.: "AM and FM Noise of BARITT Oscillators", Trans. IEEE, 1974, MTT-22, no.5, pp.517-23.

32.309. Oliver, C.J., and Pike, E.R.: "Multiplex Advantage in Detection of Optical Images in the Photon Noise Limit", Appl. Opt., 1974, 13, no.1, pp.158-61.

32.210. Srinivasan, S.K.: "Photocount Statistics of Gaussian Light of Time-Limited Spectral Profile", Phys. Lett., 1974, 47A, no.2, pp.151-2.

32.211. Budkin, L.A., and Mityugov, V.V.: "Intensity Fluctuations of Laser Radiations", Izv. VUZ Radiofiz., 1974, 17, no.2, pp.238-46.

32.212. Davidson, F., et al.: "Photon Statistics of a CW Dye Laser With and Without Internal Etalons", J. Quantum Electron. IEEE, 1974, QE-10, no.4, pp.490-12.

32.213. Palamutovoglu, O., Gardiner, J.G., and Howson, D.P.: "Image Cancelling Mixers at 2GHz", Electron. Lett., 1974, 10, no.7, pp.104-6.

32.214. Yariv, A.: "Frequency, Intensity, and Field Fluctuations in Laser Oscillators", J. Quantum Electron. IEEE, 1974, QE-10, no.6, pp.509-15.

32.215. Brown, D.C.: "Study of Noise in He-Cd^{II} Laser", Appl. Phys. Lett., 1974, 24, no.6, pp.287-9.

32.216. Janossy, M.: "Noise and Output Power in the CW 441.6-nm He-Cd^{II} Laser", Phys. Lett., 1974, 47A, no.5, pp.409-10.

32.217. Paoli, T.L.: "Intrinsic Fluctuations in the Output Intensity of Double Heterostructure Junction Lasers Operating Continuously at 300°K", Appl. Phys. Lett., 1974, 24, no.4, pp.187-90.

32.218. Oshita, S., et al.: "Output Signal-to-Noise Ratio for Amplitude-Modulated Noise-Like Lasers", Trans. IEEE, 1974, AES-10, no.2, pp.223-30.

32.219. Goedbloed, J.J., and Vlaardingerbroek, M.T.: "Noise in IMPATT-Diode Oscillators at Large Signal Levels", Trans. IEEE, 1974, ED-21, no.6, pp.342-51.

32.220. Berdnikova, V.A., Kornilov, S.A., and Uman, S.D.: "Investigation of Correlation Relations Between LF Noises of an Electron Stream and Fluctuations of Oscillations of a Reflex Klystron", Izv. VUZ Radiofiz., 1974, 17, no.5, pp.743-52.

32.221. Berdnikova, V.A., Kornilov, S.A., and Uman, S.D.: "Investigation of Flicker-Noise Sources in a Reflex Klystron", Izv. VUZ Radiofiz., 1974, 17, no.5, pp.753-63.

32.222. Salomaa, R.: "Field Fluctuations in Bistable Gas Lasers", J. Phys. A, 1974, 7, no.9, pp.1094-1116.

32.223. Lachs, G.: "Statistics for Detection of Light by Nonideal Photomultipliers", J. Quantum Electron. IEEE, 1974, QE-10, no.8, pp.590-6.

32.224. Suzuki, T., and Ieiri, S.: "Noise Due to Relaxation Oscillations in a CW Nd^{3+}: YAG Laser", Jap. J. Appl. Phys., 1974, 13, no.8, pp.1331-2.

32.225. Milushkin, G.A., and Troshin, B.I.: "Investigation of Frequency Fluctuations of a He-Ne Laser at 633 nm", Kvantovaya Elektron., 1974, 1, no.1, pp.91-5, and Sov. J. Quantum Electron., 1974, 4, no.1, pp.48-50.

32.226. Lakshminarayana, B., Rao, V.M., and Murthy, K.V.N.: "Miniature Balanced Mixers for X-Band Applications", J. Inst. Electron. Telecommun. Eng., 1974, Jan., p.45-9.

32.227. Vvedenskii, B.S., et al.: "Interferometer for Investigating Pulsations of GaAs Injection Lasers", Prib. Tekh. Eksp., 1974, 17, no.2, pp.186-8, and Instrum. Exp. Tech., 1974, 17, no.2, pp.520-2.

32.228. Seib, D.H.: "Heterodyne Detection Experiments at 118.6 micron in Ga:Ge", J. Quantum Electron. IEEE, 1974, QE-10, no.2, pp.130-1.

32.229. Andrade, O., and Rye, B.J.: "(Phase-Front) Tolerances in Optical Mixing", J. Phys. D, 1974, 7, no.2, pp.280-91.

32.230. Kolesov, Yu.I., and Listvin, V.N.: V.N.: "Sensitivity of Submillimetre Radiometers With Fabry-Perot Interferometers", Izv. VUZ Radiofiz., 1974, 17, no.1, pp.11-6.

32.231. Sugawara, F.: "Noise Reduction of Ge Bolometer Using Thermal Filter", Jap. J. Appl. Phys., 1974, 13, no.2, pp.385-6.

32.232. Il'inskii, Yu.A., and Petnikova, V.M.: "Statistics of Photocounts in Parametric Frequency Conversion", Kvantovaya Elektron., 1974, 1, no.5, pp.1133-7, and Sov. J. Quantum Electron., 1974, 4, no.5, pp.619-21.

32.233. Matsumoto, G., and Shimizu, H.: "Photoelectron Counting System Using Mini-Computer", Bull. Electrotech. Lab., 1974, 38, no.11, pp.681-90.

32.234. Zakharenko, Yu.G.: "Striation Modulation of He-Ne Laser Emission", Opt. Spekt., 1974, 37, no.5, pp.990-1, and Opt. Spectrosc., 1974, 37, no.5, pp.565-6.

33. LOW-NOISE AMPLIFIERS

33.1. Rowe, J.E., and Wadhwa, R.P.: "Noise Transformation and Cyclotron Waves in Crossed Fields", J. Appl. Phys., 1965, 36, p.9.

33.2. Howson, D.P.: "Conjugately Matched Parametric Upconverters", Proc. IEE, 1965, 112, p.447.

33.3. Arams, F.R., and Peyton, B.J.: "8-mm TW Maser (for Radiometry)", Proc. IEEE, 1965, 53, p.12.

33.4. McEvoy, J.P., Miller, D.J., and Morris, L.C.: "Wideband TW Maser Employing Magnetic Stagger Tuning", Solid-State Electron., 1965, 8, p.443.

33.5. Howson, D.P.: "Double-Pumped Parametric Converter", Int. J. Electron., 1965, 18, p.269.

33.6. Vuillaume, P.: "Study of Noise Temperature and Bandwidth of Parametric Amplifiers Using Semiconductor Diodes", Onde Electr., 1965, 45, p.74.

33.7. Ray, R.D.: "Improved Maser Performance Through Pump Modulation", Proc. IEEE, 1965, 53, p.318.

33.8. Siebecker, H.K.: "Nonreciprocal Ruby Resonance Maser", Z. Angew Phys., 1965, 18, p.270.

33.9. Daglish, H.N.: "Low-Noise, Solid-State, Microwave Amplifiers", Post Office Electr. Eng. J., 1965, 58, pt 1, p.11.

33.10. Pratt, H.J., Ince, W.J., and Sicotte, R.C.: "High-Efficiency Varactor Upper-Sideband Upconverter", Proc. IEEE, 1965, 53, p.305.

33.11. Bosch, F.: "Nonlinearity of a TW Ruby Maser", Z. Angew. Phys., 1965, 18, p.254.

33.12. Benny, A.H., and Pearson, J.D.: "High-Frequency Wideband Idler Circuit for Parametric Amplifiers", Proc. IEEE, 1965, 53, p.181.

33.13. Alfeev, V.N., and Pimenov, Yu.P.: "Theory of Multicavity Masers", Radiotekh. Elektron., 1965, 10, p.54, and Radio Eng. Electron. Phys., 1965, 10, p.43.

33.14. Penfield, P.: "Maximum Cutoff Frequency of Varactor Diodes", Proc. IEEE, 1965, 53, p.422.

33.15. Hsu, T.W., and Robson, P.N.: "Negative Absorption from Weakly Relativistic Electrons Traversing a Cuccia Coupler", Electron. Lett., 1965, 1, p.84.

33.16. Johnson, C.C., and Rouzer, L.E.: "Linewidth and Inversion Ratios of Fe^{3+}: Rutile", Proc. IEEE, 1965, 53, p.204.

33.17. Hughes, K.L., et al.: "Detuning Effects in Parametric Amplifiers", Electron. Lett., 1965, 1, p.117.

33.18. Roshal', A.S.: "Transverse Uniform-Field Cavity as Coupler to Fast Cyclotron Wave", Radiotekh. Elektron., 1965, 10, p.78, and Radio Eng. Electron. Phys., 1965, 10, p.58.

33.19. Hayashi, R., and Igarishi, T.: "Measurement of Cr^{3+}:Rutile Maser at Liquid-Nitrogen Temperature", J. Radio Res. Lab., 1965, 12, p.53.

33.20. Marcuse, D.: "Noise Performance of an Overloaded Maser Amplifier", Proc. IEEE, 1965, 53, p.687.

33.21. Alfeev, V.N., and Pimenov, Yu.P.: "Theory of Selective Multicavity Transmission Maser Amplifiers", Radiotekh. Elektron., 1965, 10, p.45, and Radio Eng. Electron. Phys., 1965, 10, p.36.

33.22. Sayakhov, F.L.: "Calculation of Noise Factor of TWT With Exponential Electron Gun", Radiotekh. Elektron., 1965, 10, p.83, and Radio Eng. Electron. Phys., 1965, 10, p.66.

33.23. Chakraborty, D., and Millward, G.F.D.: "Nitrogen-Cooled Degenerate Parametric Amplifier for Radiometric Application", Electron. Eng., 1965, 37, p.532.

33.24. Nelson, J.N., and Israelson, B.P.: "TWT With 2.7-dB Noise Factor at 12 GHz", Proc. IEEE, 1965, 53, p.548.

33.25. Morris, L.C., and Miller, D.J.: "C-Band Rutile TW Maser", J. Quantum Electron. IEEE, 1965, QE-1, p.164.

33.26. Lopukhin, V.M., and Roshal', A.S.: "Electron-Beam, Transverse-Wave, Parametric Amplifiers", Usp. Fiz. Nauk, 1965, 85, p.297, and Sov. Phys.-Usp., 1965, 8, p.117.

33.27. Rucker, C.T., Morrow, W., and Grimes, E.S.: "X-Band Parametric Amplifier With Closed-Cycle Cooling", Trans. IEEE, 1965, MTT-13, p.123.

33.28. Chung-chih, C., and Shih-zhe, H.: "Analysis of Nonlinear Characteristics and Pumped Time-Varying Parameters of Varactors", Acta Phys. Sin., 1965, 21, p.664.

33.29. Vural, B.: "Beam-Noise Reduction in High Magnetic Fields", Proc. IEEE, 1965, 53, p.510.

33.30. Alfeev, V.N., and Ratbil', E.L.: "Design of Selective Multi-Resonator Parametric Amplifiers", Elektrosvyaz, 1965, no.3, and Telecommun. Radio Eng. Pt 1, 1965, no.3, p.26.

33.31. Karlov, N.V., and Martirosyan, R.M.: "Paramagnetic-Maser Circuit With Coupled Resonators", Radiotekh. Elektron., 1965, 10, p.673, and Radio Eng. Electron. Phys., 1965, 10, p.573.

33.32. Hughes, W.E., and Richards, W.E.: "Aluminium Nitrate Powder Maser", J. Quantum Electron. IEEE, 1965, QE-1, p.221.

33.33. Tengblad, R.: "Maser Performance v. Spin Concentration in $Cr^{3+}:K_3Co(CN)_6$", Ark. Fys., 1965, 30, p.203.

33.34. Ammann, E.O.: "Broadband, Solid-State, Maser Operating at $77^{\circ}K$", Trans. IEEE, 1965, MTT-13, p.186.

33.35. Kuo-chien, F., and P'ei-Liang, S.: "Analysis of Performance of Solid-State Multicavity Masers", Acta Electron. Sin., 1965, no.2-3, p.11.

33.36. Kotzebue, K.L., and Fletcher, L.B.: "Ferrimagnetically Tuned Parametric Amplifier", Trans. IEEE, 1965, MTT-13,p.773.

33.37. Krall, A.D., and Hooper, E.T.: "Ferromagnetic Microwave Amplifier With Increased Signal Band", Trans. IEEE, 1965, MAG-1, p.382.

33.38. Rosenbaum, F.J., and Jeffers, W.Q.: "Effect of Nonuniform Population Distribution in Masers", J. Quantum Electron. IEEE, 1965, QE-1, p.375.

33.39. Shioya, H.: "Blocking Effect (via Duplexers) in Ruby Masers", Proc. IEEE, 1965, 53, p.2139.

33.40. Gurzo, V.V., Stal'makhov, V.S., and Trubetskov, D.I.: "Theory of Parametric Amplification of Cyclotron Waves in Crossed-Field, Electron-Beam, Devices", Radiotekh. Elektron., 1965, 10, p.2251, and Radio Eng. Electron. Phys., 1965, 10, p.1926.

33.41. Hayashi, R.: "Temperature Dependence of Characteristics of Cross-Relaxation Rutile Maser", J. Radio Res. Lab., 1965, 12, p.381.

33.42. Berlin, A.S., et al.: "Parametric Amplification in the 8-mm Region", Radiotekh. Elektron., 1965, 10, p.1907, and Radio Eng. Electron. Phys., 1965, 10, p.1631.

33.43. Karlov, N.V.: "Q-Factor of Active Medium in a Maser", Radiotekh. Elektron., 1966, 11, p.271, and Radio Eng. Electron. Phys., 1966, 11, no.2, pp.224-30.

33.44. Fukui, H.: "Noise Performance of Microwave Transistors", Trans. IEEE, 1966, ED-13, p.329.

33.45. Hughes, W.E., and Kremenek, C.R.: "81-GHz Zero-Field Maser", Proc. IEEE, 1966, 54, p.623.

33.46. Israelson, B.P., and Haegele, R.W.: "Low-Noise TW Amplifiers", Trans. IEEE, 1966, COM-14, p.308.

33.47. Nagai, T., et al.: "Low-Noise Receiver With Helium-Cooled Parametric Amplifier", Proc. IEEE, 1966, 54, p.882.

33.48. Bosch, B.G., and Ertelt, G.: "Noise-Matched Unidirectional Parametric Amplifier of the Four-Frequency Type", Electron. Lett., 1966, 2, p.253.

33.49. Chang, N.S., and Makimoto, T.: "Ferromagnetic Amplifier With Double Resonance", Proc. IEEE, 1966, 54, p.992.

33.50. Wallrabe, A.: "Parametric Inverting Upconverter for Microwaves", Telefunken Z., 1966, 39, p.259.

33.51. Kollberg, E.L.: "Optimum Pitch of TW Masers", Trans. IEEE, 1966, MTT-14, p.212.

33.52. Wallrabe, A.: "Bandwidth and Noise Temperature of the Parametric Inverting Upconverter for Microwaves", Telefunken Z., 1966, 39, p.253.

33.53. Vinney, D.J.: "Possible TW Parametric Amplifier Using the Gunn Effect", Electron. Lett., 1966, 2, p.357.

33.54. Nixon, W.M., and Genner, R.: "Ferric-Doped-Rutile 8-mm Maser", Electron. Lett., 1966, 2, p.406.

33.55. Grigor'yants, V.V., and Mazurov, Yu.A.: "Cavity for Tuning a Maser by Zeeman Modulation", Radiotekh. Elektron., 1966, 11, p.152, and Radio Eng. Electron. Phys., 1966, 11, p.126.

33.56. Schollmeier, G., and Roth, D.: "Properties of Cr^{3+}:Rutile for Microwave Masers", Z. Angew. Phys., 1966, 21, p.187.

33.57. Solov'ev, E.G., et al.: "TW Maser Using Cr^{3+}:Rutile and a Superconducting Magnet", Radiotekh. Elektron., 1966, 11, p.1196, and Radio Eng. Electron. Phys., 1966, 11, p.1047.

33.58. Clorfeine, A.S.: "Self-Pumped Parametric Amplification With Avalanching Diode", Proc. IEEE, 1966, 54, no.12, pp.1956-7.

33.59. Karlov, N.V., and Prokhorov, A.M.: "Regenerative TW Masers", Radiotekh. Elektron., 1966, 11, p.267, and Radio Eng. Electron. Phys., 1966, 11, p.221.

33.60. Damiano, R., and Kliphuis, J.: "Cryogenically Cooled, 4-GHz, Parametric Amplifier Using Hybrid Integrated Circuits", Proc. IEEE, 1966, 54, p.1618.

33.61. Hughes, K.L., and Pearson, J.D.: "S-Band Parametric Amplifier Using a Balanced Idler Circuit", Radio Electron. Eng., 1966, 82, p.377.

33.62. Hughes, K.L., and Pearson, J.D.: "Effect of Upper Sideband on Performance of a Parametric Amplifier", Radio Electron. Eng., 1966, 32, p.337.

33.63. Berceli, T.: "Examination of Bandwidth of Parametric Amplifiers", Hiradastechnika, 1966, 17, p.237.

33.64. Rothe, H., and Rupf, K.: "Intrinsic Noise Temperature of the Reflection Maser", J. Quantum Electron. IEEE, 1966, QE-2, p.757.

33.65. Mescheryakov, Yu.I.: "Susceptibility of a Four-Level Paramagnetic System Under Step-by-Step Conditions at Saturation", Fiz. Tver. Tela, 1966, 8, p.2430, and Sov. Phys.-Solid State, 1967, 8, p.1935.

33.66. de Coatpont, X., and Robert, A.: "TW 8-mm Maser", Electron. Lett., 1967, 3, p.5.

33.67. Aitchison, C.S., Davies, R., and Gibson, P.J.: "Simple Diode Parametric Amplifier Design for Use at S, C, and X Bands", Trans. IEEE, 1967, MTT-15, p.22.

33.68. Brenner, H.E.: "Unilateral Parametric Amplifier", Trans. IEEE, 1967, MTT-15, no.5, pp.301-6.

33.69. Brenner, H.E.: "Stabilization of Gain v. Frequency Characteristics of Parametric Amplifiers at High Input Signal Levels", Trans. IEEE, 1967, MTT-15, no.5, pp.290-4.

33.70. Jones, W.S., and Hyde, F.J.: "Two-Diode Parametric Amplifiers", Electron. Lett., 1967, 3, no.4, pp.152-3.

33.71. Bryant, P.J., and Wilson, M.G.F.: "Improved Phase Characteristic of a DC-Pumped Cyclotron-Synchronous-Wave Amplifier", Electron. Lett., 1967, 3, no.5, pp.177-8.

33.72. Bura, P.: "Self-Pumped Parametric Amplification With Avalanche Diodes", Proc. IEEE, 1967, 55, no.6, p.1076.

33.73. Garbrecht, K.: "Bandwidth of Parametric Reflection Amplifiers With Line Resonators", Frequenz, 1967, 21, no.5, pp.142-51.

33.74. James, D.S., and Hyde, F.J.: "Maximum and Resonant Gains in Parametric Amplifiers", Electron. Lett., 1967, 3, no.6, p.229.

33.75. Kempe, V.: "Double-Pumped Low-Noise Parametric Downconverter", Electron. Lett., 1967, 3, no.6, pp.274-6.

33.76. Pommereit, M.: "Experimental Results With Lower-Frequency-Pumped Parametric Amplifier", Proc. IEEE, 1967, 53, no.7, pp.1219-20.

33.77. Pavlyuchuk, V.A., and Stepukhovich, V.A.: "Possibility of Reducing Pump Frequency in the Adler Tube", Radiotekh. Elektron., 1967, 12, no.3, pp.543-5, and Radio Eng. Electron. Phys., 1967, 12, no.3.

33.78. Verkhoturov, V.N., and Trofimenko, I.T.: "Study of a Cooled Parametric Amplifier Based on a Semiconductor Diode", Radiotekh. Elektron., 1967, 12, no.4, pp.650-4, and Radio Eng. Electron. Phys., 1967, 12, no.4, pp.602-6.

33.79. Brenner, H.E.: "Large-Signal Effects in Parametric Amplifiers", Trans. IEEE, 1967, MTT-15, no.2, pp.118-9.

33.80. Saito, F.: "Operation of a Laser-Pumped Ruby Maser at $77^{\circ}K$", NEC Res. Dev., 1967, no.9, pp.80-90.

33.81. Petty, S.M., and Clauss, R.C.: "X-Band TW Maser", Trans. IEEE, 1968, MTT-16, no.1, pp.47-8.

33.82. Khan, P.J.: "Parametric Amplifier Nonresonant Gain Maximum", Proc. IEEE, 1968, 56, no.1, pp.99-100.

33.83. McNeill, M.C.: "Low-Noise Amplifier for Communications Satellite Ground Stations", Marconi Rev., 1968, 31, p.282.

33.84. Kuno, H.J., Collard, J.R., and Gobat, A.R.: "Microwave Amplification With GaAs Avalanche Diodes", Electron. Lett., 1968, 4, p.540.

33.85. Khaplanov, G.M.: "Ruby Maser With Optical Pumping", Dokl. Akad. Nauk SSSR, 1968, 179, p.562, and Sov. Phys.-Dokl., 1968, 12.

33.86. DeGruyl, J.A., Okwit, S., and Smith, J.G.: "Noise Considerations in Broadband TW Masers", Trans. IEEE, 1968, MTT-16, p.584.

33.87. Shioya, H., Hozumi, H., and Hayashi, H.: "Closed-Cycle TW Maser", NEC Res. Dev., 1968, no.11, p.84.

33.88. Josenhans, J.G.: "Liquid-Helium-Cooled 4-GHz Broadband Parametric Amplifier", Trans. IEEE, 1968, MTT-16, p.789.

33.89. Harker, K.J., and Crawford, F.W.: "Parametric Amplification of Cyclotron Harmonic Waves in Plasma", Nachr. Tech. Fachber., 1968, 35, p.364.

33.90. Lopukhin, V.M., and Mitsenko, B.K.: "Extending the Band of an Electron-Beam Parametric Amplifier Using Two Uncoupled Resonators", Vestn. Mosk. Univ. Fiz. Astron., 1968, 23, p.80.

33.91. Colles, M.J., Smith, R.C., and Stanley, C.R.: "Near-IR Parametric Devices Using LiNbO$_3$", Nachr. Tech. Fachber., 1968, 35, p.656.

33.92. Liebe, H.J.: "Amplification at 258 GHz Using a Saturated Gas Resonance", Trans. IEEE, 1968, MTT-16, p.860.

33.93. Somervuo, P., and Porra, V.: "Broadband Matching of a Parametric Amplifier Using Fano's Method", Trans. IEEE, 1968, MTT-16, p.880.

33.94. Okean, H.C., and Weingart, H.: "S-Band Integrated Parametric Amplifier Having Both Flat-Gain and Linear-Phase Response", Trans. IEEE, 1968, MTT-16, p.1057.

33.95. Miller, D.J., and Weidner, G.G.: "Wide Tuning Range S-Band Maser", Proc. IEEE, 1969, 57, p.796.

33.96. Karlov, N.V., and Shirkov, A.V.: "Paramagnetic Masers Using Different Working Substances", Radiotekh. Elektron., 1969, 14, p.315, and Radio Eng. Electron. Phys., 1969, 14, no.2, pp.269-72.

33.97. Herman, M.A.: "Influence of Line Shape on Amplification Characteristics of Multicavity Reflection Masers", Arch. Elektrotech., 1969, 18, no.4, p.635.

33.98. Bessonov, V.I., and Zhelezovskii, B.Ye.: "Coupling Coefficients Between Modes in a Three-Frequency Parametric TWT Amplifier", Radiotekh. Elektron., 1969, 14, no.3, p.465, and Radio Eng. Electron. Phys., 1969, 14, no.3, pp.402-12.

33.99. Porra, V.: "Maximum Bandwidth Performance of a Single-Varactor Parametric Amplifier With a Single Tuned Idler Circuit", Saehkoe, 1969, 42, no.5-6, p.183.

33.100. Zhelezovskii, B.Ye., et al.: "Study of a Three-Frequency Fast Space-Charge-Wave Parametric Amplifier", Izv. VUZ Radioelektron., 1969, 12, p.2407.

33.101. Goldie, H.: "Effects of Noise Contributed by TW Devices on Low-Noise Receivers", Proc. IEE, 1969, 116, p.377.

33.102. Aitchison, C.S., Corbey, C.D., and Newton, B.H.: "Self-Pumped Gunn-Effect Parametric Amplifier", Electron. Lett., 1969, 5, p.36.

33.103. Shestopalov, A.M.: "Gain of Aperiodic Parametric Amplifiers", Radiotekhnika, 1969, 24, no.1, p.67, and Telecommun. Radio Eng. Pt 2, 1969, 24, no.1.

33.104. Majewski, M., and Stachnik, A.: "Liquid-Nitrogen-Cooled Parametric Amplifier for S Band", Arch. Elektrotech., 1969, 18, no.1, p.201.

33.105. Watson, P.A., and Buther, M.E.: "Phase-Locked Degenerate Parametric Amplifier", Electron. Lett., 1969, 5, p.392.

33.106. Toyota, S., and Makimoto, T.: "Transmission-Type Esaki-Diode Amplifier Using Rectangular Waveguide", Electron. Commun. Jap., 1969, 52, p.95.

33.107. Meshcheryakov, Yu.I., et al.: "Frequency Tuning of Maser With Fully Filled Cavity", Izv. VUZ Radioelektron., 1969, 12, p.1462.

33.108. Snapp, C.P., and Hoefflinger, B.: "Degenerate Amplification With Semiconductor Diodes to a Self-Generated Avalanche-Resonance Pump", Electron. Lett., 1969, 5, p.393.

33.109. Rutulis, U.: "Low-Noise Broadband 4-GHz Parametric Amplifier", Trans. IEEE, 1969, AES-5, p.679.

33.110. Bobrov, I.N., Volkov, V.M., and Sineok, V.I.: "Dynamic Range of a Regenerative Parametric Amplifier Without a Circulator", Izv. VUZ Radioelektron., 1969, 12, no.5, p.514.

33.111. Jackolsi, S.V., Masock, J., and Ishii, T.K.: "Self-Pumped Tunnel-Diode Parametric Amplifier", Proc. IEEE, 1969, 57, p.1681.

33.112. Osborne, T.L.: "Design of Efficient Broadband Varactor Upconverters", Bell Syst. Tech. J., 1969, 48, p.1623.

33.113. Osborne, T.L., Kibler, L.U., and Snell, W.W.: "Low-Noise Receiving Downconverter", Bell Syst. Tech. J., 1969, 48, p.1651.

33.114. Houston, T.W., and Read, L.W.: "Computer-Aided Design of Broadband and Low-Noise Microwave Amplifiers", Trans. IEEE, 1969, MTT-17, p.612.

33.115. Baechtold, W., Kotyczka, W., and Strutt, M.J.O.: "Computerized Calculation of Small Signal and Noise Properties of Microwave Transistors", Trans. IEEE, 1969, MTT-17, p.614.

33.116. Dathe, G., et al.: "Parametric Amplification Using an Inverted Maser Material", J. Quantum Electron. IEEE, 1969, QE-5, p.623.

33.117. Kunger, Y., and Arams, F.: "10.6-micron CW Upconversion in Proustite Using a Nd^{3+}:YAG Laser Pump", Proc. IEEE, 1969, 57, p.1797.

33.118. Firester, A.H.: "Parametric Image Conversion. I and II", J. Appl. Phys. 1969, 40, pp.4842 and 4849.

33.119. Proshutinskii, V.I., and Smirnov, A.I.: "Efficiency of Ruby Masers for 3-cm Band", Radiotekh. Elektron., 1969, 14, no.9, p.1634, and Radio Eng. Electron. Phys., 1969, 14, no.9.

33.120. Shamfarov, Ya.L., and Cherpak, N.T.: "Experimental Study of Pulse Saturation in Ruby Maser", Radiotekh. Elektron., 1969, 14, no.9, p.1656, and Radio Eng. Electron. Phys., 1969, 14, no.9, pp.1434-7.

33.121. Sanchez, J.B.: "Construction of a Wideband Degenerate Parametric Amplifier", Electron. Fis. Apl., 1969, 12, no.47, p.179.

33.122. Ozawa, S., and Fujioka, T.: "Computer Simulation for Low-Velocity Drifting Effects on Electron-Beam Noise", Electron. Commun. Jap., 1969, 52, no.3, p.103.

33.123. Korenev, Yu.V., Ratbil', E.L., and Yan'shin, A.S.: "Design and Construction of Broadband Parametric Amplifiers Utilizing the Reactive Components of a Semiconductor Diode", Elektrosvyaz, 1969, 23, no.6, p.6, and Telecommun. Radio Eng. Pt 1, 1969, 23, no.6.

33.124. Novozhilov, O.P.: "Calculation of Equivalent Admittances and Impedances of p-n Junctions in Three-Frequency Parametric Circuits", Izv. VUZ Radioelektron., 1969, 12, no.8, p.904.

33.125. Zhelezovskii, B.Ye, and Bessonov, V.I.: "Electron-Wave Gain Diagram in a Three-Frequency Parametric Amplifier of Space-Charge Waves", Izv. VUZ Radioelektron., 1969, 12, no.9, p.1032.

33.126. Kadoo, S.: "Considerations on Bandwidth of Diode Parametric Amplifier", Tech. J. Jap. Broadcast. Corp., 1969, 21, no.3, p.39, and Electron. Commun. Jap., 1969, 52, no.3, p.108.

33.127. Standley, R.D., and Braun, F.A.: "Experimental Results on Millimetre-Wave Microstrip Downconverter", Trans. IEEE, 1970, MTT-18, p.232.

33.128. Golubentsev, A.F.: "Concept of Noise Impedance of an Electron Beam", Radiotekh. Elektron., 1970, 15, no.5, p.1016, and Radio Eng. Electron. Phys., 1970, 15, no.5, pp.896-900.

33.129. Gurzo, V.V., Kulikov, M.N., and Stal'makhov, V.S.: "Experimental Study of an M-Type Electron-Beam Parametric Amplifier With Distributed Coupling", Izv. VUZ Radioelektron., 1970, 13, no.5, p.642.

33.130. Fantom, A.E.: "Experimental Nonreciprocal Double-Pumped Parametric Converter", Int. J. Electron., 1970, 29, p.315.

33.131. Lebedev, I.V.: "Threshold Input Signals of Microwave Amplifiers", Izv. VUZ Radioelektron., 1970, 13, no.8, pp.909-15.

33.132. Hildebrand, P.: "Method for Determining the Equivalent-Circuit Elements for Varactor Diodes", Frequenz, 1970, 24, p.78.

33.133. Khan, P.J.: "Determination of Parametric-Amplifier Nonresonant Gain Maximum", J. Solid-State Circuits IEEE, 1970, SC-5, p.79.

33.134. Hughes, W.E.: "Type of Powder Maser", Proc. IEEE, 1970, 58, p.480.

33.135. Collard, J.R., and Gobat, A.R.: "High-Dynamic-Range, Low-Noise, Microwave Amplifier", Electron. Lett., 1970, 6, p.202.

33.136. Shenogin, A.A.: "Reduction of Noise Factor of a TWT Operating in the 8-mm Band", Radiotekh. Elektron., 1970, 15, no.5, p.1022, and Radio Eng. Electron. Phys., 1970, 15, no.5, pp.900-3.

33.137. Terzhov, I.I.: "Minimal Noise Factor of Electron-Beam Microwave Amplifiers", Radiotekh. Elektron., 1970, 15, p.100, and Radio Eng. Electron. Phys., 1970, 15, no.1, pp.80-7.

33.138. Kerr, A.R.: "Comparison of One- and Two-Idler Parametric Amplifiers", Trans. IEEE, 1970, MTT-18, p.277.

33.139. Kinoshita, Y., and Maeda, M.: "18-GHz Single-Tuned Parametric Amplifier With Large Gain-Bandwidth Product", Trans. IEEE, 1970, MTT-18, p.409.

33.140. Pellisier, J.P., and Mesnard, G.: "Control of Frequency of Two-Cavity Klystron for Pumping a Ferromagnetic Amplifier", C. R. Acad. Sci. (Paris), 1970, 271B, p.361.

33.141. Armstrong, J.G., Dunford, B., and Willard, J.: "Low-Noise TWT's", Pric. IEE, 1970, 117, p.285.

33.142. Bobrov, I.N., Volkov, V.M., and Sineok, V.I.: "Dynamic Range of Nonregenerative Parametric Converter", Radiotekhnika, 1970, 25, no.1, p.97, and Telecommun. Radio Eng. Pt 2, 1970, 25, no.1.

33.143. Safronov, A.N.: "Modulation Depth of a Parametric Diode Using Harmonic Pump Signal", Radiotekhnika, 1970, 25, no.3, p.68, and Telecommun. Radio Eng. Pt 2, 1970, 25, no.3.

33.144. Gorshkov, A.S., et al.: "Amplification of Short Subharmonic Pulses in a BW Parametric Amplifier", Vestn. Mosk. Univ. Fiz. Astron., 1970, 11, p.87.

33.145. Newhouse, V.L., and Freeman, J.C.: "Capacitively Coupled Meander Line for Solid-State TW Amplification", Trans. IEEE, 1970, ED-17, p.383.

33.146. Dobrowolski, J.: "Design of Transistor Microwave Amplifiers", Przegl. Telekomun., 1970, 43, no.3, p.80.

33.147. Bandilla, A., and Paul, H.: "Amplitude and Intensity Fluctuations Following Amplification by a Laser", Ann. Phys. (Leipzig), 1970, 24, no.3-4, p.119.

33.148. Malaviya, S.D., and van der Ziel, A.: "Simplified Approach to Noise in Microwave Transistors", Solid-State Electron., 1970, 13, no.12, pp.1511-8.

33.149. Kudo, T., et al.: "Design Considerations of 4-GHz Broadband Cooled Parametric Amplifier", Fujitsu Sci. Tech. J., 1970, 6, no.3, pp.1-26.

33.150. Mazurov, M.Ye., et al.: "Optimum Nonlinear Capacitance for Parametric Amplifiers", Radiotekh. Elektron., 1970, 15, no.9, and Radio Eng. Electron. Phys., 1970, 15, no.9, pp.1719-22.

33.151. Korolev, F.A., and Kurin, A.F.: "Cyclotron-Resonance Maser With Fabry-Perot Cavity", Radiotekh. Elektron., 1970, 15, no.10, and Radio Eng. Electron. Phys., 1970, 15, no.10, pp.1863-73.

33.152. Raskutin, S.A.: "Surface-Wave Amplification in Dielectric Optical Waveguides", Radiotekh. Elektron., 1970, 15, no.12, and Radio Eng. Electron. Phys., 1970, 15, no.12, pp.2231-6.

33.153. Ono, S., and Shino, T.: "Characteristics of Varactor Upconverters", Electron. Commun. Jap., 1970, 53, no.10, pp.118-25.

33.154. Kinoshita, Y., and Maeda, M.: "18-GHz Double-Tuned Parametric Amplifier", Trans. IEEE, 1970, MTT-18, no.12, pp.1114-9.

33.155. Edrich, J.: "Parametric Amplification of Millimetre Waves Using Wafer Diodes", Trans. IEEE, 1970, MTT-18, no.12, pp.1173-5.

33.156. Takahashi, S., et al.: "K-Band Cryogenically Cooled Wideband Nondegenerate Parametric Amplifier", Trans. IEEE, 1970, MTT-18, no.12, pp.1176-8.

33.157. Freyer, U.: "Noise Factor of Reflection-Type Amplifiers With IMPATT Diodes", Nachr. Tech. Z., 1971, 24, no.3, pp.143-5.

33.158. Magnoni, A., Marzocchi, G., and Riva, F.: "Parametric Multiplier-Upconverter for 13 GHz", Alta Freq., 1971, 40, no.2, pp.169-70.

33.159. Jachimovits, L.: "Tuning of the Signal-Frequency Circuit of Parametric Amplifiers", Hiradastechnika, 1971, 22, no.2, pp.33-9.

33.160. Okean, H.C., et al.: "Integrated Parametric Amplifier Module With Self-Contained Solid-State Pump Source", Trans. IEEE, 1971, MTT-19, no.5, pp.491-3.

33.161. Nikiforov, A.N., and Detinko, V.N.: "Possibility of Regenerative Parametric Amplification Without a Difference-Frequency Circuit", Izv. VUZ Radioelektron., 1971, 14, no.3, pp.346-9.

33.162. Edrich, J.: "Parametric Amplifier for 46 GHz", Proc. IEEE, 1971, 59, no.7, pp.1125-6.

33.163. Lifanov, P.S., Filatov, K.V., and Shteinshleiger, V.B.: "Broadening the Passband of Masers Operating at Elevated Temperatures", Radiotekh. Elektron., 1971, 15, no.8, and Radio Eng. Electron. Phys., 1971, 15, no.8, pp.1534-5.

33.164. Berceli, T., and Konczanszky, Gy.: "Tunnel-Diode Amplifier in the 8-GHz Band", Hiradastechnika, 1971, 22, no.5, pp.129-32.

33.165. Shchepetkin, F.V.: "Minimum Noise Factor of a Transistor Microwave Amplifier", Izv. VUZ Radioelektron., 1971, 14, no.5, pp.577-80.

33.166. Ruppli, M.: "Very-Low-Noise 4-GHz Parametric Amplifier", Rev. Tech. Thomson-CSF, 1971, 3, no.2, pp.336-56.

33.167. Frey, W., Engelmann, W.H., and Bosch, B.G.: "Unilateral TW Amplification in GaAs at Microwave Frequencies", Arch. Elektron. Ubertrag., 1971, 25, no.1, pp.1-8.

33.168. Jutzi, W.: "Tuned (GaAs) MESFET Amplifier at 16 GHz", Arch. Elektron. Ubertrag., 1971, 25, no.3, pp.159-60.

33.169. Gayda, J.P., and Herve, J.: "Optical Pumping of Solid-State Masers", Rev. Phys. Appl., 1971, 6, no.2, pp.173-5.

33.170. Haldemann, H.: "Realizibility of a Laser-Pumped Parametric Microwave Amplifier", Z. Angew Math. Phys., 1971, 22, no.3, pp.428-42.

33.171. Khan, P.J.: "Optimum Design for Low-Noise Lower-Sideband Parametric Upconverters", Trans. IEEE, 1971, ED-18, no.10, pp.924-31.

33.172. Cesone, L., Gratze, S.C., and Wisbey, P.H.: "Use of GaAs (Parametric-Amplifier) Diodes at 1-5 GHz", Antenna, 1971, 43, no.7, pp.249-52.

33.173. Manh, P.T., and Naoulo, M.: "Paramagnetic Amplification in a Beam-Plasma System With Signal Higher in Frequency than the Pump Wave", C. R. Acad. Sci. (Paris), 1971, 273B, no.12, pp.446-9.

33.174. Vokes, J.C., Dawsey, J.R., and Deadman, H.A.: "Low-Noise Room-Temperature Wideband Parametric Amplifiers", Electron. Lett., 1971, 7, no.22, pp.657-8.

33.175. Maeda, M., and Samioka, A.: "Computer-Aided Design of Parametric Amplifiers", Trans. IEEE, 1971, MTT-19, no.12, pp.916-21.

33.176. Hughes, W.E.: "Ceramic Rutile X-Band Maser", Proc. IEEE, 1971, 59, no.2, pp.296-7.

33.177. Vanke, V.A., Kryukov, S.P., and Timofeev, Yu.M.: "Noise of DC-Pumped Amplifier With Strip Beam", Izv. VUZ Radiofiz., 1971, 14, no.1, pp.141-51.

33.178. Bessonov, V.I., et al.: "Investigation of Gain and Conversion Coefficients of a Parametric Three-Frequency TWT Amplifier", Radiotekh. Elektron., 1971, 16, no.8, pp.1418-24, and Radio Eng. Electron. Phys., 1971, 16, no.8, pp.1325-30.

33.179. Kita, S., and Kanmuri, N.: "Millimetre-Wave Upconverter Using Ge Avalanche Diode", Electron. Commun. Jap., 1971, 53, no.12, pp.86-93.

33.180. Grebenchuk, V.A., et al.: "Theory of a Twistron Amplifier. I", Radiotekh. (Kharkov), 1971, no.19, pp.82-9.

33.181. Pukhov, K.K., and Solov'ev, E.G.: "Frequency Conversion in a TW Maser", Radiotekh. Elektron., 1971, 16, no.10, pp.1846-51, and Radio Eng. Electron. Phys., 1971, 16, no.10, pp.1657-61.

33.182. Morino, A., and Sugano, T.: "Conversion Loss of Schottky-Diode Microwave Downconverters", Electron. Commun. Jap., 1971, 54, no.3, pp.68-74.

33.183. Misezhnikov, G.S., et al.: "TW Maser With Broadened Passband for 5-cm Wavelength", Prib. Tekh. Eksp., 1971, 14, no.3, pp.138-9, and Instrum. Exp. Tech., 1971, 14, no.3, pp.814-5.

33.184. Shevchenko, A.K.: "Use of Ruby in Optically Pumped Masers", Vestn. Mosk. Univ. Fiz. Astron., 1971, 12, no.6, pp.726-9.

33.185. Kabanov, D.A., and Popov, A.A.: "Noise Properties of a TW Amplifier Using a Distributed Tunnel Diode", Radiotekh. Elektron., 1971, 16, no.8, pp.1442-7, and Radio Eng. Electron. Phys., 1971, 16, no.8, pp.1343-6.

33.186. Baican, R.: "Achievement of Push-Pull Condition for Ruby Maser", Rev. Roum. Phys., 1971, 16, no.10, pp.1077-82.

33.187. Chiou, W.C., and Pace, F.P.: "Parametric Image Upconversion of 10.6-micron Illuminated Objects", Appl. Phys. Lett., 1972, 20, no.1, pp.44-7.

33.188. Edrich, J.: "Low-Noise Parametric Amplifiers Tunable over One Octave", J. Solid-State Circuits IEEE, 1972, SC-7, no.1, pp.32-7.

33.189. Muller, J.: "Upper-Sideband Upconverter Using MIS Varactors", J. Solid-State Circuits IEEE, 1972, SC-7, no.1, pp.43-50.

33.190. Dickens, L.E.: "Millimetre-Wave-Pumped X-Band Uncooled Parametric Amplifier", Proc. IEEE, 1972, 60, no.3, pp.328-9.

33.191. Bessonov, V.I., et al.: "Parametric Amplification and Frequency Conversion in a Two-Section TWT", Izv. VUZ Radioelektron., 1972, 15, no.3, pp.290-5.

33.192. Levashkin, V.I., and Tarabrin, Yu.K.: "Directivity of Three-Frequency Parametric Converters", Izv. VUZ Radioelektron., 1972, 15, no.3, pp.405-6.

33.193. Buchs, J.D.: "Downconversion With Schottky Diodes", Nachr. Tech. Z., 1972, 25, no.7, pp.305-11.

33.194. Lucy, R.F.: "IR-to-Visible Parametric Upconversion", Appl. Opt., 1972, 11, no.6, pp.1329-36.

33.195. Dupraz, J., and Creac'h, M.: "Parametric Amplifiers. Performance v. Operating Frequency", Electr. Commun., 1972, 47, no.2, pp.101-7.

33.196. Chiou, W.C., et al.: "Sensitivity Measurement of a 10.6-micron Parametric Upconverter", Trans. IEEE, 1972, ED-19, no.7, p.894.

33.197. Morgan, G.B., and Snow, K.C.: "Phase Characteristics of Microwave Avalanche-Diode Amplifiers", Electron. Lett., 1972, 8, no.11, pp.280-1.

33.198. Baikov, S.S.: "Influence of Intensified Spontaneous Radiation on Amplification Processes in a TW Amplifier", Vestn. Mosk. Univ. Fiz. Astron., 1972, 13, no.2, pp.164-74.

33.199. Howson, D.P.: "Maximum Power Transfer in Parametric Circuits", Radio Electron. Eng., 1972, 42, no.7, pp.328-32.

33.200. Konarev, V.P., et al.: "Threshold Characteristics of Laser Receivers", Kvantovaya Elektron., 1972, no.7, pp.86-7, and Sov. J. Quantum Electron., 1972, 2, no.1, pp.79-80.

33.201. Cho, C.: "Inversion Ratios for Push-Pull Paramagnetic Masers", J. Korean Inst. Electron. Eng., 1972, 21, no.3, pp.53-6.

33.202. Tseng, D.Y.: "Synchronously Pulsed IR Upconversion", Appl. Phys. Lett., 1972, 21, no.8, pp.382-4.

33.203. Asai, K., and Inaba, H.: "Wideband Parametric Upconversion of IR to Visible Radiation Using Tunable-Dye-Laser Pumping", Rec. Electr. Commun. Eng. Conversaz. Tohoku Univ., 1972, 41, no.2, pp.138-45.

33.204. Aitchison, C.S., and Williams, J.G.: "Actively Compensated Parametric Amplifiers", Electron. Lett., 1972, 8, no.23, pp.567-8.

33.205. Branner, G.R., Meyer, E.R., and Scheibe, P.O.: "Broadband Parametric Amplifier Design", Trans. IEEE, 1972, MTT-20, no.2, pp.176-8.

33.206. Chramiec, J.: "Effect of Pump Noise and Interfering Signals in Parametric Amplifiers", Proc. IEEE, 1972, 60, no.1, p.149.

33.207. Baskaran, S., and Robson, P.N.: "Noise Performance of InP Reflection Amplifiers in Q-Band", Electron. Lett., 1972, 8, no.5, pp.137-8.

33.208. Archer, J.A.: "Design and Performance of Small Signal Microwave Transistors", Solid-State Electron., 1972, 15, no.3, pp.249-58.

33.209. Nekrashevich, V.B., et al.: "Maser for 8-mm Band with Miniaturized Cooling System and Noise Temperature of 35°K", Radiotekh. Elektron., 1972, 17, no.7, pp.1544-5, and Radio Eng. Electron. Phys., 1972, 17, no.7, pp.1219-20.

33.210. Abramyan, L.E., Martirosyan, R.M., and Sarkisyan, E.L.: "Double-Cavity Maser at 1.35 cm for Radio Astronomy", Izv. Akad. Nauk Arm. SSR Fiz., 1972, 7, no.6, pp.464-9.

33.211. Cherpak, N.T., and Shamfarov, Ya.L.: "Protecting a Maser from Saturation by Pulsed Pump Modulation", Radiotekh. Elektron., 1972, 17, no.8, pp.1628-31, and Radio Eng. Electron. Phys., 1972, 17, no.8, pp.1284-6.

33.212. Vendik, O.G., et al.: "Cooled Parametric Amplifier with $SrTiO_3$", Radiotekh. Elektron., 1972, 17, no.9, pp.1981-3, and Radio Eng. Electron. Phys., 1972, 17, no.9, pp.1585-6.

33.213. Algazinov, E.K., and Kitaev, Yu.I.: "Combined Amplification of Monochromatic and Noise Signals in a TWT", Radiotekh. Elektron., 1972, 17, no.10, pp.2224-7, and Radio Eng. Electron. Phys., 1972, 17, no.10, pp.1786-9.

33.214. Okajima, T., et al.: "18-GHz Paramps With Triple-Tuned Gain Characteristics", Trans. IEEE, 1972, MTT-20, no.12, pp.812-9.

33.215. Samoilo, K., and Spasov, A.: "Analysis of One-Circuit Parametric Amplifier by the Phase-Plane Method", Izv. Inst. Elektron., 1972, 6, no.10, pp.113-8.

33.216. Hulme, K.F., and Warner, J.: "Theory of Thermal Imaging Using IR-to-Visible Upconversion", Appl. Opt., 1972, 11, no.12, pp.2956-64.

33.217. Bezenkova, O.A., et al.: "Design of Pumping Circuit for a Wideband Parametric Amplifier", Radiotekh. Elektron., 1972, 17, no.2, pp.419-22, and Radio Eng. Electron. Phys., 1972, 17, no.2, pp.326-9.

33.218. Gertsenshtein, M.E., et al.: "Noise Characteristics of Reflection Parametric Amplifiers with Correction in the Signal Circuit", Radiotekh. Elektron., 1972, 17, no.4, pp.879-80, and Radio Eng. Electron. Phys., 1972, 17, no.4, pp.689-90.

33.219. Gol'din, S.M., and Repinskii, V.N.: "Phase/Amplitude Response of a Regenerative Parametric Amplifier", Izv. VUZ Radioelektron., 1972, 15, no.9, pp.1114-21.

33.220. Bura, P.: "MIC K-Band Upconverter", Trans. IEEE, 1973, MTT-21, no.3, pp.136-7.

33.221. Braddock, P.W., and Gray, K.W.: "Low-Noise Wideband InP Transferred-Electron Amplifiers", Electron. Lett., 1973, 9, no.2, pp.36-7.

33.222. Wright, J.C., et al.: "Laser-Excited Double Resonance and Efficient IR Quantum Upconversion", J. Appl. Phys., 1973, 44, no.2, pp.781-6.

33.223. Melkov, G.A., and Artyukh, N.N.: "Ferrite Amplifier Using a Dielectric Resonator", Izv. VUZ Radioelektron., 1973, 16, no.1, pp.127-8.

33.224. Tucker, R.S.: "Gain-Bandwidth Limitations of Microwave Transistor Amplifiers", Trans. IEEE, 1973, MTT-21, no.5, pp.322-7.

33.225. Voronin, E.S., et al.: "Conversion of IR to Visible Radiation in Proustite With Orthogonal Pump and Signal Beams", Opto-Electron., 1973, 5, no.3, pp.237-41.

33.226. Davis, W.A., and Khan, P.J.: "Effect of Upper-Sideband Impedance on a Lower-Sideband Upconverter", Trans. IEEE, 1973, MTT-21, no.6, pp.386-92.

33.227. Branner, G.R., and Chan, S.P.: "Technique for Synthesis of Broadband Parametric Amplifiers", Trans. IEEE, 1973, MTT-21, no.7, pp.437-44.

33.228. Frey, W., et al.: "CW Operation of GaAs TW Amplifiers for X Band", Arch. Elektron. Ubertrag., 1973, 27, no.6, pp.245-52.

33.229. Dmitrevsky, S., and Kremer, P.C.: "Double-Amplification Mode Maser", Trans. IEEE, 1973, MTT-21, no.6, pp.426-7.

33.230. Baican, R., and Demco, D.: "Inversion Ratio for Push-Pull Ruby Maser", Rev. Roum. Phys., 1973, 18, no.4, pp.523-30.

33.231. Gurski, T.R.: "High-Quantum-Efficiency IR Upconversion", Appl. Phys. Lett., 1973, 23, no.5, pp.273-5.

33.232. Riccius, H.D., and Siemsen, K.J.: "Upconversion of CO Laser Lines by Difference-Frequency Mixing in Proustite", Phys. Lett., 1973, 45A, no.5, pp.377-8.

33.233. Alcock, A.J., and Walker, A.C.: "Fast Linear (Upconversion) Detection for TE CO_2 Lasers", Appl. Phys. Lett., 1973, 23, no.8, pp.467-8.

33.234. Sicotte, R.L.: "6-GHz Broadband Varactor Upconverter", COMSAT Tech. Rev., 1973, 3, no.2, pp.387-410.

33.235. Cherpak, N.T., and Shamfarov, Ya.L.: "Pulse Inversion in Three-Level Spin System", Ukr. Fiz. Zh., 1973, 18, no.5, pp.720-5.

33.236. Gunter, J., and Lucy, R.F.: "Background Light in Proustite Parametric Upconverter", Appl. Opt., 1973, 12, no.7, pp.1400-1.

33.237. Sewell, K.G., and Volz, W.B.: "Direct IR Image Quantum Upconversion With $PrCl_3$", Appl. Phys. Lett., 1973, 23, no.2, pp.104-6.

33.238. Verber, C.M.: "IR-to-Visible Conversion in $Er^{3+}:CaF_2$. Sequential Pair Process", J. Appl. Phys., 1973, 44, no.7, pp.3263-5.

33.239. Stracca, G.B., Aspesi, F., and D'Arcangelo, T.: "Low-Noise Microwave Downconverter With Optimum Matching at Idle Frequencies", Trans. IEEE, 1973, MTT-21, no.8, pp.544-7.

33.240. Lefeuvre, S., and Hanna, V.F.: "Noise in Solid (-State) TW Amplifiers Using Coupled-Mode Analysis", Int. J. Electron., 1973, 35, no.2, pp.163-7.

33.241. Mita, Y., et al.: "Efficient IR-to-Visible Conversion in $Yb^{3+}Er^{3+}:BaY_2F_8$ by Confinement of Excitation Energy", Appl. Phys. Lett., 1973, 23, no.4, pp.173-5.

33.242. Gainer, A.V., Krivoshchekov, G.V., and Sokolovskii, R.I.: "Theory of Image Conversion in Nonlinear Optical Systems", Opt. Spekt., 1973, 34, no.2, pp.401-4, and Opt. Spectrosc., 1973, 34, no.2, pp.225-6.

33.243. Kersten, R.Th., Mockel, P., and Oberbacher, R.: "$Nd^{3+}:YAG$ Laser Amplifier of Integrated Design", Siemens Forsch. Entwick., 1973, 2, no.4, pp.239-41.

33.244. Shakhparyan, V.P., and Martirosyan, R.M.: "Application of Fe^{3+}: Andulusite for Masers", Izv. Akad. Nauk Arm. SSR Fiz., 1973, 8, no.6, pp.446-50.

33.245. Averin, Ye.V., et al.: "Degenerate Parametric Amplifier With Polarization Decoupling of Signal- and Idler-Frequency Oscillations", Radiotekh. Elektron., 1973, 18, no.5, pp.1035-45, and Radio Eng. Electron. Phys., 1973, 18, no.5, pp.757-65.

33.246. Singh, R., and Gupta, K.C.: "Effect of Nonsinusoidal Pumping on Performance of Parametric Amplifiers", J. Inst. Electron. Telecommun. Eng., 1973, 19, no.11, pp.613-6.

33.247. Mukhina, M.M., Misezhnikov, G.S., and Shteinshleiger, V.B.: "3-cm Band TW Maser for Radio Interferometer", Radiotekh. Elektron., 1973, 18, no.8, pp.1746-7, and Radio Eng. Electron. Phys., 1973, 18, no.8, pp.1280-1.

33.248. Oldendskii, V.A., and Ustimenko, V.M.: "Operating Conditions of a Large-Signal Varactor Frequency Upconverter", Izv. VUZ Radioelektron., 1973, 16, no.11, pp.104-11.

33.249. Chramiec, J., and Coumes, A.: "Balanced Microwave Parametric Amplifier", Onde Electr., 1973, 53, no.11, pp.405-11.

33.250. Borodovskii, P.A., and Bul'dygin, A.F.: "Parametric Gunn-Diode Frequency Converter", Radiotekh. Elektron., 1973, 18, no.2, pp.365-70, and Radio Eng. Electron. Phys., 1973, 18, no.2, pp.262-6.

33.251. Berlin, A.S., et al.: "Wideband Parametric Amplifier for 8-mm Wavelength", Radiotekh. Elektron., 1973, 18, no.2, pp.423-6, and Radio Eng. Electron. Phys., 1973, 18, no.2, pp.309-11.

33.252. Lau, P.K., and Watson, P.A.: "Low-Noise Room-Temperature Parametric Amplifier", Electron. Lett., 1973, 9, no.25, pp.581-2.

33.253. Grigoryan, F.A., and Berlin, A.S.: "70-GHz Parametric Amplifier for Use in Radioastronomy", Radiotekh. Elektron., 1973, 18, no.2, pp.426-8, and Radio Eng. Electron. Phys., 1973, 18, no.2, pp.311-2.

33.254. Smirnova, T.A., Cherpak, N.T., and Shamfarov, Ya.L.: "4-cm Band Ruby Maser", Radiotekh. Elektron., 1973, 18, no.2, pp.361-4, and Radio Eng. Electron. Phys., 1973, 18, no.2, pp.258-61.

33.255. Edrich, J.: "Coolable Degenerate Parametric Amplifier for Millimetre Waves", Trans. IEEE, 1974, MTT-22, no.5, pp.581-3.

33.256. Sitch, J.E.: "Computer Modelling of Low-Noise InP Amplifiers", Electron. Lett., 1974, 10, no.6, pp.74-5.

33.257. Liechti, C.A., and Tillman, R.L.: "Design and Performance of Microwave Amplifiers With GaAs Schottky-Gate FET's", Trans. IEEE, 1974, MTT-22, no.5, pp.510-7.

33.258. van Dijk, M.H.H., Hagstrom, C.E., and Kollberg, E.L.: "Dielectrically Loaded Easitron Circuit as a Slow-Wave Structure for TW-Maser Applications", Int. J. Electron., 1974, 36, no.4, pp.495-506.

33.259. Voronin, E.S., Solomatin, V.S., and Shuvalov, V.V.: "Increased Efficiency for Upconversion of 10.6-micron Radiation Into the Visible", Opto-Electron., 1974, 6, no.2, pp.189-90.

33.260. Konishi, Y., et al.: "Simplified 12-GHz Low-Noise Converter With Mounted Planar Circuit in Waveguide", Trans. IEEE, 1974, MTT-22, no.4, pp.451-4.

33.261. Fischer, H.: "Maximum Bandwidth of Nonreciprocal Parametric Amplifiers Without Circulator With Negative Feedback", Arch. Elektron. Ubertrag., 1974, 28, no.4, pp.189-91.

33.262. Stal'makhov, V.S., Shatalov, A.M., and Shchegolev, V.P.: "Effect of Distributed Attenuation in the Pump Source on Parametric Gain of Fast Cyclotron Waves in M-Type Beams", Izv. VUZ Radioelektron., 1974, 17, no.2, pp.123-5.

33.263. Tseng, D.Y.: "Real-Time Synchronously Pulsed IR Image Upconversion", Appl. Phys. Lett., 1974, 24, no.3, pp.134-6.

33.264. Estes, L.E., et al.: "Internal Thermal Noise in Optical Frequency Converters", J. Opt. Soc. Am., 1974, 64, no.3, pp.295-300.

33.265. Burg, P.: "Operation of 6GHz FET Amplifier at Reduced Ambient Temperature", Electron. Lett., 1974, 10, no.10, pp.181-2.

33.266. Vendelin, D., Archer, J.A., and Bechtel, N.G.: "Low-Noise Integrated S-Band Amplifier", Microwave J., 1974, 17, no.2, pp.47-50.

33.267. Abramyan, L.E., and Martirosyan, R.M.: "Maser at 1.35 cm for Radioastronomy", Izv. VUZ Radiofiz., 1974, 17, no.5, pp.674-8.

33.268. Stang, M.: "Cooled Parametric Amplifier for Operation at 4 GHz", Frequenz, 1974, 28, no.5, pp.123-35.

33.269. Alexander, J.W.: "Linear and Non-linear Negative-Resistance and Negative-Conductance Parametric Amplifiers", Int. J. Electron., 1974, 36, no.6, pp.779-92.

33.270. Kollberg, E.L.: "Comb-Type Slow-Wave Structure for TW Masers", Electron. Lett., 1974, 10, no.13, pp.249-50.

33.271. Thirlwell, J., McPherson, J., and Bell, R.R.: "Broadband Cryogenic Parametric Amplifier Operating at 11.6 GHz", Electron. Lett., 1974, 10, no.16, pp.329-30.

33.272. Robson, P.N.: "Low-Noise Microwave Amplifiers Using Transferred-Electron and BARITT Devices", Radio Electron. Eng., 1974, 44, no.10, pp.553-67.

33.273. Sharma, M.L.: "S-Band Low-Noise Transistor Amplifier", J. Inst. Electron. Telecommun. Eng., 1974, Jan., pp.37-41.

33.274. Cherpak, N.T., Shamfarov, Ya.L., and Smirnova, T.A.: "Paramagnetic 4-cm-Band Maser With Asymmetric Passive Circuit", Izv. VUZ Radioelektron., 1974, 17, no.11, pp.92-6.

33.275. Mania, L., and Stracca, G.B.: "Conversion Loss of Single-Ended Downconverters With Junction Diodes", Alta Freq., 1974, 43, no.12, pp.988-97.

33.276. Parish, P.T., and Chiao, R.Y.: "Amplification of Microwaves by Superconducting Microbridges in a Four-Wave Parametric Mode", Appl. Phys. Lett., 1974, 25, no.10, pp.627-9.

33.277. Kitazume, S., Kobayashi, H., and Ishihara, H.: "Up- and Down-Converters in the 60-GHz Frequency Range", NEC Res. Dev., 1974, no.34, pp.62-71.

33.278. Scott, A.C., and Chu, F.Y.F.: "Pulse Saturation in a TW Parametric Amplifier", Proc. IEEE, 1974, 62, no.12, pp.1720-1.

33.279. Egami, S.: "Design Theory for Wideband Parametric Amplifiers", Trans. IEEE, 1974, MTT-22, no.2, pp.119-25.

33.280. Stetsenko, A.I., and Shamfarov, Ya.L.: "Dispersion of an Asymmetrically Loaded Comb Delay Line for a Maser", Kvantovaya Elektron. (Kiev), 1974, no.8, pp.133-44.

33.281. Stetsenko, A.I.: "Study of ESR Spectra and Matrix Elements of $Cr^{3+}:ZnWO_4$", Kvantovaya Elektron. (Kiev), 1974, no.8, pp.144-56.

33.282. Scanlan, M.J.B.: "Microwave Parametric Amplifiers", GEC J. Sci. Technol., 1974, 41, no.4, pp.135-42.

33.283. Smilowitz, B., Irvin, G., and Kaminsky, R.: "IMPATT Pump for Low-Noise Parametric Amplifier", Trans. IEEE, 1974, MTT-22, no.12, pp.1331-4.

33.284. Molokovskii, S.I., and Tregubov, V.F.: "Motion of Modulated Beam in the Output Stage of a Klystron Employing Electrostatic Focusing", Izv. VUZ Radioelektron., 1974, 17, no.9, pp.20-4.

34. PRINCIPLES OF HOLOGRAPHY

34.1. Dooley, R.P.: "X-Band Holography", Proc. IEEE, 1965, 53, p.1733.

34.2. Bryngdahl, O.: "Nonlinear Effects in Holography", J. Opt. Soc. Am., 1968, 58, p.1325.

34.3. Fujiwara, H., and Murata, K.: "Effects of Spatial Coherence on Fourier-Transform Holography", Oyo Buturi, 1968, 37, no.9, p.834.

34.4. Aristov, V.V., et al.: "Holography Without a Reference Beam for Three-Dimensional Holograms", Dokl. Akad. Nauk SSSR, 1968, 183, no.5, p.1039, and Sov. Phys.-Dokl., 1969, 13, no.12, p.1222.

34.5. Macovski, A.: "Efficient Holography Using Temporal Modulation", Appl. Phys. Lett., 1969, 14, p.166.

34.6. Staselko, D.I., et al.: "Production of High-Quality Holograms of Three-Dimensional Diffuse Objects With Single-Mode Ruby Lasers", Opt. Spekt., 1969, 25, no.6, p.910, and Opt. Spectrosc., 1969, 25, no.6, p.505.

34.7. Iizuka, K.: "Microwave Holograms and Microwave Reconstruction", Electron. Lett., 1969, 5, p.26.

34.8. Hildebrand, B.P., and Haines, K.A.: "Holography by Scanning", Phys. Lett., 1969, 28A, p.623.

34.9. Hirth, A.: "Application of Single-Mode Ruby Laser to Ultra-Rapid Holographic Phenomena", C. R. Acad. Sci. (Paris), 1969, 268B, p.961.

34.10. Iizuka, K.: "Microwave Hologram by Photoengraving", Proc. IEEE, 1969, 57, p.813.

34.11. Augustine, C.F., et al.: "Microwave Holography Using Liquid-Crystal Area Detectors", Proc. IEEE, 1969, 57, p.1333.

34.12. Pietsch, R.: "Holographic Technique Using a Diffuser as a Beamsplitter", Am. J. Phys., 1969, 37, p.748.

34.13. Budziak, A., Musiol, K., and Starnawski, A.: "Holography With a Ruby Laser", Acta Phys. Polon., 1969, 36, no.2, p.281.

34.14. Champagne, E.B., and Massey, N.G.: "Resolution in Holography", Appl. Opt., 1969, 8, p.1879.

34.15. Mottier, F.M.: "Time-Averaged Holography With Triangular Phase Modulation of the Reference Wave", Appl. Phys. Lett., 1969, 15, p.285.

34.16. Mirande, W., and Weingartner, I.: "Holography with Partially Coherent Illumination", Phys. Lett., 1969, 28A, p.623.

34.17. Weingartner, I., Mirande, W., and Menzel, E.: "Holography With Partial Coherence. I. Fresnel Holography", Optik, 1969, 29, no.1, p.87.

34.18. Fritlzer, D., and Marom, E.: "Reduction of Bandwidth Required for High-Resolution Hologram Transmission", Appl. Opt., 1969, 8, no.6, p.1241.

34.19. Chivian, J.S., Claytor, R.N., and Eden, D.D.: "Holography at 10.6 micron", Appl. Phys. Lett., 1969, 15, p.123.

34.20. Lowenthal, S., Serres, J., and Froehly, C.: "Recording of Holograms by Spatially Noncoherent Light", C. R. Acad. Sci. (Paris), 1969, 268B, p.841.

34.21. Larson, R.W., Johansen, E.L., and Zelenka, J.S.: "Microwave Holography", Proc. IEEE, 1969, 57, p.2162.

34.22. Stroke, G.W., Furrer, F., and Lamberty, D.R.: "Deblurring of Motion-Blurred Photographs Using Extended-Range Holographic Fourier-Transform Division", Opt. Commun., 1969, 1, no.3, p.141.

34.23. Voronin, E.S., et al.: "IR Holography by Methods of Nonlinear Optics", Zh. Eksp. Teor. Fiz. Pis'ma, 1969, 10, no.4, p.172, and JETP Lett., 1969,10,no.4,p.108.

34.24. Bondarenko, M.D., Gnatovskii, A.V., and Soskin, M.S.: "Holographic Method of Converting Coherent-Light Fields", Dokl. Akad. Nauk SSSR, 1969, 187, no.3, p.538, and Sov. Phys.-Dokl., 1970, 14, no.7, p.675.

34.25. Kermisch, D.: "Wavefront Reconstruction Mechanism in Blazed Holograms", J. Opt. Soc. Am., 1970, 60, p.782.

34.26. Murata, K., Asakura, T., and Fujiwara, H.: "Effects of Spatial Coherence on Holography", Opt. Acta, 1970,17,no.1,p.5.

34.27. Booth, B.L., Jarrett, S.M., and Barker, G.C.: "Holograms Made With Pulsed Argon-Ion Lasers Operating in Various Transverse Modes", Appl. Opt., 1970, 9, p.107.

34.28. Pernik, B.J., and Reich, A.: "Holographic System for Large Time-Bandwidth Product Multichannel Spectral Analysis", Appl. Opt., 1970, 9, p.229.

34.29. King, M.C., Noll, A.M., and Berry, D.H.: "Approach to Computer-Generated Holography", Appl. Opt., 1970, 9, p.471.

34.30. Aoki, Y., and Boivin, A.: "Computer Reconstruction of Images from a Microwave Hologram", Proc. IEEE, 1970, 58, p.821.

34.31. Siebert, L.D.: "Coherence-Length Curve for Ruby Oscillator-Amplifier", Appl. Phys. Lett., 1970, 16, p.318.

34.32. Dammann, H.: "Blazed Synthetic Phase-Only Holograms", Optik, 1970, 31, p.95.

34.33. Mirande, W., Weingartner, I., and Menzel, E.: "Compensation for Aberrations in Partially Coherent Image Holography", Opt. Commun., 1970, 1, p.315.

34.34. Stroke, G.W., et al.: "Laser-Light Spatial-Domain Deconvolution of Blurred Photographs Using the General Holographic Deblurring Filter", Phys. Lett., 1970, 31A, p.341.

34.35. Gregoris, L.G., and Iizuka, K.: "Visualization of Internal Structure by Microwave Holography", Proc. IEEE, 1970, 58, p.791.

34.36. Goodman, J.W.: "Analogy Between Holography and Interferometric Image Formation", J. Opt. Soc. Am., 1970, 60, p.506.

34.37. Sukhanov, V.I., and Denisyuk, Yu.N.: "Relation Between Spatial-Frequency Spectra of a Three-Dimensional Object and Three-Dimensional Hologram", Opt. Spekt., 1970, 28, no.1, p.126, and Opt. Spectrosc., 1970, 28, no.1, p.63.

34.38. Goetz, G.G.: "Real-Time Holographic Reconstruction by Electrooptic Modulation", Appl. Phys. Lett., 1970, 17, p.63.

34.39. Iizuka, K.: "Microwave Holography by Optical Interference Holography", Appl. Phys. Lett., 1970, 17, p.99.

34.40. Budziak, A., Musiol, K., and Palasinska, B.: "Holography by Use of a Ruby Laser", Acta Phys. Pol., 1970, 38A, no.1, p.131.

34.41. Wolf, E.: "Diffraction Theory of Holography", J. Math. Phys., 1970, 11, no.8, pp.2254.

34.42. Staselko, D.I., and Denisyuk, Yu.N.: "Effect of Transverse Mode Structure of a Radiation Source on the Image Produced by a Hologram", Opt. Spekt., 1970, 28, no.2, p.323, and Opt. Spectrosc., 1970, 28, no.2, p.174.

34.43. Kakichashvili, Sh.D.: "Achromatic Reconstruction of Holograms in Transmitted White Light", Zh. Prikl. Spekt., 1970, 12, no.3, p.547.

34.44. Gabel, R.A., and Liu, B.: "Minimization of Reconstruction Errors With Computer-Generated Binary Holograms", Appl. Opt., 1970, 9, p.1180.

34.45. De, M., Biswas, S.C., and Palchowdhuri, A.: "Spectral Coherence in Holography", Indian J. Pure Appl. Phys., 1970, 8, no.1, p.8.

34.46. Macovski, A.: "Hologram Information Capacity", J. Opt. Soc. Am., 1970, 60, no.1, p.21.

34.47. Hildebrand, B.P.: "Hologram Bandwidth Reduction by Space-Time Multiplexing", J. Opt. Soc. Am., 1970, 60, p.259.

34.48. Gates, J.W.C., Hall, R.G.N., and Ross, I.N.: "Holographic Recording Using Frequency-Doubled Radiation at 530 nm", J. Phys. E, 1970, 3, p.89.

34.49. McMahon, D.H., and Caulfield, H.J.: "Technique for Producing Wide-Angle Holographic Displays", Appl. Opt., 1970, 9, p.91.

34.50. McClung, F.J., Jacobson, A.D., and Close, D.H.: "Experiments Performed With Reflected-Light Pulsed-Laser Holography System", Appl. Opt., 1970, 9, p.103.

34.51. Firester, A.H., and Heller, M.E.: "Use of Diode Lasers to Recover Holographically Stored Information", J. Quantum Electron. IEEE, 1970, QE-6, p.572.

34.52. Budziak, A.: "Holography of Phase Objects With Use of a Ruby Laser", Acta Phys. Pol., 1970, 38A, no.5, pp.805-6.

34.53. Mirande, W., and Weingartner, I.: "Recording of Inhomogeneous Mutual Coherence Functions in One Hologram by a Combination of Image and Fourier (Fresnel) Holography", Opt. Commun., 1970, 2, no.3, pp.97-100.

34.54. Som, S.C., and Lessard, R.A.: "Multiplex Fourier-Transform Holography", Opt. Commun., 1970, 2, no.3, pp.128-32.

34.55. Hirth, A.: "Holography of a Three-Dimensional Fast Moving Object Using a Ruby Laser With a Single Uncoupled Mode", Opt. Commun., 1970, 2, no.3, pp.139-41.

34.56. Clair, J.J., et al.: "Fourier Hologram Synthesis Using Birefringent Elements", Opt. Commun., 1970, 2, no.4, pp.163-4.

34.57. Farwat, N.W., and Guard, W.R.: "Holographic Imaging at 70 GHz", Proc. IEEE, 1970, 58, no.12, pp.1955-6.

34.58. Stroke, G.W.: "Image Deblurring", Opt. Spectra, 1970, 4, no.10, pp.31-2.

34.59. Aristov, V.V., et al.: "Reconstruction of Three-Dimensional Holograms Using an Extended Source", Opt. Spekt., 1970, 29, no.3, pp.604-5, and Opt. Spectrosc., 1970, 29, no.3, p.322.

34.60. Kim, K.H., et al.: "Effect of Exposure Parameters on Holographic Image Brightness", J. Korean Phys. Soc., 1970, 3, no.2, pp.39-42.

34.61. Klimenko, I.S., and Matinyan, E.G.: "Use of Random Reference Wave in Holographic Recording of Focused Images", Opt. Spekt., 1970, 29, no.6, pp.1132-7, and Opt. Spectrosc., 1970, 29, no.6, pp.603-5.

34.62. Aplin, C., Fleuret, J., and Gaggioli, N.G.: "Recent Improvements in Dynamic Holography at 1060 nm", Acta Cient., 1970, 3, no.3-4, pp.131-5.

34.63. Bykovskii, Yu.A., Elkhov, V.A., and Larkin, A.I.: "Coherence of Radiation of a Semiconductor Laser and Use in Holography", Fiz. Tekh. Poluprov., 1970, 4, no.5, p.962, and Sov. Phys.-Semicond., 1970, 4, no.5, pp.819-21.

34.64. Gabor, D.: "Laser Speckle and its Elimination (in Holograms)", IBM J. Res. Dev., 1970, 14, no.5, pp.509-14.

34.65. Buinov, G.N., Lukin, A.V., and Mustafin, K.S.: "Effect of Spatial Coherence on Characteristics of Holograms", Opt. Spekt., 1970, 28, no.5, pp.1018-21, and Opt. Spectrosc., 1970, 28, no.5, pp.550-1.

34.66. Postl, W.: "Modulating a Hologram", Proc. IEEE, 1971, 59, no.1, pp.77-8.

34.67. Lang, M., Goldmann, G., and Graf, P.: "Contribution to the Comparison of Single- and Multiple-Exposure Storage Holograms", Appl. Opt., 1971, 10, no.1, pp.168-73.

34.68. Bromley, K., et al.: "Holographic Subtraction", Appl. Opt., 1971, 10, no.1, pp.174-81.

34.69. Weigl, F.: "Generalized Technique of Two-Wavelength Nondiffuse Holographic Interferometry", Appl. Opt., 1971, 10, no.1, pp.187-92.

34.70. Varner, J.R.: "Simplified Multiple-Frequency Holographic Contouring", Appl. Opt., 1971, 10, no.1, pp.212-3.

34.71. Nishida, N., and Sakaguchi, M.: "Improvement of Nonuniformity of Reconstructed Beam Intensity from a Multiple-Exposure Hologram", Appl. Opt., 1971, 10, no.1, pp.439-40.

34.72. Matsumoto, K.: "Analysis of Holographic Multiple-Beam Interferometry", J. Opt. Soc. Am., 1971, 61, no.2, pp.176-81.

34.73. Ueda, M., and Sato, T.: "Super-resolution by Holography", J. Opt. Soc. Am., 1971, 61, no.3, pp.418-9.

34.74. Hutzler, P., Lanzl, F., and Waidelich, W.: "Extension of Spatial-Frequency Range of Fourier Holograms by Double Exposure", Opt. Commun., 1971, 2, no.8, pp.402-6.

34.75. MacGovern, A.J., and Wyant, J.C.: "Computer Generated Holograms for Testing Optical Elements", Appl. Opt., 1971, 10, no.3, pp.619-24.

34.76. Siebert, L.D.: "Holographic Coherence Length of a Pulse Laser", Appl. Opt., 1971, 10, no.3, pp.632-7.

34.77. Latta, J.N.: "Fifth-Order Hologram Aberrations", Appl. Opt., 1971, 10, no.3, pp.666-7.

34.78. Hsu, T.R., and Moyer, R.G.: "Application of Fibre Optics in Holography", Appl. Opt., 1971, 10, no.3, pp.669-70.

34.79. Lee, T.C., and Gossen, D.: "Generalized Fourier-Transform Holography and Applications", Appl. Opt., 1971, 10, no.4, pp.961-3.

34.80. Lin, L.H., and Doherty, E.T.: "Efficient and Aberration-Free Wavefront Reconstruction from Holograms Illuminated at Wavelengths Differing from the Forming Wavelength", Appl. Opt., 1971, 10, no.6, pp.1314-8.

34.81. Aleksoff, C.C.: "Temporally Modulated Holography", Appl. Opt., 1971, 10, no.6, pp.1329-41.

34.82. Porter, R.P., and Schwab, W.C.: "Electromagnetic Image Formation With Holograms of Arbitrary Shape", J. Opt. Soc. Am., 1971, 61, no.6, pp.789-96.

34.83. Lohmann, A.W.: "Real-Time Holography", Opt. Commun., 1971, 3, no.2, pp.73-6.

34.84. Shamir, J., Politch, J., and Ben Uri, J.: "Holography With 1-mW Laser", Am. J. Phys., 1971, 39, no.7, pp.840-1.

34.85. Sakusabe, T., and Kobayashi, S.: "Holography With Liquid Crystals", Jap. J. Appl. Phys., 1971, 10, no.6, pp.758-61.

34.86. Minami, M., Unno, Y., and Mizobuchi, Y.: "Holographic Image Reconstruction With an Injection Laser", Appl. Opt., 1971, 10, no.7, pp.1629-31.

34.87. Bieringer, R.J., and Ringlien, J.A.: "Diffraction-Limited Holography", Appl. Opt., 1971, 10, no.7, pp.1632-5.

34.88. Colburn, W.S., and Haines, K.A.: "Volume Hologram Formation in Photopolymer Materials", Appl. Opt., 1971, 10, no.7, pp.1636-41.

34.89. Moran, J.M.: "Compensation of Aberrations due to a Wavelength Shift in Holography", Appl. Opt., 1971, 10, no.8, pp.1909-13.

34.90. Rosenberg, R.L., and Chandross, E.A.: "Holographic Fibre Optics", Appl. Opt., 1971, 10, no.8, pp.1986-8.

34.91. Perina, J.: "Holographic Method of Deconvolution and Analytic Continuation", Czech. J. Phys., 1971, 21B, no.7, pp.731-48.

34.92. Chau, H.H., and Leppelmeier, G.W.: "Effect of Carrier-Frequency Shift on Short-Pulse Holography", J. Opt. Soc. Am., 1971, 61, no.8, pp.998-1000.

34.93. Murty, M.V.R.K., and Das, N.C.: "Theory of Certain Diffraction Gratings Produced by Holographic Method", J. Opt. Soc. Am., 1971, 61, no.8, pp.1001-6.

34.94. Zubov, V.A., et al.: "Holographic Recording of Nonstationary Processes", Zh. Eksp. Teor. Fiz. Pis'ma, 1971, 13, no.8, pp.316-7, and JETP Lett., 1971, 13, no.8, pp.443-6.

34.95. Stepanov, B.I., Ivakin, E.V., and Rubanov, A.S.: "Recording Two- and Three-Dimensional Dynamic Holograms in Transparent Substances", Dokl. Akad. Nauk SSSR, 1971, 196, no.1-3, pp.567-9, and Sov. Phys.-Dokl., 1971, 16, no.1, pp.46-8.

34.96. Brocard, R.J.: "Holograms of a Moving Body", Toute Electron., 1971, no.357, p.25.

34.97. Dainty, J.C., and Welford, W.T.: "Reduction of Speckle in Image Plane Hologram Reconstruction by Moving Pupils", Opt. Commun., 1971, 3, no.5, pp.289-94.

34.98. Caulfield, H.J.: "Speckle Averaging by Spatially Multiplexed Holograms", Opt. Commun., 1971, 3, no.5, pp.322-3.

34.99. Goodman, J.W.: "Introduction to the Principles and Applications of Holography", Proc. IEEE, 1971, 59, no.9, pp.1292-304.

34.100. Huang, T.S.: "Digital (Computer) Holography", Proc. IEEE, 1971, 59, no.9, pp.1335-46.

34.101. Farhat, N.H., and Guard, W.R.: "Millimetre-Wave Holographic Imaging of Concealed Weapons", Proc. IEEE, 1971, 59, no.9, pp.1383-4.

34.102. Upatnieks, J., and Leonard, C.D.: "Linear Wavefront Reconstruction from Non-linearly Recorded Holograms", Appl. Opt., 1971, 10, no.10, pp.2365-7.

34.103. Keneman, S.A.: "Hologram Storage in AsS_3 Thin Films", Appl. Phys. Lett., 1971, 19, no.6, pp.205-7.

34.104. Aplin, C., Fleuret, J., and Gaggioli, N.G.: "Ultrafast Colour Holography", C. R. Acad. Sci. (Paris), 1971, 273B, no.4, pp.173-6.

34.105. Magill, P.J., and Speicher, C.A.: "Photometallic Etching of Holograms", J. Electrochem. Soc., 1971, 118,no.9,pp.1514-6.

34.106. Yonezawa, S., et al.: "Computer Hologram Recording With an Electron Beam", Jap. J. Appl. Phys., 1971, 10, no.9, p.1279.

34.107. Soskin, M.S., Bondarenko, M.D., and Gnatovskii, A.V.: "Holographic Method of Amplitude-Phase Correction of Laser Beams", Zh. Eksp. Teor. Fiz. Pis'ma, 1971, 14, no.1, pp.27-32, and JETP Lett., 1971, 14, no.1, pp.17-20.

34.108. Wuerker, R.F., and Heflinger, L.O.: "Ruby-Laser Holography", Soc. Photo-Opt. Instrum. Eng. J., 1971, 9, no.4, pp.122-30.

34.109. DeBitetto, D.J., and Dalisa, A.L.: "Elimination of Flare Light in Hologram Recording of Diffuse Objects", Appl. Opt., 1971, 10, no.10, pp.2292-6.

34.110. Papi, G., Russo, V., and Sottini, S.: "Microwave Holographic Interferometry", Trans. IEEE, 1971, AP-19, no.6, pp.740-6.

34.111. Mikaelyan, A.L., et al.: "Multiple Recording of Holograms in Case of an Extended Reference-Beam Source", Kvantovaya Elektron., 1971, no.1, pp.143-5.

34.112. Richter, W., Burow, R., and Hebermehl, G.: "Synthetic Holograms", Monatsber. Dtsch. Akad. Wiss. Berlin, 1971, 13, no.2, pp.98-107.

34.113. Stroke, G.W.: "Sharpening Images by Holography", New Sci., 1971, 51, no.770, pp.671-4.

34.114. Galpern, A.D., and Denisyuk, Yu.N.: "Improving Image Quality Using a Holographic Storage Method", Opt. Spekt., 1971, 30, no.2, pp.340-4, and Opt. Spectrosc., 1971, 30, no.2, pp.184-6.

34.115. Khaikin, K.E., and Khitrova, V.S.: "Method for Computer Synthesis of Holograms", Opt. Spekt., 1971, 30, no.2, pp.375-6, and Opt. Spectrosc., 1971, 30, no.2, pp.204-5.

34.116. Winter, D.C.: "Correction of Unequal Longitudinal and Lateral Magnification in Holography", Appl. Opt., 1971, 10, no.11, pp.2551-3.

34.117. Lowenthal, S., and Braat, J.: "Subtraction of Intensites by Spatially Incoherent Fourier Holography", Appl. Opt., 1971, 10, no.11, pp.2553-4.

34.118. Politch, J., Shamir, J., and Ben Uri, J.: "Characteristics of Multibeam Holography", Opt. Laser Technol., 1971, 3, no.4, pp.226-8.

34.119. Dickinson, J., and Rinard, G.: "Resolution Criterion for Long-Wavelength Holograms", Appl. Phys. Lett., 1971, 19, no.9, pp.352-3.

34.120. Okayama, H., and Emori, Y.: "Polarization Effects in Holography", Appl. Phys. Lett., 1971, 19, no.9, pp.359-60.

34.121. Anderson, A.P.: "Basic Experiments on Optical Resolution of Sampled Microwave Holograms", Opto-Electron., 1971, 3, no.3, pp.127-30.

34.122. Asakura, T., and Nagai, S.: "Double-Pinhole Holography", Bull Res. Inst. Appl. Electr., 1971, 23, no.3-4, pp.128-31.

34.123. Aristov, V.V., and Shekhtman, V.Sh.: "Properties of Three-Dimensional Holograms", Usp. Fiz. Nauk, 1971, 14, no.3, pp.51-76, and Sov. Phys.-Usp., 1971, 14, no.3, pp.263-77.

34.124. Klimenko, I.S., and Matinyan, E.G.: "Obtaining Holograms of Focused Images in Multimode Laser Radiation", Opt. Spekt., 1971, 31, no.3, pp.471-2, and Opt. Spectrosc., 1971, 31, no.3, pp.250-1.

34.125. Shigesawa, H., et al.: "Microwave Holography by Synthetic Aperture", Proc. IEEE, 1972, 60, no.1, pp.137-9.

34.126. Waters, J.P.: "Object-Motion Compensation by Speckle-Reference-Beam Holography", Appl. Opt., 1972, 11,no.3,pp.630-6.

34.127. Torii, Y., and Sumi, M.: "Hologram Reconstruction by Incoherent Light. I", Jap. J. Appl. Phys., 1972, 11,no.5,pp.644-55.

34.128. Kronrod, M.A., et al.: "Experiments on Synthesis of Hologram Transparencies on a Digital Computer", Zh. Tekh. Fiz., 1972, 42, no.2, pp.414-8, and Sov. Phys.-Tech. Phys., 1972, 17, no.2, pp.329-32.

34.129. Swingler, D.N.: "Simple Scanning Technique for Microwave Holography", Proc. IEEE, 1972, 60, no.7, pp.918-19.

34.130. Fridman, G.Kh., and Tsvetov, Ye.R.: "Feasibility of Applying Holography to Large-Volume Scenes by Using RF Light Modulation", Radiotekh. Elektron., 1972, 16, no.9, pp.1718-9, and Radio Eng. Electron. Phys., 1972, 16, no.9, pp.1555-7.

34.131. Mikaeliane, A.I., et al.: "Possibilities of Optical Elements Design Using Phase Holograms", Appl. Opt., 1972, 11, no.9, pp.2004-6.

34.132. Givens, M.P.: "Image Location and Magnification in Holography", Am. J. Phys., 1972, 40, no.9, pp.1311-4.

34.133. Latta, J.N.: "Analysis of Multiple Hologram Optical Elements With Low Dispersion and Low Aberrations", Appl. Opt., 1972, 11, no.8, pp.1686-96.

34.134. Everett, P.N., and Cantor, A.J.: "Long-Range Holography", Appl. Opt., 1972, 11, no.8, pp.1697-707.

34.135. Johnson, R.H., and Holshouser, D.F.: "Application of a Mode-Locked Laser to Holography", Appl. Opt., 1972, 11, no.8, pp.1708-15.

34.136. Anderson, A.P., and Duvernoy, J.: "Enhancement of Optically Reconstructed Images from Microwave Holograms", Opto-Electron., 1972, 4, no.1, pp.57-62.

34.137. Papi, G., Russo, V., and Sottini, S.: "Two-Frequency Microwave Holographic Interferometry", Proc. IEEE, 1972, 60, no.8, pp.1004-5.

34.138. Amodei, J.J., and Staebler, D.L.: "Holographic Recording in LiNbO$_3$", RCA Rev., 1972, 33, no.1, pp.71-93.

34.139. Hariharan, P., and Ramprasad, B.S.: "Simplified Optical System for Holographic Subtraction", J. Phys. E, 1972, 5, no.10, pp.976-8.

34.140. Marom, E.: "Holographic Subtraction With Circularly Polarized Light", Opt. Commun., 1972, 6, no.1, pp.86-90.

34.141. Bykovskii, Yu.A., et al.: "Coherent Radiation and Holography in an Optically Inhomogeneous Medium", Zh. Tekh. Fiz., 1972, 42, no.4, pp.830-6, and Sov. Phys.-Tech. Phys., 1972, 17, no.4, pp.653-8.

34.142. Nagai, K., et al.: "Numerical Reconstruction of the Image from a Microwave Hologram", Electron. Commun. Jap., 1972, 55, no.11, pp.50-4.

34.143. Budagyan, I.F., et al.: "Holographic Versions of a Refractometer and Reflectometer", Prib. Tekh. Eksp., 1972, 15, no.6, pp.174-7, and Instrum. Exp. Tech., 1972, 15, no.6, pp.1817-20.

34.144. Ginzburg, V.M., et al.: "Use of Lateral Illumination for Holography of Small Objects", Prib. Tekh. Eksp., 1972, 15, no.6, pp.179-81, and Instrum. Exp. Tech., 1972, 15, no.6, pp.1824-6.

34.145. Bazarskii, O.V., Kotosonov, N.B., and Khlyavich, Ya.L.: "Investigation of Holographic Method of Obtaining Visible Images of Phase Objects in the Microwave Region", Radiotekh. Elektron., 1972, 17, no.8, pp.1733-4, and Radio Eng. Electron. Phys., 1972, 17, no.8, pp.1369-70.

34.146. Morozov, V.N.: "Reconstruction of Images of Transparencies With Semiconductor Laser", Kvantovaya Elektron., 1972, no.3, pp.76-8, and Sov. J. Quantum Electron., 1972, 2, no.3, pp.257-9.

34.147. Romanov, G.N., and Shakhidzhanov, S.S.: "Generation of a Quasi-Plane Wavefront by Holographic Techniques", Radiotekh. Elektron., 1972, 17, no.12, pp.2584-6, and Radio Eng. Electron. Phys., 1972, 17, no.12, pp.2071-3.

34.148. Aoki, S., Ichihara, Y., and Kikuta, S.: "X-Ray Hologram Obtained by Using Synchrotron Radiation", Jap. J. Appl. Phys., 1972, 11, no.12, p.1857.

34.149. Bondarenko, M.D., Gnatovskii, A.V., and Soskin, M.S.: "Holographic Method to Control Spatial-Angular Characteristics of Laser Radiation", Ukr. Fiz. Zh., 1972, 17, no.2, pp.1950-4.

34.150. Zuev, V.S., and Kuznetsova, T.I.: "Nonstationary Holography to Improve the Directivity of Laser Radiation", Zh. Eksp. Teor. Fiz. Pis'ma, 1972, 16, no.8, pp.466-8, and JETP Lett., 1972, 16, no.8, pp.330-2.

34.151. Iizuka, K.: "In Situ Microwave Holography", Appl. Opt., 1973, 12, no.1, pp.147-9.

34.152. Nagai, K., Aoki, Y., and Suzuki, M.: "Numerical Reconstruction of Images from a Microwave Hologram", Trans. IEEE, 1973, MTT-21, no.1, pp.13-8.

34.153. Mottier, F.M., Dandliker, R., and Ineichen, B.: "Relaxation of the Coherence Requirements in Holography", Appl. Opt., 1973, 12, no.2, pp.243-8.

34.154. Rogers, G.L.: "Holographic Recording of a Complete Closed Surface", Appl. Opt., 1973, 12, no.4, pp.886-7.

34.155. Politch, J., and Assa, A.: "Application of Longitudinal Multimode Laser Coherence Properties to Increase the Holographic Depth of Field", Opt. Commun., 1973, 7, no.3, pp.266-9.

34.156. Farhat, N.H., and Farhat, A.H.: "Double Circular Scanning in Microwave Holography", Proc. IEEE, 1973, 61, no.4, pp.509-10.

34.157. Orme, R.D., and Anderson, A.P.: "High-Resolution Microwave Holographic Technique", Proc. IEE, 1973, 12, no.4, pp.401-6.

34.158. Iin, B.J., and Collins, S.A.: "Holographic Imaging and Aberrations Due to an Incorrectly Repositioned Hologram in a System with Lenses Having Aberrations", J. Opt. Soc. Am., 1973, 63, no.5, pp.537-47.

34.159. Kato, M., and Okino, Y.: "Speckle Reduction by Double-Recorded Holograms", Appl. Opt., 1973, 12, no.6, pp.1199-201.

34.160. George, N., and Jain, A.: "Speckle Reduction Using Multiple Tones of Illumination", Appl. Opt., 1973, 12, no.6, pp.1202-12.

34.161. Forman, P.R., Humphries, S., and Peterson, R.W.: "Pulsed Holographic Interferometry at 10.6 micron", Appl. Phys. Lett., 1973, 22, no.10, pp.537-9.

34.162. Yu, F.T.S., and Wang, E.Y.: "Speckle Reduction in Holography by Random Spatial Sampling", Appl. Opt., 1973, 12, no.7, pp.1656-9.

34.163. Miler, M.: "Geometrical Constructions in Holographic Images", Jemna Mech. Opt., 1973, 18, no.3, pp.67-8.

34.164. Musiol, K.: "Holographic Diffractive Gratings", Postepy Fiz., 1973, 24, no.2, pp.209-24.

34.165. MacQuigg, D.: "Complex Wavefront Synthesis by Multiple Exposure Holography", Opt. Commun., 1973, 8, no.1, pp.76-8.

34.166. Kalestynski, A.: "Holographic Multiplication on One Exposure by Use of Multibeam Reference Field", Appl. Opt., 1973, 12, no.8, pp.1946-50.

34.167. Leith, E.N., and Chang, B.J.: "Space-Invariant Holography with Quasi-Coherent Light", Appl. Opt., 1973, 12, no.8, pp.1957-63.

34.168. Politch, J.: "Considerations in Multibeam Holography. II", Optik, 1973, 38, no.5, pp.479-501.

34.169. Gasvik, K.: "Polarizing Effects in Holographic Reconstruction", Optik, 1973, 39, no.1, pp.47-57.

34.170. Grover, C.P., and May, M.: "Interference Phenomena Produced by a Gabor Hologram Recorded with a Spherical Reference Beam", Opt. Acta, 1973, 20,no.11,pp.833-44.

34.171. Tanaka, S., et al.: "Multiple Object-Beam Holograms", Opt. Commun., 1973, 9, no.1, pp.54-7.

34.172. Welford, W.T.: "Aplanatic Hologram Lenses on Spherical Surfaces", Opt. Commun., 1973, 9, no.3, pp.268-9.

34.173. Gregoris, L.G., and Iizuka, M.: "Additive and Subtractive Microwave Holography", Appl. Opt., 1973, 12, no.11, pp.2641-8.

34.174. Deryugin, I.A., et al.: "Polarization Effects in Holography", Kvantovaya Elektron. (Kiev), 1973, no.7, pp.167-81.

34.175. Gama, M.A.M.: "Fringe Localization and Visibility in Hologram and Classical Broad-Source Interferometry", Opt. Commun., 1973, 8, no.4, pp.362-5.

34.176. Ogawa, K., and Chang, W.S.C.: "Analysis of Holographic Thin-Film Grating Coupler", Appl. Opt., 1973, 12, no.9, pp.2167-71.

34.177. Aoki, Y.: "Computer Image Reconstruction from a Microwave Hologram Using Two-Dimensional FFT of Arbitrary Sampling Number", Electron. Commun. Jap., 1973, 56, no.10, pp.86-92.

34.178. Belokrinitskii, N.S., Gnatovskii, A.V., and Danileiko, M.V.: "Holographic Compensation of Phase Distortions in Radiation of Solid-State Lasers", Kvantovaya Elektron., 1973, no.2, pp.118-21, and Sov. J. Quantum Electron., 1973, 3, no.2, pp.176-8.

34.179. Shigesawa, H., et al.: "Microwave Holography by Spherical Scanning", Electron. Commun. Jap., 1973, 56, no.3, pp.56-64.

34.180. Bondarenko, M.D., and Soskin, M.S.: "Enhancement of Axial Brightness of Laser Beams by Holographic Phase Correction", Opt. Spekt., 1973, 35, no.6, pp.1147-52, and Opt. Spectrosc., 1973, 35, no.6, pp.665-7.

34.181. Sugaya, T., and Iwamoto, A.: "Holograms Produced with Double-Heterojunction Laser Illumination", Opt. Commun., 1974, 10, no.1, pp.37-8.

34.182. Roychoudhuri, C.: "Dynamic and Multiplex Holography with Scanning Fabry-Perot Fringes", Opt. Commun., 1974, 10, no.2, pp.160-3.

34.183. Ash, E.A., et al.: "Holographic Coupler for Integrated Optics", Appl. Phys. Lett., 1974, 24, no.4, pp.207-8.

34.184. Minemoto, T.: "Holographic Deblurring Filter Made with Flying-Spot-Scanner Exposure Device", Jap. J. Appl. Phys., 1974, 13, no.6, pp.975-83.

34.185. Chow, B.S.K.: "Hologram of Large Field Depth", Opt. Commun., 1974, 11, no.3, pp.235-7.

34.186. Tsunoda, Y., and Takeda, Y.: "High Density Image-Storage Holograms by a Random Phase Sampling Method", Appl. Opt., 1974, 13, no.9, pp.2046-51.

34.187. Yamakido, K., et al.: "Resolution of Millimetre-Wave Holography", Bull. Kyushu Inst. Technol., 1974, no.28,pp.99-106.

34.188. Nikitin, V.V., Samoilov, V.D., and Semenov, G.I.: "Influence of Injection Laser Parameters on the Reconstruction of Phase Holograms", Kvantovaya Elektron., 1974, 1, no.1, pp.7-13, and Sov. J. Quantum Electron., 1974, 4, no.1, pp.1-5.

34.189. Bykovskii, Yu.A., et al.: "Single-Pulse Holography Employing Injection Lasers", Kvantovaya Elektron., 1974, 1, no.1, pp.217-9, and Sov. J. Quantum Electron., 1974, 4, no.1, p.136.

34.190. Karg, R., and Ermet, H.: "Multifrequency Quasi-Holographic Microwave Imaging System", Nachr. Tech. Z., 1974, 27, no.10, pp.369-72.

34.191. Surget, J.: "Holography Setup With Two Reference Sources", Nouv. Rev. Opt., 1974, 5, no.4, pp.201-17.

34.192. Fienup, J.P., and Goodman, J.W.: "Ways to Make Computer-Generated Colour Holograms", Nouv. Rev. Opt., 1974, 5, no.5, pp.269-75.

34.193. Kakichashvili, Sh.D.: "Method for Phase Polarization Recording of Holograms", Kvantovaya Elektron., 1974, 1, no.6, pp.1435-41, and Sov. J. Quantum Electron., 1974, 4, no.6, pp.795-8.

34.194. Som, S.C.: "Modulated Holographic Grating", Appl. Opt., 1974, 13, no.12, pp.2767-9.

34.195. Richter, A.K., and Carlson, F.P.: "Holographically Generated Lens", Appl. Opt., 1974, 13, no.12, pp.2924-30.

34.196. Burch, J.M., Forno, C., and Tanner, L.H.: "Lens-Hologram Technique for Analysing Symmetrical Aberrations of a Camera System", Opt. Laser Technol., 1974, 6, no.3, pp.109-13.

35. STORAGE AND PROCESSING

35.1. Collins, J.H., and Nielson, G.G.: "Microwave Pulse Compression With YIG Delay Line", Electron. Lett., 1965, 1, p.234.

35.2. Olson, F.A., and Yaeger, J.R.: "Microwave Delay Techniques Using YIG", Trans. IEEE, 1965, MTT-13, p.63.

35.3. Kaufman, I., and Robinson, W.A.: "Electronically Variable Delay of Microwave Pulses in YIG Rods", J. Appl. Phys., 1965, 36, p.1245.

35.4. Levitin, L.B.: "Photon Channel With Small Occupation Numbers", Probl. Peredachi Inf., 1966, 2, no.2, p.60, and Problems Inf. Transm., 1966, 2, no.2, p.48.

35.5. Hartnagel, H.L.: "Microwave Logic With Gunn Diodes", Nachr. Tech. Fachber., 1968, 35, p.461.

35.6. Mitchell, R.H., Williams, I., and Ryan, W.D.: "Delay Line and Logic Circuits Utilizing Charge-Storage Subharmonic Parametric Oscillators", Trans. IEEE, 1968, CPR-17, p.1037.

35.7. Stroke, G.W.: "Class of Optical Imaging Systems Achieving Aperture Synthesis by Lensless Fourier-Transform Holography", Phys. Lett., 1968, 28A, p.251.

35.8. Kuznetsov, V.M., and Tseitlin, V.E.: "Photographic Measurement of Laser Radiation Parameters", Izmer. Tekh. 1968, no.12, p.28, and Meas. Tech., 1968, no.12, p.1634.

35.9. Jacobs, H., et al.: "Conversion of Millimetre-Wave Images Into Visible Displays", J. Opt. Soc. Am., 1968, 58, no.2, pp.246-53.

35.10. Eliseev, P.G., and Ismailov, I.: "Observation of Memory Effects in Injection Lasers", Zh. Tekh. Fiz., 1968, 38, no.12, p.2085, and Sov. Phys.-Tech. Phys., 1969, 13, no.12, pp.1671-2.

35.11. Nikitin, V.V., et al.: "Theory of Bistable Operation of Nonuniformly Excited Semiconductor Lasers", Fiz. Tekh. Poluprov., 1968, 2, no.11, p.1662, and Sov. Phys.-Semicond., 1969, 2, no.11, pp.1382-5.

35.12. Chang, R.W.: "Photon Detector for Optical PCM System", Trans. IEEE, 1969, IT-15, p.725.

35.13. Pistoresi, D.J.: "Comparison of Error Probability and Signal/Noise Ratio Between a Coherent Heterodyne and a Photon-Limited Communication System", Appl. Opt., 1969, 8, p.1811.

35.14. Hargrove, L.E.: "Reduction of Mode-Locked-Laser Pulse Duration by Interferometric Combination of Frequency-Shifted Pulses", J. Opt. Soc. Am., 1969, 59, p.1680.

35.15. Beck, J.W.: "Size v. Energy for a Spot Written Thermomagnetically With a Gaussian Light Beam", Proc. IEEE, 1969, 57, p.1223.

35.16. Jacobs, H., Schumacher, J.D., and Register, D.: "Bulk-Semiconductor Imaging Device for Submillimetre Radiation", Trans. IEEE, 1969, ED-16, no.5, pp.419-24.

35.17. Thompson, B.J.: "Application of Lasers to Printing and Recording", Image Technol., 1969, 11, no.10, p.16.

35.18. Abrams, R.L.: "Characteristics of 10.6-micron Chirped Pulses", J. Quantum Electron. IEEE, 1969, QE-5, p.522.

35.19. Stark, H., King, M., and Arm, M.: "Linear Spatial Filtering With Crossed Ultrasonic Light Modulators", Proc. IEEE, 1969, 57, p.1456.

35.20. Solimeno, S.: "Experimental Study of Binary Photon Channel Capacity", Alta Freq., 1969, 38, no.5, p.367.

35.21. Surin, V.V., and Shevhchenko, Yu.A.: "Nonlinear Effects in Ferrite Microwave Delay Lines", Radiotekh. Elektron., 1969, 14, no.11, p.2065, and Radio Eng. Electron. Phys., 1969, 14, no.11, pp.1789-90.

35.22. Maloney, W.T.: "Acoustooptical Approaches to Radar Signal Processing", Spectrum IEEE, 1969, 6, no.10, p.40.

35.23. Klauder, J.R.: "Spectral Criterion for Mode-Locked Laser Signals", Appl. Phys. Lett., 1969, 14, p.147.

35.24. Carleton, H.R., Maloney, W.T., and Meltz, G.: "Collinear Heterodyning in Optical Processors", Proc. IEEE, 1969, 57, p.769.

35.25. Sheronov, A.A.: "Quenching Effect in Optically Coupled GaAs Injection Lasers", Fiz. Tekh. Poluprov., 1969, 3, no.3, p.368, and Sov. Phys.-Semicond., 1969, 3, no.3.

35.26. Gagliardi, R.M., and Karp, S.: "M-ary Poisson Detection and Optical Communications", Trans. IEEE, 1969, COM-17, p.208.

35.27. Hamal, K., DeMichelis, C., and Alcock, A.J.: "Schlieren Photography With Mode-Locked Laser as Light Source", Opt. Acta, 1969, 16, no.4, p.463.

35.28. Komisarova, I.I., et al.: "Two-Wavelength Holography of a Laser Spark", Phys. Lett., 1969, 29A, p.262.

35.29. Braudy, R.S.: "Laser Writing", Proc. IEEE, 1969, 57, p.1771.

35.30. Duguay, M.A., and Hansen, J.W.: "Compression of Pulses from a Mode-Locked He-Ne Laser", J. Quantum Electron. IEEE, 1969, QE-5, p.326.

35.31. Gustafson, T.K.: "Subpicosecond Pulse Generation Using the Optical Kerr Effect", J. Quantum Electron. IEEE, 1969, QE-5, p.325.

35.32. Treacy, E.B.: "Optical Pulse Compression With Diffraction Gratings", J. Quantum Electron. IEEE, 1969, QE-5, p.454.

35.33. Ripper, J.E., Paoli, T.L., and Dyment, J.C.: "Characteristics of Bistable CW GaAs Junction Lasers Operating Above the Delay-Transition Temperature", J. Quantum Electron. IEEE, 1970, QE-6, p.300.

35.34. Langdon, R.M.: "High-Capacity Holographic Memory", Marconi Rev., 1970, 33, no.177, p.113.

35.35. Payer, S.F., Moore, R.A., and Pincoffs, P.H.: "Microwave Correlator Employing YIG Delay Lines", Trans. IEEE, 1970, MTT-18, p.426.

35.36. Kock, W.E.: "Pulse Compression With Periodic Gratings and Zone-Plate Gratings", Proc. IEEE, 1970, 58, p.1395.

35.37. Haskal, H.: "Thermomagnetic Writing With Non-Gaussian Laser Beam Intensity Distributions", Proc. IEEE, 1970, 58, p.802.

35.38. Wenkoff, M.P., and Katchky, M.: "Improved Read-In Technique for Optical Delay-Line Correlators", Appl. Opt., 1970, 9, p.135.

35.39. Yuen, H.P., and Kennedy, R.S.: "Optimal Quantum Receivers for Digital Signal Detection", Proc. IEEE, 1970, 58, no.10, pp.1770-3.

35.40. Bolgiano, L.P., Lutz, B.C., and Deal, J.H.: "Simulated Communication Channel Featuring a Photoelectron Counter System", Proc. IEEE, 1970, 58, no.10, pp.1763-4.

35.41. Karp, S., O'Neill, E.L., and Gagliardi, R.M.: "Communication Theory for the Free-Space Optical Channel", Proc. IEEE, 1970, 58, no.10, pp.1611-26.

35.42. Chen, F.S.: "Modulators for Optical Communications", Proc. IEEE, 1970, 58, no.10, pp.1440-57.

35.43. Melchior, H., Fisher, M.B., and Arams, F.R.: "Photodetectors for Optical Communication Systems", Proc. IEEE, 1970, 58, no.10, pp.1466-86.

35.44. Kinsel, T.S., and De Lange, O.E.: "Wideband Optical Communication Systems. I and II", Proc. IEEE, 1970, 58, no.10, pp.1666-83 and 1683-90.

35.45. Hance, H.V., et al.: "Ultra-Wideband Laser Communications. II", Proc. IEEE, 1970, 58, no.10, pp.1714-9.

35.46. Ross, M., Green, S.I., and Brand, J.: "Short-Pulse Optical Communication Experiments", Proc. IEEE, 1970, 58, no.10, pp.1719-26.

35.47. Thompson, G.D., and Pratt, W.K.: "Optical Heterodyne Receiver Design by Nonlinear Recursive Estimation Techniques", Proc. IEEE, 1970, 58, no.10, pp.1727-32.

35.48. White, G.: "1-Gbit/s Optical PCM Communications System", Proc. IEEE, 1970, 58, no.10, pp.1779-80.

35.49. Hirano, N.: "Optical-Band Information Rate in Cascade Circuit of Attenuator and Amplifier", Electron. Commun. Jap., 1970, 53, no.3, p.23.

35.50. Konnerth, K.L., and Shah, B.R.: "Optical Transmission Utilizing Injection Light Sources", Spectrum IEEE, 1970, 7, no.9, pp.37-46.

35.51. Bass, J.C.: "Major Role for GaAs Devices in Microwave Communications", Electron. Eng., 1970, 42, no.513, pp.61-5.

35.52. Hill, B.: "Interconnection of Information Channels by Digital Light Deflectors", Nachr. Tech. Z., 1970, 23, no.11, pp.549-52.

35.53. Kozlyaev, I.P., et al.: "Use of Injection-Laser Logic Elements in Optical Communications Systems With Time Multiplexing", Radiotekh. Elektron., 1970, 15, no.4, and Radio Eng. Electron. Phys.,1970,pp.652-6.

35.54. Townsend, V.T., and Mehran, F.: "Coherent-Light Accumulator (Using Mode-Locked Pulses)", IBM Tech. Disclosure Bull., 1970, 13, no.1, p.283.

35.55. Atzeni, C., and Pantani, L.: "Optical Signal Processing Through Dual-Channel Ultrasonic Light Modulators", Proc. IEEE, 1970, 58, p.501.

35.56. Ripper, J.E., and Paoli, T.L.: "Bistable Operation of CW Junction Lasers Due to Saturable Absorbing Centres", Proc. IEEE, 1970, 58, p.178.

35.57. Romanov, A.M.: "Optimal Filter in Systems for Optical Signal Detection", Radiotekh. Elektron., 1970, 15, p.92, and Radio Eng. Electron. Phys., 1970, 15, no.1, pp.73-9.

35.58. Korado, V.A.: "Optimum Detection of Signals With Upper Bound on False-Alarm Probability", Radiotekh. Elektron., 1970, 15, no.7, and Radio Eng. Electron. Phys., 1970, 15, no.7, pp.1215-22.

35.59. Gaprindashivili, Kh.I., et al.: "Coupled Fibre Lasers (Quenching)", Radiotekh. Elektron., 1970, 15, no.7, and Radio Eng. Electron. Phys., 1970, 15, no.7, pp.1249-51.

35.60. Nakamura, S., and Yoshimoto, C.: "Rotated Pattern Detections by Superimposed Holograms", Bull. Res. Inst. Appl. Electr., 1970, 22, no.3-4, pp.144-57.

35.61. Semeoshenkov, V.N.: "Optimal Recognition of Optical Signals in Noise", Tekh. Kibern., 1970, 8, no.6, and Eng. Cybern., 1970, 8, no.6, pp.1191-5.

35.62. Kurtz, R.L., and Loh, H.Y.: "Holographic Technique for Recording a Hypervelocity Projectile With Front Surface Resolution", Appl. Opt., 1970, 9, p.1040.

35.63. Funkhouser, A.T., and Mielenz, K.D.: "High-Speed Holographic Interferometry", Appl. Opt., 1970, 9, p.1215.

35.64. Dyes, W.A., Kellen, P.F., and Klaubert, E.C.: "Velocity-Synchronized Fourier-Transform Hologram System", Appl. Opt., 1970, 9, p.1105.

35.65. Schumacher, J.D., Hofer, R.C., and Jacobs, H.: "Performance of a Single-Collector Millimetre-Wave Imaging Device", Proc. IEEE, 1970, 58, p.1390.

35.66. Nepela, D.A.: "Vesicular Image Formation for Laser Recording Systems", IBM Tech. Disclosure Bull., 1970, 13, p.203.

35.67. Harris, A.L., Chen, M., and Bernstein, H.L.: "CW-Laser Recording on Metallic Thin Film", Image Technol., 1970, 12, no.3, p.31.

35.68. Katyl, R.H., Karstadt, L.K., and Gilmore, W.F.: "Organic-Liquid CO_2-Laser Beamfinder", J. Quantum Electron. IEEE, 1970, QE-6, p.164.

35.69. Levin, B.J.: "Millimetre-Wave Image Conversion Using a Semiconductor", Proc. IEEE, 1970, 58, p.496.

35.70. Martin, F.B., and Lalanne, J.R.: "Thermal Variations of Optical Birefringence Induced in Liquids by Q-Switched Lasers", Opt. Commun., 1970, 2, no.5, pp.219-22.

35.71. Anderson, R.J.: "Real-Time Laser Photochromic Display", Proc. Soc. Inf. Disp., 1970, 11, no.4, pp.151-63.

35.72. Fallmann, W.F., Hartnagel, H.L., and Srivastava, G.P.: "Results of CW Coplanar Gunn Diodes for Pulse Processing", Phys. Status Solidi, 1970,3A,no.4,pp.K227-8.

35.73. Manning, C.D., et al.: "Optoelectronic Transfer of Information from a Rotating Shaft", Electron. Lett., 1970, 6, no.26, pp.864-6.

35.74. Barbier, M., et al.: "Pulse-Compression Technique Using Coherent Light", Geophys. Prospect., 1970,18,no.4,pp.571-80.

35.75. Tsai, C.S.: "Increase of Bragg-Diffraction Intensity Due to Acoustic Resonance", Appl. Opt., 1971,10,no.1,pp.215-8.

35.76. Kenville, R.F.: "Noise in Laser Recording", Spectrum IEEE, 1971,8,no.3,pp.50-7.

35.77. Helstrom, C.W.: "Photoelectric Detection of Coherent Light in Filtered Background Light", Trans. IEEE, 1971, AES-7, no.1, pp.210-3.

35.78. Chomat, M.: "Semipermanent Optical Memories With Holograms", Slab. Obz., 1971, 32, no.3, pp.120-5.

35.79. Akahori, H., and Sakurai, K.: "Information Search Using Superimposed Holograms", Appl. Opt., 1971, 10,no.3,pp.665-6.

35.80. Minakovic, B.: "Note on Dielectric-Filled Waveguide as a Delay Line for Pulse Compression", Trans. IEEE, 1971, MTT-19, no.6, pp.562-3.

35.81. Goldobin, I.S., et al.: "Quantum-Optical Integrated Circuits (for Computer Technology) Using GaAs", Fiz. Tekh. Poluprov., 1971, 5, no.1, pp.170-2, and Sov. Phys.-Semicond., 1971, 5, no.1, pp.146-8.

35.82. Kim, D.M., Shah, P.L., and Rabson, T.A.: "Transverse Spatial Diffusion of Phase-Dispersion Pulses", Appl. Phys. Lett., 1971, 18, no.9, pp.369-70.

35.83. Amodei, J.J., Staebler, D.L., and Stephens, A.W.: "Holographic Storage in Doped $Ba_2NaNb_5O_{15}$", Appl. Phys. Lett., 1971, 18, no.11, pp.507-9.

35.84. Amodei, J.J., and Staebler, D.L.: "Holographic Pattern Fixing in Electrooptic Crystals", Appl. Phys. Lett., 1971, 18, no.12, pp.540-2.

35.85. Sass, A.R.: "Binary-Intensity Holograms", J. Opt. Soc. Am., 1971, 61, no.7, pp.910-5.

35.86. Aristov, V.V.: "Optical Memory of Three-Dimensional Holograms", Opt. Commun., 1971, 3, no.3, pp.194-6.

35.87. Som, S.C., and Lessard, R.A.: "Technique for Holographic Multiplexing", J. Opt. Soc. Am., 1971, 61, no.9, pp.1240-5.

35.88. Lang, M.: "Diffraction Efficiency of Transmittance Storage Holograms", Opt. Commun., 1971, 3, no.4, pp.229-33.

35.89. Adem, M., and Bismuth, G.: "Holographic Recording in Thin Films of Amorphous Semiconductors", Opt. Commun., 1971, 3, no.4, pp.234-5.

35.90. Mezrich, R.S., and Schachter, H.: "Information Storage in Magnetic Holography", Israel J. Technol., 1971, 9, no.3,pp.267-73.

35.91. Mikaelyan, L.A., et al.: "Holographic Memory Device With Bulk Recording of Information", Kvantovaya Elektron., 1971, no.1, pp.79-84.

35.92. Gokgor, H.S.: "Circular TE_{01} Periodic Waveguide as Delay Line for Pulse Compression", Electron. Lett., 1971, 7, no.20, pp.607-8.

35.93. Kaminow, I.P.: "Proposed Electrooptical PCM Multiplexer-Demultiplexer", J. Quantum Electron. IEEE, 1971, QE-7, no.11, pp.533-5.

35.94. Tsujiuchi, J.: "Optical Information Processing Using Laser Light. I and II", J. Inst. Electr. Commun. Eng. Jap., 1971, 54, no.8, pp.1084-9, and no.9, pp.1276-85.

35.95. Deryugin, I.A., et al.: "Quasi-Classical Approximation for a Parametric Amplifier in Optical Information Systems", Opt. Spekt., 1971, 30, no.5, pp.952-6, and Opt. Spectrosc., 1971, 30, no.5, pp.507-9.

35.96. Desbois, J., Gires, F., and Tournois, P.: "Mode-Locked He-Ne Laser Pulse Compression and Expansion", Opt. Commun., 1971, 2, no.8, pp.370-2.

35.97. Haskell, R.E.: "Fourier Analysis Using Coherent Light", Trans. IEEE, 1971, E-14, no.3, pp.110-5.

35.98. Fuller, H.W.: "Magnetooptic Signal Processing", J. Appl. Phys., 1971, 42, no.4, p.1816.

35.99. Dahlstrom, L.: "Method for Compression of Light Pulses in a Q-Switched High-Power Laser", Opt. Commun., 1971, 4, no.4, pp.289-91.

35.100. Condas, G.A.: "Phosphor System for Converting Far-IR Laser Radiation to Visible Light", J. Quantum Electron. IEEE, 1971, QE-7, no.5, pp.202-3.

35.101. Keilman, F., and Renk, K.F.: "Thermal Image Conversion of Submillimetre-Laser Beams", Appl. Phys. Lett., 1971, 18, no.10, pp.452-4.

35.102. Dean, S.O., et al.: "Enhancement of Time Resolution in Laser Photography", Appl. Opt., 1971, 10, no.8, pp.1985-6.

35.103. Smigielski, P.: "(Schlieren) Holography of Phase Objects", Opt. Acta, 1971, 18, no.7, pp.483-506.

35.104. Aleksoff, C.C.: "Holographic Analysis and Display of Laser Modes", J. Opt. Soc. Am., 1971, 61, no.10, pp.1426-7.

35.105. Novara, M.: "Ultrafast Holographic Camera", C. R. Acad. Sci. (Paris), 1971, 273B, no.22, pp.941-3.

35.106. Vanyukov, M.P., et al.: "Q-Switched Laser for High-Speed Cinematography", Kvantovaya Elektron., 1971, no.3,pp.108-10.

35.107. Smigielski, P., et al.: "High-Speed Holography", Nouv. Rev. Opt. Appl., 1971, 2, no.4, pp.205-7.

35.108. Chen, F.S.: "Demultiplexers for High-Speed Optical PCM", J. Quantum Electron. IEEE, 1971, QE-7, no.1, pp.24-9.

35.109. Ohtsuka, Y.: "Optical Processing With Twin Ultrasonic Light Modulators", Opto-Electron., 1971, 3, no.3, pp.119-25.

35.110. Hill, B.: "Coupling of Channels by Digital Deflectors", Polytech. Tijdschr. Elektrotech., 1971, 26, no.26, pp.1030-3.

35.111. Vlad, V.I.: "Spatial-Bandwidth Reduction in Holography", Rev. Roum. Phys., 1971, 16, no.8, pp.855-64.

35.112. Nickel, G.H., et al.: "Acoustic Beamsplitter for IR Lasers", J. Quantum Electron. IEEE, 1972, QE-8, no.1, pp.20-1.

35.113. Delingat, E.: "Investigations of Modulation Behaviour of Optical Signals", Optik, 1972, 34, no.4, pp.433-41.

35.114. Gaylord, T.K., Rabson, T.A., and Tittel, F.K.: "Optically Erasable and Rewritable Solid-State Holograms", Appl. Phys. Lett., 1972, 20, no.1, pp.47-9.

35.115. Casasent, D., and Stephenson, J.: "Electrooptical Processing of Phased-Array Antenna Data", Appl. Opt., 1972, 11, no.5, pp.1269-71.

35.116. Kamal, A.K., Sharma, P.D., and Malaviya, N.: "Theory of Optimum M-ary Laser Detector", Electron. Lett., 1972, 8, no.14, pp.348-9.

35.117. Tsai, C.S., and Yao, S.K.: "Optical Bragg Diffraction by Standing Ultrasonic Waves With Application to Demultiplexing", Trans. IEEE, 1972, SU-19, no.3, pp.411-2.

35.118. Ueno, Y., and Nagura, R.: "Optical Communication System Using Envelope Modulation", Trans. IEEE, 1972, COM-20, no.5, pp.813-9.

35.119. Whitehead, D.G.: "Dynamic Data Storage at 1-Gbit/s Rate", Electron. Lett., 1972, 8, no.17, pp.441-2.

35.120. Gray, E.E.: "Laser Mass Memory System", Trans. IEEE, 1972, MAG-8, no.3, pp.416-20.

35.121. Kohashi, T.: "Solid-State IR Image Converter", Trans. IEEE, 1972, ED-19, no.1, pp.98-103.

35.122. Fuller, D.W.E., et al.: "Television Display of HCN Laser Radiation", Electron. Lett., 1972, 8, no.2, pp.44-5.

35.123. Putley, E.H., Watton, R., and Ludlow, J.H.: "Pyroelectric Thermal Imaging Devices", Ferroelectrics, 1972, 3, no.2-4, pp.263-8.

35.124. Mourou, G., et al.: "Variable Ultrafast Photographic Shutter", Appl. Phys. Lett., 1972, 20, no.11, pp.453-5.

35.125. Lesieur, J.P., Sexton, M.C., and Veron, D.: "Visualization of Modes of HCN Laser. Liquid-Crystal Camera", J. Phys. D, 1972, 5, no.7, pp.1212-7.

35.126. Fourney, M.E., and Barker, D.B.: "Picosecond Holography", Appl. Phys. Lett., 1972, 21, no.1, pp.21-3.

35.127. Meshchankin, V.M., et al.: "Visualization of Microwave Fields Through Evaporation of Water from a $CoCl_2$ Solution", Radiotekh. Elektron., 1972, 16, no.9, p.1735, and Radio Eng. Electron. Phys., 1972, 16, no.9, pp.1572-3.

35.128. von Gutfeld, R.J., and Chaudhari, P.: "Laser Writing and Erasing on Chalcogenide Films", J. Appl. Phys., 1972, 43, no.11, pp.4688-93.

35.129. Gramenopoulos, N., and Hartfield, E.D.: "Advanced Laser Image Recorder", Appl. Opt., 1972, 11, no.12, pp.2778-82.

35.130. Smigielski, P., Hirth, A., and Thery, C.: "Application of Holography to Sonic Boom Investigations", Trans. IEEE, 1972, AES-8, no.6, pp.751-6.

35.131. Szentesi, O.I.: "Acoustooptic Page Composer Based on Bragg Imaging", Proc. IEEE, 1972, 61, no.11, pp.1461-2.

35.132. Polovtseva, G.L., Dybina, A.A., and Lipatov, V.V.: "Comparison of Photographic Effect of Laser and Pseudothermal Radiations", Opt. Spekt., 1972, 33, no.2, pp.347-8, and Opt. Spectrosc., 1972, 33, no.2, p.183.

35.133. Panyakeow, S., et al.: "Photoelectric EMF Induced by Q-Switched CO_2 Laser in Semiconductors", Technol. Rep. Osaka Univ., 1972, 22, no.1053-89, pp.563-74.

35.134. Angert, N.B., et al.: "Optically Induced Inhomogeneity of Refractive Index of $LiNbO_3$", Zh. Eksp. Teor. Fiz., 1972, 62, no.5, pp.1666-72, and Sov. Phys.-JETP, 1972, 35, no.5, pp.867-9.

35.135. Stadnik, B., and Tronner, Z.: "Optical Holographic Storage in KCl and KBr Crystals at Room Temperature", Nouv. Rev. Opt. Appl., 1972, 3, no.6, pp.347-9.

35.136. Phillips, W., Amodei, J.J., and Staebler, D.L.: "Optical and Holographic Storage Properties of Transition-Metal-Doped $LiNbO_3$", RCA Rev., 1972, 33, no.1, pp.94-109.

35.137. Kock, M., and Rabe, G.: "Parallel and Time Sequential Optical Multiplex Systems for Pattern Recognition", Opt. Commun., 1972, 5, no.2, pp.73-7.

35.138. Aagard, R.L., Lee, T.C., and Chen, D.: "Advanced Optical Storage Techniques for Computers", Appl. Opt., 1972, 11, no.10, pp.2133-9.

35.139. Winzer, G., and Douklias, N.: "Improved Holographic Matched Filter Systems for Pattern Recognition Using a Correlation Method", Opt. Laser. Technol., 1972, 4, no.5, pp.222-7.

35.140. Kawachida, S., and Namekawa, T.: "Optical Matched Filter Using Raman-Nath Diffraction", J. Acoust. Soc. Jap., 1972, 28, no.7, pp.327-34.

35.141. Bieringer, R.J.: "Optical Correlation Using Diffuse Objects", Appl. Opt., 1973, 12, no.2, pp.249-54.

35.142. Pursey, H.: "Spatial Filtering Technique for Improved Holographic Character Recognition", Opt. Laser Technol., 1973, 5, no.1, pp.24-7.

35.143. Anderson, A.P., and Whitaker, A.J.T.: "Improved Optical Images Using a Multilevel Display System for Nonoptical Hologram Distributions", Opt. Laser Technol., 1973, 5, no.1, pp.28-9.

35.144. Arsenault, H.H., and Brousseau, N.: "Space Variance in Quasi-Linear Coherent Optical Processors", J. Opt. Soc. Am., 1973, 63, no.5, pp.555-8.

35.145. Hildenbrand, R.: "Code-Division Multiplexing for a Branched Glass-Fibre Communication Network", Arch. Elektron. Ubertrag., 1973, 27, no.4, pp.177-80.

35.146. Ems, S.C., Marshall, T.R., and Gucker, F.T.: "Simple Methods for High-Resolution Scanning of Laser Beams and Alignment of Small Spatial Filtered Beams", Rev. Sci. Instrum., 1973, 44, no.4, pp.475-7.

35.147. Hill, B., and Schmidt, K.P.: "Page Composer for Holographic Data Storage", Appl. Opt., 1973, 12, no.6, pp.1193-8.

35.148. Graf, P., and Kiemle, H.: "Storage of Digital Data on Holograms", Elektron Int., 1973, no.7-9, pp.135-8.

35.149. Mallick, S.: "Multiplexing With Extended Broadband Source", Opt. Commun., 1973, 7, no.4, pp.427-8.

35.150. Mehta, P.C., and Singh, M.: "Single-Exposure Holographic Multiplexing", Opt. Commun., 1973, 7, no.4, pp.394-6.

35.151. Hoff, F., Javorsky, S., and Miler, M.: "Information Storage Using Miniature Holograms", Acta Tech. CSAV, 1973, 18, no.2, pp.145-53.

35.152. Stewart, W.C., et al.: "Experimental Read-Write Holographic Memory", RCA Rev., 1973, 34, no.1, pp.3-44.

35.153. Lang, M.: "Holographic Memories With Storage Capacities Beyond 100 Mbit", Optik, 1973, 37, no.5, pp.501-15.

35.154. Nishida, N., Sakaguchi, M., and Saito, F.: "Holographic (Memory) Coding Plate", Appl. Opt., 1973, 12, no.7, pp.1663-74.

35.155. D'Auria, L., Huignard, J.P., and Spitz, E.: "Holographic Read-Write Memory and Capacity Enhancement by 3-D Storage", Trans. IEEE, 1973, MAG-9, no.2, pp.83-94.

35.156. Hubbard, W.M.: "Utilization of Optical-Frequency Carriers for Low- and Moderate-Bandwidth Channels", Bell Syst. Tech. J., 1973, 52, no.5, pp.731-65.

35.157. Eschler, H., et al.: "Laboratory Model of Holographic Read-Only Memory", Optik, 1973, 37, no.5, pp.516-27.

35.158. Ernest, J.: "Digital Telecommunications Guided by an Optical Wave", Elektron. Microelectron. Ind., 1973, no.172, pp.36-7.

35.159. Herreman, G.O.: "Laser-Calculator System Improves Encoder Plate Measurements", Hewlett-Packard J., 1973, 24, no.10, pp.16-8.

35.160. Kalestynski, A., et al.: "Holographic Directing of High-Energy IR Laser Beams", Appl. Opt., 1973, 12, no.8, pp.1749-51.

35.161. Gottlieb, M., and Conroy, J.J.: "Design and Construction of an Optoacoustic Signal Processor", Appl. Opt., 1973, 12, no.8, pp.1922-7.

35.162. Il'inskii, Yu.A., and Petnikova, V.M.: "Resolving Power of Image Conversion in Nonlinear Crystals", Izv. VUZ Radiofiz., 1973, 16, no.8, pp.1285-7.

35.163. Akatova, V.M., and Il'inskii, Yu.A.: "Influence of Inhomogeneties in a Nonlinear Crystal on Image Conversion by Sum-Frequency Generation", Kvantovaya Elektron., 1973, no.6, pp.29-34, and Sov. J. Quantum Electron., 1973, 2, no.6, pp.520-3.

35.164. Upatnieks, J., and Lewis, R.W.: "Noise Suppression in Coherent Imaging", Appl. Opt., 1973, 12, no.9, pp.2161-6.

35.165. Gara, A.D., Majkowski, R.F., and Stapleton, T.T.: "Holographic System for Automatic Surface Mapping", Appl. Opt., 1973, 12, no.9, pp.2172-9.

35.166. Tubbs, M.R.: "Reversible Holographic Recording Materials for Optical Information Storage", Opt. Laser Technol., 1973, 5, no.4, pp.155-61.

35.167. Gurevich, S.B., and Sokolov, V.K.: "Maximum Information Capacity of a Holographic System", Zh. Tekh. Fiz., 1973, 43, no.3, pp.675-8, and Sov. Phys.-Tech. Phys., 1973, 18, no.3, pp.424-6.

35.168. Goell, J.E.: "274-Mbit/s Optical-Repeater Experiment Employing GaAs Laser", Proc. IEEE, 1973, 61, no.10, pp.1504-5.

35.169. Kazaryan, R.A., Marucharyan, R.G., and Gasparyan, S.S.: "Probability of Error in a Binary Communication Channel Provided by a Polarization-Modulated Laser Beam", Kvantovaya Elektron., 1973, no.7, pp.90-5, and Sov. J. Quantum Electron., 1973, 3, no.1, pp.49-51.

35.170. Lachs, G., and Ruggieri, N.F.: "Effect of Spectral Shape on Probability of Error in Laser Binary Communication Systems", Trans. IEEE, 1973, AES-9, no.6, pp.860-3.

35.171. Polovnikov, G.G., and Bakhtin, V.D.: "(Submillimetre) Devices for Quasi-optical Paths Using a Semiconductor Diode", Prib. Tekh. Eksp., 1973, 16, no.5, pp.167-9, and Instrum. Exp. Tech., 1973, 16, no.5, pp.1497-9.

35.172. Zubov, V.A.: "Holographic Method of Extraction of a Temporal Signal from a Background of Noise", Radiotekh. Elektron., 1973, 18, no.8, pp.1584-90, and Radio Eng. Electron. Phys., 1973, 18, no.8, pp.1584-9.

35.173. Chenming, H., and Whinnery, J.R.: "Thermooptical Measurement Method", Appl. Opt., 1973, 12, no.1, pp.72-9.

35.174. Weiser, K., Gambino, R.J., and Reinhold, J.A.: "Laser-Beam Writing on Amorphous Chalcogenide Films", Appl. Phys. Lett., 1973, 22, no.1, pp.48-9.

35.175. Ishii, A., et al.: "Optical Information Retrieval System for Literature", Electron. Commun. Jap., 1973, 56, no.12, pp.115-26.

35.176. Byalik, V.L.: "Photoelectron Statistics and Calculation of Communication Link Characteristics", Radiotekhnika, 1973, 28, no.4, pp.86-7, and Telecommun. Radio Eng. Pt 2, 1973, 28, no.4, pp.103-5.

35.177. Skomorovskii, Yu.A.: "Application of Noise-Reducing Codes in Optical PCM Communication", Poluprov. Prid. Tekh. Elektrosvyazi Sb. Statei, 1973, no.12, pp.173-82.

35.178. Kawachida, S., and Namekawa, T.: "Ultrasonic Optical Matched Filters Using Bragg Reflection", Electron. Commun. Jap., 1973, 56, no.5, pp.9-14.

35.179. Gottlieb, M.: "Acoustooptic Signal Processors for Extending Radar System Capabilities", Opt. Laser Technol., 1974, 6, no.1, pp.21-7.

35.180. Morto, S., and Saita, H.: "Method of Character Reading by Optical Correlation Technique", Jap. J. Appl. Phys., 1974, 13, no.6, pp.984-8.

35.181. Braunecker, B., and Lohmann, A.W.: "Character Recognition by Digital Holography", Opt. Commun., 1974, 11, no.2, pp.141-3.

35.182. Hariharan, P., and Hegedus, Z.S.:
"Reduction of Speckle in Coherent Imaging by
Spatial-Frequency Sampling", Opt. Acta, 1974,
21, no.5, pp.345-56.

35.183. Maiorov, S.A., and Ken, L.S.:
"Realizing Arithmetic and Control Operations
by Holography", Izv. VUZ Prib., 1974, 17,
no.2, pp.53-5.

35.184. Righini, G.C., Russo, V., and
Sottini, S.: "Reflection Holographic Filters
for Compacting Optical Processors", Appl.
Opt., 1974, 13, no.5, pp.1019-22.

35.185. Dubovik, A.S., and Ushakov, L.S.:
"High-Speed Holography and Kineholography",
Zh. Nauchn. Prikl. Fotogr. Kinematogr.,
1974, 19, no.1, pp.67-76.

35.186. Novikov, A.A., Fedorov, V.B.,
and Yurchikov, B.M.: "Method of Increasing
Speed of Response in an Optical Memory
Device", Avtometriya, 1974, no.2,pp.68-73.

35.187. Lee, W.H., and Greer, M.O.:
"Matched Filter Optical Processor", Appl.
Opt., 1974, 13, no.4, pp.925-30.

35.188. Semenov, A.T., and Yakubovich,
S.D.: "Calculation of Steady-State Character-
istics of a Many-Resonator Injection Laser
(Logic System)", Kvantovaya Elektron., 1974,
1, no.1, pp.173-6, and Sov. J. Quantum
Electron., 1974, 4, no.1, pp.100-2.

35.189. Karpel'tsev, V.P., and Samoilov,
V.D.: "Reconstruction of Microholograms
Carrying Textual Information With Injection
Lasers", Kvantovaya Elektron., 1974, 1,
no.1, pp.167-9, and Sov. J. Quantum Elec-
tron., 1974, 4, no.1, p.97.

35.190. Zakharov, Yu.P., et al.: "Injection-Laser Recording of High-Resolution
Microholograms", Kvantovaya Elektron., 1974,
1, no.1, pp.176-8, and Sov. J. Quantum
Electron., 1974, 4, no.1, p.103.

35.191. Bykovskii, Yu.A., Makovkin, A.V.,
and Smirnov, V.L.: "Use of Film Waveguides
in Optoelectronic Devices for Reading Information Stored in Holograms", Kvantovaya
Elektron., 1974, 1, no.1, pp.208-11, and
Sov. J. Quantum Electron., 1974, 4, no.1,
pp.129-30.

35.192. Blume, H., Bader, T., and Luty,
F.: "Bi-Directional Holographic Information
Storage Based on Optical Reorientation of
F Centres in (NaK)Cl", Opt. Commun., 1974,
12, no.2, pp.147-51.

35.193. Golubeva, N.S., et al.: "Synchronization of Two Solid-State Lasers in
Conditions of Nonstationary Generation",
Izv. VUZ Radioelektron., 1974, 17, no.9,
pp.25-30.

35.194. Aleksakov, G.N., et al.: "Holographic Installation for Analogue Pattern
Recognition", Prib. Tekh. Eksp., 1974, 17,
no.1, pp.177-9, and Instrum. Exp. Tech.,
1974, 17, no.1, pp.203-5.

35.195. Campbell, K., Wecksung, G.W., and
Mansfield, C.R.: "Spatial Filtering by Digital Holography", Opt. Eng., 1974, 13, no.3,
pp.175-88.

35.196. Grischkowsky, D.: "Optical Pulse
Compression", Appl. Phys. Lett., 1974, 25,
no.10, pp.566-8.

35.197. Korsakov, V.V., et al.: "Reversible
Recording of Optical Information by Laser and
Electron Beams in Chalcogenide Vitreous Semiconductors", Avtometriya, 1974, no.6,
pp.24-30.

35.198. Evtikhiev, N.N., et al.: "Micro-
recording of Information by a Laser Projection Method", Kvantovaya Elektron., 1974, 1,
no.4, pp.959-62, and Sov. J. Quantum Electron., 1974, 4, no.4, pp.527-8.

35.199. Lee, G.M., et al.: "Bit-Error
Probability for Baseband and Subcarrier
Optical Communication", Trans. IEEE, 1974,
COM-21, no.13, pp.55-62.

35.200. Borisov, V.F., and Fedoseev,
P.G.: "Laser Reproducing Devices With Mecha-
nical and Combined Systems of Scanning",
Tekh. Kino Telev., 1974, no.1, pp.69-72.

35.201. Bartashevskii, Ye.L., Dolgov,
V.M., and Krasovskii, V.S.: "Visualization
of Field in a Waveguide Using Liquid-Crystal
Thermo-Indicators", Izv. VUZ Radiofiz.,
1974, 17, no.5, pp.734-8.

35.202. Voronin, E.S., et al.: "Influence
of Speckle on Resolution in IR Image Conversion", Kvantovaya Elektron., 1974, no.4,
pp.115-8, and Sov. J. Quantum Electron.,
1974, 3, no.4, pp.351-2.

35.203. Berg, A.D., Cormier, R.J., and
Courtney-Pratt, J.S.: "High-Resolution
Graphics Using a He-Cd Laser to Write on
Kalvar Film", J. Soc. Motion Pict. Telev.
Eng.,1974, 83, no.7, pp.588-99.

35.204. Mangin, J., et al.: "Visualization of Radiation at 10 micron and 337 micron
With a Pyricon Pickup Tube", Nouv. Rev. Opt.,
1974, 5, no.5, pp.305-11.

35.205. Bobrov, A.V., and Nikogosyan,
D.N.: "Recording of IR Raman-Scattered Light
by Conversion to Visible Range", Kvantovaya
Elektron., 1974, 1, no.5, pp.1242-5, and
Sov. J. Quantum Electron., 1974, 4, no.5,
pp.685-6.

35.206. Vul', V.A.: "Selective Modulation
of Light in Transverse (Image) Plane", Izv.
VUZ Radioelektron., 1974, 17, no.12, pp.92-4.

35.207. Sato, T., Nakatani, Y., and Ueda, M.: "Real-Time Display of Velocity Distribution on a Surface by Ultrasonic Laser Light Frequency Shifter and TV", Appl. Opt., 1974, 13, no.12, pp.2759-60.

35.208. Jarasiunas, K., and Vaitkus, J.: "Properties of a Laser-Induced Phase Grating in CdSe", Phys. Status Solidi, 1974, 23A, no.1, pp.K19-21.

35.209. Mirovitskii, D.I., et al.: "Integrated Holographic Multiposition Transportable Instruments", Prib. Tekh. Eksp., 1974, 17, no.3, pp.178-82, and Instrum. Exp. Tech., 1974, 17, no.3, pp.834-8.

35.210. Wang, L., and Helstrom, C.W.: "Resolution of Gaussian Images on a Photoelectrically Emissive Surface", Trans. IEEE, 1974, AES-10, no.6, pp.883-6.

35.211. Potapov, B.M., and Teleshevskii, V.I.: "Frequency/Angle Acoustooptic Converter", Prib. Tekh. Eksp., 1974, 17, no.4, pp.186-8, and Instrum. Exp. Tech., 1974, 17, no.4, pp.1170-2.

35.212. Mikaelyan, A.L., and Bobrinev, V.I.: "Holographic Memory Systems", Radiotekhnika, 1974, 29, no.5, pp.7-24, and Telecommun. Radio Eng. Pt 2, 1974, 29, no.5, pp.47-57.

Part V

MEASUREMENT PRINCIPLES AND TECHNIQUES

36. FUNDAMENTAL QUANTITIES

36.1. Schiffman, B.M., Young, L., and Larrick, R.B.: "Wire-Grid Waveguide Bolometer for Multimode Power Measurement", Trans. IEEE, 1965, MTT-13, p.427.

36.2. Engen, G.F.: "International Intercomparison of Standards for Microwave Power Measurement", Trans. IEEE,1965,MTT-13,p.713.

36.3. Berry, J.A.: "Measurement of H_{01}-Mode Loss at Q-Band in Helix Guide", Marconi Rev., 1965, 28, p.22.

36.4. Tao-Ming, C., and Tsu-Ch'uan, C.: "Measurement of Attenuation in a Rectangular Waveguide", Acta Electron. Sin., 1965, no.2-3, p.93.

36.5. Schuegraf, E.: "Rotating Standing-Wave Meter in the 5-mm Region", Frequenz, 1965, 19, p.225.

36.6. Neumann, E.G., and Dietz, K.J.: "Measurement of Reflection Factor With Guided Waves", Z. Angew. Phys., 1965, 19, p.297.

36.7. Russell, D.H.: "Unmodulated Twin-Channel Microwave Attenuation Measurement System", ISA Trans., 1965, 4, p.162.

36.8. Schanda, E., and Murnaghan, J.T.: "Method for Measurement of Plane Resistors at Microwave Frequencies", Proc. IEE, 1965, 112, p.49.

36.9. Schwelb, O., and Elliott, J.: "Precision Microwave Attenuator and Phaseshifter", Proc. IEEE, 1965, 53, p.107.

36.10. Zakaria, H., and Heaton, A.G.: "Thermoelectric Cooling Applied to the Absolute Measurement of Microwave Power", Proc. IEE, 1965, 112, p.508.

36.11. Evendorff, S.: "Microwave Peak Power Measurement", ISA J., 1965, 12, no.5, p.71.

36.12. Engen, G.F.: "Method of Transferring Calibrations Between Microwave Power Meters Having Different Input Waveguides", Int. Conv. Rec. IEEE, 1965, 13, pt 11, p.99.

36.13. Altschuler, H.M.: "Interchange of Source and Detector in Low-Power Network Measurements", Trans. IEEE, 1965,MTT-13,p.84.

36.14. Beatty, R.W.: "Measuring Impedance Through an Adaptor Without Introducing Additional Error", Proc. IEEE, 1965, 53, p.656.

36.15. de Ronde, F.C.: "Precise and Sensitive X-Band Reflectometer Providing Automatic Full-Band Display", Trans. IEEE, 1965, MTT-13, p.435.

36.16. Opricot, D.: "Direct-Reading Impedance Meter for Microwaves", Telecomunicattii, 1965, 9, no.5, p.166.

36.17. Mayo, G.: "Simplified Approach to Nodal-Shift Measurement Technique for Lossless Waveguide Discontinuities", Int. J. Electr. Eng. Educ., 1964, 4, no.2, p.155.

36.18. Clarricoats, P.J.B., and McStay, J.: "Microwave Power Measurement Using the Parallel-Pump Instability in YIG", Electron. Lett., 1966, 2, p.56.

36.19. Penfield, P.: "Electromagnetic Momentum Associated With Waveguide Modes", Proc. IEEE, 1966, 113, p.1504.

36.20. Kataoka, S.: "Application of Magnetoresistance Effect in Semiconductors to Microwave Power Measurements", Proc. IEE, 1966, 113, p.8.

36.21. Stuchly, S., and Wlodek, J.: "Measurement of Attenuation in the Microwave Region", Przegl. Elektrotech., 1966,7,p.147.

36.22. Nemoto, T., and Kawakami, T.: "National Standards for Microwave Impedance. III", Bull. Electrotech. Lab., 1966,30,p.699.

36.23. Allen, P.J.: "Radiation-Torque Experiment", Am. J. Phys., 1966, 34, p.1185.

36.24. Walliker, D.A.J.: "Resistive Thin Films in Rectangular Waveguide", Nature, 1966, 212, p.186.

36.25. Corazza, G.C., and Corzani, T.: "Measurement of Scattering Matrix of a Two-Port Microwave Junction", Electron. Lett., 1966, 2, p.238.

36.26. Wizner, W.W.: "Microwave Power Summator", Przegl. Telekomun., 1966, no.5, pp.143-5.

36.27. Kenderessy, M.: "Measurement of Low Standing-Wave Ratios With a Directional Coupler", Budavox Telecommun. Rev., 1966, no.1-2, pp.50-4.

36.28. Ely, P.C.: "Swept-Frequency Techniques", Proc. IEEE, 1967, 55, no.6, pp.991-1002.

36.29. Buckmaster, H.A., and Dering, J.C.: "Fundamental Limit to the Balance of a Microwave Bridge Containing a Dispersive Element", Trans. IEEE, 1967, IM-16, no.1, pp.13-8.

36.30. Tomiyasu, K.: "Decoupling Probe Method of Measuring High VSWR", Proc. IEEE, 1967, 55, no.6, p.1088.

36.31. Tischer, F.J.: "Slot Compensation in a Standing-Wave Meter at Millimetre Waves", Rev. Sci. Instrum., 1967, 38, no.10, pp.1481-5.

36.32. Henne, W.: "Wobble Technique in Microwave Measurements", Arch. Tech. Mess., 1967, no.378, pp.145-8.

36.33. Sakurai, K., and Nemoto, T.: "Thin-Film Bolometer Unit", Trans. IEEE, 1967, IM-16, no.3, pp.206-11.

36.34. Chandra, K., Parshad, R., and Kumar, R.C.: "Measurement of Impedance at Microwave Frequencies Using Directional Coupler and Adjustable Short-Circuit", Proc. IEE, 1967, 114, no.11, pp.1653-5.

36.35. Takeshima, M., Nakashima, S., and Miyai, Y.: "Admittance Measurements on Ge Microwave-Oscillator Diodes", Jap. J. Appl. Phys., 1967, 6, no.10, p.1254.

36.36. Levinson, D.S., and Sleven, R.L.: "Power Measurement in Multimode Waveguide", Microwave J., 1967, 10, no.11, pp.59-64.

36.37. Cheremnykh, M.A.: "Reference Impedance Samples With a Frequency-Independent SWR in Rectangular Waveguides", Izmer. Tekh., 1967, no.6, pp.42-5, and Meas. Tech., 1967, no.6, pp.699-703.

36.38. Tamaru, T.: "Methods for Precision Measurement of Microwave Impedance", Res. Electrotech. Lab., 1967, no.674, pp.1-63.

36.39. Nemoto, T.: "National Standard for Microwave Impedance. IV", Bull. Electrotech. Lab., 1967, 31, no.11, pp.1197-206.

36.40. Fitaire, M.: "Microwave Measurements of Symmetrical Two-Terminal-Pair Structures", Proc. IEEE, 1968, 56, no.4, p.777.

36.41. Lane, J.A.: "Application of Impedance Measurements in an Investigation of a Waveguide Thermistor Mount", Electron. Lett., 1968, 4, no.10, pp.186-8.

36.42. Satoda, Y., and Bodway, G.E.: "Three-Port Scattering Parameters for Microwave Transistor Measurement", J. Solid-State Circuits IEEE, 1968, SC-3, p.250.

36.43. Stumper, U.: "Influence of Higher-Order Modes in an Oversized Waveguide on Reflection-Factor Measurements", Z. Angew. Phys., 1968, 25, p.293.

36.44. Kumar, R.C.: "Microwave Terminations", J. Inst. Telecommun. Eng., 1968, 14, p.483.

36.45. Widman, D.L.: "Microwave Sweep Oscillators for the Laboratory", Electron. Wld, 1968, 80, no.5, p.49.

36.46. Herman, W., and Majewski, M.: "Measurement of Varactor Diodes by Reflection Methods", Arch. Elektrotech., 1968, 17, p.939.

36.47. Beatty, R.W.: "Effect of Realizability Conditions Upon Limits of Mismatch Error in the Calibration of Fixed Attenuators", Trans. IEEE, 1968, MTT-16, p.976.

36.48. Zwerdling, S., Smith, R.A., and Theriault, J.P.: "Fast, High-Responsivity, Bolometer for the Far-IR", Infrared Phys., 1968, 8, no.4, p.271.

36.49. Semenikhin, I.M.: "Attenuator Design Improvements", Izmer. Tekh., 1968, no.2, p.90, and Meas. Tech., 1968, no.2, p.273.

36.50. Vavrouch, D.: "Calorimeter for Measuring Output Energy of Lasers", Jemna Mech. Opt., 1968, no.9, p.287.

36.51. Lewin, L.: "Standard Mismatch. Controlled Small Reflections in Waveguides", J. Res. Natl. Bur. Stand., 1968, 72C, p.197.

36.52. Zagrodzinski, J.A.: "Microwave Measurement Setup for the 4-mm Band", Postepy Fiz., 1969, 20, no.2, p.233.

36.53. Gaskell, C.S.: "Microwave Attenuation Measurement", Des. Electron., 1969, 6, no.6, p.30.

36.54. Gechyauskas, S.I., et al.: "Transducer for SHF Transferred Pulse-Power Meter", Prib. Tekh. Eksp., 1969, 12, no.2, p.121, and Instrum. Exp. Tech., 1969, 12, no.2, p.400.

36.55. Schmidt, A.J., and Greenhow, R.C.: "Practical Laser Output Meter", J. Phys. E, 1969, 2, p.438.

36.56. Beatty, R.W., and Yates, B.C.: "Graph of Return Loss v. Frequency for Quarter-Wave Waveguide Impedance Standards", Trans. IEEE, 1969, MTT-17, p.282.

36.57. Fujisawa, K.: "Power Standard at 3 GHz", Bull. Electrotech. Lab., 1969, 33, no.4, p.100.

36.58. Hollway, D.L.: "High-Resolution Swept-Frequency Reflectometer", Trans. IEEE, 1969, MTT-17, p.185.

36.59. Stelzried, C.T., and Ollmans, D.A.: "Precision Insertion-Loss Calibration at 90 GHz", Trans. IEEE, 1969, MTT-17, p.233.

36.60. Nemoto, T., Beatty, R.W., and Fentress, G.H.: "Two-Channel Off-Null Technique for Measuring Small Changes of Attenuation", Trans. IEEE, 1969, MTT-17, p.396.

36.61. Rahman, M.H., and Gunn, M.W.: "Wave Reflections from Rotary-Vane Attenuators", Trans. IEEE, 1969, MTT-17, p.402.

36.62. Owens, R.P.: "Technique for Measurement of SWR at Low Power Level", Proc. IEE, 1969, 116, p.933.

36.63. Allen, C., et al.: "IR-to-Millimetre, Broadband, Solid-State Bolometer", Appl. Opt., 1969, 8, p.813.

36.64. Sato, T., and Sakurai, K.: "Power Measurement of Giant Pulses Using Optical Rectification", Bull. Electrotech. Lab., 1969, 33, no.12, p.37.

36.65. Nowicki, R.: "Calorimeter for Absolute and Accurate Laser CW Power Measurements", Pomiary Automat. Kontr., 1969, 15, no.4, p.165.

36.66. Kellock, H.A.: "Calorimeter for IR Laser Power Measurement", J. Phys. E, 1969, 2, p.377.

36.67. Nagatsuka, A., and Ishige, R.: "Directivity Measurements of Four-Port Directional Couplers", Bull. Electrotech. Lab., 1969, 33, no.4, p.114.

36.68. Kolomiichenko, G.N.: "Measurement of Large VSWR by a Minimum-Width Method", Izv. VUZ Radioelektron., 1969, 12,no.1,p.64.

36.69. Lienard, J.C.: "Method for Microwave Output Impedance Measurement", Rev. MBLE, 1969, 12, no.1, p.13.

36.70. Krahl-Urban, B., and Wagner, H.: "Precision Measurement of Parameters of a Dielectric Quarter-Wave Transformer", Z. Angew. Phys., 1969, 27, no.2, p.104.

36.71. Jachimovits, L.: "Measurement of Y-Circulators", Mes. Regul. Autom., 1969, 17, no.1, p.20.

36.72. Hesse, E., and Mendez, D.: "Thermistor Bolometer for Measuring Monochromatic Light", Z. Angew. Phys., 1969, 27,no.4,p.289.

36.73. Nowicki, R.: "Definition of Effective Efficiency of Laser Calorimeters", Trans. IEEE, 1969, IM-18, p.238.

36.74. Gertsenshtein, M.Ye., et al.: "Errors Due to Mismatch by Small Losses in Circulators and Reciprocal Fourpoles", Radiotekhnika, 1969, 24, no.7, p.91, and Telecommun. Radio Eng. Pt 2, 1969, 24,no.7.

36.75. Dalley, J.E.: "Computer-Aided Microwave Impedance Measurements", Trans. IEEE, 1969, MTT-17, p.572.

36.76. Bandler, J.W.: "Programme for Processing Standing-Wave Measurements", Trans. IEEE, 1969, MTT-17, p.644.

36.77. Matveev, V.I., and Drozdoz, M.M.: "Measurement of Gain and Losses in Gas Lasers", Izmer. Tekh., 1969, no.3, p.95, and Meas. Tech., 1969, no.3, p.438.

36.78. Musil, J., and Zacek, F.: "Applying a Quadrupole Microwave Probe to Dynamic Measurement of Phase and Attenuation", Slab. Obz., 1969, 30, no.9, p.395.

36.79. Drew, H.D., and Sievers, A.J.: "^{3}He-Cooled Bolometer for the Far IR", Appl. Opt., 1969, 8, p.2067.

36.80. Somlo, P.I., and Hollway, D.L.: "Microwave Locating Reflectometer", Electron. Lett., 1969, 5, p.468.

36.81. Kuz'michev, V.M., et al.: "Low-Inertia Pyroelectric Indicator of Laser Pulse Radiation", Radiotekh. Elektron., 1969, 14, no.10, p.1843, and Radio Eng. Electron. Phys., 1969, 14, no.10, pp.1591-4.

36.82. Staniforth, J.A., and Allnutt, J.E.: "Absolute Grid Attenuator", Electron. Lett., 1969, 5, p.653.

36.83. Adams, J.W., and Jarvis, S.: "Current Distribution in Barretters for Microwave Power Measurements", Trans. IEEE, 1969, MTT-17, p.778.

36.84. Firth, K., and Davies, M.B.: "Laser Power and Energy Measurement by Photoelectric Techniques", Proc. Inst. Mech. Eng., 1969, 183, pt 3D, p.34.

36.85. Balczewski, L.: "Methods of Laser Energy and Power Measurement", Postepy Fiz., 1969, 20, no.1, p.49.

36.86. Epstein, R.A., and Rockwell, R.J.: "(Phototube) Monitor for a Pulsed Laser System", Opt. Technol., 1969, 1, no.5, p.244.

36.87. Zubov, B.V., et al.: "Measuring Energy and Temporal Parameters of a Laser in the IR Region", Zh. Prikl. Spekt., 1969, 11, no.4, p.732.

36.88. Pieterse, J.D., and Versnel, W.: "Numerical Method for Obtaining the Scattering Matrix of a Microwave Two-Port from Standing-Wave Measurements", Appl. Sci. Res., 1969, 21, no.1, p.13.

36.89. Jurkus, A.: "Semi-Automatic Method for Precision Measurement of Microwave Impedance", Trans. IEEE, 1969, IM-18, p.283.

36.90. Scott, B.F.: "Laser Energy Measurements by Absolute Methods", Proc. Inst. Mech. Eng., 1969, 183, pt 3D, p.56.

36.91. Kododii, N.G., and Valitov, R.A.: "Ponderomotive Meter of Laser Output Energy", Izmer. Tekh., 1969, no.12, pp.27-30, and Meas. Tech., 1969, no.12, pp.1684-8.

36.92. Quel, E., et al.: "Simple Method for Measuring Energy of CO_2 Laser", Ann. Soc. Sci. Brux., 1969, 83, no.3, p.388.

36.93. Ruiz-Aguirre, R.D., and Gimeno, A.I.: "Method for Absolute Calibration of Microwave Attenuators", Electron. Fiz. Apl., 1969, 12, no.48, p.255.

36.94. Ueki, K., Aoki, T., and Ohi, K.: "Studies of Microwave Dry Calorimeters", Hitachi Rev., 1969, 18, no.6, p.235.

36.95. Nemoto, T.: "Measurement of Waveguide-Joint Properties", Electron. Commun. Jap., 1969, 52, no.7, p.68.

36.96. Barkhudarova, T.M., et al.: "Measurement of Intense Giant-Pulse Radiation", Electron Technol., 1969, 2, no.2-3, p.127.

36.97.   Koubarev, A.V., Oboukhov, A.S., and Ivlev, E.I.: "Equipment for Measuring Laser Energy", Electron Technol., 1969, 2, no.2-3, p.165.

36.98.   Kantor, K.: "Apparatus and Method for Determining Optical Parameters of Laser Mirrors", Electron Technol., 1969, 2, no.2-3, p.197.

36.99.   Kozelev, A.I., and Matveev, R.F.: "Pulse Measurement of Average H_{01}-Mode Total Losses in Waveguides", Izv. VUZ Radioelektron., 1969, 12, no.6, p.639.

36.100.   Jha, R.K., and Garg, V.K.: "VSWR Measurement by Cross Coupler", Int. J. Electron., 1970, 29, p.179.

36.101.   Dombi, J., et al.: "Apparatus for Measurement of Laser Pulse Energy", Acta Phys. Chem. Szeged., 1970, 16, no.1-2, p.3.

36.102.   West, E.D., and Jennings, D.A.: "Power Measurement of Large Laser Beams With a Small Dual-Cone Calorimeter", Rev. Sci. Instrum., 1970, 41, p.142.

36.103.   Mok, C.K.: "Method of Obstacle Admittance Measurement in Below-Cutoff Waveguides", Electron. Lett., 1970, 6, p.50.

36.104.   Kajfez, D.: "Numerical Data Processing of Reflection-Factor Circles", Trans. IEEE, 1970, MTT-18, p.96.

36.105.   Larson, W., Desch, R.F., and Gillard, B.F.: "Further Analysis of Off-Null v. Power-Ratio Method of Attenuation Measurement", Trans. IEEE, 1970, MTT-18, p.112.

36.106.   Nakagawa, Y., and Yoshinaga, H.: "Characteristics of High-Sensitivity Ge Bolometer", Jap. J. Appl. Phys., 1970, 9, no.1, p.125.

36.107.   Chambers, B.: "Measurement of Attenuation Coefficient of Inhomogeneous Dielectric-Filled Cables", Electron. Lett., 1970, 6, p.211.

36.108.   West, E.D., and Churney, K.L.: "Theory of Isothermal Calorimetry for Laser Power and Energy", J. Appl. Phys., 1970, 41, p.2705.

36.109.   Kendall, J.M., and Berdahl, C.M.: "Two Blackbody Radiometers of High Accuracy", Appl. Opt., 1970, 9, p.1082.

36.110.   Yakovlev, Yu.M.: "Simple Method of Determining Gain and Losses in Gas Lasers", Zh. Prikl. Spekt., 1970, 13, no.4, pp.728-9.

36.111.   Tajima, I., et al.: "Design of Thin-Film Thermoelements for High-Frequency Power Measurements", Electron. Commun. Jap., 1970, 53, no.7, pp.24-5.

36.112.   Tulip, J., and Seguin, H.J.: "Gain-Saturation Measurements in CO_2 Laser Using Fresnel Loss-Plate Technique", Can. J. Phys., 1970, 48, p.1086.

36.113.   Cheremnykh, M.A.: "Device and Method for Measuring Reflections from Flanged Joints and Short Segments of Waveguide", Izmer. Tekh., 1970, no.12, pp.54-7, and Meas. Tech., 1970, no.12, pp.1889-93.

36.114.   Bernard, P., Lengeler, H., and Vaghin, V.: "Pulse Method for Matching Mode Transformers on Disc-Loaded Waveguide", Nuclear Instrum. Methods, 1970, 82, p.319.

36.115.   Movchang, S.P., et al.: "Waveguide Gas-Discharge In-Line Power Monitor", Radiotekh. Elektron., 1970, 15, p.501, and Radio Eng. Electron. Phys., 1970, 15, no.3, pp.429-32.

36.116.   Jennings, D.A., and West, E.D.: "Laser Power Meter for Large Beams", Rev. Sci. Instrum., 1970, 41, p.565.

36.117.   Collier, R.J., and Price, D.J.: "Analogue Technique for Microstrip Parameters", Electron. Lett., 1970, 6, p.338.

36.118.   Otoshi, T.Y., et al.: "Comparisons of Waveguide Losses Calibrated by Various Techniques", Trans. IEEE, 1970, MTT-18, p.406.

36.119.   Barlow, H.E.M., and Cross, P.H.: "Radiation-Pressure Instrument for Absolute Measurements of Power at 35 GHz", Proc. IEE, 1970, 117, p.853.

36.120.   Ekinge, R., and Hedstrom, T.: "Variable Microwave Attenuator", Trans. IEEE, 1970, MTT-18, p.661.

36.121.   Uhlir, A.: "Bounds on Output VSWR of a Passive Reciprocal Two-Port from Forward Measurements", Trans. IEEE, 1970, MTT-18, p.662.

36.122.   King, R.J., and Christopherson, R.I.: "Homodyne System for Measurement of Microwave Reflection Factors", Trans. IEEE, 1970, MTT-18, p.658.

36.123.   Last, J.D., and Smith, S.T.: "Off-Line Correction of Microwave Network Analyser Reflection-Factor Errors", Electron. Lett., 1970, 6, p.641.

36.124.   Hirakawa, H., and Okumura, S.: "Direct Measurement of RF Field Using Diodes", Jap. J. Appl. Phys., 1970, 9, p.812.

36.125.   Engen, G.F.: "Evaluation of the Back-to-Back Method of Measuring Adaptor Efficiency", Trans. IEEE, 1970, IM-19, p.18.

36.126.   Rietto, G.: "Calorimetric Measurements of Reflectance of Waveguide Short-Circuits", Alta Freq., 1970, 39, p.35.

36.127.   Gimeno, A.I.: "Properties of Chrome Resistors as Microwave Attenuators", Electron. Fiz. Apl., 1970, 13, no.2, pp.93-7.

36.128.   Nemoto, T.: "Microwave Impedance Standards", Res. Electrotech. Lab., 1970, no.708, pp.1-99.

36.129. Rinehart, R.E., and Kratz, H.R.: "Microsecond-Response Bolometer for Measuring Large Thermal Fluxes", ISA Trans., 1970, 9, no.2, pp.104-12.

36.130. Rasmussen, A.L.: "Laser Energy and Power Measurement With a Double-Reflecting Plate Calorimeter", Rev. Sci. Instrum., 1970, 41, no.10, pp.1479-84.

36.131. Offenberger, A.A.: "Analysis of a Thermocouple Laser Power Meter", Appl. Opt., 1970, 9, no.11, pp.2594-7.

36.132. Napoli, L.S., and Hughes, J.J.: "Measurement Technique for Coupled Microstrip Lines", RCA Rev., 1970, 31, no.3, pp.479-98.

36.133. Barlow, H.E.M., and Wizner, W.W.: "Method for Absolute Measurements of Microwave Power", Prace PIT, 1970, 20, no.67, pp.1-12.

36.134. Gechyauskas, S.I., et al.: "Characteristics of n-Si Transducer for Measuring Microwave Pulse Power", Prib. Tekh. Eksp., 1970, no.3, pp.182-3, and Instrum. Exp. Tech., 1970, no.3, pp.844-5.

36.135. Potapov, V.A., and Maslennikova, L.P.: "Microwave Impedance Meter With Separate Measurement of Modulus and Phase", Izmer. Tekh., 1970, no.8, pp.67-70, and Meas. Tech., 1970, no.8, pp.1221-5.

36.136. Engen, G.F., and Hudson, P.A.: "International Intercomparison of Power Standards at 3 GHz", Trans. IEEE, 1971, MTT-19, no.4, pp.411-3.

36.137. Riech, V.: "Measurement Programme and Calibration Standards for Microstrip Measurements With an Automatic Network Analyser", Nachr. Tech. Z., 1971, 24, no.5, pp.255-9.

36.138. Oseki, T., and Saito, S.: "Precision Variable, Double-Prism, Attenuator for CO_2 Lasers", Appl. Opt., 1971, 10, no.1, pp.144-9.

36.139. Cook, C.C., and Allred, C.M.: "Excitation System for Piston Attenuators", Trans. IEEE, 1971, IM-20, no.1, pp.10-6.

36.140. Staniforth, J.A.: "Self-Calibrating Grid Attenuator for Use at Submillimetre Wavelengths", Proc. IEE, 1971, 118, no.2, pp.343-8.

36.141. Engen, G.F.: "Power Equations. Concept in Description and Evaluation of Microwave Systems", Trans. IEEE, 1971, IM-20, no.1, pp.49-57.

36.142. Ernst, R.L.: "Accurate Measurement of Isolation in Three-Port Circulators", RCA Rev., 1971, 32, no.1, pp.164-71.

36.143. Shaklee, K.L., and Leheny, R.F.: "Direct Determination of Optical Gain in Semiconductor Crystals", Appl. Phys. Lett., 1971, 18, no.11, pp.475-7.

36.144. Ishchenko, V.A.: "Design of Broadband Waveguide Four-Probe Impedance Meters", Izv. VUZ Radioelektron., 1971, 14, no.5, pp.572-6.

36.145. Kirkpatrick, G.R.: "Effective Stripline Device Characterization (Network Analyser)", Hewlett Packard J., 1971, 22, no.9, pp.12-6.

36.146. Edwards, J.G., and Jefferies, R.: "Power and Energy Monitor for Pulsed Lasers", J. Phys. E, 1971, 48, no.8, pp.580-4.

36.147. Zakurenko, O.E., et al.: "Compensated Power Meter for CW Laser Radiation", Izmer. Tekh., 1971, 14, no.1, pp.21-2, and Meas. Tech., 1971, 14, no.1, pp.30-2.

36.148. Little, W.E., Larson, W., and Kinder, B.J.: "Rotary-Vane Attenuator With an Optical Readout", J. Res. Natl. Bur. Stand., 1971, 75C, no.1, pp.1-5.

36.149. Otoshi, T.Y., and Stelzried, C.T.: "Precision Compact Rotary-Vane Attenuator", Trans. IEEE, 1971, MTT-19, no.11, pp.843-54.

36.150. Afonchenkov, N.G., et al.: "Helium Cryostat for a Submillimetre n-InSb Bolometer", Cryogenics, 1971, 11, no.5, pp.415-6.

36.151. Nishihara, H., Ura, K., and Terada, M.: "Measurement of Small-Signal Parameters in TWT Theory", Trans. IEEE, 1971, ED-18, no.12, pp.1155-62.

36.152. Preston, J.S.: "Radiation Thermopile for CW and Pulse Measurement", J. Phys. E, 1971, 4, no.12, pp.969-72.

36.153. Povarkov, V.I., et al.: "Differential Radiometer", Opt.-Mekh. Prom., 1971, 38, no.1, pp.39-43, and Sov. J. Opt. Technol., 1971, 38, no.1, pp.33-7.

36.154. Plotnikov, V.A., and Chastukhina, L.N.: "Measurement of Power of a He-Ne Laser", Prib. Tekh. Eksp., 1971, 14, no.4, pp.189-90, and Instrum. Exp. Tech., 1971, 14, no.4, pp.1173-4.

36.155. Nesterenko, V.M., and Morozov, B.N.: "Utilization of Optical Rectification in Measuring Laser Power Output", Kvantovaya Elektron., 1971, no.5, pp.87-92.

36.156. Bramall, K.E.: "Accurate Microwave High-Power Measurements Using a Cascaded-Coupler Method", J. Res. Natl. Bur. Stand., 1971, 75C, no.3-4, pp.185-92.

36.157. Martin, B., and Hobson, G.S.: "Technique for Standing-Wave Measurements of Pulse-Biased Semiconductor Devices", Radio Electron. Eng., 1971, 41, no.12, pp.538-40.

36.158. Engen, G.F.: "Improved Method for Microwave Power Calibration With Application to Evaluation of Connectors", J. Res. Natl. Bur. Stand., 1971, 75C, no.2, pp.89-93.

36.159. Kukush, V.D.: "Errors Due to Losses During Absolute Calibration of Ponderomotive Wattmeters", Radiotekh. (Kharkov), 1971, no.18, pp.80-4.

36.160. Serednii, V.P., Zhilkov, V.S., and Kukush, V.D.: "Standard (Ponderomotive) Setup for Reproduction of Unit of Power at 8 mm", Radiotekh. (Kharkov), 1971, no.19, pp.136-40.

36.161. Guzhva, V.G., et al.: "High-Speed Meter for Laser Pulse Energy", Radiotekh. (Kharkov), 1971, no.19, pp.140-4.

36.162. Yabe, H.: "S-Parameter Representation of Canonical Network Elements in Microwave Structures", Electron. Commun. Jap., 1971, 54, no.3, pp.7-14.

36.163. Anatychuk, L.I., et al.: "Anisotropic Radiation Detector", Opt.-Mekh. Prom., 1971, 38, no.1, pp.27-9, and Opt. Technol., 1971, 38, no.1, pp.22-3.

36.164. Jacobs, S.D.: "Simple CO_2-Laser Power Meter", Appl. Opt., 1971, 10, no.11, pp.2564-5.

36.165. Matino, H.: "Characteristic Impedance Measurements of a Coplanar Waveguide", Electron. Lett., 1971, 7, no.23, pp.678-9.

36.166. Chu, A., and Farell, E.: "Measurement of Forces in Microwaves", Rev. Electrotec., 1971, 57, no.4, pp.213-24.

36.167. Mastellari, G.A., Someda, C.G., and Valdoni, F.: "Automated Attenuation Test in Multimodal Waveguides by Resonance Return-Loss or Insertion-Loss Measurements", Trans. IEEE, 1972, MTT-20, no.1, pp.17-21.

36.168. Garver, R.V., et al.: "Errors in S_{11} Measurements Due to Residual SWR of the Measuring Equipment", Trans. IEEE, 1972, MTT-20, no.1, pp.61-9.

36.169. Somlo, P.I.: "Locating Reflectometer", Trans. IEEE, 1972, MTT-20, no.1, pp.105-12.

36.170. Chang, C.T.M.: "Application of Deschamps' Graphical Method to Measurements of Scattering Coefficients of Multiport Waveguide Junctions", Trans. IEEE, 1972, MTT-20, no.2, pp.186-7.

36.171. Beatty, R.W.: "Efficiency of Microwave 2-Ports from Reflection-Factor Measurements", Trans. IEEE, 1972, MTT-20, no.5, pp.343-4.

36.172. Davis, J.B., et al.: "3.7-4.2 GHz Computer-Operated Measurement System for Loss, Phase, Delay, and Reflection", Trans. IEEE, 1972, IM-21, no.1, pp.24-37.

36.173. Foote, W.J., and Hunter, R.D.: "Improved Gearing for Rotary-Vane Attenuators", Rev. Sci. Instrum., 1972, 43, no.7, pp.1042-3.

36.174. Clemesha, B.R.: "Laser Energy Monitor With Digital Output", J. Phys. E, 1972, 5, no.9, pp.859-61.

36.175. Preston, J.S.: "Spectral Responses of Laser Cone Calorimeters", J. Phys. E, 1972, 5, no.10, pp.1014-6.

36.176. Kalinski, J.: "Modulated Subcarrier Techniques for Microwave Attenuation Measurements in Industrial Applications", Trans. IEEE, 1972, IM-21, no.3, pp.291-3.

36.177. Knyaz'kov, B.N., Litinov, D.D., and Yanovskii, M.S.: "Quasi-Optical Reflectance Meter", Izv. VUZ Radioelektron., 1972, 5, no.1, pp.52-8.

36.178. Smathers, S.E., and Maksymonko, G.: "Calorimetric Measurement of Power from Pulsed Lasers", Trans. IEEE, 1972, IM-21, no.4, pp.430-3.

36.179. Gunn, S.R.: "Tubular Calorimeter for High-Power Laser Pulses", Rev. Sci. Instrum., 1972, 43, no.10, pp.1523-5.

36.180. Zaitsev, V.V.: "Fast-Response Wideband Microwave Power-Level Converter", Izmer. Tekh., 1972, 15, no.2, pp.59-60, and Meas. Tech., 1972, 15, no.2, pp.280-2.

36.181. Smith, R.L., et al.: "Calorimeter for High-Power CW Lasers", Trans. IEEE, 1972, IM-21, no.4, pp.434-8.

36.182. Shcherbov, V.A.: "Variable Load Standard for a Beamguide", Izv. VUZ Radioelektron., 1972, 15, no.7, pp.895-8.

36.183. Moren, R.E.: "Technique for Measuring Insertion Loss of a Waveguide", Tech. Dig., 1972, no.28, pp.33-4.

36.184. Warner, F.L., Watton, D.O., and Herman, P.: "Accurate X-Band Rotary Attenuator With Absolute Digital Angular Readout", Trans. IEEE, 1972, IM-21, no.4, pp.446-50.

36.185. Nemoto, T., Fujisawa, K., and Inoue, T.: "Bolometer Mounts for the Short-Millimetre-Wave Region", Trans. IEEE, 1972, IM-21, no.4, pp.480-3.

36.186. Dorschner, T.A.: "Measurement of Load SWR's Through Uncalibrated Lossless Two-Ports", Electron. Lett., 1972, 8, no.22, pp.544-5.

36.187. Weinert, F.K., and Weinschel, B.O.: "Microwave Peak-Power Measurement Independent of Detector Characteristics by Comparison of Video Samples", Trans. IEEE, 1972, IM-21, no.4, pp.474-9.

36.188. Okamura, S., and Okabe, Y.: "Measurement of Admittances of Microwave Oscillators With Injection Locking", Trans. IEEE, 1972, IM-21, no.4, pp.443-6.

36.189. Hume, F.R., Koide, F.K., and Dederich, D.J.: "Practical and Precise Means of Microwave Power-Meter Calibration Transfer", Trans. IEEE, 1972, IM-21, no.4, pp.457-66.

36.190. Mery, M.C., Silhouette, D., and Conard, J.: "Radiation Pressure Due to Absorption of Resonant Light", C. R. Acad. Sci. (Paris), 1972, 275B, no.19, pp.693-6.

36.191. Kim, A.Ch., et al.: "Calorimetric Wattmeter for High Power Levels", Radiotekh. (Kharkov), 1972, no.23, pp.34-7.

36.192. Gasilov, A.L., and Streletskii, R.P.: "VSWR Measurement Taking Into Account the Detector Characteristic", Izmer. Tekh., 1972, 15, no.3, p.87, and Meas. Tech., 1972, 15, no.3, pp.491-2.

36.193. Somlo, P.I.: "Display of Microwave Pulse Response via Real-Time Fourier Transform of the Transfer Function", Electron. Lett., 1972, 8, no.26, pp.631-2.

36.194. Zhilkov, V.S., and Sirotnikov, A.I.: "Error of a Two-Plate Ponderomotive Wattmeter Due to Higher Modes", Radiotekh. (Kharkov), 1972, no.22, pp.100-3.

36.195. Ivanov, K.P., Ganchev, S.I., and Tsankov, M.A.: "Slotted-Line Measurement of Insertion Loss in Three-Port Ferrite Junction Circulator", C. R. Acad. Bulg. Sci., 1972, 25, no.5, pp.601-4.

36.196. Arkhangel'skii, Yu.S., and Kolomeitsev, V.A.: "Error Analysis of SHF Power Measurement by Waveguide Calorimeters", Radiotekh. (Kharkov), 1972, no.23, pp.12-6.

36.197. Epel'man, E.G., Usov, E.V., and Fedotenkov, L.P.: "Metrological Analysis of High-Power-Level SHF Wattmeters", Radiotekh. (Kharkov), 1972, no.23, pp.17-30.

36.198. Banach, J., and Trnka, Z.: "Novel Method of Measuring and Evaluating the Division Error in 3-dB Hybrid Junctions", TESLA Electron., 1972, 5, no.3, pp.81-4.

36.199. Kim, A.Ch., Makarov, V.V., and Popov, E.I.: "Investigation of Calorimetric Loads With Dielectric Windows", Radiotekh. (Kharkov), 1972, no.23, pp.37-40.

36.200. Wizner, W.W.: "Microwave Instruments Applying Ponderomotive Forces", Pr. PIT, 1972, 22, no.76, pp.24-33.

36.201. Medresh, V.G., et al.: "Digital Meter of Laser Pulse Energy", Izmer. Tekh., 1972, 15, no.12, pp.36-7, and Meas. Tech., 1972, 15, no.12, pp.1801-3.

36.202. Falciasecca, G., and Mastellari, G.A.: "Measuring Method for Waveguide-Junction Parameters", Alta Freq., 1972, 41, no.12, pp.962-5.

36.203. Bokrinskaya, A.A., and Vuntesmeri, V.S.: "Use of Ferromagnetic Films for Measurement of Peak SHF Power", Radiotekh. (Kharkov), 1972, no.23, pp.41-4.

36.204. Balakov, V.F., et al.: "SHF Peak Wattmeter", Radiotekh. (Kharkov), 1972, no.23, pp.44-8.

36.205. Sereda, A.F., and Akulov, A.T.: "Choice of Physical Structure of Automatic Ponderomotive Wattmeters", Radiotekh. (Kharkov), 1972, no.23, pp.81-4.

36.206. Zhilkov, V.S., Serednii, V.P., and Shpagin, Yu.V.: "Investigation of a Ponderomotive-Wattmeter With Frequency Adjustment", Radiotekh. (Kharkov), 1972, no.23, pp.94-8.

36.207. Smith, R.L., and Phelan, R.J.: "Limitations of Vacuum Photodiodes for Measurement of Laser Power and Energy", Appl. Opt., 1973, 12, no.4, pp.795-8.

36.208. Raicu, D.: "Precise Measurement With Tuned Microwave Reflectometers", Rev. Roum. Phys., 1973, 18, no.2, pp.177-86.

36.209. Weber, H.P., Dunn, F.A., and Leibolt, W.N.: "Loss Measurements in Thin-Film Optical Waveguides", Appl. Opt., 1973, 12, no.4, pp.755-7.

36.210. Da Silva, E.F., and McPhon, M.K.: "Calibration of Microwave Network Analyser for Computer-Corrected S-Parameter Measurements", Electron. Lett., 1973, 9, no.6, pp.126-8.

36.211. Jacob, J.H., et al.: "Absolute Method of Measuring Energy Outputs from CO_2 Lasers", Rev. Sci. Instrum., 1973, 44, no.4, pp.471-4.

36.212. Boulanger, P., et al.: "Absolute Calorimeter for High-Power CO_2 Laser", J. Phys. E, 1973, 6, no.6, pp.559-60.

36.213. Birch, J.R., and Bradley, C.C.: "Variable-Loss Determination of HCN Laser Gain", Infrared Phys., 1973, 13, no.2, pp.99-108.

36.214. Mahlein, H.F., and Rauscher, W.: "Continuously Variable Laser-Beam Attenuator Using a Dielectric Multilayer", Optik, 1973, 38, no.2, pp.187-95.

36.215. Lang, G.: "Precision Reflectometers for Swept-Frequency Measurements", Elektron. Ind., 1973, 4, no.4, pp.61-2.

36.216. Koren, G., et al.: "Thin-Film Calorimeter for Low-Energy Laser Pulse Measurements", Appl. Phys. Lett., 1973, 23, no.2, pp.73-4.

36.217. Bishop, S.G., and Moore, W.J.: "Chalcogenide Glass Bolometers", Appl. Opt., 1973, 12, no.1, pp.80-3.

36.218. Gunn, S.R.: "Calorimetric Measurements of Laser Energy and Power", J. Phys. E, 1973, 6, no.2, pp.105-14.

36.219. De Waard, P.J.: "Measurement of Admittance of Gunn Diodes", Electron. Lett., 1973, 9, no.3, pp.59-60.

36.220. Beatty, R.W.: "Two-Port Quarter-Wave Waveguide Standards of SWR", Electron. Lett., 1973, 9, no.2, pp.24-6.

36.221. Otoshi, T.Y.: "Analytical Expression for Limits of Error in Measurement of Reflection-Factor Phase", Trans. IEEE, 1973, MTT-21, no.3, pp.151-3.

36.222. Valdoni, F.: "Method for Measurement of Coupling Between Different Modes", Note Recens. Not., 1973, 22, no.4, pp.440-9.

36.223. Jacobs, R.R.: "Technique for Gain Determinations in Pulsed CO_2 TEA Lasers", Rev. Sci. Instrum., 1973, 44, no.8, pp.1146-7.

36.224. Luther, G.: "Enhancement of Sensitivity of Microwave Admittance Measurements by Use of Matching 2-Ports", Arch. Elektron. Übertrag., 1973, 27, no.7-8, pp.297-302.

36.225. Hollway, D.L., and Somlo, P.I.: "Reduction of Errors in a Precise Microwave Attenuator Calibration System", Trans. IEEE, 1973, IM-22, no.3, pp.268-70.

36.226. Medresh, V.G., et al.: "Automated Meter for Time and Energy Parameters of Pulsed Lasers", Izmer. Tekh., 1973, 16, no.8, pp.28-30, and Meas. Tech., 1973, 16, no.8, pp.1161-3.

36.227. Egami, S.: "Swept-Frequency Impedance Indicator Using Directional Couplers", Trans. IEEE, 1973, MTT-21, no.10, pp.647-8.

36.228. de Jersey, D.E.: "System for Swept-Frequency Measurement of High Attenuations", J. Phys. E, 1973, 6, no.10,pp.964-6.

36.229. Miane, J.L.: "Q-Band Reflectometric Bridge", C. R. Acad. Sci. (Paris), 1973, 277B, no.1, pp.17-20.

36.230. Nichols, L.L., Ratcliffe, C.A., and Gordon, R.L.: "Calorimeter for Mirror Absorption Measurements", Appl. Opt., 1973, 12, no.10, pp.2232-3.

36.231. Roosen, G., and Imbert, C.: "Measurement of the Mechanical Couple Exerted by an Extraordinary Light Ray Traversing an Anisotropic Medium", C. R. Acad. Sci. (Paris), 1973, 277B, no.6, pp.135-8.

36.232. Roosen, G., and Imbert, C.: "Experiment for Measuring a Mechanical Effect of Light", C. R. Acad. Sci. (Paris), 1973, 277B, no.7, pp.147-9.

36.233. Johnson, D.C.: "Measurement of Low Absorption Coefficients in Crystals", Appl. Opt., 1973, 12, no.9, pp.2192-7.

36.234. Watt, B.E.: "Calorimeter for Picosecond Laser Pulses", Appl. Opt., 1973, 12, no.10, pp.2373-7.

36.235. Khizhnyak, N.A., et al.: "Ponderomotive Forces Acting on Two Rigidly Linked Bodies in a Rectangular Waveguide", Radiotekh. (Kharkov), 1973, no.26, pp.105-11 and 112-7.

36.236. Komisarczuk, J.: "Exact Broadband Measurement of Scattering Parameters", Bull. Acad. Pol. Sci. Tech., 1973, 21, no.6, pp.499-502.

36.237. Gould, J.W., and Rhodes, G.M.: "Computer Correction of a Microwave Network Analyser Without Accurate Frequency Measurement", Electron. Lett., 1973, 9, no.21, pp.494-5.

36.238. Petrosyan, O.G.: "Determination of the Variation of Reflection Factor of Absorptive Coatings in the Millimetre Range", Prib. Tekh. Eksp., 1973, 16, no.2, pp.141-2, and Instrum. Exp. Tech., 1973,16,no.2,pp.496-7.

36.239. Shurmer, H.V.: "Calibration Procedure for Computer-Corrected S-Parameter Characterization of Devices Mounted in Microstrip", Electron. Lett., 1973, 9, no.14, pp.323-4.

36.240. Davies, O.J., Doshi, R.B., and Nagenthiram, B.: "Correction of Microwave-Network-Analyser Measurements of Two-Port Devices", Electron. Lett., 1973, 9, no.23, pp.543-4.

36.241. Bacherikov, V.V., et al.: "Device for Measuring Laser Energy by (Heating of) a Ferromagnetic Element", Izmer. Tekh., 1973, 16, no.3, pp.69-70, and Meas. Tech., 1973, 16, no.3, pp.424-5.

36.242. Nikolaev, V.K., et al.: "High-Pulse-Energy, Large-Beam, Laser Meter", Radiotekh. (Kharkov), 1973, no.25, pp.8-14.

36.243. Starodubtsev, G.P., Nadezhkin, Yu.M., and Valitov, R.A.: "Thermal Effects in Non-Vacuum Ponderomotive Laser-Energy Meters", Radiotekh. (Kharkov), 1973, no.25, pp.14-7.

36.244. Khizhnyak, N.A., et al.: "Moment of Ponderomotive Forces Acting on Regular-Shape Bodies in a Rectangular Waveguide", Radiotekh. (Kharkov), 1973, no.25, pp.23-31.

36.245. Belyavtsev, V.B., Malyshenko, L.Ye., and Zhilkov, V.S.: "Determination of Power in Higher-Order Modes from Maximum Amplitudes of Field Components", Radiotekh. (Kharkov), 1973, no.25, pp.18-23.

36.246. Rabinovich, G.I.: "Instrument for Calibrating and Tuning Waveguide Reflectometers", Elektrosvyaz, 1973, 27, no.9, pp.49-52, and Telecommun. Radio Eng. Pt 1, 1973, 27, no.9, pp.40-3.

36.247. Dubovoi, N.D., Osokin, V.I., and Chibrikov, S.I.: "Automatic Microwave Comparison-Type Power Meter", Radiotekh. (Kharkov), 1973, no.27, pp.132-8.

36.248. Osokin, V.I., and Dubovoi, N.D.: "Microwave Power Meter Employing Periodic Comparison", Radiotekh. (Kharkov), 1973, no.27, pp.138-45.

36.249. Ohkawa, S., and Yamamoto, H.: "Multipoint Method for Precise Measurement of Reflection Factor and Scattering Matrix", Electron. Commun. Jap., 1973, 56, no.11, pp.80-7.

36.250. Kubarev, A.V., et al.: "National Standard for Units of Power and Energy of Coherent Light", Izmer. Tekh. 1973, 16, no.8, pp.3-4, and Meas. Tech., 1973, 16, no.8, pp.1113-5.

36.251. Lushchikov, I.I., et al.: "Power Meter for IR Lasers", Opt.-Mekh. Prom., 1973, 40, no.8, pp.27-9, and Sov. J. Opt. Technol., 1973, 40, no.8, pp.489-91.

36.252. Potapov, V.A., and Obidenko, L.A.: "Errors of Reflectometers", Izmer. Tekh., 1973, 16, no.8, pp.65-7, and Meas. Tech., 1973, 16, no.8, pp.1219-23.

36.253. Andresciani, R., De Leo, R., and De Sario, M.: "Determination of Open-Microstrip-Line Dispersion Characteristics by S-Parameter Measurements", Alta Freq., 1974, 43, no.3, pp.173-5.

36.254. Ekkers, J., Bauder, A., and Gunthard, H.H.: "Broadband Directional Couplers for Microwave-Microwave Double Resonance", Rev. Sci. Instrum., 1974, 45, no.2, pp.311-2.

36.255. Puntenney, D.G., et al.: "Microwave Dosimetry Using Electrochemical Effects", J. Microwave Power, 1974, 9, no.1, pp.39-45.

36.256. Hoffman, R.A.: "Apparatus for Measurement of Optical Absorptivity in Laser Mirrors", Appl. Opt., 1974, 13, no.6, pp.1405-11.

36.257. Hollway, D.L., and Somlo, P.I.: "Simple Instrument for Matching Impedances in Waveguide and Coaxial Systems", Trans. IEEE, 1974, MTT-22, no.5, pp.560-1.

36.258. Zernike, F.: "Precise Measurement of Coupling Between Optical Waveguides", Appl. Phys. Lett., 1974, 24, no.6, pp.285-6.

36.259. Batt, R.J., Doswell, A., and Harris, D.J.: "Waveguide Attenuation Measurements at 890 GHz", Electron. Lett., 1974, 10, no.9, pp.145-6.

36.260. Nayar, P.S.: "Far-IR Bolometer", Infrared Phys., 1974, 14, no.1, pp.31-6.

36.261. Mohr, F., and Sepp, G.: "Styrofoam Calorimeter for High-Energy CO_2 Lasers", Opt. Laser. Technol., 1974, 6, no.1, pp.11-2.

36.262. Grianti, F., Ottonello, P., and Troilo, M.: "Laser Torque-Measuring Device", Autom. Strum., 1974, 22, no.3, pp.115-22.

36.263. Gordy, R.S., and Rodrigue, G.P.: "Broadband Impedance Matching a Slot Radiator Using an Improved Computer-Aided Technique", Trans. IEEE, 1974, MTT-22, no.8, pp.799-801.

36.264. Ladbrooke, P.H.: "Novel Standing-Wave Meter in Microstrip", Radio Electron. Eng., 1974, 44, no.5, pp.273-80.

36.265. Vorontsov, A.N., and Zaitsev, V.V.: "Measurement of Microwave Power Using Polarization-Sensitive Wattmeter", Izv. VUZ Radioelektron., 1974, 17, no.5, pp.17-21.

36.266. Kersten, R.Th.: "Measurement of Attenuation of Optical Waves in Dielectric Film Guides", Acta Phys. Austriaca, 1974, 39, no.4, pp.385-92.

36.267. Reichelt, W.H., Stark, E.E., and Stratton, T.F.: "Gas Calorimeter for Use at 10.6 micron", Opt. Commun., 1974, 11, no.3, pp.305-8.

36.268. Somlo, P.I., and Morgan, I.G.: "Side-Arm-Switched Directional Coupler as a Single-Step Attenuator of High Long-Term Stability", Trans. IEEE, 1974, MTT-22, no.9, pp.830-5.

36.269. Soloukhin, R.I., and Yakobi, Yu.A.: "Problem of Measuring Amplification Coefficient (of a Laser)", Zh. Prikl. Mekh. Tekh. Fiz., 1974, no.3, pp.3-12.

36.270. Shindo, S.: "Tapered Waveguide Higher-Mode Measurement by a Resonant Method", Rev. Electr. Commun. Lab., 1974, 22, no.1-2, pp.139-44.

36.271. Keck, D.B.: "Spatial and Temporal Power-Transfer Measurements on a Low-Loss Optical Waveguide", Appl. Opt., 1974, 13, no.8, pp.1882-8.

36.272. Ostermayer, F.W., and Benson, W.W.: "Integrating Sphere for Measuring Scattering Loss in Optical Fibre Waveguides", Appl. Opt., 1974, 13, no.8, pp.1900-2.

36.273. Seip, B.S., and Hinderks, L.W.: "Computerized Klinger-Cavity Mode-Conversion Test Set", Trans. IEEE, 1974, MTT-22, no.10, pp.873-8.

36.274. Luskow, A.A.: "Precision Attenuators for Microwave Frequencies", Marconi Instrum., 1974, 14, no.4, pp.81-6.

36.275. Woods, D.: "Generation of Reflection-Factor Standards of Any Value by a 3-Port Coaxial Junction at Microwave Frequencies", Electron. Lett., 1974, 10, no.18, pp.379-80.

36.276. Keen, N.J.: "Milliwatt Calorimeter for the 90-140 GHz Waveguide Band", Electron. Lett., 1974, 10, no.18, pp.384-5.

36.277. Javed, A., and Goud, P.A.: "Microwave Measurements of Series Resistance of a Waveguide-Mounted Varactor Diode", Electr. Eng. Rev., 1974, 4, no.1, pp.2-6.

36.278. Barbero, J.: "Error in Impedance Measurement When the Signal is Introduced Across the Slotted-Line Probe", Trans. IEEE, 1974, MTT-22, no.10, pp.887-9.

36.279. Leszczynska, I., and Dabrowski, J.: "Measurement of Microwave Reflection Factor", Rozpr. Electrotech., 1974, 20, no.3, pp.489-501.

36.280. Bennett, W.R., and Sze, R.C.: "CW Gain Measurements in Small-Bore Argon-Ion Laser Discharges Using a Novel Modulation Technique", J. Quantum Electron. IEEE, 1974, QE-10, no.12, pp.908-10.

36.281. Russell, T.W.: "Procedure for Intercomparing Laser Power Meters", Electro-Opt. Syst. Des., 1974, 6, no.8, pp.28-33.

36.282. Cohen, R.L., et al.: "Loss Measurements in Optical Fibres. I and II", Appl. Opt., 1974, 13, no.11, pp.2518-21.

36.283. Zaganiaris, A.: "Simultaneous Measurement of Absorption and Scattering Losses in Bulk Glass and Optical Fibres by a Microcalorimetric Method", Appl. Phys. Lett., 1974, 25, no.6, pp.345-7.

36.284. Goryachev, Yu.A., Kalmyk, V.A., and Raevskii, S.B.: "Cutoff Attenuator Using Inductive Coupling Between Electrodes", Izv. VUZ Radioelektron., 1974,17,no.8,pp.42-7.

36.285. Warner, F.L., Herman, P., and Jeffs, T.: "Special Techniques for Measuring Low and High Values of Attenuation With a Modulated Subcarrier System", Trans. IEEE, 1974, IM-23, no.4, pp.381-6.

36.286. Bomer, R.F.: "Computer Controlled Attenuation Measurement System for TE_{01}-Mode Circular Waveguide at 32-110 GHz", Trans. IEEE, 1974, IM-23, no.4, pp.386-9.

36.287. Abbott, N.P., Reeves, C.J., and Orford, G.R.: "Waveguide Flow Calorimeter for Levels of 1-20 W", Trans. IEEE, 1974, IM-23, no.4, pp.414-20.

36.288. Zhilkov, V.S., Belyavtsev, V.B., and Serednii, V.P.: "Possible Application of Ponderomotive Wattmeters in Multimode Rectangular Waveguides. I and II", Radiotekh. (Kharkov), 1974, no.28, pp.139-43 and 151-3.

36.289. Matyshenko, L.E., Didyk, L.S., and Tabachnikov, V.V.: "Creation of a Multimode SHF Wattmeter", Radiotekh. (Kharkov), 1974, no.28, pp.153-9.

36.290. Hashimoto, S., Yamaguchi, M., and Asao, T.: "Microwave Power Measurements Based on Saturation of Resonant Absorption", Bull. Electrotech. Lab., 1974, 38, no.9, pp.477-85.

36.291. Kuz'michev, V.M., Latynin, Yu.M., and Priz, I.A.: "Grating Used for Measuring Laser Pulse Energy", Prib. Tekh. Eksp., 1974, 17, no.2, pp.190-3, and Instrum. Exp. Tech., 1974, 17, no.2, pp.525-8.

36.292. Engen, G.F.: "Calibration Technique for Automated Network Analysers With Application to Adaptor Evaluation", Trans. IEEE, 1974, MTT-22, no.12, pp.1255-60.

36.293. Kalmyk, V.A., and Raevskaya, O.I.: "Effect of Eccentricity of Excitation Probe on Performance of a Cutoff Attenuator", Izv. VUZ Radioelektron., 1974, 17,no.10,pp.58-62.

36.294. Uhlir, A.: "Correction for Adaptors in Microwave Measurements", Trans. IEEE, 1974, MTT-22, no.3, pp.330-2.

36.295. Akhiezer, A.N., et al.: "Standards for Unit of Power in Waveguides in Range 37.5-53.5 GHz", Izmer. Tekh., 1974, 17, no.5, pp.5-7, and Meas. Tech., 1974, 17, no.5, pp.645-8.

36.296. Komarek, E.L., and Tryon, P.V.: "Application of Power-Equation Concept and Automation Techniques to Precision Bolometer Unit Calibration", Trans. IEEE, 1974, MTT-22, no.12, pp.1260-7.

36.297. Richings, J.G.: "Accurate Experimental Method for Determining Properties of Microstrip Lines", Marconi Rev., 1974, 37, no.195, pp.209-16.

36.298. Arkhipov, R.N., et al.: "Instrument for Measuring Energy and Power of Broad Laser Beams", Prib. Tekh. Eksp., 1974, 17, no.4, pp.165-7, and Instrum. Exp. Tech., 1974, 17, no.4, pp.1142-3.

36.299. Govor, I.N., Kubarev, A.V., and Obukhov, A.S.: "Reference Grade Meter of Laser Radiation Power", Izmer. Tekh., 1974, 17, no.7, pp.50-2, and Meas. Tech., 1974, 17, no.7, pp.1059-62.

36.300. Robinson, A.M., and Nohr, M.: "Self-Contained Continuously Variable Attenuator for Pulsed CO_2-Laser Radiation", Rev. Sci. Instrum., 1974, 45, no.12, pp.1605-6.

36.301. Mekhannikov, A.I.: "Waveguide Reflectors With Reflectance Independent of Frequency", Izmer. Tekh., 1974, 17, no.7, pp.43-4, and Meas. Tech., 1974, 17, no.7, pp.1045-7.

36.302. Mekhannikov, A.I., and Perepelkin, V.A.: "Calibration of Waveguide Bolometric Transmitted-Power Transducers", Izmer. Tekh., 1974, 17, no.7, pp.46-7, and Meas. Tech., 1974, 17, no.7, pp.1051-3.

36.303. Sokolev, V.M., and Turyanskii, V.P.: "Resonance Method for Measuring Attenuation in Waveguides", Izmer. Tekh., 1974, 17, no.7, pp.44-5, and Meas. Tech., 1974, 17, no.7, pp.1048-50.

36.304. Sirotnikov, A.I., Zhilkov, V.S., and Khizhnyak, N.A.: "Ponderomotive Action of a Field on a Lossy Dielectric Ellipsoid", Radiotekh. (Kharkov), 1974, no.29, pp.144-6.

37. FREQUENCY STANDARDS AND STABILIZATION

37.1. Davidovits, P., and Stern, W.A.: "Field-Independent Optically Pumped [87]Rb Maser Oscillator", Appl. Phys. Lett., 1965, 6, p.20.

37.2. Stiglitz, M.R., and Sethares, J.C.: "Frequency Stability in Dielectric Resonators", Proc. IEEE, 1965, 53, p.311.

37.3. Rzepecka, M.: "Microwave Oscillator Frequency Stabilization by Pound's Method", Rozpr. Elektrotech., 1965, 11, p.197.

37.4. Parshad, R., et al.: "Parallel Operation of Reflex Klystrons at Locked Frequency", J. Inst. Telecommun. Eng., 1965, 11, p.402.

37.5. Davidovits, P., and Novick, R.: "Stability Considerations for a ^{87}Rb Maser", Int. Conv. Rec. IEEE, 1965, 13, pt 5, p.2.

37.6. Kester, T.: "Automatic Frequency Control for an X-Band Reflex Klystron", J. Sci. Instrum., 1965, 42, p.442.

37.7. Krupnov, A.F., and Skvortsov, V.A.: "Excitation Parameter of a Beam Maser", Izv. VUZ Radiofiz., 1965, 8, p.200, and Sov. Radiophys., 1965, 8, p.361.

37.8. Andreev, V.N., Armenskii, E.V., and Rybin, V.M.: "System of Frequency Stabilization for a Magnetron", Prib. Tekh. Eksp., 1965, no.6, p.200, and Instrum. Exp. Tech., 1965, no.6, p.1519.

37.9. Beehler, R.E., and Glaze, D.J.: "Evaluation of a Thallium Atomic Beam Frequency Standard", Trans. IEEE, 1966, IM-15, p.55.

37.10. Hakki, B.W., Beccone, J.P., and Plauski, S.E.: "Phase-Locked GaAs CW Microwave Oscillators", Trans. IEEE, 1966, ED-13, p.197.

37.11. Laine, D.C., and Smith, A.L.S.: "High-Resolution ^{14}NH$_3$ Beam Maser With Two Cascaded Cavities and Low Beam Flux", Phys. Lett., 1966, 20, p.374.

37.12. Hellwig, H.: "Investigation of Frequency Accuracy of the NH$_3$ Maser on the 3.2 Line", Z. Angew. Phys., 1966,21,p.250.

37.13. Grygorenko, V.G., et al.: "Microwave Interferometers With Amplitude Modulation", Ukr. Fiz. Zh., 1966, 11, p.491.

37.14. Smith, A.L.S., and Laine, D.C.: "Fluctuations in Amplitude of an NH$_3$ Beam Maser", Electron. Lett., 1967,3,no.2,pp.90-1.

37.15. Fabre, F., and Pescia, J.: "Frequency Stabilization of Power Klystron", C. R. Acad. Sci. (Paris), 1967, 264B, no.9, pp.680-3.

37.16. Hrvoic, I.: "Klystron Frequency Stabilization to a Resonant Cavity", Elektrotehnika, 1967, 10, no.3, pp.173-7.

37.17. Zhabotinskii, M.Ye.: "Quantum Frequency Standards", Radiotekh. Elektron., 1967, 12, no.11, pp.2023-31, and Radio Eng. Electron. Phys., 1967, 12, no.11.

37.18. Biquard, F., Crivet, P., and Septier, A.: "Frequency Stability of a Monotron Oscillator With Superconducting Cavity", Electron. Lett., 1968, 4, no.8, pp.143-4.

37.19. Shamfarov, Ya.L.: "Simple Method of Frequency Stabilization for a Microwave Oscillator by an External Cavity", Radiotekh. Elektron., 1968, 13, no.1, pp.89-94, and Radio Eng. Electron. Phys., 1968, 13, no.1.

37.20. Smith, A.L.S., and Laine, D.C.: "Further Characteristics of a Beam Maser with Two Cavities in Series", J. Phys. C, 1968, 1, no.6, pp.727-32.

37.21. Roy, A.K., and Roy, R.: "Stabilization of the Frequency of a Klystron", J. Inst. Telecommun. Eng., 1968, 14, p.319.

37.22. Hartmann, F.: "Frequency Stability of the Regenerative Rb Maser", Phys. Lett., 1968, 28A, p.193.

37.23. Hellwig, H., and Pannaci, E.: "Maser Oscillations With External Gain", J. Appl. Phys., 1968, 39, p.5496.

37.24. Ivanov, I.V.: "Automatic Temperature Stabilization in Ferroelectric Microwave Cavities", Radiotekh. Elektron., 1968, 13, p.1291, and Radio Eng. Electron. Phys., 1968, 13, no.7, pp.1120-3.

37.25. Sachkov, V.I.: "Application of U-Shaped Resonators in Cs Frequency Standards", Izmer. Tekh., 1968, no.7, p.99, and Meas. Tech., 1968, no.7, p.997.

37.26. Bazarov, Ye.N., et al.: "Short-Term Frequency Stability of a Maser Using ^{87}Rb Pairs", Radiotekh. Elektron., 1968, 13, no.11, p.2097, and Radio Eng. Electron. Phys., 1968, 13, no.11, pp.1842-3.

37.27. Bakeev, A.A., and Cherubkin, N.V.: "Effect of Frequency Drift on Operation of a Three-Mirror Interferometer", Radiotekh. Elektron., 1968, 13, no.10, p.1894, and Radio Eng. Electron. Phys., 1968, 13, no.10, pp.1662-5.

37.28. Soskin, M.S., Sal'kova, E.N., and Pogoretskii, P.P.: "Special Features of Generation Control in a Spherical Resonator of a Ruby Laser", Ukr. Fiz. Zh., 1968, 13, p.1567.

37.29. Leikin, A.Ya., and Solov'ev, V.S.: "Stabilization of Laser Frequency", Izmer. Tekh., 1968, no.8, p.19, and Meas. Tech., 1968, no.8, p.1025.

37.30. Bagaev, S.N., et al.: "Stabilization and Reproducibility of a He-Ne Laser Frequency at 633 nm", Izmer. Tekh., 1968, no.8, p.27, and Meas. Tech., 1968,no.8,p.1037.

37.31. Guisset, J.L., de Prins, J., and Detrie, R.: "Klystron Phase Stabilizing System for the 27-37 GHz Band. I", K. Dan. Vidensk. Selsk. Mat.-Fys. Skr., 1968, 54, no.10, p.1314.

37.32. Jannitti, E., and Tondello, G.: "Observations on Behaviour of a Laser Interferometer", Alta Freq., 1968, 37, p.565.

37.33. Martinot-Lagarde, P., and Coumes, C.: "Argon-Ion Laser in Single Stabilized Mode", Nachr. Tech. Fachber., 1968,35,p.695.

37.34. Korneev, N.E., and Pavlov, Yu.I.: "High-Power Single-Mode Ruby Laser", Zh. Eksp. Teor. Fiz. Pis'ma, 1968, 8, p.309, and JETP Lett., 1968, 8, p.190.

37.35. Koster, A., and Chartier, G.: "Operation of a Solid-State Laser With Non-parallel Mirrors", C. R. Acad. Sci. (Paris), 1968, 267B, p.1020.

37.36. Leschiutta, S.: "Experiments With a Commercial Cs Resonator", Alta Freq., 1968, 37, no.10, p.916.

37.37. Petrun'kin, V.Yu., et al.: "Longitudinal Mode Selection in a He-Ne Laser With a Four-Mirror T-Shaped Resonator", Zh. Tekh. Fiz., 1968, 38, no.11, p.1983, and Sov. Phys.-Tech. Phys., 1969, 13, no.11, p.1591.

37.38. Korshunov, I.P.: "AFC System for Locking Beat Frequency Between Two Lasers", Prib. Tekh. Eksp., 1969, 12, no.3, p.180, and Instrum. Exp. Tech., 1969,12,no.3,p.734.

37.39. Hercher, M.: "Tunable Single-Mode Operation of Gas Lasers Using Intracavity Tilted Etalons", Appl. Opt., 1969,8,p.1103.

37.40. Petrun'kin, V.Yu., Vysotskii, M.G., and Okunev, R.I.: "Selection of Longitudinal Modes in a Gas Laser With Annular Resonators", Zh. Tekh. Fiz., 1969, 39, no.5, p.928, and Sov. Phys.-Tech. Phys., 1969, 14, no.5, pp.694-5.

37.41. Ledneva, G.P., and Chekalinskaya, Yu.I.: "Frequency Calculation of Axial Modes and Optimal Parameters of Three- and Four-Mirror Cavities for Gas Lasers", Zh. Prikl. Spekt., 1969, 11, no.1, p.35.

37.42. Zaitsev, Yu.I., and Khurtin, L.A.: "Frequency Stabilization of He-Ne Laser by Modulation of Intensity", Prib. Tekh. Eksp., 1969, 12, no.3, p.177, and Instrum. Exp. Tech., 1969, 12, no.3, p.731.

37.43. Shul'ga, V.F.: "Analysis of Microwave Components in Pound's Circuit", Izv. VUZ Radioelektron., 1969, 12, p.1438.

37.44. Harris, S.E.: "Method to Lock an Optical Parametric Oscillator to an Atomic Transition", Appl. Phys. Lett., 1969, 14, p.335.

37.45. Smirnov, A.G., and Terent'ev, V.E.: "Kinetics of a High-Power Ruby-Laser Pulse Formed by an Ultrasonic Wave", Opt. Spekt., 1969, 27, p.81, and Opt. Spectrosc., 1969, 27, no.1, p.163.

37.46. Terent'ev, V.E.: "Control of Nd^{3+}:Glass Laser by Ultrasonic Diffraction Modulator", Opt. Spekt., 1969, 27, p.705, and Opt. Spectrosc., 1969, 27, no.4, p.384.

37.47. Adrianova, I.I., et al.: "Single Pulses from a Ruby Laser by Ultrasonic-Modulated Diffraction in a Phototropic Medium", Opt. Spekt., 1969, 27, p.968, and Opt. Spectrosc., 1969, 27, no.6, p.526.

37.48. Brady, M.M.: "Simple Temperature Compensation of Large Rectangular-Waveguide Resonators", Electron. Eng., 1969, 41, p.198.

37.49. Udelson, J.J., and Hines, R.E.: "Effects of CW Injected Signal on Pulsed Avalanche Oscillator", Proc. IEEE, 1969, 57, p.2091.

37.50. Arnold, D.H., and Hanna, D.C.: "Transverse-Mode Selection of Rotating-Mirror Q-Switched Lasers", Electron. Lett., 1969, 5, p.354.

37.51. Forsyth, J.M.: "Stabilization of Transverse-Mode Spectra in CW Ion Lasers", J. Appl. Phys., 1969, 40, p.3049.

37.52. Magyar, G.: "Mode-Selection Techniques for Solid-State Lasers", Opt. Technol., 1969, 1, no.5, p.231.

37.53. Kononchuk, G.L.: "Laser Frequency Selector", Zh. Prikl. Spekt., 1969, 11, no.4, p.735.

37.54. Golant, M.B., et al.: "Klystron 3-cm Oscillator Stabilized by a Superconducting Resonator", Prib. Tekh. Eksp., 1969, 12, no.3, p.232, and Instrum. Exp. Tech., 1969, 12, no.3, p.802.

37.55. Gerard, V.B.: "Frequency-Controlled Laser for Long-Path Interferometry", J. Phys. E, 1969, 2, p.749.

37.56. Letokhov, V.S., and Chebotaev, V.P.: "Optical Frequency Standard With Nonlinearly Absorbing Gas Cell", Zh. Eksp. Teor. Fiz. Pis'ma, 1969, 9, no.6, p.364, and JETP Lett., 1969, 9, no.6, p.215.

37.57. Gordeev, D.V., et al.: "Feasibility of Using a Resonator With Aperture in the Mirror of Argon-Ion Lasers", Radiotekh. Elektron., 1969, 14, no.9, p.1637, and Radio Eng. Electron. Phys., 1969, 14, no.9, pp.1420-2.

37.58. Danilov, A.N.: "Synchronization of Tunnel-Diode Microwave Oscillators", Radiotekhnika, 1969, 24, no.1, p.62, and Telecommun. Radio Eng. Pt 2, 1969, 24, no.1.

37.59. Bodlaj, V.: "Frequency Stabilization of He-Ne Lasers With an External Ne Absorption Cell in an Alternating Magnetic Field", Frequenz, 1969, 23, p.92.

37.60. Stone, J.L., Hartwig, W.H., and Baker, G.L.: "Automatic Tuning of a Superconducting Cavity Using Optical Feedback", J. Appl. Phys., 1969, 40, p.2015.

37.61. Mukhamedgalieva, A.F., Tatarenkov, V.M., and Titov, A.N.: "Investigation of He-Ne Laser With an Absorbing Cell", Izv. VUZ Radiofiz., 1969, 12, no.8, p.1156.

37.62. Tsetsegova, E.I.: "He-Ne Laser Frequency Stabilization at 3.39 micron", Izv. VUZ Radiofiz., 1969, 12, no.8, p.1159.

37.63. Wenk, G.J.: "Frequency Stabilization of Gas Lasers", Proc. IREE Aust., 1969, 30, no.8, p.250.

37.64. Fuller, D.W.E., Hines, J., and Compton, B.: "Short-Term Frequency Stability of HCN Maser", Electron. Lett., 1969, 5, p.448.

37.65. Bazarov, Ye.N.: "Approximate Theory of Optically Pumped Rb Maser", Radiotekh. Elektron., 1969, 14, no.6, p.1035, and Radio Eng. Electron. Phys., 1969, 14, no.6, pp.896-902.

37.66. Bazarov, Ye.N., and Gubin, V.P.: "Light Frequency Shift in a Rb Maser Using Modulated Optical Pumping", Radiotekh. Elektron., 1969, 14, no.6, p.1050, and Radio Eng. Electron. Phys., 1969, 14, no.6, pp.908-12.

37.67. Bazarov, Ye.N., and Grigor'ev, V.I.: "Frequency Shift of F_{2-1} Transition of Rb Atoms Using Pulsed Optical Pumping", Radiotekh. Elektron., 1969, 14, no.6, p.1056, and Radio Eng. Electron. Phys., 1969, 14, no.6, pp.912-8.

37.68. Bodlaj, V.: "Analysis of Frequency Stabilization of a Gas Laser With External Absorption Cell", Frequenz, 1969, 23, no.12, p.374.

37.69. Beterov, I.M., Klement'ev, V.M., and Chebotaev, V.P.: "Secondary Frequency Standard Using Mercury Laser at Microwaves", Radiotekh. Elektron., 1969, 14, no.11, p.2066, and Radio Eng. Electron. Phys., 1969, 14, no.11, pp.1790-3.

37.70. Alekseev, E.I., Bazarov, Ye.N., and Levshin, A.E.: "Frequency Shifts Due to Luminous Pumping in a Passive Frequency Standard Using ^{87}Rb", Radiotekh. Elektron., 1969, 14, no.11, p.2026, and Radio Eng. Electron. Phys., 1969, 14, no.11, pp.1750-7.

37.71. Bjorkholm, J.E., and Danielmeyer, H.G.: "Frequency Control of a Pulsed Optical Parametric Oscillator by Radiation Injection", Appl. Phys. Lett., 1969, 15, p.171.

37.72. Troitskii, Yu.V.: "Optical Resonator With Thin Absorbing Film as a Mode Selector", Opt. Spekt., 1969, 25, p.557, Opt. Spectrosc., 1969, 25, p.309, Radiotekh. Elektron., 1969, 14, no.9, p.1641, and Radio Eng. Electron. Phys., 1969, 14, no.9, pp.1423-7.

37.73. Sugawara, K.: "Wideband 4-GHz, Esaki-Diode, Injection-Locked Oscillator", Proc. IEEE, 1969, 57, p.215.

37.74. Malota, F.: "Energy Measurements in a Ruby Laser With Moving Crystal", Z. Naturforsch., 1969, 24a, p.1285.

37.75. Kinsel, T.S., et al.: "Stabilized Mode-Locked Nd^{3+}:YAG Laser", J. Quantum Electron. IEEE, 1969, QE-5, p.326.

37.76. Biverot, H.: "Modified Smith-Type Reflector for Long Laser Cavities", Opt. Commun., 1969, 1, no.4, pp.179.

37.77. Kovalenko, Ye.S., et al.: "High-Frequency Modulation of the Structure of the Space Field in a Ruby Laser", Izv. VUZ Fiz., 1969, no.1, p.81.

37.78. Lisitsyn, V.H., and Chebotaev, V.P.: "Zeeman Effect for Stabilization of Frequency of a Gas Laser With Nonlinear Absorption", Opt. Spekt., 1969, 26, no.5, p.856, and Opt. Spectrosc., 1969, 26, no.5, p.467.

37.79. Stickler, J.J.: "Injection Locking of Gunn Oscillators With Feedback Stabilization", Proc. IEEE, 1969, 57, p.1772.

37.80. Kamenets, F.F., and Feoktistov, A.A.: "Synchronization of Longitudinal and Transversal Modes in a Laser With a Passive Filter", Zh. Prikl. Spekt., 1969, 11, p.444.

37.81. Belyaev, V.P., et al.: "High-Power, Single-Frequency, Argon-Ion Laser", J. Quantum Electron. IEEE, 1969, QE-5, p.589.

37.82. Donin, V.I., Troitskii, Yu.V., and Goldina, N.D.: "Single-Mode Generation and Lamb Dip in an Argon-Ion Laser", Opt. Spekt., 1969, 26, no.1, p.118, and Opt. Spectrosc., 1969, 26, no.1.

37.83. Eliseev, P.G., et al.: "Injection Semiconductor Laser With Compound Resonator", Zh. Eksp. Teor. Fiz. Pis'ma, 1969, 9, no.10, p.594, and JETP Lett., 1969, 9, no.10, p.362.

37.84. Phillip-Rutz, E.M., and Edmonds, H.D.: "Diffraction-Limited GaAs Laser with External Resonator", Appl. Opt., 1969, 8, p.1859.

37.85. Korneev, N.E., Pavlov, Yu.I., and Folomeev, A.V.: "Power of a High-Coherence Ruby Laser With Diffractive Dispersion", Zh. Tekh. Fiz., 1969, 39, p.2250, and Sov. Phys.-Tech. Phys., 1970, 14, no.12, pp.1699-700.

37.86. Bykovskii, Yu.A., et al.: "Stabilization of Frequency of a GaAs Injection Laser by an External Fabry-Perot Resonator", Zh. Eksp. Teor. Fiz., 1969, 57, no.4, p.1109, and Sov. Phys.-JETP, 1970, 30, no.4, pp.605-6.

37.87. Belousova, I.M., et al.: "Selection of Longitudinal Modes in a Gas Laser With a Three-Mirror Resonator", Zh. Tekh. Fiz., 1969, 39, p.2223, and Sov. Phys.-Tech. Phys., 1970, 14, no.12, pp.1676-7.

37.88. Mungall, A.G., and Daams, H.: "Cs Frequency Standard Cavity Design", Metrologia, 1970, 6, no.2, p.60.

37.89. Danielmeyer, H.G.: "Stabilized Efficient Single-Frequency Nd^{3+}:YAG Laser", J. Quantum Electron. IEEE, 1970, QE-6, p.101.

37.90. Vanier, J., et al.: "85Rb Maser Oscillator", J. Appl. Phys., 1970,41,p.3188.

37.91. Samson, A.M., and Ry'bakov, V.A.: "Selection of Filters for Obtaining Sustained Oscillations of Laser Intensity", Zh. Prikl. Spekt., 1970, 12, no.4, p.641.

37.92. Laine, D.C., and Bardo, W.S.: "Anomalous Saturation Effect in an NH_3 Beam Maser", J. Phys. B, 1970, 3, p.L23.

37.93. Bykovskii, Yu.A., et al.: "Use of Fabry-Perot Resonator in Stabilization of the Frequency of an Injection Laser", Fiz. Tekh. Poluprov., 1970, 4, no.4, p.685, and Sov. Phys.-Semicond., 1970, 4, no.4, pp.580-3.

37.94. Le Floch, A., Frere, P., and Brun, P.: "Frequency Stabilization of a Gas Laser Using the Magnetic Lamb Dip", Appl. Phys. Lett., 1970, 17, p.40.

37.95. Skolnick, M.L.: "Use of Plasma-Tube Impedance Variations to Frequency Stabilize a CO_2 Laser", J. Quantum Electron. IEEE, 1970, QE-6, p.139.

37.96. Kooi, P.S., and Walsh, D.: "Novel Technique for Improving Frequency Stability of Gunn Oscillators", Electron. Lett., 1970, 6, p.85.

37.97. Stern, W.A., and Novick, R.: "Field-Independent Optically Pumped 85Rb Maser Oscillator", Appl. Phys. Lett., 1970, 17, p.216.

37.98. Birnbaum, M., Wendzikowski, P.H., and Fincher, C.L.: "CW Nonspiking Single-Mode Ruby Lasers", Appl. Phys. Lett., 1970, 16, p.436.

37.99. Goldobin, I.S.: "Microwave Synchronization of Pulsed Emission of a Semiconductor Layer", Fiz. Tekh. Poluprov., 1970, 4, no.6, p.1201, and Sov. Phys.-Semicond., 1970, 4, no.6, p.1021.

37.100. Kobayashi, T., and Matsuo, Y.: "Single-Frequency Oscillation Using Two Coupled Cavities Incorporating a Fabry-Perot Electrooptic Modulator", Appl. Phys. Lett., 1970, 16, p.217.

37.101. Tsvirko, Yu.A., and Ivanchenko, I.A.: "Mode Tuning and Hysteresis Behaviour of Gunn-Effect Cavity-Controlled Generator", Electron. Lett., 1970, 6, p.9.

37.102. Poizner, B.N.: "Synchronization of a Single-Frequency Gas Laser by a Small Harmonic Signal", Izv. VUZ Fiz., 1970, no.7, p.158.

37.103. Mamedov, K.Ya.: "Frequency Stability of Magnetrons", Izv. VUZ Radioelektron., 1970, 13, no.8, pp.1019-22.

37.104. Milovskii, N.D.: "Stability of Regenerative TW Laser Amplifier", Izv. VUZ Radiofiz., 1970, 13, p.258.

37.105. Malota, F.: "Regular Relaxation Oscillations of a Ruby Laser With Movable Reflectors", Z. Naturforsch., 1970, 25a, no.6, p.916.

37.106. Kornienko, L.S., et al.: "Single-Mode Ruby Ring Laser", Zh. Eksp. Teor. Fiz., 1970, 58, no.2, p.541, and Sov. Phys.-JETP, 1970, 31, no.2, pp.290-1.

37.107. Kupka, M., and Kupka, J.: "Hurwitz Criteria Optimalization of Reflectances of Mirrors in a High-Selective Resonator", Optik, 1970, 31, p.15.

37.108. Kornienko, L.S., et al.: "Ruby Laser with Optical Delay Line Inside the Resonator", Zh. Eksp. Teor. Fiz. Pis'ma, 1970, 11, no.12, p.585, and JETP Lett., 1970, 11, no.12, p.404.

37.109. Hanna, D.C.: "Increasing Laser Brightness by Transverse Mode Selection. I and II", Opt. Laser Technol., 1970, 2, no.3, p.122, and no.4, pp.175-8.

37.110. Rigrod, W.W.: "Selectivity of Open-Ended Interferometric Resonators", J. Quantum Electron. IEEE, 1970, QE-6, p.9.

37.111. Troitskii, Yu.V.: "Comparison of Methods of Resonator Mode Selection", Zh. Prikl. Spekt., 1970, 12, no.3, p.425.

37.112. Tako, T., et al.: "Frequency Stabilization of 3.39-micron Laser on CH_4 Line", Jap. J. Appl. Phys., 1970, 9, no.12, p.1535.

37.113. Basov, N.G., Danileiko, M.V., and Nikitin, V.V.: "Investigation of the CH_4 Line for Stabilization of He-Ne Lasers at 3.39 micron", Zh. Eksp. Teor. Fiz. Pis'ma, 1970, 12, no.2, pp.95-7, and JETP Lett., 1970, 12, no.2, pp.66-8.

37.114. Voitovich, A.P., and Smirnov, A.Ya.: "Nonlinear Effects in He-Ne Lasers With Absorption Cell in a Magnetic Field", Zh. Prikl. Spekt., 1970, 13, no.1, pp.33-9.

37.115. Hayes, R.E., et al.: "Rb Atomic Frequency Source Having Small Size and Fast Warmup", Electron. Lett., 1970, 6, no.23, pp.734-5.

37.116. Nagano, S.: "Stabilization of Microwave Solid-State Oscillators by a Reaction Cavity", Electron. Commun. Jap., 1970, 53, no.3, p.23.

37.117. Krupnov, A.F., and Gershtein, L.I.: "Frequency Stabilization of Submillimetre Generators by Quartz Crystal", Prib. Tekh. Eksp., 1970, no.1, pp.159-60, and Instrum. Exp. Tech., 1970, no.1, pp.180-1.

37.118. Kuznetsov, V.M.: "Gas-Laser Automatic-Frequency-Stabilization System", Prib. Tekh. Eksp., 1970, no.1, pp.189-91, and Instrum. Exp. Tech., 1970, no.1, pp.218-20.

37.119. Milovskii, N.D.: "Stability of a Single-Frequency TW Laser", Phys. Lett., 1970, 33A, no.8, pp.492-3.

37.120. Ito, Y., Komizo, H., and Sasagawa, S.: "Cavity-Stabilized X-Band Gunn Oscillator", Trans. IEEE, 1970, MTT-18, no.11, pp.890-7.

37.121. Kukushkin, A.V., and Nasonov, V.S.: "Phase Fluctuations of a TWT Generator With Superconducting Resonator", Cryogenics, 1970, 10, no.6, pp.503-4.

37.122. Bonch-Bruevich, A.M., et al.: "Passive Q-Switching and Stabilization of Ruby-Laser Frequency With Molecular Rubidium Vapour", Zh. Eksp. Teor. Fiz. Pis'ma, 1970, 12, no.7, pp.354-6, and JETP Lett., 1970, 12, no.7, pp.242-3.

37.123. Bodlaj, V.: "Frequency Stabilization of He-Ne Laser With External Ne-Absorption Tube in an Alternating Magnetic Field", Opto-Electron., 1970, 2, no.4, pp.221-6.

37.124. Bakhert, Kh.Yu., et al.: "Spectral Investigation of Injection-Laser Vibration Synchronization", Zh. Prikl. Spekt., 1970, 13, no.2, pp.232-7.

37.125. Eichler, H., and Weisemann, W.: "Mode-Quenching in a He-Ne Laser by a Coupled Optical Resonator", Z. Angew. Phys., 1970, 30, no.4, pp.283-5.

37.126. Miura, S., and Furuhama, Y.: "Heterodyne Experiment With Stable CO_2 Laser", Rev. Radio Res. Lab., 1970, 16, no.85, pp.365-9.

37.127. Koshelyaevskii, N.B., et al.: "Stabilization of Frequency of a He-Ne Laser", Izmer. Tekh., 1970, no.8, pp.38-40, and Meas. Tech., 1970, no.8, pp.1170-3.

37.128. Ginodman, V.V., et al.: "Klystron Oscillator Stabilized by a Superconducting Resonator", Prib. Tekh. Eksp., 1970, no.4, pp.154-5, and Instrum. Exp. Tech., 1970, no.4, pp.1136-7.

37.129. Saprykin, E.G., and Yudin, R.N.: "Frequency Stabilization of a Two-Mode Laser", Zh. Prikl. Spekt., 1970, 13, no.6, pp.1072-3.

37.130. Zory, P.: "Degenerate Vernier Interferometric Laser Resonators", Appl. Phys. Lett., 1970, 17, p.77.

37.131. Galaktionova, N.M., et al.: "Selection of a Single Longitudinal Mode in Solid-State Lasers", Opt. Spekt., 1970, 28, no.4, pp.751-8, and Opt. Spectrosc., 1970, 28, no.4, pp.404-8.

37.132. Kimura, T., and Otsuka, K.: "Response of a CW Nd^{3+}:YAG Laser to Sinusoidal Cavity Perturbations", J. Quantum Electron. IEEE, 1970, QE-6, no.12, pp.764-9.

37.133. Zakharov, M.I., and Troitskii, Yu.V.: "Design of an Optical Cavity With an Absorbing Film", Radiotekh. Elektron., 1970, 15, no.12, and Radio Eng. Electron. Phys., 1970, 15, no.12, pp.2344-6.

37.134. Malinin, Yu.N., and Pol'skii, Yu.E.: "Locking of Gas Lasers in the FM Mode of Operation", Radiotekh. Elektron., 1970, 15, no.12, and Radio Eng. Electron. Phys., 1970, 15, no.12, pp.2284-8.

37.135. Bakumenko, V.M., and Valitov, R.A.: "Frequency Stabilization and Instability Measurement of He-Ne Lasers at 633 nm", Radiotekh. Elektron., 1970, 15, no.12, and Radio Eng. Electron. Phys., 1970, 15, no.12, pp.2289-93.

37.136. Moskienko, M.V., et al.: "Frequency Stabilization of Wideband Submillimetre BW Tube", Prib. Tekh. Eksp., 1970, no.5, pp.131-2, and Instrum. Exp. Tech., 1970, no.5, pp.1397-8.

37.137. Strakhovskii, G.M., et al.: "Frequency Stabilization of 633-nm He-Ne Laser With Internal Absorptive Cell", Izmer. Tekh., 1970, no.12, pp.25-8, and Meas. Tech., 1970, no.12, pp.1839-43.

37.138. Iga, K., and Fukuyo, H.: "Optical-Pumping and -Detection Mechanism in Rb Atomic Oscillator", Electron. Commun. Jap., 1970, 53, no.1, pp.54-9.

37.139. Bazarov, Ye.N., and Gubin, V.P.: "Short-Term Frequency Instability of Optically Pumped Rb Maser", Radiotekh. Elektron., 1970, 15, no.10, pp.1860-7, and Radio Eng. Electron. Phys., 1970, 15, no.10, pp.1860-7.

37.140. Ivanov, I.V., and Semenova, T.G.: "Effect of Automatic Thermal Stabilization of Resonance in Ferroelectric Microwave Cavities", Radiotekh. Elektron., 1970, 15, no.12, and Radio Eng. Electron. Phys., 1970, 15, no.12, pp.2330-2.

37.141. Kutin, V.N., and Troshin, B.I.: "Laser (Smith-Fox) Ring Interferometer With Special Selective Characteristics", Opt. Spekt., 1970, 29, no.2, pp.371-3, and Opt. Spectrosc., 1970, 29, no.2, pp.197-8.

37.142. Belyaev, V.P., et al.: "Single-Frequency (Smith-Fox) Argon-Ion Laser With High Output", Zh. Prikl. Spekt., 1970, 13, no.2, pp.223-6.

37.143. Kravchenko, V.I., and Tarabrov, V.V.: "Adjustable Single-Frequency TW Laser With Nd^{3+}:Glass", Zh. Prikl. Spekt., 1970, 13, no.4, pp.719-21.

37.144. Yoshino, T.: "Single-Frequency Output from an Internal-Mirror He-Ne Laser Utilizing the Polarization Properties", Appl. Opt., 1971, 10, no.1, p.221.

37.145. Terekin, D.K., Andreeva, E.Yu., and Fridrikhov, S.A.: "Single-Frequency He-Ne Laser in a Magnetic Field", Zh. Prikl. Spekt., 1971, 14, no.1, pp.53-9.

37.146. Tschiedel, W., et al.: "Interference Arrangement Using Multimode Lasers", Z. Angew. Phys., 1971, 31, no.1, pp.15-21.

37.147. Boyne, H.S.: "Laser Frequency-Stabilization Techniques and Applications", Trans. IEEE, 1971, IM-20, no.1, pp.19-22.

37.148. Michaelides, M., and Stephenson, I.M.: "Injection Locking of Microwave Solid-State Oscillators", Proc. IEEE, 1971, 59, no.2, pp.319-21.

37.149. Vyse, B., and Gissing, J.M.: "Spectra of Short, Locked, Magnetron Pulses", Trans. IEEE, 1971, ED-18, no.3, pp.221-3.

37.150. Gerhardt, H., Bodecker, V., and Welling, H.: "Frequency-Stabilized Nd^{3+}:YAG Laser", Z. Angew. Phys., 1971, 31, no.1, pp.11-5.

37.151. Jimenez, J.J., Sudraud, P., and Septier, A.: "Frequency Stabilization of a Reflex-Klystron Oscillator by a Superconducting Cavity", Electron. Lett., 1971, 7, no.7, pp.153-4.

37.152. Goud, P.A.: "Cavity Frequency Stabilization with (Temperature Stabilized) Compound Tuning", Microwave J., 1971, 14, no.3, pp.55-6.

37.153. Bruce, C.F.: "Stable Single-Frequency He-Ne Laser", Appl. Opt., 1971, 10, no.4, pp.880-3.

37.154. Goldberg, M.W., and Yusek, R.: "Doppler Jitter Stabilization of CO_2 Laser", Appl. Phys. Lett., 1971, 18, no.4, pp.135-7.

37.155. Maischberger, K.: "Long-Term Frequency Stabilization of Composite-Cavity Argon Laser", J. Quantum Electron. IEEE, 1971, QE-7, no.6, pp.250-2.

37.156. Bodlaj, V.: "Frequency Stabilization of He-Ne Laser With External Absorption Cell in an Alternating Magnetic Field", Z. Angew. Phys., 1971, 31, no.2, pp.97-105, and Phys. Lett., 1971, 35A, no.5, pp.381-2.

37.157. Culshaw, W., and Kannelaud, J.: "Two-Component-Mode Filters for Optimum Single-Frequency Operation of Nd^{3+}:YAG Lasers", J. Quantum Electron. IEEE, 1971, QE-7, no.8, pp.381-7.

37.158. Bradley, C.C., and Knight, D.J.E.: "Frequency Locking HCN Laser to a Molecular Absorption Line", Electron. Lett., 1971, 7, no.13, pp.381-2.

37.159. Sasnett, M.W., and Reynolds, R.S.: "Single-Wavelength CO_2 Laser Without Dispersive Elements", J. Quantum Electron. IEEE, 1971, QE-7, no.7, pp.372-3.

37.160. Leikin, A.Ya., Solov'ev, V.S., and Moskienko, N.V.: "Stabilization of Laser Frequency by Extremum Storage", Kvantovaya Elektron., 1971, no.4, pp.95-7.

37.161. Gorlanov, A.V., Lyubimov, V.V., and Petrov, V.F.: "Quasi-CW Nd^{3+} Laser (With Mercury Mirror)", Kvantovaya Elektron., 1971, no.4, pp.116-7.

37.162. Vasil'eva, N.N., and Ovechkin, A.P.: "Methods of Adjusting Resonators With Spherical Mirrors", Izmer. Tekh., 1971, 14, no.5, pp.28-9, and Meas. Tech., 1971, 14, no.5, pp.704-6.

37.163. Galutva, G.V., and Ryazantsev, A.I.: "Gas-Laser Resonator Stabilized by Thermal Compensation", Kvantovaya Elektron., 1971, no.2, pp.32-9.

37.164. Zaks, V.S.: "Frequency Shifts in Optically Pumped Standards", Radiotekh. Elektron., 1971, 16, no.8, pp.1515-8, and Radio Eng. Electron. Phys., 1971, 16, no.8, pp.1403-5.

37.165. Morozov, V.A., et al.: "Automatic Frequency Control of Two He-Ne Lasers and Fluctuations of Difference Frequency", Prib. Tekh. Eksp., 1971, 14, no.5, pp.192-4, and Instrum. Exp. Tech., 1971,14,no.5,pp.1489-91.

37.166. Escudero, J.L., Guerra, J.M., and Sancho, J.: "Microwave Maser Between Paschen-Back Levels of Ground-State Atomic Sodium", Opt. Pura Apl., 1971, 4, no.3, pp.122-4.

37.167. Novikov, M.A.: "Use of Polarizing Interferometer for Laser Frequency Selection", Radiotekh. Elektron., 1971, 16, no.10, pp.1992-3, and Radio Eng. Electron. Phys., 1971, 16, no.10, pp.1781-3.

37.168. Maloney, P.J., and Smith, P.W.: "Measurements of Mode-Competition Discriminant in a Single-Frequency Argon-Ion Ring Laser", J. Quantum Electron. IEEE, 1971, QE-8, no.9, pp.744-9.

37.169. Bardeche, G., Hauden, D., and Uebersfeld, J.: "Polarization and Radioelectric Emission in a Two-Cavity NH_3 Maser", C. R. Acad. Sci. (Paris), 1971, 272B, no.19, pp.1147-50.

37.170. Knight, D.J.E.: "Phase-Locking Millimetre-Wave Klystrons for Measurement of Submillimetre-Laser Frequencies", Electron. Lett., 1971, 7, no.13, pp.383-4.

37.171. Laine, D.C., and Smart, G.D.S.: "Ammonia Beam Masers Employing Parallel-Plate and Conical Types of Open Resonator", J. Phys. D, 1971, 4, no.8, pp.L23-5.

37.172. Bokhonov, A.F., et al.: "Regular Pulsations in Ruby Lasers by Use of Filters", Zh. Prikl. Spekt., 1971, 14, no.6, pp.994-9.

37.173. Letokhov, V.S.: "Problems of Laser Frequency Stabilization by Nonlinear Saturation Absorption Resonance in Gases", Comments At. Mol. Phys., 1971, 2, no.6, pp.181-94.

37.174. Hercher, M., and Pike, H.A.: "Single-Mode Operation of a CW Tunable Dye Laser", Opt. Commun., 1971, 3,no.5,pp.346-8.

37.175. Kohiyama, K., and Momma, K.: "Type of Frequency-Stabilized Gunn-Oscillator", Proc. IEEE, 1971, 59, no.10, pp.1532-3.

37.176. Ikenoue, J., and Nakajima, M.: "Synchronization Phenomena of Microwave Solid-State Oscillators", J. Inst. Electron. Commun. Eng. Jap., 1971, 54, no.5,pp.693-9.

37.177. Ivanov, I.V., Karyagin, S.N., and Semenov, T.G.: "Effect of Temperature Stabilization in a Microwave Resonator With a Ferroelectric Nonlinear Element", Vestn. Mosk. Univ. Fiz. Astron., 1971, no.3, pp.298-304.

37.178. Papadopoulos, G.D.: "Phase Stability of Two Independently Locked Local Oscillators Near 22 GHz", Proc. IEEE, 1971, 59, no.11, pp.1620-2.

37.179. Troitskii, Yu.V., and Khyuppenen, V.P.: "Resetting and Stabilization of Frequency of Optical Single-Mode Lasers With Selective Frequency Dumping", Avtometriya, 1971, no.1, pp.52-6.

37.180. Zakharov, M.I., and Troitskii, Yu.V.: "Frequency Properties of Optical Resonator With Attenuating Films", Avtometriya, 1971, no.1, pp.111-3.

37.181. Mikhal'tsova, I.A., et al.: "Single-Frequency Stabilized He-Ne Laser", Avtometriya, 1971, no.1, pp.10-5.

37.182. Lokhmatov, A.I., and Khanov, C.A.: "Frequency Stabilizing System for Gas Laser Using Lamb Dip", Avtometriya, 1971, no.1, pp.16-20.

37.183. Troitskii, Yu.V.: "Analysis of Single-Frequency Gas Laser Using Q-Factor Mode Selection", Avtometriya, 1971, no.1, pp.102-9.

37.184. Atutov, S.N., Saprykin, E.G., and Yudin, R.N.: "Influence of Magnetic Field of Attenuation Cell on Power of Single-Frequency He-Ne Laser", Avtometriya, 1971, no.1, pp.114-5.

37.185. Belousov, P.Ya., and Lokhmatov, A.I.: "Stabilized He-Cd Laser for Measuring Instruments", Avtometriya, 1971, no.1, pp.120-1.

37.186. Blaney, T.G., et al.: "Phase Locking Klystrons to Far-IR Lasers via Harmonic Mixing", Phys. Lett., 1971, 36A, no.4, pp.285-6.

37.187. Siegman, A.E.: "Stabilizing Output With Unstable Resonators", Laser Focus, 1971, 7, no.5, pp.42-7.

37.188. Akerman, D., et al.: "Methods for Mode Selection in Injection Lasers", Kvantovaya Elektron., 1971, no.1, pp.85-90.

37.189. Locke, E.V., Hella, R., and Westra, L.: "Performance of an Unstable Oscillator on a 30-kW CW Gasdynamic Laser", J. Quantum Electron. IEEE, 1971, QE-7, no.12, pp.581-3.

37.190. Moss, G.E.: "High-Power Single-Mode He-Ne Laser", Appl. Opt., 1971, 10, no.11, pp.2565-6.

37.191. Letokhov, V.S., and Pavlik, B.D.: "Gas Laser With Nonlinear Absorption Under Quasi-TW Conditions", Kvantovaya Elektron., 1971, no.1, pp.53-63.

37.192. Bardo, W.S., and Laine, D.C.: "Oscillation Transient in a Molecular Q-Switched NH₃ Maser", J. Phys. D, 1971, 4, no.11, pp.L42-4.

37.193. Laine, D.C., and Sweeting, R.C.: "Operation of Molecular-Beam Masers With Electret Focusers", J. Phys. D, 1971, 4, no.11, pp.L44-6.

37.194. Krawczyk, M.: "Circuit for Automatic Frequency Control of Microwave Oscillators With Microwave Discriminators", Rozpr. Electrotech., 1971, 17, no.3, pp.541-61.

37.195. Gubin, M.A., Popov, A.I., and Protsenko, E.D.: "Contrast Enhancement of Resonances of a He-Ne Laser With an Absorption Cell", Kvantovaya Elektron., 1971, no.3, pp.99-102.

37.196. Borisova, M.S., and Yasinskii, V.M.: "Interaction of Modes in a TW Argon Laser", Opt. Spekt., 1971, 31, no.3, pp.433-5, and Opt. Spectrosc., 1971, 31, no.3, pp.231-2.

37.197. Kitaeva, V.F., et al.: "Spectral Characteristics of a Single-Frequency Argon Laser With an Absorbing Film", Kvantovaya Elektron., 1971, 1, no.6, pp.91-4.

37.198. Burmakin, V.A., et al.: "Method for Absolute Frequency Stabilization of an Argon-Ion Laser With a Magnetic Field", Radiotekh. Elektron., 1971, 16, no.7, pp.1292-6, and Radio Eng. Electron. Phys., 1971, 16, no.7, pp.1228-31.

37.199. Lamalle, B., et al.: "Frequency Stabilization of a Gunn Oscillator by Phase-Locked Crystal Control", Electron. Microelectron. Ind., 1971, no.148, pp.59-62.

37.200. Mikaelyan, A.L., et al.: "Single-Mode Ruby Laser With a Ring Resonator", Kvantovaya Elektron., 1971, no.1, pp.136-9.

37.201. Stepanov, D.P.: "Dispersion Characteristics at 633 nm of a He-Ne Laser in a Magnetic Field", Izv. VUZ Radiofiz., 1971, 14, no.9, pp.1348-52.

37.202. Hochuli, U.E., and Haldemann, P.: "Relative Frequency Stability of Stable He-Ne Gas-Laser Structures", J. Quantum Electron. IEEE, 1971, QE-7, no.12, pp.573-5.

37.203. Kornienko, L.S., et al.: "Ring Solid-State Laser With Selector in the Feedback Loop", Vestn. Mosk. Univ. Fiz. Astron., 1971, no.4, pp.486-8.

37.204. Basov, N.G., Belenov, E.M., et al.: "Power Resonances and Frequency Stabilization of a Gas Laser With a Nonlinear Absorption Cell", Kvantovaya Elektron., 1971, no.1, pp.42-52.

37.205. Weiss, J.A., and Goldberg, L.S.: "Single-Longitudinal-Mode Operation of a TE CO_2 Laser", J. Quantum Electron. IEEE, 1971, QE-8, no.9, pp.757-8.

37.206. Troitskii, Yu.V.: "Continuous Variation of Spectral Linewidth in a Gas Laser", Opt. Spekt., 1971, 31, no.6, pp.1031-2, and Opt. Spectrosc., 1971, 31, no.6, p.558.

37.207. Fukuyo, H., et al.: "Pure-Ne Laser and Fundamental Characteristics", Bull. Tokyo Inst. Technol., 1971, no.107, pp.67-73.

37.208. Tapkov, A.N., Dyubko, S.F., and Schmidt, V.V.: "Monochromaticity of a Sub-millimetre Laser at 337 micron", Radiotekh. Elektron., 1971, 16, no.1, and Radio Eng. Electron. Phys., 1971, 16, no.1, pp.181-2.

37.209. Gershenzon, Ye.M., and Levites, A.A.: "Oscillations and Frequency Stabilization in a Multicircuit Microwave Oscillator With Delayed Feedback", Radiotekh. Elektron., 1971, 16, no.2, pp.331-9, and Radio Eng. Electron. Phys., 1971, 16, no.2, pp.285-91.

37.210. Beterov, I.M., Lisitsyn, V.N., and Chebotaev, V.P.: "Saturation and Mode Selection in He-Ne Lasers. I", Opt. Spekt., 1971, 30, no.5, pp.932-9, and Opt. Spectrosc., 1971, 30, no.5, pp.497-501.

37.211. Voitovich, A.P., et al.: "Frequency Selection in a He-Ne Laser With an Absorbing Neon Cell Inside the Cavity", Opt. Spekt., 1971, 30, no.5, pp.940-6, and Opt. Spectrosc., 1971, 30, no.5, pp.501-4.

37.212. Feldmann, J.: "Temperature Stabilization of a Cavity Resonator by a Gas-Pressure-Controlled Membrane", Nachr. Tech. Z., 1971, 24, no.11, pp.580-4.

37.213. Barchukov, A.I., Konev, Yu.B., and Prokhorov, A.M.: "Use of Quasioptical Lines to Create High-Power Single-Mode CO_2 Lasers", Dokl. Akad. Nauk SSSR, 1971, 197, no.5, pp.74-5, and Sov. Phys.-Dokl., 1971, 16, no.5, pp.388-9.

37.214. McElroy, J.H., Schiffner, G., and Reynolds, R.S.: "Temperature-Dependent Etalon Effects in Laser Systems", Appl. Opt., 1971, 10, no.9, pp.2065-9.

37.215. Bashkin, A.S., et al.: "Stabilization of the Oscillation Frequency of a Gas Laser by Comparison With a Radio-Frequency Standard", Kvantovaya Elektron., 1971, no.2, pp.40-8.

37.216. Troitskii, Yu.V.: "Effect of Method of Introducing Loss on Competition of Two Modes in a Gas Laser", Opt. Spekt., 1971, 31, no.1, pp.158-60, and Opt. Spectrosc., 1971, 31, no.1, pp.82-3.

37.217. Komizo, H., Sasagawa, S., and Oya, T.: "Cavity Stabilized X-Band Gunn Oscillator", Fujitsu Sci. Tech. J., 1971, 7, no.4, pp.91-108.

37.218. Matyugin, Yu.A., Troshin, B.I., and Chebotaev, V.P.: "Method for Stabilizing He-Ne Laser Frequency Using the Lorentzian Absorption Profile in an External Gas Cell", Opt. Spekt., 1971, 31, no.1, pp.111-5, and Opt. Spectrosc., 1971, 31, no.1, pp.56-8.

37.219. Troitskii, Yu.V.: "Strong Coupling Between Modes and Frequency Pulling in He-Ne Laser", Izv. VUZ Radiofiz., 1971, 14, no.12, pp.1795-800.

37.220. Spencer, M.B., and Lamb, W.E.: "Laser (Locking) With a Transmitting Window", Phys. Rev., 1972, 5A, no.2, pp.884-92.

37.221. Spencer, M.B., and Lamb, W.E.: "Theory of Two Coupled Lasers", Phys. Rev., 1972, 5A, no.2, pp.893-8.

37.222. Corcoran, V.J., and Smith, W.T.: "Laser (Stabilization) Millimetre-Wave Techniques", Appl. Opt., 1972, 11, no.2, pp.269-72.

37.223. Rowley, W.R.C., and Wilson, D.C.: "Optical Coupling Effects in Frequency-Stabilized Lasers", Appl. Opt., 1972, 11, no.2, pp.475-6.

37.224. Clarke, J.: "Simple Stabilized Microwave Source", Trans. IEEE, 1972, IM-21, no.1, pp.83-4.

37.225. Benard, J., et al.: "Frequency-Stability Improvement in a Klystron Stabilized by a Superconducting Cavity", Electron. Lett., 1972, 8, no.5, pp.117-8.

37.226. Smith, P.W.: "Mode Selection in Lasers", Proc. IEEE, 1972, 60, no.4, pp.422-40.

37.227. Gunderson, M., Lloyd, H.B., and Poarch, B.W.: "Mirror Mount for Long-Wave Lasers", Rev. Sci. Instrum., 1972, 43, no.2, pp.333-4.

37.228. Hohimer, J.P., Kelly, R.C., and Tittel, F.K.: "Frequency Stabilization of a High-Power Argon-Ion Laser", Appl. Opt., 1972, 11, no.3, pp.626-9.

37.229. Greenstein, H.: "Theory of a Gas Laser With Internal Absorption Cell", J. Appl. Phys., 1972, 43, no.4, pp.1732-50.

37.230. Wang, C.C.: "Frequency Locking of Laser Oscillators by Injected Signal", J. Appl. Phys., 1972, 43, no.1, pp.158-9.

37.231. Tanaka, K.: "Method for Stabilizing Frequency of an Unmodulated Laser Output", J. Opt. Soc. Am., 1972, 62, no.1, pp.24-9.

37.232. Reilly, J.P.: "Single-Mode Operation of High-Power Pulsed N_2-CO_2 Laser", J. Quantum Electron. IEEE, 1972, QE-8, no.2, pp.136-9.

37.233. Bykovskii, Yu.A., et al.: "Coherence of a Pulsed Single-Mode Injection Laser", Dokl. Akad. Nauk SSSR, 1972, 203, no.5, pp.1027-9, and Sov. Phys.-Dokl., 1972, 17, no.4, pp.359-61.

37.234. Pinard, J., and Young, J.F.: "Interferometric Stabilization of Optical Parametric Oscillator", Opt. Commun., 1972, 4, no.5, pp.425-7.

37.235. Bodlaj, V.: "Neon Absorption Tube in an Alternating Magnetic Field Inside Laser Cavity as Frequency Standard", Opt. Commun., 1972, 6, no.1, pp.12-4.

37.236. Ryan, T.J., et al.: "Molecular-(I_2)-Beam-Stabilized Argon Laser", Appl. Phys. Lett., 1972, 21, no.7, pp.320-2.

37.237. Eremin, V.I., Kolosov, V.A., and Norinskii, L.V.: "Powerful Single-Pulse, Single-Mode, Nd^{3+} Laser With Frequency Stabilization", Prib. Tekh. Eksp., 1972, 15, no.1, pp.170-1, and Instrum. Exp. Tech., 1972, 15, no.1, pp.196-7.

37.238. Antipov, B.A., Pyrsikova, P.D., and Sapozhnikova, V.A.: "Tunable Single-Frequency Laser for 3.39 micron", Opt. Spekt., 1972, 33, no.5, pp.954-7, and Opt. Spectrosc., 1972, 33, no.5, pp.522-4.

37.239. Revenko, V.I., and Timofeev, V.B.: "Tunable Dye, Single-Frequency Laser With High Monochromaticity and Stability", Prib. Tekh. Eksp., 1972, 15, no.6, pp.168-9, and Instrum. Exp. Tech., 1972, 15, no.6, pp.1809-10.

37.240. Nagano, S., and Ohnaka, S.: "Highly Stabilized K-Band Gunn Oscillator", Trans. IEEE, 1972, MTT-20, no.2, pp.174-6.

37.241. Ammann, E.O., and Yarborough, J.M.: "Mode-Selection Technique for Continuously Pumped Repetitively-Q-Switched Lasers", Appl. Phys. Lett., 1972, 20, no.3, pp.117-20.

37.242. Hellwig, H., et al.: "Stability of Methane-Stabilized He-Ne Lasers", J. Appl. Phys., 1972, 43, no.2, pp.450-2.

37.243. Frerking, M.E.: "Ruggedized Rb Frequency Standard", Proc. IEEE, 1972, 60, no.5, pp.628-9.

37.244. Buczek, C.J., and Freiberg, R.J.: "Hybrid Injection Locking of Higher-Power CO_2 Lasers", J. Quantum Electron. IEEE, 1972, QE-8, no.7, pp.641-50.

37.245. Llewellyn-Jones, D.T., and James, M.D.: "Stabilization of HCN Laser", J. Phys. E, 1972, 5, no.5, pp.468-72.

37.246. Balochin, Y., et al.: "Monochromatic TEA CO_2 Laser", C. R. Acad. Sci. (Paris), 1972, 274B, no.24, pp.1322-5.

37.247. Bennett, W.R.: "Role of Hole Burning in Self-Stabilization of Gas-Laser Frequencies", Comments At. Mol. Phys., 1972, 3, no.3, pp.63-8.

37.248. Stein, S.R., and Turneaure, J.P.: "Superconducting-Cavity-Stabilized Oscillator", Electron. Lett., 1972,8,no.13,pp.321-3.

37.249. Iga, K., and Fukuyo, H.: "Frequency Stabilization of Pure Neon Laser", Bull. Tokyo Inst. Technol., 1972, no.112, pp.43-7.

37.250. Wallard, A.J.: "Frequency Stabilization of He-Ne Laser by Saturated Absorption in Iodine Vapour", J. Phys. E, 1972, 5, no.9, pp.926-30.

37.251. Basov, N.G., Gubin, M.A., et al.: "Frequency Stabilization of a Gas Laser Using Mode-Interaction Effects", Zh. Eksp. Teor. Fiz. Pis'ma, 1972, 15, no.9, pp.525-8, and JETP Lett., 1972, 15, no.9, pp.371-3.

37.252. Basov, N.G., Belenov, E.M., et al.: "Stabilization of Ring-Laser Frequency", Zh. Eksp. Teor. Fiz. Pis'ma, 1972, 15, no.12, pp.466-8.

37.253. Lokhmatov, A.I., et al.: "Determination of Wavelength of a He-Cd Laser from Centre of Lamb Dip", Opt. Spekt., 1972, 32, no.1, pp.223-5, and Opt. Spectrosc., 1972, 32, no.1, pp.116-7.

37.254. Troitskii, Yu.V.: "Single-Frequency Operation of the He-Ne Laser", Zh. Tekh. Fiz., 1972, 42, no.2, pp.395-7, and Sov. Phys.-Tech. Phys., 1972, 17, no.2, pp.314-6.

37.255. Ryan, T.J., et al.: "High-Resolution Laser-Excited Spectrum of I_2 Molecular Beam", Q. Progr. Rep. Res. Lab. Electron. MIT, 1972, no.104, pp.140-2.

37.256. Cirkel, H.J., and Schafer, F.P.: "Passive Nonreciprocal Element for TW (Ruby) Ring Laser", Opt. Commun., 1972, 5, no.3, pp.183-6.

37.257. Ohtsuka, Y.: "Mode-Selective Output-Coupling Aperture in IR Gas Laser", Mem. Fac. Eng. Hokkaido Univ., 1972, 13, no.2, pp.133-45.

37.258. Dutu, C.A.: "Frequency Stabilization by Phase Control of a Single-Frequency/Single-Mode CO_2 Laser", Stud. Cercet. Fiz., 1972, 24, no.5, pp.535-44.

37.259. Clobes, A.R., and Brienza, M.J.: "Single-Frequency TW Nd^{3+}:YAG Laser", Appl. Phys. Lett., 1972, 21, no.6, pp.265-7.

37.260. Bagaev, S.N., et al.: "Investigation of the Frequency Stability of Gas Lasers at Various Wavelengths", Opt. Spekt., 1972, 32, no.4, pp.802-8, and Opt. Spectrosc., 1972, 32, no.4, pp.422-5.

37.261. Bua, D.P., Fradin, D.W., and Bass, M.: "Simple Technique for Longitudinal Mode Selection", J. Quantum Electron. IEEE, 1972, QE-8, no.12, pp.916-7.

37.262. Malyshev, G.F.: "Stabilization of a Single-Frequency He-Ne Laser", Avtometriya, 1972, no.5, pp.86-93.

37.263. Kostin, N.N., Khodovoi, V.A., and Chigir, N.A.: "Passive Q-Switching and Frequency Stabilization of a Nd^{3+}:Glass Laser Using Molecular Cs Vapour", Opt. Spekt., 1972, 32, no.3, pp.585-8, and Opt. Spectrosc., 1972, 32, no.3, pp.310-1.

37.264. Chester, A.N.: "Mode Selectivity and Mirror Misalignment Effects in Unstable Laser Resonators", Appl. Opt., 1972, 11, no.11, pp.2584-90.

37.265. Bazarov, Ye.N., Biketov, V.D., and Gubin, V.P.: "Short-Term Instability of an Optically Pumped Rb Maser", Radiotekh. Elektron., 1972, 17, no.4, pp.887-8, and Radio Eng. Electron. Phys., 1972, 17, no.4, pp.697-9.

37.266. Leikin, A.Ya., Solov'ev, V.S., and Moskienko, N.V.: "Problems in the Calculation of Frequency Stabilization of Lasers Using the Lamb Dip", Radiotekh. (Kharkov), 1972, no.22, pp.104-11.

37.267. Beterov, I.M., et al.: "High-Stability Gas Laser Based on Nonlinear Absorption. I-IV", Avtometriya, 1972, no.5, pp.59-70 and 71-85, and no.6, pp.55-63 and pp.64-8.

37.268. Kinsel, T.S.: "Stabilized Mode-Locked Nd^{3+}:YAG Laser Using Electronic Feedback", J. Quantum Electron. IEEE, 1972, QE-9, no.1, pp.3-8.

37.269. Yakobi, Yu.A.: "Phase Relations in a Michelson Laser Interferometer", Radiotekh. Elektron., 1972, 17, no.4, pp.787-93, and Radio Eng. Electron. Phys., 1972, 17, no.4, pp.615-9.

37.270. Bhawalkar, D.D., and Nair, L.G.: "Proposed Method of Spike Suppression in Solid-State Lasers", Opto-Electron., 1972, 4, no.3, pp.225-34.

37.271. Zhupan, Yu.Yu., Zaika, V.V., and Kravchenko, V.I.: "Generation Spectra of Ruby Laser With Frequency Scanning", Ukr. Fiz. Zh., 1972, 17, no.11, pp.1803-8.

37.272. Hanna, D.C., Luther-Davies, B., and Smith, R.C.: "Single Longitudinal Mode Selection of High-Power Actively Q-Switched Lasers", Opto-Electron., 1972, 4, no.3, pp.249-56.

37.273. Bakumenko, V.M., and Valitov, R.A.: "Swept Frequency Lasers", Radiotekh. Elektron., 1972, 17, no.1, pp.90-3, and Radio Eng. Electron. Phys., 1972, 17, no.1, pp.72-4.

37.274. Vasilenko, L.S., et al.: "Frequency Stabilization of a CO_2 Laser", Opt. Spekt., 1972, 32, no.6, pp.1123-9, and Opt. Spectrosc., 1972, 32, no.6, pp.609-12.

37.275. Antsiferov, V.V., et al.: "Spikeless Oscillation of a Ruby Laser With Frequency Selection and Tuning", Opt. Spekt., 1972, 32, no.6, pp.1159-62, and Opt. Spectrosc., 1972, 32, no.6, pp.628-30.

37.276. Anan'ev, Yu.A., et al.: "Control of Radiation of Lasers With Telescopic Resonators by Signal Injection", Kvantovaya Elektron., 1972, no.2, pp.85-8, and Sov. J. Quantum Electron., 1972, 2, no.2, pp.157-9.

37.277. Lemaire, J., Houriez, J., and Lapauw, J.M.: "Frequency Stabilization of CO_2 or N_2O Laser by Rapid Sampling Method", Rev. Phys. Appl., 1972, 7, no.4, pp.323-8.

37.278. Le Floch, A., and Frere, P.: "Frequency and Intensity Stabilization of High-Power Monomode Laser by Magnetic Lamb Dip", Rev. Phys. Appl., 1972, 7, no.4, pp.409-12.

37.279. Avtonomov, V.P., et al.: "Separation of Rotational Lines of a CO_2 Laser With Film Selector in Resonator", Kvantovaya Elektron., 1972, no.3, pp.112-5, and Sov. J. Quantum Electron., 1972, 2, no.3, pp.300-2.

37.280. Bogdanov, A.A., et al.: "Frequency Stability of a Q-Switched Ruby Laser", Opt. Spekt., 1972, 33, no.2, pp.352-3, and Opt. Spectrosc., 1972, 33, no.2, pp.186-7.

37.281. Bikmukhametov, K.A., Klement'ev, V.M., and Chebotaev, V.P.: "Investigation of Stability of Oscillation Frequency of a Hg Laser at 1530 nm", Kvantovaya Elektron., 1972, pp.74-6, and Sov. J. Quantum Electron., 1972, 2, no.3, pp.254-6.

37.282. Gurevich, G.L., Ingel', L.Kh., and Khanin, Ya.I.: "Influence of Nonlinear Lens on Stability of Steady-State Laser Emission", Kvantovaya Elektron., 1972, no.3, pp.45-52, and Sov. J. Quantum Electron., 1972, 2, no.3, pp.230-5.

37.283. Koshelyaevskii, N.B., Tatarenkov, V.M., and Titov, A.N.: "Frequency Meters Based on He-Ne Lasers", Izmer. Tekh. 1972, 15, no.11, pp.32-5, and Meas. Tech., 1972, 15, no.11, pp.1644-8.

37.284. Abashev, Yu.G., Voronin, G.F., and Valitov, R.A.: "Velocity Distribution in an Atomic Beam", Izmer. Tekh. 1972, 15, no.11, pp.53-5, and Meas. Tech., 1972, 15, no.11, pp.1679-81.

37.285. Kuzovkova, T.A., Nilov, E.V., and Chertkov, A.A.: "Obtaining Quasistationary Lasing With Ruby or Nd^{3+}:Glass", Prib. Tekh. Eksp., 1972, 15, no.5, pp.191-3, and Instrum. Exp. Tech., 1973, 15, no.5, pp.1515-7.

37.286. Letokhov, V.S., and Pavlik, B.D.: "Frequency Fluctuations in a Gas Laser With Nonlinear Absorption", Kvantovaya Elektron., 1972, no.4, pp.32-9, and Sov. J. Quantum Electron., 1973, 2, no.4, pp.324-8.

37.287. Marugin, A.M., and Ovchinnikov, V.M.: "Mode Selection in a Resonator With Electrooptic and Piezooptic Control", Kvantovaya Elektron., 1972, no.4, pp.104-5, and Sov. J. Quantum Electron., 1973, 2, no.4, pp.378-9.

37.288. Hanes, G.R., Baird, K.M., and DeRemigis, J.: "Stability, Reproducibility, and Absolute Wavelength of a 633-nm He-Ne Laser Stabilized to an Iodine Hyperfine Component", Appl. Opt., 1973, 12, no.7, pp.1600-5.

37.289. Spieweck, F.: "$^{127}I_2$ Absorption by Three Kr^{II}-Laser Lines", Metrologia, 1973, 9, no.1, pp.24-5.

37.290. Knight, D.J.E., and Brown, E.C.: "Frequency Stability of Millimetre-Band Reflex Klystrons With Various Cooling Techniques", Electron. Lett., 1973, 9, no.7, pp.163-5.

37.291. Tohma, K.: "Analysis of Frequency Locking of Organic Lasers by Faraday Filter", J. Radio Res. Lab., 1973, 19, no.101, pp.267-321, and Electron. Commun. Jap., 1973, 56, no.2, pp.124-31.

37.292. Milovskii, N.D.: "Single-Frequency Laser Stability", Izv. VUZ Radiofiz., 1973, 16, no.4, pp.537-44.

37.293. Spieweck, F.: "Single-Mode Operation of Argon-Ion Laser With Intracavity I_2 Absorption Cell", Appl. Phys., 1973, 1, no.4, pp.233-4.

37.294. Marowsky, G.: "Reliable Single-Mode (Michelson-Type) Operation of a Flash-Pumped Dye Laser", Rev. Sci. Instrum., 1973, 44, no.7, pp.890-2.

37.295. Busca, G., Tetu, M., and Vanier, J.: "Light Shift and Broadening in ^{87}Rb Maser", Can. J. Phys., 1973, 51, no.13, pp.1379-87.

37.296. Tetu, M., Busca, G., and Vanier, J.: "Short-Term Frequency Stability of ^{87}Rb Maser", Trans. IEEE, 1973, IM-22, no.3, pp.250-7.

37.297. Junghans, J., and Stemme, R.: "Electrooptic Modulator to Influence Relaxation Spikes in Solid-State Lasers", Z. Angew. Math. Phys., 1973, 24, no.3, p.451.

37.298. Annabi, M., and Gillet, D.: "Adiabatic Passage on Electric Dipole Transitions of a Molecular Beam in an Inhomogeneous Magnetic Field", C. R. Acad. Sci. (Paris), 1973, 277B, no.18, pp.515-7.

37.299. Lefrere, P.R., and Laine, D.C.: "Beat Mode Phase Shifts in a Molecular Beam Zeeman Maser Operated With Cavities in Series", Phys. Lett., 1973, 45A, no.5, pp.405-6.

37.300. Hill, K.O.: "Lowest-Order Mode Selection in a Laser Interferometer", Can. J. Phys., 1973, 51, no.6, pp.624-8.

37.301. Smith, P.W., and Maloney, P.J.: "Self-Stabilized 3.5-micron Waveguide He-Xe Laser", Appl. Phys. Lett., 1973, 22, no.12, pp.667-9.

37.302. Wellegehausen, B., and Guttner, A.: "Relative Frequency Stability Measurements of Single-Mode Lasers", Z. Naturforsch., 1973, 28a, no.6, pp.968-72.

37.303. Mumola, P.B.: "Dye-Laser (Etalon) Tuning With Pellicles", J. Appl. Phys., 1973, 44, no.7, pp.3198-9.

37.304. Shimoda, K.: "Ultimate Stability of Methane-Stabilized Lasers", Jap. J. Appl. Phys., 1973, 12, no.8, pp.1222-6.

37.305. Bagaev, S.N., and Dmitriev, A.K.: "Use of Sharp Resonances in Methane for He-Ne Laser Frequency Stabilization at 3.39 micron", Opt. Spekt., 1973, 34, no.2, pp.337-42, and Opt. Spectrosc., 1973, 34, no.2, pp.191-3.

37.306. Batarchukova, N.R., Glosman, Ts.I., and Kartashev, A.I.: "Absorption of He-Ne Laser Radiation by an Iodine Molecular Beam", Opt. Spekt., 1973, 34, no.2, pp.413-4, and Opt. Spectrosc., 1973, 34, no.2, pp.233-4.

37.307. Shank, C.V., and Klein, M.B.: "Frequency Locking of a CW Dye Laser Near Atomic Absorption Lines in a Gas Discharge", Appl. Phys. Lett., 1973, 23, no.3, pp.156-7.

37.308. Wallard, A.J.: "Frequency Stabilization of Gas Lasers", J. Phys. E, 1973, 6, no.9, pp.793-807.

37.309. Tschirnich, J.: "Measurement of the Wavelength Stability of Lasers", Feingeraete Tech., 1973, 22, no.7, pp.299-301.

37.310. Bourdet, G., Orszag, A., and de Valence, Y.: "Utilization of Impedance Variation of the Plasma of a CO_2 Laser for Frequency Stabilization on the Lamb Dip", C. R. Acad. Sci. (Paris), 1973, 277B, no.9, pp.207-9.

37.311. Schiek, B.: "Stabilization Factor of a Cavity-Controlled Microwave Oscillator With Several Output Ports", Arch. Elektron. Ubertrag., 1973, 37, no.11, pp.490-1.

37.312. Jimenez, J.J., et al.: "Frequency Standards Using Superconducting Cavities", Electron. Fis. Apl., 1973, 16, no.2, pp.79-84.

37.313. Galutin, V.Z., et al.: "Frequency Stability of a Gas Laser Operating in a Two-Mode Domain", Izv. VUZ Radioelektron., 1973, 16, no.9, pp.90-3.

37.314. Salamon, T., et al.: "Single-Mode Operation in a Hollow-Cathode Transverse Discharge He-Cd Ion Laser", Phys. Lett., 1973, 46A, no.1, pp.17-8.

37.315. Vysotskii, V.G., et al.: "Longitudinal Mode Selection and Frequency Stabilization in the He-Ne Ring Laser", Zh. Tekh. Fiz., 1973, 43, no.4, pp.881-3, and Sov. Phys.-Tech. Phys., 1973, no.4, pp.560-1.

37.316. Hyatt, R.C., Mueller, L.F., and Osterdock, T.N.: "High-Performance Beam Tube for Cs Frequency Standards", Hewlett-Packard J., 1973, 25, no.1, pp.14-23.

37.317. Schweitzer, W.G., et al.: "Description, Performance, and Wavelengths of Iodine-Stabilized Lasers", Appl. Opt., 1973, 12, no.12, pp.2927-38.

37.318. Matsumura, K., Mushiake, Y., and Nakajima, N.: "Generation of Nonunidirectionally Polarized Modes by a Gas Laser With Mode Selector", Electron. Commun. Jap., 1973, 56, no.1, pp.85-90.

37.319. Wiesemann, W.: "Longitudinal Mode Selection in Lasers With Three-Mirror Reflectors", Appl. Opt., 1973, 12, no.12, pp.2909-12.

37.320. Zakharov, M.I., and Troitskii, Yu.V.: "Design of an Optical Resonator With a Thin-Film Anisotropic Mode Selector", Radiotekh. Elektron., 1973, 18, no.2, pp.394-8, and Radio Eng. Electron. Phys., 1973, 18, no.2, pp.280-5.

37.321. Gnatovskii, A.V., Danileiko, M.V., and Shpak, M.T.: "Ring-Type Optical Frequency Standard", Kvantovaya Elektron., 1973, no.2, pp.122-3, and Sov. J. Quantum Electron., 1973, 3, no.2, pp.179-80.

37.322. Goldevskii, A.P., Lopasov, V.P., and Makogon, M.M.: "Frequency Scanning and Stabilization of Parameters of a Ruby Laser", Kvantovaya Elektron., 1973, no.2, pp.681-71, and Sov. J. Quantum Electron., 1973, 3, no.2, pp.130-1.

37.323. Leikin, A.Ya., Solov'ev, V.S., and Fisher, A.M.: "Stabilizing the Frequency of a He-Ne Laser at 633 nm With Respect to Generation at 3.39 micron", Izmer. Tekh., 1973, 16, no.9, pp.29-30, and Meas. Tech., 1973, 16, no.9, pp.1321-2.

37.324. Ivanov, V.A., et al.: "Method for Frequency Stabilization of a CO_2 Laser", Kvantovaya Elektron., 1973, no.3, pp.133-4, and Sov. J. Quantum Electron., 1973, 3, no.3, pp.272-3.

37.325. Spieweck, F.: "Wavelength Stabilization of an Ar^{II} Laser with an External $^{129}I_2$ Absorption Cell", Appl. Phys., 1974, 3, no.5, pp.429-30.

37.326. Voitovich, A.P., and Smirnov, A.Ya.: "Autostabilization of Intermode Pulse Frequencies in a Gas Laser", Zh. Prikl. Spekt., 1974, 20, no.3, pp.510-2.

37.327. Popov, L.N., and Pivovarov, B.L.: "Effects of Frequency Modulation on Multimode Lasers", Izv. VUZ Fiz., 1974, no.5, pp.114-5.

37.328. Baird, K.M., and Hanes, G.R.: "Stabilization of Wavelengths from Gas Lasers", Rep. Progr. Phys., 1974, 37, no.7, pp.927-50.

37.329. Marowsky, G.: "Single-Mode Dye Ring Laser With Output Coupler Using Frustrated Total Internal Reflection", Z. Naturforsch., 1974, 29a, no.3, pp.536-8.

37.330. Mikhnov, S.A., Matyushkov, V.E., and Andreichev, V.A.: "Increased Working Stability of Giant-Pulse Lasers With Aid of a Telescopic System", Zh. Prikl. Spekt., 1974, 20, no.1, pp.150-2.

37.331. Atutov, S.N., et al.: "Stabilization of a He-Ne Laser With Internal Mirrors in a Variable Magnetic Field", Avtometriya, 1974, no.1, pp.83-8.

37.332. Brillet, A., Cerez, P., and Clergeot, H.: "Frequency Stabilization of He-Ne Lasers by Saturated Absorption", J. Quantum Electron. IEEE, 1974, QE-10, no.6, pp.526-8.

37.333. Troitskii, Yu.V.: "Calculation of Heat Dissipation in a Film-Type Mode Selector in a Single-Frequency Laser", Avtometriya, 1974, no.1, pp.75-80.

37.334. Bespalova, M.P., et al.: "Frequency Characteristics of ^{87}Rb Pair Maser", Izv. VUZ Radiofiz., 1974, 17, no.5, pp.767-70.

37.335. Kompanets, O.N., et al.: "Stabilization of the Emission Frequency of a CO_2 Laser by an External Nonlinearly Absorbing SF_6 Cell", Kvantovaya Elektron., 1974, no.4, pp.28-34, and Sov. J. Quantum Electron., 1974, 3, no.4, pp.293-6.

37.336. Gnatovskii, A.V., et al.: "Reproducibility of Frequency of a Laser Stabilized by an Absorbing Gas", Zh. Eksp. Teor. Fiz. Pis'ma, 1974, 19, no.6, pp.368-71, and JETP Lett., 1974, 19, no.6.

37.337. Blair, D.P.: "Frequency Offset of a Stabilized Laser Due to Modulation Distortion", Appl. Phys. Lett., 1974, 25, no.1, pp.71-3.

37.338. Helmcke, J., and Bayer-Helms, F.: "Stabilizing $^3He-^{22}Ne$ Lasers by Saturated Absorption in $^{129}I_2$", Metrologia, 1974, 10, no.2, pp.69-71.

37.339. Marowsky, G., and Tittel, F.K.: "Single-Mode Operation of a Tunable CW Dye Laser", Appl. Phys., 1974, 5, no.2, pp.181-2.

37.340. Luybimov, V.V., et al.: "Laser Spectrum Selection With Telescopic Resonators", Opt. Spektr., 1974, 36, no.4, pp.806-8, and Opt. Spectrosc., 1974, 36, no.4, pp.469-70.

37.341. Ivanov, V.A., et al.: "Confocal Interferometer With Extra-Axial Beam Input", Izmer. Tekh., 1974, 17, no.1, pp.65-6, and Meas. Tech., 1974, 17, no.1, pp.107-9.

37.342. Bogdanovich, B.Yu., Lyashenko, O.N., and Milovanov, O.S.: "Synchronization of Three Microwave Self-Running Oscillators", Izv. VUZ Radioelektron., 1974, 17, no.8, pp.56-62.

37.343. Avtonomov, V.P., et al.: "Active Stabilization of a CO_2 Laser With Diffraction Selector", Kvantovaya Elektron., 1974, 1, no.2, pp.456-8, and Sov. J. Quantum Electron., 1974, 4, no.2, pp.263-4.

37.344. Dushechkin, G.A., Solov'ev, V.S., and Fisher, A.M.: "Stabilization of Radiation at 633 nm and 3.39 micron With a Methane Absorption Line", Radiotekhnika (Kharkov), 1974, no.28, pp.144-8.

37.345. Letokhov, V.S., and Chebotaev, V.P.: "Quantum Optical Frequency Standards", Kvantovaya Elektron., 1974, 1, no.2, pp.245-67, and Sov. J. Quantum Electron., 1974, 4, no.2, pp.137-48.

37.346. Clairon, A., and Henry, L.: "Stabilization of CO_2 Lasers by Saturated Absorption", C. R. Acad Sci. (Paris), 1974, 279B, no.16, pp.419-22.

37.347. Mende, F., et al.: "Highly Stable Microwave Oscillator for 3-cm Band", Radiotekhnika, 1974, 29, no.3, pp.71-4, and Telecommun. Radio Eng. Pt 2, 1974, 29, no.3, pp.111-3.

37.348. Beterov, I.M., et al.: "High-Stability (He-Ne) Gas Laser Based on Nonlinear Absorption", Avtometriya, 1974, no.6, pp.53-64.

37.349. Matyshev, G.F., and Troitskii, Yu.V.: "Frequency Stabilization of Single-Mode He-Ne Laser With Diffraction Selector", Avtometriya, 1974, no.6, pp.71-6.

37.350. Hochuli, U.E., Haldemann, P., and Li, H.A.: "Factors Influencing Relative Frequency Stability of He-Ne Laser Structures", Rev. Sci. Instrum., 1974, 45, no.11, pp.1378-81.

37.351. Krivoshchekov, G.V., et al.: "Quasistationary Ruby-Laser Emission With External Signal", Avtometriya, 1974, no.6, pp.64-71.

37.352. Hammond, C.R., et al.: "Single Longitudinal Mode Operation of TE CO_2 Laser", J. Phys. E, 1974, 7, no.1, pp.45-8.

37.353. Wallard, A.J., and Wilson, D.C.: "Digital Frequency Tripling Technique for Use in Control System of a Frequency-Stabilized Gas Laser", J. Phys. E, 1974, 7, no.3, pp.161-3.

37.354. Bobrik, V.I., Kolomnikov, Yu.D., and Mogilnitskii, B.S.: "Production of Narrow Resonances in a Long Iodine Absorption Cell", Opt. Spekt., 1974, 37, no.3, p.606, and Opt. Spectrosc., 1974, 37, no.3, p.343.

37.355. Matsuda, I., et al.: "Frequency Stability of Rb Standard", Bull. Tokyo Inst. Technol., 1974, no.125, pp.9-15.

37.356. Jechart, E.: "Miniaturized Rb Frequency Standard", Elektronik, 1974, 23, no.12, pp.457-60.

37.357. Tanaka, K., Sakurai, T., and Kurosawa, T.: "Frequency Stabilization of He-Ne Laser by Saturated Absorption in Iodine", Trans. Soc. Instrum. Control. Eng., 1974, 10, no.6, pp.669-74.

37.358. Nussmeier, T.A., and Abrams, R.L.: "Stark-Cell Stabilization of CO_2 Laser", Appl. Phys. Lett., 1974, 25, no.10, pp.615-7.

37.359. Kapralov, V.P., and Bulygin, A.S.: "Use of Dispersion of a Laser to Stabilize Frequency", Opt. Spekt., 1974, 37, no.5, pp.993-4, and Opt. Spectrosc., 1974, 37, no.5, pp.568-9.

37.360. Bobrik, V.I., et al.: "Stabilization of Laser Frequencies", Izmer. Tekh., 1974, 17, no.8, pp.67-8, and Meas. Tech., 1974, 17, no.8, pp.1252-3.

37.361. Zubarev, I.G., and Mikhailov, S.I.: "External-Signal Control of Emission Spectrum of a Q-Switched Nd^{3+} Laser", Kvantovaya Elektron., 1974, 1, no.3, pp.625-8, and Sov. J. Quantum Electron., 1974, 4, no.3, pp.348-50.

38. WAVELENGTH AND FREQUENCY

38.1. Adlem, I., et al.: "Absolute Method of Measuring Acceleration Due to Gravity Using Doppler Radar", Electron. Lett., 1965, 1, p.63.

38.2. Collier, J.R., and Tai, C.T.: "Guided Waves in Moving Media", Trans. IEEE, 1965, MTT-13, p.441.

38.3. Mednikov, O.I.: "Parametric Divider in the Microwave Band", Radiotekh. Elektron., 1965, 10, p.1159, and Radio Eng. Electron. Phys., 1965, 10, no.6, pp.997-8.

38.4. Lisitano, G.: "Automatic Phase-Measuring System for 8-mm Carrier Wave and its 4-mm Harmonic", Rev. Sci. Instrum., 1965, 36, p.364.

38.5. Ernst, W.P.: "Technique for Measuring Phase Modulation or Rapid Phase Changes of a Microwave Signal", Trans. IEEE, 1965, MTT-13, p.70.

38.6. Ellerbruch, D.A.: "Evaluation of a Microwave Phase-Measurement System", J. Res. Natl. Bur. Stand., 1965, 69C, p.55.

38.7. Marshall, J.A.: "Technique for Making Ultra-Precise Measurements of Microwave Frequency Stability", Hewlett-Packard J., 1965, 17, no.3, p.1.

38.8. Du, L.J., and Compton, R.T.: "Cutoff Phenomena for Guided Waves in Moving Media", Trans. IEEE, 1966, MTT-14, p.358.

38.9. Grauling, C.H., and Healey, D.J.: "Instrumentation for Measurement of the Short-Term Frequency Stability of Microwave Sources", Proc. IEEE, 1966, 54, p.249.

38.10. Gardner, A.L.: "Null Method of Measuring Microwave Phase Shifts", Rev. Sci. Instrum., 1966, 37, p.23.

38.11. Shapiro, I.I.: "Testing General Relativity With Radar", Phys. Rev., 1966, 141, p.1219.

38.12. Rasch, P.J., and Duval, J.F.: "High-Speed Microwave Frequency Synthesizer", Microwave J., 1966, 9, no.6, p.97.

38.13. Hellwig, H.: "Measurement of Frequencies and Spectra in the Microwave Region With the Aid of the NH_3 Maser", Nachr. Tech. Z., 1966, 19, no.9, pp.558-62.

38.14. Schulten, G., and Stoll, J.P.: "High-Precision Wideband Wavemeter for Millimetre Waves", Philips Res. Rep., 1967, 22, no.3, pp.309-14.

38.15. Mergerian, D., and Bozanic, D.A.: "System for Measuring Frequencies Up to 345 GHz", Rev. Sci. Instrum., 1967, 38, no.11, pp.1662-3.

38.16. Frenkel, L., and Pollack, M.A.: "Methods for Measurement of Frequency and Stability of Lasers in the Far-IR Region", Nachr. Tech. Fachber., 1968, 35, p.757.

38.17. Gehre, O., Mayer, H.M., and Tutter, M.: "Experiments on Fresnel Drag of 3-cm Waves by Low-Pressure Discharge", Phys. Lett., 1968, 28A, p.35.

38.18. Landa, P.S., et al.: "Study of Properties of Ring Gas Lasers Using Electronic Modelling", Radiotekh. Elektron., 1968, 13, no.11, pp.2026, and Radio Eng. Electron. Phys., 1968, 13, no.11, pp.1777-83.

38.19. Tuong, T.T.: "Effect of Rotational Movement on Electromagnetic Cavities", Ann. Inst. Poincare, 1968, 9A, p.303.

38.20. Wood, L.E., and Thompson, M.C.: "Simultaneous Measurement of Transit Period of Radio and Optical Signals", Appl. Opt., 1968, 7, p.1955.

38.21. Ogorodniichuk, L.D.: "SHF Phase Meter With Two Modulators", Izv. VUZ Radioelektron., 1968, 11, p.370.

38.22. Kofanov, V.L.: "Evaluation of Phase-Difference Measurement Error Due to Coupling Between Channels", Izv. VUZ Radioelektron., 1968, 11, p.373.

38.23. James, A.V.: "High-Accuracy Microwave Phase Standard", Trans. IEEE, 1968, MTT-16, p.944.

38.24. Young, L., and Bahr, A.J.: "Proposal for a Microwave Rotation Sensor", Proc. IEEE, 1968, 56, p.2076.

38.25. Kalinin, N.A.: "Absolute Measurements of the 633-nm Wavelength of a He-Ne Laser With Internal Absorption Cell", Izmer. Tekh., 1968, no.12, p.27, and Meas. Tech., 1968, no.12, p.1632.

38.26. Golubev, Yu.M., et al.: "Optimum Composition Ratio in He-Ne Ring Laser", Zh. Tekh. Fiz., 1968, 38, no.11, p.1990, and Sov. Phys.-Tech. Phys., 1969, 13, no.11, p.1598.

38.27. Kumar, R.C.: "X-Band Wavemeter", J. Inst. Telecommun. Eng., 1969, 15, no.5, p.352.

38.28. Schilder, D.: "Wavemeter With Optical-Type Resonator for Wavelength of 8-10 mm", Nachr. Tech. Z., 1969, 19, no.8, p.292.

38.29. Knight, D.J.E.: "Harmonic Coincidences Between CW-Gas-Laser Lines for Extending Frequency Measurements to the 10-micron Region", Opto-Electron., 1969, 1, p.161.

38.30. Wolinski, W., et al.: "Measurement of Laser Radiation Spectrum With Fabry-Perot Scanning Interferometer", Przegl. Elektron., 1969, 10, no.6, p.281.

38.31. Frenkel, F., and Sullivan, T.: "Frequency Measurements in the Far-IR Region", Trans. IEEE, 1969, MTT-17, p.281.

38.32. Bolgarfalvy, K.: "Linearization of Cavity Resonators for Wavelength Measurements", Mes. Regul. Autom., 1969, 17, no.1, p.11.

38.33. Corcoran, V.J.: "Locking a HCN Laser to a Microwave Frequency Standard", J. Quantum Electron. IEEE, 1969, QE-5, p.424.

38.34. Whitney, C.: "Contributions to Theory of Ring Lasers", Phys. Rev., 1969, 181, p.535.

38.35. Whitney, C.: "Ring-Laser Mode Coupling", Phys. Rev., 1969, 181, p.542.

38.36. Vesnitskii, A.I.: "Propagation in a Waveguide With Movable Walls", Izv. VUZ Radiofiz., 1969, 12, no.6, p.935.

38.37. Kajfez, D.: "Three-Phase Separator for Circular Polarization", Trans. IEEE, 1969, MTT-17, p.726.

38.38. Neumann, J.: "Electronic Interpolation Method for Light-Wave Interference", Zeiss Mitt., 1969, 5, no.1-2, p.5.

38.39. Irwin, C.G.: "Microwave Phase Testing", Western Electr. Eng., 1969, 13, no.3, p.47.

38.40.   Nitka, E.F., and Ishii, T.K.: "Microwave Ferrite (Magnetostriction) Acceleration Sensors", Trans. IEEE, 1969, ED-16, p.845.

38.41.   Deily, G.R., and Herbstritt, R.L.: "Octave-Band Microwave Phase Comparator", Trans. IEEE, 1969, IM-18, p.290.

38.42.   Zeiger, S.G., Fradkin, E.E., and Filatov, P.P.: "Single-Mode Generation in a Ring Gas Laser", Opt. Spekt., 1969, 26, no.4, p.622, and Opt. Spectrosc., 1969, 26, no.4, p.340.

38.43.   Frenkel, L.: "Methods for Measurement of Frequency and Stability of Submillimetre Lasers", Electron Technol., 1969, 2, no.2-3, p.47.

38.44.   Leikin, A.J., and Solovjov, V.S.: "Determination of Emission Wavelength Stability of a Gas Laser", Electron Technol., 1969, 2, no.2-3, p.77.

38.45.   Kawakami, T., and Ishige, R.: "Phase-Shift Standard at 10 GHz", Bull. Electrotech. Lab., 1969, 33, no.4, p.50.

38.46.   Prishivalko, A.P., Rubanov, V.S., and Katova, A.I.: "Effect of Rotating Discharge Tube Around the Longitudinal Axis Upon Polarization, Q-Factor, and Radiation Frequency of a Ring Laser", Zh. Prikl. Spekt., 1969, 11, no.3, p.425.

38.47.   Zborovskii, V.A., and Kulikov, V.N.: "Investigating the Frequency Characteristics of a Ring Laser", Zh. Prikl. Spekt., 1969, 11, no.4, p.730.

38.48.   Baserova, I.V., et al.: "Reduction of Phase-Shift Measurement Errors at Microwaves in Wave-Interference Systems", Izv. VUZ Radioelektron., 1969, 12, p.1380.

38.49.   Troshin, B.I.: "Spectral Composition of Radiation of a Gas Ring Laser", Opt. Spekt., 1969, 27, p.51, and Opt. Spectrosc., 1969, 27, p.107.

38.50.   Rybakov, B.V., et al.: "Polarization Characteristics of Radiation from a Ring Laser with Circularly Anisotropic Resonator", Opt. Spekt., 1969, 27, p.55, and Opt. Spectrosc., 1969, 27, p.113.

38.51.   Kamyshan, V.V., and Valitov, R.A.: "Heterodyne Method of Frequency Measurement in Millimetre Range", Prib. Tekh. Eksp., 1969, 12, no.4, p.106, and Instrum. Exp. Tech., 1969, 12, no.4, p.925.

38.52.   Korzhenevich, I.M., and Ratner, A.M.: "Spatial Distribution and Field Spectrum in Ring Resonator", Radiotekh. Elektron., 1969, 14, no.9, p.1676, and Radio Eng. Electron. Phys., 1969, 14, no.9, pp.1451-3.

38.53.   Dashchuk, M.: "Transverse Doppler Effect With Laser Light in a Reflection System", Proc. IEEE, 1969, 57, p.2148.

38.54.   Dandliker, R., and Tschudi, T.: "Coupled (Fabry-Perot) Resonators. Dependence of Mutual Interaction on Separation", Appl. Opt., 1969, 8, p.1119.

38.55.   Schmidt, V.V., et al.: "Frequency Measurement of a Gas Laser Operating Near 0.3 mm", Radiotekh. Elektron., 1969, 14, no.9, p.1708, and Radio Eng. Electron. Phys., 1969, 14, no.9, pp.1487-8.

38.56.   Fukai, I., Suzuki, M., and Kazama, T.: "Reflection and Transmission of Plane Waves by an Anisotropic Moving Medium", Electron. Commun. Jap., 1969, 52, no.10, pp.61-8.

38.57.   Kleiman, A.S., et al.: "Measurement of Short-Term Instability of Laser Frequency on the 3.3 $^{15}NH_3$ Line", Izv. VUZ Radiofiz., 1969, 12, no.8, p.1165.

38.58.   Gehre, O., Mayer, H.M., and Tutter, M.: "Interferometer for Measurement of Fresnel Drag on Microwaves by a Drifting Electron Plasma", Trans. IEEE, 1969, IM-18, p.194.

38.59.   Balakhanov, V.Ya., Zhivotov, V.K., and Titov, A.V.: "Combined Fabry-Perot Interferometers in the Microwave Region", Zh. Prikl. Spekt., 1969, 11, no.1, p.161.

38.60.   Hetherington, A., Burrell, G.J., and Moss, T.S.: "Properties of He-Ne Ring Lasers at 3.39 micron", Infrared Phys., 1969, 9, p.109.

38.61.   Andronova, I.A., and Bershtein, I.L.: "Experimental Investigation of Inequality of Optical Paths of Opposite Waves in a 3.39-micron Ring Laser", Zh. Eksp. Teor. Fiz., 1969, 57, no.1, p.100, and Sov. Phys.-JETP, 1970, 30, no.1, pp.58-62.

38.62.   Volkov, A.M., and Kiselev, V.A.: "Proper Frequencies of a Rotating Ring Resonator", Zh. Eksp. Teor. Fiz., 1969, 57, no.4, p.1353, and Sov. Phys.-JETP, 1970, 30, no.4.

38.63.   Basov, N.G., Belenov, E.M., et al.: "Ring Laser With Nonlinear Absorbing Cell", Zh. Eksp. Teor. Fiz., 1969, 57, no.6, p.1991, and Sov. Phys.-JETP, 1970, 30, no.6.

38.64.   Rybakov, B.V., et al.: "Amplitude and Frequency Characteristics of a Ring Laser", Zh. Eksp. Teor. Fiz., 1969, 57, no.4, p.1184, and Sov. Phys.-JETP, 1970, 30, no.4, pp.646-50.

38.65.   Fradkin, E.E.: "Effect of Distribution of Losses on Oppositely Travelling Waves in a Ring Gas Laser", Opt. Spekt., 1970, 28, no.2, p.422, and Opt. Spectrosc., 1970, 28, no.2, p.227.

38.66.   Gus'kov, N.A.: "Wavelength Measurement in Optical Range Using a Fabry-Perot Interferometer", Radiotekh. Elektron., 1970, 15, no.5, p.1109, and Radio Eng. Electron. Phys., 1970, 15, no.5, pp.940-3.

38.67. Fradkin, E.E., and Khayutin, L.M.: "Competition of Oppositely Travelling Waves in a Ring Gas Laser in Longitudinal Magnetic Field", Opt. Spekt., 1970, 28, no.1, p.89, and Opt. Spectrosc., 1970, 28, no.1, p.45.

38.68. Landa, P.S., and Lariontsev, E.G.: "Beating and Synchronization States of Backward Waves in Rotating Ring Gas Laser", Radiotekh. Elektron., 1970, 15, no.6, p.1214, and Radio Eng. Electron. Phys., 1970, 15, no.6, pp.1027-38.

38.69. Faxvog, F.R., et al.: "Measured Pulse Velocity Greater than Light in Neon Absorption Cell", Appl. Phys. Lett., 1970, 17, p.192.

38.70. Kruglik, G.S., et al.: "Frequency Characteristic of Ring Laser Near Parametric Resonance", Zh. Prikl. Spekt., 1970, 13, no.5, pp.913-4.

38.71. Miyashita, T., and Ikenoue, J.: "Unidirectional Oscillation of Internally Loss-Modulated Ring Laser", Jap. J. Appl. Phys., 1970, 9, no.12, pp.1547-8.

38.72. Volkov, A.M., and Kiselev, V.A.: "Rotating Ring Resonator With Nonreciprocal Element", Opt. Spekt., 1970, 29, no.2, pp.365-70, and Opt. Spectrosc., 1970, 29, no.2, pp.194-7.

38.73. Kutin, V.N., and Troshin, B.I.: "Laser Ring Interferometer With Special Selective Characteristics", Opt. Spekt., 1970, 29, no.2, pp.371-3, and Opt. Spectrosc., 1970, 29, no.2, pp.197-8.

38.74. Bondarev, V.A., et al.: "(Direct) Frequency Measurement of a HCN Laser", Izmer. Tekh., 1970, 13, no.11, pp.1628-31, and Meas. Tech., 1970, 13, no.11, pp.5-8.

38.75. Volkov, A.M., and Skrotskii, G.V.: "Certain Effects Arising in the Locking Zone of a Ring Laser", Opt. Spekt., 1970, 29, no.5, pp.965-9, and Opt. Spectrosc., 1970, 29, no.5, pp.512-4.

38.76. Vetkin, V.A., and Khromykh, A.M.: "Combination Interaction of Two Longitudinal Modes in a Ring Laser", Opt. Spekt., 1970, 29, no.4, pp.765-71, and Opt. Spectrosc., 1970, 29, no.4, pp.407-11.

38.77. Blazhnov, B.A., et al.: "Amplitude and Phase Characteristics of a Ring Laser", Zh. Prikl. Spekt., 1970, 13, no.6,pp.96-101.

38.78. Sunduchkov, K.S.: "Accurate Phase Measurements on Low-Level Microwave Signals", Radiotekhnika, 1970, 25, no.11, and Telecommun. Radio Eng. Pt 2, 1970, 25, no.11, pp.127-30.

38.79. Van Welzenis, R.G., and Daub, D.: "Spectrum Analysis of Nanosecond Pulses", Trans. IEEE, 1970, MTT-18, p.280.

38.80. Aronowitz, F.: "Loss Lock-In in the Ring Laser", J. Appl. Phys., 1970,41,p.2453.

38.81. Kowalski, G.: "Measurement of Phase Characteristics of Tapered Slow-Wave Structures", Arch. Elektrotech., 1970, 19, p.3.

38.82. Bidikhov, S.A., et al.: "Self-Oscillatory States in a Ring-Type Gas Laser", Radiotekh. Elektron., 1970, 15, p.529, and Radio Eng. Electron. Phys., 1970, 15, no.3, pp.450-7.

38.83. Dryagin, Yu.A.: "Method of Measuring Wavelength by a High-Quality Fabry-Perot Resonator", Izv. VUZ Radiofiz., 1970, 13, p.141.

38.84. Kruglik, G.S., et al.: "Parametric Resonance in a Ring Laser", Zh. Prikl. Spekt., 1970, 12, no.3, p.432.

38.85. Dubovets, V.G., and Prishivalko, A.P.: "Investigating Polarization, Frequency, and Losses, in the Radiation of a Ring Laser With Anisotropic Plate", Zh. Prikl. Spekt., 1970, 12, no.4, p.647.

38.86. Landa, P.S.: "Fluctuations in Ring Lasers", Zh. Eksp. Teor. Fiz., 1970, 58, no.5, p.1651, and Sov. Phys.-JETP, 1970, 31, no.5, pp.886-90.

38.87. Massey, G.A., and Siegman, A.E.: "Fresnel-Drag Technique for Determining the Spin-Axis Rotation of a Spherical Rotor", J. Quantum Electron. IEEE, 1970, QE-6, p.496.

38.88. Washwell, E.R., and Cuff, K.F.: "High-Speed Electrooptic Spectral (Interferometric) Scanning", Appl. Opt., 1970,9,p.1911.

38.89. Klimontovich, Yu.L., and Landa, P.S.: "Sources of Natural Fluctuations in Ring Lasers", Zh. Eksp. Teor. Fiz., 1970, 58, no.4, p.1367, and Sov. Phys.-JETP, 1970, 31, no.4, pp.733-7.

38.90. Schacter, H., and Chi, C.: "Resonant Modes in Laser Cavities Immersed in a Homogeneous Moving Medium", Opt. Acta, 1970, 17, no.11, pp.801-9.

38.91. Sokoloff, D.R., et al.: "Extension of Laser Harmonic-Frequency Mixing Into the 5-micron Region", Appl. Phys. Lett., 1970, 17, no.6, pp.257-9.

38.92. Basov, N.G., Belenov, E.M., et al.: "Intense Power Resonances of a Ring Laser with an Absorbing Cell", Zh. Eksp. Teor. Fiz. Pis'ma, 1970, 12, no.3, pp.145-7, and JETP Lett., 1970, 12, no.3, pp.101-2.

38.93. Rozanov, N.N.: "Noise in the Locking Band of a Ring Laser", Opt. Spekt., 1970, 28, no.4, pp.740-3, and Opt. Spectrosc., 1970, 28, no.4, pp.398-400.

38.94. Gut'man, G.B., Rolich, V.I., and Filatov, Yu.V.: "Frequency Characteristics of a Ring Gas Laser", Zh. Prikl. Spekt., 1970, 13, no.4, pp.722-5.

38.95. Aronowitz, F.: "Lock-In and Intensity-Phase Interaction in the Ring Laser", J. Appl. Phys., 1970, 41, p.130.

38.96. Zhelnov, B.L., and Smirnov, V.S.: "Effect of Collisions on the Oscillation Conditions of a Ring Gas Laser", Opt. Spekt., 1970, 28, no.4, pp.747-50, and Opt. Spectrosc., 1970, 28, no.4, pp.402-3.

38.97. Landa, P.S., and Slin'ko, E.F.: "Frequency Characteristics of a Ring Laser With a Vibrating Support", Vestn. Mosk. Univ. Fiz. Astron., 1970, no.4, pp.400-5.

38.98. Adonina, A.I., et al.: "Prismatic (Fabry-Perot) Absorption-Type Wavemeter", Izv. VUZ Radioelektron., 1970, 13, no.8, pp.993-9.

38.99. Hobson, G.S.: "Measurement of Variation of Frequency With Ambient Temperature of Microwave Semiconductor Oscillators", J. Phys. E, 1970, 3, no.10, pp.801-5.

38.100. Bauer, R.K.: "Further Development of an Accurate-Method for Measurements of the Polarization of Light", J. Phys. E, 1970, 3, no.12, pp.965-8.

38.101. Roland, J.J., and Lamarre, J.M.: "Coupling Phenomena in Laser-Gyro and Angular Measurements", Rev. Phys. Appl., 1970, 5, no.5, pp.757-42.

38.102. Evenson, K.M., Wells, J.S., and Matarrese, L.M.: "Absolute Frequency Measurements of CO_2 CW Laser at 28 THz", Appl. Phys. Lett., 1970, 16, p.251.

38.103. Zhelnov, B.L., Smirnov, V.S., and Fadeev, A.P.: "Instability of Unidirectional Radiation in a Ring Laser", Opt. Spekt., 1970, 28, no.4, pp.744-6, and Opt. Spectrosc., 1970, 28, no.4, pp.400-1.

38.104. Fradkin, E.E., and Khayutin, L.M.: "Theory of a Ring Gas Laser in a Magnetic Field", Zh. Eksp. Teor. Fiz., 1970, 59, no.5, pp.1634-44, and Sov. Phys.-JETP, 1971, 32, no.5, pp.891-6.

38.105. Watkins, L.S., and Smith, R.C.: "Operation of a Circularly Polarized Ring Laser", J. Quantum Electron. IEEE, 1971, QE-7, no.2, pp.59-62.

38.106. Jacobs, G.B.: "CO_2 Laser Gyro", Appl. Opt., 1971, 10, no.1, pp.219-20.

38.107. Peterson, R.W., and Jahoda, F.C.: "Method for Absolute Calibration of Electrooptic (Phase) Modulators", Rev. Sci. Instrum., 1971, 42, no.4, pp.532-3.

38.108. Basov, N.G., Belenov, E.M., et al.: "Investigation of Power Resonances of a Ring Laser With Nonlinear Absorbing Cell", Zh. Eksp. Teor. Fiz., 1971, 60, no.1, pp.117-23, and Sov. Phys.-JETP, 1971, 33, no.1, pp.66-9.

38.109. Bershtein, I.L.: "Fluctuations in Ring-Laser Radiation", Izv. VUZ Radiofiz., 1971, 14, no.2, pp.252-62.

38.110. Letokhov, V.S., and Pavlik, B.D.: "Frequency Effects in a Ring Gas Laser With Nonlinear Absorption", Izv. VUZ Radiofiz., 1971, 14, no.2, pp.244-51.

38.111. Ke, B., and Hiyama, T.: "Simple Wavelength-Scanning (Filter Wedge) Monitor for Tunable Dye Laser", Rev. Sci. Instrum., 1971, 42, no.3, pp.395-6.

38.112. Cleveland, F.H., and Kernweis, N.P.: "Technique for Measuring Phase at Millimetre Wavelengths", Trans. IEEE, 1971, MTT-19, no.4, pp.406-10.

38.113. Grudina, N.A., and Ishchenko, V.A.: "Meter for Measuring Instantaneous Phase Differences of RF Pulses", Izv. VUZ Radioelektron., 1971, 14, no.4, pp.460-3.

38.114. Kruglik, G.S., and Kutsak, A.A.: "Multimode Effect on Beat Regime of a Ring Laser", Zh. Prikl. Spekt., 1971, 14, no.1, pp.59-64.

38.115. Chamberlain, J., and Gebbie, H.A.: "Use of Phase Modulation in Submillimetre-Wave Interferometers", Appl. Opt., 1971, 10, no.5, pp.1184-5.

38.116. Chamberlain, J.: "Phase Modulation in Submillimetre-Wave Interferometers. I-III", Infrared Phys., 1971, 11, no.1, pp.25-55, 57-73, and 75-84.

38.117. Tschudi, T., and Dandliker, R.: "Reflection Properties of an Active Three-Mirror Resonator", Z. Angew. Phys., 1971, 31, no.2, pp.105-11.

38.118. Bykovskii, Yu.A., et al.: "Method for Investigating Semiconductor Lasers With a Fabry-Perot Resonator", Fiz. Tekh. Poluprov., 1971, 5, no.3, pp.498-501, and Sov. Phys.-Semicond., 1971, 5, no.3, pp.435-8.

38.119. Andronova, I.A., and Bershtein, I.L.: "Experimental Investigation of Feedback Effect Upon Ring-Laser Operation", Izv. VUZ Radiofiz., 1971, 14, no.5,pp.698-705.

38.120. Miyashita, T., Mori, H., and Ikenoue, J.: "Theory of Multimode Ring Laser", Jap. J. Appl. Phys., 1971, 10,no.8,pp.1051-9.

38.121. Whitney, C.: "Modulation of Ring-Laser Mode Coupling", Trans. IEEE, 1971, AES-7, no.5, pp.914-21.

38.122. Zimokosov, G.A., et al.: "(Interferometric) Wavelength Measurement on Lasers", Izmer. Tekh., 1971, 14, no.6, pp.42-3, and Meas. Tech., 1971, 14, no.6, pp.868-71.

38.123. Bakalyar, A.I., and Usol'tsev, I.F.: "Influence of Axial Magnetic Field on Beat Frequency in Ring Laser With Linear Polarization", Kvantovaya Elektron., 1971, no.4, pp.91-4.

38.124. Keiter, R.C.: "Fully Calibrated Solid-State Microwave Spectrum Analyser", Hewlett-Packard J., 1971, 23, no.1, pp.4-9.

38.125. Fradkin, E.E.: "Diffraction Splitting of Frequencies in Ring Gas Laser. I", Opt. Spekt., 1971, 31, no.6, pp.952-60, and Opt. Spectrosc., 1971, 31, no.6, pp.514-8.

38.126. Bailey, A.G.: "Relative-Phase Measurement at Q Band", Proc. IEE, 1971, 118, no.10, pp.1345-50.

38.127. Volkov, A.M., Izmestev, A.A., and Skrotskii, G.V.: "Rotating Ring Laser in an Arbitrary Gravitational Field", Opt. Spekt., 1971, 30, no.4, pp.762-6, and Opt. Spectrosc., 1971, 30, no.4, pp.411-3.

38.128. Mel'Tsin, A.L.: "Frequency Stabilization of Ring Laser With an External Absorption Chamber", Zh. Prikl. Spekt., 1971, 15, no.2, pp.214-8.

38.129. Lyon, D.L., and George, E.V.: "TEA CO_2 Ring Laser", Q. Progr. Rep., 1971, no.101, pp.55-60.

38.130. Klochan, E.L., and Landa, P.S.: "Frequency Characteristics of a Ring Laser With Natural Fluctuations", Izv. VUZ Radiofiz., 1971, 14, no.10, pp.1518-25.

38.131. Apanasevich, P.A., and Zhovna, G.I.: "Nonlinear Link Between Modes of Ring Lasers", Zh. Prikl. Spekt., 1971, 15, no.4, pp.622-9.

38.132. Bogdanov, V.V., and Mynbaev, D.K.: "Lock-In Band of a Gas Ring Laser", Opt. Spekt., 1971, 31, no.1, pp.101-2, and Opt. Spectrosc., 1971, 31, no.1, pp.51-2.

38.133. Bilger, M.R., and Zavodny, A.T.: "Fresnel Drag in a Ring Laser. Measurement of the Dispersive Term", Phys. Rev., 1972, 5, no.2, pp.591-9.

38.134. Ulrich, R., and Weber, H.P.: "Unidirectional Thin-Film Ring Laser", Appl. Phys. Lett., 1972, 20, no.1, pp.38-40.

38.135. Evenson, K.M., et al.: "Extension of Absolute Frequency Measurements to the CW He-Ne Laser at 88 THz", Appl. Phys. Lett., 1972, 20, no.3, pp.133-4.

38.136. Mygind, J.: "Stabilized Heterodyne Microwave Interferometer With High Resolution", J. Phys. E, 1972, 5, no.2, pp.186-9.

38.137. Kafri, O., Kimel, S., and Shamir, J.: "Description of Laser Pulses Using Fourier Analysis", J. Quantum Electron. IEEE, 1972, QE-8, no.3, pp.295-301.

38.138. Wax, S.I., and Chodorow, M.: "Phase Modulation of a Ring-Laser Gyro. I and II", J. Quantum Electron. IEEE, 1972, QE-8, no.3, pp.343-52 and 352-61.

38.139. Aronowitz, F.: "Single-Isotope Laser Gyro", Appl. Opt., 1972, 11, no.2, pp.405-12.

38.140. Arthur, L., et al.: "Microwave Phase Measurements Associated With the Goos-Hanchen Shift", Can. J. Phys., 1972, 50, no.1, pp.52-6.

38.141. Goddard, N.E.: "Instantaneous Frequency-Measuring Receivers", Trans. IEEE, 1972, MTT-20, no.4, pp.292-3.

38.142. Lim, T.K., and Garside, B.K.: "(Ring) Laser Pulses in a System With Discrete Loss", J. Quantum Electron. IEEE, 1972, QE-8, no.5, pp.454-5.

38.143. Takata, K.: "Stability of TW Oscillation With Arbitrary Intensity in Ring Lasers", Jap. J. Appl. Phys., 1972, 11, no.5, pp.699-709.

38.144. Schade, W.J.: "Accurate Measurements for Cyanic Laser Wavelengths Using Resonator Interferometry", Opt. Commun., 1972, 4, no.5, pp.399-403.

38.145. McDonald, D.G., et al.: "Four-Hundredth-Order Harmonic Mixing of Microwave and IR Radiations Using a Josephson Junction and Maser", Appl. Phys. Lett., 1972, 20, no.8, pp.296-9.

38.146. Andronova, I.A., and Khandokhin, P.A.: "Influence of Magnetic Field on Characteristics of a Ring Laser at 3.39 micron", Izv. VUZ Radiofiz., 1972, 15, no.5, pp.703-12.

38.147. Zhelnov, B.L., and Smirnov, G.I.: "Gas Ring Laser With Optically Active Element", Opt. Spekt., 1972, 32, no.2, pp.388-91, and Opt. Spectrosc., 1972, 32, no.2, pp.202-4.

38.148. Pandit, L., and Bhawalkar, D.D.: "Simple Scanning Spherical-Mirror Fabry-Perot Interferometer", Indian J. Pure Appl. Phys., 1972, 10, no.6, pp.487-8.

38.149. Kovacs, K.P.: "Deforming-Light-Path Ring-Laser Experiments", J. Opt. Soc. Am., 1972, 62, no.11, pp.1264-7.

38.150. Evenson, K.M., et al.: "Speed of Light from Direct Frequency and Wavelength Measurements of Methane-Stabilized Laser", Phys. Rev. Lett., 1972, 29, no.19, pp.1346-9.

38.151. Ogorodniichuk, L.D., and Ogorodniichuk, N.D.: "Effect of Noise on Accuracy of Microwave Phasemeters", Izv. VUZ Radioelektron., 1972, 15, no.7, pp.915-20.

38.152. Luk'yanov, D.P., Rogachev, A.F., and Samokhin, V.S.: "Nature of Losses Introduced in a Ring Resonator by a Nonreciprocal Faraday Phaseshifter", Opt. Spekt., 1972, 32, no.4, pp.814-8, and Opt. Spectrosc., 1972, 32, no.4, pp.428-30.

38.153. Voswinkel, G.: "Automatic Heterodyne Method for Measurement of Frequencies Up to 18 GHz", Elektron. J., 1972, 7, no.11, pp.70-4.

38.154. Gol'dort, V.G., Puchkov, V.N., and Toropov, A.K.: "Spectrometers for Analysis of Radiation from He-Ne Lasers", Prib. Tekh. Eksp., 1972, 15, no.2, pp.171-3, and Instrum. Exp. Tech., 1972, 15, no.2, pp.506-8.

38.155. Baird, K.M., Riccius, H.D., and Siemsen, K.J.: "CO$_2$-Laser Wavelengths and the Velocity of Light", Opt. Commun., 1972, 6, no.2, pp.91-5.

38.156. Gomenyuk, A.S., and Ratner, Ye.S.: "Measuring the Spread Function of IR Spectrometers by Means of Gas Lasers", Opt.-Mekh. Prom., 1972, 39, no.3, p.58, and Sov. J. Opt. Technol., 1972, 39, no.3, pp.177-8.

38.157. Koronkevich, V.P., and Lenkova, G.A.: "Application of Laser Interferometers in Accurate Measurement", Avtometriya, 1972, no.6, pp.69-75.

38.158. Somlo, P.I., Hollway, D.L., and Morgan, I.G.: "Absolute Calibration of Periodic Microwave Phaseshifters", Trans. IEEE, 1972, MTT-20, no.8, pp.532-7.

38.159. Fradkin, E.E.: "Diffraction Splitting of Frequencies in a Ring Gas Laser. II", Opt. Spekt., 1972, 32, no.1, pp.132-42, and Opt. Spectrosc., 1972, 32, no.1, pp.65-70.

38.160. Bradley, C.C., Edwards, G., and Knight, D.J.E.: "Absolute Measurement of Submillimetre Laser Frequencies", Radio Electron. Eng., 1972, 42, no.7, pp.321-7.

38.161. Pestov, E.G., and Kruglik, G.S.: "Polarizing Effect of Decrease of Competition by Counter Waves in Ring Lasers", Zh. Prikl. Spekt., 1972, 16, no.2, pp.985-90.

38.162. Landa, P.S.: "Peculiarities of a Ring Laser With Mixture of Active Gas Isotopes", Opt. Spekt., 1972, 32, no.2, pp.383-7, and Opt. Spectrosc., 1972, 32, no.2, pp.200-2.

38.163. Vetkin, V.A., and Khromykh, A.M.: "Competition Between Longitudinal Modes in a Ring Laser With Anisotropic Resonator", Kvantovaya Elektron., 1972, no.3, pp.59-68, and Sov. J. Quantum Electron., 1972, 2, no.3, pp.240-6.

38.164. Kravtsov, N.V.: "Exact Frequency Measurements in the Optical Range", Zh. Prikl. Spekt., 1972, 17, no.2, pp.368-70.

38.165. Fradkin, E.E.: "Backward Diffraction Scattering of Oppositely Travelling Waves by Mirrors of a Ring Resonator", Opt. Spekt., 1972, 33, no.4, pp.716-9, and Opt. Spectrosc., 1972, 33, no.4, pp.395-7.

38.166. Orlov, A.I., Orlov, L.N., and Rubanov, V.S.: "Generation of Oppositely Travelling Waves Polarized in Different Planes in a Ring Laser", Opt. Spekt., 1972, 33, no.4, pp.729-32, and Opt. Spectrosc., 1972, 33, no.4, pp.401-3.

38.167. Korshunov, V.A., Kuznetsova, T.I., and Malyutin, A.A.: "Time Characteristics of a Ring Laser With Bleachable Filter", Kvantovaya Elektron., 1972, no.3, pp.69-72, and Sov. J. Quantum Electron., 1972, 2, no.3, pp.247-50.

38.168. Kruglik, G.S., and Pestov, E.G.: "General Method for Calculation of Frequency of Beats in Single-Mode Laser", Kvantovaya Elektron., 1972, no.5, pp.22-9, and Sov. J. Quantum Electron., 1973, 2, no.5, pp.411-5.

38.169. Blaney, T.G., et al.: "Absolute Frequency Measurement of a Lamb-Dip-Stabilized Water-Vapour Laser at 10.7 THz", Phys. Lett., 1973, 43A, no.5, pp.471-2.

38.170. Checcacci, P.F., Falciai, R., and Scheggi, A.M.: "Microwave Four-Mirror Ring Resonator", Trans. IEEE, 1973, MTT-21, no.5, pp.361-2.

38.171. Molchanov, V.Ya., and Shrotskii, G.V.: "Natural Polarization Conditions of a Rotating Circular Resonator", Zh. Prikl. Spekt., 1973, 18, no.3, pp.485-7.

38.172. Markelov, N.A., Rogachev, V.A., and Turkin, A.A.: "Experimental Study of Difference-Frequency Fluctuations of a Ring Laser", Izv. VUZ Radiofiz., 1973, 16, no.4, pp.545-51.

38.173. Evenson, K.M., et al.: "(Measurement of) Accurate Frequencies of Molecular Transitions Used in Laser Stabilization", Appl. Phys. Lett., 1973, 22, no.4, pp.192-5.

38.174. Barger, R.L.: "(Measurement of) Wavelength of the 3.39-micron Laser-Saturated Absorption Line of Methane", Appl. Phys. Lett., 1973, 22, no.4, pp.196-9.

38.175. Apanasevich, P.A., and Zhovna, G.I.: "Calculation of Generation of Ring Lasers in the Locking Region", Zh. Prikl. Spekt., 1973, 18, no.1, pp.32-7.

38.176. Chekalinskaya, Yu.I., and Ledneva, G.P.: "Frequency-Polarization Characteristics of a Three-Mirror Ring Resonator With Anisotropic Elements", Zh. Prikl. Spekt., 1973, 18, no.2, pp.219-26.

38.177. Manankova, A.V., and Borisov, V.V.: "Pulse Signal Reflection from a Moving Mirror", Izv. VUZ Radiofiz., 1973, 16, no.1, pp.312-6.

38.178. Blaney, T.G., and Knight, D.J.E.: "Experiments Using a Superconducting Point-Contact Harmonic Mixer Near 1 THz", J. Phys. D, 1973, 6, no.8, pp.936-52.

38.179. Pachev, Kh., Sabotinov, N.V., and Blagoev, K.D.: "Laboratory-Type Laser Gyroscope", Elektro Prom. Prib., 1973, 8, no.2, pp.61-3.

38.180. Wells, J.S.: "Stabilized HCN Laser for IR Frequency Synthesis", Trans. IEEE, 1973, IM-22, no.2, pp.113-8.

38.181. Klochan, E.L., et al.: "Generation Regimes of Solid-State Ring Laser", Zh. Eksp. Teor. Fiz. Pis'ma, 1973, 17, no.8, pp.405-9, and JETP Lett., 1973, 17, no.8, pp.189-92.

38.182. Roland, J.J., and Lamarre, J.M.: "Periodic Faraday Bias and Lock-In Phenomena in a Laser Gyro", Appl. Opt., 1973, 12, no.7, pp.1460-7.

38.183. Prussak, W., and Gierszal, H.: "Fabry-Perot Resonator Wavementer for K Band", Postepy Fiz., 1973,24,no.3,pp.357-9.

38.184. Zeiger, S.G.: "Three-Mode Oscillation in a Ring Laser", Opt. Spekt., 1973, 34, no.1, pp.133-40, and Opt. Spectrosc., 1973, 34, no.1, pp.72-6.

38.185. Kozhevnikov, N.M., et al.: "Effect of External Magnetic Field on Beat Frequency in a Ring Laser With Nonreciprocal Phaseshifter", Zh. Tekh. Fiz., 1973, 43, no.2, pp.349-52, and Sov. Phys.-Tech. Phys., 1973, 18, no.2, pp.225-6.

38.186. Mochalov, A.V., and Mynbaev, D.K.: "Frequency Dependence of Locking in a Ring Laser", Zh. Tekh. Fiz., 1973, 43, no.3, pp.674-5, and Sov. Phys.-Tech. Phys., 1973, 18, no.3, pp.422-3.

38.187. Bennett, S.J., and Rowley, W.R.C.: "Automatic Compensation for Changes of Signal Level in Laser Interferometry", J. Phys. E, 1973, 6, no.10, pp.963-4.

38.188. Sakurai, T.: "Preliminary Experiments With an He-Ne Ring Laser for Angular-Velocity Measurement", Trans. Soc. Instrum. Control Eng., 1973, 9, no.2, pp.137-43.

38.189. Dal Pozzo, P., et al.: "Unstable Ring Resonator", J. Quantum Electron. IEEE, 1973, QE-9, no.11, pp.1061-3.

38.190. Kulukov, Yu.Yu., et al.: "Heterodyne Spectrum Analyser for 1.1-1.7 mm Range", Izv. VUZ Radiofiz., 1973, 16, no.9, pp.1442-3.

38.191. Busca, G., Tetu, M., and Vanier, J.: "Light-Shift Effects in the ^{87}Rb Maser", Appl. Phys. Lett., 1973, 23, no.7,pp.395-6.

38.192. Vershinina, L.N.: "Phase Measurements at Submillimetre Wavelengths", Prib. Tekh. Eksp., 1973, 16, no.2, pp.138-40, and Instrum. Exp. Tech., 1973, 16, no.2,pp.493-5.

38.193. Basov, N.G., Belenov, E.M., et al.: "Frequency Reproducibility of a Stabilized Ring Laser", Dokl. Akad. Nauk SSSR, 1973, 210, no.1-3, pp.1-3, and Sov. Phys.-Dokl., 1973, 18, no.5, pp.316-7.

38.194. Biswas, P.K.: "Laser Amplification by Gravitational Field", Nuovo Cim., 1973, 18B, no.2, pp.345-53.

38.195. Menegozzi, L.N., and Lamb, W.E.: "Theory of a Ring Laser", Phys. Rev., 1973, 8, no.4, pp.2103-25.

38.196. Petrun'kin, V.Yu., et al.: "Influence of Faraday Effect on Beat Frequency of a He-Ne Ring Laser", Zh. Tekh. Fiz., 1973, 18, no.5, pp.702-3.

38.197. Arutyunyan, A.G., Tunkin, V.G., and Chirkin, A.S.: "High-Luminosity, High-Resolution, Interferometer for Measurement of Coherence of Optical Radiation", Kvantovaya Elektron., 1973, no.7, pp.111-3, and Sov. J. Quantum Electron., 1973, 3, no.1, pp.63-4.

38.198. Vanin, N.V., Migulin, A.V., and Sukhorukov, A.P.: "Deflection of a Light Beam by a Moving Liquid", Zh. Tekh. Fiz., 1973, 43, no.5, pp.1102-4, and Sov. Phys.-Tech. Phys., 1973, 18, no.5, pp.704-5.

38.199. Birulin, A.I., Kurenev, Yu.P., and Okhrimenko, N.I.: "Analysis of the Output Signal of a Ring Laser", Opt.-Mekh. Prom., 1973, 40, no.4, pp.15-7, and Sov. J. Opt. Technol., 1973, 40, no.4, pp.221-3.

38.200. Paneseikin, Yu.V., Rybkin, L.L., and Stepanyan, R.G.: "Method of Measuring Microwave Signal Phase", Radiotekhnika, 1973, 28, no.3, pp.108-11, and Telecommun. Radio Eng. Pt 2, 1973, 28, no.3, pp.138-41.

38.201. Szczypka, Z.: "Calibration of a Q-Band Two-Cavity High-Accuracy Wavemeter", Pr. PIT, 1973, 23, no.77, pp.25-31.

38.202. Gudkov, Yu.P., and Rozanov, N.N.: "Theory of Locking in a Ring Gas Laser. I and II", Opt. Spekt., 1973, 35, no.4, pp.736-45 and no.5, pp.919-28, and Opt. Spectrosc., 1973, 35, no.4, pp.426-31 and no.5, pp.532-6.

38.203. Burnashev, M.N., and Filatov, Yu.V.: "Nonreciprocity of Oppositely Travelling Waves in a Ring Gas Laser at 633 nm by Diaphragming", Opt. Spekt., 1973, 35, no.5, pp.992-4, and Opt. Spectrosc., 1973, 35, no.5, pp.577-8.

38.204. Solomakha, D.A.: "System Errors in Measuring Laser Wavelength by Fabry-Perot Interferometer", Izmer. Tekh., 1973, 16, no.8, pp.32-3, and Meas. Tech., 1973, 16, no.8, pp.1167-8.

38.205. Bryzzhev, L.D., Miroshnichenko, O.N., and Khimchenko, V.P.: "Comparing Wavelengths of a Lamp and Laser by an Interferometer With Movable Reflector", Izmer. Tekh., 1973, 16, no.6, p.93, and Meas. Tech., 1973, 16, no.6, pp.946-7.

38.206. Newburgh, R.G.: "Motional Effects in Retardation Plates and Mode Locking in Ring Lasers", Appl. Opt., 1973, 12, no.1, pp.116-9.

38.207. Voronov, V.I., and Pol'skii, Yu.E.: "Mode Locking in Laser With Ring Cavity", Radiotekh. Elektron., 1973, 18, no.7, pp.1434-9, and Radio Eng. Electron. Phys., 1973, 18, no.7, pp.1055-8.

38.208. Sokolov, V.A., and Fradkin, E.E.: "Two-Mode Operation in a Ring Laser", Zh. Tekh. Fiz., 1973, 43, no.11, pp.2367-74, and Sov. Phys.-Tech. Phys., 1974, 18, no.11, pp.1494-8.

38.209. Burnashev, M.N., and Filatov, Yu.V.: "Dependence of Frequency Characteristic of Ring Laser on Gain", Zh. Tekh. Fiz., 1973, 43, no.11, pp.2364-6, and Sov. Phys.-Tech. Phys., 1974, 18, no.11, pp.1492-3.

38.210. Galkin, S.L., et al.: "Effect of Axial Magnetic Field on Beat Frequency in a Synchronized He-Ne Ring Laser", Zh. Tekh. Fiz., 1973, 43, no.9, pp.1995-8, and Sov. Phys.-Tech. Phys., 1974, 18, no.9, pp.1257-8.

38.211. Hanson, D.R., and Sargent, M.: "Theory of a Zeeman Ring Laser", Phys. Rev., 1974, 9A, no.1, pp.466-80.

38.212. Gut'man, G.B.: "Measuring the Absolute Value of the Gravitational Acceleration (With He-Ne Laser)", Avtometriya, 1974, no.2, pp.98-100.

38.213. Andronova, I.A.: "Backscattering in Ring Resonators", Izv. VUZ Radiofiz., 1974, 17, no.5, pp.775-7.

38.214. Boitsov, V.F., Murina, T.A., and Fradkin, E.E.: "Frequency Splitting of Oppositely Travelling Waves in a Ring Laser with a Gaussian Diaphragm", Opt. Spekt., 1974, 36, no.3, pp.539-45, and Opt. Spectrosc., 1974, 36, no.3, pp.311-14.

38.215. Sokolov, V.A., and Fradkin, E.E.: "Unsynchronized Two-Mode Lasing in a Ring Gas Laser", Opt. Spekt., 1974, 36, no.3, pp.603-5, and Opt. Spectrosc., 1974, 36, no.3, pp.348-9.

38.216. Zborovskii, V.A., and Fradkin, E.E.: "Nonlinear Interaction Between Opposite Waves With Various Polarizations in a Ring Laser", Zh. Eksp. Teor. Fiz., 1974, 66, no.4, pp.1219-28, and Sov. Phys.-JETP, 1974, 39, no.4.

38.217. Kuhlke, D., and Dietel, W.: "Homogeneous Broadening and Nonlinear Absorption in a Ring Gas Laser", Phys. Lett., 1974, 48A, no.6, pp.441-2.

38.218. Andronova, I.A., and Kazarin, Yu.K.: "Experimental Investigation of Scattering in a Ring Resonator", Izv. VUZ Radiofiz., 1974, 17, no.9, pp.1287-90.

38.219. Pankratova, T.F.: "Natural Frequencies and Polarizations of a Ring Resonator", Opt. Spekt., 1974, 36, no.5, pp.969-74, and Opt. Spectrosc., 1974, 36, no.5, pp.569-71.

38.220. Guseva, T.V., and Fradkin, E.E.: "Diffraction Splitting of the Frequencies of Oppositely Travelling Waves in a Ring Gas Laser", Opt. Spekt., 1974, 36, no.5, pp.975-81, and Opt. Spectrosc., 1974, 36, no.5, pp.572-5.

38.221. Faxvog, F.R., and Gara, A.D.: "Travelling-Wave Gas Laser", Appl. Phys. Lett., 1974, 25, no.5, pp.306-8.

38.222. Whitford, B.G., et al.: "Frequency Measurements and Techniques in the 30-THz Region", Trans. IEEE, 1974, IM-23, no.4, pp.535-9.

38.223. Markelov, N.A.: "Frequency Characteristic of a Ring He-Ne Laser", Izv. VUZ Radiofiz., 1974, 17, no.11, pp.1642-8.

38.224. Burcev, P.: "Time Delay of Radar Wave in a Rotating System", Phys. Lett., 1974, 47A, no.5, pp.365-6.

38.225. Eng, R.S., et al.: "Determination of Absolute Frequencies of $^{12}C^{16}O$ and $^{13}C^{16}O$ Laser Lines", Appl. Phys. Lett., 1974, 24, no.5, pp.231-3.

38.226. Andronova, I.A., Bershtein, I.L., and Markelov, N.A.: "Experimental Determination of Zero of Phase Characteristic of Ring Laser at 3.39 micron", Kvantovaya Elektron., 1974, 1, no.3, pp.645-52, and Sov. J. Quantum Electron., 1974, 4, no.3, pp.360-4.

38.227. Geller, V.M., and Grif, G.I.: "Determination of Absolute Oscillation Frequency of the $3s_2-2p_4$ Transition in He-Ne Laser", Kvantovaya Elektron., 1974, 1, no.8, pp.1883-5.

38.228. Blazhnov, B.A.: "Frequency Synchronization Region for Colliding Waves in a Ring Laser During Perimeter Reorganization", Zh. Prikl. Spekt., 1974, 21, no.6, pp.990-6.

38.229. Solomakha, D.A.: "Equipment for Absolute Measurements of Wavelengths of Lasers", Izmer. Tekh., 1974, 17, no.8, pp.71-2, and Meas. Tech., 1974, 17, no.8, pp.1258-60.

38.230. Batarchukova, N.R., et al.: "Measuring the Wavelength of He-Ne Lasers Stabilized by the Lamp Dip", Izmer. Tekh., 1974, 17, no.7, p.83, and Meas. Tech., 1974, 17, no.7, pp.1116-7.

39. MATERIALS PARAMETERS

39.1. Roussy, G.: "Fixed-Frequency Measurement of Complex Permittivity of Solids by Resonant Cavities", J. Phys. (Paris), 1965, 26, p.64A.

39.2. Seifert, F.: "Electrodeless Microwave Measurements of Semiconductor Resistivity and Faraday Effect by Cavity-Wall Replacement", Proc. IEEE, 1965, 53, p.752.

39.3. Ruzicka, J.: "Measurement of Large Q-Factors of Cavity Resonators by the Decrement Method", Slab. Obz., 1965, 26, p.347.

39.4. Redhardt, A.: "Q Measurements on Reentrant Resonators", Z. Angew. Phys., 1965, 19, p.310.

39.5. Imanov, L.M., and Zul'fugarzade, K.E.: "Measuring the Temperature Dependence of Complex Permittivity at Microwave Frequencies", Prib. Tekh. Eksp., 1965, no.4, p.192, and Instrum. Exp. Tech., 1965, no.4, p.939.

39.6. Waldron, R.A., and Maxwell, S.P.: "Measurement of Material Properties by Stripline Cavity", Trans. IEEE, 1965, MTT-13, p.711.

39.7. Fatuzzo, E., and Mason, P.R.: "Precision Measurement of Microwave Permittivity. Liquids", J. Appl. Phys., 1965, 36, p.427.

39.8. Gunn, M.W., and Brown, J.: "Measurement of Semiconductor Properties in a Slotted-Waveguide Structure", Proc. IEE, 1965, 112, p.463.

39.9. Staniforth, J.A., Bennett, R.G., and Calderwood, J.H.: "Dielectric Measurement of High-Loss Liquids in Rectangular Waveguides", Electron. Lett., 1965, 1, p.4.

39.10. Verweel, J.: "Determination of Microwave Permeability and Permittivity in Cylindrical Cavities", Philips Res. Rep., 1965, 20, p.404.

39.11. Brydon, G.M., and Hepplestone, D.J.: "Microwave Measurement of Permittivity and Loss Tangent Over the Temperature Range 20-700°C", Proc. IEE, 1965, 112, p.421.

39.12. Newell, A.C., and Baird, R.C.: "Absolute Determination of Refractive Indices of Gases at 47.7 GHz", J. Appl. Phys., 1965, 36, p.3751.

39.13. Grigulis, Yu.K., and Aboltyn'sh, E.E.: "Contactless Two-Parameter Microwave Measurement on Semiconductors", Latv. PSR Zinat. Akad. Vestis, 1965, no.2, p.46.

39.14. Berecz, E., and Gellen, G.: "Problems in Measurement of Dielectric Properties of High-Conductivity Liquids and Solutions at Microwave Frequencies", Magy. Fiz. Foly., 1965, 13, p.399.

39.15. Seifert, F.: "Microwave Measurements of Conductivity of Low-Resistance Semiconductors", Arch. Elektr. Ubertrag., 1965, 19, p.492.

39.16. Vadnjal, M.: "Microwave Measurement of Permittivity of Low-Loss Dielectrics", Alta Freq., 1965, 34, p.204.

39.17. Larrabee, R.D.: "(Measurement of) Microwave Impedance of Semiconductor Posts in Waveguide. I", J. Appl. Phys., 1965, 36, p.1597.

39.18. Larrabee, R.D., Woodard, D.W., and Hicinbothem, W.A.: "Two-Position Probe Method for Microwave Determination of Electrical Properties of InSb. II", J. Appl. Phys., 1965, 36, p.1659.

39.19. Coumes, A., and Pic, E.: "Use of Tee (Junction) for Measurement of Microwave Permittivity of Low-Loss Liquids", Onde Electr., 1965, 45, p.1056.

39.20. Fatuzzo, E., and Mason, P.R.: "Method of Measurement of Microwave Dielectric Properties of Weakly Polar Liquids", J. Sci. Instrum., 1965, 42, p.37.

39.21. Mansingh, A.: "Dielectric Relaxation in Organic Liquids. Measurement at Microwaves", J. Inst. Telecommun. Eng., 1965, 11, p.433.

39.22. Kraeft, W.D.: "Measurement Method for Complex Permittivity of Concentrated Ion Solutions Using Cavity Resonators. I", Z. Phys. Chem., 1965, 230, p.368.

39.23. Brodwin, M.E., and Lu, P.S.: "Precise Cavity Technique for Measuring Low Resistivity", Proc. IEEE, 1965, 53, p.1942.

39.24. Aron, C.P., and Watkins, J.: "Measurement of Dielectric Properties of Low-Loss Ceramics at Microwave Frequencies", Proc. IEE, 1965, 112, p.1252.

39.25. Zal'tsman, E.B.: "Measuring Loss Tangent of Dielectrics by Resonator Transmission", Prib. Tekh. Eksp., 1966, 9, no.6, p.101, and Instrum. Exp. Tech., 1966, 9, no.6, p.1408.

39.26. Doucet, Y., Morabin, A., and Tete, A.: "Apparatus for Measuring Complex Permittivity of High-Loss Liquids", C. R. Acad. Sci. (Paris), 1966, 262B, p.752.

39.27. Roussy, G., and Felden, M.: "Sensitive Method for Measuring Complex Permittivity With Microwave Resonator", Trans. IEEE, 1966, MTT-14, p.171.

39.28. Steele, C.W.: "Nonresonant Perturbation Theory", Trans. IEEE, 1966, MTT-14, p.70.

39.29. Mallory, K.B., and Miller, R.H.: "Nonresonant Perturbation Measurements", Trans. IEEE, 1966, MTT-14, p.99.

39.30. Ralston, G.: "Simplified Measurement of Q of Microwave Transmission Cavities", Proc. IEEE, 1966, 54, p.311.

39.31. Stuchly, S., and Kraszewski, A.: "Microwave Measurement Method for Moisture Content in Solids and Liquids. I", Automat. Kontrol., 1966, 12, no.2, p.51.

39.32. Budzinski, W.V., and Garfunkel, M.P.: "Magnetic-Field-Induced Anisotropy of the Superconducting Energy Gap Determined by Microwave Absorption", Phys. Rev. Lett., 1966, 16, p.1100.

39.33. Champlin, K.S., and Glover, G.H.: "Influence of Waveguide Contact on Measured Complex Permittivity of Semiconductors", J. Appl. Phys., 1966, 37, p.2355.

39.34. Freedman, N.J.: "Swept-Field System for Rapid Measurements on Microwave Ferrite Devices", Mullard Tech. Commun., 1966, 8, p.271.

39.35. Tamm, K., and Schneider, M.: "Determination of Complex Permittivity of Electrolytes at Microwave Frequencies", Z. Angew. Phys., 1966, 20, p.544.

39.36. Helberg, H.W., and Wartenberg, B.: "Measurement of Permittivity and Permeability in the Microwave Region by Resonators", Z. Angew. Phys., 1966, 20, p.505.

39.37. Reboul, J.P., and Caillon, P.: "Measurement of Permittivity at Hyperfrequencies by Resonant Methods", C. R. Acad. Sci. (Paris), 1966, 262B, p.903.

39.38. Champlin, K.S., and Glover, G.H.: "Gap Effect in Measurement of Large Permittivities", Trans. IEEE, 1966, MTT-14, p.397.

39.39. Gelas, J., and Lestrade, J.C.: "Permittivity Measurements in Liquids at 3-cm Wavelength", Onde Electr., 1966, 46, p.989.

39.40. Dianov, E.M., and Irisova, N.A.: "Determination of Absorption Coefficients of Solids in the Short-Millimetre Range", Zh. Prikl. Spekt., 1966, 5, p.251.

39.41. Bondarenko, R.N., and Gladyshev, G.I.: "Measurement of Permittivity of Liquids Within a Cavity", Radiotekh. Elektron., 1966, 11, p.149, and Radio Eng. Electron. Phys., 1966, 11, p.123.

39.42. Rupke, H.D.: "Coupling of Microwave Resonators for Measurement on Dielectrics", Arch. Elektr. Ubertrag., 1966, 20, p.617.

39.43. Charru, A., and Bretenoux, A.: "Measurement of Permittivity of Absorbent Media at 35 GHz by Reflecto-Polarimetry in Free Space", C. R. Acad. Sci. (Paris), 1966, 263B, p.45.

39.44. Hanna, F.F., and Bishai, A.M.: "Measurement of Dielectric Properties of Samples of Egyptian Soil by Waveguide Method", Slab. Obz., 1966, 27, p.13.

39.45. Fukumitsu, O.: "Location Invariant Method for Measuring Permittivity at Microwave Frequencies", J. Inst. Electr. Commun. Eng. Jap., 1966, 49, p.15.

39.46. Schunzel, M., and Stockhausen, M.: "Measurement of Complex Permittivity of Liquids by Free-Space Reflection of Millimetre Waves", Z. Angew. Phys., 1966, 21, p.508.

39.47. Wagner, H.: "Measurement of Microwave Complex Permittivity of Te", Z. Phys. 1966, 193, p.218.

39.48. Seifert, F.: "Measurement of Microwave Faraday Effect in Low-Resistance Semiconductors", Arch. Elektr. Ubertrag., 1966, 20, p.169.

39.49. Dube, D.C., Parshad, R., and Yadav, R.S.: "Determination of Permittivity of Liquid Air at Microwave Frequencies", Indian J. Pure Appl. Phys., 1966, 4, p.428.

39.50. Vinogradov, E.A., et al.: "Measurement of Dielectric Characteristics of Liquid Nitrogen at 2.3 mm", Zh. Tekh. Fiz., 1966, 36, p.1319, and Sov. Phys.-Tech. Phys., 1967, 11, no.7, p.983.

39.51. Hamaguchi, C., Kono, T., and Inuishi, Y.: "Microwave Measurement of Differential Negative Conductivity in n-GaAs", Phys. Lett., 1967, 24A, no.10, pp.500-1.

39.52. Rueggeberg, W.: "Microwave Properties of Adhesively Bonded Aluminium Foil Seams", Trans. IEEE, 1967, MTT-15, no.2, pp.117-8.

39.53. Aron, C.P.: "Effect of Degenerate E_{11n} Mode in H_{01n} Mode Cavity on Measurement of Complex Permittivity", Proc. IEEE, 1967, 114, no.8, pp.1030-4.

39.54. Wagner, H.: "Microwave Measurements on Semiconductor Discs Without Rigid Mounting in the Waveguide", Arch. Elektr. Ubertrag., 1967, 21, no.6, pp.321-5.

39.55. Lizuka, K., and Sugimoto, T.: "Measurement of the Dielectric Properties of Soft Materials With High Loss and High Permittivity in a Parallel-Plate Region", Proc. IEE, 1967, 114, no.9, pp.1219-22.

39.56. Bodi, A., and Baican, R.: "Measurement of Complex Permittivity and Permeability, and Nonreciprocal Attenuation for Ferrites at Microwave Frequencies", Rev. Roum. Phys., 1967, 12, no.3, pp.305-10.

39.57. Summerhill, S.: "Microwaves in the Measurement of Moisture", Instrum. Rev., 1967, 14, no.10, pp.419-22.

39.58. Breeden, K.H., and Sheppard, A.P.: "Submillimetre-Wave Dielectric Measurements", Microwave J., 1967, 10, no.12, pp.59-64.

39.59. Tateno, H., and Kataoka, S.: "Measurements of Conductivity and Permittivity of Semiconductors at Microwave Frequencies", Bull. Electrotech. Lab., 1967, 31, no.3, pp.371-80.

39.60. Griffin, D.W.: "Wideband Balancing of Bimodal Cavities Used for Microwave Measurements on Gyrotropic Media", Br. J. Appl. Phys., 1967, 18, no.12, pp.1743-51.

39.61. Grant, E.H., and Shack, R.: "Complex Permittivity Measurements at 8.6 mm Over the Temperature Range 1-60°C", Br. J. Appl. Phys., 1967, 18, no.12, pp.1807-14.

39.62. Amrhein, E.M., Roder, H., and Muller, F.H.: "Cavity Resonator for Measurement of Complex Permittivity at 32 GHz between 4.2°K and 300°K", Z. Angew. Phys., 1967, 24, no.1, pp.18-20.

39.63. Zal'tsman, E.B., et al.: "Equipment for Measuring Parameters of Heated Solid Dielectrics at 8-mm", Izmer. Tekh., 1967, no.8, pp.54-6, and Meas. Tech., 1967, no.8, pp.966-8.

39.64. Hanna, F.F., and Abdel-Nour, K.N.: "Reflection of Microwaves from Buildings and Determination of Brickwork Permittivity", Slab. Obz., 1968, 29, no.4, pp.231-2.

39.65. Casini, G., and Fagioli, O.: "Microwave Refractometer for Gases", Alta Freq., 1968, 37, no.1, pp.64-9.

39.66. Amilac, J.L., Boutan, L., and Matheau, J.C.: "Measurement of Permittivity of Low-Loss Liquids and Solids in an Overmoded Waveguide", C. R. Acad. Sci. (Paris), 1968, 266B, no.24, pp.1441-4.

39.67. Iwasaki, H.: "Reflectance of a Thin Rod of High Permittivity in a Waveguide at 9.3 GHz", J. Radio Res. Lab., 1968, 15, no.82, p.295.

39.68. Wolff, I.: "Dynamical Modes in a Ferrite-Filled Coaxial Cavity for Measuring Materials Parameters at Millimetre Wavelengths", Z. Angew. Phys., 1968, 25, p.286.

39.69. Brydon, G.M.: "High-Temperature Microwave Permittivity Measurements", Component Technol., 1968, 3, no.2, p.21.

39.70. Thunqvist, D.E.O., and Sollbrand, S.G.: "Two Methods for Measuring High Permittivity at Microwave Frequencies", Trans. IEEE, 1968, IM-17, p.170.

39.71. Goulon, J., Roussy, G., and Rivail, J.L.: "Michelson's Interferometer in an Overmoded Waveguide. Permittivity Measurements on Liquids", Rev. Phys. Appl., 1968, 3, p.231.

39.72. Leibrecht, K.: "Method for Measurement of Q-Factor of Cavities", Rev. Sci. Instrum., 1968, 39, p.1919.

39.73. Horton, J.B., and Burdick, C.A.: "Measurement of Materials With Large Permittivity", Trans. IEEE, 1968, MTT-16, p.873.

39.74. Collins, J.H., et al.: "Measurements of Magnetoelastic Constants by Microwave Techniques", Int. J. Electron., 1968, 24, p.453.

39.75. Naumenko, I.G., and Petinov, V.I.: "Measurement of Cavity Q by Resonance-Curve Derivative", Prib. Tekh. Eksp., 1968, no.2, p.139, and Instrum. Exp. Tech., 1968, no.2, p.388.

39.76. Yamanaka, H.: "Dielectric Measurements in the Microwave Region", Res. Electrotech. Lab., 1968, no.683, p.1.

39.77. Toki, M., and Murakami, I.: "Measurement of Properties of Materials by Reflection Factor", Bull. Fac. Eng. Yokohama Natl. Univ., 1968, 17, March, p.55.

39.78. Ura, K., and Takaoka, A.: "Buildup of Oscillations in Microwave Resonant Cavity Excited by Swept-Frequency Signal", Technol. Rep. Osaka Univ., 1968, 18, p.197.

39.79. Weissman, I.: "Niobium TM_{010}-Mode Cavity With High Electric Field and Q", Appl. Phys. Lett., 1968, 13, p.390.

39.80. Petrov, A.S., and Tyul'kov, G.I.: "Apparatus for Measuring Resonator Q-Factors in the Microwave Region", Prib. Tekh. Eksp., 1968, no.6, p.126, and Instrum. Exp. Tech., 1968, no.6, p.1414.

39.81. Yamanaka, H.: "Study of Q Measurement by Decrement Method", Electron. Commun. Jap., 1969, 54, p.113.

39.82. Champlin, K.S., Glover, G.H., and Holm, J.D.: "Bulk Microwave Conductivity of Semiconductors Determined from TE_{01}-Mode Reflectivity of Boule Surface", Trans. IEEE, 1969, IM-18, p.105.

39.83. Botsco, R.J.: "Nondestructive Testing of Plastics With Microwaves", Mater. Eval., 1969, 27, no.6, p.25a.

39.84. Milewski, A.: "Q Measurement of Microwave Resonant Cavities With Use of Double-Klystron Modulation", Przegl. Elektron., 1969, 10, no.10, p.498.

39.85. Mikhailov, Yu.A., and Khramov, V.A.: "Application of AFC for Measuring Parameters of High-Q Cavities", Otbor Peredacha Inf., 1969, 69, no.20, p.113.

39.86. Richter, K.R.: "Measurement of Short-Term Change of Cavity Q and Resonant Frequency", Trans. IEEE, 1969, MTT-17, p.339.

39.87. Hall, R.B.: "Perturbation of Microwave Cavities by Lossy Dielectrics and Plasmas", J. Appl. Phys., 1969, 40, p.30.

39.88. Brady, M.M.: "Loss Measurements of Wet Textiles at 9 GHz", J. Microwave Power, 1969, 3, no.4, p.194.

39.89. Miane, J.L.: "Reflectance of a Dielectric Sheet in a Waveguide. Measurement of Complex Permittivities", Rev. Phys. Appl., 1969, 4, no.1, p.42.

39.90. Charru, A., Bretenoux, A., and Sarremejean, A.: "Measurement of Permittivity Up to 1200°C by Reflecto-Polarimetry at 35 GHz", Rev. Phys. Appl., 1969, 4, no.1, p.37.

39.91. Yamanaka, H., and Hashiba, K.: "Microwave Measurements of (Thin) Fan-Shaped Dielectrics", Bull. Electrotech. Lab., 1969, 33, no.4, p.65.

39.92. Yamanaka, H.: "Effect of Decay Period of Trapezoidal Modulating Wave on Q-Factor Measurement by Decrement Method", Bull. Electrotech. Lab., 1969, 33, no.4, p.73.

39.93. Julke, L.: "Cavity Resonator Method for Determination of Permittivity and Permeability", Hochfreq. Elektroak., 1969, 78, no.2, p.70.

39.94. Dryagin, Yu.A., and Chukhvichev, A.N.: "Measuring the Parameters of Solid Dielectrics in the Millimetre Band by a Resonance Method", Izv. VUZ Radiofiz., 1969, 12, no.8, p.1245.

39.95. Breeden, K.H.: "Error Analysis for Waveguide-Bridge Permittivity Measurements at Millimetre Wavelengths", Trans. IEEE, 1969, IM-18, p.203.

39.96. Tinga, W.R.: "Dielectric Properties of Douglas Fir at 2.45 GHz", J. Microwave Power, 1969, 4, no.3, pp.162-4.

39.97. Buts, V.A.: "Theory of Resonators With a Ferrite", Izv. VUZ Radiofiz., 1969, 12, no.4, p.628.

39.98. Grigas, I.P., and Shugurov, V.K.: "Determination of Microwave Reflection Factor and Permittivity of Thin Cylindrical Samples of Dielectrics and Semiconductors", Izv. VUZ Radiofiz., 1969, 12, no.2, p.307.

39.99. Buckmaster, H.A., and Kloza, M.J.: "Fundamental Limit to Balance of a Microwave Bridge Containing a Dispersive Element", Trans. IEEE, 1969, IM-18, p.237.

39.100. Bakhtin, V.D., et al.: "Measurement of Permittivity by Amplitude Ratio of Waves Reflected From and Transmitted Through the Dielectric", Izv. VUZ Radioelektron., 1969, 12, no.6, p.643.

39.101. Hughes, J.J., Napoli, L.S., and Reichert, W.F.: "Novel Technique for Measuring the Q-Factor of Thin-Film Lumped Elements at Microwave Frequencies", Electron. Lett., 1969, 5, p.535.

39.102. Cutnell, J.D., et al.: "Dielectric Permittivity Measurements at Centimetre Wavelengths", Rev. Sci. Instrum., 1969, 40, p.908.

39.103. Zeeck, E., and Voitlander, J.: "Microwave Equipment for Measurements of Powdered Semiconductors", Z. Angew. Phys., 1969, 28, no.3, p.137.

39.104. Chakravarti, A.N., Biswas, S.N., and Rakshit, S.: "Method for Measuring Electrical Conductivity in the Active Region of a GaAs Junction Laser", Int. J. Electron., 1969, 27, p.397.

39.105. Brodwin, M.E., and Lu, P.S.: "Cryogenic Microwave Cavity for Semiconductor Diagnostics", Trans. IEEE, 1969, IM-18, p.208.

39.106. Haniotis, Z., and Gunthard, H.H.: "Frequency Shift and Q Deterioration of TE_{011} Cylindrical Cavity by Coaxial Loading With Dielectric", Z. Angew. Math. Phys., 1969, 20, no.5, p.771.

39.107. Swiderski, J., and Mroziewicz, B.: "Application of GaAs Junction Lasers in Investigation of Properties of Semiconductor Material", Electron Technol., 1969, 2, no.1, p.77.

39.108. Soumpasis, D., and Luders, K.: "Superconducting Nb-Ti Microwave Cavity", J. Appl. Phys., 1970, 41, p.2475.

39.109. Olyphant, M., and Ball, J.H.: "Stripline Methods for Dielectric Measurements at Microwave Frequencies", Trans. IEEE, 1970, EI-5, p.26.

39.110. Grant, E.H., and Sheppard, R.J.: "Measurement of Permittivity of High-Loss Liquids at Microwave Frequencies Using an Unmatched Phaseshifter", J. Phys. D, 1970, 3, p.84.

39.111. Bahl, I.J., and Gupta, K.C.: "Measurement of Parameters of an Artificial Dielectric Using a Partially Filled Parallel-Plate Waveguide", Int. J. Electron., 1970, 28, p.173.

39.112. Lindberg, K.: "Microwave Moisture Meters for Paper and Pulp Industry", Meas. Control, 1970, 3, no.3, p.T33.

39.113. Braginskii, V.B., Manukin, A.B., and Tikhonov, M.Yu.: "Investigation of Dissipative Ponderomotive Effects of EM Radiation", Zh. Eksp. Teor. Fiz., 1970, 58, no.5, p.1549, and Sov. Phys.-JETP, 1970, 31, no.5, pp.829-30.

39.114. Bilenko, D.I., Lun'kov, A.E., and Yazikov, V.N.: "Measurement of Complex Reflection Factor of Semiconductors in the Millimetre Range", Izv. VUZ Radiofiz., 1970, 13, no.3, p.453.

39.115. Gehre, O.: "Microwave Interferometer for Wideband Measurements of Complex Transmission Factors", Trans. IEEE, 1970, IM-19, p.14.

39.116. Stuchly, S.: "Measurement Method for Dielectric Properties of Granulated Substances in the Microwave Range", Arch. Elektrotech., 1970, 19, no.2, p.291.

39.117. Horikx, C.M.: "Resonant Cavities for Dielectric Measurements", J. Phys. E, 1970, 3, no.11, pp.871-4.

39.118. Batsco, R.: "Microwave Moisture Measurement", Instrum. Control Syst., 1970, 43, no.5, p.116.

39.119. Friedburg, H., and Szecsi, L.: "Superconducting Coupling System for Microwave Resonators", Z. Angew. Phys., 1970, 29, no.4, p.252.

39.120. Harrison, A.W., Hansen, C., and Will, D.W.: "Simple Digital Near-IR Spectrometer", Appl. Opt., 1970, 9, p.1610.

39.121. Faucheron, G.: "Determination of Complex Permittivity of Dielectric Samples Inserted in a Waveguide", Ann. Telecommun., 1970, 25, no.7-8, pp.248-58.

39.122. Maier, H.G.: "Method for Measurement of Complex Permittivities of Foils and Thin Plates in the Microwave Band", Frequenz, 1970, 24, no.10, pp.303-7.

39.123. Brain, M.: "Measurement and Control of Moisture in Wheat", Meas. Control, 1970, 3, no.11, pp.T181-7.

39.124. Burtovoi, D.P., et al.: "Use of Open Cylindrical Below-Cutoff Cavity for Study of Dielectric Properties", Izv. VUZ Radioelektron., 1970, 13, no.9, pp.1085-91.

39.125. Hansen, N.P.: "Determination of Submillimetre Harmonics Using a Lamellar Grating Fourier Spectrometer", Rev. Sci. Instrum., 1970, 41, no.11, pp.1678-80.

39.126. Battaglia, A., and Gozzini, A.:A "Experimental Study of Confocal Fabry-Perot Microwave Resonators", Nuovo Cim., 1970, 69, no.2, pp.121-51.

39.127. Deutsch, J., and Jung, H.J.: "Measurement of the Effective Permittivity of Microstrip Lines in the Range 2-12 GHz", Nachr. Tech. Z., 1970, 23, no.12,pp.613-9.

39.128. Dem'yanov, A.A., and Meriakri, V.V.: "Automatic Measurement Method of Moisture Content in Petroleum", Avtometriya, 1970, no.5, pp.70-5.

39.129. Ivashka, V.L., et al.: "Determination of Specific Surface Resistance of Thin Films Using Microwave Fields", Litov. Fiz. Sb., 1970, 10, no.3, pp.385-90.

39.130. Ashley, J.R., and Palka, F.M.: "Modulation Method for Measurement of Microwave Oscillator Q-Factor", Trans. IEEE, 1970, MTT-18, no.11, pp.1002-4.

39.131. Bosisio, R.G., Giroux, M., and Couderc, D.: "Paper-Sheet Moisture Measurements by Microwave Phase Perturbation Techniques", J. Microwave Power, 1970, 5, no.1, pp.25-34.

39.132. Beran, Z.: "Application of Microwaves to Moisture Measurement", TESLA Electron., 1970, 3, no.4, pp.117-22.

39.133. Apletalin, V.N., Meriakri, V.V., and Chigryay, Ye.Ye.: "Measurement of Absorbing and Reflecting Properties of Water at 0.8-2.0 mm", Radiotekh. Elektron., 1970, 15, no.7, and Radio Eng. Electron. Phys., 1970, 15, no.7, pp.1286-8.

39.134. Danilyuk, Yu.L., Koleda, F.A., and Trubitsyha, O.N.: "Fabry-Perot Interferometer for Submillimetre (Dielectric) Measurements", Prib. Tekh. Eksp., 1970, no.6, pp.149-51, and Instrum. Exp. Tech., 1970, no.6, pp.1717-9.

39.135. Yuba, Y.: "Measurement of Q-Factor of Fabry-Perot Resonators at Millimetre Wavelengths", Mem. Fac. Ind. Arts Kyoto Tech. Univ., 1970, 19, pp.28-43.

39.136. Lonngren, K.E., Russell, J.P., and Meyers, B.L.: "Use of Microwaves to Evaluate Temporal Hydration of Cement Paste", J. Microwave Power, 1970, 5, no.1, pp.23-4.

39.137. Kumagai, H., et al.: "Measurement of Q of a Superconducting Resonant Cavity", J. Fac. Eng. Univ. Tokyo, 1970, ser.A, no.8, pp.36-7.

39.138. Kozlov, G.V.: "Apparatus for Investigating the Birefringence of Anisotropic Media at Short Millimetre Wavelengths", Prib. Tekh. Eksp., 1970, 13, no.6, pp.147-8, and Instrum. Exp. Tech., 1970, 13,no.6,pp.1714-6.

39.139. Al'tshuler, Yu.G., Kats, L.I., and Revzin, R.M.: "Measurement of the Refractive Index of Dielectrics in the Submillimetre Range", Prib. Tekh. Eksp., 1970, 13, no.6, pp.145-6, and Instrum. Exp. Tech., 1970, 13, no.6, pp.1712-3.

39.140. Badian, L., and Milewski, A.: "Perturbation Method for Investigation of Dielectrics in the Microwave Band", Elektronika, 1971, no.2, pp.72-5.

39.141. Lin, T., and Wolff, I.: "Method for Measuring Dielectric and Magnetic Losses in Gyrotropic Ferrites", Arch. Elektron. Übertrag., 1971, 25, no.1, pp.9-16.

39.142. Dohi, T., and Suzuki, T.: "Attainment of High-Resolution Holographic Fourier-Transform Spectroscopy", Appl. Opt., 1971, 10, no.5, pp.1137-40.

39.143. Wapoli, L.S., and Hughes, J.J.: "Simple Technique for Accurate Determination of Permittivity of MIC Substrates", Trans. IEEE, 1971, MTT-19, no.7, pp.664-5.

39.144. Ahluwalia, H.P.S., Boerner, W.M., and Hamid, M.A.K.: "Microwave Test Chamber for Measuring Permittivity of Thin Layers", Proc. IREE Aust., 1971, 32, no.6,pp.253-69.

39.145. Ermet, H.: "Measurements of Ferrites at Millimetre Wavelengths Using Quasi-Optical Methods", Frequenz, 1971, 25, no.6, pp.171-7.

39.146. Dix, G., Helberg, H.W., and Wartenberg, B.: "Microwave (Resonator) Apparatus for Measurement of Anisotropic Permittivity", Phys. Status Solidi, 1971, 5, no.3, pp.633-6.

39.147. Bottreau, A., and Marzat, Cl.: "Reflection Interferometer Converted to an Open-Branch Type for Measurement of Complex Permittivities in the 8-mm Band", Rev. Phys. Appl., 1971, 6, no.2, pp.211-2.

39.148. Pradoux, D., et al.: "Propagation in a Cylindrical Waveguide Filled With an Anisotropic Medium. Permittivity Determination", Rev. Phys. Appl., 1971, 6, no.2, pp.219-23.

39.149. Bassett, H.L.: "Free-Space Focused Microwave System to Determine the Complex Permittivity of Materials to Temperatures Exceeding 2000°C", Rev. Sci. Instrum., 1971, 42, no.2, pp.200-4.

39.150. Molchanov, V.I., and Poplavko, Yu.M.: "Measurement of Dielectric Properties of Ferroelectrics at Microwaves Using a Symmetric Stripline", Izv. VUZ Radioelektron., 1971, 14, no.2, pp.158-62.

39.151. Vernon, R.J., and Dorschmer, T.A.: "Improved Interferometric Polarization Analyser for Measuring the Microwave Magneto-Kerr Effect in Semiconductors", Trans. IEEE, 1971, MTT-19, no.3, pp.287-94.

39.152. Lazebuyi, B.V., and Shul'ga, V.F.: "Dynamic Method of Measuring Cavity Characteristics", Izv. VUZ Radioelektron., 1971, 14, no.1, pp.99-102.

39.153. Charru, A., Bretenoux, A., and Duhau, A.: "Measurement by Reflectometry at 35 GHz of Permittivity of Anisotropic Absorbing Media", Rev. Phys. Appl., 1971, 6, no.2, pp.199-201.

39.154. Jager, D., and Rabus, W.: "Determination of Thin-Film Material Properties in a Stripline Resonator", Z. Angew. Phys., 1971, 31, no.5-6, pp.271-4.

39.155. Lopez, P., Gourdon, J.C., and Pescia, J.: "Method for Measuring Q-Factor of Cavity Resonator Using Microwave Amplitude Modulation", C. R. Acad. Sci. (Paris), 1971, 273B, no.5, pp.239-42.

39.156. Stephenson, I.M., and Easter, B.: "Resonant Techniques for Establishing Equivalent Circuits of Small Discontinuities in Microstrip", Electron. Lett., 1971, 17, no.19, pp.582-4.

39.157. Bhagyalakshmi, K.K., and Kataria, B.K.: "Use of Precision Slotted Line in Dielectric Measurements at Higher Frequency Ranges", J. Inst. Telecommun. Eng., 1971, 17, no.4, pp.131-4.

39.158. Buzin, I.M.: "Dynamic Method of Measuring Q-Factor of Microwave Resonators", Prib. Tekh. Eksp., 1971, 14, no.1, pp.160-1, and Instrum. Exp. Tech., 1971, 14, no.1, pp.188-9.

39.159. Cullen, A.L., and Yu, P.K.: "Accurate Measurement of Permittivity by Open Resonator", Proc. Roy. Soc., 1971, 325A, no.1536, pp.493-509.

39.160. Baican, R.: "Investigations on Semiconductors by Microwave Methods", Stud. Cercet. Fiz., 1971, 23, no.9, pp.1015-31.

39.161. Kovbasa, A.P., and Shelamov, G.N.: "Measurements of Parameters of Dielectrics Using Waveguide-Resonance Method", Izmer. Tekh., 1971, 14, no.3, pp.58-9, and Meas. Tech., 1971, 14, no.3, pp.447-9.

39.162. Apletalin, V.N., et al.: "Quasi-Optic Technique of Measuring Complex Permittivity", Radiotekh. Elektron., 1971, 16, no.1, and Radio Eng. Electron. Phys., 1971, 16, no.1, pp.152-5.

39.163. Veszely, Gy.: "Microwave Diagnostics of Semiconductor Crystals", Acta Tech. Acad. Sci. Hung., 1971, 70, no.3-4, pp.331-42.

39.164. Gertsenshtein, M.Ye., et al.: "Cold (Q-Factor) Measurements on Idler Circuit of a Parametric Amplifier", Radiotekh. Elektron., 1971, 16, no.3, and Radio Eng. Electron. Phys., 1971, 16, no.3, pp.475-80.

39.165. Kazantsev, Yu.N., and Udalov, V.V.: "Measurement of Attenuation in Gas-Dielectric Waveguides at Submillimetre Wavelengths", Radiotekh. Elektron., 1971, 16, no.3, and Radio Eng. Electron. Phys., 1971, 16, no.3, pp.544-6.

39.166. Gazimuddin, M.: "TM_{010}-Mode Resonator. Temperature Variation of Resonant Frequency, Resonance Curves, and Analysis", Pak. J. Sci. Ind. Res., 1971, 14, no.4-5, pp.328-32.

39.167. Compy, E.M., and Hansen, U.J.: "(Resonator) Apparatus for Microwave Studies at High Pressures and Low Temperatures", Rev. Sci. Instrum., 1971, 42, no.8, pp.1215-7.

39.168. Burroughs, W.J., and Harries, J.E.: "Use of Apodization in Submillimetre Fourier-Transform Spectroscopy", Infrared Phys., 1971, 11, no.2, pp.99-108.

39.169. Kolosovskii, O.A.: "Refractive Index of Gas-Discharge Medium in a CO_2 Laser", Kvantovaya Elektron., 1971, no.4, pp.107-9.

39.170. Aleshechkin, V.N., et al.: "Investigation of Solids at Submillimetre Wavelengths", Prib. Tekh. Eksp., 1971, 14, no.4, pp.150-1, and Instrum. Exp. Tech., 1971, 14, no.4, pp.1123-5.

39.171. Kozlov, G.V.: "Measurement of the Refractive Indices of Dielectrics at Millimetre Wavelengths", Prib. Tekh. Eksp., 1971, 14, no.4, pp.152-4, and Instrum. Exp. Tech., 1971, 14, no.4, pp.1126-8.

39.172. Kliger, M., and Zagrodzinski, J.A.: "Measurements of High-Permittivity Materials in the Microwave Region", Arch. Elektron. Ubertrag., 1972, 26, no.5, pp.243-4.

39.173. Ungurs, I.: "Effect of Curvilinearity of Electromagnetic Wavefront on Measurement of Impedance of Conducting Media", Latv. PSR Zinat. Akad. Vestis, Fiz. Teh. Ser., 1972, no.2, pp.69-78.

39.174. Lynch, A.C., and Ayers, S.: "Measurement of Small Dielectric Loss at Microwave Frequencies", Proc. IEE, 1972, 119, no.6, pp.767-70.

39.175. Kutik, M., and Potmesil, J.:
"Measuring Loss of Optical Devices Within
a Gas-Laser Resonator", Slab. Obz., 1972,
33, no.1, pp.14-20.

39.176. Strebel, B.: "Measuring the Q of
a Microwave Resonator Coupled With a Drifting InSb Plasma", Nachr. Tech. Z., 1972,
25, no.2, pp.69-71.

39.177. Menmet, K., McPhun, M.K., and
Michie, D.F.: "Simple Resonator Method for
Measuring Dispersion of Microstrip", Elec-
tron. Lett., 1972, 8, no.6, pp.165-6.

39.178. Wenger, N.C., and Smetana, J.:
"Hydrogen-Density Measurements Using an Open-
Ended Microwave Cavity", Trans. IEEE, 1972,
IM-21, no.2, pp.105-14.

39.179. Groll, H., and Wiesbeck, W.:
"Measurement of Permittivity of Substrates
of Microwave Striplines", Nachr. Tech. Z.,
1971, 25, no.6, pp.265-9.

39.180. Karimov, N.N., and Shakov, Kh.K.:
"Measurement of Glass/Resin Ratio in a Fibre-
glass Sheet", Izv. VUZ Elektromekh., 1972,
no.3, pp.333-6.

39.181. Eugene, G., and Mollet, B.: "Automatic Measurement of Microwave-Cavity Parameters Using Stable Sampled Control Loops",
Electron. Lett., 1972, 8, no.17, pp.434-6.

39.182. Galwas, B.: "Wobbulation Method
of Electron Admittance Measurement for a
Reflex Klystron", Arch. Elektrotech., 1972,
21, no.1, pp.159-68.

39.183. Frey, W.: "Microwave Measurement
of Surface Conductivity and Permittivity of
Thin Layers in an E_{010} Resonator", Electron.
Lett., 1972, 8, no.19, pp.486-8.

39.184. Kent, M.: "Use of Stripline Configurations in Microwave Moisture Measurement", J. Microwave Power, 1972, 7, no.3,
pp.185-93.

39.185. Gupta, H.M.: "Microwave Measurement of Complex Permittivity of Semiconductor Using Partially Loaded Waveguides", J.
Inst. Telecommun. Eng., 1972, 18, no.2,
pp.100-5.

39.186. Picherti, F., and Easthope, J.P.:
"Measurement of Variation of Emissivity
With Temperature at 35 GHz. Permittivity
Determination", Rev. Sci. Instrum., 1972,
43, no.11, pp.1710-1.

39.187. Otoshi, T.Y.: "Precision Reflect-
ance-Loss Measurements of Perforated-Plate
Mesh Materials by a Waveguide Technique",
Trans. IEEE, 1972, IM-21, no.4, pp.451-7.

39.188. Deutsch, J., and Jung, H.J.:
"Measurement of Permittivity of Ceramics
(for Microstrip Substrates)", Nachr. Tech.
Z., 1972, 25, no.10, pp.463-4.

39.189. Chamberlain, J.: "Measurements in
the Submillimetre-Wave Region", Trans. IEEE,
1972, IM-21, no.4, pp.438-42.

39.190. Cullen, A.L., Nagenthiram, P.,
and Williams, A.D.: "Improvement in Open-
Resonator Permittivity Measurement", Electron. Lett., 1972, 8, no.23, pp.577-9.

39.191. Badian, L., Makarewicz, W., and
Milewski, A.: "Precision Equipment for
Microwave Measurement of Complex Permittiv-
ity", Elektronika, 1972, no.10, pp.425-7.

39.192. Davidson, J.J., and Watkins, J.:
"Measurements on Alumina and Glasses Using
a TM_{020}-Mode Resonant Cavity", Proc. IEE,
1972, 119, no.12, pp.1759-63.

39.193. Moskalenko, N.I., Zotov, O.V.,
and Dugin, V.P.: "Absorption of Emission by
CO_2, NH_3, and Water Vapour from CO_2 Laser",
Zh. Prikl. Spekt., 1972, 17, no.5, pp.881-4.

39.194. Uhlir, A.: "Automatic Microwave
Q Measurements for Determination of Small
Attenuations", Trans. IEEE, 1972, MTT-20,
no.1, pp.38-41.

39.195. Eldumiati, I.I., and Haddad, G.I.:
"Cavity Perturbation Techniques for Measurement of Microwave Conductivity and Dielectric
Constant of Bulk Semiconductors", Trans. IEEE,
1972, MTT-20, no.2, pp.126-32.

39.196. Hanfling, J., and Bott, L.: "Measurement of Dielectric Materials Using a Cutoff Circular-Waveguide Cavity", Trans. IEEE,
1972, MTT-20, no.3, pp.233-5.

39.197. Bourgoin, D., Volf, E., and Joly,
M.: "Microwave Complex-Permittivity Measurement in Strongly Polar Liquids", J. Phys. D,
1972, 5, no.3, pp.589-600.

39.198. Kuznetsov, V.A., and Shchuka, A.A.:
"Use of Laser Probe for Investigating Sorp-
tion Properties of Surfaces", Prib. Tekh.
Eksp., 1972, 15, no.3, pp.171-3, and Instrum.
Exp. Tech., 1972, 15, no.3, pp.817-9.

39.199. Saha, A.R., Das, A.K., and
Banerjee, P.K.: "Microwave Studies of c-Axis
Resistivity of Pyrolytic Graphite", Indian
J. Phys., 1972, 46, no.12, pp.537-46.

39.200. Kornev, Yu.V., and Sysoev, V.Ya.:
"Measurements of Losses and Phase Shifts in
Striplines With Thin Ferromagnetic Films",
Izmer. Tekh., 1972, 15, no.8, p.89, and
Meas. Tech., 1972, 15, no.8, pp.1273-4.

39.201. Kolosovskii, O.A., and Ustimenko,
L.N.: "Measurement of Temperature Coefficient
of Refractive Index of IR Materials Using a
CO_2 Laser", Opt. Spekt., 1972, 33, no.4,
pp.781-2, and Opt. Spectrosc., 1972, 33,
no.4, pp.430-1.

39.202. Ohkawa, S., Saito, T., and
Yamamoto, H.: "Multipoint Method for Measurement of Complex Permittivity in Microwave
Range", Electron. Commun. Jap., 1972, 55,
no.7, pp.54-61.

39.203. Aleksandrov, E.B., and Kulyasov, V.N.: "Faraday Effect in ^{136}Xe Discharge", Opt. Spekt., 1972, 33, no.5, pp.1010-1, and Opt. Spectrosc., 1972, 33, no.5, pp.557-8.

39.204. Al'tshuler, Yu.G., Yershov, V.V., and Kats, L.I.: "Experimental Investigation of Propagation in Waveguide Partially Filled with n-InSb in a Transverse Magnetic Field", Radiotekh. Elektron., 1972, 17, no.8, pp.1737-9, and Radio Eng. Electron. Phys., 1972, 17, no.8, pp.1373-5.

39.205. Butlin, R.S., and McPhun, M.K.: "Surface-Resistance Measurements of Thin Conducting Films at 10 GHz", Electron. Lett., 1972, 8, no.26, pp.637-9.

39.206. Baranov, L.I., Gamanyuk, V.B., and Usanov, D.A.: "Determining the Conductivity and Permittivity of Semiconductors at Microwave Frequencies", Radiotekh. Elektron., 1972, 17, no.2, pp.426-8, and Radio Eng. Electron. Phys., 1972, 17, no.2, pp.332-4.

39.207. Balkhanov, V.Ya., Zhivotov, V.K., and Titov, A.V.: "Use of Holographic Fourier Spectroscopy for Analysis of Microwave Radiation", Prib. Tekh. Eksp., 1972, 15, no.3, pp.146-50, and Instrum. Exp. Tech., 1972, 15, no.3, pp.783-6.

39.208. Howell, J.Q.: "Quick Accurate Method to Measure the Permittivity of Microwave Integrated-Circuit Substrates", Trans. IEEE, 1973, MTT-21, no.3, pp.142-3.

39.209. Kent, M.: "Rapid Method Suitable for Liquids and Powders for Determination of Permittivity in the Microwave Region", Electron. Lett., 1973, 9, no.2, pp.39-40.

39.210. Lunazzi, J.J., and Garavaglia, M.: "Fabry-Perot Laser Interferometry to Measure Refractive Index or Thickness of Transparent Materials", J. Phys. E, 1973, 6, no.3, pp.237-40.

39.211. Pananakakis, G., Gabalda, F., and Boudouris, G.: "Experimental Study of Fields in Resonators", Ann. Telecommun., 1973, 28, no.1-2, pp.33-46.

39.212. Hoffmann, R.K., Kurzweg, J.P., and Mutzig, J.P.: "Measurement of Permittivity and Q-Factor of Slot Lines on Ceramic Substrate at 1-18 GHz", Frequenz, 1973, 27, no.2, pp.32-40.

39.213. Linzer, M., and Stokesberry, D.P.: "Frequency-Lock Method for Measurement of Q-Factors of Reflection and Transmission Resonators", Trans. IEEE, 1973, IM-22, no.1, pp.61-77.

39.214. Stumper, U.: "TE$_{01}$-Cavity Method for Determination of Complex Permittivity of Low-Loss Liquids at Millimetre Wavelengths", Rev. Sci. Instrum., 1973, 44, no.2, pp.165-9.

39.215. Helszajn, J.: "Microwave Measurement Techniques for Below-Resonance Junction Circulators", Trans. IEEE, 1973, MTT-21, no.5, pp.347-51.

39.216. Garault, Y., and Fenelon, J.P.: "Iris Impedance Measurements of the Dispersion Characteristics of an Open Periodic Structure", C. R. Acad. Sci. (Paris), 1973, 276B, no.9, pp.327-30.

39.217. Kohanzadeh, Y., Ma, K.W., and Whinnery, J.R.: "Measurement of Refractive-Index Change with Temperature Using Thermal Self-Phase Modulation", Appl. Opt., 1973, 12, no.7, pp.1584-7.

39.218. Gerlach, H.: "Method for Measuring Electrooptic Constants at High Frequencies", Appl. Phys., 1973, 1, no.5, pp.279-83.

39.219. Rzepecka, M.A.: "Cavity Perturbation Method for Routine Permittivity Measurement", J. Microwave Power, 1973, 8, no.1, pp.3-11.

39.220. Nelson, S.O., Stetson, L.E., and Schlaphoff, G.W.: "Computer Programme for Short-Circuited-Waveguide, Dielectric-Properties, Measurements on High- and Low-Loss Materials", J. Microwave Power, 1973, 8, no.1, pp.13-22.

39.221. Couderc, D., Giroux, M., and Bosisio, R.G.: "Dynamic High-Temperature Microwave Complex-Permittivity Measurements on Samples Heated via Microwave Absorption", J. Microwave Power, 1973, 8, no.1, pp.69-82.

39.222. Kraszewski, A.: "Measurement of Water Content by Microwave Method", Rozpr. Electrotech., 1973, 19, no.1, pp.137-63.

39.223. Berliner, M.A., Lelyanov, B.N., and Ivanov, V.A.: "Use of Cole-Cole Diagrams in SHF Hygrometry", Izv. VUZ Prib., 1973, 16, no.4, pp.101-6.

39.224. Moreau, R.: "Microwave Measurement of Anisotropy Factor for Si and Ge", C. R. Acad. Sci. (Paris), 1973, 277B, no.1, pp.25-8.

39.225. Balakhanov, V.Ya., Zhivotov, V.K., and Krotov, M.F.: "Multiple-Beam (Fourier) Holographic Spectroscopy", Dokl. Akad. Nauk SSSR, 1973, 208, no.4-6, pp.805-9, and Sov. Phys.-Dokl., 1973, 18, no.2, pp.123-4.

39.226. Noll, E.D.: "Measuring the Index of Refraction of Liquids With a Laser. I and II", Phys. Teach., 1973, 11, no.5, pp.307-8 and p.309.

39.227. Kent, M.: "Use of Stripline Configuration in Microwave Moisture Measurements. II", J. Microwave Power, 1973, 8, no.2, pp.189-94.

39.228. van Loon, R., and Finsy, R.: "Measurement of Complex Permittivity of Liquids at Frequencies of 5-40 GHz", Rev. Sci. Instrum., 1973, 44, no.9, pp.1204-8.

39.229. Patel, B.S.: "Determination of Gain, Saturation Intensity, and Internal Losses of a Laser Using an Intracavity Rotatable Reflector", J. Quantum Electron. IEEE, 1973, QE-9, no.12, pp.1150-1.

39.230. Meriakri, V.V., and Ushatkin, E.F.: "Measurement of Refractive Index at Submillimetre Wavelengths", Prib. Tekh. Eksp., 1973, 16, no.2, pp.143-5, and Instrum. Exp. Tech., 1973, 16, no.2,pp.498-500.

39.231. Watanabe, K., and Takao, I.: "Bridge Method for Simultaneous Measurements of Coupling Coefficient and Loaded Q of Single-Ended Cavity", Rev. Sci. Instrum., 1973, 44, no.11, pp.1625-7.

39.232. Olver, A.D., Clarricoats, P.J.B., and Chong, S.L.: "Experimental Determination of Attenuation in Corrugated Circular Waveguides", Electron. Lett., 1973, 9, no.18, pp.424-6.

39.233. Bilenko, D.I., Lun'ko, A.E., and Akincheva, N.S.: "Method of Solving Inverse Problem in Measurement of Parameters of Two-Layer Structures Using Microwave Methods", Radiotekh. Elektron., 1973, 18, no.7, pp.1496-9, and Radio Eng. Electron. Phys., 1973, 18, no.7, pp.1104-7.

39.234. Kawamura, M., and Sasamori, E.: "Measurement of High-Permittivity, Low-Loss, Dielectrics at 50 GHz Using Ghost-Mode Techniques", Electron. Commun. Jap., 1973, 56, no.6, pp.61-6.

39.235. Vystavkin, A.N., et al.: "Spectroradiometer for the Submillimetre Range Having an n-InSb Detector", Prib. Tekh. Eksp., 1973, 16, no.4, pp.164-7, and Instrum. Exp. Tech., 1973, 19,no.4,pp.1180-3.

39.236. Galaktionova, N.M., et al.: "Feasibility of a Supersensitive Laser Meter for Artificial Anisotropy and Faraday Rotation", Zh. Eksp. Teor. Fiz. Pis'ma, 1973, 18, no.8, pp.507-10, and JETP Lett., 1973, 18, no.8, pp.298-9.

39.237. Presby, H.M.: "(Measurement by) Optical Dispersion of Unclad Fibres Over Limited Wavelengths", Appl. Opt., 1974, 13, no.3, pp.465-7.

39.238. Druelle, Y., Citerne, J., and Raczy, L.: "Method for Measuring Dielectric Properties of Liquids With a Microslotline", Electron. Lett., 1974, 10, no.8, pp.117-8.

39.239. McRee, D.I.: "Determination of the Absorption of Microwave Radiation by a Biological Specimen in a 2.45-GHz Field", Health Phys., 1974, 26, no.5, pp.385-90.

39.240. Kaiser, P., and Ashe, H.W.: "Measurement of Spectral Total and Scattering Losses in Unclad Optical Fibres", J. Opt. Soc. Am., 1974, 64, no.4, pp.469-74.

39.241. Adonina, A.I., et al.: "Quasi-Optical Two-Channel Interferometer", Izv. VUZ Radioelektron., 1974, 17, no.5, p.3-9.

39.242. Stuchly, S.S., Rzepecka, M.A., and Iskander, M.F.: "Permittivity Measurements at Microwave Frequencies Using Lumped Elements", Trans. IEEE, 1974, IM-23, no.1, pp.56-62.

39.243. Petru, F., and Krsek, J.: "Reflectometer for Measurements on Dielectric Mirrors", Opt. Acta, 1974, 21, no.4, pp.293-314.

39.244. Rudduck, R.C., and Yu, C.L.: "Circular-Waveguide Method for Measuring Reflection Properties of Absorber Panels", Trans. IEEE, 1974, AP-22, no.2, pp.235-40.

39.245. Bahl, I.J., and Gupta, H.M.: "Microwave Measurement of Permittivity of Liquids and Solids Using Partially Loaded Slotted Waveguide", Trans. IEEE, 1974, MTT-22, no.1, pp.52-4.

39.246. Das Gupta, C.: "Microwave Measurement of Complex Permittivity Over a Wide Range of Values by a Waveguide-Resonator Method", Trans. IEEE, 1974, MTT-22, no.4, pp.365-72.

39.247. Weir, W.B.: "Automatic Measurement of Complex Permittivity and Permeability at Microwave Frequencies", Proc. IEEE, 1974, 62, no.1, pp.33-6.

39.248. Castle, G.S.P., and Roberts, J.: "Microwave Instrument for Continuous Monitoring of the Water Content of Crude Oil", Proc. IEEE, 1974, 62, no.1, pp.103-8.

39.249. Ajmera, R.C., et al.: "Microwave Measurements With Active Systems", Proc. IEEE, 1974, 62, no.1, pp.118-27.

39.250. Zlunitsyn, E.S., Zykov, A.I., and Kushmir, V.A.: "Method for Studying Parameters of Superconducting Resonators", Cryogenics, 1974, 14, no.1, pp.45-7.

39.251. Wolff, I.: "(Technique for) Measuring Effective Permittivity of Even and Odd Modes on Coupled Microstrip Lines", Nachr. Tech. Z., 1974, 27, no.1, pp.30-4.

39.252. Roy, P.K., and Datta, A.N.: "Application of Quarter-Wave Transformers for Precise Measurement of Complex Microwave Conductivity of Semiconductors", Trans. IEEE, 1974, MTT-22, no.2, pp.144-6.

39.253. Teague, J.R., and Rice, R.R.: "Dielectric-Loss-Tangent Measurements for Electrooptic Modulator Crystals", Rev. Sci. Instrum., 1974, 45, no.5, pp.710-1.

39.254. van Loon, R., and Finsy, R.: "Measurement of Complex Permittivity of Liquids at 60-150 GHz", Rev. Sci. Instrum., 1974, 45, no.4, pp.523-5.

39.255. Wyslouzil, W., and Van Koughnett, A.L.: "Attenuation-Based Microwave Moisture Gauge for Sheet Materials", J. Microwave Power, 1974, 9, no.2, pp.91-8.

39.256. Bussey, H.E., Morris, D., and Zal'tsman, E.B.: "International Comparison of Complex Permittivity Measurement at 9 GHz", Trans. IEEE, 1974, IM-23, no.3, pp.235-9.

39.257. Kuprenyuk, V.I., and Sherstobitov, V.E.: "Simple Method for Measuring Reflectance of Metallic Mirrors at 10.6 micron", Zh. Prikl. Spekt., 1974, 20, no.5, pp.962-8.

39.258. Martin, W.E.: "Refractive-Index Profile Measurements of Diffused Optical Waveguides", Appl. Opt., 1974, 13, no.9, pp.2112-6.

39.259. Dusoiu, N.: "Cell for Microwave Permittivity Measurements on Molten Salts", Stud. Cercet. Fiz., 1974, 26,no.5,pp.563-7.

39.260. Nagenthiram, P., and Cullen, A.L.: "Microwave Barrel Resonator for Permittivity Measurements on Dielectric Rods", Proc. IEEE, 1974, 62, no.11, pp.1613-4.

39.261. Bliss, E.S., Speck, D.R., and Simmons, W.W.: "Direct Interferometric Measurements of Nonlinear-Refractive-Index Coefficient in Laser Materials", Appl. Phys. Lett., 1974, 25, no.12, pp.728-30.

39.262. Garelis, E.: "Initial Heating Rate of a Droplet in a Spherical Microwave Cavity", Phys. Fluids, 1974, 17, no.11, pp.2002-8.

39.263. Tereshchenko, A.I., and Moronenko, V.I.: "Investigation of Dielectric Parameters With a Semicoaxial Resonator", Prib. Tekh. Eksp., 1974, 17, no.2, pp.141-2, and Instrum. Exp. Tech., 1974, 17, no.2, pp.463-4.

39.264. Borukhov, M.Yu., et al.: "Microwave Properties of Dielectrics Studied in Optical Furnaces", Geliotekhnika, 1974, 10, no.3, pp.78-9, and Appl. Sol. Energy, 1974, 10, no.3-4, pp.59-60.

39.265. Milward, R.C.: "Commercial Fourier Spectrometers for Submillimetre Wavelengths", Trans. IEEE, 1974, MTT-22, no.12, pp.1018-23.

39.266. Parker, T.J.: "Technique for Dispersive-Reflection Spectroscopy in Far-IR", Trans. IEEE, 1974, MTT-22, no.12, pp.1032-6.

39.267. M'Baye, K.: "Resonant-Ring Interferometric Device for X-Band Measurement of Permittivity", Rev. Gen. Electr., 1974, 83, no.9, pp.553-9.

39.268. Zal'tsman, E.B., Kiselev, O.F., and Poyarkova, V.E.: "High-Precision Equipment for Reproducing Unit Permittivity of Solid Dielectrics at 9.5 GHz", Izmer. Tekh., 1974, 17, no.7, pp.17-8, and Meas. Tech., 1974, 17, no.7, pp.1001-3.

39.269. Matushkin, N.I., and Burtovoi, D.P.: "Differential (Microwave) Measurements of Losses of Local Absorbents", Radiotekh. (Kharkov), 1974, no.29, pp.153-7.

39.270. Dobrovol'skii, I.F., et al.: "Method of Measuring Material Parameters in a Radial Line", Izmer. Tekh., 1974, 17, no.8, pp.58-9, and Meas. Tech., 1974, 17, no.8, pp.1235-8.

39.271. Czerlinski, G., and Bracokova, V.: "Coaxial Alignment of Laser Beams for Perturbation Experiments", Appl. Opt., 1974, 13, no.7, pp.1639-45.

39.272. Antsiferov, V.V., and Folin, K.G.: "Time Behaviour of the Spectrum of a Ruby Laser With Spherical Mirrors in a Quasi-stationary Operating Mode", Avtometriya, 1974, no.6, pp.103-4.

40. RESONANCE SPECTROMETRY

40.1. Buckmaster, H.A., and Dering, J.C.: "Experimental Sensitivity Study of a 9-GHz ESR Spectrometer", Can. J. Phys., 1965, 43, p.1088.

40.2. Rinehart, E.A., Legan, R.L., and Lin, C.C.: "Microwave Spectrograph for Linewidth Measurements", Rev. Sci. Instrum., 1965, 36, p.511.

40.3. Nag, B.R., and Engineer, M.H.: "Experimental Observation of Faraday Rotation in Artificial Dielectrics", J. Appl. Phys., 1965, 36, p.3388.

40.4. Tsung-tang, S., et al.: "Characteristics of an ESR Spectrometer With Frequency and Magnetic-Field Modulation", Acta Phys. Sin., 1965, 21, p.866.

40.5. Place, H.: "Measurement of Spin- and Lattice-Relaxation Periods by Nonlinearities of Ferromagnetic Resonance of a Disc", C. R. Acad. Sci. (Paris), 1965, 260B, p.2177.

40.6. Schanda, E., and Kint, L.V.D.: "Rapid Linewidth Measurement of Microwave Ferrites", Proc. IEEE, 1965, 53, p.189.

40.7. Arndt, R.: "Analytical Line Shapes for Lorentzian Signals Broadened by Modulation", J. Appl. Phys., 1965, 36, p.2522.

40.8. Andresen, S.G., and de Prins, J.: "Possible Stark-Field Millimetre-Wave Interaction Structure", Proc. IEEE, 1965, 53, p.511.

40.9. Danilyuk, Yu.L., Pakhol'chik, P.L., and Koleda, F.A.: "Double-Rotation Goniometer for Microwave Spectroscopy", Prib. Tekh. Eksp., 1965, no.1, p.213, and Instrum. Exp. Tech., 1965, no.1, p.221.

40.10. Griffiths, D.J., and Glattli, H.: "Optical Faraday Rotation Studies of Paramagnetic Resonance and Relaxation", Can. J. Phys., 1965 43, p.2361.

40.11. Tobler, H.J., Bauder, A., and Gunthard, H.H.: "Distortion of Line Shapes in Stark-Modulated Microwave Spectrometers", J. Sci. Instrum., 1965, 42, p.236.

40.12. Tobler, H.J., et al.: "Electrical and Mechanical Properties of an Improved Microwave Stark Cell", J. Sci. Instrum., 1965, 42, p.420.

40.13. Cox, A.P., Flynn, G.W., and Wilson, E.B.: "Microwave Double-Resonance Experiments", J. Chem. Phys., 1965, 42, p.3094.

40.14. Kaplan, R.: "Method for Observation of Cyclotron Resonance at Millimetre Wavelengths", Solid-State Commun., 1965, 3, no.2, p.35.

40.15. Dianov, E.M., Irisova, N.A., and Karlov, N.V.: "Use of Dielectric Waveguides in Millimetric Spectroscopy", Prib. Tekh. Eksp., 1965, 8, no.4, p.144, and Instrum. Exp. Tech., 1965, 8, no.4, p.885.

40.16. Strigutskii, V.P.: "Cavity for ESR Study of Aqueous Solutions at 3-cm Wavelength", Prib. Tekh. Eksp., 1966, 9, no.6, p.202, and Instrum. Exp. Tech., 1966, 9, no.6, p.1521.

40.17. Srivastava, C.M.: "Wall Effect in Ferrimagnetic Resonance Experiments", Br. J. Appl. Phys., 1966, 17, p.1173.

40.18. Ernst, R.R., and Anderson, W.A.: "Sensitivity Enhancement in Magnetic Resonance", Rev. Sci. Instrum., 1966, 36, pp.1689 and 1696.

40.19. Ernst, R.R., and Anderson, W.A.: "Application of Fourier-Transform Spectroscopy to Magnetic Resonance", Rev. Sci. Instrum., 1966, 37, p.93.

40.20. Schulten, G.: "Resonators for Millimetre Waves and Use in Observation of Gas Resonances", Frequenz, 1966, 20, p.10.

40.21. Macke, B., Messelyn, J., and Wertheimer, R.: "Application of Phase Stabilization in the Microwave Measurement of a Phenomenon with Double-Resonance Spectroscopy", Onde Electr., 1966, 46, p.123.

40.22. Garif'yanov, N.S., and Kharakhash'yan, E.G.: "Simple Special-Purpose Cryostat for ESR at Helium Temperatures", Prib. Tekh. Eksp., 1966, 9, no.1, p.225, and Instrum. Exp. Tech., 1966, 9, no.1, p.245.

40.23. Goodings, J.M., and Sugden, T.M.: "Zeeman-Modulated Microwave Spectrometer Suitable for Study of Free Radicals", J. Sci. Instrum., 1966, 43, p.692.

40.24. Urban, W.: "ESR Spectrometer With Continuously Variable Frequency and High Sensitivity", Z. Angew. Phys., 1966, 20, p.215.

40.25. Strauch, R.G., et al.: "Millimetre Electric-Resonance Spectroscopy", Proc. IEEE, 1966, 54, p.506.

40.26. Decailliot, M., and Uebersfeld, J.: "ESR Spectrometer With High Sensitivity Permitting Simultaneous Observation of Absorption and Dispersion", C. R. Acad. Sci. (Paris), 1966, 262B, p.141.

40.27. Gilbert, J., and Vaillancourt, R.M.: "Saturation Effect Spectrometer", Proc. IEEE, 1966, 54, p.514.

40.28. Zwarts, C.M.G., and van Ormondt, D.: "Simple Sensitive X-Band Heterodyne Spectrometer for ESR Measurements", J. Sci. Instrum., 1966, 43, p.317.

40.29. Woods, R.C., Ronn, A.M., and Wilson, E.B.: "Double-Resonance Modulated Microwave Spectrometer", Rev. Sci. Instrum., 1966, 37, p.927.

40.30. Radford, H.E.: "Free-Radical Microwave Absorption Meter", Rev. Sci. Instrum., 1966, 37, p.790.

40.31. Faulkner, E.A., and Whippey, P.W.: "ESR Spectrometer Using RF Amplitude Modulation", Proc. IEE, 1966, 113, p.1159.

40.32. Shvets, A.D., et al.: "Low-Temperature Apparatus for Studying ESR at 8-mm Wavelength", Cryogenics, 1966, 6, p.174.

40.33. Schmid, P.E., and Gunthard, H.H.: "K-Band Heterodyne ESR Spectrometer", Z. Angew. Math. Phys., 1966, 17, p.404.

40.34. Nakamura, S., and Kusumoto, H.: "Variable Coupling Scheme for X-Band ESR Cavities", Rev. Sci. Instrum., 1966, 37, p.1740.

40.35. Rorke, D.: "Lengthened H_{102} Cavity for ESR Studies of Aqueous Samples", J. Sci. Instrum., 1966, 43, p.396.

40.36. Buckmaster, H.A., and Dering, J.C.: "Application of Phase-Lock Microwave Frequency Stabilizers to EPR Spectrometers", J. Sci. Instrum., 1966, 43, p.554.

40.37. Collier, R.J., and Wilmshurst, T.H.: "Improvements to the Ammonia-Maser ESR Spectrometer", Phys. Lett., 1966, 23, p.333.

40.38. Aseltine, C.L., and Kim, Y.W.: "Microwave Resonant Cavity for EPR Work Near 77°K", Rev. Sci. Instrum., 1966, 37, p.1270.

40.39. Antipin, A.A.: "Cavity for 10-cm EPR Investigations", Prib. Tekh. Eksp., 1967, 10, no.2, p.98, and Instrum. Exp. Tech., 1967, 10, no.2, p.366.

40.40. Denisov, Yu.N., and Kalinichenko, V.V.: "Wideband Absorption Chamber for Detection of Centimetric EPR Signals", Prib. Tekh. Eksp., 1967, 10, no.3, p.152, and Instrum. Exp. Tech., 1967, 10, no.3, p.702.

40.41. Buckmaster, H.A., and Dering, J.C.: "9-GHz, Single-Klystron, EPR Spectrometer Using Heterodyne Demodulation", Can. J. Phys., 1967, 45, p.107.

40.42. Pearlman, M.R., and Webb, R.H.: "Characteristics of TW Helices in ESR Spectrometers", Rev. Sci. Instrum., 1967, 38, no.9, pp.1264-7.

40.43. Britt, C.O.: "Solid-State Microwave Spectrometer", Rev. Sci. Instrum., 1967, 38, no.10, pp.1496-501.

40.44. Smetana, Z., and Dusek, J.: "Thermal Effects of Absorbed Microwave Power in Measurement of Ferromagnetic Resonance", Czech. J. Phys., 1968, 18B, no.3, pp.389-92.

40.45. Waldron, R.A.: "Errors Due to the Uncertainty Principle in Swept-Frequency Cavity Measurements of Properties of Materials", Trans. IEEE, 1968, MTT-16, no.5, pp.314-5.

40.46. Kobayashi, T.: "Analysis of Signal With Fabry-Perot Interferometer Stark Cell", Bull. Electrotech. Lab., 1968, 32, no.2, pp.55-64.

40.47. Legrand, J., et al.: "Frequency Marking for Microwave Spectra", Rev. Phys. Appl., 1968, 3, no.2, pp.199-202.

40.48. Bady, I.: "Errors in Measurement of Linewidth of Single-Crystal Ferrites Due to Nonideal Microwave Components", Trans. IEEE, 1968, MAG-4, p.716.

40.49. Toda, M.: "Voltage Induction Method for Microwave Susceptibility Measurements", Trans. IEEE, 1968, MTT-16, p.828.

40.50. Yakovlev, Yu.M., et al.: "Resonator Method of Measuring the Parameters of Ferrites", Izv. VUZ Radioelektron., 1968, 11, p.834.

40.51. Gurin, E.I., et al.: "Spectroscopy of Gas Laser Frequencies", Izmer. Tekh., 1968, no.8, p.22, and Meas. Tech., 1968, no.8, p.1030.

40.52. Grugor'yants, V.V.: "Population Measurement of Metastable Level of Laser Active Material", Radiotekh. Elektron., 1968, 13, no.12, p.2186, and Radio Eng. Electron. Phys., 1968, 13, no.12, pp.1917-21.

40.53. Culshaw, W.: "Double Resonance in Gas Laser", J. Quantum Electron. IEEE, 1968, QE-4, p.979.

40.54. Hertz, J.H.: "Application of Combined Resonators in Laser Spectroscopy", Monatsber. Dtsch. Akad. Wiss. Berlin, 1968, 10, no.7, p.478.

40.55. Belan, V.R., et al.: "Fluorescence Method of Determining Laser Parameters", Nachr. Tech. Fachber., 1968, 35, p.782, and Electron Technol., 1969, 2, no.2-3, p.175.

40.56. Volosov, V.D.: "Arrangement for Studying Spectral Characteristics of Laser Radiation", Zh. Tekh. Fiz., 1968, 38, p.1769, and Sov. Phys.-Tech. Phys., 1969, 13, p.1429.

40.57. Kalinski, J.: "Stabilization of Signal Level in Microwave Measurement Setups With Double-Modulation Method", Rozpr. Elektrotech., 1969, 15, p.635.

40.58. Trukhanenko, E.M., and Ishchenko, P.I.: "Direct Method of Measuring Amplification Contour of He-Ne Gas Mixture", Zh. Prikl. Spekt., 1969, 11, p.940.

40.59. Gheorghiu, O.C.: "Cavity Resonator for Study of Electronic Susceptibility of an Ionized Gas", Rev. Roum. Phys., 1969, 14, p.61.

40.60. Strauch, R.G.: "Technique for Measurement of HCN-Laser Linewidth", Electron. Lett., 1969, 5, p.246.

40.61. Johnson, L.F., Dillon, J.F., and Remeika, J.P.: "Optical Studies of Ho^{3+} Ions in YGaG and YIG", J. Appl. Phys., 1969, 40, p.1499.

40.62. Siahatgar, S., and Hochuli, U.E.: "Display of 852.1-nm Line of Cs Utilizing a Swept GaAs Laser", J. Quantum Electron. IEEE, 1969, QE-5, p.295.

40.63. Kalliomaki, K., and Kalliomaki, P.L.: "Methods of Measuring Cyclotron Resonance at Microwave Frequencies", Saehkoe, 1969, 42, no.7-8, p.218.

40.64. Fork, R.L., Dienes, A., and Kluver, J.W.: "Effects of Combined RF and Optical Fields on a Laser Medium", J. Quantum Electron. IEEE, 1969, QE-5, p.607.

40.65. Liu, C.S., Cherrington, B.E., and Verdeyen, J.T.: "Dispersion Effects in a High-Gain 3.39-micron He-Ne Laser", J. Appl. Phys., 1969, 40, p.3556.

40.66. Miller, P.A., Verdeyen, J.T., and Cherrington, B.E.: "Measurement of Dispersion at 632.8 nm and 640.1 nm Due to Neon $1s_5$ Metastable Atoms in a He-Ne Discharge", J. Quantum Electron. IEEE, 1969, QE-5, p.473.

40.67. Levinson, G.R., et al.: "Measurement of 00^01 Population Levels of CO_2 Molecules", Radiotekh. Elektron., 1969, 14, no.4, p.682, and Radio Eng. Electron. Phys., 1969, 14, no.4.

40.68. Okada, F., and Chino, M.: "Measurements of Tensor Permeability of Planar Ferrites at Millimetre Wavelengths", Mem. Def. Acad., 1969, 9, no.1, p.135.

40.69. Audion, C., et al.: "Double-Resonance Method for Determination of Level Populations", J. Quantum Electron. IEEE, 1969, QE-5, p.431.

40.70. Kohanzadeh, Y., and Auston, D.H.: "Measurement of Low Absorption Coefficients Using the Beat Frequency Shift Between Transverse Modes of a Laser", J. Quantum Electron. IEEE, 1970, QE-6, p.475.

40.71. Carroll, T.O.: "Double-Resonance Spectroscopy in Gas Lasers", J. Quantum Electron. IEEE, 1970, QE-6, p.512.

40.72. Johnston, T.F.: "Measurement of Lower-Level Population of the Argon-Ion Laser Using Laser-Induced Population Changes", Appl. Phys. Lett., 1970, 17, p.161.

40.73. Courtney, W.E., and Temme, D.H.: "Spin-Wave Linewidth Measurements With Low-Power RF Sources", Trans. IEEE, 1970, MTT-18, p.510.

40.74. Vance, M.E.: "Saturation and Excited-State Absorption in Nd^{3+}:Glass", J. Quantum Electron. IEEE, 1970,QE-6,p.249.

40.75. Dienes, A., and Sosnowski, T.P.: "Magnetic-Field Dip Measurements on the He-Cd Laser", Appl. Phys. Lett., 1970, 16, p.512.

40.76. Nicoll, F.H.: "Apparatus for Studying Electron-Beam-Pumped Semiconductor Lasers", Rev. Sci. Instrum., 1970, 41, no.8, pp.1175-8.

40.77. Moruzzi, G.: "Technique for Optical Pumping in Vacuum UV", Opt. Commun., 1970, 2, no.6, pp.279-81.

40.78. Kaminskii, A.A.: "High-Temperature Spectroscopic Investigation of Stimulated Emission from Lasers Based on Crystals Activated With Nd^{3+} Ions", Phys. Status Solidi, 1970, 1, no.3, pp.573-89.

40.79. Baican, R.: "Use of Maser as Preamplifier in an ESR Spectrometer", Stud. Cercet. Fiz., 1970, 22, no.9, pp.1027-33.

40.80. Krupnov, A.F., and Gershtein, L.I.: "Submillimetre Gas Spectrometer With Quartz Frequency Readout", Prib. Tekh. Eksp., 1970, 13, no.5, pp.128-30, and Instrum. Exp. Tech., 1970, 13, no.5, pp.1394-6.

40.81. Kozlov, Yu.I., and Rodkin, E.A.: "Experimental Device for Studying EPR Under Uniaxial Pressure", Prib. Tekh. Eksp., 1970, 13, no.5, p.246, and Instrum. Exp. Tech., 1970, 13, no.5, pp.1535-6.

40.82. Poehler, T.O., and Apel, J.R.: "Cyclotron Resonance in Solid-State Plasma", Appl. Tech. Dig., 1970, 9, no.5, pp.2-12.

40.83. Petutin, A.I., et al.: "Submillimetre-Band (ESR) Spectrometer", Prib. Tekh. Eksp., 1970, 13, no.4, pp.163-5, and Instrum. Exp. Tech., 1970, 13, no.4, pp.1146-9.

40.84. Parekh, S.V.: "Ferromagnetic Resonance of Spherical Ferrite Samples", Int. J. Electron., 1971, 30, no.6, pp.585-7.

40.85. Kaminskii, A.A., and Vylegzhanin, D.N.: "Stimulated-Emission Investigations of Effects of Electron-Phonon Interaction in Crystals Activated With Nd^{3+} Ions", J. Quantum Electron. IEEE, 1971, QE-7,no.7,pp.329-38.

40.86. Oppenheim, U.P., and Melman, P.: "Excited-State Absorption of SF_6 at 28.1 THz", J. Quantum Electron. IEEE, 1971, QE-7, no.8, pp.426-7.

40.87. Grieneisen, H.P., Kurnit, N.A., and Szoke, A.: "Fluorescence Induced by Coherent Optical Pulses", Opt. Commun., 1971, 3, no.4, pp.259-63.

40.88. Gorshkov, V.I., et al.: "Investigation of Vibrational Relaxation in the HCl Chemical Laser", Appl. Opt., 1971, 10, no.8, pp.1781-5.

40.89. Mikhnenko, G.A., et al.: "Experimental Study of Collision Broadening of the 633-nm Line of a He-Ne Laser", Opt. Spekt., 1971, 30, no.1, pp.124-32, and Opt. Spectrosc., 1971, 30, no.1, pp.65-9.

40.90. Standley, K.J., Stevens, R., and Storey, B.E.: "Technique for Measuring Permeability Tensor and Permittivity of Ferrites and Garnets", J. Phys. E, 1971, 4, no.2, pp.111-4.

40.91. Hardin, J., and Uebersfeld, J.: "NH_3 Maser Used for EPR", Rev. Phys. Appl., 1971, 6, no.2, pp.165-72.

40.92. Smith, A.M., and Netterfield, R.P.: "Production of Frequency Markers for Microwave Spectroscopy", J. Phys. E, 1971, 48, no.8, pp.618-9.

40.93. Nill, K.W., et al.: "IR Spectroscopy of CO Using a Tunable Pb(SSe) Diode Laser", Appl. Phys. Lett., 1971, 19, no.4, pp.79-82.

40.94. Hoeft, J., et al.: "Microwave Spectroscopy at High Temperatures. II", Z. Angew. Phys., 1971, 31, no.5-6, pp.337-42.

40.95. Ginodman, V.B., et al.: "Use of a Klystron Oscillator, Stabilized by a Superconducting Resonator, in an EPR Spectrometer With RF Modulator", Cryogenics, 1971, 11, no.4, pp.304-6.

40.96. Chakravarti, A.N., and Parui, D.P.: "Proposed Method for Determining Characteristic Energies of Exponential Band Tails in GaAs Junction Lasers", Int. J. Electron., 1971, 31, no.4, pp.359-64.

40.97. Hultzsch, R.: "Correlations Between Active and Passive Parameters of Solid-State Laser Media. II", Phys. Status Solidi, 1971, 7A, no.1, pp.13-46.

40.98. Bykovskii, Yu.A., et al.: "Pulsed Semiconductor Laser as a High-Resolution Spectroscope", Opt. Spekt., 1971, 30, no.3, pp.508-10, and Opt. Spectrosc., 1971, 30, no.3, pp.277-8.

40.99. Collier, F., et al.: "Amplification Cross Section of 1052-nm Transition of Nd^{3+}: $POCl_3 \cdot SnCl_4$ by Three Different Methods", J. Quantum Electron. IEEE, 1971, QE-7, no.11, pp.519-22.

40.100. Zuev, V.E., et al.: "Use of High-Speed Laser Spectroscopy to Study the Absorption Spectrum of Atmospheric Gases", Appl. Opt., 1971, 10, no.11, pp.2452-5.

40.101. White, K.O., et al.: "High-Resolution Laser Atmospheric-Transmission Measurement", Appl. Phys. Lett., 1971, 19, no.10, pp.381-2.

40.102. Sze, R.C., Antropov, Y.T., and Bennett, W.R.: "Lorentz Width Measurements on Argon-Ion-Laser Transitions", Appl. Opt., 1972, 11, no.1, pp.197-8.

40.103. Kido, G., Nagasaka, K., and Narita, S.: "Technique for Measurement of Far-IR Cyclotron Resonance in Intense Pulsed Magnetic Field", Jap. J. Appl. Phys., 1972, 11, no.2, pp.237-46.

40.104. Blum, F.A., et al.: "Measurement of the Gain Line Shape of a Gas Laser Using a Tunable Semiconductor Laser", Appl. Phys. Lett., 1972, 20, no.10, pp.377-9.

40.105. Kwok, M.A., Cross, E.F., and Jacobs, T.A.: "Spectroscopy of Pulsed HF Chemical Lasers Using an IR Vidicon Camera Tube", Rev. Sci. Instrum., 1972, 43, no.7, pp.1043-5.

40.106. Djeu, N., and Searles, S.K.: "Method of Measuring Temperature, Inversion Ratio, and Pressure-Broadened Linewidth in a CW Molecular Laser", J. Quantum Electron. IEEE, 1972, QE-8, no.10, pp.811-3.

40.107. Lee, P.H.: "Introduction to Optical Saturated-Absorption (Techniques)", Trans. IEEE, 1972, IM-21, no.4, pp.384-7.

40.108. Levenson, M.D., Flytzanis, C., and Bloembergen, N.: "Interference of Resonant and Nonresonant Three-Wave Mixing in Diamond", Phys. Rev., 1972, 6B, no.10, pp.3962-5.

40.109. Fetterman, H.R., et al.: "Identification of Donor Species in High-Purity GaAs Using Optically Pumped Submillimetre Lasers", Appl. Phys. Lett., 1972, 21, no.9, pp.434-6.

40.110. White, K.O., and Schleusener, S.A.: "Coincidence of Er^{3+}:YAG Laser Emission With Methane Absorption at 1645.1 nm", Appl. Phys. Lett., 1972, 21, no.9, pp.419-20.

40.111. Gibbs, H.M., and Slusher, R.E.: "Sharp-Line Self-Induced Transparency", Phys. Rev., 1972, 6A, no.6, pp.2326-34.

40.112. Zakharov, S.M., and Manykin, E.A.: "Photon-Echo Polarization in Ruby", Opt. Spekt., 1972, 32, no.4, pp.717-23, and Opt. Spectrosc., 1972, 32, no.4, pp.378-81.

40.113. Gadomskii, O.N., and Solovarov, N.K.: "Concentration Dependence of Photon-Echo Intensity", Phys. Lett., 1972, 42A, no.3, pp.219-20.

40.114. Becker, K.H., Haaks, D., and Tatarczyk, T.: "Monitoring of Radicals by a Tunable Dye Laser", Z. Naturforsch., 1972, 27a, no.10, pp.1519-20.

40.115. Frohlich, D.: "Two-Photon Absorption v. Raman Effect", Commun. Solid State Phys., 1972, 4, no.6, pp.179-82.

40.116. Osswald, R.: "Microwave Cavities Designed for Paraelectric Resonance at 10, 35, and 70 GHz", J. Phys. E, 1972, 5, no.12, pp.1155-6.

40.117. Morgenshtern, Z.L., and Neustruev, V.B.: "(Measurement of) Luminescence Efficiency of Ruby With Resonant Excitation", Opt. Spekt., 1972, 32, no.5, pp.953-8, and Opt. Spectrosc., 1972, 32, no.5, pp.510-3.

40.118. Freund, I.: "Nonlinear (Two-Photon) X-Ray Spectroscopy", Opt. Commun., 1972, 6, no.4, pp.421-3.

40.119. Mikaberidze, A.A., Ochkin, V.N., and Sobolev, N.N.: "Measurement of Vibrational Temperatures in a CO Laser", Zh. Tekh. Fiz., 1972, 42, no.12, pp.2550-5, and Sov. Phys.-Tech. Phys., 1973, 17, no.12, pp.1982-5.

40.120. Rice, D.K.: "Absorption Measurements of CO-Laser Radiation by Water Vapour", Appl. Opt., 1973, 12, no.2, pp.218-25.

40.121. Menzies, R.T.: "Measurement of Ne 3.39-micron Transition Decay Rates Using Laser Faraday Rotation", Phys. Lett., 1973, 43A, no.3, pp.209-10.

40.122. Alferov, G.N., et al.: "Direct Determination of Fluorescence Energy Yields of Rhodamine 6G Solutions Using an Ar^{II} Laser", Zh. Prikl. Spekt., 1973, 18, no.2, pp.316-9.

40.123. Chackerian, C., and Weisbach, M.F.: "Amplified Laser Absorption. Detection of NO", J. Opt. Soc. Am., 1973, 63, no.3, pp.342-5.

40.124. Czarnecki, S., and Krasinski, J.: "Measurements of Naphthalene Triplet-Level Population by Paramagnetic Properties", Lett. Nuovo Cim., 1973, 6, no.12, pp.473-8.

40.125. Youmans, D.G., Hackel, L.A., and Ezekiel, S.: "High-Resolution Spectroscopy of I_2 Using Laser-Molecular-Beam Techniques", J. Appl. Phys., 1973, 44, no.5, pp.2319-21.

40.126. Haroche, S., Paisner, J.A., and Schawlow, A.L.: "Hyperfine Quantum Beats Observed in Cs Vapour Under Pulsed Dye-Laser Excitation", Phys. Rev. Lett., 1973, 30, no.20, pp.948-51.

40.127. Blanaru, D.L.: "Spectral-Line Absorption Measurement Using Optical Cavities", J. Appl. Phys., 1973, 44, no.6, pp.2735-45.

40.128. Hartig, W., and Walther, H.: "High-Resolution Spectroscopy With, and Frequency Stabilization of, a CW Dye Laser", Appl. Phys., 1973, 1, no.3, pp.171-4.

40.129. Christridis, T.C., and Heineken, F.W.: "Microwave Cavity for High-Power ENDOR Spectroscopy", J. Phys. D, 1973, 6, no.5, pp.432-4.

40.130. Gamo, H., Ostrem, J.S., and Chuang, S.S.: "Determination of Line-Shape Parameters of High-Gain Laser Transitions Based on Line-Narrowing Measurements", J. Appl. Phys., 1973, 44, no.6, pp.2750-5.

40.131. Reid, J., et al.: "Vibrational Relaxation Measurements in CO_2 Employing an Incremental TEA-Laser Gain Technique", J. Quantum Electron. IEEE, 1973, QE-9, no.6, pp.602-4.

40.132. Bean, B.L., and Izatt, J.R.: "Verification of the Kramers-Kronig Relations in Optically Pumped Ruby", J. Opt. Soc. Am., 1973, 63, no.7, pp.832-9.

40.133. Schuda, F., Hercher, M., and Stroud, C.R.: "Direct Optical Measurement of Sodium Hyperfine Structure Using a CW Dye Laser", Appl. Phys. Lett., 1973, 22, no.8, pp.360-2.

40.134. Grischkowsky, D.: "Adiabatic Following and Slow Optical Pulse Propagation in Rb Vapour", Phys. Rev., 1973, 7A, no.6, pp.2096-102.

40.135. Linford, G.J.: "Experimental Studies of a Zeeman-Tuned Xenon Laser Differential Absorption Apparatus", Appl. Opt., 1973, 12, no.6, pp.1130-9.

40.136. Beterov, I.M., Chebotaev, V.P., and Provorov, A.S.: "High Precision Spectroscopy of SF_6 With CW High-Pressure Tunable CO_2 Laser", Opt. Commun., 1973, 7, no.4, pp.410-1.

40.137. Rages, K.A., and Sawyer, R.E.: "Properties of Microwave Cavities Containing Magnetic Resonant Samples", Rev. Sci. Instrum., 1973, 44, no.7, pp.830-4.

40.138. Dunn, M.H., and Ross, J.N.: "Investigation of Population Inversions by Perturbation Spectroscopy", Phys. Lett., 1973, 44A, no.4, pp.247-8.

40.139. Sevast'yanov, B.K., and Orekhova, V.P.: "Determination of the Population of a Metastable Level in Cr^{3+}-Activated Crystals", Zh. Prikl. Spekt., 1973, 18, no.4, pp.641-7.

40.140. Pine, A.S., Glassbrenner, G.J., and Kafalas, J.A.: "Pressure-Tuned, GaAs-Diode-Laser, Absorption Spectroscopy of Xenon Hyperfine Structure", J. Quantum Electron. IEEE, 1973, QE-9, no.8, pp.800-7.

40.141. Lange, W., et al.: "High-Resolution Fluorescence Spectroscopy by CW Dye Laser", Opt. Commun., 1973, 8, no.2, pp.157-9.

40.142. Nill, K.W., Strauss, A.J., and Blum, F.A.: "Tunable CW (PbCd)S Diode Lasers for 3.5 micron. Ultrahigh-Resolution Spectroscopy", Appl. Phys. Lett., 1973, 22, no.12, pp.677-9.

40.143. Atkinson, G.H., Laufer, A.H., and Kurylo, M.J.: "Detection of Free Radicals by an Intracavity Dye Laser Technique", J. Chem. Phys., 1973, 59, no.1, pp.350-4.

40.144. Petersen, F.R., et al.: "Rotational Constants for $^{12}C^{16}O_2$ from Beats Between Lamb-Dip-Stabilized Lasers", Phys. Rev. Lett., 1973, 31, no.9, pp.573-6.

40.145. Bonch-Bruevich, A.M., Razumova, T.K., and Rubanova, G.M.: "Induced Absorption of Polymethine Dyes", Opt. Spekt., 1973, 34, no.2, pp.305-11, and Opt. Spectrosc., 1973, 34, no.2, pp.172-5.

40.146. Andreev, R.B., Volosov, V.D., and Kalintsev, A.G.: "Nonlinear Controlled-Dispersion Spectrograph", Opt. Spekt., 1973, 34, no.1, pp.186-7, and Opt. Spectrosc., 1973, 34, no.1, pp.102-3.

40.147. Lis, L.: "Study of Population Changes of Cd^{II} s and p Levels Induced by 441.6-nm Laser Action", Acta Phys. Pol., 1973, 44A, no.2, pp.173-6.

40.148. Ivakin, E.V., Petrovich, I.P., and Rubanov, A.S.: "Self-Diffraction of Emission Caused by Absorption in Excited Levels", Zh. Prikl. Spekt., 1973, 18, no.6, pp.1003-6.

40.149. Doughty, J.R., Jack, J.L., and O'Pray, J.E.: "Absorption of CO_2 Laser Radiation by Carbonyl Fluoride", J. Appl. Phys., 1973, 44, no.9, pp.4065-6.

40.150. Dienes, A., and Madden, M.: "Study of Excitation Transfer in Dye Mixtures by Measurement of Gain Spectra", J. Appl. Phys., 1973, 44, no.9, pp.4164-4.

40.151. Osipov, A.S., et al.: "Ring Resonator for Spectral Analysis of CO_2-Laser Radiation", Prib. Tekh. Eksp., 1973, 16, no.1, pp.186-7, and Instrum. Exp. Tech., 1973, 16, no.1, pp.225-7.

40.152. Yamashita, M., and Kashiwagi, H.: "Triplet-State ESR of Rhodamine 6G During Laser Irradiation", J. Chem. Phys., 1973, 59, no.4, pp.2156-7.

40.153. Andra, H.J., et al.: "Doppler-Tuned Beam-Laser Spectroscopy", Nucl. Instrum. Methods, 1973, 110, no.7, pp.453-7.

40.154. Schuda, F., and Stroud, C.R.: "CW Velocity Selection by Dye Laser Optical Pumping", Opt. Commun., 1973, 9, no.1, pp.14-6.

40.155. Jones, H., and Eyer, A.: "Collisional Transfer of Energy in NH_3. Triple Resonance Experiments", Z. Naturforsch., 1973, 28a, no.10, pp.1703-6.

40.156. Borde, C., et al.: "Experimental Observation of Saturated Dispersion Phenomenon in Iodine at 514.5 nm", C. R. Acad. Sci. (Paris), 1973, 277B, no.14, pp.381-3.

40.157. Green, W.H., and Hancock, J.K.: "Laser-Excited, Vibrational-Energy-Transfer, Studies of HF, CO, and NO", J. Quantum Electron. IEEE, 1973, QE-9, no.1, pp.50-8.

40.158. Entschladen, H., and Severin, H.: "Permeability Tensor of Ferrites Measured With Rods in a Circular Cylindrical Waveguide", Arch. Elektron. Übertrag., 1973, 27, no.4, pp.153-61.

40.159. Yabuzaki, T., and Ogawa, T.: "Double-Resonance Phenomena of Ne in Spontaneous Emission from Lower Laser Level", J. Phys. Soc. Jap., 1973, 34, no.3, pp.769-76.

40.160. Klein, M.B., Shank, C.V., and Dienes, A.: "Detection of Small Laser Gains in He-Se Discharge Using Dye-Laser Assisted Oscillation", Opt. Commun., 1973, 7, no.3, pp.178-80.

40.161. Kreiner, W.A., Romheld, M., and Rudolph, H.D.: "IR-Microwave Double Resonance Experiments With NH_3", Z. Naturforsch., 1973, 28a, no.10, pp.1707-11.

40.162. Kopvillem, U.Kh., Smolyakov, B.P., and Sharipov, R.Z.: "Polarized Echo and Possible Detection in the Submillimetre and Optical Ranges", Izv. Akad. Nauk SSSR, Ser. Fiz., 1973, 37, no.10, pp.2240-3.

40.163. Green, J.M., Hohimer, J.P., and Tittel, F.K.: "High-Resolution CW-Dye-Laser Spectrometer", Opt. Commun., 1973, 9, no.4, pp.407-11.

40.164. Lingel, C., et al.: "Automated System for Measuring Gain in Organic Dyes", Appl. Opt., 1973, 12, no.12, pp.2939-41.

40.165. Nagibarov, V.R., et al.: "Method of Investigating Resonance Media by Modulation of Laser Pulses", Izv. VUZ Fiz., 1973, no.12, pp.154-6,

40.166. Letokhov, V.S.: "Possibilities for Laser Spectroscopy Inside the Doppler Line in the Optical and Gamma Ranges", Laser Unconv. Opt. J., 1973, no.46, pp.3-27.

40.167. Gershenzon, Ye.M., Gol'tsman, G.N., and Ptitsina, N.G.: "Submillimetre Spectroscopy of Semiconductors", Zh. Eksp. Teor. Fiz., 1973, 64, no.2, pp.587-98, and Sov. Phys.-JETP, 1973, 37, no.2, pp.299-304.

40.168. Batarchukova, N.R., and Naidenov, A.S.: "Interferometer Method of Calibrating Spectrometers With Lasers", Izmer. Tekh., 1973, 16, no.8, p.28, and Meas. Tech., 1973, 16, no.8, pp.1159-60.

40.169. Browne, P.G., and Dunn, M.H.: "Perturbation Spectroscopy of the He-Cd Laser Discharge", J. Phys. B, 1974, 7, no.10, pp.1113-21.

40.170. Godlevskii, A.P., and Lopasov, V.P. "Ruby Laser With a Multimode Optical Vessel in Resonator for Study of Weak Absorption Lines", Zh. Prikl. Spekt., 1974, 20, no.2, pp.299-301.

40.171. Kleinschmidt, J., Tottleben, W., and Rentsch, S.: "Measurements of Two-Photon Absorption in Cavity of an Organic Dye Laser", Exp. Tech. Phys., 1974, 22, no.3, pp.191-5.

40.172. Weber, W.H., et al.: "High-Resolution Stark Spectroscopy in the v_4 Band of NH_3 Using a Thin-Film Diode Laser", Appl. Opt., 1974, 13, no.6, pp.1431-4.

40.173. Uehara, H., Tanimoto, M., and Ijuuin, Y.: "Stark-Sweep Microwave Cavity Spectrometer for Zeeman Effect Studies", Chem. Phys. Lett., 1974, 26, no.4, pp.578-81.

40.174. Wu, F.Y., Grove, R.E., and Ezekiel, S.: "CW Dye Laser for Ultrahigh-Resolution Spectroscopy", Appl. Phys. Lett., 1974, 25, no.1, pp.73-5.

40.175. Bykovskii, Yu.A., et al.: "Observation of Hyperfine Structure of Selective Reflection from ^{133}Cs Vapour Using a Semiconductor Laser", Kvantovaya Elektron., 1974, 1, no.1, pp.146-9, and Sov. J. Quantum Electron., 1974, 4, no.1, pp.82-3.

40.176. Preier, H., and Riedel, W.: "NO Spectroscopy by Pulsed Pb(SSe) Diode Lasers", J. Appl. Phys., 1974, 45, no.9, pp.3955-8.

40.177. Yakushev, A.I.: "Method for Measuring Difference Between Populations of the Zeeman Sublevels of the $2s_2-2p_4$ Ne Transition", Kvantovaya Elektron., 1974, 1, no.4, pp.963-4, and Sov. J. Quantum Electron., 1974, 4, no.4, pp.529-30.

40.178. Woods, R.C., and Dixon, T.A.: "Computer Controlled Microwave Spectrometer System", Rev. Sci. Instrum., 1974, 45, no.9, pp.1122-6.

40.179. Weyssenhoff, H.V., and Rehling, U.: "Intracavity Enhancement of Absorption Spectra in a Dye Laser", Z. Naturforsch., 1974, 29a, no.2, pp.256-60.

40.180. Johansson, B., et al.: "Stripline Resonator for ESR", Rev. Sci. Instrum., 1974, 45, no.11, pp.1445-7.

40.181. Gast, J.: "Performance of Amplitude Fourier Spectrometer for Far-IR Materials", Trans. IEEE, 1974, MTT-22, no.12, pp.1026-7.

40.182. Lowndes, R.P.: "High-Pressure Far-IR Spectroscopy of Ionic Solids", Trans. IEEE, 1974, MTT-22, no.12, pp.1076-80.

40.183. Von Ortenberg, M.: "Submillimetre-Laser Magnetospectroscopy in Tellurium by Nernst Effect", Trans. IEEE, 1974, MTT-22, no.12, pp.1081-5.

40.184. Shimizu, F.: "Laser Stark Spectrometer With High Sensitivity", J. Fac. Eng. Univ. Tokyo, 1974, no.12, pp.40-1. See also "Q-Switching by Stark Effect of Ammonia", Appl. Phys. Lett., 1970, 16, p.368.

40.185. Nikitenko, A.G., and Solov'ev, M.V.: "Spectrometer for Investigating Mode Composition of CO_2-Laser Radiation", Izmer. Tekh., 1974, 17, no.8, pp.72-4, and Meas. Tech., 1974, 17, no.8, pp.1261-4.

41. RADIATION SCATTERING

41.1. Wort, D.J.H., and Heald, M.A.: "Scattering of Microwaves from Moving Inhomogeneities in Laboratory Plasma", Plasma Phys., 1965, 7, p.79.

41.2. Stern, R.A., and Tzoar, N.: "Incoherent Microwave Scattering from Resonant Plasma Oscillations", Phys. Rev. Lett., 1965, 15, p.485.

41.3. Chen, Y.G., Leheny, R.F., and Marshall, T.C.: "Combination Scattering of Microwaves from Space-Charge Waves in Laboratory Magnetoplasma", Phys. Rev. Lett., 1965, 15, p.184.

41.4. Shiobara, S.: "Electron-Density Measurement by Microwave Scattering from Underdense Cylindrical Plasma", J. Phys. Soc. Jap., 1966, 21, p.1380.

41.5. Perepelkin, N.F.: "Raman Scattering of Microwaves by Plasma Oscillations", Zh. Eksp. Teor. Fiz. Pis'ma, 1966, 3, p.258, and JETP Lett., 1966, 3, p.165.

41.6. Rice, D.K., and Wada, J.Y.: "Measurements of Microwaves Scattered from Moving Density-Modulated Plasma", Electron. Lett., 1966, 2, p.385.

41.7. Hotston, E.S.: "Simple 4-mm Receiver for Plasma Scattering Experiments", J. Sci. Instrum., 1967, 44, p.47.

41.8. Akhmanov, S.A., and Bol'shov, M.A.: "Observation of Stimulated Brillouin Scattering With Laser Optical Harmonics", Opt. Spekt., 1968, 25, p.432, and Opt. Spectrosc., 1968, 25, p.233.

41.9. Planner, A.: "Experiments on Influence of Liquids on Power and Spatial Pattern of a Raman Laser", Acta Phys. Polon., 1969, 36, no.2, p.287.

41.10. Barak, S., Rokni, M., and Yatsiv, S.: "Induced v. Parametric Scattering Processes in Potassium Vapour", J. Quantum Electron. IEEE, 1969, QE-5, p.448.

41.11. Derkacheva, L.D., and Krymova, A.I.: "Four-Photon Resonant Parametric Interaction in Lasers Using Dye Solutions", Zh. Eksp. Teor. Fiz. Pis'ma, 1969, 9, no.10, p.564, and JETP Lett., 1969, 9, no.10, p.343.

41.12. Avizonis, P.V., and Heinlich, R.M.: "Pulse Shape of Stimulated Raman Emission from an Oscillator Cavity", J. Appl. Phys., 1969, 40, p.3650.

41.13. Ishida, A., and Inuishi, Y.: "Acoustic Domains in Semiconductor CdS Revealed by Brillouin Scattering Measurements", J. Phys. Soc. Jap., 1969, 26, p.957.

41.14. Auth, D.C., Mayer, W.G., and Thaler, W.J.: "Light Diffraction Technique for Measuring Dielectric Constants at Microwave Frequencies", Proc. IEEE, 1969,57,p.96.

41.15. Tabata, N., and Mori, M.: "Determination of Fine Particle Size by Laser-Beam Scattering", Mitsubishi Denki Lab. Rep., 1969, 10, no.1, p.15.

41.16. Korolev, F.A., et al.: "Excitation of SRS in Liquids Upon Q-Switching of a Laser by the Investigated Substance", Zh. Eksp. Teor. Fiz. Pis'ma, 1970, 11, no.6, p.295, and JETP Lett., 1970, 11, no.6, p.193.

41.17. Kao, K.C.: "Coherent-Light Scattering Measurements on Single and Cladded Optical Glass Fibres", Radio Electron. Eng., 1970, 39, no.2, p.105.

41.18. Allwood, R.L., et al.: "Tunable Stimulated IR Spin-Flip Magneto-Raman Scattering With Dielectric-Coated Cavities", J. Phys. C, 1970, 3, no.11, pp.L186-90.

41.19. Carman, R.L., et al.: "Theory of Stokes Pulse Shapes in Transient SRS", Phys. Rev., 1970, 2A, p.60.

41.20. Aussenegg, F.: "SRS Excited by 530-nm Light", Opt. Commun., 1970, 2, no.6, pp.295-7.

41.21. Inoue, K.: "Study on Stimulated Brillouin Scattering by Varying both Phonon Lifetime and Pulse Duration of Laser", Jap. J. Appl. Phys., 1970, 9, no.11, pp.1347-55.

41.22. Hordvik, A., and Collins, R.J.: "Time Behaviour of Stimulated Raman Scattering", J. Quantum Electron. IEEE, 1970, QE-6, p.254.

41.23. Key, P.Y., Harrison, R.G., and Little, V.I.: "Bragg Reflection from a Phase Grating Induced by Nonlinear Optical Effects in Liquids", J. Quantum Electron. IEEE, 1970, QE-6, p.641.

41.24. Guthart, H., and Graf, K.A.: "Scattering from Turbulent Plasma", Radio Sci., 1970, 5, no.7, p.1099.

41.25. Slobodnik, A.J.: "Microwave Acoustic Surface-Wave Investigations Using Light Deflection", Proc. IEEE, 1970, 58, p.488.

41.26. Hughes, A.J., Maines, J.D., and O'Shaughnessy, J.: "Strain Measurements in the CdS Electroacoustic Oscillator", J. Phys. D, 1970, 3, p.751.

41.27. Shaw, E.D., and Patel, C.K.N.: "Stimulated anti-Stokes Spin-Flip Raman Scattering in InSb", Appl. Phys. Lett., 1971, 18, no.6, pp.215-8.

41.28. Patel, C.K.N.: "Stimulated Second-Stokes Spin-Flip Raman Scattering in InSb", Appl. Phys. Lett., 1971, 18, no.7, pp.274-6.

41.29. Smith, R.A.: "Lasers and Light Scattering", Proc. Roy. Soc., 1971, 323A, No.1554, pp.305-20.

41.30. Colles, M.J.: "Ultrashort Pulse Formation in Short-Pulse-Stimulated Raman Oscillator", Appl. Phys. Lett., 1971, 19, no.2, pp.23-5.

41.31. Ito, H., and Inaba, H.: "Measurement of Four-Photon Optical Parametric Fluorescence in Nonlinear Solids", Rec. Electr. Commun. Eng. Conversaz. Tohoku Univ., 1971, 40, no.1, pp.27-33.

41.32. Graf, K.A., Guthart, H., and Douglas, D.G.: "Scattering from Turbulent Laboratory Plasma at 31 GHz", Radio Sci., 1971, 6, no.7, pp.737-52.

41.33. Gorelik, V.S., et al.: "Use of Argon-Ion Laser for Raman Light Scattering", Prib. Tekh. Eksp., 1971, 14, no.2, pp.205-7, and Instrum. Exp. Tech., 1971, 14, no.2, pp.566-8.

41.34. Zaitsev, V.P., et al.: "He-^{114}Cd Laser for Exciting Scattered-Light Spectra", Prib. Tekh. Eksp., 1971, 14, no.3, pp.189-90, and Instrum. Exp. Tech., 1971, 14, no.3, pp.878-9.

41.35. Bozhkov, A.I.: "Stimulated Light Scattering on Liquid Surface With Full Fresnel Reflection", Izv. VUZ Radiofiz., 1972, 15, no.2, pp.233-41.

41.36. Apanasevich, P.A., et al.: "Properties of SRS in a Laser Resonator", Zh. Prikl. Spekt., 1972, 16, no.2, pp.256-61.

41.37. Reinhold, I., and Maier, M.: "Gain Measurements of Stimulated Raman Scattering Using a Tunable Dye Laser", Opt. Commun., 1972, 5, no.1, pp.31-4.

41.38. Uzgiris, E.E.: "Electrophoresis of Particles and Biological Cells Measured by the Doppler Shift of Scattered Laser Light", Opt. Commun., 1972, 6, no.1, pp.55-7.

41.39. Sacchi, C.A., Svelto, O., and Zaraga, F.: "Stimulated Scattering in the Wing of the Rayleigh Line in CS$_2$ With Picosecond Excitation", Opt. Commun., 1972, 6, no.1, pp.71-4.

41.40. Saikan, S.: "Transient SRS in Binary Mixed Liquids", Opt. Commun., 1972, 6, no.1, pp.77-80.

41.41. Welsch, D.: "Theory of SRS in Molecular Crystals", Wiss. Z. Friedrich-Schiller Univ. Jena, 1972, 21, no.1, pp.125-30.

41.42. Schmidt, W., and Appt, W.: "Tunable Stimulated Raman Emission Generated by a Dye Laser", Z. Naturforsch., 1972, 27a, no.8-9, pp.1372-4.

41.43. Getmantsev, G.G., and Tokarev, Yu.V.: "Depression of Power Spectrum of Compton Radiation from Relativistic Particles in a Rarefied Plasma", Astrophys. Lett., 1972, 12, no.1, pp.57-60.

41.44. Duclos, J., and Ripoche, J.: "Stroboscopic Visualization of Ultrasonic Waves in Liquid Media", C. R. Acad. Sci. (Paris), 1972, 275B, no.15, pp.525-8.

41.45. Steudel, H.: "Stimulated Raman Emission With Ultrashort Light Pulses", Exp. Tech. Phys., 1972, 20, no.5, pp.409-15.

41.46. Kondilenko, I.I., Korotkov, P.A., and Maly, V.I.: "Nonlinear High-Order Effects in SRS Spectrum", Phys. Lett., 1972, 42A, no.1, pp.72-4.

41.47. Dolino, G.: "Effects of Domain Shapes on Second-Harmonic Scattering in Triglycine Sulphate", Phys. Rev., 1972, 6B, no.10, pp.4025-35.

41.48. Kudyavtseva, A.D., et al.: "Investigation of Self-Focusing in the Case of SRS", Kvantovaya Elektron., 1972, no.7, pp.73-5, and Sov. J. Quantum Electron., 1972, 2, no.1, pp.63-5.

41.49. Kovalev, V.I., et al.: "Gain and Linewidths in Stimulated Brillouin Scattering in Gases", Kvantovaya Elektron., 1972, no.7, pp.78-80, and Sov. J. Quantum Electron., 1972, 2, no.1, pp.69-71.

41.50. Rawson, E.G.: "Measurement of the Angular Distribution of Light Scattered from a Glass-Fibre Optical Waveguide", Appl. Opt., 1972, 11, no.11, pp.2477-81.

41.51. Parkash, V., and Jaseja, T.S.: "Stimulated Raman Emission from Inorganic Liquids", Indian J. Pure Appl. Phys., 1972, 10, no.6, pp.427-32.

41.52. Neely, G.O., Nelson, L.Y., and Harvey, A.B.: "Modification of a Commercial Argon-Ion Laser for Enhancement of Gas-Phase Raman Scattering", Appl. Spectrosc., 1972, 26, no.5, pp.553-5.

41.53. Mooradian, A., et al.: "Electric-Field-Induced Transient Spin-Flip Raman Laser Pulses in InSb", Appl. Phys. Lett., 1972, 21, no.10, pp.482-4.

41.54. Macheleidt, G.: "Anisotropic Effect in Crystals of Brillouin Scattering. I and II", Exp. Tech. Phys., 1972, 20, no.6, pp.539-44 and 545-51.

41.55. Arbatskaya, A.N., Prokhorov, K.A., and Suschinskii, M.M.: "Investigation of Angular Distribution of First Stokes Component for SRS", Zh. Eksp. Teor. Fiz., 1972, 62, no.3, pp.872-8, and Sov. Phys.-JETP, 1972, 35, no.3, pp.462-5.

41.56. Gadomskii, O.N., and Nagibarov, V.R.: "Stimulated Coherent (Scattering) Processes", Zh. Eksp. Teor. Fiz., 1972, 62, no.3, pp.896-900,and Sov.Phys.-JETP,35,p.475.

41.57. Stolen, R.H., Ippen, E.P., and Tynes, A.R.: "Raman Oscillation in Glass Optical Waveguide", Appl. Phys. Lett., 1972, 20, no.2, pp.62-4.

41.58. Grabe, M.: "Nonstationary Surface Roughness Measurement by Light Scattering", Proc. IEEE, 1972, 60, no.3, pp.339-40.

41.59. Franklin, W., and Sengupta, P.: "Laser-Stimulated Atomic Migration", J. Quantum Electron. IEEE, 1972, QE-8, no.4, pp.393-400.

41.60. Wherett, B.S., and Firth, W.J.: "Spin Saturation and Pump Depletion in CW Spin-Flip Raman Oscillation", J. Quantum Electron. IEEE, 1972, QE-8, no.12, pp.865-8.

41.61. Firth, W.J.: "Power Output of a Pulsed Raman Laser With Saturable Excitation", J. Quantum Electron. IEEE, 1972, QE-8, no.12, pp.869-72.

41.62. Grasyuk, A.Z., Zubarev, I.G., and Suyazov, N.V.: "Influence of Spectral Linewidth of Exciting Radiation on the Gain in Stimulated Scattering", Zh. Eksp. Teor. Fiz. Pis'ma, 1972, 16, no.4, pp.237-40.

41.63. Korolev, F.A., Gulyaeva, L.S., and Sokolova, E.Yu.: "Excitation of SRS in Nitrogen Using a Cavity", Opt. Spekt., 1972, 32, no.3, pp.518-21, and Opt. Spectrosc., 1972, 32, no.3, pp.271-2.

41.64. Dennis, R.B., et al.: "Stimulated Spin-Flip Raman Scattering. Magnetically Tunable IR Laser. I", Proc. Roy. Soc., 1972, 331A, no.1585, pp.203-36.

41.65. Scarlet, R.I.: "Diffractive Scattering of Picosecond Light Pulses in Absorbing Liquids", Phys. Rev., 1972, 6A, no.6, pp.2281-91.

41.66. Meyer, J.: "Plasma Density Fluctuations in the Presence of Optical Mixing of Light Beams", Phys. Rev., 1972, 6A, no.6, pp.2291-7.

41.67. Bairamov, B.Kh., et al.: "(Investigation of) Self-Focusing of Argon-Laser Radiation in $Bi_{12}GeO_{20}$ by Raman Scattering", Fiz. Tver. Tela, 1972, 14, no.5, pp.1374-83, and Sov. Phys.-Solid State, 1972, 14, no.5, pp.1181-8.

41.68. Ippen, E.P., and Stolen, R.H.: "Stimulated Brillouin Scattering in Optical Fibres", Appl. Phys. Lett., 1972, 21, no.11, pp.539-41.

41.69. Emel'yanov, V.I., and Klimontovich, Yu.L.: "Theory of Parametric Scattering of Light by Polaritons", Zh. Eksp. Teor. Fiz., 1972, 35, no.2, pp.778-88, and Sov. Phys.-JETP, 1972, 35, no.2, pp.411-6.

41.70. Kats, A.V., and Maslov, V.V.: "Stimulated Scattering from a Highly Conducting Surface", Zh. Eksp. Teor. Fiz., 1972, 35, no.2, pp.496-504, and Sov. Phys.-JETP, 1972, 35, no.2, pp.264-8.

41.71. Silin, V.A.: "Stimulated Raman Scattering of Microwaves in a Layer of Collisionless Plasma", Zh. Eksp. Teor. Fiz. Pis'ma, 1972, 16, no.3, pp.153-7, and JETP Lett., 1972, 16, no.3, pp.105-8.

41.72. Akhmanov, S.A., et al.: "Combined Effects of Molecular Relaxation and Medium Dispersion in SRS of Ultrashort Light Pulses", Zh. Eksp. Teor. Fiz., 1972, 35, no.2, pp.525-40, and Sov. Phys.-JETP, 1972, 35, no.2, pp.279-86.

41.73. Smith, R.G.: "Optical Power Handling Capacity of Low-Loss Optical Fibres as Determined by Stimulated Raman and Brillouin Scattering", Appl. Opt., 1972, 11, no.11, pp.2489-94.

41.74. Bhatnagar, R., Bhawalkar, D.D., and Pant, H.C.: "PTM Q-Switched Laser Utilizing SRS", Indian J. Pure Appl. Phys., 1972, 10, no.7, pp.569-70.

41.75. Terao, K., and Tatsumi, H.: "Measurement of Flame Temperatures by Means of Laser Radiation", Jap. J. Appl. Phys., 1972, 11, no.12, p.1856.

41.76. Wherrett, B.S.: "Stokes/ anti-Stokes Spin-Flip Cavity Oscillation", Opt. Commun., 1972, 6, no.4, pp.402-6.

41.77. Sacchi, C.A.: "Stimulated Low-Frequency Raman Lines in Liquid Benzene", Opt. Commun., 1972, 6, no.4, pp.418-20.

41.78. Lugovoi, V.N., and Strel'tsov, V.N.: "Stimulated Brillouin Emission in an Optical Resonator", Zh. Eksp. Teor. Fiz., 1972, 62, no.4, pp.1312-20, and Sov. Phys.-JETP, 1972, 35, no.4, pp.692-5.

41.79. Sorokin, S.A.: "Absolute Instability of SRS", Kvantovaya Elektron., 1972, no.2, pp.98-101, and Sov. J. Quantum Electron., 1972, 2, no.2, pp.172-4.

41.80. Dlab, J., and Schork, L.: "Scattering of Laser Light. Nondestructive Testing of Plastics", Siemens Forsch.-Entwick., 1972, 1, no.4, pp.376-9.

41.81. Korobkin, V.V., et al.: "Ultrashort SRS Pulses and Multifocus Structure of Light Beams", Zh. Eksp. Teor. Fiz. Pis'ma, 1972, 16, no.11, pp.595-9, and JETP Lett., 1972, 16, no.11, pp.419-22.

41.82. Klepsvik, J.O.: "Two-Photon Absorption in Rhodamine 6G", Phys. Norv., 1972, 6, no.3-4, p.203.

41.83. Apanasevich, P.A., and Afanas'ev, A.A.: "Four-Photon Stimulated Scattering of Light in Resonant Media", Opt. Spekt., 1972, 33, no.2, pp.300-7, and Opt. Spectrosc., 1972, 33, no.2, pp.160-3.

41.84. Levleva, L.D., Karagodova, T.Ya., and Kovner, M.A.: "Zeeman Effect and Electronic Stimulated Raman Scattering", Zh. Eksp. Teor. Fiz., 1972, 62, no.5, pp.1681-5, and Sov. Phys.-JETP, 1972, 35,no.5,pp.874-6.

41.85. Dneprovskii, V.S., et al.: "Stimulated Two- and Three-Photon Raman Scattering in Water", Izv. Akad. Nauk. Arm. SSR Fiz., 1972, 7, no.5, pp.348-53.

41.86. Tsikin, B.G., and Dubrovskii, V.A.: "Possibility of Storing Laser Radiation Scattered by an Electron Beam", Radiotekh. Elektron., 1972, 17, no.7, pp.1433-8, and Radio Eng. Electron. Phys., 1972, 17, no.7, pp.1123-7.

41.87. Lee, S.P., Tscharnuter, W., and Chu, B.: "Calibration of Optical Self-Beating Spectrometer by Polystyrene Latex Spheres and Confirmation of Stokes-Einstein Formula", J. Polym. Sci., 1972, 10, no.12, pp.2453-9.

41.88. Chihara, K., et al.: "Two-Photon Absorption in Anthracene", Sci. Rep. Kanazawa Univ., 1972, 17, no.2, pp.17-21.

41.89. Korolev, F.A., Odintsov, V.I., and Sokolova, E.Yu.: "Spectral Width and Structure of Lines of Induced Raman Scattering in Methane and Nitrogen Under Excitation in a Cavity", Opt. Spekt., 1972, 33, no.6, pp.1093-8, and Opt. Spectrosc., 1972, 33, no.6, pp.600-3.

4 .90. Popovichev, V.I., Ragul'skii, V.V., and Faizullov, F.S.: "Switching of Resonator Q-Factor by Stimulated Brillouin Scattering", Kvantovaya Elektron., 1972, no.5, pp.126-9, and Sov. J. Quantum Electron., 1973, 2, no.5, pp.496-8.

41.91. Galeev, A.A., and Syunyaev, R.A.: "Plasma Effects in Stimulated Compton Interaction Between Matter and Radiation", Zh. Eksp. Teor. Fiz., 1972, 63, no.4, pp.1266-81, and Sov. Phys.-JETP, 1973, 36, no.4, pp.669-76.

41.92. Akaneev, B.A., et al.: "Stimulated Raman Scattering in SF_6", Kvantovaya Elektron., 1972, no.5, pp.88-90, and Sov. J. Quantum Electron., 1973, 2, no.5, pp.457-8.

41.93. Kyzylasov, Yu.I., Starunov, V.S., and Fabelinskii, I.L.: "Stimulated Entropy Scattering", Zh. Eksp. Teor. Fiz., 1972, 63, no.2, pp.407-20, and Sov. Phys.-JETP, 1973, 36, no.2, pp.216-22.

41.94. Peregudov, G.V., Ragozin, E.N., and Chirkov, V.A.: "Energy and Time Characteristics of SRS in Powder Medium at Various Temperatures", Zh. Eksp. Teor. Fiz., 1972, 63, no.2, pp.421-30, and Sov. Phys.-JETP, 1973, 36, no.2, pp.223-7.

41.95. Delalande, C., and Mysyrowicz, A.: "Laser Induced Two-Photon Decay in Pr^{3+}: $LaAlO_3$", Opt. Commun., 1973, 7, no.1, pp.10-2.

41.96. Sokolovskaya, A.I., et al.: "Angular Distribution of Stimulated Raman Light in Liquid Nitrogen", Zh. Prikl. Spekt., 1973, 18, no.1, pp.122-6.

41.97. Stolen, R.H., and Ippen, E.P.: "Raman Gain in Glass Optical Waveguides", Appl. Phys. Lett., 1973, 22, no.6, pp.276-81.

41.98. Kleinman, D.A., Miller, R.C., and Nordland, W.A.: "Two-Photon Absorption of Nd^{3+}-Laser Radiation in GaAs", Appl. Phys. Lett., 1973, 23, no.5, pp.243-4.

41.99. Koppel, D.E., and Schaefer, D.W.: "Scaled Photocount Correlation of non-Gaussian Scattered Light", Appl. Phys. Lett., 1973, 22, no.1, pp.36-7.

41.100. Offenberger, A.A., and Burnett, N.H.: "CO_2-Laser-Induced Compton Effect in Underdense Hydrogen Plasma", Phys. Lett., 1973, 42A, no.7, pp.527-8.

41.101. Peterson, L.M., and Wiggins, T.A.: "Forward Stimulated Thermal Rayleigh Scattering", J. Opt. Soc. Am., 1973, 63, no.1, pp.13-6.

41.102. Sorokin, P.P., and Lankard, J.R.: "Efficient Parametric Conversion in Cs Vapour Irradiated by 347-nm Mode-Locked Pulses", J. Quantum Electron. IEEE, 1973, QE-9, no.2, pp.227-30.

41.103. Yamada, M., et al.: "Brillouin-Scattering Study of Propagating Acousto-electric Domains in Semiconducting CdS", Phys. Rev., 1973, 7, no.6, pp.2682-92.

41.104. Gelbert, U., and Many, A.: "Brillouin-Scattering Studies of Acoustoelectric Gain and Lattice Attenuation in Semiconducting CdS", Phys. Rev., 1973, 7,no.6,pp.2713-26.

41.105. Wood, R.A., et al.: "Quantum Oscillations and Pump Depletion Effects in an Efficient High-Power Tunable Spin-Flip Laser", J. Phys. C, 1973, 6, no.6, pp.L144-9.

41.106. Carlsten, J.L., and McIlrath, T.J.: "Observations of Stimulated anti-Stokes Radiation in Barium Vapour", J. Phys. B, 1973, 6, no.4, pp.L80-5.

41.107. Barocchi, F., Vallauri, R., and Zoppi, M.: "Memory Effects Associated With Bulk Viscosity on the Spectrum of Stimulated Brillouin Scattering", Nuov. Cim., 1973, 14B, no.1, pp.39-51.

41.108. Sukhatme, V.P., and Wolff, P.A.: "Stimulated Compton Scattering as a Radiation Source", J. Appl. Phys., 1973, 44, no.5, pp.2331-4.

41.109. Karmenyan, K.V., and Chilingaryan, Yu.S.: "Nonstationary Stimulated Scattering by Polaritons in LiIO3", Zh. Eksp. Teor. Fiz. Pis'ma, 1973, 17, no.2, pp.106-10, and JETP Lett., 1973, 17, no.2, pp.73-6.

41.110. Kaye, W.: "Low-Angle Laser Light Scattering", Anal. Chem., 1973, 45, no.2, pp.221A-5.

41.111. Krasyuk, I.K., Pashinin, P.P., and Prokhorov, A.M.: "Role of Stimulated Compton Scattering in the Interaction of Laser Radiation With a Superdense Plasma", Zh. Eksp. Teor. Fiz. Pis'ma, 1973, 17, no.2, pp.130-2, and JETP Lett., 1973, 17, no.2, pp.92-4.

41.112. Hensler, D., et al.: "Laser Specular Reflectometer for Ceramic Surface Diagnostics", Am. Ceram. Soc. Bull., 1973, 52, no.2, pp.191-4.

41.113. Bershtein, I.L., and Stepanov, S.P.: "Detection and Measurement of Small Back-Scattering of Laser Radiation", Izv. VUZ Radiofiz., 1973, 16, no.4, pp.531-6.

41.114. Brincourt, G., Cacioli, R., and Millet, Ch.: "Multiphoton Ionization of Cs Atoms by a Ruby Laser", Opt. Commun., 1973, 7, no.4, pp.384-5.

41.115. DeSilets, C.S., and Patel, C.K.N.: "Characteristics of a Low-Field Spin-Flip Raman Laser", Appl. Phys. Lett., 1973, 22, no.10, pp.543-5.

41.116. Butylkin, V.S., et al.: "Spatially Bounded Phase Capture and Axial anti-Stokes Radiation in SRS in Gases", Zh. Eksp. Teor. Fiz. Pis'ma, 1973, 17, no.8, pp.400-5, and JETP Lett., 1973, 17, no.8, pp.285-9.

41.117. Srivastava, G.P., Nath, R., and Gupta, S.C.: "Absorption Cross Section and Spin Analysis in the Three-Photon Process", Opt. Acta, 1973, 20, no.7, pp.565-76.

41.118. Srivastava, M.K., and Crow, R.W.: "Raman Susceptibility Measurements and Stimulated Raman Effect in KDP", Opt. Commun., 1973, 8, no.1, pp.82-4.

41.119. Stewart, P.: "Phase Velocity Effects in Nonlinear Compton Scattering", Astron. Astrophys., 1973, 25, no.3, pp.457-60.

41.120. Wood, R.A., et al.: "High-Power Pulsed Operation of a Tunable Spin-Flip Raman Laser at Low Magnetic Field", Opt. Commun., 1973, 8, no.3, pp.248-50.

41.121. Brueck, S.R.J., and Mooradian, A.: "Spontaneous Spin-Flip Raman Linewidth and Nonlinear Processes in InSb", Opt. Commun., 1973, 8, no.3, pp.263-6.

41.122. Schehl, R.R.: "Portable (Scattering) Monitor for Respirable Dust Utilizing a Gas Laser", J. Phys. E, 1973, 6,no.8,pp.732-4.

41.123. Karger, A.M., English, R.P., and Smith, R.J.D.: "Laser Raman Spectrometer for Process Control", Appl. Opt., 1973, 12, no.9, pp.2083-7.

41.124. Fuhrmann, H., et al.: "Investigation of Intensity Dependence of Frequency Displacement in Stimulated Brillouin Scattering", Exp. Tech. Phys., 1973, 21, no.4, pp.349-54.

41.125. Sukhorukov, A.P., and Shchednova, A.K.: "SRS of Phase-Modulated Light Pulses", Opt. Spekt., 1973, 34, no.2, pp.351-5, and Opt. Spectrosc., 1973, 34, no.2, pp.198-200.

41.126. Lugovoi, V.N., and Strel'tsov, V.N.: "Stimulated Raman and Brillouin Radiations in a Laser Resonator", Opt. Acta, 1973, 20, no.3, pp.165-75.

41.127. Herrmann, J.: "Influence of Population Change on SRS by Short Light Pulses", Phys. Lett., 1973, 43A, no.2, pp.133-4.

41.128. Strel'tsov, V.N.: "Stimulated Scattering of Light by Plasmons at Large Gradients of Medium Density", Zh. Eksp. Teor. Fiz. Pis'ma, 1973, 18, no.8, pp.532-5, and JETP Lett., 1973, 18, no.8, pp.314-5.

41.129. Harrison, R.G., Key, P.Y., and Little, V.I.: "Stimulated Scattering and Induced Bragg Reflection of Light in Liquid Media. I and II", Proc. Roy. Soc., 1973, 334A, no.1597, pp.193-214 and 215-29.

41.130. Korolev, F.A., Vokhnik, O.M., and Odintsov, V.I.: "Mode-Locked Pulses and Stimulated Brillouin Scattering in an Optical Resonator", Zh. Eksp. Teor. Fiz. Pis'ma, 1973, 18, no.1, pp.58-61, and JETP Lett., 1973, 18, no.1, pp.32-3.

41.131. Regnier, P.R., and Taran, J.P.E.: "Measuring Gas Concentrations by Stimulated anti-Stokes Scattering", Appl. Phys. Lett., 1973, 23, no.5, pp.240-2.

41.132. Hughes, J.L.: "Scattering of Light by Light in a Vacuum Free of the Influence of Any Field", Phys. Lett., 1973, 46B, no.2, pp.211-3.

41.133. Kestenbaum, A.: "Light Intensity Modulation in a Thermal Medium", Appl. Opt., 1973, 12, no.10, pp.2378-80.

41.134. Kotsubanov, V.D., Leikin, A.Ya., and Pavlichenko, O.S.: "Increase in Efficiency of Lasers in Experiments Involving Light Scattering in Plasmas", Fiz. Plazmy, 1973, no.4, pp.208-12.

41.135. Firth, W.J.: "Theory of Spin-Flip Raman Amplification in InSb", Opt. Commun., 1973, 9, no.1, pp.84-8.

41.136. Aref'ev, I.M., et al.: "Brillouin Spectra in the Critical Mixture of Nitroethane/Isooctane", Opt. Commun., 1973, 9, no.1, pp.69-73.

41.137. Kondilenko, I.I., et al.: "Effect of Spontaneous Lines on the Spectral Composition of Stimulated Raman Scattering", Opt. Spekt., 1973, 34, no.3, pp.475-8, and Opt. Spectrosc., 1973, 34, no.3, pp.271-3.

41.138. Guerra, M.A., Brueck, S.R.J., and Mooradian, A.: "Gradient-Field, Permanent-Magnet, Spin-Flip Laser", J. Quantum Electron. IEEE, 1973, QE-9, no.12, pp.1157-9.

41.139. Aussenegg, F., Deserno, U., and Scherr, D.: "Collinear Emission of Second-Order Stokes Radiation in SRS" Z. Naturforsch., 1973, 28a, no.10, pp.1654-9.

41.140. Hickman, R.S., and Liang, L.: "Intracavity Laser Raman Spectroscopy Using a Commercial Laser", Appl. Spectrosc., 1973, 27, no.6, pp.425-7.

41.141. Markhviladze, T.M., and Shelepin, L.A.: "Superscattering and Stimulated Raman Scattering", Izv. Akad. Nauk SSR, Ser. Fiz., 1973, 37, no.10, pp.2190-3.

41.142. Fijnaut, H.M., and Vrij, A.: "Laser Beat Spectroscopy of Thin, Free, Liquid Films", Nature, 1973, 246, no.155, pp.118-9.

41.143. Galushkin, M.G., Davydov, V.V., and Yukov, E.A.: "Transverse Pumping of a Raman Laser", Kvantovaya Elektron., 1973, no.7, pp.114-5, and Sov. J. Quantum Electron., 1973, 3, no.1, pp.65-6.

41.144. Leiderer, P., Berberich, P., and Hunklinger, S.: "Measurement of Hypersonic Attenuation by Stimulated Brillouin Scattering", Rev. Sci. Instrum., 1973, 44, no.11, pp.1610-2.

41.145. Montero, S., and Orza, J.M.: "Simple Device for Raman Laser Spectrography", An. Fis., 1973, 69, no.4-6, pp.103-16.

41.146. Akhmanov, S.A., and D'yakov, Yu.E.: "Saturation Effects in SRS and Resonant Absorption of a Strong Non-Monochromatic Field", Zh. Eksp. Teor. Fiz. Pis'ma, 1973, 18, no.8, pp.519-22, and JETP Lett., 1973, 18, no.8, pp.305-7.

41.147. Dobele, H.F., and Kirsch, K.: "Detection of Weak Thomson Scattered Radiation from a Magnetized Arc Using a CW Argon-Ion Laser", Phys. Lett., 1974, 46A, no.5, pp.352-4.

41.148. Daree, K., and Kaiser, W.: "Transient Stimulated Scattering With High Conversion of Laser Into Scattered Light", Opt. Commun., 1974, 10, no.1, pp.63-7.

41.149. Sparks, M.: "Stimulated Raman and Brillouin Scattering. Parametric-Instability Explanation of Anomalies", Phys. Rev. Lett., 1974, 32, no.9, pp.450-3.

41.150. Thuy, C.D., et al.: "Effect of Self-Focusing on Inverse Raman Scattering", Exp. Tech. Phys., 1974, 22, no.2, pp.111-7.

41.151. Hsieh, C.T., Foltz, N.D., and Cho, C.W.: "Production of Stimulated Raman Lines in H_2 Gas with a Focused Laser Beam", J. Opt. Soc. Am., 1974, 64, no.2, pp.202-5.

41.152. Korobkin, V.V., et al.: "Self-Focusing of Ultrashort SRS Pulses", Phys. Lett., 1974, 47A, no.5, pp.381-2.

41.153. Herrmann, J., et al.: "SRS of Ultrashort Light Pulses by Polaritons in $LiIO_3$", Exp. Tech. Phys., 1974, 22, no.2, pp.97-110.

41.154. Stith, J.H., et al.: "Hypersound Speeds in CS_2, Acetone, and Benzene (by Brillouin Scattering)", J. Acoust. Soc. Am., 1974, 55, no.4, pp.785-9.

41.155. Levy, Y., et al.: "Raman Scattering of Thin Films as Waveguide", Opt. Commun., 1974, 11, no.1, pp.66-9.

41.156. Werncke, W., et al.: "Investigation of Inverse Raman Scattering Using Intracavity Spectroscopy", Opt. Commun., 1974, 11, no.2, pp.159-63.

41.157. Amy, J.W., et al.: "Programmed Cell Instrument for Digital Raman Difference Spectroscopy", Appl. Spectrosc., 1974, 28, no.3, pp.262-9.

41.158. Little, V.I., et al.: "Scattering of Light from Light-Induced Periodic Structures", Contemp. Phys., 1974, 15, no.3, pp.271-98.

41.159. Bloom, G.H., and Keeler, R.N.: "Stimulated Brillouin Scattering in Shock-Compressed Fluids", J. Appl. Phys., 1974, 45, no.3, pp.1200-7.

41.160. Apollonov, V.V., Barchukov, A.I., and Konyukhov, V.K.: "Measurement of Scattering of Mirrors in a CO_2 Laser", Kvantovaya Elektron., 1974, no.4, pp.103-5, and Sov. J. Quantum Electron., 1974, 3, no.4, pp.341-2.

41.161. Popovichev, V.I., Ragul'skii, V.V., and Faizullov, F.S.: "Stimulated Brillouin Scattering With a Broad Exciting Spectrum", Zh. Eksp. Teor. Fiz. Pis'ma, 1974, 19, no.6, pp.350-5, and JETP Lett., 1974, 19, no.6.

41.162. Dakin, J.P.: "Simplified Photometer for Rapid Measurement of Total Scattering Attenuation of Fibre Optical Waveguides", Opt. Commun., 1974, 12, no.1, pp.83-8.

41.163. Sprangle, P., and Granatstein, V.L.: "Stimulated Cyclotron Resonance Scattering and Production of Powerful Submillimetre Radiation", Appl. Phys. Lett., 1974, 25, no.7, pp.377-9.

41.164. Kato, D.: "Nondestructive Technique for Precise Evaluation of (Scattering of) Optical Waveguides", Appl. Phys. Lett., 1974, 25, no.7, pp.406-8.

41.165. Anderson, D., and Wilhelmsson, H.: "Simple Approach to Stimulated Backscattering in Inhomogeneous Plasmas", Phys. Lett., 1974, 50A, no.5, pp.383-4.

41.166. Ott, E., Manheimer, W.M., and Klein, H.H.: "Stimulated Compton Scattering and Self-Focusing in the Outer Regions of a Laser-Fusion Plasma", Phys. Fluids, 1974, 17, no.9, pp.1757-61.

41.167. Welsch, D.: "Linewidths in a Solid-State anti-Stokes Raman Oscillator", Phys. Lett., 1974, 47A, no.6, pp.487-8.

42. PLASMAS AND TIME RESOLUTION

42.1. Nag, B.R., and Engineer, M.H.: "Measurement of Microwave Hall Mobility of Semiconductors", Int. J. Electron., 1965, 18, p.529.

42.2. Shaw, T.M., Brooks, G.H., and Gunton, R.C.: "Bakeable Microwave Cavity for Measurement of Electron Loss Rates in Photoionized Nitric Oxide", Rev. Sci. Instrum., 1965, 36, p.478.

42.3. Golden, K.E.: "Plasma Simulation With Artificial Dielectric in a Horn Geometry", Trans. IEEE, 1965, AP-13, p.587.

42.4. Pescia, J.: "Measurement of Very Short Spin-Relaxation Periods", Ann. Phys. (Paris), 1965, 10, p.389.

42.5. Squire, P.T.: "Experimental Study of 2:1 Harmonic Cross Relaxation in Ruby", Proc. Phys. Soc., 1965, 86, p.573.

42.6. Abazadze, Yu.V., and Solov'ev, E.G.: "Measuring Group Velocity in Maser Delay Line", Prib. Tekh. Eksp., 1965, 8, no.6, p.124, and Instrum. Exp. Tech., 1965, 8, no.6, p.1433.

42.7. McLane, C.K., et al.: "Electron Density Measurements in the Magnetically Confined Arc", J. Appl. Phys., 1965, 36, p.337.

42.8. Maloney, W.T.: "Investigation of Hot Opaque Arc Plasmas by Microwave Cavity Techniques", J. Appl. Phys., 1965,36,p.703.

42.9. Valakhanov, V.Ya., Rusanov, V.D., and Striganov, A.R.: "Multiple-Beam Interferometer for Plasma Diagnostics", Zh. Tekh. Fiz., 1965, 35, p.127, and Sov. Phys.-Tech. Phys., 1965, 10, no.1, pp.96-9.

42.10. Andreev, S.I., and Sokolov, B.M.: "Investigation of Deionization of Plasma at Atmospheric Pressure by Microwaves", Zh. Tekh. Fiz., 1965, 35, p.101, and Sov. Phys.-Tech. Phys., 1965, 10, no.1, pp.75-80.

42.11. Hotston, E.S., and Seidl, M.: "Microwave Interferometer for Measurement of Small Phase Angles", J. Sci. Instrum., 1965, 42, p.225.

42.12. Green, A.H.: "Flame Attenuation Analysis Using Millimetre Techniques", Trans. IEEE, 1965, MIL-9, p.172.

42.13. Makios, W., and Munterbruch, H.: "Microwave Interferometry of Electromagnetically Generated Shock Waves", Z. Naturforsch., 1965, 20a, p.870.

42.14. Noon, J.H., Holt, E.H., and Reynolds, J.F.: "X-Band Waveguide Cell for Study of Microwave Propagation Through Magnetoplasma", Rev. Sci. Instrum., 1965, 36, p.622.

42.15. Buckmaster, H.A.: "Microwave Plasma Densitometers", Rev. Sci. Instrum., 1965, 36, p.711.

42.16. Edwards, D.H., and Lawrence, T.R.: "Ionization Measurements in Detonation Waves", Proc. Roy. Soc., 1965, 285, p.415.

42.17. Pilar, J., and Sicha, M.: "Verification of Microwave Method of Measuring Small Changes in Concentration of Electrons in Striated Plasma", Czech. J. Phys., 1965, 15B, p.399.

42.18. Shiobara, S.: "Measurement of Low-Electron-Density Plasma by Microwave Free-Space Method", Jap. J. Appl. Phys., 1965, 4, p.513.

42.19. Bromer, H.H., and Dobler, F.: "Electron Density Determinations With a Microwave Bridge in Weakly Ionized Plasmas", Z. Naturforsch., 1965, 20a, p.599.

42.20. Takeda, S., and Funahashi, A.: "Microwave Reflection Measurement of Plasmas Produced by Shock Waves", J. Phys. Soc. Jap., 1965, 20, p.1090.

42.21. Marzat, Cl., and Bottreau, A.: "Design of an Open-Branch Interferometer for 35 GHz", Onde Electr., 1965, 45, p.485.

42.22. Thomassen, K.I.: "Microwave Plasma-Density Measurements", J. Appl. Phys., 1965, 36, p.3642.

42.23. Balakhanov, V.Ya., Rusanov, V.D., and Striganov, A.R.: "Determination of Plasma Parameters by Multibeam Interferometer", At. Energ., 1965, 18, p.515.

42.24. Studnicka, J.: "Measurement of Stationary Striations by Microwave Method", Br. J. Appl. Phys., 1965, 16, p.1739.

42.25. Iannuzzi, M., and Enriques, L.: "Microwave Study of Thermally Ionized Plasmas", Appl. Phys. Lett., 1965, 7, p.47.

42.26. Anicin, B.A.: "Electron Density Profiles in Cylindrical Plasmas from Microwave Refraction Data", J. Res. Natl. Bur. Stand., 1965, 69D, p.721.

42.27. Markwardt, G.E.: "Investigation of Plasma Clot Motion by Microwaves", Magn. Gidrodin., 1965, no.4, p.27.

42.28. Dushin, L.A., et al.: "Plasma Investigation by Interferometer and Microwave-Cutoff Methods", Ukr. Fiz. Zh., 1965, 10, p.977.

42.29. Beerwold, H., and Hartwig, H.: "Fabry-Perot Resonator for Microwave Plasma Diagnostics", Z. Angew. Phys., 1965, 19, p.545.

42.30. Prinzler, M.: "Microwave Diagnostics of Cold-Cathode Arcs", Monatsber. Dtsch. Akad. Wiss. Berlin, 1965, 7, p.264.

42.31. Basu, J., and Dutta, C.: "Microwave Determination of Electron-Density Distribution in a Plasma", J. Inst. Telecommun. Eng., 1965, 11, p.510.

42.32. Lubin, M.J.: "Transverse Microwave Cavity Technique for Sensitive Measurement of Transient Electron Density Distributions", Rev. Sci. Instrum., 1966, 37, p.1034.

42.33. Parker, A.J., Laine, D.C., and Ingram, D.J.E.: "ESR Spectrometer for Studies of Short-Lived Species Using Sampling Techniques", J. Sci. Instrum., 1966, 43, p.688.

42.34. Unland, M.L., and Flygare, W.H.: "Direct Measurement of Rotational Relaxation", J. Chem. Phys., 1966, 45, p.2421.

42.35. Haussler, P., and Welles, S.J.: "Determination of Relaxation Periods in Cyclotron Resonance in Copper", Phys. Rev., 1966, 152, p.675.

42.36. Hermansdorfer, H.: "Measurement of High Reflection Densities With Microwave Reflection Probe on a Linear z-Pinch", Z. Naturforsch., 1966, 21a, p.1471.

42.37. Lutz, B.C., Lindsay, P.D., and Roemer, L.E.: "Microwave Bridge Measurement of Plasma Frequency", Rev. Sci. Instrum., 1966, 37, p.168.

42.38. Takeda, S., and Masumi, M.: "Extension of Measurable Density Ranges in Microwave Plasma Diagnostics", Jap. J. Appl. Phys., 1966, 5, p.1100.

42.39. Musil, J.: "Measurement of Radial Distribution of Electron Density in Plasma Cylinder With a Multibeam Microwave Interferometer", Czech. J. Phys., 1966, 16B, p.782.

42.40. Cronson, H.M.: "Spatial Variations of Plasma Electron Temperature in a Standing Wave at Microwave Frequencies", Phys. Fluids, 1966, 9, p.581.

42.41. Fischer, M.: "Exact Determination of Electron Density Distribution in a Finite Anisotropic Plasma", Ann. Phys. (Leipzig), 1966, 17, p.7.

42.42. Estin, A.J., and Anderson, M.M.: "Time-Resolved Microwave Interferometry as Diagnostic Tool for Decaying Plasma Afterglows", Rev. Sci. Instrum., 1966, 37, p.468.

42.43. Dushin, L.A., et al.: "Measurement of Plasma Density Distribution by Refraction of a Microwave Beam", Zh. Tekh. Fiz., 1966, 36, p.304, and Sov. Phys.-Tech. Phys., 1966, 11, no.2, pp.220-5.

42.44. Lonngren, K.E.: "Interpretation of Multiple Reflections in Microwave Interferometers Used in Plasma Diagnostics", Jap. J. Appl. Phys., 1966, 5, p.223.

42.45. Ferendeci, A.M.: "Microwave Cavity Q-Sampler for Plasma Diagnostics", Rev. Sci. Instrum., 1966, 37, p.1089.

42.46. Bachynski, M.P., Osborne, F.J.F., and Gibbs, B.W.: "Measurements of Anisotropic Plasmas Using a Turnstile Multiple-Probe Polarimeter", Can. J. Phys., 1966, 44, p.1649.

42.47. Ipatov, V.A., and Kalmykov, S.G.: "Microwave Diagnostics in the Investigation of Shock-Wave Structure", Zh. Tekh. Fiz., 1966, 36, p.981, and Sov. Phys.-Tech. Phys., 1966, 11, no.6, pp.726-30.

42.48. Dushin, L.A., et al.: "Investigation of Spatial Distribution of Plasma Density by Refraction of a Microwave Beam With Several Frequency Components", Zh. Tekh. Fiz., 1966, 36, p.1842, and Sov. Phys.-Tech. Phys., 1967, 11, no.10, pp.1372-8.

42.49. Hayami, R.A., and Kelley, K.J.: "Open Microwave Resonators for Ionized Wake Measurements", Trans. IEEE, 1967, AES-3, p.339.

42.50. Shankowski, A.E., and Kharadly, M.M.Z.: "Microwave Refraction Technique for Measuring Electron-Density Distribution in a Transient Plasma Column", Electron. Lett., 1967, 3, no.7, pp.335-6.

42.51. Miyoshi, Y., et al.: "Interferometer Measurements at 2 mm of Plasma Parameters in Controlled Fusion Experiment", Bull. Nagoya Inst. Technol., 1967, 19, pp.269-76.

42.52. Zakrzewski, A.: "Determination of Plasma Parameters from Measurements of Reflection Factor in a Waveguide", Rozpr. Elektrotech., 1967, 13, no.3, pp.403-21.

42.53. Graf, K.A., and Jassby, D.L.: "Measurements of Dipole Antenna Impedance in an Isotropic Laboratory Plasma", Trans. IEEE, 1967, AP-15, no.5, pp.681-8.

42.54. Cronin, J.C., and Sexton, M.C.: "Optimization of an X-Band Interferometer for Plasma Diagnostics", Int. J. Electron., 1967, 22, no.6, pp.581-5.

42.55. Polman, J.: "Sensitive 4-mm Lecher-Wire Interferometer for Electron-Concentration Measurements in Low-Density Plasmas", Rev. Sci. Instrum., 1967, 38, no.11, pp.1631-3.

42.56. Lederman, S., and Dawson, E.F.: "Microwave Technique for Measurement of Electron Density and Ionization Time", Phys. Fluids, 1967, 10, no.12, pp.2570-8.

42.57. Harris, J.H., and Balfour, D.: "Microwave Cavity Studies of Ionization Decay in Cs Vapour", J. Phys. D, 1968, 1, no.4, pp.409-23.

42.58. Chaffin, R.J., and Beyer, J.B.: "Diagnostics of an Anisotropic Plasma With a Microwave Fabry-Perot Resonator", Trans. IEEE, 1968, MTT-16, p.878.

42.59. Caron, P.R., and Russo, F.: "Technique for Measuring Electron Density of Dense, Thick, Steady-State Plasmas", Trans. IEEE, 1968, AP-16, p.611.

42.60. Brand, G.F., and Hooker, C.A.: "Microwave Measurements of Plasma Temperature", Nucl. Fusion, 1968, 8, p.272.

42.61. Kaitmazov, S.D., et al.: "Characteristics of a Mode-Locked Laser", Dokl. Akad. Nauk SSSR, 1968, 180, p.1331, and Sov. Phys.-Dokl., 1968, 13, p.591.

42.62. Lukac, P.: "Determination of Electron Density by Cylindrical TM_{010} Microwave Cavity", J. Phys. D, 1968, 1, p.1495.

42.63. Vlachos, M.A.W., and Hsuan, H.C.S.: "Profile Measurements of Plasma Columns Using Microwave Resonant Cavities", J. Appl. Phys., 1968, 39, p.5009.

42.64. Burkley, C.J., and Sexton, M.C.: "Measurement of Plasma Electron Distributions Using Microwave Cavities", J. Appl. Phys., 1968, 39, p.5013.

42.65. Musil, J.: "Measurement of Mean Electron Density in a Plasma Cylinder", Cesk. Casopis Fys., 1968, 18A, no.1, pp.7-10.

42.66. Musil, J., and Zacek, F.: "Influence of Multiple Reflections on the Accuracy of Microwave-Interferometer Measurements of Plasma Electron Density", Cesk. Casopis Fys., 1968, 18A, no.1, pp.1-6.

42.67. Aro, T.O., and Walsh, D.: "Measurement of Plasma Temperature Using a Waveguide Probe", Phys. Fluids, 1968, 11, no.5, pp.1070-5.

42.68. Okamoto, Y., and Fujita, J.: "Comparison of 35-GHz Interferometry", Jap. J. Appl. Phys., 1969, 8, p.281.

42.69. Moma, Yu.A., and Nevskii, M.V.: "Apparatus for Studying Semiconductor Lasers", Prib. Tekh. Eksp., 1969, 12, no.2, p.184, and Instrum. Exp. Tech., 1969, 12, no.2, p.469.

42.70. Yamanaka, H.: "Variable-Characteristic-Impedance Component for Time-Domain Reflectometry", Bull. Electrotech. Lab., 1969, 33, no.4, p.88.

42.71. Gunn, J.B.: "Spectrum and Width of Mode-Locked Laser Pulses", J. Quantum Electron. IEEE, 1969, QE-5, p.513.

42.72. van der Sijde, B., and van Run, L.P.M.: "Improvement of Phase Stabilization of a Simple (Plasma) Microwave Interferometer", J. Phys. E, 1969, 2, p.584.

42.73. Schaller, W.: "Measurement of Very Low Electron Density Using Laser Interferometry", Nachr. Tech. Z., 1969, 22, no.5, p.265.

42.74. Wallington, J.R., and Beynon, J.D.E.: "Sensitive Microwave Interferometer for Plasma Diagnostics", J. Plasma Phys., 1969, 3, pt 3, p.371.

42.75. Seshadri, S.R.: "Plasma Diagnostics With a Fabry-Perot Resonator", Proc. IEEE, 1969, 57, p.1187.

42.76. Vendik, O.G., and Foulds, K.W.H.: "Hybrid-T Measurement of Transient Impedance Changes", J. Phys. E, 1969, 2, p.269.

42.77. Bhagavat, G.K., and Vanvari, N.: "Microwave Power Absorption (by Cavity) in Plasma", Indian J. Pure Appl. Phys., 1969, 7, no.12, p.778.

42.78. Vauge, Ch., and Delpech, J.F.: "Measurement of Parameters Characterizing the Plasma of a He-Ne Laser", Onde Electr., 1969, 49, p.544.

42.79. Levinson, G.R., et al.: "Measurement of Lifetime of Oscillatory Levels in CO_2 Molecules", Radiotekh. Elektron., 1969, 14, no.4, p.675, and Radio Eng. Electron. Phys., 1969, 14, no.4, pp.580-5.

42.80. Hubbard, W.M.: "Measurement of Duration of Picosecond Pulses by Beat-Frequency Detection", J. Quantum Electron. IEEE, 1969, QE-5, p.326.

42.81. Brousseau, M., and Schuttler, R.: "Microwave Techniques for Measuring Carrier Lifetime and Mobility in Semiconductors", Solid-State Electron., 1969, 12, p.417.

42.82. Shapiro, S.I., and Duguay, M.A.: "Observation of Subpicosecond Components in Mode-Locked Nd^{3+}:Glass Laser", Phys. Lett., 1969, 28A, p.698.

42.83. Bridgett, K.A., and King, T.A.: "Excitation and Relaxation Effects on Laser Action in Helium", J. Phys. B, 1969, 2, p.902.

42.84. Klose, J.Z.: "Transition Probabilities and Mean Lives of the $2s_2$ Laser Level in Ne^I", J. Quant. Spectrosc., 1969, 9, p.881.

42.85. Kano, T., and Heaton, A.G.: "High-Resolution Laser Interferometer for Measurement of Electron Density in Transient Plasma", Electron. Lett., 1969, 5, p.413.

42.86. Harrache, R.J.: "Determination of Ultrashort Pulse Widths by Two-Photon Fluorescence Patterns", Appl. Phys. Lett., 1969, 14, p.148.

42.87. Vasil'eva, A.N., et al.: "Plasma Diagnostics in a Pinch-Discharge Laser", Zh. Tekh. Fiz., 1969, 39, no.2, p.341, and Sov. Phys.-Tech. Phys., 1969, 14, no.2, pp.246-50.

42.88. Gos'kov, P.I.: "Measurement of Electrodynamic Characteristics by Combined-Probe Method", Izv. VUZ Fiz., 1969, no.2, p.144.

42.89. Raptis, A.C., and Lonngren, K.E.: "Microwave Cavities for Plasma Diagnostics", J. Microwave Power, 1969, 4, no.3, pp.182-7.

42.90. Shimba, M.: "Attenuation Measurement of Millimetre Waveguide by Pulse Reflection Method", Electron. Commun. Jap., 1969, 52, no.12, pp.53-8.

42.91. Bradley, D.J., New, G.H.C., and Caughey, S.J.: "Subpicosecond Structure in Mode-Locked Nd^{3+}:Glass Lasers", Phys. Lett., 1969, 30A, p.78.

42.92. Wood, L.E., Grady, T.K., and Thompson, M.C.: "Technique for Measurement of Photomultiplier Transit-Time Variation", Appl. Opt , 1969, 8, p.2143.

42.93. Maly'shev, V.I., Markin, A.S., and Sychev, A.A.: "Determining the Rise Period of a Giant Pulse in a Laser", Zh. Prikl. Spekt., 1969, 10, no.2, p.248.

42.94. Slama, L.: "Measurement of Characteristics of a Very Dense Plasma Created by a Laser", Electron Technol., 1969, 2, no.2-3, p.29.

42.95. Korolev, F.A., et al.: "Measurement of the Lifetimes of Low Working Levels in an Argon Laser", Radiotekh. Elektron., 1969, 14, no.8, p.1519, and Radio Eng. Electron. Phys., 1969, 14, no.8, pp.1318-20.

42.96. Bradley, D.J., et al.: "Picosecond Pulses from Mode-Locked Dye Lasers", Phys. Lett., 1969, 30A, p.535.

42.97. Kon'kov, I.D., Rovinskii, R.E., and Cheburkin, N.V.: "Dependence of Lifetime and Rate of Population of Ionized Argon Level on Discharge Region", Radiotekh. Elektron., 1969, 14, no.11, p.2069, and Radio Eng. Electron. Phys., 1969, 14, no.11, pp.1793-6.

42.98. Banys, T., et al.: "Investigation of Ge Conductivity Relaxation in High Microwave Fields", Phys. Status Solidi, 1969, 36, p.755.

42.99. Kasuya, K., Nakai, S., and Yamanaka, C.: "Study of Ionizing Shock Waves by Millimetre-Wave Techniques", Trans. IEE, Jap., 1969, 89, no.4, p.78.

42.100. Anson, M., and Smith, R.C.: "Experimental Upper Limit to the Nonradiative Relaxation Period Between the $^4T_2$ and $^2T_1$ States of Ruby", J. Quantum Electron. IEEE, 1970, QE-6, p.268.

42.101. Sasaki, T., et al.: "Investigation of Pulse Structure of Mode-Locked Nd^{3+}:Glass Laser Using Two-Photon Fluorescence", Jap. J. Appl. Phys., 1970, 9, p.228.

42.102. Klein, M.B.: "Measurement of Upper-Laser-Level Lifetime in the He-Cd Laser by Fast Cavity-Dumping Techniques", Appl. Phys. Lett., 1970, 16, p.509.

42.103. Kent, G., and Thomas, D.: "Studies of Cavity Resonators Containing Magnetoplasmas", J. Appl. Phys., 1970, 41, no.12, pp.4945-53.

42.104. Smith, P.W.: "Pulse Velocity in a Resonant Absorber", J. Quantum Electron. IEEE, 1970, QE-6, p.416.

42.105. Bradley, D.J., Morrow, T., and Petty, M.S.: "Quenching of Two-Photon Fluorescence of a Mode-Locked Ruby Laser", Opt. Commun., 1970, 2, no.1, p.1.

42.106. Bradley, D.J., New, G.H.C., and Caughey, S.J.: "Amplitude and Phase Structure of Picosecond Pulses from Nd^{3+}:Glass Lasers", Phys. Lett., 1970, 32A, p.313.

42.107. Treacy, E.B.: "Direct Demonstration of Picosecond-Pulse Frequency Sweep", Appl. Phys. Lett., 1970, 17, p.14.

42.108. Burnham, D.C.: "Picosecond-Pulse Measurement Employing Nonlinear Photoelectric Effect", Appl. Phys. Lett., 1970, 17, p.45.

42.109. Klein, M.B.: "Time-Resolved Temperature Measurements in the Pulsed Argon-Ion Laser", Appl. Phys. Lett., 1970, 17, p.29.

42.110. Takeda, S., and Yasuda, A.: "Measurement of High Electron Densities by TM_{010} Cavity", Jap. J. Appl. Phys., 1970, 9, p.806.

42.111. Nagibarov, V.R., and Samartsev, V.V.: "Space-Time Superposition of Giant-Pulse Structures for Generation of a Light Echo", Ukr. Fiz. Zh., 1970, 15, no.8, p.1386.

42.112. Rowe, H.E., and Li, T.: "Theory of Two-Photon Measurement of Laser Output", J. Quantum Electron. IEEE, 1970, QE-6, p.49.

42.113. Siegman, A.E., and Kuizenga, D.J.: "Proposed Method for Measuring Picosecond-Pulse Widths and Shapes in CW Mode-Locked Lasers", J. Quantum Electron. IEEE, 1970, QE-6, p.212.

42.114. Kannewurf, C.R., and Motamedi, M.E.: "Pulsed GaAs Laser for Carrier-Lifetime Measurements", Trans. IEEE, 1970, IM-19, p.6.

42.115. Shimba, M., and Kikushima, M.: "Accurate Delay-Time Measuring Equipment for 50 GHz", Trans. IEEE, 1970, IM-19, p.9.

42.116. Bradley, D.J., New, G.H.C., and Caughey, S.J.: "Relationship Between Saturable Absorber Cell Length and Pulse Duration in Passively Mode-Locked Lasers", Opt. Commun., 1970, 2, no.1, p.41.

42.117. Uhlhorn, R.W., and Holshouser, D.F.: "Cross-Correlation Detection of Subnanosecond Optical Pulses", J. Quantum Electron. IEEE, 1970, QE-6, no.12, pp.775-82.

42.118. von der Linde, D., Bernecker, O., and Kaiser, W.: "Experimental Investigation of Single Picosecond Pulses", Opt. Commun., 1970, 2, no.4, pp.149-52.

42.119. Smith, A.W., and Landon, A.J.: "Mode-Locked Laser Pulse Shapes", Appl. Phys. Lett., 1970, 17, no.8, pp.340-3.

42.120. Bhagavat, G.K., and Vanvari, N.: "Effect of External DC Electric Field on (Cavity) Absorption of Microwaves in Plasma", Indian J. Pure Appl. Phys., 1970, 8, no.7, p.383.

42.121. Pike, H.A., and Hercher, M.: "Basis for Picosecond Structure in Mode-Locked Laser Pulses", J. Appl. Phys., 1970, 41, no.11, pp.4562-5.

42.122. Fried, Z.: "Structure of Picosecond Pulses", Phys. Lett., 1970, 33A, no.2, pp.62-3.

42.123. Hazan, J.P., et al.: "Real-Time Oscilloscope Observation of an Ultrafast Photodiode Response to Mode-Locked Laser Pulses", J. Quantum Electron. IEEE, 1970, QE-6, no.11, pp.744-5.

42.124. Duguay, M.A., and Hansen, J.W.: "Study of Nd^{3+}:Glass Laser Emission", J. Quantum Electron. IEEE, 1970, QE-6, no.11, pp.725-43.

42.125. Malley, M.M., and Rentzepis, P.M.: "Picosecond Time-Resolved Stimulated Light Emission", Chem. Phys. Lett., 1970, 7, no.1, pp.57-60.

42.126. Braslau, N., and Hauge, P.S.: "Microwave Measurement of Velocity/Field Characteristic of GaAs", Trans. IEEE, 1970, ED-17, no.8, pp.616-22.

42.127. Shimba, M., Kikushima, M., and Tamura, Y.: "Delay-Period Measuring Equipment for Waveguides in the 50-GHz Region", Rev. Electr. Commun. Lab., 1970, 18, no.7-8, pp.545-56.

42.128. Sharapov, L.I., Vakser, I.Kh., and Lanin, I.M.: "Equipment for 3-cm Waves With Pulse Duration of 1ns", Prib. Tekh. Eksp., 1970, 13, no.3, pp.179-81, and Instrum. Exp. Tech., 1970, 13, no.3, pp.841-3.

42.129. Glebovich, G.V., and Gur'eva, I.S.: "Analysis of Profiles of Nonuniform Transmission Lines Using Pulse Methods", Izv. VUZ Radioelektron., 1970, 13, no.11, pp.1376-8.

42.130. Keilmann, F.: "Plasma Diagnostics by Focused Laser Beams", Laser, 1970, 2, no.4, pp.55-9.

42.131. Fujimoto, T., Ogata, Y., and Fukuda, K.: "Measurement of Electron Density and Temperature in a Pulsed Argon-Ion Laser", Mem. Fac. Eng. Kyoto Univ., 1970, 32, no.2, pp.236-48.

42.132. Auston, D.H.: "Measurement of Picosecond Pulse Shape and Background Level", Appl. Phys. Lett., 1971, 18, no.6, pp.49-51.

42.133. Grigor'yants, V.V., et al.: "Measuring the Relaxation Period of a Laser Level of Nd^{3+} in Glass", Zh. Prikl. Spekt., 1971, 14, no.1, pp.73-7.

42.134. Papayoanou, A., and Gumeiner, I.M.: "Interferometric Measurements of Time-Dependent Electron Density in the Xe Pinched-Plasma Laser", J. Appl. Phys., 1971, 42, no.5, pp.1914-6.

42.135. Zory, P.S., and Lynch, G.W.: "Plasma Refractive-Index Measurements by He-Ne Vernier Laser", Proc. IEEE, 1971, 59, no.4, pp.684-9.

42.136. Verevkin, Yu.K., et al.: "Measurement of Ultrashort Pulse Durations", Izv. VUZ Radiofiz., 1971, 14, no.6, pp.840-4.

42.137. Kozelev, A.I.: "(Pulse) Measurement of Losses on Conversion of TE_{01} to TE_{02} Mode in Waveguides With Random Discontinuities", Izv. VUZ Radioelektron, 1971, 14, no.2, pp.213-6.

42.138. Glover, G.H.: "Error in Microwave Measurement of Velocity/Field Characteristic of n-GaAs Due to Energy Relaxation Effects", Appl. Phys. Lett., 1971, 18, no.7, pp.290-1.

42.139. Hubbard, W.M.: "Method for Measurement of Duration of Picosecond Pulses by Beat-Frequency Detection of Laser Output", Bell Syst. Tech. J., 1971, 50, no.1, pp.1-21.

42.140. Patel, C.K.N.: "Measurement of Subnanosecond Laser Pulses at 10.6 micron", Appl. Phys. Lett., 1971, 18, no.1, pp.25-8.

42.141. Ventrice, C.A.: "Microwave Propagation in Presence of Finite Amplitude Helical Instability", Phys. Fluids, 1971, 14, no.1, pp.192-4.

42.142. Kirdyashev, K.P., and Zaikina, A.N.: "Microwave Probing of a Plasma-Beam Discharge Subjected to an External RF Field", Radiotekh. Elektron., 1971, 16, no.4, pp.634-6, and Radio Eng. Electron. Phys., 1971, 16, no.4, pp.697-9.

42.143. Gloge, D., and Lee, T.P.: "Study of a Self (Picosecond) Pulsing GaAs Laser by Intensity Correlation in $LiIO_3$", J. Appl. Phys., 1971, 42, no.1, pp.307-9.

42.144. Cummins, W.F.: "Stabilized 70-GHz Plasma Interferometer", Rev. Sci. Instrum., 1970, 41, p.234.

42.145. Comte, G., Daujard, G., and Vuillaume, G.: "Pulse Measurements on Circular Millimetric Waveguides", Cables Transm., 1971, 25, no.1, pp.88-112.

42.146. Chakravarti, A.N., and Parui, D.P.: "Method for Determining Thickness of Active Region in GaAs Junction Lasers", Int. J. Electron., 1971, 30, no.2, pp.180-4.

42.147. Chakravarti, A.N., and Parui, D.P.: "Method for Determining the Recombination Constant in GaAs Junction Lasers", Indian J. Pure Appl. Phys., 1971, 9, no.3, pp.196-8.

42.148. Afromowitz, M.A., and DiDomenico, M.: "Measurement of Free-Carrier Lifetimes in GaP by Photoinduced Modulation of IR Absorption", J. Appl. Phys., 1971, 42, no.8, pp.3205-8.

42.149. Hayase, K., and Okuda, T.: "Measurement of Electron Density and Temperature in Slow Theta-Pinch Plasma by Laser Scattering", Trans. IEE Jap,, 1971, 90, no.1-2, pp.113-20.

42.150. Archambault, Y.: "Spherical Plasma Diffraction of Microwaves. Measurement of High Electron Densities", Rev. Phys. Appl., 1971, 6, no.2, pp.195-6.

42.151. Jansen, G.: "Continuous Measurement of Plasma Electron Density by Cavity Resonators", Z. Angew. Phys., 1971, 31, no.5-6, pp.296-300.

42.152. Akhmanov, S.A., and Chirkin, A.S.: "Two-Photon Fluorescence Technique of Ultrashort Laser Pulse Measurement", Opto-Electron., 1971, 3, no.2, pp.111-6.

42.153. Makios, V., and Thomas, R.E.: "Measurement of Minority-Carrier Lifetime in Si at Microwave Frequencies Using Microstrip Techniques", Electron. Lett., 1971, 7, no.17, pp.496-7.

42.154. Treacy, E.B.: "Measurement and Interpretation of Dynamic Spectrograms of Picosecond Light Pulses", J. Appl. Phys., 1971, 42, no.10, pp.3848-58.

42.155. Dahlstrom, L.: "Measurement of Ultrashort Laser Pulses by the Optical Kerr Effect", Opt. Commun., 1971, 3, no.6, pp.399-403.

42.156. Bradley, D.J., et al.: "Direct Measurement of Duration of Dye-Laser Picosecond Pulses", Opt. Commun., 1971, 3, no.6, pp.426-8.

42.157. Dolgov-Savel'ev, G.G., et al.: "Vibrational-Rotational Transitions in HF Chemical Laser", Zh. Eksp. Teor. Fiz., 1971, 61, no.1, pp.64-71, and Sov. Phys.-JETP, 1972, 34, no.1.

42.158. Nicolson, A.M., et al.: "Applications of Time-Domain Metrology to Automation of Broadband Microwave Measurements", Trans. IEEE, 1972, MTT-20, no.1, pp.3-9.

42.159. Fein, M.E., et al.: "Numerical Method for Calibrating Microwave Cavities for Plasma Diagnostics. I and II", Trans. IEEE, 1972, MTT-20, no.1, pp.22-30 and 83.

42.160. Uhlir, A.: "Frequency Domain Characterization of Microwave Delay Lines", Trans. IEEE, 1972, MTT-20, no.1, pp.51-3.

42.161. Iizuka, K.: "Subtractive Microwave Holography and Application to Plasma Studies", Appl. Phys. Lett., 1972, 20, no.1, pp.27-9.

42.162. Lin, C.H., and Gustafson, T.K.: "Optical Pulsewidth Measurement Using Self-Phase Modulation", J. Quantum Electron. IEEE, 1972, QE-8, no.4, pp.429-30.

42.163. von der Linde, D.: "Experimental Study of Single Picosecond Light Pulses", J. Quantum Electron. IEEE, 1972, QE-8, no.3, pp.328-38.

42.164. Biron, E.J.: "Time-Domain Reflectometry for Testing Waveguide and Antenna Systems", Western Electr. J., 1972, 16, no.1, pp.20-3.

42.165. Steward, K.W.F.: "Frequency Response Measuring Equipment", Marconi Rev., 1972, 35, no.184, pp.54-74.

42.166. Granek, H., Freed, C., and Haus, H.A.: "Experiment on Cross Relaxation in CO_2", J. Quantum Electron. IEEE, 1972, QE-8, no.4, pp.404-14.

42.167. Kuznetsova, T.I.: "Measurements of Time Radiation Characteristics Based on Multiphoton Processes in Opposite Light Beams", Izv. VUZ Radiofiz., 1972, 15, no.2, pp.227-32.

42.168. Ripper, J.E.: "Measurement of Spontaneous Carrier Lifetime from Stimulated-Emission Delays in Semiconductor Lasers", J. Appl. Phys., 1972, 43, no.4, pp.1762-3.

42.169. Attwood, D.: "Microwave Scattering from Underdense and Overdense Turbulent Plasmas", Phys. Fluids, 1972, 15, no.5, pp.942-4.

42.170. Glover, G.H.: "Microwave Measurement of the Velocity/Field Characteristic of n-InP", Appl. Phys. Lett., 1972, 20, no.6, pp.224-5.

42.171. Busca, G.L., and Bergeron, M.: "Variation of Pulse Duration of a Mode-Locked Nd^{3+}:Glass Laser With Cavity Length and Dye Transmittance", Can. J. Phys., 1972, 50, no.4, pp.407-9.

42.172. O'Neil, R.W., et al.: "TEA Laser Medium Diagnostics", Appl. Phys. Lett., 1972, 20, no.11, pp.461-3.

42.173. Ohmi, T., Hasuo, S., and Hori, S.: "Observation of Transient Behaviour of Picosecond Laser Pulses", J. Appl. Phys., 1972, 43, no.9, pp.3773-5.

42.174. Reintjes, J., and Carman, R.L.: "Direct Observation of Orientation Kerr Effect in Self-Focusing of Picosecond Pulses", Phys. Rev. Lett., 1972, 28, no.26, pp.1697-700.

42.175. Shimba, M., Yamaguchi, K., and Kondoh, K.: "(Pulse) Method of Measuring Millimetre Waveguide Attenuation", Trans. IEEE, 1972, IM-21, no.3, pp.215-9.

42.176. Gibson, A.F., et al.: "Optical Bridge for Assessment of Mode-Locked CO_2 Lasers", J. Phys. D, 1972,5,no.10,pp.1800-6.

42.177. Dorman, F.H., and McTaggart, F.K.: "(Measurement of) Electron Density and Temperature in Microwave Plasmas at High Pressures", J. Microwave Power, 1972, 7, no.3, pp.181-4.

42.178. Nemes, G.: "Techniques for Detection and Measurement of Picosecond Light Pulses", Stud. Cercet. Fiz., 1972, 24, no.6, pp.727-40.

42.179. Ricard, D., Lowdermilk, W.H., and Ducuing, J.: "Direct Observation of Vibrational Relaxation of Dye Molecules in Solution", Chem. Phys. Lett., 1972, 16, no.3, pp.617-21.

42.180. New, G.H.C.: "Ultrashort Pulse Measurements", Alta Freq., 1972, 41, no.10, pp.718-25.

42.181. Fanchenko, S.D., and Frolov, B.A.: "Picosecond Structure of the Emission of a Laser With a Nonlinear Absorber", Zh. Eksp. Teor. Fiz. Pis'ma, 1972, 16, no.3, pp.147-50, and JETP Lett., 1972, 16, no.3, pp.101-4.

42.182. Ku, R.T., et al.: "Plasma Diagnostics With a Tuned GaAs Laser Diode", J. Appl. Phys., 1972, 43, no.11, pp.4579-86.

42.183. Jassby, D.L., and Marhic, M.E.: "Far-IR Plasma Reflectometry and Instability Detection", J. Appl. Phys., 1972, 43, no.11, pp.4586-90.

42.184. Pichamuthu, J.P., Hassler, J.C., and Coleman, P.D.: "Gas-Temperature Measurement in Pulsed H_2O Laser Discharges", J. Appl. Phys., 1972, 43, no.11, pp.4562-5.

42.185. Erdmann, T.A., Figger, H., and Walther, H.: "Lifetime Measurements With a Tunable Flash-Pumped Dye Laser", Opt. Commun., 1972, 6, no.2, pp.166-8.

42.186. Petrov, G.D., Petryakov, A.I., and Samarskii, P.A.: "Submillimetre Laser Interferometry of a Carbon-Arc Plasma", Teplofiz. Vys. Temp., 1972, 10, no.1, pp.181-2, and High Temp., 1972, 10, no.1, pp.154-5.

42.187. Voitovich, A.P., Komar, V.A., and Smirnov, A.Ya.: "Possibility of Measuring Electron Concentrations in a Plasma Using a Gas Laser with Nonlinear Absorbing Cell", Zh. Prikl. Spekt., 1972, 17,no.4,pp.705-7.

42.188. Cirkel, H.J., Ringwelski, L., and Schafer, F.P.: "Fluorescence Lifetime Measurements Using Two-Photon Absorption". Z. Phys. Chem. Frankfurt, 1972, 81, no.1-4, pp.158-62.

42.189. Davydov, Yu.T., and Merkishin, G.V.: "(Pulsed) Measurement of Multiplication Factor and Current of an Avalanche Photodiode With Guard Ring", Fiz. Tekh. Poluprov., 1972, 6, no.10, pp.2072-3, and Sov. Phys.-Semicond., 1972, 6, no.10, p.1765.

42.190. Mak, A.A., et al.: "Measurement of Relaxation Rates in Gases Activated With Nd^{3+} Ions", Opt. Spekt., 1972, 33, no.4, pp.689-97, and Opt. Spectrosc., 1972, 33, no.4, pp.381-5.

40.191. Zaritskii, A.R., et al.: "Measurement of Polarization of Radiation Reflected Backwards from a Laser-Heated Plasma", Kvantovaya Elektron., 1972, no.2, pp.89-90, and Sov. J. Quantum Electron., 1972,2,no.2,p.162.

42.192. Alfano, R.R., and Shapiro, S.L.: "Direct Measurement of Vibrational Decay of Dye Molecules in the Excited State", Opt. Commun., 1972, 6, no.2, pp.98-100.

42.193. Kurbatov, L.N., Nikitin, V.V., and Sharin, A.I.: "Unit for Investigating the Temporal Characteristics of Injection Lasers", Prib. Tekh. Eksp., 1972, 15, no.3, pp.203-5, and Instrum. Exp. Tech., 1972, 15, no.3, pp.857-60.

42.194. Dewhurst, R.J., et al.: "Direct Measurement of Pulse Broadening in the Second Harmonic of a Mode-Locked Nd^{3+}:Glass Laser", Opt. Commun., 1972, 6, no.4,pp.356-9.

42.195. Wollrab, J.E., and Rasmussen, R.L.: "Microwave Resonator for Studying (Lifetimes of) Chemical Laser Reactions", Rev. Sci. Instrum., 1973, 44, no.2,pp.177-9.

42.196. McGeoch, M.W.: "Measurement of Ultrashort Pulses in Ruby Laser", Opt. Commun., 1973, 7, no.2, pp.116-20.

42.197. Varma, C.A.G.O., and Rentzepis, P.M.: "Time-Resolved Picosecond Spectroscopy and Light-Gate Method", Chem. Phys. Lett., 1973, 19, no.2, pp.162-5.

42.198. Bancroft, J.C., and Johnston, R.H.: "Microwave Measurements by Fourier Analysis of Network Pulse Response", Proc. IEEE, 1973, 61, no.4, pp.472-3.

42.199. Desbois, J., Tournois, P., and Gires, F.: "Approach to Picosecond-Laser-Pulse Analysis, Shaping, and Coding", J. Quantum Electron. IEEE, 1973, QE-9, no.2, pp.213-8.

42.200. Busch, G.E., Jones, R.P., and Rentzepis, P.M.: "Picosecond Spectroscopy Using a Picosecond (Self-Focusing) Continuum", Chem. Phys. Lett., 1973, 18, no.2, pp.178-85.

42.201. Tsinsli, P.E., et al.: "Photon-Counting Method for Investigating Dyes With Short Quenching Periods", Izv. Akad. Nauk SSSR, Ser. Fiz., 1973, 37, no.2, pp.391-5.

42.202. Sackett, P.B., Hordvik, A., and Schlossberg, H.: "Measurement of v-v Energy Transfer Rate from CO (v=2) Using Tunable Parametric-Oscillator Excitation", Appl. Phys. Lett., 1973, 22, no.8, pp.367-8.

42.203. Konjevic, R., Jovicic, J., and Konjevic, N.: "Time-Resolved Spectroscopy of Rhodamine-6G Dye Laser", Fizika, 1973, 5, no.1, pp.17-26.

42.204. Alekseev, A.I., and Yevseev, I.V.: "(Relaxation Time by) Photon Echo on a Homogeneously Broadened Line", Phys. Lett., 1973, 43A, no.5, pp.465-6.

42.205. Reintjes, J.F., and McGroddy, J.C.: "Indirect Two-Photon Transitions in Si at 1060 nm", Phys. Rev. Lett., 1973, 30, no.19, pp.901-3.

42.206. Baumhacker, H., Fill, E., and Schmid, W.: "Detection of Short CO_2-Laser Pulses Using Optical Kerr Effect", Phys. Lett., 1973, 44A, no.1, pp.3-4.

42.207. Varma, C.A.G.O., and Rentzepis, P.M.: "Time Resolution and Characteristics of a Broadband Picosecond Continuum and Light Gate", J. Chem. Phys., 1973, 58, no.12, pp.5237-46.

42.208. Billman, K.W., and Stallcop, J.R.: "Measurement of Density and Temperature of a Hydrogen Plasma Using an Argon Laser", Appl. Phys. Lett., 1973, 22, no.11,pp.565-7.

42.209. Schreiber, P.W., Hunter, A.M., and Smith, D.R.: "Determination of Plasma Electron Density from Refraction Measurements", Plasma Phys., 1973, 15, no.7, pp.635-46.

42.210. Chakravarti, A.N., and Parui, D.P.: "Determination of the Diffusivity-Mobility Ratio in Degenerate Semiconductors from Linewidth Measurements in Laser Diodes", Czech. J. Phys., 1973, 23B, no.5,pp.548-50.

42.211. Leheny, R.F., Nahory, R.E., and Pollack, M.A.: "Millimetre-Wave Determination of Photoinjected Free-Carrier Concentrations in Highly Excited GaAs", Phys. Rev., 1973, 8B, no.2, pp.620-3.

42.212. Gornik, W., et al.: "Lifetime Measurements Using Stepwise Excitation by Two Pulsed Dye Lasers", Appl. Phys., 1973, 1, no.5, pp.285-6.

42.213. Richardson, M.C.: "Investigation of a Mode-Locked Nd^{3+}:Glass Laser With a Picosecond Streak Camera", J. Quantum Electron. IEEE, 1973, QE-9, no.7, pp.768-72.

42.214. Arthurs, E.G., Bradley, D.J., and Roddie, A.G.: "(Measurement of) Buildup of Picosecond Pulses in Passively Mode-Locked Dye Lasers", Appl. Phys. Lett., 1973, 23, no.2, pp.88-9.

42.215. Malz, D., Pohler, M., and Staupendahl, G.: "Multiphoton Absorption in Te by Irradiation of CO_2 Laser Pulses", Phys. Status Solidi, 1973, 58, no.1, pp.K35-8.

42.216. Davis, C.C., and King, T.A.: "Decay Rates of IR Emitting Laser Levels in Xe^I", J. Quant. Spectrosc., 1973, 13, no.9, pp.825-35.

42.217. Grasyuk, A.Z., et al.: "Dependence of Two-Photon Absorption in GaAs on Light-Pulse Duration", Zh. Eksp. Teor. Fiz. Pis'ma, 1973, 17, no.10, pp.584-7, and JETP Lett., 1973, 17, no.10, pp.416-8.

42.218. Sasaki, T., and Yamanaka, C.: "Measurement of Picosecond Pulse Duration Using Optical Kerr Effect", Technol. Rep. Osaka Univ., 1973, 23, no.1090-1120,pp.205-13.

42.219. Alimpiev, S.S., and Karlov, N.V.: "Photon Echo in the Gases BCl_3 and SF_6", Zh. Eksp. Teor. Fiz., 1973, 36, no.2, pp.482-90, and Sov. Phys.-JETP, 1973, 36,no.2,pp.255-9.

42.220. Green, J.M., Collins, G.J., and Webb, C.E.: "Collisional Excitation and Destruction of Excited Zn^{II} Levels in a Helium Afterglow", J. Phys. B, 1973, 6, no.8, pp.1551-61.

42.221. Stark, E.E.: "Measurement of the 10^00-02^00 Relaxation Rate in CO_2", Appl. Phys. Lett., 1973, 23, no.6, pp.335-7.

42.222. Smith, I.W.M., and Wittig, C.: "Vibrational Energy Transfer in CO at Low Temperatures", J. Chem. Soc. Faraday Trans. II, 1973, 69, pt 7, pp.939-51.

42.223. Ferrario, A.: "Measurements of Upper- and Lower-Level Lifetime in He-Se Lasers", Opt. Commun., 1973, 8,no.4,pp.333-5.

42.224. Zlenko, A.A., Sychugov, V.A., and Shipulo, G.P.: "Measurement of Relaxation Period of Nd^{3+}:YAG", Kvantovaya Elektron., 1973, no.5, pp.103-6, and Sov. J. Quantum Electron., 1973, 2, no.5, pp.474-6.

42.225. Lin, C., and Dienes, A.: "Direct Measurement of Radiationless Internal Conversion Rate of Excited Singlet States of Laser Dye Molecules", Opt. Commun., 1973, 9, no.1, pp.21-4.

42.226. Bartlett, D.V., and Brand, G.F.: "Microwave Plasma Measurements With a Gunn-Effect Oscillator", J. Phys. E, 1973, 6, no.12, pp.1213-5.

42.227. Pilipovich, V.A., et al.: "Measurement of Shape of a Laser Pulse", Prib. Tekh. Eksp., 1973, 16, no.2, pp.177-9, and Instrum. Exp. Tech. 1973, 16, no.2, pp.541-3.

42.228. Alimpiev, S.S., and Karlov, N.V.: "Experimental Methods of Observation of Coherent Interaction of Pulsed IR Radiation With Molecular Gases", Izv. Akad. Nauk SSSR, Ser. Fiz., 1973, 37, no.10, pp.2022-31.

42.229. Reintjes, J., Carman, R.L., and Shimizu, F.: "Study of Self-Focusing and Self-Phase-Modulation in the Picosecond Time Regime", Phys. Rev., 1973, 8A,no.3,pp.1486-503.

42.230. Nechaev, S.Yu., and Ponomarev, Yu.N.: "Change in Form of a Powerful Light Pulse in a Resonance-Absorbing Medium", Izv. VUZ Fiz., 1973, no.12, pp.148-50.

42.231. Lin, C., and Dienes, A.: "Study of Excitation Transfer in Laser Dye Mixtures by Direct Measurement of Fluorescence Lifetime", J. Appl. Phys., 1973, 44, no.11, pp.5050-2.

42.232. Borisevich, N.A., and Gruzinskii, V.V.: "Time-Resolved Spectral Characteristics of Lasing Solutions of Complex Molecules", Acta Phys. Chem. Szeged., 1973, 19, no.4, pp.327-44.

42.233. Appelt, J., and Sadowski, M.: "Application of Laser Interferometers to Measurements of Electron Concentration in Plasma", Nukleonika, 1973, 18,no.7,pp.277-97.

42.234. Fried, S.S., Wilson, J., and Taylor, R.L.: "Measurement of Temperature Dependence of Vibrational Relaxation Rate of HF", J. Quantum Electron. IEEE, 1973, QE-9, no.1, pp.59-64.

42.235. Dneprovskii, V.S., Koshchug, D.G., and Khattatov, V.U.: "Investigation of Intensity and Duration of Ultrashort Pulses Emitted by a Mode-Locked Nd^3+ Laser", Kvantovaya Elektron., 1973, no.2, pp.84-6, and Sov. J. Quantum Electron., 1973, 3, no.2, pp.144-5.

42.236. Pirozhkov, V.A., et al.: "(Apparent) Acceleration of Laser Pulses Due to a Phase Memory", Kvantovaya Elektron., 1973, no.2, pp.115-6, and Sov. J. Quantum Electron., 1973, 3, no.2, pp.172-3.

42.237. Johnson, L.C., and Chu, T.K.: "Measurement of Electron Density Evolution and Beam Self-Focusing in a Laser-Produced Plasma", Phys. Rev. Lett., 1974, 32, no.10, pp.517-20.

42.238. Kuwahara, C., Matsuura, K., and Miyahara, A.: "Phase-Shift Measurement With Microwave Interferometer for Plasma Diagnostics", Jap. J. Appl. Phys., 1974, 13, no.2, pp.318-26.

42.239. Kuwahara, C., Matsuura, K., and Miyahara, A.: "Density Measurement of Small-Diameter Plasma by Microwave Interferometer", Jap. J. Appl. Phys., 1974, 13,no.2,pp.327-33.

42.240. Seguin, H.J., Tulip, J., and McKen, D.: "(Measurement of) UV Photoionization in TEA (CO_2) Lasers", J. Quantum Electron. IEEE, 1974, QE-10, no.3, pp.311-9.

42.241. Bradley, D.J., and New, G.H.C.: "Ultrashort Pulse Measurements", Proc. IEEE, 1974, 62, no.3, pp.313-45.

42.242. Watanabe, S., Chihara, M., and Ogura, I.: "Decay Rate Measurements of Upper Laser Levels in He-Ne and He-Se Lasers", Jap. J. Appl. Phys., 1974, 13, no.1, pp.164-9.

42.243. Granek, H.: "Observation of Diffusion as an Effective Vibrational Relaxation Rate in CO_2", J. Quantum Electron. IEEE, 1974, QE-10, no.3, pp.320-5.

42.244. Tomov, I.V.: "Ultrashort Pulse Measurements Using a Nonlinear Mirror", Opt. Commun., 1974, 10, no.2, pp.154-6.

42.245. Lee, C.H., and Jayaraman, S.: "Measurement of Ultrashort Optical Pulses by Two-Photon Conductivity (in Semiconductors)", Opto-Electron., 1974, 6, no.1, pp.115-20.

42.246. Bebelaar, D.: "Time-Resolved Molecular Spectroscopy Using High-Power Solid-State Lasers in Pulse Transmission Mode", Chem. Phys., 1974, 3, no.2, pp.205-16.

42.247. Craig, A.D., et al.: "Measurement of Long-Wavelength Turbulence in a Collisionless Shock by Scattering of Radiation from a CO_2 Laser", Phys. Rev. Lett., 1974, 32, no.18, pp.975-8.

42.248. Turner, R.: "Electron-Density Measurement in HCN Laser Using Faraday Mode-Splitting Technique", Appl. Opt., 1974, 13, no.4, pp.968-73.

42.249. Alfano, R.R., and Zawadzkas, G.A.: "Observation of Backward-SRS Generated by Picosecond Laser Pulses in Liquids", Phys. Rev., 1974, 9A, no.2, pp.822-4.

42.250. Clerc, M., Jones, R.P., and Rentzepis, P.M.: "Picosecond Time Resolution of SRS of Alcohols", Chem. Phys. Lett., 1974, 26, no.2, pp.167-73.

42.251. Sahar, E., and Wider, I.: "Absorption Cross Sections of First Excited Singlet State of Laser Dyes at 337.1 nm", J. Quantum Electron. IEEE, 1974, QE-10, no.8, pp.612-4.

42.252. Aung, H., Nagai, K., and Katayama, M.: "Cooling Effects of CO_2, CO_2-N_2, and CO_2-He, Gases by Absorption of Q-Switched Laser Radiations", J. Phys. Soc. Jap., 1974, 37, no.1, pp.186-92.

42.253. Sam, C.L.: "Experimental Study of Linear-Growth Region of Ultrashort-Pulse Generation in a Mode-Locked Nd^3+:Glass Laser", Appl. Phys. Lett., 1974, 24, no.12, pp.631-3.

42.254. Peterson, L.M., Arnold, C.B., and Lindquist, G.H.: "Pulsed HF Chemical-Laser Linewidth Measurements Using Time-Resolvable Bleachable Absorption", Appl. Phys. Lett., 1974, 24, no.12, pp.615-7.

42.255. Jones, P.A.: "Thomson-Scattering Plasma Diagnostics Using a Four-Pulse Ruby Laser", J. Phys. E, 1974, 7, no.9, pp.704-6.

42.256. Pappalardo, R., and Lempicki, A.: "Method for Detecting Transient Gain (or Loss) in Pulsed Gas Discharges", J. Quantum Electron. IEEE, 1974, QE-10,no.10,pp.816-8.

42.257. Lytle, F.E., and Kelsey, M.S.: "Cavity-Dumped Argon-Ion Laser as Excitation Source in Time-Resolved Fluorimetry", Anal. Chem., 1974, 46, no.7, pp.855-60.

42.258. Chin, S.L.: "Theory of the Superposition of a Train of Short Laser Pulses in a Dye Medium", Opt. Commun., 1974, 12, no.1, pp.1-4.

42.259. Adamski, W.: "Accuracy of Measurement Using Microwave (Frequency-Domain) Reflectometer", Rozpr. Electrotech., 1974, 20, no.3, pp.471-88.

42.260. Kitazima, I.: "Relaxation-Period Measurement of CO_2 Laser by Q-Switching Technique", J. Appl. Phys., 1974, 45, no.11, pp.4961-3.

42.261. Nichols, D.B., Wrolstad, K.H., and McClure, J.D.: "Time-Resolved Spectroscopy of a Pulsed HF Laser With Well-Defined Initial Conditions", J. Appl. Phys., 1974, 45, no.12, pp.5360-6.

42.262. Klimov, A.V.: "Instrument for Measurement of Intensity of Short Light Pulses", Prib. Tekh. Eksp., 1974, 17, no.2, pp.189-90, and Instrum. Exp. Tech., 1974, 17, no.2, pp.523-4.

42.263. Yasuda, A., and Takeda, S.: "Simultaneous Measurement of Transient Plasma Density and its Radial Profile by a Laser Interferometer", Oyo Buturi, 1974, 43, no.9, pp.919-24.

42.264. Bennett, W.R., Carlin, D.B., and Collins, G.J.: "Picosecond Time-Interval Measurement and Intensity Correlations Using the Two-Quantum Photoelectric Effect", J. Quantum Electron. IEEE, 1974, QE-10, no.1, pp.97-9.

42.265. Andreeva, L.I., et al.: "Measurement of Time Characteristics of Photomultipliers by GaAs Injection Laser", Prib. Tekh. Eksp., 1974, 17, no.3, pp.159-61, and Instrum. Exp. Tech., 1974, 17, no.3, pp.809-12.

42.266. Gehre, O.: "Heterodyne Detection for Measurements in Collective HCN-Laser Scattering from Thermonuclear Plasmas", Trans. IEEE, 1974, MTT-22, no.12, pp.1061-4.

42.267. Brossier, P., and Blanken, R.A.: "Interferometry at 337 micron on a Tokamak Plasma", Trans. IEEE, 1974, MTT-22, no.12, pp.1053-6.

42.268. Datskevich, N.P., et al.: "Space-Time Characteristics of Pulses Emitted from a Double-Discharge CO_2 Laser", Kvantovaya Elektron., 1974, 1, no.6, pp.1416-9, and Sov. J. Quantum Electron., 1974, 4, no.6, pp.783-5.

43. NOISE AND FLUCTUATIONS

43.1. Stelzried, C.T.: "Temperature Calibration of Microwave Thermal Noise Sources", Trans. IEEE, 1965, MTT-13, p.128.

43.2. Daglish, H.N., and Carter, J.W.: "Cooled Terminations for Use as 4-GHz, Microwave-Standard, Noise Sources", Proc. IEE, 1965, 112, p.705.

43.3. St. Michel, H., and Prinzler, H.: "Noise Generators in the Microwave Range. I", Nachr. Tech., 1965, 15, no.1, p.33.

43.4. Penzias, A.A.: "Helium-Cooled Reference Noise Source in 4-GHz Waveguide", Rev. Sci. Instrum., 1965, 36, p.68.

43.5. Kloza, M.J.: "Heterodyne Method of Microwave Noise Measurements", Trans. IEEE, 1965, MTT-13, p.882.

43.6. Gambling, W.A., Nudd, G.R., and Ryley, J.E.: "Measurement of Noise Parameters of an Electron Beam", Proc. IEE, 1965, 112, p.1695.

43.7. Kuypers, W., and Vlaardingerbroek, M.T.: "Measurement of Electron-Beam Noise", Philips Res. Rep., 1965, 20, p.349.

43.8. Grangeon, M.J.: "Measurement of Maser Noise Temperature and Stability", Onde Electr., 1965, 45, p.53.

43.9. Fox, A.J., Mansell, J.R., and Phillips, J.L.: "Measurement of Noise Parameters in a Low-Noise BW Amplifier", Proc. IEEE, 1965, 53, p.2113.

43.10. Majewski, S.: "Measurement of Noise in TWT's", Pr. PIE, 1966, 7, no.1,p.19.

43.11. Baertsch, R.D.: "LF Measurements in Si Avalanche Photodiodes", Trans. IEEE, 1966, ED-13, no.3, pp.383-4.

43.12. Low, F.J.: "Thermal Detection Radiometry at Short Millimetre Wavelengths", Proc. IEEE, 1966, 54, no.4, pp.477-84.

43.13. Harris, I.A.: "Dependence of Receiver Noise-Temperature Measurement on Source Impedance", Electron. Lett., 1966, 2, no.4, pp.130-1.

43.14. Pascalar, K.G., and Jordan, P.R.: "Tunnel-Diode K-Band Radiometer", Proc. IEEE, 1966, 54, no.3, pp.442-3.

43.15. Howell, T.F., and Field, C.: "Coaxial Terminations as Low-Temperature Noise Reference Sources", Electron. Lett., 1966, 2, no.6, pp.198-9.

43.16. Miller, C.K.S., Daywitt, W.C., and Arthur, M.G.: "Noise Standards, Measurements, and Receiver Noise Definitions", Proc. IEEE, 1967, 55, no.6, pp.865-77.

43.17. Singer, A., Ulrich, R., and Naess, E.: "Thermal Calibrators in Millimetre-Wave Radiometry", Proc. IEEE, 1967, 55, no.6, pp.1094-6.

43.18. Faris, J.J.: "Sensitivity of a Correlation Radiometer", J. Res. Natl. Bur. Stand., 1967, 71C, no.2, pp.153-70.

43.19. Wait, D.F.: "Sensitivity of the Dicke Radiometer", J. Res. Natl. Bur. Stand., 1967, 71C, no.2, pp.127-52.

43.20. Taylor, H.P.: "Radiometer Equation", Microwave J., 1967,10,no.6,pp.39-42.

43.21. Musztacs, I.: "Measurement of Noise of Microwave Oscillator Tubes", Hirad-astechnika, 1967, 18, no.8, pp.258-62.

43.22. Hollway, D.L., and Somlo, P.I.: "Stable Broadband Variable Noise Source for Microwave Radiometry", Electron. Lett., 1968, 4, no.2, pp.24-5.

43.23. Schanda, E.: "Measuring Emissivity With a Microwave Radiometer", Arch. Elektr. Ubertrag., 1968, 22, no.3, pp.133-40.

43.24. Gambling, W.A., and Kitching, D.M.: "Construction of a Sensitive Microwave Noise Spectrometer", Proc. IEE, 1968, 115, no.5, pp.615-21.

43.25. Golubentsev, A.F., and Minkin, L.M.: "Measurement of the Noise S and Pi Invariants of an Electron Beam Using an O-Type Beam Amplifier", Radiotekh. Elektron., 1968, 13, no.6, pp.1139-40, and Radio Eng. Electron. Phys., 1968, 13, no.6.

43.26. Scherer, E.: "Investigations of the Noise Spectra of Avalanche Oscillators", Trans. IEEE, 1968, MTT-16, p.779.

43.27. Kenney, J.M.: "Simultaneous Measurement of Gain and Noise Using Only Noise Generators", Trans. IEEE, 1968, MTT-16,p.601.

43.28. Wait, D.F., and Nemoto, T.: "Measurement of the Noise Temperature of a Mismatched Noise Source", Trans. IEEE, 1968, MTT-16, p.668.

43.29. Otoshi, T.Y.: "Effect of Mismatched Components on Microwave Noise-Temperature Calibrations", Trans. IEEE, 1968, MTT-16, p.673.

43.30. Mukaihata, T.: "Analysis of Noise Generation in N-Cascaded Mismatched Two-Port Networks", Trans. IEEE, 1968, MTT-16, p.697.

43.31. Ashley, J.R., Searles, C.B., and Palka, F.M.: "Measurement of Oscillator Noise at Microwave Frequencies", Trans. IEEE, 1968, MTT-16, p.751.

43.32. Stelzried, C.T.: "Microwave Thermal Noise Standards", Trans. IEEE, 1968, MTT-16, p.644.

43.33. Sann, K.H.: "Measurement of Near-Carrier Noise in Microwave Amplifiers", Trans. IEEE, 1968, MTT-16, p.759.

43.34. Ondria, J.G.: "Microwave System for Measurements of AM and FM Noise Spectra", Trans. IEEE, 1968, MTT-16, p.765.

43.35. Bates, C., and Ettenberg, M.: "Measurement of the Probability Density Function of a Microwave Noise Generator", Trans. IEEE, 1968, MTT-16, p.791.

43.36. Gottfried, A.H., and Tangredi, J.J.: "Utility and Measurement of Voltage Probability Density of Wideband Microwave Noise", Trans. IEEE, 1968, MTT-16, p.793.

43.37. Olson, K.W.: "Band-Transition-Noise Measurements of Gas-Discharge Sources", Trans. IEEE, 1968, MTT-16, p.795.

43.38. Seidel, B.L., and Stelzried, C.T.: "Radiometric Method for Measuring Insertion Loss of Radome Materials", Trans. IEEE, 1968, MTT-16, p.623.

43.39. Trembath, C.L., et al.: "Low-Temperature Microwave Noise Standard", Trans. IEEE, 1968, MTT-16, p.707.

43.40. Artem'ev, V.V.: "Measurement of Time Correlation of Photons", Radiotekh. Elektron., 1968, 13, no.10, p.1848, and Radio Eng. Electron. Phys., 1968, 13, no.10, pp.1616-20.

43.41. Copinpath, A., Ono, S., and Hartnagel, H.L.: "Measurement of Noise Along Electrostatically Focused Electron Beams", J. Phys. Chem. Solids, 1968, 29, no.12,p.936.

43.42. Lindstrom, J.I., and Nilsson, O.: "Cross Modulation in TWT's", Int. J. Electron., 1969, 26, p.205.

43.43. Rainal, A.J.: "Phase Principle for Measuring Antenna Temperature", Proc. IEEE, 1969, 57, p.1677.

43.44. Chang, R.F., et al.: "Correlations in Light from a Laser at Threshold", Phys. Rev., 1969, 178, p.612.

43.45. Ninomiya, K., and Okoshi, T.: "Measurement of Noise Parameters S and Pi of an Electron Beam Using a Sealed-Off Tube", Electron. Commun. Jap., 1969, 52, no.2, p.68.

43.46. Zaitsev, Yu.I.: "Intensity Fluctuations of He-Ne Laser Radiation at 633 nm", Izv. VUZ Radiofiz., 1969, 12, no.1, p.60.

43.47. Andronova, I.A.: "Intensity Fluctuations of a Single-Frequency Gas Laser at 3.39 micron", Zh. Eksp. Teor. Fiz., 1969, 56, no.2, p.417, and Sov. Phys.-JETP, 1969, 29, no.2.

43.48. Kerecman, A.J., et al.: "Photon-Induced Current Changes in a CO_2 Laser Amplifier", J. Quantum Electron. IEEE, 1969, QE-5, p.474.

43.49.   Stelzried, C.T., and Otoshi, T.Y.: "Radiometric Evaluation of Antenna-Fed Component Losses", Trans. IEEE, 1969, IM-18, p.172.

43.50.   Nemoto, T.: "Measurement of Noise Temperature of a Mismatched Noise Source", Bull. Electrotech. Lab., 1969, 33,no.4,p.44.

43.51.   Arrathoon, R., and Siegman, A.E.: "Measurements of Quantum Phase Noise in He-Ne Laser", J. Appl. Phys., 1969,40,p.910.

43.52.   Waksberg, A., and Wood, J.: "Noise Spectrum for He-Ne Laser Under Various Discharge Conditions", Rev. Sci. Instrum., 1969, 40, p.1306.

43.53.   Yokoshima, I.: "Microwave Low-Noise Measurements at Liquid-Nitrogen Temperature", Bull. Electrotech. Lab., 1969, 33, no.4, p.23.

43.54.   Davidson, F.: "Measurement of Photon Correlations in a Laser Beam Near Threshold With Time-to-Amplitude Converter Techniques", Phys. Rev., 1969, 185, p.446.

43.55.   Pertsev, A.N., et al.: "Method of Measuring Counting Characteristics of Single-Electron Photomultipliers", Zh. Prikl. Spekt., 1970, 12, no.5, p.952.

43.56.   Kuznetsov, Ye.P., and Ogurok, N.D.D.: "Experimental Investigation of Modulation Noise in a He-Ne TW Amplifier at 3.39 micron", Radiotekh. Elektron., 1970, 15, p.629, and Radio Eng. Electron. Phys., 1970, 15, no.3, pp.543-5.

43.57.   Boileau, E.: "Experiment for Determining the Second-Order Coherence Function of a Single-Mode Laser", Opt. Commun., 1970, 2, no.2, pp.49-50.

43.58.   Meltzer, D., and Mandel, L.: "Time-Dependent Photoelectric Counting Statistics for a Q-Switched Laser Near Threshold", Phys. Rev. Lett., 1970, 25, no.17, pp.1151-4.

43.59.   Straub, W.F.: "Measuring Excess Noise of Microwave Devices", Instrum. Control Syst., 1970, 43, no.3, p.107.

43.60.   Sweet, A.S., and Mackenzie, L.A.: "FM Noise of a CW Gunn Oscillator", Proc. IEEE, 1970, 58, p.822.

43.61.   Telegin, G.G., Ugozhaev, V.D., and Folin, K.G.: "Transients and Statistical Effects in a He-Ne Laser Near Threshold", Opt. Spektr., 1970, 28, no.2, p.353, and Opt. Spectrosc., 1970, 28, no.2, p.189.

43.62.   Kotyczka, W., and Strutt, M.J.O.: "Noise Measurements of Si Planar Microwave Transistors in the Range 4-8 GHz", Electron. Lett., 1970, 6, p.478.

43.63.   Shabel'nikov, A.V.: "Study of Phase Fluctuations of Optical Waves Using Diffraction Gratings", Radiotekh. Elektron., 1970, 15, no.5, p.1077, and Radio Eng. Electron. Phys., 1970, 15, no.5, pp.904-6.

43.64.   Epifanov, V.P., and Petrashko, G.A.: "Use of Gas Lasers for Measuring Frequency Characteristics of Photodetectors", Radiotekh. Elektron., 1970, 15, no.6, p.1317, and Radio Eng. Electron. Phys., 1970, 15, no.6, pp.1127-8.

43.65.   Okoshi, T., et al.: "Long-Line Method for Measuring Microwave Oscillator Noise", Electron. Commun. Jap., 1970, 53, no.8, pp.80-5.

43.66.   Meltzer, D., Davis, W., and Mandel, L.: "Measurements of Photoelectric Counting Distributions for Laser Near Threshold", Appl. Phys. Lett., 1970, 17, no.6, pp.242-5.

43.67.   Jakeman, E., et al.: "Intensity Fluctuation Distribution of Laser Light", J. Phys. A, 1970, 3, no.6, pp.152-5.

43.68.   Picard, R.H., and Schweitzer, P.: "Theory of Intensity-Correlation Measurements on Imperfectly Mode-Locked Lasers", Phys. Rev., 1970, 1A, p.1803.

43.69.   Lax, M., and Zwanziger, M.: "Exact Photocount Distributions for Lasers Near Threshold", Phys. Rev. Lett., 1970, 24,p.937.

43.70.   Mathieu, E., and Keller, H.J.: "Intensity Correlation Functions by a Non-Q-Switched Laser Measured by SHG", J. Appl. Phys., 1970, 41, p.1560.

43.71.   Gudnov, V.M., et al.: "Anomolous LF Noise of a Maser", Radiotekh. Elektron., 1970, 15, p.632, and Radio Eng. Electron. Phys., 1970, 15, no.3, pp.545-7.

43.72.   Zaitsev, Yu.I.: "Natural Two-Mode Laser Intensity and Frequency Fluctuations", Izv. VUZ Radiofiz., 1970, 13, no.6,pp.898-903.

43.73.   Hamilton, C.H.: "Signal Generator for Testing Microwave Links", Marconi Instrum., 1970, 12, no.7, pp.130-2.

43.74.   Jordan, P.R.: "Microwave Noise Standard Having Output Temperature of 40-370°K", Rev. Sci. Instrum., 1970, 41, no.11, pp.1649-51.

43.75.   Haken, H.: "Quantum Fluctuations in Nonlinear Optics", Opto-Electron., 1970, 2, no.3, pp.161-7.

43.76.   Clunie, D.M., and Tearle, C.A.: "FM Noise Measurements on Si IMPATT Oscillators", Electron. Lett., 1971, 7, no.2, pp.39-40.

43.77.   Harris, I.A.: "Zero-Point Fluctuations and Thermal-Noise Standards", Electron. Lett., 1971, 7, no.7, pp.148-9.

43.78.   Buckmaster, H.A., and Rathie, R.S.: "Noise Spectrum Measurements from 10 Hz to 1 MHz Using a Tunable Switching Radiometer. I", Can. J. Phys., 1971, 49, no.7, pp.849-52.

43.79.   Steward, K.W.F., and Cooper, D.: "Intra-Spectral Noise Measurement in Microwave Pulsed Amplifiers", Marconi Instrum., 1971, 13, no.1, pp.11-5.

43.80. Keen, N.J.: "Avalanche Diodes as Transfer Noise Standards for Microwave Radiometers", Radio Electron. Eng., 1971, 41, no.3, pp.133-6.

43.81. Strutt, M.J.O.: "(Measurement of) Current and Photon Noise of Semiconductor Injection Lasers", Sci. Electr., 1971, 17, no.1, pp.15-36.

43.82. Kim, D.M., Shah, P.L., and Rabson, T.A.: "Correlation Measurements of Fluctuating Mode-Locked Laser Pulses", Phys. Lett., 1971, 35A, no.4, pp.260-2.

43.83. Boileau, E., et al.: "Study of Frequency Fluctuations in Monomode Lasers", Rev. Phys. Appl., 1971, 6, no.1, pp.23-30.

43.84. Deryugin, I.A., and Mirzaev, A.T.: "Initial Distribution (Measurement) of Photons in Laser Emission", Ukr. Fiz. Zh., 1971, 16, no.5, pp.858-60.

43.85. Matsui, T., Hara, K., and Kobayashi, T.: "Measurement of FM Noise on Phase-Locked Microwave Oscillator Signal", Bull. Electrotech. Lab., 1971, 35, no.6, pp.18-28.

43.86. Kunzi, K., and Magun, A.: "Statistical Gain Fluctuations of Microwave Amplifiers Measured With a Dicke Radiometer", Z. Angew. Math. Phys., 1971, 22, no.3, pp.404-12.

43.87. Falk, J., and Yarborough, J.M.: "Detection of Room-Temperature Blackbody Radiation by Parametric Upconversion", Appl. Phys. Lett., 1971, 19, no.3, pp.68-70.

43.88. Trembath, C.L., Foote, W.J., and Wait, D.F.: "Liquid-Nitrogen-Cooled Microwave Noise Standard", Rev. Sci. Instrum., 1971, 42, no.8, pp.1261-2.

43.89. Ovsepyan, Zh.M., and Moskalyuk, V.A.: "Measurement of Noise in Fast Transverse-Electron-Beam Waves", Izv. VUZ Radioelektron., 1971, 14, no.9, pp.1095-8.

43.90. Gordon, A.C., and Gibbins, C.J.: "Simple Overall Absolute Calibrator for Millimetre-Wavelength Horn-Radiometer Systems", Electron. Lett., 1972, 8, no.3, pp.59-60.

43.91. Yanhoutte, J.C.: "Study and Achievement of a Blackbody Standard", Electron. Microelectron. Ind., 1972, no.158, pp.35-8.

43.92. Bartos, J.: "Methods of Measuring Noise in Microwave Generators", Slab. Obz., 1972, 33, no.3, pp.129-36.

43.93. Gerhardt, H., Welling, H., and Guttner, A.: "Measurement of Laser Linewidth Due to Quantum Phase and Amplitude Noise", Z. Phys., 1972, 253, no.2, pp.113-26.

43.94. Fikart, J.L., and Goud, P.A.: "Direct-Detection Noise-Measuring System", Trans. IEEE, 1972, IM-21, no.3, pp.219-24.

43.95. Fikart, J.L., Nigrin, J., and Goud, P.A.: "Accuracy of AM and FM Noise Measurements With Carrier Suppression and Phase Detector", Trans. IEEE, 1972, MTT-20, no.10, pp.702-3.

43.96. Varecha, K.: "Accuracy of Measuring the Noise Factor of Microwave Two-Ports", Slab. Obz., 1972, 33, no.9, pp.427-31.

43.97. Chen, S.H., et al.: "Method for Clipped Intensity Correlation Measurement", J. Phys. A, 1972, 5, no.11, pp.1619-23.

43.98. Kato, Y., Ojima, T., and Miki, C.: "Measurements of Equivalent Temperature of Liquid-Nitrogen-Cooled Noise Source", Bull. Electrotech. Lab., 1972, 36, no.7, pp.559-65.

43.99. Kato, Y., and Yokoshima, I.: "Method of Measuring Equivalent Temperature of a Coaxial Noise Source With a Waveguide Standard", Bull. Electrotech. Lab., 1972, 36, no.11, pp.745-52.

43.100. Keen, N.J., Haddad, G.G., and Hills, D.L.: "Waveguide Reference Termination for Absolute Noise Calibration at 2.7 GHz", J. Phys. E, 1973, 6, no.10, pp.979-80.

43.101. Aslanyan, A.M., and Gulyan, A.G.: "Measurement of Antenna Noise Temperature", Izv. Akad. Nauk Arm. SSR Fiz., 1973, 8, no.2, pp.148-55.

43.102. Hardy, W.N.: "Precision Temperature Reference for Microwave Radiometry", Trans. IEEE, 1973, MTT-21, no.3, pp.149-50.

43.103. Dushin, L.A., Skibenko, A.I., and Fomin, I.P.: "Measurement on a Fluctuating Plasma by Modulation of a Microwave Signal", Zh. Tekh. Fiz., 1973, 43, no.2, pp.317-22, and Sov. Phys.-Tech. Phys., 1973, 18, no.2, pp.207-10.

43.104. Noll, K.L.: "Measurement of Small Microwave Noise by Frequency Switching", Arch. Elektron. Ubertrag., 1973, 27, no.10, pp.443-6.

43.105. Vertii, A.A., and Shestopalov, V.P.: "Visualization of Amplitude-Phase Structure in the Submillimetre Region", Prib. Tekh. Eksp., 1973, 16, no.2, pp.145-6, and Instrum. Exp. Tech., 1973, 16, no.2, pp.501-2.

43.106. Andrianov, V.I., et al.: "Radiometric Method Applied to Antenna Measurements", Antenny, 1973, no.18, pp.3-17.

43.107. Yeshtokin, V.N., and Tager, A.S.: "Problem of Measuring AM Noise of Avalanche Drift Diode Oscillators", Radiotekh. Elektron., 1973, 18, no.6, pp.1257-60, and Radio Eng. Electron. Phys., 1973, 18, no.6, pp.920-2.

43.108. Kollberg, E.L.: "Measurements of Noise in Low-Noise Receiving Systems", Trans. IEEE, 1974, IM-23, no.3, pp.226-32.

43.109. Cohn-Sfetcu, S., and Buckmaster, H.A.: "Method to Measure the Factor-of-Merit of Microwave Detector Diodes", Trans. IEEE, 1974, IM-23, no.1, pp.102-3.

43.110. Cheung, W.N.: "Measurement of FM Noise from Gunn Oscillators Using a Microwave Interferometer", Int. J. Electron., 1974, 37, no.6, pp.809-15.

43.111. Carli, B.: "Design of Blackbody Reference Standard for Submillimetre Region", Trans. IEEE, 1974, MTT-22, no.12, pp.1094-9.

43.112. Vvedenskii, B.S., et al.: "Interferometer for Investigating Pulsations of GaAs Injection Lasers", Prib. Tekh. Eksp., 1974, 17, no.2, pp.186-8, and Instrum. Exp. Tech., 1974, 17, no.2, pp.520-2.

44. RADIATION PARAMETERS

44.1. Martsafei, V.V.: "Application of Collimating Devices for Scatter Measurements at Microwave Frequencies", Radiotekh. Elektron., 1965, 10, p.561, and Radio Eng. Electron. Phys., 1965, 10, p.478.

44.2. Smith, A.G., and Bryant, D.J.: "Accuracy Requirements in RCS Measurements from a Systems Viewpoint", Proc. IEEE, 1965, 53, p.1159.

44.3. Freeny, C.C.: "Target Support Parameters Associated With Radar Reflectivity Measurements", Proc. IEEE, 1965, 53, p.929.

44.4. Tseitlin, V.B., and Kinber, B.Ye.: "Measurement of Directive Gain of Horn Antennas at a Short Distance", Radiotekh. Elektron., 1965, 10, p.14, and Radio Eng. Electron. Phys., 1965, 10, p.10.

44.5. Hogg, D.C., and Wilson, R.W.: "Precise Measurement of Gain of a Large Horn-Reflector Antenna", Bell Syst. Tech. J., 1965, 44, p.1019.

44.6. Carswell, A.I.: "Measurements of the Longitudinal Component of the Field at the Focus of a Coherent Beam", Phys. Rev. Lett., 1965, 15, p.647.

44.7. Morimoto, M.: "Method of Calibration for Large Aerial Arrays", Electron. Lett., 1965, 1, p.192.

44.8. Trainer, R.F., and Young, W.M.: "Research on an Experimental Corrected Spherical-Reflector Antennas Using Cosmic Radio Sources", Microwave J., 1965, 8, no.1, p.69.

44.9. Jacobs, E., and King, H.E.: "Millimetre-Wave, 0.8-mrad-Beamwidth, Antenna. Measurement", Int. Conv. Rec. IEEE, 1965, 13, pt 5, p.92.

44.10. Tseitlin, N.M.: "Methods of Radio Astronomy in Investigation of Antennas", Radiotekh. Elektron., 1965, 10, p.1363, and Radio Eng. Electron. Phys., 1965, 10, p.1175.

44.11. Cann, A.J.: "CW-Equivalent RCS Measurements With a Pulse", Proc. IEEE, 1965, 53, p.1644.

44.12. Emerson, W.H., and Sefton, H.B.: "Improved Design for Indoor (Measurement) Ranges", Proc. IEEE, 1965, 53, p.1079.

44.13. Kinber, B.Ye., and Tseitlin, V.B.: "Measurement of Antenna Parameters in the Field of a Plane Collimated Wave", Radiotekh. Elektron., 1965, 10, p.1190, and Radio Eng. Electron. Phys., 1965, 10, p.1021.

44.14. Tseitlin, V.B.: "Measurement of Sidelobe Levels and Antenna Phase Patterns in the Near Zone", Radiotekh. Elektron., 1965, 10, p.1127, and Radio Eng. Electron. Phys., 1965, 10, p.963.

44.15. Padgitt, R.D.: "Angular Calibration Using Celestial Radio Sources", Bell Lab. Rec., 1965, 43, p.2.

44.16. Carswell, A.J.: "Microwave Scattering Measurements in the Rayleigh Region Using a Focused-Beam System", Can. J. Phys., 1965, 43, p.1962.

44.17. Schetne, K.H., and Mount, W.V.: "Full-Scale Bistatic RCS Measurement Method", Proc. IEEE, 1965, 53, p.1083.

44.18. Bachman, C.G.: "Developments in RCS Measurement Techniques", Proc. IEEE, 1965, 53, p.962.

44.19. Olin, I.D., and Queen, F.D.: "Dynamic Measurement of Radar Cross Section", Proc. IEEE, 1965, 53, p.954.

44.20. Huynen, J.: "Measurement of Target Scattering Matrix", Proc. IEEE, 1965, 53, p.936.

44.21. Youyoumjian, R.G., and Peters, L.: "Range Requirements in RCS Measurements", Proc. IEEE, 1965, 53, p.920.

44.22. Blacksmith, P., Hiatt, R.E., and Mack, R.B.: "Introduction to RCS Measurements", Proc. IEEE, 1965, 53, p.901.

44.23. Iizuka, K.: "Technique for Measuring Mutual Admittance of Antennas", Trans. IEEE, 1965, AP-13, p.469.

44.24. Blakey, J.R.: "Feasibility of Aerial-Pattern Determination by Processing the Near-Field Signal", Electron. Lett., 1965, 1, p.298.

44.25. Debski, T.R., and Hannan, P.W.: "Complex Mutual Coupling Measured in a Large Phased-Array Aerial", Microwave J., 1965, 8, no.6, p.93.

44.26. Poirier, J.L., Rotman, W., and Shore, R.A.: "Microwave Measurements of Partially Coherent Fields", Appl. Opt., 1965, 4, p.1321.

44.27. King, R.J.: "Amplitude and Phase Measuring System Using a Small Modulated Scatterer", Microwave J., 1965, 8, no.3, p.95.

44.28. Levinson, D.S., and Rubinstein, I.: "Technique for Measuring Individual Modes Propagating in Overmoded Waveguide", Trans. IEEE, 1966, MTT-14, p.310.

44.29. Arteza, A., and Dovan, J.A.: "Bistatic-Radar Method for Determination of Permeability and Permittivity for a Smooth Spherical Target", Radio Sci., 1966,1,p.995.

44.30. Smith, P.G.: "Measurement of the Complete Far-Field Pattern of Large Antennas by Radio-Star Sources", Trans. IEEE, 1966, AP-14, p.6.

44.31. Fry, C.R.: "Impedance Measurements on Zigzag Aerial Structures", Electron. Lett., 1966, 2, p.101.

44.32. Fitzgerrell, R.G.: "Swept-Frequency Antenna-Gain Measurements", Trans. IEEE, 1966, AP-14, p.173.

44.33. Stankevich, K.S., and Tseitlin, N.M.: "Effect of Feed Displacement from Focus on Accuracy of Antenna Measurements", Radiotekh. Elektron., 1966, 11, p.451, and Radio Eng. Electron. Phys., 1966,11,p.380.

44.34. Blakey, J.R.: "Errors in Near-Field Measurements and Effect on Computed Radiation Patterns", Electron. Lett., 1966, 2, p.299.

44.35. Dickstein, H.D.: "Near-Field Phase Measurements Using Standard Microwave Equipment", Microwave J., 1966, 9, no.2, p.55.

44.36. Bale, F.V., Gourlay, J.A., and Meadows, R.W.: "Measuring Shape of Large Reflectors by a Simple Radio Method", Electron. Lett., 1966, 2, p.252.

44.37. Puttock, M.J., and Minnett, H.C.: "Instrument for Rapid Measurement of Surface Deformations of a 64-m Radiotelescope", Proc. IEE, 1966, 113, p.1723.

44.38. Golota, S.I.: "Microwave Instrument for Control of Water Content and Density of Materials", Prib. Sist. Upr., 1967, no.1, pp.15-18.

44.39. Lacombat, M.: "Application of Lasers to Metrology", Tech. Mod., 1967, 59, no.4, pp.35-8.

44.40. Schede, R.W.: "(Laser) Interferometers as Integral Parts of Machine Tools", Trans. IEEE, 1967, IGA-3, no.4, p.328.

44.41. Redman, J.D.: "Holographic Velocity Measurement", J. Sci. Instrum., 1967, 44, no.12, pp.1033-4.

44.42. Campbell, J.W.: "Extending the Laser Feedback Interferometer", Instrum. Control Syst., 1967, 40, no.11, pp.75-80.

44.43. Robertson, C.W.: "Industrial Measurement With Lasers", Ind. Electron., 1968, 6, no.1, pp.11-15.

44.44. Broussaud, G.: "Application of Coherent Optics to Shape Definition", Rev. Gen. Electr., 1967, 76, no.10, pp.1261-71.

44.45. Kovalev, V.P.: "Measurement of Scattering Fields at Microwaves", Radiotekh. Elektron., 1968, 13, p.1679, and Radio Eng. Electron. Phys., 1968, 13, no.9, pp.1463-6.

44.46. Kurochkin, A.P.: "Optical Modelling of Microwave Antennas", Radiotekh. Elektron., 1968, 13, pp.1169 and 1347, and Radio Eng. Electron. Phys., 1968, 13, no.7, pp.1020-4 and no.8, pp.1171-7.

44.47. Burrows, M.L.: "Surface Tolerance of a Radar Calibration Sphere", Trans. IEEE, 1968, AP-6, p.718.

44.48. D'yachenko, A.A., and Shushpanov, O.Ye.: "Use of Holograms in Analysis of Wavebeams in Quasi-Optical Transmission Lines", Radiotekh. Elektron., 1968, 13, no.11, p.2067, and Radio Eng. Electron. Phys., 1968, 13, no.11, pp.1811-3.

44.49. Gel'freikh, G.B., and Korzhavin, A.N.: "Optical Modelling of the Patterns of Microwave Antennas With Varying Profile Reflectors", Radiotekh. Elektron., 1968, 13, p.1176, and Radio Eng. Electron. Phys., 1968, 13, no.7, pp.1025-35.

44.50. Tsujiuchi, J., Takeya, N., and Matsuda, K.: "Measurement of Deformation by Use of Holography", Oyo Buturi, 1968, 37, no.9, p.877.

44.51. Nassenstein, H.: "Holography and Interference Studies With Inhomogeneous Surface Waves", Phys. Lett., 1968, 28A, p.249.

44.52. Cornbleet, S.: "Determination of the Aperture Field of an Antenna by a Beam-Displacement Measurement", Proc. IEE, 1968, 115, p.1398.

44.53. Flower, R.A.: "Laser Instruments for Measurements", Mech. Eng., 1968, 90, no.10, p.27.

44.54. Yevseev, V.I., and Deryagin, V.N.: "Automatic Recording of Spatial Distribution of Radiation from p-n Junction Diodes", Opt.-Mekh. Prom., 1968, 35, no.2, p.33, and Sov. J. Opt. Technol., 1968, 35, no.2, p.178.

44.55. Buckley, E.F., and Niles, G.E.: "Evaluation of Anechoic Chambers for Microwaves" Electrotechnik, 1969, 51, no.20,p.12.

44.56. Stankevich, K.S.: "Measurement of Small Antennas Using Two Blackbodies of Different Temperatures", Radiotekh. Elektron., 1969, 14, no.3, p.528, and Radio Eng. Electron. Phys., 1969, 14, no.3.

44.57. Huber, F.R., and Schiller, M.: "Hyperbolic Technique for Contactless Measurement of the Surfaces of Large Parabolic Antennas", Nachr. Tech. Z., 1969, 22, no.6, p.333.

44.58. Barkalov, S.S., and Deryagin, V.N.: "Study of Modulation Phase Distribution Over the Emitting Surface of the p-n Junction of a Semiconductor Laser", Opt.-Mekh. Prom., 1969, 36, no.7-8, p.40, and Sov. J. Opt. Technol., 1969, 36, no.4, p.487.

44.59. Buchtemann, W., and Hohn, D.H.: "Measurements of Coherence of Pulsed GaAs Laser Diodes", Optik, 1969, 29,no.4,p.401.

44.60. Smith, R.L.: "Demonstration of Equality of Phases of Laser-Cavity Modes by an Oscilloscope Trace", Phys. Lett., 1969, 30A, p.132.

44.61. Antsiferov, V.V., et al.: "Experimental Investigation of Spectral, Angular, and Time Characteristics of TW Ruby Laser", Zh. Tekh. Fiz., 1969, 39, no.5, p.931, and Sov. Phys.-Tech. Phys., 1969, 14, no.5, pp.696-8.

44.62. Chatterjee, S.K., et al.: "Measurement of Back-Scattering Cross Sections of Metallic Bodies of Revolution at X-Band", J. Inst. Eng., 1969, 49, no.9, p.87.

44.63. Rowley, W.R.C., and Wilson, D.C.: "Design Tolerances in Laser Measurement Systems", Proc. Inst. Mech. Eng., 1969, 183, pt 3D, p.29.

44.64. Korpel, A., and Whitman, R.L.: "Visualization of a Coherent-Light Field by Heterodyning with a Scanning Laser Beam", Appl. Opt., 1969, 8, p.1577.

44.65. Tipton, H.: "Laser Measuring Techniques in Machine-Tool Applications", Proc. Inst. Mech. Eng., 1969, 183, pt 3D, p.1.

44.66. Bakhtadze, Sh.N., et al.: "Optoelectronic Methods and Means for Automation of Metrological Work", Izmer. Tekh., 1969, no.5, p.3, and Meas. Tech., 1969,no.5,p.597.

44.67. Chuang, K.C., and Mueller, R.K.: "Detection of Strain by Coherent Optical Techniques", Mater. Eval., 1969, 27, no.4, p.76.

44.68. Zverev, V.A., et al.: "Measurement of Antenna Radiation Patterns by Optical Modelling in Noncoherent Light", Izv. VUZ Radiofiz., 1969, 12, p.1829.

44.69. Buckley, E.F., and Niles, G.E.: "Anechoic Antenna Measuring Chamber for Microwaves", Elektrotechnik, 1969, 51,no.17,p.14.

44.70. Froehly, C., et al.: "Study of Small Displacements of Opaque Objects and of Optical Distortion in Solid Lasers by Holographic Interferometry", Opt. Acta, 1969, 16, no.3, p.343.

44.71. Tirro, S., and Grego, F.: "Measurement of Gain Using Radio Stars", Elettronica Telecom., 1969, 18, no.5, p.158.

44.72. Heflinger, L.O., and Wuerker, R.F.: "Holographic Contouring via Multifrequency Lasers", Appl. Phys. Lett., 1969, 15, p.28.

44.73. Zarghamee, M.S.: "Prediction of Antenna Tolerance from Efficiency Measurements", Trans. IEEE, 1969, AP-17, p.354.

44.74. Sheridan, K.V.: "Use of Atomic Frequency Standards for Phase Calibration of Large Antenna Arrays", Electron. Lett., 1969, 5, p.363.

44.75. Knott, E.F.: "Laboratory Method of Measuring Phase of a Backscatter Signal", Electron. Lett., 1969, 5, p.667.

44.76. Rode, D.L.: "Method for Determining Frequency Response of a Laser", Rev. Sci. Instrum., 1969, 40, p.506.

44.77. Holtz, E.: "Indication for Location of Axis of a Laser Beam", Messtechnik, 1969, 77, p.263.

44.78. Willet, C.S.: "Simple Technique for Detection of Oscillation at 3.39 micron in a He-Ne Laser Operating at 633 nm", J. Quantum Electron. IEEE, 1969, QE-5, p.524.

44.79. Arnold, D.H., and Hanna, D.C.: "Measurement of Diffraction Loss in a Solid-State Laser", Appl. Opt., 1969, 8, p.2146.

44.80. Eichler, H., and Wiesemann, W.: "Measurement of Length Shifts Down to 0.1 pm With a Two-Mode Laser", Z. Angew. Phys., 1969, 28, no.3, p.129.

44.81. Voitekhovich, A.V., et al.: "Features of Transverse Intensity Redistribution of He-Ne Laser Emission", Opt. Spekt., 1969, 26, no.4, p.662, and Opt. Spectrosc., 1969, 26, no.4, p.363.

44.82. Blaszczak, Z.: "Portable He-Ne Laser for Alignment of Optical Systems", Postepy Fiz., 1969, 20, no.4, p.489.

44.83. Lohmann, H.D., and Tholl, H.: "Holographic Investigations of Vibrating Piezoelectric Quartz Plates", Z. Naturforsch., 1969, 24a, no.11, p.1806.

44.84. D'mitrenko, D.A., Romanychev, A.A., and Tseitlin, N.M.: "Measurement of Antenna Parameters Using a Blackbody Disc in the Fresnel Region", Radiotekh. Elektron., 1969, 14, no.12, p.2108, and Radio Eng. Electron. Phys., 1969, 14, no.12, pp.1825-31.

44.85. Gustyr, L.Ya., and Dontsova, V.V.: "Measurement of Divergence of Radiation of Gas Lasers Operating in TEM_{nm} Modes", Opt. Spekt., 1969, 25, no.6, p.958, and Opt. Spectrosc., 1969, 25, no.6, p.530.

44.86. Matsuo, M.M., Yuba, Y., and Yamane, K.: "Bistatic RCS Measurements by Pendulum Method", Trans. IEEE, 1970, AP-18, p.83.

44.87. Ennos, A.E.: "Holographic Techniques in Engineering Metrology", Proc. Inst. Mech. Eng., 1969, 183, pt 3D, p.5.

44.88. Welford, W.T.: "Fringe Visibility and Localization in Hologram Interferometry", Opt. Commun., 1969, 1, no.3, p.123.

44.89. Fontain, J., and Roux, G.: "Beam Alignment for a Linear Accelerator", Onde Electr., 1969, 49, p.1216.

44.90. Pavlenko, M.F., et al.: "Measurement of Radiation Patterns of Parabolic Antennas in Laboratory Conditions", Izv. VUZ Radioelektron., 1969, 12, p.1394.

44.91. Rowley, W.R.C.: "Signal Strength in Two-Beam Interferometers With Laser Illumination", Opt. Acta, 1969, 16, p.159.

44.92. Iijima, K., Tsuzuki, Y., and Hirose, Y.: "Holographic Method of Vibration Analysis", Bull. Fac. Eng. Yokohama Natl. Univ., 1969, 18, March, p.91.

44.93. Mirovitskii, D.I., et al.: "Optical Modelling of Reflection and Scattering of Microwaves", Radiotekh. Elektron., 1970, 15, p.38, and Radio Eng. Electron. Phys., 1970, 15, no.1, pp.30-8.

44.94. Kerns, D.M.: "Correction of Near-Field Antenna Measurements Made With an Arbitrary but Known Measuring Antenna", Electron. Lett., 1970, 6, p.346.

44.95. Kerns, D.M.: "Method of Gain Measurement Using Two Identical Antennas", Electron. Lett., 1970, 6, p.348.

44.96. Baird, R.C., et al.: "Experimental Results in Near-Field Antenna Measurements", Electron. Lett., 1970, 6, p.349.

44.97. Weingartner, I.: "Measurement of Mutual Coherence Functions by Image Holography", J. Opt. Soc. Am., 1970, 60, p.572.

44.98. Hanfling, J.D.: "Aperture Fields of Paraboloidal Reflectors by Stereographic Mapping of Feed Polarization", Trans. IEEE, 1970, AP-18, p.392.

44.99. Rzepecka, M.: "(Sheet-Thickness) Measurement by Microwave Quasi-Closed Resonators", Arch. Elektrotech., 1970, 19, no.4, pp.647-55.

44.100. Cullen, A.L., and Kumar, A.: "Absolute Determination of Extinction Cross Sections by an Open Resonator", Proc. Roy. Soc., 1970, 315A, p.217.

44.101. Rice, P.L., Noble, J.L., and Thompson, W.I.: "Idealized Pencil-Beam-Antenna Patterns for Use in Interference Studies", Trans. IEEE, 1970, COM-18, p.27.

44.102. Hockley, B.S., and Butters, J.N.: "Holography as a Routine Method of Vibration Analysis", J. Mech. Eng. Sci., 1970, 12, no.7, p.37.

44.103. Fryer, P.A.: "Scanning Technique for Allowing Whole Vibration Cycles to be Stored on One Hologram", Appl. Opt., 1970, 9, p.1216.

44.104. Kalestynski, A., and Zardecki, A.: "Diffraction Measurements of Laser-Beam Intensity Distribution", Acta. Phys. Polon., 1970, 37A, p.437.

44.105. Menzel, R., and Shofner, F.M.: "Investigation of Fraunhofer Holography for Velocimetry Applications", Appl. Opt., 1970, 9, p.2073.

44.106. Anderson, A.P., and Swingler, D.N.: "Location of an Irregularity in a Microwave Array by Optical Signal Processing of its Microwave Hologram", Electron. Lett., 1970, 6, p.577.

44.107. Kakichashvili, Sh.D.: "Focused Integrated-Image Holography of Long Objects", Opt.-Mekh. Prom., 1970, 37, no.10, pp.15-8, and Sov. J. Opt. Technol., 1970, 37, no.10, pp.641-3.

44.108. Ginzburg, V.M., et al.: "Holographic Interferometry in the Microwave Region", Radiotekh. Elektron., 1970, 15, no.12, pp.2342-3, and Radio Eng. Electron. Phys., 1970, 15, no.12.

44.109. Engelhard, E., and Spieweck, F.: "Ion Laser for Metrological Applications", Z. Naturforsch., 1970, 25a, no.1, p.156.

44.110. Bennett, S.J., and Gates, J.W.C.: "Design of Detector Arrays for Laser Alignment Systems", J. Phys. E, 1970, 3, p.65.

44.111. Belousova, I.M., et al.: "Investigation of Spectrum of a Laser Employed as a Detector of a Doppler-Shifted Signal", Zh. Eksp. Teor. Fiz., 1970, 58, no.2, p.394, and Sov. Phys.-JETP, 1970, 31, no.2, pp.209-15.

44.112. Tsuruta, T., and Itoh, Y.: "Holographic Interferometry for Rotating Subject", Appl. Phys. Lett., 1970, 17, p.85.

44.113. Magyar, G.: "Measurements of the Spatial Coherence of a Giant Pulse Laser", Opto-Electron., 1970, 2, no.2, pp.68-72.

44.114. Hirth, A.: "Measurement and Interpretation of the Coherence Length of a Single-Mode Pulsed Ruby Laser", C. R. Acad. Sci. (Paris), 1970, 271B, no.16, pp.853-6.

44.115. Teichman, M.: "Precision Phase-Centre Measurements of Horn Antennas", Trans. IEEE, 1970, AP-18, no.5, pp.689-90.

44.117. Rude, A.F.: "Two-Frequency Laser Interferometer for Control and Calibration of Machine Tools", Electron. Ind., 1970, no.138, pp.685-9.

44.118. Ando, S., Taniguchi, J., and Okada, K.: "Laser Profiling Machine for Contour Measurement", Laser, 1970, 2, no.2, pp.9-12.

44.119. Tursunov, A.T.: "Spatial Structure of Radiation from a Moving-Medium Laser", Dokl. Akad. Nauk SSSR, 1970, 192, no.1-3, pp.538-40, and Sov. Phys.-Dokl., 1970, 15, no.5, pp.487-9.

44.120. Pyshkin, O.S., et al.: "Experimental Studies of Radiation Characteristics of a Ruby Laser with a Lens Resonator", Ukr. Fiz. Zh., 1970, 15, no.10, pp.1667-73.

44.121. Hager, H.: "Standard-Gain Horn Antennas", News Rohde Schwarz, 1970, 10, no.42, pp.32-3.

44.122. Carey, G., and Hickman, P.A.: "Laser Alignment Techniques", ISA Trans., 1970, 9, no.3, pp.222-8.

44.123. Ohly, J.K.: "Frequency-Stabilized Laser for Precision (Length) Interferometry", Mess. Pruef., 1970, 6, no.9, pp.723-7.

44.124. Young, M.: "Spatial Coherence (Measurement) of Gas-Laser Light", Opt. Commun., 1970, 2, no.6, pp.253-4.

44.125. Fercher, A.F., and Torge, R.: "Holographic Test Glasses", Opt. Laser Technol., 1970, 2, no.4, pp.200-1.

44.126. Kulczyk, W.K., and Davis, Q.V.: "Laser Measurements of Vibrations of Rotating Objects", Opto-Electron., 1970, 2, no.3, pp.177-9.

44.127. Lowe, B.A., and Pearson, H.E.: "Measurements on Goonhilly 25-m Antenna. II", Post Office Electr. Eng. J., 1970, 63, pt 1, pp.20-3.

44.128. Petrushkin, A.A., et al.: "Device for Studying Fields in Open Resonators for Millimetre Wavelengths", Prib. Tekh. Eksp., 1970, no.2, pp.147-9, and Instrum. Exp. Tech., 1970, no.2, pp.481-4.

44.129. Iuzuka, K., and Gregoris, L.G.: "Application of Microwave Holography in the Study of the Field from a Radiating Source", Appl. Phys. Lett., 1970, 17, no.12, pp.509-12.

44.130. Testa, P., and Viti, M.: "Optical Method of Registration of Rapidly Variable Liquid Levels", Acqua, 1970, 48, no.4, pp.91-102.

44.131. Ivanov, E.I., and Chaika, M.P.: "Observation of Interference Beats in Spontaneous Emission from He-Ne Laser", Opt. Spekt., 1970, 29, no.2, pp.124-7, and Opt. Spectrosc., 1970, 29, no.2, pp.66-8.

44.132. Helmberger, T.: "Electrooptical Light-Speed Measurements by He-Ne Lasers", Laser, 1970, 2, no.3, pp.29-32.

44.133. Korshunov, I.P.: "Apparatus for Analysing the Field at the Output of a Quasi-Optical Lens Line", Radiotekh. Elektron., 1970, 15, no.7, and Radio Eng. Electron. Phys., 1970, 15, no.7, pp.1255-60.

44.134. Lamb, C.B., and Tuffy, E.P.: "Measurement of Bracings of Ship's Deck by Laser-Alignment System", Laser, 1970, 2, no.4, pp.15-6.

44.135. Nazarova, L.G.: "Measurement of Laser Coherence by Young's Method", Opt. Spekt., 1970, 29, no.4, pp.757-60, and Opt. Spectrosc., 1970, 29, no.4, pp.403-5.

44.136. Hirose, Y., Tsuzuki, Y., and Iijima, K.: "Measurement of Contour Vibrations of Quartz Plates by Holographic Technique", Electron. Commun. Jap., 1970, 53, no.6, pp.49-54.

44.137. Vlasov, B.I., et al.: "Investigation of Diffraction Fields of Plane Irregularities in a Waveguide Using Metal/Semiconductor Film Structures", Izv. VUZ Radiofiz., 1970, 13, no.10, pp.1532-40.

44.138. White, J.R.: "Measuring the Microwave Field in a Cavity", J. Microwave Power, 1970, 5, no.2, pp.145-7.

44.139. Asakura, T.: "Spatial Coherence of Laser Light Passed Through Rotating Ground Glass", Opto-Electron., 1970, 2, no.3, pp.115-23.

44.140. Miyamoto, T.: "Measurement of Beam Parameters and Diffraction Field of a Laser With a Hologram", Appl. Opt., 1971, 10, no.1, pp.161-7.

44.141. Newell, A.C., and Kerns, D.M.: "Determination of Both Polarization and Power Gain of Antennas by a Generalized Three-Antenna Method", Electron. Lett., 1971, 7, no.3, pp.68-70.

44.142. Birand, T., and Marini, A.: "Correction of Errors in Antenna Far-Field Radiation-Pattern Measurements", Electron. Lett., 1971, 7, no.3, pp.86-7.

44.143. Midgley, J.A., and Sander, R.F.: "High-Resolution Laser Homodyne Interferometer", Electron. Lett., 1971, 7, no.5-6, pp.117-8.

44.144. Vikram, C.S., and Sirohi, R.S.: "Holographic Images of Objects Moving With Constant Acceleration", Appl. Opt., 1971, 10, no.3, pp.672-3.

44.145. Butters, J.N., and Leendertz, J.A.: "Speckle Pattern and Holographic Techniques in Engineering Metrology", Opt. Laser Technol., 1971, 3, no.1, pp.26-30.

44.146. King, R.J., and Hustig, C.H.: "Microwave Surface Impedance Measurements of a Dielectric Wedge on a Perfect Conductor", Can. J. Phys., 1971, 49, no.7, pp.820-30.

44.147. Vallese, L.M.: "Measurement of the Beam Parameters of a Laser", Appl. Opt., 1971, 10, no.4, pp.959-60.

44.148. Kurbatov, L.N., and Shakhidzhanov, S.S.: "Method for Determination of Optical Constants of a p-n Junction Laser", Fiz. Tekh. Poluprov. 1971, 5, no.2, pp.251-5, and Sov. Phys.-Semicond., 1971, 5, no.2,

44.149. Floyd, R.P., and Collins, D.J.: "Holographic Determination of Simple Translation and Rotation", Am. J. Phys., 1971, 39, no.4, pp.359-62.

44.150. Abramson, N.: "Moire Patterns and Hologram Interferometry", Natl. Phys. Sci., 1971, 231, no.20, pp.65-7.

44.151. Nelle, G.: "Length Measurement by Laser", Werkstattechnik, 1971, 61, no.3, pp.137-44.

44.152. Deitz, P.H., and Evans, J.M.: "Holographic Method of Measuring Scintillation Effects", Appl. Opt., 1971, 10, no.5, pp.1080-2.

44.153. Burchett, O.J., and Irwin, J.L.: "Using Laser Holography for Nondestructive Testing", Mech. Eng., 1971, 93, no.3, pp.27-33.

44.154. Birand, T., and Marincic, A.: "Determination of Antenna Far-Field Radiation Patterns from Near-Field Measurements", Electron. Lett., 1971, 7, no.11, pp.296-8.

44.155. Binks, S.D.: "Laser-Operated Scanning Rod Gauge", Meas. Control, 1971, 4, no.4, pp.149-53.

44.156. Weber, H.: "Laser as Measuring Instrument. I and II", Mess. Pruef., 1971, 7, no.1, pp.17-21, and no.2, pp.57-61.

44.157. Knopf, A.: "Radar Target Simulation by Control of the Phasefront in the Receiving Field", Nachr. Tech. Z., 1971, 24, no.5, pp.244-5.

44.158. Yansen, D.E., Reynolds, G.O., and Cronin, D.J.: "Optical Synthetic-Aperture Analogues of Two Radio Interferometers", Opt. Acta, 1971, 18, no.3, pp.167-80.

44.159. Liska, D.J.: "Electric-Field Measurement in Klystron Cavities", Trans. IEEE, 1971, ED-18, no.7, pp.450-3.

44.160. Tschudi, T.: "Holography in Measurement Technology", Mess. Pruf., 1971, 7, no.4, pp.147-53.

44.161. Stuchly, S.S., Hamid, M.A.K., and Andres, A.: "Microwave Surface-Level Monitor", Trans. IEEE, 1971, IECI-18, no.3, pp.85-92.

44.162. Matsumoto, K.: "Laser Applications in the Manufacture of Lenses", Laser Angew. Strahlentech., 1971, 3, no.1, pp.24-6.

44.163. Kersch, L.A.: "Advanced Concepts of Holographic Nondestructive Testing", Mater. Eval., 1971, 29, no.6, pp.125-9.

44.164. Mustafin, K.S., and Seleznev, V.G.: "Three-Beam Holographic Interferometry", Opt. Spekt., 1971, 30, no.1, pp.154-8, and Opt. Spectrosc., 1971, 30, no.1, pp.80-2.

44.165. Vikram, C.S., and Sirohi, R.S.: "Time-Average Holography of Objects Vibrating Sinusoidally With Uniform Slow Drift", Phys. Lett., 1971, 35A, no.6, pp.460-1.

44.166. Bates, R.H.T., and Napier, P.J.: "Determining Antenna-Pattern Phase from Holographic Measurement", Proc. IREE Aust., 1971, 32, no.4, pp.164-6.

44.167. Czyz, Z.: "Five-Antenna Method for Determination of Polarization and Gain", Pr. PIT, 1971, 21, no.71, pp.1-14.

44.168. Dey, K.K., and Singh, V.: "Microwave Antenna Measurements at Reduced Ranges by On-Axis Defocus of the Feed", Indian J. Pure Appl. Phys., 1971, 9, no.3, pp.179-82.

44.169. Landry, M., and Chasse, Y.: "Measurement of Field Intensity in Focal Region of a Wide-Angle Paraboloid", Trans. IEEE, 1971, AP-19, no.4, pp.539-43.

44.170. Wolf, D., and Rassow, B.: "Method to Determine the Focusing of Optical Systems With Inaccessible Image Plane", Optik, 1971, 33, no.6, pp.597-9.

44.171. Mizuno, H., and Tanaka, S.: "Application of Nematic Liquid Crystals to Control the Coherence of Laser Beams", Opt. Commun., 1971, 3, no.5, pp.320-1.

44.172. Kuwada, H., Fujita, T., and Sugihara, K.: "Investigation of Deformation of the Surface by Holographic Interferometry", Kogaku Giyutsu, 1971, 4, no.3, pp.3-9.

44.173. Schmidt, W., and Fercher, A.F.: "Holographic Generation of Depth Contours Using a Flash-Pumped Dye Laser", Opt. Commun., 1971, 3, no.5, pp.363-5.

44.174. Klejman, H.: "Laser Interferometry (in Metrology)", Pomiary Autom. Kontr., 1971, 17, no.7, pp.292-4.

44.175. Dudderar, T.D., and O'Reagan, R.O.: "Laser Holography and Interferometry in Materials Research", Mater. Res. Stand., 1971, 11, no.9, pp.8-15.

44.176. Felber, C.K., and Massialas, F.G.: "Design Features of Holographic Apparatus", Mater. Res. Stand., 1971, 11, no.9, pp.19-21.

44.177. Sampson, R.C.: "Structural Measurements With Holographic Interferometry", Mater. Res. Stand., 1971, 11, no.9, pp.26-31.

44.178. Varner, J.R.: "Holographic Contouring Techniques Applicable to Mechanical Testing", Mater. Res. Stand., 1971, 11, no.9, pp.31-5.

44.179. Roszhart, T.V., Pearson, D.J., and Bohm, J.R.: "Pulsed-Ruby Holographic Instrumentation for Materials Testing", Mater. Res. Stand., 1971, 11, no.9, pp.36-43.

44.180. Graf, W., et al.: "Sun as Test Source for Boresight Calibration of Microwave Antennas", Trans. IEEE, 1971, AP-19, no.5, pp.606-12.

44.181. Dobyrn, V.V., et al.: "Combination of an Interference Field and Sinusoidal Holographic Diffraction Gratings", Opt. Spekt., 1971, 30, no.3, pp.550-5, and Opt. Spectrosc., 1971, 30, no.3, pp.297-300.

44.182. Bates, R.H.T., and Napier, P.J.: "Holographic Approach to Radiation-Pattern Measurement. I and II", Int. J. Eng. Sci., 1971, 9, no.11, pp.1047-60 and no.12, p.1193.

44.183. Vasilu, V., et al.: "He-Ne Laser Alignment System", Stud. Cercet. Fiz., 1971, 23, no.10, pp.1241-6.

44.184. Butters, J.N., and Leendertz, J.A.: "Holographic Techniques Applied to Engineering Measurement", Meas. Control, 1971, 4, no.12, pp.349-54.

44.185. Wort, D.J.H.: "Microwave Interferometry as Nondestructive Method of Measuring Dynamic Clearances", Non-Destruct. Test., 1971, 4, no.6, pp.380-1.

44.186. West, P.: "Fourier-Transform Optics for Micro-Gauging", Microscope, 1971, 19, no.4, p.422.

44.187. Sen, D.: "Use of He-Ne Lasers for Alignment", Proc. Indian Natl. Sci. Acad., 1971, 37A, no.3, pp.257-65.

44.188. Barrier, B.: "Vernier Laser Interferometers", Inter Electron., 1971, 26, no.26, pp.33-4.

44.189. Bodlaj, V.: "He-Ne Laser as Secondary Length Standard", Laser Angew. Strahlentech., 1971, 3, no.2, pp.21-8.

44.190. Shibayama, K., and Uchiyama, H.: "Measurement of Three-Dimensional Displacements by Hologram Interferometry", Appl. Opt., 1971, 10, no.9, pp.2150-4.

44.191. Bartlett, J., and Adams, R.J.: "Holographic Technique for Sizing Particles in Moving Aerosols", Microscope, 1971, 19, no.4, p.424.

44.192. Worthington, R.: "Measurement of Mode Cutoff Wavelengths of Optical Fibres", J. Phys. E, 1971, 4, no.12, pp.1052-4.

44.193. Rabouille, B.: "Measuring Small Radar Cross Sections", Rev. Tech. Thomson-CSF, 1971, 3, no.4, pp.737-50.

44.194. Bogomolov, A.S., Vlasov, N.G., and Solov'ev, E.G.: "Method for Time Averaging in Holographic Interferometry of Non-Periodically Moving Bodies", Opt. Spekt., 1971, 31, no.3, pp.481-2, and Opt. Spectrosc., 1971, 31, no.3, pp.256-7.

44.195. Laszlo, T.I., and Cioara, F.: "Automatic Apparatus for Tracing Antenna Radiation Patterns", Bul. Inst. Politeh. Iasi, 1971, 17, pt III, no.1-2, pp.31-4.

44.196. Bakhrakh, L.D., et al.: "Holographic Determination of Antenna Pattern With Source in Fresnel Region", Dokl. Akad. Nauk SSSR, 1971, 201, no.3, pp.580-2, and Sov. Phys.-Dokl., 1971, 24, no.3.

44.197. Billman, K.W., Leonard, E.T., and Yafee, M.A.: "True Vertical Laser", Appl. Opt., 1971, 10, no.2, pp.422-5.

44.198. Burkhardt, D.: "Use of Synchronized Signal Generators in Contactless (Microwave) Measurement of Parabolic Reflectors", Microwave J., 1971, 14, no.1, p.24.

44.199. Pospisil, J.: "Holographic Methods for Measurement of Optical Transfer Function of Objectives", Cesk. Cas. Fis., 1971, 21, no.6, pp.603-13.

44.200. Goldberg, J.L., and O'Toole, K.M.: "Holography and Potential Application to Production Engineering", J. Inst. Eng. Aust., 1971, 43, no.10-11, pp.8-12.

44.201. Baldwin, R.R., Gordon, G.B., and Rude, A.F.: "Remote Laser Interferometry", Hewlett-Packard J., 1971, 23, no.4, pp.14-20.

44.202. Luizov, A.V., and Fedorova, N.S.: "Investigation of Coherence of Nd^{3+}:Glass Lasers", Opt.-Mekh. Prom., 1971, 38, no.8, pp.15-9, and Sov. J. Opt. Technol., 1971, 38, no.8, pp.466-9.

44.203. Seleznev, V.G., Sobolev, N.D., and Yakovlev, V.V.: "Instrument for Measuring Displacements by Holographic Interferometry", Zavod. Lab., 1971, 37, no.8, pp.979-80, and Ind. Lab., 1971, 37, no.8, pp.1257-8.

44.204. Goggin, W.R., and Paquin, R.A.: "He-Ne Laser for Thermal Expansion Measurements", Image Technol., 1971, 13, no.6, pp.19-22.

44.205. Lamare, M., and Simon, J.: "Laser Interferometer for Optical Systems Control", Nouv. Rev. Opt. Appl., 1971, 2, no.6, pp.317-20.

44.206. Mirovitskii, D.I., et al.: "Analysis of Cartographic Radiation Patterns in Optical Modelling of Antennas", Radiotekh. Elektron., 1971, 16, no.10, pp.1946-9, and Radio Eng. Electron. Phys., 1971, 16, no.10, pp.1737-41.

44.207. Linnik, V.P., et al.: "Interferometer for Studying Wavefront of a Laser Beam", Opt.-Mekh. Prom., 1971, 38, no.11, pp.27-9, and Sov. J. Opt. Technol., 1971, 38, no.11, pp.671-3.

44.208. Toyonaga, T., Miyake, M., and Hori, M.: "Antennas for Measurement of Microwave Fields by Modulated-Light Scattering Technique", Electron. Commun. Jap., 1971, 54, no.11, pp.46-53.

44.209. Dijk, J., et al.: "Digitized Antenna Measurements", Trans. IEEE, 1972, MTT-20, no.1, pp.48-51.

44.210. Dale, C.H., and Howland, A.R.: "Automated Test Equipment for Phased-Array Modules", Trans. IEEE, 1972, MTT-20, no.1, pp.10-7.

44.211. Stetson, K.A.: "Fringes of Hologram Interferometry for Simple Nonlinear Oscillations", J. Opt. Soc. Am., 1972, 62, no.2, pp.297-8.

44.212. Tiziani, H.J.: "Optical Methods for Vibration Analysis of a Tuning Fork", Optik, 1972, 34, no.4, pp.442-55.

44.213. Skinner, D.R., and Whitcher, R.E.: "Measurement of the Radius of a High-Power Laser Beam Near Focus of a Lens", J. Phys. E, 1972, 5, no.3, pp.273-8.

44.214. Whitton, D.T.N.: "Laser Technique for Precise (Distance) Measurements", Electron. Power, 1972, 18, Feb., pp.46-8.

44.215. Saifi, M.A., and Stolen, R.H.: "Far-IR Interference Technique for Determining Epitaxial Si Thickness", J. Appl. Phys., 1972, 43, no.3, pp.1171-7.

44.216. Ribbens, W.B.: "Surface-Roughness Measurement by Holographic Interferometry", Appl. Opt., 1972, 11, no.4, pp.807-10.

44.217. Kopf, U.: "Coherent-Optical Method for Contactless Measurements of Local Displacements and Vibrations", Optik, 1972, 35, no.2, pp.144-51.

44.218. Meneely, C.T., She, C.Y., and Edwards, D.F.: "Measurement of Flow and Turbulence of Free Jet by Laser Photon-Correlation Spectroscopy", Opt. Commun., 1972, 6, no.4, pp.380-2.

44.219. Vorob'eva, N.N., et al.: "Optical Doppler Meter for Measuring Velocity of Working Surfaces", Zh. Prikl. Spekt., 1972, 17, no.6, pp.988-91.

44.220. Tanaka, K.: "Experimental Results of Diffraction Field of a Wave Beam from a Circular Aperture", Opt. Commun., 1972, 6, no.3, pp.267-9.

44.221. Tanimura, Y., and Yaminoto, A.: "Measuring Machine Leadscrews With Laser Interferometer", Bull. Jap. Soc. Precis. Eng., 1972, 6, no.2, pp.45-50.

44.222. Franz, C.M.: "Laser Optical-Lever System for Measuring Pitch and Yaw of a Ground-Launched Rocket", Image Technol., 1972, 14, no.6, pp.30-8.

44.223. Harden, D.T., and Schwemmer, K.H.: "Laser Eases Turbine-Generator Alignment", Electr. Wld, 1972, 178, no.12, pp.30-1.

44.224. Arutyunyan, A.A., et al.: "Measurement of Antenna Far-Field Pattern by Machine Reconstruction of the Microwave Hologram in the Aperture", Izv. Akad. Nauk Arm. SSR Fiz., 1972, 7, no.5, pp.373-6.

44.225. Joy, E.B., and Paris, D.T.: "Spatial Sampling and Filtering in Near-Field Measurements", Trans. IEEE, 1972, AP-20, no.3, pp.253-61.

44.226. Moyer, R.G., and Gilespie, G.E.: "Displacement Measurement by Holographic Interferometry", AECL Res. Dev. Eng., 1972, no.2, pp.12-4.

44.227. Coultas, F.W.: "Non-Contacting Measuring Using Mini-Radars", Chart. Mech. Eng., 1972, 19, no.3, pp.62-6.

44.228. Smeets, G.: "Laser Interferometer for Measurements on Transient Phase Objects", Trans. IEEE, 1972, AES-8, no.2, pp.186-90.

44.229. Bhatnagar, R., and Bhawalkar, D.D.: "Measurement of Focal Length of Pump-Induced Lenses in Laser Media", J. Quantum Electron. IEEE, 1972, QE-8, no.6, pp.497-9.

44.230. Karger, A.M., and Holeman, J.M.: "Microscopic Holography of Small Parts", Appl. Opt., 1972, 11, no.7, pp.1646-7.

44.231. Kopf, U.: "Application of Speckling for Measuring the Deflection of Laser Light by Phase Objects", Opt. Commun., 1970, 5, no.5, pp.347-50.

44.232. Valot-Degueurce, M.O., and Bainier, C.: "Evaluation of Laser Coherence Factor by Superposition of Two Systems of Stationary Waves", Opt. Commun., 1972, 5, no.5, pp.367-9.

44.233. D'Orazio, R.J., and George, N.: "Detection of Mode-Locked Laser Signals", Opt. Commun., 1972, 5, no.5, pp.407-9.

44.234. Cheremiskin, I.V., and Chekhlova, T.K.: "Phase-Distribution Measurement of Field on Laser Output Mirror", Opt. Spekt., 1972, 32, no.1, pp.160-2, and Opt. Spectrosc., 1972, 32, no.1, pp.80-1.

44.235. Hansche, B.D., and Murphy, C.G.: "Strain Measurements by Holometry", ISA Trans., 1972, 11, no.1, pp.1-14.

44.236. Tiziani, H.J.: "Use of Laser Speckle to Measure Small Tilts of Rough Surfaces", Opt. Commun., 1972, 5, no.4, pp.271-6.

44.237. Popela, B.: "Influence of the Atmosphere on Measurements Made With He-Ne Laser Interferometer", Opt. Acta, 1972, 19, no.7, pp.605-12.

44.238. Cowles, P.R., and Parker, E.A.: "Antenna Radiation-Pattern Measurements Over Long Test Ranges", J. Phys. E, 1972, 5, no.9, pp.857-9.

44.239. Young, D.D., et al.: "Holographic Interferometry Measurement of Thermal Coefficients of Nd^{3+}:YAG and Nd^{3+}:YALO", J. Quantum Electron. IEEE, 1972, QE-8, no.8, pp.720-1.

44.240. Majkowski, R.F., and Gara, A.D.: "Accuracy of Holographic Images", Appl. Opt., 1972, 11, no.8, pp.1867-9.

44.241. Keller, R., Salathe, R., and Tschudi, T.: "Interferometric Measurement of Elongation of a Pulsed Diode Laser", J. Quantum Electron. IEEE, 1972, QE-8, no.10, pp.783-7.

44.242. Schwider, J.: "(Testing) Absolute Flatness by Combination of a Standard With a Compensating Hologram", Opt. Commun., 1972, 6, no.1, pp.58-62.

44.243. Larionov, N.P., Lukin, A.V., and Mustofin, K.S.: "Artificial Hologram as an Optical Compensator", Opt. Spekt., 1972, 32, no.2, pp.396-9, and Opt. Spectrosc., 1972, 32, no.2, pp.206-7.

44.244. Sakurai, Y.: "Measurement of Distance Using a Laser", J. Soc. Instrum. Control Eng., 1972, 11, no.7, pp.617-23.

44.245. Birch, K.G., and Green, F.J.: "Application of Computer-Generated Holograms to Testing Optical Elements", J. Phys. D, 1972, 5, no.11, pp.1982-92.

44.246. Pirlet, R., and Noel, Y.: "Measurement of Thickness of Bloom Slabs by Laser Gauge", Energie, 1972, 200, no.3, pp.145-57.

44.247. Kindl, H.: "Construction Machines Guided by Laser", Umsch. Wiss. Tech., 1972, 72, no.20, pp.664-5.

44.248. Erdmann, J.C., and Jahoda, J.A.: "Remotely Sensing Strain-Rate Meter Based on Doppler Shift of Laser Light", Rev. Sci. Instrum., 1972, 43, no.11, pp.1643-7.

44.249. Afanas'eva, A.L., et al.: "Polychromatic Holographic Interferometry With Recording in a Three-Dimensional Medium", Opt. Spekt., 1972, 32, no.3, pp.589-91, and Opt. Spectrosc., 1972, 32, no.3, pp.312-3.

44.250. Rosenberger, D.: "Alignment With Laser Beams", Alta Freq., 1972, 41, no.10, pp.780-6.

44.251. Arnautov, G.P., et al.: "Accurate Laser Gravimeter", Avtometriya, 1972, no.5, pp.29-38.

44.252. Wyant, J.C., and Bennett, V.P.: "Using Computer-Generated Holograms to Test Aspheric Wavefronts", Appl. Opt., 1972, 11, no.12, pp.2833-9.

44.253. Farmer, W.M.: "Measurement of Particle Size, Number Density, and Velocity Using a Laser Interferometer", Appl. Opt., 1972, 11, no.11, pp.2603-12.

44.254. Barker, L.M., and Hollenbach, R.E.: "Laser Interferometer for Measuring High Velocities of a Reflecting Surface", J. Appl. Phys., 1972, 43, no.11, pp.4669-75.

44.255. Vlasov, N.G., Skrotskii, G.V., and Solov'ev, E.G.: "Investigation of Coherence With Aid of Diffraction Shearing Interferometer", Kvantovaya Elektron., 1972, no.3, pp.84-6, and Sov. J. Quantum Electron., 1972, 2, no.3, pp.266-8.

44.256. Nemoto, S., and Makimoto, T.: "Measurement of Focal Length of Dielectric Lens Using Field Distribution in the Image Space", Electron. Commun. Jap., 1972, 55, no.4, pp.58-65.

44.257. Gubarev, V.Ya., et al.: "Measurement of Small Displacements", Izv. VUZ Mashinostr., 1972, no.9, pp.190-1.

44.258. von Willesen, F.K., and Glantschnig, F.: "Laser Interferometer for Industrial Measurement of Length", Tec. Regul. Mando Autom., 1972, no.26, pp.65-80.

44.259. Barchukov, A.I., Lyubin, A.A., and Terin, V.S.: "Measurement of Radius of Curvature of Laser Beam by an Interferometric Method", Kvantovaya Elektron., 1972, no.4, pp.99-101, and Sov. J. Quantum Electron., 1973, 2, no.4, pp.374-5.

44.260. Pernick, B.J.: "Self-Consistent and Direct-Reading Homodyne Measurement Technique", Appl. Opt., 1973, 12, no.3, pp.607-10.

44.261. Baars, J.W.M.: "Measurement of Large Antennas With Cosmic Radio Sources", Trans. IEEE, 1973, AP-21, no.4, pp.461-74.

44.262. van Ligten, R.F.: "Speckle Reduction by Simulation of Partially Coherent Object Illumination in Holography", Appl. Opt., 1973, 12, no.2, pp.255-65.

44.263. Evans, J.D.: "Analysis of He-Ne Laser Surface Reflections from Off-Axis Parabolic Mirror", Appl. Opt., 1973, 12, no.2, pp.212-7.

44.264. Vysokosov, E.P., et al.: "Methods of Measuring the Beamspread of Lasers", Izmer. Tekh., 1973, 16, no.5, pp.32-6, and Meas. Tech., 1973, 16, no.5, pp.681-6.

44.265. Tilford, C.R.: "Fringe-Counting Laser Interferometer Manometer", Rev. Sci. Instrum., 1973, 44, no.2, pp.180-2.

44.266. Grianti, F., Ottonello, P., and Troilo, M.: "Laser Torque Meter", Autom. Strum., 1973, 21, no.1, pp.26-31.

44.267. Bartolotta, C.S., and Pernick, B.J.: "Holographic Nondestructive Evaluation of Interference-Fit Fasteners", Appl. Opt., 1973, 12, no.4, pp.885-6.

44.268. Dabergerova, L.: "Contactless Interference Control of Spherical Surfaces", Jemna Mech. Opt., 1973, 18, no.1, pp.10-3.

44.269. Hockley, B.S.: "Holographic Visualization of Large Amplitude Vibration Using Reference-Beam Phase Modulation", J. Phys. E, 1973, 6, no.4, pp.377-80.

44.270. Kawase, S., et al.: "Vibration Measurement by Holograms", Opt. Commun., 1973, 7, no.1, pp.6-9.

44.271. Fercher, A.F.: "Testing Optical Mirrors by Holography", Umsch. Wiss. Tech., 1973, 73, no.9, pp.270-4.

44.272. Michael, H.: "Deformation Measurements of Diffusely Reflecting Objects With the Aid of Holography", Appl. Opt., 1973, 12, no.6, pp.1111-3.

44.273. Velzel, C.H.F.: "Contours of Equal In-Plane Displacement in Holographic Interferometry", Opt. Commun., 1973, 7, no.4, pp.302-4.

44.274. Faulde, M., et al.: "Optical Testing by Synthetic Holograms and Partial Lens Compensation", Opt. Commun., 1973, 7, no.4, pp.363-5.

44.275. Kupper, F.P., and van Dijk, C.A.: "Reference Fringes in Holographic Interferometry", Opt. Laser Technol., 1973, 5, no.2, pp.69-74.

44.276. Ginzburg, V.M., et al.: "Holographic Interferometry for Observation of Solution State During Crystal Growth", Kristallogr., 1973, 17, no.5, pp.1012-4, and Sov. Phys.-Cryst., 1973, 17, no.5, pp.889-91.

44.277. Doi, Y., and Kawabe, S.: "Recognition of 3D Objects by Laser Beam Section", Trans. Soc. Instrum. Control Eng., 1973, 9, no.1, pp.16-21.

44.278. Levine, J.: "Ultra-Sensitive Laser Interferometers and Application to Problems of Geophysical Interest", Philos. Trans., 1973, 274A, no.1239, pp.279-84.

44.279. Manzoni, G., and Marchesini, C.: "60-m Laser Strainmeter", Philos. Trans., 1973, 274A, no.1239, p.285.

44.280. Oh, L.L.: "Accurate Boresight Measurements of Large Antennas and Radomes", Trans. IEEE, 1973, AP-21, no.4, pp.567-9.

44.281. Zambuto, M.H., and Fischer, W.K.: "Shifted Reference Holographic Interferometry", Appl. Opt., 1973, 12, no.7, pp.1651-5.

44.282. Cain, D.A., et al.: "Holography of Large Objects in a Turbulent Atmosphere With a CW Laser", Appl. Phys. Lett., 1973, 23, no.1, pp.37-8.

44.283. Hung, Y.Y., et al.: "Surface Displacement Measurements by Holographic Interferometry", Opt. Commun., 1973, 8, no.1, pp.48-51.

44.284. Hsiao, C.C.: "Laser Diffraction of Brittle Polymer Under Tension", Appl. Phys. Lett., 1973, 23, no.1, pp.20-1.

44.285. Waddell, P., and Fagan, W.: "Speckle-Reference-Beam Holography for Real-Time Visualization of Vibration Patterns", Proc. Br. Acoust. Soc., 1973, 2, no.1, pp.1-4.

44.286. Ludwig, A.C., and Norman, R.A.: "Method for Calculating Correction Factors for Near-Field Gain Measurements", Trans. IEEE, 1973, AP-21, no.5, pp.623-8.

44.287. Gliddon, C.W., and Carson, C.T.: "Antenna Radiation-Pattern Measurement Using Model Aircraft", Trans. IEEE, 1973, AP-21, no.5, pp.700-2.

44.288. Burrell, G.A., and Jamieson, A.R.: "Antenna-Radiation-Pattern Measurement Using Time-to-Frequency Transformation", Trans. IEEE, 1973, AP-21, no.5, pp.702-4.

44.289. Belozerov, A.F., et al.: "Hologram Study of Gas Flow in a Ballistic Wind Tunnel", Zh. Tekh. Fiz., 1973, 43, no.4, pp.777-81, and Sov. Phys.-Tech. Phys., 1973, 18, no.4, pp.488-90.

44.290. Fuhlrott, H., and Korf, H.: "Double-Beam He-Ne Laser for Interferometrically Measuring Displacements in the Nanometre Range", Appl. Opt., 1973, 12, no.8, pp.1741-3.

44.291. Sciammarella, C.A., and Gilbert, J.A.: "Strain Analysis of a Disc Subjected to Diametral Compression by Holographic Interferometry", Appl. Opt., 1973, 12, no.8, pp.1951-6.

44.292. Apostol, D., and Nicolau, S.: "Measurement of Small Angular Movement", Stud. Cercet. Fiz., 1973, 25, no.6, pp.765-6.

44.293. Bauer, W.: "Reflectance Measurements of Flat Surfaces", J. Appl. Phys., 1973, 44, no.8, pp.3694-6.

44.294. Vikram, C.S., and Bhatnagar, G.S.: "Application of Holographic Subtraction to Time-Average Interferometry of Vibrating Objects", Appl. Opt., 1973, 12, no.10, pp.2239-40.

44.295. Vikram, C.S.: "Time-Average Holography of Objects Vibrating Sinusoidally and Moving With Constant Acceleration", Opt. Commun., 1973, 8, no.4, pp.355-7.

44.296. Yokozeki, S., Okuyama, H., and Banda, S.: "Speckle Noise Reduction by Composition of Diffused Fraunhofer Holograms", Opt. Commun., 1973, 8, no.4, pp.358-61.

44.297. Albe, F., Smigielski, P., and Fagot, H.: "Application of Double Exposure Holographic Interferometry to Study of Ceramic Deformation Under Projectile Impact", Opt. Commun., 1973, 8, no.4, pp.369-71.

44.298. Bjelkhagen, H.: "Simplified Interpretation of Interference Fringes Obtained by Time-Average Holography", Opt. Laser Technol., 1973, 5, no.4, pp.172-5.

44.299. Larionov, N.P., Lukin, A.V., and Mustafin, K.S.: "Holographic Inspection of Rough Surfaces in Reflected Light", Opt.-Mekh. Prom., 1973, 40, no.1, pp.66-7, and Sov. J. Opt. Technol., 1973, 40, no.1, pp.61-2.

44.300. Filippov, A.V., and Andreev, V.A.: "Experimental Determination of the Field Parameters in Sectoral-Horn Aperture Using a Passive Probe", Izv. VUZ Radiofiz., 1973, 16, no.8, pp.1265-70.

44.301. Gerhardt, H., Welling, H., and Frolich, D.: "Ideal Laser Amplifier as a Phase Measuring System of a Microscopic Radiation Field", Appl. Phys., 1973, 2, no.2, pp.91-3.

44.302. Staselko, D.I., Voronin, V.B., and Smirnov, A.G.: "Holographic Method for Measuring Spatial Coherence Functions", Opt. Spekt., 1973, 34, no.3, pp.561-6, and Opt. Spectrosc., 1973, 34, no.3, pp.320-3.

44.303. Bellani, V.F., and Sona, A.: "Technique for Measurement of Displacements by Double-Exposure Holographic Interferometry", Proc. Br. Acoust. Soc., 1973, 2, no.2, pp.1-4.

44.304. Elsworth, Y., and James, J.F.: "Optical Screw With a Pitch of One Wavelength", J. Phys. E, 1973, 6,no.11,pp.1134-6.

44.305. Henshaw, P.O., and Ezekiel, S.: "High-Resolution Holographic Contour Generation Using a Pulsed Multicolour Ion Laser", Appl. Opt., 1973, 12, no.11, pp.2550-2.

44.306. Der, V.K., Holloway, O.C., and Fourney, W.L.: "Four-Exposure Holographic Moire Technique", Appl. Opt., 1973, 12, no.11, pp.2552-4.

44.307. Hecht, N.L., et al.: "Quantitative Theory for Predicting Fringe Pattern Formation in Holographic Interferometry", Appl. Opt., 1973, 12, no.11, pp.2665-76.

44.308. Cathey, W.T., Hadwin, J.F., and Pace, J.D.: "Imaging Through Turbulent Water Using Speckle Reference Holography", Appl. Opt., 1973, 12, no.11, pp.2683-5.

44.309. Bogomolov, A.S., Vlasov, N.G., and Solov'ev, E.G.: "Investigation of Thermal Deformations by Holographic Interferometry", Prib. Tekh. Eksp., 1973, 16, no.2, pp.185-7, and Instrum. Exp. Tech., 1973, 16, no.2, pp.551-2.

44.310. Tanimura, Y., and Nara, J.: "Pitch Signals of Screw and Rack Detected by Holographic Interferometry", J. Jap. Soc. Precis. Eng., 1973, 39, no.5,pp.541-6.

44.311. Cannistraci, P., Migliardo, P., and Wanderlingh, F.: "Holographic Control of Surface Modifications", Opt. Laser Technol., 1973, 5, no.5, pp.210-5.

44.312. Vikram, C.S.: "Coherent Moire Technique for Time-Average Holography of Objects Vibrating Sinusoidally With Uniform Slow Drift", Phys. Lett., 1973, 45A, no.5, p.426.

44.313. Al'kaev, M.I., et al.: "Laser Interferometer System for Measuring Displacement Based on a Small Digital Computer", Avtometriya, 1973, no.3, pp.52-9.

44.314. Solodkin, Yu.N.: "Holographic Interferometer as a Measuring Instrument", Avtometriya, 1973, no.5, pp.64-8.

44.315. Brodskii, B.I.: "Preliminary Results of Determinations of Gravitational Acceleration by Free-Falling-Body Method", Vrash. Pril. Deform. Zemli, 1973, no.5, pp.10-1.

44.316. Johnson, R.C., Ecker, H.A., and Hollis, J.S.: "Determination of Antenna Patterns from a Near-Field Measurements", Proc. IEEE, 1973, 61, no.12, pp.1668-94.

44.317. Vasil'ev, A.M., et al.: "Investigation of Vibrational Characteristics of Bodies by Holographic Interferometry", Avtometriya, 1973, no.5, pp.59-62.

44.318. Nagata, K., Umehara, T., and Nishiwaki, J.: "Determination of RMS Roughness and Correlation Length by Spatial Coherence Function", Jap. J. Appl. Phys., 1973, 12, no.11, pp.1693-8.

44.319. Vlasov, N.G., Smirnova, S.N., and Presnyakov, Yu.P.: "Extraction of the Components of the Deformation Vector in Interference Measurements", Zh. Tekh. Fiz., 1973, 43, no.5, and Sov. Phys.-Tech. Phys., 1973, 18, no.5, pp.706-7.

44.320. Chomat, M., and Miler, M.: "Application of Holography to the Analysis of Mechanical Vibrations of Electronic Components", TESLA Electron., 1973, 6, no.3, pp.83-93.

44.321. Brutti, C., and Branca, F.P.: "Measurement of Mechanical Vibration Using Modulation of Laser Light", Ingegnere, 1973, 48, no.6, pp.321-32.

44.322. Zemskov, G.G., and Makukhin, V.P.: "Automatic Inspection of Dimensions of Shaped Workpieces", Mekh. Avtom. Proiz., 1973, no.9, pp.15-9.

44.323. Mintrop, H., and Roth, P.: "Use of Microwaves for Inspection of Internal Geometry of Fine Steel Tubes", Mess. Pruef., 1973, no.10, pp.635-7.

44.324. Zastrogin, Yu.F.: "Optical Non-contacting Doppler Methods of Mechanical-Vibration Measurement", Izmer. Tekh., 1973, 16, no.3, pp.35-7, and Meas. Tech., 1973, 16, no.3, pp.365-8.

44.325. Dorogaya, L.N., et al.: "Simple Method for Measuring the Divergence Angle of Lasers", Izmer. Tech., 1973, 16, no.4, pp.30-1, and Meas. Tech., 1973, 16, no.4, pp.520-2.

44.326. Lukin, A.V., Mustafin, K.S., and Rafikov, R.A.: "Inspecting the Profile of Aspherical Surfaces With Aid of One-Dimensional Synthesized Holograms", Opt.-Mekh. Prom., 1973, 40, no.6, p.67, and Sov. J. Opt. Technol., 1973, 40, no.6, pp.398-9.

44.327. Yamamoto, A., et al.: "Construction of Screw Lead Measuring Machine Using Laser Interferometer", Bull. Jap. Soc. Precis. Eng., 1973, 7, no.3, pp.91-2.

44.328. Caradot, J.C.: "Laser in Dimensional Metrology", Mes. Regul. Autom., 1973, 38, no.11, pp.37-42.

44.329. Hariharan, P.: "Application of Holographic Subtraction to Time-Average Interferometry of Vibrating Objects", Appl. Opt., 1973, 12, no.1, pp.143-6.

44.330. Garault, Y., and Fenelon, J.P.: "Determination of Dispersion Characteristics of Open Periodic Structure Using Variable Length Cavity With a Single Iris", C. R. Acad. Sci. (Paris), 1973, 276B, no.11, pp.409-12.

44.331. Tober, G., Anderson, R.C., and Sherndin, O.H.: "Laser Instrument for Detecting Water-Ripple Slopes", Appl. Opt., 1973, 12, no.4, pp.788-94.

44.332. Matsumoto, T., Iwata, K., and Nagata, R.: "Measuring Accuracy of Three-Dimensional Displacements in Holographic Interferometry", Appl. Opt., 1973, 12, no.5, pp.961-7.

44.333. Dalton, B.L.: "Microwave Noncontact Measurement and Instrumentation in the Steel Industry", J. Microwave Power, 1973, 8, no.3-4, pp.235-44.

44.334. Soga, H.: "Microwave Thickness Gauge", J. Microwave Power, 1973, 8, no.3-4, pp.253-66.

44.335. Kozyrev, Yu.I.: "Use of Laser to Determine Contour of a Crack in Transparent Materials", Zavod. Lab., 1973, 39, no.8, pp.1006-7, and Ind. Lab., 1973, 39, no.8, pp.1322-3.

44.336. Levites, A.F., and Teleshevskii, V.I.: "Heterodyne Laser Interferometer With Acoustooptic Modulator", Prib. Tekh. Eksp., 1973, 16, no.6, pp.139-40, and Instrum. Exp. Tech., 1973, 16, no.6, pp.1776-8.

44.337. Sharma, M.G.: "Background Cancellation in Radar-Cross-Section Measurements by Screw Tuner in Main Line", J. Inst. Eng., 1973, 54, pt ET2, pp.66-8.

44.338. Nishiwaki, A., and Mikami, T.: "Measurement of Rebounding of Concrete Piles= With He-Cd Laser", Mem. Chubu Inst. Technol., 1973, 9A, pp.27-31.

44.339. Beketova, A.K., and Yanichkin, V.I.: "Use of Laser-Beam Diffraction for Placing Elements of a System on a Straight Line", Opt.-Mekh. Prom., 1973, 40, no.8, pp.47-9, and Sov. J. Opt. Technol., 1973, 40, no.8, pp.506-7.

44.340. Meshchankin, V.M.: "Correlation Analysis of SHF Holograms", Radiotekh. Elektron., 1973, 18, no.6, pp.1159-64, and Radio Eng. Electron. Phys., 1973, 18, no.6, pp.841-4.

44.341. Osmolovskaya, E.P., and Lodi, M.N.: "Use of Lasers for Measuring Width of Ribbons", Izmer. Tekh., 1973, 16, no.6, pp.30-1, and Meas. Tech., 1973, 16, no.6, pp.839-40.

44.342. Tanimura, Y., and Nara, J.: "Pitch Measurement of Objects Having Periodic Shapes by Holographic Interference. I", Bull. Jap. Soc. Precis. Eng., 1973, 7, no.4, pp.109-14.

44.343. Zlatin, N.A., et al.: "Laser Differential Interferometer", Zh. Tekh. Fiz., 1973, 43, no.9, pp.1961-4, and Sov. Phys.-Tech. Phys., 1974, 18, no.9, pp.1235-7.

44.344. Wachutka, H., Ewers, W.M., and Barwinkel, K.: "Measurement of Surface Contours Using Holographically Stored Interference Patterns", Optik, 1974, 40, no.1, pp.69-79.

44.345. Nakajima, T.: "Detection of Small Amplitude Vibrations by Modulated Reference Holography", Jap. J. Appl. Phys., 1974, 13, no.3, pp.471-83.

44.346. Longrigg, P.: "Laser Precision Liquid Level Sensor", Trans. IEEE, 1974, IECI-21, no.2, pp.88-96.

44.347. Jones, R., and Bijl, D.: "Holographic Interferometric Study of Effects Associated With a Technique for Measuring Poisson's Ratio", J. Phys. E, 1974, 7, no.5, pp.357-8.

44.348. Burch, J.M.: "Holographic Measurement of Displacement and Strain", J. Strain Anal., 1974, 9, no.1, pp.1-3.

44.349. Jones, R.: "Strain Distribution and Elastic-Constant Measurement Using Holographic and Speckle-Pattern Interferometry", J. Strain Anal., 1974, 9, no.1, pp.4-9.

44.350. Archbold, E., and Ennos, A.E.: "Application of Holography and Speckle Photography to the Measurement of Displacement and Strain", J. Strain Anal., 1974, 9, no.1, pp.10-16.

44.351. Robertson, E.R., and King, W.: "Technique of Holographic Interferometry Applied to the Study of Transient Stresses", J. Strain Anal., 1974, 9, no.1, pp.44-9.

44.352. Farmer, W.M.: "Observation of Large Particles With a Laser Interferometer", Appl. Opt., 1974, 13, no.3, pp.610-22.

44.353. Boersch, H., et al.: "Measurement of Length Shifts down to 10^{-3} pm With a Three-Mode Laser", J. Quantum Electron. IEEE, 1974, QE-10, no.6, pp.501-4.

44.354. Blows, L.G., and Tanner, L.H.: "(Spatial-Frequency) Method for Measurement of Fluid Surface Velocities Using Particles and a Light Source", J. Phys. E, 1974, 7, no.5, pp.402-5.

44.355. Joynes, G.M.S., and Davis, Q.V.: "Determination of Structural Vibrations Using a Modulated Laser", Acustica, 1974, 30, no.5, pp.267-70.

44.356. Nakatani, N., Kawata, K., and Yamada, T.: "Flow Visualization by Improved Double-Exposure Method in Holography", Opt. Laser Technol., 1974, 6, no.2, pp.82-3.

44.357. Vikram, C.S., and Bhatnagar, G.S.: "Application of Holographic Addition to Time-Average Hologram Interferometry of Constant Velocity Motion", Appl. Opt., 1974, 13, no.4, pp.720-1.

44.358. Ohtsuka, Y., and Sasaki, I.: "Laser Heterodyne Measurement of Small Arbitrary Displacements", Opt. Commun., 1974, 10, no.4, pp.362-5.

44.359. Thomason, W.H., and Macomber, J.D.: "(Laser) Apparatus for Simultaneous Measurement of Reflectance, Absorptivity, and Transmittance", Rev. Sci. Instrum., 1974, 45, no.2, pp.264-9.

44.360. New, B.M.: "Versatile Electro-optic Alignment System for Field Applications", Appl. Opt., 1974, 13, no.4, pp.937-41.

44.361. Mathews, N.A., and Stachera, H.: "Vibrating-Dipole Technique for Measuring Millimetre-Wave Fields in Free Space", Trans. IEEE, 1974, MTT-22, no.2, pp.103-10.

44.362. Mathews, N.A., and Stachera, H.: "Automatic System for Simultaneous Measurement of Amplitude and Phase of Millimetre-Wave Fields", Trans. IEEE, 1974, MTT-22, no.2, pp.140-2.

44.363. Rzepecka, M.A., and Stuchly, S.S.: "Microwave System for Measurement of Diameter of Thin Dielectric Fibres", Trans. IEEE, 1974, IM-23, no.1, pp.100-1.

44.364. Wang, T.K., Tsui, P.C., and Chang, I.C.H.: "Optical-Parallelism Measurement of Ruby Laser Rods", Appl. Opt., 1974, 13, no.6, pp.1379-83.

44.365. Svetlik, J.: "Simple Methods for Measurement of Laser Beam Parameters", Appl. Opt., 1974, 13, no.6, pp.1276-8.

44.366. Sato, T., Ogawa, H., and Ueda, M.: "Contour Generation of Vibrating Object by Weighted Subtraction of Holograms", Appl. Opt., 1974, 13, no.6, pp.1280-2.

44.367. Bellani, V.F., and Sona, A.: "Measurement of Three-Dimensional Displacements by Scanning a Double-Exposure Hologram", Appl. Opt., 1974, 13, no.6, pp.1337-41.

44.368. Hariharan, P., and Hegedus, Z.S.: "Measurement of Symmetrical and Antisymmetrical Deformations by Hologram Interferometry", Opt. Commun., 1974, 11, no.2, pp.127-31.

44.369. Smith, P.L.: "Measurement of Antenna Gain in Radar Systems", Microwave J., 1974, 17, no.4, pp.37-40.

44.370. Ruiz, H.J., Williams, C.S., and Padovani, F.A.: "Silicon-Slice Analyser Using a He-Ne Laser", J. Electrochem. Soc., 1974, 121, no.5, pp.689-92.

44.371. Akimovich, I.N.: "Application of Lasers to Crystallographic Orientation of Ruby", Ukr. Fiz. Zh., 1974, 19, no.6, pp.1031-4.

44.372. Burch, J.M., Forno, C., and Tanner, L.H.: "Lens-Hologram Technique for Analysing Symmetrical Aberrations of a Camera System", Opt. Laser Technol., 1974, 6, no.3, pp.109-13.

44.373. Hoff, D., and Monich, G.: "Probe for Direct Measurement of Energy Flow in the Near Field of Transmitting and Receiving Antennas", Nachr. Tech. Z., 1974, 27, no.8, pp.313-8.

44.374. Fouere, J.C.: "Holographic Interferometers for Optical Testing", Opt. Laser Technol., 1974, 6, no.4, pp.181-3.

44.375. Suminov, V.M., Gol'dberg, M.M., and Grebnev, A.A.: "Automatic Television-Laser Measuring Devices", Prib. Sist. Upr., 1974, no.4, pp.40-2.

44.376. Hanlon, J., and Aiken, S.: "Alignment Technique for Unstable Resonators", Appl. Opt., 1974, 13, no.11, p.2461.

44.377. Filippov, A.V., Kulagin, V.A., and Andreev, V.A.: "Measurement of Field Parameters Across the Face of a Pyramidal Horn", Izv. VUZ Radioelektron., 1974, 17, no.8, pp.98-100.

44.378. Clarricoats, P.J.B., Olver, A.D., and Parini, C.: "Radiation Method for Measurement of Mode-Conversion Levels", Electron. Lett., 1974, 10, no.25-26, pp.525-6.

44.379. Hsu, T.R.: "Large-Deformation Measurements by Real-Time Holographic Interferometry", Exp. Mech., 1974, 14, no.10, pp.408-11.

44.380. Pedersen, H.M., Lokberg, O.J., and Forre, B.M.: "Holographic Vibration Measurement Using a TV Speckle Interferometer", Opt. Commun., 1974, 12, no.4, pp.421-6.

44.381. Kutik, M.: "Alignment and Levelling With a Gas Laser", TESLA Electron., 1974, 7, no.3, pp.79-84.

44.382. Grange, J.: "Laser Interferometry and Application in Microelectronics", Electron. Microelectron. Ind., 1974, no.197, pp.87-9.

44.383. Mogil'nyi, A.G., Popov, V.D., and Rybakov, B.V.: "Method for Measuring Polarization and Spatial Characteristics of Light Scattered by Laser Mirrors", Kvantovaya Elektron., 1974, 1, no.5, pp.1279-81, and Sov. J. Quantum Electron., 1974, 4, no.5, pp.713-4.

44.384. Uyemura, T., et al.: "Application of Holography to Precise Measurement", J. Jap. Soc. Precis. Eng., 1974, 40, no.9, pp.729-38.

44.385. Vasiliu, V., Ristici, M., and Blaj, V.: "Alignment System With He-Ne Laser", Stud. Cercet. Fiz., 1974, 26, no.7, pp.801-5.

Part VI

APPLICATIONS

45. THERMAL AND INDUSTRIAL ASPECTS

45.1. Brown, W.C., Heenan, N.I., and Mims, J.R.: "Experimental Microwave-Powered Helicopter", Int. Conv. Rec. IEEE, 1965, 13, pt 5, p.225.

45.2. Garnier, R.C., and Ishii, T.K.: "Microwave Motor Utilizing Double Antenna and Double Coil", Proc. IEEE, 1965,53,p.178.

45.3. Voss, W.A.G.: "Factors Affecting Operation of High-Power Microwave Heating for the Lumber Industry", Trans. IEEE, 1966, IAG-2, p.234.

45.4. Steinoff, R.W., and Ishii, T.K.: "Pickup Antennas for Waveguide Motors", Trans. IEEE, 1966, MTT-14, p.438.

45.5. Loewenstern, W., and Dunn, D.A.: "Cylindrical Waveguides as Power Transmission Medium", Proc. IEEE, 1966, 54, p.955.

45.6. Ishii, T.K., Garnier, R.C., and Steinoff, R.W.: "Electric Motor Energized by Microwaves in a Waveguide", Proc. Natl. Electron. Conf., 1966, 22, p.259.

45.7. Dougal, A.A., and Friedrich, O.M.: "Interaction of Laser Beams With Ionized Gases", ISA Trans., 1968, 7, p.344.

45.8. Farkas, Gy., Kertesz, S., and Naray, Zs.: "Discrimination of Laser-Induced Nonlinear Photoelectric Effect from Thermionic Emission by Time-Response Measurements", Phys. Lett., 1968, 28A, p.190.

45.9. Kato, T., and Yamaguchi, T.: "Laser Machining", NEC Res. Dev., 1968, no.12,p.57.

45.10. Akhmanov, S.A., et al.: "Thermal Self-Actions of Laser Beams", J. Quantum Electron. IEEE, 1968, QE-4, p.568.

45.11. Peppers, N.A., et al.: "Q-Switched Ruby Laser for Emission Microspectroscopic Analysis", Anal. Chem., 1968,40,p.1178.

45.12. Savchenko, M.M., and Stepanov, V.K.: "Structure of Laser Spark Image", Zh. Eksp. Teor. Fiz. Pis'ma, 1968, 8, no.9, p.458, and JETP Lett., 1968, 8, no.9, p.281.

45.13. Williams, J.L.R., and Reynolds, G.A.: "Pyrylium Salts as Q-Switches", J. Appl. Phys., 1968, 39, p.5327.

45.14. Hora, H.: "Self-Focusing of Laser Beams in a Plasma by Ponderomotive Forces", Z. Phys., 1969, 226, no.2, p.156.

45.15. Vladimirov, V.I.: "Acoustooptical Domain Generation in CdS Under Intense Laser Irradiation", Opto-Electron., 1969, 1, p.209.

45.16. Kaczmarek, F.: "Flux Densities and Electric- and Magnetic-Field Strengths of Laser Beams", Postepy Fiz., 1969, 20, no.2, p.201.

45.17. Vanyukov, M.P., et al.: "Effectiveness of Various Polymethine Pigments Used in Passive Shutters of Nd^{3+} Lasers", Zh. Prikl. Spekt., 1969, 10, no.5, p.732.

45.18. Gryaznov, Yu.M., et al.: "Stable Passive Shutter for Nd^{3+} Laser", Zh. Prikl. Spekt., 1969, 10, no.5, p.739.

45.19. Vasishht, R.C., and Cote, W.A.: "Microwave Redry Veneer", J. Microwave Power, 1969, 4, no.3, pp.158-61.

45.20. Moriarty, J.J., and Brown, W.C.: "Toroidal Microwave Discharge Heating of Gas", J. Microwave Power, 1969, 3, p.180.

45.21. Kruzhilin, Yu.I., et al.: "Mechanism of Degradation of Injection Lasers at High Levels of Excitation", Izv. VUZ Radioelektron., 1969, 12, no.10, p.1124.

45.22. Rowe, T.J., and Moule, D.J.: "Laser Machining of Photolithographic Masks in Thin Metallic Films", Proc. Inst. Mech. Eng., 1969, 183, pt 3D, p.13.

45.23. Wheeler, C.B., and Troughton, J.: "Measurement of Plasma Temperature by Absorption of Resonant Laser Radiation", Plasma Phys., 1969, 11, p.391.

45.24. Beatrice, E.S., et al.: "Electric-Spark Cross-Excitation in Laser Microprobe-Emission Spectroscopy for Samples of 10-25 micron Diameter", Appl. Spect., 1969,23,p.257.

45.25. Beatrice, E.S., and Glick, D.: "Direct-Reading Polychromator for Emission Spectroscopy", Appl. Spect., 1969, 23, p.260.

45.26. Lamb, G.L., and Kinney, R.B.: "Evaporation of Mist by an Intense Laser Beam", J. Appl. Phys., 1969, 40, p.416.

45.27. Carman, R.L., et al.: "Transient and Steady-State Thermal Self-Focusing", Appl. Phys. Lett., 1969, 14, p.136.

45.28. Dewey, R.J.: "Microwave Motor", Proc. IEEE, 1969, 57, p.248.

45.29. Pack, J.L., George, T.V., and Engelhardt, A.G.: "Microwave Diagnostics of Laser-Produced Aluminium Plasmas", Phys. Fluids, 1969, 12, p.469.

45.30. Bradley, D.J., et al.: "Simple, Laser-Triggered, Spark Gap for Kilovolt Pulses of Accurate Variable Timing", Opto-Electron., 1969, 1, p.62.

45.31. Dovger, L.S., et al.: "Characteristics of Phototropic Shutters and Effect on Laser Parameters", Opt. Spekt., 1969, 25, p.617, and Opt. Spectrosc., 1969,25,p.346.

45.32. Hirosawa, H., Tanimoto, M., and Sekiguchi, T.: "Plasma Generation by Laser Beam Irradiation on a Single Solidified Gas Particle", Trans. IEE Jap., 1969, 89, no.1, p.68.

45.33. Hokanson, J.L.: "Laser Machining Thin-Film Electrode Arrays on Quartz-Crystal Substrates", J. Appl. Phys., 1969, 40,p.3157.

45.34. Krul, L.: "Power Transport at Microwave Frequencies", Elektrotechniek, 1969, 47, no.18, p.417.

45.35. Klewe, R.C., Quigley, M.B.C., and Tozer, B.A.: "(Laser-Trigger) Technique for Studying Nonuniform-Field Breakdown Phenomena", Appl. Phys. Lett., 1969, 15, p.155.

45.36. Skinner, D.R.: "Thermal Defocusing of Brief Laser Pulses", Opt. Commun., 1969, 1, no.2, p.57.

45.37. Huang, H.F.: "Microwave Apparatus for Rapid Heating of Threadlines", J. Microwave Power, 1969, 4, no.4, pp.288-93.

45.38. Eason, H.O.: "Microwave Formation and Heating of Plasma for Controlled-Fusion Research", J. Microwave Power, 1969, 4, no.2, pp.88-99.

45.39. Novikov, N.P., et al.: "Experimental Study of Effect of Laser Beams on Transparent Dielectrics", Mekh. Polim., 1969, 5, no.5, pp.827-35, and Polym. Mech., 1969, 5, no.5, pp.734-42.

45.40. Akhmanov, S.A., et al.: "Self-Focusing of Radiation from a CW Gas Laser", Zh. Eksp. Teor. Fiz., 1969, 57, no.1, p.16, and Sov. Phys.-JETP, 1970, 30,no.1,pp.9-12.

45.41. Pikhtin, A.N., Papov, V.A., and Yas'kov, D.A.: "Use of Laser Beam in Production of Ohmic Contacts With Semiconductors", Fiz. Tekh. Poluprov., 1969, 3, no.11, p.1646, and Sov. Phys.-Semicond., 1970, 3, no.11, pp.1383-5.

45.42. Tomlinson, W.J., et al.: "Photo-induced Refractive-Index Increase in Lucite", Appl. Phys. Lett., 1970, 16, p.486.

45.43. Khan, S.H., and Walsh, D.: "Effect of Various Gases on Formative Periods of Laser-Triggered Spark Gap", Electron. Lett., 1970, 6, p.551.

45.44. Waters, R.L., and Weiner, M.J.: "Resistor Trimming and Micromachining With Nd^{3+}:YAG Laser", Solid-State Technol., 1970, 13, no.4, p.43.

45.45. Ready, J.F.: "Selecting a Laser for Material Working", Laser Focus, 1970, 6, no.3, p.38.

45.46. Sobra, K., and Maloch, J.: "Laser Beam as Auxiliary Ionization Factor", Cesk. Casopis Fys., 1970, 20A, no.2, p.123.

45.47. Zverev, G.M., Levchuk, E.A., and Maldutis, E.K.: "Thermal Self-Focusing of Laser Radiation in KDP and ADP", Zh. Eksp. Teor. Fiz., 1970, 58, no.5, p.1487, and Sov. Phys.-JETP, 1970, 31, no.5, pp.794-5.

45.48. Arakelyan, V.S., et al.: "Irradiation and Dissociation of Gaseous BCl_3 by CO_2-Laser Radiation", Radiotekh. Elektron., 1970, 15, p.634, and Radio Eng. Electron. Phys., 1970, 15, no.3, pp.547-9.

45.49. Alcock, A.J., Richardson, M.C., and DeMichelis, C.: "Breakdown and Self-Focusing Effects in Gases Produced by Single-Mode Ruby Laser", J. Quantum Electron. IEEE, 1970, QE-6, p.622.

45.50. Vali, W., et al.: "Radiation Pressure Effects of Focused Picosecond Light Pulses", J. Quantum Electron. IEEE, 1970, QE-6, p.649.

45.51. Buzhinskii, I.M., et al.: "Filter Glasses for Lasers", Zh. Prikl. Spekt., 1970, 12, no.6, p.1007.

45.52. Hopper. R.W., and Uhlmann, L.R.: "Mechanism of Inclusion Damage in Laser Glass", J. Appl. Phys., 1970, 41, p.4023.

45.53. Kenemuth, J.R.: "Thermal Blooming of a 10.6-micron Laser Beam in CO_2", Appl. Phys. Lett., 1970, 17, p.220.

45.54. Smith, D.C., and Gebhardt, F.G.: "Saturation of Self-Induced Thermal Distortion of Laser Radiation in a Wind", Appl. Phys. Lett., 1970, 16, p.275.

45.55. Lalanne, J.R.: "Simple Hydrodynamic Model Accounting for Intermediate Stages of Temporal Development of Laser-Induced Plasmas", J. Quantum Electron. IEEE, 1970, QE-6, no.12, pp.770-5.

45.56. Jungling, K.C., et al.: "Reduction of CO_2-Laser-Induced Thermal Lensing in CS_2 With a Steady Electric Field", J. Quantum Electron. IEEE, 1970, QE-6, no.11, pp.669-72.

45.57. Longfellow, J.: "Production of an Extensible (Rubber) Matrix by Laser Drilling", Rev. Sci. Instrum., 1970, 41,no.10,pp.1485-6.

45.58. Ready, J.F.: "Using the Laser in Production. Design Implications", Mech. Eng. 1970, 92, no.9, pp.18-24.

45.59. Harris, K.D.: "Laser Drilling System With Numerical Control", Elektro-Anz., 1970, 23, no.22, pp.441-2.

45.60. Woodyard, D.: "Experiments With Laser Machining", Engineering, 1970, 210, no.5443, p.249.

45.61. Brown, W.C.: "Receiving Antenna and Microwave Power Rectification (for Space Power System)", J. Microwave Power, 1970, 5, no.4, pp.279-92.

45.62. Weber, H.: "Plasma Production by Laser-Nuclear Fusion", Naturwiss. Rdsch., 1970, 23, no.11, pp.461-7.

45.63. Volkova, N.V.: "Structural Characteristics of Laser Damage in Perspex", Mekh. Polim., 1970, 6, no.5, pp.944-5, and Polym. Mech., 1970, 6, no.5, pp.826-7.

45.64. Agostini, P., et al.: "Multiphoton Ionization of Rare Gases at 1060 nm and 530 nm", J. Quantum Electron. IEEE, 1970, QE-6, no.12, pp.782-8.

45.65. Qureshi, M.S., and Nichols, K.G.: "Laser Machining System for Making Integrated-Circuit Masks", Radio Electron. Eng., 1970, 40, no.5, pp.233-40.

45.66. DeMichelis, C.: "Gas Breakdown Produced by Train of Mode-Locked Pulses", Opt. Commun., 1970, 2, no.6, pp.255-6.

45.67. Khan, S.H., and Walsh, D.: "Laser-Triggered Spark Gap", Electr. Eng. Rev., 1970, 1, no.2, pp.10-20.

45.68. Sharlai, S.F.: "Phototropic Film Shutter for Passive Q-Modulation of a Ruby Laser", Zh. Prikl. Spekt., 1970, 13, no.4, pp.730-2.

45.69. Livingston, P.M.: "Thermally Induced Modifications of a High-Power CW Laser Beam", Appl. Opt., 1971, 10, no.2, pp.426-36.

45.70. Waxler, R.M.: "Laser-Glass Composition and Possibility of Eliminating Electrostrictive (Damage) Effects", J. Quantum Electron. IEEE, 1971, QE-7, no.4, pp.166-7.

45.71. Ippen, E.P., Shank, C.V., and Dienes, A.: "Rapid Photobleaching of Organic Laser Dyes in Continuously Operated Devices", J. Quantum Electron. IEEE, 1971, QE-7, no.4, pp.178-9.

45.72. Ryzhakova, S.I., and Tomov, I.V.: "Thermal Nonlinear Rotation of Plane of Polarization of Laser Radiation", Vestn. Mosk. Univ. Fiz. Astron., 1971, no.2, pp.218-20.

45.73. Chu, T.S.: "Maximum Power Transmission Between Two Reflector Antennas in the Fresnel Zone", Bell Syst. Tech. J., 1971, 50, no.4, pp.1407-20.

45.74. Bliss, E.S.: "Pulse-Duration Dependence of Laser Damage Mechanisms", Opto-Electron., 1971, 3, no.2, pp.99-108.

45.75. Winogradoff, N.N., Haller, W.K., and Hockey, B.J.: "Cross-Sectional Energy Distributions in Gigawatt Laser Pulses", Opto-Electron., 1971, 3, no.3, pp.145-8.

45.76. Babenko, V.A., Malyshev, V.I., and Sychev, A.A.: "Method of Reducing the Relaxation Period of a Passive Nd^{3+}:Glass Laser (Dye) Shutter", Zh. Eksp. Teor. Fiz. Pis'ma, 1971, 14, no.8, pp.461-5, and JETP Lett., 1971, 14, no.8, pp.314-7.

45.77. Weber, H.P.: "Two-Photon Absorption Laws for Coherent and Incoherent Radiation", J. Quantum Electron. IEEE, 1971, QE-7, no.5, pp.189-95.

45.78. Guenther, A.H., and Bettis, J.R.: "Review of Laser-Triggered Switching", Proc. IEEE, 1971, 59, no.4, pp.689-97.

45.79. Young, P.A.: "Thermal Runaway in Ge Laser Windows", Appl. Opt., 1971, 10, no.3, pp.638-43.

45.80. Cockayne, B., and Gasson, D.B.: "Machining of Oxides Using Gas Lasers", J. Mater. Sci., 1971, 6, no.2, pp.126-9.

45.81. Siekman, J.G.: "Materials Processing With a CO_2 Laser", Elektrotechnik, 1971, 49, no.7, pp.299-304.

45.82. Brandt, G., Spengelis, B., and Van Hulle, J.: "Cutting With a CO_2 Laser", Schweissen Schneiden, 1971, 23, no.2, pp.56-9.

45.83. Steinhauer, L.C., and Ahlstrom, H.G.: "Propagation of (Optical) Coherent Radiation in a Cylindrical Plasma Column", Phys. Fluids, 1971, 14, no.6, pp.1109-14.

45.84. Belland, P., DeMichelis, C., and Mattioli, M.: "Holographic Interferometry of Laser-Produced Plasmas Using Picosecond Pulses", Opt. Commun., 1971, 3, no.1, pp.7-8.

45.85. Feinleib, J., et al.: "Rapid Reversible Light-Induced Crystallization of Amorphous Semiconductors", Appl. Phys. Lett., 1971, 18, no.6, pp.254-7.

45.86. Witkowski, S.: "Production of Fusion Plasmas by Lasers", Elektro-Tech. Z., 1971, 92, no.5, pp.273-7.

45.87. Stitch, M.L., and Weiner, M.J.: "Laser Appliances for Superfinishing Work", Werk. Betr., 1971, 104, no.4, pp.209-12.

45.88. Dabby, F.W., and Paek, U.C.: "CW Self-Induced Frequency Modulation and Switching of a Laser Beam in Liquids", Appl. Phys. Lett., 1971, 18, no.10, pp.430-2.

45.89. Scott, R.H., Jackson, P.F.S., and Strasheim, A.: "Laser-Source Mass Spectroscopy for Analysis of Geological Material", Nature, 1971, 232, no.5313, pp.623-4.

45.90. Rusbuildt, D.: "High-Power Pulsed CO_2 Lasers in Plasma Physics", Elektrotech. Z., 1971, 92A, no.8, pp.475-80.

45.91. Qureshi, M.S., and Nichols, K.G.: "Machining Integrated-Circuit Masks With He-Ne/ CO_2 Laser System", Israel J. Technol., 1971, 9, no.3, pp.249-55.

45.92. Maydan, D.: "Micromachining and Image Recording on Thin Films by Laser Beams", Bell Syst. Tech. J., 1971, 50, no.6, pp.1761-89.

45.93. Christmas, T.M., and Ley, J.M.: "Laser-Induced Damage in XDP Crystals", Electron. Lett., 1971, 7, no.18, pp.544-6.

45.94. Bradley, L.P., and Davies, T.J.: "Laser-Controlled Switching", J. Quantum Electron. IEEE, 1971, QE-7, no.9, p.464.

45.95. Capua, R.: "Ruby-Laser (Multiple) Microwelding Machine", Israel J. Technol., 1971, 9, no.3, pp.239-43.

45.96. Okada, M., and Ieiri, S.: "Influences of Self-Induced Thermal Effects on Phase Matching in Nonlinear Optical Crystals", J. Quantum Electron. IEEE, 1971, QE-7, no.12, pp.560-3.

45.97. Burns, F.B.: "Use of Sealed-Off CO_2 Laser for Calibrating Film Resistors", Ind. Ital. Elettrotec., 1971, 24, no.8, pp.657-61.

45.98. Schweisheimer, W.: "Cutting Clothes by Laser", Polytech. Tijdschr. Werk., 1971, 26, no.23, pp.974-5.

45.99. Gebhardt, F.G., and Smith, D.C.: "Self-Induced Thermal Distortion in Near Field for Laser Beam in Moving Medium", J. Quantum Electron. IEEE, 1971, QE-7, no.2, pp.63-73.

45.100. Takehana, A., et al.: "Automatic Laser Trimming System for Thin-Film Resistors", Natl. Tech. Rep., 1971, 17, no.6, pp.718-24.

45.101. Mikhailov-Teplyakov, V.A.: "Device for Programmable Cutting of Textile Materials", Mekh. Autom. Proiz., 1971, no.9, pp.21-2.

45.102. Kazakevich, V.I., et al.: "Simulating Thermal Breakdown in a Transistor by Heating With a Laser Beam", Radiotekh. Elektron., 1971, 16, no.3, and Radio Eng. Electron. Phys., 1971, 16, no.3, pp.567-9.

45.103. Bol'shov, V.F., et al.: "Laser Unit for Cutting a Glass Rod", Kvantovaya Elektron., 1971, no.6, pp.84-6.

45.104. Tulip, J., Manes, K.R., and Seguin, H.J.: "Air-Breakdown Characteristics Within a TEA Laser Cavity", Opto-Electron., 1971, 3, no.3, pp.131-6.

45.105. Ashmarin, I.I., et al.: "Investigation of Gaseous Breakdown With Laser Flare Method of Pulse Holography", Zh. Tekh. Fiz., 1971, 41, no.11, pp.2369-77, and Sov. Phys.-Tech. Phys., 1972, 16, no.11, pp.1881-7.

45.106. Draggoo, V.G., et al.: "Effects of Laser Mode Structure on Damage in Quartz", J. Quantum Electron. IEEE, 1972, QE-8, no.2, pp.54-7.

45.107. Dewhurst, R.J., Pert, G.J., and Ramsden, S.A.: "Picosecond Triggering of a Laser-Triggered Spark Gap", J. Phys. D, 1972, 5, no.1, pp.97-103.

45.108. Ahmad, S.R.: "Evidence for Reverse Photoelectrons in Laser-Induced Current", J. Appl. Phys., 1972, 43, no.1, pp.244-5.

45.109. Dabby, F.W., and Paek, U.C.: "High-Intensity Laser-Induced Vaporization and Explosion of Solid Material", J. Quantum Electron. IEEE, 1972, QE-8, no.2, pp.106-11.

45.110. Epshtein, E.M.: "Thermal Semiconductor Instability in a Laser Beam", Izv. VUZ Radiofiz., 1972, 15, no.1, pp.33-7.

45.111. Ulrich, P.B., Hayes, J.N., and Aitken, A.H.: "Comparison of Wave-Optics Computer Model With Nonlinear (Thermal Blooming) Laser Propagation Experiments", J. Opt. Soc. Am., 1972, 62, no.2, pp.298-9.

45.112. Hugenschmidt, M., Vollrath, K., and Hirth, A.: "Schlieren Diagnostics of Laser-Produced Xe Plasma", Appl. Opt., 1972, 11, no.2, pp.339-44.

45.113. Duley, W.W., and Gonsalves, J.N.: "Industrial Applications of CO_2 Lasers", Can. Res. Rev., 1972, 5, no.1, pp.25-9.

45.114. Hahn, Yu.H.: "Optical Coating for High-Power Lasers", Electro-Opt. Syst. Des., 1972, 4, no.1, pp.18-9.

45.115. Schneider, R., Walther, H., and Woste, L.: "Atomic-Beam Deflection by Light of a Tunable Dye Laser", Opt. Commun., 1972, 5, no.5, pp.337-40.

45.116. Beer, D., and Weber, J.: "Photobleaching of Organic Laser Dyes", Opt. Commun., 1972, 5, no.4, pp.307-9.

45.117. Weber, H.: "Nuclear Fusion With Lasers", Bull. Assoc. Suisse Electr., 1972, 63, no.4, pp.192-7.

45.118. Hanna, D.C., et al.: "Q-Switched Laser Damage of IR Nonlinear Materials", J. Quantum Electron. IEEE, 1972, QE-8, no.3, pp.317-24.

45.119. Bass, M., and Barrett, H.H.: "Avalanche Breakdown and Probabilistic Nature of Laser-Induced Damage", J. Quantum Electron. IEEE, 1972, QE-8, no.3, pp.338-43.

45.120. Giuliani, J.F.: "Saturable Absorption and Q-Switching in a Triphenylmethane Dye", J. Appl. Phys., 1972, 43, no.3, p.1290.

45.121. Glass, A.J., and Guenther, A.H.: "Damage in Laser Materials", Appl. Opt., 1972, 11, no.4, pp.832-40.

45.122. Spiller, E.: "Saturable (Absorber) Optical Resonator", J. Appl. Phys., 1972, 43, no.4, pp.1673-81.

45.123. Pert, G.J.: "Role of Excited States in Multiphoton Ionization", J. Quantum Electron. IEEE, 1972, QE-8, no.7, pp.623-31.

45.124. Hollier, R.A., and Macomber, J.D.: "Light Source Responsible for Deterioration of Cryptocyanine Q-Switches", Appl. Opt., 1972, 11, no.6, pp.1360-4.

45.125. Penzkofer, A.: "(Saturable-Absorber) Generation of Picosecond Light Pulses", Appl. Phys. Lett., 1972, 20, no.9, pp.531-4.

45.126. Eleccion, M.: "Materials Processing With Lasers", Spectrum IEEE, 1972, 9, no.4, pp.62-72.

45.127. Karlov, N.V., et al.: "Effect of Plasma Mirror in the Breakdown of Air in a CO_2-Laser Cavity", Zh. Eksp. Teor. Fiz. Pis'ma, 1972, 16, no.2, pp.95-8, and JETP Lett., 1972, 16, no.2, pp.65-7.

45.128. Avotin, S.S., et al.: "Change of Electrical Resistance of Laser-Irradiated Beryllium", Zh. Eksp. Teor. Fiz., 1972, 62, no.1, pp.288-93, and Sov. Phys.-JETP, 1972, 35, no.1, pp.155-7.

45.129. Belozerov, S.A., et al.: "Breakdown of Transparent Dielectrics by Radiation from Mode-Locked Lasers", Zh. Eksp. Teor. Fiz., 1972, 62, no.1, pp.294-9, and Sov. Phys.-JETP, 1972, 35, no.1, pp.158-60.

45.130. Zverev, G.M., et al.: "Damage to the Surface of $LiNbO_3$ by Light", Zh. Eksp. Teor. Fiz., 1972, 62, no.1, pp.307-12, and Sov. Phys.-JETP, 1972, 35, no.1, pp.165-7.

45.131. Anan'in, O.B., et al.: "Fracture of Nonlinear Crystals by Ruby-Laser Radiation", Zh. Tekh. Fiz., 1972, 42, no.4, pp.837-40, and Sov. Phys.-Tech. Phys., 1972, 17, no.4, pp.659-61.

45.132. Lisitsa, M.P., and Fekeshgazi, I.V.: "Analogies in the Laser-Induced Destruction of Transparent Glass", Zh. Tekh. Fiz., 1972, 42, no.4, pp.895-6, and Sov. Phys.-Tech. Phys., 1972, 17, no.4, pp.708-9.

45.133. Kuznetsov, A.Ye., Orlov, A.A., and Ulyakov, P.I.: "Pulsating Conditions in Evaporation of Materials by CO_2-Laser Radiation", Kvantovaya Elektron., 1972, no.7, pp.57-60, and Sov. J. Quantum Electron., 1972, 2, no.1, pp.44-6.

45.134. King, S.R., Hartwick, T.S., and Chase, A.B.: "Optical Damage in KTN", Appl. Phys. Lett., 1972, 21, no.7, pp.312-4.

45.135. Gibson, A.F., et al.: "Absorption Saturation in Ge, Si, and GaAs, at 10.6 micron", Appl. Phys. Lett., 1972, 21, no.8, pp.356-7.

45.136. Crisp, M.D., Boling, N.L., and Dube, G.: "Importance of Fresnel Reflection in Laser Surface Damage of Transparent Dielectrics", Appl. Phys. Lett.; 1972, 21, no.8, pp.364-6.

45.137. Penzkofer, A., and Kaiser, W.: "Nonlinear Loss in Nd^{3+}:Glass", Appl. Phys. Lett., 1972, 21, no.9, pp.427-30.

45.138. Kerr, E.L.: "Electric Stress and Laser Surface Damage", J. Quantum Electron. IEEE, 1972, QE-8, no.8, pp.723-4.

45.139. Drexhage, K.H.: "Q-Switch Compounds for IR Lasers", J. Quantum Electron. IEEE, 1972, QE-8, no.9, p.759.

45.140. Olsen, J.N., Jones, E.D., and Gobeli, G.W.: "Picosecond Laser-Produced CD_2 Plasmas", J. Appl. Phys., 1972, 43, no.10, pp.3991-9.

45.141. Armstrong, J.J., and Gaddy, O.L.: "Saturation Behaviour of SF_6 at High Pressure and Laser Intensity", J. Quantum Electron. IEEE, 1972, QE-8, no.10, pp.797-802.

45.142. Field, J.E., and Zafar, M.A.: "Effect of Surface Films on Laser Damage in Glasses", J. Phys. D, 1972, 5, no.11, pp.2105-14.

45.143. Haller, W.K., and Simmons, J.H.: "Laser Self-Focusing Damage Tracks in Glass", J. Res. Natl. Bur. Stand., 1972, 76A, no.4, pp.337-45.

45.144. Tanno, N., Yokoto, K., and Inaba, H.: "Two-Photon Self-Induced Transparency of Different-Frequency Optical Short Pulses in Potassium", Phys. Rev. Lett., 1972, 29, no.18, pp.1211-4.

45.145. Makhlin, A.N., and Skrotskii, G.V.: "Behaviour of an Intense Light Beam in a Nonideal Gas", Kvantovaya Elektron., 1972, no.7, pp.56-7, and Sov. J. Quantum Electron., 1972, 2, no.1, pp.42-3.

45.146. Litvinov, V.F., et al.: "Transmission of Light Pulses by a Two-Component Semiconductor", Kvantovaya Elektron., 1972, no.7, pp.89-92, and Sov. J. Quantum Electron., 1972, 2, no.1, pp.83-6.

45.147. Saum, K.A., and Koopman, D.W.: "Discharges Guided by Laser-Induced Rarefaction Channels", Phys. Fluids, 1972, 15, no.11, pp.2077-9.

45.148. Smith, D.M., and Wiggins, T.A.: "Sound Speeds and Laser-Induced Damage in Polystyrene", Appl. Opt., 1972, 11, no.11, pp.2680-3.

45.149. O'Keefe, J.D., and Skeen, C.H.: "Laser-Induced Stress-Wave and Impulse Augmentation", Appl. Phys. Lett., 1972, 21, no.10, pp.464-6.

45.150. Boling, N.L., and Dube, G.: "Morphological Asymmetry in Laser Damage of Transparent Dielectric Surfaces", Appl. Phys. Lett., 1972, 21, no.10, pp.487-9.

45.151. Yamanaka, C., et al.: "Anomalous Heating of a Plasma by a Laser", Phys. Rev., 1972, 6A, no.6, pp.2335-42.

45.152. Aseev, G.I., and Kats, M.L.: "Mechanisms for the Laser Destruction of Alkali-Halide Crystals", Fiz. Tver. Tela, 1972, 14, no.5, pp.1303-7, and Sov. Phys.-Solid State, 1972, 14, no.5, pp.1122-5.

45.153. Grieneisen, H.P., et al.: "Observation of Transparency of a Resonant Medium to Zero-Degree Optical Pulses", Appl. Phys. Lett., 1972, 21, no.11, pp.559-62.

45.154. Geguzin, Ya.E., Emets, A.K., and Boiko, Yu.I.: "Reduction in the Optical Strength of Transparent Solids Containing Macroscopic Defects", Fiz. Tver. Tela, 1972, 14, no.5, pp.1565-6, and Sov. Phys.-Solid State, 1972, 14, no.5, p.1350.

45.155. Arkhipov, Yu.V., et al.: "Energy Balance and Dynamics of Damage of Transparent Dielectrics by Laser Radiation", Fiz. Tver. Tela, 1972, 14, no.6, pp.1756-60, and Sov. Phys.-Solid State, 1972, 14, no.6, p.1510.

45.156. Kuznetsov, A.Ye., et al.: "Destruction of Reflecting Dielectric Coatings by Laser Radiation", Opt.-Mekhan. Prom., 1972, 39, no.3, pp.39-42, and Sov. J. Opt. Technol., 1972, 39, no.3, pp.158-60.

45.157. Grabiner, F.R., Siebert, D.R., and Flynn, G.W.: "Laser-Induced Time-Dependent Thermal-Lensing Studies of Vibrational Relaxation", Chem. Phys. Lett., 1972, 17, no.2, pp.189-94.

45.158. Aleshkevich, V.A., et al.: "Aberrations and Extreme Divergences of CW Laser Radiation in Defocusing Media", Zh. Eksp. Teor. Fiz., 1972, 35, no.2, pp.551-61, and Sov. Phys.-JETP, 1972, 35, no.2, pp.292-7.

45.159. Andreev, S.I., et al.: "Time Taken for a Laser Pulse to Make a Hole in a Metal Film", Zh. Tekh. Fiz., 1972, 42, no.4, pp.893-5, and Sov. Phys.-Tech. Phys., 1972, 17, no.4, pp.705-7.

45.160. Petrov, A.A., et al.: "Possible Use of a Laser Torch as Light Source for Spectral-Isotopic Method", Zh. Prikl. Spekt., 1972, 17, no.3, pp.391-3.

45.161. Mikhnov, S.A., et al.: "Possible Use of an Organic Laser for Spectral Analysis", Zh. Prikl. Spekt., 1972, 17, no.3, pp.394-8.

45.162. Anderson, L.C., Fraser, H.R., and Jemkins, J.B.: "Laser Micromachining of Green Ceramic Dielectrics", Am. Ceram. Soc. Bull., 1972, 51, no.8, p.652.

45.163. Levine, J.S.: "Possibility of Generating Laser Plasmas by Photoionization", Phys. Lett., 1972, 42A, no.2, pp.173-5.

45.164. Johnson, R.L., and O'Keefe, J.D.: "Thermal Runaway in Semiconductor Laser Windows", Appl. Opt., 1972, 11, no.12, pp.2926-32.

45.165. Stabnikov, M.V., and Tombak, M.Sh.: "Holograms of Spark Discharges Produced by Nanosecond Electrical Pulses", Zh. Tekh. Fiz., 1972, 17, no.5, pp.1073-5, and Sov. Phys.-Tech. Phys., 1972, 17, no.5, pp.852-4.

45.166. Gonsalves, J.N., and Duley, W.W.: "Cutting Thin Metal Sheets With a CW CO_2 Laser", J. Appl. Phys., 1972, 43, no.11, pp.4684-7.

45.167. Jassby, D.L.: "IR-Laser Heating of Dense Arc Plasmas", Phys. Fluids, 1972, 15, no.12, pp.2442-4.

45.168. Ahmad, K.: "RF Burnout in X-Band Schottky Mixers", J. Appl. Phys., 1972, 43, no.11, pp.4826-7.

45.169. Petrun'kin, V.Yu., et al.: "High-Voltage Nanosecond Pulse Generator Triggered by Laser Radiation", Prib. Tekh. Eksp., 1972, 15, no.2, pp.178-80, and Instrum. Exp. Tech., 1972, 15, no.2, pp.515-7.

45.170. Bukatii, V.I., and Pogodaev, V.A.: "IR Radiation Vaporization of a Water Drop", Laser Unconv. Opt. J., 1972, no.40, pp.3-6.

45.171. Crisp, M.D.: "Laser-Induced Damage of Totally Internal Reflecting Surfaces", Opt. Commun., 1972, 6, no.2, pp.213-5.

45.172. Eichler, H., and Stahl, H.: "Thermal Excitation of Ultrasonic Waves by Laser Light", Opt. Commun., 1972, 6, no.3, pp.239-41.

45.173. Meyer, I., and Timm, U.: "Generation of Spark in Air by a Train of Mode-Locked Laser Pulses", Opt. Commun., 1972, 6, no.4, pp.339-41.

45.174. Alcock, A.J., Kato, K., and Richardson, M.C.: "Features of Laser-Induced Gas Breakdown in the UV", Opt. Commun., 1972, 6, no.4, pp.342-4.

45.175. Burnett, N.H., Kerr, R.D., and Offenberger, A.A.: "High-Intensity CO_2 Laser Plasma Interaction", Opt. Commun., 1972, 6, no.4, pp.372-6.

45.176. Bonch-Bruevich, A.M., et al.: "Action of Single-Pulse Ruby Laser Radiation on a Mercury-Lamp Plasma", Opt. Spekt., 1972, 32, no.6, pp.1171-5, and Opt. Spectrosc., 1972, 32, no.6, pp.635-7.

45.177. Usami, A., and Matsuoka, Y.: "Damage on Semiconductor Surfaces by Q-Switched Laser Bombardment", Bull. Nagoya Inst. Technol., 1972, 24, pp.211-6.

45.178. Velichko, O.A., et al.: "Pulsed Laser Beam Welding for Integrated Circuits", Avtom. Svarka, 1972, 25, no.8, pp.50-1, and Autom. Weld., 1972, 25, no.8, pp.48-50.

45.179. Howe, J.T.: "Lateral Expansion of Laser-Supported Detonation Wave in a Gas", AIAA J., 1972, 10, no.12, pp.1710-1.

45.180. Feiock, F.D., and Goodwin, L.K.: "Calculation of Laser-Induced Stresses in Water", J. Appl. Phys., 1972, 43, no.12, pp.5061-4.

45.181. Anisimov, S.I.: "Transition of Hydrogen Into the Metallic State in a Compression Wave Induced by a Laser Pulse", Zh. Eksp. Teor. Fiz. Pis'ma, 1972, 16, no.10, pp.570-2, and JETP Lett., 1972, 16, no.10, pp.404-6.

45.182. Askar'yan, G.A., Kaitmazov, S.D., and Medvedev, A.A.: "Light Flash from Shock Wave of a Laser Spark. Effect of Strong External Magnetic Field", Zh. Eksp. Teor. Fiz., 1972, 62, no.3, pp.918-23, and Sov. Phys.-JETP, 1972, 35, no.3, pp.487-9.

45.183. Weick, W.W.: "Laser Generation of Conductor Patterns", J. Quantum Electron. IEEE, 1972, QE-8, no.2, pp.126-31.

45.184. Peak, V.C., and Gagliano, F.P.: "Thermal Analysis of Laser-Drilling Process", J. Quantum Electron. IEEE, 1972, QE-8, no.2, pp.112-9.

45.185. Kocher, E., et al.: "Dynamics of Laser Processing in Transparent Media", J. Quantum Electron. IEEE, 1972, QE-8, no.2, pp.120-5.

45.186. Thomas, S.W., and Coleman, L.W.: "Laser-Triggered Avalanche-Transistor Voltage Generator for Picosecond Streak Camera", Appl. Phys. Lett., 1972, 20, no.2, pp.83-4.

45.187. Akulenok, E.M., et al.: "Mechanics of Damage to Ruby Crystal by Laser Radiation", Zh. Eksp. Teor. Fiz. Pis'ma, 1972, 16, no.6, pp.336-9, and JETP Lett., 1972, 16, no.6, pp.238-40.

45.188. Moeckel, W.E.: "Propulstion by Impinging Laser Beams", J. Spacecr. Rockets, 1972, 9, no.12, pp.942-4.

45.189. Meyer, R.T., and Lynch, A.W.: "Mechanism of Particle Emission from Graphite During Pulsed Laser Heating", High Temp. Sci., 1972, 4, no.4, pp.283-9.

45.190. Fanchenko, S.D., and Sholin, G.V.: "Possible Mechanisms of Turbulent Heating of a Plasma by Ultrashort Pulses of Laser Radiation", Dokl. Akad. Nauk SSSR, 1972, 201, no.6, pp.1090-3, and Sov. Phys.-Dokl., 1972, 17, no.6, pp.572-5.

45.191. Mushiake, Y., et al.: "Generation of Radially Polarized Optical Beam Mode by Laser", Proc. IEEE, 1972, 60, no.9, pp.1107-9.

45.192. Shatilov, A.V., Gusev, G.P., and Dvornikov, G.D.: "Self-Focusing Thresholds of Nanosecond Radiation in Optical Glasses", Opt.-Mekh. Prim., 1972, 39, no.4, pp.18-20, and Sov. J. Opt. Technol., 1972, 39, no.4, pp.203-4.

45.193. Yamanaka, C., et al.: "Plasma Generation and Heating by Lasers", Kvantovaya Elektron., 1972, no.2, pp.45-52, and Sov. J. Quantum Electron., 1972, 2, no.2, pp.127-32.

45.194. Zverev, G.M., et al.: "Laser-Induced Damage to Surface of $LiNbO_3$ and $LiTaO_3$ Crystals", Kvantovaya Elektron., 1972, no.2, pp.94-6, and Sov. J. Quantum Electron., 1972, 2, no.2, pp.167-9.

45.195. Nagao, Y., et al.: "Improvement of Damage Thresholds of Glass Lasers", Electr. Eng. Jap., 1972, 92, no.5, pp.138-42.

45.196. Offenberger, A.A., and Burnett, N.H.: "CO_2-Laser-Induced Gas Breakdown in Hydrogen", J. Appl. Phys., 1972, 43, no.12, pp.4977-80.

45.197. Kimura, T., Yanai, H., and Kamiyama, M.: "Optical Triggering of Gunn-Effect Devices", Electron. Commun. Jap., 1972, 55, no.10, pp.86-92.

45.198. Volkova, N.V., and Tsirul'nik, P.N.: "Effect of Soluble Impurity on Resistance of LiF to Destruction by Light", Opt.-Mekh. Prom., 1972, 39, no.12, pp.35-6, and Sov. J. Opt. Technol., 1972, 39, no.12, pp.755-6.

45.199. Kobayashi, A., et al.: "Drilling of Nonmetals With Ruby Laser", Bull. Jap. Soc. Precis. Eng., 1972, 6, no.3, pp.73-8.

45.200. Bunkin, F.B., Kazakov, A.E., and Fedorov, M.V.: "Interaction of Intense Optical Radiation With Free Electrons", Usp. Fiz. Nauk, 1972, 107, no.3-4, pp.559-93, and Sov. Phys.-Usp., 1973, 15, no.4, pp.416-35.

45.201. Danileiko, Yu.K., et al.: "Role of Absorbing Impurities in Laser-Induced Damage of Transparent Dielectrics", Zh. Eksp. Teor. Fiz., 1972, 63, no.3, pp.1030-5, and Sov. Phys.-JETP, 1973, 36, no.3, pp.541-3.

45.202. Lisitsa, M.P., and Fekeshgazi, I.V.: "Nature of Damage Caused by Laser Radiation on Surface or in Bulk of Transparent Glasses", Kvantovaya Elektron., 1972, 2, no.5, pp.86-8, and Sov. J. Quantum Electron., 1973, 2, no.5, pp.454-6.

45.203. Basov, N.G., Belenov, E.M., et al.: "Optical Breakdown of Compressed Gases by CO_2 Laser Radiation", Zh. Eksp. Teor. Fiz., 1972, 63, no.6, pp.2010-4, and Sov. Phys.-JETP, 1973, 36, no.6, pp.1061-3.

45.204. Fersman, I.A., and Khazov, L.D.: "Mechanism of Damage to Surfaces of Transparent Dielectrics by Short Light Pulses", Kvantovaya Elektron., 1972, no.4, pp.25-31, and Sov. J. Quantum Electron., 1973, 2, no.4, pp.319-23.

45.205. Zuev, V.E., et al.: "Effect of Heating Water Droplets by Optical Radiation", Dokl. Akad. Nauk SSSR, 1972, 205, no.5, pp.1069-72, and Sov. Phys.-Dokl., 1973, 17, no.8, pp.765-8.

45.206. Gulyaeva, A.S., et al.: "Change in Photoluminescence of GaAs in Regions Damaged by Laser Beam", Dokl. Akad. Nauk SSSR, 1972, 205, no.4, pp.815-7, and Sov. Phys.-Dokl., 1973, 17, no.8, pp.780-2.

45.207. Bubunov, M.M., et al.: "Change in Sign of Thermal Lens With Change in Thermo-optical Constant of Glass", Dokl. Akad. Nauk SSSR, 1972, 205, no.1-3, pp.556-9, and Sov. Phys.-Dokl., 1973, 17, no.7, pp.682-4.

45.208. Ashkin, A., and Dziedzic, J.M.: "Radiation Pressure on a Free Liquid Surface", Phys. Rev. Lett., 1973, 30, no.4, pp.139-42.

45.209. Clarke, J.S., Fisher, H.N., and Mason, R.J.: "Laser-Driven Implosion of Spherical DT Targets to Thermonuclear Burn Conditions", Phys. Rev. Lett., 1973, 30, no.3, pp.89-92.

45.210. Withers, P.B., and Wilshaw, T.R.: "Fracture of Transparent Brittle Solids Due to Stress Pulses Generated by Q-Switched Ruby Laser", J. Phys. D, 1973, 6, no.3, pp.322-36.

45.211. Siegrist, M., Adam, B., and Kneubuhl, F.K.: "Interaction of TEA-CO_2-Laser Pulses With Metals Enhanced by Liquid Layers", Phys. Lett., 1973, 42A, no.5, pp.352-4.

45.212. Metz, S.A.: "Impulse Loading of Targets by Subnanosecond Laser Pulses", Appl. Phys. Lett., 1973, 22, no.5,pp.211-3.

45.213. Pirozhkov, V.A., et al.: "Deformation of Laser Pulses in Resonant Media", Phys. Lett., 1973, 43A, no.1, pp.31-2.

45.214. Kato, K., et al.: "Laser Induced Gas Breakdown in the UV", Oyo Buturi, 1973, 42, no.7, pp.684-9.

45.215. Bodner, S.E., Chapline, G.F., and DeGroot, J.: "Anomalous Ion Heating in a Laser-Heated Plasma", Plasma Phys., 1973, 15, no.1, pp.21-7.

45.216. Kafalas, P., and Ferdinand, A.P.: "Fog Droplet Vaporization and Fragmentation by 10.6-micron Laser Pulse", Appl. Opt., 1973, 12, no.1, pp.29-33.

45.217. Aitken, A.H., Hayes, J.N., and Ulrich, P.B.: "Thermal Blooming of Pulsed Focused Gaussian Laser Beams", Appl. Opt., 1973, 12, no.2, pp.193-7.

45.218. Buser, R.G., and Rohde, R.S.: "Severe Self-Induced Beam Distortion in Laboratory-Simulated Laser Propagation at 10.6 micron", Appl. Opt., 1973, 12, no.2, pp.205-11.

45.219. Moody, C.D.: "Effects of CW Power on Pulsed Gas-Breakdown Threshold in Argon at 10.6 micron", Appl. Phys. Lett., 1973, 22, no.1, pp.31-2.

45.220. Robinson, A.M.: "Laser-Induced Gas Breakdown Initiated by UV Photoionization", Appl. Phys. Lett., 1973, 22, no.1, pp.33-5.

45.221. Fradin, D.W.: "Comparison of Laser-Induced Surface and Bulk Damage", Appl. Phys. Lett., 1973, 22, no.4, pp.157-9.

45.222. Fradin, D.W.: "Electron Avalanche Breakdown Induced by Ruby Laser", Appl. Phys. Lett., 1973, 22, no.5, pp.206-8.

45.223. Sam, C.L.: "Laser Damage of GaAs and ZnTe at 1060 nm", Appl. Opt., 1973, 12, no.4, pp.878-9.

45.224. Prokhorov, A.M., et al.: "Metal Evaporation Under Powerful Optical Radiation", J. Quantum Electron. IEEE, 1973, QE-9, no.5, pp.503-10.

45.225. Moore, M.: "Time-Dependent, Ray-Geometrical, Treatment of Interference Fringes Within a Thermally Self-Defocused Laser Beam", Opto-Electron., 1973, 5, no.2, pp.189-200.

45.226. Bendow, B., and Gianino, P.D.: "Optics of Thermal Lensing in Solids", Appl. Opt., 1973, 12, no.4, pp.710-8.

45.227. Ghatak, A.K., and Sharma, S.K.: "Thermal Self-Defocusing of Laser Beams", Appl. Phys. Lett., 1973, 22, no.4, pp.141-2.

45.228. Leupold, D., Voigt, B., and Konig, R.: "Nature of Hole Burning in Saturable Organic Absorbers", Exp. Tech. Phys., 1973, 21, no.1, pp.41-4.

45.229. van Hulsteyn, D.B., and Anderson, L.: "Quasimodo. Device for Suspending Small Solid Targets at the Focus of a High-Power Laser", Rev. Sci. Instrum., 1973, 44, no.4, pp.453-6.

45.230. Boling, N.L., Crisp, M.D., and Dube, G.: "Laser-Induced Surface Damage", Appl. Opt., 1973, 12, no.4, pp.650-60.

45.231. Bloembergen, N.: "Role of Cracks, Pores, and Absorbing Inclusions on Laser Induced Damage Threshold at Surfaces of Transparent Dielectrics", Appl. Opt., 1973, 12, no.4, pp.661-4.

45.232. Austin, R.R., et al.: "Effects of Structure, Composition, and Stress, on Laser-Damage Threshold of Single and Multiple Films", Appl. Opt., 1973, 12, no.4, pp.665-76.

45.233. Bliss, E.S., Milam, D., and Bradbury, R.A.: "Dielectric-Mirror Damage by Laser Radiation Over a Range of Pulse Durations and Beam Radii", Appl. Opt., 1973, 12, no.4, pp.677-89.

45.234. Bass, M., and Barrett, H.H.: "Laser-Induced Damage Probability at 1060 nm and 684 nm", Appl. Opt., 1973, 12, no.4, pp.690-9.

45.235. Fradin, D.W., Yablonovitch, E., and Bass, M.: "Confirmation of an Electron Avalanche Causing Laser-Induced Bulk Damage at 1060 nm", Appl. Opt., 1973, 12, no.4, pp.700-9.

45.236. Kafalas, P., and Herrmann, J.: "Dynamics and Energetics of the Explosive Vaporization of Fog Droplets by a 10.6-micron Laser Pulse", Appl. Opt., 1973, 12, no.4, pp.772-5.

45.237. Bissonnette, L.R.: "Thermally Induced Nonlinear Propagation of a Laser Beam in an Absorbing Fluid", Appl. Opt., 1973, 12, no.4, pp.719-28.

45.238. Kamecke, W.: "Laser Trimming of Thick Film Resistors", Int. Elektron. Rdsch., 1973, 27, no.3, pp.63-6.

45.239. Guenther, A.H., et al.: "Pulsed Interferometric Holography of Laser-Produced Air Breakdown", Opt. Laser Technol., 1973, 5, no.1, pp.20-3.

45.240. Bendow, B., et al.: "Theory of Thermally Induced Interference and Lensing in Transparent Materials", Opt. Commun., 1973, 7, no.3, pp.219-24.

45.241. Fleck, J.A., and Layne, C.: "Study of Self-Focusing Damage in a High-Power Nd^{3+}:Glass Amplifier", Appl. Phys. Lett., 1973, 22, no.9, pp.467-9.

45.242. Canavan, G.H., and Nielsen, P.E.: "Focal-Spot-Size Dependence of Gas Breakdown Induced by Particulate Ionization", Appl. Phys. Lett., 1973, 22, no.8, pp.409-10.

45.243. Muzii, L., Stagni, L., and Vitali, G.: "Experiments on Reflectivity of Laser-Irradiated Solid Surfaces", Nuovo Cim., 1973, 14B, no.2, pp.173-89.

45.244. Mirkin, L.I.: "Formation of Oriented Structures by Action of a Laser Beam on Metals", Dokl. Akad. Nauk SSSR, 1973, 206, no.6, pp.1339-41, and Sov. Phys.-Dokl., 1973, 17, no.10, pp.1026-7.

45.245. Mourou, G., et al.: "Kinetics of Bleaching in Polymethine Cyanine Dyes", J. Quantum Electron. IEEE, 1973, QE-9, no.7, pp.745-8.

45.246. Weber, J.: "Study of the Influence of Triplet Quencher on Photobleaching of Rhodamine-6G", Opt. Commun., 1973, 7, no.4, pp.420-2.

45.247. Fox, J.A., and Barr, D.N.: "Laser-Induced Shock Effects in Plexiglas and Aluminium", Appl. Phys. Lett., 1973, 22, no.11, pp.594-6.

45.248. Gurevich, G.L., and Murav'ev, V.A.: "Action of Laser Radiation on Thin Films", Fiz. Khim. Obrab. Mater., 1973, no.1, pp.3-8.

45.249. Libenson, M.I., and Nikitin, M.N.: "Diffusion of Atoms from a Film Into a Substrate Under the Action of Laser Radiation", Fiz. Khim. Obrab. Mater., 1973, no.1,pp.9-14.

45.250. Mirkin, L.I.: "Contact Melting at a Ferrite-Graphite Boundary Under the Influence of Laser Light", Fiz. Khim. Obrab. Mater., 1973, no.1, pp.143-5.

45.251. Clark, A.F., Moulder, J.C., and Reed, R.P.: "Ability of a CO_2 Laser to Assist Ice Breakers", Appl. Opt., 1973, 12, no.6, pp.1103-4.

45.252. Abramyan, E.A., et al.: "Investigation of the Distribution of Oxygen in Deformed Niobium by a Laser Method", Fiz. Khim. Obrab. Mater., 1973, no.1, p.146.

45.253. Razdobreev, A.A., and Bukatii, V.I.: "Technique of Studying Combustion and Burning of Metal Particles by a Laser", Izv. VUZ Fiz., 1973, no.4, pp.155-7.

45.254. Fradin, D.W., Bloembergen, N., and Letellier, J.P.: "Dependence of Laser-Induced Breakdown Field on Pulse Duration", Appl. Phys. Lett., 1973, 22, no.12,pp.635-7.

45.255. Kleiman, H., and O'Neil, R.W.: "Thermal Blooming of Pulsed Laser Radiation", Appl. Phys. Lett., 1973, 23, no.1, pp.43-4.

45.256. Papirov, I.I., et al.: "Deformation of Be Crystals by Laser Radiation", Fiz. Khim. Obrab. Mater., 1973,no.2,pp.147-8.

45.257. Mourou, G., et al.: "Observation of Hole-Burning in a Solution of Cryptocyanine in Methanol", Opt. Commun., 1973, 8, no.1, pp.56-9.

45.258. Smith, K.W., and Allen, L.: "Incoherent Bleaching and Self-Induced Transparency", Opt. Commun., 1973, 8,no.2,pp.166-70.

45.259. Chang, C.S., and Stehle, P.: "Theory of Resonant Multiphoton Ionization", Phys. Rev. Lett., 1973, 30, no.26, pp.1283-5.

45.260. Lencioni, D.E.: "Effect of Dust on 10.6-micron Laser-Induced Air Breakdown", Appl. Phys. Lett., 1973, 23, no.1, pp.12-4.

45.261. Siegrist, M., and Kneubuhl, F.K.: "Shock and Compression by TEA-CO_2-Laser Pulses Drastically Enhanced by Liquid Layers", Appl. Phys., 1973, 2, no.1, pp.43-4.

45.262. Siegrist, M., Kaech, G., and Kneubuhl, F.K.: "Formation of Periodic Wave Structure on the Dry Surface of a Solid by TEA-CO_2-Laser Pulses", Appl. Phys., 1973, 2, no.1, pp.45-6.

45.263. Sparks, M., and Duthler, C.J.: "Theory of IR Absorption and Material Failure in Crystals Containing Inclusions", J. Appl. Phys., 1973, 44, no.7, pp.3038-45.

45.264. Veiko, V.P., et al.: "Thermochemical Action of Laser Radiation", Dokl. Akad. Nauk SSSR, 1973, 208, no.3, pp.587-90, and Sov. Phys.-Dokl., 1973, 18, no.1, pp.83-5.

45.265. Bendow, B., and Gianino, P.D.: "Thermal Lensing of Laser Beams in Optically Transmitting Materials. I and II", Appl. Phys., 1973, 2, no.1, pp.1-10 and no.2, pp.71-90.

45.266. Agranat, B.B., et al.: "Increasing the Optical Strength of a Liquid", Zh. Eksp. Teor. Fiz. Pis'ma, 1973, 17, no.9, pp.501-4, and JETP Lett., 1973, 17, no.9, pp.361-2.

45.267. Bazhenov, S.N., et al.: "Programmed Scanning Laser System for Photomask Synthesis", Izv. VUZ Prib., 1973, 16, no.5,pp.121-5.

45.268. Bartoli, F., et al.: "Laser Damage to Triglycine Sulphate", J. Appl. Phys., 1973, 44, no.8, pp.3713-20.

45.269. von der Linde, D., and Rodgers, K.F.: "Recovery Period of Saturable Absorbers for 1060 nm", J. Quantum Electron. IEEE, 1973, QE-9, no.9, pp.960-1.

45.270. von Gutfeld, R.J.: "Laser-Induced Anisotropic Thermoelectric Voltages in Thin Films", Appl. Phys. Lett., 1973, 23, no.4, pp.206-8.

45.271. Reichert, J.D., Wagner, W.G., and Chen, W.Y.: "Instabilities of Intense Laser Beams in Air", J. Appl. Phys., 1973, 44, no.8, pp.3641-6.

45.272. Reichert, J.D., Wagner, W.G., and Chen, W.Y.: "Propagation of Intense Gaussian Laser Pulse in Air", J. Appl. Phys., 1973, 44, no.8, pp.3647-58.

45.273. Stegman, R.L., Schriempf, J.T., and Hettche, L.R.: "Experimental Studies of Laser-Supported Absorption Waves With 5-ms Pulses of 10.6-micron Radiation", J. Appl. Phys., 1973, 44, no.8, pp.3675-81.

45.274. Ammann, E.O., and Wintemute, J.D.: "Damage to ZnS Thin Films from 1060-nm Laser Radiation", J. Opt. Soc. Am., 1973, 63, no.8, pp.965-71.

45.275. Ciura, A.I., Mihailescu, I.N., and Popescu, I.M.: "Laser-Produced Plasmas on Solid Targets", Stud. Cercet. Fiz., 1973, 25, no.3, pp.341-58.

45.276. Fradin, D.W.: "Measurement of Self-Focusing Parameters Using Intrinsic Optical Damage", J. Quantum Electron. IEEE, 1973, QE-9, no.9, pp.954-6.

45.277. Peterson, G.E., et al.: "Control of Laser Damage in $LiNbO_3$", J. Am. Ceram. Soc., 1973, 56, no.5, pp.278-82.

45.278. Shank, C.V., and Schmidt, R.V.: "Optical Technique for Producing 100-nm Periodic Surface Structures", Appl. Phys. Lett., 1973, 23, no.3, pp.154-5.

45.279. David, C.D., and Clark, W.M.: "Self-Induced Transparency With CO_2 Laser Pulses in Ammonia Gas", Appl. Phys. Lett., 1973, 23, no.6, pp.306-8.

45.280. Cornish, W.D., Wong, K.Y.W., and Young, L.: "Ellipsometric Investigation of Radiation Damage in $LiNbO_3$", Am. Ceram. Soc. Bull., 1973, 52, no.4, p.372.

45.281. Brukner, F., et al.: "Self-Induced Transparency in Semiconductor by Single-Photon Excitation by Ultrashort Laser Pulse", Zh. Eksp. Teor. Fiz. Pis'ma, 1973, 18, no.1, pp.27-30, and JETP Lett., 1973, 18, no.1, pp.14-5.

45.282. Bass, M., and Fradin, D.W.: "Surface and Bulk Laser-Damage Statistics and Identification of Intrinsic Breakdown Processes", J. Quantum Electron. IEEE, 1973, QE-9, no.9, pp.890-6.

45.283. Strickland, D.M., Bettis, J.R., and Guenther, A.H.: "Low-Power Laser-Triggered Switching at Voltages Above 500 kV", Rev. Sci. Instrum., 1973, 44, no.8, pp.1121-2.

45.284. Guccione, S.A.: "Pulse Burnout of Microwave Mixer Diodes", Trans. IEEE, 1973, R-22, no.4, pp.196-207.

45.285. Sparks, M.: "Short-Pulse Operation of IR Windows Without Thermal Defocusing", Appl. Opt., 1973, 12, no.9, pp.2033-5.

45.286. Pridmore-Brown, D.C.: "Absorption Saturation Effects on High-Power CO_2 Laser Beam Transmission", Appl. Opt., 1973, 12, no.9, pp.2188-91.

45.287. Gurevich, G.L.: "Theory of Breakdown of Thin Films by Laser Radiation", Fiz. Khim. Obrab. Mater., 1973, no.3, pp.5-11.

45.288. Gulyaeva, A.S., et al.: "Change of Structure of GaAs With Laser Radiation", Fiz. Khim. Obrab. Mater., 1973, no.3, pp.17-21.

45.289. Girard, G., and Michon, M.: "Transmission of Kodak 9740 Dye Solution Under Picosecond Pulses", J. Quantum Electron. IEEE, 1973, QE-9, no.10, pp.979-84.

45.290. Mirkin, L.I.: "Dissociation of Semiconducting Compounds Under the Influence of a Laser Beam", Izv. Akad. Nauk SSSR Neorg. Mater., 1973, 9, no.1, pp.125-6, and Inorg. Mater., 1973, 9, no.1, pp.109-10.

45.291. Hettche, L.R., Schriempf, J.T., and Stegman, R.L.: "Impulse Reaction Resulting from In-Air Irradiation of Aluminium by a Pulsed CO_2 Laser", J. Appl. Phys., 1973, 44, no.9, pp.4079-85.

45.292. Boni, A.A., and Su, F.Y.: "Metal-Oxide Absorption Coefficients for Use in Intense Laser Interaction With Solids", J. Appl. Phys., 1973, 44, no.9, pp.4086-94.

45.293. Sodha, M.S., et al.: "Cross-Focusing of Two Coaxial Laser Beams in a Dielectric", Radio Sci., 1973, 8, no.6, pp.559-62.

45.294. Stegman, R.L., Schriempf, J.T., and Hettche, L.R.: "Experimental Studies of Laser-Supported Absorption Waves With 5-ms Pulses of 10.6-micron Radiation", Rep. NRL Prog., 1973, Feb., pp.25-35.

45.295. Vlasov, R.A., et al.: "Impact Ionization Mechanism in the Optical Breakdown of Transparent Dielectrics", Fiz. Tver. Tela, 1973, 15, no.2, pp.444-8, and Sov. Phys.-Solid State, 1973, 15, no.2, pp.317-9.

45.296. Rubinshtein, A.I., and Fain, V.M.: "Theory of Avalanche Ionization in Transparent Dielectrics Under the Action of a Strong Electromagnetic Field", Fiz. Tver. Tela, 1973, 15, no.2, pp.470-8, and Sov. Phys.-Solid State, 1973, 15, no.2, pp.332-6.

45.297. Belyaev, L.M., et al.: "Focused-Laser-Beam Damage Mechanism for CsI Crystals", Kristallogr., 1973, 18, no.2, pp.334-8, and Sov. Phys.-Crystallogr., 1973, 18, no.2, pp.207-8.

45.298. Bloembergen, N.: "Influence of Electron Plasma Formation on Superbroadening in Light Filaments", Opt. Commun., 1973, 8, no.4, pp.285-8.

45.299. Beaulieu, A.J.: "Rapid Balancing of Gyroscopes with TEA CO_2 Laser", Electro-Opt. Syst. Des., 1973, 5, no.5, pp.36-7.

45.300. Duley, W.W., and Young, W.A.: "Kinetic Effects in Drilling With the CO_2 Laser", J. Appl. Phys., 1973, 44, no.9, pp.4236-7.

45.301. Yablonovitch, E.: "Spectral Broadening in Light Transmitted Through a Rapidly Growing Plasma", Phys. Rev. Lett., 1973, 31, no.14, pp.877-9.

45.302. Britton, B.: "Laser Trimming of Thick-Film Resistors", Electron. Equip. News, 1973, 15, no.6, pp.52-4.

45.303. Lee, C.T.: "Self-Induced Transparency of an Extremely Short Pulse", Opt. Commun., 1973, 9, no.1, pp.1-3.

45.304. Fox, J.A., and Barr, D.N.: "Laser-Induced Stress Waves in Aluminium", Appl. Opt., 1973, 12, no.11, pp.2547-8.

45.305. Andriesh, A.M., Kats, M.S., and Fekeshgazi, I.V.: "Surface Breakdown in Vitreous As_2S_3 Under Laser Irradiation", Fiz. Khim. Obrab. Mater., 1973, no.4, pp.14-8.

45.306. Gurevich, G.L., and Murav'ev, V.A.: "Influence of Temperature Coefficient of Reflection on Heating of Thin Films by Laser Radiation", Fiz. Khim. Obrab. Mater., 1973, no.4, pp.26-9.

45.307. Aleksandrov, V.I., Solov'ev, A.G., and Ulyakov, P.I.: "Space-Time Distribution of Laser Radiation With 1-ms Duration", Fiz. Khim. Obrab. Mater., 1973, no.4, pp.30-3.

45.308. Hora, H.: "Estimates for Efficient Production of Antihydrogen by Lasers of Very High Intensity", Opto-Electron., 1973, 5, no.6, pp.491-501.

45.309. Zubarev, I.G., et al.: "Anomalous High Values of Transparency of GaAs Under the Interaction of Picosecond Light Pulses", Izv. Akad. Nauk SSSR, Ser. Fiz., 1973, 37, no.10, pp.2099-103.

45.310. George. Y.H., and Moore, F.K.: "Nearly Spherical Constant-Power Detonation Waves as Driven by Focused Radiation", J. Fluid Mech., 1973, 61, pt 3, pp.481-98.

45.311. Pirri, A.N.: "Theory for Momentum Transfer to a Surface With a High-Power Laser", Phys. Fluids, 1973, 16, no.9, pp.1435-40.

45.312. Volod'kina, V.L., et al.: "Heating an Oxidized Metal With a CO_2-Laser Beam", Dokl. Akad. Nauk SSSR, 1973, 210, no.1-3, pp.66-9, and Sov. Phys.-Dokl., 1973, 18, no.5, pp.335-7.

45.313. Anisimov, S.I., and Makshantsev, B.I.: "Role of Absorbing Inclusions in the Optical Breakdown of Transparent Media", Fiz. Tver. Tela, 1973, 15, no.4, pp.1090-5, and Sov. Phys.-Solid State, 1973, 15, no.4, pp.743-5.

45.314. Emmony, D.C., Howson, R.P., and Willis, L.J.: "Laser Mirror Damage in Ge at 10.6 micron", Appl. Phys. Lett., 1973, 23, no.11, pp.598-600.

45.315. Fradin, D.W., and Bass, M.: "Effects of Lattice Disorder on Intrinsic Optical Damage Fields of Solids", Appl. Phys. Lett., 1973, 23, no.11, pp.604-6.

45.316. DeShazer, L.G., Newnam, B.E., and Leung, K.M.: "Role of Coating Defects in Laser-Induced Damage to Dielectric Thin Films", Appl. Phys. Lett., 1973, 23, no.11, pp.607-9.

45.317. Karasev, I.G., and Kirillov, V.M.: "Possible Improvements in the Effectiveness of the Laser Processing of Metals", Fiz. Khim. Obrab. Mater., 1973, no.5, pp.3-9.

45.318. Pogodaev, V.A., and Chistyakov, L.K.: "Formation and Behaviour of Vapour Bubbles in a Drop Under the Action of a Laser Pulse", Izv. VUZ Fiz., 1973, no.12, pp.137-9.

45.319. O'Keefe, J.D., Skeen, C.H., and York, C.M.: "Laser-Induced Deformation Modes in Thin Metal Targets", J. Appl. Phys., 1973, 44, no.10, pp.4622-6.

45.320. Alexander, J.C., and Nurmikko, A.V.: "Excitation of Thin Elastic Membranes by Momentum of Laser Light", Opt. Commun., 1973, 9, no.4, pp.404-6.

45.321. Winogradoff, N.N.: "Laser-Induced Damage and Water Content of Vitreous Quartz", Opt. Commun., 1973, 9, no.4, pp.417-9.

45.322. Zhiryakov, B.M., et al.: "Relationships Governing Ejection of Matter from the Zone of Interaction With Laser Radiation", Kvantovaya Elektron., 1973, no.7, pp.119-21, and Sov. J. Quantum Electron., 1973, 3, no.1, pp.70-1.

45.323. Tsuya, H., and Fujino, Y.: "Optical Damage in $LiTaO_3$", Jap. J. Appl. Phys., 1973, 12, no.12, pp.1896-903.

45.324. Hsu, T.R.: "Application of Laser Beam Technique to the Improvement of Metal Strength", J. Test. Eval., 1973, 1, no.6, pp.457-8.

45.325. Milam, D., Bradbury, R.A., and Bass, M.: "Laser Damage Threshold for Dielectric Coatings as Determined by Inclusions", Appl. Phys. Lett., 1973, 23, no.12, pp.654-7.

45.326. Boiling, N.L., and Dube, G.: "Laser-Induced Inclusion Damage of Surfaces of Transparent Dielectrics", Appl. Phys. Lett., 1973, 23, no.12, pp.658-60.

45.327. Lysikov, Yu.I.: "Evaporation in Vacuum of Thin Metallic Films Heated by Laser Radiation", Fiz. Khim. Obrab. Mater., 1973, no.6, pp.67-71.

45.328. Neusser, H.J., and Puell, H.: "Laser-Induced Breakdown in a High-Pressure Gas Jet", Z. Naturforsch., 1973, 28a, no.2, pp.264-72.

45.329. Stamper, J.A., and Tidman, D.A.: "Magnetic Field Generation Due to Radiation Pressure in a Laser-Produced Plasma", Phys. Fluids, 1973, 16, no.11, pp.2024-5.

45.330. Brueckner, K.A., and Jorna, S.: "Stimulated Backward Brillouin Scattering in Laser-Heated Plasmas", Phys. Fluids, 1973, 16, no.12, pp.2350-1.

45.331. Lisitsa, M.P., and Fekeshgazi, I.V.: "Changes in Energy and Shape of a Laser Pulse Damaging an Area of Glass", Kvantovaya Elektron. (Kiev), 1973, no.7, pp.64-70.

45.332. Lisitsa, M.P., and Fekeshgazi, I.V.: "Effect of Atmospheric Pressure on Laser Damage Process at Transparent Glass Surfaces", Kvantovaya Elektron., 1973, no.7, pp.71-6.

45.333. Bedair, S.M., and Smith, H.P.: "Laser Deposition and Ordering Kinetics of Aluminium on (111) Silicon Surface", Surf. Sci., 1973, 40, no.2, pp.419-22.

45.334. Barinov, V.V., and Sorokin, S.A.: "Explosion of Water Drops Under Action of Optical Radiation", Kvantovaya Elektron., 1973, no.2, pp.8-11, and Sov. J. Quantum Electron., 1973, 3, no.2, pp.89-92.

45.335. Shatilov, A.V., et al.: "Effect of Microinhomogeneous Structure of Glasses on Threshold of Self-Focusing Radiation", Opt.-Mekh. Prom., 1973, 40, no.7, pp.48-50, and Sov. J. Opt. Technol., 1973, 40, no.7, pp.444-6.

45.336. Bystrova, T.V., Librovich, V.B., and Lisitsyn, V.I.: "Elements of Combustion Theory in Gas-Laser Cutting of Metals", Fiz. Goreniya Vzryva, 1973, 9, no.5, pp.725-32.

45.337. Niino, M., Toda, S., and Egusa, T.: "Experimental Investigation of Nucleation and Growth of a Single Bubble Using Laser-Beam Heating", Heat Transfer-Jap. Res., 1973, 2, no.4, pp.26-36.

45.338. Basov, N.G., Belenov, E.M., et al.: "Stimulation of Chemical Reactions by Laser Radiation", Zh. Eksp. Teor. Fiz., 1973, 64, no.2, pp.485-97, and Sov. Phys.-JETP, 1973, 37, no.2, pp.247-52.

45.339. Holzinger, G., Kosanke, K., and Menz, W.: "Printing of Part Numbers Using High-Power Laser Beam", Opt. Laser Technol., 1973, 5, no.6, pp.256-65.

45.340. Takamoto, K., and Nakayama, S.: "Temperature Distributions in Thin Metal Films Irradiated by a Gaussian Laser Beam", Rev. Electr. Commun. Lab., 1973, 21, no.9-10, pp.647-53.

45.341. Saifi, M.A., and Paek, U.C.: "Scribing Glass With Pulsed and Q-Switched CO_2 Laser", Am. Ceram. Soc. Bull., 1973, 52, no.11, pp.838-41.

45.342. Rykalin, N.N., and Uglov, A.A.: "Calculation of Heating of Metals by CO_2-Laser Radiation", Svar. Proizvod., 1973, no.7, no.1-3, and Weld. Prod., 1973, 20, no.7, pp.1-4.

45.343. Begunov, A.N., et al.: "Effect of Polishing Mixture on Threshold of Surface Damage of KDP Crystals by Optical Radiation", Opt.-Mekh. Prom., 1973, 40, no.9, p.66, and Sov. J. Opt. Technol., 1973, 40, no.9, pp.598-9.

45.344. Askar'yan, G.A., et al.: "Effect of Powerful Laser Beam on Surface of Water With Thin Liquid Film", Zh. Eksp. Teor. Fiz. Pis'ma, 1973, 18, no.11, pp.665-7, and JETP Lett., 1973, 18, no.11, pp.389-90.

45.345. Lokhov, Yu.N., Mospanov, V.S., and Fiveiskii, Yu.D.: "Optical Strength of the Surface of a Transparent Dielectric", Kvantovaya Elektron., 1973, no.2, pp.71-4, and Sov. J. Quantum Electron., 1973, 3, no.2, pp.132-3.

45.346. Bakeev, A.A., et al.: "Role of Plasma Jet in the Energy Balance of the Interaction Between Laser Radiation and Matter", Kvantovaya Elektron., 1973, no.2, pp.77-80, and Sov. J. Quantum Electron., 1973, 3, no.2, pp.138-40.

45.347. Machulka, G.A., and Muratova, L.P.: "Self-Channelling of a Laser Beam in Opaque Solids", Kvantovaya Elektron., 1973, no.2, pp.93-6, and Sov. J. Quantum Electron., 1973, 3, no.2, pp.152-4.

45.348. Vakulenko, V.M., et al.: "(Nd^{3+}: Laser) Apparatus for Drilling Holes (in Dies)", Kvantovaya Elektron., 1973, no.2, pp.99-102, and Sov. J. Quantum Electron., 1973, 3, no.2, pp.158-60.

45.349. Levinson, G.R., and Smilga, V.I.: "Mechanism of Damage to Thin Metal Films by Focused Laser Radiation", Kvantovaya Elektron., 1973, no.3, pp.72-8, and Sov. J. Quantum Electron., 1973, 3, no.3, pp.220-3.

45.350. Denus, S., et al.: "Generation of Fusion Neutrons in Plasma Produced by a Strong Laser Pulse", Bull. Acad. Pol. Sci., 1973, 21, no.11, pp.937-46.

45.351. Koopman, D.W., and Sanm, K.A.: "Formation and Guiding of High-Velocity Electrical Streamers by Laser-Induced Ionization", J. Appl. Phys., 1973, 44, no.12, pp.5328-36.

45.352. Lowder, J.E., and Kleiman, H.: "Long-Pulse Breakdown With 10.6-micron Laser Radiation", J. Appl. Phys., 1973, 44, no.12, pp.5504-5.

45.353. Kask, N.E., et al.: "Thermal Mechanisms Involved in Destruction of Optical Glass by Laser Radiation", Dokl. Akad. Nauk SSSR, 1973, 211, no.4-6, pp.1317-9, and Sov. Phys.-Dokl., 1974, 18, no.8, pp.550-1.

45.354. Gagliano, F.P., and Paek, U.C.: "Observation of Laser-Induced Explosion of Solid Materials and Correlation With Theory", Appl. Opt., 1974, 13, no.2, pp.214-9.

45.355. Duthler, C.J.: "Explanation of Laser-Damage Cone-Shaped Surface Pits", Appl. Phys. Lett., 1974, 24, no.1, pp.5-7.

45.356. Pan, Y.L., Simpson, J.R., and Bernhardt, A.F.: "(Breakdown) Limitation for Nanosecond CO_2 Pulse Amplification", Appl. Phys. Lett., 1974, 24, no.2, pp.87-90.

45.357. Bar-Isaac, C., Korn, U., and Shtrikman, S.: "Thermoelectric Temperature Measurements in Laser Pulsed Heating of Metals", Appl. Phys., 1974, 3,no.4,pp.285-90.

45.358. Duley, W.W., and Young, W.A.: "Study of Stress and Defect Structures Adjacent to Laser-Drilled Holes in Fused Quartz", J. Phys. D, 1974, 7,no.7,pp.937-9.

45.359. Pogorzelski, S.: "Power Transmission Between Antennas in Fresnel Region", Rozpr. Electrotech., 1974,20,no.1,pp.105-22.

45.360. Bell, C.E., and Maccabee, B.S.: "Shock-Wave Generation in Air and in Water by CO_2 TEA Laser Radiation", Appl. Opt., 1974, 13, no.3, pp.605-9.

45.361. O'Neil, R.W., Kleiman, H., and Lowder, J.E.: "Observation of Hydrodynamic Effects on Thermal Blooming", Appl. Phys. Lett., 1974, 24, no.3, pp.118-20.

45.362. Lowder, J.E., and Pettingill, L.C.: "Measurement of CO_2-Laser Generated (Surface) Impulse and Pressure", Appl. Phys. Lett., 1974, 24, no.4, pp.204-7.

45.363. Pogodaev, V.A., et al.: "Problems of the Explosive Regime in Evaporation of Water Drops", Izv. VUZ Fiz., 1974, no.3, pp.56-60.

45.364. Alexandrescu, R., Cojocaru, E., and Velenlescu, V.G.: "Nonlinear Thermal Effects by Laser Beam Irradiation", Rev. Roum. Phys., 1974, 19, no.2, pp.167-76.

45.365. Chik, K.P.: "Laser Damage in Glass Due to a Metal Film", Thin Solid Films", 1974, 21, no.2, pp.S27-30.

45.366. Zhbankov, R.G., et al.: "Effect of Monochromatic Emission on Changes of the Physical Structure of Polymers", Zh. Prikl. Spekt., 1974, 20, no.2, pp.317-9.

45.367. Douglas-Hamilton, D.H., Hoag, E.D., and Seitz, J.R.M.: "Diamond as a High-Power-Laser Window", J. Opt. Soc. Am., 1974, 64, no.1, pp.36-8.

45.368. Crisp, M.D.: "Laser-Induced Surface Damage of Transparent Dielectrics", J. Quantum Electron. IEEE, 1974, QE-10, no.1, pp.57-62.

45.369. Bar-Isaac, C., and Korn, U.: "Moving Heat Source Dynamics in Laser Drilling Processes", Appl. Phys., 1974, 3, no.1, pp.45-54.

45.370. Bloembergen, N.: "Laser-Induced Electric Breakdown in Solids", J. Quantum Electron. IEEE, 1974, QE-10, no.3, pp.375-86.

45.371. Schumacher, B.W.: "Quasi-Adiabatic Melting and Vaporization Due to Radiation Beam of High Power Density", Optik, 1974, 39, no.5, pp.558-80.

45.372. Casperson, L.W., and Shekhani, M.S.: "Air Breakdown in Radial Mode Focusing Element", Appl. Opt., 1974, 13,no.1,pp.104-8.

45.373. Pohl, D.W., et al.: "Laser-Induced Phase Transition in Surface of SmS Crystals", Appl. Opt., 1974, 13, no.1, pp.95-7.

45.374. Yamanaka, C., et al.: "Brillouin Backscattering and Parametric Double Resonance in Laser-Produced Plasma", Phys. Rev. Lett., 1974, 32, no.19, pp.1038-41.

45.375. Raamot, J., and Zaleckas, V.J.: "Laser Pattern Generation Using X-Y Beam Deflection", Appl. Opt., 1974, 13, no.5, pp.1179-83.

45.376. Skolnik, L.H., Bendow, B., and Cross, E.F.: "IR Vidicon Technique for Measuring Thermal Lensing from Laser Windows", Appl. Opt., 1974, 13, no.4, pp.726-9.

45.377. Bendow, B., Skolnik, L.H., and Cross, E.F.: "Investigations of Laser-Induced Thermal Lensing and Interference from IR-Transmitting Materials", Appl. Opt., 1974, 13, no.4, pp.29-31.

45.378. O'Keefe, J.D., and Johnson, R.L.: "Optical Response of High-Power Laser Windows. Ultrashort Pulse Regime", Appl. Opt., 1974, 13, no.5, pp.1141-6.

45.379. Fried, D.L.: "Absence of Thermal Blooming for a Uniformly Illuminated Square-Aperture High-Power Laser Transmitter", Appl. Opt., 1974, 13, no.5, pp.989-91.

45.380. Fox, J.A.: "Effect of Pulse Shaping on Laser-Induced Spallation", Appl. Phys. Lett., 1974, 24, no.7, pp.340-3.

45.381. Humphries, S.: "Propagation of Laser Radiation Through a Long Plasma Column", Plasma Phys., 1974, 16,no.7,pp.623-34.

45.382. Elliott, C.J.: "Degradation of Laser Medium Due to Ionization by Laser Pulse", J. Appl. Phys., 1974, 45, no.1, pp.345-9.

45.383. Ivanov, L.I., et al.: "Investigation of Surface of Ge Crystals Irradiated by Q-Switched Lasers", Fiz. Khim. Obrab. Mater., 1974, no.1, pp.30-2.

45.384. Siebert, D.R., Grabiner, F.R., and Flynn, G.W.: "Time-Resolved Thermal Lensing Studies of Laser-Induced Translational Energy Fluctuations in CD_4, SO_2, and OCS", J. Chem. Phys., 1974, 60, no.4, pp.1564-74.

45.385. Hughes, J.L., and Strachan, J.D.: "Versatile Laser-CTR Pulse-Tailoring System Utilizing Rear Excitation of Pellets", Nucl. Fusion, 1974, 14, no.2, pp.292-6.

45.386. Chun, K.R.: "Surface Heating of Metallic Mirrors in High-Power Laser Cavities", Trans. ASME, 1974, 96C, no.1, pp.43-7.

45.387. Bukzdorf, N.V., et al.: "Laser-Induced Explosion of a Spherical Drop", Izv. VUZ Fiz., 1974, no.5, pp.36-40.

45.388. Chan, C.H., and Moody, C.D.: "Solutions of the Classical Boltzmann Equation for He and Ne Gas Breakdown (at 10.6 micron)", J. Appl. Phys., 1974, 45, no.3, pp.1105-11.

45.389. Hoffman, C.G.: "Laser-Target Interactions", J. Appl. Phys., 1974, 45, no.5, pp.2125-8.

45.390. Maher, W.E., Hall, R.B., and Johnson, R.R.: "Experimental Study of Ignition and Propagation of Laser-Supported Detonation Waves", J. Appl. Phys., 1974, 45, no.5, pp.2138-45.

45.391. Asahara, Y., and Izumitani, T.: "Effect of Irradiation of As_6Se_4 Glass", J. Non-Cryst. Solids, 1974, 15, no.2, pp.343-6.

45.392. Bernhardt, H.J., and Hultzsch, R.: "Ruby-Laser-Induced Changes of the Absorption Spectrum of Additively Coloured (LaEr):CaF_2", Phys. Status Solidi, 1974, 23, no.2, pp.K183-6.

45.393. Paek, U.C.: "Laser Drawing of Optical Fibres", Appl. Opt., 1974, 13, no.6, pp.1383-6.

45.394. Baev, V.M., Savchenko, A.N., and Sviridenkov, E.A.: "Investigation of Ruby Breakdown by a Train of Single Ultrashort Pulses", Zh. Eksp. Teor. Fiz., 1974, 66, no.3, pp.913-9, and Sov. Phys.-JETP, 1974, 39, no.3.

45.395. Ashkin, A., and Dziedzic, J.M.: "Stability of Optical Levitation by Radiation Pressure", Appl. Phys. Lett., 1974, 24, no.12, pp.586-8.

45.396. Szedny, A., and Wyrebski, W.: "Application of Pulse Lasers for Artificially Formed Clouds Investigation", Acta Geophys. Pol., 1974, 22, no.1, pp.65-73.

45.397. Annenkov, V.D., et al.: "Investigation of the Structure of Steel and Cast Iron in the Area of Influence of CW CO_2-Laser Radiation", Fiz. Khim. Obrab. Mater., 1974, no.2, pp.38-42.

45.398. Goryachev, N.S., et al.: "Investigation of the Increase in Hardness of Steels Exposed to Laser Radiation", Fiz. Khim. Obrab. Mater., 1974, no.2, pp.43-9.

45.399. Uglov, A.A., Kokora, A.N., and Orekhov, M.V.: "Movement of the Melt During the Piercing of a Hole in a Metallic Plate by a Laser Beam", Fiz. Khim. Obrab. Mater., 1974, no.2, pp.32-7.

45.400. Schwenn, U., and Sigel, R.: "Continuous Droplet Source for Plasma Production With Pulsed Lasers", J. Phys. E, 1974, 7, no.9, pp.715-8.

45.401. Soures, J., Kumpan, S., and Hoose, J.: "High-Power Nd^{3+}:Glass Laser for Fusion Applications", Appl. Opt., 1974, 13, no.9, pp.2081-94.

45.402. Schildbach, K., and Basting, D.: "Dye Laser Triggered Spark Gap", Rev. Sci. Instrum., 1974, 45, no.8, pp.1015-6.

45.403. Bar-Isaac, C., et al.: "Thermal Structure of the Evaporation Front in Laser Drilling Processes", Appl. Phys., 1974, 5, no.2, pp.121-5.

45.404. Hora, H.: "Nuclear Fusion by Lasers With Improved Nonlinear Compression of the Plasma", Laser Elektro-Opt., 1974, 6, no.1, pp.24-7.

45.405. Yoshida, K., et al.: "High-Power Glass Laser System (for Nuclear Fusion)", Technol. Rep. Osaka Univ., 1974, 24, no.1155-90, pp.83-93.

45.406. Demidov, B.A., et al.: "High-Voltage Water Spark Gap With Laser Firing", Prib. Tekh. Eksp., 1974, 17, no.1, pp.120-2, and Instrum. Exp. Tech., 1974, 17, no.1, pp.131-3.

45.407. Ohmori, Y., Yasojima, Y., and Inuishi, Y.: "Control of Optical Damage in Reduced $LiNbO_3$ by Applied Electric Field", Appl. Phys. Lett., 1974, 25, no.12, pp.716-7.

45.408. Sviridov, A.N., Tropikhin, Yu.D., and Kamenskii, A.G.: "Technological Laser Installation With Programme Control", Prib. Tekh. Eksp., 1974, 17, no.2, pp.260-1, and Instrum. Exp. Tech., 1974, 17, no.2, pp.606-8.

45.409. Little, V.I., Selden, A.C., and Stamatakis, T.: "Observation of Wedge Fringes on CO_2 Laser Optical Components", J. Phys. E, 1974, 7, no.12, pp.962-3.

45.410. Ghosh, A.K.: "Review on High-Power-Laser Damage to Materials. II", RCA Rev., 1974, 35, no.2, pp.279-319.

45.411. Cremosnik, G.: "Nuclear Fusion by Lasers", Tech. Rundsch., 1974, 66, no.10, pp.29-31.

45.412. Soileau, M.J., and Wang, V.: "Improved Damage Thresholds for Metal Mirrors", Appl. Opt., 1974, 13,no.6,pp.1286-8.

45.413. Fox, J.A.: "Effect of Water and Paint Coatings on Laser-Irradiated Targets", Appl. Phys. Lett., 1974, 24, no.10,pp.461-4.

45.414. Fradin, D.W., and Bua, D.P.: "Laser-Induced Damage in ZnSe", Appl. Phys. Lett., 1974, 24, no.11, pp.555-7.

45.415. Danileiko, Yu.K., et al.: "Optical Properties and Laser-Induced Destruction of Ideal Single-Crystal Ruby Surfaces", Fiz. Tver. Tela, 1974, 16, no.6, pp.1725-7, and Sov. Phys.-Solid State, 1974, 16, no.6, pp.1121-2.

46. BIOLOGICAL AND MEDICAL ASPECTS

46.1. Mintz, M., and Heimer, G.: "Techniques for Microwave Radiation Hazard Monitoring", Trans. IEEE, 1965, EMC-7, p.179.

46.2. Ketcham, A.S., and Minton, J.P.: "Laser Irradiation as a Clinical Tool in Cancer Therapy", Federation Proc., 1965, 24, p.159.

46.3. Minton, J.P., et al.: "Application of Pulsed, High-Energy, Laser Radiation to Multiple Inter-Abdominal Tumour Implants in Experimental Animals", Surgery, 1965, 58, p.12.

46.4. Minton, J.P., et al.: "Effect of Nd^{3+}-Laser Radiation on Two Experimental Malignant Tumour Systems", Surg. Gynec. Obstet., 1965, 120, p.481.

46.5. Campbell, C.J., et al.: "Clinical Studies in Laser Photocoagulation", Arch. Ophthal., 1965, 74, p.57.

46.6. Amy, R.L., and Storb, R.: "Selective Mitochondrial Damage by Ruby Laser Microbeam", Science, 1965, 150, p.757.

46.7. Saks, N.M., Zuzolo, R.C., and Kopac, M.J.: "Microsurgery of Living Cells by Ruby-Laser Irradiation", Ann. N.Y. Acad. Sci., 1965, 122, p.695.

46.8. Kinersly, T., et al.: "Laser Effects on Tissues and Materials Related to Dentistry", J. Am. Dental Assoc., 1965, 10, p.593.

46.9. Goldman, L., et al.: "Effect of Laser-Beam Impacts on Teeth", J. Am. Dental Assoc., 1965, 70, p.601.

46.10. Taylor, R., et al.: "Effects of Laser Radiation on Teeth, Dental Pulp, and Oral Mucosa of Experimental Animals", Oral Surg. Med. Pathol., 1965, 19, p.776.

46.11. Goldman, L., et al.: "Effect of Radiation from a 10-MW, Q-Switched, Laser on a Tattoo of a Man", J. Invest. Derm., 1965, 44, p.69.

46.12. Seto, T., and Pomerat, C.M.: "In Vitro Study of Somatic Chromosomes in Newts. Genus Taricha", Copeia, 1965, no.4, p.415.

46.13. Minton, J.P., and Ketcham, A.S.: "Utilization of Cyclophosphamide to Potentiate Tumour Destruction by Laser Energy", Surg. Forum, 1965, 16, p.113.

46.14. Davis, C.O., Smith, O., and Olander, J.: "Microwave Processing of Potato Chips. I-III", Potato Chipper, 1965, 25, Oct., p.38, Nov., p.72, and Dec., p.78.

46.15. Mullins, F., et al.: "Effect of High-Energy Laser Pulses on Primate Liver", Surg. Gynec. Obstet., 1966, 122, p.727.

46.16. Hume, R., Ketcham, A.S., and Minton, J.P.: "Light-Absorption Characteristics of Tumour Tissue at Ruby and Nd^{3+} Laser Wavelengths", J. Surg. Res., 1966, 6, p.531.

46.17. Spalter, H.F., et al.: "Prophylactic Photocoagulation of Recurrent Toxoplasmic Retinochoroiditis", Arch. Ophthal., 1966, 75, p.21.

46.18. Campbell, C.J., et al.: "Threshold of Retina to Damage by Laser Energy", Arch. Ophthal., 1966, 76, p.437.

46.19. Storb, R., et al.: "Electron-Microscope Study of Vitally Stained Single Cells Irradiated With Ruby-Laser Microbeam", J. Cell Biol., 1966, 33, p.11.

46.20. Gordon, T.E.: "Effects of Laser Impacts on Extracted Teeth", J. Dental Res., 1966, 45, p.372.

46.21. Jones, A.E., and McCartney, A.J.: "Ruby-Laser Effects on the Monkey Eye", Invest. Ophthal., 1966, 5, p.474.

46.22. Stern, R.H., et al.: "Laser Effect on In Vitro Enamel Permeability and Solubility", J. Am. Dental Assoc., 1966,73,p.838.

46.23. Goldman, L., Siler, V.E., and Blaney, D.: "Laser Therapy of Melanomas", Surg. Gynec. Obstet., 1967, 124, p.49.

46.24. Amy, R.L., et al.: "Ruby-Laser Micro-Irradiation of Single Tissue Culture Cells Vitally Stained With Janus Green B. I-III", Exp. Cell Res., 1967, 45, pp.361, 374, and 381.

46.25. Zweng, H.C., et al.: "Experimental Q-Switched Ruby Laser Retinal Damage", Arch. Ophthal., 1967, 78, p.634.

46.26. Goldman, L., et al.: "Laser Treatment of Tattoos", J. Am. Med. Assoc., 1967, 201, p.841.

46.27. Yules, R.B., et al.: "Effect of Q-Switched Ruby-Laser Radiation on Dermal Tattoo Pigment in Man", Arch. Surg., 1967, 95, p.179.

46.28. Wilson, R., Goldman, L., and Brech, F.: "Calcinosis Cutis With Laser Microprobe Analysis", Arch. Derm., 1967, 95, p.490.

46.29. Riggle, G.C., and Hoye, R.C.: "Effects of Laser Energy on Living Tissue", Int. Conv. Rec. IEEE, 1967, 15, pt 9, pp.94-6.

46.30. Kinersly, T., et al.: "Tooth Transillumination With Laser Radiation", Med. Biol. Illus., 1967, 17, p.5.

46.31. Hoye, R.C., et al.: "Potentiation of Laser Oncolysis With Pretreatment X-Irradiation", Radiat. Res., 1967, 32, p.112.

46.32. Wise, G.N., et al.: "Photocoagulation of Vascular Lesions of the Macula", Am. J. Ophthal., 1968, 66, p.452.

46.33. Rounds, D.E., Olson, R.S., and Johnson, F.M.: "Effect of Laser on Cellular Respiration", Z. Zellforsch., 1968, 87, p.193.

46.34. Solomon, H., et al.: "Histopathology of Laser Treatment of Portwine Lesions", J. Invest. Derm., 1968, 50, p.141.

46.35. Mullins, F., Jennings, B., and McClusky, L.: "Liver Resection With CW CO_2 Laser. Experimental Observations", Am. Surg., 1968, 34, p.717.

46.36. Gordon, T.E., et al.: "Laser Blockage or Delay of Cell Division at Prophase in Human Leukocyte Cultures", J. Dental Res., 1968, 47, p.171.

46.37. Cohen, E., Klein, E., and Fine, S.: "Effects of Laser Irradiation on Serologic Properties of Human gamma-Globulin", Life Sci., 1968, 7, p.569.

46.38. Tsou, K.C.: "Cytochemical Method for Selective Intracellular Damage With Ruby Laser", Life Sci., 1968, 7, p.10.

46.39. Zweng, H.C., Little, H.L., and Peabody, R.R.: "Laser Photocoagulation of Macula Lesions", Trans. Am. Acad. Ophthal. Otol., 1968, 72, p.377.

46.40. Landers, M.B., Kreiger, A., and Neidlinger, R.: "Ocular Hazards of Lasers", Laser Focus, 1968, 4, no.21, p.40.

46.41. Campbell, C.J., et al.: "Ocular Effects Produced by Experimental Lasers. I-IV", Am. J. Ophthal., 1968, 66, pp.459, 604, and 614, and 1969, 67, p.671.

46.42. Siler, V.E.: "Status of Laser in Medicine", Postgrad. Med., 1969, 46, pp.82-6.

46.43. Griffin, J.L., Stein, M.S., and Stowell, R.E.: "Laser Microscope Irradiation of Polysarum Polycephalum", J. Cell Biol., 1969, 40, p.108.

46.44. McKinnell, R.G., Mims, M.F., and Reed, L.A.: "Laser Ablation of Maternal Chromosomes in Eggs of Rana Pipiens", Z. Zellforsch., 1969, 93, p.30.

46.45. Goldman, L., et al.: "Preliminary Investigation of Fat Embolization from Pulsed Ruby-Laser Impacts of Bone", Nature, 1969, 221, p.361.

46.46. Harris, K.D.: "Aspects of Laser Safety", Proc. Inst. Mech. Eng., 1969, 183, pt 3D, p.43.

46.47. Lelieveld, H.L.M.: "Microwave Oven as Tool in Microbiology", Lab. Pract., 1969, 18, p.165.

46.48. Glick, D.: "Cytochemical Analysis by Laser Microprobe-Emission Spectroscopy", Ann. N. Y. Acad. Sci., 1969, 157, p.265.

46.49. Janes, D.E., et al.: "Effect of 2.45-GHz Radiation on Protein Synthesis and on Chromosomes in Chinese Hamsters", Non-Ioniz. Radiat., 1969, 1, no.3, pp.125-30.

46.50. Schwan, H.P.: "Effect of Microwave Radiation on Tissue", Non-Ioniz. Radiat., 1969, 1, no.1, pp.23-31.

46.51. Vucicevic, Z.M., et al.: "Cytochemical Approach to Laser Coagulation of Ciliary Body", Bibl. Ophthalmol., 1969, no.79, pp.467-78.

46.52. Berns, M.W., Olson, R.S., and Rounds, D.E.: "Effects of Laser Micro-Irradiation on Chromosomes", Exp. Cell Res., 1969, 56, p.292.

46.53. Mester, E., et al.: "Effect of Laser Irradiation on Catalase Activity of Leukocytes", Biol. Kozlem., 1969, 17, no.1, pp.11-6.

46.54. Lawrence, J.C.: "Effect of Pulsed X-Band Radiation on Skin Metabolism", Non-Ioniz. Radiat., 1969, 1, no.2, pp.80-4.

46.55. Edgerton, M.T., and McKnelly, L.O.: "Coherent Light in Biomedical Research", Plast. Reconstr. Surg., 1969, 43, pp.269-76.

46.56. Simakov, Yu.G., Poluektova, M.L., and Popov, V.V.: "Decreased Ca Content in Crystalline Lens Affected by Laser Radiation", Dokl. Akad. Nauk SSSR, 1969, 188, no.6, pp.1387-9.

46.57. Watts, G.K.: "Ruby-Laser Damage and Pigmentation of the Iris", Exp. Eye Res., 1969, 8, no.4, pp.470-6.

46.58. Kolar, J., Babicky, A., and Blabla, J.: "Effect of Laser on Bones", Experientia, 1969, 25, no.4, pp.365-6.

46.59. Webb, S.J., and Booth, A.D.: "Absorption of Microwaves by Microorganisms", Nature, 1969, 222, no.5199, pp.1199-200.

46.60. Bilbro, M.: "Safety Precautions for Ships' Radar", Pr. PIT, 1969,no.63,p.23.

46.61. Solem, D.L., et al.: "Measurements of Radiation Fields Near Microwave Ovens", Non-Ioniz. Radiat., 1969, 1, no.2, p.88.

46.62. Hamid, M.A.K., and Boulanger, R.J.: "Control of Moisture and Insect Infestations of Grain by Microwaves", J. Microwave Power, 1969, 4, no.1, pp.11-8.

46.63. May, K.N.: "Application of Microwaves to Preparation of Poultry", J. Microwave Power, 1969, 4, no.2, pp.54-9.

46.64. Bosisio, R.G., and Barthakur, N.: "Microwave Protection of Plants from Cold", J. Microwave Power, 1969, 4, no.3,pp.190-3.

46.65. Hamid, M.A.K., et al.: "Microwave Pasteurization of Raw Milk", J. Microwave Power, 1969, 4, no.4, pp.272-5.

46.66. Sluce, P.M.: "Determining Energy Distribution in a Microwave Oven", Non-Ioniz. Radiat., 1969, 1, p.131.

46.67. Fletcher, K., and Woods, D.: "Thin-Film Spherical Bolometer for Measurement of Hazardous Fields at 0.4-40 GHz", Non-Ioniz. Radiat., 1969, 1, no.2, p.57.

46.68. Taleff, M.E., et al.: "Laser Coagulation of Retina Using Argon-Ion Laser", Am. J. Ophthal., 1969, 67, p.666.

46.69. Birenbaum, L., et al.: "Effect of Microwaves on the Eye", Trans. IEEE, 1969, BME-16, p.7.

46.70. Berns, M.W., Olson, R.S., and Rounds, D.E.: "In Vitro Production of Chromosomal Lesions With Argon-Laser Microbeam", Nature, 1969, 221, p.74.

46.71. Lohmann, A.W.: "Possibility of Recording the Retinal After-Image on a Hologram", Opt. Commun., 1970, 1, no.6, p.303.

46.72. Deitz, P.H.: "(Laser) Safety Considerations in Outdoor Applications", Laser Focus, 1970, 6, no.6, pp.40-3.

46.73. Baillie, H.D.: "Thermal and Non-Thermal Cataractogenesis by Microwaves", Non-Ioniz. Radiat., 1970, 1, no.4,pp.159-63.

46.74. Baillie, H.D., Heaton, A.G., and Pal, D.K.: "Dissipation of Microwaves as Heat in the Eye", Non-Ioniz. Radiat., 1970, 1, no.4, pp.164-8.

46.75. Stellar, S., Polanyi, T.G., and Bredemeier, H.C.: "Studies With CO_2 Laser as Neurosurgical Instrument", Med. Biol. Eng., 1970, 8, no.6, pp.549-58.

46.76. Watts, G.K.: "Argon-Laser Irradiation of the Iris", Non-Ioniz. Radiat., 1970, 1, no.4, pp.155-8.

4646.77. Rogers, S.J., and King, R.S.: "Radio Hazards in Microwave Bands", Non-Ioniz. Radiat., 1970, 1, no.4, pp.178-89.

46.78. Ismailov, E.Sh.: "Effect of Microwaves on Erythrocyte Permeability for K and Na Ions. I-IV", Vop. Fiziol., 1970, no.4, pp.94-5, 96-7, 98-100, and 101-3.

46.79. Arfors, K.E., et al.: "Counteraction of Platelet Activity Using Laser-Induced Endothelial Trauma", Thromb. Diath. Haemorrh., 1970, no.42, pp.315-9.

46.80. Jesionowski, M.: "Effect of Laser Radiation on Dental Pulp in Rabbits", Pol. Med. J., 1970, 9, pp.468-74.

46.81. Cleary, S.F.: "Biological Effects of Microwave Radiation", Crit. Rev. Environ. Control, 1970, 1, pp.257-306.

46.82. Miyagawa, K., and Nishi, N.: "Microwave Cooking. Ascorbic Acid Retention in Vegetables", Osaka Shiritsu Daigaku Kaseigakubu Kiyo, 1970, 18, pp.15-8.

46.83. Berns, M.W., et al.: "Laser Photosensitization and Metabolic Inhibition of Tissue Culture Cells", Life Sci., 1970, 9, no.18, pp.1061-9.

46.84. Koldaev, V.M.: "Correlation Between Survival of Rats After Microwave Irradiation and Oxidative Processes in Muscle Tissues", Byull. Eksp. Biol. Med., 1970, 70, no.11, pp.69-70.

46.85. Makoc, Z., et al.: "Effect of Caffeine on Rat Resistance to Microwave-Generated Hyperthermia", Vojen. Zdrav. Listy., 1970, 39, no.5, pp.186-90.

46.86. Dietrich, W.C., et al.: "Microwaves for Blanching Corn-on-the-Cob", Food Technol., 1970, 24, no.3, pp.87-90.

46.87. Gordon, T.E., Waldron, C.A., and Gordon, L.S.: "Laser Effect on DMBA-Induced Dyskeratoses of Hamster Cheek Pouch", Cancer, 1970, 25, no.4, pp.851-7.

46.88. Kavetskii, R.E., et al.: "Intensification of Antineoplastic Action of Laser Radiation", Potol. Fiziol. Eksp. Ter., 1970, 14, no.3, pp.12-7.

46.89. Simakov, Yu.G., Poluektova, M.L., and Popov, V.V.: "Change in Lead Content in Lens Damaged by Laser Radiation", Biofizika, 1970, 15, no.3, pp.554-6.

46.90. Simakov, Yu.G., Poluektova, Yu.G., and Popov, V.V.: "Effect of Laser Radiation on Lipid Content in Lens of Rana Temporaria", Izv. Akad. Nauk SSSR, Ser. Biol., 1970, no.4, pp.609-10.

46.91. Berns, M.W., et al.: "Enzyme Inactivation With 265-nm Laser Radiation", Science, 1970, 169, no.3951, pp.1215-7.

46.92. Carney, S.A., Laurence, J.C., and Richetts, C.R.: "Effect of X-Band Radiation on Guinea-Pig Skin in Tissue Culture", Br. J. Ind. Med., 1970, 27, no.1, pp.72-6.

46.93. Gordon, T.E., and Smith, D.L.:
"Laser in the Dental (-Equipment) Laboratory", Laser Focus, 1970, 6, no.6, pp.37-9.

46.94. Jull, E.V.: "Bird Feathers as Dielectric-Rod Antennas", Bull. Radio Electr.
Eng. Div., 1970, 20, no.2, pp.28-9.

46.95. Fujii, Y.: "Eye Damage from Laser",
Electron. Commun. Jap., 1970, 53, no.7,
pp.973-6.

46.96. Michaelson, S.M.: "Pathophysiolo-
gical Aspects of Microwave Irradiation. I
and II", Non-Ioniz. Radiat., 1970, 1, no.4,
pp.169-76, and 1971, 2, no.1, pp.27-38.

46.97. Forster, G., and Muuss, H.: "Experiments Involving Laser Beams for Effect-
ing Joints in Skin and Bone Formations",
Schweissen Schneiden, 1971,23,no.2,pp.59-60.

46.98. Moore, R.L., et al.: "Comparison
of Microwave (Hazard) Power-Density Meters",
Non-Ioniz. Radiat., 1971, 2, no.1, pp.11-4.

46.99. Crapuchettes, P.W.: "Instrumentation for Microwave Leakage", Non-Ioniz.
Radiat., 1971, 2, no.1, pp.15-9.

46.100. Wing, R.W., and Alexander, J.C.:
"Heating of Soyabean Meals by Microwave
Radiation", Nutr. Rep. Int., 1971, 4, no.6,
pp.387-96.

46.101. Sidorik, E.P., and Danko, M.I.:
"Free-Radical Oxidative Processes in Liver
Under Effect of Nd^{3+}-Laser Radiation",
Dopov. Akad. Nauk Ukr. RSR, 1971, 33B,
no.11, pp.1001-5.

46.102. Cleary, S.F., and Hamrick, P.E.:
"Laser-Induced Retinal Damage", Non-Ioniz.
Radiat., 1971, 2, no.1, pp.1-10.

46.103. Vahl, J.: "Analysis of Laser Irradiated Dental Enamel and Filling Material", Laser Angew. Strahlentech., 1971, 3,
no.1, pp.41-4.

46.104. Nealeigh, R.C., et al.: "Effect
of Microwaves on Y-Maze Learning in the
White Rat", J. Microwave Power, 1971, 6,
no.1, pp.49-54.

46.105. Pronin, V.R.: "Recommendations
for Drawing Up Safety Rules for Lasers",
Kvantovaya Elektron., 1971, no.2, pp.87-91.

46.106. Kenyon, E.M., et al.: "Continuous Processing of Food Pouches With Microwaves", J. Food Sci., 1971, 36, no.2,
pp.289-93.

46.107. Rouse, A.H., and Moore, E.L.:
"Microwave Processing of Citrus Salad Gels
in Plastic Container", Proc. Fla. State
Hort. Soc., 1971, 84, pp.241-4.

46.108. Floyd, R.A., Keyhani, E., and
Chance, B.: "Membrane Structure and Function
(Ruby-Laser Pulses)", Arch. Biochem. Bio-
phys., 1971, 146, no.2, pp.627-34 and 618-26.

46.109. Berns, M.W., and Floyd, A.D.:
"Chromosomal Microdissection by Laser", Exp.
Cell Res., 1971, 67, no.2, pp.305-10.

46.110. Mester, E., et al.: "Use of AET
to Prevent Laser-Induced Changes in Intesti-
nal Mucosa", Kiserl. Orvostud., 1971, 23,
no.4, pp.398-402.

46.111. Mester, E., et al.: "Effect of
Laser Radiation on Haemoglobin Synthesis In
Vitro", Kiserl. Orvostud., 1971, 23, no.5,
pp.449-54.

46.112. Faitel'berg-Blank, V.R., and
Kutsenko, P.Ya.: "Effect of Microwaves on
Resorptive Activity of Alimentary Apparatus",
Fiziol. Zh., 1971, 17, no.6, pp.813-9.

46.113. O'Brien, C.K., et al.: "Histo-
pathologic Changes in Rat Liver After 2.45-
GHz Radiation", TIT J. Life Sci., 1971, 1,
no.1, pp.1-8.

46.114. Schiffman, R.F., et al.: "Application of Microwaves to Doughnut Production",
Food Technol., 1971, 25, no.7, pp.718-22.

46.115. Tota, J.G., Mester, E., and
Bertha, I.: "Temperature Changes in Blood
In Vitro After Ruby-Laser Irradiation",
Biol. Kozlem., 1971, 19, no.1, pp.9-16.

46.116. Schmidt, M.J., Schmidt, D.E.,
and Robison, G.A.: "Microwave Irradiation
as Means of Tissue Fixation", Science, 1971,
173, no.4002, pp.1142-3.

46.117. Chen, S.C., et al.: "Blanching of
White Potatoes by Microwaves", J. Food Sci.,
1971, 36, no.5, pp.742-3.

46.118. Zhokhov, V.P., et al.: "Biochemical Changes in Eye Tissue by Laser Radiation", Oftal'mol. Zh., 1971, 26, no.4,
pp.273-7.

46.119. Koldaev, V.M.: "Effect of Anti-
oxidant on Protein Metabolism After Microwave Irradiation", Vap. Kurortol., 1971, 36,
no.3, pp.246-8.

46.120. Sliney, D.H., Bason, F.C., and
Freasier, B.C.: "Measurement of (Laser) UV,
Visible, and IR Radiation (for Ocular
Hazards)", J. Am. Ind. Hyg. Assoc., 1971,
32, no.7, pp.415-31.

46.121. Sharp, J.C., and Paperiello,
C.J.: "Effect of Microwave Exposure on Thy-
idine Uptake in Albino Rats", Radiat. Res.,
1971, 45, no.2, pp.434-9.

46.122. Rubin, L.B., et al.: "Action of
Ruby Laser on Pigment Apparatus of Photosyn-
thesizing Organisms", Zh. Prikl. Spekt.,
1971, 14, no.1, pp.78-81.

46.123. Berns, M.W., et al.: "Chromosome
Lesions Produced With Argon Laser Without
Dye Sensitization", Science, 1971, 171,
no.3974, pp.903-4.

46.124. Matsin, S., Rounds, D.E., and
Olson, R.S.: "Effect of 265-nm Laser Radiation on Deoxyribonucleic Acid", Life Sci.,
1971, 10, no.4, pp.217-21.

46.125. Ismailov, E.Sh.: "Mechanism of Action of Microwaves on Erythrocyte Permeability for K and Na Ions", Biol. Nauki, 1971, 14, no.3, pp.58-60.

46.126. May, J.F., Rounds, D.E., and Cone, C.D.: "Intracellular Transfer of Toxic Components After Laser Irradiation", J. Natl. Cancer Inst., 1971, 46, no.3, pp.655-63.

46.127. Baranki, S.: "Effect of Microwave Irradiation on Central Nervous System of Rabbits and Guinea Pigs", Am. J. Phys. Med., 1972, 51, no.4, pp.182-91.

46.128. Kantola, S.: "Laser-Induced Effects on Tooth Structure. V", Acta Odontol. Scand., 1972, 30, no.4, pp.475-84.

46.129. Szwarc, B., Hamerski, W., and Wachowiak, B.: "Histochemical Studies on Effect of Laser Energy on Retinal Enzymes", Klin. Oczna, 1972, 42, no.5, pp.1161-4.

46.130. Sidorik, E.P., Baglei, E.A., and Danko, M.I.: "Antioxidation Activity of Liver Lipids During Exposure to Nd^{3+}-Laser Irradiation", Visn. Akad. Nauk Ukr. RSR, 1972, no.40, pp.41-3.

46.131. Milroy, W.C., and Michaelson, S.M.: "Thyroid Pathophysiology of Microwave Radiation", Aerosp. Med., 1972, 43, no.10, pp.1126-31.

46.132. Schmidt, M.J., Schmidt, D.E., and Robison, G.A.: "Cyclic AMP in Rat Brain. Microwave Irradiation for Tissue Fixation", Adv. Cyclic Nucleotide Res., 1972, 1, pp.425-34.

46.133. Boyanov, B., Kulikov, I., and Aleksieva, K.: "Effect of Laser Radiation Upon Stomatological Materials", Stomatologiya, 1972, 54, no.5, pp.399-403.

46.134. Olaru, M., and Csath, Z.: "Bacteriological Control of Pharmaceutical Emulsions by Laser Radiation Scattering", Rev. Med., 1972, 18, no.4, pp.468-71.

46.135. Dubin, S.B.: "Measurement of Diffusion Coefficients (in Enzymology) by Laser Light Scattering", Methods Enzymol., 1972, 26C, pp.119-74.

46.136. Zyss, R., and Boczynski, E.: "Morphological Changes in Cells of Corti's Organ Following Exposure to Microwaves", Otolaryngol. Pol., 1972, 26, no.4,pp.399-406.

46.137. Boczynski, E., and Zyss, R.: "Changes in Enzyme Activity in Cells of Corti's Organ in Guinea Pigs After Long-Term Exposure to Microwaves", Otolaryngol. Pol., 1972, 26, no.4, pp.407-13.

46.138. Krendeleva, T.E., et al.: "Selective Damage of Pea Chloroplasts by Ruby Laser Radiation", Nature, 1972, 24,no.102,pp.223-4.

46.139. Hamerski, W.: "Effect of Laser Energy on Cornea", Klin. Oczna, 1972, 42, 1a, pp.323-8.

46.140. Mester, E., et al.: "Effect of Laser Irradiation on Phagocytosis of Leukocytes Treates With Acridine Orange", Kiserl. Orvostud., 1972, 24, no.4, pp.391-6.

46.141. Sakharov, V.N.: "UV and Local Laser Irradiation Study of a Living Cell", Usp. Sovrem. Biol., 1972, 73,no.2,pp.231-49.

46.142. Bark, K.R., et al.: "Killing and Preserving Nematodes in Soil Samples With Chemicals and Microwave Energy", J. Nematol., 1972, 4, no.2, pp.75-9.

46.143. Zubkova, S.M., Zhuravlev, A.I., and Abramov, V.M.: "Effect of Microwaves on Intensity of Luminescence of Bacteria", Tr. Mosk. Obshchest. Ispyt. Prir., 1972, 39, pp.116-20.

46.144. Mester, E., et al.: "Laser-Induced Modifications of Phagocytosis in Rat Leukocytes", Kiserl. Orvostud., 1972, 24, no.2, pp.150-5.

46.145. Rosen, C.G.: "Effect of Microwaves on Food and Related Materials", Food Technol., 1972, 26, no.7, pp.36-40.

46.146. Michaelson, S.M.: "Biological Effects of Microwave Exposure. Overview", J. Microwave Power, 1972, 6, no.3, pp.259-67.

46.147. Koldaev, V.M.: "Effect of Microwaves on Rats Subjected to Action of Gaseous Media", Patol. Fiziol. Eksp. Ter., 1972, no.2, pp.71-3.

46.148. Jaszagi-Nagy, E., Mester, E., and Jolan, G.T.: "Biological Effects of Laser Irradiance", Kiserl, Orvostud., 1972, 24, no.3, pp.225-32.

46.149. Vozelj, M., Rajver, I., and Vrenko, E.: "Laser-Induced Excitation of Fluorescein Isothiocyanate in Immunofluorescence", Experientia, 1972, 28, no.9, pp.1098-9.

46.150. Aref, M.M., Noel, J.G., and Miller, H.: "Inactivation of alpha-Amylase in Wheat Flour With Microwaves", J. Microwave Power, 1972, 7, no.3, pp.215-21.

46.151. Schmidt, M.J., et al.: "Cyclic AMP in Brain Areas", Brain Res., 1972, 42, no.2, pp.465-77.

46.152. Schmidt, D.E., et al.: "Use of Microwaves in Determination of Acetylcholine in Rat Brain", Brain Res., 1972, 38, no.2, pp.377-89.

46.153. Vilenskaya, R.L., et al.: "Induction of Colicin Synthesis With Millimetre Waves", Byull. Eksp. Biol. Med., 1972, 73, no.4, pp.52-4.

46.154. Krendeleva, T.E., et al.: "Action of Ruby Laser on Primary Biochemical Photosynthesis Reactions", Biokhimiya, 1972, 37, no.1, pp.158-62.

46.155. Liben, W., Hochheimer, B.F., and Patz, A.: "Argon-Laser Photocoagulator", APL Tech. Dig., 1972, 11, no.3, pp.2-14.

46.156. Ovchinnikov, B.V., and Gan'kovskaya, V.A.: "Eye Protection in Laser Operation", Opt.-Mekh. Prom., 1972, 39, no.6, pp.37-8, and Sov. J. Opt. Technol., 1972, 39, no.6, pp.348-9.

46.157. Hennessy, R.T., and Leibowitz, H.W.: "Laser Optometer (Eye Refraction) Incorporating the Badal Principle", Behav. Res. Method, 1972, 4, no.5, pp.237-9.

46.158. Mohon, N., and Rodemann, A.: "Laser Speckle for Determining Ametropia and Accomodation Response of the Eye", Appl. Opt., 1973, 12, no.4, pp.783-7.

46.159. Breitwieser, P., et al.: "CO$_2$-Laser as a Surgical Instrument in Experimental Urology", Biomed. Tech., 1973, 18, no.1, pp.6-13.

46.160. Goldman, L., et al.: "Studies in Laser Safety of High-Output Systems. II. TEA CO$_2$ Laser Impacts", Opt. Laser Technol., 1973, 5, no.2, pp.58-9.

46.161. Eleccion, M.: "Laser Hazards", Spectrum IEEE, 1973, 10, no.8, pp.32-8.

46.162. Mahlein, H.F.: "Scientific Foundations of Laser Safety Regulations", Atomkernenergie, 1973, 21, no.4, pp.262-8.

46.163. Rubin, L.B., Khoklov, R.V., and Pashchenko, V.Z.: "Use of Lasers in Biophotophysical Investigations", Tr. Mosk. Obshchest. Ispyt. Prir., 1973,49,pp.258-64.

46.164. Mester, E., and Jaszsagi-Nagy, E.: "Effect of Laser Radiation on Wound Healing Collagen Synthesis", Stud. Biophys., 1973, 35, no.3, pp.227-30.

46.165. Putenney, D.G., Born, G.S., and Vetter, R.J.: "Effect of Microwaves on Photosensitizing Agents in the Albino Rabbit", Environ. Lett., 1973, 5,no.1,pp.21-7.

46.166. Alekaev, N.S.: "Effect of (Microwave) Heat Treatment on Food Values of Rice Proteins", Vop. Pitan., 1973,no.3,pp.68-73.

46.167. Schiller, E.A., Pratt, D.E., and Reber, E.F.: "Lipid Changes in Egg Yolks and Cakes Baked in Microwave Ovens", J. Am. Diet. Assoc., 1973, 62, no.5, pp.529-33.

46.168. Parker, L.N.: "Thyroid Suppression and Adrenomedullary Activation by Low-Intensity Microwave Radiation", Am. J. Physiol., 1973, 224, no.6, pp.1388-90.

46.169. Nelson, S.R.: "Effects of Microwave Irradiation on Enzymes and Metabolites in Mouse Brain", Radiat. Res., 1973, 55, no.1, pp.153-9.

46.170. Stavinoka, W.B., Weintraub, S.T., and Modak, A.T.: "Microwave Heating to Inactivate Cholinesterase in Rat Brain", J. Neurochem., 1973, 20, no.2, pp.361-71.

46.171. Rabinowitz, J.R.: "Mechanism of Biomolecular Absorption of Microwave Radiation", Trans. IEEE, 1973, MTT-21, no.12, pp.850-1.

46.172. Kantola, S.: "Laser-Induced Effects on Tooth Structure. VI and VII", Acta Odontol. Scand., 1973, 31, no.6, pp.369-79 and 381-6.

46.173. Drapeau, A.J., and Thank, L.V.: "Laser Application to Study of Photosynthetic Mechanism. I-III", Annee Biol., 1973, 12, no.5-6, pp.193-201 and 202-8, and no.11-12, pp.525-33.

46.174. Ehlers, G., and Florian, H.J.: "Cytophotometric Study of Carcinogenic Effect of Ruby Laser Light", Hautarzt, 1973, 24, no.10, pp.423-8.

46.175. Kozlov, A.P., et al.: "Antitumour Effect of Laser Radiation", Acta Radiol., 1973, 12, no.3, pp.241-56.

46.176. Swartz, H.M., et al.: "Survey of Clinical Application of ESR", Ann. N. Y. Acad. Sci., 1973, 222, pp.989-1009.

46.177. Maksimova, L.I.: "Effect of Microwaves on Catecholamine Metabolism in Healthy Rabbits", Vop. Kurotol., 1973,no.6,pp.490-3.

46.178. Ivanov, A.V., Ganago, A.O., and Rubin, L.B.: "Mechanism of Selective Damage of Pea Chloroplasts After Ruby-Laser Irradiation", Biofizika, 1973, 18, no.6, pp.1117-9.

46.179. Matolin, S.: "Regulation of Embryogenesis in Pyrrhocoris Apterus With Sterilants and Laser Beam", Acta Entomol. Bohemoslov., 1973, 70, no.4, pp.225-37.

46.180. Zeman, G., et al.: "Aminobutyric Acid Metabolism in Rats Following Microwave Exposure", J. Microwave Power, 1973, 8, no.3-4, pp.213-6.

46.181. Hamid, M.A.K., and Badour, S.S.: "Effect of Microwaves on Green Algae", J. Microwave Power, 1973, 8, no.3-4,pp.267-73.

46.182. Hamrick, P.E.: "Thermal Denaturation of DNA Exposed to 2.45-GHz CW Radiation", Radiat. Res., 1973, 56,no.2,pp.400-4.

46.183. Shishlo, M.A., Korolev, Yu.N., and Popov, V.I.: "Oxidative and Energy Processes in Animal Liver After Exposure to Microwaves", Vop. Kurortol., 1973, no.5, pp.414-20.

46.184. Mayers, C.P., and Habeshane, J.A.: "Depression of Phagocytosis by Microwave Radiation", Int. J. Radiat. Biol., 1973, 24, no.5, pp.449-61.

46.185. Holst, G.C.: "Proper Selection and Testing of Laser Protective Materials", Am. J. Optom., 1973, 50, no.6, pp.466-83.

46.186. Hannah, S., Hartnagel, H., and Kennair, J.T.: "Continuous Monitor of Dangerous Levels of Microwave Power", Electron. Lett., 1974, 10, no.14, pp.274-6.

46.187. Gandhi, O.P.: "Polarization and Frequency Effects on Whole Animal Absorption of RF Energy", Proc. IEEE, 1974, 62, no.8, pp.1171-5.

46.188. Edrich, J., and Hardee, P.C.: "Thermography (of Human Body) at Millimetre Wavelengths", Proc. IEEE, 1974, 62, no.8, pp.1184-6.

46.189. Lin, J.C.: "Cavity-Backed Slot Radiator for Microwave Biological-Effect Research", J. Microwave Power, 1974, 9, no.2, pp.63-7.

46.190. Elder, R.L.: "Laser Protective Eyewear", Appl. Opt., 1974, 13, no.4, p.725.

46.191. Ruegg, W., Willi, W., and Helfenstein, W.: "Mossbauer and Laser Interferometry Studies of the Inner Ear of Cats", Helv. Phys. Acta, 1974, 47, no.1, p.61.

46.192. McAfee, R.D., Cazenavette, L.L., and Shubert, H.A.: "Thermistor Probe Error (in NaCl Solutions) in X-Band Fields", J. Microwave Power, 1974, 9, no.3, pp.177-80.

46.193. Moskalik, K.G., Skachkov, A.P., and Il'chenko, A.M.: "Effect of Laser Radiation on Reactivity of Animals With Tumours", Patol. Fiziol. Eksp. Ter., 1974, no.1, pp.76-8.

46.194. Recht, P., and Jolivet, A.: "Radioprotection and Utilization of Laser Sources", J. Belge Radiol., 1974, 57, no.2, pp.117-28.

46.195. Nicholls, D.M., Retryshyn, R., and Warner, L.: "Laser Irradiation Induces Increased Activity of Liver Elongation Factor", Radiat. Res., 1974, 60, no.1, pp.98-107.

46.196. Couch, R., and Gangstad, E.O.: "Response of Water Hyacinth to Laser Radiation", Weed Sci., 1974, 22, no.5, pp.450-3.

46.197. Butcher, S.G., and Butcher, L.L.: "Acetylcholine and Choline Levels in Rat Corpus Striatum After Microwave Irradiation", Proc. West. Pharmocol. Soc., 1974,17,pp.37-9.

46.198. Stelmachow, J.: "Effects (on Chicken Embryos) of Repeated Exposure to Microwaves", Lek. Wojsk., 1974, 50, no.3, pp.190-4.

46.199. Wolbarsht, M.L.: "Laser Safety", Laser Unconv. Opt. J., 1974,no.53,pp.3-13.

46.200. Sliney, D.H.: "Laser Safety Update. LDI Report", Electro-Opt. Syst. Des., 1974, 6, no.11, pp.28-30.

46.201. Corker, G.A., and Sharpe, S.A.: "Kinetics of Photo-Induced EPR Signal in Whole-Cell Rhodospirillum Rubrum", Photochem. Photobiol., 1974, 19, no.6, pp.443-55.

46.202. Faitel'berg-Blank, V.R., and Shenkerman, E.D.: "Methionine Absorption in Gastrointestinal Tract of Chicks Under Effect of Microwaves", Fiziol. Zh., 1974, 20, no.3, pp.379-85.

46.203. Faitel'berg-Blank, V.R., and Kharim, M.P.: "Penetrability of Sinus Frontalis Mucosa Under Effect of Microwaves", Fiziol. Zh., 1974, 20, no.1, pp.100-7.

46.204. Belkhode, M.L., Muc, A.M., and Johnson, D.L.: "Effects of 2.8-GHz Radiation on Three Human Serum Enzymes", J. Microwave Power, 1974, 9, no.1, pp.23-9.

46.205. Kolaev, V.M.: "Effect of Ephedrine and Cordiamine on Outcome of Microwave Injury to Mice", Byull. Eksp. Biol. Med., 1974, 77, no.3, pp.79-81.

46.206. Metaxas, A.C.: "Design of TM_{010} Resonant Cavity as Heating Device at 2.45 GHz", J. Microwave Power, 1974, 9, no.2, pp.123-8.

46.207. Belkhode, M.L., Johnson, D.L., and Muc, A.M.: "Effects of Microwave Radiation on Activity of Glucose in Human Blood", Health Phys., 1974, 26, no.1, pp.45-51.

46.208. Luskow, A.A.: "Precision Attenuators for Microwave Frequencies", Marconi Instrum., 1974, 14, no.4, pp.81-6.

47. ELECTRON ACCELERATORS

47.1. Jarvis, T.R., Saxon, G., and Crowley-Milling, M.C.: "Experimental Observations of Pulse Shortening in a Linac Waveguide", Proc. IEE, 1965, 112, p.1795.

47.2. Purser, K.H., et al.: "Properties of Inclined-Fields Acceleration Tubes", Rev. Sci. Instrum., 1965, 36, p.453.

47.3. Bjorkholm, J.E., and Hyneman, R.E.: "Analysis of TM_{11}-Mode Beam-Blowup in a Linac", Trans. IEEE, 1965, ED-12, p.281.

47.4. Rogenhagen, H., and Haberstock, F.: "Calculation of Electron Tracks in the Microtron", Z. Angew. Phys., 1965, 19, p.5.

47.5. Ostrovskii, E.K.: "Obtaining Short Electron Bunches in a Linac", Zh. Tekh. Fiz., 1965, 35, p.290, and Sov. Phys.-Tech. Phys., 1965, 10, no.2, pp.232-4.

47.6. Makhnenko, L.A., Zykov, A.I., and Kramskoi, G.D.: "Calculation of the Field in a TW Linac", Zh. Tekh. Fiz., 1965, 35, p.496, and Sov. Phys.-Tech. Phys., 1965, 10, no.3, pp.384-8.

47.7. Sells, V., Froelich, H., and Brannen, E.: "Energy Spectrum of the Electron Beam in a Racetrack Microton", J. Appl. Phys., 1965, 36, p.3264.

47.8. Zykov, A.I., Makhnenko, L.A., and Kramskoi, G.D.: "Determination of Equivalent Coefficient of Reflection for the Iris Waveguide of a Linac", Zh. Tekh. Fiz., 1965, 35, p.508, and Sov. Phys.-Tech. Phys., 1965, 10, no.3, pp.393-5.

47.9. Zykov, A.I., and Makhnenko, L.A.: "Calculation of Phase Velocity in the Iris Waveguide of a Linac", Zh. Tekh. Fiz., 1965, 35, p.489, and Sov. Phys.-Tech. Phys., 1965, 10, no.3, pp.379-83.

47.10. Makhnenko, L.A., Pakhomov, V.I., and Stepanov, K.N.: "High-Frequency Focusing in a Linac", Zh. Tekh. Fiz., 1965, 35, p.618, and Sov. Phys.-Tech. Phys., 1965, 10, no.4, pp.486-9.

47.11. Zykov, A.I., and Makhnenko, L.A.: "Calculation of the Group Velocity and Damping of Waves in the Iris Waveguide of a Linac", Zh. Tekh. Fiz., 1965, 35, p.502, and Sov. Phys.-Tech. Phys., 1965, 10, no.3, pp.389-92.

47.12. Haimson, J.: "Absorption and Generation of Power in Electron Linac Systems", Nucl. Instrum. Methods, 1965, 33, p.93.

47.13. Wilson, P.B., and Schwettman, H.A.: "Superconducting Accelerators", Trans. IEEE, 1965, NS-12, p.1045.

47.14. Morris, A.J., and Martin-Vegue, C.A.: "Comparison of Triodes and Klystrons for Particle Accelerator Applications", Trans. IEEE, 1965, NS-12, p.96.

47.15. Ruden, T.E.: "Amplitron as a High-Power, Efficient, RF Power Source for Long-Pulse, High-Resolution, Linacs", Trans. IEEE, 1965, NS-12, p.169.

47.16. Haimson, J.: "Injector and Waveguide Design Parameters for a High-Energy, Electron-Positron, Linac", Trans. IEEE, 1965, NS-12, p.499.

47.17. Leiss, J.E.: "Beam Loading in Linacs", Trans. IEEE, 1965, NS-12, p.566.

47.18. Wheeler, G.W., and Giordano, S.: "RF Structures for Linacs", Trans. IEEE, 1965, NS-12, p.110.

47.19. Giordano, S.: "RF Accelerating Structures", Trans. IEEE, 1965, NS-12, p.213.

47.20. Haimson, J.: "High-Current TW Electron Linacs", Trans. IEEE, 1965, NS-12, p.996.

47.21. Winter, S.D.: "Waveguide Resonant-Ring Electron Accelerator", Trans. IEEE, 1965, NS-12, p.494.

47.22. Kanter, B.Z.: "Stability of the Parameters of Accelerated Beams in Microtrons", Prib. Tekh. Eksp., 1965, no.3, p.34, and Instrum. Exp. Tech., 1965, no.3, p.496.

47.23. Haimson, J.: "Electron Bunching in TW Linacs", Nuclear Instrum. Methods, 1966, 39, p.13.

47.24. Hirel, R.: "Study of a System of Bunching of Electrons at the Input of a Linac and of the Harmonic Coefficients of the Resonant Cavities", Ann. Phys. (Paris), 1965, 10, p.623.

47.25. Luccio, A.: "Proposal for a High-Intensity Microtron", Nuovo Cim., 1966, 42, p.376.

47.26. Zverev, B.V., and Sobenin, N.P.: "Tuning Circular Iris Waveguides of Linacs by Resonance Method", Prib. Tekh. Eksp., 1965, no.5, p.26, and Instrum. Exp. Tech., 1966, no.5, p.1039.

47.27. Kolomenskii, A.A., and Lebedev, A.N.: "Quasilinear Acceleration of Particles by a Transverse Electromagnetic Wave", Zh. Eksp. Teor. Fiz., 1966, 50, p.1101, and Sov. Phys.-JETP, 1966, 23, no.4.

47.28. Combe, R., and Caillaud, A.: "Superconducting Cavities for Microtrons", C. R. Acad. Sci. (Paris), 1966, 263B, p.131.

47.29. Bernard, J., Combe, R., and Morignot, P.: "Dispersion Formula for Waveguides Loaded With Irises With Rounded Internal Edges", C. R. Acad. Sci. (Paris), 1966, 263B, p.1010.

47.30. Verbitskii, I.L.: "Finite Motion of Charged Particles in the Field of a Travelling Wave", Zh. Tekh. Fiz., 1966, 36, no.6, pp.1065-74, and Sov. Phys.-Tech. Phys., 1966, 11, no.6.

47.31. Anan'ev, V.D., et al.: "30-MeV Microtron Injector for Pulsed Fast-Neutron Reactor", Plasma Phys., 1966, 8, no.6, pp.711-5.

47.32. Miller, D.B.: "Experimental X-Band Electron-Cyclotron-Resonance Plasma Accelerator", Trans. IEEE, 1966, MTT-14, no.3, pp.162-4.

47.33. Aleksandrov, I.A., Vagin, V.A., and Kotov, V.I.: "Waves With Complex Propagation Coefficients in an Iris Waveguide. I and II", Zh. Tekh. Fiz., 1966, 36, pp.1995 and 2002, and Sov. Phys.-Tech. Phys., 1967, 11, no.11, pp.1486-90 and 1491-7.

47.34. Grishaev, I.A., and Shenderovich, A.M.: "Loading of an Electron Linac With a Beam in the Transition Regime", Zh. Tekh. Fiz., 1966, 36, no.11, pp.2013-6, and Sov. Phys.-Tech. Phys., 1967, 11, no.11.

47.35. Vishnyakov, V.A., Grishaev, I.A., and Zykov, A.I.: "Influence of Beam Loading on Current Pulse Duration in a Multisection Linac", Zh. Tekh. Fiz., 1966, 36, no.11, p.201, and Sov. Phys.-Tech. Phys., 1967, 11, no.11.

47.36. Dem'yanenko, G.K., Kolesnikov, L.Ya., and Miroshnichenko, I.I.: "Radial Shaping of a Beam in an Electron Linac", Zh. Tekh. Fiz., 1967, 37, no.7, pp.1219-24, and Sov. Phys.-Tech. Phys., 1967, 12, no.7.

47.37. Gallagher, W.J.: "Design of TW Electron Linacs", Trans. IEEE, 1967, NS-14, no.3, pp.282-5.

47.38. Speciale, R.A.: "Accelerating Systems With Ring-Resonator Configuration", Nuclear Instrum. Methods, 1967, 55, no.2, pp.205-37.

47.39. Rosander, S.: "Two-Frequency Microtron", Nuclear Instrum. Methods, 1967, 56, no.1, pp.154-6.

47.40. Babic, H., and Sedlacek, M.: "Method for Stabilizing Particle Orbits in the Race-Track Microtron", Nuclear Instrum. Methods, 1967, 56, no.1, pp.170-2.

47.41. Nunan, C.S.: "Microwave Electron Accelerators in the Medium Energy Range", Trans. IEEE, 1967, NS-14, no.3, pp.1174-85.

47.42. Marcou, J., Papiernik, A., and Wartski, L.: "Experimental Study of the Transient Response of an Iris-Loaded Accelerator Cavity", C. R. Acad. Sci. (Paris), 1967, 265B, no.3, pp.172-5.

47.43. Gos'kov, P.I.: "Investigation of Double-Period Waveguide Delay System (for Iron-Free Synchrotrons)", Izv. VUZ Radiofiz., 1967, no.5, pp.152-4.

47.44. Wiik, B.H., and Wilson, P.B.: "Design of a High-Energy, High-Duty-Cycle, Racetrack Microtron", Nuclear Instrum. Methods, 1967, 56, no.2, pp.197-208.

47.45. Kreindel', Yu.E., et al.: "Microtron With Plasma Injector", Zh. Eksp. Teor. Fiz. Pis'ma, 1968, 8, p.361, and JETP Lett., 1968, 8, p.223.

47.46. Nakamura, M.: "Effect of Nonuniformity in Disc-Loaded Waveguide of Electron Linacs", Jap. J. Appl. Phys., 1968, 7, no.2, pp.156-62.

47.47. Davies, J.B., and Goldsmith, B.J.: "Analysis of Mode Propagation and Pulse Shortening in Electron Linacs", Philips Res. Rep., 1968, 23, no.2, pp.207-32.

47.48. Takeda, Y., and Matsui, I.: "Laser Linac With Grating", Nuclear Instrum. Methods, 1968, 62, no.3, pp.306-10.

47.49. Hirel, R.: "Application of Superconductivity to Particle Accelerators", Onde Electr., 1968, 48, no.6, pp.595-600.

47.50. Passow, C.: "Problems in the Construction of Superconducting Linacs", Elektro-Tech. Z., 1968, 89, no.14, pp.341-6.

47.51. Gos'kov, P.I.: "Application of H-Shaped Retarding Systems in Waveguide Synchrotrons", Izv. VUZ Fiz., 1968, no.3, pp.148-50.

47.52. Knapp, E.A., Knapp, B.C., and Potter, J.M.: "Standing-Wave, High-Energy, Linac Structures", Rev. Sci. Instrum., 1968, 39, no.7, pp.979-91.

47.53. Gos'kov, P.I.: "Sectionalizing the Accelerating Systems of Waveguide Synchrotrons", Izv. VUZ Fiz., 1968, no.9, p.139.

47.54. Snedkov, B.A., and Galkov, V.A.: "Harmonic Composition of the Current in a Resonator Buncher, Allowing for Space Charge", Zh. Tekh. Fiz., 1968, 38, p.1317, and Sov. Phys.-Tech. Phys., 1969, 13, no.8, pp.1078-80.

47.55. Vishnyakov, V.A., et al.: "Current-Pulse Shortening in a Double-Section Linac", Zh. Tekh. Fiz., 1968, 38, no.1, pp.133-8, and Sov. Phys.-Tech. Phys., 1969, 13, no.1.

47.56. Vishnyakov, V.A., and Ostrovskii, E.K.: "Dynamics of Particles in a Waveguide Bunching System", Zh. Tekh. Fiz., 1968, 38, no.1, pp.139-42, and Sov. Phys.-Tech. Phys., 1969, 13, no.1.

47.57. Zykov, A.I., et al.: "Investigation of Effect of Current-Pulse Shortening in Multisection Linacs", Zh. Tekh. Fiz., 1968, 38, no.1, pp.129-32, and Sov. Phys.-Tech. Phys., 1969, 13, no.1.

47.58. Vishnyakov, V.A., et al.: "Increasing (by EH_{11}-Mode Elimination) the Limiting Current in a Linac", Zh. Tekh. Fiz., 1969, 39, no.2, p.321, and Sov. Phys.-Tech. Phys., 1969, 14, no.2.

47.59. Gallagher, W.J.: "Standing-Wave Operation of Electron Linacs", Trans. IEEE, 1969, NS-16, p.321.

47.60. Norris, N.J., and Hanst, R.K.: "Velocity-Modulation System for Enhancement of 50-ps Radiation Pulses", Trans. IEEE, 1969, NS-16, p.323.

47.61. Jameson, R.A., Hoffert, W.J., and Morris, D.I.: "Microwave Instrumentation for Accelerator Systems", Trans. IEEE, 1969, NS-16, p.367.

47.62. Lewin, J.D., Smith, P.F., and Spurway, A.H.: "Work on Superconducting Synchrotrons", Trans. IEEE, 1969, NS-16, p.715.

47.63. Toda, T., Irie, K., and Nemoto, Y.: "Measurement of Phase Bunching in Electron Linacs by RF Deflector", Jap. J. Appl. Phys., 1969, 8, no.12, p.1535.

47.64. Kaminskaya, R.G., et al.: "One-Resonator Electron Accelerator", Prib. Tekh. Eksp., 1969, no.2, p.23, and Instrum. Exp. Tech., 1969, no.2, p.291.

47.65. Grizhko, V.M., et al.: "Excitation of Hybrid Modes of a Relativistic Electron Beam in an Iris Waveguide", Zh. Tekh. Fiz., 1969, 39, no.9, p.1646, and Sov. Phys.-Tech. Phys., 1970, 14, no.9.

47.66. Reich, H., Lons, K., and Feist, H.: "Influence of Cavity Gap Length on Phase Stability of Microtrons", Z. Angew. Phys., 1970, 28, no.4, p.244.

47.67. Stapleton, R.E., Gritzo, L.A., and Venable, D.: "Beam Loading in a High-Current Standing-Wave Electron Accelerator", J. Appl. Phys., 1970, 41, p.82.

47.68. Bornard, P., Lengeler, H., and Vagin, V.A.: "Experimental Study on a Circularly Polarized Deflecting Mode in a Disc-Loaded Waveguide", Nuclear Instrum. Methods, 1970, 82, p.321.

47.69. Ostrovskii, E.K., and Zykov, A.I.: "Measurement of Phase Characteristics of Buncher in an Electron Linac", Prib. Tekh. Eksp., 1970, no.3, pp.48-50, and Instrum. Exp. Tech., 1970, no.3, pp.683-5.

47.70. Kul'man, V.G., et al.: "Accelerating Structures With Ring-Coupling Resonators", Prib. Tekh. Eksp., 1970, no.4, pp.56-61, and Instrum. Exp. Tech., 1970, no.4, pp.1020-4.

47.71. Haag, H.W., and Senkowski, E.: "9-MeV Microtron for C-Band Operation", Z. Angew. Phys., 1970, 30, no.2-3, pp.201-7.

47.72. Fast, R.W., et al.: "Superconducting Two-Tesla Bending Magnet", Part. Accel., 1970, 1, no.3, pp.265-8.

47.73. Wilson, P.B., et al.: "Superconducting Accelerator Research and Development", Part. Accel., 1970, 1, no.3, pp.223-38.

47.74. Allen, J.S., et al.: "Design of 600-MeV Microtron Using a Superconducting Linac", Part. Accel., 1970, 1, no.3, pp.239-45.

47.75. Haag, H.W.: "Capture of Field Electrons Into the Phase-Stable Acceleration of a C-Band Microtron", Z. Angew. Phys., 1970, 30, no.4, pp.299-304.

47.76. Geissler, W., and Haag, H.W.: "Calculation of Phase-Acceptance Ranges of a C-Band Microtron", Z. Angew. Phys., 1970, 30, no.4, pp.304-9.

47.77. Melekhin, V.N., and Luganskii, L.B.: "Power Instability in a Microtron", Zh. Tekh. Fiz., 1970, 40, no.11, pp.2465-7, and Sov. Phys.-Tech. Phys., 1971, 15, no.11.

47.78. Pivit, E.: "High-Power Circulator for Electron Synchrotron", Int. Elektron. Rdsch., 1971, 25, no.4, pp.101-3.

47.79. De Mutti, M., Mancini, E., and Manuzio, G.: "Stability Conditions for Spatial Focusing in a Racetrack Microtron", Nuclear Instrum. Methods, 1971, 93, no.1, pp.93-7.

47.80. Allen, J.S., et al.: "Superconducting Accelerator Section for 600-MeV Microtron", J. Appl. Phys., 1971, 42, no.1, p.106.

47.81. Whitham, K.: "12-MeV S-Band Standing-Wave (Accelerator) Guide", Trans. IEEE, 1971, NS-18, no.3, pp.542-4.

47.82. Seryapin, V.G., Patrenin, V.A., and Limasov, Yu.M.: "Microtron for Radiation Physics of Semiconductors", Prib. Tekh. Eksp., 1971, 14, no.2, pp.21-3, and Instrum. Exp. Tech., 1971, 14, no.2, pp.350-2.

47.83. Zhulinskii, S.F., Luk'yanenko, E.A., and Mirzoyan, A.R.: "AFC in a Microtron", Prib. Tekh. Eksp., 1971, 14, no.2, pp.24-5, and Instrum. Exp. Tech., 1971, 14, no.2, pp.353-4.

47.84. Fowkes, W.R., and Wilson, P.B.: "Application of TW Resonators to Superconducting Linacs", Trans. IEEE, 1971, NS-18, no.3, pp.173-5.

47.85. Hanson, A.O.: "Performance of a Superconducting Linac", Trans. IEEE, 1971, NS-18, no.3, pp.149-52.

47.86. Halama, H.J.: "Superconducting-Nb S-Band Cavities", Trans. IEEE, 1971, NS-18, no.3, pp.188-92.

47.87. Bronca, G., et al.: "Pulsed Superconducting Magnet Work", Trans. IEEE, 1971, NS-18, no.3, pp.636-8.

47.88. Bomko, V.A., Klyucharev, A.P., and Pudyak, B.I.: "Linac With Variable Energy", At. Energ., 1971, 31, no.2, pp.123-6.

47.89. Alaux, A., and Audet, J.C.: "Limitations and Losses of Superconducting Microtrons", Can. J. Phys., 1972, 50, no.3, pp.298-300.

47.90. Irie, K.: "Q-Factor Effects of Accelerating Structure in TW Linacs", Jap. J. Appl. Phys., 1972, 11, no.7, p.1056.

47.91. Berzin, A.K., et al.: "Compact, Transportable, Electron Accelerators", Prib. Tekh. Eksp., 1972, 15, no.6, p.227, and Instrum. Exp. Tech., 1972, 15, no.6, p.1886.

47.92. Serov, V.L., and Baryshev, A.I.: "Interaction of High-Current Single Bunch With Accelerating Resonator", Izv. Akad. Nauk Arm. SSR Fiz., 1972, 7, no.6, pp.406-12.

47.93. Takaoka, A., and Ura, K.: "Experiments on Superconducting Cavity for Electron Linacs", Technol. Rep. Osaka Univ., 1973, 23, no.1090-120, pp.223-33.

47.94. Zhulinskii, S.F., and Mirzoyan, A.R.: "Determination of Path Parameters Ensuring Stable Operation of a Microtron", Prib. Tekh. Eksp., 1973, 16, no.5, pp.24-5, and Instrum. Exp. Tech., 1973, 16, no.5, pp.1315-7.

47.95. Csonka, P.L.: "Particle Acceleration by Template-Modified Coherent Light", Part. Accel., 1973, 5, no.3, pp.129-54.

47.96. Haag, H.W.L "Optimum Coupling of a Microtron Accelerator Cavity in the Case of Variable Electron Beam Loading", J. Phys. E, 1973, 6, no.1, pp.79-81.

47.97. Kaw, P.K., and Kulsrud, R.M.: "Relativistic Acceleration of Charged Particles by Superintense Laser Beams", Phys. Fluids, 1973, 16, no.2, pp.321-8.

47.98. Septier, A., and Boussoukaya, M.: "Conditions for Observation of Beam Instabilities in a Superconducting Accelerator With a Single TM_{010} Cavity", Nuclear Instrum. Methods, 1973, 107, no.3, pp.437-43.

47.99. Alaux, A., and Audet, J.C.: "Superconducting Microtron Losses, Refrigeration Energy, and Orbit Current", Can. J. Phys., 1973, 51, no.5, pp.510-2.

47.100. Konrad, A.: "Linac Cavity-Field Calculation by Finite-Element Method", Trans. IEEE, 1973, NS-20, no.1, pp.802-8.

47.101. Aleshin, V.M., et al.: "Stabilization of the Acceleration Mode in a Microtron", Prib. Tekh. Eksp., 1973, 16, no.6, pp.19-21, and Instrum. Exp. Tech., 1973, 16, no.6, pp.1630-1.

47.102. Asgekar, V.B., et al.: "Beam Current Losses in Race-Track Microtron", Indian J. Pure Appl. Phys., 1974, 12, no.3, pp.237-8.

47.103. Moiseev, M.A., and Nusinovich, G.S.: "Theory of Multimode Generation in a Gyromonotron", Izv. VUZ Radiofiz., 1974, 17, no.11, pp.1709-17.

47.104. Biller, E.Z., et al.: "Installation for Investigation of Superconducting Electron Accelerators", Prib. Tekh. Eksp., 1974, 17, no.4, pp.15-7, and Instrum. Exp. Tech., 1974, 17, no.4, pp.940-3.

48. GUIDED-WAVE COMMUNICATION

48.1. Waldron, R.A., and Bowe, D.J.: "Normal Modes of Helix Waveguide", Marconi Rev., 1965, 28, p.29.

48.2. Bowe, D.J., and Waldron, R.A.: "Coupling Between Modes in Helix Waveguide", Marconi Rev., 1965, 28, p.65.

48.3. Westcott, B.: "Statistics of Coupling Into Unwanted Modes in a Long Multimode Transmission System", Marconi Rev., 1965, 28, p.89.

48.4. Steier, W.H.: "Attenuation of the Holmdel Helix Waveguide in the 100-125 GHz Band", Bell Syst. Tech. J., 1965, 44, p.899.

48.5. Ito, M.: "Dispersion of Very Short Microwave Pulses in Waveguide", Trans. IEEE, 1965, MTT-13, p.357.

48.6. Torgow, E.N.: "Equalization of Waveguide Delay Distortion", Trans. IEEE, 1965, MTT-13, p.756.

48.7. Peaudecerf, M., and Lefeuvre, S.: "Experimental Study of Propagation in a Rectangular Oversized Waveguide", C. R. Acad. Sci. (Paris), 1965, 261B, p.2177.

48.8. Kindermann, H.P.: "Optimized TE_{01}-Mode Bends With Dielectric Layer", Arch. Elektr. Ubertrag., 1965, 19, p.699.

48.9. Hanckel, W., and Stockhausen, M.: "Change of Mode at the Sloping Surface of a Dielectric in a Rectangular Overmoded Waveguide", Z. Angew. Phys., 1965, 19, p.138.

48.10. Young, D.T.: "Measured TE_{01}-Mode Attenuation in Helix Waveguide With Controlled Straightness Deviations", Bell Syst. Tech. J., 1965, 44, p.273.

48.11. Krul, L.: "Waveguides With Distributed Reflection", PTT Bedr., 1965, 14, p.34.

48.12. Semenov, V.V.: "Fast Waves in a Helix Waveguide With Dielectric Coating and a Metallic Jacket", Radiotekh. Elektron., 1965, 10, p.2240, and Radio Eng. Electron. Phys., 1965, 10, p.1913.

48.13. Sedlmair, S.: "Filters With Thin Metallized Film for H_{01} Long-Haul Waveguides", Frequenz, 1966, 20, p.372.

48.14. Vidallon, C.: "Guided Propagation of Pulses With Very Short Duration", C. R. Acad. Sci. (Paris), 1966, 262B, p.1036.

48.15. Tang, C.C.H.: "Mode Conversion in Tapered Waveguides At and Near Cutoff", Trans. IEEE, 1966, MTT-14, p.233.

48.16. Barlow, H.E.M.: "Field Distribution at Bends in Circular H_{0n} and Cylindrical Surface Waveguides", Proc. IEE, 1966, 113, p.1913.

48.17. Chen, Y.M.: "Wave Propagation in Large Waveguides Containing Random Media", Radio Sci., 1966, 1, p.697.

48.18. Tang, C.C.H.: "Wave Propagation and Mode Conversion in a Helically Corrugated Multimode Circular Waveguide", Trans. IEEE, 1966, MTT-14, p.275.

48.19. Eigner, H.: "Problem in TE_{01} Long-Distance Communication Technique", Elektrotech. Maschin., 1966, 83, p.303.

48.20. Hosono, T., Yoshida, S., and Namiki, T.: "Calculation of Attenuation Coefficient of Helix Waveguide", J. Inst. Electr. Commun. Eng. Jap., 1966, 49, no.3, p.17.

48.21. Comte, G.: "Propagation of TE_{01} Mode Through Elbows in Helix Waveguides", Cables Transm., 1966, 20, p.177.

48.22. Conklin, G.E.: "Observed 50-60 GHz Attenuation for Circular Electric Wave in Dielectric-Coated Cylindrical Waveguide Bend", Bell Syst. Tech. J., 1966, 45, p.723.

48.23. Yoshida, S.: "Unwanted Mode Propagation in Helix Waveguide", J. Inst. Electr. Commun. Eng. Jap., 1966, 49, p.943.

48.24. Jaumann, A.: "Delay Equalizer for Long-Distance Waveguide", Nachr. Tech. Z., 1966, 19, p.451.

48.25. Meriakri, V.V.: "Tunable Multimode Waveguide Coupler", Prib. Tekh. Eksp., 1966, no.2, p.204, and Instrum. Exp. Tech., 1966, no.2, p.479.

48.26. Averbakh, V.S., et al.: "Experimental Study of a Reflecting Beamguide", Radiotekh. Elektron., 1966, 11, no.4, pp.750-2, and Radio Eng. Electron. Phys., 1966, 11, no.4.

48.27. Larsen, H., et al.: "Construction, Testing, and Transmission Properties of an H_{01} Long-Haul Waveguide Using Aluminium Tubes With Dielectric Lining. I and II", Frequenz, 1967, 21, no.11, pp.344-55, and no.12, pp.381-4.

48.28. Kerzhentseva, N.P.: "Reflection Factor Due to Axial Bends in a Multimode Waveguide", Radiotekh. Elektron., 1967, 12, no.8, pp.1378-85, and Radio Eng. Electron. Phys., 1967, 12, no.8, pp.1284-91.

48.29. Ruckdeschel, H.: "Measurement of Group Delay of an H_{01}-Mode Waveguide Link for 32-39 GHz", Nachr. Tech. Z., 1967, 20, no.11, pp.635-9.

48.30. Yamauchi, N.: "Characteristics of Bent Circular Waveguide Near Cutoff of Unwanted Modes", Bull. Naqoyd Inst. Technol., 1967, 19, pp.263-8.

48.31. Vaganov, R.B., and Matveev, R.F.: "Theoretical Study of Extended Slightly Irregular Beamguides", Radiotekh. Elektron., 1968, 13, no.2, pp.232-42, and Radio Eng. Electron. Phys., 1968, 13, no.2.

48.32. Morita, N., and Nakanishi, Y.: "Circumferential Gap in TE_{01}-Mode Multimode Circular Waveguide", Trans. IEEE, 1968, MTT-16, no.3, pp.183-8.

48.33. Korshunova, E.N., and Korshunov, I.P.: "Results of Measurements of Total Losses in Lenslike Quasi-Optical Lines", Izv. VUZ Radioelektron., 1968, 11, p.378.

48.34. Webster, J.R.: "Gamma-Radiation Effects on Fibre-Optic Guides", Opt. Spectra, 1968, 2, p.59.

48.35. Pruzhanovskii, V.A.: "Metal Optical Waveguides", Zh. Tekh. Fiz., 1968, 38, p.1035, and Sov. Phys.-Tech. Phys., 1968, 13, no.6, pp.784-9.

48.36. Shutpanov, O.Ye.: "Quasi-Optical Line Consisting of Imperfect Correctors", Radiotekh. Elektron., 1968, 13, p.1950, and Radio Eng. Electron. Phys., 1968, 13, no.11, pp.1710-5.

48.37. Kurtz, C.N., and Streifer, W.: "Guided Waves in Inhomogeneous Focusing Media. II", Trans. IEEE, 1969, MTT-17, p.250.

48.38. Gloge, D., and Steier, W.H.: "Experimental Simulation of a Multiple-Beam Optical Guide", Bell Syst. Tech. J., 1969, 48, p.1445.

48.39. Daly, J.C.: "Step Correction of Misaligned Beam Waveguides", Bell Syst. Tech. J., 1969, 48, p.1909.

48.40. Prache, P.M.: "Circular Waveguides With Internal Dielectric Linings and External Metallic Tube", Cables Transm., 1969, 23, p.3.

48.41. Sugimoto, S., Kaneko, H., and Kuroda, T.: "Guided Millimetre-Wave Communication System. II", NEC Res. Dev., 1969, no.14, p.5.

48.42. Gallawa, R.L., et al.: "Surface-Wave Transmission Line and Use in Communicating With High-Speed (Railroad) Vehicles", Trans. IEEE, 1969, COM-17, p.518.

48.43. Marcatili, E.A.J.: "Bends in Optical Dielectric Guides", Bell Syst. Tech. J., 1969, 48, p.2103.

48.44. Sawa, S., and Kumagai, N.: "Curved Glass Lens", Electron. Commun. Jap., 1969, 52, no.3, p.114.

48.45. Noda, K., and Miyauchi, K.: "Experimental Guided Millimetre-Wave Transmission System", Jap. Telecommun. Rev., 1969, 11, no.3, p.156.

48.46. Tutubalin, V.N.: "Investigation of Mathematical Models for Waveguides With Random Irregularities", Teor. Veroyatn. Primen., 1969, 14, no.4, pp.547-66, and Theory Probab. Appl., 1969, 14, no.4.

48.47. Suematsu, Y., and Nagashima, H.: "Light-Beam Instability in Sequential Lens Guide With Fourth-Order Aberration", Electron. Commun. Jap., 1969, 52, no.12, pp.76-82.

48.48. Ishida, N.: "Pseudo Transmission Line for Millimetre-Wave Communication Using Cutoff Tapered Waveguides", Electron. Commun. Jap., 1969, 52, no.12, pp.125-33.

48.49. Iga, K., et al.: "Propagation of Light Beam in Gas Flow Having a Transverse Temperature Difference", Electron. Commun. Jap., 1969, 52, no.3, p.81.

48.50. Vaganov, R.B.: "Damping of Beam Swinging in Iris Waveguide", Izv. VUZ Radiofiz., 1969, 12, no.10, p.1546.

48.51. Daglish, H.N.: "Light Scattering in Selected Optical Glasses", Glass Technol., 1970, 11, no.2, p.30.

48.52. Kapany, N.S., and Sawatari, T.: "Thermally Induced Beat Phenomenon in Coupled Optical Waveguides", J. Opt. Soc. Am., 1970, 60, p.135.

48.53. Gloge, D.: "Crosstalk in Multiple-Beam Waveguides", Bell Syst. Tech. J., 1970, 49, p.55.

48.54. Chan, K.B., et al.: "Propagation Characteristics of an Optical Waveguide With a Diffused-Core Boundary", Electron. Lett., 1970, 6, no.23, pp.748-9.

48.55. Gloge, D.: "Optical Waveguide Transmission", Proc. IEEE, 1970, 58, no.10, pp.1513-22.

48.56. Taheri, S.H.: "Mode Filtering in Long Multimode Transmission Lines With Random Coupling", Proc. IEE, 1970, 117, p.1615.

48.57. Klapka, J.L.: "Influence of Wall Losses on Energy-Flow-Centre Velocity of Pulses in Waveguides", Trans. IEEE, 1970, MTT-18, p.689.

48.58. Schulz-DuBois, E.O.: "Sommerfeld Pre- and Post-Cursors in the Context of Waveguide Transients", Trans. IEEE, 1970, MTT-18, p.455.

48.59. Kogelnik, H., and Sosnowski, T.P.: "Holographic Thin-Film Couplers", Bell Syst. Tech. J., 1970, 49, no.7, pp.1602-8.

48.60. Marie, P.: "Guided Propagation of Coherent Light in a Spiral Waveguide", Ann. Telecommun., 1970, 25, no.9-10,pp.320-4.

48.61. Sawa, S., and Kumagai, N.: "Response of Gaussian Beam Mismatched Into a Periodic Waveguide Consisting of Lenslike Media", Electron. Commun. Jap., 1970, 53, no.3, p.23.

48.62. Mullins, J.H.: "Using Computers in Designing Millimetre Waveguide Systems", Bell Lab. Rec., 1970, 48, no.10, pp.293-8.

48.63. Kapron, F.P., Keck, D.B., and Maurer, R.D.: "Radiation Losses in Glass Optical Waveguides", Appl. Phys. Lett., 1970, 17, no.10, pp.423-5.

48.64. Arnaud, J.A.: "Nonorthogonal (Deformed) Optical Waveguides and Resonators", Bell Syst. Tech. J., 1970, 49, no.9, pp.2311-48.

48.65. Voitovich, N.N., and Semenov, V.V.: "Quasi-Optical Iterative Feeders", Radiotekh. Elektron., 1970, 15, no.4, and Radio Eng. Electron. Phys., 1970, 15,no.4,pp.588-94.

48.66. Kaznacheev, Yu.I., and Minenko, V.K.: "Power Losses in Steel Waveguides of Circular Cross Section", Izv. VUZ Radioelektron., 1970, 13, no.12, pp.1416-22.

48.67. Suematsu, Y., Shimizu, T., and Kitano, T.: "Beam Waves Along a Complex-Permittivity Lenslike Medium", Electron. Commun. Jap., 1970, 53, no.12, pp.77-85.

48.68. Padanyi, G.: "Theory of Light Propagation by Fibre-Optical Methods", Finommechanika, 1970, 9, no.11, pp.353-6.

48.69. Miyauchi, K., Matsuda, S., and Shimada, S.: "Millimetre-Wave Guided Transmission. I and II", J. Inst. Electr. Commun. Eng. Jap., 1970, 53, pp.1544-7 and 1548-52.

48.70. Sushi, N., Kurauchi, N., and Tanaka, S.: "Millimetre-Wave Guided Transmission. III", J. Inst. Electr. Commun. Eng. Jap., 1970, 53, no.11, pp.1553-6.

48.71. Suzuki, N., and Shimada, S.: "Cut-off Filter for Millimetre-Wave Communicator System", Rev. Electr. Commun. Lab., 1970, 18, no.11-12, pp.781-95.

48.72. Barchukov, A.I., Konev, Yu.B., and Prokhorov, A.M.: "Accuracy of Alignment of Mirror Lines", Radiotekh. Elektron., 1970, 15, no.10, and Radio Eng. Electron. Phys., 1970, 15, no.10, pp.1920-1.

48.73. Yamamoto, S., and Makimoto, T.: "Equivalence Properties of Two-Dimensional Distributed-Parameter Systems. Lenslike Media", Electron. Commun. Jap., 1970, 53, no.2, pp.88-95.

48.74. Korshunov, I.P.: "(Phase Correction of) Laser Beam Partially Covered by a Shield", Radiotekh. Elektron., 1970, 15, no.10, and Radio Eng. Electron. Phys., 1970, 15, no.10, pp.1900-3.

48.75. Suematsu, Y., and Nagashima, H.: "Transmission of Optical Waveguide With Temperature Difference Between Centre and Periphery", Electron. Commun. Jap., 1970, 53, no.6, pp.110-6.

48.76. Kazantsev, Yu.N.: "Attenuation of Eigenwaves in an Overmoded Circular Guide With Finite Dielectric Coating", Radiotekh. Elektron., 1970, 15, p.207, and Radio Eng. Electron. Phys., 1970, 15, no.1, pp.179-82.

48.77. Hatakeyama, K., and Kitta, R.: "Analysis of Reflection Characteristics of Rectangular Waveguides", KDD Tech. J., 1970, no.65, pp.30-6.

48.78. Kapron, F.P., and Keck, D.B.: "Pulse Transmission Through a Dielectric Optical Waveguide", Appl. Opt., 1971, 10, no.7, pp.1519-23.

48.79. Falciasecca, G., Someda, C.G., and Valdoni, F.: "Wall Impedances and Application to Long-Distance Waveguides", Alta Freq., 1971, 40, no.5, pp.426-34.

48.80. Brayer, M., and Yhuel, J.: "Spectral Representation of Propagation in Long-Distance Circular Waveguides. I and II", Ann. Telecommun., 1971, 26, no.5-6, pp.215-30, and no.7-8, pp.279-302.

48.81. Carlin, J.W., and D'Agostino, P.: "Low-Loss Modes in Dielectric-Lined Waveguide", Bell Syst. Tech. J., 1971, 50, no.5, pp.1631-8.

48.82. Van Bladel, J.: "Mode Coupling Through Wall Losses in a Waveguide", Electron. Lett., 1971, 7, no.8, pp.178-80.

48.83. Cohen, L.G.: "Measured Attenuation and Depolarization of Light Transmitted Along Glass Fibres", Bell Syst. Tech. J., 1971, 50, no.1, pp.23-42.

48.84. Zvatitskii, V.A.: "Regular Beamguides of the Second Kind", Radiotekh. Elektron. 1971, 15, no.8, and Radio Eng. Electron. Phys., 1971, 15, no.8, pp.1378-85.

48.85. Tien, P.K., and Martin, R.J.: "Experiments on Light Waves in a Thin Tapered Film With Coupler", Appl. Phys. Lett., 1971, 18, no.9, pp.398-401.

48.86. Personick, S.D.: "Time Dispersion in Dielectric Waveguides", Bell Syst. Tech. J., 1971, 50, no.3, pp.843-59.

48.87. Harris, J.H., Giarusso, D.P., and Shubert, R.: "Optical Waveguide Scattering and Griffith's Microcracks", Proc. IEEE, 1971, 59, no.7, pp.1123-4.

48.88. Stern, J.R., and Dyott, R.B.: "Off-Axis Launching Into Fibre-Optical Waveguide", Electron. Lett., 1971, 7, no.2, pp.52-3.

48.89. Dyott, R.B., and Stern, J.R.: "Group Delay in Glass-Fibre Waveguide", Electron. Lett., 1971, 7, no.3, pp.82-4.

48.90. Carlin, J.W.: "Relation for Loss Characteristics of Circular-Electric and -Magnetic Modes in Dielectric-Lined Waveguide", Bell Syst. Tech. J., 1971, 50, no.5, pp.1639-43.

48.91. Chown, M., and Kao, K.C.: "Wideband Fibre-Waveguide Communication Systems for Optical Frequencies", Electr. Commun., 1971, 46, no.2, pp.118-24.

48.92. Kerzhentseva, N.P.: "Mode Conversion in Waveguides With Smoothly Varying Wall Impedance", Radiotekh. Elektron., 1971, 16, no.1, and Radio Eng. Electron. Phys., 1971, 16, no.1, pp.24-31.

48.93. Yoshikiyo, H., and Hirano, J.: "Transmission Bandwidth of Optical Beam Waveguide", Electr. Commun. Lab. Tech. J., 1971, 20, no.9, pp.2029-41.

48.94. Borner, M.: "Optical Communication Transmission System With Glassfibre Waveguides", Wiss. Ber. AEG-Telefunken, 1971, 44, no.2, pp.41-5.

48.95. Krumpholz, O.: "Monomode Glassfibre Optical Waveguides", Wiss. Ber. AEG-Telefunken, 1971, 44, no.2, pp.64-70.

48.96. Jacobsen, A.: "Problems in the Production of Dielectric Lightguides", Wiss. Ber. AEG-Telefunken, 1971, 44, no.2, pp.71-3.

48.97. Arnaud, J.A.: "Modes of Propagation of Optical Beams in Helical Gas Lenses", Proc. IEEE, 1971, 59, no.9, pp.1378-9.

48.98. Hirano, J., and Okawara, C.: "Hybrid Optical Transmission Systems", Electr. Commun. Lab. Tech. J., 1971, 20, no.10, pp.2357-60.

48.99. Tutubalin, V.N.: "Design of Waveguides With Random Inhomogeneities", Radiotekh. Elektron., 1971, 16, no.8, pp.1352-60, and Radio Eng. Electron. Phys., 1971, 16, no.8, pp.1274-80.

48.100. Kumagai, N., and Sawa, S.: "Analogies Between Lenslike and Circular-TE$_{01}$ Waveguides", Electron. Commun. Jap., 1971, 54, no.1, pp.92-7.

48.101. Nagashima, H., Suematsu, Y., and Yonezawa, N.: "Limit of Beam Capacity Caused by Irregularity of Focuser in Multiple-Beam Waveguide", Electron. Commun. Jap., 1971, 54, no.3, pp.61-7.

48.102. Matveev, R.F.: "Multibeam Transmission Along an Optical Beamguide", Radiotekh. Elektron., 1971, 16, no.10, pp.1950-3, and Radio Eng. Electron. Phys., 1971, 16, no.10, pp.1741-4.

48.103. Sawa, S., and Kumagai, N.: "Analysis of Optical Beamguide Consisting of Tapered Lenslike Medium", Electron. Commun. Jap., 1971, 54, no.4, pp.110-7.

48.104. Kondoh, K.: "Method of Transmission-Characteristics Estimation of Millimetre-Wave Long-Distance Circular Waveguide", Electron. Commun. Jap., 1971, 54, no.7, pp.55-63.

48.105. Suematsu, Y., and Furuya, K.: "Vector-Wave Solution of Light Beam Propagating Along Lenslike Medium", Electron. Commun. Jap., 1971, 54, no.7, pp.39-46.

48.106. Suematsu, Y., et al.: "Reduction of Temperature Difference in Shielding Pipes for Light-Beam Transmission", Electron. Commun. Jap., 1971, 54, no.7, pp.47-54.

48.107. Sawa, S.: "Design Theory for Circular Bend of Optical Waveguide Consisting of Lenslike Medium", Electron. Commun. Jap., 1971, 54, no.10, pp.68-72.

48.108. Vershinina, L.N., and Shevchenko, V.V.: "Quasi-Optical Channels at Submillimetre Wavelengths", Prib. Tekh. Eksp., 1971, 14, no.4, pp.147-9, and Instrum. Exp. Tech., 1971, 14, no.4, pp.1120-2.

48.109. Marcuse, D.: "Crosstalk Caused by Scattering in Slab Waveguides", Bell Syst. Tech. J., 1971, 50, no.6, pp.1817-31.

48.110. Marcuse, D.: "Coupling of Degenerate Modes in Two Parallel Dielectric Waveguides", Bell Syst. Tech. J., 1971, 50, no.6, pp.1791-816.

48.111. Gloge, D.: "Weakly Guiding Fibres", Appl. Opt., 1971, 10, no.10, pp.2252-8.

48.112. Gambling, W.A., Payne, D.N., and Sunak, H.R.D.: "Pulse Dispersion in Glass Fibres", Electron. Lett., 1971, 7, no.18, pp.549-50.

48.113. Mink, J.W.: "Assessment of Optical Iris-Beam Waveguides", Electron. Lett., 1971, 7, no.18, pp.527-8.

48.114. Kita, H., et al.: "Light-Focusing Glass Fibres and Rods", J. Am. Ceram. Soc., 1971, 54, no.7, pp.321-6.

48.115. Dyott, R.B., and Stern, J.R.: "Effects of Multiple Scattering in Optical-Fibre Transmission Line", Electron. Lett., 1971, 7, no.20, pp.624-5.

48.116. Bisbee, D.L.: "Measurements of Loss Due to Offsets and End Separations of Optical Fibres", Bell Syst. Tech. J., 1971, 50, no.10, pp.3159-68.

48.117. Marcuse, D.: "Attenuation of Un-wanted Cladding Modes", Bell Syst. Tech. J., 1971, 50, no.8, pp.2565-83.

48.118. Gloge, D., et al.: "Picosecond Pulse Distortion in Optical Fibres", J. Quantum Electron. IEEE, 1972, QE-8, no.2, pp.217-21.

48.119. Kapron, F.P., Borrelli, N.F., and Keck, D.B.: "Birefringence in Dielectric Optical Waveguides", J. Quantum Electron. IEEE, 1972, QE-8, no.2, pp.222-5.

48.120. Arnaud, J.A., Hogg, D.C., and Ruscio, J.T.: "Focusing of 52-GHz Beams by Cylindrical Mirrors", Trans. IEEE, 1972, MTT-20, no.5, pp.344-5.

48.121. Marcuse, D.: "Derivation of Coup-led Power Equations", Bell Syst. Tech. J., 1972, 51, no.1, pp.229-37.

48.122. Stone, J.: "Optical Transmission in Liquid-Core Quartz Fibres", Appl. Phys. Lett., 1972, 20, no.7, pp.239-40.

48.123. Rich, T.C., and Pinnow, D.A.: "Total Optical Attenuation in Bulk Fused Silica", Appl. Phys. Lett., 1972, 20, no.7, pp.264-6.

48.124. Snyder, A.W.: "Optical Fibre With Nonuniform Refractive Index", Electron. Lett., 1972, 8, no.7, pp.183-4.

48.125. Dakin, J.P., et al.: "Launching Into Glass-Fibre Optical Waveguides", Opt. Commun., 1972, 4, no.5, pp.354-7.

48.126. Ihaya, A., Furuta, H., and Noda, H.: "Thin-Film Optical Directional Coupler", Proc. IEEE, 1972, 60, no.4, pp.470-1.

48.127. Rowe, H.E., and Young, D.T.: "Transmission Distortion in Multimode Random Waveguides", Trans. IEEE, 1972, MTT-20, no.6, pp.349-65.

48.128. Schlosser, W.O.: "Delay Distortion in Weakly Guiding Optical Fibres Due to El-liptic Deformation of Boundary", Bell Syst. Tech. J., 1972, 51, no.2, pp.487-92.

48.129. Pearson, A.D., and French, W.G.: "Low-Loss Glass Fibres for Optical Trans-mission", Bell Lab. Rec., 1972, 50, no.4, pp.103-9.

48.130. Arnaud, J.A.: "Modes in Helical Gas Lenses", Appl. Opt., 1972, 11, no.11, pp.2514-21.

48.131. Schicketanz, D., and Schubert, J.: "Coupling Losses Between Laser Diodes and Multimode Glass Fibres", Siemens Forsch. Entwick., 1972, 1, no.4, pp.329-31.

48.132. Matveev, R.F.: "Effect of Air Turbulence on Ray Oscillation in Beamguides", Radiotekh. Elektron., 1972, 17, no.5, pp.1073-5, and Radio Eng. Electron. Phys., 1972, 17, no.5, pp.833-6.

48.133. Ishida, N., and Yokoyama, H.: "1.7-GHz Delay Equalizers for Millimetre-Wave Communication", Electron. Commun. Jap., 1972, 55, no.4, pp.87-94.

48.134. Marcuse, D.: "Power Distribution and Radiation Losses in Multimode Dielectric-Slab Waveguides", Bell Syst. Tech. J., 1972, 51, no.2, pp.429-54.

48.135. Cohen, L.G.: "Power Coupling from GaAs Injection Lasers Into Optical Fibres", Bell Syst. Tech. J., 1972, 51, no.3, pp.573-94.

48.136. Rowbotham, T.R., and Johns, P.B.: "Waveguide Analysis by Random Walks", Electron. Lett., 1972, 8, no.10, pp.251-3.

48.137. Suematsu, Y., and Furuya, K.: "Propagation Mode and Scattering Loss of a Two-Dimensional Dielectric Waveguide With Gradual Distribution of Refractive Index", Trans. IEEE, 1972, MTT-20, no.8, pp.524-31.

48.138. Imai, T.: "Semiconductor Devices for Millimetre-Wave Repeaters", J. Inst. Telev. Eng. Jap., 1972, 26, no.2, pp.111-47.

48.139. Kapron, F.P., Maurer, R.D., and Teter, M.P.: "Theory of Backscattering Ef-fects in Waveguides", Appl. Opt., 1972, 11, no.6, pp.1352-6.

48.140. Comte, G., and Trezeguet, J.P.: "Negotiation of Bends by TE_{01} Mode in Cir-cular Waveguides", Cables Transm., 1972, 26, no.2, pp.166-82.

48.141. Vuorinen, P.A.: "Mode-Conversion Reduction in Curved Circular Waveguide", J. Microwave Power, 1972, 7, no.2, pp.99-108.

48.142. Yamaguchi, K., and Kondoh, K.: "Improvement of Peak Loss Characteristics in Uniformly Dielectric-Lined Waveguide", Electron. Commun. Jap., 1972, 53, no.12, pp.19-26.

48.143. Miyauchi, K., et al.: "Character-istics of an Experimental Millimetre-Wave Guided Transmission System", Trans. IEEE, 1972, COM-20, no.5, pp.808-13.

48.144. Keck, D.B., and Tynes, A.R.: "Spectral Response of Lowloss Optical Wave-guides", Appl. Opt., 1972, 11, no.7, pp.1502-6.

48.145. Marcuse, D.: "Pulse Propagation in a Two-Mode Guide", Bell Syst. Tech. J., 1972, 51, no.8, pp.1785-91.

48.146. Rosman, G.: "Variation of Pulse Delay With Launch Angle in a Liquid-Filled Fibre", Electron. Lett., 1972, 8, no.18, pp.455-6.

48.147. Schicketanz, D., and Schubert, J.: "Coupling Losses Between Laser Diodes and Multimode Glass Fibres", Opt. Commun., 1972, 5, no.4, pp.291-2.

48.148. Volkov, V.I., et al.: "Pulsating Wave Beams in Quasi-Optical Beamguides", Radiotekh. Elektron., 1972, 16, no.9, pp.1618-22, and Radio Eng. Electron. Phys., 1972, 16, no.9, pp.1477-81.

48.149. Marcuse, D.: "Higher-Order Scattering Losses in Dielectric Waveguides", Bell Syst. Tech. J., 1972, 51,no.8,pp.1801-17.

48.150. Janssen, W., and Odemar, N.: "Tolerance in Waveguides for Long-Distance Transmission", Frequenz, 1972, 26, no.9, pp.258-65.

48.151. Suematsu, Y., et al.: "Fundamental-Mode Selection for Thin-Film Asymmetric Lightguides", Appl. Phys. Lett., 1972, 21, no.6, pp.291-3.

48.152. Borner, M., et al.: "Detachable Connector for Monomode Glass-Fibre Light Guides", Arch. Elektron. Ubertrag., 1972, 26, no.6, pp.288-9.

48.153. Marcuse, D.: "Pulse Propagation in Multimode Dielectric Waveguide", Bell Syst. Tech. J., 1972, 51, no.6,pp.1199-232.

48.154. Herlent, Y.: "Telecommunication by Circular Waveguides", Toute Electron., 1972, no.368, pp.11-16.

48.155. Nishizawa, J., and Otsuka, A.: "Solid-State Self-Focusing Surface Waveguide Microguide", Appl. Phys. Lett., 1972, 21, no.2, pp.48-50.

48.156. Payne, D.N., and Gambling, W.A.: "Lowloss Liquid-Core Fibre Waveguide", Electron. Lett., 1972, 8, no.15,pp.374-6.

48.157. Garmire, E., and Stoll, H.: "Propagation Losses in Metal-Film-Substrate Optical Waveguides", J. Quantum Electron. IEEE, 1972, QE-8, no.10, pp.763-6.

48.158. Gloge, D., Chinnock, E.L., and Lee, T.P.: "Self-Pulsing GaAs Laser for Fibre-Dispersion Measurements", J. Quantum Electron. IEEE, 1972, QE-8, no.11,pp.844-6.

48.159. Gloge, D., and Chinnock, E.L.: "Fibre-Dispersion Measurements Using a Mode-Locked Kr Laser", J. Quantum Electron. IEEE, 1972, QE-8, no.11, pp.852-4.

48.160. Marcuse, D.: "Higher-Order Loss Processes and Loss Penalty of Multimode Operation", Bell Syst. Tech. J., 1972, 51, no.8, pp.1819-36.

48.161. Gloge, D.: "Optical Power Flow in Multimode Fibres", Bell Syst. Tech. J., 1972, 51, no.8, pp.1767-83.

48.162. Gloge, D.: "Bending Loss in Multimode Fibres With Graded and Ungraded Core Index", Appl. Opt., 1972, 11, no.11,pp.2506-13.

48.163. Sattarov, D.K., et al.: "Light Transmission of Individual Components of a Flux Leaving a Fibre-Optical Element", Opt. Spekt., 1972, 33, no.1, pp.159-64, and Opt. Spectrosc., 1972, 33, no.1, pp.86-8.

48.164. Krumpholz, O.: "Refractive-Index Profile in a Glass-Fibre Lightguide", Wiss. Ber. AEG-Telefunken, 1972, 45, no.1-2,pp.1-8.

48.165. Shcherbov, V.A., and Litvinov, D.D.: "Matching of Beamguide Loads", Izv. VUZ Radiofiz., 1972, 15, no.9, pp.1175-8.

48.166. Corzani, C.E., and Stracca, G.B.: "Channelling for Circular Waveguide Transmission Systems", Note Recens. Not., 1972, 21, no.4, pp.335-52.

48.167. Carli, E., Corzani, T., and Stracca, G.B.: "Commutating Branching Filters for Circular Waveguide Communication Systems", Note Recens. Not., 1972, 21, no.4, pp.353-400.

48.168. Gambling, W.A., Payne, D.N., and Matsumura, H.: "Gigahertz Bandwidths in Multimode, Liquid-Core, Optical-Fibre Waveguides", Opt. Commun., 1972,6,no.4,pp.317-22.

48.169. Dakin, J.P., and Gambling, W.A.: "Angular Distribution of Light Scattering in Bulk Glass and Fibre Waveguides", Opt. Commun., 1972, 6, no.3, pp.235-8.

48.170. Svalov, S.I., Popov, M.M., and Lisienko, E.G.: "Use of Waveguide as Group Protection Conductor for a Traction Power-Supply Line", Autom. Telemekh. Suyaz, 1972, no.10, pp.26-8.

48.171. Kazakova, N.A., and Percikov, M.V.: "Scattering of Waves at Step in Circular Multimode Waveguide", Radiotekh. Elektron., 1972, 17, no.8, pp.1573-80, and Radio Eng. Electron. Phys., 1972, 17, no.8, pp.1239-45.

48.172. Sawa, S.: "Analysis of Optical Wave Beam Propagating Along Lenslike Medium With General Taper", Electron. Commun. Jap., 1972, 55, no.12, pp.97-103.

48.173. Korshunov, I.P.: "Test-Range Quasi-Optic Lens Line", Radiotekh. Elektron., 1972, 17, no.12, pp.2597-601, and Radio Eng. Electron. Phys., 1972, 17, no.12, pp.2084-7.

48.174. Lapta, S.I.: "Propagation in a Circular Waveguide With Spirally Conducting Elements", Izv. VUZ Radiofiz., 1972, 15, no.12, pp.1919-25.

48.175. Gloge, D., Chinnock, E.L., and Koizumi, K.: "Study of Pulse Distortion in Selfoc Fibres", Electron. Lett., 1972, 8, no.21, pp.526-7.

48.176. Gloge, D., et al.: "Dispersion in a Lowloss Multimode Fibre Measured at Three Wavelengths", Electron. Lett., 1972, 8, no.21, pp.527-9.

48.177. Hubbard, W.M.: "Double-Reverse Scatter Interference in Optical-Fibre Communication", Appl. Opt., 1972, 11, no.11, pp.2495-501.

48.178. Papanicolaou, G.C. McLaughlin, D., and Burridge, R.: "Stochastic Gaussian Beam (in Random Focusing Medium)", J. Math. Phys., 1973, 14, no.1, pp.84-9.

48.179. Ruscio, J.T.: "Focusing of 104-GHz Beams by Cylindrical Mirrors", Trans. IEEE, 1973, MTT-21, no.3, pp.154-5.

48.180. Unger, H.G.: "Circular-Electric Guide With Minimum Loss and Elastic Flexibility", Arch. Elektron. Ubertrag., 1973, 27, no.1, pp.49-50.

48.181. Schicketanz, D.: "Pulse Broadening in Multimode Fibres Excited by GaAs Lasers", Electron. Lett., 1973, 9, no.1, pp.5-6.

48.182. Maxia, V., Murgia, M., and Testa, F.: "Angular Distribution of Light Transmitted by a Cylindrical Guide", Appl. Opt., 1973, 12, no.1, pp.98-102.

48.183. Kato, D.: "Fused-Silica-Core Glass Fibre as Lowloss Optical Waveguide", Appl. Phys. Lett., 1973, 22, no.1, pp.3-4.

48.184. Snyder, A.W., Pask, C., and Mitchell, D.J.: "Light-Acceptance Property of an Optical Fibre", J. Opt. Soc. Am., 1973, 63, no.1, pp.59-64.

48.185. Matsuhara, M.: "Analysis of Modes in Lenslike Media", J. Opt. Soc. Am., 1973, 63, no.2, pp.135-8.

48.186. Marcuse, D.: "Coupling Coefficients for Imperfect Asymmetric Slab Waveguides", Bell Syst. Tech. J., 1973, 52, no.1, pp.63-82.

48.187. Brayer, M., and Yhuel, J.: "Study on Curvature on Long-Distance Waveguides. II and III", Ann. Telecommun., 1973, 28, no.3-4, pp.143-76, and no.5-6, pp.273-82.

48.188. Lit, J.W.Y., and Van Rooy, D.L.: "Hybrid Lens Beamguide", Appl. Opt., 1973, 12, no.4, pp.749-54.

48.189. Kondoh, K.: "Error Probability Characteristics of Millimetre-Wave Transmission Line", Electron. Commun. Jap., 1973, 56, no.2, pp.64-71.

48.190. Yoneyama, T., and Nishida, S.: "Effects of Random Surface Irregularities of Lenses on Wave-Beam Transmission", Rep. Res. Inst. Electr. Commun. Tohoku Univ., 1973, 25, no.2, pp.67-77.

48.191. Marcuse, D.: "Scattering Losses Caused by the Support Structure of an Uncladded Fibre", Bell Syst. Tech. J., 1973, 52, no.2, pp.205-17.

48.192. Kaiser, P., Marcatili, E.A.J., and Miller, S.E.: "Novel Optical Fibre", Bell Syst. Tech. J., 1973, 52, no.2, pp.265-9.

48.193. Young, M.: "Geometrical Theory of Multimode Optical Fibre-to-Fibre Connectors", Opt. Commun., 1973, 7, no.3, pp.253-5.

48.194. Ghatak, A.K., Malik, D.P.S., and Goyal, I.C.: "Wave Propagation Through a Gas Lens", Opt. Acta, 1973, 20, no.4, pp.303-11.

48.195. Carlin, J.W., and D'Agostino, P.: "Normal Modes in Overmoded Dielectric-Lined Circular Waveguide", Bell Syst. Tech. J., 1973, 52, no.4, pp.453-86.

48.196. Carlin, J.W., and Maione, A.: "Experimental Verification of Lowloss TM Modes in Dielectric-Lined Waveguide", Bell Syst. Tech. J., 1973, 52, no.4, pp.487-96.

48.197. Kirchhoff, H.: "Wave Propagation in Gradient Fibres With Curvature", Arch. Elektron. Ubertrag., 1973, 27, no.4, pp.161-7.

48.198. Marcuse, D., and Mammel, W.L.: "Tube Waveguide for Optical Transmission", Bell Syst. Tech. J., 1973, 52, no.3, pp.423-35.

48.199. Someda, C.G.: "Simple, Low-Loss, Joints Between Single-Mode Optical Fibres", Bell Syst. Tech. J., 1973, 52, no.4, pp.583-96.

48.200. Van Uitert, L.G., et al.: "Borosilicate Glasses for Optical-Fibre Waveguides", Mater. Res. Bull., 1973, 8, no.4, pp.469-76.

48.201. Kato, D.: "Light Coupling from a Stripe-Geometry GaAs Diode Laser Into an Optical Fibre With Spherical End", J. Appl. Phys., 1973, 44, no.6, pp.2756-8.

48.202. Smith, L.: "Dispersion Minimization of Dielectric Waveguides", Appl. Opt., 1973, 12, no.7, pp.1592-9.

48.203. Mar'in, V.I., Meriakri, V.V., and Lagerev, L.I.: "Quality Control of Circular Waveguide Sections", Izv. VUZ Radioelektron., 1973, 16, no.5, pp.51-4.

48.204. Tacke, M., and Ulrich, R.: "Submillimetre Waveguiding on Thin Dielectric Films", Opt. Commun., 1973, 8, no.3, pp.234-8.

48.205. Tsandoulas, G.N.: "Bandwidth Enhancement in Dielectric-Lined Circular Waveguides", Trans. IEEE, 1973, MTT-21, no.10, pp.651-4.

48.206. Marcuse, D.: "Coupled-Mode Theory of Round Optical Fibres", Bell Syst. Tech. J., 1973, 52, no.6, pp.817-42.

48.207. Marcuse, D.: "Effect of grad-n^2 Term on Modes of an Optical Square-Law Medium", J. Quantum Electron. IEEE, 1973, QE-9, no.9, pp.958-60.

48.208. Marcuse, D.: "TE Modes of Graded-Index Slab Waveguides", J. Quantum Electron. IEEE, 1973, QE-9, no.10, pp.1000-6.

48.209. Smithgall, D.H., and Dabby, F.W.: "Graded-Index Planar-Dielectric Waveguides", J. Quantum Electron. IEEE, 1973, QE-9, no.10, pp.1023-8.

48.210. Gambling, W.A., and Matsumura, H.: "Pulse Dispersion in a Lenslike Medium", Opto-Electron., 1973, 5, no.5, pp.429-37.

48.211. Costa, B., Di Vita, P., and Selt, C.: "Group Velocity of Modes and Pulse Distortion in Dielectric Optical Waveguides", Opto-Electron., 1973, 5, no.5, pp.439-56.

48.212. Bouillie, R., Steiner, K.H., and Treheux, M.: "Pulse Broadening in Dielectric Multimode Waveguides", Opto-Electron., 1973, 5, no.5, pp.457-77.

48.213. Personick, S.D.: "Receiver Design for Digital Fibre-Optic Communication Systems. I and II", Bell Syst. Tech. J., 1973, 52, no.6, pp.843-74 and 875-86.

48.214. Sodha, M.S., Chakravarti, A.K., and Gautama, G.D.: "Propagation of Optical Pulses Through Cladded Fibres. Modified Theory", Appl. Opt., 1973, 12, no.10, pp.2482-5.

48.215. Burrus, C.A., et al.: "Pulse Dispersion and Refractive Index Profiles of Low-Noise Multimode Optical Fibres", Proc. IEEE, 1973, 61, no.10, pp.1498-9.

48.216. Bernardi, P., and Falciasecca, G.: "TE_{01}-Mode Attenuation in a Circular Waveguide Link With Continuous Imperfections", Alta Freq., 1973, 42, no.8, pp.368-73.

48.217. Abram, R.A., and Rees, J.: "Mode Conversion in an Imperfect Waveguide", J. Phys. A, 1973, 6, no.11, pp.1693-708.

48.218. Gloge, D., and Marcatili, E.A.J.: "Impulse Response of Fibres With Ring Shaped Parabolic-Index Distribution", Bell Syst. Tech. J., 1973, 52, no.7, pp.1161-8.

48.219. Marcuse, D.: "Impulse Response of an Optical Fibre With Parabolic Index Profile", Bell Syst. Tech. J., 1973, 52, no.7, pp.1169-74.

48.220. Derosier, R.M., and Stone, J.: "Low-Loss Splices in Optical Fibres", Bell Syst. Tech. J., 1973, 52, no.7,pp.1229-35.

48.221. Gambling, W.A., and Matsumura, H.: "Pulse Response of a Graded-Index Optical Fibre", Electron. Lett., 1973, 9, no.15, pp.336-8.

48.222. Personick, S.D.: "Baseband Linearity and Equalization in Fibre-Optic Digital Communication Systems", Bell Syst. Tech. J., 1973, 52, no.7, pp.1175-94.

48.223. Marcuse, D.: "Losses and Impulse Response of a Parabolic Index Fibre With Random Bends", Bell Syst. Tech. J., 1973, 52, no.8, pp.1423-37.

48.224. Cook, J.S., Mammel, W.L., and Grow, R.J.: "Effect of Misalignments on Coupling Efficiency of Single-Mode Optical-Fibre Butt Joints", Bell Syst. Tech. J., 1973, 52, no.8, pp.1439-48.

48.225. Yip, G.L., Martucci, J., and Farnell, G.W.: "Scattering Loss in a Cladded-Fibre Optical Waveguide", Electron. Lett., 1973, 9, no.13, pp.293-5.

48.226. Dyott, R.B., Day, C.R., and Brain, M.C.: "Glass-Fibre Waveguide With Triangular Core", Electron. Lett., 1973, 9, no.13, pp.288-90.

48.227. Spence, J.E., and Daly, J.C.: "Beam Stability Control in Optical Waveguides Subject to Statistical Variations", Trans. IEEE, 1973, COM-21, no.12, pp.1409-14.

48.228. Flad, E.: "Waveform Distortion of High Speed Pulses Due to Discontinuities in Oversized Transmission Lines", Arch. Elektrotech., 1973, 55, no.6, pp.320-9.

48.229. Gloge, D., and Marcatili, E.A.J.: "Multimode Theory of Graded-Core Fibres", Bell Syst. Tech. J., 1973, 52, no.9, pp.1563-78.

48.230. Gloge, D., et al.: "Optical-Fibre End Preparation for Low-Loss Splices", Bell Syst. Tech. J., 1973, 52, no.9, pp.1579-88.

48.231. Marcuse, D.: "Cutoff Condition of Optical Fibres", J. Opt. Soc. Am., 1973, 63, no.11, pp.1369-71.

48.232. Frazier, J.F.: "Propagation in Jacketed Optical Waveguide Glass Fibres", Polytech. Tijdschr. Elektrotech., 1973, 28, no.26, pp.861-3.

48.233. Gambling, W.A., Payne, D.N., and Matsumura, H.: "Effect of Loss on Propagation in Multimode Fibres", Radio Electron. Eng., 1973, 43, no.11, pp.683-8.

48.234. Vard'ya, V.P., et al.: "Quasi-Optical Test Line for Investigating Propagation of Laser Beams Over Long Distances", Radiotekh. Elektron., 1973, 18, no.2, pp.391-4, and Radio Eng. Electron. Phys., 1973, 18, no.2, pp.278-80.

48.235. Bernardi, P., Falciasecca, G., and Valdoni, F.: "Experimental Evaluation of Effects Due to Random Ellipticity in Circular Waveguide", Note Recens. Not., 1973, 22, no.6, pp.643-55.

48.236. Nashima, H., Yonezawa, N., and Suematsu, Y.: "Computer Simulation on Beam Transmission in Optical Waveguides Between Tokyo and Osaka", Electr. Eng. Jap., 1973, 93, no.2, pp.124-9.

48.237. Korshunov, I.P.: "Investigation of Displacements of Light Beam in Underground Lens Line", Radiotekh. Elektron., 1973, 18, no.6, pp.1267-9, and Radio Eng. Electron. Phys., 1973, 18, no.6, pp.928-30.

48.238.   Yamamoto, M., et al.: "Optical Coupler With Control Function and Application", Mem. Fac. Eng. Osaka City Univ., 1973, 14, pp.87-100.

48.239.   Kazantsev, Yu.N.: "Smooth Bend of a Gas-Dielectric Waveguide", Radiotekh. Elektron., 1973, 18, no.7, pp.1329-35, and Radio Eng. Electron. Phys., 1973, 18, no.7, pp.981-5.

48.240.   Sawa, S., Ono, K., and Kumagai, N.: "Design Considerations for Circular Bends of Optical Lens Waveguides", Electron. Commun. Jap., 1973, 56, no.7, pp.103-7.

48.241.   Henderson, D.M.: "Dispersion and Equalization in Fibre-Optic Communication Systems", Bell Syst. Tech. J., 1973, 52, no.10, pp.1867-76.

48.242.   Brzhechko, L.V.: "Numerical Method of Calculating Mode Conversion at Discontinuities in Multimode Waveguides of Arbitrary Cross Section", Radiotekh. (Kharkov), 1973, no.25, pp.101-5.

48.243.   Guttmann, J., and Krumpholz, O.: "Theoretical and Experimental Investigations of Coupling of Two Glass-Fibre Lightguides", Wiss. Ber. AEG-Telefunken, 1973, 46, no.1, pp.8-15.

48.244.   Davies, D.E.N., and Kingsley, S.: "Method of Phase-Modulating Signals in Optical Fibres", Electron. Lett., 1974, 10, no.2, pp.21-2.

48.245.   Miller, S.E.: "Delay Distortion in Generalized Lenslike Media", Bell Syst. Tech. J., 1974, 53, no.2, pp.177-93.

48.246.   Marcuse, D.: "Losses and Impulse Response in Parabolic-Index Fibres With Square Cross Section", Bell Syst. Tech. J., 1974, 53, no.2, pp.195-215.

48.247.   Ikeda, M.: "Propagation Characteristics of Multimode Fibres With Graded Core Index", J. Quantum Electron. IEEE, 1974, QE-10, no.3, pp.362-71.

48.248.   Chakravarti, A.K., Gautama, G.D., and Rattan, I.: "Propagation of FM Pulses in Waveguides", Int. J. Electron., 1974, 36, no.4, pp.461-4.

48.249.   Cohen, L.G., and Schneider, M.V.: "Microlenses for Coupling Junction Lasers to Optical Fibres", Appl. Opt., 1974, 13, no.1, pp.89-94.

48.250.   Timmermann, C.C.: "Mode Distribution and Impulse Response of General Graded Multimode Fibres", Arch. Elektron. Übertrag., 1974, 28, no.4, pp.186-8.

48.251.   Gambling, W.A., et al.: "(Delay in) Optical Fibres and the Goos-Hanchen Shift", Electron. Lett., 1974, 10, no.7, pp.99-101.

48.252.   Gambling, W.A., Payne, D.N., and Matsumura, H.: "Pulse Dispersion for Single-Mode Operation of Multimode Cladded Optical Fibres", Electron. Lett., 1974, 10, no.9, pp.148-9.

48.253.   Stracca, G.B.: "Considerations on Optimization of Transmission Systems in Circular Waveguide", Alta Freq., 1974, 43, no.2, pp.90-101.

48.254.   Stracca, G.B.: "Characteristics of Circular Waveguides Used for Long-Distance Communications", Alta Freq., 1974, 43, no.2, pp.102-16.

48.255.   Bouillie, R., et al.: "Ray Delay in Gradient Waveguides With Arbitrary Symmetric Refractive Profile", Appl. Opt., 1974, 13, no.5, pp.1045-9.

48.256.   Marcuse, D.: "Rayleigh Scattering and the Impulse Response of Optical Fibres", Bell Syst. Tech. J., 1974,53,no.4,pp.705-15.

48.257.   Newns, G.R., Beales, K.J., and Duncan, W.J.: "Lowloss Glass for Optical Transmission", Electron. Lett., 1974, 10, no.10, pp.201-2.

48.258.   Goell, J.E.: "Optical Repeater With High Impedance Input Amplifier", Bell Syst. Tech. J., 1974, 53, no.4, pp.629-43.

48.259.   Miyashita, T., et al.: "Eccentric-Core Glass Optical Waveguide", J. Appl. Phys., 1974, 45, no.2, pp.808-9.

48.260.   Personick, S.D.: "Optimal Trade-Off of Mode-Mixing, Optical Filtering, and Index Difference, in Digital Fibre-Optic Communication Systems", Bell Syst. Tech. J., 1974, 53, no.5, pp.785-800.

48.261.   Beales, K.J., et al.: "Materials and Fibre for Optical Transmission Systems", Post Office Electr. Eng. J., 1974, 67, pt 2, pp.80-7.

48.262.   French, W.G., et al.: "Optical Waveguides With Very Low Losses", Bell Syst. Tech. J., 1974, 53, no.5, pp.951-4.

48.263.   Treheux, M., Bouillie, R., and Steiner, K.H.: "Filter Theory Applied to Multimode Fibre Optics", Ann. Telecommun., 1974, 29, no.5-6, pp.209-18.

48.264.   Cozannet, A., Treheux, M., and Bouillie, R.: "Propagation of Light Through Optical Fibres Composed of Graded Index Material", Ann. Telecommun., 1974, 29, no.5-6, pp.219-26.

48.265.   Black, P.W., et al.: "Measurements on Waveguide Properties of GeO_2-SiO_2-Cored Optical Fibres", Electron. Lett., 1974, 10, no.12, pp.239-40.

48.266.   Payne, D.N., and Gambling, W.A.: "Silica-Based Low-Loss Optical Fibre", Electron. Lett., 1974, 10, no.15, pp.289-90.

48.267. Brackett, C.A.: "Efficiency of Coupling Light from Stripe-Geometry GaAs Lasers Into Multimode Optical Fibres", J. Appl. Phys., 1974, 45, no.6, pp.2636-7.

48.268. Sunak, H.R.D., and Gambling, W.A.: "Picosecond-Pulse Dispersion in Cladded Glass Fibre", Opt. Commun., 1974, 11, no.3, pp.277-81.

48.269. Vassell, M.O.: "Calculation of Propagating Modes in a Graded-Index Optical Fibre", Opto-Electron., 1974, 6, no.4, pp.271-86.

48.270. Tasker, G.W., and French, W.G.: "Low-Loss Optical Waveguides With Pure Fused-SiO$_2$ Cores", Proc. IEEE, 1974, 62, no.9, pp.1283-4.

48.271. Boisrobert, C., and Leboutet, A.: "Use of a PIN Photodetector in a Fibre-Optic Transmission System", Ann. Telecommun., 1974, 29, no.5-6, pp.227-34.

48.272. Kaiser, P., and Astle, H.W.: "Low-Loss Single-Material Fibres Made from Pure Fused Silica", Bell Syst. Tech. J., 1974, 53, no.6, pp.1021-39.

48.273. Payne, D.N., and Gambling, W.A.: "Preparation of Water-Free, Silica-Based, Optical-Fibre Waveguide", Electron. Lett., 1974, 10, no.16, pp.335-6.

48.274. Kuchikyah, L.M.: "Diffraction on the Entrance-Face Plane of an Optical Fibre", Opt. Spekt., 1974, 36, no.3, pp.600-1, and Opt. Spectrosc., 1974, 36, no.3, pp.346-7.

48.275. Schnitger, H., and Unger, H.G.: "Circular Waveguide System for Trunk Communication", Trans. IEEE, 1974, COM-22, no.9, pp.1374-7.

48.276. Standley, R.D.: "Fibre-Ribbon Optical Transmission Lines", Bell Syst. Tech. J., 1974, 53, no.6, pp.1183-5.

48.277. Ohnsorge, H., and Schenkel, K.D.: "Integrated (Glass-Fibre) Communication System With Fully Decentralized Switching", Trans. IEEE, 1974, COM-22, no.9, pp.1292-6.

48.278. Davies, W.S., and Kidd, G.P.: "Bandwidth Results for Liquid-Core Optical Fibres", Electron. Lett., 1974, 10, no.19, pp.406-8.

48.279. Weidel, E.: "Light Coupling from a Junction Laser Into a Monomode Fibre With a Glass Cylindrical Lens", Opt. Commun., 1974, 12, no.1, pp.93-7.

48.280. Yakubov, A.F.: "Stability of Light Ray Distribution in a Thermohydrodynamic Lightguide", Vestsi Akad. Navuk BSSR, 1974, no.3, pp.92-5.

48.281. Hashimoto, K., Kondoh, K., and Shimada, S.: "Wave-Coupled TE$_{02}$-Mode Filters", Rev. Electr. Commun. Lab., 1974, 22, no.1-2, pp.130-8.

48.282. Sushi, N., et al.: "Waveguide Line Installed in Steel Conduit", Rev. Electr. Commun. Lab., 1974, 22, no.1-2, pp.1-19.

48.283. Yamaguchi, K., Nihei, F., and Nagahama, N.: "Design Consideration and Characteristics of Circular Waveguides and Components", Rev. Electr. Commun. Lab., 1974, 22, no.1-2, pp.20-38.

48.284. Kuribayashi, M., et al.: "Transmission Characteristics of Waveguide Line Protected by Steel Conduit", Rev. Electr. Commun. Lab., 1974, 22, no.1-2, pp.51-62.

48.285. Denger, L.A.: "Coupled-Wave Equations With Three or More Elements", J. Appl. Phys., 1974, 45, no.8, pp.3394-5.

48.286. Colvin, J.: "Launching Efficiency for a Light-Emitting Diode Into Optical Fibres", Opto-Electron., 1974, 6, no.5, pp.387-92.

48.287. Steiner, K.H.: "Ray Delay in Gradient Waveguides With Asymmetric Transverse Refractive-Index Profiles", Opto-Electron., 1974, 6, no.5, pp.401-9.

48.288. Midwinter, J.E., and Reeve, M.H.: "Technique for Study of Mode Cutoffs in Multimode Optical Fibres", Opto-Electron., 1974, 6, no.5, pp.411-6.

48.289. Arnaud, J.A.: "Transverse Coupling in Fibre Optics. III", Bell Syst. Tech. J., 1974, 53, no.7, pp.1379-94.

48.290. Arnaud, J.A.: "Pulse Spreading in Multimode, Planar, Optical Fibres", Bell Syst. Tech. J., 1974, 53, no.8, pp.1599-618.

48.291. Ostermayer, F.W., and Pinnow, D.A.: "Optimum Refractive-Index Difference for Graded-Index Fibres Resulting from Concentration-Fluctuation Scattering", Bell Syst. Tech. J., 1974, 53, no.7, pp.1395-402.

48.292. Marcuse, D.: "Theory of the Single-Material Fibre", Bell Syst. Tech. J., 1974, 53, no.7, pp.1619-41.

48.293. Arnaud, J.A.: "Theory of the Single-Material Helicoidal Fibre", Bell Syst. Tech. J., 1974, 53, no.8, pp.1643-56.

48.294. Hudson, M.C., and Thiel, F.L.: "Star Coupler. Unique Interconnection Component for Multimode Optical Waveguide Communications Systems", Appl. Opt., 1974, 13, no.11, pp.2540-5.

48.295. Da, B.: "French System of Long-Distance Communication by Circular Waveguide", Ann. Telecommun., 1974, 29, no.9-10, pp.338-59.

48.296. Dupuis, P., and Verdot, G.: "(Error) Measurements on an Experimental Repeater for Circular Waveguide", Ann. Telecommun., 1974, 29, no.9-10, pp.418-25.

48.297. Martin, M.J.: "Group-Delay Equalization on a Circular-Waveguide Link", Ann. Telecommun., 1974, 29, no.9-10, pp.426-42.

48.298. Gisin, B.V.: "Influence of Non-linear Optical Effects on Propagation in an Optical Fibre", Kvantovaya Elektron., 1974, 1, no.4, pp.968-9, and Sov. J. Quantum Electron., 1974, 4, no.4, p.534.

48.299. Luskinovich, P.N., and Shomorovskii, Yu.A.: "Propagation of a Semiconductor-Laser Pulse in an Optical Fibre", Kvantovaya Elektron., 1974, 1, no.6, pp.1460-3, and Sov. J. Quantum Electron., 1974, 4, no.6, pp.811-2.

48.300. Brayer, M., and Leguen, J.C.: "Long-Distance Transmission of Digital Signals in Circular Waveguide", Ann. Telecommun., 1974, 29, no.11-12, pp.587-628.

48.301. Kanmuri, N., et al.: "Low-Noise Downconverter and High-Efficiency Upconverter for Transmitter-Receiver Applications in the 60-86-GHz Region", Trans. IEEE, 1974, MTT-22, no.12, pp.1286-90.

48.302. Morrison, J.A.: "Average Output Power of Incident Wave Randomly Coupled to a Reflected Wave", Trans. IEEE, 1974, MTT-22, no.2, pp.126-30.

48.303. Clarricoats, P.J.B., Olver, A.D., and Al-Hariri, A.M.B.: "Dielectric Waveguides for Millimetre-Wave Transmission", Electron. Lett., 1974, 10, no.1, pp.1-2.

48.304. Snyder, A.W., and Mitchell, D.J.: "Bending Losses of Multimode Optical Fibres", Electron. Lett., 1974, 10, no.1, pp.11-2.

48.305. Popov, Yu.M., and Shuikin, N.N.: "Calculation of the Matching of an Injection Laser to a Dielectric Waveguide", Kvantovaya Elektron., 1974, 1, no.8, pp.1780-4.

48.306. DiDomenico, M.: "Review of Fibre Optical Transmission Systems", Opt. Eng., 1974, 13, no.5, pp.423-8.

48.307. Hashimoto, K.: "Circular-TE$_{0n}$ Mode Filters for Millimetre-Wave Communication", Electron. Commun. Jap., 1974, 57, no.1, pp.102-11.

48.308. Wanselow, R.D., and Taggart, D.A.: "Circularly Polarized Equalizer Networks", Trans. IEEE, 1974, MTT-22, no.1, pp.63-6.

48.309. Matsuhara, M., and Hill, K.O.: "Optical Waveguide Bandreject Filters", Appl. Opt., 1974, 13, no.12, pp.2886-8.

49. TERRESTRIAL PROPAGATION

49.1. de Wolf, D.A.: "Point-to-Point Propagation Through an Intermediate Layer of Random Anisotropic Irregularities", Trans. IEEE, 1965, AP-13, p.48.

49.2. Herman, B.M.: "Multiple-Scatter Effects in Radar Return from Large Hail", J. Geophys. Res., 1965, 70, p.1215.

49.3. Mjolsness, R.C., and Petschek, A.G.: "Radar Attenuation Due to Delayed Gamma Rays from High-Altitude Nuclear Explosions", J. Geophys. Res., 1965, 70, p.2619.

49.4. Murty, B.V.R., Roy, A.K., and Biswas, K.R.: "Radar Echo Intensities Below Bright Band", J. Atmos. Sci., 1965, 22, p.91.

49.5. Deirmendjian, D.: "Complete Scattering Parameters of Polydispersed Hydrometeors in the 1-100 mm Wavelength Range", J. Res. Natl. Bur. Stand., 1965, 69D, p.893.

49.6. Hay, D.R., and Naito, K.: "Investigation of Clear-Air Stratification With Radar and Elevated Instruments", J. Res. Natl. Bur. Stand., 1965, 69D, p.877.

49.7. Smith, P.L.: "Inferring the Refractive-Index Structure of the Troposphere from Scattering Experiments", J. Res. Natl. Bur. Stand., 1965, 69D, p.881.

49.8. Dutton, E.J., and Bean, B.R.: "Bi-Exponential Nature of Tropospheric Gaseous Absorption of Radio Waves", J. Res. Natl. Bur. Stand., 1965, 69D, p.885.

49.9. Fung, A.K., Moore, R.K., and Parkins, B.E.: "Notes on Backscattering and Depolarization by Gently Undulating Surfaces", J. Geophys. Res., 1965, 70, pp.1559 and 1563.

49.10. Raghavan, S.: "Explanation of Anomalous Radar Propagation Following Thunderstorm Activity", Indian J. Meteorol. Geophys., 1965, 16, p.91.

49.11. Thompson, M.C., and Grant, W.B.: "Noise Tests of an Airborne Microwave Refractometer System", Rev. Sci. Instrum., 1965, 36, p.758.

49.12. Gjessing, D.T.: "Microwave Propagation Over Long Distances in the Troposphere. Fluctuations in the Refractive Index and Wind Velocity", Telektronikk, 1965, no.1-2, p.27.

49.13. Siderman, J.A., Peterson, A.C., and Robbiani, R.L.: "Army Atmospheric Sounding System", Trans. IEEE, 1965, MIL-9, p.153.

49.14. Nicolis, J.: "Wavelength Dependence of Back-Reflected Energy from Small Ice Particles During the Melting Process", Proc. IEEE, 1965, 53, p.551.

49.15. Gilmer, R.O., McGavin, R.E., and Bean, B.R.: "Response of Microwave Refractometer Cavities to Atmospheric Variations", J. Res. Natl. Bur. Stand., 1965, 69D, p.1213.

49.16.   Croom, D.L.: "Microwave Studies of the Stratosphere and Mesosphere", Proc. Roy. Soc., 1965, 288A, p.556.

49.17.   Zhevakin, C.A., and Naumov, A.P.: "Absorption of Millimetre Waves in Atmospheric Oxygen", Radiotekh. Elektron., 1965, 10, p.987, and Radio Eng. Electron. Phys., 1965, 10, p.844.

49.18.   Takeya, Y., Okumoto, T., and Tatebe, W.: "Chaff Method for Upper-Atmospheric Wind Measurement", Mem. Fac. Eng. Osaka City Univ., 1965, 5, p.69.

49.19.   Legg, A.J.: "Propagation Measurements at 11 GHz Over a 35-km Near-Optical Path Involving Diffraction at Two Obstacles", Electron. Lett., 1965, 1, p.285.

49.20.   Funakawa, K.: "Measurement of Total Cross Sections of Water Drops at 5-mm Wavelength Utilizing the Shadow Theorem", J. Radio Res. Lab., 1965, 12, p.111.

49.21.   Volynets, L.M., Markovich, M.L., and Muchnik, V.M.: "Results of Rainfall Measurements by Radar With a Correction for Distance", Izv. Akad. Nauk SSSR, Fiz. Atmos. Okeana, 1966, 11, p.617.

49.22.   Rider, G.C.: "Propagation Measurements of Q-Band Rainfall Attenuation", Marconi Rev., 1966, 29, p.24.

49.23.   Burroughs, W.J., Pyatt, C.C., and Gebbie, H.A.: "Transmission of Submillimetre Waves in Fog", Nature, 1966, 212, pp.387-8.

49.24.   Harrold, T.W.: "Measurement of Horizontal Convergence in Precipitation Using Doppler Radar", Q. J. Roy. Meteorol. Soc., 1966, 92, p.31.

49.25.   Hardy, K.R., Atlas, D., and Glover, K.M.: "Multi-Wavelength Backscatter from the Clear Atmosphere", J. Geophys. Res., 1966, 71, p.1537.

49.26.   Carlson, A.B., and Waterman, A.T.: "Microwave Propagation Over Mountain Diffraction Paths", Trans. IEEE, 1966, AP-14, p.489.

49.27.   Bull, G.: "Power Spectra of Atmospheric Refractive Index from Microwave Refractometer Measurements", J. Atmos. Terr. Phys., 1966, 28, p.513.

49.28.   Hoffman, L.A., Wintroub, H.J., and Garber, W.A.: "Propagation Observations at 3.2 mm", Proc. IEEE, 1966, 54, p.449.

49.29.   Voss, R.A.: "Variability of Microwave Transmission Time Over Tropospheric Paths", Trans. IEEE, 1966, AP-14, p.403.

49.30.   Oguchi, T.: "Scattering and Absorption of Millimetre Waves Due to Melting Ice Spheres", Proc. IEEE, 1966, 54, p.883.

49.31.   Shimabukuro, F.I.: "Propagation Through the Atmosphere at 3.3 mm", Trans. IEEE, 1966, AP-14, p.228.

49.32.   Singh, J.K.: "Attenuation of 3-cm Waves Due to Artificial Rain", Def. Sci. J., 1966, 16, p.105.

49.33.   Minervin, V.E., and Shupyaatzkii, A.B.: "Investigation of the Phase State of Clouds by Radar", Izv. Akad. Nauk SSSR, Fiz. Atmos. Okeana, 1966, 2, p.933.

49.34.   Fowler, C.S., Champion, R.J.B., and Tyler, J.N.: "Three-Cavity Refractometer and Associated Telemetry Equipment", Radio Electron. Eng., 1966, 32, p.186.

49.35.   Heywood, B.: "Measurement of Phase Variation on a Microwave Path", N. Z. Eng., 1966, 21, p.36.

49.36.   Kieburtz, R.B., and Fantera, I.A.: "Angle-of-Arrival Measurements Performed Over a 296-km Troposcatter Path", Radio Sci., 1966, 1, p.1245.

49.37.   Lammers, U.: "Measurement of Backscatter from Water Droplets and Effect of Shape at Millimetre Wavelengths", Nachr. Tech. Z., 1966, 19, p.591.

49.38.   Morgunov, L.N.: "Received-Signal Fluctuations at a Vehicle Antenna in Microwave Urban Communication System", Elektrosvyaz, 1966, 20, no.10, and Telecommun. Radio Eng. Pt 1, 1966, 20, no.10, pp.43-8.

49.39.   Rakshit, D.K., and De, A.C.: "Radar Estimation of Rainfall from Convective Clouds", Indian J. Meteorol. Geophys., 1966, 17, no.2, p.257.

49.40.   Stankevich, K.S.: "Absorption of Centimetre Waves by Molecular Oxygen in the Atmosphere", Radiotekh. Elektron., 1966, 11, p.445, and Radio Eng. Electron. Phys., 1966, 11, p.375.

49.41.   Ikegami, F., et al.: "Experimental Studies on Atmospheric Ducts and Microwave Fading", Rev. Electr. Commun. Lab., 1966, 14, p.505.

49.42.   Peter, T.V.: "Polarization of Radar Returns from Rain", Trans. S. Afr. IEE, 1966, 57, p.243.

49.43.   Lhermitte, R.M.: "Application of Pulse Doppler Radar Technique to Meteorology", Bull. Am. Meteorol. Soc., 1966, 47, p.703.

49.44.   Barrett, A.H., Kuiper, J.W., and Lenoir, W.B.: "Observations of Microwave Emission by Molecular Oxygen in the Terrestrial Atmosphere", J. Geophys. Res., 1966, 71, p.4723.

49.45.   Harrold, T.W.: "Attenuation of 8.6-mm Radiation in Rain", Proc. IEE, 1967, 114, p.201.

49.46.   Crane, R.K.: "Coherent Pulse Transmission Through Rain", Trans. IEEE, 1967, AP-15, p.252.

49.47.   Stogryn, A.: "Apparent Temperature of the Sea at Microwave Frequencies", Trans. IEEE, 1967, AP-15, p.278.

49.48. Etcheverry, R.D., et al.: "Measurements of Spatial Coherence in 3.2-mm Horizontal Transmission", Trans. IEEE, 1967, AP-15, no.1, pp.136-41.

49.49. Lane, J.A., Gordon-Smith, A.C., and Zavody, A.M.: "Absorption and Scintillation Effects at 3 mm on a Short Line-of-Sight Radio Link", Electron. Lett., 1967, 3, no.5, pp.185-6.

49.50. Blevis, B.C., Dohoo, R.M., and McCormick, K.S.: "Measurements of Rainfall Attenuation at 8 GHz and 15 GHz", Trans. IEEE, 1967, AP-15, no.3, pp.394-403.

49.51. Weibel, G.E., and Dressel, H.O.: "Propagation Studies in Millimetre-Wave Links", Proc. IEEE, 1967, 55,no.4,pp.497-513.

49.52. Kuhn, U.: "Propagation Investigations at 8.1 GHz on a Line-of-Sight Link", Tech. Mitt. RFZ, 1967, 11, no.2, pp.65-70.

49.53. Krasyuk, N.P.: "Scattering of Microwaves from a Region With Nonuniform Distribution of Particles", Radiotekhnika, 1967, 22, no.3, pp.34-41, and Telecommun. Radio Eng. Pt 2, 1967, 22, no.3.

49.54. Hearson, L.T.: "Unusual Propagation Factors in Point-to-Point Microwave System Performance", Trans. IEEE, 1967, COM-15, no.4, pp.615-25.

49.55. Kundu, M.M., and De, A.C.: "Radar Study of Line-Type Echoes", Indian J. Meteorol. Geophys., 1967, 18,no.2,pp.247-54.

49.56. Caton, W.M., Welch, W.J., and Silver, S.: "Absorption and Emission in the 8-mm Region by Ozone in the Upper Atmosphere", J. Geophys. Res., 1967, 72, no.24, pp.6137-8.

49.57. Kirdyashev, K.P.: "Variations of Radio Emission in Cloudy Atmosphere at Microwave Frequencies", Radiotekh. Elektron., 1967, 12, no.12, pp.2099-107, and Radio Eng. Electron. Phys., 1967, 12, no.12.

49.58. Picherit, F.: "Measurement of Atmospheric Absorption at 35 GHz Using Solar Radiation", C. R. Acad. Sci. (Paris), 1968, 266B, no.12, pp.784-6.

49.59. Atlas, D., Naito, K., and Carbone, R.E.: "Bistatic Microwave Probing of a Refractively Perturbed Clear Atmosphere", J. Atmos. Sci., 1968, 25, no.2, pp.257-68.

49.60. Crawford, A.B., Hogg, D.C., and Kummer, W.H.: "Relationship Between Cross-Path Winds and Fading Rates in Microwave Propagation Beyond the Horizon", Proc. IEEE, 1968, 56, p.1758.

49.61. Altshuler, E.E., et al.: "Troposcatter Propagation Experiment at 15.7 GHz over a 500-km Path", Proc. IEEE, 1968, 56, p.1729.

49.62. Evans, G.C., and Shaw, A.H.: "Measurements of the Propagation Component of Radar Tracking Noise at 3 cm", Proc. IEE, 1968, 115, p.1431.

49.63. Plechkov, V.M.: "Radiowave Absorption in the Atmosphere at 1.8-2.7 cm", Izv. VUZ Radiofiz., 1968, 11, p.1435.

49.64. Eklund, F., and Wickerts, S.: "Wavelength Dependence of Microwave Propagation Far Beyond the Horizon", Radio Sci., 1968, 3, p.1066.

49.65. Chang, S.Y., and Lester, J.D.: "Performance of a 300-GHz Radiometer for Atmospheric Attenuation Measurements", Trans. IEEE, 1968, AP-16, p.588.

49.66. Davydova, I.N.: "Computation of Laser Radiation Backscattered Along Oblique Paths", Opt.-Mekh. Prom., 1968, 35, no.2, p.6, and Sov. J. Opt. Technol., 1968, 35, p.77.

49.67. Dubost, G., and Thuilleaux, J.M.: "Influence of Diameter of Rain Drops on Scattering at 33 GHz", Ann. Telecommun., 1968, 22, no.9-10, p.249.

49.68. Robert, A., and Charvier, H.: "Microwave Radiometer for Meteorology", Ann. Radioelectr., 1968, 23, p.189.

48.69. Sloss, P.W., and Atlas, D.: "Wind-Shear and Reflectivity-Gradient Effects on Doppler Radar Spectra", J. Atmos. Sci., 1968, 25, p.1080.

49.70. Cato, J.E., et al.: "Propagation in the Earth's Atmosphere", Trans. IEEE, 1969, AP-17, p.110.

49.71. Mitchell, A.J.M., Fitzsimons, T.K., and Rider, G.C.: "Tropospheric Scatter Propagation Experiments", Radio Electron. Eng., 1969, 37, p.16.

49.72. Tyabotov, A.E., et al.: "Study of Optical Characteristics of the Atmosphere With a Laser", Izv. Akad. Nauk SSSR, Fiz. Atmos. Okeana, 1969, 5, no.2, p.192.

49.73. Staeca, G.B.: "Propagation Tests at 11 GHz and 18 GHz on Two Unequal Paths", Alta Freq., 1969, 38, no.5, p.345.

49.74. King, R.W.: "Design of Model Propagation Path to Study Obstacle Diffraction", Electron. Lett., 1969, 5, p.234.

49.75. Mondloch, A.J.: "Overwater Propagation of Millimetre Waves", Trans. IEEE, 1969, AP-17, p.82.

49.76. Gorelik, A.G., and Logunov, V.F.: "Determination of Vertical Motions in Rain by Doppler Radar", Izv. Akad. Nauk SSSR, Fiz. Atmos. Okeana, 1969, 5, no.5, p.543.

49.77. Zuev, V.E., Sosmin, A.V., and Khmelevtsov, S.S.: "Transparency of Atmospheric Surface Layer for IR Lasers", Izv. Akad. Nauk SSSR, Fiz. Atmos. Okeana, 1969, 5, no.2, p.201.

49.78. Fitzmaurice, M.W., Bufton, J.L., and Minott, P.O.: "Wavelength Dependence of Laser-Beam Scintillation", J. Opt. Soc. Am., 1969, 59, p.7.

49.79. Moreland, J.P., and Collins, S.A.: "Optical Heterodyne Detection of a Randomly Distorted Signal Beam", J. Opt. Soc. Am., 1969, 59, p.10.

49.80. Hardy, K.R., and Katz, I.: "Probing the Clear Atmosphere With High-Power, High-Resolution, Radars", Proc. IEEE, 1969, 57, p.468.

49.81. Schneider, A.: "Oversea Radar Propagation Within a Surface Duct", Trans. IEEE. 1969, AP-17, p.254.

49.82. Malyshenko, Yu.I.: "Measurement of Absorption of Water Vapour in the 1.3-mm Window", Radiotekh. Elektron., 1969, 14, no.3, p.522, and Radio Eng. Electron. Phys., 1969, 14, no.3, pp.447-8.

49.83. Artem'ev, A.V.: "Distortion of Coherence by Atmospheric Turbulence", Radiotekh. Elektron., 1969, 14, no.3, p.544, and Radio Eng. Electron. Phys., 1969, 14, no.3, pp.469-71.

49.84. Majumdar, S.C.: "Tropospheric Radio-Refractivity Over India", Radio Electron. Eng., 1969, 38, no.2, p.99.

49.85. Rosner, R.D.: "Performance of an Optical Heterodyne Receiver for Various Receiving Apertures", Trans. IEEE, 1969, AP-17, p.324.

49.86. Semplak, R.A., and Turrin, R.H.: "Measurements of Attenuation by Rainfall at 18.5 GHz", Bell Syst. Tech. J., 1969, 48, p.1767.

49.87. Naumov, A.P., and Stankevich, V.S.: "Millimetre-Wave Attenuation in Rain", Izv. VUZ Radiofiz., 1969, 12, no.2, p.181.

49.88. Plechkov, V.M.: "Atmospheric Absorption in the Region of Water-Vapour Rotational Resonance at 1.35 cm", Izv. VUZ Radiofiz., 1969, 12, no.2, p.185.

49.89. Gracheva, M.E., and Gurvich, A.S.: "Averaging Effect of Receiving Aperture on Light-Intensity Fluctuations", Izv. VUZ Radiofiz., 1969, 12, no.2, p.253.

49.90. Luchinin, A.G., and Savel'ev, V.A.: "Propagation of a Sinusoidal-Modulated Light Beam in Scattering Medium", Izv. VUZ Radiofiz., 1969, 12, no.2, p.256.

49.91. McCoy, J.H., Rensch, D.B., and Long, R.K.: "Water-Vapour Continuum Absorption of CO_2-Laser Radiation Near 10 micron", Appl. Opt., 1969, 8, p.1471.

49.92. Orszag, A., and Pourney, J.C.: "Evaluation of Atmospheric Density by Optical Radar", Ann. Geophys., 1969, 25, no.2, p.567.

49.93. Birch, J.R., and Burroughs, W.J.: "Observation of Atmospheric Absorption Using Submillimetre Maser Sources", Infrared Phys., 1969, 9, no.2, p.75.

49.94. Liebe, H.J.: "Calculated Tropospheric Dispersion and Absorption Due to 22-GHz Water-Vapour Line", Trans. IEEE, 1969, AP-17, p.621.

49.95. Stotskii, A.A.: "Phase-Difference Measurement of Centimetre Waves Propagated in the Atmosphere Close to Earth", Radiotekh. Elektron., 1969, 14, no.9, p.1547, and Radio Eng. Electron. Phys., 1969, 14, no.9, pp.1343-8.

49.96. Hohn, D.H.: "Atmospheric Propagation of a Laser Beam", Optik, 1969, 30, no.2, p.161.

49.97. Hogg, D.C.: "Statistics on Attenuation of Microwaves by Intense Rain", Bell Syst. Tech. J., 1969, 48, p.2949.

49.98. Ruthroff, C.L.: "Microwave Attenuation and Rain Gauge Measurements", Proc. IEEE, 1969, 57, p.1235.

49.99. Jeske, H.: "Fine-Scale Structure of Refractive Index Over Sea Within Height Interval 50-2400 m", Z. Geophys., 1969, 35, no.5, p.529.

49.100. Sievering, H.C., and Semonin, R.G.: "Laser Radar Coherence Considerations", J. Opt. Soc. Am., 1969, 59, p.1679.

49.101. Yakoto, K., and Ota, K.: "Random Fluctuations of 633-nm Laser Beam Transmitted Through the Atmosphere", Electron. Commun. Jap., 1969, 52, no.7, p.53.

49.102. Laussade, J.P., Yariv, A., and Comly, J.C.: "Optical Communication Through Random-Atmosphere Turbulence", Appl. Opt., 1969, 8, p.1607.

49.103. Whaley, T.W., and Fannin, B.M.: "Characteristics of Propagation Near the 183-GHz H_2O Line", Trans. IEEE, 1969, AP-17, p.682.

49.104. Foster, P.R.: "Atmospheric Effects at 9 mm", Trans. IEEE, 1969, AP-17, p.684.

49.105. Kerr, J.R., Titterton, P.J., and Brown, C.M.: "Atmospheric Distortion of Short Laser Pulses", Appl. Opt., 1969, 8, p.2233.

49.106. Gruss, R.: "Transmission by Laser Beams in the Atmosphere", Laser, 1969, 1, no.4, p.23.

49.107. Yamanaka, C., et al.: "Meteorological Laser Radar", Electron Technol., 1969, 2, no.2-3, p.209.

49.108. Ulaby, F.T., and Straiton, A.W.: "Atmospheric Attenuation Studies in the 183-325 GHz Region", Trans. IEEE, 1969, AP-17, p.337.

49.109. Gracheva, M.E., et al.: "Fluctu-ations of Intensity of a Focused Laser Beam Propagated in the Atmosphere", Radiotekh. Elektron., 1970, 15, no.6, p.1290, and Radio Eng. Electron. Phys., 1970, 15, no.6, pp.1093-5.

49.110. Imai, M., Kikuchi, S., and Matsumoto, T.: "Mode Conversion of Gaussian Light Beams Propagating Through a Random Medium", Radio Sci., 1970, 5, no.7, p.1009.

49.111. Gray, D.A.: "Transit-Time Varia-tions in Line-of-Sight Tropospheric Propaga-tion Paths", Bell Syst. Tech. J., 1970, 49, p.1059.

49.112. Turner, D.J.W., and Turner, D.: "Attenuation Due to Rainfall on a 24-km Link at 11 GHz, 18 GHz, and 36 GHz", Elec-tron. Lett., 1970, 6, p.297.

49.113. Roche, J.F., et al.: "Radio Propa-gation at 27-40 GHz", Trans. IEEE, 1970, AP-18, p.452.

49.114. Vignali, J.A.: "Overwater Line-of-Sight Fade and Diversity Measurements at 37 GHz", Trans. IEEE, 1970, AP-18, p.463.

49.115. Ulaby, F.T., and Straiton, A.W.: "Atmospheric Absorption Between 150 GHz and 350 GHz", Trans. IEEE, 1970, AP-18, p.479.

49.116. Bertolotti, M., et al.: "Optical Processing of the Phase Correlation Induced by a Turbulent Medium in a Laser Beam", Appl. Opt., 1970, 9, p.962.

49.117. S'edin, V.Ya., et al.: "Intensity Fluctuations in Pulsed Laser Beam Propagating in the Atmosphere at Distances Up to 9.8 km", Izv. VUZ Radiofiz., 1970, 13, p.44.

49.118. Gracheva, M.E., Gurvich, A.S., and Kallistratova, M.A.: "Measurement of Dispersion of Strong Intensity Fluctuations of Laser Radiation in the Atmosphere", Izv. VUZ Radiofiz., 1970, 13, p.56.

49.119. Wait, J.R.: "Oblique Reflection of a Plane Wave from a Striated Impedance Surface", J. Math. Phys., 1970, 11, p.1437.

49.120. Gracheva, M.E., Gurvich, A.S., and Kallistratova, M.A.: "Measurements of Mean Level of a Light Wave Propagating in a Turbulent Atmosphere", Izv. VUZ Radiofiz., 1970, 13, p.50.

49.121. Gel'fer, E.I., Filatova, E.I., and Cheremukhin, A.M.: "Intensity of a Focused Laser Beam Passed Through a Turbulent Atmosphere", Izv. VUZ Radiofiz., 1970, 13, p.271.

49.122. Farrow, J.B., and Gibson, A.F.: "Influence of the Atmosphere on Optical Sys-tems", Opt. Acta, 1970, 17, p.317.

49.123. Guinard, N.W., and Daley, J.C.: "Experimental Study of a Sea Clutter Model", Proc. IEEE, 1970, 58, p.543.

49.124. Venkiteshwaran, S.P., and Narayan, V.S.: "Refractive Index Over India for Micro-wave Propagation", J. Sci. Ind. Res., 1970, 29, no.2, p.76.

49.125. Straiton, A.W., Bailey, C.R., and Vogel, W.: "Amplitude Variations of 15-GHz Waves Transmitted Through Clear Air and Through Rain", Radio Sci., 1970,5,no.3,p.551.

49.126. Semplak, R.A.: "Effect of Oblate Raindrops on Attenuation at 30.9 GHz", Radio Sci., 1970, 5, no.3, p.559.

49.127. Rosenberg, V.I.: "Radar Scattering of Centimetric Radiation by Flaky Hail", Izv. Akad. Nauk SSSR, Fiz. Atmos. Okeana, 1970, 6, no.2, p.168.

49.128. Thompson, M.C., and Janes, H.B.: "Measurements of Phase-Front Distortion on an Elevated Line-of-Sight Path", Trans. IEEE, 1970, AES-6, p.645.

49.129. Valenzuela, G.R., and Laing, M.B.: "Doppler Spectra of Radar Sea Echo", J. Geophys. Res., 1970, 75, p.551.

49.130. Sancer, M.I., and Varvatsis, A.D.: "Saturation Calculations for Light Propaga-tion in the Turbulent Atmosphere", J. Opt. Soc. Am., 1970, 60, p.654.

49.131. Ho, T.L.: "Coherence Degradation of Gaussian Beams in a Turbulent Atmosphere", J. Opt. Soc. Am., 1970, 60, p.667.

49.132. Kazaryan, R.A., et al.: "Measure-ment of the Average Structural Characteris-tic of Atmospheric Refractive Index", Proc. IEEE, 1970, 58, no.10, pp.1546-7.

49.133. Lerner, R.M., and Holland, A.E.: "Optical Scatter Channel", Proc. IEEE, 1970, 58, no.10, pp.1547-63.

49.134. Bucher, E.A., Lerner, R.M., and Niessen, C.W.: "Propagation of Light Pulses Through Clouds", Proc. IEEE, 1970, 58, no.10, pp.1564-7.

49.135. Gilder, J.R., and Yao, K.: "Trans-ient Response of Multiple Scattered Laser Radiation", Proc. IEEE, 1970, 58, no.10, pp.1764-6.

49.136. Shimabukuro, F.I., and Epstein, E.E.: "Attenuation and Emission of the At-mosphere at 3.3 mm", Trans. IEEE, 1970, AP-18, p.485.

49.137. Consortini, A., and Ronchi, L.: "Gaussian Beams in Turbulent Media", Appl. Opt., 1970, 9, p.125.

49.138. Reber, E.E., Mitchell, R.L., and Carter, C.J.: "Attenuation at 5 mm in a Variable Atmosphere", Trans. IEEE, 1970, AP-18, p.472.

49.139. Basart, J.P., Miley, G.K., and Clark, B.G.: "Phase Measurements With an Interferometer Baseline of 11.3 km", Trans. IEEE, 1970, AP-18, p.375.

49.140. Janes, H.B., et al.: "Comparison of Simultaneous Line-of-Sight Signals at 9.6 GHz and 34.5 GHz", Trans. IEEE, 1970, AP-18, p.447.

49.141. Semplak, R.A.: "Influence of Heavy Rainfall on Attenuation at 18.5 GHz and 30.9 GHz", Trans. IEEE, 1970, AP-18, p.507.

49.142. Godard, S.: "Propagation of Microwaves Through Precipitation", Trans. IEEE, 1970, AP-18, p.530.

49.143. Ippolito, L.J.: "Millimetre-Wave Propagation Measurements from a Satellite", Trans. IEEE, 1970, AP-18, p.535.

49.144. Brookner, E.: "Multipath Dispersion for Atmospheric Laser Channel", Proc. IEEE, 1970, 58, no.10, pp.1767-9.

49.145. Diament, P., and Teich, M.C.: "Photodetection of Low-Level Radiation Through the Turbulent Atmosphere", J. Opt. Soc. Am., 1970, 60, no.11, pp.1489-94.

49.146. Yokoi, H., Yamada, M., and Satoh, T.: "Attenuation and Scintillation of Microwaves Originating from Space Sources", Electron. Commun. Jap., 1970, 53, no.5, pp.14-15.

49.147. King, H.E., et al.: "Terrain-Backscatter Measurements at 40-90 GHz", Trans. IEEE, 1970, AP-18, no.6, pp.780-4.

49.148. Bouricius, G.M.B., and Clifford, S.F.: "Atmospherically Induced Phase Fluctuations in an Optical Signal", J. Opt. Soc. Am., 1970, 60, no.11, pp.1484-9.

49.149. Clark, J.R., and Karp, S.: "Approximations for Lognormally Fading Optical Signals", Proc. IEEE, 1970, 58, no.12, pp.1964-5.

49.150. Laussade, J.P., and Yariv, A.: "Theory of Optical Propagation Through Random Atmospheric Turbulence", Radio Sci., 1970, 5, no.8-9, pp.1119-26.

49.151. Colavito, C.: "Statistical Study of Fading in Line-of-Sight Microwave Links", Alta Freq., 1970, 39, no.11, pp.964-73.

49.152. Setzer, D.E.: "Computed Transmission Through Rain at Microwave and Visible Frequencies", Bell Syst. Tech. J., 1970, 49, no.8, pp.1873-92.

49.153. Schotland, R.M., and Reiss, N.M.: "Double-Scattering Computations for a Bistatic Pulsed Laser Radar", J. Geophys. Res., 1970, 75, no.36, pp.7581-7.

49.154. D'mitrenko, D.A., et al.: "Measurement of Absorption in the Atmosphere at 5.2-20 cm", Izv. VUZ Radiofiz., 1970, 13, no.12, pp.1761-8.

49.155. Sokolov, A.V., and Sukhonin, Ye.V.: "Attenuation of Submillimetre Waves in Rain", Radiotekh. Elektron., 1970, 15, no.12, and Radio Eng. Electron. Phys., 1970, 15, no.12, pp.2167-71.

49.156. Babkin, Yu.S., et al.: "Attenuation of 0.96-mm Radiation in Snow", Radiotekh. Elektron., 1970, 15, no.12, and Radio Eng. Electron. Phys., 1970, 15, no.12, pp.2171-4.

49.157. Finkel'shteyn, M.I.: "Optimum Pulse Form in Radar Sounding of Sea Ice", Radiotekh. Elektron., 1970, 15, no.12, and Radio Eng. Electron. Phys., 1970, 15, no.12, pp.2179-82.

49.158. Feyzulin, Z.I.: "Amplitude and Phase Fluctuations of a Confined Wave Beam Propagating in a Randomly Inhomogeneous Medium", Radiotekh. Elektron., 1970, 15, no.7, and Radio Eng. Electron. Phys., 1970, 15, no.7, pp.1189-95.

49.159. Miura, S., Furuhama, Y., and Fukushima, M.: "Problems of CO_2-Laser Propagation Through the Atmosphere", Rev. Radio Res. Lab., 1970, 16, no.87, pp.577-84.

49.160. Kawecki, A.: "Deformation of Radar Display of Atmospheric-Precipitation Distribution Resulting from Finite Pulse Dimensions and Wave Attenuation", Pr. PIT, 1970, 20, no.68, pp.7-25.

49.161. Falcone, V.J.: "Atmospheric Attenuation of Microwaves", J. Microwave Power, 1970, 5, no.4, pp.269-78.

49.162. Rosenberg, V.I.: "Radar Characteristics of Rain for Submillimetre Waves", Radiotekh. Elektron., 1970, 15, no.12, and Radio Eng. Electron. Phys., 1970, 15, no.12, pp.2157-63.

49.163. Babkin, Yu.S., et al.: "Measurement of Attenuation in Rain Over 1-km Path at 0.96 mm", Radiotekh. Elektron., 1970, 15, no.12, and Radio Eng. Electron. Phys., 1970, 15, no.12, pp.2164-6.

49.164. Sokolov, A.V.: "Attenuation of Visible and IR Radiation in Rain and Snow", Radiotekh. Elektron., 1970, 15, no.12, and Radio Eng. Electron. Phys., 1970, 15, no.12, pp.2175-8.

49.165. Qureshi, M.A.A., and Shearman, E.D.R.: "Scattering of Waves from Rough Surfaces", Electr. Eng. Rev., 1970, 1, no.2, pp.2-9.

49.166. D'Auria, G., and Solimini, D.: "Coherence Properties of a 9-GHz Propagation Path Near the Ground", Radio Sci., 1970, 5, no.12, pp.1387-95.

49.167. Kawecki, A.: "Problems of Measurements of Atmospheric Propagation", Pr. PIT, 1970, 20, no.67, pp.13-26.

49.168. Andreev, G.A., et al.: "Two-Dimensional Fluctuations of Pulsed Optical Radiation Propagated in the Lower Atmosphere", Izv. VUZ Radiofiz., 1971, 14, no.2, pp.276-84.

49.169. Brookner, E.: "Log-Amplitude Fluctuations of a Laser Beam", J. Opt. Soc. Am., 1971, 61, no.5, p.641.

49.170. Wrixon, G.T.: "Measurements of Atmospheric Attenuation on an Earth-Space Path at 90 GHz Using a Sun Tracker", Bell Syst. Tech. J., 1971, 50, no.1, pp.103-14.

49.171. Hogge, C.B., and Visinsky, W.L.: "Laser Beam Probing of Jet-Exhaust Turbulence", Appl. Opt., 1971, 10, no.4,pp.889-92.

49.172. Setzer, D.E.: "Anisotropic Scattering Due to Rain at Radio-Relay Frequencies", Bell Syst. Tech. J., 1971, 50, no.3, pp.861-8.

49.173. Vigants, A.: "Number and Duration of Fades at 6 GHz and 4 GHz", Bell Syst. Tech. J., 1971, 50, no.3, pp.815-41.

49.174. Selden, A.C.: "(Saturation Absorption) Transmission of an Optical Signal", Electron. Lett., 1971, 7, no.11, pp.287-8.

49.175. Llewellyn-Jones, D.T., and Zavody, A.M.: "Rainfall Attenuation at 110 GHz and 890 GHz", Electron. Lett., 1971, 7, no.12, pp.321-2.

49.176. Kuhn, U., and Nielsen, G.: "Fading Behaviour of 56-km Sea Path and Measurement of Subrefraction at 7 GHz", Tech. Mitt. RFZ, 1971, 15, no.1, pp.31-8.

49.177. Saunders, M.J.: "Cross Polarization at 18 GHz and 30 GHz Due to Rain", Trans. IEEE, 1971, AP-19, no.2, pp.273-7.

49.178. Barnum, J.R.: "High-Frequency Backscatter from Terrain With Cement-Block Walls", Trans. IEEE, 1971, AP-19, no.3, pp.343-7.

49.179. Schuler, C.J., Pike, C.T., and Miranda, H.A.: "Dye-Laser Probing of the Atmosphere Using Resonant Scattering", Appl. Opt., 1971, 10, no.7, pp.1689-90.

49.180. Artem'ev, A.V., and Gurvich, A.S.: "Experimental (Atmospheric Turbulence) Study of Coherence Function Spectra", Izv. VUZ Radiofiz., 1971, 14, no.5, pp.734-8.

49.181. Sirkis, M.D.: "Contribution of Water Vapour to Index-of-Refraction Structure at Microwaves", Trans. IEEE, 1971, AP-19, no.4, pp.572-4.

49.182. Skerjanec, R.E., and Samson, C.A.: "Rain Attenuation Measurements in Mississipi at 10 GHz and 14.43 GHz", Trans. IEEE, 1971, AP-19, no.4, pp.575-8.

49.183. Akeyama, A., and Shimada, S.: "Millimetre-Wave Propagation", Electron. Commun. Jap., 1971, 54, no.5, pp.729-34.

49.184. Vorob'ev, V.V.: "Average Intensity of Light Beam in a Weakly Nonlinear Turbulent Atmosphere", Izv. VUZ Radiofiz., 1971, 14, no.6, pp.865-75.

49.185. Vlasova, T.G., Markus, F.A., and Cheremukhin, A.M.: "Measurement of the Coherence Function of Light Beam Propagated in the Atmosphere", Izv. VUZ Radiofiz., 1971, 14, no.6, pp.876-9.

49.186. Il'ich, G.K., et al.: "Parameters of Modulated Light Signals in a Scattering Medium", Izv. Akad. Nauk SSSR, Fiz., Atmos. Okeana, 1971, 7, no.6, pp.674-7.

49.187. Borisov, V.A., and Markov, I.M.: "Method and Apparatus for Measuring (Optical) Atmospheric Transmission", Opt.-Mekh. Prom., 1971, 37, no.9, pp.11-5, and Sov. J. Opt. Technol., 1971, 37, no.9, pp.578-81.

49.188. Platt, C.M.R., and Gambling, D.J.: "Laser Radar Reflections and Downward IR Flux Enhancement Near Small Cumulus Clouds", Nature, 1971, 232, no.5307,pp.182-5.

49.189. Zaromb, S., et al.: "Lidar Instrument for Spectroscopic Analysis of Air Pollution", Laser Angew. Strahlentech., 1971, 3, no.1, pp.21-3.

49.190. Lhermitte, R.M.: "Probing of Atmospheric Motion by Airborne Pulse-Doppler Radar", J. Appl. Meteorol., 1971, 10, no.2, pp.234-46.

49.191. Lane, J.A., Ashwell, G.E., and Dagnall, A.: "Results of Lidar Probing of the Troposphere", Atmos. Environ., 1971, 5, no.1, pp.49-54.

49.192. Battan, L.J.: "Radar Attenuation by Wet Ice Spheres", J. Appl. Meteorol., 1971, 10, no.2, pp.247-52.

49.193. Boucher, R.J., and Ottersten, H.: "Doppler Radar Observation of Wind Structure in Snow", J. Appl. Meteorol., 1971, 10, no.2, pp.228-33.

49.194. Cooney, J.A.: "Comparisons of Water-Vapour Profiles Obtained by Radiosonde and Laser Backscatter", J. Appl. Meteorol., 1971, 10, no.2, pp.301-8.

49.195. Allen, J.R.: "Measurement of Cloud Emissivity in the 3-8 micron Waveband", J. Appl. Meteorol., 1971, 10, no.2,pp.260-5.

49.196. Gel'fer, E.I., Gurvich, A.S., and Cheremukhin, A.M.: "Intensity Distribution in Focal Plane of Light Beam Passed Through a Turbulent Atmospheric Layer", Izv. VUZ Radiofiz., 1971, 14, no.8, pp.1208-11.

49.197. Imai, M.: "Mode Conversion of Gaussian Light Beams in the Atmosphere", Proc. IEEE, 1971, 59, no.9, pp.1379-80.

49.198. Clifford, S.F., et al.: "Phase Variations in Atmospheric Optical Propagation", J. Opt. Soc. Am., 1971, 61, no.10, pp.1279-84.

49.199. Schappert, G.T.: "Technique for Measuring Visibility", Appl. Opt., 1971, 10, no.10, pp.2325-8.

49.200. Clarricoats, P.J.B., et al.: "Scattering (Noise) from a Turbulent Rocket-Exhaust Jet Illuminated by a Focused Microwave Beam", Electron. Lett., 1971, 17, no.19, pp.597-600.

49.201. Bullington, K.: "Phase and Amplitude Variations in Multipath Fading of Microwave Signals", Bell Syst. Tech. J., 1971, 50, no.6, pp.2039-53.

49.202. Vilar, E., and Mathews, P.A.: "Measurement of Phase Fluctuations on Millimetric Propagation", Electron. Lett., 1971, 7, no.18, pp.566-8.

49.203. Doviak, R.J., Goldhirsch, J., and Miller, A.: "Simultaneous Bistatic and Monostatic Detection of Tropospheric Layers", Trans. IEEE, 1971, AP-19, no.5, pp.714-6.

49.204. Kallistratova, M.A., and Pokasov, V.V.: "Defocusing and Shift Fluctuations of Focused Laser Beam in the Atmosphere", Izv. VUZ Radiofiz., 1971, 14, no.8, pp.1200-7.

49.205. Clifford, S.F.: "Temporal-Frequency Spectra for a Spherical Wave Propagating Through Atmospheric Turbulence", J. Opt. Soc. Am., 1971, 61, no.10, pp.1285-92.

49.206. Laing, M.B.: "Upwind-Downwind Dependence of Doppler Spectra of Radar Sea Echo", Trans. IEEE, 1971, AP-19, no.5, pp.712-4.

49.207. Plass, G.N., and Kattawar, G.W.: "Reflection of Light Pulses from Clouds", Appl. Opt., 1971, 10, no.10, pp.2304-10.

49.208. Williams, H., Wilson, A.S., and Blake, C.C.: "Scattering (Noise) from a Turbulent Rocket Exhaust Jet Illuminated by a Plane (Microwave) Beam", Electron. Lett., 1971, 17, no.19, pp.595-7.

49.209. Kanevskii, M.B.: "Parameter Fluctuations of a Normal Wave at Superrefraction", Izv. VUZ Radiofiz., 1971, 14, no.9, pp.1392-9.

49.210. Klyatskin, V.I., and Tatarskii, V.I.: "Method of Successive Approximations for Propagation in a Medium With Random Large-Scale Inhomogeneities", Izv. VUZ Radiofiz., 1971, 14, no.9, pp.1400-15.

49.211. Petrishchev, V.A.: "Application of the Moment's Method to Propagation of Partially Coherent Light", Izv. VUZ Radiofiz., 1971, 14, no.9, pp.1416-26.

49.212. Gochelashvily, K.S., and Shishov, V.I.: "Multiple Scattering of Light in a Turbulent Medium", Opt. Acta, 1971, 18, no.10, pp.767-77.

49.213. Morita, K., and Yoshida, F.: "Light-Wave Attenuation in Propagation Through the Atmosphere", Rev. Electr. Commun. Lab., 1971, 19, no.5-6, pp.714-25.

49.214. Harries, J.E., and Burroughs, W.J.: "Measurements of Submillimetre Radiation Emitted by the Stratosphere", Q. J. Roy. Meteorol. Soc., 1971, 94, no.414. pp.519-36.

49.215. Kildal, H., and Byer, R.L.: "Comparison of Laser Methods for Remote Detection of Atmospheric Pollutants", Proc. IEEE, 1971, 59, no.12, pp.1644-63.

49.216. Yura, H.T.: "Atmospheric-Turbulence-Induced Laser Beamspread", Appl. Opt., 1971, 10, no.12, pp.2771-3.

49.217. Chiba, T.: "Spot Dancing of a Laser Beam Propagated Through Turbulent Atmosphere", Appl. Opt., 1971, 10, no.11, pp.2456-61.

49.218. Battesti, J., Boithias, L., and Misme, P.: "Determination of Attenuation Due to Rain Above 10 GHz", Ann. Telecommun., 1971, 26, no.11-12, pp.439-44.

49.219. Lefrancois, G.: "Theoretical Model of Equivalent Precipitation Over a Radio Path (above 10 GHz)", Ann. Telecommun., 1971, 26, no.11-12, pp.445-53.

49.220. Gaudio, R., and Boccardo, P.: "Propagation Tests at 7 GHz and 11 GHz", Elettron. Telecomun., 1971, 20, no.5, pp.173-83.

49.221. Dunlop, A.J., and Stachera, H.S.: "Polarization of Electromagnetic Waves Scattered by Water Drops", Electron. Lett., 1971, 7, no.3, pp.87-8.

49.222. Atlas, D., and Srivastava, R.C.: "Method for Radar Turbulence Detection", Trans. IEEE, 1971, AES-7, no.1, pp.179-87.

49.223. Gel'fer, E.I., et al.: "Method of Measuring Intensity-Centre Displacement of a Light Beam Passed Through the Atmosphere", Izv. VUZ Radiofiz., 1971, 14, no.12, pp.1838-42.

49.224. Kinpara, A.: "Attenuation at 10 GHz Due to Precipitation With High-Elevation Propagation", J. Inst. Telev. Eng. Jap., 1971, 25, no.9, pp.706-14.

49.225. Morita, K., and Higuti, I.: "Statistical Studies on Attenuation Due to Rain", Rev. Electr. Commun. Lab., 1971, 19, no.7-8, pp.798-842.

49.226. Semplak, R.A.: "Dual-Frequency Measurements of Rain-Induced Microwave Attenuation on a 2.6-km Propagation Path", Bell Syst. Tech. J., 1971, 50, no.8, pp.2599-606.

49.227. Vaytsel', V.I.: "Optical Heterodyning Through a Turbulent Atmosphere", Radiotekh. Elektron., 1971, 16, no.2, pp.439-41, and Radio Eng. Electron. Phys., 1971, 16, no.2, pp.374-6.

49.228. Murayama, M.: "Laser-Beam Propagation Through Falling Snow", Res. Rep. Nagaoka Tech. Coll., 1971, 7,no.4,pp.279.85.

49.229. Akeyama, A., and Satou, R.: "Measurements of Millimetre-Wave Attenuation Due to Rain", Electr. Commun. Lab. Tech. J., 1971, 20, no.11, pp.2415-27.

49.230. Kwon, K.Y.: "Transmission Characteristics of Laser Light Communication in Water and Atmospheric Media", J. Korean Inst. Electron. Eng., 1971, 8, no.4, pp.168-75.

49.231. Bisyarin, V.P., Bisyarin, I.P., and Sokolov, A.V.: "Attenuation of 10.6-micron Laser Radiation in Artificial and Real Fog", Radiotekh. Elektron., 1971, 16, no.10, pp.1758-64, and Radio Eng. Electron. Phys., 1971, 16, no.10, pp.1589-94.

49.232. Bisyarin, V.P., et al.: "Attenuation of 10.6-micron and 633-nm Laser Radiation in Atmospheric Precipitation", Radiotekh. Elektron., 1971, 16, no.10,pp.1765-9, and Radio Eng. Electron. Phys., 1971, 16, no.10, pp.1594-7.

49.233. Fijisawa, A., and Nakao, S.: "Attenuation Characteristics of Laser Light in Water", Electron. Commun. Jap., 1971, 54, no.7, pp.77-83.

49.234. Chiba, T.: "Beamspread of Laser Light Propagating Through the Atmosphere", Electron. Commun. Jap., 1971, 54, no.11, pp.90-5.

49.235. Nishitsuji, A.: "Anomalous Attenuation of Millimetre Waves Due to Snowfall", Electron. Commun. Jap., 1971, 54, no.11, pp.27-33.

49.236. Nishitsuji, A., et al.: "Millimetre-Wave Propagation Test of Snowfall", Monogr. Res. Inst. Appl. Electr., 1971, no.19, pp.1-20.

49.237. Nishitsuji, A., and Matsumoto, A.: "Problems on Millimetre-Wave Attenuation in Rainfall", Monogr. Res. Inst. Appl. Electr., 1971, no.19, pp.93-105.

49.238. Nishitsuji, A., Hirayama, M., and Matsumoto, A.: "Problems on Millimetre-Wave Attenuation in Snowfall", Monogr. Res. Inst. Appl. Electr., 1971, no.19, pp.79-81.

49.239. Maher, W.E.: "Laser Beam Propagation Model With Hydrodynamic Treatment of Transmission Medium", Appl. Opt., 1972, 11, no.2, pp.249-56.

49.240. Hayes, J.N., Ulrich, P.B., and Aitken, A.H.: "Effects of the Atmosphere on Propagation of 10.6-micron Laser Beams", Appl. Opt., 1972, 11, no.2, pp.257-60.

49.241. Milton, J.E., Anderson, R.C., and Browell, E.V.: "Lidar Reflectance of Fair-Weather Cumulus Clouds at 903 nm", Appl. Opt., 1972, 11, no.3, pp.697-8.

49.242. Breig, E.L.: "Limitations on Atmospheric Thermal Effects for High-Power CO_2-Laser Beams", J. Opt. Soc. Am., 1972, 26, no.4, pp.518-28.

49.243. Norbury, J.R., and White, W.J.K.: "Attenuation at 35.8 GHz Due to Rainfall", Electron. Lett., 1972, 8, no.4, pp.91-2.

49.244. Llewellyn-Jones, D.T., and Zavody, A.M.: "Attenuation Due to Rain at 110 GHz", Electron. Lett., 1972, 8, no.4, p.97.

49.245. Gebhardt, F.G., and Smith, D.C.: "Diffraction of Self-Induced Thermal Distortion of a Laser Beam in a Crosswind", Appl. Opt., 1972, 11, no.2, pp.244-8.

49.246. Lees, M.L.: "High-Resolution Measurement of Microwave Refraction on Short Tropospheric Paths", Trans. IEEE, 1972, AP-20, no.2, pp.176-81.

49.247. Wulfsberg, K.N., and Altshuler, E.E.: "Rain Attenuation at 15 GHz and 35 GHz", Trans. IEEE, 1972, AP-20,no.2,pp.181-7.

49.248. Wheelon, A.D.: "Backscattering by Turbulent Irregularities", Proc. IEEE, 1972, 60, no.3, pp.252-65.

49.249. Collett, E., and Alferness, R.: "Depolarization of a Laser Beam in a Turbulent Medium", J. Opt. Soc. Am., 1972, 26, no.4, pp.529-33.

49.250. Barnett, W.T.: "Multipath Propagation at 4 GHz, 6 GHz, and 11 GHz", Bell Syst. Tech. J., 1972, 51, no.2, pp.321-61.

49.251. Croom, D.L., Davies, P.G., and Powell, R.J.: "Anomalies in Attenuation and Emission by Rain at 37 GHz", Electron. Lett., 1972, 8, no.8, pp.189-91.

49.252. Albertin, F., and Querzola, B.: "Propagation of Laser Beams Through the Atmosphere. I and II", Alta Freq., 1972, 41, no.3, pp.121-33 and no.5, pp.350-73.

49.253. Torrieri, D.J., and Taylor, L.S.: "Irradiance Fluctuations in Optical Transmission Through the Atmosphere", J. Opt. Soc. Am., 1972, 62, no.1, pp.145-7.

49.254. Kanevskii, M.B.: "Influence of Absorption on Amplitude Fluctuations of Submillimetre Waves in the Atmosphere", Izv. VUZ Radiofiz., 1972, 15, no.12, pp.1939-40.

49.255. Weinman, J.A., and Shipley, S.T.: "Effects of Multiple Scattering on Laser Pulses Transmitted Through Clouds", J. Geophys. Res., 1972, 77, no.36, pp.7123-8.

49.256. Yegorov, Yu.P.: "Effect of Atmospheric Fluctuations on Heterodyne Reception of Two-Frequency Optical Signals", Radiotekh. Elektron., 1972, 17, no.6, pp.1312-5, and Radio Eng. Electron. Phys., 1972, 17, no.6, pp.1020-4.

49.257. Richter, K.R.: "Enhanced Microwave Absorption in the Lower Atmosphere of Venus", Radio Sci., 1972, 7, no.4, pp.443-7.

49.258. Nishitsuji, A., et al.: "Calculation of Propagation Loss of He-Ne Laser Light Due to Snowfall", Bull. Res. Inst. Appl. Electr., 1972, 24, no.3, pp.100-17.

49.259. Reagan, J.A., and Herman, B.M.: "Three Optical Methods for Remotely Measuring Aerosol Size Distributions", AIAA J., 1972, 10, no.11, pp.1401-7.

49.260. Yeshenko, S.D., and Lande, B.Sh.: "Radar Mapping of Sea Surface", Radiotekh. Elektron., 1971, 17, no.8, pp.1590-7, and Radio Eng. Electron. Phys., 1972, 17, no.8, pp.1253-8.

49.261. Krasyuk, N.P., Lande, B.Sh., and Megretskaya, I.I.: "Effect of Radar Resolution on Spectral Width of Signals Scattered by Sea Surface", Radiotekh. Elektron., 1972, 17, no.10, pp.2182-4, and Radio Eng. Electron. Phys., 1972, 17, no.10, pp.1743-5.

49.262. Arsen'yan, T.I., et al.: "Interferometric Investigation of Phase Fluctuations of Coherent Light in the Atmosphere", Izv. VUZ Radiofiz., 1972, 15, no.8, pp.1228-32.

49.263. Kerr, J.R., and Eiss, R.: "Transmitter Aperture and Focus Effects of (Atmospheric) Scintillations", J. Opt. Soc. Am., 1972, 62, no.5, pp.682-4.

49.264. Lawrence, R.S.: "Irradiance Fluctuations in Optical Transmission Through the Atmosphere", J. Opt. Soc. Am., 1972, 62, no.5, p.701.

49.265. McCormick, K.S., and Maynard, L.A.: "Measurements of SHF Tropospheric Fading Along Earth-Space Paths at Low Elevation", Electron. Lett., 1972, 8, no.10, pp.274-6.

49.266. Watson, P.A., and Arbabi, M.: "Rainfall Crosspolarization of Linearly and Circularly Polarized Waves at Microwave Frequencies", Electron. Lett., 1972, 8, no.11, pp.283-5.

49.267. Joseph, R.I., and Smith, G.D.: "Propagation in an Evaporation Duct. Analytic Models", Radio Sci., 1972, 7, no.4, pp.433-41.

49.268. Zuev, V.E., Sosnin, A.V., and Khmelevstov, S.S.: "Attenuation of Ruby-Laser Radiation in the Lower Layer of the Atmosphere Due to Temperature Changes of its Wavelength", Zh. Prikl. Spekt., 1972, 17, no.2, pp.361-3.

49.269. Consortini, A., et al.: "Influence of Beam Shape on Phase Fluctuations of Waves Propagated Through a Turbulent Medium", Appl. Opt., 1972, 11, no.5, pp.1229-33.

49.270. Mironov, V.L., and Khmelevtsov, S.S.: "Laser Beam Broadening in the Turbulent Atmosphere Along Sloping Propagation Paths", Izv. VUZ Radiofiz., 1972, 15, no.5,pp.743-50.

49.271. Yura, H.T.: "Mutual Coherence Function of a Finite Optical Beam Propagating in a Turbulent Medium", Appl. Opt., 1972, 11, no.6, pp.1399-406.

49.272. Trunk, G.V.: "Radar Properties of non-Rayleigh Sea Clutter", Trans. IEEE, 1972, AES-8, no.2, pp.196-204.

49.273. Gebhardt, F.G.: "Self-Induced Thermal Distortion Effects on Target Image Quality", Appl. Opt., 1972, 11, no.6, pp.1419-23.

49.274. Nishitsuji, A.: "Calculation of Millimetre-Wave Attenuation in Snowfall", Electron. Commun. Jap., 1972, 54, no.1, pp.74-81.

49.275. Gochelashvily, K.S., and Shishov, V.I.: "Focused Irradiance Fluctuations Beyond a Layer of Turbulent Atmosphere", Opt. Acta, 1972, 19, no.4, pp.327-32.

49.276. Mironov, V.L., and Patrushev, G.Ya.: "Field Fluctuations of a Laser Beam Propagated in the Turbulent Atmosphere", Izv. VUZ Radiofiz., 1972, 15, no.6,pp.865-72.

49.277. Chiba, T., and Sugiura, Y.: "Spot Dancing of Laser Beam in Atmospheric Propagation", Electron. Commun. Jap., 1972, 54, no.2, pp.89-96.

49.278. Gambling, D.J., and Bartusek, K.: "Lidar Observations of Tropospheric Aerosols", Atmos. Environ., 1972, 6, no.3, pp.181-90.

49.279. Fernald, F.G., Herman, B.M., and Reagan, J.A.: "Determination of Aerosol Height Distributions by Lidar", J. Appl. Meteorol., 1972, 11, no.3, pp.482-9.

49.280. Carlson, P.E., and Marshall, J.S.: "Measurement of Snowfall by (Microwave) Radar", J. Appl. Meteorol., 1972, 11, no.3, pp.494-500.

49.281. Akiyama, T., and Sakagami, S.: "Simplified Method of Calculating Microwave Diffraction Loss Over Spherical Earth", Rev. Electr. Commun. Lab., 1972, 20, no.1-2, pp.47-52.

49.282. Hirono, M., et al.: "Measurements of Atmospheric Transmission at Ruby-Laser Wavelength in Fukuoka", Mem. Fac. Sci. Kyushu Univ., 1972, 4B, no.4, pp.111-7.

49.283. Hirono, M., and Uchino, O.: "Measuring the Vertical Distribution of Molecular Oxygen and Temperature Variations in the Upper Atmosphere by Two-Wavelength Lidar", Mem. Fac. Sci. Kyushu Univ., 1972, 4, no.4, pp.119-28.

49.284. Kapitanov, V.A., et al.: "Spectrum Shape of a Radar Echo from Precipitation", Izv. Akad. Nauk SSSR Fiz. Atmos. Okeana, 1972, 8, no.9, pp.963-72.

49.285. Lopez, M.E., and Vickers, W.W.: "Refraction Correction of Rocket-Tracking Radar Inputs in Near Real Time", J. Atmos. Sci., 1972, 29, no.5, pp.893-9.

49.286. Grams, G.W., et al.: "Complex Index of Refraction of Airborne Fly Ash Determined by Laser Radar at 13 km", J. Atmos. Sci., 1972, 29, no.5, pp.900-5.

49.287. Chiba, T., and Sugiura, Y.: "Characteristics of Propagation of a Laser Beam Through the Turbulent Atmosphere. I", NHK Tech. J., 1972, 24, no.4, pp.41-50.

49.288. Hoekstra, P., and Spanogle, D.: "Backscatter from Snow and Ice Surfaces at Near Incident Angles", Trans. IEEE, 1972, AP-20, no.6, pp.788-90.

49.289. Perina, J.: "Photon-Counting Statistics of Light Passing Through an Inhomogeneous Random Medium", Czech. J. Phys., 1972, 22B, no.11, pp.1075-84.

49.290. Vorob'ev, V.V.: "Influence of Heating of a Turbulent Atmosphere by a Light Beam on the Intensity Fluctuations", Kvantovaya Elektron., 1972, no.7, pp.5-13, and Sov. J. Quantum Electron., 1972, 2, no.1, pp.1-7.

49.291. Vartanyan, E.S., et al.: "Measurement of Angular Divergence and Refraction of a Laser Beam in the Ground Layer of the Atmosphere", Kvantovaya Elektron., 1972, no.7, pp.60-2, and Sov. J. Quantum Electron., 1972, 2, no.1, pp.47-8.

49.292. Ryadov, V.Ya., and Furashov, N.I.: "Spectrum of Waver-Vapour Absorption in the 1.1-1.5 mm Band", Izv. VUZ Radiofiz., 1972, 15, no.10, pp.1469-74.

49.293. Ryadov, V.Ya., and Furashov, N.I.: "Atmospheric Absorption in the Window at 0.73 mm", Izv. VUZ Radiofiz., 1972, 15, no.10, pp.1475-85.

49.294. Harrison, H., Herbet, J., and Waggoner, A.P.: "Mie-Theory Computations of Lidar and Nephelometric Scattering Parameters for Power-Law Aerosols", Appl. Opt., 1972, 11, no.12, pp.2880-5.

49.295. Perina, J., and Perinova, V.: "Photon-Counting Statistics of Superposition of Coherent and Chaotic Light Propagated Through the Turbulent Atmosphere", Czech. J. Phys., 1972, 22B, no.11, pp.1085-94.

49.296. Sodha, M.S., et al.: "Propagation of a Gaussian Beam in Planar and Cylindrically Inhomogeneous Media", Opt. Acta, 1972, 19, no.11, pp.941-50.

49.297. Werner, C.: "Lidar Measurements of Atmospheric Aerosol as a Function of Relative Humidity", Opto-Electron., 1972, 4, no.2, pp.125-32.

49.298. Nomura, T.: "Attenuation of Signals in the 12-GHz Band Due to Rainfall", ABU Tech. Rev., 1972, no.23, pp.23-30.

49.299. Nakahara, S., et al.: "Detection of SO_2 in Stack Plume by Laser-Raman Radar", Opto-Electron., 1972, 4, no.2, pp.169-77.

49.300. Proctor, T.D.: "Use of a Gas Laser for Sizing Single Particles of Airborne Dust", J. Phys. E, 1972, 5, no.12, pp.1226-9.

49.301. Collis, R.T.H., and Uthe, E.E.: "Mie Scattering Techniques for Air Pollution Measurement With Lasers", Opto-Electron., 1972, 4, no.2, pp.87-99.

49.302. Inaba, H., and Kobayas, T.: "Laser-Raman Scattering Methods for Remote Detection and Analysis of Atmospheric Pollution", Opto-Electron., 1972, 4, no.2, pp.101-23.

49.303. Measures, R.M., and Pilon, G.: "Study of Tunable Laser Techniques for Remote Mapping of Specific Gaseous Constituents of the Atmosphere", Opto-Electron., 1972, 4, no.2, pp.141-53.

49.304. Smith, W.M.H.: "Method for Detection of Raman Scattering from Atmospheric Pollutants", Opto-Electron., 1972, 4, no.2, pp.161-7.

49.305. Menzies, R.T.: "Remote Sensing With IR Heterodyne Radiometers", Opto-Electron., 1972, 4, no.2, pp.179-86.

49.306. Seals, R.K.: "Atmospheric Windows for HF-Laser Radiation in Range 2.7-3.2 micron", Appl. Opt., 1972, 11, no.12, pp.2979-80.

49.307. Waggoner, A.P., Ahlqvist, N.C., and Charlson, R.J.: "Measurement of Aerosol Total Scatter/Backscatter Ratio", Appl. Opt., 1972, 11, no.12, pp.2886-9.

49.308. Bonczyk, P.A., and Ultee, C.J.: "NO Detection by Zeeman Effect and CO Laser", Opt. Commun., 1972, 6, no.2, pp.196-8.

49.309. Kuhn, U.: "Propagation Measurements at 11 GHz Within the Radio Horizon", Tech. Mitt. RFZ, 1972, 16, no.4, pp.114-8.

49.310. Lukin, V.P., Pokasov, V.V., and Khmelevtsov, S.S.: "Time Characteristics of Phase Fluctuations of Optical Waves Propagating in the Lower Atmosphere", Izv. VUZ Radiofiz., 1972, 15, no.12, pp.1861-6.

49.311. Zagorodnikov, A.A.: "Spectrum of a Radar Signal Scattered by the Sea Surface", Radiotekh. Elektron., 1972, 17, no.3, pp.477-87, and Radio Eng. Electron. Phys., 1972, 17, no.3, pp.369-77.

49.312. Gurvich, A.S., and Pokasov, V.V.: "Spectrum of Fluctuations of Laser Radiation in a Turbulent Atmosphere", Izv. Akad. Nauk SSSR, Fiz. Atmos. Okeana, 1972, 8, no.8, pp.878-9.

49.313. Kazakov, L.Ya.: "Reflectivity of Surfaces in the Optical Range", Radiotekh. Elektron., 1972, 17, no.6, pp.1309-11, and Radio Eng. Electron. Phys., 1972, 17, no.6, pp.1018-20.

49.314. Furuhama, Y., and Fukushima, M.: "Measurement of Log-Irradiance Fluctuations of He-Ne Laser in the Atmosphere", J. Radio Res. Lab., 1972, 19, no.100, pp.197-212.

49.315. Polishchuk, Yu.M., and Rybakov, B.S.: "Experimental Investigation of a Gaussian Model for the Field Below the Horizon at Centimetre Wavelengths", Radiotekh. Elektron., 1972, 17, no.6, p.1191, and Radio Eng. Electron. Phys., 1972, 17, no.6, pp.925-30.

49.316. Gochelashvily, K.S.: "Focused Laser Irradiance Fluctuations in a Turbulent Medium", Opt. Acta, 1973, 20, no.3, pp.193-206.

49.317. Edenhofer, P., Franklin, J.N., and Papas, C.H.: "Inversion Method in (Atmospheric) Wave Propagation", Trans. IEEE, 1973, AP-21, no.2, pp.260-3.

49.318. Ulaby, F.T.: "Absorption in the 220-GHz Atmospheric Window", Trans. IEEE, 1973, AP-21, no.2, pp.266-9.

49.319. Morris, G.J.: "Airborne Laser-Beam Scintillation Measurements at High Altitudes", J. Opt. Soc. Am., 1973, 63, no.3, pp.263-70.

49.320. Watson, P.A.: "Rainfall Cross-polarization at Microwave Frequencies", Proc. IEE, 1973, 120, no.4, pp.413-8.

49.321. Gel'fer, E.I., Kon, A.I., and Cheremukhin, A.M.: "Correlation of Centre-of-Intensity Shift of a Focused Light Beam in the Turbulent Atmosphere", Izv. VUZ Radiofiz., 1973, 16, no.1, pp.245-53.

49.322. Yura, H.T.: "Propagation of Finite Laser Beams in Sea Water", Appl. Opt., 1973, 12, no.1, pp.108-15.

49.323. Mason, J.B., and Lindberg, J.D.: "Laser Beam Behaviour on a Long High Path", Appl. Opt., 1973, 12, no.2, pp.187-90.

49.324. Brinkworth, B.J.: "Pulsed-Lidar Reflectance of Clouds", Appl. Opt., 1973, 12, no.2, pp.427-8.

49.325. Ulrich, P.B., and Wallace, J.: "Propagation Characteristics of Collimated Pulsed Laser Beams Through an Absorbing Atmosphere", J. Opt. Soc. Am., 1973, 63, no.1, pp.8-12.

49.326. Kerr, J.R., and Dunphy, J.R.: "Experimental Effects of Finite Transmitter Apertures on Scintillations", J. Opt. Soc. Am., 1973, 63, no.1, pp.1-8.

49.327. Sica, L.: "Interferometric Observations of Kinetic (Atmospheric) Cooling", Appl. Phys. Lett., 1973, 22, no.8, pp.396-8.

49.328. Arbabi, M., and Watson, P.A.: "Slant-Path Microwave Propagation Through Distorted Raindrops", Electron. Lett., 1973, 9, no.8-9, pp.187-8.

49.329. Lin, S.H.: "Statistical Behaviour of Rain Attenuation", Bell Syst. Tech. J., 1973, 52, no.4, pp.557-81.

49.330. Chiba, T., and Sugiura, Y.: "Characteristics of Propagation of a Laser Beam Through the Turbulent Atmosphere. II", NHK Tech. J., 1973, 25, no.1, pp.42-50.

49.331. Mandics, P.A., Lee, R.W., and Waterman, A.T.: "Spectra of Short-Term Fluctuations of Line-of-Sight Signals", Radio Sci., 1973, 8, no.3, pp.185-201.

49.332. Morrison, J.A., Cross, M.J., and Chu, T.S.: "Rain-Induced Differential Attenuation and Phase Shift at Microwave Frequencies", Bell Syst. Tech. J., 1973, 52, no.4, pp.599-604.

49.333. Collett, E., Alferness, R., and Forbes, T.: "Log-Intensity Correlations of a Laser Beam in a Turbulent Medium", Appl. Opt., 1973, 12, no.5, pp.1067-70.

49.334. Waldteufel, P.: "Attenuation of Microwaves by Rain", Ann. Telecommun., 1973, 28, no.5-6, pp.255-72.

49.335. Mendes, J.A.C.S.: "Attenuation Measurements at 11 GHz Under Rain Conditions", Tecnica, 1973, 35, no.419, pp.171-9.

49.336. Hogg, D.C.: "Intensity and Extent of Rain on Earth-Space Paths", Nature, 1973, 243, no.5406, pp.337-8.

49.337. Luder, J., Radermacher, K., and Schlotterbeck, W.: "Effect of Rainfall on Attenuation in the 15-GHz Band", Tech. Mitt. AEG-Telefunken, 1973, 63, no.2, pp.55-7.

49.338. Atlas, D., Srivastava, R.C., and Sekhon, R.S.: "Doppler-Radar Characteristics of Precipitation at Vertical Incidence", Rev. Geophys. Space Phys., 1973, 11, no.1, pp.1-35.

49.339. Samson, C.A., and Kirby, R.S.: "Observed 7-8 GHz Signal Attenuation During Rainfall", Trans. IEEE, 1973, COM-21, no.7, pp.862-3.

49.340. Weil, T.A.: "Atmospheric Lens Effect. Loss for Radar Range", Trans. IEEE, 1973, AES-9, no.1, pp.51-4.

49.341. Byer, R.L., and Garbuny, M.: "Pollutant Detection by Absorption Using Mie Scattering and Topographic Targets as Retroreflectors", Appl. Opt., 1973, 12, no.7, pp.1495-505.

49.342. Harries, J.E., et al.: "Measurements of Submillimetre Stratospheric Emission from a Balloon Platform", Infrared Phys., 1973, 13, no.2, pp.149-55.

49.343. Shaw, G.E., Reagan, J.A., and Herman, B.M.: "Investigations of Atmospheric Extinction Using Solar Radiation and a Multiple-Wavelength Radiometer", J. Appl. Meteorol., 1973, 12, no.2, pp.374-80.

49.344. Taur, R.R.: "Ionospheric Scintillation at 4-6 GHz", COMSAT Tech. Rev., 1973, 3, no.1, pp.145-63.

49.345. Cianos, N., and Waterman, A.T.: "Angle and Doppler Measurements of the Quasi-Coherent and Incoherent Components of Microwave Transhorizon Signals", Trans. IEEE, 1973, AP-21, no.5, pp.746-50.

49.346. Akeyama, A.: "Measurements of Millimetre-Wave Attenuation Due to Rain", Rev. Electr. Commun. Lab., 1973, 21, no.1-2, pp.87-93.

49.347. Kato, S.: "Estimation of Microwave Attenuation Due to Water Vapour Along a Slanting Path", Rev. Electr. Commun. Lab., 1973, 21, no.1-2, pp.94-101.

49.348. Bramley, E.N., and Cherry, S.M.: "Microwave Scattering by Tall Buildings", Proc. IEE, 1973, 120, no.8, pp.833-42.

49.349. Weissman, D.E.: "Two-Frequency Radar Interferometry Applied to Measurement of Ocean Wave Height", Trans. IEEE, 1973, AP-21, no.5, pp.649-56.

49.350. Miura, S., et al.: "Underwater Transmission Characteristics of Laser Beam", Rev. Radio Res. Lab., 1973, 19, no.100, pp.31-9.

49.351. Lutomirski, R.F., and Buser, R.G.: "Mutual Coherence Function of a Finite Optical Beam and Application to Coherent Detection", Appl. Opt., 1973, 12, no.9, pp.2153-60.

49.352. Bucher, E.A.: "Computer Simulation of Light Pulse Propagation for Communication Through Thick Clouds", Appl. Opt., 1973, 12, no.10, pp.2391-400.

49.353. Bucher, E.A., and Lerner, R.M.: "Experiments on Light-Pulse Propagation Through Atmospheric Clouds", Appl. Opt., 1973, 12, no.10, pp.2401-14.

49.354. Dirmans, D., and Doviak, R.J.: "Pulsed-Doppler Velocity Isotach Displays of Storm Winds in Real Time", J. Appl. Meteorol., 1973, 12, no.4, pp.694-7.

49.355. Brown, R.T.: "Lidar for Meteorological Application", J. Appl. Meteorol., 1973, 12, no.4, pp.698-708.

49.356. Perina, J., Perinova, V., and Horak, R.: "Evolution of Photon Statistics of Light Propagating Through a Random Medium. I and II", Czech. J. Phys., 1973, 23B, no.10, pp.975-93.

49.357. Zavody, A.M.: "Effect of Scattering on Radiometer Measurements of Attenuation in Rain", Electron. Lett., 1973, 9, no.15, pp.328-9.

49.358. Khmelevstov, S.S., and Tsvyk, R.Sh.: "Intensity Fluctuations and Angle of Arrival of Finite Collimated Light Beams in a Turbulent Atmosphere", Izv. VUZ Fiz., 1973, no.9, pp.108-12.

49.359. Flavin, R.K., et al.: "11-GHz Radiometer for Rain-Attenuation Studies", Aust. Telecommun. Res., 1973, 7, no.1, pp.10-9.

49.360. Craig, E.R., and Jenkinson, G.F.: "Study of Tropical Rain Attenuation at 11 GHz Using a Solar Radiometer", Aust. Telecommun. Res., 1973, 7, no.1, pp.3-9.

49.361. Cleverley, M.E.: "Observation of Small-Scale Structure of Intense Rain by Scattering of Microwaves", Electron. Lett., 1973, 9, no.22, pp.535-6.

49.362. Fuks, I.M.: "Fluctuations of a Turning Point in a Stratified Inhomogeneous Medium", Izv. VUZ Radiofiz., 1973, 16, no.10, pp.1558-67.

49.363. Lund, I.A.: "Model for Estimating Joint Probabilities of Cloud-Free Lines-of-Sight Through the Atmosphere", J. Appl. Meteorol., 1973, 12, no.6, pp.1040-3.

49.364. Evans, B.G., and Thompson, P.T.: "Use of Cancellation Techniques in the Measurement of Atmospheric Crosspolarization", Electron. Lett., 1973, 9, no.19, pp.447-8.

49.365. Rosengren, L.G.: "Analysis of a Long-Distance System Measuring the Concentration of Atmospheric Gaseous Pollutants", Trans. IEEE, 1973, AES-9, no.5, pp.725-31.

49.366. White, K.O., and Linberg, J.D.: "Measuring the Transmittance of Atmospheric Dust in KBr Pellets Using a CO_2 Laser", Appl. Opt., 1973, 12, no.11, pp.2544-5.

49.367. Carbone, R.E., et al.: "Dual-Wavelength Radar Hail Detection", Bull. Am. Meteorol. Soc., 1973, 54, no.9, pp.921-4.

49.368. Mitrik, L.M.: "Use of Microwave Radiometry for Spectral Investigations of the Earth's Upper Atmosphere", Izv. Akad. Nauk SSSR, Fiz. Atmos. Okeana, 1973, 9, no.10, pp.1092-6.

49.369. Pilon, R.O., and Purves, C.G.: "Radar Imagery of Oil Slicks", Trans. IEEE, 1973, AES-9, no.5, pp.630-6.

49.370. Subramanian, M., O'Brien, K.C., and Puglis, P.J.: "Phase-Dispersion Characteristics During Fade in a Microwave Line-of-Sight Channel", Bell Syst. Tech. J., 1973, 52, no.10, pp.1877-902.

49.371. Morrison, J.A., and Chu, T.S.: "Perturbation Calculations of Rain-Induced Differential Attenuation and Differential Phase Shift at Microwave Frequencies", Bell Syst. Tech. J., 1973, 52, no.10, pp.1907-13.

49.372. Gibbins, C.J.: "Nomograms for Estimating Clear-Sky Zenith Atmospheric Attenuation in Range 80-130 GHz", Electron. Lett., 1973, 9, no.26, pp.605-7.

49.373. Vardanyan, A.S., et al.: "Measurements of 0.98-1.60 mm Atmospheric Absorption by Radioastronomical Means", Radiotekh. Elektron., 1973, 18, no.2, pp.217-20, and Radio Eng. Electron. Phys., 1973, 18, no.2, pp.163-5.

49.374. Neisser, J.: "Fading Statistics of Received Signals Over Tropospheric Microwave Paths", Z. Elektr. Inf. Energietech., 1973, 3, no.4, pp.197-208.

49.375. Raidt, H., and Hohn, D.H.: "Transmission of a GaAs Laser Beam Through the Atmosphere", Appl. Opt., 1973, 12, no.1, pp.103-7.

49.376. Hasegawa, T., Sato, H., and Sakurada, T.: "Study on Propagation of CO_2-Laser Beams Through the Atmosphere", Mem. Def. Acad., 1973, 13, no.2, pp.119-29.

49.377. Lukin, V.P., et al.: "Fluctuation of Phase Modulation of Optical Carriers Propagating in Turbulent Atmosphere", Radiotekh. Elektron., 1973, 18, no.3, pp.502-7, and Radio Eng. Electron. Phys., 1973, 18, no.3, pp.370-4.

49.378. Kucerovsky, Z., et al.: "Characteristics of Laser System for Atmospheric Absorption and Air-Pollution Experiments", J. Appl. Meteorol., 1973, 12, no.8, pp.1387-93.

49.379. Furuhama, Y., et al.: "Propagation Characteristics of Laser Waves Through the Turbulent Atmosphere", Electron. Commun. Jap., 1973, 56, no.4, pp.50-6.

49.380. Gusev, V.D., Polykakov, B.I., and Fadeev, V.V.: "Measurement of Fluctuations Spectra of Laser Radiation Propagating in a Turbulent Atmosphere", Radiotekh. Elektron., 1973, 18, no.5, pp.934-9, and Radio Eng. Electron. Phys., 1973, 18, no.5, pp.677-81.

49.381. Gellwachs, J.: "NO_2 Lidar Comparison. Fluorescence v. Backscattered Differential Absorption", Appl. Opt., 1973, 12, no.12, pp.2812-3.

49.382. Grenier, P., Langlet, A., and Talureav, B.: "Photometer for Submillimetre Measurement (of Sky Background)", Appl. Opt., 1973, 12, no.12, pp.2863-8.

49.383. Gollub, J.P., Chabay, I., and Flygare, W.H.: "Optical Heterodyne Measurement of Cloud Droplet Size Distributions", Appl. Opt., 1973, 12, no.12, pp.2838-42.

49.384. Lang, R.H., and Minott, P.O.: "Determination of Atmospheric Structure Function by Single Coherent Detector", Appl. Opt., 1973, 12, no.12, pp.2843-7.

49.385. Sica, L.: "Three-Beam Interferometer for the Observation of Kinetic Cooling in Air", Appl. Opt., 1973, 12, no.12, pp.2848-54.

49.386. Nakajima, S., Abe, T., and Saito, Y.: "Propagation of Laser Light in Snowfall", Electron. Commun. Jap., 1973, 56, no.8, pp.79-84.

49.387. Semplak, R.A.: "Effect of Rain on Circular Polarization at 18 GHz", Bell Syst. Tech. J., 1973, 52, no.6, pp.1029-31.

49.388. Bukatii, V.I., et al.: "Dynamics of Clearing a Low-Absorbing Turbid Medium in an Intense Light Field", Opt. Spekt., 1973, 35, no.4, pp.720-3, and Opt. Spectrosc., 1973, 35, no.4, pp.418-9.

49.389. Andrianov, V.V., Armand, N.A., and Vetrov, V.I.: "Refraction of Centimetre Waves in the Boundary Layer of the Atmosphere", Radiotekh. Elektron., 1973, 18, no.4, pp.673-80, and Radio Eng. Electron. Phys., 1973, 18, no.4, pp.487-92.

49.390. Armand, N.A., et al.: "Fluctuation of Waves Propagated Through a Turbulent Atmosphere Near the Oxygen 5-mm Absorption Line", Radiotekh. Elektron., 1973, 18, no.4, pp.680-7, and Radio Eng. Electron. Phys., 1973, 18, no.4, pp.492-7.

49.391. Inoue, T.: "Crosspolarization Characteristics for Oversea Line-of-Sight Path", Electron. Commun. Jap., 1973, 56, no.9, pp.98-105.

49.392. Kobayasi, T., and Inaba, H.: "Raman-Laser Radar for Remote Measurement of Atmospheric Molecular Constituents", Electron. Commun. Jap., 1973, 56, no.2, pp.101-9.

49.393. Seger, G.: "Application of Laser Light Sources for Measurement of Air Pollution by Small Particles", Messtechnik, 1973, 81, no.5, pp.142-8.

49.394. Nakahara, S., et al.: "Laser Radar for Monitoring Stack Effluents", Mitsubishi Electr. Eng., 1973, no.6, pp.20-9.

49.395. Kamal, A.K., and Malaviya, N.: "Remote Probing of Atmospheric Turbulence", J. Inst. Eng., 1973, 53, pt ET5, pp.173-5.

49.396. Kelsall, D.: "Optical Seeing Through the Atmosphere by Interferometric Technique", J. Opt. Soc. Am., 1973, 63, no.11, pp.1472-84.

49.397. Platt, C.M.R.: "Lidar and Radiometric Observations of Cirrus Clouds", J. Atmos. Sci., 1973, 30, no.6, pp.1191-204.

49.398. Gerkhen-Gubanov, G.V.: "Effect of Atmospheric Dust on Error of a Laser Distance Meter", Izv. VUZ Prib., 1973, 16, no.11, pp.111-3.

49.399. Tomiyasu, K.: "Remote Sensing of Earth by Microwaves", Proc. IEEE, 1974, 62, no.1, pp.86-92.

49.400. Matthews, D.B., and Cole, R.S.: "Use of Open Spherical Resonator as Sampling Cavity in Microwave Refractometer", J. Phys. E, 1974, 7, no.2, pp.110-4.

49.401. Gracheva, M.E., et al.: "Distribution of Probabilities of Laser-Beam Intensity Fluctuations in the Atmosphere", Izv. VUZ Radiofiz., 1974, 17, no.1, pp.105-12.

49.402. Semplak, R.A.: "Simultaneous Measurements of Depolarization by Rain Using Linear and Circular Polarizations at 18 GHz", Bell Syst. Tech. J., 1974, 53, no.2, pp.400-4.

49.403. Barrick, D.E.: "Wind Dependence of Quasi-Specular Microwave Sea Scatter", Trans. IEEE, 1974, AP-22, no.1, pp.135-6.

49.404. Rosenbaum, S., and Bowles, L.W.: "Clutter Return from Vegetated Areas", Trans. IEEE, 1974, AP-22, no.2, pp.211-20.

49.405. Fruchtenicht, H.W.: "Notes on Duct Influences on Line-of-Sight Propagation", Trans. IEEE, 1974, AP-22, no.2, pp.279-86.

49.406. Wang, J.Y.: "IR Atmospheric Transmission of Laser Radiation", Appl. Opt., 1974, 13, no.1, pp.56-62.

49.407. Mironov, V.L., and Nosov, V.V.: "Influence on Spatial Correlation of Random Displacements of Light Beams", Izv. VUZ Radiofiz., 1974, 17, no.2, pp.247-51.

49.408. Banakh, V.A., Krekov, G.M., and Moronov, V.L.: "Beamspread and Spatial Correlation of Intensity of Wave Beams Propagating in a Turbulent Atmosphere", Izv. VUZ Radiofiz., 1974, 17, no.2, pp.252-60.

49.409. Haikonen, T., and Luomaranta, R.: "Open Resonator for Measurement of Rain Attenuation Above 10 GHz", Saehkoe, 1974, 47, no.3, pp.137-40.

49.410. Ulaby, F.T.: "Radar Measurement of Soil Moisture Content", Trans. IEEE, 1974, AP-22, no.2, pp.241-9.

49.411. Ohtsuka, Y., and Sasaki, I.: "Measurement of Phase Fluctuations of Laser Beam Propagating Through a Turbulent Atmosphere", Appl. Phys., 1974, 3, no.1, pp.15-20.

49.412. Rothe, K.W., Brinkmann, U., and Walther, H.: "Applications of Tunable Dye Lasers to Air-Pollution Detection", Appl. Phys., 1974, 3, no.2, pp.115-9.

49.413. Lin, S.H.: "Occurrence of Very Heavy Rain on a (4-GHz) 42-km Path", Trans. IEEE, 1974, COM-22, no.5, pp.708-10.

49.414. Gardner, C.S., and Plonus, M.A.: "Optical Pulses in Atmospheric Turbulence", J. Opt. Soc. Am., 1974, 64, no.1, pp.68-77.

49.415. Crane, R.K.: "Bistatic Scatter from Rain", Trans. IEEE, 1974, AP-22, no.2, pp.296-304.

49.416. Crane, R.K.: "Propagation Through a Simulated Rain Environment", Trans. IEEE, 1974, AP-22, no.2, pp.305-12.

49.417. Rotheram, S.: "Microwave Propagation in Evaporation Duct", Marconi Rev., 1974, 37, no.192, pp.18-40.

49.418. Hall, M.P.M., and Dowling, G.R.: "Effect of Rain on 4-GHz Troposcatter Radio Paths", Electron. Lett., 1974, 10, no.11, pp.210-2.

49.419. Semplak, R.A.: "Measurement of Rain-Induced Polarization Rotation at 30.9 GHz", Radio Sci., 1974, 9, no.4, pp.425-9.

49.420. Lombardini, P.P.: "Radiation from a Microwave Source in the Intermediate (Earth) Zone", Radio Sci., 1974, 9, no.4, pp.431-7.

49.421. Gel'fer, E.I., et al.: "Measurement of Two-Dimensional Intensity Correlation Function in a Focused Light Beam (in a Turbulent Atmosphere)", Izv. VUZ Radiofiz., 1974, 17, no.5, pp.710-3.

49.422. Fante, R.L.: "Mutual Coherence Function and Frequency Spectrum of a Laser Beam Propagating Through Atmospheric Turbulence", J. Opt. Soc. Am., 1974, 64, no.5, pp.592-8.

49.423. Misme, P., Benoit-Guyot, G.: "Simplification of Theoretical Calculation of Attenuation Due to Rainfall", Ann. Telecommun., 1974, 29, no.3-4, pp.132-8.

49.424. Meyerhoff, H.J., Buige, A., and Robertson, E.A.: "15.3-GHz Precipitation Attenuation Measurements Using a Transportable Earth Station", COMSAT Tech. Rev., 1974, 4, no.1, pp.169-86.

49.425. Taur, R.R.: "Rain Depolarization. Theory and Experiment", COMSAT Tech. Rev., 1974, 4, no.1, pp.187-90.

49.426. Gibbins, C.J.: "Tropospheric Emission and Attenuation Statistics at 110 GHz", Electron. Lett., 1974, 10, no.12, pp.241-2.

49.427. Willis, D.M.: "Phase Variations at Millimetre Wavelengths on an Earth-Space Path Through Model Atmosphere", Electron. Lett., 1974, 10, no.14, pp.281-2.

49.428. King, R.J., et al.: "Experimental Data for Groundwave Propagation Over Cylindrical Surfaces", Trans. IEEE, 1974, AP-22, no.4, pp.551-6.

49.429. Inoue, T., and Akiyama, T.: "Propagation Characteristics on Line-of-Sight Oversea Paths", Trans. IEEE, 1974, AP-22, no.4, pp.557-65.

49.430. Woo, R., and Ishimaru, A.: "Effects of Turbulence in a Planetary Atmosphere on Radio Occultation", Trans. IEEE, 1974, AP-22, no.4, pp.566-73.

49.431. Straiton, A.W., Fannin, B.M., and Perry, J.W.: "Measurements of Index of Refraction and Signal Loss Due to Ice-Fog Medium at 97 GHz Using a Fabry-Perot Resonator", Trans. IEEE, 1974, AP-22, no.4, pp.613-6.

49.432. Okamoto, H., et al.: "Underwater Transmission Characteristics of a Laser Beam", J. Radio Res. Lab., 1974, 21, no.103, pp.19-38.

49.433. Woodman, R.F., and Guillen, A.: "Radar Observations of Winds and Turbulence in the Stratosphere and Mesosphere", J. Atmos. Sci., 1974, 31, no.2, pp.493-505.

49.434. Schotland, R.M.: "Errors in Lidar Measurement of Atmospheric Gases by Differential Absorption", J. Appl. Meteorol., 1974, 13, no.1, pp.71-7.

49.435. DeLong, H.P.: "Air-Pollution Field Studies With a Raman Lidar", Opt. Eng., 1974, 13, no.1, pp.5-9.

49.436. Leonard, D.A.: "Single-Ended Atmospheric Transmissometer", Opt. Eng., 1974, 13, no.1, pp.10-14.

49.437. Fante, R.L.: "Intensity of a Focused Beam in a Turbulent Medium", Proc. IEEE, 1974, 62, no.10, pp.1400-2.

49.438. Wenzel, H.G.: "Calculation of Microwave Path for Distance Measurement Over the Sea", Z. Vermessungswes., 1974, 99, no.5, pp.215-9.

49.439. Anfossi, D., Bacci, P., and Longhetto, A.: "Application of Lidar Technique to Study of Nocturnal Radiation Inversion", Atmos. Environ., 1974, 8, no.6, pp.537-41.

49.440. Cooper, D.W., Davis, J.W., and Byer, R.L.: "Measurements of Depolarization by Dry and Humidified Salt Aerosols Using a Lidar Analogue", J. Aerosol Sci., 1974, 5, no.2, pp.117-23.

49.441. Morrison, J.A., and Cross, M.J.: "Scattering of a Plane Wave by Axisymmetric Raindrops", Bell Syst. Tech. J., 1974, 53, no.6, pp.955-1019.

49.442. Stankevich, K.S.: "Absorption by a Dry Atmosphere of Submillimetre Waves", Izv. VUZ Radiofiz., 1974, 17, no.5, pp.764-6.

49.443. Long, M.W.: "Two-Scatterer Theory of Sea Echo", Trans. IEEE, 1974, AP-22, no.5, pp.662-6.

49.444. Tomiyasu, K.: "Note on Specular Ocean Surface Radar Cross Section", J. Geophys. Res., 1974, 79, no.21, p.3101.

49.445. Gorelik, A.G., and Logunov, V.F.: "Determination of Vertical Velocities and Rain Microstructure by Doppler Spectrum and Intensity of Reflected Signal", Izv. Akad. Nauk SSSR, Fiz. Atmos. Okeana, 1974, 10, no.7, pp.742-51.

49.446. Chaevskii, E.V.: "Wave Propagation in a Medium With Random Irregularities Strongly Extended Along the Direction of Propagation", Izv. VUZ Radiofiz., 1974, 17, no.6, pp.886-95.

49.447. Kannan, D.: "Wave Propagation in One-Dimensional Random Media", J. Math. Phys. Sci., 1974, 8, no.3, pp.201-18.

49.448. de Bettencourt, J.T.: "Statistics of Millimetre-Wave Rainfall Attenuation", J. Rech. Atmos., 1974, 8, no.102, pp.89-119.

49.449. Oguchi, T., and Hosoya, Y.: "Differential Attenuation and Phase Shift of Millimetre Waves Due to Rain", J. Rech. Atmos., 1974, 8, no.1-2, pp.121-8.

49.450. Evans, B.G., and Thompson, P.T.: "Crosspolarization Due to Precipitation at 11.6 GHz", J. Rech. Atmos., 1974, 8, no.1-2, pp.129-36.

49.451. Wallace, J., Itskan, I., and Camm, J.: "Irradiance Tailoring as a Method of Reducing Thermal Blooming in an Absorbing Medium", J. Opt. Soc. Am., 1974, 64, no.8, pp.1123-8.

49.452. Davies, P.G., and Croom, D.L.: "Diversity Measurements of Attenuation at 37 GHz With Solar-Tracking Radiometers", Electron. Lett., 1974, 10, no.23, pp.482-3.

49.453. Harden, B.N., Norbury, J.R., and White, W.J.K.: "Model of Intense Convective Rain Cells for Estimating Attenuation on Terrestrial Millimetric Links", Electron. Lett., 1974, 10, no.23, pp.483-4.

49.454. Walsh, E.J.: "Analysis of Experimental Radar Altimeter Data", Radio Sci., 1974, 9, no.8-9, pp.711-22.

49.455. Srivastava, R.C., and Atlas, D.: "Effect of Finite Radar Pulse Volume on Turbulence Measurements", J. Appl. Meteorol., 1974, 13, no.4, pp.472-80.

49.456. Consortini, A., and Ronchi, L.: "Laser Propagation Through Atmospheric Turbulence", Alta Freq., 1974, 43, no.10, pp.769-72.

49.457. Chu, T.S.: "Rain-Induced Crosspolarization at Centimetre and Millimetre Wavelengths", Bell Syst. Tech. J., 1974, 53, no.8, pp.1557-79.

49.458. Bodtmann, W.F., and Ruthroff, C.L.: "Rain Attenuation on Short Paths", Bell Syst. Tech. J., 1974, 53, no.7, pp.1329-49.

49.459. Andrianov, V.V., Armand, N.A., and Rakitin, B.V.: "Measurement of Radiation Patterns of Microwaves Propagating in the Earth-Bound Layer of the Atmosphere", Izv. VUZ Radiofiz., 1974, 17, no.10, pp.1478-85.

49.460. Shapiro, J.H.: "Optimum Adaptive Imaging Through Atmospheric Turbulence", Appl. Opt., 1974, 13, no.11, pp.2609-13.

49.461. Shapiro, J.H.: "Normal-Mode Approach to Wave Propagation in the Turbulent Atmosphere", Appl. Opt., 1974, 13, no.11, pp.2614-9.

49.462. Seals, R.K.: "Analysis of Tunable Laser Heterodyne Radiometry. Remote Sensing of Atmospheric Gases", AIAA J., 1974, 12, no.8, pp.1118-22.

49.463. Konenenko, L.G., and Muchnik, V.M.: "Accuracy of Rainfall Measurement by Single-Wave Radar", Meteorol. Gidrol., 1974, no.9, pp.107-10.

49.464. Fante, R.L.: "Numerical Evaluation of Mutual Coherence Function of a Laser Beam in Atmospheric Turbulence", Proc. IEEE, 1974, 62, no.11, pp.1604-6.

49.465. Inoue, T., Sakagami, S., and Ogawa, M.: "Oversea Propagation Characteristics at 18-22 GHz", Electr. Commun. Lab. J., 1974, 23, no.6, pp.1267-81.

49.466. Morita, K., et al.: "Propagation Characteristics Due to Rain at 20 GHz Band", Rev. Electr. Commun. Lab., 1974, 22, no.7-8, pp.619-32.

49.467. Valenzuela, G.R.: "Effect of Capillarity and Resonant Interactions on the Second-Order Doppler Spectrum of Radar Sea Echo", J. Geophys. Res., 1974, 79, no.33, pp.5031-7.

49.468. Lugomer, S., and Stipancic, M.: "Atmospheric Diagnostics by Method of Laser Spectroscopy", Elektrotehnika, 1974, 17, no.6, pp.346-9.

49.469. Fuks, I.M.: "Correlation of Frequency-Spaced Signal Fluctuations in a Randomly Inhomogeneous Medium", Izv. VUZ Radiofiz., 1974, 17, no.11, pp.1665-70.

49.470. Krekov, G.M., and Titov, G.A.: "Spatial Energy Structure of Light Haze Near an Optical Communication Channel", Izv. VUZ Radiofiz., 1974, 17, no.11, pp.1678-83.

49.471. Agarwal, D.C.: "Optical Wave Propagation Through Turbulent Atmosphere", J. Inst. Eng., 1974, 54, pt ET3, pp.81-2.

49.472. Antipov, A.B., and Ponomarev, Yu.N.: "Investigation of Weak Absorption Lines of Gases With a Laser Spectrophone", Kvantovaya Elektron., 1974, 1, no.6, pp.1345-9, and Sov. J. Quantum Electron., 1974, 4, no.6, pp.740-2.

49.473. Kerhardt, J., Kourimski, J., and Kuhn, U.: "Measurement for Assessing Interference in the 3.5 GHz Band by Tropospheric Long-Range Propagation", Tech. Mitt. RFZ, 1974, 18, no.3, pp.65-70.

49.474. Spencer, D.J., Denault, G.C., and Takimoto, H.H.: "Atmospheric Gas Absorption at DF-Laser Wavelengths", Appl. Opt., 1974, 13, no.12, pp.2855-68.

49.475. Corsi, S., et al.: "Atmospheric Noise in the Far-IR", Trans. IEEE, 1974, MTT-22, no.12, pp.1036-44.

49.476. Gochelashvily, K.S., Pevgov, V.G., and Shishov, V.I.: "Saturation of Intensity Fluctuations of Laser Radiation at Large Distances in a Turbulent Atmosphere", Kvantovaya Elektron., 1974, 1, no.5, pp.1156-65, and Sov. J. Quantum Electron., 1974, 4, no.5, pp.632-7.

49.477. Krekov, G.M., Krekova, M.M., and Khmelevtsov, S.S.: "Time Transformation of a Lidar Signal in the Brightened Zone of an Optical Channel", Izv. VUZ Fiz., 1974, no.11, pp.72-8.

49.478. Menzies, R.T., and Chahine, M.T.: "Remote Atmospheric Sensing With Airborne Laser Absorption Spectrometer", Appl. Opt., 1974, 13, no.12, pp.2840-9.

49.479. Prade, B.: "Polarization Effects in Random (Atmosphers) Optics", Rev. Cethedec, 1974, 11, no.40, pp.1-2.

49.480. Herrmann, H., et al.: "Lidar Measurements of Atmospheric Visibility", Alta Freq., 1974, 43, no.9, pp.732-5.

49.481. Stanchev, K.I., Petrov, R.D., and Boev, P.V.: "Possibility of Using 10-cm Radar Equipment to Indicate Hail-Bearing Clouds", C. R. Acad. Bulg. Sci., 1974, 27, no.10, pp.1367-70.

49.482. Kitamura, S., et al.: "Laser Radar System for Observing Atmospheric Conditions", Technol. Rep. Osaka Univ., 1974, 24, no.1191-1229, pp.545-55.

50. PRACTICAL ANTENNA FEATURES

50.1. Seshadri, S.R.: "Radiation from an Electric Dipole in a Plasma Column", Proc. IEE, 1965, 112, p.249.

50.2. Davis, M.: "Design Considerations for a Steerable Satellite Phased-Array Antenna", Int. Conv. Rec. IEEE, 1965, 13, pt 2, p.114.

50.3. Denison, E., and Rogers, G.L.: "Aerial Noise Temperatures at 5.65 GHz", Proc. IEE, 1965, 12, p.1075.

50.4. Penzias, A.A., and Wilson, R.W.: "Measurement of Excess Antenna Temperature at 4.08 GHz", Astrophys. J., 1965, 142, p.419.

50.5. Stranak, F.: "Periscope Antenna With Phase Equalization", Slab. Obz., 1965, 26, p.279.

50.6. Lo, Y.T., and Lee, S.W.: "Optimization of Signal-to-Noise Ratio and Gain of Arbitrary Antenna Arrays", Proc. IEEE, 1965, 53, p.655.

50.7. Foldes, P., et al.: "Cassegrainian Feed for Wideband Satellite Communications", RCA Rev., 1965, 26, p.369.

50.8. Takada, M., and Shinji, M.: "Diffractor Grating for 11-GHz Microwave Systems", Trans. IEEE, 1965, AP-13, p.532.

50.9. Cooper, D.N.: "Circularly Polarized Monopulse Feed", Proc. IEEE, 1965,53,p.1252.

50.10. Pomot, C., et al.: "Van Atta Type Reflector", C. R. Acad. Sci. (Paris), 1965, 260, p.1889.

50.11. Matveenko, L.I., Kardashev, N.S., and Sholomitskii, G.B.: "Large-Baseline Radio Interferometer", Izv. VUZ Radiofiz., 1965, 8, p.651, and Sov. Radiophys., 1965, 8, p.461.

50.12. Giger, A.J., and Turrin, R.H.: "Triply-Folded Horn Reflector for Satellite Communications", Bell Syst. Tech. J., 1965, 44, p.1229.

50.13. Blevis, B.C.: "Losses Due to Rain on Radomes and Antenna Reflecting Surfaces", Trans. IEEE, 1965, AP-13, p.175.

50.14. Mathis, H.F., and Mathis, R.F.: "Synthesizing Air With a Radome Sandwich", Trans. IEEE, 1965, MTT-13, p.708.

50.15. Kay, A.F.: "Electrical Design of Metal Space-Frame Radomes", Trans. IEEE, 1965, AP-13, p.188.

50.16. Webster, A.J.: "Wind Torques on Rotating Radar Aerials", Marconi Rev., 1965, 28, p.147.

50.17. Hirst, H., and McKee, K.E.: "Wind Forces on Parabolic Antennas", Microwave J., 1965, 8, no.11, p.43.

50.18. Tolbert, C.W., Straiton, A.W., and Krause, L.C.: "Millimetre-Wave 5-m Antenna System", Trans. IEEE, 1965, AP-13, p.225.

50.19. de Marchin, P., and Tyras, G.: "Radiation from an Infinite Axial Slot on a Circular Cylinder Clad With Magnetoplasma", J. Res. Natl. Bur. Stand., 1965, 69D, p.529.

50.20. Galejs, J.: "Self and Mutual Admittances of Waveguides Radiating Into Plasma Layers", J. Res. Natl. Bur. Stand., 1965, 69D, p.179.

50.21. Hasserjian, G.: "Fields of a Curved Plasma Layer Excited by a Slot", Trans. IEEE, 1965, AP-13, p.339.

50.22. Chen, H.C., and Cheng, D.K.: "Radiation from an Axially Slotted Anisotropic Plasma-Clad Cylinder", Trans. IEEE, 1965, AP-13, p.395.

50.23. Galejs, J.: "Admittance of a Waveguide Radiating Into Stratified Plasma", Trans. IEEE, 1965, AP-13, p.64.

50.24. Villeneuve, A.T.: "Admittance of Waveguide Radiating Into Plasma Environment", Trans. IEEE, 1965, AP-13, p.115.

50.25. Smith, T.M., and Golden, K.E.: "Radiation Patterns of a Slot Covered by a Simulated Plasma Sheet", Trans. IEEE, 1965, AP-13, p.285.

50.26. Fujimoto, K.: "Noise Performance of Amplifier-Antenna Systems", Proc. IEEE, 1965, 53, p.1671.

50.27. Waldman, A., and Wooley, G.J.: "Noise Temperature of a Phased-Array Receiver", Microwave J., 1966, 9, no.9, p.89.

50.28. Schmitt, H.J., Harrison, C.W., and Williams, C.S.: "Response of Thin Cylindrical Antennas to Pulse Excitation", Trans. IEEE, 1966, AP-14, p.120.

50.29. Wait, J.R.: "Radiation from a Spherical-Aperture Antenna in a Compressible Plasma", Trans. IEEE, 1966, AP-14, p.360.

50.30. Ruze, J.: "Antenna Tolerance Theory. Review", Proc. IEEE, 1966, 54, p.633.

50.31. Cheng, D.K., and Tseng, F.I.: "Gain-Optimization Principle for Arrays Responding to Quasi-Monochromatic Periodic Signals", Trans. IEEE, 1966, AP-14, p.250.

50.32. Esepkina, N.A., et al.: "Effect of Atmosphere on Characteristics of Radiotelescopes of Maximally Large Dimensions", Radiotekh. Elektron., 1966, 11, p.1405, and Radio Eng. Electron. Phys., 1966, 11, p.1222.

50.33. Tricoles, C.: "Ray Tracing for Predicting Properties of a Small, Axially Symmetric, Missile Radome", Trans. IEEE, 1966, AP-14, p.244.

50.34. Braude, B.V., et al.: "Optical Methods of Surface Control for Adjustment of Pencil-Beam Radiotelescopes", Radiotekh. Elektron., 1966, 11, p.1499, and Radio Eng. Electron. Phys., 1966, 11, p.1302.

50.35. Zhuravlev, V.S., Petrovskii, A.A., and Pogrebnyi, B.P.: "Universal Radiotelescope With 15-m Reflector", Astron. Zh., 1966, 43, p.220.

50.36. Lee, R.W., and Waterman, A.T.: "Large Antenna Array for Millimetre-Wave Propagation Studies", Proc. IEEE, 1966, 54, p.454.

50.37. Hyltin, T.M.: "MIC's in Phased-Array Radars", Int. Conv. Rec. IEEE, 1966, 14, pt 5, p.105.

50.38. Messiaen, A.M., and Vandenplas, P.E.: "Theory and Experiments of the Enhanced Radiation from a Plasma-Coated Antenna", Electron. Lett., 1967, 3, no.1, pp.26-7.

50.39. Hoerner, S.: "Design of Large Steerable Antennas", Astron. J., 1967, 72, no.1, pp.35-47.

50.40. Samuel, I.J.: "Chilbolton Steerable Aerial", AEI Eng., 1967, 7, no.1, pp.6-15.

50.41. McKenzie, J.F.: "Dipole Radiation in Moving Media", Proc. Phys. Soc., 1967, 91, pt 3, pp.537-51.

50.42. Reed, H.H.: "Noise Curves for High-Gain Antennas", Microwaves, 1967, 6, no.4, pp.46-9.

50.43. Drabowitch, S.: "Optimization of a Cold Antenna by Use of Multimode Sources", Onde Electr., 1967, 47, no.2, pp.281-93.

50.44. Mandel, P., Roger, G., and Tocquec, Y.: "High-Gain, Low-Noise, Antenna for Space Communications", Onde Electr., 1967, 47, no.2, pp.258-65.

50.45. Wunsch, A.D.: "Current Distribution on a Dipole Antenna in a Warm Plasma", Electron. Lett., 1967, 3, no.7, pp.320-1.

50.46. Oh, L.L., and Lee, H.F.: "Radant. Integrated Radome-Antenna System", Microwave J., 1967, 10, no.8, pp.50-6.

50.47. Kristal, R., and Shizume, P.: "Antenna-Plasma Interaction in a Conical Geometry", Trans. IEEE, 1967, AP-15, no.5, pp.710-2.

50.48. Graff, P.: "Method for Solving the Problem of an Antenna in a Plasma", Ann. Telecommun., 1967, 22, no.1-2, pp.3-16.

50.49. Zarghamee, M.S.: "Antenna Tolerance Theory", Trans. IEEE, 1967, AP-15, no.6, pp.777-81.

50.50. McKenzie, J.F.: "Effect of Motion of a Strongly Magnetized Plasma on Radiation by a Finite Dipole", J. Appl. Phys., 1967, 38, no.13, pp.5249-55.

50.51. Profera, C.E., and Yorinks, L.H.: "Improved Cassegrain Monopulse Feed System", RCA Rev., 1967, 28, no.4, pp.620-33.

50.52. Zucker, H.: "Gain of Antennas With Random Surface Deviations", Bell Syst. Tech. J., 1968, 47, p.1637.

50.53. Becker, K.D.: "Disturbance of Field of an Antenna by a Surrounding Dielectric", Arch. Elektr. Ubertrag., 1968, 22, p.548.

50.54. Vetter, H.: "Studies of Centimetric Antennas Excited by Comparatively Short Pulses", Nachr. Tech., 1968, 18, p.416.

50.55. Jurkiewicz, R.: "Procedure for Fixing Tolerances of Waveguide Slot Antennas", Pr. PIT, 1968, 18, no.62, p.15.

50.56. Miller, E.K.: "Infinite Cylindrical Antenna in a Uniaxial Compressible Plasma", Can. J. Phys., 1968, 46, p.2846.

50.57. Dijk, J., Jeuken, M.E.J., and Maanders, E.J.: "Antenna Noise Temperature", Proc. IEEE, 1968, 115, p.1403.

50.58. Mayhan, J.W., Caldecott, R., and Bohley, P.: "Antenna Impedance in a Reentry Environment", Trans. IEEE, 1968, AP-16, p.573.

50.59. Bachynski, P., and Gibbs, B.W.: "Antenna Pattern Distortion by an Isotropic Plasma Slab", Trans. IEEE, 1968, AP-16, p.583.

50.60. Bohme, J.F.: "Model for Random (Antenna) Noise in Space", Elektr. Ubertrag., 1968, 22, p.585.

50.61. Galejs, J.: "Antenna Impedances in a Cold Plasma With Perpendicular Static Magnetic Field", Trans. IEEE, 1968, AP-16, p.728.

50.62. Nguyen, D.T.: "Linear Antennas in a Warm Plasma Driven from a Coaxial Line", Electron. Lett., 1968, 4, p.475.

50.63. Polishchuk, Yu.M.: "Antenna-Gain Losses Calculated from Phase-Difference Measurements at S Band", Radiotekh. Elektron., 1968, 13, no.10, p.1871, and Radio Eng. Electron. Phys., 1968, 13, no.10, pp.1636-7.

50.64. Nicotra, G.: "Flat and Curved Passive Reflectors in Periscope System", Note Recens. Not., 1968, 17, p.822.

50.65. Howard, D.D.: "Contour Pattern Analysis of a Monopulse Radar Cassegrain Antenna", Microwave J., 1968, 11, no.12, p.61.

50.66. Ottl, H., and Thomanek, L.: "Monopulse Antenna of a Ground Station for Satellite Transmission", Nachr. Tech. Z., 1968, 21, pp.631-799.

50.67. Golden, K.E., and Stewart, G.E.: "Self and Mutual Admittances of Rectangular-Slot Antennas in Inhomogeneous Plasma", Trans. IEEE, 1969, AP-17, p.763.

50.68. Mirovitskii, D.I., et al.: "Laws Governing Radiation Patterns of Multielement Large Antenna Systems", Radiotekh. Elektron., 1969, 14, no.3, p.530, and Radio Eng. Electron. Phys., 1969, 14, no.3.

50.69. Foster, P.R., Flett, A.M., and Howie, I.H.: "Atmospheric Noise at 33.5 GHz", Nature, 1969, 221, p.160.

50.70. Walker, G.B.: "Superconducting Antennas", J. Appl. Phys., 1969, 40, p.2035.

50.71. Horne, M.R., and Barrett, N.T.: "Structural Setting and Operation of Large Paraboloidal Antennas for Optimum Performance", Int. J. Mech. Sci., 1969, 11, no.1, p.87.

50.72. Calvez, C.A., and Casey, K.F.: "Normalized Signal-to-Noise Ratios of Aperture Antennas in Random Signal and Noise Fields", Trans. IEEE, 1969, AP-17, p.232.

50.73. Meinke, H.H.: "Noise Matching of Transistorized Receiving Antennas", Nachr. Tech. Z., 1969, 22, no.6, p.319.

50.74. Pierrot, R.: "Optimization of Airborne Radomes", Rev. Tech. Thomson-CSF, 1969, 1, no.4, p.597.

50.75. De Vito, G.: "Parabolic-Segment Antenna for High-Capacity Microwave Links", Elettronica Telecom., 1969, 18, no.2, p.38.

50.76. Crawford, A.B., and Turrin, R.H.: "Packaged Antenna for Short-Hop Microwave Links", Bell Syst. Tech. J., 1969, 48, p.1605.

50.77. Jaeger, L.G., and Harris, P.J.: "Distortion of Paraboloidal Antenna Structures", Eng. J., 1969, 52, p.41.

50.78. Woo, R., and Ishimaru, A.: "Radiation from a Circularly Polarized Antenna Through the Ionized Wake of a Mars-Entry Capsule", Trans. IEEE, 1969, AP-17, p.488.

50.79. Kuz'mina, G.A.: "Signal-to-Noise Ratio in Multielement Active and Passive Antennas", Izv. VUZ Radiofiz., 1969, 12, no.8, p.1181.

50.80. Deryugin, L.N., and Chekan, A.V.: "Ultimate Resolving Power of Multibeam Spectrum Analysers", Dokl. Akad. Nauk SSSR, 1969, 184, no.4, p.807, and Sov. Phys.-Dokl., 1969, 14, no.4, p.123.

50.81. Russo, A.J.: "Radiation Pattern of Open-Ended Waveguide Covered by Plasma Layers", Trans. IEEE, 1969, AP-17, p.672.

50.82. Moller, W.: "Modern Radar Antennas. Construction Aspects", Elektro-Tech. Z., 1969, 51, no.24, p.22.

50.83. Nadenenko, B.S.: "Antenna System for Centimetre Tropospheric Links", Elektrosvyaz, 1969, 23, no.10, and Telecommun. Radio Eng. Pt 1, 1969, 23, no.10, p.12.

50.84. Nadenenko, B.S.: "Tropospheric Link Antenna", Elektrosvyaz, 1969, 23, no.4, and Telecommun. Radio Eng. Pt 1, 1969, 23, no.4, p.11.

50.85. Rankin, J.B., Devane, M.E., and Rosenthal, M.L.: "Multifunction Single-Package Antenna System for Spin-Stabilized Near-Synchronous Satellite", Trans. IEEE, 1969, AP-17, p.435.

50.86. Gross, A.A., et al.: "Improvement of Noise Protection and Back-to-Front Ratio of Reflector Antennas", Elektrosvyaz, 1969, 23, no.12, and Telecommun. Radio Eng. Pt 1, 1969, 23, no.12, p.1.

50.87. Croswell, W.F., and Cockrell, C.R.: "Omnidirectional Microwave Antenna for Use on Spacecraft", Trans. IEEE, 1969, AP-17, p.459.

50.88. Chow, Y.L.: "Comparison of Correlation Array Configurations for Radio Astronomy", Trans. IEEE, 1970, AP-18, p.567.

50.89. Paris, D.T.: "Computer-Aided Radome Analysis", Trans. IEEE, 1970, AP-18, p.7.

50.90. Shifrin, Ya.S., and Maslov, A.F.: "Efficiency of Self-Focusing Antennas", Radiotekh. Elektron., 1970, 15, no.2, p.378, and Radio Eng. Electron. Phys., 1970, 15, no.2.

50.91. Otsu, Y.: "Measurement of Sky-Noise Temperature at 16 GHz and 35 GHz", Rev. Radio Res. Lab., 1970, 16, no.85, pp.379-94.

50.92. Waksberg, A.: "Dual-Scan Acquisition Technique for Laser Communication System", Trans. IEEE, 1970, AES-6, p.407.

50.93. Checcacci, P.F., Russo, V., and Scheggi, A.M.: "Holographic Antennas", Trans. IEEE, 1970, AP-18, no.6, pp.811-3.

50.94. Disk, J., Groothuis, H.H.H., and Maanders, E.J.: "Improvements in Antenna-Noise-Temperature Calculations", Trans. IEEE, 1970, AP-18, no.5, pp.690-2.

50.95. Yokoi, H., et al.: "7-m Antenna for Space-Earth Millimetric Propagation Test", KDD Tech. J., 1970, no.65, pp.11-9.

50.96. Lin, C.C., and Chen, K.M.: "Effect of Electroacoustic Wave on Radiation of a Plasma-Coated Spherical Antenna", Trans. IEEE, 1970, AP-18, no.6, pp.831-4.

50.97. Alekseev, V.A., et al.: "Designing Radio Interferometers With Independent Reception", Izv. VUZ Radiofiz., 1970, 13, p.5.

50.98. Hongo, K., and Ohta, M.: "Radiation Field from Electric Dipole in an Anisotropic and Compressible Plasma", Trans. IEEE, 1970, AP-18, p.294.

50.99. Korolev, A.N.: "Linear, Circular, and Crossed Receiving Arrays for Maximum Signal-to-Noise Ratio", Radiotekh. Elektron., 1970, 15, p.166, and Radio Eng. Electron. Phys., 1970, 15, no.1, pp.135-9.

50.100. De Vito, G.: "Feed System for Satellite-Communication Earth Station", Alta Freq., 1970, 39, p.182.

50.101. Lewis, T.S., and Hutchings, H.S.: "Synthetic Aperture at Optical Frequencies", Proc. IEEE, 1970, 58, p.587.

50.102. Keeney, J.: "Observations on Microwave Emission from Colliding Charged Water Drops", J. Geophys. Res., 1970, 75, p.1123.

50.103. Jarvis, E.G.: "Antenna-Noise-Temperature Variations Due to Atmospheric Changes at 3.95 GHz, 11.75 GHz, and 17 GHz", Electron. Lett., 1970, 6, p.254.

50.104. Mayhan, J.W.: "Calculation of Effective Temperature of Planar Antennas in a Plasma Environment", Trans. IEEE, 1970, AP-18, p.136.

50.105. Barrick, D.E.: "Technique for Short-Pulse Compensation in Phased-Array Scanning", Proc. IEEE, 1970, 58, p.1133.

50.106. Tseitlin, N.M.: "Aperture Synthesis in Radio Astronomy. Survey", Radiotekh. Elektron., 1970, 15, p.427, and Radio Eng. Electron. Phys., 1970, 15, no.3, pp.369-87.

50.107. Slater, R.H.: "Radiation Pattern of Imperfect Paraboloidal Reflectors", Electron. Lett., 1970, 6, no.25, pp.796-8.

50.108. Stotskii, A.A., and Shivris, O.N.: "Adjustment and Mounting of Variable Profile Antenna", Izv. Gl. Astron. Obs. Pulkove, 1970, no.185, pp.236-41.

50.109. Wang, J.J.H.: "Tolerance of Aperture Antennas", Proc. IEEE, 1971, 59, no.1, pp.108-9.

50.110. Lin, C.C., and Chen, K.M.: "Radiation from a Spherical Antenna Covered by a Layer of Lossy Hot Plasma", Proc. IEE, 1971, 118, no.1, pp.36-42.

50.111. Fante, R.L.: "Admittance of an Aperture Antenna Radiating Into a Warm Plasma", Trans. IEEE, 1971, AP-19, no.1, pp.150-1.

50.112. Kovner, M.S., Lapidus, V.A., and Lupanov, G.A.: "Radiation from Electric and Magnetic Dipoles Placed in a Cavity Formed by a Plasma Layer", Izv. VUZ Radiofiz., 1971, 14, no.1, pp.28-35.

50.113. Bertossa, S.: "Statistical Determination of the Accuracy of the Main Reflecting Surface of a Space-Communications Antenna", Alta Freq., 1971, 40, no.2, pp.186-93.

50.114. Brennan, L.E., Pugh, E.L., and Reed, I.S.: "Control-Loop Noise in Adaptive Array Antennas", Trans. IEEE, 1971, AES-7, no.2, pp.254-62.

50.115. Casey, K.F.: "Radiation from a Slot Antenna in a Ground Plane Coated With a Moving Plasma Sheath", Trans. IEEE, 1971, AP-19, no.3, pp.401-5.

50.116. Glasman, V.N., et al.: "Nonsteerable Radio Telescope for Millimetre Wavelengths", Izv. VUZ Radiofiz., 1971, 14, no.5, pp.663-72.

50.117. Esepkina, N.A.: "Polarization Characteristics of Radiotelescopes", Izv. VUZ Radiofiz., 1971, 14, no.5, pp.673-9.

50.118. Bell, T.F., and Wang, T.N.C.: "Radiation Resistance of a Small Filamentary Loop Antenna in a Cold Multicomponent Magnetoplasma", Trans. IEEE, 1971, AP-19, no.4, pp.517-22.

50.119. Chu, T.S.: "Focal-Plane Distribution of an Imperfect Paraboloid", Trans. IEEE, 1971, AP-19, no.4, pp.550-2.

50.120. Drabowitch, S., Daveau, B., and Beguerie, H.: "Design for an Experimental Space-Communication Antenna in the Range 10-35 GHz", Onde Electr., 1971, 51, no.6, pp.502-8.

50.121. Barton, D.K.: "Interferometer Phase (Fluctuations) Measurement. Comparison With Theory", Trans. IEEE, 1971, AP-19, no.4, pp.566-9.

50.122. Ishiguro, M.: "Image Correction in High-Resolution Radio Interferometer", Proc. Res. Inst. Atmos. Nagoya Univ., 1971, 18, pp.73-88.

50.123. Meeks, M.L., and Ruze, J.: "Evaluation of the Haystack Antenna and Radome", Trans. IEEE, 1971, AP-19, no.6, pp.723-8.

50.124. Hinder, R., and Ryle, M.: "Atmospheric Limitations to the Angular Resolution of Aperture Synthesis Radiotelescopes", Mon. Not. Roy. Astron. Soc., 1971, 154, no.2, pp.229-53.

50.125. Novotny, Za.: "Noise Parameters of Antennas for Space Links", Slab. Obz., 1971, 32, no.12, pp.541-50.

50.126. Dugin, N.A., et al.: "Use of Two-Element Interferometer With Fixed Base for Obtaining Knife-Edge Radiation Patterns", Radiotekh. Elektron., 1971, 16, no.6, pp.918-28, and Radio Eng. Electron. Phys., 1971, 16, no.6, pp.921-35.

50.127. van Heuven, J.H.C.: "PIN Switching Diodes in Phaseshifters for Electronically Scanned Antenna Arrays", Philips Tech. Rev., 1971, 32, no.9-12, pp.405-12.

50.128. Yokoi, H., and Fukumuro, H.: "Low-Sidelobe Paraboloids With Microwave Absorber", Electron. Commun. Jap., 1971, 54, no.11, pp.34-9.

50.129. Otsu, Y.: "Measurement of Sky-Noise Temperature at 16 GHz and 35 GHz", J. Radio Res. Lab., 1971, 18, no.96, pp.87-111.

50.130. Dorge, G.: "Luneberg Lens Antenna for Satellite Microwave Links", Int. Elektron. Rdsch., 1971, 25, no.3, pp.64-6.

50.131. Cardot, C.: "Analytical Optimization of Cassegrain Antennas of Revolution", Ann. Telecommun., 1971, 26, no.1-2, pp.37-48.

50.132. Fried, D.L., and Yura, H.T.: "(Optical) Telescope Performance Reciprocity for Propagation in a Turbulent Medium", J. Opt. Soc. Am., 1972, 62, no.4, pp.600-2.

50.133. Lee, S.W., and Fong, T.T.: "Electromagnetic-Wave Scattering from an Active Corrugated Structure", J. Appl. Phys., 1972, 43, no.2, pp.388-96.

50.134. Singh, K.P., Shukla, P.K., and Misra, K.D.: "Radiation Resistance of a Cylindrical Antenna in Weakly Ionized Plasma", Int. J. Electron., 1972, 32, no.2, pp.147-52.

50.135. Jeuken, M.E.J., Knoben, M.H.M., and Wellington, K.J.: "Dual-Frequency, Dual-Polarized, Feed for Radio Astronomy", Nachr. Tech. Z., 1972, 25, no.8, pp.374-6.

50.136. Vu, T.B.: "Corrugated Waveguide for Monopulse Feed", Int. J. Electron., 1972, 33, no.4, pp.477-80.

50.137. Sanyal, G.S.: "Noise Considerations in Space Communication Antennas", J. Inst. Telecommun. Eng., 1972, 18, no.9, pp.437-48.

50.138. Stotskii, A.A.: "Aberrations of the Main Mirror of the Variable Profile Antenna and Scanning by Shift of Primary Feed", Izv. Gl. Astron. Obs. Pulkove, 1972, no.188, pp.63-76.

50.139. Spitkovskii, V.M.: "Phase Errors at the Aperture of a Paraboloid With Feed Shifted from the Focus", Izv. Gl. Astron. Obs. Pulkove, 1972, no.188, pp.77-82.

50.140. Gel'freikh, G.B.: "Radioastronomical Method of Adjustment of Variable Profile Antennas", Izv. Gl. Astron. Obs. Pulkove, 1972, no.188, pp.139-48.

50.141. Buloshnikov, A.M., and Savin,
M.G.: "Radiation from a Dipole-Type Antenna
in a Gyrotropic Medium", Kosm. Issled.,
1972, 19, no.5, pp.789-91.

50.142. Danilov, Yu.N.: "Thickness Cal-
culation of a Pointed Monolithic Antenna
Radome for Least Angular Errors", Izv. VUZ
Radioelektron. 1973, 16, no.9, pp.100-2.

50.143. Kuznetsov, V.D.: "Sidelobe Level
of Large Two-Mirror Parabolic Antennas",
Radiotekhnika, 1973, 28, no.9, pp.36-43,
and Telecommun. Radio Eng. Pt 2, 1973, 28,
no.9, pp.76-81.

50.144. Vu, T.B.: "Low-Noise Dual-Hybrid-
Mode Horn", Int. J. Electron., 1973, 34,
no.3, pp.391-400.

50.145. Vu, T.B.: "Corrugated Horn as
High-Performance Monopulse Feed", Int. J.
Electron., 1973, 34, no.4, pp.433-44.

50.146. Zahm, C.L.: "Comparison of Opti-
mal and Conventional Arrays (for Signal/
Noise Performance)", Trans. IEEE, 1973,
AP-21, no.3, pp.379-80.

50.147. Dowsett, P.H.: "Crosspolarization
in Radomes. Programme for Computation",
Trans. IEEE, 1973, AES-9, no.3, pp.421-33.

50.148. Hugli, P.: "Antenna Radomes of
Polyurethane Hard Foam", Tech. Mitt. PTT,
1973, 51, no.6, pp.242-7.

50.149. Shiau, Y.: "Limitation of Axial
Gain of Large Antennas Under Partial Coherent Illumination", Proc. IEEE, 1973, 61,
no.8, pp.1159-60.

50.150. Weigand, R.M.: "Performance of a
Water-Repellant Radome Coating in an Airport
Surveillance Radar", Proc. IEEE, 1973, 61,
no.8, pp.1167-8.

50.151. Akaba, K.: "Reflector Surface of
6-m Millimetre-Wave Telescope", Tokyo Astron. Obs. Rep., 1973, 16, no.3, pp.471-99.

50.152. Akaba, K.: "Pointing Accuracy of
6-m Millimetre-Wave Telescope", Tokyo Astron. Obs. Rep., 1973, 16, no.3, pp.500-21.

50.153. Cogdell, J.R., and Davis, J.H.:
"Separating Aberrant Effects from Random
Scattering Effects in Radiotelescopes",
Proc. IEEE, 1973, 61, no.9, pp.1344-5.

50.154. Hizal, A.: "Radiation from a
Spherical Antenna Covered by an Inhomogeneous Over-Dense Plasma Layer", J. Phys. D,
1973, 6, no.16, pp.1843-9.

50.155. Burnside, W.D., Marhefka, R.J.,
and Lu, C.L.: "Roll-Plane Analysis of On-
Aircraft Antennas", Trans. IEEE, 1973, AP-21,
no.6, pp.780-6.

50.156. Vu, T.B., and Hien, N.V.: "High-
Performance Monopulse Feed", Trans. IEEE,
1973, AP-21, no.6, pp.855-7.

50.157. Kuhne, E., and Neske, H.K.: "Behaviour of Microwave Antennas Covered by Ice",
Robotron. Tech. Commun., 1973, 15, no.10,
pp.1-17.

50.158. Calla, O.P.N., and Khola, R.K.:
"Using Mesh as the Reflector Surface for
Microwave Antennas", J. Inst. Electron. Tele-
commun. Eng., 1973, 19, no.8, pp.425-7.

50.159. Budiansky, J.H.: "Phased Arrays
for ECM", Microwaves, 1973, 12, no.10, pp.56-59.

50.160. Vinichenko, Yu.P., et al.: "Matching a Phased-Array Antenna to Space", Radiotekh. Elektron., 1973, 18, no.6, pp.1137-44,
and Radio Eng. Electron. Phys., 1973, 18,
no.6, pp.830-5.

50.161. Hance, H.V., and Fried, D.L.:
"Experimental Test of Optical Antenna-Gain
Reciprocity", J. Opt. Soc. Am., 1973, 63,
no.8, pp.1015-6.

50.162. De, D.K., et al.: "Fabrication
of Gooseneck for Line-of-Sight Microwave
Communication Antennas", J. Inst. Electron.
Telecommun. Eng., 1973, 19, no.8, pp.464-6.

50.163. Vil'kotskii, M.A., Kaplun, V.A.,
and Kravchenko, I.T.: "Radiation Characteristics of a Directive Antenna Under a
Slotted Difracting Radome", Antenny, 1973,
no.18, pp.67-80.

50.164. Bogomolov, A.F., and Poperechenko,
B.A.: "Efficiency Improvement of 64-128 m
Millimetre-Wave Radiotelescopes", Izv. VUZ
Radiofiz., 1973, 16, no.12, pp.1893-7.

50.165. Christopher, E.J.: "Electrically
Scanned TACAN Antenna", Trans. IEEE, 1974,
AP-22, no.1, pp.12-6.

50.166. Shestag, L.N.: "Cylindrical Array
for TACAN System", Trans. IEEE, 1974, AP-22,
no.1, pp.17-25.

50.167. Wong, N.S., Tang, R., and Barber,
E.E.: "Multielement High-Power Monopulse Feed
With Low Sidelobes and High Aperture Efficiency", Trans. IEEE, 1974, AP-22, no.3,
pp.402-7.

50.168. Steinberg, B.D.: "Effects of Relative Source Strength and Signal/Noise Ratio
on Angular Resolution of Antennas", Proc.
IEEE, 1974, 62, no.6, pp.758-62.

50.169. Smirnov, E.P., and Yablochkin,
N.A.: "Radiation from a Circular Waveguide
in Anisotropic Plasma", Izv. VUZ Radiofiz.,
1974, 17, no.7, pp.944-9.

50.170. Brennan, L.E., Reed, I.S., and
Swerling, P.: "Adaptive Arrays", Microwave
J., 1974, 17, no.5, pp.43-6.

50.171. Scudder, R.M., and Sheppard, W.H.:
"Phased Array Antenna", Microwave J., 1974,
17, no.5, pp.51-5.

50.172. Heppenheimer, T.A.: "Holographic Structural Control for Large Space Reflectors and Radiotelescopes", J. Spacecr. Rockets, 1974, 11, no.7, pp.536-8.

50.173. Pelton, E.L., and Munk, B.A.: "Streamlined Metallic Radome", Trans. IEEE, 1974, AP-22, no.6, pp.799-803.

50.174. Dunphy, J.R., and Kerr, J.R.: "Atmospheric Beam-Wander Cancellation by Fast-Tracking Transmitter", J. Opt. Soc. Am., 1974, 64, no.7, pp.1015-6.

50.175. Shimada, S., Koyama, M., and Shinji, M.: "Focused-Beam-Feed Cassegrain Antenna for Experimental Earth Station", Rev. Electr. Commun. Lab., 1974, 22, no.5-6, pp.547-57.

50.176. Holley, A.E., DuFort, E.C., and Dell-Imagine, R.A.: "Electronically Scanned Beacon Antenna", Trans. IEEE, 1974, AP-22, no.1, pp.3-12.

50.177. Ogawa, Y., and Hongo, K.: "Radiation Into Anisotropic Plasma from Flanged Parallel-Plate Waveguide", J. Appl. Phys., 1974, 45, no.6, pp.2493-6.

50.178. Franceschetti, G., and Papas, C.H.: "Pulsed Antennas", Trans. IEEE, 1974, AP-22, no.5, pp.651-61.

50.179. Dixon, R.S.: "Isotropic Frequency-Independent Antenna System for Simultaneous Omnidirectional Doppler-Shift Removal", Trans. IEEE, 1974, AP-22, no.5, pp.707-9.

50.180. Rusch, W.V.T.: "Double-Aperture Blocking by Two Wavelength-Sized Feed-Support Struts", Electron. Lett., 1974, 10, no.15, pp.296-7.

50.181. Sanyal, G.S.: "Antenna Techniques for Space Systems", J. Inst. Electron. Telecommun. Eng., 1974, 20, no.6, pp.255-63.

50.182. Pasupathy, S., and Venetsanopoulos, A.N.: "Optimum Active Array Processing Structure and Space-Time Factorability", Trans. IEEE, 1974, AES-10, no.6, pp.770-8.

50.183. Hansen, R.C.: "Segmented Synthetic-Aperture Radar", Trans. IEEE, 1974, AES-10, no.6, pp.800-4.

50.184. Reed, I.S., Mallett, J.D., and Brennan, L.E.: "Rapid Convergence Rate in Adaptive Arrays", Trans. IEEE, 1974, AES-10, no.6, pp.853-63.

50.185. Beckman, J.E.: "Optimal Design of Optics for Submillimetre Astronomy", Trans. IEEE, 1974, MTT-22, no.12, pp.1113-5.

51. RADIO-LINK COMMUNICATION

51.1. Houssin, J.P.: "Sources of Noise in (Satellite) Communications Receivers", Onde Electr., 1965, 45, p.15.

51.2. Karlov, N.V., et al.: "Influence of Mismatch of Antenna Feeder Channels on Frequency Characteristics of Cavity Masers", Radiotekh. Elektron., 1965, 10, p.40, and Radio Eng. Electron. Phys., 1965, 10, p.32.

51.3. Lelliott, S.R., and Thurlow, E.W.: "Path Testing for Microwave Links", Post Office Electr. Eng. J., 1965, 58, pt 1, p.26.

51.4. Bray, W.J.: "Improved Performance of Goonhilly, Satellite-Communication, Earth-Station Aerial System", Electron. Lett., 1965, 1, p.108.

51.5. Pearson, K.W.: "Method for Prediction of Fading Performance of a Multisection Microwave Link", Proc. IEE, 1965, 112, p.1291.

51.6. Patrick, W.S.: "Troposcatter (Link) for Tactical Communications", Trans. IEEE, 1965, MIL-9, p.137.

51.7. Davidson, D., et al.: "Link for Multichannel Telephony and Television in the 6-GHz Range", Electr. Commun., 1965, 40, pp.173, 184, 200, and 209.

51.8. Ensslin, G., and Maier, H.G.: "Filters and Circulators for 6-GHz Wideband Communication System", Nachr. Tech. Z., 1965, 18, p.369.

51.9. Surenian, D.: "Results of Angle-Diversity System Tests", Trans. IEEE, 1965, COM-13, p.208.

51.10. Nakagami, M., and Kaneku, S.: "Characteristics of Common-Phase-Control Space Diversity", Rev. Electr. Commun. Lab., 1965, 13, p.75.

51.11. Makino, H., and Morita, K.: "Space-Diversity Reception and Transmission Systems for Line-of-Sight Microwave Links", Rev. Electr. Commun. Lab., 1965, 13, p.111.

51.12. Cook, J.S., and Giger, A.J.: "All-Weather, Earth-Station, Satellite Communications Antennas", Bell Syst. Tech. J., 1965, 44, p.1225.

51.13. Gillitzer, E.: "Antennas for a 6-GHz Wideband Communication System", Nachr. Tech. Z., 1965, 18, p.479.

51.14. Okumura, Y., and Nakamura, S.: "Space-Diversity Effects Measured on a Long Mountain Diffraction Path", Rev. Electr. Commun. Lab., 1965, 13, p.183.

51.15. Boithias, L., and Battesti, J.:
"Transhorizon Microwave Links of High
Quality", Ann. Telecommun., 1965, 20, p.138.

51.16. Kato, S., and Ohashi, K.: "Antenna
Systems for Microwave Links", J. Inst.
Electr. Commun. Eng. Jap., 1965, 48, p.614.

51.17. Sasaki, T., and Uda, H.: "Ground
Antenna for Space Communication", J. Inst.
Electr. Commun. Eng. Jap., 1965, 48, p.632.

51.18. Pistilli, A.: "Elements in Waveguides of Reduced Profile in Microwave Radio
Links", Alta Freq., 1965, 34, p.847.

51.19. Takeshita, S.: "Effects of Frost
on Microwave Passive-Reflector Efficiency",
J. Inst. Electr. Commun. Eng. Jap., 1965,
48, p.712.

51.20. Tewson, M.B.: "Planning Microwave
Links", GEC Telecommun., 1966, no.34, p.32.

51.21. Chen, C.C.: "Turnstile Antenna
for Space Communications", Int. Conv. Rec.
IEEE, 1966, 14, pt 4, p.120.

51.22. Shinn, D.H.: "Earth-Terminal
Aerials for Satellite Communication Systems",
Point-to-Point Telecommun., 1966, 10, p.40.

51.23. Bunin, D.A.: "Operational Reliability of Railway Microwave Links", Avtom.
Telemekh. Svyaz, 1966, no.1, p.12.

51.24. Abraham, L.G.: "Reliability of
Microwave Links", Trans. IEEE, 1966, COM-14,
p.805.

51.25. Yoshida, S., Yonemitsu, H., and
Takahashi, S.: "Satellite Communications by
TW Maser", Toshiba Rev., 1966, no.26, p.39.

51.26. Rao, M.S.V.G.: "Tropospheric-Scatter Communication Systems", J. Inst. Eng.,
1967, 47, no.12, pt ET3, p.329.

51.27. Chakrabarti, N.B., and Datta,
A.K.: "Diversity Combining Using Carrier
Lock and Sideband Lock Techniques. II",
Indian J. Phys., 1967, 41, no.2, pp.87-98.

51.28. Hogg, D.C.: "Path Diversity in
Propagation of Millimetre Waves Through
Rain", Trans. IEEE, 1967, AP-15, no.3,
pp.410-6.

51.29. Mansfeld, W.: "11-GHz Directional
Communication System With High Reliability",
Nachr. Tech., 1967, 17, no.5, pp.167-73.

51.30. Dimeff, J., Gunter, W.D., and
Hruby, R.J.: "Spectral Dependence of Deep-
Space Communications Capability", Spectrum
IEEE, 1967, 4, no.9, pp.98-104.

51.31. Makino, H., and Morita, K.: "De-
sign of Space-Diversity Systems for Line-of-
Sight Microwave Links", Int. Conv. Rec.
IEEE, 1967, 15, pt 2, pp.15-6.

51.32. Lumb, D.R.: "Study of Codes for
Deep-Space Telemetry", Int. Conv. Rec. IEEE,
1967, 15, pt 11, pp.130-6.

51.33. Tamas, F., and Andras, R.: "Studies
of Medium-Speed Data Transmission on a Microwave Link", Hiradstechnika, 1967, 18, no.4,
pp.113-7.

51.34. Valkenburgh, L.A.: "Influence of
High-Voltage Circuits and Pylons on Microwave
Links", PTT-Bedr., 1968, 16, p.10.

51.35. Wanless, G.: "Microwave-Link Characteristics for Harbour-Surveillance System",
Radio Electron. Eng., 1968, 36, p.153.

51.36. Bruntrup, H., Scherner, U., and
Roth, D.: "Helium-Cooled Parametric Amplifier
for Satellite Ground Stations", Nachr. Tech.
Fachber., 1968, 35, p.558.

51.37. Aitchison, C.S., et al.: "Low-
Noise Wideband Parametric Amplifier for Satellite Communications", Nachr. Tech. Fachber., 1968, 35, p.564.

51.38. Fairley, D.O.: "Noise Considerations for Solid-State Microwave Sources in
High-Capacity FM Links", Telecommunications,
1968, 2, p.11.

51.39. Niemeyer, M.: "Tunnel-Diode Amplifier for Communications Satellites in the
5.5-6.5 GHz Band", Nachr. Tech. Fachber.,
1968, 35, p.584.

51.40. Vendelin, G.V.: "K-Band Integrated
Receiver Front End", J. Solid-State Circuits
IEEE, 1968, SC-3, p.255.

51.41. Boithias, L., and Battesti, J.:
"Protection Against Fading in Line-of-Sight
Microwave Links", Ann. Telecommun., 1968,
22, no.9-10, p.230.

51.42. Zherebtsov, B.V., and Visel', A.A.:
"Regeneration of Short Microwave Pulses by
a Parametric Generator", Radiotekh. Elektron.,
1968, 13, p.1623, and Radio Eng. Electron.
Phys., 1968, 13, no.9, pp.1414-7.

51.43. Yasuda, S., and Ohrui, R.: "6-GHz,
Compact, Plug-In Replacement-Type TWT for
Microwave Links", NEC Res. Dev., 1968,
no.11, p.59.

51.44. Mizukaga, Y., et al.: "High-Power
Transmitting Tubes for Satellite-Communication Earth Stations", Toshiba Rev., 1968,
no.38, p.22.

51.45. Yonemitsu, H., et al.: "Low-Noise
Preamplifiers in Satellite Communications
Earth-Terminal", Toshiba Rev., 1968, no.38,
p.40.

51.46. Flack, M., and Whittaker, A.:
"Selecting Microwave (-Link) Equipment",
Teleph. Eng. Management, 1968,72,no.19,p.35.

51.47. Iwai, F., and Miyakawa, T.:
"Solid-State Microwave Heterodyne Link",
Fujitsu Sci. Tech. J., 1968, 4, p.53.

51.48. Kube, E.: "Communications via the
Atmosphere by Light Beams", Nachr. Tech.,
1969, 19, p.201.

51.49. Deerkoski, L.F.: "Measured Gain-Improvement Characteristics of a Space-Diversity Antenna System", Trans. IEEE, 1969, AES-5, p.872.

51.50. Banerjee, K., Parikh, P., and Nath. R.: "He-Ne Laser Communication", J. Inst. Telecommun. Eng., 1969, 15, no.5, p.364.

51.51. Holmes, J.F., and Ishimaru, A.: "Relativistic Communications Effects Associated With Moving Space Antennas", Trans. IEEE, 1969, AP-17, p.484.

51.52. Anderson, E.W.: "Performance Measurement of Tropospheric Scatter Systems", Point-to-Point Telecommun., 1969, 13, p.219.

51.53. Komarovich, V.F., and Lebedinskii, Ye.V.: "Diversity Reception Under Random Interference Conditions", Elektrosvyaz, 1969, 23, no.1, p.16, and Telecommun. Radio Eng. Pt 1, 1969, 23, no.1.

51.54. Colavito, C.: "Design of Tropospheric-Scatter Link", Alta Freq., 1969, 38, no.10, p.808.

51.55. Gilbert, E.N.: "Mobile-Radio Diversity Reception", Bell Syst. Tech. J., 1969, 48, p.2473.

51.56. Kalinin, A.I.: "Statistical Distribution of Thermal Noise Power in Tropospheric Link Channels", Elektrosvyaz, 1969, 23, no.3, p.1, and Telecommun. Radio Eng. Pt 1, 1969, 23, no.3, p.1.

51.57. Witkover, R.L.: "Pulse-Width-Modulated Optical Data Link", Rev. Sci. Instrum., 1969, 40, p.469.

51.58. Harger, R.O.: "Maximum-Likelihood and Optimized-Coherent-Heterodyne Receivers for Strongly Scattered Gaussian Fields", Opt. Acta, 1969, 16, no.6, p.745.

51.59. Vaizburg, G.M., and Nemirovskii, A.S.: "Additive Systems for Diversity Reception Using Tracking Heterodynes", Elektrosvyaz, 1969, 23, no.11, p.8, and Telecommun. Radio Eng. Pt 1, 1969, 23, no.11, p.7.

51.60. Romanov, A.M., and Lebed'ko, E.G.: "Generalized Matching Criterion for Electrnic Photodetector Optical Communication System", Opt.-Mekh. Prom., 1969, 36, no.3, p.60, and Opt. Technol., 1969, 36, no.3, p.369.

51.61. Khambaty, M.B., et al.: "GaAs Injection Laser in a Communication System", Indian J. Pure Appl. Phys., 1969, 7, no.1, p.29.

51.62. Nakamura, Y., Watanabe, T., and Nara, T.: "2700-Channel Microwave Link", Jap. Telecommun. Rev., 1969, 11, no.3, p.175.

51.63. Mocker, H.W.: "10.6-micron Optical Heterodyne Communication System", Appl. Opt., 1969, 8, p.677.

51.64. Viddeleer, R.: "Influence of Power-Line Pylons on Sidelobe Suppression of Antennas Used for Microwave Links", PTT-Bedr., 1969, 16, no.2, p.108.

51.65. Lake, H., and Roche, J.F.: "Reliability of 11-GHz Communication System in a Tropical Environment", Telecommunications, 1969, 3, no.1, p.15.

51.66. McGavin, R.E., Dougherty, H.T., and Emmanuel, C.B.: "Microwave Space- and Frequency-Diversity Performance Under Adverse Conditions", Trans. IEEE, 1970, COM-18, p.261.

51.67. Takita, S., and Aoki, S.: "Solid-State 4-GHz Link", Jap. Telecommun. Rev., 1970, 12, no.1, p.34.

51.68. Schlisser, G.: "Semiconductor-Laser Military Communications", Laser Focus, 1970, 6, no.5, p.32.

51.69. Ebihara, I., and Yano, T.: "18-GHz TV Relay System", Jap. Telecommun. Rev., 1970, 12, no.2, p.143.

51.70. Mayo, W.T.: "Spatial Filtering Properties of the Reference Beam in Optical Heterodyne Receiver", Appl. Opt., 1970, 9, p.1159.

51.71. De Lange, O.E.: "Optical Communications Experiments", Appl. Opt., 1970, 9, p.1167.

51.72. Borsuk, G.M., and Thaler, W.J.: "FM Laser Communication System in Turbulent Media", J. Opt. Soc. Am., 1970, 60, p.1245.

51.73. Cook, J.S.: "Deep-Space Communications", Bell Lab. Rec., 1970, 48, p.213.

51.74. Levin, B.R., and Ginzburg, S.A.: "Diversity Reception in the Presence of Lognormal Fluctuations", Elektrosvyaz, 1970, 24, no.1, and Telecommun. Radio Eng. Pt 1, 1970, 24, no.1, p.51.

51.75. Kwiatkowski, W., Arthanayabe, T., and Knight, V.H.: "Efficient High-Level Upconverter for Microwave Link", Electron. Lett., 1970, 6, p.625.

51.76. Moxon, J.: "Practical Problems in Planning Microwave Links", GEC-AEI Telecommun., 1970, no.38, p.30.

51.77. Carroll, W., and Poronnik, K.: "IR, Solid-State, Optical Communication System", Proc. IREE Aust., 1970, 31, p.212.

51.78. Baker, A.E., and Brice, P.J.: "(Atmospheric) Propagation as a Factor in Design of Microwave Links", Post Office Electr. Eng. J., 1970, 62, pt 4, p.239.

51.79. Gerdine, M.A., and Lenzing, H.F.: "Reduction of Delay Distortion in a Horn Reflector With Overmoded-Waveguide Feeder", Trans. IEEE, 1970, COM-18, p.21.

51.80. Ruthroff, C.L.: "Rain Attenuation and Radio Path Design", Bell Syst. Tech. J., 1970, 49, p.121.

51.81. Eliseev, P.G., Ismailov, I., and Fedorov, Yu.F.: "Injection Lasers for Multichannel Optical Communication", J. Quantum Electron. IEEE, 1970, QE-6, p.38.

51.82. Drufuca, G., and Paraboni, A.: "Theoretical Consideration of Rain Attenuation Parameters Affecting Millimetre Links", Alta Freq., 1970, 39, no.5, p.115e.

51.83. Hoversten, E.V.: "Communication Theory for the Turbulent Atmosphere", Proc. IEEE, 1970, 58, no.10, pp.1626-50.

51.84. Delaloye, B.: "Problems Relating to Choice of Site for a Ground Station for Satellite Communications", Bull. Ass. Suisse Electr., 1970, 61, no.18, pp.826-33.

51.85. Whitmer, R.F., et al.: "Ultrawide-Band Laser Communications. I", Proc. IEEE, 1970, 58, no.10, pp.1710-4.

51.86. McIntyre, C.M., et al.: "Components and Technology in Laser Space Communications", Proc. IEEE, 1970, 58, no.10, pp.1491-503.

51.87. Chatterjee, U.K., Menon, M.V.G., and Pant, H.C.: "Daytime Optical Communication Using a GaAs Laser", Indian J. Pure Appl. Phys., 1970, 8, no.8, pp.488-92.

51.88. Kopeika, N.S., and Bordogna, J.: "Background Noise in Optical Communication Systems", Proc. IEEE, 1970, 58, no.10, pp.1571-7.

51.89. Kerr, J.R., et al.: "Atmospheric Optical Communications Systems", Proc. IEEE, 1970, 58, no.10, pp.1691-709.

51.90. Pratt, W.K., Stokes, L.S., and Hinckley, R.: "Optimization of Optical Communication Systems", Proc. IEEE, 1970, 58, no.10, pp.1737-41.

51.91. Ekberg, J.: "Transmission Calculations for Optical Links", State Inst. Tech. Res. Rep., 1970, no.157, pp.5-29.

51.92. Yokoi, H., Yamada, M., and Satoh, T.: "Atmospheric Attenuation and Scintillation in Satellite Communications", KDD Tech. J., 1970, no.65, pp.1-10.

51.93. Lawrence, R.S., and Strohbehn, J.W.: "Survey of Clear-Air Propagation Effects in Optical Communications", Proc. IEEE, 1970, 58, no.10, pp.1523-45.

51.94. Gusler, L.T., and Hogg, D.C.: "Coupling Between Satellite and Terrestrial Communications Links Due to Scattering by Rain", Bell Syst. Tech. J., 1970, 49, no.7, pp.1491-511.

51.95. Vigants, A.: "Number of Fades in Space-Diversity Reception", Bell Syst. Tech. J., 1970, 49, no.7, pp.1513-30.

51.96. Valenzula, E., et al.: "Rocketsonde Transmission-Link Characteristics", Trans. IEEE, 1970, AES-6, no.6, pp.843-6.

51.97. Yeh, Y.S.: "Analysis of Adaptive Retransmission Arrays in a Fading Environment", Bell Syst. Tech. J., 1970, 49, no.8, pp.1811-25.

51.98. Kinkuchi, Y., and Matsumoto, S.: "Short-Haul (Communications) Link at 15 GHz", Jap. Telecommun. Rev., 1970, 12, no.4, pp.267-72.

51.99. Yano, T., and Matsumoto, S.: "Transportable Large-Capacity Microwave System", Jap. Telecommun. Rev., 1970, 12, no.4, pp.290-3.

51.100. Barnett, W.T.: "Microwave Line-of-Sight Propagation With and Without Frequency Diversity", Bell Syst. Tech. J., 1970, 49, no.8, pp.1827-71.

51.101. Vammen, C.M., and McCormick, F.L.: "Millimetre-Wave Satellite Propagation Experiment", Trans. IEEE, 1970, AES-6, no.6, pp.825-31.

51.102. Gowrishankar, S.M.: "Microwave Communication on Indian Railways", Electron. Today, 1970, 3, no.10, pp.12-6.

51.103. Masuda, T., et al.: "Experimental High-Speed PCM/AM Optical Communication System Using Mode-Locked He-Ne Laser", NEC Res. Dev., 1970, no.19, pp.1-14.

51.104. Cox, R.D.: "Measurements of Waveguide-Component Mixing Products in 6-GHz Frequency-Diversity Systems", Trans. IEEE, 1970, COM-18, p.33.

51.105. Morita, K., and Higuti, I.: "Estimation of Differential Fading in Microwave Links", Rev. Electr. Commun. Lab., 1970, 20, no.3-4, pp.263-70.

51.106. Mori, Y.: "Earth Station and Communication Satellite", J. Inst. Electr. Commun. Eng. Jap., 1970, 53, no.11, pp.1579-83.

51.107. Nomura, T., and Yasuda, Y.: "Deep-Space Communication", J. Inst. Electr. Commun. Eng. Jap., 1970, 53, no.11, pp.1587-91.

51.108. Hirano, J.: "Optical Communication System", J. Inst. Electr. Commun. Eng. Jap., 1970, 53, no.11, pp.1557-63.

51.109. Turner, D.: "Microwave Links and the Weather", Post Office Telecommun. J., 1970-1, 22, no.4, pp.13-5.

51.110. Timoshenko, V.G.: "Short-Range Millimetre-Wave Link", Izv. VUZ Radioelektron., 1971, 14, no.1, pp.34-9.

51.111. Pacini, G.P., Gaudio, R., and Rossi-Doria, F.: "Experimental Investigation of Man-Made Noise in the 12-GHz Band", Alta Freq., 1971, 40, no.2, pp.132-9.

51.112. Setzer, D.E.: "Influence of Scattered Radiation on Narrow-Field (Optical) Communication Links", Appl. Opt., 1971, 10, no.1, pp.109-13.

51.113. Crane, R.K.: "Propagation Phenomena Affecting Satellite-Communication Microwave Links", Proc. IEEE, 1971, 59, no.2, pp.173-88.

51.114. Kazaryan, R.A., et al.: "Multi-channel Laser Communication Link", Trans. IEEE, 1971, AES-7, no.1, pp.111-4.

51.115. Ippolito, L.J.: "Effects of Precipitation on 15.3 GHz and 31.65 GHz Earth-Satellite Communications", Proc. IEEE, 1971, 59, no.2, pp.189-205.

51.116. May, A.S., and Pagones, M.J.: "Model for Computation of Interference to Radio-Relay Systems from Geostationary Satellites", Bell Syst. Tech. J., 1971, 50, no.1, pp.81-102.

51.117. Ito, Y., Yokouchi, H., and Komizo, H.: "Integrated Circuits for 6-GHz Upconverter and 4-GHz Mixer", Fujitsu Sci. Tech. J., 1971, 7, no.1, pp.1-18.

51.118. Kuhn, U.: "Measurement of Fading Behaviour of a 11-GHz Radio Relay With Horizontal and Vertical Polarization and Depolarization", Nachr. Tech., 1971, 21, no.1, pp.13-4.

51.119. Rowe, R.P.: "Modular Tropospheric Scatter Equipment", Point-to-Point Commun., 1971, 15, no.2, pp.80-8.

51.120. Takahashi, S., et al.: "Millimetre-Wave (Satellite) Parametric Amplifier", Toshiba Rev., 1971, no.55, pp.30-5.

51.121. Evans, B.G., et al.: "Effects of Precipitation on Parabolic Antennas Employing Linear Orthogonal Polarizations at 11 GHz", Electron. Lett., 1971, 7, no.13, pp.375-7.

51.122. Parsons, J.D., and Ratcliff, P.A.: "Self-Phasing Antenna Array for FM Communication Links", Electron. Lett., 1971, 7, no.13, pp.380-1.

51.123. Hartnagel, H.: "Pulse Communication Using Gunn Diodes and Heterojunction Lasers", Arch. Elektron. Ubertrag., 1971, 25, no.1, p.51.

51.124. Hirano, J., Okawara, C., and Ito, T.: "Optical PCM Transmission Experiments Through the Atmosphere", Jap. Telecommun. Rev., 1971, 13, no.2, pp.94-6.

51.125. D'Ambrosio, A.: "Parametric Amplifier for Satellite Communication Earth Stations", Alta Freq., 1971, 40, no.6, pp.534-43.

51.126. Babler, G.M.: "Scintillation Effects at 4 GHz and 6 GHz on a Line-of-Sight Link", Trans. IEEE, 1971, AP-19, no.4, pp.574-5.

51.127. Sugahara, H., et al.: "Horn-Reflector Antenna and Feeder System Using 4 GHz, 6 GHz, and 7 GHz Bands", Rev. Electr. Commun. Lab., 1971, 19, no.4, pp.455-91.

51.128. Noda, K., et al.: "Experimental Millimetre-Wave PCM-AM Repeater", Rev. Electr. Commun. Lab., 1971, 19, no.4, pp.506-54.

51.129. Yano, T., and Matsumoto, S.: "Transportable (Link) Equipment for Emergency Operation in the 4 GHz and 6 GHz Bands", Jap. Telecommun. Rev., 1971, 13, no.3, pp.135-9.

51.130. Shapiro, J.H.: "Optimal Power Transfer Through Atmospheric Turbulence", Trans. IEEE, 1971, COM-19, no.4, pp.410-4.

51.131. Sharma, K.J.: "Aspects of Optical Communication Techniques", Israel J. Technol., 1971, 9, no.3, pp.257-62.

51.132. Dworkin, L.U., and Schwartz, M.: "Application of Information Feedback to an AM-Laser Communication System", Trans. IEEE, 1971, COM-19, no.5, pp.618-27.

51.133. Boithias, L., and Battesti, J.: "Reliability of Beamed Radio Links", Ann. Telecommun., 1971, 26, no.9-10, pp.381-92.

51.134. Flaherty, J.M.: "Preventing Polarization Fading in Microwave Systems", Electron. Des., 1971, 19, no.17, pp.68-9.

51.135. Chu, T.S.: "Restoring the Orthogonality of Two Polarizations in Radio-Communication Systems. I", Bell Syst. Tech. J., 1971, 50, no.9, pp.3063-9.

51.136. Wittkop, C.F.: "Microwave (Bands) for Railroad Communications", Communications, 1971, Oct., pp.22-6.

51.137. Thomas, D.T.: "Crosspolarization Distortion in Microwave Transmission Due to Rain", Radio Sci., 1971, 6, no.10, pp.833-9.

51.138. Gaignebet, J., and Goebbels, A.: "Space Telemetry by Laser Stations", Rech. Spat., 1971, 10, no.5, pp.6-9.

51.139. Kaye, D.N.: "Powerful (GaAs) Laser Communicator in a 1.5-kg Binocular Package", Electron. Des., 1971, 19, no.5, p.32.

51.140. Vakulenko, A.M., et al.: "Optical (GaAs) Telephone System", Kvantovaya Elektron., 1971, no.4, pp.134-6.

51.141. Monsen, P.: "Experimental Angle-Diversity Troposcatter System", Trans. IEEE, 1972, COM-20, no.2, pp.242-7.

51.142. Jovanovic, A., Schweizer, J., and Steffen, W.: "Optical Information Transmission", Bull. Assoc. Suisse Electr., 1972, 63, no.4, pp.198-201.

51.143. Sanchez-Cordoves, J.: "Microwave-Link Interference", Metal. Electr., 1972, 36, no.412, pp.243-7.

51.144. Babler, G.M.: "Study of Frequency-Selective Fading for Microwave Line-of-Sight Narrowband Radio Channel", Bell Syst. Tech. J., 1972, 51, no.3, pp.731-57.

51.145. Motoki, T., Sugiura, Y., and Chiba, T.: "Automatic Acquisition and Tracking System for Laser Communication", NHK Tech. J., 1972, 24, no.3, pp.12-20, and Trans. IEEE, 1972, COM-20, no.5, pp.847-51.

51.146. Inatomi, T., et al.: "IF-Combined, Space-Diversity, System", Fujitsu Sci. Tech. J., 1972, 8, no.1, pp.91-116.

51.147. Malek, K., and Rzewuski, M.: "Speech IR Link", Arch. Elektrotech., 1972, 21, no.1, pp.211-28.

51.148. Dworkin, L.U., and Schwartz, M.: "Information-Feedback Approach Applied to Polarization-Modulated Laser Communication System", Trans. IEEE, 1972, COM-20, no.3, pp.419-23.

51.149. Villard, O.G., Lomasney, J.M., and Kawachika, N.M.: "Mode-Averaging Diversity Combiner", Trans. IEEE, 1972, AP-20, no.4, pp.463-9.

51.150. Reudink, D.O.: "Comparison of Transmission at X-Band in Suburban and Urban Areas", Trans. IEEE, 1972, AP-20, no.4, pp.470-3.

51.151. Goeldner, J., et al.: "Transistorized Microwave Links", Tech. Mitt. AEG-Telefunken, 1972, 62, no.1, pp.1-8.

51.152. Tsao, G.K.H.: "Tropospheric Effects on Design of Line-of-Sight Space-Diversity System", Electron. Lett., 1972, 8, no.15, pp.388-9.

51.153. Snyder, D.L., and Rhodes, I.B.: "Phase- and Frequency-Tracking Accuracy in Direct-Detection Optical Communication Systems", Trans. IEEE, 1972, COM-10, no.6, pp.1139-42.

51.154. Quercia, P.: "Lunar Radio Communication", Antenna, 1972, 44, no.9, pp.320-1.

51.155. Tada, K., Sada, S., and Kamiya, K.: "Line-of-Sight Microwave Link Between Okinawa and Mainland", Jap. Telecommun. Rev., 1972, 14, no.4, pp.211-20.

51.156. Troitskii, V.N.: "Efficiency of Angle-Diversity Reception in Long-Distance Tropospheric Links", Elektrosvyaz, 1972, 26, no.9, pp.20-6, and Telecommun. Radio Eng. Pt 1, 1972, 26, no.9, pp.17-23.

51.157. Zargar'yants, M.N., et al.: "Use of Semiconductor Lasers in Compact Communication Systems", Kvantovaya Elektron., 1972, no.3, pp.101-3, and Sov. J. Quantum Electron., 1972, 2, no.3, pp.286-7.

51.158. Peyton, B.J., et al.: "High Sensitivity for IR-Laser Communications", J. Quantum Electron. IEEE, 1972, QE-8, no.2, pp.252-62.

51.159. Forster, D.C., Goodwin, F.E., and Bridges, W.B.: "Wideband Laser Communications in Space", J. Quantum Electron. IEEE, 1972, QE-8, no.2, pp.263-72.

51.160. Mathe, L., and Csernoch, J.: "Noise Problems in Microwave Equipments of Low and Medium Channel Capacity", Budavox Telecommun. Rev., 1972, no.3-4, pp.28-48.

51.161. Ueno, Y., and Kajitani, M.: "Semiconductor-Laser Communication System Using Differential PPM", Electron. Commun. Jap., 1972, 5, no.12, pp.83-9.

51.162. Makkaveev, V.I., and Morozov, D.N.: "Influence of Laser Intensity Fluctuations on Noise Immunity of a Binary Data Transmission System", Probl. Peredachi Inf., 1972, 8, no.2, pp.115-9, and Probl. Inf. Transm., 1972, 8, no.2, pp.175-8.

51.163. Mainchits, E.A., and Yampol'skii, V.G.: "Microwave Link Antennas", Elektrosvyaz, 1972, 26, no.17, pp.21-8, and Telecommun. Radio Eng. Pt 1, 1972, 26, no.11, pp.11-6.

51.164. Englisch, W.: "TV Transmission With CO_2 Laser", Umsch. Wiss. Tech., 1972, 72, no.24, pp.798-9.

51.165. Gochelashvily, K.S.: "Problem of Noise in Laser Communication Links", Radiotekh. Elektron., 1972, 17, no.5, pp.1093-4, and Radio Eng. Electron. Phys., 1972, 17, no.5, pp.852-4.

51.166. Khinrikus, Kh.V., and Afinogenov, V.N.: "Depolarization of Laser Emission in an Optical Channel", Izv. VUZ Radioelektron., 1972, 15, no.12, pp.1501-6.

51.167. Blair, R.H.: "Effects of Thermal Noise in FM Microwave Digital Link", Aust. Telecommun. Res., 1972, 6, no.2, pp.8-12.

51.168. Fried, D.L.: "Statistics of Laser Beam Fade Induced by Pointing Jitter", Appl. Opt., 1973, 12, no.2, pp.422-3.

51.169. Wulfsberg, K.N.: "Path Diversity for Millimetre-Wave Earth-Satellite Links", Radio Sci., 1973, 8, no.1, pp.1-5.

51.170. Babler, G.M.: "Selectively Faded Non- and Space-Diversity Narrowband Microwave Channels", Bell Syst. Tech. J., 1973, 52, no.2, pp.239-61.

51.171. Daino, B., Galeotti, M., and Sette, D.: "Error Probability of Binary Optical Communications in Turbulent Atmosphere", Alta Freq., 1973, 42, no.2, pp.80-3.

51.172. Rhomberg, B.: "13-GHz PCM Link", Electr. Commun., 1973, 48, no.1-2, pp.163-8.

51.173. Debois, P.G., Liekens, A., and Quaghebeur, G.: "Broadband Solid-State Link for 6-7 GHz", Electr. Commun., 1973, 48, no.1-2, pp.155-62.

51.174. Fussgaenger, K.: "CO_2 Laser Communication Through an Urban Atmosphere", Siemens Forsch. Entwick., 1973, 2, no.2, pp.105-12.

51.175. Coackley, B.: "Measurement Techniques for Microwave Communications", Electron. Equip. News, 1973, 15, no.4, pp.63-6.

51.176. Epstein, N., and Gregg, W.D.: "Line-of-Sight PCM Performance at 15.3 GHz", Trans. IEEE, 1973, AP-21, no.1, pp.78-83.

51.177. Tada, K., Oguchi, M., and Kikuchi, T.: "11-GHz Solid-State Short-Haul Link", Jap. Telecommun. Rev., 1973, 15, no.12, pp.138-42.

51.178. Petasne, A., et al.: "Transmission of Video- and Audio-Frequency Signals on He-Ne Laser Carrier", Rev. Telegr. Electron., 1973, no.722, pp.16-9.

51.179. Tycz, M., Fitzmaurice, M.W., and Premo, D.A.: "Optical Communication System Performance With Tracking-Error-Induced Signal Fading", Trans. IEEE, 1973, COM-21, no.9, pp.1069-72.

51.180. Okamura, S., Aoyama, K., and Oimatsu, K.: "Microwave Communication-Circuit Components", Rep. Univ. Electro-Commun., 1973, 24, no.1, pp.17-21.

51.181. Kalinin, A.I., and Shkud, M.A.: "Method of Reducing Depth of Interference Minima in Radio Links", Elektrosvyaz, 1973, 27, no.8, pp.19-25, and Telecommun. Radio Eng. Pt 1, 1973, 27, no.8, pp.15-20.

51.182. Rider, G.C., and Gouch, M.W.: "Case for Angle Diversity in Troposcatter Systems", Point-to-Point Commun., 1973, 17, no.3, pp.108-18.

51.183. Anderson, E.W.: "Application of Angle Diversity to Tropospheric Scatter Links", Point-to-Point Commun., 1973, 17, no.3, pp.130-8.

51.184. Rosenberg, S., and Teich, M.C.: "Photocounting Array Receivers for Optical Communication Through Lognormal Atmospheric Channel. I-III", Appl. Opt., 1973, 12, no.11, pp.2616-24 and 2625-35, and Trans. IEEE, 1973, IT-19, no.6, pp.807-9.

51.185. Lau, P.K., and Watson, P.A.: "Low-Noise, Room-Temperature, Parametric Amplifier (for Digital Links)", Electron. Lett., 1973, 9, no.25, pp.581-2.

51.186. Dmitrachenko, V.M., et al.: "Millimetre-Wave PCM Links", Elektrosvyaz, 1973, 27, no.2, pp.1-6, and Telecommun. Radio Eng. Pt 1, 1973, 27, no.2, pp.1-5.

51.187. Gusyatinskii, I.A., and Nemirovskii, A.S.: "System for Interference Fading Prevention in Tropospheric Links", Elektrosvyaz, 1973, 27, no.2, pp.7-12, and Telecommun. Radio Eng. Pt 1, 1973, 27,no.2,pp.6-9.

51.188. Shchelkunov, K.N.: "Systems for Message Transmission Over Parallel Optical Channels", Radiotekhnika, 1973, 28, no.2, pp.92-3, and Telecommun. Radio Eng. Pt 2, 1973, 28, no.2, pp.124-6.

51.189. Braude, V.B.: "Spectral Analysis of Pulsed Signals as Applied to the Synchonization of PCM Laser Links", Radiotekhnika, 1973, 28, no.3, pp.7-11, and Telecommun. Radio Eng. Pt 2, 1973, 28, no.3, pp.71-4.

51.190. Bremenson, C.: "High-Capacity Analogue and Digital Microwave Links", Rev. Tech. Thomson-CSF, 1973, 5, no.3, pp.641-68.

51.191. Daino, B., Galeotti, M., and Sette, D.: "Error-Rate Measurements in an Atmospheric Twin-Channel Optical Link", J. Quantum Electron. IEEE, 1974, QE-10, no.1, pp.86-7.

51.192. Spindler, W.: "Microwave Link Test Setup", Frequenz, 1974, 28, no.2, pp.44-51.

51.193. Lum, Y.F.: "Performance of a Long Troposcatter Link in the South-East-Asia Equatorial Region", Trans. IEEE, 1974, COM-22, no.5, pp.703-8.

51.194. Noesen, P., Winzeler, H.R., and Bodmer, F.: "Semiconductor Oscillator for 11-GHz Beamed Links of Large Channel Capacity", Bull. Assoc. Suisse Electr., 1974, 65, no.11, pp.813-20.

51.195. Yamamoto, H., Kuramoto, M., and Kamei, T.: "Experimental 20-GHz Multi-Hop Digital Link", Jap. Telecommun. Rev., 1974, 16, no.2, pp.124-32.

51.196. Webb, P.R.W.: "Military Satellite Communications Using Small Earth Terminals", Trans. IEEE, 1974, AES-10, no.3, pp.306-18.

51.197. Yoshikawa, T., and Kuramoto, M.: "20-GHz High-Speed Digital Communication Experiment During Heavy Rainfall", Proc. IEEE, 1974, 62, no.7, pp.1036-7.

51.198. Grimm, E.: "Reliability of Optical Systems for Atmospheric Transmission of Information", Feingeraete Tech., 1974, 23, no.8, pp.360-4.

51.199. Allnutt, J.E., and Hall, J.E.: "Site-Diversity Advantage for Satellite Communication at 11.6 GHz", Electron. Lett., 1974, 10, no.25-26, pp.527-8.

51.200. Yurlov, F.F., and Remeshkov, Yu.I.: "Technical and Economic Efficiency of Diversity Reception of Uninterrupted Data", Izv. VUZ Radioelektron., 1974, 17, no.9, pp.72-6.

51.201. Yamamoto, H., and Kohiyama, K.: "Construction and Overall Performance of Experimental 20-GHz Digital Repeater", Rev. Electr. Commun. Lab., 1974, 22, no.7-8, pp.571-8.

51.202. Kohiyama, K., et al.: "Experimental 20-GHz Digital Link Transmitter Design and Characteristics", Rev. Electr. Commun. Lab., 1974, 22, no.7-8, pp.579-81.

51.203. Shinji, M., et al.: "Antenna and Multiplexers for Experimental 20-GHz Digital Link", Rev. Electr. Commun. Lab., 1974, 22, no.7-8, pp.650-2.

51.204. Frolov, O.P.: "Determination of Excitation of Troposcatter Volume from Instantaneous Antenna Patterns", Radiotekhnika, 1974, 29, no.2, pp.94-5, and Telecommun. Radio Eng. Pt 2, 1974, 29, no.2, pp.124-5.

51.205. Rau, B.S.: "Microwave Network Planning", J. Inst. Electron. Telecommun. Eng., 1974, 20, no.6, pp.234-8.

51.206. Gupta, B.C.: "Radio Relays for Telecommunication Services in India", J. Inst. Electron. Telecommun. Eng., 1974, 20, no.6, pp.243-7.

51.207. Saran, S.: "Microwave Network Planning on Railways", J. Inst. Electron. Telecommun. Eng., 1974, 20, no.6, pp.248-9.

51.208. Sharma, V.T.V.: "Long-Path, Line-of-Sight, Microwave Communication Links", J. Inst. Telecommun. Eng., 1974, 20, no.6, pp.282-6.

51.209. Borisov, N.N., and Frolov, O.P.: "Alignment of Antennas Operating on Tropospheric Links", Elektrosvyaz, 1974, 28, no.2, pp.40-2, and Telecommun. Radio Eng. Pt 1, 1974, 28, no.2, pp.29-31.

51.210. Dworkin, L.U., et al.: "Tactical Optical Communications Set", Opt. Eng., 1974, 13, no.5, pp.401-8.

51.211. Braun, L.D.: "Characteristics of a Communications Satellite Transponder", Microwave J., 1974, 17, no.12, pp.45-7.

51.212. Gjessing, D.T., and McCormick, K.S.: "Prediction of Parameters of Long-Distance Tropospheric Communication Links", Trans. IEEE, 1974, COM-22, no.9, pp.1325-31.

51.213. Ince, A.N.: "Design, Testing, and Operation of X-Band Satellite Communication System", Trans. IEEE, 1974, COM-22, no.9, pp.1338-53.

51.214. Price, T.E.: "Atmospheric IR Communications Link", Electron. Eng., 1974, 46, no.560, pp.57-9.

51.215. Pao, Y.H., Allen, J.W., and Claspy, P.C.: "Wideband He-Ne Laser Communication System", Opt. Eng., 1974, 13, no.5, pp.383-8.

51.216. Ross, M.: "Optical Communications in Space", Opt. Eng., 1974, 13, no.5, pp.374-82.

51.217. Kincaid, B.E., Cowen, S., and Campbell, D.: "Applications of GaAs Lasers and Si Avalanche Detectors to Optical Communications", Opt. Eng., 1974, 13, no.5, pp.389-95.

51.218. Ishida, N.: "4-GHz-Band Delay Equalizer Using a Meander Line", Rev. Electr. Commun. Lab., 1974, 22, no.1-2, pp.101-13.

52. RADAR TECHNIQUES

52.1. Waters, P.L.: "Frequency-Diversity Performance of a Ground Surveillance Radar", Electron. Lett., 1965, 1, p.282.

52.2. Barton, D.K.: "Radar System Performance Charts", Trans. IEEE, 1965, MIL-9, p.255.

52.3. Hall, W.M., and Barton, D.K.: "Antenna-Pattern Loss Factor for Scanning Radars", Proc. IEEE, 1965, 53, p.1257.

52.4. Robbiani, R.L.: "High-Performance Weather Radar", Trans. IEEE, 1965, AES-1, p.185.

52.5. Vincent, N.D., and Lynn, P.A.: "Assessment of Site Effects on Radar Polar Diagrams", Marconi Rev., 1965, 28, p.111.

52.6. Pelchat, G.M.: "Effects of Receiver and Antenna Noise on the Performance of a Conical-Scan Tracking System", Microwave J., 1965, 8, no.2, p.37.

52.7. Terrington, D.G.: "Development of Secondary Surveillance Radar for Air-Traffic Control", Proc. IEE, 1965, 112, p.861.

52.8. Sleeper, G.B., and Hazen, W.: "Low-Cost Solution to Automatic Tracking Requirements", Microwave J., 1965, 8, no.4, p.76.

52.9. Seppen, J.M.G.: "Performance Calculations for High-Definition, Q-Band, River Radar", Radio Electron. Eng., 1965, 29, p.263.

52.10. Musal, H.M., et al.: "Millimetre Radar Instrumentation", Trans. IEEE, 1965, EAS-1, p.225.

52.11. Alongi, A.V., Kell, R.E., and Newton, D.J.: "High-Resolution, X-Band, FM CW Radar for Cross-Section Measurements", Proc. IEEE, 1965, 53, p.1072.

52.12. Ishii, T.K.: "Error of Doppler Radar in Target-Speed Determination for Traffic Control", Trans. IEEE, 1965, MTT-13, p.389.

52.13. Helgostam, L.F., and Ronnerstam, B.: "Ground-Clutter Calculation for Airborne Doppler Radars", Trans. IEEE, 1965, MIL-9, p.294.

52.14. Indiresan, P.V.: "Method for Improving the Accuracy of Tracking Radar", J. Inst. Telecommun. Eng., 1965, 11, p.483.

52.15. Venugopalan, B.S., Nayar, V.K.K., and Murthy, S.S.R.: "Antenna System for a Fire-Control Radar", J. Inst. Telecommun. Eng., 1965, 11, p.479.

52.16. Rihaczek, A.W.: "Optimum Filters for Signal Detection in Clutter", Trans. IEEE, 1965, AES-1, p.297.

52.17. Ruze, J., Sheftman, F.I., and Cahlander, D.A.: "Radar Ground-Clutter Shields", Proc. IEEE, 1966, 54, p.1171.

52.18. Rihaczek, A.W., and King, D.D.: "Radar Performance Prediction from Test-Range Measurements", Trans. IEEE, 1966, AES-2, p.462.

52.19. Parker, C.V., and Cazenavette, L.L.: "Artificial Beam-Sharpening Technique for Radar Beacon Systems", Trans. IEEE, 1966, AES-2, p.278.

52.20. Ray, H.K.: "Improving Radar Range and Angle Detection With Frequency Agility", Microwave J., 1966, 9, no.5, p.63.

52.21. Mavroides, W.G., Dennett, L.G., and Dorr, L.: "3-D Radar Based on Phase-in-Space Principle", Trans. IEEE, 1966, AES-2, p.323.

52.22. Withers, M.J.: "Matched Filter for FM CW Radar Systems", Proc. IEE, 1966, 113, p.405.

52.23. Blinn, J.C., and Campbell, J.P.: "Microwave Radiometric Sensing for Air Navigation", Trans. IEEE, 1966, AES-2, no.5, pp.585-90.

52.24. Knop, C.M., and Lichtenwalter, G.: "Radar Echo Reduction Using Circulators", Trans. IEEE, 1966, AP-14, p.789.

52.25. Rainal, A.J.: "Monopulse Radars Excited by Gaussian Signals", Trans. IEEE, 1966, AES-2, p.337.

52.26. MacPhie, R.H.: "Phase-Switched Radar System Giving Improved Control of Directional Pattern", Radio Electron. Eng., 1966, 3, no.2, p.81.

52.27. McGillem, C.D.: "Monopulse Radar Air-to-Ground Ranging", Trans. IEEE, 1966, AES-2, p.303.

52.28. Kirkpatrick, G.M.: "Use of Airborne Monopulse Radar as a Low-Approach Aid", Trans. IEEE, 1966, AES-2, p.353.

52.29. Thorne, T.G.: "Multi-Mode Weather Radar", J. Inst. Navig., 1966, 19, p.235.

52.30. Sekiguchi, T., and Goto, N.: "Beacon Radar System Using a Rapidly Scanned Antenna", Trans. IEEE, 1966, AES-2, p.461.

52.31. Knop, C.M.: "Radar Return from a Perfectly Reflecting Target in the Presence of a Second Medium", Proc. IEEE, 1966, 54, p.783.

52.32. Voles, R.: "Frequency Correlation of Clutter", Proc. IEEE, 1966, 54, p.881.

52.33. Whitfield, G.R.: "Optimization of a Surveillance Radar", Proc. IEE, 1966, 113, p.1277.

52.34. Colliver, D.J., and Hilsum, C.: "Miniature Radar Set Using a Transferred-Electron Oscillator", J. Sci. Instrum., 1966, 43, p.513.

52.35. Cornbleet, S.: "Use for the Fourth Output of a Monopulse Network", Electron. Lett., 1966, 2, p.83.

52.36. Croney, J.: "Improved Radar Visibility of Small Targets in Sea Clutter", Radio Electron. Eng., 1966, 32, p.135.

52.37. Hajovsky, R.G., Deam, A.P., and LaGrone, A.H.: "Radar Reflections from Insects in the Lower Atmosphere", Trans. IEEE, 1966, AP-14, p.224.

52.38. Long, M.W.: "Backscattering for Circular Polarization", Electron. Lett., 1966, 2, p.341.

52.39. Slattery, B.R.: "Use of Mills-Cross Receiving Arrays in Radar Systems", Proc. IEE, 1966, 113, p.1712.

52.40. Glover, K.M., et al.: "Radar Observations of Insects in Free Flight", Science, 1966, 154, p.967.

52.41. Takeshima, T.: "Slot-Array Antenna for Monopulse Tracking Radar", Microwave J., 1966, 9, no.12, p.63.

52.42. Hair, T., and Agar, W.O.: "Precision CW-Radar Ranging Technique", Marconi Rev., 1966, 29, p.191.

52.43. McGinn, J.W.: "Thermal Noise in Amplitude Comparison Monopulse Systems", Trans. IEEE, 1966, AES-2, no.5, pp.550-6.

52.44. Peter, T.V.: "Polarization of Radar Returns from Rain", Trans. S. Afr. IEE, 1966, 57, pt 11, pp.243-51.

52.45. Shenoy, R.P.: "Performance of Doppler Radar in the Presence of Random Fading", Trans. IEEE, 1966, AES-2, no.6, pp.676-80.

52.46. Painter, J.H.: "Designing Pseudo-random Coded Ranging Systems", Trans. IEEE, 1967, AES-3, no.1, pp.14-27.

52.47. Kownacki, S.: "Simulation of Radar Range and of Doppler Effect by a Stationary Target", Trans. IEEE, 1967, AES-3, no.1, pp.148-9.

52.48. Rihaczek, A.W.: "Radar Resolution of Moving Targets", Trans. IEEE, 1967, IT-13, no.1, pp.51-6.

52.49. Wong, J.L., Reed, I.S., and Kaprielian, Z.A.: "Model for Radar Echo from a Random Collection of Rotating Dipole Scatterers", Trans. IEEE, 1967, AES-3, no.2, pp.171-8.

52.50. Harrison, R.N.: "Selective Moving-Target Indication", Ind. Electron., 1967, 5, no.4, pp.146-8.

52.51. Moll, J.W.: "Radar Echoes from Aircraft", Trans. IEEE, 1967, AES-3, no.3, pp.574-7.

52.52. Phillips, C.S.E.: "Computer-Controlled Adaptive Radar", Proc. IEE, 1967, 114, no.7, pp.869-73.

52.53. Rihaczek, A.W., and Mitchell, R.L.: "Radar Waveforms for Suppression of Extended Clutter", Trans. IEEE, 1967, AES-3, no.3, pp.510-7.

52.54. Rihaczek, A.W.: "Delay/Doppler Ambiguity Function for Wideband Signals", Trans. IEEE, 1967, AES-3, no.3, pp.705-11.

52.55. Scholefield, P.H.R.: "Statistical Aspects of Ideal Radar Targets", Proc. IEEE, 1967, 55, no.4, pp.587-9.

52.56. Kownacki, S.: "Screening Effect of a Chaff Cloud", Trans. IEEE, 1967, AES-3, no.4, pp.731-4.

52.57. Beckmann, P.: "Estimation of the Number of Unresolvable Targets Producing a Single Radar Return", Radio Sci., 1967, 2, no.9, pp.955-60.

52.58. Erteza, A., and Lenhert, D.H.: "Field Theory of Depolarization of Radar Backscatter", Radio Sci., 1967, 2, no.9, pp.979-90.

52.59. Schaffer, H.: "Radar Receiver as an Optimum Filter and Correlator", Nach. Tech. Z., 1967, 20, no.4, pp.218-26.

52.60. Lipman, M.A.: "Useful Property of the Generalized Chirp-Signal Ambiguity Function", Proc. IEEE, 1967, 55, no.7, pp.1241-2.

52.61. Lizon, A.: "Doppler Radar Frequency-Modulated by a Sine Wave", Rozpr. Elektrotech., 1967, 13, no.3, pp.477-93.

52.62. Backmark, N., Krim, J.E.V., and Sellberg, F.: "Frequency-Agile Radar", Philips Tech. Rev., 1967, 28, no.11, pp.323-8.

52.63. Becker, J.E., and Millet, R.E.: "Double-Slot Radar Fence for Increased Clutter Suppression", Trans. IEEE, 1968, AP-16, no.1, pp.103-8.

52.64. Waters, W.M., and Eikenberg, A.F.: "Design of a Large-Aperture Radar for Target Imaging", Trans. IEEE, 1968, AES-4, p.886.

52.65. Mishra, S.R., and Mahajan, K.B.: "Staggered Modulation in Beam-Riding Guidance", J. Inst. Telecommun. Eng. 1968, 14, p.189.

52.66. Mayer, R.: "Sensitivity of a Pulsed-Doppler Tracking Radar", Nachr. Tech. Z., 1968, 21, p.457.

52.67. Mitchell, R.L., and Rihaczek, A.W.: "Clutter Suppression Properties of Weighted Pulse Trains", Trans. IEEE, 1968, AES-4, p.822.

52.68. Peebles, P.Z., and Berkowitz, R.S.: "Multiple-Target Monopulse Radar Processing Techniques", Trans. IEEE, 1968, AES-4, p.845.

52.69. Gawron, T.: "Estimation of Angular Position in Amplitude Monopulse Radar", Pr. PIT, 1968, 18, no.61, p.1.

52.70. Bertagna, J.C.: "Secondary Interrogator and Responder Radars", Onde Electr., 1968, 48, p.1105.

52.71. Kahrilas, P.J.: "Design of Electronic-Scanning Radar Systems", Proc. IEEE, 1968, 56, p.1763.

52.72. Grasso, G., and Guarguaglini, P.F.: "Improvement Factor for Limited Coherent MTI", Proc. IEEE, 1968, 56, p.2064.

52.73. Fung, A.K.: "Radar Determination of the Properties of a Randomly Rough Surface", Proc. IEEE, 1968, 56, p.2163.

52.74. Szyszkiewicz, J.: "Angle Measurement With Amplitude Monopulse Radar in Presence of Passive Interference", Pr. PIT, 1968, 18, no.61, p.13.

52.75. Kleiber, P., and Reinhard, L.: "Fire-Control Radar for Ships", Albiswerk Ber., 1968, 20, no.2, p.79.

52.76. Shamfarov, Ya.L.: "Possibility of Using Transients for Protecting Masers in Pulsed Radars", Radiotekh. Elektron., 1968, 13, p.2019, and Radio Eng. Electron. Phys., 1968, 13.

52.77. Blake, L.V.: "Radar Range-Height-Angle Charts", Microwave J., 1968, 11, no.10, p.49.

52.78. Gorelik, A.G., and Uglova, L.N.: "Radar Characteristics of the Reflections from Angels", Izv. Akad. Nauk SSSR, Fiz. Atmos. Okeana, 1968, 4, p.1235.

52.79. Neuvy, J., and Schifrine, J.: "Special Slide Rule for Practical Determination of Radar Range", Onde Electr., 1968, 48, p.1030.

52.80. Khaytun, F.I., and Nepogodin, I.A.: "Transmission and Reception of Optical Pulses", Opt-Mekh. Prom., 1968, 35, no.5, p.1, and Sov. J. Opt. Technol., 1968, 35, no.5, p.285.

52.81. Strauss, H.J.: "Control of Electronically Scanned Radar", Sperry Rand Eng. Rev., 1968, 21, no.4, p.35.

52.82. Milner, C.J., and Shell, G.S.: "Superregenerative Microwave Doppler MTI (for Road Traffic)", Trans. IEEE, 1968, VT-17, p.13.

52.83. Vigneri, R.J.: "Scene Imaging Using Laser Radiation", Opt. Spectra, 1968, 2, no.5, p.25.

52.84. Hannan, W.J.: "Laser Mobility Aid for the Blind", RCA Tech. Notes, 1968, no.21, p.1.

52.85. Amemiya, Y., et al.: "(Railroad) Obstacle Detection Using Beamguide With Discrete Confocal Reflectors", Hitachi Rev., 1968, 17, p.479.

52.86. Broderick, R.F., and Hayre, H.S.: "Doppler Return from Random Rough Surface", Trans. IEEE, 1969, AES-5, p.441.

52.87. Moore, R.K., Waite, W.P., and Rouse, J.W.: "Panchromatic and Polypanchromatic Radar", Proc. IEEE, 1969, 57, p.590.

52.88. Navara, P.: "Use of Laser for Measuring Position of Earth Satellites", Cesk. Cas. Fys., 1969, 19A, p.153.

52.89. Teich, M.C.: "Homodyne Detection of IR Radiation from Moving Diffuse Target", Proc. IEEE, 1969, 57, p.786.

52.90. Hansen, J.P.: "Profiling With a GaAs Laser Diode Radar", Proc. IEEE, 1969, 57, p.854.

52.91. Bartling, J.Q.: "Use of Alford-Gold (Coding) Effect as Ranging Technique", Proc. IEEE, 1969, 57, p.1335.

52.92. Honda, H.: "Electronic Boresight Shift in Spaceborne Monopulse Systems", RCA Rev., 1969, 30, p.150.

52.93. Fric, V., and Hornak, T.: "Evaluation of the Frequency Stability of a Microwave Source in a Radar Transmitter With Fixed Target Suppression", Slab. Obz., 1969, 30, p.254.

52.294. Reid, M.S.: "Millimetre-Wave Pseudorandom-Coded Meteorological Radar", Trans. IEEE, 1969, GE-7, p.146.

52.295. Barton, D.K.: "Simple Procedures for Radar Detection Calculations", Trans. IEEE, 1969, AES-5, p.837.

52.296. Swamy, R.S.: "Triple Pulse Doppler Radar", Int. J. Electron., 1969, 26, no.1, p.85.

52.297. Chieco, G.A.: "Use of Amplitrons in High-Power Radar Systems With Frequency Agility", Int. J. Electron., 1969,26,p.237.

52.98. Panov, D.N., and P'yavchenko, T.A.: "Digital Distance-Measuring Tracking Radar", Izv. VUZ Radioelektron., 1969, 12, no.5, p.480.

52.99. Levanon, N., and Stremler, F.G.: "Accurate Pulse-Radar Altimeter for Meteorological Balloons", Proc. IEEE, 1969, 57, p.1680.

52.100. Bottenberg, H.: "Range Resolution and Bandwidth of Phased-Array Antennas", Frequenz, 1969, 23, no.9, p.262.

52.101. Graf, K.A., and Guthart, H.: "Velocity Effects in Synthetic Aperture", Trans. IEEE, 1969, AP-17, p.541.

52.102. Croney, J., Woroncow, A., and Salt, H.: "Detection of Surface Targets in Sea Clutter by Shipborne Radar", Electron. Lett., 1969, 5, p.619.

52.103. Kuriger, W.L.: "Laser Doppler Velocimeter Employing a Scanning Interferometer", Proc. IEEE, 1969, 57, p.2161.

52.104. Jackson, D.A., and Paul, D.M.: "Assessment of the Argon-Ion Laser as a Suitable Source of Low-Frequency Heterodyne Anemometry Experiments", J. Phys. E, 1969, 2, p.1077.

52.105. Nagano, S., et al.: "Self-Excited Microwave Mixer With a Gunn Diode and Applications to Doppler Radar", Electron. Commun. Jap., 1969, 52, no.3, p.112.

52.106. Zhuravlev, A.K., and Suslov, N.A.: "Statistical Characteristics of Signals at Output of a Monopulse Radar Receiver", Radiotekh. Elektron., 1969, 14, no.12, p.2253, and Radio Eng. Electron. Phys., 1969, 14, no.12, pp.1945-8.

52.107. Courtney, J.E., and Halpern, H.M.: "Parallel Processing for Phased-Array Radars", RCA Rev., 1969, 30, p.709.

52.108. Simpson, R.E.: "Single-Antenna, X-Band, Doppler-Shift Apparatus", Am. J. Phys., 1969, 37, no.7, p.744.

52.109. Nanjo, M., Isawa, Y., and Yamanaka, C.: "Multi-Coincidence Method of Photon Counting for Optical Tracking", Bull. Electrotech. Lab., 1969, 33, no.2, p.18.

52.110. Carpentier, M.H.: "Application to Radar of Pseudorandom Codes", Onde Electr., 1969, 49, p.836.

52.111. Mikhailov, V.F.: "Effect of Plasma Shroud Surrounding a Vehicle on Operating Distance of Radar", Izv. VUZ Radioelektron., 1969, 12, no.10, p.1238.

52.112. Zabka, M.: "Radar Distance-Measuring System With Binary Phase Modulation", Slab. Obz., 1969, 30, no.12, p.564.

52.113. Peebles, P.Z.: "Signal Processors and Accuracy of Three-Beam Monopulse Tracking Radar", Trans. IEEE, 1969, AES-5, p.52.

52.114. Croney, J., and Woroncow, A.: "Radar Polarization Comparisons in Sea-Clutter Suppression by Decorrelation and CFAR Receivers", Radio Electron. Eng., 1969, 38, no.4, p.187.

52.115. Malota, F.: "Satellite Location and Tracking by Laser Radar", Laser, 1969, 1, no.4, pp.49-52.

52.116. Adrianova, I.I., Nesterova, Z.V., and Popov, Yu.V.: "Modulation Method for Phase Detection in an Electrooptic Rangefinder", Opt.-Mekh. Prom., 1969, 36, no.5, and Sov. J. Opt. Technol., 1969, 36, no.5, pp.662-5.

52.117. Clancy, T.M., and Wilmot, D.W.: "Estimation and Reduction of Aerosol Backscatter in Optical Systems", Proc. IEEE, 1969, 57, p.205.

52.118. Ussisoo, I.: "Correction Problems in Electronic Distance Measurements", Tellus, 1969, 21, no.4, p.549.

52.119. Dodington, S.H.: "Developments of TACAN Navigation System", Electr. Commun., 1969, 44, no.4, p.316.

52.120. Popp, H.: "Solid-State VOR. Ommirage Navigation Aids", Electr. Commun., 1969, 44, no.4, p.322.

52.121. Willoughby, E.O.: "Navigation Without False Courses", Electron. Eng., 1969, 41, no.502, p.61.

52.122. Willoughby, E.O.: "Landing, Take-Off, and En-Route, Guidance of Aircraft Using ILS", Proc. IREE Aust., 1969, 30, no.9, p.298, and Electron. Eng., 1970, 42, no.504, p.66.

52.123. Levanon, N.: "Balloon-Borne Radio Altimeter", Trans. IEEE, 1970, GE, p.19.

52.124. Ishii, T.K.: "Analysis of Target-Speed Determination With Doppler Radar", Trans. IEEE, 1970, IM-19, p.85.

52.125. Johansen, E.L.: "Synthetic-Array Radar Image of a Flat Plate", Trans. IEEE, 1970, AES-6, p.395.

52.126. Varakin, L.Ye.: "Effect of Acceleration on Reception of Radar Signals", Radiotekhnika, 1970, 25, no.3, p.13, and Telecommun. Radio Eng. Pt 2, 1970, 25, no.3.

52.127. Alberti, C.: "Developments in Radar for Marine Navigation", Alta Freq., 1970, 74, p.8.

52.128. Williams, W.P.: "Electronic (Radar) Systems for Berthing Big Tankers", Electron. Aust., 1970, 31, no.10, p.29.

52.129. Serafin, R., and Kazel, S.: "Side-looking Radiometry", Proc. IEEE, 1970, 58, p.819.

52.130. Veret, C.: "Influence of Reflecting Surface Characteristics on a Laser Rangefinder", Radio Electron. Eng., 1970,39,p.201.

52.131. Huffaker, R.M., Jelalian, A.V., and Thomson, J.A.L.: "Laser Doppler System for Detection of Aircraft Trailing Vortices", Proc. IEEE, 1970, 58, p.322.

52.132. Dubnishchev, Yu.N., Senin, A.G., and Sobolev, V.S.: "Evaluating Potential Applications of Laser Doppler Velocimeter in Terms of Accuracy", Avtometriya, 1970, no.5, pp.47-50.

52.133. Mazumder, M.K.: "Laser Doppler Velocity Measurement Without Directional Ambiguity by Using Frequency-Shifted Incident Beams", Appl. Phys. Lett., 1970, 16, p.462.

52.134. Lind, G.: "Measurement of Sea-Clutter Correlation With Agile- and Fixed-Frequency Radars", Philips Telecommun. Rev., 1970, 29, no.1, p.32.

52.135. Waite, W.P., and MacDonald, H.C.: "Snowfield Mapping With K-Band Radar", Remote Sensing Envir., 1970, 1, no.2, p.143.

52.136. Hunter, J.A.: "Unique Digital Doppler Ambiguity-Resolution Technique", Trans. IEEE, 1970, AES-6, p.716.

52.137. Howell, J.M.: "Tracking Performance of a Monopulse Radar in the Presence of Multiple Targets", Trans. IEEE, 1970, AES-6, p.718.

52.138. Shirman, Ya.D.: "Optimum Detection of a Radar Target in a Cloud of Passive Reflectors", Radiotekh. Elektron., 1970, 15, no.5, p.934, and Radio Eng. Electron., Phys., 1970, 15, no.5, pp.799-806.

52.139. Kock, W.E.: "Passive (Cooperative) Hologram Radar", Proc. IEEE, 1970,58,p.1297.

52.140. Belen'skii, Ya.E., et al.: "Errors in Velocity-Vector Measurement of Moving Objects in Two-Point Ranging Using Doppler Effect", Otbor Peredacha Inf. 1970, no.24,p.130.

52.141. Davy, B., and Phillips, C.S.E.: "Pulsed AM/FM CW Radar", Proc. IEEE, 1970, 117, p.485.

52.142. Ward, R.B., and Strubel, F.L.: "Precision Earth-Satellite Range Measurements Using (Noise-Modulation) Delay-Lock Techniques", Trans. IEEE, 1970, IM-19, p.117.

52.143. Porcello, L.J.: "Turbulence-Induced Phase Errors in Synthetic-Aperture Radars", Trans. IEEE, 1970, AES-6, p.636.

52.144. Trunk, G.V., and George, S.F.: "Detection of Targets in Non-Gaussian Sea Clutter", Trans. IEEE, 1970, AES-6, p.620.

52.145. Salwen, H.C.: "Error Analysis of Optical Range-Measurement Systems", Proc. IEEE, 1970, 58, no.10, pp.1741-5.

52.146. Cooper, D.C.: "Estimation of Loss of Echoing Area With Very-High-Resolution Radars", Radio Electron. Eng., 1970, 40, no.4, pp.159-64.

52.147. Leith, E.N., Friesen, A.A., and Funkhouser, A.T.: "Optical Simulation of Radar Ambiguities", Trans. IEEE, 1970, AES-6, no.6, pp.832-40.

52.148. Donati, S., and Sona, A.: "Optical Range Gating to Extend Visibility in Fog", Alta Freq., 1970, 39, p.202.

52.149. Heffes, H., and Horing, S.: "Optimal Allocation of Tracking Pulses for an Array Radar", Trans. IEEE, 1970, AC-15, p.81.

52.150. Young, G.O., and Howard, J.E.: "Antenna Processing for Surface Target Direction", Trans. IEEE, 1970, AP-18, p.335.

52.151. Ormsby, J.F.A., et al.: "Analytic Coherent Radar Techniques for Target Mapping", Trans. IEEE, 1970, AES-6, p.295.

52.152. Daly, R.T.: "Optical-Radar Pilot Warning Indicator", Proc. IEEE, 1970, 58, p.456.

52.153. Kamenov, P., and Kinova, L.: "Mossbauer Speedometer", Annu. Univ. Sofia Fac. Phys., 1970, 64-5, pp.343-8.

52.154. Aggarwal, G.K.: "Phased-Array Radar for Automatic Random Search", Int. J. Electron., 1970, 28, p.365.

52.155. Takenouchi, T., et al.: "Satellite Ranging With a Laser", Hitachi Rev., 1970, 19, no.4, pp.153-64.

52.156. Watrasiewicz, B.M.: "Improved Signal/Noise Ratio in Laser Velocimeter", J. Phys. E, 1970, 3, no.10, p.823.

52.157. Fuller, K.L., and Lambell, A.J.: "Traffic-Flow Analysis by Radar", Philips Tech. Rev., 1970, 31, no.1, pp.17-22.

52.158. Kock, W.E.: "Holographic Amplitude Pulse Compression for Synthetic-Aperture Radar", Proc. IEEE, 1970, 58, no.10, pp.1773-4.

52.159. Lewis, T.S., and Hutchings, H.S.: "Synthetic-Aperture (Radar) at 10.6 micron", Proc. IEEE, 1970, 58, no.10, pp.1781-2.

52.160. Kock, W.E.: "Synthetic Endfire Hologram Radar", Proc. IEEE, 1970, 58, no.11, pp.1858-9.

52.161. Kock, W.E.: "Holographic Techniques in CW Bistatic Radars", Proc. IEEE, 1970, 58, no.11, pp.1683-4.

52.162. Taylor, M., Keough, R.A., and Moeller, A.W.: "Beam Broadening of a Monopulse Tracking Antenna by Feed Defocusing", Trans. IEEE, 1970, AP-18, no.5, pp.622-7.

52.163. Nuttall, T.W.B., and Hartland, B.J.: "Automation in Marine Radar. Collision Warning and Course Tracking", J. Sci. Technol., 1970, 37, no.3, pp.94-102.

52.164. Laharrague, P., Durand, J.M., and le Bihan, A.: "Measurement Method of Missile Speed Using Laser and Fabry-Perot Interferometer", Onde Electr., 1970, 50, no.9, pp.804-11.

52.165. Harris, J.: "Interrogating Flow Fields With Microwave and Laser Radars", Meas. Control, 1970, 3, no.11, p.T188.

52.166. de Munck, J.C.: "Electromagnetic Distance Measurement", Tijdschr. Ned. Elektron. Radiogenoot., 1970, 35, no.9, pp.ET129-33.

52.167. Lewis, D.J., and Buck, G.J.: "Phase-Monopulse Tracking Errors Due to Mutual Scattering Effects in Phase-Scanned Arrays", Radio Sci., 1970, 5, no.8-9, pp.1215-20.

52.168. McAlister, E.D., and McLeish, W.: "Radiometric System for Airborne Measurement of Total Heat Flow from the Sea", Appl. Opt., 1970, 9, no.12, pp.2697-705.

52.169. Deichl, K., and Sigl, R.: "Distance Measurements Using Tellurometers of Different Wavelengths", Messtechnik, 1970, 78, no.8, pp.162-7.

52.170. Blythe, J.H.: "Separation of Radars on Common Frequencies by PRF Discrimination", J. Sci. Technol., 1970, 37, no.4, pp.157-62.

52.171. Drufuca, G., and Giorgetti, P.: "Radar Location Experiments in Snow", Alta Freq., 1970, 39, no.10, pp.858-63.

52.172. Kroszczynski, J.: "Two-Frequency MTI System", Pr. PIT, 1970, 20, no.66, pp.1-8, and Radio Electron. Eng., 1970, 39, p.172.

52.173. Kroszczynski, J.: "Efficiency of Two-Frequency MTI System", Pr. PIT, 1970, 20, no.68, pp.1-5.

52.174. Stotskii, A.A., and Solovjev, V.M.: "(Microwave) Phase Comparator for Distances", Izv. Gl. Astron. Obs. Pulkove, 1970, no.185, pp.242-7.

52.175. Nathanson, F.E., and Reilly, J.P.: "Frequency Agility for Radar Target Detection and Tracking", Appl. Tech. Dig., 1970, 9, no.6, pp.2-8.

52.176. Gruzdev, Yu.P., and Safronov, I.N.: "Method for Testing Sensitivity of Optical Rangefinders", Opt.-Mekh. Prom., 1970, 37, no.5, p.67, and Sov. J. Opt. Technol., 1970, 37, no.5, pp.335-6.

52.177. Kyaytun, F.I., and Nepogodin, I.A.: "Threshold and Power Relations for Optical Radars", Opt.-Mekh. Prom., 1970, 37, no.8, pp.13-4, and Sov. J. Opt. Technol., 1970, 37, no.8, pp.504-6.

52.178. Atzeni, C., Cionini, A., and Masotti, L.: "Influence of Symmetry of Power Spectrum on Ambiguity Function of a Radar Signal", Alta Freq., 1970, 39, no.12, pp.1097-100.

52.179. Atlas, D., Harris, F.I., and Richter, J.H.: "Measurement of Point-Target Speeds With Incoherent (Microwave) Non-Tracking Radar", J. Geophys. Res., 1970, 75, no.36, pp.7588-95.

52.180. Kock, W.E.: "Holographic (Synthetic-Aperture) Method for Increasing Gain of Ground-to-Air Radars", Proc. IEEE, 1971, 59, no.3, pp.426-7.

52.181. Kroszczynski, J.: "Efficiency of Two-Frequency MTI System", Radio Electron. Eng., 1971, 41, no.2, pp.77-80.

52.182. Fuller, K.L.: "FM CW Short-Range High-Definition Radar", Wireless Wld, 1971, 77, no.1425, pp.110-3.

52.183. Bullock, L.G., Oeh, G.R., and Sparagna, J.J.: "Analysis of Wideband Microwave Monopulse Direction-Finding Techniques", Trans. IEEE, 1971, AES-7, no.1, pp.188-203.

52.184. Porter, R.P.: "Radar Imaging System Using the Object to Provide the Reference Signal", Proc. IEEE, 1971, 59, no.2, pp.307-8.

52.185. Samuelsson, H.: "Infrared Systems. I and II", Trans. IEEE, 1971, AES-7, no.1, pp.27-33 and 34-41.

52.186. Paul, D.M., and Jackson, D.A.: "Rapid Velocity Sensor Using a Static Confocal Fabry-Perot and a Single-Frequency Argon-Ion Laser", J. Phys. E, 1971, 4, no.3, pp.170-2.

52.187. Jackson, D.A., and Paul, D.M.: "Measurement of Supersonic Velocity and Turbulence by Laser Anemometry", J. Phys. E, 1971, 4, no.3, pp.173-7.

52.188. Moss, S.J.: "Performance of Laser Ranging System in Satellite Tracking", Trans. IEEE, 1971, GE-9, no.1, pp.1-9.

52.189. Buhring, W., and Wirth, W.D.: "Detection of Moving Targets Out of Clutter With Doppler-Filter System", Nachr. Tech. Z., 1971, 24, no.2, pp.72-6.

52.190. Bures, K.J., and Stremler, F.G.: "Synthetic-Aperture Bistatic Radar for Mapping Tropospheric Radio Scatterers", Proc. IEEE, 1971, 59, no.4, pp.715-6.

52.191. Jackson, J.S., et al.: "IR Techniques for Surveillance and Crime Deterrence", Trans. IEEE, 1971, IECI-18, no.2, pp.38-42.

52.192. Rehmann, G.: "IR System Counts and Measures (Traffic) to ±1 km/h", Elektrotechnik, 1971, 53, no.1-2, pp.22-3.

52.193. Rogers, P.J., and Eccles, P.J.: "Bistatic Radar Equation for Randomly Distributed Targets", Proc. IEEE, 1971, 59, no.6, pp.1019-21.

52.194. Watkins, C.D.: "High-Power Radar for Meteorological Studies in Clear Air", Proc. IEEE, 1971, 118, no.3-4, pp.519-28.

52.195. Lanz, O., Johnson, C.C., and Morikawa, S.: "Directional Laser Doppler Velocimeter", Appl. Opt., 1971, 10, no.4, pp.884-8.

52.196. Schanda, E.: "Graphs for Radiometer Applications", Ingenieur, 1971, 83, no.17, pp.51-5, and Tijdschr. Ned. Elektron. Radiogenoot., 1971, 36, no.4, pp.51-5.

52.197. Meier, P.J., Okean, H.C., and Sard, E.W.: "Integrated X-Band Sweeping Heterodyne Receiver", Trans. IEEE, 1971, MTT-19, no.7, pp.600-9.

52.198. Osterwalder, J.M.: "K-Band Electronically Tunable Monopulse Receiver", Trans. IEEE, 1971, MTT-19, no.7, pp.627-33.

52.199. Vigneri, R.J.: "Dual-Frequency Laser Ranging", Instrum. Control Syst., 1971, 44, no.4, pp.117-8.

52.200. Ostrovityanov, R.V., and Basalov, F.A.: "Correlation Function and Excursions of Angular (Glint) Noise", Radiotekh. Elektron., 1971, 15, no.8, and Radio Eng. Electron. Phys., 1971, 15, no.8, pp.1511-3.

52.201. Bichara, M.R.E.: "Gunn Diode as a Component for Portable Radar", Ann. Telecommun., 1971, 26, no.1-2, pp.2-14.

52.202. Ackerman, S.: "Calibration, Precision, and Efficiency, of Optical Rangefinders", Appl. Opt., 1971, 10, no.5, pp.1051-6.

52.203. Marini, J.W.: "Effect of Satellite Spin on Two-Way Doppler Range-Gate Measurements", Trans. IEEE, 1971, AES-7, no.2, pp.316-20.

52.204. Dannemann, H., and Hanle, E.: "Simulation of Radar Clutter Signals With Given Correlation", Arch. Elektron. Ubertrag., 1971, 25, no.6, pp.262-6.

52.205. Watanabe, M., et al.: "Japanese 3D Radar for Air-Traffic Control", Electronics, 1971, 44, no.13, pp.68-72.

52.206. Greated, C.A.: "Resolution and Backscattering Optical Geometry of Laser Doppler Systems", J. Phys. E, 1971, 48, no.8, pp.585-8.

52.207. Ewell, G.W.: "Monopulse Antenna Performance When Receiving Nonuniform Waves", Trans. IEEE, 1971, AES-7, no.3, pp.561-4.

52.208. Bramanti, M., Calamia, M., and Franceschetti, G.: "Locating Low-Level Targets by Radar", Alta Freq., 1971, 40, no.5, pp.442-51.

52.209. Trunk, G.V.: "Further Results on Detection of Targets in Non-Gaussian Sea Clutter", Trans. IEEE, 1971, AES-7, no.3, pp.553-6.

52.210. Grasso, G., and Guarguaglini, P.F.: "Behaviour of Noncoherent MTI in Nonlinear Conditions", Alta Freq., 1971, 40, no.5, pp.452-9.

52.211. Sims, R.J., and Graf, E.R.: "Reduction of Radar Glint by Diversity Techniques", Trans. IEEE, 1971, AP-19, no.4, pp.462-8.

52.212. Raney, R.K.: "Synthetic-Aperture Imaging Radar and Moving Targets", Trans. IEEE, 1971, AES-7, no.3, pp.499-505.

52.213. Manning, R.: "Theoretical Analysis of Laser Flowmeters", Opto-Electron., 1973, 3, no.2, pp.93-7.

52.214. Durst, F.: "Laser Doppler-Shift Anemometry", Tecnica, 1971, 33, no.408, pp.431-46.

52.215. Brayton, D.B., and Goethert, W.H.: "Dual-Scatter Laser Doppler Velocity-Measuring Technique", ISA Trans., 1971, 10, no.1, pp.40-50.

52.216. Teich, M.C., and Rosenberg, S.: "N-Fold Joint Photocounting Distribution for Modulated Laser Radiation. Turbulent Atmosphere", Opto-Electron., 1971, 3, no.2, pp.63-76.

52.217. Kobayasi, T., Aruga, T., and Inaba, H.: "Analytical Study on Resonance-Scattering Radar Using Frequency-Tunable Lasers", Rec. Electr. Commun. Eng. Conversaz. Tohoku Univ., 1971, 40, no.1, pp.33-9.

52.218. Leith, E.N.: "Quasi-Holographic Techniques in the Microwave Region", Proc. IEEE, 1971, 59, no.9, pp.1305-18.

52.219. Yurchenko, Yu.S.: "Effect of Fluctuation Noise on Monopulse Tracking Radar", Izv. VUZ Radioelektron., 1971, 14, no.8, pp.894-901.

52.220. Nakazawa, N.: "Electrooptical Rangefinder With Accuracy of ±10 mm at 1 km", J. Electr. Eng., 1971, no.56, pp.30-6.

52.221. Morikawa, S., Lanz, O., and Johnson, C.C.: "Laser-Doppler Measurements of Localized Pulsatile Fluid Velocity", Trans. IEEE, 1971, BME-18, no.6, pp.416-20.

52.222. Farmer, W.M., and Brayton, D.B.: "Analysis of Atmospheric Laser Doppler Velocimeters", Appl. Opt., 1971, 10, no.10, pp.2319-24.

52.223. Hamel, K., Danicek, T., and Novotny, A.: "Experimental Satellite Laser Radar", Czech. J. Phys., 1971, 21B, no.10, pp.1118-20.

52.224. Jones, P., Mickle, E.A., and Swetnam, G.D.: "Homing Radar Tracking Accuracy Improvement With Frequency Diversity", J. Spacecr. Rockets, 1971, 8, no.9, pp.983-9.

52.225. Kulczyk, W.K.: "Laser Vibration-Measuring Instrument Based on the Doppler Effect", Laser Angew. Strahlentech., 1971, 3, no.2, pp.44-5.

52.226. Kacprzak, K.: "Radar Equipment for Checking Vehicle Speed", Pomiary Autom. Kontr., 1971, 17, no.10, pp.446-8.

52.227. Stevens, M.C.: "Precision Secondary Radar", Proc. IEE, 1971, 118, no.12, pp.1729-35.

52.228. Lizon, A.: "Doppler Radar (Ship) Velocimeter", Pr. PIT, 1971, 21, no.72, pp.37-50.

52.229. Khaytun, F.I., and Safronov, I.N.: "Comparison of Pulse and Doppler Techniques for Measuring Velocity of Moving Targets", Opt.-Mekh. Prom., 1971, 38, no.1, pp.14-5, and Sov. J. Opt. Technol., 1971, 38, no.1, pp.11-3.

52.230. Knyazev, L.V., and Uglova, L.N.: "Determination of Wind-Speed Derivatives in Precipitation by a Radar Method", Izv. Akad. Nauk SSSR, Fiz. Atmos. Okeana, 1971, 7, no.9, pp.1002-7.

52.231. Berger, H.: "Angular Accuracy of Amplitude-Monopulse Off-Boresight Radar", Proc. IEEE, 1971, 59, no.3, pp.411-2.

52.232. Dubnishchev, Yu.N., and Kovshov, Yu.M.: "Laser Doppler Velocimeter Nonsensitive to Geometry of Incident Ray", Avtometriya, 1971, no.3, pp.87-90.

52.233. Vasilenko, Yu.G., et al.: "Laser Doppler Velocimeter Using Fabry-Perot Interferometer", Avtometriya, 1971, no.3, pp.90-2.

52.234. Kokurin, Yu.L., et al.: "Laser Radar Location of Retroflector Mounted on Space Vehicle", Kosm. Issled., 1971, 9, no.6, pp.912-9, and Cosmic Res., 1971, 9, no.6, pp.841-7.

52.235. Earp. C.W., Overbury, F.G., and Sothcott, P.: "Doppler Scanning Guidance System", Electr. Commun., 1971, 46, no.4, pp.253-70.

52.236. Nirasawa, T.: "Shipboard Collision Avoidance Systems Utilizing Marine Radars", J. Soc. Instrum. Control Eng., 1971, 10, no.11, pp.789-800.

52.237. Berkowitz, R.S.: "Information Derivable from Monopulse Radar Measurements of Two Unresolved Targets", Trans. IEEE, 1971, AES-7, no.5, pp.1011-3.

52.238. van der Spek, G.A.: "Detection of a Distributed Target", Trans. IEEE, 1971, AES-7, no.5, pp.922-31.

52.239. Kaszerman, P.: "Frequency of Pulse Coincidence Given n Radars of Different Pulse Durations and PRF's", Trans. IEEE, 1971, AES-7, no.5, pp.1013-4.

52.240. Honeycutt, T.E., and Otto, W.F.: "FM CW Radar Range Measurement With CO_2 Laser", J. Quantum Electron. IEEE, 1972, QE-8, no.2, pp.91-2.

52.241. Poultney, S.K.: "Capabilities of High-PRF Laser Radars for Measurement of Atmosphere Above 30 km", J. Atmos. Terr. Phys., 1972, 34, no.2, pp.339-42.

52.242. Hermet, P.: "Design of a (Laser) Rangefinder for Military Purposes", Appl. Opt., 1972, 11, no.2, pp.273-6.

52.243. Cooke, C.R.: "Automatic Laser Tracking and Ranging System", Appl. Opt., 1972, 11, no.2, pp.277-84.

52.244. Flom, T.: "Spaceborne Laser Radar", Appl. Opt., 1972, 11, no.2, pp.291-9.

52.245. Lehr, C.G., et al.: "Transportable Lunar-Ranging System", Appl. Opt., 1972, 11, no.2, pp.300-4.

52.246. Hanle, E.: "Models for Radar Clutter and Comparison With Measurement Results", Arch. Elektron. Ubertrag., 1972, 26, no.4, pp.159-64.

52.247. Bossel, H.H., Hiller, W.J., and Meier, G.E.A.: "Self-Aligning Comparison Beam Methods for One-, Two-, and Three-Dimensional Optical Velocity Measurements", J. Phys. E, 1972, 5, no.9, pp.897-900.

52.248. Rattman, W., and Smith, T.: "Lasers for Depth Sounding and Underwater Viewing", Hydrospace, 1972, 5, no.1, pp.57-9.

52.249. Weber, P.: "Adaptation of Lidar to Aircraft and Missile Rangefinding", Rech. Aerosp., 1972, no.1, pp.23-35.

52.250. Boothroyd, R.G.: "Tracer Behaviour in Laser Anemometry for Turbulent Flow", Opt. Laser Technol., 1972, 4, no.2, pp.87-90.

52.251. Holtum, A.G., and Yang, R.F.H.: "Generalized Radar Equations and Doppler Shifts in the Fresnel Region", Proc. IEEE, 1972, 61, no.6, pp.748-9.

52.252. Coffrey, D.W., and Norris, V.J.: "Nd^{3+}:YAG Laser Target Designators and Rangefinders", Appl. Opt., 1972, 11, no.5, pp.1013-8.

52.253. Wang, C.P.: "Instantaneous Turbulence Velocity Measurement by Laser Doppler Velocimeter", Appl. Phys. Lett., 1972, 20, no.9, pp.339-41.

52.254. Kampstra, A.J.: "Survey of Anti-Clutter Measures for Long-Range Surveillance Radars", Ingenieur, 1972, 84,no.15,pp.39-50.

52.255. Salinger, S.N., and Wangsness, D.: "Target-Handling Capacity of a Phased-Array Tracking Radar", Trans. IEEE, 1972, AES-8, no.1, pp.43-50.

52.256. Larson, R.W., Zelenka, J.S., and Johansen, E.L.: "Microwave Hologram Radar System", Trans. IEEE, 1972, AES-8, no.2, pp.208-17.

52.257. Durrani, T.S.: "Noise Analysis for Laser Doppler Velocimeter Systems", Trans. IEEE, 1972, COM-20, no.3, pp.296-307.

52.258. Giles, T.G., and Saw, J.E.: "Doppler Radar Using Gunn Oscillator for Small Rotating Objects", Mullard Tech. Commun., 1972, 12, no.114, pp.114-9.

52.259. Eggins, P.L., and Jackson, D.A.: "Laser Doppler Velocity Measurements in Supersonic Flow Without Artificial Seeding", Phys. Lett., 1972, 42A, no.2, pp.122-4.

52.260. Brandewie, R.A., and Davis, W.C.: "Parametric Study of a 10.6-micron Laser Radar", Appl. Opt., 1972, 11, no.7, pp.1526-33.

52.261. Shefer, J., and Kaplan, G.S.: "X-Band Electronic Fence Automates Vehicle Location", Commun. Equip. Syst. Des., 1972, no.2, pp.3-4.

52.262. Daricek, T., et al.: "Laser Distance Measurement of Satellites", Elektrotech. Cas., 1972, 23, no.6, pp.321-30.

52.263. Steiner, K.: "Doppler Radar Using Solid-State Devices", Tech. Mitt. AEG-Telefunken, 1972, 62, no.1, pp.39-41.

52.264. Logan, S.E.: "Laser Velocimeter for Reynolds Stress and Other Turbulence Measurements", AIAA J., 1972, 10, no.7, pp.933-5.

52.265. Bossel, H.H., Hiller, W.J., and Meier, G.E.A.: "Noise-Cancelling Signal Difference Method for Optical Velocity Measurements", J. Phys. E, 1972, 5,no.9,pp.893-6.

52.266. Domaratskii, A.N., et al.: "Effect of Concentration of Scattering Particles on Correlation Period of a Laser Doppler Velocimeter", Avtometriya, 1972, no.5, pp.122-5.

52.267. Barill, G.A., Zhuravel', F.A., and Sobolev, V.S.: "Evaluation of Systematic Error in Laser Doppler Velocimeters by Computer Simulation", Avtometriya, 1972, no.6, pp.98-100.

52.268. Vasilenko, Yu.G., et al.: "Differential Laser Doppler Velocimeter Using a Fabry-Perot Interferometer", Opt. Spekt., 1972, 33, no.1, pp.170-2, and Opt. Spectrosc., 1972, 33, no.1, pp.93-5.

52.269. Singh, R.: "Optimization of Slope of Difference-Mode Radiation Pattern in Monopulse Radar", Proc. IEE, 1972, 119, no.9, pp.1278-9.

52.270. Sullivan, J.P., and Ezekiel, S.: "Two-Dimensional Laser Doppler Velocimeter", Q. Progr. Rep. Res. Lab. Electron. MIT, 1972, no.104, pp.142-5.

52.271. Farmer, W.M., Hornkohl, J.O., and Brayton, D.B.: "Analysis of Atmospheric Laser Doppler Velocimeter Methods", Opt. Eng., 1972, 11, no.1, pp.24-30.

52.272. Cook, C.S., Bethke, G.W., and Conner, W.D.: "Remote Measurement of Smoke Plume Transmittance Using Lidar", Appl. Opt., 1972, 11, no.8, pp.1742-8.

52.273. Harris, J.: "Flow Measurement Using Microwave Radar Techniques", Powder Technol., 1972, 6, no.2, pp.85-9.

52.274. Berger, H.: "Optimum Squint Angles of Amplitude Monopulse Radar and Beacon Tracking Systems", Trans. IEEE, 1972, AES-8, no.4, pp.545-7.

52.275. Kazel, S.: "Multilateration (Mapping) Radar", Proc. IEEE, 1972, 60, no.10, pp.1238-9.

52.276. Allen, R.J., and Evans, W.E.: "Laser Radar for Mapping Aerosol Structure", Rev. Sci. Instrum., 1972, 43, no.10, pp.1422-32.

52.277. Abbiss, J.B., et al.: "Laser Anemometry in an Unseeded Supersonic Wind Tunnel by Photon Correlation Spectroscopy of Backscattered Light", J. Phys. D, 1972, 5, no.11, pp.L100-2.

52.278. MacIntyre, F.: "Instantaneous Velocity Correlations from a Single Dopplermeter", Appl. Phys. Lett., 1972, 21, no.9, pp.430-2.

52.279. Kock, W.E.: "Extending Maximum Range of Synthetic Aperture (Hologram) Systems", Proc. IEEE, 1972, 61,no.11,pp.1459-60.

52.280. Hughes, A.J., O'Shaughnessy, J., and Pike, E.R.: "FM CW Radar Range Measurement at 10 micron", J. Quantum Electron. IEEE, 1972, QE-8, no.12, pp.909-10.

52.281. Castruccio, P.A.: "Principles of Synthetic-Aperture Radar", Aerotec. Missile Spazio, 1972, 51, no.4, pp.235-47.

52.282. Hartley-Smith, A.: "Ground and Angel Clutter in Radar Systems. Two-Beam Solution", GEC J. Sci. Technol., 1972, 39, no.4, pp.173-80.

52.283. Galutin, V.Z., Zenkevich, S.S., and Skibarko, A.P.: "Operating Feature of an FM Rangefinder Using a Gas Laser", Izv. VUZ Radioelektron., 1972, 15, no.12, pp.1421-7.

52.284. Dessus, B., Napie, P., and Perneker, L.: "Laser Flow Velocimeter", Houille Blanche, 1972, no.8, pp.685-94.

52.285. Steinbach, M., and Neubert, R.: "Measuring Positions of Satellites With the Aid of Laser Pulses", Jena Rev., 1972, 17, no.7, pp.331-6.

52.286. Waksberg, A.: "FM Laser Noise Effects on Optical Doppler-Radar Systems", Trans. IEEE, 1972, AES-8, no.6, pp.791-9.

52.287. Lind, G.: "Approximate Formula for Glint Improvement With Frequency Agility", Trans. IEEE, 1972, AES-8, no.6, pp.854-5.

52.288. Harger, R.O.: "Radar Altimeter Optimization for Geodesy Over the Sea", Trans. IEEE, 1972, AES-8, no.6, pp.728-42.

52.289. Chadwick, R.B.: "Measurement of Distributed Targets With Random-Signal Radar", Trans. IEEE, 1972, AES-8, no.6, pp.743-50.

52.290. Deryagin, V.N., Marasin, L.E., and Popov, Yu.V.: "Compact Pulsed Semiconductor Laser Rangefinder", Opt.-Mekh. Prom., 1972, 39, no.7, pp.23-6, and Sov. J. Opt. Technol., 1972, 39, no.7, pp.400-3.

52.291. Iida, T.: "Automatic Tracking of an Aircraft by Radar", Electron. Commun. Jap., 1972, 55, no.7, pp.99-107.

52.292. Saito, S., and Takeda, N.: "Laser Distance-Measurement System by Comparing Phase of Modulating Signal", Electron. Commun. Jap., 1972, 55, no.6, pp.86-90.

52.293. Yoder, P.R., and Friedman, I.: "Low-Light-Level Objective Lens With Integral Laser Channel", Opt. Eng., 1972, 11, no.6, pp.127-30.

52.294. Baburin, V.I., Zakhar'ev, L.N., and Lemanskii, A.A.: "Effect of Different Antennas of a Doppler Velocimeter on Accuracy", Radiotekhnika, 1972, 27, no.12, pp.94-6, and Telecommun. Radio Eng. Pt 2, 1972, 27, no.12, pp.112-4.

52.295. Dannemann, H.: "Optimal Search Procedures for Phased-Array Radar", Nach. Tech. Z., 1973, 26, no.3, pp.118-22.

52.296. Benda, O.: "Method for Measuring the Spatial Spectral Density of Refractive Index in Turbulent Air Flow Using Laser Correlation Technique", Proc. IEEE, 1973, 61, no.5, pp.684-5.

52.297. Brown, G.S.: "Analysis of Four- and Five-Horn-Fed Cassegrainian Reflectors", Trans. IEEE, 1973, AP-21, no.3, pp.382-4.

52.298. Brayton, D.B., Kalb, H.T., and Crosswy, F.L.: "Two-Component Dual-Scatter Laser Doppler Velocimeter With Frequency-Burst Signal Readout", Appl. Opt., 1973, 12, no.6, pp.1145-56.

52.299. Barill, G.A., et al.: "Use of Laser Doppler Velocimeters for Observation of Turbulent Flow", Zh. Prikl. Mekh. Tekh. Fiz., 1973, no.1, pp.110-20.

52.300. Hughes, A.J., and Pike, E.R.: "Remote Measurement of Wind Speed by Laser Doppler Systems", Appl. Opt., 1973, 12, no.3, pp.597-601.

52.301. Lading, L.: "Analysis of Signal-to-Noise Ratio of Laser Doppler Velocimeter", Opto-Electron., 1973, 5, no.2, pp.175-87.

52.302. Krasnenko, N.P., and Glazov, G.N.: "Amplitude Tracking Accuracy Using Phased Arrays", Izv. VUZ Radioelektron., 1973, 16, no.2, pp.23-7.

52.303. Ohtsuka, Y.: "Frequency-Shifted Laser Interferometer for Study of Small Dynamic Motions", Opt. Commun., 1973, 7, no.3, pp.244-7.

52.304. Manning, R.: "Symmetric Transforms for Laser Velocimeter", J. Phys. D, 1973, 6, no.10, pp.1173-87.

52.305. Politch, J.: "Distance Measurement With an Unmodulated CW Laser", Optik, 1973, 38, no.2, pp.138-46.

52.306. Golden, K.E., et al.: "Laser Ranging System With 10-mm Resolution", Appl. Opt., 1973, 12, no.7, pp.1447-53.

52.307. Tanner, L.H.: "Particle Timing Laser Velocity Meter", Opt. Laser Technol., 1973, 5, no.3, pp.108-10.

52.308. Manning, R.: "Optical Design for Laser Flowmeter", Opt. Laser Technol., 1973, 5, no.3, pp.114-8.

52.309. Hallermeier, R.J.: "Laser Doppler Velocimeter for Studying Gravity Waves in Shallow Water", Appl. Opt., 1973, 12, no.2, pp.294-300.

52.310. Wang, C.P.: "Effect of Doppler Ambiguity on Measurement of Turbulence Spectra by Laser Velocimeter", Appl. Phys. Lett., 1973, 22, no.4, pp.154-6.

52.311. D'Orazio, R.J.: "Matched-Filter Detection of Mode-Locked Laser Signals", Appl. Opt., 1973, 12, no.10, pp.2367-72.

52.312. Orloff, K.L., and Logan, S.E.: "Confocal Backscatter Laser Velocimeter With On-Axis Sensitivity", Appl. Opt., 1973, 12, no.10, pp.2477-81.

52.313. George, W.K., and Berman, N.S.: "Ambiguity in Laser Doppler Velocimeters", Appl. Phys. Lett., 1973, 23, no.5, pp.222-3.

52.314. Arbenz, K.: "Optimum Cross-Coupled Tracker for Pulse-Doppler Radar", Trans. IEEE, 1973, AES-9, no.4, pp.495-8.

52.315. Macphie, R.H.: "Circular Synthetic Radar", Trans. IEEE, 1973, AES-9, no.4, pp.608-11.

52.316. Kulczyk, W.K., and Davis, Q.V.: "Laser Doppler Instrument for Measurement of Vibration of Moving Turbine Blades", Proc. IEE, 1973, 120, no.9, pp.1017-23.

52.317. Kulczyk, W.K.: "Signal/Noise Ratio in Laser Doppler Systems", Proc. IEE, 1973, 120, no.9, pp.1024-9.

52.318. Wisler, D.C., and Mossey, P.W.: "Measurements Within a Compressor Rotor Passage Using Laser Doppler Velocimeter", Trans. ASME, 1973, 95A, no.2, pp.91-6.

52.319. Durrani, T.S., and Greated, C.A.: "Frequency-Domain Analysis of Laser Doppler Signals for Turbulence Parameters", Proc. IEE, 1973, 120, no.8, pp.913-8.

52.320. McAulay, R.J.: "Effect of Staggered PRF's on MTI Signal Detection" Trans. IEEE, 1973, AES-9, no.4, pp.615-8.

52.321. Ovsyannikov, V.A., and Romanov, A.M.: "Signal/Noise Ratio in Optical Radar Systems", Opt.-Mekh. Prom., 1973, 40, no.1, pp.11-3, and Sov. J. Opt. Technol., 1973, 40, no.1, pp.8-10.

52.322. Vereshchaka, A.I., Popov, Yu.V., and Smirnov, V.P.: "Optical Rangefinder Employing a CO_2 Laser", Opt.-Mekh. Prom., 1973, 40, no.1, pp.63-4, and Sov. J. Opt. Technol., 1973, 40, no.1, pp.58-9.

52.323. Riegl, J.: "Short-Distance Rangefinding by Optical Pulse Radar", Nachr. Tech. Z., 1973, 26, no.9, pp.435-40.

52.324. Dubnishchev, Yu.N., et al.: "Measurement of Velocity of a Body by Optical Doppler Effect", Opt. Spekt., 1973, 34, no.3, pp.587-8, and Opt. Spectrosc., 1973, 34, no.3, p.335.

52.325. Fritsche, R., and Mesch, F.: "Non-Contact Speed Measurement. Comparison of Optical Systems", Meas. Control., 1973, 6, no.7, pp.293-300.

52.326. Moore, C.E., and Elpi, J.D.: "Optimum Collision Avoidance for Merchant Ships", Trans. IEEE, 1973, IA-9, no.6, pp.640-9.

52.327. Schuck, W.D.: "Thermal Effects of IMPATT Diodes Used as Self-Detecting Oscillators in Doppler Radar", Nachr. Tech. Z., 1973, 26, no.11, pp.517-9.

52.328. Riley, J.R.: "Angular and Temporal Variations in the Radar Cross Section of Insects", Proc. IEE, 1973, 120, no.10, pp.1229-32.

52.329. Riegl, J.: "Optical Miniature Pulse-Radar Rangefinders for Short Distances", Nachr. Tech. Z., 1973, 26, no.11, pp.481-6.

52.330. Shimizu, M., and Ueno, Y.: "Tracking and Ranging Systems Using Semiconductor Lasers and Application to Large-Scale Ship's Guiding Apparatus", J. Soc. Instrum. Control Eng., 1973, 12, no.8, pp.688-95.

52.331. Farmer, W.M., and Hornkohl, J.O.: "Two-Component, Self-Aligning, Laser Vector Velocimeter", Appl. Opt., 1973, 12, no.11, pp.2636-40.

52.332. Wirth, W.D.: "Fine Resolution in Azimuth (by Synthetic Array) for Rectilinear Moving Targets", Nachr. Tech. Z., 1973, 26, no.12, pp.539-41.

52.333. Harger, R.O.: "Synthetic-Aperture Radar System Design for Random Field Classification", Trans. IEEE, 1973, AES-9, no.5, pp.732-40.

52.334. Davies, D.E.N., and Makridis, H.: "Two-Frequency Secondary Radar Incorporating Passive Transponders", Electron. Lett., 1973, 9, no.25, pp.592-3.

52.335. Singer, R.A., and Sea, R.G.: "Optimizing Surveillance-System Tracking and Data Correlation Performance in Dense Multi-target Environments", Trans. IEEE, 1973, AC-18, no.6, pp.571-82.

52.336. Prinsen, P.J.A.: "Elimination of Blind Velocities of MTI Radar by Modulating the Interpulse Period", Trans. IEEE, 1973, AES-9, no.5, pp.714-24.

52.337. Holmstrom, F.R., et al.: "Microwave Anticipatory Crash Sensor for Automobiles", Trans. IEEE, 1973, VT-22, no.2, pp.46-54.

52.338. Hsiao, J.K., and Kretschmer, F.F.: "Design of a Staggered-PRF MTI Filter", Radio Electron. Eng., 1973, 43, no.11, pp.689-93.

52.339. Lee, B.: "Estimating Aircraft Velocity Directly from Airborne Range Measurements", Navigation, 1973, 20, no.1, pp.29-40.

52.340. Klement'sev, V.M., and Potap'ev, S.V.: "Laser Seismostation", Geofiz. Appar., 1973, no.51, pp.67-9.

52.341. Rihaczek, A.W.: "Choice of Burst Number for Blind-Speed Avoidance", Trans. IEEE, 1973, AES-9, no.5, pp.778-80.

52.342. Vinogradov, Ye.M.,et al.: "Modelling of EMC of Radar Equipment", Izv. VUZ Radioelektron., 1973, 16, no.11, pp.115-7.

52.343. Kapitanov, V.A., Mel'nichuk, Yu.V., and Chernikov, A.A.: "Spectra of Radar Signals Reflected from Forests at Centimetre Wavelengths", Radiotekh. Elektron., 1973, 18, no.9, pp.1816-25, and Radio Eng. Electron. Phys., 1973, 18, no.9, pp.1330-8.

52.344. Kroszczynski, J.: "Autocoherent MTI System", Pr. PIT, 1973,23,no.78,pp.9-14.

52.345. Shakhgil'dyan, V.V., and Burd-zeyko, B.P.: "Statistical Dynamics of a Rangefinder With Variable PRF", Radiotekh. Elektron., 1973, 18, no.6, pp.1172-80, and Radio Eng. Electron. Phys., 1973, 18, no.6, pp.850-7.

52.346. Suzuki, T., and Arai, I.: "Doppler Radar System Capable of Detecting a Stationary Target", Rep. Univ. Electro-Commun., 1973, 24, no.2, pp.235-41.

52.347. Volkonskii, V.B., et al.: "Laser Rangefinder With Microwave Modulation and Frequency Conversion in the Photodetector", Opt.-Mekh. Prom., 1973, 40, no.10, pp.22-4, and Sov. J. Opt. Technol., 1973, 40, no.10, pp.616-8.

52.348. Kroszczynski, J., and Granato-wickz, D.: "Anti-Collision Automobile Radar", Przegl. Telekomun., 1973, 46, no.11-12, pp.364-8.

52.349. Heinz, K.J.: "Accuracy of Target Angle Measurements of Dual-Beam Radars", Siemens Forsch. Entwick., 1973, 2, no.6, pp.333-7.

52.350. Kolacny, P., and Lipinsky, M.v.R.: "Quantization Losses During Digital Radar Signals", Siemens Forsch. Entwick., 1973, 2, no.6, pp.338-44.

52.351. Kolacny, P.: "Elimination of Blind Speeds in Pulse Doppler Radars by Staggered PRF", Siemens Forsch. Entwick., 1973, 2, no.6, pp.345-9.

52.352. Richter, J.H., and Jensen, D.R.: "RCS Measurements of Insects", Proc. IEEE, 1973, 61, no.1, pp.143-4.

52.353. Brennan, L.E., and Reed, L.S.: "Theory of Adaptive Radar", Trans. IEEE, 1973, AES-9, no.2, pp.237-52.

52.354. Roy, R.: "Velocity Adaptive MTI Filter", Trans. IEEE, 1973, AES-9, no.2, pp.324-6.

52.355. Ollendorff, F.: "Radar (Detection) of Targets Moving With High Velocities", Arch. Elektrotech., 1973, 55, no.3, pp.123-30.

52.356. Oldengarm, J., van Krieken, A.H., and Raterink, H.J.: "Laser Doppler Veloci-meter With Optical Frequency Shifting", Opt. Laser Technol., 1973, 5, no.6, pp.249-52.

52.357. Einav, S., and Lee, S.L.: "Meas-urement of Velocity Distribution in Two-Phase Suspension Flows by Laser-Doppler Technique", Rev. Sci. Instrum., 1973, 44, no.10, pp.1478-80.

52.358. Rihaczek, A.W.: "Systematic Ap-proach to Blind-Speed Elimination", Trans. IEEE, 1973, AES-9, no.6, pp.940-7.

52.359. Arora, R.K., and Kodali, V.P.: "Optimal Microwave System Parameters for Long-Range Radars", J. Inst. Electron. Tele-commun. Eng., 1973, 19, no.8, pp.428-37.

52.360. Basalov, F.A., and Ostrovityanov, R.V.: "Effects of Signal Polarization on Range Noise of a Complex Target", Radiotekh. Elektron., 1973, 18, no.4, pp.866-7, and Radio Eng. Electron. Phys., 1973, 18, no.4, pp.632-3.

52.361. Khlopov, G.I.: "Resolution and Power Losses in a Millimetre-Wave Imaging System", Radiotekh. (Kharkov), 1973, no.27, pp.7-16.

52.362. Rinkevichyus, B.S., and Yamina, G.M.: "Effect of Particle Size on Signal Magnitude in an Optical Doppler Velocimeter", Radiotekh. Elektron., 1973, 18, no.7, pp.1353-8, and Radio Eng. Electron. Phys., 1973, 18, no.7, pp.1000-4.

52.363. Durrani, T.S., and Greated, C.A.: "Statistical Analysis and Computer Simula-tion of Laser Doppler Velocimeters", Trans. IEEE, 1973, IM-22, no.1, pp.23-34.

52.364. Yu'vev, A.N.: "Accuracy of Esti-mating Target Angle by Maximum Likelihood Method in a Correlation Noise Background", Radiotekhnika, 1973, 28, no.3, pp.12-7, and Telecommun. Radio Eng. Pt 2, 1973, 28, no.3, pp.75-8.

52.365. Papoff, A., Romani, G., and Matitti, T.: "Rendezvous and Docking Radar System Study", J. Br. Interplanet. Soc., 1973, 26, no.10, pp.606-21.

52.366. Bulgakov, B.M., et al.: "Laser Meter for Measuring the Parameters of Solid (Dust) Particles", Prib. Tekh. Eksp., 1973, 16, no.3, p.270, and Instrum. Exp. Tech., 1973, 16, no.3, p.981.

52.367. Jernqvist, L.F., and Johansson, T.G.: "Laser Doppler Anemometer for Measure-ment of Arbitrary Velocity Component in Highly Turbulent Fluid Flows", J. Phys. E, 1974, 7, no.4, pp.246-7.

52.368. Schmugge, T., et al.: "Remote Sensing of Soil Moisture With Microwave Radiometers", J. Geophys. Res., 1974, 79, no.2, pp.317-23.

52.369. Gouesbet, G., and Ledoux, M.: "Study of the Laminar Boundary Layer on a Cylinder by Laser Doppler Anemometry", C. R. Acad. Sci. (Paris), 1974, 278B, no.9, pp.315-8.

52.370. Sullivan, J.P., and Ezekiel, S.: "Two-Component Laser Doppler Velocimeter for Periodic Flow Fields", J. Phys. E, 1974, 7, no.4, pp.272-4.

52.371. Wang, C.P., and Snyder, D.: "Laser Doppler Velocimetry. Experimental Study", Appl. Opt., 1974, 13, no.1, pp.98-103.

52.372. Lowenthal, S., Lapierre, J., and Phalippon, D.: "Optical Analysis of Vibra-tion Modes of Elastic Bars", J. Appl. Phys., 1974, 45, no.1, pp.30-1.

52.373. Nygren, T., and Sjolund, A.: "Sensitivity of Doppler Radar With Self-Detecting Diode Oscillators", Trans. IEEE, 1974, MTT-22, no.5, pp.494-8.

52.374. Cugiani, C., and Piazzo, C.: "S-Band Interferential Ranging System", Alta Freq., 1974, 43, no.4, pp.207-10.

52.375. Shefer, J., et al.: "Clutter-Free Radar for Cars", Wireless Wld, 1974, 80, no.1461, pp.117-22 and no.1462, pp.199-202.

52.376. Tompkins, W.R., Monti, R., and Intaglietta, M.: "Velocity Measurement by Self-Tracking Correlator", Rev. Sci. Instrum., 1974, 45, no.5, pp.647-9.

52.377. Lizon, A.: "Analysis of Disturbing Signal Influence Upon Velocity Measurement by Doppler Method", Rozpr. Elektrotech., 1974, 20, no.2, pp.389-96.

52.378. Farhat, N.H., and Wang, P.C.: "Holographic Imaging With Object Synthesized Apertures", Trans. IEEE, 1974, MTT-22, no.5, pp.531-5.

52.379. Landry, M.J.: "Beam Characteristics of a Lidar System", Appl. Opt., 1974, 13, no.1, pp.63-73.

52.380. Hardy, W.N., Gray, K.W., and Love, A.W.: "S-Band Radiometer Design With High Absolute Precision", Trans. IEEE, 1974, MTT-22, no.4, pp.382-90.

52.381. Silverberg, E.C.: "Operation and Performance of a Lunar Laser Ranging Station", Appl. Opt., 1974, 13, no.3,pp.565-74.

52.382. Skolnik, M.I.: "Empirical Formula for RCS of Ships at Grazing Incidence", Trans. IEEE, 1974, AES-10, no.2, pp.292-3.

52.383. Steinberg, B.D.: "Angular Resolution in Microwave Radar", Proc. IEEE, 1974, 62, no.4, pp.519-20.

52.384. Pernick, B.J.: "Distortions of Synthetic Aperture Radar Imagery Due to Target Motion", Appl. Opt., 1974, 13, no.3, pp.471-2.

52.385. Vasilenko, Yu.G., et al.: "Application of Laser Doppler Measurements of Velocity to Boundary Layers in a Stream", Avtometriya, 1974, no.1, pp.71-5.

52.386. Iten, P.D., and Mastner, J.: "Laser Doppler Velocimeter", Electron. Anz., 1974, 6, no.2, pp.31-2.

52.387. Kirmse, R.: "Use of Laser Doppler Anemometers for Flow-Rate Measurement", Arch. Tech. Mess., 1974, no.458, pp.41-6.

52.388. Mishina, H., Asakura, T., and Nagai, S.: "Laser Doppler Microscope", Opt. Commun., 1974, 11, no.1, pp.99-102.

52.389. Yang, B.T., and Meroney, R.N.: "Portable Laser Light-Scattering Probe for Turbulent Diffusion Studies", Rev. Sci. Instrum., 1974, 45, no.2, pp.210-5.

52.390. Hendry, A., and McCormick, G.C.: "Deterioration of Circular-Polarization Clutter Cancellation in Anisotropic Precipitation Scatter", Electron. Lett., 1974, 10, no.10, pp.165-6.

52.391. Flock, W.L., and Green, J.L.: "Detection and Identification of Birds in Flight Using Coherent and Noncoherent Radars", Proc. IEEE, 1974, 62, no.6, pp.745-53.

52.392. Nicholls, L.A.: "Rapid Determination of Glint Spectral Parameters for Symmetrical Two-Complex-Source Targets", Trans. IEEE, 1974, AES-10, no.2, pp.288-90.

52.393. Barton, D.K.: "Low-Angle Radar Tracking", Proc. IEEE, 1974, 62, no.6, pp.687-704.

52.394. Mitchell, R.L.: "Models of Extended Targets and Coherent Radar Images", Proc. IEEE, 1974, 62, no.6, pp.754-8.

52.395. Lobarev, A.S., and Nizhegorodtseva, I.I.: "Doppler Effect in the SHF Region", Izv. VUZ Fiz., 1974, no.4, pp.144-6.

52.396. van Roessel, J.W., and de Godoy, R.C.: "Side-Looking Airborne Radar (Mapping) Mosaics", Photogramm. Eng., 1974, 40, no.5, pp.583-95.

52.397. Derenyi, E.E.: "Side-Looking Airborne Radar Geometric Test", Photogramm. Eng., 1974, 40, no.5, pp.597-604.

52.398. Gobert, J.F.: "Analysis and Suppression of Dynamic Effects of Scanning Caused by Phased-Array MTI Radar Operation", Arch. Elektron. Ubertrag., 1974, 28, no.7-8, pp.314-20.

52.399. Eliasson, B., and Dandliker, R.: "Theoretical Analysis of Laser Doppler Flowmeters", Opt. Acta, 1974, 21, no.2,pp.119-49.

52.400. Lank, G.W.: "Estimation of Clutter Cross Section", Trans. IEEE, 1974, AES-10, no.1, pp.152-3.

52.401. Chadwick, R.B.: "Clutter Reduction Techniques for Random Signal Radars", Trans. IEEE, 1974, AES-10, no.1, pp.156-60.

52.402. Vannicola, V.: "Detection of Slow Fluctuating Targets With Frequency-Diversity Channels", Trans. IEEE, 1974, AES-10, no.1, pp.43-52.

52.403. Lee, C.K.: "Tunnel-Diode Oscillator as Self-Excited Mixer for Moving Targets", J. Korean Inst. Electron. Eng., 1974, 11, no.1, pp.40-6.

52.404. Fitzgerald, R.J.: "Effect of Range-Doppler Coupling on Chirp Radar Tracking Accuracy", Trans. IEEE, 1974, AES-10, no.4, pp.528-32.

52.405. Arora, R.K., and Agrawal, V.D.: "Frequency Spread Associated With Fast Electronic Scanning", Proc. IEEE, 1974, 62, no.8, pp.1175-6.

52.406. Bates, C.J.: "Experimental Pipe Flow Analysis Using a Laser Doppler Anemometer", DISA Inf., 1974, no.16, pp.5-10.

52.407. Enander, B., and Larson, G.: "Microwave Radiometric Measurements of the Temperature Inside a Body", Electron. Lett., 1974, 10, no.15, p.317.

52.408. Ohtsuka, Y., and Ozawa, T.: "Three-Dimensional Laser Doppler Scalar Velocimeter With Directional Nonambiguity", Jap. J. Appl. Phys., 1974, 13, no.9, pp.1435-9.

52.409. Wiesbeck, W.: "Precipitation Echoes in Millimetre-Wave CW Radars", Arch. Elektron. Ubertrag., 1974, 28, no.9,pp.371-4.

52.410. Stevens, J.E., and Nagy, L.L.: "Diplex Doppler Radar for Automotive Obstacle Detection", Trans. IEEE, 1974, VT-23, no.2, pp.34-44.

52.411. Constant, J.N.: "Microwave Automatic Vehicle Identification System", Trans. IEEE, 1974, VT-23, no.2, pp.44-54.

52.412. Zieleniewski, J.: "Requirements for a Doppler Satellite Navigation System", J. Br. Interplanet. Soc., 1974, 27, no.8, pp.602-11.

52.413. Ballantyne, A., Blackmore, C.S., and Rizzo, J.E.: "Frequency Shifting for Laser Anemometers by Scattering", Opt. Laser Technol., 1974, 6, no.4, pp.170-3.

52.414. Hoge, F.E.: "Integrated Laser/ Radar Satellite Ranging and Tracking System", Appl. Opt., 1974, 13, no.10, pp.2352-8.

52.415. Pallavicino, L.: "Microwave Landing Systems", Radio Ind., 1974, 41, no.7, pp.21-4 and no.8, pp.34-40.

52.416. Sobolev, V.S.: "Phase-Noise Spectrum at the Outlet of a Laser Doppler Flowmeter", Avtometriya, 1974, no.6, pp.111-2.

52.417. Sobolev, V.S., and Timokhin, S.A.: "Selection of Scattering-Particle Concentration in Determining Flow Rate With a Laser Doppler Meter", Avtometriya, 1974, no.6, pp.112-4.

52.418. Wardrop, B.: "Performance of MTI Systems Used With PRF Stagger", Marconi Rev., 1974, 37, no.195, pp.217-32.

52.419. Lyapin, K.K., Polyanskii, V.A., and Shishkin, I.F.: "Selection of Targets in Presence of Reflections from the Sea", Radiotekhnika, 1974, 29, no.2, p.76, and Telecommun. Radio Eng. Pt 2, 1974, 29,no.2,p.111.

52.420. Vasilenko, Yu.G., et al.: "Laser Doppler Velocimeters With Frequency Shift", Avtometriya, 1974, no.6, pp.83-90.

52.421. Agapov, G.A., Lakin, Yu.G., and Sizov, V.V.: "High-Sensitivity Laser Manometer-Flowmeter for Corrosive Media", Kvantovaya Elektron., 1974, 1, no.5, pp.1281-3, and Sov. J. Quantum Electron., 1974, 4, no.5, pp.715-6.

52.422. Loomis, J.M.: "Frequency-Agility Processing to Reduce Radar Glint Pointing Error", Trans. IEEE, 1974, AES-10, no.6, pp.811-20.

52.423. White, W.D.: "Low-Angle Radar Tracking in the Presence of Multipath", Trans. IEEE, 1974, AES-10, no.6, pp.835-52.

52.424. Ral'nikov, V.I., and Kharchenko, I.P.: "Mathematical Model for EMC of Radar Installations", Izv. VUZ Radioelektron., 1974, 17, no.12, pp.72-6.

52.425. Howard, D.D., et al.: "Experimental Results of Complex Indicated Angle Technique for Multipath Correction", Trans. IEEE, 1974, AES-10, no.6, pp.779-87.

52.426. Mansharamani, N.: "Compact Time-Measuring Unit for Laser Rangefinder", J. Phys. E, 1974, 7, no.12, pp.966-8.

52.427. Lutomirski, R.F.: "Phase-Difference and Angle-of-Arrival Fluctuations in Tracking a Moving Point Source", Appl. Opt., 1974, 13, no.12, pp.2869-73.

52.428. Vincent, R.P.: "Microwave Aircraft Digital Guidance Equipment. I", Philips Tech. Rev., 1974, 34, no.9, pp.225-41.

52.429. Munoz, R.M., Mocker, H.W., and Koehler, L.O.: "Airborne Laser Doppler Velocimeter", Appl. Opt., 1974, 13, no.12, pp.2890-8.

52.430. Calamia, M., et al.: "Radar Tracking of Low-Altitude Targets", Trans. IEEE, 1974, AES-10, no.4, pp.539-44.

52.431. Hildebrand, B.P.: "Statistics of Focused and Defocused Radar Maps", Trans. IEEE, 1974, AES-10, no.5, pp.615-21.

52.432. Lutte, N.P., and Thomas, H.W.: "Frequency-Domain Approaches to MTI Filters With Staggered PRF's",Proc. IEE, 1974, 121, no.9, pp.954-6.

52.433. Donati, S.: "Optoelectronic Techniques for Navigation Aids in Poor Weather", Alta Freq., 1974, 43, no.9, pp.725-8.

52.434. Minnett, H.C., and O'Keefe, H.B.: "INTERSCAN. Microwave Approach and Landing Guidance System for International Civil Aviation", J. AEU, 1974, 7, no.2, pp.15-20.

52.435. Mishina, H., et al.: "Beat-Signal Analysing System in a Laser Doppler Microscope", Bull. Res. Inst. Appl. Electr., 1974, 26, no.1-2, pp.51-66.

52.436. Murty, Y.S.N., and Rao, V.N.: "Analysis of Radar Clutter Environment", Electro-Technol., 1974, 18, no.2, pp.39-47.

52.437. Mizumachi, M., and Sakai, Y.: "Digital Detector for Surveillance Radar", Electron. Commun. Jap., 1974, 57, no.1, pp.79-88.

52.438. Gol'dman, R.S., and Titov, M.S.: "Recognition of Objects on Basis of Reflected Signals", Tekh. Kibern., 1974,12,no.2,p.197, and Eng. Cybern., 1974,12,no.2,pp.154-9.

52.439. Szajnowski, W.: "Correlative Properties of Sub-Sequences of Binary Psuedo-Random Sequences", Pr. PIT, 1974, 24, no.81, pp.22-30.

52.440. Turkel'taub, R.M.: "Spectrum of a Coherent Sequence of Stochastic Pulses", Radiotekhnika, 1974, 29, no.5, pp.79-81, and Telecommun. Radio Eng. Pt 2, 1974, 29, no.5, pp.101-2.

52.441. Gadre, M.N., Subramanian, A.K., and Taneja, V.S.: "Scheme for Continuously Tunable Stable Local Oscillator for Radar Systems", Electro-Technol., 1974, 18, no.1, pp.27-9.

52.442. Kailasam, T.K., et al.: "Radar Principles of Distance Measurement", Electro-Technol., 1974, 18, no.2, pp.57-63.

52.443. Mellon, D.W., and Daniels, W.D.: "Pulse-Compression Distance-Measuring Equipment Using Surface-Acoustic-Wave Devices", Trans. IEEE, 1974, MTT-22, no.12, pp.1308-12.

52.444. Marukawa, T., Morinaga, N., and Namekawa, T.: "Accurate System of FM CW Radar for Approaching Target Using Phase Detection", Technol. Rep. Osaka Univ., 1974, 24, no.1191-1229, pp.635-53.

52.445. Carletti, U., et al.: "Target Tracking", Alta Freq., 1974, 43, no.9, p.708.

53. ASTRONOMY TECHNIQUES

53.1. Howard, H.T., and Erickson, W.C.: "Ground-Based Bistatic Radar Astronomy", J. Geophys. Res., 1965, 70, p.1270.

53.2. Pollack, J.B., and Sagan, C.: "Microwave Phase Effect of Venus", Icarus, 1965, 4, p.62.

53.3. Kaydanovskii, N.L., and Smirnova, N.A.: "Resolution Limits of Radiotelescopes and Interferometers by Propagation in Space and the Earth's Atmosphere", Radiotekh. Elektron., 1965, 10, p.1574, and Radio Eng. Electron. Phys., 1965, 10, p.1355.

53.4. Schell, A.C.: "Enhancing the Angular Resolution of Incoherent Sources", Radio Electron. Eng., 1965, 29, p.21.

53.5. Beckman, P.: "Radar Backscatter from Surface of the Moon", J. Geophys. Res., 1965, 70, p.2345.

53.6. Nash, R.T., O'Donnell, S.R., and Fitch, L.: "Time-Sharing Radiometers", Proc. IEEE, 1965, 53, p.308.

53.7. Vinokur, M.: "Optimization of the Detection of a Sine Wave of Known Period in the Presence of Noise. Application to Radio Astronomy", Ann. Astrophys., 1965, 28, p.412.

53.8. Simon, M.: "Solar Observations at 3.2-mm", Astrophys. J., 1965, 141, p.1513.

53.9. Broten, N.W., et al.: "Observations at 6 cm With the Australian 64-m Radiotelescope", Aust. J. Phys., 1965, 18, p.85.

53.10. Pettingill, G.H., and Dyce, R.B.: "Radar Determination of the Rotation of the Planet Mercury", Nature, 1965, 206, p.1240.

53.11. Yaplee, B.S., et al.: "Mean Distance to the Moon as Determined by Radar", Bull. Astron., 1965, 25, p.81.

53.12. Matveenko, L.I., et al.: "Application of Masers to Radio-Astronomical Investigation at 8 cm", Dokl. Akad. Nauk SSSR, 1965, 161, p.810, and Sov. Phys.-Dokl., 1965, 11.

53.13. Weaver, H., et al.: "Observations of a Strong Unidentified Microwave Line and of Emission from the OH Molecule", Nature, 1965, 208, p.29.

53.14. Berge, G.L.: "Interferometric Study of Jupiter at 10 cm and 21 cm", J. Res. Natl. Bur. Stand., 1965, 69D, p.1552.

53.15. Sagan, C., and Pollack, J.B.: "Analysis of Microwave Observations of Venus", J. Res. Natl. Bur. Stand., 1965, 69D, p.1584.

53.16. Muhleman, D.O.: "Radar Scattering from Venus and Mercury at 12.5 cm", J. Res. Natl. Bur. Stand., 1965, 69D, p.1632.

53.17. Evans, J.V.: "Radar Studies of the Moon", J. Res. Natl. Bur. Stand., 1965, 69D, p.1639.

53.18. Drake, F.D.: "Search for the 1.36-cm Water-Vapour Line in Venus", J. Res. Natl. Bur. Stand., 1965, 69D, p.1577.

53.19. Gibson, J.E.: "Observations of Jupiter at 8.6 mm", J. Res. Natl. Bur. Stand., 1965, 69D, p.1560.

53.20. Mezger, P.G.: "Polarization of Thermal Radiation of the Moon at 14.5 GHz", J. Res. Natl. Bur. Stand., 1965, 69D, p.1612.

53.21. Salomonovich, A.E.: "Radio Emission from the Planets Mercury, Mars, and Saturn, at 8 mm", J. Res. Natl. Bur. Stand., 1965, 69D, p.1576.

53.22. Gardner, F.F.: "Polarization of 6-cm Radiation from Crab Nebula", Aust. J. Phys., 1965, 18, p.385.

53.23. Kamenskaya, S.A., et al.: "Radio Eclipses of the Moon at Millimetre Wavelengths", Izv. VUZ Radiofiz., 1965, 8, p.219, and Sov. Radiophys., 1965, 8, p.155.

53.24. Kuz'min, A.D.: "Measurements of the Brightness Temperature of the Illuminated Side of Venus at 10.6 cm", Astron. Zh., 1965, 42, p.1281, and Sov. Astron.-AJ, 1965, 42.

53.25. Gibson, J.E., and Corbett, H.H.: "Radiation of Venus at the 13.5-mm Water-Vapour Line", J. Res. Natl. Bur. Stand., 1965, 69D, p.1577.

53.26. Pettengill, G.H.: "Radar Studies of Planetary Surfaces", J. Res. Natl. Bur. Stand., 1965, 69D, p.1619.

53.27. Tansworthe, R.C.: "Precision Planetry Range-Tracking Radar", Trans. IEEE, 1965, SET-11, p.78.

53.28. Kotel'nikov, V.A.: "Radar Observations of Venus", Dokl. Akad. Nauk SSSR, 1965, 163, p.50, and Sov. Phys.-Dokl., 1965, 11.

53.29. Katz, I.: "Wavelength Dependence of Radar Reflectivity of the Earth and the Moon", J. Geophys. Res., 1966, 71, p.361.

53.30. Epstein, E.E.: "Mercury. Anomalous Absence from the 3.4-mm Emission of Variation With Phase", Science, 1966, 151, p.445.

53.31. King, H.E., Jacobs, E., and Stacey, J.M.: "Lunar Eclipse Observations at 3.2 mm", Trans. IEEE, 1966, AP-14, p.82.

53.32. Gee, S.: "Bistatic-Radar Measurements of Interplanetary Plasma Streams", J. Geophys. Res., 1966, 71, p.2353.

53.33. Meyer, J.W.: "Radar Astronomy at Submillimetre Wavelengths", Proc. IEEE, 1966, 54, p.484.

53.34. Coates, R.J.: "Solar Observations at Millimetre Wavelengths", Proc. IEEE, 1966, 54, p.471.

53.35. Thaddeus, P., and Clauser, J.F.: "Cosmic Radiation at 2.63 mm from Observations of Interstellar CN", Phys. Rev. Lett., 1966, 16, p.819.

53.36. Varshalovich, D.A.: "Coherent Amplification of Radio Emission in a Cosmic Medium", Zh. Eksp. Teor. Fiz. Pis'ma, 1966, 4, no.5, pp.180-2, and JETP Lett., 1966, 4, no.5, pp.124-5.

53.37. Maxwell, A., and Rinehart, R.: "Flux Densities of Radio Sources at 5 GHz", Astron. J., 1966, 71, p.927.

53.38. Hughes, M.P.: "Planetary Observations at 6 cm", Planet. Space Sci., 1966, 14, p.1017.

53.39. Welch, W.J., Thornton, D.D., and Lohman, R.: "Observations of Jupiter, Saturn, and Mercury at 1.53 cm", Astrophys. J., 1966, 46, p.799.

53.40. Kellerman, K.I.: "Thermal Radio Emission from Mercury, Venus, Mars, Saturn, and Uranus", Icarus, 1966, 5, p.478.

53.41. Evans, J.V., et al.: "Radar Observations of Venus at 3.8 cm", Astron. J., 1966, 71, p.902.

53.42. Pistol'kors, A.A.: "Method for Determination of Spectrum of an Extraterrestrial Source", Radiotekh. Elektron., 1966, 11, no.5, and Radio Eng. Electron. Phys., 1966, 11, no.5, pp.673-7.

53.43. Dickel, J.R.: "Measurement of Temperature of Venus at 3.75 cm for a Full Cycle of Planetary Phase Angles", Icarus, 1966, 5, p.305.

53.44. Seielstad, G.A.: "Distributions of 10.6-cm Linearly Polarized Radiation Over Eight Extragalactic Radio Sources", Astrophys. J., 1967, 147, p.24.

53.45. Charru, A., Picherit, F., and Lissayou, J.: "First Experiments at the Bordeaux Faculty of Sciences With a Radiotelescope for the 35-GHz Band", Ann. Astrophys., 1967, 30, p.119.

53.46. Ko, H.C.: "Coherence Theory of Radio-Astronomical Measurements", Trans. IEEE, 1967, AP-15, no.1, pp.10-20.

53.47. Clapp, R.E., and Maxwell, J.C.: "Complex-Correlation Radiometer", Trans. IEEE, 1967, AP-15, no.2, pp.286-91.

53.48. Hagfors, T.: "Study of the Depolarization of Lunar Radar Echoes", Radio Sci., 1967, 2, no.5, pp.445-65.

53.49. Frenkel, L.: "Microwave Method for Planetary Atmosphere Research", Rev. Sci. Instrum., 1967, 38, no.4, pp.557-8.

53.50. Ko, H.C.: "Theory of Tensor Aperture Synthesis", Trans. IEEE, 1967, AP-15, no.1, pp.188-9.

53.51. Dyce, R.B., and Pettingill, G.H.: "Radar Determination of the Rotations of Venus and Mercury", Astron. J., 1967, 72, no.3, pp.351-9.

53.52. Sagan, C., Pollack, J.B., and Goldstein, R.M.: "Radar Doppler Spectroscopy of Mars. I", Astron. J., 1967, 72, no.1, pp.20-34.

52.53. Epstein, E.E.: "Small-Scale Distribution at 3.4 mm of the Reported $3^{O}K$ Background Radiation", Astrophys. J., 1967, 148, no.3, pp.L157-9.

53.54. Naumov, A.P., and Kanevskii, M.B.: "Venus Radiation Spectrum at Millimetre Wavelengths", Izv. VUZ Radiofiz., 1967, 10, no.8, pp.1058-69.

53.55. Akabane, K., et al.: "High-Resolution Observations of the Galactic Radio Sources at 4.17 GHz", J. Radio Res. Lab., 1967, 14, no.1, pp.88-95.

53.56. Kronberg, P.P., and Conway, R.G.: "Interferometer of 425-m Baseline Consisting of Two-Fully Steerable Radiotelescopes", Mon. Notes Roy. Astron. Soc., 1967, 135, no.2, pp.199-206.

53.57. Feix, G.: "Solar Radiation in the Millimetre Range", Nachr. Tech., 1967, 20, no.8, pp.429-34.

53.58. Castelli, J.P., and Strauss, F.M.: "Correlation Between Sudden Cosmic Noise Absorption and Solar Radio Bursts Observed at Five Microwave Frequencies", Nature, 1967, 216, 25 Nov., pp.776-7.

53.59. Ewing, M.S., Burke, B.F., and Staelin, D.H.: "Cosmic Background Measurement at 9.24 mm", Phys. Rev. Lett., 1967, 19, no.21, pp.1251-3.

53.60. Shapiro, I.I.: "Theory of the Radar Determination of Planetary Rotations", Astron. J., 1967, 72, no.10, pp.1309-23.

53.61. Hagfors, T., Nanni, B., and Stone, K.: "Aperture Synthesis in Radar Astronomy and Applications to Lunar and Planetary Studies", Radio Sci., 1968, 3, no.5, pp.491-509.

53.62. Rainal, A.J.: "Phase Principle for Measuring Location or Spectral Shape of a Discrete Radio Source", Bell Syst. Tech. J., 1968, 47, no.3, pp.415-28.

53.63. Stankevich, K.S.: "Influence of Terrestrial Tides on the Operation of Interferometers With Long Base", Radiotekh. Elektron., 1968, 13, no.3, p.529, and Radio Eng. Electron. Phys., 1968, 13, no.3.

53.64. Feix, G.: "Radioastronomical Interferometry With Long Baseline", Nachr. Tech. Z., 1968, 21, no.5, pp.263-6.

53.65. Penzias, A.A.: "Measurement of Cosmic Microwave Background Radiation", Trans. IEEE, 1968, MTT-16, p.606.

53.66. Johnson, W.A.: "Performance of a 3.3-mm Radiometer", Trans. IEEE, 1968, MTT-16, p.619.

53.67. Stankevich, K.S.: "Limitation of Maximum Sensitivity of Radiotelescopes by Fluctuations of Thermal Emission by the Atmosphere", Radiotekh. Elektron., 1968, 13, p.1570, and Radio Eng. Electron. Phys., 1968, 13, no.9, pp.1366-72.

53.68. Laffineur, M., and Koutchmy, S.: "Millimetre-Wave Radiotelescope at Sineis Observatory", Onde Electr., 1969, 49, p.246.

53.69. Berlin, A.B., et al.: "Radiometer for 4 cm Based on Tunnel-Diode and Parametric Amplifiers", Prib. Exp. Tech., 1969, 12, no.3, p.146, and Instrum. Exp. Tech., 1969, 12, no.3, p.695.

53.70. Bakhrakh, L.D., et al.: "Improvement of Operation Efficiency of Radiotelescope at 8 mm", Izv. VUZ Radiofiz., 1969, 12, no.8, p.1115.

53.71. Alekseev, V.A.: "Composite Interferometer With Independent Local Oscillators for Studying Radiation Sources", Radiotekh. Elektron., 1969, 14, no.6, p.1091, and Radio Eng. Electron. Phys., 1969, 14,no.6,pp.940-2.

53.72. Yefanov, V.A., et al.: "Radiometer for Q-Band Using Parametric Amplifier", Radiotekh. Elektron., 1970, 15, p.627, and Radio Eng. Electron. Phys., 1970, 15, no.3, pp.541-2.

53.73. Charru, A., et al.: "Study and Synchronous-Oscillators Realization of a Radioastronomic Interferometer at 35 GHz", Onde Electr., 1970, 50, no.6, p.517.

53.74. Sodin, L.G.: "Method of Statistical Estimations in Solving Brightness Reconstruction in Radioastronomy", Izv. VUZ Radiofiz., 1971, 14, no.5, pp.739-47.

53.75. Buhl, D., and Snyder, L.E.: "Microwave Receivers for Molecular-Line Radio Astronomy", Nature, 1971, 232, no.34,pp.161-3.

53.76. Fourikis, N.: "Microwave Solar Spectrograph", Proc. IREE Aust., 1971, 32, no.10, pp.361-6.

53.77. Papadopoulos, G.C., and Brke, B.F.: "K-Band Radio Interferometer", Radio Sci., 1972, 7, no.6, pp.667-74.

53.78. Oki, K., et al.: "17-GHz Interferometer", Tokyo Astron. Obs. Rep., 1972, 16, no.1, pp.241-58.

53.79. Timofeev, Yu.M., Gruzdeva, M.A., and Pokrovskii, O.M.: "Determining the Temperature Profile in Venus' Atmosphere from Thermal Emission", Kosm. Issled., 1972, 19, no.5, pp.751-9, and Cosmic Res., 1972, 10, no.5, pp.680-7.

53.80. Krupenio, N.N., and Cherkasov, V.V.: "Radio Scattering Characteristics of the Moon's Surface at Luna Landing Sites", Kosm. Issled., 1972, 19, no.5, pp.794-6, and Cosmic Res., 1972, 10, no.5, pp.722-4.

53.81. Shapiro, I.I., et al.: "Lunar Topography Determination by Radar", Science, 1972, 178, no.4064, pp.939-48 and 977-80.

53.82. Pampaloni, P., and Tofani, G.: "Radiometer at 8.4 mm for Radioastronomy", Mem. Soc. Astron. Ital., 1972, 43, no.3, pp.523-34.

53.83. Dent, W.A.: "Flux-Density Scale for Microwave Frequencies", Astrophys. J., 1972, 177, no.1, pp.93-9.

53.84. Rudnitskii, G.M., and Strel'nitskii, V.S.: "Constraints on Cosmic Maser Intensity and Possibility of Detecting OH and H_2O Sources", Astron. Zh., 1972, 49, no.6, pp.1323-5, and Sov. Astron.-AJ, 1973, 16, no.6, pp.1049-50.

53.85. Vardanyan, A.S., et al.: "Submillimetre Telescope With an n-InSb Detector", Astron. Zh., 1972, 49, no.5, pp.986-9, and Soviet Astron.-AJ, 1973, 16, no.5, pp.806-8.

53.86. Gay, J., and Journet, A.: "Heterodyne Detection of Blackbody Radiation", Appl. Phys. Lett., 1973, 22, no.9, pp.448-9.

53.87. Linsky, J.L.: "Moon as a Proposed Radiometric Standard for Microwave and IR Observations of Extended Sources", Astrophys. J., 1973, 25, no.216, pp.163-204.

53.88. Parker, M.N., and Tyler, G.L.: "Bistatic-Radar Estimation of Surface-Slope Probability Distributions With Applications to the Moon", Radio Sci., 1973, 8, no.3, pp.177-84.

53.89. Bracewell, R.N., and Thompson, A.R.: "Main Beam and Ringlobes of an East-West Rotation-Synthesis Array", Astrophys. J., 1973, 182, no.1, pp.77-94.

53.90. Korchak, A.: "Possibility of Measuring Gravitational Redshift by Earth Satellites", Comments Astrophys. Space Phys., 1973, 5, no.2, pp.37-42.

53.91. Woo, R.: "Remote Sensing of Turbulence Characteristics of a Planetary Atmosphere by Radio Occultation of a Space Probe", Radio Sci., 1973, 8, no.2, pp.103-8.

53.92. Shapiro, I.I., et al.: "Venus. Radar Determination of Gravity Potential", Science. 1973, 179, no.4072, pp.473-6.

53.93. Bell, M.B., Covington, A.E., and Kennedy, W.A.G.: "Polarization Interferometer for 2.8-GHz Solar Noise Studies With 0.15-mrad Fan Beam", Sol. Phys., 1973, 28, no.1, pp.123-36.

53.94. Linsky, J.L.: "Recalibration of the Quiet-Sun Millimetre Spectrum Using the Moon as an Absolute Radiometric Standard", Sol. Phys., 1973, 28, no.2, pp.409-18.

53.95. Kaufmann, P.: "Brazilian Radiotelescope for Millimetre Wavelengths", Sky Telesc., 1973, 45, no.3, pp.144-5.

53.96. Calame, O.: "Determination of Lunar Libration by Laser Moon Ranging", Astron. Astrophys., 1973, 22, no.1, pp.75-80.

53.97. Abbot, R.I., et al.: "Laser Observations of the Moon", Astron. J., 1973, 78, no.8, pp.784-93.

53.98. Ponsonby, J.E.B., and Morison, I.: "Sampling Phase Rotator for Removal of Doppler Shift in Lunar Radar", Electron. Lett., 1973, 9, no.18, pp.409-10.

53.99. Davis, J.H., and Bout, P.V.: "Intensity Calibration of Interstellar CO Line at 2.6 mm", Astrophys. Lett., 1973, 15, no.1, pp.43-7.

53.100. Standish, E.M.: "Figure of Mars and Effect on Radar Ranging", Astron. Astrophys., 1973, 26, no.3, pp.463-6.

53.101. Delannoy, J., Lacroix, J., and Blum, E.J.: "8-mm Interferometer for Solar Radio Astronomy at Bordeaux", Proc. IEEE, 1973, 61, no.9, pp.1282-4.

53.102. Shukla, P.K.: "Synchrotron Radiation from Magnetoplasma (of Solar Corona)", Int. J. Electron., 1973, 35, no.4, pp.533-9.

53.103. Berlin, A.B., et al.: "4-cm Radiometer With Nitrogen-Cooled Parametric Amplifier", Izv. VUZ Radiofiz., 1973, 16, no.9, pp.1444-7.

53.104. Kislyakov, A.G., and Shvetsov, A.A.: "Limiting Sensitivity of Radiotelescopes for Investigations of Inhomogeneities in Radiation Distribution", Izv. VUZ Radiofiz., 1973, 16, no.12, pp.1846-52.

53.105. Sazonov, V.V., and Karavaev, V.V.: "Theory of Measurements of Positions of Noise Radio Sources Using Interferometer With Aperture Synthesis", Izv. VUZ Radiofiz., 1973, 16, no.12, pp.1861-6.

53.106. Stutzman, W.L., and Ko, H.C.: "Measurement of Antenna Beamwidth Using Extraterrestrial Radio Sources", Trans. IEEE, 1974, AP-22, no.3, pp.493-5.

53.107. Ade, P.A.R., Rather, J.D.G., and Clegg, P.E.: "Limits to Solar Limb Darkening at 1.4 mm Derived from Antenna-Beam Parameters", Astrophys. J., 1974, 187, no.2, pp.389-92.

53.108. Burenin, A.V., and Krupnov, A.K.: "Limiting Parameters of Radiotelescopes", Izv. VUZ Radiofiz., 1974, 17, no.8, pp.1242-4.

53.109. Akabane, K., Miyaji, T., and Chikada, Y.: "Reflector Beam Switching for Millimetre-Wave Telescope", Tokyo Astron. Bull., 1974, no.229, pp.2639-47.

53.110. Akabane, K., Chikada, Y., and Miyazawa, K.: "Phaseshifter Beam Switching for Millimetre-Wave Telescope", Tokyo Astron. Bull., 1974, no.232, pp.2675-85.

53.111. van de Stadt, H.: "Heterodyne Detection at 3.39 micron for Astronomical Purposes", Astron. Astrophys., 1974, 36, no.3, pp.341-8.

53.112. Vystavkin, A.N., et al.: "Submillimetre Spectroradiometers With n-InSb Detectors", Trans. IEEE, 1974, MTT-22, no.12, pp.1041-6.

Chan, A.K., 16.426
Chan, C.H., 45.388
Chan, C.K., 25.373, 27.669
Chan, K.B., 12.148, 48.54
Chan, S.P., 33.227
Chan, W.K., 28.149
Chance, B., 46.108
Chandezon, J., 12.283
Chandra, K., 14.37, 18.72,
  .36.34
Chandra, N., 27.613, 32.62
Chandra, S., 21.241
Chandross, E.A., 21.176,
  34.90
Chang, B.J., 34.167
Chang, C.S., 20.239, 45.259
Chang, C.T.M., 10.304,
  10.351, 12.221, 36.170
Chang, D.B., 6.177
Chang, D.C., 16.329
Chang, I.C., 27.92
Chang, I.C.H., 44.364
Chang, K., 10.346
Chang, K.K.N., 25.134
Chang, M.S., 21.272
Chang, N.C., 22.792
Chang, N.S., 4.60, 25.215,
  33.49
Chang, R.F., 43.44
Chang, R.K., 20.68
Chang, R.W., 35.12
Chang, S.Y., 49.65
Chang, T.M., 10.347
Chang, T.V., 22.709
Chang, T.Y., 22.43, 22.172,
  22.187, 22.273, 22.382,
  22.433, 22.551, 22.817,
  22.936, 24.342, 27.314
Chang, V.W.H., 16.312
Chang, W.S.C., 17.39,
  17.120, 30.223, 34.176
Channin, D.J., 13.76, 30.255,
  30.312, 30.343
Chao, C., 25.317
Chao, G., 10.163, 14.139
Chapline, G.F., 45.215
Chapovskii, P.I., 22.517,
  27.668
Charan, S., 24.379, 24.390
Charap. S.H., 12.313
Charles, C., 22.555
Charlson, R.J., 49.307
Charlton, D.A., 14.47,
  17.122
Charlton, G.G., 16.145
Charlton, R., 25.372
Charru, A., 39.43, 39.90,
  39.153, 53.45, 53.73
Charschan, S.S., 7.73

Chartier, G., 21.45, 26.278,
  27.19, 37.55
Charvier, H., 49.68
Chase, A.B., 45.134
Chase, I.L., 10.186
Chashchin, S.P., 23.40,
  23.82, 23.101, 23.163
Chasse, Y., 44.169
Chastukhina, L.N., 36.154
Chatterjee, J.S., 16.179
Chatterjee, R., 3.4, 11.33,
  12.41, 16.167, 16.196,
  16.216, 16.462, 16.464,
  16.465
Chatterjee, S.K., 3.4, 9.25,
  11.33, 12.40, 12.41,
  12.76, 12.205, 12.206,
  12.294, 16.216, 16.417,
  44.62
Chatterjee, U.K., 17.171,
  51.87
Chattopadhyay, D., 26.68,
  30.117
Chau, H.H., 34.92
Chaudhari, P., 35.128
Chaudhuri, M., 16.362
Chaudhuri, S.K., 10.354
Chavka, G.G., 11.215
Chawla, B.R., 15.73
Chebotaev, V.P., 20.33,
  22.349, 22.496, 22.646,
  27.304, 27.478, 27.519,
  27.626, 30.46, 37.56,
  37.69, 37.78, 37.210,
  37.218, 37.345, 40.136
Chebotarev, G.N., 18.191
Chebotarev, V.I., 16.138
Cheburkin, N.V., 20.41,
  42.97
Chebykhin, N.N., 24.549
Checcacci, P.F., 8.134, 9.51,
  9.53, 9.66, 11.49, 11.78,
  11.247, 11.264, 38.170,
  50.93
Chechenina, E.P., 20.54,
  20.69, 22.324, 27.533
Chegis, I.L., 12.209
Chekalinskaya, Yu.I., 20.54,
  20.69, 20.250, 22.324,
  27.533, 37.41, 38.176
Chekan, A.V., 50.80
Chekhlova, T.K., 24.545,
  44.234
Chelishchev, N.N., 14.211,
  14.249
Chemla, D.S., 6.30, 6.103,
  6.199, 6.213, 6.222, 6.265,
  26.155
Chen, B.V., 26.358

Chen, C.A., 16.455
Chen, C.C., 7.71, 7.75, 7.85,
  16.441, 16.501, 29.156,
  51.21
Chen, C.J., 22.679, 22.741,
  22.840, 22.917
Chen, D., 5.37, 12.341,
  14.315, 35.138
Chen, F.S., 30.403, 35.42,
  35.108
Chen, H.C., 14.291,
  50.22
Chen, J.M., 6.145,
  6.164
Chen, K.M., 9.9, 16.69,
  16.175, 16.307, 50.96
  50.110
Chen, L.W., 11.221,
  11.227
Chen, M., 35.67
Chen, M.H., 3.143, 16.445,
  16.509
Chen, M.Y., 12.24
Chen, S.C., 46.117
Chen, S.H., 5.75, 43.97
Chen, S.N.C., 16.291
Chen, T.S., 28.24
Chen, W.H., 10.173
Chen, W.Y., 6.143, 45.271,
  45.272
Chen, Y.G., 41.3
Chen, Y.M., 48.17
Chenausky, P.P., 24.416,
  27.307
Cheng, D., 27.407
Cheng, D.K., 13.29, 16.120,
  16.294, 16.455, 29.78,
  50.22
Cheng, Y.C., 17.108
Cheming, H., 35.173
Cheo, P.K., 5.86, 9.40,
  12.252, 22.43, 24.47,
  30.296, 30.420
Cheremiskin, I.V., 24.545,
  44.234
Cheremnykh, M.A., 36.37,
  36.113
Cheremukhin, A.M., 49.121,
  49.185, 49.196, 49.321
Cherkasov, E.M., 22.8, 22.52,
  22.59
Cherkasov, V.V., 53.80
Cherne, Kh.I., 10.153, 11.77
Chernenko, A.A., 27.668
Chernikov, A.A., 52.343
Chernikov, V.A., 22.508
Chernoch, J.P., 21.441,
  21.538
Chernov, V.A., 27.378,
  27.482, 27.492

Hasson, V., 24.531
Hasted, J.B., 1.1
Hasty, T.E., 1.2, 13.8, 25.5
Hasuo, S., 27.639, 42.173
Hatakeyama, K., 48.77
Hatch, A.J., 15.85
Hatcher, D.M., 16.47
Hatsuda, T., 2.50, 2.78,
    2.93, 2.99
Hattori, S., 22.50, 22.78,
    22.79, 22.116, 22.310,
    22.603, 22.826, 31.201
Hattori, T., 1.84
Hauden, D., 37.169
Haueisen, D.C., 6.142
Haug, C., 20.139, 27.589
Haug, H., 23.78
Hauge, P.S., 42.126
Haupt, G., 10.355
Haus, H.A., 24.71, 30.128,
    32.81, 42.166
Haussler, P., 42.35
Haussuhl, S., 6.115
Haveisen, D.C., 6.178
Havey, M.E., 22.254
Hawke, R.S., 13.15
Hawkins, D.C., 10.265,
    16.373, 16.473
Hawkins, P.W., 25.336
Hawrylo, F.Z., 23.68, 23.111,
    23.175, 23.265, 25.79
Hay, D.R., 49.6
Hayami, R.A., 42.49
Hayase, K., 42.149
Hayashi, H., 17.8, 25.369,
    33.87
Hayashi, I., 23.6, 23.48,
    23.64, 23.72, 23.121,
    23.155, 23.332, 24.105
Hayashi, R., 21.234, 33.19,
    33.41
Hayashi, T., 31.117
Hayashi, Y., 14.247
Haydl, W.H., 25.7
Hayes, C.L., 7.95
Hayes, J.N., 45.111, 45.217,
    49.240
Hayes, R.E., 37.115
Hayre, H.S., 16.332,
    52.86
Hazan, J.P., 42.123
Hazen, W., 52.8
Heald, M.A., 5.51, 41.1
Healey, D.J., 32.11, 38.9
Healy, J.J., 22.489, 22.562
Hearson, L.T., 49.54
Heaton, A.G., 36.10, 42.85,
    46.74
Heathcote, V.A., 19.32

Hebel, L.C., 4.7
Hebermehl, G., 34.112
Hebner, R.E., 24.248
Hecht, D.L., 21.247
Hecht, N.L., 44.307
Hechtel, J.R., 18.81, 18.129
Hecken, R.P., 8.97, 8.111
Heckscher, H., 23.247,
    23.263
Hedstrom, T., 36.120
Hedvall, P., 15.54
Heeks, J.S., 25.5, 25.244
Heenan, N.I., 45.1
Heer, C.V., 22.108
Heeren, R.G., 3.94
Heffes, H., 52.149
Heflinger, L.O., 34.108,
    44.72
Hegedus, Z.S., 35.182, 44.368
Heidt, R.C., 29.33
Heimer, G., 46.1
Heinard, W.G., 16.30
Heineken, F.W., 40.129
Heinlich, R.M., 41.12
Heinz, K.J., 52.349
Heinz, W.W., 13.7, 25.292,
    30.1
Heising, S.J., 30.164, 30.199
Helberg, H.W., 10.29, 39.36,
    39.146
Held, H., 31.185
Helfenstein, W., 46.191
Helfrich, W., 5.52
Helgostam, L.F., 52.13
Hella, R., 37.189
Heller, G.S., 14.12, 14.90
Heller, H.J., 11.233
Heller, M., 30.414
Heller, M.E., 9.69, 27.293,
    34.51
Hellman, M.E., 2.27
Hellwig, H., 37.12, 37.23,
    37.242, 38.13
Helmberger, T., 44.132
Helmcke, J., 37.338
Helmstrom, C.W., 35.77,
    35.210
Helszajn, J., 4.43, 14.20,
    14.25, 14.52, 14.65, 14.77,
    14.134, 14.156, 14.160,
    14.164, 14.216, 14.227,
    14.230, 14.244, 14.268,
    14.296, 28.184, 28.200,
    30.6, 39.215
Helsztynski, J., 30.268
Henaff, J., 30.187
Henderson, D.B., 22.754,
    22.755
Henderson, D.M., 30.121,
    30.235, 48.241

Hendry, A., 52.390
Heng, T.M.S., 17.44, 25.394
Henk, T., 25.384
Henke, H., 8.131
Henne, W., 36.32
Hennessy, R.T., 46.157
Henningsen, T., 26.122
Henoch, B.T., 30.152
Henry, A., 22.39, 22.806
Henry, L., 22.39, 22.113,
    37.346
Henry, R., 13.69
Henshall, G.D., 27.617
Henshaw, P.O., 44.305
Hensler, D., 41.112
Hepner, G., 5.93
Heppinstall, R., 18.108
Heppenheimer, T.A., 50.172
Hepplestone, D.J., 39.11
Herbet, H., 32.82
Herbet, J., 49.294
Herbst, H., 32.155
Herbst, R.L., 26.183, 26.359
Herbstritt, R.L., 38.41
Herceg, J.E., 22.65
Hercher, M., 21.202, 26.214,
    37.39, 37.174, 40.133,
    42.121
Herczfeld, P.R., 31.168
Hering, K.H., 14.47, 14.235
Herlent, Y., 48.154
Herman, B.M., 49.2, 49.259,
    49.279, 49.343
Herman, I.P., 22.376
Herman, M.A., 33.97
Herman, P., 36.184, 36.285
Herman, W., 36.46
Hermann, F., 6.84, 26.280
Hermann, G., 27.599
Hermann, J.P., 6.62, 6.130,
    6.159
Hermansdorfer, H., 42.36
Hermet, P., 52.242
Herndon, M., 19.153
Hernqvist, K.G., 22.182,
    22.354, 22.383, 24.32,
    24.99, 24.397, 32.36
Herreman, G.O., 35.159
Herrmann, H., 49.480
Herrmann, J., 41.127, 41.153,
    45.236
Hershberger, W.D., 13.19,
    13.23
Hershenov, B., 12.13, 17.11
Hertz, J.H., 40.54
Herve, J., 33.169
Herziger, G., 21.466, 22.10,
    22.688, 24.100
Herzog, D.G., 23.56

Maes, J.P., 30.300

Maffett, A.L., 7.13, 7.14, 7.56

Magalhaes, F.M., 25.191

Magar, R., 21.118

Magarshack, J., 25.65, 29.31

Magda, A.N., 29.13

Magde, D., 21.479

Magdich, L.N., 20.37, 27.245, 30.345

Magee, C.J., 23.78

Magerl, G., 30.432

Magill, E.G., 29.16

Magill, P.J., 34.105

Magnante, P.C., 26.37, 30.55

Magnoni, A., 33.158

Magoulas, P., 16.346

Maguire, E.A., 4.18

Magun, A., 32.79, 43.86

Magyar, G., 21.102, 21.284, 24.72, 26.127, 27.344, 37.52, 44.113

Magyary, K., 29.67

Mah, S.Q., 22.931

Mahajan, K.B., 52.65

Mahan, A.I., 12.218

Mahapatra, S., 25.319

Maher, A.T., 31.34

Maher, W.E., 45.390, 49.239

Mahlein, H.F., 9.67, 9.68, 9.83, 21.168, 24.108, 27.93, 36.214, 46.162

Mahler, R.J., 31.114

Mahmoud, S.F., 3.141

Mahr, H., 6.178, 22.813

Maier, H.G., 39.122, 51.8

Maier, M., 6.233, 41.37

Mailloux, R.J., 10.145, 16.207, 16.247, 29.132

Mainchits, E.A., 51.163

Maines, J.D., 41.26

Maione, A., 48.196

Maiorov, S.A., 35.183

Maischberger, K., 37.155

Maiti, J.N., 15.118

Maitland, A., 22.67, 22.608, 24.185, 24.289, 24.373

Majewski, M., 25.231, 33.104, 36.46

Majewski, S., 30.24, 43.10

Majkowski, R.F., 35.165, 44.240

Major, L.B., 9.79

Majumdar, S.C., 49.84

Mak, A.A., 11.190, 20.278, 21.122, 21.159, 24.91, 24.434, 24.554, 27.14, 27.15, 27.23, 27.85, 27.494, 27.688, 27.695, 27.706, 30.53, 30.151, 42.190

Makarewicz, W., 39.191

Makarov, A.I., 20.211

Makarov, A.K., 27.332

Makarov, T.V., 8.51

Makarov, V.N., 19.98

Makarov, V.V., 36.199

Makeeva, G.S., 13.131

Makhlin, A.N., 6.75, 45.145

Makhnenko, L.A., 47.6, 47.8, 47.9, 47.11, 47.10

Makhnev, V.P., 27.370

Makhorin, V.I., 27.505

Maki, K., 1.62

Makimoto, T., 7.76, 11.38, 13.128, 13.141, 14.302, 17.74, 30.141, 30.258, 33.49, 33.106, 44.256, 48.73

Makino, H., 51.11, 51.31

Makino, Y., 21.178

Makios, V., 2.66, 2.101, 24.247, 24.276, 42.153

Makios, W., 42.13

Makita, Y., 27.379, 27.473

Makkaveev, V.I., 51.162

Makoc, Z., 46.85

Makogon, M.M., 21.446, 26.55, 26.73, 27.91, 27.117, 27.472, 30.390, 37.322

Makosch, G., 22.10

Makovkin, A., 35.191

Makridis, H., 52.334

Makshantsev, B.I., 45.313

Maksimov, V.I., 14.147

Maksimova, L.I., 46.177

Maksimenkov, P.P., 30.292

Maksymouko, G., 36.178

Makukhin, V.P., 44.322

Malacara, D., 22.111, 24.188

Malakhov, A.N., 20.218, 32.90

Malan, O.G., 22.425, 22.472, 22.498, 22.661, 22.760

Malanchenko, V.P., 10.274

Malanin, Yu.N., 11.187

Malaviya, N., 35.116, 49.395

Malaviya, S.D., 33.148

Maldutis, E.K., 45.47

Malek, K., 51.147

Malherbe, C.W., 30.16

Malik, D.P.S., 48.194

Malinin, Yu.N., 37.134

Malinov, I.A., 11.3

Mal'ishev, V.I., 20.57

Malissin, R., 29.72

Malivanchuk, V.I., 12.110

Mallet, J.D., 50.184

Malley, M.M., 42.125

Mallick, A.K., 12.308

Mallick, S., 35.149

Mallik, A., 24.379

Mallory, K.B., 39.29

Mallozzi, P.J., 9.36

Malmstadt, H.V., 30.433

Maloch, J., 45.46

Maloney, E.D., 18.205

Maloney, P.J., 24.407, 37.168, 37.301

Maloney, W.T., 35.22, 35.24, 42.8

Maloratskii, L.G., 10.153, 10.283, 11.77

Malota, F., 27.131, 30.131, 37.74, 37.105, 52.115

Malovickho, A.A., 10.288

Maltese, U., 28.121

Mal'tsev, A.A., 27.383

Mal'tsev, V.P., 12.161, 12.199, 16.254

Malushkov, G.D., 7.41

Malvezzi, A.M., 22.811

Maly, V.I., 41.46

Malyakin, A.K., 28.110

Malysh, A.G., 24.200

Malyshenko, L.Ye., 36.245

Malyshenko, Yu.I., 49.82

Malyshev, B.N., 21.205

Malyshev, G.F., 37.262

Malyshev, V.A., 13.167, 18.33, 18.44, 18.99, 20.95, 25.146

Malyshev, V.I., 27.50, 27.103, 27.178, 27.220, 30.76, 42.93, 45.76

Malyutenko, V.K., 31.30

Malyutin, A.A., 27.71, 27.126, 38.167

Malyutin, N.D., 2.69, 2.100

Malz, D., 42.215

Mamedov, K.Ya., 37.103

Mammel, W.L., 8.127, 48.198, 48.224

Mamonov, S.K., 24.65

Manankova, A.V., 38.177

Manarini, A.M.M., 16.100

Mancini, E., 47.79

Mandel, L., 20.120, 20.192, 32.113, 32.156, 32.205, 43.58, 43.66

Mandel, P., 20.204, 20.297, 20.333, 50.44

Mandelberg, H.I., 21.535

Mandel'shtam, M.Ya., 11.23

Mandeville, G.D., 30.162

Mandics, P.A., 49.331

Manenkov, A.A., 13.21, 27.707

Manenkov, A.B., 11.114, 11.127, 12.79, 12.358

Manes, K.R., 22.493, 24.385, 45.104

Miyata, N., 4.15

Miyauchi, K., 48.45, 48.69, 48.143

Miyazaki, M., 25.153

Miyazaki, Y., 11.156, 12.249, 16.142

Miyazawa, K., 53.110

Miyazoe, Y., 21.142, 21.274, 21.457, 21.473, 21.493, 27.187, 27.587

Miyoshi T., 2.84

Miyoshi, Y., 42.51

Mizeraczyk, J., 22.835

Mizobuchi, A., 14.252

Mizobuchi, Y., 34.86

Mizukaga, Y., 51.44

Mizumachi, M., 52.437

Mizuno, H., 25.35, 44.171

Mizuno, J., 6.65

Mizuno, K., 18.131, 18.189, 27.676

Mizusawa, M., 16.474, 16.484

Mizushima, Y., 13.88

Mjolsness, R.C., 49.3

Mkrtchyan, M.M., 22.565

Mochalov, A.V., 38.186

Mockel, P., 21.296, 24.503, 27.230, 27.450, 33.243

Mocker, H.W., 51.63, 52.429

Modak, A.T., 46.170

Model', A.M., 10.46, 28.19, 28.21, 28.195

Modenov, V.P., 14.278

Modi, J.K., 31.116

Moeckel, W.E., 45.188

Moeller, A.W., 52.162

Moeller, C.E., 21.161

Moeny, W.M., 22.405

Moffatt, D.L., 7.8

Mogilnitskii, B.S., 37.354

Mogil'nyi, A.G., 44.383

Mohan, M., 6.234

Mohan, R., 25.283

Mohn, E., 24.197, 27.57

Mohon, N., 46.158

Mohr, C., 10.109

Mohr, F., 36.261

Mohr, W.B., 21.535

Mohsen, A., 3.66, 3.70, 12.142

Moiseev, M.A., 47.103

Mok, C.K., 28.130, 28.164, 28.216, 36.103

Mokhamed, M.O., 12.190

Molchanov, A.G., 23.42

Molchanov, M.I., 24.43, 24.396

Molchanov, V.I., 39.150

Molchanov, V.Ya., 11.141, 38.171

Moldavskaya, V.M., 31.90

Moldovan, P.K., 11.46

Moles, M., 1.86

Molinari, V.G., 15.106

Moll, J.W., 52.51

Mollenauer, L.F., 21.480

Moller, E., 10.245

Moller, W., 50.82

Mollet, B., 39.181

Molochev, V.I., 23.136

Molodkin, V.A., 3.51

Molokovskii, S.I., 18.211, 33.284

Molyavko, V.I., 12.100

Moma, Yu.A., 42.69

Momma, K., 37.175

Monaco, V.A., 10.176

Monaghan, S.R., 14.232

Monahan, B.C., 31.82

Moncur, K., 24.515

Mondloch, A.J., 49.75

Monich, G., 44.373

Moniz, W.B., 21.320

Monosov, Ya.A., 5.16

Monroe, J.W., 25.190, 25.337

Monsen, P., 51.141

Monson, D.J., 22.110

Montero, S., 41.145

Montgomery, J.P., 3.81, 8.98, 10.272

Montgomery, P.C., 26.99

Monti, R., 52.376

Moody, C.D., 45.219, 45.388

Mooradian, A., 23.174, 26.166, 26.329, 41.53, 41.121, 41.138

Moore, C.E., 52.326

Moore, E.L., 46.107

Moore, F.K., 45.310

Moore, I.J.H.S., 14.149, 30.323

Moore, J.S., 6.235

Moore, L., 1.1

Moore, M., 30.300, 45.225

Moore, R.A., 14.300, 30.14, 35.35

Moore, R.K., 49.9, 52.87

Moore, R.L., 46.98

Moore, W.J., 23.54, 36.217

Moore, W.S., 28.203

Morabin, A., 39.26

Morales, A., 39.26

Moran, J.M., 24.125, 34.89

Morantz, D.J., 22.38

Morawski, T., 11.173, 30.400

Moreau, R., 39.224

Moreland, J., 7.20

Moreland, J.P., 49.79

Moren, R.E., 36.183

Morenkov, A.D., 32.63

Moreno, J.B., 22.847

Moresco, M., 15.76

Morey, W.W., 21.307

Morgan, D.J., 32.121

Morgan, G.B., 25.344, 33.197

Morgan, I.G., 36.268, 38.158

Morgan, J.R., 25.189

Morgan, P.D., 30.181

Morgenshtern, Z.L., 21.8, 40.117

Morgun, Yu.F., 27.444

Morgunov, L.N., 49.38

Mori, H., 10.230, 38.120

Mori, K., 21.442

Mori, M., 41.15

Mori, S., 10.81

Mori, T.T., 32.54

Mori, Y., 51.106

Moriarty, J.J., 45.20

Morignot, P., 47.29

Morikawa, S., 52.195, 52.221

Morimoto, M., 44.7

Morinaga, N., 52.444

Morino, A., 33.182

Morisaki, H., 26.18

Morison, I., 53.98

Morita, K., 49.213, 49.225, 49.466, 51.11, 51.31, 51.105

Morita, N., 48.32

Morita, T., 15.46

Moritani, A., 13.31

Moriyama, M., 26.6

Moroney, W., 25.53

Moroney, W.J., 31.106

Morosin, B., 6.109

Morotani, T., 4.54

Morozov, B.N., 36.155

Morozov, D.N., 24.319, 51.162

Morozov, V.A., 37.165

Morozov, V.N., 20.59, 23.334, 27.640, 34.146

Morozov, V.V., 24.10

Morreal, J.A., 27.38

Morris, A.J., 47.14

Morris, D., 39.256

Morris, D.I., 47.61

Morris, G.E., 22.99

Morris, G.J., 49.319

Morris, L.C., 33.4, 33.25

Morrison, C.A., 21.392

Morrison, J.A., 13.35, 48.302, 49.332, 49.371, 49.441

Morrison, R.W., 24.247, 24.275, 24.276

Morrow, T., 42.105

Morrow, W., 33.27

Smirnov, E.P., 50.169
Smirnov, G.I., 38.147
Smirnov, G.T., 11.191
Smirnov, V.A., 27.427,
    27.698
Smirnov, V.L., 35.191
Smirnov, V.P., 12.20, 12.181,
    52.322
Smirnov, V.S., 14.210, 14.212,
    21.393, 21.449, 21.451,
    22.511, 27.100, 38.96,
    38.103
Smirnova, A.V., 14.263
Smirnova, N.A., 53.3
Smirnova, S.N., 44.319
Smirnova, T.A., 12.525,
    33.254, 33.274
Smith, A.G., 44.2
Smith, A.L.S., 22.788,
    22.802, 22.828, 24.517,
    37.11, 37.14, 37.20
Smith, A.M., 40.92
Smith, A.W., 26.103, 27.188,
    32.98, 42.119
Smith, B., 24.341
Smith, B.H., 32.10
Smith, D.C., 9.21, 22.820,
    24.68, 24.137, 24.199,
    24.274, 30.198, 45.54,
    45.99, 49.245
Smith, D.J., 21.114
Smith, D.L., 27.323, 46.93
Smith, D.M., 45.148
Smith, D.R., 42.209
Smith, D.S., 52.566
Smith, F.G., 13.71
Smith, G.D., 49.267
Smith, G.E., 4.7
Smith, H.P., 45.333
Smith, I.W.M., 22.716,
    42.222
Smith, J.G., 33.86
Smith, J.L., 24.133
Smith, K.W., 45.258
Smith, L., 48.202
Smith, M.A., 22.119
Smith, N.S., 22.932
Smith, O., 46.14
Smith, P.F., 47.62
Smith, P.G., 44.30
Smith, P.L., 44.369, 49.7
Smith, P.W., 22.386, 22.411,
    22.453, 22.881, 24.198,
    24.407, 27.5, 30.77, 30.100,
    30.129, 30.230, 37.168,
    37.226, 37.301, 42.104
Smith, R.A., 36.48, 41.29
Smith, R.C., 22.365, 26.80,
    26.95, 26.126, 26.255,

Smith, R.C. (cont.), 26.264,
    26.308, 27.310, 33.91,
    37.272, 38.105, 42.100
Smith, R.G., 26.79, 26.86,
    26.106, 26.129, 26.247,
    29.128, 41.73
Smith, R.J.D., 41.123
Smith, R.L., 36.181, 36.207,
    44.60
Smith, R.M., 2.39
Smith, S.D., 26.230
Smith, S.T., 36.123
Smith, S.W., 20.31
Smith, T., 52.248
Smith, T.I., 13.154
Smith, T.M., 16.159, 50.25
Smith, W.A., 32.160, 32.180
Smith, W.D., 9.47
Smith, W.F., 8.2
Smith, W.M.H., 49.304
Smith, W.T., 24.19, 37.222
Smithers, M.E., 20.231,
    32.174
Smithgall, D.H., 48.209
Smits, V.C., 16.199
Smoczynski, L., 14.92
Smolinsky, G., 12.277
Smol'skaya, T.I., 21.225,
    24.268, 27.52
Smolyakov, B.P., 40.162
Smorchkova, S.A., 7.48
Smorgonskii, A.V., 18.186,
    19.149
Smorgonskii, V.Ya., 3.8,
    3.22, 3.55, 3.74, 3.87,
    3.88, 3.90, 3.95, 3.112,
    3.133, 3.134, 3.136
Snapp, C.P., 25.257, 26.131,
    33.108
Snavely, B.B., 21.22, 21.39,
    21.78
Snedkov, B.A., 18.2, 47.54
Sneider, J., 5.49
Snell, W.L., 24.168
Snell, W.W., 28.147, 33.113
Snelling, D.R., 22.558
Snider, D.M., 25.150
Sniekers, J.P.F., 16.402
Snigirev, O.V., 25.391
Snow, K.C., 33.197
Snurnikova, G.K., 10.262,
    12.135, 12.198, 12.229
Snyder, A.W., 10.171, 12.14,
    12.104, 12.108, 12.128,
    12.132, 12.138, 12.140,
    12.169, 12.178, 12.233,
    12.300, 12.319, 12.337,
    12.348, 48.124, 48.184,
    48.304

Snyder, D., 52.371
Snyder, D.L., 51.153
Snyder, H.H., 3.130
Snyder, L.C., 4.6
Snyder, L.E., 53.75
Snyder, R.V., 28.104, 28.269
Sobel, H., 10.205, 28.144
Sobel'man, I.I., 22.531,
    22.773
Sobenin, N.P., 47.26
Sobol, G.A., 31.2
Sobol, H., 17.3, 17.33
Sobolev, G.L., 19.23, 19.24,
    19.93, 19.137, 19.141,
    32.133
Sobolev, N.D., 44.203
Sobolev, N.N., 22.256,
    22.402, 22.738, 24.208,
    27.217, 40.119
Sobolev, V.S., 52.132, 52.267,
    52.416, 52.417
Sobolev, V.V., 6.106
Sobra, K., 45.46
Socci, R.J., 25.40
Sochor, V., 20.185
Soderholm, L.G., 24.182
Soderman, D.A., 31.160
Sodha, M.S., 6.44, 6.102,
    6.114, 6.175, 6.208, 8.115,
    15.120, 15.125, 26.23,
    26.42, 26.58, 31.60,
    45.293, 48.214, 49.296
Sodin, L.G., 53.74
Soffer, B.H., 21.28, 21.148,
    21.254
Soga, H., 44.334
Soileau, M.J., 45.412
Sokolev, V.M., 36.303
Sokoloff, D.R., 38.91
Sokoloski, M.M., 1.9
Sokolov, A.K., 30.282
Sokolov, A.V., 49.155, 49.164,
    49.231
Sokolov, B.M., 42.10
Sokolov, D.V., 19.18
Sokolov, S.A., 22.831, 24.523
Sokolov, V.A., 38.208, 38.215
Sokolov, V.K., 35.167
Sokolova, E.Yu., 41.63, 41.89
Sokolova, L.I., 19.2
Sokolovskaya, A.I., 41.96
Sokolovskii, I.I., 30.257,
    30.278, 30.299
Sokolovskii, R.I., 6.74,
    6.189, 26.132, 26.239,
    32.149, 33.242
Sokolowski, T., 8.2
Solamaa, R., 27.382
Solbakken, K., 17.42, 30.104,
    30.146

Zhilkov, V.S., 36.160,
   36.194, 36.206, 36.245,
   36.288, 36.304
Zhironkin, A.V., 10.376,
   10.379
Zhiryakov, B.M., 27.696,
   45.322
Zhitkova, M.B., 21.221,
   24.313
Zhivotov, V.K., 38.59,
   39.207, 39.225
Zhokhov, V.P., 46.118
Zhovna, G.I., 38.131,
   38.175
Zhukovskii, V.V., 27.33
Zhulinskii, S.F., 47.83,
   47.94
Zhupan, Yu.Ya., 37.271
Zhurakhovskii, V.A., 14.312,
   18.70
Zhurav, S.M., 8.124
Zhuravel', F.A., 52.267
Zhuravle, V.A., 31.129
Zhuravlev, A.I., 46.143
Zhuravlev, A.K., 52.106
Zhuravlev, V.S., 50.35
Zhuravleva, L.N., 21.512
Zich, R., 16.3, 16.140,
   29.66
Zieleniewski, J., 52.412
Ziermann, A., 10.319, 21.76,
   21.84
Zil'berberg, V.V., 6.88
Zilli, E., 15.76
Zimmer, H., 17.7
Zimmerl, O.F., 25.165

Zimmerman, J., 27.554
Zimokosov, G.A., 38.122
Ziolkowski, F.P., 16.84
Zipf, E.C., 30.110
Zitelli, L.T., 18.82
Zito, G., 18.36, 18.59
Zlatin, N.A., 44.343
Zlenko, A.A., 12.345, 42.224
Zlunitsyn, E.S., 13.132,
   39.250
Zocher, E., 16.249
Zohta, Y., 1.57
Zolotov, E.M., 2.139, 12.373,
   12.374
Zook, J.D., 7.81
Zoppi, M., 41.107
Zorkin, A.F., 3.156, 12.310
Zoroofchi, J., 27.618
Zory, P., 24.271, 24.360,
   37.130
Zory, P.S., 23.295, 42.135
Zorya, V.D., 10.377
Zotov, O.V., 39.193
Zritskii, A.R., 26.274
Zubarev, I.G., 21.216,
   26.368, 27.354, 27.637,
   27.708, 37.361, 41.62,
   45.309
Zubarev, T.N., 20.49, 20.193,
   27.426, 27.463, 30.282
Zubkov, V.I., 5.16, 30.208
Zubkova, S.M., 46.132
Zubov, B.V., 36.87
Zubov, V.A., 34.94, 35.172
Zubovskii, V.P., 12.251
Zubrinov, I.I., 5.55

Zucker, F.J., 16.289
Zucker, H., 11.101, 16.300,
   16.376, 29.115, 50.52
Zucker, J., 31.3
Zucker, M.S., 8.80
Zuev, M.G., 24.401
Zuev, V.E., 40.100, 45.205,
   49.77, 49.268
Zuev, V.S., 34.150
Zul'fugarzade, K.E., 39.5
Zun'kova, Z.E., 22.532
Zurcher, L.A., 16.35
Zuryk, J.A., 22.794
Zusman, M.I., 30.239, 30.374
Zuzolo, R.C., 46.7
Zvatitskii, V.A., 48.84
Zverev, B.V., 47.26
Zverev, G.M., 6.149, 21.246,
   45.47, 45.130, 45.194
Zverev, V.A., 24.414, 29.125,
   44.68
Zwanziger, M., 32.181, 43.69
Zwarts, C.M.G., 40.28
Zweng, H.C., 46.25, 46.39
Zwerdling, S., 36.48
Zwick, S.A., 22.729
Zyatitskii, V.A., 9.34,
   12.193
Zybin, M.I., 21.227
Zykov, A.I., 11.183, 13.126,
   13.132, 39.250, 47.6, 47.8,
   47.9, 47.11, 47.35, 47.57,
   47.69
Zyss, R., 46.136, 46.137